Practice
"Work the Problems"

Feature	Description	Benefit	Page
"Assess Your Understanding" contains a variety of problems at the end of each section.			
'Are You Prepared?' Problems	These assess your retention of the prerequisite material you'll need. Answers are given at the end of the section exercises. This feature is related to the Preparing for This Section feature.	Do you always remember what you've learned? Working these problems is the best way to find out. If you get one wrong, you'll know exactly what you need to review and where to review it!	280 & 284
Concepts and Vocabulary	These Fill-in-the-Blank and True/False items assess your understanding of key definitions and concepts in the current section.	Learning math is more than memorization – it's about discovering connections. These problems help you understand the 'big ideas' before diving into skill building.	280
Skill Building	Correlated to section examples, these problems provide straightforward practice.	It's important to dig in and develop your skills. These problems provide you with ample practice to do so.	280-283
Applications and Extensions	These problems allow you to apply your skills to real-world problems. These problems also allow you to extend concepts learned in the section.	You will see that the material learned within the section has many uses in everyday life.	283-284
Discussion and Writing	"Discussion and Writing" problems are colored red. These support class discussion, verbalization of mathematical ideas, and writing and research projects.	To verbalize an idea, or to describe it clearly in writing, shows real understanding. These problems nurture that understanding. They're challenging but you'll get out what you put in.	284
Now Work Problems	Many examples refer you to a related homework problem. These related problems are marked by a pencil and yellow numbers.	If you get stuck while working problems, look for the closest Now Work problem and refer back to the related example to see if it helps.	273 & 281
Graphing Calculator	These optional problems require the use of a graphing utility, and are marked by a special icon and green numbers.	Your instructor will usually provide guidance on whether or not to do these problems. If so, these problems help to verify and visualize your analytical results.	324

Review
"Study for Quizzes and Tests"

Feature	Description	Benefit	Page
Chapter Reviews at the end of each chapter contain...			
"Things to Know"	A detailed list of important theorems, formulas, and definitions from the chapter.	Review these and you'll know the most important material in the chapter!	353
"You Should be Able To..."	Contains a complete list of objectives by section, with corresponding practice exercises.	Do the recommended exercises and you'll have mastery over the key material. If you get something wrong, review the suggested page numbers and try again.	354-355
Review Exercises	These provide comprehensive review and practice of key skills, matched to the Learning Objectives for each section.	Practice makes perfect. These problems combine exercises from all sections, giving you a comprehensive review in one place.	355-358
Chapter Test	About 15–20 problems that can be taken as a Chapter Test. Be sure to take the Chapter Test under test conditions—no notes!	Be prepared. Take the sample practice test. This will get you ready for your instructor's test. If you get a problem wrong, watch the Chapter Test Prep video.	359
Chapter Projects	The Chapter Project applies what you've learned in the chapter. Additional projects are available on the Instructor's Resource Center (IRC).	The Project gives you an opportunity to apply what you've learned in the chapter to solve a problem related to the opening article. If your instructor allows, these make excellent opportunities to work in a group, which is often the best way of learning math.	359-360
Cumulative Review	These problem sets appear at the end of each chapter, beginning with Chapter 2. They combine problems from previous chapters, providing an ongoing cumulative review.	These are really important. They will ensure that you are not forgetting anything as you go. These will go a long way toward keeping you constantly primed for quizzes and tests.	360

COLLEGE ALGEBRA

COLLEGE ALGEBRA

Concepts Through Functions

Michael Sullivan

Chicago State University

Michael Sullivan, III

Joliet Junior College

PEARSON

Prentice
Hall

Upper Saddle River, New Jersey 07458

Editor-in-Chief: Sally Yagan
Acquisitions Editor: Adam Jaworski
Project Manager: Dawn Murrin
Executive Managing Editor: Kathleen Schiaparelli
Senior Managing Editor: Linda Mihatov Behrens
Production Editor: Bob Walters, Prepress Management, Inc.
Media Production Editor: Jenelle Woodrup
Assistant Managing Editor, Science and Math Print
 Supplements: Karen Bosch
Manufacturing Buyer: Maura Zaldivar
Manufacturing Manager: Alexis Heydt-Long
Executive Marketing Manager: Halee Dinsey
Art Director: Heather Scott
Interior Designer: Judith Matz-Coniglio
Cover Designer: John Christiana
Creative Director: Juan R. Lõpez
Director of Creative Services: Paul Belfanti
Composition: ESM Inhouse Formatting
 Manager, Composition: Allyson Graesser
 Assistant Manager, Formatting: Clara Bartunek
 Electronic Production Specialist: Karen Noferi

Electronic Page Makeup: Karen Noferi, Karen Stephens, and
 Julita Nazario
Director, Image Resource Center: Melinda Reo
Manager, Rights and Permissions: Zina Arbia
Manager, Visual Research: Beth Brenzel
Cover Image Specialist: Karen Sanatar
Image Permission Coordinator: Joanne Dippel
Photo Researcher: Melinda Alexander
Cover Image: Royalty Free/CORBIS
Art Studio: Artworks
 Senior Managing Editor, Art Production
 and Management: Patricia Burns
 Manager, Production Technolgies: Matthew Haas
 Managing Editor, Art Management: Abigail Bass
 Manager, Illustration Production: Sean Hogan
 Assistant Manager, Illustration Production: Ronda C. Whitson
 Art Editor: Tom Benfatti
 Illustrations: Stacy Smith, Scott Wieber, Nathan Storck,
 Ryan Currier, Mark Landis, Audrey Simonetti,
 David Lynch
 Quality Assurance: Panela Taylor, Timoethy Nguyen,
 Cathy Shelley

© 2007 Pearson Education, Inc.
Pearson Prentice Hall
Pearson Education, Inc.
Upper Saddle River, New Jersey 07458

Printed in the United States of America
10 9 8 7 6 5 4 3 2 1

ISBN: 0-13-227982-7 (Instructor's Edition)
ISBN: 0-13-187478-0 (Student Edition)

Pearson Education LTD., London
Pearson Education Australia PTY, Limited, Sydney
Pearson Education Singapore, Pte. Ltd
Pearson Education North Asia Ltd. Hong Kong
Pearson Education Canada, Ltd., Toronto
Pearson Educación de Mexico, S.A. de C.V.
Pearson Education – Japan, Tokyo
Pearson Education Malaysia, Pte. Ltd

In Memory of Mary . . .

Wife and Mother



Contents

Chapter 7 Sequences; Induction; The Binomial Theorem 495

Chapter 8 Counting and Probability 535

Appendix A Review A1

Three Distinct Series

Students have different goals, learning styles, and levels of preparation. Instructors have different teaching philosophies, styles, and techniques. Rather than write one series to fit all, the Sullivans have written three distinct series. All share the same goal—to develop a high level of mathematical understanding and an appreciation for the way mathematics can describe the world around us. The manner of reaching that goal, however, differs from series to series.

Contemporary Series, Seventh Edition

The Contemporary Series is the most traditional in approach, yet modern in its treatment of precalculus mathematics. Graphing utility coverage is optional and can be included or excluded at the discretion of the instructor: *College Algebra, College Algebra Essentials, Algebra & Trigonometry, Trigonometry, Precalculus, Precalculus Essentials*

Enhanced with Graphing Utilities Series, Fourth Edition

This series provides a more thorough integration of graphing utilities into topics, allowing students to explore mathematical concepts and foreshadow ideas usually studied in later courses. Using technology, the approach to solving certain problems differs from the Contemporary series, while the emphasis on understanding concepts and building strong skills does not: *College Algebra, College Algebra Essentials, Algebra & Trigonometry, Trigonometry, Precalculus, Precalculus Essentials.*

Concepts through Functions Series, First Edition

This series differs from the others, utilizing a functions approach that serves as the organizing principle tying concepts together. Functions are introduced early in various formats. This approach supports the Rule of Four, which states that functions are represented symbolically, numerically, graphically, and verbally. Each chapter introduces a new type of function and then develops all concepts pertaining to that particular function. The solutions of equations and inequalities, instead of being developed as stand-alone topics, are developed in the context of the underlying functions. Graphing utility coverage is optional and can be included or excluded at the discretion of the instructor: *College Algebra; Precalculus, with a Unit Circle Approach to Trigonometry; Precalculus, with a Right Triangle Approach to Trigonometry.*

Preface to the Instructor

As professors at both an urban university and a community college, Michael Sullivan and Michael Sullivan, III are aware of the varied needs of College Algebra students, ranging from those who have little mathematical background and a fear of mathematics courses, to those having a strong mathematical education and a high level of motivation. For some of your students, this will be their last course in mathematics, whereas others will further their mathematical education. This text is written for both groups.

As a teacher, and as an author of precalculus, engineering calculus, finite mathematics, and business calculus texts, Michael Sullivan understands what students must know if they are to be focused and successful in upper level math courses. However, as a father of four, he also understands the realities of college life. His co-author and son, Michael Sullivan, III, believes in the value of technology as a tool for learning that enhances understanding without sacrificing math skills. Together, both authors have taken great pains to ensure that the text contains solid, student-friendly examples and problems, as well as a clear and seamless writing style.

A tremendous benefit of authoring a successful series is the broad-based feedback we receive from teachers and students. Many of the innovative ideas in this series are in response to their thoughtful comments and suggestions. We are sincerely grateful for this support and hope that we have been able to take these ideas and create a new and invigorating approach to the subject. We continue to encourage you to share with us your experiences teaching from this text.

About this Book

This book utilizes a functions approach to College Algebra. Functions are introduced early (Chapter 1) in various formats: maps, tables, equations, and graphs. This approach suppports the Rule of Four; that functions are represented symbolically, numerically, graphically, and verbally. The solutions of equations and inequalities, instead of being stand-alone topics, are studied in the context of the underlying functions. At the same time, properties of these functions are listed and applications and models involving them are developed.

Linear and quadratic functions form the basis of Chapter 2, polynomial and rational functions are studied in Chapter 3, and exponential and logarithmic functions are discussed in Chapter 4. The Table of Contents on page vii shows the detail of the material in these chapters as well as a listing of the remaining topics found in this book.

Features of this Book

Rather than provide a list of features here, we have placed that information on the inside front cover of the book. This places features in their proper context, as building blocks of an overall learning system that has been carefully crafted over the years to help students get the most out of the time they put into studying.

Please take the time to review the features, and to discuss them with your students at the beginning of your course. Our experience has been that when students utilize these features, they are more successful in the course.

Using this Book Effectively and Efficiently with Your Syllabus

To meet the varied needs of diverse syllabi, this book contains more content than is likely to be covered in a typical College Algebra course. As the chart illustrates, this book has been organized with flexibility of use in mind. Even within a given chapter, certain sections are optional and can be omitted without loss of continuity. (See the detail following the flow chart).

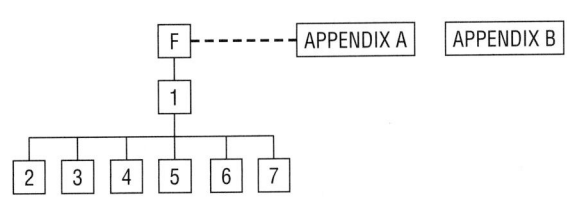

Chapter F Foundations

Quick coverage of this chapter, which is mainly review material, will enable you to get to Chapter 1, "Functions and Their Graphs," earlier.

Chapter 1 Functions and Their Graphs

Perhaps the most important chapter. Section 1.6 is optional.

Chapter 2 Linear and Quadratic Functions

Topic selection depends on your syllabus. Section 2.2, 2.6, and 2.7 may be omitted without a loss of continuity.

Chapter 3 Polynomial and Rational Functions

Topic selection depends on your syllabus. Section 3.6 is optional.

Chapter 4 Exponential and Logarithmic Functions

Sections 4.1–4.6 follow in sequence. Sections 4.7–4.9 are optional.

Chapter 5 Conics

Sections 5.2–5.4 follow in sequence.

Chapter 6 Systems of Equations and Inequalities

Sections 6.2–6.7 may be covered in any order. Section 6.8 requires Section 6.7.

Chapter 7 Sequences; Induction; The Binomial Theorem

There are three independent parts: Sections 7.1–7.3, Section 7.4, and Section 7.5.

Chapter 8 Counting and Probability

The sections follow in sequence.

Appendix A Review

Sections A.1–A.9 consist of review material, which may be covered at the start of a course or used as a just-in-time review. Specific references to this material occur throughout the text to assist in the review process. Coverage of Sections A.10 and A.11 will depend on your syllabus.

Appendix B Graphing Utilities

Reference is made to these sections at the appropriate place in the text.

Acknowledgments

Textbooks are written by authors, but evolve from an idea to final form through the efforts of many people.

We would like to extend thanks to the following individuals for their important assistance and encouragement in the preparation of this edition: From Prentice Hall: Halee Dinsey for her innovative marketing skills; Adam Jaworski for his substantial contributions, ideas, and enthusiasm; Patrice Jones, who remains a huge fan and supporter; Dawn Murrin for her talent and skill in getting supplements out; Bob Walters, who continues to amaze us with his organizational skill as Production Editor; Sally Yagan for her continued support and genuine interest; and the Prentice Hall Sales Team, for their continued confidence in our books. We also extend our thanks to Anna Maria Mendiola and Perian Herring for their accuracy check of page proof and to Kathleen Miranda, Karla Neal, and Roxanne Byrne for checking our answers. Special thanks to Celeste Hernandez for her help in preparing chapter projects, and to Kevin Bodden for his work in preparing the chapter tests.

Finally, our grateful thanks to the dedicated users and reviewers of our books, whose collective insights form the backbone of each series. We apologize for any omissions:

James Africh, College of DuPage
Steve Agronsky, Cal Poly State University
Grant Alexander, Joliet Junior College
Dave Anderson, South Suburban College
Joby Milo Anthony, University of CentralFlorida
James E. Arnold, University of Wisconsin-Milwaukee
Adel Arshaghi, Center for Educational Merit
Carolyn Autray, University of West Georgia
Agnes Azzolino, Middlesex County College
Wilson P Banks, Illinois State University
Sudeshna Basu, Howard University
Dale R. Bedgood, East Texas State University
Beth Beno, South Suburban College
Carolyn Bernath, Tallahassee Community College
William H. Beyer, University of Akron

Annette Blackwelder, Florida State University
Richelle Blair, Lakeland Community College
Kevin Bodden, Lewis and Clark College
Bob Bradshaw, Ohlone College
Trudy Bratten, Grossmont College
Tim Bremer, Broome Community College
Joanne Brunner, Joliet Junior College
Warren Burch, Brevard Community College
Mary Butler, Lincoln Public Schools
Jim Butterbach, Joliet Junior College
William J. Cable, University of Wisconsin-Stevens Point
Lois Calamia, Brookdale Community College
Jim Campbell, Lincoln Public Schools
Roger Carlsen, Moraine Valley Community College
Elena Catoiu, Joliet Junior College

Mathews Chakkanakuzhi, Palomar College
John Collado, South Suburban College
Nelson Collins, Joliet Junior College
Jim Cooper, Joliet Junior College
Denise Corbett, East Carolina University
Theodore C. Coskey, South Seattle Community College
Donna Costello, Plano Senior High School
Paul Crittenden, University of Nebraska at Lincoln
John Davenport, East Texas State University
Faye Dang, Joliet Junior College
Antonio David, Del Mar College
Duane E. Deal, Ball State University
Timothy Deis, University of Wisconsin-Platteville
Vivian Dennis, Eastfield College
Deborah Dillon, R. L. Turner High School

Guesna Dohrman, Tallahassee Community College
Karen R. Dougan, University of Florida
Louise Dyson, Clark College
Paul D. East, Lexington Community College
Don Edmondson, University of Texas-Austin
Jason Eltrevoog, Joliet Junior College
Erica Egizio, Joliet Junior College
Christopher Ennis, University of Minnesota
Ralph Esparza, Jr., Richland College
Garret J. Etgen, University of Houston
Pete Falzone, Pensacola Junior College
W.A. Ferguson, University of Illinois-Urbana/Champaign
Iris B. Fetta, Clemson University
Mason Flake, student at Edison Community College
Timothy W. Flood, Pittsburg State University
Merle Friel, Humboldt State University
Richard A. Fritz, Moraine Valley Community College
Dewey Furness, Ricke College
Randy Gallaher, Lewis and Clark College
Tina Garn, University of Arizona
Dawit Getachew, Chicago State University
Wayne Gibson, Rancho Santiago College
Robert Gill, University of Minnesota Duluth
Sudhir Kumar Goel, Valdosta State University
Joan Goliday, Sante Fe Community College
Frederic Gooding, Goucher College
Sue Graupner, Lincoln Public Schools
Jennifer L. Grimsley, University of Charleston
Ken Gurganus, University of North Carolina
James E. Hall, University of Wisconsin-Madison
Judy Hall, West Virginia University
Edward R. Hancock, DeVry Institute of Technology
Julia Hassett, DeVry Institute-Dupage
Christopher Hay-Jahans, University of South Dakota
Michah Heibel, Lincoln Public Schools
LaRae Helliwell, San Jose City College
Brother Herron, Brother Rice High School
Robert Hoburg, Western Connecticut State University
Lynda Hollingsworth, Northwest Missouri State University
Charla Holzbog, Denison High School
Lee Hruby, Naperville North High School
Miles Hubbard, St. Cloud State University
Kim Hughes, California State College-San Bernardino
Ron Jamison, Brigham Young University
Richard A. Jensen, Manatee Community College
Sandra G. Johnson, St. Cloud State University
Tuesday Johnson, New Mexico State University
Moana H. Karsteter, Tallahassee Community College
Donna Katula, Joliet Junior College
Arthur Kaufman, College of Staten Island
Thomas Kearns, North Kentucky University
Shelia Kellenbarger, Lincoln Public Schools
Lynne Kowski, Raritan Valley Community College
Keith Kuchar, Manatee Community College
Tor Kwembe, Chicago State University
Linda J. Kyle, Tarrant Country Jr. College
H.E. Lacey, Texas A & M University

Harriet Lamm, Coastal Bend College
James Lapp, Fort Lewis College
Matt Larson, Lincoln Public Schools
Christopher Lattin, Oakton Community College
Julia Ledet, Lousiana State University
Adele LeGere, Oakton Community College
Kevin Leith, University of Houston
JoAnn Lewin, Edison College
Jeff Lewis, Johnson County Community College
Janice C. Lyon, Tallahassee Community College
Jean McArthur, Joliet Junior College
Virginia McCarthy, Iowa State University
Karla McCavit, Albion College
Tom McCollow, DeVry Institute of Technology
Marilyn McCollum, North Carolina State University
Jill McGowan, Howard University
Will McGowant, Howard University
Angela McNulty, Joliet Junior College
Laurence Maher, North Texas State University
Jay A. Malmstrom, Oklahoma City Community College
Rebecca Mann, Apollo High School
Sherry Martina, Naperville North High School
Alec Matheson, Lamar University
Nancy Matthews, University of Oklahoma
James Maxwell, Oklahoma State University-Stillwater
Marsha May, Midwestern State University
Judy Meckley, Joliet Junior College
David Meel, Bowling Green State University
Carolyn Meitler, Concordia University
Samia Metwali, Erie Community College
Rich Meyers, Joliet Junior College
Eldon Miller, University of Mississippi
James Miller, West Virginia University
Michael Miller, Iowa State University
Kathleen Miranda, SUNY at Old Westbury
Thomas Monaghan, Naperville North High School
Craig Morse, Naperville North High School
Samad Mortabit, Metropolitan State University
Pat Mower, Washburn University
A. Muhundan, Manatee Community College
Jane Murphy, Middlesex Community College
Richard Nadel, Florida International University
Gabriel Nagy, Kansas State University
Bill Naegele, South Suburban College
Karla Neal, Lousiana State University
Lawrence E. Newman, Holyoke Community College
James Nymann, University of Texas-El Paso
Mark Omodt, Anoka-Ramsey Community College
Seth F. Oppenheimer, Mississippi State University
Linda Padilla, Joliet Junior College
E. James Peake, Iowa State University
Kelly Pearson, Murray State University
Philip Pina, Florida Atlantic University
Michael Prophet, University of Northern Iowa
Neal C. Raber, University of Akron
Thomas Radin, San Joaquin Delta College

Ken A. Rager, Metropolitan State College
Kenneth D. Reeves, San Antonio College
Elsi Reinhardt, Truckee Meadows Community College
Jane Ringwald, Iowa State University
Stephen Rodi, Austin Community College
William Rogge, Lincoln Northeast High School
Howard L. Rolf, Baylor University
Mike Rosenthal, Florida International University
Phoebe Rouse, Lousiana State University
Edward Rozema, University of Tennessee at Chattanooga
Dennis C. Runde, Manatee Community College
Alan Saleski, Loyola University of Chicago
Susan Sandmeyer, Jamestown Community College
Linda Schmidt, Greenville Technical College
A.K. Shamma, University of West Florida
Martin Sherry, Lower Columbia College
Tatrana Shubin, San Jose State University
Anita Sikes, Delgado Community College
Timothy Sipka, Alma College
Lori Smellegar, Manatee Community College
Gayle Smith, Loyola Blakefield
John Spellman, Southwest Texas State University
Rajalakshmi Sriram, Okaloosa-Walton Community College
Becky Stamper, Western Kentucky University
Judy Staver, Florida Community College-South
Neil Stephens, Hinsdale South High School
Patrick Stevens, Joliet Junior College
Christopher Terry, Augusta State University
Diane Tesar, South Suburban College
Tommy Thompson, Brookhaven College
Martha K. Tietze, Shawnee Mission Northwest High School
Richard J. Tondra, Iowa State University
Marvel Townsend, University of Florida
Jim Trudnowski, Carroll College
Robert Tuskey, Joliet Junior College
Richard G. Vinson, University of South Alabama
Mary Voxman, University of Idaho
Jennifer Walsh, Daytona Beach Community College
Donna Wandke, Naperville North High School
Darlene Whitkenack, Northern Illinois University
Christine Wilson, West Virginia University
Brad Wind, Florida International University
Mary Wolyniak, Broome Community College
Canton Woods, Auburn University
Tamara S. Worner, Wayne State College
Terri Wright, New Hampshire Community Technical College, Manchester
George Zazi, Chicago State University
Steve Zuro, Joliet Junior College

Michael Sullivan
Michael Sullivan, III

STUDENT RESOURCES	INSTRUCTOR RESOURCES
Student Study-Pak	**Instructor Resource Distribution**

Everything a student needs to succeed in one place. The Student Study Pack can be packaged with the book, or is available for purchase standalone. Study-Pak contains:

- *Student Solutions Manual*
 Fully worked solutions to odd-numbered exercises.

- *Pearson Tutor Center*
 Tutors provide one-on-one tutoring for any problem with an answer at the back of the book. Students access the Tutor Center via toll-free phone, fax, or email. Available only to college students in the U.S. and Canada.

- *CD Lecture Series*
 A comprehensive set of CD-ROMs, tied to the textbook, containing short video clips of an instructor working key book examples.

- *Algebra Review*
 Four chapters of Intermediate Algebra review. Perfect for a slower-paced course or for individual review.

All instructor resources can be downloaded from the web site, www.prenhall.com. Select "Browse our catalog," then, click on "Mathematics;" select your course and choose your text. Under "Resources," on the left side, select "instructor" and choose the supplement you need to download. You will be required to run through a one time registration before you can complete this process.

- *TestGen*
 Easily create tests from textbook section objectives. Questions are algorithmically generated allowing for unlimited versions. Edit problems or create your own.

- *Test Item File*
 A printed test bank derived from TestGen.

- *PowerPoint Lecture Slides*
 Fully editable slides that follow the textbook. Project in class or post to a website in an online course.

- *Instructor Solutions Manual*
 Fully worked solutions to all textbook exercises and chapter projects.

- *Mini Lecture Notes*
 These include additional examples and helpful teaching tips, by section.

Instructor's Edition

Provides answers to *all* exercises in the back of the text.

MathXL®

MathXL® is a powerful online homework, tutorial, and assessment system that accompanies your textbook. Instructors can create, edit, and assign online homework and tests using algorithmically generated exercises correlated at the objective level to the textbook. Student work is tracked in an online gradebook. Students can take chapter tests and receive personalized study plans based on their results. The study plan diagnoses weaknesses and links students to tutorial exercises for objectives they need to study. Students can also access video clips from selected exercises. MathXL® is available to qualified adopters. For more information, visit our website at **www.mathxl.com**, or contact your Prentice Hall sales representative for a demonstration.

MyMathLab

MyMathLab is a text-specific, customizable online course for your textbooks. MyMathLab is powered by CourseCompass™— Pearson Education's online teaching and learning environment—and by MathXL®—our online homework, tutorial, and assessment system. MyMathLab gives you the tools you need to deliver all or a portion of your course online, whether your students are in a lab setting or working from home.

 MyMathLab provides a rich and flexible set of course materials, featuring free-response exercises that are algorithmically generated for unlimited practice. Students can use online tools such as video lectures and a multimedia textbook to improve their performance. Instructors can use MyMathLab's homework and test managers to select and assign online exercises correlated to the textbook, and can import TestGen tests for added flexibility. The online gradebook—designed specifically for mathematics—automatically tracks students' homework and test results and gives the instructor control over how to calculate final grades. MyMathLab is available to qualified adopters. For more information, visit our website at **www.mymathlab.com** or contact your Prentice Hall sales representative for a product demonstration.

List of Applications

Photo Credits

To The Student

As you begin, you may feel anxious about the number of theorems, definitions, procedures, and equations. You may wonder if you can learn it all in time. Don't worry, your concerns are normal. This textbook was written with you in mind. If you attend class, work hard, and read and study this book, you will build the knowledge and skills you need to be successful. Here's how you can use the book to your benefit.

Read Carefully

When you get busy, it's easy to skip reading and go right to the problems. Don't…the book has a large number of examples and clear explanations to help you break down the mathematics into easy-to-understand steps. Reading will provide you with a clearer understanding, beyond simple memorization. Read sections to be covered before class (not after) so you can ask questions about anything you didn't understand. You'll be amazed at how much more you'll get out of class if you do this.

Getting the Most From This Book

We've provided a brief guide to getting the most from this book (refer to the "Prepare for Class," "Practice," and "Review" pages on the inside front cover of this book). Spend 15 minutes reviewing the guide and familiarizing yourself with the features by flipping to the page numbers provided. Then, as you read, use them. This is the best way to make the most of your textbook.

Do not hesitate to contact us, through Prentice Hall, with any questions, suggestions or comments that would improve this text. We look forward to hearing from you, and best of luck with all your studies.

Best Wishes!

Michael Sullivan
Michael Sullivan, III

F

Foundations:
A Prelude to Functions

A LOOK BACK Appendix A reviews skills from Intermediate Algebra.

A LOOK AHEAD Here we make the connection between algebra and geometry through the rectangular coordinate system. The idea of using a system of rectangular coordinates dates back to ancient times, when such a system was used for surveying and city planning. Apollonius of Perga, in 200 BC, used a form of rectangular coordinates in his work on conics, although this use does not stand out as clearly as it does in modern treatments. Sporadic use of rectangular coordinates continued until the 1600s. By that time, algebra had developed sufficiently so that René Descartes (1596–1650) and Pierre de Fermat (1601–1665) could take the crucial step, which was the use of rectangular coordinates to translate geometry problems into algebra problems, and vice versa. This step was important for two reasons. First, it allowed both geometers and algebraists to gain new insights into their subjects, which previously had been regarded as separate, but now were seen to be connected in many important ways. Second, these insights made the development of calculus possible, which greatly enlarged the number of areas in which mathematics could be applied and made possible a much deeper understanding of these areas.

F.1 The Distance and Midpoint Formulas

PREPARING FOR THIS SECTION *Before getting started, review the following:*

• Algebra Essentials (Appendix A, Section A.1, pp. A1–A9) • Geometry Essentials (Appendix A, Section A.2, pp. A11–A16)

Now work the 'Are You Prepared?' problems on page 6.

OBJECTIVES 1 Use the Distance Formula
2 Use the Midpoint Formula

Figure 1

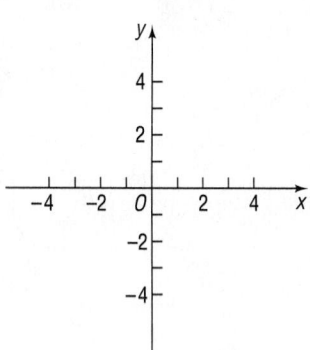

We locate a point on the real number line by assigning it a single real number, called the *coordinate of the point*. For work in a two-dimensional plane, we locate points by using two numbers.

We begin with two real number lines located in the same plane: one horizontal and the other vertical. We call the horizontal line the **x-axis**, the vertical line the **y-axis**, and the point of intersection the **origin O**. See Figure 1. We assign coordinates to every point on these number lines using a convenient scale. In mathematics, we usually use the same scale on each axis; in applications, a different scale is often used on each axis.

The origin O has a value of 0 on both the x-axis and y-axis. Points on the x-axis to the right of O are associated with positive real numbers, and those to the left of O are associated with negative real numbers. Points on the y-axis above O are associated with positive real numbers, and those below O are associated with negative real numbers. In Figure 1, the x-axis and y-axis are labeled as x and y, respectively, and we have used an arrow at the end of each axis to denote the positive direction.

Figure 2

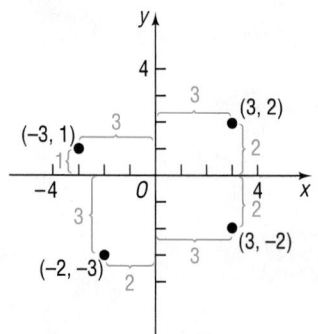

The coordinate system described here is called a **rectangular** or **Cartesian**[*] **coordinate system**. The plane formed by the x-axis and y-axis is sometimes called the **xy-plane**, and the x-axis and y-axis are referred to as the **coordinate axes**.

Any point P in the xy-plane can be located by using an **ordered pair** (x, y) of real numbers. Let x denote the signed distance of P from the y-axis (*signed* means that, if P is to the right of the y-axis, then $x > 0$, and if P is to the left of the y-axis, then $x < 0$); and let y denote the signed distance of P from the x-axis. The ordered pair (x, y), also called the **coordinates** of P, then gives us enough information to locate the point P in the plane.

For example, to locate the point whose coordinates are $(-3, 1)$, go 3 units along the x-axis to the left of O and then go straight up 1 unit. We **plot** this point by placing a dot at this location. See Figure 2, in which the points with coordinates $(-3, 1)$, $(-2, -3)$, $(3, -2)$, and $(3, 2)$ are plotted.

The origin has coordinates $(0, 0)$. Any point on the x-axis has coordinates of the form $(x, 0)$, and any point on the y-axis has coordinates of the form $(0, y)$.

If (x, y) are the coordinates of a point P, then x is called the **x-coordinate**, or **abscissa**, of P and y is the **y-coordinate**, or **ordinate**, of P. We identify the point P by its coordinates (x, y) by writing $P = (x, y)$. Usually, we will simply say "the point (x, y)" rather than "the point whose coordinates are (x, y)."

Figure 3

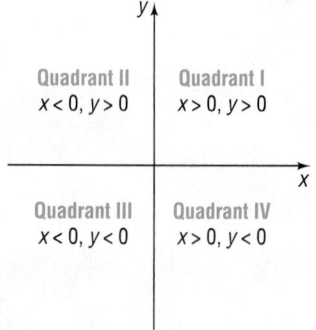

The coordinate axes divide the xy-plane into four sections called **quadrants**, as shown in Figure 3. In quadrant I, both the x-coordinate and the y-coordinate of all points are positive; in quadrant II, x is negative and y is positive; in quadrant III, both x and y are negative; and in quadrant IV, x is positive and y is negative. Points on the coordinate axes belong to no quadrant.

NOW WORK PROBLEM 11.

Figure 4

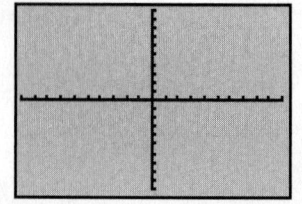

COMMENT: On a graphing calculator, you can set the scale on each axis. Once this has been done, you obtain the **viewing rectangle**. See Figure 4 for a typical viewing rectangle. You should now read Section B.1, *The Viewing Rectangle*, in Appendix B.

[*]Named after René Descartes (1596–1650), a French mathematician, philosopher, and theologian.

1 Use the Distance Formula

If the same units of measurement, such as inches, centimeters, and so on, are used for both the *x*-axis and *y*-axis, then all distances in the *xy*-plane can be measured using this unit of measurement.

EXAMPLE 1	**Finding the Distance between Two Points**

Find the distance *d* between the points $(1, 3)$ and $(5, 6)$.

Solution First we plot the points $(1, 3)$ and $(5, 6)$ and connect them with a straight line. See Figure 5(a). We are looking for the length *d*. We begin by drawing a horizontal line from $(1, 3)$ to $(5, 3)$ and a vertical line from $(5, 3)$ to $(5, 6)$, forming a right triangle, as shown in Figure 5(b). One leg of the triangle is of length 4 (since $|5 - 1| = 4$), and the other is of length 3 (since $|6 - 3| = 3$). By the Pythagorean Theorem, the square of the distance *d* that we seek is

$$d^2 = 4^2 + 3^2 = 16 + 9 = 25$$
$$d = \sqrt{25} = 5$$

Figure 5

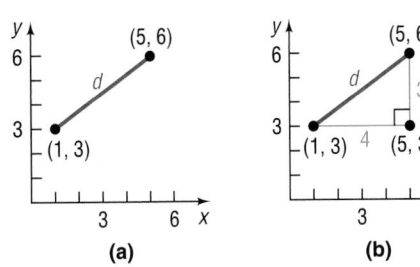

(a) (b)

The **distance formula** provides a straightforward method for computing the distance between two points.

Theorem

In Words

To compute the distance between two points, find the difference of the x-coordinates, square it, and add this to the square of the difference of the y-coordinates. The square root of this sum is the distance.

Distance Formula

The distance between two points $P_1 = (x_1, y_1)$ and $P_2 = (x_2, y_2)$, denoted by $d(P_1, P_2)$, is

$$d(P_1, P_2) = \sqrt{(x_2 - x_1)^2 + (y_2 - y_1)^2} \qquad \textbf{(1)}$$

Proof of the Distance Formula Let (x_1, y_1) denote the coordinates of point P_1 and let (x_2, y_2) denote the coordinates of point P_2. Assume that the line joining P_1 and P_2 is neither horizontal nor vertical. Refer to Figure 6(a). The coordinates of P_3 are (x_2, y_1). The horizontal distance from P_1 to P_3 is the absolute value of the difference of the *x*-coordinates, $|x_2 - x_1|$. The vertical distance from P_3 to P_2 is the absolute value of the difference of the *y*-coordinates, $|y_2 - y_1|$. See Figure 6(b). The distance $d(P_1, P_2)$ that we seek is the length of the hypotenuse of the right triangle, so, by the Pythagorean Theorem, it follows that

$$[d(P_1, P_2)]^2 = |x_2 - x_1|^2 + |y_2 - y_1|^2$$
$$= (x_2 - x_1)^2 + (y_2 - y_1)^2$$
$$d(P_1, P_2) = \sqrt{(x_2 - x_1)^2 + (y_2 - y_1)^2}$$

Figure 6

(a)

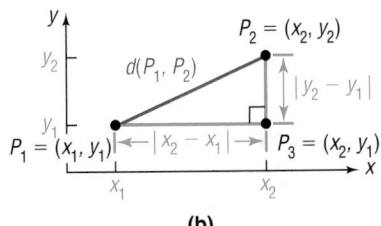

(b)

Now, if the line joining P_1 and P_2 is horizontal, then the y-coordinate of P_1 equals the y-coordinate of P_2; that is, $y_1 = y_2$. Refer to Figure 7(a). In this case, the distance formula (1) still works, because for $y_1 = y_2$, it reduces to

$$d(P_1, P_2) = \sqrt{(x_2 - x_1)^2 + 0^2} = \sqrt{(x_2 - x_1)^2} = |x_2 - x_1|$$

Figure 7

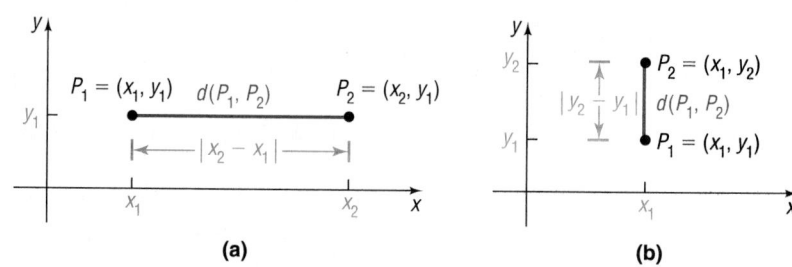

(a) (b)

A similar argument holds if the line joining P_1 and P_2 is vertical. See Figure 7(b). The distance formula is valid in all cases. ∎

EXAMPLE 2 **Using the Distance Formula**

Find the distance d between the points $(-4, 5)$ and $(3, 2)$.

Solution Using the distance formula, equation (1), the solution is obtained as follows:

$$d = \sqrt{[3 - (-4)]^2 + (2 - 5)^2} = \sqrt{7^2 + (-3)^2}$$
$$= \sqrt{49 + 9} = \sqrt{58} \approx 7.62$$ ∎

═══ **NOW WORK PROBLEMS 15 AND 19.**

The distance between two points $P_1 = (x_1, y_1)$ and $P_2 = (x_2, y_2)$ is never a negative number. Furthermore, the distance between two points is 0 only when the points are identical, that is, when $x_1 = x_2$ and $y_1 = y_2$. Also, because $(x_2 - x_1)^2 = (x_1 - x_2)^2$ and $(y_2 - y_1)^2 = (y_1 - y_2)^2$, it makes no difference whether the distance is computed from P_1 to P_2 or from P_2 to P_1; that is, $d(P_1, P_2) = d(P_2, P_1)$.

The introduction to this chapter mentioned that rectangular coordinates enable us to translate geometry problems into algebra problems, and vice versa. The next example shows how algebra (the distance formula) can be used to solve geometry problems.

EXAMPLE 3 **Using Algebra to Solve Geometry Problems**

Consider the three points $A = (-2, 1)$, $B = (2, 3)$, and $C = (3, 1)$.
(a) Plot each point and form the triangle ABC.
(b) Find the length of each side of the triangle.
(c) Verify that the triangle is a right triangle.
(d) Find the area of the triangle.

Solution (a) Points A, B, C and triangle ABC are plotted in Figure 8.

(b) We use the distance formula, equation (1).

$$d(A, B) = \sqrt{[2 - (-2)]^2 + (3 - 1)^2} = \sqrt{16 + 4} = \sqrt{20} = 2\sqrt{5}$$
$$d(B, C) = \sqrt{(3 - 2)^2 + (1 - 3)^2} = \sqrt{1 + 4} = \sqrt{5}$$
$$d(A, C) = \sqrt{[3 - (-2)]^2 + (1 - 1)^2} = \sqrt{25 + 0} = 5$$

Figure 8

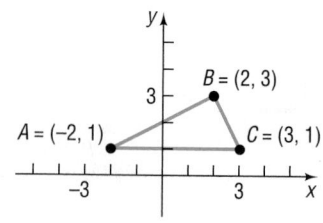

(c) To show that the triangle is a right triangle, we need to show that the sum of the squares of the lengths of two of the sides equals the square of the length of the third side. (Why is this sufficient?) Looking at Figure 8, it seems reasonable to conjecture that the right angle is at vertex B. We shall check to see whether

$$[d(A, B)]^2 + [d(B, C)]^2 = [d(A, C)]^2$$

We find that

$$[d(A, B)]^2 + [d(B, C)]^2 = (2\sqrt{5})^2 + (\sqrt{5})^2$$
$$= 20 + 5 = 25 = [d(A, C)]^2$$

so it follows from the converse of the Pythagorean Theorem that triangle ABC is a right triangle.

(d) Because the right angle is at vertex B, the sides AB and BC form the base and height of the triangle. Its area is

$$\text{Area} = \frac{1}{2}(\text{Base})(\text{Height}) = \frac{1}{2}(2\sqrt{5})(\sqrt{5}) = 5 \text{ square units} \quad \blacksquare$$

NOW WORK PROBLEM 29.

2 Use the Midpoint Formula

Figure 9

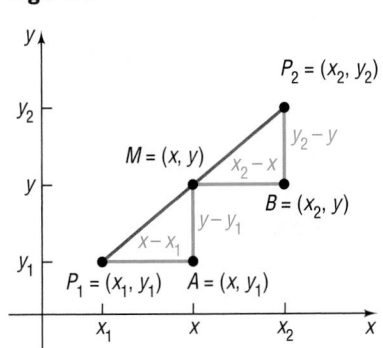

We now derive a formula for the coordinates of the **midpoint of a line segment**. Let $P_1 = (x_1, y_1)$ and $P_2 = (x_2, y_2)$ be the endpoints of a line segment, and let $M = (x, y)$ be the point on the line segment that is the same distance from P_1 as it is from P_2. See Figure 9. The triangles P_1AM and MBP_2 are congruent. [Do you see why? Angle AP_1M = angle BMP_2,* angle P_1MA = angle MP_2B, and $d(P_1, M) = d(M, P_2)$ is given. So, we have angle–side–angle.] Hence, corresponding sides are equal in length. That is,

$$x - x_1 = x_2 - x \quad \text{and} \quad y - y_1 = y_2 - y$$
$$2x = x_1 + x_2 \qquad\qquad 2y = y_1 + y_2$$
$$x = \frac{x_1 + x_2}{2} \qquad\qquad y = \frac{y_1 + y_2}{2}$$

Theorem

In Words
To find the midpoint of a line segment, average the x-coordinates and average the y-coordinates of the endpoints.

Midpoint Formula

The midpoint $M = (x, y)$ of the line segment from $P_1 = (x_1, y_1)$ to $P_2 = (x_2, y_2)$ is

$$M = (x, y) = \left(\frac{x_1 + x_2}{2}, \frac{y_1 + y_2}{2}\right) \tag{2}$$

EXAMPLE 4 **Finding the Midpoint of a Line Segment**

Find the midpoint of a line segment from $P_1 = (-5, 5)$ to $P_2 = (3, 1)$. Plot the points P_1 and P_2 and their midpoint. Check your answer.

Solution We apply the midpoint formula (2) using $x_1 = -5$, $y_1 = 5$, $x_2 = 3$, and $y_2 = 1$. Then the coordinates (x, y) of the midpoint M are

$$x = \frac{x_1 + x_2}{2} = \frac{-5 + 3}{2} = -1 \quad \text{and} \quad y = \frac{y_1 + y_2}{2} = \frac{5 + 1}{2} = 3$$

Figure 10

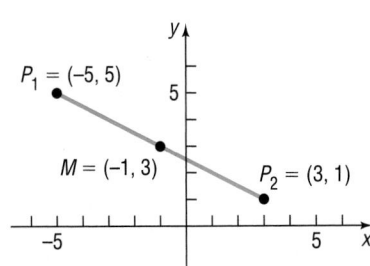

That is, $M = (-1, 3)$. See Figure 10.

*A postulate from geometry states that the transversal $\overline{P_1P_2}$ forms congruent corresponding angles with the parallel line segments $\overline{P_1A}$ and \overline{MB}.

✔ CHECK: Because M is the midpoint, we check the answer by verifying that $d(P_1, M) = d(M, P_2)$:

$$d(P_1, M) = \sqrt{[-1 - (-5)]^2 + (3 - 5)^2} = \sqrt{16 + 4} = \sqrt{20}$$

$$d(M, P_2) = \sqrt{[3 - (-1)]^2 + (1 - 3)^2} = \sqrt{16 + 4} = \sqrt{20}$$ ∎

NOW WORK PROBLEM 35.

F.1 Assess Your Understanding

'Are You Prepared?'

Answers are given at the end of these exercises. If you get a wrong answer, read the pages listed in red.

1. On the real number line the origin is assigned the number _____. (p. A4)

2. If -3 and 5 are the coordinates of two points on the real number line, the distance between these points is _____. (p. A6)

3. If 3 and 4 are the legs of a right triangle, the hypotenuse is _____. (pp. A11–A12)

4. Use the converse of the Pythagorean Theorem to show that a triangle whose sides are of lengths $11, 60$, and 61 is a right triangle. (p. A12)

Concepts and Vocabulary

5. If (x, y) are the coordinates of a point P in the xy-plane, then x is called the _____ of P and y is the _____ of P.

6. The coordinate axes divide the xy-plane into four sections called _____.

7. If three distinct points P, Q, and R all lie on a line and if $d(P, Q) = d(Q, R)$, then Q is called the _____ of the line segment from P to R.

8. *True or False:* The distance between two points is sometimes a negative number.

9. *True or False:* The point $(-1, 4)$ lies in quadrant IV of the Cartesian plane.

10. *True or False:* The midpoint of a line segment is found by averaging the x-coordinates and averaging the y-coordinates of the endpoints.

Skill Building

In Problems 11 and 12, plot each point in the xy-plane. Tell in which quadrant or on what coordinate axis each point lies.

11. (a) $A = (-3, 2)$
 (b) $B = (6, 0)$
 (c) $C = (-2, -2)$
 (d) $D = (6, 5)$
 (e) $E = (0, -3)$
 (f) $F = (6, -3)$

12. (a) $A = (1, 4)$
 (b) $B = (-3, -4)$
 (c) $C = (-3, 4)$
 (d) $D = (4, 1)$
 (e) $E = (0, 1)$
 (f) $F = (-3, 0)$

13. Plot the points $(2, 0)$, $(2, -3)$, $(2, 4)$, $(2, 1)$, and $(2, -1)$. Describe the set of all points of the form $(2, y)$, where y is a real number.

14. Plot the points $(0, 3)$, $(1, 3)$, $(-2, 3)$, $(5, 3)$ and $(-4, 3)$. Describe the set of all points of the form $(x, 3)$, where x is a real number.

In Problems 15–28, find the distance $d(P_1, P_2)$ between the points P_1 and P_2.

15.

16.

17.

18.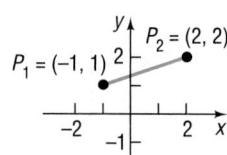

19. $P_1 = (3, -4)$; $P_2 = (5, 4)$

20. $P_1 = (-1, 0)$; $P_2 = (2, 4)$

21. $P_1 = (-3, 2)$; $P_2 = (6, 0)$

22. $P_1 = (2, -3)$; $P_2 = (4, 2)$

23. $P_1 = (4, -3)$; $P_2 = (6, 4)$

24. $P_1 = (-4, -3)$; $P_2 = (6, 2)$

25. $P_1 = (-0.2, 0.3)$; $P_2 = (2.3, 1.1)$

26. $P_1 = (1.2, 2.3)$; $P_2 = (-0.3, 1.1)$

27. $P_1 = (a, b)$; $P_2 = (0, 0)$

28. $P_1 = (a, a)$; $P_2 = (0, 0)$

In Problems 29–34, plot each point and form the triangle ABC. Verify that the triangle is a right triangle. Find its area.

29. $A = (-2, 5)$; $B = (1, 3)$; $C = (-1, 0)$

30. $A = (-2, 5)$; $B = (12, 3)$; $C = (10, -11)$

31. $A = (-5, 3)$; $B = (6, 0)$; $C = (5, 5)$

32. $A = (-6, 3)$; $B = (3, -5)$; $C = (-1, 5)$

33. $A = (4, -3)$; $B = (0, -3)$; $C = (4, 2)$

34. $A = (4, -3)$; $B = (4, 1)$; $C = (2, 1)$

In Problems 35–44, find the midpoint of the line segment joining the points P_1 and P_2.

35. $P_1 = (3, -4)$; $P_2 = (5, 4)$

36. $P_1 = (-2, 0)$; $P_2 = (2, 4)$

37. $P_1 = (-3, 2)$; $P_2 = (6, 0)$

38. $P_1 = (2, -3)$; $P_2 = (4, 2)$

39. $P_1 = (4, -3)$; $P_2 = (6, 1)$

40. $P_1 = (-4, -3)$; $P_2 = (2, 2)$

41. $P_1 = (-0.2, 0.3)$; $P_2 = (2.3, 1.1)$

42. $P_1 = (1.2, 2.3)$; $P_2 = (-0.3, 1.1)$

43. $P_1 = (a, b)$; $P_2 = (0, 0)$

44. $P_1 = (a, a)$; $P_2 = (0, 0)$

Applications and Extensions

45. Find all points having an x-coordinate of 2 whose distance from the point $(-2, -1)$ is 5.

46. Find all points having a y-coordinate of -3 whose distance from the point $(1, 2)$ is 13.

47. Find all points on the x-axis that are 5 units from the point $(4, -3)$.

48. Find all points on the y-axis that are 5 units from the point $(4, 4)$.

49. Geometry The **medians** of a triangle are the line segments from each vertex to the midpoint of the opposite side (see the figure). Find the lengths of the medians of the triangle with vertices at $A = (0, 0)$, $B = (6, 0)$, and $C = (4, 4)$.

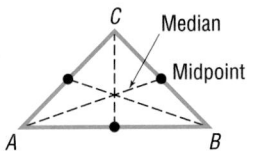

50. Geometry An **equilateral triangle** is one in which all three sides are of equal length. If two vertices of an equilateral triangle are $(0, 4)$ and $(0, 0)$, find the third vertex. How many of these triangles are possible?

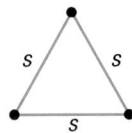

51. Geometry Find the midpoint of each diagonal of a square with side of length s. Draw the conclusion that the diagonals of a square intersect at their midpoints.

[**Hint:** Use $(0, 0)$, $(0, s)$, $(s, 0)$, and (s, s) as the vertices of the square.]

52. Geometry Verify that the points $(0, 0)$, $(a, 0)$, and $\left(\dfrac{a}{2}, \dfrac{\sqrt{3}a}{2}\right)$ are the vertices of an equilateral triangle. Then show that the midpoints of the three sides are the vertices of a second equilateral triangle (refer to Problem 50).

*In Problems 53–56, find the length of each side of the triangle determined by the three points P_1, P_2, and P_3. State whether the triangle is an isosceles triangle, a right triangle, neither of these, or both. (An **isosceles triangle** is one in which at least two of the sides are of equal length.)*

53. $P_1 = (2, 1)$; $P_2 = (-4, 1)$; $P_3 = (-4, -3)$

54. $P_1 = (-1, 4)$; $P_2 = (6, 2)$; $P_3 = (4, -5)$

55. $P_1 = (-2, -1)$; $P_2 = (0, 7)$; $P_3 = (3, 2)$

56. $P_1 = (7, 2)$; $P_2 = (-4, 0)$; $P_3 = (4, 6)$

57. Baseball A major league baseball "diamond" is actually a square, 90 feet on a side (see the figure). What is the distance directly from home plate to second base (the diagonal of the square)?

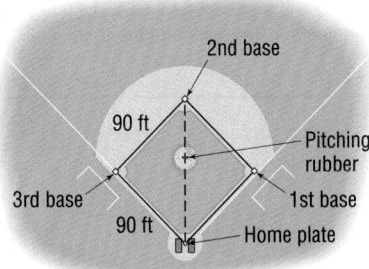

58. Little League Baseball The layout of a Little League playing field is a square, 60 feet on a side.* How far is it directly from home plate to second base (the diagonal of the square)?
*Source: *Little League Baseball, Official Regulations and Playing Rules,* 2005.

59. Baseball Refer to Problem 57. Overlay a rectangular coordinate system on a major league baseball diamond so that the origin is at home plate, the positive x-axis lies in the direction from home plate to first base, and the positive y-axis lies in the direction from home plate to third base.

(a) What are the coordinates of first base, second base, and third base? Use feet as the unit of measurement.

(b) If the right fielder is located at $(310, 15)$, how far is it from there to second base?

(c) If the center fielder is located at $(300, 300)$, how far is it from there to third base?

60. Little League Baseball Refer to Problem 58. Overlay a rectangular coordinate system on a Little League baseball diamond so that the origin is at home plate, the positive x-axis lies in the direction from home plate to first base, and the positive y-axis lies in the direction from home plate to third base.
(a) What are the coordinates of first base, second base, and third base? Use feet as the unit of measurement.
(b) If the right fielder is located at $(180, 20)$, how far is it from there to second base?
(c) If the center fielder is located at $(220, 220)$, how far is it from there to third base?

61. A Dodge Neon and a Mack truck leave an intersection at the same time. The Neon heads east at an average speed of 30 miles per hour, while the truck heads south at an average speed of 40 miles per hour. Find an expression for their distance apart d (in miles) at the end of t hours.

62. A hot-air balloon, headed due east at an average speed of 15 miles per hour and at a constant altitude of 100 feet, passes over an intersection (see the figure). Find an expression for the distance d (measured in feet) from the balloon to the intersection t seconds later.

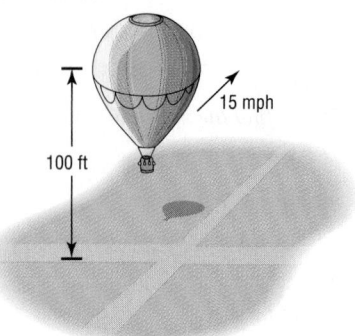

15 mph

100 ft

'Are You Prepared?' Answers

1. 0 **2.** 8 **3.** 5 **4.** $11^2 + 60^2 = 61^2$

F.2 Graphs of Equations in Two Variables; Intercepts; Symmetry

PREPARING FOR THIS SECTION *Before getting started, review the following:*

• Solving Equations (Appendix A, Section A.8, pp. A60–A64)

✏ Now work the 'Are You Prepared?' problems on page 15.

OBJECTIVES
1 Graph Equations by Plotting Points
2 Find Intercepts from a Graph
3 Find Intercepts from an Equation
4 Test an Equation for Symmetry with Respect to the x-Axis, the y-Axis, and the Origin
5 Know How to Graph Key Equations

1 **Graph Equations by Plotting Points**

An **equation in two variables**, say x and y, is a statement in which two expressions involving x and y are equal. The expressions are called the **sides** of the equation. Since an equation is a statement, it may be true or false, depending on the value of the variables. Any values of x and y that result in a true statement are said to **satisfy** the equation.

For example, the following are all equations in two variables x and y:

$$x^2 + y^2 = 5 \qquad 2x - y = 6 \qquad y = 2x + 5 \qquad x^2 = y$$

The first of these, $x^2 + y^2 = 5$, is satisfied for $x = 1, y = 2$, since $1^2 + 2^2 = 1 + 4 = 5$. Other choices of x and y, such as $x = -1, y = -2$, also satisfy this equation. It is not satisfied for $x = 2$ and $y = 3$, since $2^2 + 3^2 = 4 + 9 = 13 \neq 5$.

The **graph of an equation in two variables** x and y consists of the set of points in the xy-plane whose coordinates (x, y) satisfy the equation.

EXAMPLE 1	**Determining Whether a Point Is on the Graph of an Equation**

Determine if the following points are on the graph of the equation $2x - y = 6$.

(a) $(2, 3)$ (b) $(2, -2)$

Solution (a) For the point $(2, 3)$, we check to see if $x = 2$, $y = 3$ satisfies the equation $2x - y = 6$.

$$2x - y = 2(2) - 3 = 4 - 3 = 1 \neq 6$$

The equation is not satisfied, so the point $(2, 3)$ is not on the graph.

(b) For the point $(2, -2)$, we have

$$2x - y = 2(2) - (-2) = 4 + 2 = 6$$

The equation is satisfied, so the point $(2, -2)$ is on the graph. ■

NOW WORK PROBLEM **11.**

| EXAMPLE 2 | **Graphing an Equation by Plotting Points** |

Graph the equation: $y = 2x + 5$

Solution We want to find all points (x, y) that satisfy the equation. To locate some of these points (and thus get an idea of the pattern of the graph), we assign some numbers to x and find corresponding values for y.

Figure 11

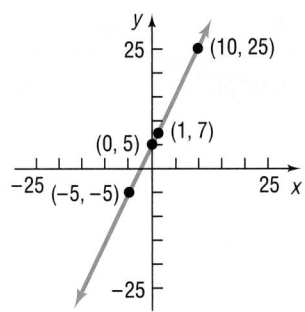

If	Then	Point on Graph
$x = 0$	$y = 2(0) + 5 = 5$	$(0, 5)$
$x = 1$	$y = 2(1) + 5 = 7$	$(1, 7)$
$x = -5$	$y = 2(-5) + 5 = -5$	$(-5, -5)$
$x = 10$	$y = 2(10) + 5 = 25$	$(10, 25)$

By plotting these points and then connecting them, we obtain the graph of the equation (a *line*), as shown in Figure 11. ■

| EXAMPLE 3 | **Graphing an Equation by Plotting Points** |

Graph the equation: $y = x^2$

Solution Table 1 provides several points on the graph. In Figure 12 we plot these points and connect them with a smooth curve to obtain the graph (a *parabola*).

Table 1

x	$y = x^2$	(x, y)
-4	16	$(-4, 16)$
-3	9	$(-3, 9)$
-2	4	$(-2, 4)$
-1	1	$(-1, 1)$
0	0	$(0, 0)$
1	1	$(1, 1)$
2	4	$(2, 4)$
3	9	$(3, 9)$
4	16	$(4, 16)$

Figure 12

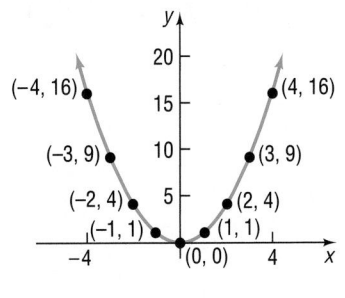

■

The graphs of the equations shown in Figures 11 and 12 do not show all points. For example, in Figure 11, the point $(20, 45)$ is a part of the graph of $y = 2x + 5$, but it is not shown. Since the graph of $y = 2x + 5$ could be extended out as far as we please, we use arrows to indicate that the pattern shown continues. It is important when illustrating a graph to present enough of the graph so that any viewer of the illustration will "see" the rest of it as an obvious continuation of what is actually there. This is referred to as a **complete graph**.

One way to obtain a complete graph of an equation is to plot a sufficient number of points on the graph until a pattern becomes evident. Then these points are connected with a smooth curve following the suggested pattern. But how many points are sufficient? Sometimes knowledge about the equation tells us. For example, we will learn in the next section that, if an equation is of the form $y = mx + b$, then its graph is a line. In this case, only two points are needed to obtain the graph.

One purpose of this book is to investigate the properties of equations in order to decide whether a graph is complete. Sometimes we shall graph equations by plotting points. Shortly, we shall investigate various techniques that will enable us to graph an equation without plotting so many points.

 COMMENT: *Another way to obtain the graph of an equation is to use a graphing utility. Read Section B.2, Using a Graphing Utility to Graph Equations, in Appendix B.* ■

Two techniques that sometimes reduce the number of points required to graph an equation involve finding *intercepts* and checking for *symmetry*.

2 Find Intercepts from a Graph

The points, if any, at which a graph crosses or touches the coordinate axes are called the **intercepts**. See Figure 13. The x-coordinate of a point at which the graph crosses or touches the x-axis is an **x-intercept**, and the y-coordinate of a point at which the graph crosses or touches the y-axis is a **y-intercept**. For a graph to be complete, all its intercepts must be displayed.

Figure 13

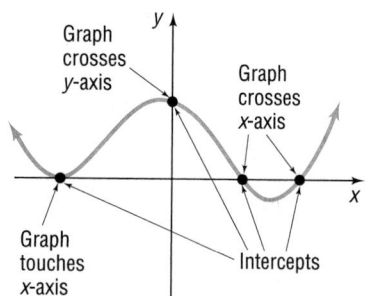

| EXAMPLE 4 | **Finding Intercepts from a Graph** |

Find the intercepts of the graph in Figure 14. What are its x-intercepts? What are its y-intercepts?

Solution The intercepts of the graph are the points

$$(-3, 0), \quad (0, 3), \quad \left(\frac{3}{2}, 0\right), \quad \left(0, -\frac{4}{3}\right), \quad (0, -3.5), \quad (4.5, 0)$$

The x-intercepts are $-3, \dfrac{3}{2}$, and 4.5; the y-intercepts are $-3.5, -\dfrac{4}{3}$, and 3. ■

Figure 14

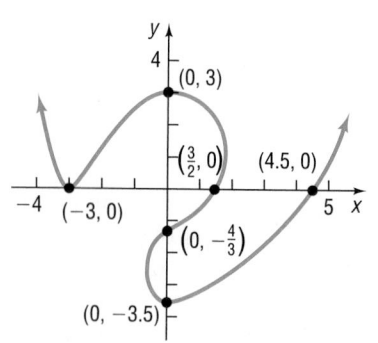

In Example 4, you should notice the following usage: If we do not specify the type of intercept (x- versus y-), then we report the intercept as an ordered pair. However, if we specify the type of intercept, then we only report the coordinate of the intercept. For x-intercepts, we report the x-coordinate of the intercept; for y-intercepts, we report the y-coordinate of the intercept.

NOW WORK PROBLEM 39(a).

3 Find Intercepts from an Equation

The intercepts of a graph can be found from its equation by using the fact that points on the x-axis have y-coordinates equal to 0 and points on the y-axis have x-coordinates equal to 0.

> **Procedure for Finding Intercepts**
> **1.** To find the x-intercept(s), if any, of the graph of an equation, let $y = 0$ in the equation and solve for x.
> **2.** To find the y-intercept(s), if any, of the graph of an equation, let $x = 0$ in the equation and solve for y.

Because the x-intercepts of the graph of an equation are those x-values for which $y = 0$, they are also called the **zeros** (or **roots**) of the equation.

EXAMPLE 5 **Finding Intercepts from an Equation**

Find the x-intercept(s) and the y-intercept(s) of the graph of $y = x^2 - 4$.

Solution To find the x-intercept(s), we let $y = 0$ and obtain the equation

$$x^2 - 4 = 0$$
$$(x + 2)(x - 2) = 0 \qquad \text{Factor.}$$
$$x + 2 = 0 \quad \text{or} \quad x - 2 = 0 \qquad \text{Zero-Product Property}$$
$$x = -2 \quad \text{or} \quad x = 2 \qquad \text{Solve.}$$

The equation has two solutions, -2 and 2. The x-intercepts (or zeros) are -2 and 2.
To find the y-intercept(s), we let $x = 0$ in the equation.

$$y = x^2 - 4$$
$$= 0^2 - 4 = -4$$

The y-intercept is -4. ■

 NOW WORK PROBLEM 21.

COMMENT: For many equations, finding intercepts may not be so easy. In such cases, a graphing utility can be used. Read the first part of Section B.3, *Using a Graphing Utility to Locate Intercepts and Check for Symmetry*, in Appendix B to find out how a graphing utility locates intercepts. ■

4 Test an Equation for Symmetry with Respect to the x-Axis, the y-Axis, and the Origin

We just saw the role that intercepts play in obtaining key points on the graph of an equation. Another helpful tool for graphing equations involves symmetry, particularly symmetry with respect to the x-axis, the y-axis, and the origin.

> A graph is said to be **symmetric with respect to the x-axis** if, for every point (x, y) on the graph, the point $(x, -y)$ is also on the graph.

Figure 15
Symmetry with respect to the x-axis

Figure 15 illustrates the definition. When a graph is symmetric with respect to the x-axis, notice that the part of the graph above the x-axis is a reflection or mirror image of the part below it, and vice versa.

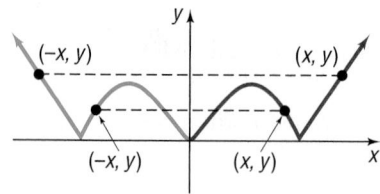

| EXAMPLE 6 | Points Symmetric with Respect to the *x*-Axis |

If a graph is symmetric with respect to the *x*-axis and the point $(3, 2)$ is on the graph, then the point $(3, -2)$ is also on the graph. ∎

Figure 16
Symmetry with respect to the *y*-axis

> A graph is said to be **symmetric with respect to the *y*-axis** if, for every point (x, y) on the graph, the point $(-x, y)$ is also on the graph.

Figure 16 illustrates the definition. When a graph is symmetric with respect to the *y*-axis, notice that the part of the graph to the right of the *y*-axis is a reflection of the part to the left of it, and vice versa.

| EXAMPLE 7 | Points Symmetric with Respect to the *y*-Axis |

If a graph is symmetric with respect to the *y*-axis and the point $(5, 8)$ is on the graph, then the point $(-5, 8)$ is also on the graph. ∎

Figure 17
Symmetry with respect to the origin

> A graph is said to be **symmetric with respect to the origin** if, for every point (x, y) on the graph, the point $(-x, -y)$ is also on the graph.

Figure 17 illustrates the definition. Notice that symmetry with respect to the origin may be viewed in two ways:

1. As a reflection about the *y*-axis, followed by a reflection about the *x*-axis
2. As a projection along a line through the origin so that the distances from the origin are equal

| EXAMPLE 8 | Points Symmetric with Respect to the Origin |

If a graph is symmetric with respect to the origin and the point $(4, 2)$ is on the graph, then the point $(-4, -2)$ is also on the graph. ∎

NOW WORK PROBLEMS 29 AND 39(b).

When the graph of an equation is symmetric with respect to a coordinate axis or the origin, the number of points that you need to plot in order to see the pattern is reduced. For example, if the graph of an equation is symmetric with respect to the *y*-axis, then, once points to the right of the *y*-axis are plotted, an equal number of points on the graph can be obtained by reflecting them about the *y*-axis. Because of this, before we graph an equation, we first want to determine whether it has any symmetry. The following tests are used for this purpose.

Tests for Symmetry

To test the graph of an equation for symmetry with respect to the

x-Axis	Replace *y* by $-y$ in the equation. If an equivalent equation results, the graph of the equation is symmetric with respect to the *x*-axis.
y-Axis	Replace *x* by $-x$ in the equation. If an equivalent equation results, the graph of the equation is symmetric with respect to the *y*-axis.
Origin	Replace *x* by $-x$ and *y* by $-y$ in the equation. If an equivalent equation results, the graph of the equation is symmetric with respect to the origin.

EXAMPLE 9	**Testing an Equation for Symmetry**

Test $y = \dfrac{4x^2}{x^2 + 1}$ for symmetry.

Solution

x-Axis: To test for symmetry with respect to the *x*-axis, replace *y* by $-y$. Since $-y = \dfrac{4x^2}{x^2 + 1}$ is not equivalent to $y = \dfrac{4x^2}{x^2 + 1}$, the graph of the equation is not symmetric with respect to the *x*-axis.

y-Axis: To test for symmetry with respect to the *y*-axis, replace *x* by $-x$. Since $y = \dfrac{4(-x)^2}{(-x)^2 + 1} = \dfrac{4x^2}{x^2 + 1}$ is equivalent to $y = \dfrac{4x^2}{x^2 + 1}$, the graph of the equation is symmetric with respect to the *y*-axis.

Origin: To test for symmetry with respect to the origin, replace *x* by $-x$ and *y* by $-y$.

$$-y = \frac{4(-x)^2}{(-x)^2 + 1} \qquad \text{Replace x by } -x \text{ and y by } -y.$$

$$-y = \frac{4x^2}{x^2 + 1} \qquad \text{Simplify.}$$

$$y = -\frac{4x^2}{x^2 + 1} \qquad \text{Multiply both sides by } -1.$$

Since the result is not equivalent to the original equation, the graph of the equation $y = \dfrac{4x^2}{x^2 + 1}$ is not symmetric with respect to the origin. ∎

Figure 18

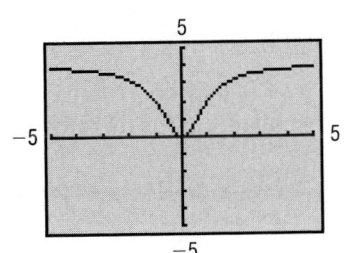

─── **Seeing the Concept** ───

Figure 18 shows the graph of $y = \dfrac{4x^2}{x^2 + 1}$ using a graphing utility. Do you see the symmetry with respect to the *y*-axis?

NOW WORK PROBLEM 59.

5 Know How to Graph Key Equations

The next three examples use intercepts, symmetry, and point plotting to obtain the graphs of key equations. It is important to know the graphs of these key equations because we use them later. The first of these is $y = x^3$.

EXAMPLE 10	**Graphing the Equation $y = x^3$ by Finding Intercepts and Checking for Symmetry**

Graph the equation $y = x^3$ by plotting points. Find any intercepts and check for symmetry first.

Solution

First, we seek the intercepts. When $x = 0$, then $y = 0$; and when $y = 0$, then $x = 0$. The origin $(0, 0)$ is the only intercept. Now we test for symmetry.

x-Axis: Replace *y* by $-y$. Since $-y = x^3$ is not equivalent to $y = x^3$, the graph is not symmetric with respect to the *x*-axis.

y-Axis: Replace *x* by $-x$. Since $y = (-x)^3 = -x^3$ is not equivalent to $y = x^3$, the graph is not symmetric with respect to the *y*-axis.

Origin: Replace *x* by $-x$ and *y* by $-y$. Since $-y = (-x)^3 = -x^3$ is equivalent to $y = x^3$ (multiply both sides by -1), the graph is symmetric with respect to the origin.

To graph $y = x^3$, we use the equation to obtain several points on the graph. Because of the symmetry, we only need to locate points on the graph for which $x \geq 0$. See Table 2. Since $(1, 1)$ is on the graph, and the graph is symmetric with respect to the origin, the point $(-1, -1)$ is also on the graph. Figure 19 shows the graph.

Table 2

x	$y = x^3$	(x, y)
0	0	$(0, 0)$
1	1	$(1, 1)$
2	8	$(2, 8)$
3	27	$(3, 27)$

Figure 19

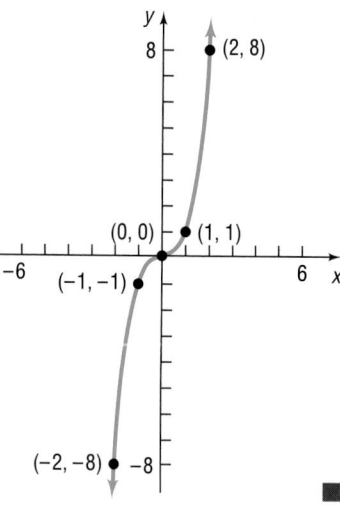

EXAMPLE 11

Graphing the Equation $x = y^2$

Graph the equation $x = y^2$. Find any intercepts and check for symmetry first.

Solution The lone intercept is $(0, 0)$. The graph is symmetric with respect to the x-axis. (Do you see why? Replace y by $-y$.) Figure 20 shows the graph. ■

If we restrict y so that $y \geq 0$, the equation $x = y^2$, $y \geq 0$, may be written equivalently as $y = \sqrt{x}$. The portion of the graph of $x = y^2$ in quadrant I is therefore the graph of $y = \sqrt{x}$. See Figure 21.

Figure 20

Figure 21

Figure 22

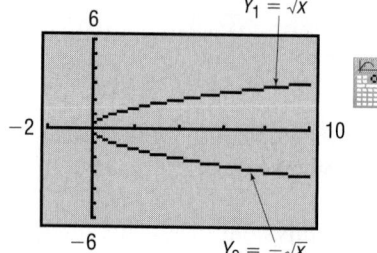

COMMENT: To see the graph of the equation $x = y^2$ on a graphing calculator, you will need to graph two equations: $Y_1 = \sqrt{x}$ and $Y_2 = -\sqrt{x}$. We discuss why in Chapter 1. See Figure 22. ■

EXAMPLE 12

Graphing the Equation $y = \dfrac{1}{x}$

Graph the equation $y = \dfrac{1}{x}$. Find any intercepts and check for symmetry first.

Solution We check for intercepts first. If we let $x = 0$, we obtain 0 in the denominator, which makes y undefined. We conclude that there is no y-intercept. If we let $y = 0$, we get the equation $\dfrac{1}{x} = 0$, which has no solution. We conclude that there is no x-intercept. The graph of $y = \dfrac{1}{x}$ does not cross or touch the coordinate axes.

Table 3

x	$y = \dfrac{1}{x}$	(x, y)
$\dfrac{1}{10}$	10	$\left(\dfrac{1}{10}, 10\right)$
$\dfrac{1}{3}$	3	$\left(\dfrac{1}{3}, 3\right)$
$\dfrac{1}{2}$	2	$\left(\dfrac{1}{2}, 2\right)$
1	1	$(1, 1)$
2	$\dfrac{1}{2}$	$\left(2, \dfrac{1}{2}\right)$
3	$\dfrac{1}{3}$	$\left(3, \dfrac{1}{3}\right)$
10	$\dfrac{1}{10}$	$\left(10, \dfrac{1}{10}\right)$

Next we check for symmetry:

x-Axis Replacing y by $-y$ yields $-y = \dfrac{1}{x}$, which is not equivalent to $y = \dfrac{1}{x}$.

y-Axis Replacing x by $-x$ yields $y = \dfrac{1}{-x} = -\dfrac{1}{x}$, which is not equivalent to $y = \dfrac{1}{x}$.

Origin Replacing x by $-x$ and y by $-y$ yields $-y = -\dfrac{1}{x}$, which is equivalent to $y = \dfrac{1}{x}$. The graph is symmetric with respect to the origin.

Now, we set up Table 3, listing several points on the graph. Because of the symmetry with respect to the origin, we use only positive values of x. From Table 3 we infer that if x is a large and positive number then $y = \dfrac{1}{x}$ is a positive number close to 0. We also infer that if x is a positive number close to 0, then $y = \dfrac{1}{x}$ is a large and positive number. Armed with this information, we can graph the equation.

Figure 23 illustrates some of these points and the graph of $y = \dfrac{1}{x}$. Observe how the absence of intercepts and the existence of symmetry with respect to the origin were utilized. ∎

Figure 23

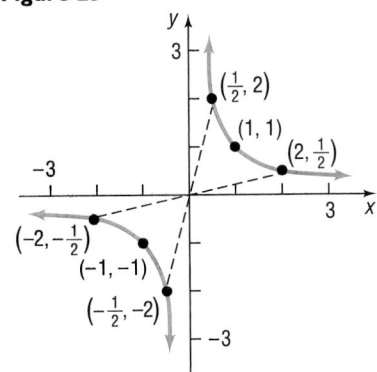

COMMENT: Refer to Example 3 in Appendix B, Section B.3, for the graph of $y = \dfrac{1}{x}$ using a graphing utility.

F.2 Assess Your Understanding

'Are You Prepared?'

Answers are given at the end of these exercises. If you get a wrong answer, read the pages listed in red.

1. Solve the equation $2(x + 3) - 1 = -7$. (pp. A61–A63) **2.** Solve the equation $x^2 - 9 = 0$. (p. A64)

Concepts and Vocabulary

3. The points, if any, at which a graph crosses or touches the co-ordinate axes are called _____.

4. Because the x-intercepts of the graph of an equation are those x-values for which $y = 0$, they are also called _____ or _____.

5. If for every point (x, y) on the graph of an equation the point $(-x, y)$ is also on the graph, then the graph is symmetric with respect to the _____.

6. If the graph of an equation is symmetric with respect to the y-axis and -4 is an x-intercept of this graph, then _____ is also an x-intercept.

7. If the graph of an equation is symmetric with respect to the origin and $(3, -4)$ is a point on the graph, then _____ is also a point on the graph.

8. *True or False:* To find the y-intercepts of the graph of an equation, let $x = 0$ and solve for y.

9. *True or False:* The y-coordinate of a point at which the graph crosses or touches the x-axis is an x-intercept.

10. *True or False:* If a graph is symmetric with respect to the x-axis, then it cannot be symmetric with respect to the y-axis.

Skill Building

In Problems 11–16, tell whether the given points are on the graph of the equation.

11. Equation: $y = x^4 - \sqrt{x}$
 Points: $(0, 0); (1, 1); (-1, 0)$

12. Equation: $y = x^3 - 2\sqrt{x}$
 Points: $(0, 0); (1, 1); (1, -1)$

13. Equation: $y^2 = x^2 + 9$
 Points: $(0, 3); (3, 0); (-3, 0)$

14. Equation: $y^3 = x + 1$
 Points: $(1, 2); (0, 1); (-1, 0)$

15. Equation: $x^2 + y^2 = 4$
 Points: $(0, 2); (-2, 2); \left(\sqrt{2}, \sqrt{2}\right)$

16. Equation: $x^2 + 4y^2 = 4$
 Points: $(0, 1); (2, 0); \left(2, \dfrac{1}{2}\right)$

In Problems 17–28, find the intercepts and graph each equation by plotting points. Be sure to label the intercepts.

17. $y = x + 2$

18. $y = x - 6$

19. $y = 2x + 8$

20. $y = 3x - 9$

21. $y = x^2 - 1$

22. $y = x^2 - 9$

23. $y = -x^2 + 4$

24. $y = -x^2 + 1$

25. $2x + 3y = 6$

26. $5x + 2y = 10$

27. $9x^2 + 4y = 36$

28. $4x^2 + y = 4$

In Problems 29–38, plot each point. Then plot the point that is symmetric to it with respect to (a) the x-axis; (b) the y-axis; (c) the origin.

29. $(3, 4)$

30. $(5, 3)$

31. $(-2, 1)$

32. $(4, -2)$

33. $(5, -2)$

34. $(-1, -1)$

35. $(-3, -4)$

36. $(4, 0)$

37. $(0, -3)$

38. $(-3, 0)$

In Problems 39–50, the graph of an equation is given. (a) Find the intercepts. (b) Indicate whether the graph is symmetric with respect to the x-axis, the y-axis, or the origin.

39.

40.

41.

42.

43.

44.

45.

46.

47.

48.

49.

50.
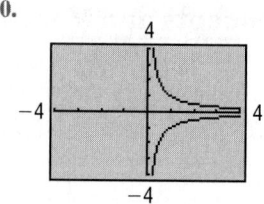

In Problems 51–54, draw a complete graph so that it has the type of symmetry indicated.

51. *y*-axis

52. *x*-axis

53. Origin

54. *y*-axis

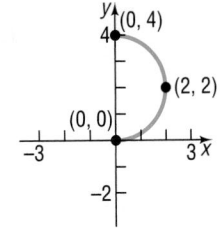

In Problems 55–70, list the intercepts and test for symmetry.

55. $y^2 = x + 4$

56. $y^2 = x + 9$

57. $y = \sqrt[3]{x}$

58. $y = \sqrt[5]{x}$

59. $x^2 + y - 9 = 0$

60. $x^2 - y - 4 = 0$

61. $9x^2 + 4y^2 = 36$

62. $4x^2 + y^2 = 4$

63. $y = x^3 - 27$

64. $y = x^4 - 1$

65. $y = x^2 - 3x - 4$

66. $y = x^2 + 4$

67. $y = \dfrac{3x}{x^2 + 9}$

68. $y = \dfrac{x^2 - 4}{2x}$

69. $y = \dfrac{-x^3}{x^2 - 9}$

70. $y = \dfrac{x^4 + 1}{2x^5}$

In Problems 71–74, draw a quick sketch of each equation.

71. $y = x^3$

72. $x = y^2$

73. $y = \sqrt{x}$

74. $y = \dfrac{1}{x}$

75. If $(3, b)$ is a point on the graph of $y = 4x + 1$, what is *b*?

76. If $(-2, b)$ is a point on the graph of $2x + 3y = 2$, what is *b*?

77. If $(a, 4)$ is a point on the graph of $y = x^2 + 3x$, what is *a*?

78. If $(a, -5)$ is a point on the graph of $y = x^2 + 6x$, what is *a*?

Applications and Extensions

79. Given that the point $(1, 2)$ is on the graph of an equation that is symmetric with respect to the origin, what other point is on the graph?

80. If the graph of an equation is symmetric with respect to the *y*-axis and 6 is an *x*-intercept of this graph, name another *x*-intercept.

81. If the graph of an equation is symmetric with respect to the origin and -4 is an *x*-intercept of this graph, name another *x*-intercept.

82. If the graph of an equation is symmetric with respect to the *x*-axis and 2 is a *y*-intercept, name another *y*-intercept.

Discussion and Writing

In Problem 83, use a graphing utility.

83. (a) Graph $y = \sqrt{x^2}$, $y = x$, $y = |x|$, and $y = (\sqrt{x})^2$, noting which graphs are the same.
(b) Explain why the graphs of $y = \sqrt{x^2}$ and $y = |x|$ are the same.
(c) Explain why the graphs of $y = x$ and $y = (\sqrt{x})^2$ are not the same.
(d) Explain why the graphs of $y = \sqrt{x^2}$ and $y = x$ are not the same.

84. Explain what is meant by a complete graph.

85. Draw a graph of an equation that contains two *x*-intercepts; at one the graph crosses the *x*-axis, and at the other the graph touches the *x*-axis.

86. Make up an equation with the intercepts $(2, 0)$, $(4, 0)$, and $(0, 1)$. Compare your equation with a friend's equation. Comment on any similarities.

87. Draw a graph that contains the points $(-2, -1)$, $(0, 1)$, $(1, 3)$, and $(3, 5)$. Compare your graph with those of other students. Are most of the graphs almost straight lines? How many are "curved"? Discuss the various ways that these points might be connected.

88. An equation is being tested for symmetry with respect to the *x*-axis, the *y*-axis, and the origin. Explain why, if two of these symmetries are present, the remaining one must also be present.

'Are You Prepared?' Answers

1. $\{-6\}$ **2.** $\{-3, 3\}$

F.3 Lines

OBJECTIVES
1 Calculate and Interpret the Slope of a Line
2 Graph Lines Given a Point and the Slope
3 Find the Equation of a Vertical Line
4 Use the Point–Slope Form of a Line; Identify Horizontal Lines
5 Find the Equation of a Line Given Two Points
6 Write the Equation of a Line in Slope–Intercept Form
7 Identify the Slope and y-Intercept of a Line from Its Equation
8 Graph Lines Written in General Form Using Intercepts
9 Find Equations of Parallel Lines
10 Find Equations of Perpendicular Lines

In this section we study a certain type of equation that contains two variables, called a *linear equation*, and its graph, a *line*.

1 Calculate and Interpret the Slope of a Line

Figure 24

Consider the staircase illustrated in Figure 24. Each step contains exactly the same horizontal **run** and the same vertical **rise**. The ratio of the rise to the run, called the *slope*, is a numerical measure of the steepness of the staircase. For example, if the run is increased and the rise remains the same, the staircase becomes less steep. If the run is kept the same, but the rise is increased, the staircase becomes more steep. This important characteristic of a line is best defined using rectangular coordinates.

Let $P = (x_1, y_1)$ and $Q = (x_2, y_2)$ be two distinct points. If $x_1 \neq x_2$, the **slope m** of the nonvertical line L containing P and Q is defined by the formula

$$m = \frac{y_2 - y_1}{x_2 - x_1} \qquad x_1 \neq x_2 \tag{1}$$

If $x_1 = x_2$, L is a **vertical line** and the slope m of L is **undefined** (since this results in division by 0).

Figure 25(a) provides an illustration of the slope of a nonvertical line; Figure 25(b) illustrates a vertical line.

Figure 25

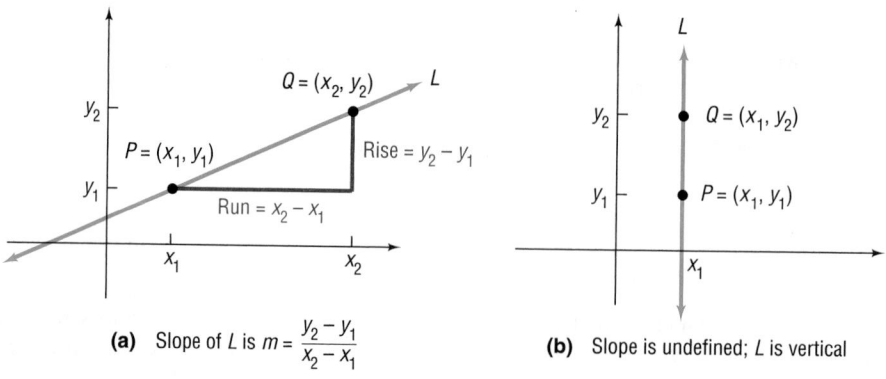

(a) Slope of L is $m = \dfrac{y_2 - y_1}{x_2 - x_1}$

(b) Slope is undefined; L is vertical

As Figure 25(a) illustrates, the slope m of a nonvertical line may be viewed as

$$m = \frac{y_2 - y_1}{x_2 - x_1} = \frac{\text{Rise}}{\text{Run}}$$

We can also express the slope m of a nonvertical line as

$$m = \frac{y_2 - y_1}{x_2 - x_1} = \frac{\text{Change in } y}{\text{Change in } x} = \frac{\Delta y}{\Delta x}$$

That is, the slope m of a nonvertical line L measures the amount that y changes as x changes from x_1 to x_2. This is called the **average rate of change** of y with respect to x.

Two comments about computing the slope of a nonvertical line may prove helpful:

1. Any two distinct points on the line can be used to compute the slope of the line. (See Figure 26 for justification.)

Figure 26
Triangles ABC and PQR are similar (equal angles). Hence, ratios of corresponding sides are proportional so that

Slope using P and Q = $\dfrac{y_2 - y_1}{x_2 - x_1}$

$= \dfrac{d(B, C)}{d(A, C)}$ = Slope using A and B

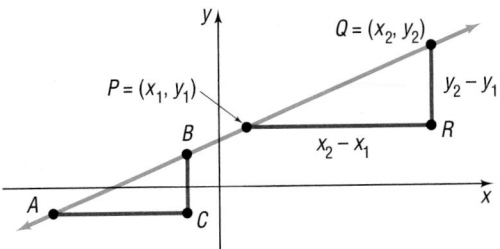

2. The slope of a line may be computed from $P = (x_1, y_1)$ to $Q = (x_2, y_2)$ or from Q to P because

$$\frac{y_2 - y_1}{x_2 - x_1} = \frac{y_1 - y_2}{x_1 - x_2}$$

EXAMPLE 1	**Finding and Interpreting the Slope of a Line Containing Two Points**

The slope m of the line containing the points $(1, 2)$ and $(5, -3)$ may be computed as

$$m = \frac{-3 - 2}{5 - 1} = \frac{-5}{4} = -\frac{5}{4} \quad \text{or as} \quad m = \frac{2 - (-3)}{1 - 5} = \frac{5}{-4} = -\frac{5}{4}$$

For every 4-unit change in x, y will change by -5 units. That is, if x increases by 4 units, then y will decrease by 5 units. The average rate of change of y with respect to x is $-\dfrac{5}{4}$. ∎

 NOW WORK PROBLEMS 11 AND 17.

To get a better idea of the meaning of the slope m of a line L, consider the following example.

EXAMPLE 2	**Finding the Slopes of Various Lines Containing the Same Point (2, 3)**

Compute the slopes of the lines L_1, L_2, L_3, and L_4 containing the following pairs of points. Graph all four lines on the same set of coordinate axes.

$$L_1: \quad P = (2, 3) \qquad Q_1 = (-1, -2)$$
$$L_2: \quad P = (2, 3) \qquad Q_2 = (3, -1)$$
$$L_3: \quad P = (2, 3) \qquad Q_3 = (5, 3)$$
$$L_4: \quad P = (2, 3) \qquad Q_4 = (2, 5)$$

Figure 27

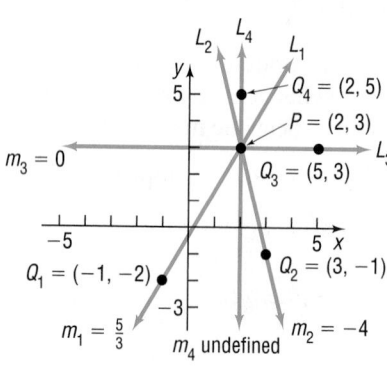

$$m_1 = \frac{5}{3}$$
$$m_4 \text{ undefined}$$
$$m_2 = -4$$

Solution Let m_1, m_2, m_3, and m_4 denote the slopes of the lines L_1, L_2, L_3, and L_4, respectively. Then

$$m_1 = \frac{-2 - 3}{-1 - 2} = \frac{-5}{-3} = \frac{5}{3}$$

A rise of 5 divided by a run of 3

$$m_2 = \frac{-1 - 3}{3 - 2} = \frac{-4}{1} = -4$$

$$m_3 = \frac{3 - 3}{5 - 2} = \frac{0}{3} = 0$$

m_4 is undefined because $x_1 = x_2 = 2$

The graphs of these lines are given in Figure 27.

Figure 27 illustrates the following facts:

1. When the slope of a line is positive, the line slants upward from left to right (L_1).
2. When the slope of a line is negative, the line slants downward from left to right (L_2).
3. When the slope is 0, the line is horizontal (L_3).
4. When the slope is undefined, the line is vertical (L_4).

—— **Seeing the Concept** ——

On the same square screen, graph the following equations:

Figure 28

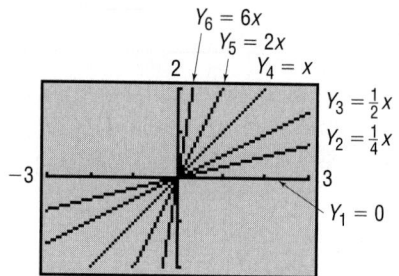

$Y_1 = 0$	*Slope of line is 0.*
$Y_2 = \frac{1}{4}x$	*Slope of line is $\frac{1}{4}$.*
$Y_3 = \frac{1}{2}x$	*Slope of line is $\frac{1}{2}$.*
$Y_4 = x$	*Slope of line is 1.*
$Y_5 = 2x$	*Slope of line is 2.*
$Y_6 = 6x$	*Slope of line is 6.*

See Figure 28.

—— **Seeing the Concept** ——

On the same square screen, graph the following equations:

Figure 29

$Y_1 = 0$	*Slope of line is 0.*
$Y_2 = -\frac{1}{4}x$	*Slope of line is $-\frac{1}{4}$.*
$Y_3 = -\frac{1}{2}x$	*Slope of line is $-\frac{1}{2}$.*
$Y_4 = -x$	*Slope of line is -1.*
$Y_5 = -2x$	*Slope of line is -2.*
$Y_6 = -6x$	*Slope of line is -6.*

See Figure 29.

Figures 28 and 29 illustrate that the closer the line is to the vertical position, the greater the magnitude of the slope.

2 Graph Lines Given a Point and the Slope

The next example illustrates how the slope of a line can be used to graph the line.

| EXAMPLE 3 | Graphing a Line Given a Point and a Slope |

Draw a graph of the line that contains the point $(3, 2)$ and has a slope of:

(a) $\dfrac{3}{4}$ (b) $-\dfrac{4}{5}$

Solution (a) Slope $= \dfrac{\text{Rise}}{\text{Run}}$. The fact that the slope is $\dfrac{3}{4}$ means that for every horizontal movement (run) of 4 units to the right there will be a vertical movement (rise) of 3 units. If we start at the given point $(3, 2)$ and move 4 units to the right and 3 units up, we reach the point $(7, 5)$. By drawing the line through this point and the point $(3, 2)$, we have the graph. See Figure 30.

Figure 30

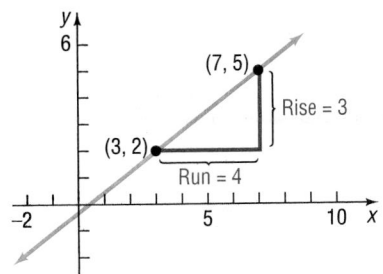

(b) The fact that the slope is

$$-\frac{4}{5} = \frac{-4}{5} = \frac{\text{Rise}}{\text{Run}}$$

means that for every horizontal movement of 5 units to the right there will be a corresponding vertical movement of -4 units (a downward movement). If we start at the given point $(3, 2)$ and move 5 units to the right and then 4 units down, we arrive at the point $(8, -2)$. By drawing the line through these points, we have the graph. See Figure 31.

Alternatively, we can set

$$-\frac{4}{5} = \frac{4}{-5} = \frac{\text{Rise}}{\text{Run}}$$

Figure 31

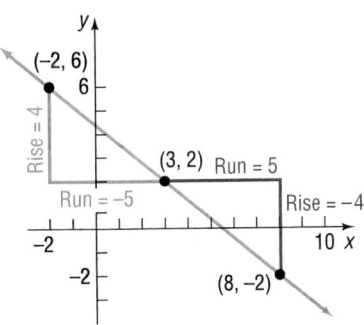

so that for every horizontal movement of -5 units (a movement to the left) there will be a corresponding vertical movement of 4 units (upward). This approach brings us to the point $(-2, 6)$, which is also on the graph shown in Figure 31. ∎

✏ NOW WORK PROBLEM 23.

3 Find the Equation of a Vertical Line

Now that we have discussed the slope of a line, we are ready to derive equations of lines. As we shall see, there are several forms of the equation of a line. Let's start with an example.

| EXAMPLE 4 | Graphing a Line |

Graph the equation: $x = 3$

Solution To graph $x = 3$, recall that we are looking for all points (x, y) in the plane for which $x = 3$. No matter what y-coordinate is used, the corresponding x-coordinate always equals 3. Consequently, the graph of the equation $x = 3$ is a vertical line with x-intercept 3 and undefined slope. See Figure 32.

Figure 32

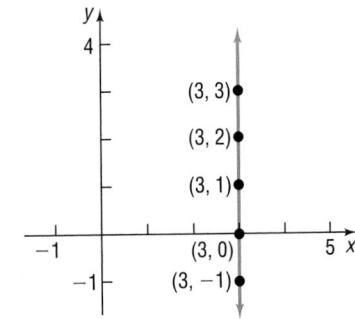

As suggested by Example 3, we have the following result:

Theorem

Equation of a Vertical Line

A vertical line is given by an equation of the form

$$x = a$$

where a is the x-intercept.

 COMMENT: To graph an equation using a graphing utility, we need to express the equation in the form y = {expression in x}. But x = 3 cannot be put in this form. To overcome this, most graphing utilities have special commands for drawing vertical lines. LINE, PLOT, and VERT are among the more common ones. Consult your manual to determine the correct methodology for your graphing utility. ∎

Figure 33

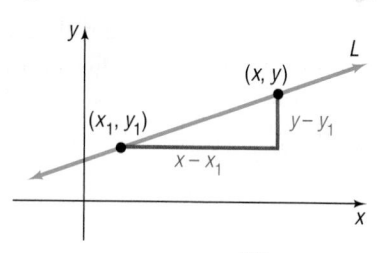

4 **Use the Point–Slope Form of a Line; Identify Horizontal Lines**

Now let L be a nonvertical line with slope m and containing the point (x_1, y_1). See Figure 33. For any other point (x, y) on L, we have

$$m = \frac{y - y_1}{x - x_1} \quad \text{or} \quad y - y_1 = m(x - x_1)$$

Theorem

Point–Slope Form of an Equation of a Line

An equation of a nonvertical line with slope m that contains the point (x_1, y_1) is

$$y - y_1 = m(x - x_1) \qquad \textbf{(2)}$$

EXAMPLE 5

Using the Point–Slope Form of a Line

An equation of the line with slope 4 and containing the point $(1, 2)$ can be found by using the point–slope form with $m = 4$, $x_1 = 1$, and $y_1 = 2$.

$$y - y_1 = m(x - x_1)$$
$$y - 2 = 4(x - 1) \qquad m = 4, x_1 = 1, y_1 = 2$$
$$y = 4x - 2$$

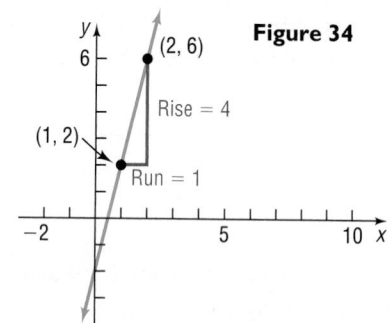

Figure 34

See Figure 34 for the graph. ∎

EXAMPLE 6

Finding the Equation of a Horizontal Line

Find an equation of the horizontal line containing the point $(3, 2)$.

Solution

Because all the y-values are equal on a horizontal line, the slope of a horizontal line is 0. To get an equation, we use the point–slope form with $m = 0$, $x_1 = 3$, and $y_1 = 2$.

$$y - y_1 = m(x - x_1)$$
$$y - 2 = 0 \cdot (x - 3) \qquad m = 0, x_1 = 3, \text{ and } y_1 = 2$$
$$y - 2 = 0$$
$$y = 2$$

Figure 35

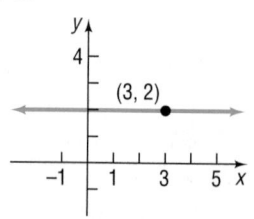

See Figure 35 for the graph. ∎

As suggested by Example 6, we have the following result:

Theorem

Equation of a Horizontal Line

A horizontal line is given by an equation of the form

$$y = b$$

where b is the y-intercept.

5 Find the Equation of a Line Given Two Points

We use the slope formula and the point–slope form of a line to find the equation given two points.

EXAMPLE 7 **Finding an Equation of a Line Given Two Points**

Find an equation of the line L containing the points $(2, 3)$ and $(-4, 5)$. Graph the line L.

Solution We first compute the slope of the line.

$$m = \frac{5 - 3}{-4 - 2} = \frac{2}{-6} = -\frac{1}{3}$$

Figure 36

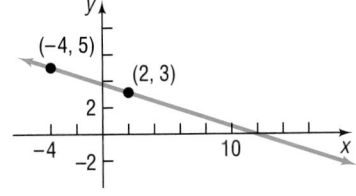

We use the point $(2, 3)$ and the slope $m = -\dfrac{1}{3}$ to get the point–slope form of the equation of the line.

$$y - 3 = -\frac{1}{3}(x - 2)$$

See Figure 36 for the graph. ∎

In the solution to Example 7, we could have used the other point, $(-4, 5)$, instead of the point $(2, 3)$. The equation that results, although it looks different, is equivalent to the equation that we obtained in the example. (Try it for yourself.)

NOW WORK PROBLEM **37**.

6 Write the Equation of a Line in Slope–Intercept Form

Another useful equation of a line is obtained when the slope m and y-intercept b are known. In this event, we know both the slope m of the line and a point $(0, b)$ on the line; then we may use the point–slope form, equation (2), to obtain the following equation:

$$y - b = m(x - 0) \quad \text{or} \quad y = mx + b$$

Theorem

Slope–Intercept Form of an Equation of a Line

An equation of a line L with slope m and y-intercept b is

$$y = mx + b \tag{3}$$

NOW WORK PROBLEM **45** (EXPRESS ANSWER IN SLOPE–INTERCEPT FORM).

Figure 37
$y = mx + 2$

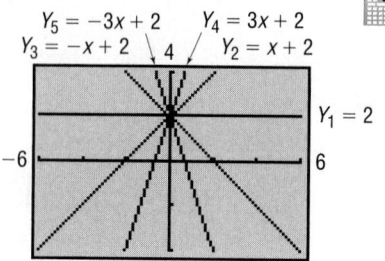

To see the role that the slope m plays, graph the following lines on the same square screen.

$$Y_1 = 2$$
$$Y_2 = x + 2$$
$$Y_3 = -x + 2$$
$$Y_4 = 3x + 2$$
$$Y_5 = -3x + 2$$

See Figure 37. What do you conclude about the lines $y = mx + 2$?

Figure 38
$y = 2x + b$

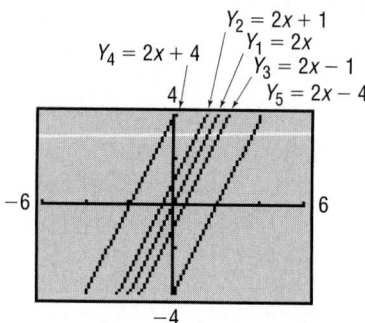

To see the role of the y-intercept b, graph the following lines on the same square screen.

$$Y_1 = 2x$$
$$Y_2 = 2x + 1$$
$$Y_3 = 2x - 1$$
$$Y_4 = 2x + 4$$
$$Y_5 = 2x - 4$$

See Figure 38. What do you conclude about the lines $y = 2x + b$?

7 Identify the Slope and y-Intercept of a Line from Its Equation

When the equation of a line is written in slope–intercept form, it is easy to find the slope m and y-intercept b of the line. For example, suppose that the equation of a line is

$$y = -2x + 3$$

Compare it to $y = mx + b$.

$$y = -2x + 3$$
$$\quad\quad\uparrow\quad\quad\uparrow$$
$$y = \quad mx + b$$

The slope of this line is -2 and its y-intercept is 3.

NOW WORK PROBLEM **71.**

EXAMPLE 8 | **Finding the Slope and y-Intercept**

Find the slope m and y-intercept b of the equation $2x + 4y = 8$. Graph the equation.

Solution

To obtain the slope and y-intercept, we write the equation in slope–intercept form by solving for y.

$$2x + 4y = 8$$
$$4y = -2x + 8$$
$$y = -\frac{1}{2}x + 2 \quad\quad y = mx + b$$

Figure 39
$2x + 4y = 8$

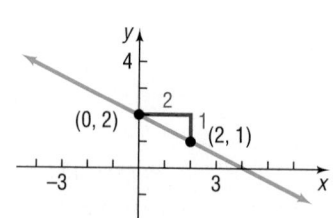

The coefficient of x, $-\frac{1}{2}$, is the slope, and the y-intercept is 2. We can graph the line using the fact that the y-intercept is 2 and the slope is $-\frac{1}{2}$. Then, starting at the point $(0, 2)$, go to the right 2 units and then down 1 unit to the point $(2, 1)$. See Figure 39. ∎

NOW WORK PROBLEM **77.**

8 Graph Lines Written in General Form Using Intercepts

The form of the equation of the line in Example 8, $2x + 4y = 8$, is called the *general form*.

> The equation of a line L is in **general form*** when it is written as
>
> $$Ax + By = C \qquad (4)$$
>
> where A, B, and C are real numbers and A and B are not both 0.

When we want to graph an equation that is written in general form, we can solve the equation for y and write the equation in slope–intercept form as we did in Example 8. Another approach to graphing the equation would be to find its intercepts. Remember, the intercepts of the graph of an equation are the points where the graph crosses or touches a coordinate axis.

EXAMPLE 9	**Graphing an Equation in General Form Using Its Intercepts**

Graph the equation $2x + 4y = 8$ by finding its intercepts.

Solution To obtain the x-intercept, let $y = 0$ in the equation and solve for x.

$$2x + 4y = 8$$
$$2x + 4(0) = 8 \qquad \text{Let } y = 0.$$
$$2x = 8$$
$$x = 4 \qquad \text{Divide both sides by 2.}$$

The x-intercept is 4 and the point $(4, 0)$ is on the graph of the equation.
To obtain the y-intercept, let $x = 0$ in the equation and solve for y.

$$2x + 4y = 8$$
$$2(0) + 4y = 8 \qquad \text{Let } x = 0.$$
$$4y = 8$$
$$y = 2 \qquad \text{Divide both sides by 4.}$$

Figure 40

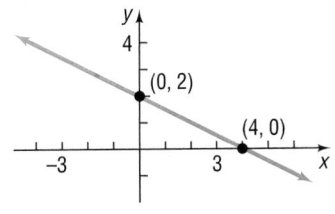

The y-intercept is 2 and the point $(0, 2)$ is on the graph of the equation.
We plot the points $(4, 0)$ and $(0, 2)$ in a Cartesian plane and draw a line through the points. See Figure 40. ■

NOW WORK PROBLEM 91.

Every line has an equation that is equivalent to an equation written in general form. For example, a vertical line whose equation is

$$x = a$$

can be written in the general form

$$1 \cdot x + 0 \cdot y = a \qquad A = 1, B = 0, C = a$$

A horizontal line whose equation is

$$y = b$$

can be written in the general form

$$0 \cdot x + 1 \cdot y = b \qquad A = 0, B = 1, C = b$$

*Some books use the term **standard form**.

Lines that are neither vertical nor horizontal have general equations of the form

$$Ax + By = C \qquad A \neq 0 \text{ and } B \neq 0$$

Because the equation of every line can be written in general form, any equation equivalent to equation (4) is called a **linear equation**.

9 Find Equations of Parallel Lines

When two lines (in the plane) do not intersect (that is, they have no points in common), they are said to be **parallel**. Look at Figure 41. There we have drawn two lines and have constructed two right triangles by drawing sides parallel to the coordinate axes. These lines are parallel if and only if the right triangles are similar. (Do you see why? Two angles are equal.) And the triangles are similar if and only if the ratios of corresponding sides are equal.

Figure 41
The distinct lines are parallel if and only if their slopes are equal.

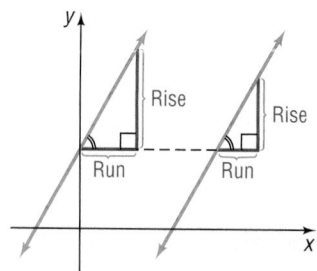

This suggests the following result:

Theorem

> **Criterion for Parallel Lines**
>
> Two nonvertical lines are parallel if and only if their slopes are equal and they have different y-intercepts.

The use of the words "if and only if" in the preceding theorem means that actually two statements are being made, one the converse of the other.

If two nonvertical lines are parallel, then their slopes are equal and they have different y-intercepts.

If two nonvertical lines have equal slopes and they have different y-intercepts, then they are parallel.

EXAMPLE 10 **Showing That Two Lines Are Parallel**

Show that the lines given by the following equations are parallel:

$$L_1: \quad 2x + 3y = 6, \qquad L_2: \quad 4x + 6y = 0$$

Solution To determine whether these lines have equal slopes and different y-intercepts, we write each equation in slope–intercept form:

Figure 42
Parallel lines

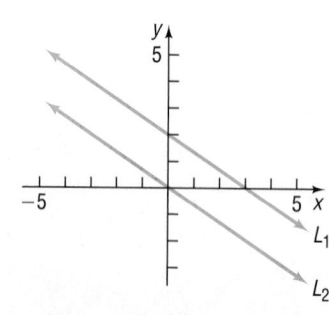

$$
\begin{array}{ll}
L_1: \quad 2x + 3y = 6 & \qquad L_2: \quad 4x + 6y = 0 \\
 3y = -2x + 6 & \qquad 6y = -4x \\
 y = -\dfrac{2}{3}x + 2 & \qquad y = -\dfrac{2}{3}x
\end{array}
$$

Slope $= -\dfrac{2}{3}$; y-intercept $= 2$ Slope $= -\dfrac{2}{3}$; y-intercept $= 0$

Because these lines have the same slope, $-\dfrac{2}{3}$, but different y-intercepts, the lines are parallel. See Figure 42. ∎

EXAMPLE 11	**Finding a Line That Is Parallel to a Given Line**

Find an equation for the line that contains the point $(2, -3)$ and is parallel to the line $2x + y = 6$.

Solution Since the two lines are to be parallel, the slope of the line that we seek equals the slope of the line $2x + y = 6$. We begin by writing the equation of the line $2x + y = 6$ in slope–intercept form.

$$2x + y = 6$$
$$y = -2x + 6$$

Figure 43

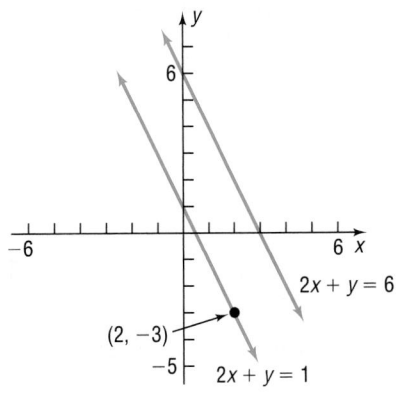

The slope is -2. Since the line that we seek contains the point $(2, -3)$, we use the point–slope form to obtain

$$y - y_1 = m(x - x_1) \quad \text{Point–slope form}$$
$$y - (-3) = -2(x - 2) \quad m = -2, x_1 = 2, y_1 = -3$$
$$y + 3 = -2x + 4 \quad \text{Simplify}$$
$$y = -2x + 1 \quad \text{Slope–intercept form}$$
$$2x + y = 1 \quad \text{General form}$$

This line is parallel to the line $2x + y = 6$ and contains the point $(2, -3)$. See Figure 43. ∎

NOW WORK PROBLEM 59.

10 Find Equations of Perpendicular Lines

When two lines intersect at a right angle (90°), they are said to be **perpendicular**. See Figure 44.

Figure 44
Perpendicular lines

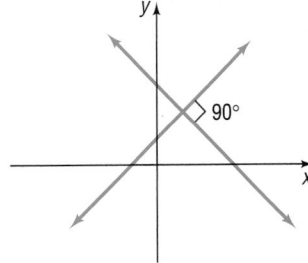

The following result gives a condition, in terms of their slopes, for two lines to be perpendicular.

Theorem

Criterion for Perpendicular Lines

Two nonvertical lines are perpendicular if and only if the product of their slopes is -1.

Here we shall prove the "only if" part of the statement:

If two nonvertical lines are perpendicular, then the product of their slopes is -1.

You are asked to prove the "if" part of the theorem; that is:

If two nonvertical lines have slopes whose product is -1, then the lines are perpendicular.

Figure 45

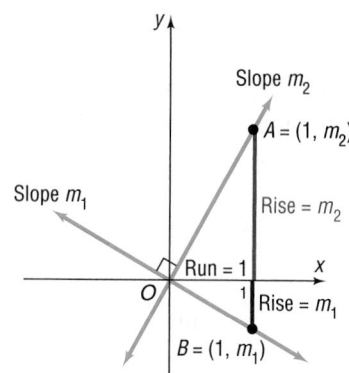

Proof Let m_1 and m_2 denote the slopes of the two lines. There is no loss in generality (that is, neither the angle nor the slopes are affected) if we situate the lines so that they meet at the origin. See Figure 45. The point $A = (1, m_2)$ is on the line having slope m_2, and the point $B = (1, m_1)$ is on the line having slope m_1. (Do you see why this must be true?)

Suppose that the lines are perpendicular. Then triangle OAB is a right triangle. As a result of the Pythagorean Theorem, it follows that

$$[d(O, A)]^2 + [d(O, B)]^2 = [d(A, B)]^2 \qquad \textbf{(5)}$$

By the distance formula, we can write the squares of these distances as

$$[d(O, A)]^2 = (1 - 0)^2 + (m_2 - 0)^2 = 1 + m_2^2$$

$$[d(O, B)]^2 = (1 - 0)^2 + (m_1 - 0)^2 = 1 + m_1^2$$

$$[d(A, B)]^2 = (1 - 1)^2 + (m_2 - m_1)^2 = m_2^2 - 2m_1m_2 + m_1^2$$

Using these facts in equation (5), we get

$$(1 + m_2^2) + (1 + m_1^2) = m_2^2 - 2m_1m_2 + m_1^2$$

which, upon simplification, can be written as

$$m_1m_2 = -1$$

If the lines are perpendicular, the product of their slopes is -1. ∎

You may find it easier to remember the condition for two nonvertical lines to be perpendicular by observing that the equality $m_1m_2 = -1$ means that m_1 and m_2 are negative reciprocals of each other; that is, either $m_1 = -\dfrac{1}{m_2}$ or $m_2 = -\dfrac{1}{m_1}$.

EXAMPLE 12 | **Finding the Slope of a Line Perpendicular to Another Line**

If a line has slope $\dfrac{3}{2}$, any line having slope $-\dfrac{2}{3}$ is perpendicular to it. ∎

EXAMPLE 13 | **Finding the Equation of a Line Perpendicular to a Given Line**

Find an equation of the line that contains the point $(1, -2)$ and is perpendicular to the line $x + 3y = 6$. Graph the two lines.

Solution

We first write the equation of the given line in slope–intercept form to find its slope.

$$x + 3y = 6$$
$$3y = -x + 6 \qquad \textit{Proceed to solve for y.}$$
$$y = -\frac{1}{3}x + 2 \qquad \textit{Place in the form } y = mx + b.$$

Figure 46

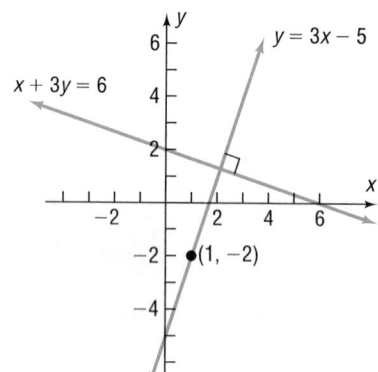

The given line has slope $-\dfrac{1}{3}$. Any line perpendicular to this line will have slope 3.

Because we require the point $(1, -2)$ to be on this line with slope 3, we use the point–slope form of the equation of a line.

$$y - y_1 = m(x - x_1) \qquad \textit{Point–slope form}$$
$$y - (-2) = 3(x - 1) \qquad \textit{m = 3, } x_1 = 1, y_1 = -2.$$

To obtain other forms of the equation, we proceed as follows:

$$y + 2 = 3(x - 1)$$
$$y + 2 = 3x - 3 \qquad \textit{Simplify}$$
$$y = 3x - 5 \qquad \textit{Slope–intercept form}$$
$$3x - y = 5 \qquad \textit{General form}$$

Figure 46 shows the graphs. ∎

━━━━ **NOW WORK PROBLEM 65.**

F.3 Assess Your Understanding

Concepts and Vocabulary

1. The slope of a vertical line is _____; the slope of a horizontal line is _____.

2. For the line $2x + 3y = 6$, the x-intercept is _____ and the y-intercept is _____.

3. A horizontal line is given by an equation of the form _____, where b is the _____.

4. *True or False:* Vertical lines have an undefined slope.

5. *True or False:* The slope of the line $2y = 3x + 5$ is 3.

6. *True or False:* The point $(1, 2)$ is on the line $2x + y = 4$.

7. Two nonvertical lines have slopes m_1 and m_2, respectively. The lines are parallel if _____ and the _____ are unequal; the lines are perpendicular if _____.

8. The lines $y = 2x + 3$ and $y = ax + 5$ are parallel if $a = $ _____.

9. The lines $y = 2x - 1$ and $y = ax + 2$ are perpendicular if $a = $ _____.

10. *True or False:* Perpendicular lines have slopes that are reciprocals of one another.

Skill Building

In Problems 11–14, (a) find the slope of the line and (b) interpret the slope.

11.
12.
13.
14.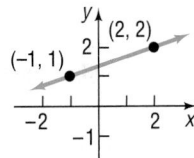

In Problems 15–22, plot each pair of points and determine the slope of the line containing them. Graph the line.

15. $(2, 3)$; $(4, 0)$
16. $(4, 2)$; $(3, 4)$
17. $(-2, 3)$; $(2, 1)$
18. $(-1, 1)$; $(2, 3)$

19. $(-3, -1)$; $(2, -1)$
20. $(4, 2)$; $(-5, 2)$
21. $(-1, 2)$; $(-1, -2)$
22. $(2, 0)$; $(2, 2)$

In Problems 23–30, graph the line containing the point P and having slope m.

23. $P = (1, 2)$; $m = 3$
24. $P = (2, 1)$; $m = 4$
25. $P = (2, 4)$; $m = -\dfrac{3}{4}$
26. $P = (1, 3)$; $m = -\dfrac{2}{5}$

27. $P = (-1, 3)$; $m = 0$
28. $P = (2, -4)$; $m = 0$
29. $P = (0, 3)$; slope undefined
30. $P = (-2, 0)$; slope undefined

In Problems 31–36, the slope and a point on a line are given. Use this information to locate three additional points on the line. Answers may vary.

[**Hint:** *It is not necessary to find the equation of the line. See Example 2.*]

31. Slope 4; point $(1, 2)$
32. Slope 2; point $(-2, 3)$
33. Slope $-\dfrac{3}{2}$; point $(2, -4)$

34. Slope $\dfrac{4}{3}$; point $(-3, 2)$
35. Slope -2; point $(-2, -3)$
36. Slope -1; point $(4, 1)$

In Problems 37–44, find an equation of the line L.

37.
38.
39.
40.

41.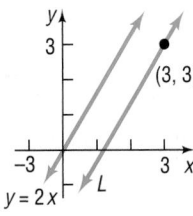

$y = 2x$

L is parallel to $y = 2x$

42.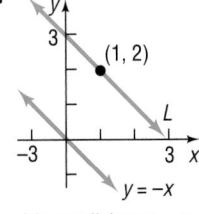

$y = -x$

L is parallel to $y = -x$

43.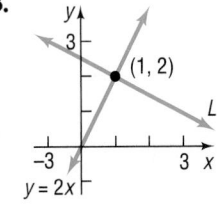

$y = 2x$

L is perpendicular to $y = 2x$

44.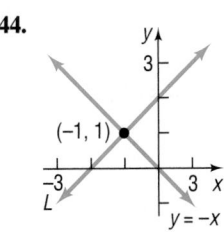

$y = -x$

L is perpendicular to $y = -x$

In Problems 45–70, find an equation for the line with the given properties. Express your answer using either the general form or the slope–intercept form of the equation of a line, whichever you prefer.

45. Slope $= 3$; containing the point $(-2, 3)$

46. Slope $= 2$; containing the point $(4, -3)$

47. Slope $= -\dfrac{2}{3}$; containing the point $(1, -1)$

48. Slope $= \dfrac{1}{2}$; containing the point $(3, 1)$

49. Containing the points $(1, 3)$ and $(-1, 2)$

50. Containing the points $(-3, 4)$ and $(2, 5)$

51. Slope $= -3$; y-intercept $= 3$

52. Slope $= -2$; y-intercept $= -2$

53. x-intercept $= 2$; y-intercept $= -1$

54. x-intercept $= -4$; y-intercept $= 4$

55. Slope undefined; containing the point $(2, 4)$

56. Slope undefined; containing the point $(3, 8)$

57. Horizontal; containing the point $(-3, 2)$

58. Vertical; containing the point $(4, -5)$

59. Parallel to the line $y = 2x$; containing the point $(-1, 2)$

60. Parallel to the line $y = -3x$; containing the point $(-1, 2)$

61. Parallel to the line $2x - y = -2$; containing the point $(0, 0)$

62. Parallel to the line $x - 2y = -5$; containing the point $(0, 0)$

63. Parallel to the line $x = 5$; containing the point $(4, 2)$

64. Parallel to the line $y = 5$; containing the point $(4, 2)$

65. Perpendicular to the line $y = \dfrac{1}{2}x + 4$; containing the point $(1, -2)$

66. Perpendicular to the line $y = 2x - 3$; containing the point $(1, -2)$

67. Perpendicular to the line $2x + y = 2$; containing the point $(-3, 0)$

68. Perpendicular to the line $x - 2y = -5$; containing the point $(0, 4)$

69. Perpendicular to the line $x = 8$; containing the point $(3, 4)$

70. Perpendicular to the line $y = 8$; containing the point $(3, 4)$

In Problems 71–90, find the slope and y-intercept of each line. Graph the line.

71. $y = 2x + 3$

72. $y = -3x + 4$

73. $\dfrac{1}{2}y = x - 1$

74. $\dfrac{1}{3}x + y = 2$

75. $y = \dfrac{1}{2}x + 2$

76. $y = 2x + \dfrac{1}{2}$

77. $x + 2y = 4$

78. $-x + 3y = 6$

79. $2x - 3y = 6$

80. $3x + 2y = 6$

81. $x + y = 1$

82. $x - y = 2$

83. $x = -4$

84. $y = -1$

85. $y = 5$

86. $x = 2$

87. $y - x = 0$

88. $x + y = 0$

89. $2y - 3x = 0$

90. $3x + 2y = 0$

In Problems 91–100, (a) find the intercepts of the graph of each equation and (b) graph the equation.

91. $2x + 3y = 6$

92. $3x - 2y = 6$

93. $-4x + 5y = 40$

94. $6x - 4y = 24$

95. $7x + 2y = 21$

96. $5x + 3y = 18$

97. $\dfrac{1}{2}x + \dfrac{1}{3}y = 1$

98. $x - \dfrac{2}{3}y = 4$

99. $0.2x - 0.5y = 1$

100. $-0.3x + 0.4y = 1.2$

101. Find an equation of the x-axis.

102. Find an equation of the y-axis.

In Problems 103–106, the equations of two lines are given. Determine if the lines are parallel, perpendicular, or neither.

103. $y = 2x - 3$
 $y = 2x + 4$

104. $y = \dfrac{1}{2}x - 3$
 $y = -2x + 4$

105. $y = 4x + 5$
 $y = -4x + 2$

106. $y = -2x + 3$
 $y = -\dfrac{1}{2}x + 2$

In Problems 107–110, write an equation of each line. Express your answer using either the general form or the slope–intercept form of the equation of a line, whichever you prefer.

107.

108.

109.

110.
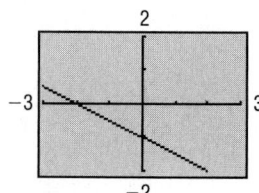

Applications and Extensions

111. Geometry Use slopes to show that the triangle whose vertices are $(-2, 5)$, $(1, 3)$, and $(-1, 0)$ is a right triangle.

112. Geometry Use slopes to show that the quadrilateral whose vertices are $(1, -1)$, $(4, 1)$, $(2, 2)$, and $(5, 4)$ is a parallelogram.

113. Geometry Use slopes to show that the quadrilateral whose vertices are $(-1, 0)$, $(2, 3)$, $(1, -2)$, and $(4, 1)$ is a rectangle.

114. Geometry Use slopes and the distance formula to show that the quadrilateral whose vertices are $(0, 0)$, $(1, 3)$, $(4, 2)$, and $(3, -1)$ is a square.

115. Truck Rentals A truck rental company rents a moving truck for one day by charging $29 plus $0.07 per mile. Write a linear equation that relates the cost C, in dollars, of renting the truck to the number x of miles driven. What is the cost of renting the truck if the truck is driven 110 miles? 230 miles?

116. Cost Equation The **fixed costs** of operating a business are the costs incurred regardless of the level of production. Fixed costs include rent, fixed salaries, and costs of buying machinery. The **variable costs** of operating a business are the costs that change with the level of output. Variable costs include raw materials, hourly wages, and electricity. Suppose that a manufacturer of jeans has fixed costs of $500 and variable costs of $8 for each pair of jeans manufactured. Write a linear equation that relates the cost C, in dollars, of manufacturing the jeans to the number x of jeans manufactured. What is the cost of manufacturing 400 pairs of jeans? 740 pairs?

117. Cost of Sunday Home Delivery The cost to the *Chicago Tribune* for Sunday home delivery is approximately $0.53 per newspaper with fixed costs of $1,070,000. Write an equation that relates the cost C and the number x of copies delivered.

SOURCE: *Chicago Tribune*, 2002.

118. Wages of a Car Salesperson Dan receives $375 per week for selling new and used cars at a car dealership in Oak Lawn, Illinois. In addition, he receives 5% of the profit on any sales that he generates. Write an equation that relates Dan's weekly salary S when he has sales that generate a profit of x dollars.

119. Electricity Rates in Illinois Commonwealth Edison Company supplies electricity to residential customers for a monthly customer charge of $7.58 plus 8.275 cents per kilowatt-hour for up to 400 kilowatt-hours.

(a) Write an equation that relates the monthly charge C, in dollars, to the number x of kilowatt-hours used in a month, $0 \le x \le 400$.

(b) Graph this equation.

(c) What is the monthly charge for using 100 kilowatt-hours?

(d) What is the monthly charge for using 300 kilowatt-hours?

(e) Interpret the slope of the line.

SOURCE: Commonwealth Edison Company, June, 2005.

120. Electricity Rates in Florida Florida Power & Light Company supplies electricity to residential customers for a monthly customer charge of $5.25 plus 8.35 cents per kilowatt-hour for up to 750 kilowatt-hours.

(a) Write an equation that relates the monthly charge C, in dollars, to the number x of kilowatt-hours used in a month, $0 \le x \le 750$.

(b) Graph this equation.

(c) What is the monthly charge for using 200 kilowatt-hours?

(d) What is the monthly charge for using 500 kilowatt-hours?

(e) Interpret the slope of the line.

SOURCE: Florida Power & Light Company, June, 2005.

121. Measuring Temperature The relationship between Celsius ($°C$) and Fahrenheit ($°F$) degrees of measuring temperature is linear. Find an equation relating $°C$ and $°F$ if $0°C$ corresponds to $32°F$ and $100°C$ corresponds to $212°F$. Use the equation to find the Celsius measure of $70°F$.

122. Measuring Temperature The Kelvin (K) scale for measuring temperature is obtained by adding 273 to the Celsius temperature.

(a) Write an equation relating K and $°C$.

(b) Write an equation relating K and $°F$ (see Problem 121).

123. Product Promotion A cereal company finds that the number of people who will buy one of its products in the first month that it is introduced is linearly related to the amount of money it spends on advertising. If it spends $40,000 on advertising, then 100,000 boxes of cereal will be sold, and if it spends $60,000, then 200,000 boxes will be sold.

(a) Write an equation describing the relation between the amount A spent on advertising and the number x of boxes sold.

(b) How much advertising is needed to sell 300,000 boxes of cereal?

(c) Interpret the slope.

124. Show that the line containing the points (a, b) and (b, a), $a \ne b$, is perpendicular to the line $y = x$. Also show that the midpoint of (a, b) and (b, a) lies on the line $y = x$.

125. The equation $2x - y = C$ defines a **family of lines**, one line for each value of C. On one set of coordinate axes, graph the members of the family when $C = -4$, $C = 0$, and $C = 2$. Can you draw a conclusion from the graph about each member of the family?

126. Prove that if two nonvertical lines have slopes whose product is -1 then the lines are perpendicular. [**Hint:** Refer to Figure 45 and use the converse of the Pythagorean Theorem.]

Discussion and Writing

127. Which of the following equations might have the graph shown? (More than one answer is possible.)
(a) $2x + 3y = 6$
(b) $-2x + 3y = 6$
(c) $3x - 4y = -12$
(d) $x - y = 1$
(e) $x - y = -1$
(f) $y = 3x - 5$
(g) $y = 2x + 3$
(h) $y = -3x + 3$

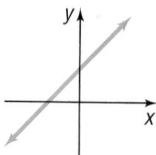

128. Which of the following equations might have the graph shown? (More than one answer is possible.)
(a) $2x + 3y = 6$
(b) $2x - 3y = 6$
(c) $3x + 4y = 12$
(d) $x - y = 1$
(e) $x - y = -1$
(f) $y = -2x - 1$
(g) $y = -\dfrac{1}{2}x + 10$
(h) $y = x + 4$

129. The figure shows the graph of two parallel lines. Which of the following pairs of equations might have such a graph?
(a) $x - 2y = 3$
$x + 2y = 7$
(b) $x + y = 2$
$x + y = -1$
(c) $x - y = -2$
$x - y = 1$
(d) $x - y = -2$
$2x - 2y = -4$
(e) $x + 2y = 2$
$x + 2y = -1$

130. The figure shows the graph of two perpendicular lines. Which of the following pairs of equations might have such a graph?
(a) $y - 2x = 2$
$y + 2x = -1$
(b) $y - 2x = 0$
$2y + x = 0$
(c) $2y - x = 2$
$2y + x = -2$
(d) $y - 2x = 2$
$x + 2y = -1$
(e) $2x + y = -2$
$2y + x = -2$

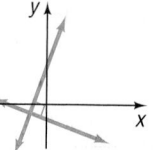

131. The accepted symbol used to denote the slope of a line is the letter m. Investigate the origin of this symbolism. Begin by consulting a French dictionary and looking up the French word *monter*. Write a brief essay on your findings.

132. The term *grade* is used to describe the inclination of a road. How does this term relate to the notion of slope of a line? Is a 4% grade very steep? Investigate the grades of some mountainous roads and determine their slopes. Write a brief essay on your findings.

133. Carpentry Carpenters use the term *pitch* to describe the steepness of staircases and roofs. How does pitch relate to slope? Investigate typical pitches used for stairs and for roofs. Write a brief essay on your findings.

134. Can the equation of every line be written in slope–intercept form? Why?

135. Does every line have exactly one x-intercept and one y-intercept? Are there any lines that have no intercepts?

136. What can you say about two lines that have equal slopes and equal y-intercepts?

137. What can you say about two lines with the same x-intercept and the same y-intercept? Assume that the x-intercept is not 0.

138. If two lines have the same slope, but different x-intercepts, can they have the same y-intercept?

139. If two lines have the same y-intercept, but different slopes, can they have the same x-intercept?

140. Which form of the equation of a line do you prefer to use? Justify your position with an example that shows that your choice is better than another. Have reasons.

F.4 Circles

PREPARING FOR THIS SECTION *Before getting started, review the following:*

• Completing the Square (Appendix A, Section A.4, pp. A35–A36)

Now work the 'Are You Prepared?' problems on page 36.

OBJECTIVES 1 Write the Standard Form of the Equation of a Circle
2 Graph a Circle
3 Work with the General Form of the Equation of a Circle

1 **Write the Standard Form of the Equation of a Circle**

One advantage of a coordinate system is that it enables us to translate a geometric statement into an algebraic statement, and vice versa. Consider, for example, the following geometric statement that defines a circle.

> A **circle** is a set of points in the xy-plane that are a fixed distance r from a fixed point (h, k). The fixed distance r is called the **radius**, and the fixed point (h, k) is called the **center** of the circle.

Figure 47

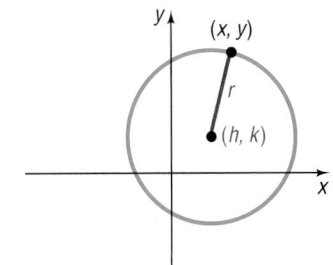

Figure 47 shows the graph of a circle. To find the equation, we let (x, y) represent the coordinates of any point on a circle with radius r and center (h, k). Then the distance between the points (x, y) and (h, k) must always equal r. That is, by the distance formula

$$\sqrt{(x - h)^2 + (y - k)^2} = r$$

or, equivalently,

$$(x - h)^2 + (y - k)^2 = r^2$$

The **standard form of an equation of a circle** with radius r and center (h, k) is

$$(x - h)^2 + (y - k)^2 = r^2 \qquad \textbf{(1)}$$

The standard form of an equation of a circle of radius r with center at the origin $(0, 0)$ is

$$x^2 + y^2 = r^2$$

Figure 48
Unit circle $x^2 + y^2 = 1$

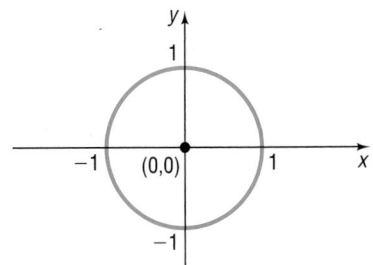

If the radius $r = 1$, the circle whose center is at the origin is called the **unit circle** and has the equation

$$x^2 + y^2 = 1$$

See Figure 48.

| EXAMPLE 1 | Writing the Standard Form of the Equation of a Circle |

Write the standard form of the equation of the circle with radius 5 and center $(-3, 6)$.

Solution Using the form of equation (1) and substituting the values $r = 5$, $h = -3$, and $k = 6$, we have

$$(x - h)^2 + (y - k)^2 = r^2$$
$$(x + 3)^2 + (y - 6)^2 = 25$$

■

 NOW WORK PROBLEM 7.

2 Graph a Circle

The graph of any equation in the form of equation (1) is that of a circle with radius r and center (h, k).

| EXAMPLE 2 | Graphing a Circle |

Graph the equation: $(x + 3)^2 + (y - 2)^2 = 16$

Figure 49

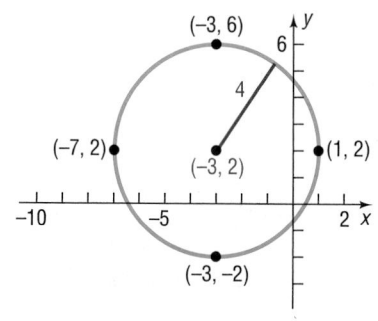

Solution The graph of the equation is a circle. To graph the equation, we first compare the given equation to the standard form of the equation of a circle. The comparison yields information about the circle.

$$(x + 3)^2 + (y - 2)^2 = 16$$
$$(x - (-3))^2 + (y - 2)^2 = 4^2$$
$$\uparrow \qquad\qquad \uparrow \qquad \uparrow$$
$$(x - h)^2 + (y - k)^2 = r^2$$

We see that $h = -3$, $k = 2$, and $r = 4$. The circle has center $(-3, 2)$ and a radius of 4 units. To graph this circle, we first plot the center $(-3, 2)$. Since the radius is 4, we can locate four points on the circle by plotting points 4 units to the left, to the right, up, and down from the center. These four points can then be used as guides to obtain the graph. See Figure 49. ■

 NOW WORK PROBLEMS 23(a) AND (b).

3 Work with the General Form of the Equation of a Circle

If we eliminate the parentheses from the standard form of the equation of the circle given in Example 2, we get

$$(x + 3)^2 + (y - 2)^2 = 16$$
$$x^2 + 6x + 9 + y^2 - 4y + 4 = 16$$

which we find, upon simplifying, is equivalent to

$$x^2 + y^2 + 6x - 4y - 3 = 0 \qquad\qquad \textbf{(2)}$$

It can be shown that any equation of the form

$$x^2 + y^2 + ax + by + c = 0$$

has a graph that is a circle, or a point, or has no graph at all. For example, the graph of the equation $x^2 + y^2 = 0$ is the single point $(0, 0)$. The equation $x^2 + y^2 + 5 = 0$, or $x^2 + y^2 = -5$, has no graph, because sums of squares of real numbers are never negative.

When its graph is a circle, the equation

$$x^2 + y^2 + ax + by + c = 0$$

is referred to as the **general form of the equation of a circle**.

✏️ **NOW WORK PROBLEM 35.**

If an equation of a circle is in the general form, we use the method of completing the square to put the equation in standard form so that we can identify its center and radius.

EXAMPLE 3 | **Graphing a Circle Whose Equation Is in General Form**

Graph the equation $x^2 + y^2 + 4x - 6y + 12 = 0$

Solution We complete the square in both x and y to put the equation in standard form. Group the expression involving x, group the expression involving y, and put the constant on the right side of the equation. The result is

$$(x^2 + 4x) + (y^2 - 6y) = -12$$

Next, complete the square of each expression in parentheses. Remember that any number added on the left side of the equation must also be added on the right.

Figure 50

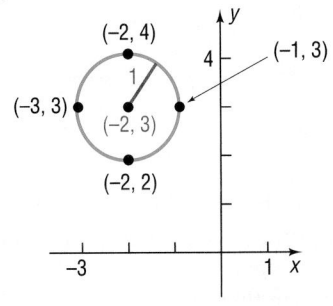

$$(x^2 + 4x + 4) + (y^2 - 6y + 9) = -12 + 4 + 9$$

$$\left(\frac{4}{2}\right)^2 = 4 \qquad \left(\frac{-6}{2}\right)^2 = 9$$

$$(x + 2)^2 + (y - 3)^2 = 1 \quad \text{Factor.}$$

We recognize this equation as the standard form of the equation of a circle with radius 1 and center $(-2, 3)$.

To graph the equation use the center $(-2, 3)$ and the radius 1. See Figure 50. ■

✏️ **NOW WORK PROBLEM 27.**

COMMENT: Example 4 requires information from Section B.5, *Square Screens*, in Appendix B. ▪

EXAMPLE 4 | **Using a Graphing Utility to Graph a Circle**

Graph the equation: $x^2 + y^2 = 4$

Solution This is the equation of a circle with center at the origin and radius 2. To graph this equation, we must first solve for y.

Figure 51

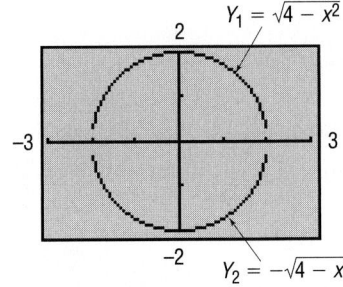

$$x^2 + y^2 = 4$$
$$y^2 = 4 - x^2 \qquad \text{Subtract } x^2 \text{ from each side.}$$
$$y = \pm\sqrt{4 - x^2} \qquad \begin{array}{l}\text{Apply the Square Root}\\\text{Method to solve for } y.\end{array}$$

There are two equations to graph: first, we graph $Y_1 = \sqrt{4 - x^2}$ and then $Y_2 = -\sqrt{4 - x^2}$ on the same square screen. (Your circle will appear oval if you do not use a square screen.) See Figure 51. ■

F.4 Assess Your Understanding

Are You Prepared?

Answers are given at the end of these exercises. If you get a wrong answer, read the pages listed in red.

1. To complete the square of $x^2 + 10x$, you would (add/subtract) the number ____. (pp. A35–A36)

2. Complete the square of $x^2 - 7x$. (pp. A35–A36)

Concepts and Vocabulary

3. *True or False:* Every equation of the form
$$x^2 + y^2 + ax + by + c = 0$$
has a circle as its graph.

4. For a circle, the ____ is the distance from the center to any point on the circle.

5. *True or False:* The radius of the circle $x^2 + y^2 = 9$ is 3.

6. *True or False:* The center of the circle
$$(x + 3)^2 + (y - 2)^2 = 13$$
is $(3, -2)$.

Skill Building

In Problems 5–8, find the center and radius of each circle. Write the standard form of the equation.

7.

8.

9.

10.
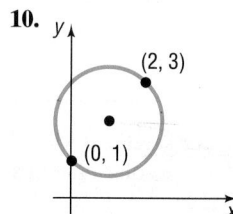

In Problems 11–20, write the standard form of the equation and the general form of the equation of each circle of radius r and center (h, k). Graph each circle.

11. $r = 2$; $(h, k) = (0, 0)$

12. $r = 3$; $(h, k) = (0, 0)$

13. $r = 2$; $(h, k) = (0, 2)$

14. $r = 3$; $(h, k) = (1, 0)$

15. $r = 5$; $(h, k) = (4, -3)$

16. $r = 4$; $(h, k) = (2, -3)$

17. $r = 4$; $(h, k) = (-2, 1)$

18. $r = 7$; $(h, k) = (-5, -2)$

19. $r = \dfrac{1}{2}$; $(h, k) = \left(\dfrac{1}{2}, 0\right)$

20. $r = \dfrac{1}{2}$; $(h, k) = \left(0, -\dfrac{1}{2}\right)$

In Problems 21–34, (a) find the center (h, k) and radius r of each circle; (b) graph each circle.

21. $x^2 + y^2 = 4$

22. $x^2 + (y - 1)^2 = 1$

23. $2(x - 3)^2 + 2y^2 = 8$

24. $3(x + 1)^2 + 3(y - 1)^2 = 6$

25. $x^2 + y^2 - 2x - 4y - 4 = 0$

26. $x^2 + y^2 + 4x + 2y - 20 = 0$

27. $x^2 + y^2 + 4x - 4y - 1 = 0$

28. $x^2 + y^2 - 6x + 2y + 9 = 0$

29. $x^2 + y^2 - x + 2y + 1 = 0$

30. $x^2 + y^2 + x + y - \dfrac{1}{2} = 0$

31. $2x^2 + 2y^2 - 12x + 8y - 24 = 0$

32. $2x^2 + 2y^2 + 8x + 7 = 0$

33. $2x^2 + 4x + 2y^2 = 0$

34. $3x^2 + 3y^2 - 12y = 0$

In Problems 35–40, find the general form of the equation of each circle.

35. Center at the origin and containing the point $(-2, 3)$

36. Center $(1, 0)$ and containing the point $(-3, 2)$

37. Center $(2, 3)$ and tangent to the x-axis

38. Center $(-3, 1)$ and tangent to the y-axis

39. With endpoints of a diameter at $(1, 4)$ and $(-3, 2)$

40. With endpoints of a diameter at $(4, 3)$ and $(0, 1)$

In Problems 41–44, match each graph with the correct equation.

(a) $(x - 3)^2 + (y + 3)^2 = 9$ (b) $(x + 1)^2 + (y - 2)^2 = 4$ (c) $(x - 1)^2 + (y + 2)^2 = 4$ (d) $(x + 3)^2 + (y - 3)^2 = 9$

41.

42.

43.

44.
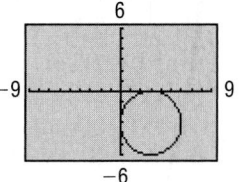

Applications and Extensions

45. Find the area of the square in the figure.

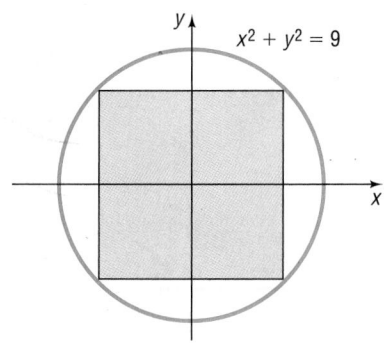

46. Find the area of the shaded region in the figure assuming the quadrilateral inside the circle is a square.

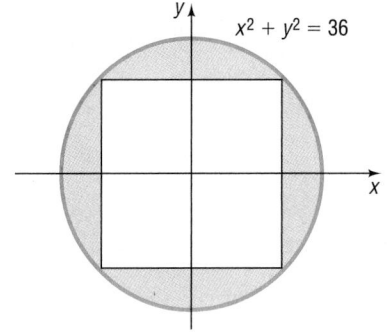

47. Weather Satellites Earth is represented on a map of a portion of the solar system so that its surface is the circle with equation $x^2 + y^2 + 2x + 4y - 4091 = 0$. A weather satellite circles 0.6 unit above Earth with the center of its circular orbit at the center of Earth. Find the equation for the orbit of the satellite on this map.

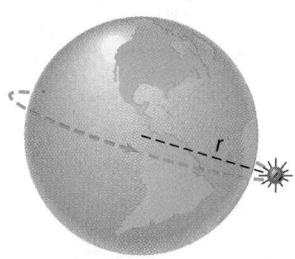

48. The **tangent line** to a circle may be defined as the line that intersects the circle in a single point, called the **point of tangency** (see the figure).

If the equation of the circle is $x^2 + y^2 = r^2$ and the equation of the tangent line is $y = mx + b$, show that:
(a) $r^2(1 + m^2) = b^2$

[**Hint:** The quadratic equation $x^2 + (mx + b)^2 = r^2$ has exactly one solution.]

(b) The point of tangency is $\left(\dfrac{-r^2 m}{b}, \dfrac{r^2}{b} \right)$.

(c) The tangent line is perpendicular to the line containing the center of the circle and the point of tangency.

49. The Greek method for finding the equation of the tangent line to a circle used the fact that at any point on a circle the lines containing the center and the tangent line are perpendicular (see Problem 48). Use this method to find an equation of the tangent line to the circle $x^2 + y^2 = 9$ at the point $(1, 2\sqrt{2})$.

50. Use the Greek method described in Problem 49 to find an equation of the tangent line to the circle $x^2 + y^2 - 4x + 6y + 4 = 0$ at the point $(3, 2\sqrt{2} - 3)$.

51. Refer to Problem 48. The line $x - 2y + 4 = 0$ is tangent to a circle at $(0, 2)$. The line $y = 2x - 7$ is tangent to the same circle at $(3, -1)$. Find the center of the circle.

52. Find an equation of the line containing the centers of the two circles

$$x^2 + y^2 - 4x + 6y + 4 = 0 \qquad \text{and}$$
$$x^2 + y^2 + 6x + 4y + 9 = 0$$

53. If a circle of radius 2 is made to roll along the x-axis, what is an equation for the path of the center of the circle?

54. If the circumference of a circle is 6π, what is its radius?

Discussion and Writing

55. Which of the following equations might have the graph shown? (More than one answer is possible.)
(a) $(x - 2)^2 + (y + 3)^2 = 13$
(b) $(x - 2)^2 + (y - 2)^2 = 8$
(c) $(x - 2)^2 + (y - 3)^2 = 13$
(d) $(x + 2)^2 + (y - 2)^2 = 8$
(e) $x^2 + y^2 - 4x - 9y = 0$
(f) $x^2 + y^2 + 4x - 2y = 0$
(g) $x^2 + y^2 - 9x - 4y = 0$
(h) $x^2 + y^2 - 4x - 4y = 4$

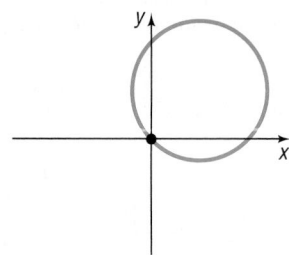

56. Which of the following equations might have the graph shown? (More than one answer is possible.)
(a) $(x - 2)^2 + y^2 = 3$
(b) $(x + 2)^2 + y^2 = 3$
(c) $x^2 + (y - 2)^2 = 3$
(d) $(x + 2)^2 + y^2 = 4$
(e) $x^2 + y^2 + 10x + 16 = 0$
(f) $x^2 + y^2 + 10x - 2y = 1$
(g) $x^2 + y^2 + 9x + 10 = 0$
(h) $x^2 + y^2 - 9x - 10 = 0$

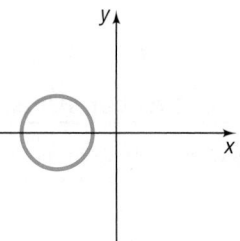

57. Explain how the center and radius of a circle can be used to graph a circle.

Are You Prepared? Answers

1. add; 25 **2.** $x^2 - 7x + \dfrac{49}{4}$.

1 Functions and Their Graphs

A LOOK BACK So far, we have developed techniques for graphing equations containing two variables.

A LOOK AHEAD In this chapter, we look at a special type of equation involving two variables called a *function*. This chapter deals with what a function is, how to graph functions, properties of functions, and how functions are used in applications. The word function apparently was introduced by René Descartes in 1637. For him, a function simply meant any positive integral power of a variable *x*. Gottfried Wilhelm Leibniz (1646–1716), who always emphasized the geometric side of mathematics, used the word function to denote any quantity associated with a curve, such as the coordinates of a point on the curve. Leonhard Euler (1707–1783) employed the word to mean any equation or formula involving variables and constants. His idea of a function is similar to the one most often seen in courses that precede calculus. Later, the use of functions in investigating heat flow equations led to a very broad definition, due to Lejeune Dirichlet (1805–1859), which describes a function as a correspondence between two sets. It is his definition that we use here.

U.S. Consumers Make Wireless Top Choice

July 11, 2005—WASHINGTON, D.C., CTIA-The Wireless Association, President and CEO Steve Largent highlights the Federal Communications Commission (FCC) report on local telephone competition, which indicates for the first time there are more wireless subscribers in the United States than wireline service lines. According to the FCC report, as of December 31, 2004 there were 181.1 million wireless subscribers in the U.S., compared to a combined 178 million incumbent local exchange carrier switched access lines and competitive local exchange carrier switched access lines. Also according to the report, there was a 15% increase in wireless subscribers for the 12-month period ending December 31, 2004.

"This significant milestone reflects consumers' approval of a vibrant, ultracompetitive industry that lets us communicate how we want, where we want and with whom we want. Americans are on the go now more than ever and we like the versatility, mobility, and affordability unique to wireless communication," said Largent.

"Wireless is so popular because it lets people communicate on their terms. And it's not just about talking anymore. You can browse the web, take pictures and video, download music, play games or conduct business. Wireless can satisfy a lot of different communication needs and desires and that's extremely popular to millions of consumers."

"Why has wireless taken off as it has? Because public policy got it right. It let consumers set the marketplace by choosing their preferred means of communication. The more of wireless they get, the more they like and the more they want, and the industry is responding with new products and services that continue giving Americans even more choices. That's why it's so important to have a consistent regulatory framework that recognizes the unique qualities of wireless communication and allows wireless to keep delivering the innovative offerings consumers want and expect."

SOURCE: www.CTIA.org

—*See Chapter Project 1.*

1.1 Functions

PREPARING FOR THIS SECTION *Before getting started, review the following:*

• Intervals (Appendix A, Section A.9, pp. A66–A67) • Solving Inequalities (Appendix A, Section A.9, pp. A69–A70)
• Evaluating Algebraic Expressions, Domain of a Variable
 (Appendix A, Section A.1, pp. A3–A4)

Now work the 'Are You Prepared?' problems on page 51.

OBJECTIVES 1 Determine Whether a Relation Represents a Function
 2 Find the Value of a Function
 3 Find the Domain of a Function
 4 Form the Sum, Difference, Product, and Quotient of Two Functions

1 Determine Whether a Relation Represents a Function

We often see situations where one variable is somehow linked to the value of another variable. For example, an individual's level of education is linked to annual income. Engine size is linked to gas mileage. When the value of one variable is related to the value of a second variable, we have a *relation*. A **relation** is a correspondence between two sets. If x and y are two elements in these sets and if a relation exists between x and y, then we say that x **corresponds** to y or that y **depends on** x, and we write $x \rightarrow y$.

We have a number of ways to express relations between two sets. For example, the equation $y = 3x - 1$ shows a relation between x and y. It says that if we take some number x, multiply it by 3, and then subtract 1, we obtain the corresponding value of y. In this sense, x serves as the **input** to the relation and y is the **output** of the relation. We can also express this relation as a graph as shown in Figure 1.

Not only can relations be expressed through an equation or graph, but we can also express relations through a technique called *mapping*. A **map** illustrates a relation by using a set of inputs and drawing arrows to the corresponding element in the set of outputs. Ordered pairs can be used to represent $x \rightarrow y$ as (x, y). We illustrate these two concepts in Example 1.

Figure 1

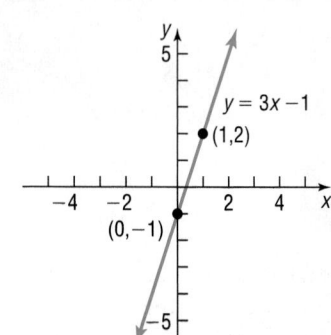

EXAMPLE 1 **Maps and Ordered Pairs as Relations**

Figure 2 shows a relation between states and the number of representatives each has in the House of Representatives. The relation might be named "number of representatives."

Figure 2

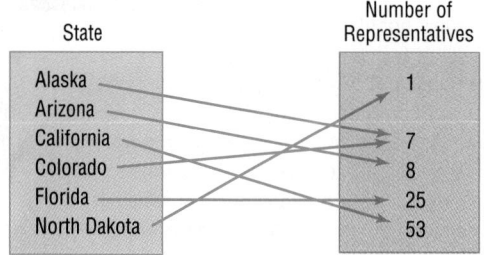

In this relation, Alaska corresponds to 7, Arizona corresponds to 8, and so on. Using ordered pairs, this relation would be expressed as

{(Alaska, 7), (Arizona, 8), (California, 53), (Colorado, 7), (Florida, 25), (North Dakota, 1)} ∎

We now present one of the most important concepts in algebra—the *function*. A function is a special type of relation. To understand the idea behind a function, let's

revisit the relation presented in Example 1. If we were to ask, "How many representatives does Alaska have?," you would respond 7. In other words, each input *state* corresponds to a single output *number of representatives*.

Let's consider a second relation where we have a correspondence between four people and their phone numbers. See Figure 3. Notice that Maureen has two telephone numbers. If asked, "What is Maureen's phone number?", you cannot assign a single number to her.

Figure 3

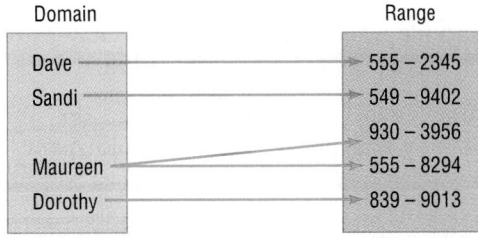

Let's look at one more relation. Figure 4 is a relation that shows a correspondence between *animals* and *life expectancy*. If asked to determine the life expectancy of a dog, we would all respond "11 years." If asked to determine the life expectancy of a rabbit, we would all respond "7 years."

Figure 4

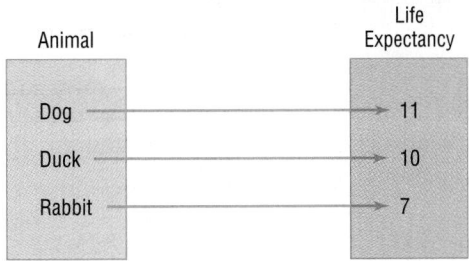

Notice that the relations presented in Figures 2 and 4 have something in common. What is it? The common link between these two relations is that each input corresponds to exactly one output. This leads to the definition of a *function*.

> Let X and Y be two nonempty sets.[*] A **function** from X into Y is a relation that associates with each element of X exactly one element of Y.

The set X is called the **domain** of the function. For each element x in X, the corresponding element y in Y is called the **value** of the function at x, or the **image** of x. The set of all images of the elements in the domain is called the **range** of the function. See Figure 5.

Figure 5

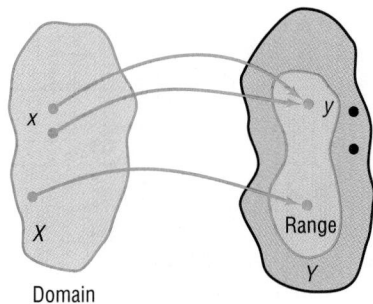

[*]The sets X and Y will usually be sets of real numbers, in which case a (real) function results. The two sets can also be sets of complex numbers, and then we have defined a complex function. In the broad definition (due to Lejeune Dirichlet), X and Y can be any two sets.

Since there may be some elements in Y that are not the image of some x in X, it follows that the range of a function may be a subset of Y, as shown in Figure 5.

Not all relations between two sets are functions. The next example shows how to determine whether a relation is a function or not.

EXAMPLE 2 | **Determining Whether a Relation Represents a Function**

Determine whether the following relations represent functions. If the relation is a function, then state its domain and range.

(a) See Figure 6. For this relation, the domain represents the level of education and the range represents the unemployment rate.

Figure 6
[SOURCE: *Statistical Abstract of the United States, 2003*]

(b) See Figure 7. For this relation, the domain represents the number of calories in a sandwich from a fast-food restaurant and the range represents the fat content (in grams).

Figure 7
[SOURCE: Each company's Web site]

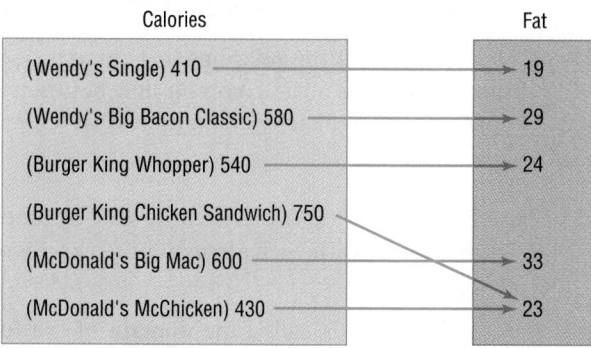

(c) See Figure 8. For this relation, the domain represents the weight (in carats) of pear-cut diamonds and the range represents the price (in dollars).

Figure 8
[SOURCE: diamonds.com]

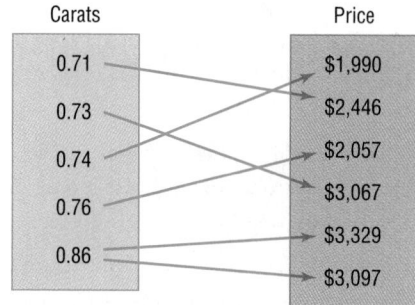

Solution

(a) The relation is a function because each element in the domain corresponds to exactly one element in the range. The domain of the relation is {No High School Diploma, High School Diploma, Some College, College Graduate} and the range of the relation is {7.7%, 5.9%, 5.4%, 3.4%}.

(b) The relation is a function because each element in the domain corresponds to exactly one element in the range. Notice that it is okay for more than one element in the domain to correspond to the same element in the range (McDonald's and Burger King's chicken sandwich both have 23 grams of fat). The domain of the relation is {410, 580, 540, 750, 600, 430}. The range of the relation is {19, 29, 24, 33, 23}.

(c) The relation is not a function because each element in the domain does not correspond to exactly one element in the range. If a 0.86 carat diamond is chosen from the domain, a single price cannot be assigned to it. ∎

═══► **NOW WORK PROBLEM 15.**

In Words

For a function, no input has more than one output.

In Words

For a function, the domain is the set of inputs, and the range is the set of outputs.

The idea behind a function is its predictability. If the input is known, we can use the function to determine the output. With "nonfunctions," we don't have this predictability. Look back at Figure 7. The inputs are {410, 580, 540, 750, 600, 430}. The correspondence is *number of fat grams* and the outputs are {19, 29, 24, 33, 23}. If asked, "How many fat grams in a 410-calorie sandwich?", we can use the correspondence to answer "19". Now consider Figure 8. If asked, "What is the price of a 0.86 carat diamond?", we could not give a single response because two outputs result from the single input "0.86". For this reason, the relation in Figure 8 is not a function.

We may also think of a function as a set of ordered pairs (x, y) in which no ordered pairs have the same first element and different second elements. The set of all first elements x is the domain of the function, and the set of all second elements y is its range. Each element x in the domain corresponds to exactly one element y in the range.

| EXAMPLE 3 | **Determining Whether a Relation Represents a Function** |

Determine whether each relation represents a function. If it is a function, state the domain and range.

(a) $\{(1, 4), (2, 5), (3, 6), (4, 7)\}$

(b) $\{(1, 4), (2, 4), (3, 5), (6, 10)\}$

(c) $\{(-3, 9), (-2, 4), (0, 0), (1, 1), (-3, 8)\}$

Solution

(a) This relation is a function because there are no ordered pairs with the same first element and different second elements. The domain of this function is $\{1, 2, 3, 4\}$, and its range is $\{4, 5, 6, 7\}$.

(b) This relation is a function because there are no ordered pairs with the same first element and different second elements. The domain of this function is $\{1, 2, 3, 6\}$, and its range is $\{4, 5, 10\}$.

(c) This relation is not a function because there are two ordered pairs, $(-3, 9)$ and $(-3, 8)$, that have the same first element and different second elements. ∎

In Example 3(b), notice that 1 and 2 in the domain each have the same image in the range. This does not violate the definition of a function; two different first elements can have the same second element. A violation of the definition occurs when two ordered pairs have the same first element and different second elements, as in Example 3(c).

> ──── **NOW WORK PROBLEM 19.**

Up to now we have shown how to identify when a relation is a function for relations defined by mappings (Example 2) and ordered pairs (Example 3). We know that relations can also be expressed as equations. We discuss next the circumstances under which equations are functions.

To determine whether an equation, where y depends on x, is a function, it is often easiest to solve the equation for y. If any value of x in the domain corresponds to more than one y, the equation does not define a function; otherwise, it does define a function.

EXAMPLE 4	**Determining Whether an Equation Is a Function**

Determine if the equation $y = 2x - 5$ defines y as a function of x.

Solution The equation tells us to take an input x, multiply it by 2, and then subtract 5. For any input x, these operations yield only one output y. For example, if $x = 1$, then $y = 2(1) - 5 = -3$. If $x = 3$, then $y = 2(3) - 5 = 1$. For this reason, the equation is a function. ∎

EXAMPLE 5	**Determining Whether an Equation Is a Function**

Determine if the equation $x^2 + y^2 = 1$ defines y as a function of x.

Solution To determine whether the equation $x^2 + y^2 = 1$, which defines the unit circle, is a function, we need to solve the equation for y.

$$x^2 + y^2 = 1$$
$$y^2 = 1 - x^2$$
$$y = \pm\sqrt{1 - x^2}$$

For values of x between -1 and 1, two values of y result. For example, if $x = 0$, then $y = \pm 1$, so two different outputs result from the same input. This means that the equation $x^2 + y^2 = 1$ does not define a function. ∎

> ──── **NOW WORK PROBLEM 33.**

2 Find the Value of a Function

Functions are often denoted by letters such as f, F, g, G, and others. If f is a function, then for each number x in its domain the corresponding image in the range is designated by the symbol $f(x)$, read as "f of x" or as "f at x." We refer to $f(x)$ as the **value of f at the number x**; $f(x)$ is the number that results when x is given and the function f is applied; $f(x)$ is the output corresponding to x or the image of x; $f(x)$ does *not* mean "f times x." For example, the function given in Example 4 may be written as $y = f(x) = 2x - 5$. Then $f\left(\dfrac{3}{2}\right) = -2$.

Figure 9 illustrates some other functions. Notice that, in every function illustrated, for each x in the domain there is one value in the range.

Figure 9

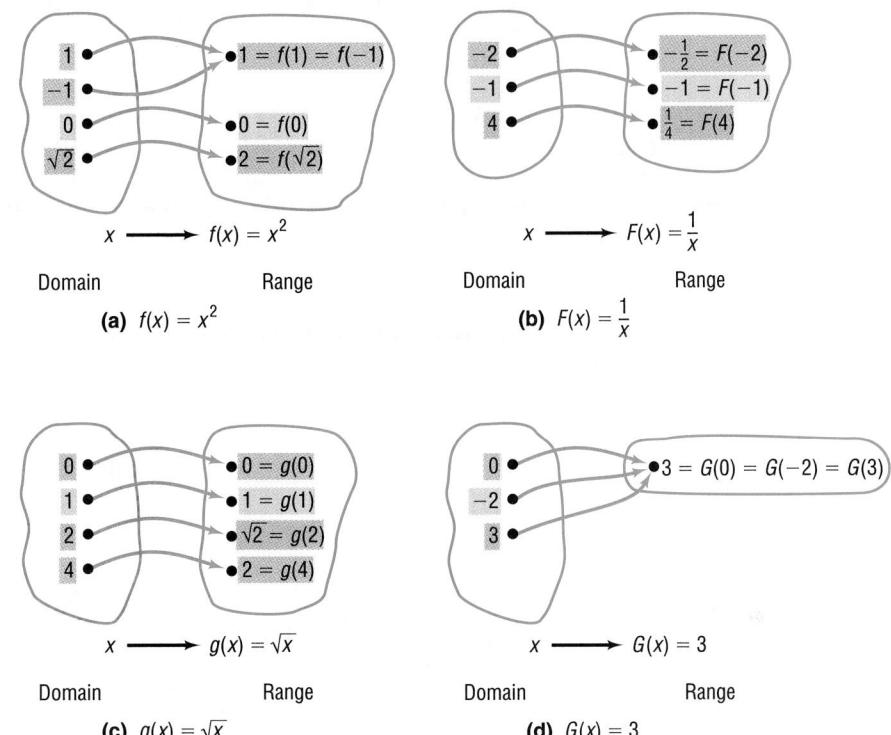

(a) $f(x) = x^2$

(b) $F(x) = \dfrac{1}{x}$

(c) $g(x) = \sqrt{x}$

(d) $G(x) = 3$

Sometimes it is helpful to think of a function f as a machine that receives as input a number from the domain, manipulates it, and outputs the value. See Figure 10.

Figure 10

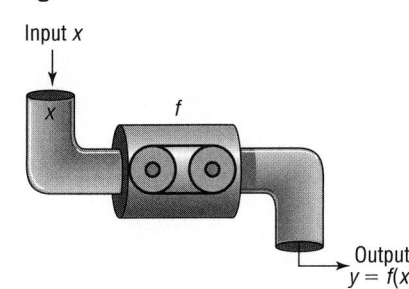

Input x

Output $y = f(x)$

The restrictions on this input/output machine are as follows:

1. It only accepts numbers from the domain of the function.

2. For each input, there is exactly one output (which may be repeated for different inputs).

For a function $y = f(x)$, the variable x is called the **independent variable**, because it can be assigned any of the permissible numbers from the domain. The variable y is called the **dependent variable**, because its value depends on x.

Any symbol can be used to represent the independent and dependent variables. For example, if f is the *cube function*, then f can be given by $f(x) = x^3$ or $f(t) = t^3$ or $f(z) = z^3$. All three functions are the same. Each tells us to cube the independent variable to get the output. In practice, the symbols used for the independent and dependent variables are based on common usage, such as using C for cost in business.

The independent variable is also called the **argument** of the function. Thinking of the independent variable as an argument can sometimes make it easier to find the value of a function. For example, if f is the function defined by $f(x) = x^3$, then f tells us to cube the argument. Thus, $f(2)$ means to cube 2, $f(a)$ means to cube the number a, and $f(x + h)$ means to cube the quantity $x + h$.

| EXAMPLE 6 | **Finding Values of a Function** |

For the function f defined by $f(x) = 2x^2 - 3x$, evaluate

(a) $f(3)$ (b) $f(x) + f(3)$ (c) $3f(x)$

(d) $f(-x)$ (e) $-f(x)$ (f) $f(3x)$

(g) $f(x + 3)$ (h) $\dfrac{f(x + h) - f(x)}{h}$ $h \neq 0$

Solution (a) We substitute 3 for x in the equation for f, $f(x) = 2x^2 - 3x$, to get

$$f(3) = 2(3)^2 - 3(3) = 18 - 9 = 9$$

(b) $f(x) + f(3) = (2x^2 - 3x) + (9) = 2x^2 - 3x + 9$

(c) We multiply the equation for f by 3.

$$3f(x) = 3(2x^2 - 3x) = 6x^2 - 9x$$

(d) We substitute $-x$ for x in the equation for f and simplify.
$$f(-x) = 2(-x)^2 - 3(-x) = 2x^2 + 3x$$

(e) $-f(x) = -(2x^2 - 3x) = -2x^2 + 3x$

(f) We substitute $3x$ for x in the equation for f and simplify.

$$f(3x) = 2(3x)^2 - 3(3x) = 2(9x^2) - 9x = 18x^2 - 9x$$

(g) We substitute $x + 3$ for x in the equation for f and simplify.

$$f(x + 3) = 2(x + 3)^2 - 3(x + 3) \qquad \text{Notice the use of parentheses here.}$$

$$= 2(x^2 + 6x + 9) - 3x - 9$$

$$= 2x^2 + 12x + 18 - 3x - 9$$

$$= 2x^2 + 9x + 9$$

(h) $\dfrac{f(x + h) - f(x)}{h} = \dfrac{[2(x + h)^2 - 3(x + h)] - [2x^2 - 3x]}{h}$

\uparrow
$f(x + h) = 2(x + h)^2 - 3(x + h)$

$$= \dfrac{2(x^2 + 2xh + h^2) - 3x - 3h - 2x^2 + 3x}{h} \qquad \text{Simplify.}$$

$$= \dfrac{2x^2 + 4xh + 2h^2 - 3h - 2x^2}{h} \qquad \text{Distribute and combine like terms.}$$

$$= \dfrac{4xh + 2h^2 - 3h}{h} \qquad \text{Combine like terms.}$$

$$= \dfrac{h(4x + 2h - 3)}{h} \qquad \text{Factor out } h.$$

$$= 4x + 2h - 3 \qquad \text{Divide out the } h\text{'s.}$$

■

Notice in this example that $f(x + 3) \neq f(x) + f(3), f(-x) \neq -f(x),$ and $3f(x) \neq f(3x)$.

 The expression in part (h) is called the **difference quotient** of f, an important expression in calculus.

NOW WORK PROBLEMS 39 AND 73.

Most calculators have special keys that enable you to find the value of certain commonly used functions. For example, you should be able to find the square function $f(x) = x^2$, the square root function $f(x) = \sqrt{x}$, the reciprocal function $f(x) = \dfrac{1}{x} = x^{-1}$, and many others that will be discussed later in this book (such as $\ln x$ and $\log x$). Verify the results of Example 7, which follows, on your calculator.

EXAMPLE 7

Finding Values of a Function on a Calculator

(a) $f(x) = x^2 \quad f(1.234) = 1.234^2 = 1.522756$

(b) $F(x) = \dfrac{1}{x} \quad F(1.234) = \dfrac{1}{1.234} \approx 0.8103727715$

(c) $g(x) = \sqrt{x} \quad g(1.234) = \sqrt{1.234} \approx 1.110855526$ ∎

 COMMENT: Graphing calculators can be used to evaluate any function that you wish. Figure 11 shows the result obtained in Example 6(a) on a TI-84 Plus graphing calculator with the function to be evaluated, $f(x) = 2x^2 - 3x$, in Y_1.

Figure 11

The numbers at which the value of a function f equals zero, that is, those numbers for which $f(x) = 0$, are called **zeros of the function f.** For example, the zeros of the function $f(x) = x^2 - 4$ are -2 and 2 since the solutions of the equation $f(x) = 0$ are

$$f(x) = 0$$
$$x^2 - 4 = 0$$
$$(x + 2)(x - 2) = 0$$
$$x = -2 \text{ or } x = 2$$

Implicit Form of a Function

In general, when a function f is defined by an equation in x and y, we say that the function f is given **implicitly**. If it is possible to solve the equation for y in terms of x, then we write $y = f(x)$ and say that the function is given **explicitly**. For example,

Implicit Form	**Explicit Form**
$3x + y = 5$	$y = f(x) = -3x + 5$
$x^2 - y = 6$	$y = f(x) = x^2 - 6$
$xy = 4$	$y = f(x) = \dfrac{4}{x}$

COMMENT: The explicit form of a function is the form required by a graphing calculator. ∎

We list next a summary of some important facts to remember about a function f.

Summary

Important Facts About Functions

(a) For each x in a domain of a function f, there is exactly one image $f(x)$ in the range; however, an element in the range can result from more than one x in the domain.

(b) f is the symbol that we use to denote the function. It is symbolic of the equation that we use to get from an x in the domain to $f(x)$ in the range.

(c) If $y = f(x)$, then x is called the independent variable or argument of f, and y is called the dependent variable or the value of f at x.

3 Find the Domain of a Function

Often the domain of a function f is not specified; instead, only the equation defining the function is given. In such cases, we agree that the **domain of f** is the largest set of real numbers for which the value $f(x)$ is a real number. The domain of a function f is the same as the domain of the variable x in the expression $f(x)$.

EXAMPLE 8 **Finding the Domain of a Function**

Find the domain of each of the following functions:

(a) $f(x) = x^2 + 5x$ (b) $g(x) = \dfrac{3x}{x^2 - 4}$ (c) $h(t) = \sqrt{4 - 3t}$

Solution (a) The function tells us to square a number and then add five times the number. Since these operations can be performed on any real number, we conclude that the domain of f is the set of all real numbers.

(b) The function g tells us to divide $3x$ by $x^2 - 4$. Since division by 0 is not defined, the denominator $x^2 - 4$ can never be 0, so x can never equal -2 or 2. The domain of the function g is $\{x | x \neq -2, x \neq 2\}$.

(c) The function h tells us to take the square root of $4 - 3t$. But only nonnegative numbers have real square roots, so the expression under the square root must be nonnegative (greater than or equal to zero). This requires that

$$4 - 3t \geq 0$$
$$-3t \geq -4$$
$$t \leq \frac{4}{3}$$

The domain of h is $\left\{t | t \leq \dfrac{4}{3}\right\}$ or the interval $\left(-\infty, \dfrac{4}{3}\right]$. ∎

NOW WORK PROBLEM 51.

If x is in the domain of a function f, we shall say that **f is defined at x**, or **$f(x)$ exists**. If x is not in the domain of f, we say that **f is not defined at x**, or **$f(x)$ does not exist**. For example, if $f(x) = \dfrac{x}{x^2 - 1}$, then $f(0)$ exists, but $f(1)$ and $f(-1)$ do not exist. (Do you see why?)

We have not said much about finding the range of a function. The reason is that when a function is defined by an equation it is often difficult to find the range.[*] Therefore, we shall usually be content to find just the domain of a function when only the rule for the function is given. We shall express the domain of a function using inequalities, interval notation, set notation, or words, whichever is most convenient.

When we use functions in applications, the domain may be restricted by physical or geometric considerations. For example, the domain of the function f defined

[*]In Section 4.2 we discuss a way to find the range for a special class of functions.

by $f(x) = x^2$ is the set of all real numbers. However, if f is used to obtain the area of a square when the length x of a side is known, then we must restrict the domain of f to the positive real numbers, since the length of a side can never be 0 or negative.

EXAMPLE 9	**Finding the Domain in an Application**

Express the area of a circle as a function of its radius. Find the domain.

Solution See Figure 12. We know that the formula for the area A of a circle of radius r is $A = \pi r^2$. If we use r to represent the independent variable and A to represent the dependent variable, the function expressing this relationship is

Figure 12

$$A(r) = \pi r^2$$

In this setting, the domain is $\{r | r > 0\}$. (Do you see why?) ∎

Observe in the solution to Example 9 that we used the symbol A in two ways: It is used to name the function, and it is used to symbolize the dependent variable. This double use is common in applications and should not cause any difficulty.

 NOW WORK PROBLEM 85.

4 Form the Sum, Difference, Product, and Quotient of Two Functions

Next we introduce some operations on functions. We shall see that functions, like numbers, can be added, subtracted, multiplied, and divided. For example, if $f(x) = x^2 + 9$ and $g(x) = 3x + 5$, then

$$f(x) + g(x) = (x^2 + 9) + (3x + 5) = x^2 + 3x + 14$$

The new function $y = x^2 + 3x + 14$ is called the *sum function* $f + g$. Similarly,

$$f(x) \cdot g(x) = (x^2 + 9)(3x + 5) = 3x^3 + 5x^2 + 27x + 45$$

The new function $y = 3x^3 + 5x^2 + 27x + 45$ is called the *product function* $f \cdot g$.
 The general definitions are given next.

If f and g are functions:

The **sum** $f + g$ is the function defined by

$$(f + g)(x) = f(x) + g(x)$$

The domain of $f + g$ consists of the numbers x that are in the domains of both f and g.

The **difference** $f - g$ is the function defined by

$$(f - g)(x) = f(x) - g(x)$$

The domain of $f - g$ consists of the numbers x that are in the domains of both f and g.

The product $f \cdot g$ is the function defined by

$$(f \cdot g)(x) = f(x) \cdot g(x)$$

The domain of $f \cdot g$ consists of the numbers x that are in the domains of both f and g.

> The **quotient** $\dfrac{f}{g}$ is the function defined by
>
> $$\left(\dfrac{f}{g}\right)(x) = \dfrac{f(x)}{g(x)} \qquad g(x) \neq 0$$

The domain of $\dfrac{f}{g}$ consists of the numbers x for which $g(x) \neq 0$ that are in the domains of both f and g.

EXAMPLE 10 **Operations on Functions**

Let f and g be two functions defined as

$$f(x) = \dfrac{1}{x+2} \quad \text{and} \quad g(x) = \dfrac{x}{x-1}$$

Find the following, and determine the domain in each case.

(a) $(f+g)(x)$ (b) $(f-g)(x)$ (c) $(f \cdot g)(x)$ (d) $\left(\dfrac{f}{g}\right)(x)$

Solution The domain of f is $\{x | x \neq -2\}$ and the domain of g is $\{x | x \neq 1\}$.

(a) $(f+g)(x) = f(x) + g(x) = \dfrac{1}{x+2} + \dfrac{x}{x-1} = \dfrac{x-1}{(x+2)(x-1)} + \dfrac{x(x+2)}{(x+2)(x-1)} = \dfrac{x^2 + 3x - 1}{(x+2)(x-1)}$

The domain of $f+g$ consists of those numbers x that are in the domains of both f and g. Therefore, the domain of $f+g$ is $\{x | x \neq -2, x \neq 1\}$.

(b) $(f-g)(x) = f(x) - g(x) = \dfrac{1}{x+2} - \dfrac{x}{x-1} = \dfrac{x-1}{(x+2)(x-1)} - \dfrac{x(x+2)}{(x+2)(x-1)} = \dfrac{-(x^2 + x + 1)}{(x+2)(x-1)}$

The domain of $f-g$ consists of those numbers x that are in the domains of both f and g. Therefore, the domain of $f-g$ is $\{x | x \neq -2, x \neq 1\}$.

(c) $(f \cdot g)(x) = f(x) \cdot g(x) = \dfrac{1}{x+2} \cdot \dfrac{x}{x-1} = \dfrac{x}{(x+2)(x-1)}$

The domain of $f \cdot g$ consists of those numbers x that are in the domains of both f and g. Therefore, the domain of $f \cdot g$ is $\{x | x \neq -2, x \neq 1\}$.

(d) $\left(\dfrac{f}{g}\right)(x) = \dfrac{f(x)}{g(x)} = \dfrac{\dfrac{1}{x+2}}{\dfrac{x}{x-1}} = \dfrac{1}{x+2} \cdot \dfrac{x-1}{x} = \dfrac{x-1}{x(x+2)}$

The domain of $\dfrac{f}{g}$ consists of the numbers x for which $g(x) \neq 0$ that are in the domains of both f and g. Since $g(x) = 0$ when $x = 0$, we exclude 0 as well as -2 and 1 from the domain. The domain of $\dfrac{f}{g}$ is $\{x | x \neq -2, x \neq 0, x \neq 1\}$. ∎

NOW WORK PROBLEM **61**.

In calculus, it is sometimes helpful to view a complicated function as the sum, difference, product, or quotient of simpler functions. For example,

$F(x) = x^2 + \sqrt{x}$ is the sum of $f(x) = x^2$ and $g(x) = \sqrt{x}$.

$H(x) = \dfrac{x^2 - 1}{x^2 + 1}$ is the quotient of $f(x) = x^2 - 1$ and $g(x) = x^2 + 1$.

Summary

We list here some of the important vocabulary introduced in this section, with a brief description of each term.

Function	A relation between two sets of real numbers so that each number x in the first set, the domain, has corresponding to it exactly one number y in the second set. A set of ordered pairs (x, y) or $(x, f(x))$ in which no first element is paired with two different second elements. The range is the set of y values of the function that are the images of the x values in the domain. A function f may be defined implicitly by an equation involving x and y or explicitly by writing $y = f(x)$. Numbers for which $f(x) = 0$ are the zeros of f.
Unspecified domain	If a function f is defined by an equation and no domain is specified, then the domain will be taken to be the largest set of real numbers for which the equation defines a real number.
Function notation	$y = f(x)$ f is a symbol for the function. x is the independent variable or argument. y is the dependent variable. $f(x)$ is the value of the function at x, or the image of x.

1.1 Assess Your Understanding

'Are You Prepared?'

Answers are given at the end of these exercises. If you get a wrong answer, read the pages listed in red.

1. The inequality $-1 < x < 3$ can be written in interval notation as _____. (pp. A66–A67)

2. If $x = -2$, the value of the expression $3x^2 - 5x + \dfrac{1}{x}$ is _____. (p. A3)

3. The domain of the variable in the expression $\dfrac{x - 3}{x + 4}$ is _____. (pp. A3–A4)

4. Solve the inequality: $3 - 2x > 5$. Graph the solution set. (pp. A69–A70)

Concepts and Vocabulary

5. If f is a function defined by the equation $y = f(x)$, then x is called the _____ variable and y is the _____ variable.

6. The set of all images of the elements in the domain of a function is called the _____.

7. If the domain of f is all real numbers in the interval $[0, 7]$ and the domain of g is all real numbers in the interval $[-2, 5]$, the domain of $f + g$ is all real numbers in the interval _____.

8. The domain of $\dfrac{f}{g}$ consists of numbers x for which $g(x)$ _____ 0 that are in the domains of both _____ and _____.

9. If $f(x) = x + 1$ and $g(x) = x^3$, then _____ $= x^3 - (x + 1)$.

10. *True or False:* Every relation is a function.

11. *True or False:* The domain of $(f \cdot g)(x)$ consists of the numbers x that are in the domains of both f and g.

12. *True or False:* The independent variable is sometimes referred to as the argument of the function.

13. *True or False:* If no domain is specified for a function f, then the domain of f is taken to be the set of real numbers.

14. *True or False:* The domain of the function $f(x) = \dfrac{x^2 - 4}{x}$ is $\{x \,|\, x \neq \pm 2\}$.

Skill Building

In Problems 15–26, determine whether each relation represents a function. For each function, state the domain and range.

15.
Person Birthday
Elvis, Colleen, Kaleigh, Marissa → Jan. 8, Mar. 15, Sept. 17

16.
Father Daughter
Bob, John, Chuck → Beth, Diane, Linda, Marcia

17.

18.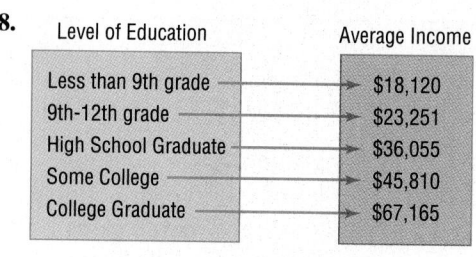

19. $\{(2, 6), (-3, 6), (4, 9), (2, 10)\}$ **20.** $\{(-2, 5), (-1, 3), (3, 7), (4, 12)\}$ **21.** $\{(1, 3), (2, 3), (3, 3), (4, 3)\}$

22. $\{(0, -2), (1, 3), (2, 3), (3, 7)\}$ **23.** $\{(-2, 4), (-2, 6), (0, 3), (3, 7)\}$ **24.** $\{(-4, 4), (-3, 3), (-2, 2), (-1, 1), (-4, 0)\}$

25. $\{(-2, 4), (-1, 1), (0, 0), (1, 1)\}$ **26.** $\{(-2, 16), (-1, 4), (0, 3), (1, 4)\}$

In Problems 27–38, determine whether the equation defines y as a function of x.

27. $y = x^2$ **28.** $y = x^3$ **29.** $y = \dfrac{1}{x}$ **30.** $y = |x|$

31. $y^2 = 4 - x^2$ **32.** $y = \pm\sqrt{1 - 2x}$ **33.** $x = y^2$ **34.** $x + y^2 = 1$

35. $y = 2x^2 - 3x + 4$ **36.** $y = \dfrac{3x - 1}{x + 2}$ **37.** $2x^2 + 3y^2 = 1$ **38.** $x^2 - 4y^2 = 1$

In Problems 39–46, find the following values for each function:

(a) $f(0)$ (b) $f(1)$ (c) $f(-1)$ (d) $f(-x)$ (e) $-f(x)$ (f) $f(x + 1)$ (g) $f(2x)$ (h) $f(x + h)$

39. $f(x) = 3x^2 + 2x - 4$ **40.** $f(x) = -2x^2 + x - 1$ **41.** $f(x) = \dfrac{x}{x^2 + 1}$ **42.** $f(x) = \dfrac{x^2 - 1}{x + 4}$

43. $f(x) = |x| + 4$ **44.** $f(x) = \sqrt{x^2 + x}$ **45.** $f(x) = \dfrac{2x + 1}{3x - 5}$ **46.** $f(x) = 1 - \dfrac{1}{(x + 2)^2}$

In Problems 47–60, find the domain of each function.

47. $f(x) = -5x + 4$ **48.** $f(x) = x^2 + 2$ **49.** $f(x) = \dfrac{x}{x^2 + 1}$ **50.** $f(x) = \dfrac{x^2}{x^2 + 1}$

51. $g(x) = \dfrac{x}{x^2 - 16}$ **52.** $h(x) = \dfrac{2x}{x^2 - 4}$ **53.** $F(x) = \dfrac{x - 2}{x^3 + x}$ **54.** $G(x) = \dfrac{x + 4}{x^3 - 4x}$

55. $h(x) = \sqrt{3x - 12}$ **56.** $G(x) = \sqrt{1 - x}$ **57.** $f(x) = \dfrac{4}{\sqrt{x - 9}}$ **58.** $f(x) = \dfrac{x}{\sqrt{x - 4}}$

59. $p(x) = \sqrt{\dfrac{2}{x - 1}}$ **60.** $q(x) = \sqrt{-x - 2}$

In Problems 61–70, for the given functions f and g, find the following functions and state the domain of each.

(a) $(f + g)(x)$ (b) $(f - g)(x)$ (c) $(f \cdot g)(x)$ (d) $\left(\dfrac{f}{g}\right)(x)$

(e) $(f + g)(3)$ (f) $(f - g)(4)$ (g) $(f \cdot g)(2)$ (h) $\left(\dfrac{f}{g}\right)(1)$

61. $f(x) = 3x + 4$; $g(x) = 2x - 3$ **62.** $f(x) = 2x + 1$; $g(x) = 3x - 2$

63. $f(x) = x - 1$; $g(x) = 2x^2$ **64.** $f(x) = 2x^2 + 3$; $g(x) = 4x^3 + 1$

65. $f(x) = \sqrt{x}$; $g(x) = 3x - 5$ **66.** $f(x) = |x|$; $g(x) = x$

67. $f(x) = 1 + \dfrac{1}{x}$; $g(x) = \dfrac{1}{x}$ **68.** $f(x) = \sqrt{x - 1}$; $g(x) = \sqrt{4 - x}$

69. $f(x) = \dfrac{2x + 3}{3x - 2}$; $g(x) = \dfrac{4x}{3x - 2}$ **70.** $f(x) = \sqrt{x + 1}$; $g(x) = \dfrac{2}{x}$

71. Given $f(x) = 3x + 1$ and $(f + g)(x) = 6 - \dfrac{1}{2}x$, find the function g.

72. Given $f(x) = \dfrac{1}{x}$ and $\left(\dfrac{f}{g}\right)(x) = \dfrac{x + 1}{x^2 - x}$, find the function g.

In Problems 73–78, find the difference quotient of f, that is, find $\dfrac{f(x + h) - f(x)}{h}$, $h \neq 0$, for each function. Be sure to simplify.

73. $f(x) = 4x + 3$ **74.** $f(x) = -3x + 1$ **75.** $f(x) = x^2 - x + 4$

76. $f(x) = x^2 + 5x - 1$ **77.** $f(x) = x^3 - 2$ **78.** $f(x) = \dfrac{1}{x + 3}$

Applications and Extensions

79. If $f(x) = 2x^3 + Ax^2 + 4x - 5$ and $f(2) = 5$, what is the value of A?

80. If $f(x) = 3x^2 - Bx + 4$ and $f(-1) = 12$, what is the value of B?

81. If $f(x) = \dfrac{3x + 8}{2x - A}$ and $f(0) = 2$, what is the value of A?

82. If $f(x) = \dfrac{2x - B}{3x + 4}$ and $f(2) = \dfrac{1}{2}$, what is the value of B?

83. If $f(x) = \dfrac{2x - A}{x - 3}$ and $f(4) = 0$, what is the value of A? Where is f not defined?

84. If $f(x) = \dfrac{x - B}{x - A}$, $f(2) = 0$, and $f(1)$ is undefined, what are the values of A and B?

85. Geometry Express the area A of a rectangle as a function of the length x if the length of the rectangle is twice its width.

86. Geometry Express the area A of an isosceles right triangle as a function of the length x of one of the two equal sides.

87. Constructing Functions Express the gross salary G of a person who earns \$10 per hour as a function of the number x of hours worked.

88. Constructing Functions Tiffany, a commissioned salesperson, earns \$100 base pay plus \$10 per item sold. Express her gross salary G as a function of the number x of items sold.

89. Population as a Function of Age The function

$$P(a) = 0.015a^2 - 4.962a + 290.580$$

represents the population P (in millions) of Americans in 2005 that are a years of age or older.
[SOURCE: U.S. Census Bureau]
(a) Identify the dependent and independent variable.
(b) Evaluate $P(20)$. Provide a verbal explanation of the meaning of $P(20)$.
(c) Evaluate $P(0)$. Provide a verbal explanation of the meaning of $P(0)$.

90. Number of Rooms The function

$$N(r) = -1.44r^2 + 14.52r - 14.96$$

represents the number N of housing units (in millions) in 2005 that have r rooms, where $1 \le r \le 9$.
[SOURCE: U.S. Census Bureau]
(a) Identify the dependent and independent variable.
(b) Evaluate $N(3)$. Provide a verbal explanation of the meaning of $N(3)$.
(c) Why is it unreasonable to evaluate $N(0)$?

91. Effect of Gravity on Earth If a rock falls from a height of 20 meters on Earth, the height H (in meters) after x seconds is approximately

$$H(x) = 20 - 4.9x^2$$

(a) What is the height of the rock when $x = 1$ second? $x = 1.1$ seconds? $x = 1.2$ seconds? $x = 1.3$ seconds?
(b) When is the height of the rock 15 meters? When is it 10 meters? When is it 5 meters?
(c) When does the rock strike the ground?

92. Effect of Gravity on Jupiter If a rock falls from a height of 20 meters on the planet Jupiter, its height H (in meters) after x seconds is approximately

$$H(x) = 20 - 13x^2$$

(a) What is the height of the rock when $x = 1$ second? $x = 1.1$ seconds? $x = 1.2$ seconds?
(b) When is the height of the rock 15 meters? When is it 10 meters? When is it 5 meters?
(c) When does the rock strike the ground?

93. Cost of Trans-Atlantic Travel A Boeing 747 crosses the Atlantic Ocean (3000 miles) with an airspeed of 500 miles per hour. The cost C (in dollars) per passenger is given by

$$C(x) = 100 + \frac{x}{10} + \frac{36{,}000}{x}$$

where x is the ground speed (airspeed \pm wind).
(a) What is the cost per passenger for quiescent (no wind) conditions?
(b) What is the cost per passenger with a head wind of 50 miles per hour?
(c) What is the cost per passenger with a tail wind of 100 miles per hour?
(d) What is the cost per passenger with a head wind of 100 miles per hour?

94. Cross-sectional Area The cross-sectional area of a beam cut from a log with radius 1 foot is given by the function $A(x) = 4x\sqrt{1 - x^2}$, where x represents the length, in feet, of half the base of the beam. See the figure. Determine the cross-sectional area of the beam if the length of half the base of the beam is as follows:
(a) One-third of a foot
(b) One-half of a foot
(c) Two-thirds of a foot

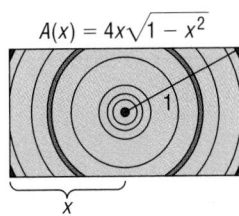

$$A(x) = 4x\sqrt{1 - x^2}$$

95. Economics The **participation rate** is the number of people in the labor force divided by the civilian population (excludes military). Let $L(x)$ represent the size of the labor force in year x and $P(x)$ represent the civilian population in year x. Determine a function that represents the participation rate R as a function of x.

96. Crimes Suppose that $V(x)$ represents the number of violent crimes committed in year x and $P(x)$ represents the number of property crimes committed in year x. Determine a function T that represents the combined total of violent crimes and property crimes in year x.

97. Health Care Suppose that $P(x)$ represents the percentage of income spent on health care in year x and $I(x)$ represents income in year x. Determine a function H that represents total health care expenditures in year x.

98. Income Tax Suppose that $I(x)$ represents the income of an individual in year x before taxes and $T(x)$ represents the individual's tax bill in year x. Determine a function N that represents the individual's net income (income after taxes) in year x.

99. Profit Function Suppose that the revenue R, in dollars, from selling x cell phones, in hundreds, is $R(x) = -1.2x^2 + 220x$. The cost C, in dollars, of selling x cell phones is $C(x) = 0.05x^3 - 2x^2 + 65x + 500$.
(a) Find the profit function, $P(x) = R(x) - C(x)$.
(b) Find the profit if $x = 15$ hundred cell phones are sold.
(c) Interpret $P(15)$.

100. Profit Function Suppose that the revenue R, in dollars, from selling x clocks is $R(x) = 30x$. The cost C, in dollars, of selling x clocks is $C(x) = 0.1x^2 + 7x + 400$.
(a) Find the profit function, $P(x) = R(x) - C(x)$.
(b) Find the profit if $x = 30$ clocks are sold.
(c) Interpret $P(30)$.

101. Some functions f have the property that $f(a + b) = f(a) + f(b)$ for all real numbers a and b. Which of the following functions have this property?
(a) $h(x) = 2x$ (b) $g(x) = x^2$
(c) $F(x) = 5x - 2$ (d) $G(x) = \dfrac{1}{x}$

Discussion and Writing

102. Are the functions $f(x) = x - 1$ and $g(x) = \dfrac{x^2 - 1}{x + 1}$ the same? Explain.

103. Investigate when, historically, the use of the function notation $y = f(x)$ first appeared.

'Are You Prepared?' Answers

1. $(-1, 3)$ **2.** 21.5 **3.** $\{x \mid x \neq -4\}$ **4.** $\{x \mid x < -1\}$

1.2 The Graph of a Function

PREPARING FOR THIS SECTION *Before getting started, review the following:*

- Graphs of Equations (Foundations, Section F.2, pp. 8–10)
- Intercepts (Foundations, Section F.2, pp. 10–11)

✏ Now work the 'Are You Prepared?' problems on page 59.

OBJECTIVES 1 Identify the Graph of a Function
 2 Obtain Information from or about the Graph of a Function

In applications, a graph often demonstrates more clearly the relationship between two variables than, say, an equation or table would. For example, Table 1 on page 55 shows the average price of gasoline, adjusted for inflation, for the years 1978–2004. If we plot these data and then connect the points, we obtain Figure 13.

Figure 13
Price of Gasoline Adjusted
for Inflation, 1978–2004

Table 1

Year	Price	Year	Price
1978	1.5207	1992	1.3460
1979	1.9263	1993	1.4622
1980	2.4197	1994	1.4166
1981	2.4258	1995	1.4111
1982	2.1136	1996	1.4804
1983	1.8210	1997	1.4697
1984	1.7732	1998	1.2685
1985	1.6998	1999	1.4629
1986	1.3043	2000	1.7469
1987	1.2831	2001	1.6821
1988	1.2674	2002	1.5339
1989	1.3134	2003	1.8308
1990	1.4160	2004	2.0047
1991	1.4490		

SOURCE: *Statistical Abstract of the United States.*

COMMENT:
When we select a viewing window to graph a function on a calculator, the values of Xmin, Xmax give the domain that we wish to view, while Ymin, Ymax give the range that we wish to view. These settings usually do not represent the actual domain and range of the function. ■

We can see from the graph that the price of gasoline (adjusted for inflation) rose rapidly from 1978 to 1981 and was falling from 2000 to 2002. The graph also shows that the highest price occurred in 1981. To learn information such as this from an equation requires that some calculations be made.

Look again at Figure 13. The graph shows that for each date on the horizontal axis there is only one price on the vertical axis. The graph represents a function, although the exact rule for getting from date to price is not given.

When a function is defined by an equation in x and y, the **graph of the function** is the graph of the equation, that is, the set of points (x, y) in the xy-plane that satisfies the equation.

1 Identify the Graph of a Function

Not every collection of points in the xy-plane represents the graph of a function. Remember, for a function, each number x in the domain has exactly one image y in the range. This means that the graph of a function cannot contain two points with the same x-coordinate and different y-coordinates. Therefore, the graph of a function must satisfy the following **vertical-line test**.

Theorem

Vertical-line Test
A set of points in the xy-plane is the graph of a function if and only if every vertical line intersects the graph in at most one point.

In other words, if any vertical line intersects a graph at more than one point, the graph is not the graph of a function.

EXAMPLE 1 **Identifying the Graph of a Function**

Which of the graphs in Figure 14 are graphs of functions?

Figure 14

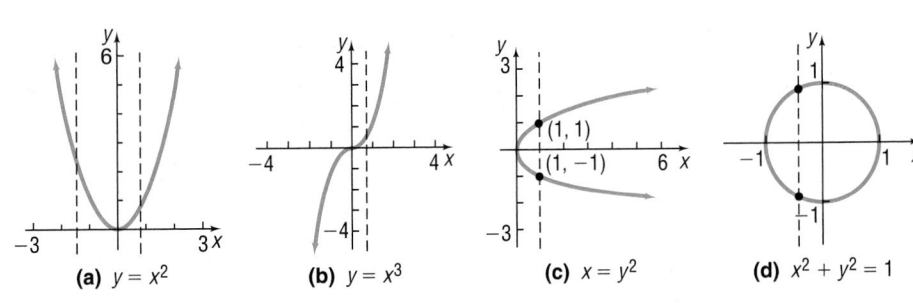

(a) $y = x^2$ (b) $y = x^3$ (c) $x = y^2$ (d) $x^2 + y^2 = 1$

Solution The graphs in Figures 14(a) and 14(b) are graphs of functions, because every vertical line intersects each graph in at most one point. The graphs in Figures 14(c) and 14(d) are not graphs of functions, because there is a vertical line that intersects each graph in more than one point. ∎

NOW WORK PROBLEM 15.

2 Obtain Information from or about the Graph of a Function

If (x, y) is a point on the graph of a function f, then y is the value of f at x; that is, $y = f(x)$. The next example illustrates how to obtain information about a function if its graph is given.

EXAMPLE 2 **Obtaining Information from the Graph of a Function**

Figure 15

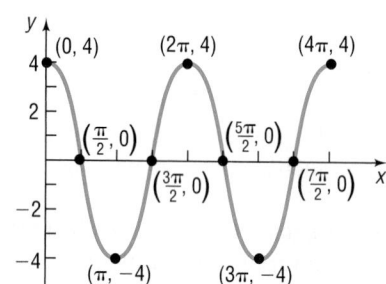

Let f be the function whose graph is given in Figure 15. (The graph of f might represent the distance that the bob of a pendulum is from its *at-rest* position. Negative values of y mean that the pendulum is to the left of the at-rest position, and positive values of y mean that the pendulum is to the right of the at-rest position.)

(a) What are $f(0)$, $f\left(\dfrac{3\pi}{2}\right)$, and $f(3\pi)$?

(b) What is the domain of f?

(c) What is the range of f?

(d) List the intercepts. (Recall that these are the points, if any, where the graph crosses or touches the coordinate axes.)

(e) What are the zeros of f?

(f) How often does the line $y = 2$ intersect the graph?

(g) For what values of x does $f(x) = -4$?

(h) For what values of x is $f(x) > 0$?

Solution
(a) Since $(0, 4)$ is on the graph of f, the y-coordinate 4 is the value of f at the x-coordinate 0; that is, $f(0) = 4$. In a similar way, we find that when $x = \dfrac{3\pi}{2}$ then $y = 0$, so $f\left(\dfrac{3\pi}{2}\right) = 0$. When $x = 3\pi$, then $y = -4$, so $f(3\pi) = -4$.

(b) To determine the domain of f, we notice that the points on the graph of f have x-coordinates between 0 and 4π, inclusive; and for each number x between 0 and 4π, there is a point $(x, f(x))$ on the graph. The domain of f is $\{x | 0 \le x \le 4\pi\}$ or the interval $[0, 4\pi]$.

(c) The points on the graph all have y-coordinates between -4 and 4, inclusive; and for each such number y, there is at least one number x in the domain. The range of f is $\{y | -4 \le y \le 4\}$ or the interval $[-4, 4]$.

(d) The intercepts are

$$(0, 4), \left(\frac{\pi}{2}, 0\right), \left(\frac{3\pi}{2}, 0\right), \left(\frac{5\pi}{2}, 0\right), \quad \text{and} \quad \left(\frac{7\pi}{2}, 0\right)$$

(e) Recall that the zeros of function f are those numbers at which $f(x) = 0$. Look at the intercepts listed in part (d). Each x-intercept is a number for which $f(x) = 0$. The zeros of f are the same as the x-intercepts of the graph of f. That is, $\dfrac{\pi}{2}, \dfrac{3\pi}{2}, \dfrac{5\pi}{2}$, and $\dfrac{7\pi}{2}$ are zeros of f.

(f) If we draw the horizontal line $y = 2$ on the graph in Figure 15, we find that it intersects the graph four times.

(g) Since $(\pi, -4)$ and $(3\pi, -4)$ are the only points on the graph for which $y = f(x) = -4$, we have $f(x) = -4$ when $x = \pi$ and $x = 3\pi$.

(h) To determine where $f(x) > 0$, we look at Figure 15 and determine the x-values for which the y-coordinate is positive. This occurs on the intervals $\left[0, \dfrac{\pi}{2}\right), \left(\dfrac{3\pi}{2}, \dfrac{5\pi}{2}\right)$, and $\left(\dfrac{7\pi}{2}, 4\pi\right]$. Using inequality notation, $f(x) > 0$ for

$$0 \le x < \dfrac{\pi}{2}, \dfrac{3\pi}{2} < x < \dfrac{5\pi}{2}, \text{ and } \dfrac{7\pi}{2} < x \le 4\pi.$$ ∎

When the graph of a function is given, its domain may be viewed as the shadow created by the graph on the x-axis by vertical beams of light. Its range can be viewed as the shadow created by the graph on the y-axis by horizontal beams of light. Try this technique with the graph given in Figure 15.

NOW WORK PROBLEMS 9 AND 13.

EXAMPLE 3

Obtaining Information about the Graph of a Function

Consider the function: $f(x) = \dfrac{x + 1}{x + 2}$

(a) Is the point $\left(1, \dfrac{1}{2}\right)$ on the graph of f?

(b) If $x = 2$, what is $f(x)$? What point is on the graph of f?

(c) If $f(x) = 2$, what is x? What point is on the graph of f?

(d) What are the zeros of f? What point is on the graph of f?

Solution

(a) When $x = 1$, then

$$f(x) = \dfrac{x + 1}{x + 2}$$

$$f(1) = \dfrac{1 + 1}{1 + 2} = \dfrac{2}{3}$$

The point $\left(1, \dfrac{2}{3}\right)$ is on the graph of f; the point $\left(1, \dfrac{1}{2}\right)$ is not.

(b) If $x = 2$, then

$$f(x) = \dfrac{x + 1}{x + 2}$$

$$f(2) = \dfrac{2 + 1}{2 + 2} = \dfrac{3}{4}$$

The point $\left(2, \dfrac{3}{4}\right)$ is on the graph of f.

(c) If $f(x) = 2$, then

$$f(x) = 2$$
$$\dfrac{x + 1}{x + 2} = 2$$
$$x + 1 = 2(x + 2) \quad \text{Multiply both sides by } x + 2.$$
$$x + 1 = 2x + 4 \quad \text{Remove parentheses.}$$
$$x = -3 \quad \text{Solve for } x.$$

If $f(x) = 2$, then $x = -3$. The point $(-3, 2)$ is on the graph of f.

(d) The zeros of f are those numbers for which $f(x) = 0$. The only solution of the equation $f(x) = 0$ or $\dfrac{x + 1}{x + 2} = 0$ is $x = -1$ So, -1 is the only zero of f. Since $f(-1) = 0$, the point $(-1, 0)$ is on the graph of f. ∎

 NOW WORK PROBLEM 25.

EXAMPLE 4	**Average Cost Function**

The average cost \overline{C} of manufacturing x computers per day is given by the function

$$\overline{C}(x) = 0.56x^2 - 34.39x + 1212.57 + \frac{20{,}000}{x}$$

Determine the average cost of manufacturing:

(a) 30 computers in a day
(b) 40 computers in a day
(c) 50 computers in a day
(d) Graph the function $\overline{C} = \overline{C}(x), 0 < x \le 80$.
(e) Create a TABLE with TblStart $= 1$ and ΔTbl $= 1$. Which value of x minimizes the average cost?

Solution

(a) The average cost of manufacturing $x = 30$ computers is

$$\overline{C}(30) = 0.56(30)^2 - 34.39(30) + 1212.57 + \frac{20{,}000}{30} = \$1351.54$$

(b) The average cost of manufacturing $x = 40$ computers is

$$\overline{C}(40) = 0.56(40)^2 - 34.39(40) + 1212.57 + \frac{20{,}000}{40} = \$1232.97$$

(c) The average cost of manufacturing $x = 50$ computers is

$$\overline{C}(50) = 0.56(50)^2 - 34.39(50) + 1212.57 + \frac{20{,}000}{50} = \$1293.07$$

(d) See Figure 16 for the graph of $\overline{C} = \overline{C}(x)$.
(e) With the function $\overline{C} = \overline{C}(x)$ in Y_1, we create Table 2. We scroll down until we find a value of x for which Y_1 is smallest. Table 3 shows that manufacturing $x = 41$ computers minimizes the average cost at $\$1231.74$ per computer.

Figure 16

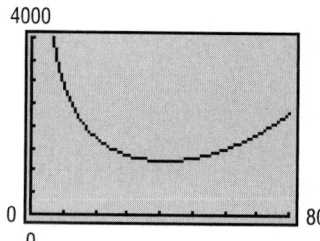

Table 2

X	Y1	
1	21179	
2	11146	
3	7781.1	
4	6084	
5	5054.6	
6	4359.7	
7	3856.4	

Y1◻.56X²−34.39X...

Table 3

X	Y1	
38	1240.7	
39	1235.9	
40	1233	
41	▓▓▓▓	
42	1232.2	
43	1234.4	
44	1238.1	

Y1=1231.74487805

∎

 NOW WORK PROBLEM 29.

Summary

Graph of a function	The collection of points (x, y) that satisfies the equation $y = f(x)$.
Vertical Line Test	A collection of points is the graph of a function provided that every vertical line intersects the graph in at most one point.

1.2 Assess Your Understanding

'Are You Prepared?'

Answers are given at the end of these exercises. If you get a wrong answer, read the pages listed in red.

1. The intercepts of the equation $x^2 + 4y^2 = 16$ are _____. (pp. 10–11)

2. *True or False:* The point $(-2, -6)$ is on the graph of the equation $x = 2y - 2$. (pp. 8–9)

Concepts and Vocabulary

3. A set of points in the xy-plane is the graph of a function if and only if every _____ line intersects the graph in at most one point.

4. If the point $(5, -3)$ is a point on the graph of f, then $f(___) = ___$.

5. Find a so that the point $(-1, 2)$ is on the graph of $f(x) = ax^2 + 4$.

6. *True or False:* A function can have more than one y-intercept.

7. *True or False:* The graph of a function $y = f(x)$ always crosses the y-axis.

8. *True or False:* The y-intercept of the graph of the function $y = f(x)$, whose domain is all real numbers, is $f(0)$.

Skill Building

9. Use the given graph of the function f to answer parts (a)–(o).

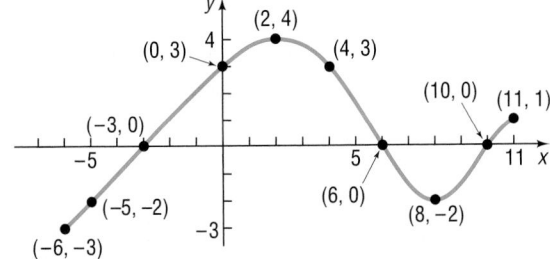

(a) Find $f(0)$ and $f(-6)$.
(b) Find $f(6)$ and $f(11)$.
(c) Is $f(3)$ positive or negative?
(d) Is $f(-4)$ positive or negative?
(e) For what values of x is $f(x) = 0$?
(f) For what values of x is $f(x) > 0$?
(g) What is the domain of f?
(h) What is the range of f?
(i) What are the x-intercepts?
(j) What is the y-intercept?
(k) How often does the line $y = \dfrac{1}{2}$ intersect the graph?
(l) How often does the line $x = 5$ intersect the graph?
(m) For what values of x does $f(x) = 3$?
(n) For what values of x does $f(x) = -2$?
(o) What are the zeros of f?

10. Use the given graph of the function f to answer parts (a)–(o).

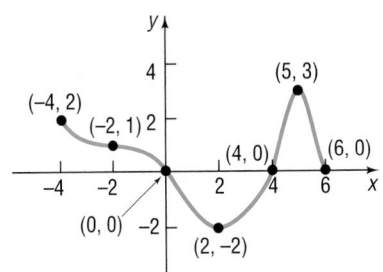

(a) Find $f(0)$ and $f(6)$.
(b) Find $f(2)$ and $f(-2)$.
(c) Is $f(3)$ positive or negative?
(d) Is $f(-1)$ positive or negative?
(e) For what values of x is $f(x) = 0$?
(f) For what values of x is $f(x) < 0$?
(g) What is the domain of f?
(h) What is the range of f?
(i) What are the x-intercepts?
(j) What is the y-intercept?
(k) How often does the line $y = -1$ intersect the graph?
(l) How often does the line $x = 1$ intersect the graph?
(m) For what value of x does $f(x) = 3$?
(n) For what value of x does $f(x) = -2$?
(o) What are the zeros of f?

In Problems 11–22, determine whether the graph is that of a function by using the vertical-line test. If it is, use the graph to find:
 (a) *The domain and range*
 (b) *The intercepts, if any*
 (c) *Any symmetry with respect to the x-axis, the y-axis, or the origin*

11.

12.

13.

14.

15.

16.

17.

18.

19.

20.

21.

22.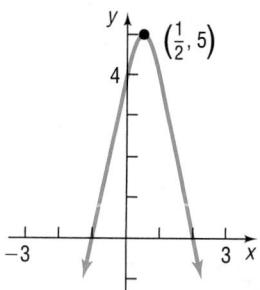

In Problems 23–28, answer the questions about the given function.

23. $f(x) = 2x^2 - x - 1$
 (a) Is the point $(-1, 2)$ on the graph of f?
 (b) If $x = -2$, what is $f(x)$? What point is on the graph of f?
 (c) If $f(x) = -1$, what is x? What point(s) are on the graph of f?
 (d) What is the domain of f?
 (e) List the x-intercepts, if any, of the graph of f.
 (f) List the y-intercept, if there is one, of the graph of f.
 (g) What are the zeros of f?

24. $f(x) = -3x^2 + 5x$
 (a) Is the point $(-1, 2)$ on the graph of f?
 (b) If $x = -2$, what is $f(x)$? What point is on the graph of f?
 (c) If $f(x) = -2$, what is x? What point(s) are on the graph of f?
 (d) What is the domain of f?
 (e) List the x-intercepts, if any, of the graph of f.
 (f) List the y-intercept, if there is one, of the graph of f.
 (g) What are the zeros of f?

25. $f(x) = \dfrac{x + 2}{x - 6}$
 (a) Is the point $(3, 14)$ on the graph of f?
 (b) If $x = 4$, what is $f(x)$? What point is on the graph of f?
 (c) If $f(x) = 2$, what is x? What point(s) are on the graph of f?
 (d) What is the domain of f?
 (e) List the x-intercepts, if any, of the graph of f.
 (f) List the y-intercept, if there is one, of the graph of f.
 (g) What are the zeros of f?

26. $f(x) = \dfrac{x^2 + 2}{x + 4}$
 (a) Is the point $\left(1, \dfrac{3}{5}\right)$ on the graph of f?

(b) If $x = 0$, what is $f(x)$? What point is on the graph of f?
(c) If $f(x) = \dfrac{1}{2}$, what is x? What point(s) are on the graph of f?
(d) What is the domain of f?
(e) List the x-intercepts, if any, of the graph of f.
(f) List the y-intercept, if there is one, of the graph of f.
(g) What are the zeros of f?

27. $f(x) = \dfrac{2x^2}{x^4 + 1}$
 (a) Is the point $(-1, 1)$ on the graph of f?
 (b) If $x = 2$, what is $f(x)$? What point is on the graph of f?
 (c) If $f(x) = 1$, what is x? What point(s) are on the graph of f?
 (d) What is the domain of f?
 (e) List the x-intercepts, if any, of the graph of f.
 (f) List the y-intercept, if there is one, of the graph of f.
 (g) What are the zeros of f?

28. $f(x) = \dfrac{2x}{x - 2}$
 (a) Is the point $\left(\dfrac{1}{2}, -\dfrac{2}{3}\right)$ on the graph of f?
 (b) If $x = 4$, what is $f(x)$? What point is on the graph of f?
 (c) If $f(x) = 1$, what is x? What point(s) are on the graph of f?
 (d) What is the domain of f?
 (e) List the x-intercepts, if any, of the graph of f.
 (f) List the y-intercept, if there is one, of the graph of f.
 (g) What are the zeros of f?

Applications and Extensions

29. Motion of a Golf Ball A golf ball is hit with an initial velocity of 130 feet per second at an inclination of 45° to the horizontal. In physics, it is established that the height h of the golf ball is given by the function

$$h(x) = \frac{-32x^2}{130^2} + x$$

where x is the horizontal distance that the golf ball has traveled.

(a) Determine the height of the golf ball after it has traveled 100 feet.
(b) What is the height after it has traveled 300 feet?
(c) What is the height after it has traveled 500 feet?
(d) How far was the golf ball hit?
(e) Graph the function $h = h(x)$.
(f) Use a graphing utility to determine the distance that the ball has traveled when the height of the ball is 90 feet.
(g) Create a TABLE with TblStart = 0 and ΔTbl = 25. To the nearest 25 feet, how far does the ball travel before it reaches a maximum height? What is the maximum height?
(h) Adjust the value of ΔTbl until you determine the distance, to within 1 foot, that the ball travels before it reaches a maximum height.

30. Cross-sectional Area The cross-sectional area of a beam cut from a log with radius 1 foot is given by the function $A(x) = 4x\sqrt{1 - x^2}$, where x represents the length, in feet, of half the base of the beam. See the figure.

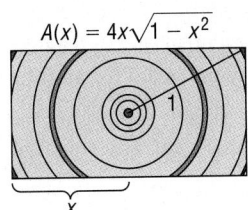

$$A(x) = 4x\sqrt{1 - x^2}$$

(a) Find the domain of A.
(b) Graph the function $A = A(x)$.

(c) Create a TABLE with TblStart = 0 and ΔTbl = 0.1 for $0 \le x \le 1$. Which value of x maximizes the cross-sectional area? What should be the length of the base of the beam to maximize the cross-sectional area?

31. Cost of Trans-Atlantic Travel A Boeing 747 crosses the Atlantic Ocean (3000 miles) with an airspeed of 500 miles per hour. The cost C (in dollars) per passenger is given by

$$C(x) = 100 + \frac{x}{10} + \frac{36,000}{x}$$

where x is the ground speed (airspeed ± wind).
(a) Graph the function $C = C(x)$.
(b) Create a TABLE with TblStart = 0 and ΔTbl = 50.
(c) To the nearest 50 miles per hour, what ground speed minimizes the cost per passenger?

32. Effect of Elevation on Weight If an object weighs m pounds at sea level, then its weight W (in pounds) at a height of h miles above sea level is given approximately by

$$W(h) = m\left(\frac{4000}{4000 + h}\right)^2$$

(a) If Amy weighs 120 pounds at sea level, how much will she weigh on Pike's Peak, which is 14,110 feet above sea level?
(b) Use a graphing utility to graph the function $W = W(h)$. Use $m = 120$ pounds.
(c) Create a Table with TblStart = 0 and ΔTbl = 0.5 to see how the weight W varies as h changes from 0 to 5 miles.
(d) At what height will Amy weigh 119.95 pounds?
(e) Does your answer to part (d) seem reasonable? Explain.

33. The graph of two functions, f and g, is shown below. Use the graph to answer parts (a)–(f).

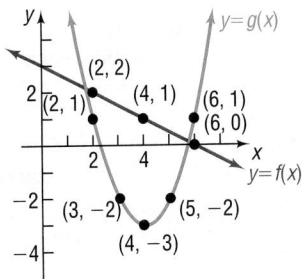

(a) $(f + g)(2)$ (b) $(f + g)(4)$
(c) $(f - g)(6)$ (d) $(g - f)(6)$
(e) $(f \cdot g)(2)$ (f) $\left(\dfrac{f}{g}\right)(4)$

Discussion and Writing

34. Describe how you would proceed to find the domain and range of a function if you were given its graph. How would your strategy change if you were given the equation defining the function instead of its graph?

35. How many x-intercepts can the graph of a function have? How many y-intercepts can the graph of a function have?

36. Is a graph that consists of a single point the graph of a function? Can you write the equation of such a function?

37. Match each of the following functions with the graphs that best describes the situation.
 (a) The cost of building a house as a function of its square footage
 (b) The height of an egg dropped from a 300-foot building as a function of time
 (c) The height of a human as a function of time
 (d) The demand for Big Macs as a function of price
 (e) The height of a child on a swing as a function of time

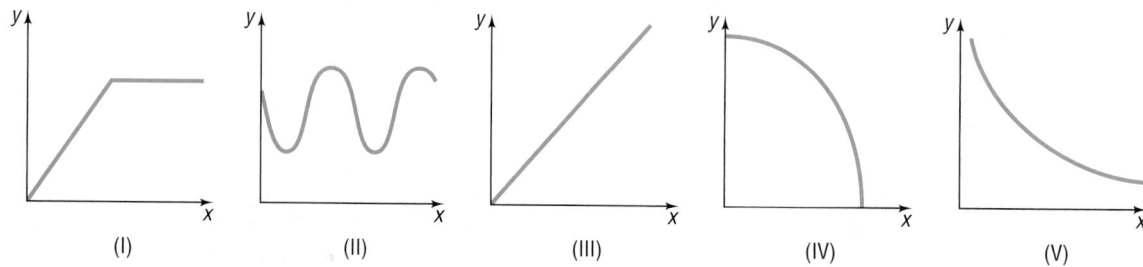

(I) (II) (III) (IV) (V)

38. Match each of the following functions with the graph that best describes the situation.
 (a) The temperature of a bowl of soup as a function of time
 (b) The number of hours of daylight per day over a 2-year period
 (c) The population of Florida as a function of time
 (d) The distance of a car traveling at a constant velocity as a function of time
 (e) The height of a golf ball hit with a 7-iron as a function of time

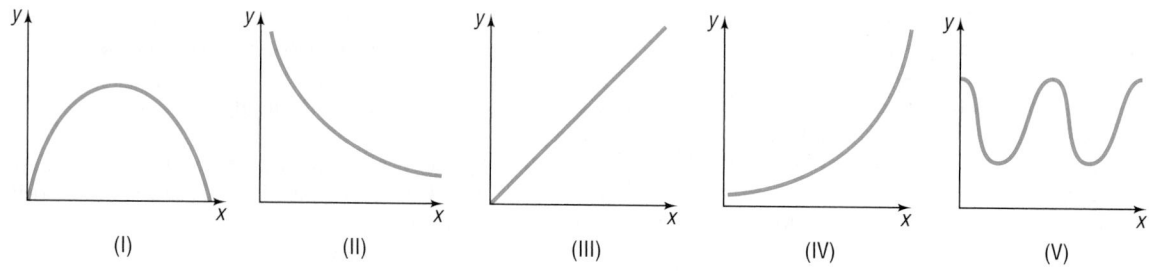

(I) (II) (III) (IV) (V)

39. Consider the following scenario: Barbara decides to take a walk. She leaves home, walks 2 blocks in 5 minutes at a constant speed, and realizes that she forgot to lock the door. So Barbara runs home in 1 minute. While at her doorstep, it takes her 1 minute to find her keys and lock the door. Barbara walks 5 blocks in 15 minutes and then decides to jog home. It takes her 7 minutes to get home. Draw a graph of Barbara's distance from home (in blocks) as a function of time.

40. Consider the following scenario: Jayne enjoys riding her bicycle through the woods. At the forest preserve, she gets on her bicycle and rides up a 2000-foot incline in 10 minutes. She then travels down the incline in 3 minutes. The next 5000 feet is level terrain and she covers the distance in 20 minutes. She rests for 15 minutes. Jayne then travels 10,000 feet in 30 minutes. Draw a graph of Jayne's distance traveled (in feet) as a function of time.

41. The following sketch represents the distance d (in miles) that Kevin is from home as a function of time t (in hours). Answer the questions based on the graph. In parts (a)–(g), how many hours elapsed and how far was Kevin from home during this time?

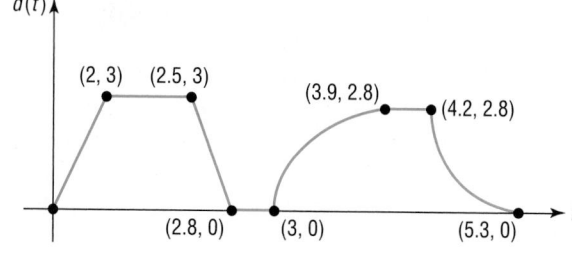

 (a) From $t = 0$ to $t = 2$
 (b) From $t = 2$ to $t = 2.5$
 (c) From $t = 2.5$ to $t = 2.8$
 (d) From $t = 2.8$ to $t = 3$
 (e) From $t = 3$ to $t = 3.9$
 (f) From $t = 3.9$ to $t = 4.2$
 (g) From $t = 4.2$ to $t = 5.3$
 (h) What is the farthest distance that Kevin is from home?
 (i) How many times did Kevin return home?

42. The following sketch represents the speed v (in miles per hour) of Michael's car as a function of time t (in minutes).

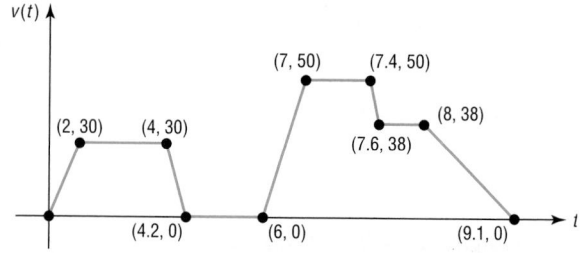

 (a) Over what interval of time is Michael traveling fastest?
 (b) Over what interval(s) of time is Michael's speed zero?
 (c) What is Michael's speed between 0 and 2 minutes?
 (d) What is Michael's speed between 4.2 and 6 minutes?
 (e) What is Michael's speed between 7 and 7.4 minutes?
 (f) When is Michael's speed constant?

43. Draw the graph of a function whose domain is $\{x \mid -3 \le x \le 8, \ x \ne 5\}$ and whose range is $\{y \mid -1 \le y \le 2, \ y \ne 0\}$. What point(s) in the rectangle $-3 \le x \le 8, -1 \le y \le 2$ cannot be on the graph? Compare your graph with those of other students. What differences do you see?

44. Is there a function whose graph is symmetric with respect to the x-axis? Explain.

'Are You Prepared?' Answers

1. $(-4, 0), (4, 0), (0, -2), (0, 2)$ **2.** False

1.3 Properties of Functions

PREPARING FOR THIS SECTION *Before getting started, review the following:*

- Intervals (Appendix A, Section A.9, pp. A66–A67)
- Intercepts (Foundations, Section F.2, pp. 10–11)
- Slope of a Line (Foundations, Section F.3, pp. 18–20)
- Point–Slope Form of a Line (Foundations, Section F.3, p. 22)
- Symmetry (Foundations, Section F.2, pp. 11–13)

Now work the 'Are You Prepared?' problems on page 71.

OBJECTIVES
1. Determine Even and Odd Functions from a Graph
2. Identify Even and Odd Functions from the Equation
3. Use a Graph to Determine Where a Function Is Increasing, Decreasing, or Constant
4. Use a Graph to Locate Local Maxima and Local Minima
5. Use a Graphing Utility to Approximate Local Maxima and Local Minima and to Determine Where a Function Is Increasing or Decreasing
6. Find the Average Rate of Change of a Function

It is easiest to obtain the graph of a function $y = f(x)$ by knowing certain properties that the function has and the impact of these properties on the way that the graph will look.

1 Determine Even and Odd Functions from a Graph

The words *even* and *odd*, when applied to a function f, describe the symmetry that exists for the graph of the function.

A function f is even if and only if, whenever the point (x, y) is on the graph of f then the point $(-x, y)$ is also on the graph. Using function notation, we define an even function as follows:

A function f is **even** if, for every number x in its domain, the number $-x$ is also in the domain and

$$f(-x) = f(x)$$

A function f is odd if and only if, whenever the point (x, y) is on the graph of f then the point $(-x, -y)$ is also on the graph. Using function notation, we define an odd function as follows:

A function f is **odd** if, for every number x in its domain, the number $-x$ is also in the domain and

$$f(-x) = -f(x)$$

Refer to page 12, where the tests for symmetry are listed. The following results are then evident.

Theorem

> A function is even if and only if its graph is symmetric with respect to the *y*-axis. A function is odd if and only if its graph is symmetric with respect to the origin.

EXAMPLE 1 **Determining Even and Odd Functions from the Graph**

Determine whether each graph given in Figure 17 is the graph of an even function, an odd function, or a function that is neither even nor odd.

Figure 17

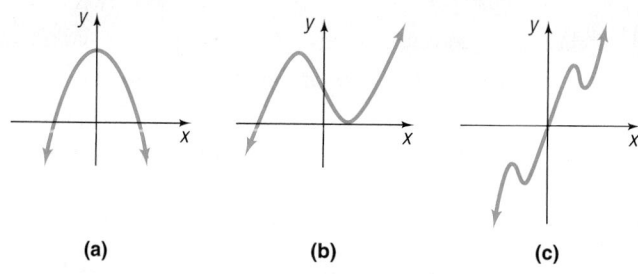

(a) (b) (c)

Solution (a) The graph in Figure 17(a) is that of an even function, because the graph is symmetric with respect to the *y*-axis.

(b) The function whose graph is given in Figure 17(b) is neither even nor odd, because the graph is neither symmetric with respect to the *y*-axis nor symmetric with respect to the origin.

(c) The function whose graph is given in Figure 17(c) is odd, because its graph is symmetric with respect to the origin. ∎

NOW WORK PROBLEMS 21(a), (b), AND (d).

2 Identify Even and Odd Functions from the Equation

In the next example, we use algebraic techniques to verify whether a given function is even, odd, or neither.

EXAMPLE 2 **Identifying Even and Odd Functions Algebraically**

Determine whether each of the following functions is even, odd, or neither. Then determine whether the graph is symmetric with respect to the *y*-axis or with respect to the origin.

(a) $f(x) = x^2 - 5$ (b) $g(x) = x^3 - 1$

(c) $h(x) = 5x^3 - x$ (d) $F(x) = |x|$

Solution (a) To determine whether f is even, odd, or neither, we replace x by $-x$ in $f(x) = x^2 - 5$. Then

$$f(-x) = (-x)^2 - 5 = x^2 - 5 = f(x)$$

Since $f(-x) = f(x)$, we conclude that f is an even function, and the graph is symmetric with respect to the *y*-axis.

(b) We replace x by $-x$ in $g(x) = x^3 - 1$. Then

$$g(-x) = (-x)^3 - 1 = -x^3 - 1$$

Since $g(-x) \neq g(x)$ and $g(-x) \neq -g(x) = -(x^3 - 1) = -x^3 + 1$, we conclude that g is neither even nor odd. The graph is not symmetric with respect to the *y*-axis nor is it symmetric with respect to the origin.

(c) We replace x by $-x$ in $h(x) = 5x^3 - x$. Then

$$h(-x) = 5(-x)^3 - (-x) = -5x^3 + x = -(5x^3 - x) = -h(x)$$

Since $h(-x) = -h(x)$, h is an odd function, and the graph of h is symmetric with respect to the origin.

(d) We replace x by $-x$ in $F(x) = |x|$. Then

$$F(-x) = |-x| = |-1| \cdot |x| = |x| = F(x)$$

Since $F(-x) = F(x)$, F is an even function, and the graph of F is symmetric with respect to the y-axis. ∎

 NOW WORK PROBLEM 33.

3 Use a Graph to Determine Where a Function Is Increasing, Decreasing, or Constant

Consider the graph given in Figure 18. If you look from left to right along the graph of the function, you will notice that parts of the graph are rising, parts are falling, and parts are horizontal. In such cases, the function is described as *increasing*, *decreasing*, or *constant*, respectively.

Figure 18

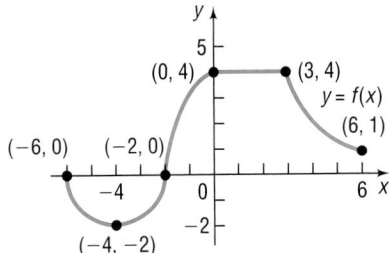

EXAMPLE 3

Determining Where a Function Is Increasing, Decreasing, or Constant from Its Graph

Where is the function in Figure 18 increasing? Where is it decreasing? Where is it constant?

Solution

WARNING
We describe the behavior of a graph in terms of its x-values. Do not say the graph in Figure 18 is increasing from the point $(-4, -2)$ to $(0, 4)$. Rather, say it is increasing on the interval $(-4, 0)$. ∎

To answer the question of where a function is increasing, where it is decreasing, and where it is constant, we use strict inequalities involving the independent variable x, or we use open intervals* of x-coordinates. The graph in Figure 18 is rising (increasing) from the point $(-4, -2)$ to the point $(0, 4)$, so we conclude that it is increasing on the open interval $(-4, 0)$ or for $-4 < x < 0$. The graph is falling (decreasing) from the point $(-6, 0)$ to the point $(-4, -2)$ and from the point $(3, 4)$ to the point $(6, 1)$. We conclude that the graph is decreasing on the open intervals $(-6, -4)$ and $(3, 6)$ or for $-6 < x < -4$ and $3 < x < 6$. The graph is constant on the open interval $(0, 3)$ or for $0 < x < 3$. ∎

More precise definitions follow:

> A function f is **increasing** on an open interval I if, for any choice of x_1 and x_2 in I, with $x_1 < x_2$, we have $f(x_1) < f(x_2)$.

> A function f is **decreasing** on an open interval I if, for any choice of x_1 and x_2 in I, with $x_1 < x_2$, we have $f(x_1) > f(x_2)$.

> A function f is **constant** on an open interval I if, for all choices of x in I, the values $f(x)$ are equal.

*The open interval (a, b) consists of all real numbers x for which $a < x < b$.

Figure 19 illustrates the definitions. The graph of an increasing function goes up from left to right, the graph of a decreasing function goes down from left to right, and the graph of a constant function remains at a fixed height.

Figure 19

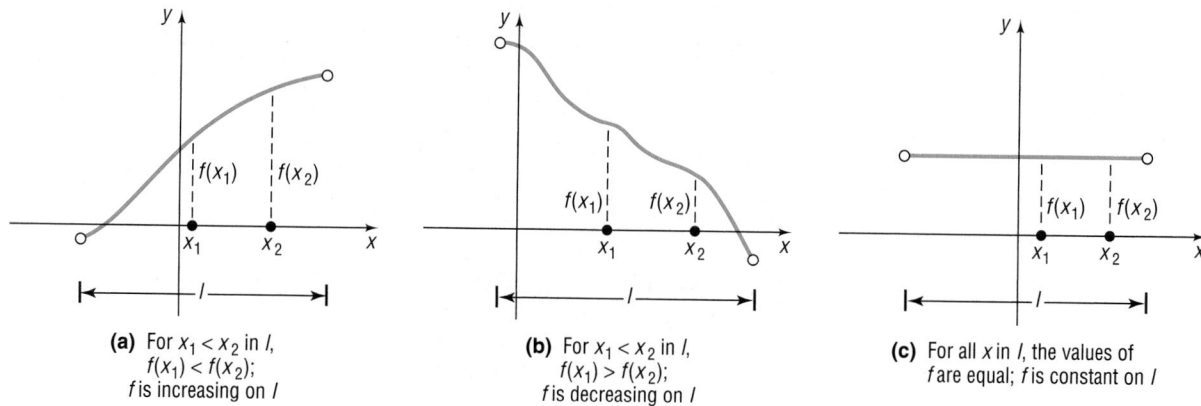

(a) For $x_1 < x_2$ in I,
$f(x_1) < f(x_2)$;
f is increasing on I

(b) For $x_1 < x_2$ in I,
$f(x_1) > f(x_2)$;
f is decreasing on I

(c) For all x in I, the values of
f are equal; f is constant on I

NOW WORK PROBLEMS 11, 13, 15, AND 21(c).

4 Use a Graph to Locate Local Maxima and Local Minima

When the graph of a function is increasing to the left of $x = c$ and decreasing to the right of $x = c$, then at c the value of f is largest. This value is called a *local maximum* of f. See Figure 20(a).

When the graph of a function is decreasing to the left of $x = c$ and is increasing to the right of $x = c$, then at c the value of f is the smallest. This value is called a *local minimum* of f. See Figure 20(b).

Figure 20

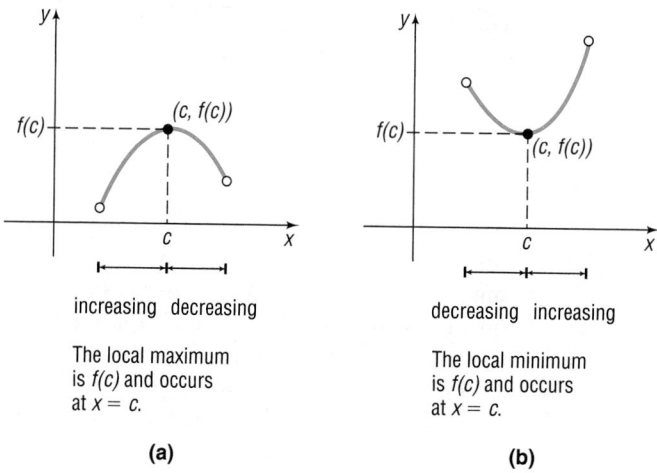

increasing decreasing

The local maximum
is $f(c)$ and occurs
at $x = c$.

(a)

decreasing increasing

The local minimum
is $f(c)$ and occurs
at $x = c$.

(b)

A function f has a **local maximum** at c if there is an open interval I containing c so that, for all x in I, $f(x) \leq f(c)$. We call $f(c)$ a **local maximum of f**.

A function f has a **local minimum** at c if there is an open interval I containing c so that, for all x in I, $f(x) \geq f(c)$. We call $f(c)$ a **local minimum of f**.

If f has a local maximum at c, then the value of f at c is greater than or equal to the values of f near c. If f has a local minimum at c, then the value of f at c is less than or equal to the values of f near c. The word *local* is used to suggest that it is only near c that the value $f(c)$ is largest or smallest.

| **EXAMPLE 4** | **Finding Local Maxima and Local Minima from the Graph of a Function and Determining Where the Function Is Increasing, Decreasing, or Constant** |

Figure 21

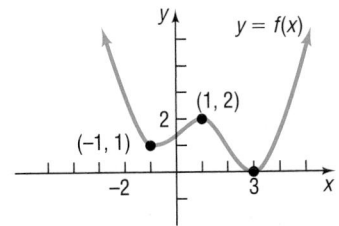

Figure 21 shows the graph of a function f.

(a) At what number(s), if any, does f have a local maximum?
(b) What are the local maxima?
(c) At what number(s), if any, does f have a local minimum?
(d) What are the local minima?
(e) List the intervals on which f is increasing. List the intervals on which f is decreasing.

Solution The domain of f is the set of real numbers.

(a) f has a local maximum at 1, since for all x close to 1, $x \neq 1$, we have $f(x) \leq f(1)$.
(b) The local maximum is $f(1) = 2$.
(c) f has a local minimum at -1 and at 3.
(d) The local minima are $f(-1) = 1$ and $f(3) = 0$.
(e) The function whose graph is given in Figure 21 is increasing for all values of x between -1 and 1 and for all values of x greater than 3. That is, the function is increasing on the intervals $(-1, 1)$ and $(3, \infty)$ or for $-1 < x < 1$ and $x > 3$. The function is decreasing for all values of x less than -1 and for all values of x between 1 and 3. That is, the function is decreasing on the intervals $(-\infty, -1)$ and $(1, 3)$ or for $x < -1$ and $1 < x < 3$. ■

WARNING
The y-value is the local maximum or local minimum and it occurs at some x-value. In Figure 21, we say the local maximum is 2 and occurs at x = 1. ■

━━ **NOW WORK PROBLEMS 17 AND 19.**

5 Use a Graphing Utility to Approximate Local Maxima and Local Minima and to Determine Where a Function Is Increasing or Decreasing

To locate the exact value at which a function f has a local maximum or a local minimum usually requires calculus. However, a graphing utility may be used to approximate these values by using the MAXIMUM and MINIMUM features.

| **EXAMPLE 5** | **Using a Graphing Utility to Approximate Local Maxima and Minima and to Determine Where a Function Is Increasing or Decreasing** |

(a) Use a graphing utility to graph $f(x) = 6x^3 - 12x + 5$ for $-2 < x < 2$. Approximate where f has a local maximum and where f has a local minimum.

(b) Determine where f is increasing and where it is decreasing.

Solution (a) Graphing utilities have a feature that finds the maximum or minimum point of a graph within a given interval. Graph the function f for $-2 < x < 2$. See Figure 22(a). Using MAXIMUM, we find that the local maximum is 11.53 and it occurs at $x = -0.82$, rounded to two decimal places. See Figure 22(b). Using MINIMUM, we find that the local minimum is -1.53 and it occurs at $x = 0.82$, rounded to two decimal places.

Figure 22

(a)

(b)

(b) Looking at Figures 22(a) and (b), we see that the graph of f is increasing from $x = -2$ to $x = -0.82$ and from $x = 0.82$ to $x = 2$, so f is increasing on the intervals $(-2, -0.82)$ and $(0.82, 2)$ or for $-2 < x < -0.82$ and $0.82 < x < 2$. The graph is decreasing from $x = -0.82$ to $x = 0.82$, so f is decreasing on the interval $(-0.82, 0.82)$ or for $-0.82 < x < 0.82$. ∎

NOW WORK PROBLEM 45.

6 Find the Average Rate of Change of a Function

In the Foundations chapter, Section F.3, we said that the slope of a line could be interpreted as the average rate of change. Often we are interested in the rate at which functions change. To find the average rate of change of a function between any two points on its graph, we calculate the slope of the line containing the two points.

> If c is in the domain of a function $y = f(x)$, the **average rate of change of f** from c to x is defined as
>
> $$\text{Average rate of change} = \frac{\Delta y}{\Delta x} = \frac{f(x) - f(c)}{x - c} \qquad x \neq c \qquad \textbf{(1)}$$

In calculus, this expression is called the **difference quotient** of f at c.

The symbol Δy in (1) is the "change in y" and Δx is the "change in x." The average rate of change of f is the change in y divided by the change in x.

| EXAMPLE 6 | Finding the Average Rate of Change |

Find the average rate of change of $f(x) = 3x^2$:

(a) From 1 to 3 (b) From 1 to 5 (c) From 1 to 7

Solution

(a) The average rate of change of $f(x) = 3x^2$ from 1 to 3 is

$$\frac{\Delta y}{\Delta x} = \frac{f(3) - f(1)}{3 - 1} = \frac{27 - 3}{3 - 1} = \frac{24}{2} = 12$$

(b) The average rate of change of $f(x) = 3x^2$ from 1 to 5 is

$$\frac{\Delta y}{\Delta x} = \frac{f(5) - f(1)}{5 - 1} = \frac{75 - 3}{5 - 1} = \frac{72}{4} = 18$$

(c) The average rate of change of $f(x) = 3x^2$ from 1 to 7 is

$$\frac{\Delta y}{\Delta x} = \frac{f(7) - f(1)}{7 - 1} = \frac{147 - 3}{7 - 1} = \frac{144}{6} = 24$$ ∎

See Figure 23 for a graph of $f(x) = 3x^2$. The function f is increasing for $x > 0$. The fact that the average rates of change are getting larger indicates that the graph is getting steeper, that is, it is increasing at an increasing rate.

NOW WORK PROBLEM 53.

Figure 23

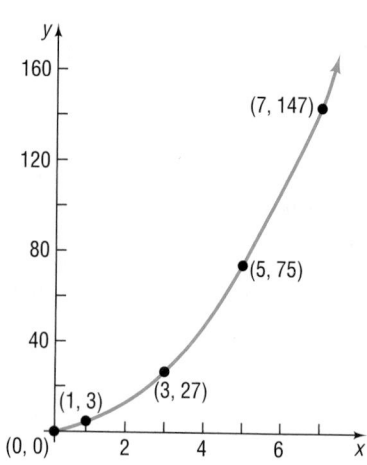

| EXAMPLE 7 | Average Rate of Change of a Population |

A strain of E-coli Beu 397-recA441 is placed into a nutrient broth at 30° Celsius and allowed to grow. The data shown in Table 4 are collected. The population is measured in grams and the time in hours. Since population P depends on time t

and each input corresponds to exactly one output, we can say that population is a function of time, so $P(t)$ represents the population at time t. For example, $P(2.5) = 0.18$. Figure 24 shows a graph of the function.

(a) Find the average rate of change of the population from 0 to 2.5 hours.

(b) Find the average rate of change of the population from 4.5 to 6 hours.

(c) What is happening to the average rate of change as time passes?

Table 4

Time (hours), t	Population (grams), P
0	0.09
2.5	0.18
3.5	0.26
4.5	0.35
6	0.50

Figure 24

Solution (a) Average rate of change $= \dfrac{\text{Population at 2.5 hours} - \text{Population at 0 hours}}{2.5 - 0}$

$$= \frac{P(2.5) - P(0)}{2.5 - 0}$$

$$= \frac{0.18 - 0.09}{2.5}$$

$$= 0.036 \text{ gram per hour}$$

On average, the population is increasing at a rate of 0.036 gram per hour from 0 to 2.5 hours.

(b) Average rate of change $= \dfrac{\text{Population at 6 hours} - \text{Population at 4.5 hours}}{6 - 4.5}$

$$= \frac{P(6) - P(4.5)}{6 - 4.5}$$

$$= \frac{0.5 - 0.35}{1.5}$$

$$= 0.1 \text{ gram per hour}$$

On average, the population is increasing at a rate of 0.1 gram per hour from 4.5 to 6 hours.

(c) Not only is the size of the population increasing over time, but the rate of increase is also increasing. We say that the population is increasing at an increasing rate. ∎

NOW WORK PROBLEM 69.

The Secant Line

The average rate of change of a function has an important geometric interpretation. Look at the graph of $y = f(x)$ in Figure 25. We have labeled two points on the graph: $(c, f(c))$ and $(x, f(x))$. The line containing these two points is called the **secant line**; its slope is

$$m_{\text{sec}} = \frac{f(x) - f(c)}{x - c}$$

Figure 25

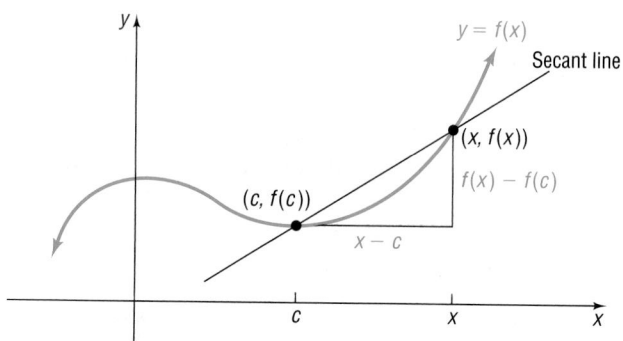

Theorem

Slope of the Secant Line

The average rate of change of a function equals the slope of the secant line containing two points on its graph.

EXAMPLE 8

Finding the Equation of a Secant Line

Suppose that $g(x) = 3x^2 - 2x + 3$.

(a) Find the average rate of change of g from -2 to x.

(b) Use the result of part (a) to find the average rate of change of g from -2 to 1. Interpret this result.

(c) Find an equation of the secant line containing $(-2, g(-2))$ and $(1, g(1))$.

Solution

(a) The average rate of change of $g(x) = 3x^2 - 2x + 3$ from -2 to x is

$$\text{Average rate of change} = \frac{g(x) - g(-2)}{x - (-2)}$$

$$= \frac{(3x^2 - 2x + 3) - 19}{x + 2} \qquad \text{\color{gray} $g(-2) = 3(-2)^2 - 2(-2) + 3 = 19$}$$

$$= \frac{3x^2 - 2x - 16}{x + 2}$$

$$= \frac{(3x - 8)(x + 2)}{x + 2} \qquad \text{\color{gray} Factor.}$$

$$= 3x - 8 \qquad \text{\color{gray} Divide out like factors.}$$

(b) The average rate of change of g from -2 to x is given by $3x - 8$. Therefore, the average rate of change of g from -2 to 1 is $3(1) - 8 = -5$. The slope of the secant line joining $(-2, g(-2))$ and $(1, g(1))$ is $m_{\text{sec}} = -5$.

(c) We use the point–slope form to find an equation of the secant line.

$$y - y_1 = m_{\text{sec}}(x - x_1) \qquad \text{\color{gray} Point–slope form of the secant line}$$

$$y - 4 = -5(x - 1) \qquad \text{\color{gray} $x_1 = 1, y_1 = g(1) = 4, m_{\text{sec}} = -5$}$$

$$y - 4 = -5x + 5 \qquad \text{\color{gray} Simplify}$$

$$y = -5x + 9 \qquad \text{\color{gray} Slope–intercept form of the secant line}$$

■

NOW WORK PROBLEM 59.

1.3 Assess Your Understanding

'Are You Prepared?'

Answers are given at the end of these exercises. If you get a wrong answer, read the pages listed in red.

1. The interval $(2, 5)$ can be written as the inequality _____. (pp. A66–A67)

2. The slope of the line containing the points $(-2, 3)$ and $(3, 8)$ is _____. (p. 18)

3. Test the equation $y = 5x^2 - 1$ for symmetry with respect to the *x*-axis, the *y*-axis, and the origin. (pp. 11–13)

4. Write the point–slope form of the line with slope 5 containing the point $(3, -2)$. (p. 22)

5. The intercepts of the equation $y = x^2 - 9$ are _____. (pp. 10–11)

Concepts and Vocabulary

6. A function f is _____ on an open interval I if, for any choice of x_1 and x_2 in I, with $x_1 < x_2$, we have $f(x_1) < f(x_2)$.

7. A(n) _____ function f is one for which $f(-x) = f(x)$ for every x in the domain of f; (an) _____ function f is one for which $f(-x) = -f(x)$ for every x in the domain of f.

8. *True or False:* A function f is decreasing on an open interval I if, for any choice of x_1 and x_2 in I, with $x_1 < x_2$, we have $f(x_1) > f(x_2)$.

9. *True or False:* A function f has a local maximum at c if there is an open interval I containing c so that, for all x in I, $f(x) \leq f(c)$.

10. *True or False:* Even functions have graphs that are symmetric with respect to the origin.

Skill Building

In Problems 11–20, use the graph of the function f given.

11. Is f increasing on the interval $(-8, -2)$?

12. Is f decreasing on the interval $(-8, -4)$?

13. Is f increasing on the interval $(2, 10)$?

14. Is f decreasing on the interval $(2, 5)$?

15. List the interval(s) on which f is increasing.

16. List the interval(s) on which f is decreasing.

17. Is there a local maximum at 2? If yes, what is it?

18. Is there a local maximum at 5? If yes, what is it?

19. List the numbers at which f has a local maximum. What are these local maxima?

20. List the numbers at which f has a local minimum. What are these local minima?

In Problems 21–28, the graph of a function is given. Use the graph to find:
 (a) The intercepts, if any
 (b) The domain and range
 (c) The intervals on which it is increasing, decreasing, or constant
 (d) Whether it is even, odd, or neither

21.

22.

23.

24.

25.

26.

27.

28.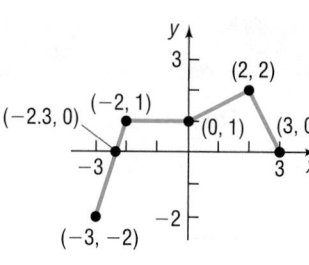

In Problems 29–32, the graph of a function f is given. Use the graph to find:
 (a) *The numbers, if any, at which f has a local maximum. What are these local maxima?*
 (b) *The numbers, if any, at which f has a local minimum. What are these local minima?*

29.

30.

31.

32.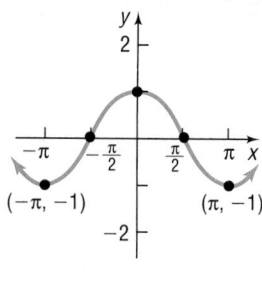

In Problems 33–44, determine algebraically whether each function is even, odd, or neither.

33. $f(x) = 4x^3$

34. $f(x) = 2x^4 - x^2$

35. $g(x) = -3x^2 - 5$

36. $h(x) = 3x^3 + 5$

37. $F(x) = \sqrt[3]{x}$

38. $G(x) = \sqrt{x}$

39. $f(x) = x + |x|$

40. $f(x) = \sqrt[3]{2x^2 + 1}$

41. $g(x) = \dfrac{1}{x^2}$

42. $h(x) = \dfrac{x}{x^2 - 1}$

43. $h(x) = \dfrac{-x^3}{3x^2 - 9}$

44. $F(x) = \dfrac{2x}{|x|}$

In Problems 45–52, use a graphing utility to graph each function over the indicated interval and approximate any local maxima and local minima. Determine where the function is increasing and where it is decreasing. Round answers to two decimal places.

45. $f(x) = x^3 - 3x + 2$ $(-2, 2)$

46. $f(x) = x^3 - 3x^2 + 5$ $(-1, 3)$

47. $f(x) = x^5 - x^3$ $(-2, 2)$

48. $f(x) = x^4 - x^2$ $(-2, 2)$

49. $f(x) = -0.2x^3 - 0.6x^2 + 4x - 6$ $(-6, 4)$

50. $f(x) = -0.4x^3 + 0.6x^2 + 3x - 2$ $(-4, 5)$

51. $f(x) = 0.25x^4 + 0.3x^3 - 0.9x^2 + 3$ $(-3, 2)$

52. $f(x) = -0.4x^4 - 0.5x^3 + 0.8x^2 - 2$ $(-3, 2)$

53. Find the average rate of change of $f(x) = -2x^2 + 4$
 (a) From 0 to 2
 (b) From 1 to 3
 (c) From 1 to 4

54. Find the average rate of change of $f(x) = -x^3 + 1$
 (a) From 0 to 2
 (b) From 1 to 3
 (c) From −1 to 1

55. Find the average rate of change of $g(x) = x^3 - 2x + 1$
 (a) From −3 to −2
 (b) From −1 to 1
 (c) From 1 to 3

56. Find the average rate of change of $h(x) = x^2 - 2x + 3$
 (a) From −1 to 1
 (b) From 0 to 2
 (c) From 2 to 5

57. $f(x) = 5x - 2$
 (a) Find the average rate of change from 1 to x.
 (b) Use the result of part (a) to find the average rate of change of f from 1 to 3. Interpret this result.
 (c) Find an equation of the secant line containing $(1, f(1))$ and $(3, f(3))$.

58. $f(x) = -4x + 1$
 (a) Find the average rate of change from 2 to x.
 (b) Use the result of part (a) to find the average rate of change of f from 2 to 5. Interpret this result.
 (c) Find an equation of the secant line containing $(2, f(2))$ and $(5, f(5))$.

59. $g(x) = x^2 - 2$
 (a) Find the average rate of change from −2 to x.
 (b) Use the result of part (a) to find the average rate of change of g from −2 to 1. Interpret this result.

(c) Find an equation of the secant line containing $(-2, g(-2))$ and $(1, g(1))$.

60. $g(x) = x^2 + 1$
 (a) Find the average rate of change from -1 to x.
 (b) Use the result of part (a) to find the average rate of change of g from -1 to 2. Interpret this result.
 (c) Find an equation of the secant line containing $(-1, g(-1))$ and $(2, g(2))$.

61. $h(x) = x^2 - 2x$
 (a) Find the average rate of change from 2 to x.

(b) Use the result of part (a) to find the average rate of change of h from 2 to 4. Interpret this result.
(c) Find an equation of the secant line containing $(2, h(2))$ and $(4, h(4))$.

62. $h(x) = -2x^2 + x$
 (a) Find the average rate of change from 0 to x.
 (b) Use the result of part (a) to find the average rate of change of h from 0 to 3. Interpret this result.
 (c) Find an equation of the secant line containing $(0, h(0))$ and $(3, h(3))$.

Applications and Extensions

63. Constructing an Open Box An open box with a square base is to be made from a square piece of cardboard 24 inches on a side by cutting out a square from each corner and turning up the sides (see the figure).

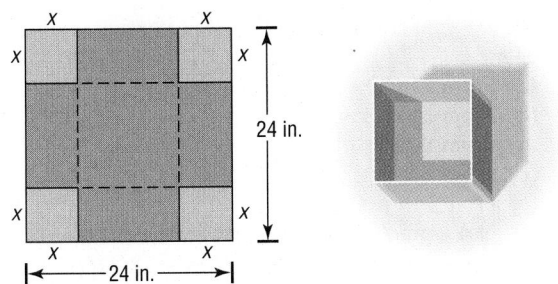

(a) Express the volume V of the box as a function of the length x of the side of the square cut from each corner.
(b) What is the volume if a 3-inch square is cut out?
(c) What is the volume if a 10-inch square is cut out?
(d) Graph $V = V(x)$. For what value of x is V largest?

64. Constructing an Open Box An open box with a square base is required to have a volume of 10 cubic feet.
(a) Express the amount A of material used to make such a box as a function of the length x of a side of the square base.
(b) How much material is required for a base 1 foot by 1 foot?
(c) How much material is required for a base 2 feet by 2 feet?
(d) Graph $A = A(x)$. For what value of x is A smallest?

65. Maximum Height of a Ball The height s of a ball (in feet) thrown with an initial velocity of 80 feet per second from an initial height of 6 feet is given as a function of the time t (in seconds) by

$$s(t) = -16t^2 + 80t + 6$$

(a) Graph $s = s(t)$.
(b) Determine the time at which height is maximum.
(c) What is the maximum height?

66. Maximum Height of a Ball On July 1, 2004, the Cassini probe became the first spacecraft to orbit the planet Saturn. Although Saturn is about 764 times the size of Earth, it has a very similar gravitational force. The height s of an object thrown upward from Saturn's surface with an ini-

tial velocity of 100 feet per second is given as a function of time t (in seconds) by $s(t) = -17.28t^2 + 100t$.
(a) Use a graphing utility to graph $s = s(t)$.
(b) Determine the time at which height is a maximum.
(c) What is the maximum height?
(d) The same object thrown from the surface of Earth would have a height given by $s(t) = -16t^2 + 100t$. Determine the maximum height of the object on Earth and compare this to your result from part (c).

67. Minimum Average Cost The average cost per hour in dollars of producing x riding lawn mowers is given by

$$\overline{C}(x) = 0.3x^2 + 21x - 251 + \frac{2500}{x}$$

(a) Graph $\overline{C} = \overline{C}(x)$
(b) Determine the number of riding lawn mowers to produce in order to minimize average cost.
(c) What is the minimum average cost?

68. Medicine Concentration The concentration C of a medication in the bloodstream t hours after being administered is given by

$$C(t) = -0.002x^4 + 0.039t^3 - 0.285t^2 + 0.766t + 0.085.$$

(a) After how many hours will the concentration be highest?
(b) A woman nursing a child must wait until the concentration is below 0.5 before she can feed him. After taking the medication, how long must she wait before feeding her child?

69. Revenue from Selling Bikes The following data represent the total revenue R, in dollars, that would be received from selling x bicycles at Tunney's Bicycle Shop.

Number of Bicycles, x	Total Revenue, R (Dollars)
0	0
25	28,000
60	45,000
102	53,400
150	59,160
190	62,360
223	64,835
249	66,525

(a) Plot the points (x, R) in a rectangular coordinate system.
(b) Draw a line through the points $(0, 0)$ and $(25, 28000)$ on the graph found in part (a).
(c) Find the average rate of change of revenue from 0 to 25 bicycles.
(d) Interpret the average rate of change found in part (c).
(e) Draw a line through the points $(190, 62360)$ and $(223, 64835)$ on the graph found in part (a).
(f) Find the average rate of change of revenue from 190 to 223 bicycles.
(g) Interpret the average rate of change found in part (f).
(h) What is happening to the average rate of change of revenue as the number of bicycles increases?

70. **Cost of Manufacturing Bikes** The following data represent the monthly cost C, in dollars, of producing bicycles at Tunney's Bicycle Shop.

Number of Bicycles, x	Total Cost of Production, C (Dollars)
0	24,000
25	27,750
60	31,500
102	35,250
150	39,000
190	42,750
223	46,500
249	50,250

(a) Plot the points (x, C) in a rectangular coordinate system.
(b) Draw a line through the points $(0, 24000)$ and $(25, 27750)$ on the graph found in part (a).
(c) Find the average rate of change of the cost from 0 to 25 bicycles.

(d) Interpret the average rate of change found in part (c).
(e) Draw a line through the points $(190, 42750)$ and $(223, 46500)$ on the graph found in part (a).
(f) Find the average rate of change of the cost from 190 to 223 bicycles.
(g) Interpret the average rate of change found in part (f).
(h) What is happening to the average rate of change of cost as the number of bicycles increases?

71. For the function $f(x) = x^2$, compute each average rate of change:
(a) From 0 to 1
(b) From 0 to 0.5
(c) From 0 to 0.1
(d) From 0 to 0.01
(e) From 0 to 0.001
(f) Use a graphing utility to graph each of the secant lines along with f.
(g) What do you think is happening to the secant lines?
(h) What is happening to the slopes of the secant lines? Is there some number that they are getting closer to? What is that number?

72. For the function $f(x) = x^2$, compute each average rate of change:
(a) From 1 to 2
(b) From 1 to 1.5
(c) From 1 to 1.1
(d) From 1 to 1.01
(e) From 1 to 1.001
(f) Use a graphing utility to graph each of the secant lines along with f.
(g) What do you think is happening to the secant lines?
(h) What is happening to the slopes of the secant lines? Is there some number that they are getting closer to? What is that number?

Problems 73–80 require the following discussion of a secant line. The slope of the secant line containing the two points $(x, f(x))$ and $(x + h, f(x + h))$ on the graph of a function $y = f(x)$ may be given as

$$m_{\text{sec}} = \frac{f(x + h) - f(x)}{(x + h) - x} = \frac{f(x + h) - f(x)}{h} \qquad h \neq 0$$

*In calculus, this expression is called the **difference quotient of** f.*
(a) *Express the slope of the secant line of each function in terms of x and h. Be sure to simplify your answer.*
(b) *Find m_{sec} for $h = 0.5, 0.1$, and 0.01 at $x = 1$. What value does m_{sec} approach as h approaches 0?*
(c) *Find the equation for the secant line at $x = 1$ with $h = 0.01$.*
(d) *Use a graphing utility to graph f and the secant line found in part (c) on the same viewing window.*

73. $f(x) = 2x + 5$
74. $f(x) = -3x + 2$
75. $f(x) = x^2 + 2x$
76. $f(x) = 2x^2 + x$
77. $f(x) = 2x^2 - 3x + 1$
78. $f(x) = -x^2 + 3x - 2$
79. $f(x) = \dfrac{1}{x}$
80. $f(x) = \dfrac{1}{x^2}$

Concepts and Vocabulary

4. The graph of $f(x) = mx + b$ is decreasing if m is _____ than zero.

5. When functions are defined by more than one equation they are called _____ functions.

6. *True or False:* The cube function is odd and is increasing on the interval $(-\infty, \infty)$.

7. *True or False:* The cube root function is odd and is decreasing on the interval $(-\infty, \infty)$.

8. *True or False:* The domain and the range of the reciprocal function are the set of all real numbers.

Skill Building

In Problems 9–16, match each graph to its function.

A.	Constant function	B.	Linear function	C.	Square function
D.	Cube function	E.	Square root function	F.	Reciprocal function
G.	Absolute value function	H.	Cube root function		

9.

10.

11.

12.

13. **14.** **15.** **16.**

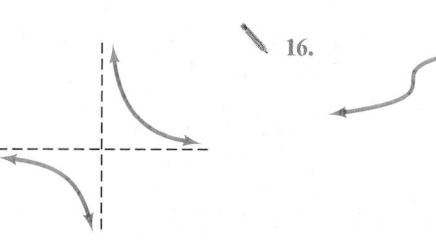

In Problems 17–24, sketch the graph of each function. Be sure to label three points on the graph.

17. $f(x) = x$

18. $f(x) = x^2$

19. $f(x) = x^3$

20. $f(x) = \sqrt{x}$

21. $f(x) = \dfrac{1}{x}$

22. $f(x) = |x|$

23. $f(x) = \sqrt[3]{x}$

24. $f(x) = 3$

25. If $f(x) = \begin{cases} x^2 & \text{if } x < 0 \\ 2 & \text{if } x = 0 \\ 2x + 1 & \text{if } x > 0 \end{cases}$

find: (a) $f(-2)$ (b) $f(0)$ (c) $f(2)$

26. If $f(x) = \begin{cases} x^3 & \text{if } x < 0 \\ 3x + 2 & \text{if } x \geq 0 \end{cases}$

find: (a) $f(-1)$ (b) $f(0)$ (c) $f(1)$

27. If $f(x) = \text{int}(2x)$, find (a) $f(1.2)$, (b) $f(1.6)$, (c) $f(-1.8)$.

28. If $f(x) = \text{int}\left(\dfrac{x}{2}\right)$, find (a) $f(1.2)$, (b) $f(1.6)$, (c) $f(-1.8)$.

In Problems 29–40:
(a) *Find the domain of each function.*
(b) *Locate any intercepts.*
(c) *Graph each function.*
(d) *Based on the graph, find the range.*

29. $f(x) = \begin{cases} 2x & \text{if } x \neq 0 \\ 1 & \text{if } x = 0 \end{cases}$

30. $f(x) = \begin{cases} 3x & \text{if } x \neq 0 \\ 4 & \text{if } x = 0 \end{cases}$

31. $f(x) = \begin{cases} -2x + 3 & x < 1 \\ 3x - 2 & x \geq 1 \end{cases}$

32. $f(x) = \begin{cases} x + 3 & \text{if } x < -2 \\ -2x - 3 & \text{if } x \geq -2 \end{cases}$

33. $f(x) = \begin{cases} x + 3 & \text{if } -2 \leq x < 1 \\ 5 & \text{if } x = 1 \\ -x + 2 & \text{if } x > 1 \end{cases}$

34. $f(x) = \begin{cases} 2x + 5 & \text{if } -3 \leq x < 0 \\ -3 & \text{if } x = 0 \\ -5x & \text{if } x > 0 \end{cases}$

35. $f(x) = \begin{cases} 1 + x & \text{if } x < 0 \\ x^2 & \text{if } x \geq 0 \end{cases}$

36. $f(x) = \begin{cases} \dfrac{1}{x} & \text{if } x < 0 \\ \sqrt[3]{x} & \text{if } x \geq 0 \end{cases}$

37. $f(x) = \begin{cases} |x| & \text{if } -2 \leq x < 0 \\ 1 & \text{if } x = 0 \\ x^3 & \text{if } x > 0 \end{cases}$

38. $f(x) = \begin{cases} 3 + x & \text{if } -3 \leq x < 0 \\ 3 & \text{if } x = 0 \\ \sqrt{x} & \text{if } x > 0 \end{cases}$

39. $f(x) = 2\,\text{int}(x)$

40. $f(x) = \text{int}(2x)$

When $x = 2$, the equation for f is $f(x) = x^2$. So

$$f(2) = 2^2 = 4$$

Figure 40

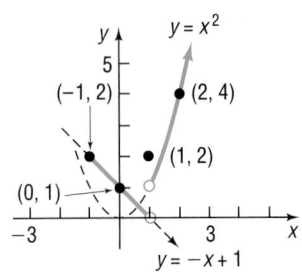

(b) To find the domain of f, we look at its definition. We conclude that the domain of f is $\{x \mid x \geq -1\}$, or the interval $[-1, \infty)$.

(c) To graph f, we graph "each piece." First we graph the line $y = -x + 1$ and keep only the part for which $-1 \leq x < 1$. Then we plot the point $(1, 2)$ because, when $x = 1$, $f(x) = 2$. Finally, we graph the parabola $y = x^2$ and keep only the part for which $x > 1$. See Figure 40.

(d) From the graph, we conclude that the range of f is $\{y \mid y > 0\}$, or the interval $(0, \infty)$. ■

⬛▬▬▬▶ NOW WORK PROBLEM 29.

EXAMPLE 4

Cost of Electricity

In May 2005, Commonwealth Edison Company supplied electricity to residences for a monthly customer charge of \$7.58 plus 8.275¢ per kilowatt-hour (kWhr) for the first 400 kWhr supplied in the month and 6.208¢ per kWhr for all usage over 400 kWhr in the month.

(a) What is the charge for using 300 kWhr in a month?

(b) What is the charge for using 700 kWhr in a month?

(c) If C is the monthly charge for x kWhr, express C as a function of x.

SOURCE: Commonwealth Edison Co., Chicago, Illinois, 2005.

Solution

(a) For 300 kWhr, the charge is \$7.58 plus 8.275¢ = \$0.08275 per kWhr. That is,

$$\text{Charge} = \$7.58 + \$0.08275(300) = \$32.41$$

(b) For 700 kWhr, the charge is \$7.58 plus 8.275¢ per kWhr for the first 400 kWhr plus 6.208¢ per kWhr for the 300 kWhr in excess of 400. That is,

$$\text{Charge} = \$7.58 + \$0.08275(400) + \$0.06208(300) = \$59.30$$

(c) If $0 \leq x \leq 400$, the monthly charge C (in dollars) can be found by multiplying x times \$0.08275 and adding the monthly customer charge of \$7.58. So, if $0 \leq x \leq 400$, then $C(x) = 0.08275x + 7.58$.

For $x > 400$, the charge is $0.08275(400) + 7.58 + 0.06208(x - 400)$, since $x - 400$ equals the usage in excess of 400 kWhr, which costs \$0.06208 per kWhr. That is, if $x > 400$, then

$$C(x) = 0.08275(400) + 7.58 + 0.06208(x - 400)$$
$$= 40.68 + 0.06208(x - 400)$$
$$= 0.06208x + 15.848$$

The rule for computing C follows two equations:

$$C(x) = \begin{cases} 0.08275x + 7.58 & \text{if } 0 \leq x \leq 400 \\ 0.06208x + 15.848 & \text{if } x > 400 \end{cases}$$

See Figure 41 for the graph. ■

Figure 41

1.4 Assess Your Understanding

'Are You Prepared?'

Answers are given at the end of these exercises. If you get a wrong answer, read the pages listed in red.

1. Sketch the graph of $y = \sqrt{x}$. (p. 14)

2. Sketch the graph of $y = \dfrac{1}{x}$. (p. 14)

3. List the intercepts of the equation $y = x^3 - 8$. (pp. 10–11)

called a *discontinuity*; that is, at integer values, the graph suddenly "steps" from one value to another without taking on any of the intermediate values. For example, to the immediate left of $x = 3$, the y-coordinates of the points on the graph are 2, and to the immediate right of $x = 3$, the y-coordinates of the points on the graph are 3.

COMMENT: When graphing a function using a graphing utility, you can choose either the **connected mode**, in which points plotted on the screen are connected, making the graph appear without any breaks, or the **dot mode**, in which only the points plotted appear. When graphing the greatest integer function with a graphing utility, it is necessary to be in the **dot mode**. This is to prevent the utility from "connecting the dots" when **f(x)** changes from one integer value to the next. See Figure 39.

Figure 39
$f(x) = \text{int}(x)$

(a) Connected mode **(b)** Dot mode

The functions that we have discussed so far are basic. Whenever you encounter one of them, you should see a mental picture of its graph. For example, if you encounter the function $f(x) = x^2$, you should see in your mind's eye a picture like Figure 32.

NOW WORK PROBLEMS 9 THROUGH 16.

2 Graph Piecewise-defined Functions

Sometimes a function is defined differently on different parts of its domain. For example, the absolute value function $f(x) = |x|$ is actually defined by two equations: $f(x) = x$ if $x \geq 0$ and $f(x) = -x$ if $x < 0$. For convenience, we generally combine these equations into one expression as

$$f(x) = |x| = \begin{cases} x & \text{if } x \geq 0 \\ -x & \text{if } x < 0 \end{cases}$$

When functions are defined by more than one equation, they are called **piecewise-defined** functions.

Let's look at another example of a piecewise-defined function.

EXAMPLE 3 **Analyzing a Piecewise-defined Function**

The function f is defined as

$$f(x) = \begin{cases} -x + 1 & \text{if } -1 \leq x < 1 \\ 2 & \text{if } x = 1 \\ x^2 & \text{if } x > 1 \end{cases}$$

(a) Find $f(0)$, $f(1)$, and $f(2)$. (b) Determine the domain of f.

(c) Graph f. (d) Use the graph to find the range of f.

Solution (a) To find $f(0)$, we observe that when $x = 0$ the equation for f is given by $f(x) = -x + 1$. So we have

$$f(0) = -0 + 1 = 1$$

When $x = 1$, the equation for f is $f(x) = 2$. Thus,

$$f(1) = 2$$

Figure 36
Reciprocal Function

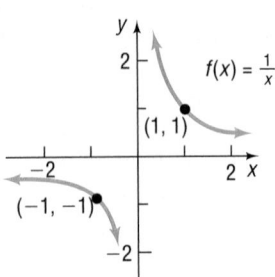

Reciprocal Function

$$f(x) = \frac{1}{x}$$

Refer to Example 12, page 14, for a discussion of the equation $y = \frac{1}{x}$. See Figure 36.

The domain and the range of the **reciprocal function** are the set of all nonzero real numbers. The graph has no intercepts. The reciprocal function is decreasing on the intervals $(-\infty, 0)$ and $(0, \infty)$ and is an odd function.

Figure 37
Absolute Value Function

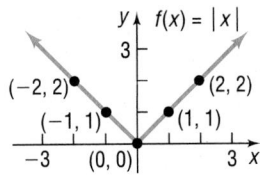

Absolute Value Function

$$f(x) = |x|$$

See Figure 37.

The domain of the **absolute value function** is the set of all real numbers; its range is the set of nonnegative real numbers. The intercept of the graph is at $(0, 0)$. If $x \geq 0$, then $f(x) = x$, and the graph of f is part of the line $y = x$; if $x < 0$, then $f(x) = -x$, and the graph of f is part of the line $y = -x$. The absolute value function is an even function; it is decreasing on the interval $(-\infty, 0)$ and increasing on the interval $(0, \infty)$.

The notation $\text{int}(x)$ stands for the largest integer less than or equal to x. For example,

$$\text{int}(1) = 1, \quad \text{int}(2.5) = 2, \quad \text{int}\left(\frac{1}{2}\right) = 0, \quad \text{int}\left(-\frac{3}{4}\right) = -1, \quad \text{int}(\pi) = 3$$

This type of correspondence occurs frequently enough in mathematics that we give it a name.

Greatest Integer Function

$$f(x) = \text{int}(x)^* = \text{greatest integer less than or equal to } x$$

Table 7

	$y = f(x)$	
x	$= \text{int}(x)$	(x, y)
-1	-1	$(-1, -1)$
$-\dfrac{1}{2}$	-1	$\left(-\dfrac{1}{2}, -1\right)$
$-\dfrac{1}{4}$	-1	$\left(-\dfrac{1}{4}, -1\right)$
0	0	$(0, 0)$
$\dfrac{1}{4}$	0	$\left(\dfrac{1}{4}, 0\right)$
$\dfrac{1}{2}$	0	$\left(\dfrac{1}{2}, 0\right)$
$\dfrac{3}{4}$	0	$\left(\dfrac{3}{4}, 0\right)$

We obtain the graph of $f(x) = \text{int}(x)$ by plotting several points. See Table 7. For values of x, $-1 \leq x < 0$, the value of $f(x) = \text{int}(x)$ is -1; for values of x, $0 \leq x < 1$, the value of f is 0. See Figure 38 for the graph.

Figure 38
Greatest Integer Function

The domain of the **greatest integer function** is the set of all real numbers; its range is the set of integers. The y-intercept of the graph is 0. The x-intercepts lie in the interval $[0, 1)$. The greatest integer function is neither even nor odd. It is constant on every interval of the form $[k, k + 1)$, for k an integer. In Figure 38, we use a solid dot to indicate, for example, that at $x = 1$ the value of f is $f(1) = 1$; we use an open circle to illustrate that the function does not assume the value of 0 at $x = 1$.

From the graph of the greatest integer function, we can see why it also called a **step function**. At $x = 0$, $x = \pm 1$, $x = \pm 2$, and so on, this function exhibits what is

*Some books use the notation $f(x) = [x]$ instead of $\text{int}(x)$.

A **constant function** is a special linear function ($m = 0$). Its domain is the set of all real numbers; its range is the set consisting of a single number b. Its graph is a horizontal line whose y-intercept is b. The constant function is an even function whose graph is constant over its domain.

Identity Function

$$f(x) = x$$

See Figure 31.

The **identity function** is also a special linear function. Its domain and range are the set of all real numbers. Its graph is a line whose slope is $m = 1$ and whose y-intercept is 0. The line consists of all points for which the x-coordinate equals the y-coordinate. The identity function is an odd function that is increasing over its domain. Note that the graph bisects quadrants I and III.

Square Function

$$f(x) = x^2$$

See Figure 32.

The domain of the **square function** f is the set of all real numbers; its range is the set of nonnegative real numbers. The graph of this function is a parabola whose intercept is at $(0, 0)$. The square function is an even function that is decreasing on the interval $(-\infty, 0)$ and increasing on the interval $(0, \infty)$.

Cube Function

$$f(x) = x^3$$

See Figure 33.

The domain and the range of the **cube function** are the set of all real numbers. The intercept of the graph is at $(0, 0)$. The cube function is odd and is increasing on the interval $(-\infty, \infty)$.

Square Root Function

$$f(x) = \sqrt{x}$$

See Figure 34.

The domain and the range of the **square root function** are the set of nonnegative real numbers. The intercept of the graph is at $(0, 0)$. The square root function is neither even nor odd and is increasing on the interval $(0, \infty)$.

Cube Root Function

$$f(x) = \sqrt[3]{x}$$

See Figure 35.

The domain and the range of the **cube root function** are the set of all real numbers. The intercept of the graph is at $(0, 0)$. The cube root function is an odd function that is increasing on the interval $(-\infty, \infty)$.

Figure 31
Identity Function

Figure 32
Square Function

Figure 33
Cube Function

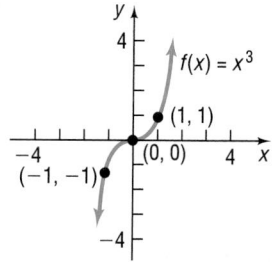

Figure 34
Square Root Function

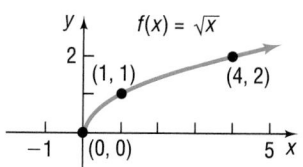

Figure 35
Cube Root Function

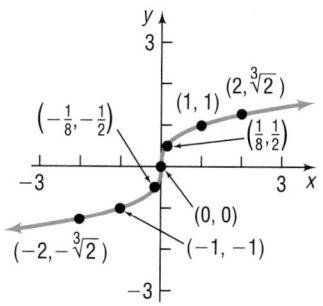

(c) We use the function to form Table 6 and obtain some points on the graph. Because of the symmetry with respect to the y-axis, we only need to find points (x, y) for which $x \geq 0$. Figure 28 shows the graph of $f(x) = |x|$.

Table 6

| x | $y = f(x) = |x|$ | (x, y) |
|---|---|---|
| 0 | 0 | $(0, 0)$ |
| 1 | 1 | $(1, 1)$ |
| 2 | 2 | $(2, 2)$ |
| 3 | 3 | $(3, 3)$ |

Figure 28

From the results of Example 2 and Figure 28, we have the following properties of the absolute value function.

Properties of $f(x) = |x|$

1. The x-intercept of the graph of $f(x) = |x|$ is 0. The y-intercept of the graph of $f(x) = |x|$ is also 0.
2. The function is even.
3. It is decreasing on the interval $(-\infty, 0)$.
 It is increasing on the interval $(0, \infty)$.
4. It has a local minimum of 0 at $x = 0$.

───── **Seeing the Concept** ─────

Graph $y = |x|$ on a square screen and compare what you see with Figure 28. Note that some graphing calculators use abs(x) for absolute value.

1 **Graph the Functions Listed in the Library of Functions**

We now provide a summary of the key functions that we have encountered. In going through this list, pay special attention to the properties of each function, particularly to the shape of each graph. Knowing these graphs will lay the foundation for later graphing techniques.

Figure 29
Linear Function

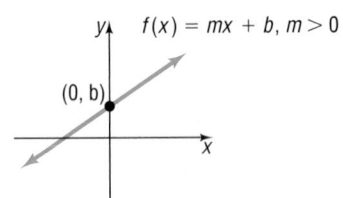

Linear Function

$$f(x) = mx + b \qquad m \text{ and } b \text{ are real numbers}$$

See Figure 29.

The domain of a **linear function** is the set of all real numbers. The graph of this function is a nonvertical line with slope m and y-intercept b. A linear function is increasing if $m > 0$, decreasing if $m < 0$, and constant if $m = 0$.

Figure 30
Constant Function

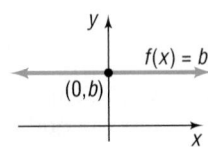

Constant Function

$$f(x) = b \qquad b \text{ is a real number}$$

See Figure 30.

In Problems 41–44, the graph of a piecewise-defined function is given. Write a definition for each function.

41.

42.

43.

44.
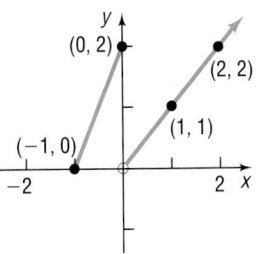

Applications and Extensions

45. Cell Phone Service Sprint PCS offers a monthly cellular phone plan for $35. It includes 300 anytime minutes plus $0.40 per minute for additional minutes. The following function is used to compute the monthly cost for a subscriber:

$$C(x) = \begin{cases} 35 & \text{if } 0 < x \le 300 \\ 0.40x - 85 & \text{if } x > 300 \end{cases}$$

where x is the number of anytime minutes used. Compute the monthly cost of the cellular phone for use of the following anytime minutes:
(a) 200 (b) 365 (c) 301
SOURCE: Sprint PCS.

46. Parking at O'Hare International Airport The short-term (no more than 24 hours) parking fee F (in dollars) for parking x hours at O'Hare International Airport's main parking garage can be modeled by the function

$$F(x) = \begin{cases} 3 & \text{if } 0 < x \le 3 \\ 5\,\text{int}(x+1) + 1 & \text{if } 3 < x < 9 \\ 50 & \text{if } 9 \le x \le 24 \end{cases}$$

Determine the fee for parking in the short-term parking garage for
(a) 2 hours (b) 7 hours
(c) 15 hours (d) 8 hours and 24 minutes
SOURCE: O'Hare International Airport.

47. Cost of Natural Gas In May 2005, the Peoples Gas Company had the following rate schedule for natural gas usage in single-family residences:

Monthly service charge	$9.45
Per therm service charge	
1st 50 therms	$0.36375/therm
Over 50 therms	$0.11445/therm
Gas charge	$0.7958/therm

(a) What is the charge for using 50 therms in a month?
(b) What is the charge for using 500 therms in a month?
(c) Construct a function that relates the monthly charge C for x therms of gas.
(d) Graph this function.
SOURCE: The Peoples Gas Company, Chicago, Illinois, 2005.

48. Cost of Natural Gas In May 2005, Nicor Gas had the following rate schedule for natural gas usage in single-family residences:

Monthly customer charge	$6.45
Distribution charge	
1st 20 therms	$0.2012/therm
Next 30 therms	$0.1117/therm
Over 50 therms	$0.0374/therm
Gas supply charge	$0.7727/therm

(a) What is the charge for using 40 therms in a month?
(b) What is the charge for using 202 therms in a month?
(c) Construct a function that gives the monthly charge C for x therms of gas.
(d) Graph this function.
SOURCE: Nicor Gas, Aurora, Illinois, 2005.

49. Federal Income Tax Two 2005 Tax Rate Schedules are given in the accompanying table. If x equals taxable income and y equals the tax due, construct a function $y = f(x)$ for Schedule X.

REVISED 2005 TAX RATE SCHEDULES											
Schedule X— Single						**Schedule Y-1— Married Filing Jointly or Qualifying Widow(er)**					
If Taxable Income		The Tax Is				If Taxable Income		The Tax Is			
Is over	But Not over	This Amount		Plus This %	Of the Excess over	Is over	But Not over	This Amount		Plus This %	Of the Excess over
$0	7,300	0.00	+	10%	$0	$0	$14,600	0.00	+	10%	$0
7,300	29,700	$730.00	+	15%	7,300	14,600	$59,400	$1,460.00	+	15%	$14,600
29,700	71,950	$4090.00	+	25%	29,700	59,400	$119,950	$8,180.00	+	25%	$59,400
71,950	150,150	$14,652.50	+	28%	71,950	119,950	$182,800	$23,317.50	+	28%	$119,950
150,150	326,450	$36,548.50	+	33%	150,150	182,850	$326,450	$40,915.50	+	33%	$182,800
326,450		$92,727.50	+	35%	326,450	326,450	—	$88,320.00	+	35%	$326,450

SOURCE: Internal Revenue Service

50. Federal Income Tax Refer to the revised 2005 tax rate schedules. If x equals taxable income and y equals the tax due, construct a function $y = f(x)$ for Schedule Y-1.

51. Cost of Transporting Goods A trucking company transports goods between Chicago and New York, a distance of 960 miles. The company's policy is to charge, for each pound, $0.50 per mile for the first 100 miles, $0.40 per mile for the next 300 miles, $0.25 per mile for the next 400 miles, and no charge for the remaining 160 miles.
 (a) Graph the relationship between the cost of transportation in dollars and mileage over the entire 960-mile route.
 (b) Find the cost as a function of mileage for hauls between 100 and 400 miles from Chicago.
 (c) Find the cost as a function of mileage for hauls between 400 and 800 miles from Chicago.

52. Car Rental Costs An economy car rented in Florida from National Car Rental® on a weekly basis costs $95 per week. Extra days cost $24 per day until the day rate exceeds the weekly rate, in which case the weekly rate applies. Also, any part of a day used counts as a full day. Find the cost C of renting an economy car as a piecewise-defined function of the number x of days used, where $7 \leq x \leq 14$. Graph this function.

53. Minimum Payments for Credit Cards Holders of credit cards issued by banks, department stores, oil companies, and so on, receive bills each month that state minimum amounts that must be paid by a certain due date. The minimum due depends on the total amount owed. One such credit card company uses the following rules: For a bill of less than $10, the entire amount is due. For a bill of at least $10 but less than $500, the minimum due is $10. A minimum of $30 is due on a bill of at least $500 but less than $1000, a minimum of $50 is due on a bill of at least $1000 but less than $1500, and a minimum of $70 is due on bills of $1500 or more. Find the function f that describes the minimum payment due on a bill of x dollars. Graph f.

54. Interest Payments for Credit Cards Refer to Problem 53. The card holder may pay any amount between the minimum due and the total owed. The organization issuing the card charges the card holder interest of 1.5% per month for the first $1000 owed and 1% per month on any unpaid balance over $1000. Find the function g that gives the amount of interest charged per month on a balance of x dollars. Graph g.

55. Wind Chill The wind chill factor represents the equivalent air temperature at a standard wind speed that would produce the same heat loss as the given temperature and wind speed. One formula for computing the equivalent temperature is

$$W = \begin{cases} t & 0 \leq v < 1.79 \\ 33 - \dfrac{(10.45 + 10\sqrt{v} - v)(33 - t)}{22.04} & 1.79 \leq v \leq 20 \\ 33 - 1.5958(33 - t) & v > 20 \end{cases}$$

where v represents the wind speed (in meters per second) and t represents the air temperature (°C). Compute the wind chill for the following:
 (a) An air temperature of 10°C and a wind speed of 1 meter per second (m/sec)
 (b) An air temperature of 10°C and a wind speed of 5 m/sec
 (c) An air temperature of 10°C and a wind speed of 15 m/sec
 (d) An air temperature of 10°C and a wind speed of 25 m/sec
 (e) Explain the physical meaning of the equation corresponding to $0 \leq v < 1.79$.
 (f) Explain the physical meaning of the equation corresponding to $v > 20$.

56. Wind Chill Redo Problem 55(a)–(d) for an air temperature of −10°C.

Discussion and Writing

In Problems 57–64, use a graphing utility.

57. Exploration Graph $y = x^2$. Then on the same screen graph $y = x^2 + 2$, followed by $y = x^2 + 4$, followed by $y = x^2 - 2$. What pattern do you observe? Can you predict the graph of $y = x^2 - 4$? Of $y = x^2 + 5$?

58. Exploration Graph $y = x^2$. Then on the same screen graph $y = (x - 2)^2$, followed by $y = (x - 4)^2$, followed by $y = (x + 2)^2$. What pattern do you observe? Can you predict the graph of $y = (x + 4)^2$? Of $y = (x - 5)^2$?

59. Exploration Graph $y = |x|$. Then on the same screen graph $y = 2|x|$, followed by $y = 4|x|$, followed by $y = \frac{1}{2}|x|$. What pattern do you observe? Can you predict the graph of $y = \frac{1}{4}|x|$? Of $y = 5|x|$?

60. Exploration Graph $y = x^2$. Then on the same screen graph $y = -x^2$. What pattern do you observe? Now try $y = |x|$ and $y = -|x|$. What do you conclude?

61. Exploration Graph $y = \sqrt{x}$. Then on the same screen graph $y = \sqrt{-x}$. What pattern do you observe? Now try $y = 2x + 1$ and $y = 2(-x) + 1$. What do you conclude?

62. Exploration Graph $y = x^3$. Then on the same screen graph $y = (x - 1)^3 + 2$. Could you have predicted the result?

63. Exploration Graph $y = x^2$, $y = x^4$, and $y = x^6$ on the same screen. What do you notice is the same about each graph? What do you notice that is different?

64. Exploration Graph $y = x^3$, $y = x^5$, and $y = x^7$ on the same screen. What do you notice is the same about each graph? What do you notice that is different?

65. Consider the equation

$$y = \begin{cases} 1 & \text{if } x \text{ is rational} \\ 0 & \text{if } x \text{ is irrational} \end{cases}$$

Is this a function? What is its domain? What is its range? What is its y-intercept, if any? What are its x-intercepts, if any? Is it even, odd, or neither? How would you describe its graph?

66. Define some functions that pass through $(0, 0)$ and $(1, 1)$ and are increasing for $x \geq 0$. Begin your list with $y = \sqrt{x}$, $y = x$, and $y = x^2$. Can you propose a general result about such functions?

'Are You Prepared' Answers

1.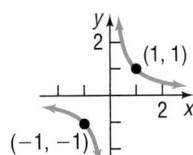

2.

3. $(0, -8), (2, 0)$

1.5 Graphing Techniques: Transformations

OBJECTIVES **1** Graph Functions Using Vertical and Horizontal Shifts

 2 Graph Functions Using Compressions and Stretches

 3 Graph Functions Using Reflections about the x-Axis and the y-Axis

At this stage, if you were asked to graph any of the functions defined by $y = x$, $y = x^2$, $y = x^3$, $y = \sqrt{x}$, $y = \sqrt[3]{x}$, $y = |x|$, or $y = \dfrac{1}{x}$, your response should be, "Yes, I recognize these functions and know the general shapes of their graphs." (If this is not your answer, review the previous section, Figures 31 through 37.)

Sometimes we are asked to graph a function that is "almost" like one that we already know how to graph. In this section, we look at some of these functions and develop techniques for graphing them. Collectively, these techniques are referred to as **transformations.**

1 Graph Functions Using Vertical and Horizontal Shifts

| EXAMPLE 1 | **Vertical Shift Up** |

Use the graph of $f(x) = x^2$ to obtain the graph of $g(x) = x^2 + 3$.

Solution We begin by obtaining some points on the graphs of f and g. For example, when $x = 0$, then $y = f(0) = 0$ and $y = g(0) = 3$. When $x = 1$, then $y = f(1) = 1$ and $y = g(1) = 4$. Table 8 lists these and a few other points on each graph. Notice that each y-coordinate of a point on the graph of g is 3 units larger than the y-coordinate of the corresponding point on the graph of f. We conclude that the graph of g is identical to that of f, except that it is shifted vertically up 3 units. See Figure 36.

Table 8

x	$y = f(x)$ $= x^2$	$y = g(x)$ $= x^2 + 3$
-2	4	7
-1	1	4
0	0	3
1	1	4
2	4	7

Figure 42

 ——— **Exploration** ———

On the same screen, graph each of the following functions:

$$Y_1 = x^2$$
$$Y_2 = x^2 + 2$$
$$Y_3 = x^2 - 2$$

Figure 43

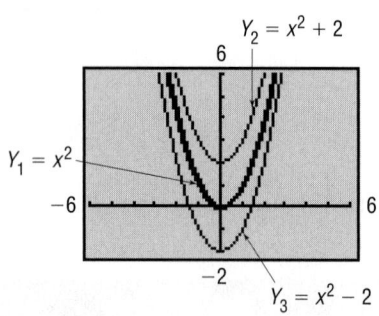

$Y_2 = x^2 + 2$

$Y_1 = x^2$

$Y_3 = x^2 - 2$

Figure 43 illustrates the graphs. You should have observed a general pattern. With $Y_1 = x^2$ on the screen, the graph of $Y_2 = x^2 + 2$ is identical to that of $Y_1 = x^2$, except that it is shifted vertically up 2 units. The graph of $Y_3 = x^2 - 2$ is identical to that of $Y_1 = x^2$, except that it is shifted vertically down 2 units.

We are led to the following conclusion:

> If a positive real number k is added to the right side of a function $y = f(x)$, the graph of the new function $y = f(x) + k$ is the graph of f **shifted vertically up** k units.

> If a positive real number k is subtracted from the right side of a function $y = f(x)$, the graph of the new function $y = f(x) - k$ is the graph of f **shifted vertically down** k units.

Let's look at another example.

EXAMPLE 2 | **Vertical Shift Down**

Use the graph of $f(x) = x^2$ to obtain the graph of $g(x) = x^2 - 4$.

Solution Table 9 lists some points on the graphs of f and g. Notice that each y-coordinate of g is 4 units less than the corresponding y-coordinate of f. The graph of g is identical to that of f, except that it is shifted down 4 units. See Figure 44.

Table 9

x	$y = f(x)$ $= x^2$	$y = g(x)$ $= x^2 - 4$
−2	4	0
−1	1	−3
0	0	−4
1	1	−3
2	4	0

Figure 44

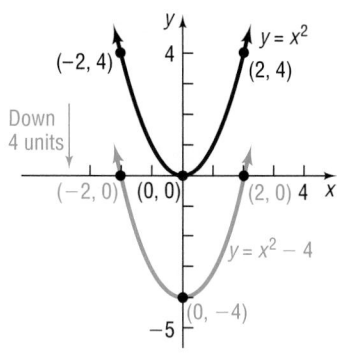

NOW WORK PROBLEM 35.

EXAMPLE 3 | **Horizontal Shift to the Right**

Use the graph of $f(x) = x^2$ to obtain the graph of $g(x) = (x - 2)^2$.

Solution The function $g(x) = (x - 2)^2$ is basically a square function. Table 10 lists some points on the graphs of f and g. Note that when $f(x) = 0$ then $x = 0$, and when $g(x) = 0$, then $x = 2$. Also, when $f(x) = 4$, then $x = -2$ or 2, and when $g(x) = 4$, then $x = 0$ or 4. We conclude that the graph of g is identical to that of f, except that it is shifted 2 units to the right. See Figure 45.

Table 10

x	$y = f(x)$ $= x^2$	$y = g(x)$ $= (x-2)^2$
−2	4	16
0	0	4
2	4	0
4	16	4

Figure 45

Figure 46

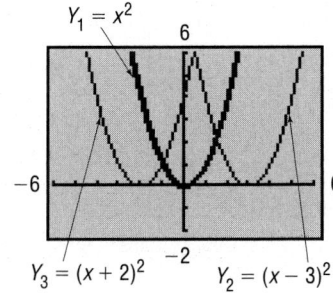

Exploration

On the same screen, graph each of the following functions:

$$Y_1 = x^2$$
$$Y_2 = (x-3)^2$$
$$Y_3 = (x+2)^2$$

Result Figure 46 illustrates the graphs.

You should have observed the following pattern. With the graph of $Y_1 = x^2$ on the screen, the graph of $Y_2 = (x-3)^2$ is identical to that of $Y_1 = x^2$, except that it is shifted horizontally to the right 3 units. The graph of $Y_3 = (x+2)^2$ is identical to that of $Y_1 = x^2$, except that it is shifted horizontally to the left 2 units.

We are led to the following conclusion.

> If the argument x of a function f is replaced by $x - h, h > 0$, the graph of the new function $y = f(x - h)$ is the graph of f **shifted horizontally right** h units. If the argument x of a function f is relaced by $x + h, h > 0$, the graph of the new function $y = f(x + h)$ is the graph of f **shifted horizontally left** h units.

EXAMPLE 4 **Horizontal Shift to the Left**

Use the graph of $f(x) = x^2$ to obtain the graph of $g(x) = (x + 4)^2$.

Solution Again, the function $g(x) = (x + 4)^2$ is basically a square function. Its graph is the same as that of f, except that it is shifted horizontally 4 units to the left. See Figure 47.

Figure 47

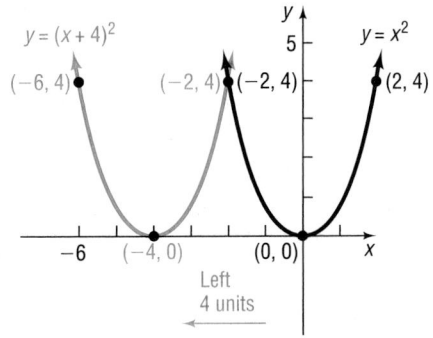

NOW WORK PROBLEM 39.

Vertical and horizontal shifts are sometimes combined.

EXAMPLE 5	**Combining Vertical and Horizontal Shifts**

Graph the function: $f(x) = (x + 3)^2 - 5$

Solution We graph f in steps. First, we note that the rule for f is basically a square function, so we begin with the graph of $y = x^2$ as shown in Figure 48(a). Next, to get the graph of $y = (x + 3)^2$, we shift the graph of $y = x^2$ horizontally 3 units to the left. See Figure 48(b). Finally, to get the graph of $y = (x + 3)^2 - 5$, we shift the graph of $y = (x + 3)^2$ vertically down 5 units. See Figure 48(c). Note the points plotted on each graph. Using key points can be helpful in keeping track of the transformation that has taken place.

Figure 48

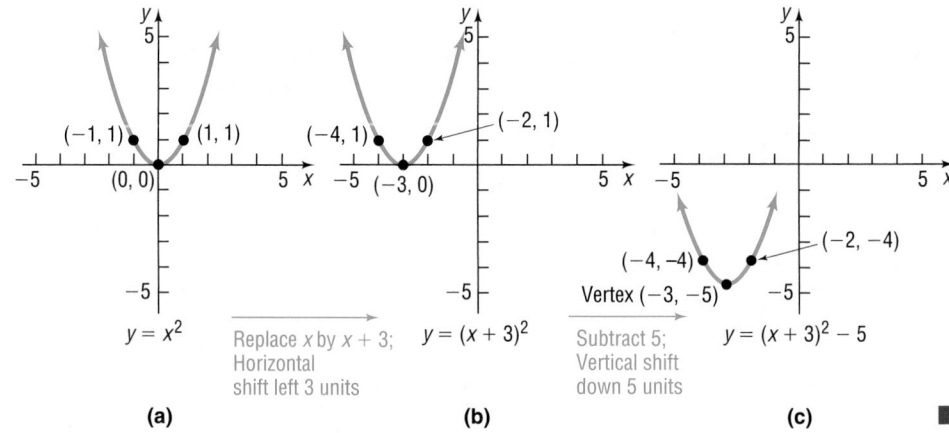

(a) (b) (c)

✔ CHECK: Graph $Y_1 = f(x) = (x + 3)^2 - 5$ and compare the graph to Figure 48(c).

In Example 5, if the vertical shift had been done first, followed by the horizontal shift, the final graph would have been the same. Try it for yourself.

NOW WORK PROBLEM **41**.

2 Graphing Functions Using Compressions and Stretches

EXAMPLE 6	**Vertical Stretch**

Use the graph of $f(x) = |x|$ to obtain the graph of $g(x) = 2|x|$.

Solution To see the relationship between the graphs of f and g, we form Table 11, listing points on each graph. For each x, the y-coordinate of a point on the graph of g is 2 times as large as the corresponding y-coordinate on the graph of f. The graph of $f(x) = |x|$ is vertically stretched by a factor of 2 to obtain the graph of $g(x) = 2|x|$. [for example, $(1, 1)$ is on the graph of f, but $(1, 2)$ is on the graph of g.] See Figure 49.

Table 10

| x | $y = f(x)$ $= |x|$ | $y = g(x)$ $= 2|x|$ |
|---|---|---|
| -2 | 2 | 4 |
| -1 | 1 | 2 |
| 0 | 0 | 0 |
| 1 | 1 | 2 |
| 2 | 2 | 4 |

Figure 49

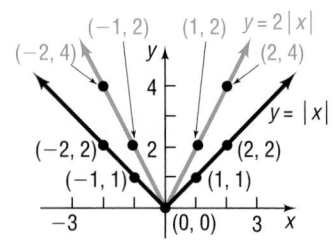

| EXAMPLE 7 | Vertical Compression |

Use the graph of $f(x) = |x|$ to obtain the graph of $g(x) = \frac{1}{2}|x|$.

Solution For each x, the y-coordinate of a point on the graph of g is $\frac{1}{2}$ as large as the corresponding y-coordinate on the graph of f. The graph of $f(x) = |x|$ is vertically compressed by a factor of $\frac{1}{2}$ to obtain the graph of $g(x) = \frac{1}{2}|x|$ [for example, $(2, 2)$ is on the graph of f, but $(2, 1)$ is on the graph of g.]. See Table 12 and Figure 50.

Table 12

x	$y = f(x)$ $= \lvert x\rvert$	$y = g(x)$ $= \frac{1}{2}\lvert x\rvert$
-2	2	1
-1	1	$\frac{1}{2}$
0	0	0
1	1	$\frac{1}{2}$
2	2	1

Figure 50

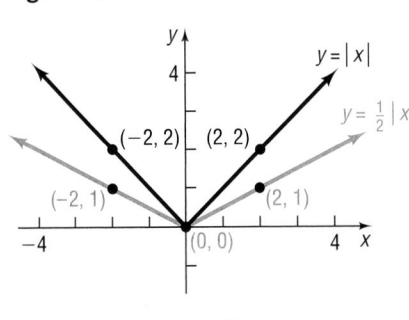

> When the right side of a function $y = f(x)$ is multiplied by a positive number a, the graph of the new function $y = af(x)$ is obtained by multiplying each y-coordinate on the graph of $y = f(x)$ by a. The new graph is a **vertically compressed** (if $0 < a < 1$) or a **vertically stretched** (if $a > 1$) version of the graph of $y = f(x)$.

 NOW WORK PROBLEM 43.

What happens if the argument x of a function $y = f(x)$ is multiplied by a positive number a, creating a new function $y = f(ax)$? To find the answer, we look at the following Exploration.

Exploration

On the same screen, graph each of the following functions:

$$Y_1 = f(x) = x^2$$
$$Y_2 = f(2x) = (2x)^2$$
$$Y_3 = f\left(\frac{1}{2}x\right) = \left(\frac{1}{2}x\right)^2$$

RESULT You should have obtained the graphs shown in Figure 51. The graph of $Y_2 = (2x)^2$ is the graph of $Y_1 = x^2$ compressed horizontally. Look at Table 13.

Figure 51

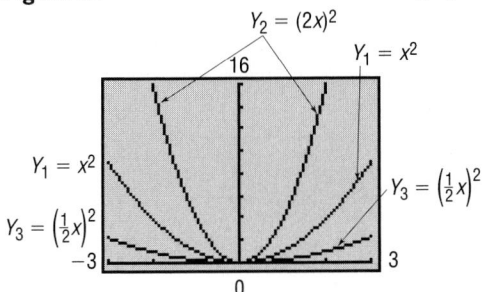

Table 13

X	Y₁	Y₂
0	0	0
.5	.25	1
1	1	4
2	4	16
4	16	64
8	64	256
16	256	1024

Y₂■(2X)²

X	Y₁	Y₃
0	0	0
.5	.25	.0625
1	1	.25
2	4	1
4	16	4
8	64	16
16	256	64

Y₃■(X/2)²

(a) (b)

Notice that $(1, 1)$, $(2, 4)$, $(4, 16)$, and $(16, 256)$ are points on the graph of $Y_1 = x^2$. Also, $(0.5, 1)$, $(1, 4)$, $(2, 16)$, and $(8, 256)$ are points on the graph of $Y_2 = (2x)^2$. For equal y-coordinates,

the x-coordinate on the graph of Y_2 is $\frac{1}{2}$ the x-coordinate on Y_1. The graph of $Y_2 = (2x)^2$ is obtained by multiplying the x-coordinate of each point on the graph of $Y_1 = x^2$ by $\frac{1}{2}$.

The graph of $Y_3 = \left(\frac{1}{2}x\right)^2$ is the graph of $Y_1 = x^2$ stretched horizontally. Look at Table 13(b). Notice that $(0.5, 0.25)$, $(1, 1)$, $(2, 4)$, and $(4, 16)$ are points on the graph of $Y_1 = x^2$. Also, $(1, 0.25)$, $(2, 1)$, $(4, 4)$, and $(8, 16)$ are points on the graph of $Y_3 = \left(\frac{1}{2}x\right)^2$. For equal y-coordinates, the x-coordinate on the graph of Y_3 is 2 times the x-coordinate on Y_1. The graph of $Y_3 = \left(\frac{1}{2}x\right)^2$ is obtained by multiplying the x-coordinate of each point on the graph of $Y_1 = x^2$ by a factor of 2.

If the argument x of a function $y = f(x)$ is multiplied by a positive number a, the graph of the new function $y = f(ax)$ is obtained by multiplying each x-coordinate of $y = f(x)$ by $\frac{1}{a}$. A **horizontal compression** results if $a > 1$, and a **horizontal stretch** occurs if $0 < a < 1$.

Let's look at an example.

| **EXAMPLE 8** | **Graphing Using Stretches and Compressions** |

The graph of $y = f(x)$ is given in Figure 52. Use this graph to find the graphs of:

(a) $y = 3f(x)$ (b) $y = f(3x)$

Figure 52

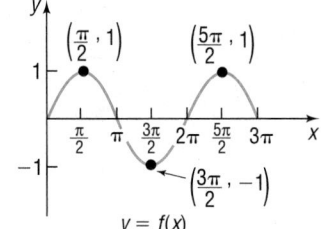

$y = f(x)$

Solution (a) The graph of $y = 3f(x)$ is obtained by multiplying each y-coordinate of $y = f(x)$ by a factor of 3. See Figure 53(a).

(b) The graph of $y = f(3x)$ is obtained from the graph of $y = f(x)$ by multiplying each x-coordinate of $y = f(x)$ by a factor of $\frac{1}{3}$. See Figure 53(b).

Figure 53

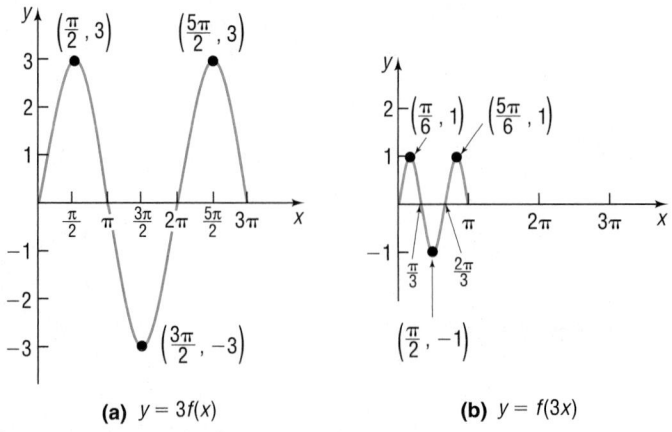

(a) $y = 3f(x)$ **(b)** $y = f(3x)$

NOW WORK PROBLEMS 65(e) AND (g).

3 **Graph Functions Using Reflections about the x-Axis and the y-Axis**

EXAMPLE 9 **Reflection about the x-Axis**

Graph the function: $f(x) = -x^2$

Solution We begin with the graph of $y = x^2$, as shown in black in Figure 54. For each point (x, y) on the graph of $y = x^2$, the point $(x, -y)$ is on the graph of $y = -x^2$, as indicated in Table 14. We can draw the graph of $y = -x^2$ by reflecting the graph of $y = x^2$ about the x-axis. See Figure 54.

Figure 54

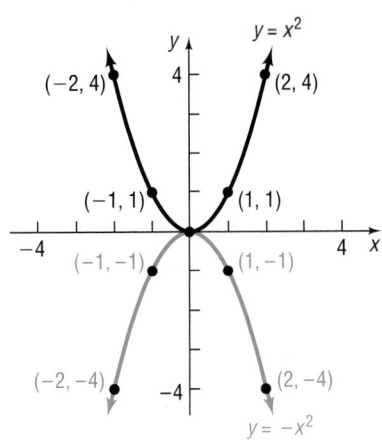

Table 14	x	$y = x^2$	$y = -x^2$
	-2	4	-4
	-1	1	-1
	0	0	0
	1	1	-1
	2	4	-4

■

> When the right side of the function $y = f(x)$ is multiplied by -1, the graph of the new function $y = -f(x)$ is the **reflection about the x-axis** of the graph of the function $y = f(x)$.

 NOW WORK PROBLEM 47.

EXAMPLE 10 **Reflection about the y-Axis**

Graph the function: $f(x) = \sqrt{-x}$

Solution First, notice that the domain of f consists of all real numbers x for which $-x \geq 0$ or, equivalently, $x \leq 0$. To get the graph of $f(x) = \sqrt{-x}$, we begin with the graph of $y = \sqrt{x}$, as shown in Figure 55. For each point (x, y) on the graph of $y = \sqrt{x}$, the point $(-x, y)$ is on the graph of $y = \sqrt{-x}$. We obtain the graph of $y = \sqrt{-x}$ by reflecting the graph of $y = \sqrt{x}$ about the y-axis. See Figure 55.

Figure 55

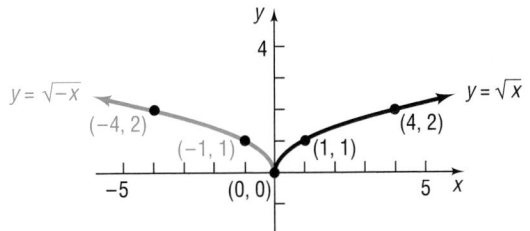

■

> When the graph of the function $y = f(x)$ is known, the graph of the new function $y = f(-x)$ is the **reflection about the y-axis** of the graph of the function $y = f(x)$.

Summary of Graphing Techniques

Table 15	To Graph:	Draw the Graph of f and:	Functional Change to f(x)
	Vertical shifts		
	$y = f(x) + k, \quad k > 0$	Raise the graph of f by k units.	Add k to $f(x)$.
	$y = f(x) - k, \quad k > 0$	Lower the graph of f by k units.	Subtract k from $f(x)$.
	Horizontal shifts		
	$y = f(x + h), \quad h > 0$	Shift the graph of f to the left h units.	Replace x by $x + h$.
	$y = f(x - h), \quad h > 0$	Shift the graph of f to the right h units.	Replace x by $x - h$.
	Compressing or stretching		
	$y = af(x), \quad a > 0$	Multiply each y-coordinate of $y = f(x)$ by a. Stretch the graph of f vertically if $a > 1$. Compress the graph of f vertically if $0 < a < 1$.	Multiply $f(x)$ by a.
	$y = f(ax), \quad a > 0$	Multiply each x-coordinate of $y = f(x)$ by $\frac{1}{a}$. Stretch the graph of f horizontally if $0 < a < 1$. Compress the graph of f horizontally if $a > 1$.	Replace x by ax.
	Reflection about the x-axis		
	$y = -f(x)$	Reflect the graph of f about the x-axis.	Multiply $f(x)$ by -1.
	Reflection about the y-axis		
	$y = f(-x)$	Reflect the graph of f about the y-axis.	Replace x by $-x$.

The examples that follow combine some of the procedures outlined in this section to get the required graph.

EXAMPLE 11 | **Determining the Function Obtained from a Series of Transformations**

Find the function that is finally graphed after the following three transformations are applied to the graph of $y = |x|$.

1. Shift left 2 units.
2. Shift up 3 units.
3. Reflect about the y-axis.

Solution
1. Shift left 2 units: Replace x by $x + 2$. $y = |x + 2|$
2. Shift up 3 units: Add 3. $y = |x + 2| + 3$
3. Reflect about the y-axis: Replace x by $-x$. $y = |-x + 2| + 3$ ∎

━━━━► **NOW WORK PROBLEM 27.**

EXAMPLE 12 | **Combining Graphing Procedures**

Graph the function: $f(x) = \dfrac{3}{x - 2} + 1$

Solution We use the following steps to obtain the graph of f:

STEP 1: $y = \dfrac{1}{x}$ Reciprocal function.

STEP 2: $y = \dfrac{3}{x}$ Multiply by 3; vertical stretch of the graph of $y = \dfrac{1}{x}$ by a factor of 3.

STEP 3: $y = \dfrac{3}{x - 2}$ Replace x by x − 2; horizontal shift to the right 2 units.

STEP 4: $y = \dfrac{3}{x - 2} + 1$ Add 1; vertical shift up 1 unit.

See Figure 56.

Figure 56

(a) $y = \dfrac{1}{x}$ Multiply by 3; Vertical stretch

(b) $y = \dfrac{3}{x}$ Replace x by x − 2; Horizontal shift right 2 units

(c) $y = \dfrac{3}{x-2}$ Add 1; Vertical shift up 1 unit

(d) $y = \dfrac{3}{x-2} + 1$

Other ordering of the steps shown in Example 12 would also result in the graph of f. For example, try this one:

STEP 1: $y = \dfrac{1}{x}$ Reciprocal function

STEP 2: $y = \dfrac{1}{x - 2}$ Replace x by x − 2; horizontal shift to the right 2 units.

STEP 3: $y = \dfrac{3}{x - 2}$ Multiply by 3; vertical stretch of the graph of $y = \dfrac{1}{x - 2}$ by a factor of 3.

STEP 4: $y = \dfrac{3}{x - 2} + 1$ Add 1; vertical shift up 1 unit.

EXAMPLE 13 **Combining Graphing Procedures**

Graph the function: $f(x) = \sqrt{1 - x} + 2$

Solution We use the following steps to get the graph of $y = \sqrt{1 - x} + 2$:

STEP 1: $y = \sqrt{x}$ Square root function

STEP 2: $y = \sqrt{x + 1}$ Replace x by x + 1; horizontal shift left 1 unit.

STEP 3: $y = \sqrt{-x + 1} = \sqrt{1 - x}$ Replace x by −x; reflect about y-axis.

STEP 4: $y = \sqrt{1 - x} + 2$ Add 2; vertical shift up 2 units.

See Figure 57.

Figure 57

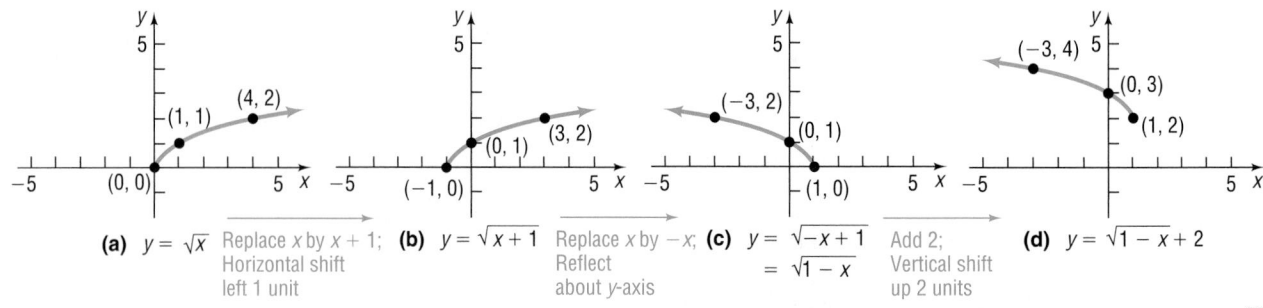

(a) $y = \sqrt{x}$ Replace x by $x + 1$;
Horizontal shift
left 1 unit

(b) $y = \sqrt{x + 1}$ Replace x by $-x$;
Reflect
about y-axis

(c) $y = \sqrt{-x + 1}$ Add 2;
$= \sqrt{1 - x}$ Vertical shift
up 2 units

(d) $y = \sqrt{1 - x} + 2$

NOW WORK PROBLEM 57.

1.5 Assess Your Understanding

Concepts and Vocabulary

1. Suppose that the graph of a function f is known. Then the graph of $y = f(x - 2)$ may be obtained by a(n) _____ shift of the graph of f to the _____ a distance of 2 units.

2. Suppose that the graph of a function f is known. Then the graph of $y = f(-x)$ may be obtained by a reflection about the _____-axis of the graph of the function $y = f(x)$.

3. Suppose that the x-intercepts of the graph of $y = f(x)$ are $-2, 1,$ and 5. The x-intercepts of $y = f(x + 3)$ are _____, _____, and _____.

4. *True or False:* The graph of $y = -f(x)$ is the reflection about the x-axis of the graph of $y = f(x)$.

5. *True or False:* To obtain the graph of $y = f(x + 2) - 3$, shift the graph of $y = f(x)$ horizontally to the right 2 units and vertically down 3 units.

6. *True or False:* Suppose that the x-intercepts of the graph of $y = f(x)$ are -3 and 2. Then the x-intercepts of the graph of $y = 2f(x)$ are -3 and 2.

Skill Building

In Problems 7–18, match each graph to one of the following functions.

A. $y = x^2 + 2$

B. $y = -x^2 + 2$

C. $y = |x| + 2$

D. $y = -|x| + 2$

E. $y = (x - 2)^2$

F. $y = -(x + 2)^2$

G. $y = |x - 2|$

H. $y = -|x + 2|$

I. $y = 2x^2$

J. $y = -2x^2$

K. $y = 2|x|$

L. $y = -2|x|$

7.

8.

9.

10.

11.

12.

13.

14.

15. **16.** **17.** **18.**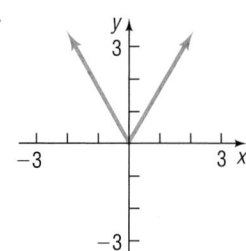

In Problems 19–26, write the function whose graph is the graph of $y = x^3$, but is:

19. Shifted to the right 4 units

20. Shifted to the left 4 units

21. Shifted up 4 units

22. Shifted down 4 units

23. Reflected about the y-axis

24. Reflected about the x-axis

25. Vertically stretched by a factor of 4

26. Horizontally stretched by a factor of 4

In Problems 27–30, find the function that is finally graphed after the following transformations are applied to the graph of $y = \sqrt{x}$.

27. (1) Shift up 2 units
(2) Reflect about the x-axis
(3) Reflect about the y-axis

28. (1) Reflect about the x-axis
(2) Shift right 3 units
(3) Shift down 2 units

29. (1) Reflect about the x-axis
(2) Shift up 2 units
(3) Shift left 3 units

30. (1) Shift up 2 units
(2) Reflect about the y-axis
(3) Shift left 3 units

31. If $(3, 0)$ is a point on the graph of $y = f(x)$, which of the following points must be on the graph of $y = -f(x)$?
(a) $(0, 3)$ (b) $(0, -3)$
(c) $(3, 0)$ (d) $(-3, 0)$

32. If $(3, 0)$ is a point on the graph of $y = f(x)$, which of the following points must be on the graph of $y = f(-x)$?
(a) $(0, 3)$ (b) $(0, -3)$
(c) $(3, 0)$ (d) $(-3, 0)$

33. If $(0, 3)$ is a point on the graph of $y = f(x)$, which of the following points must be on the graph of $y = 2f(x)$?
(a) $(0, 3)$ (b) $(0, 2)$
(c) $(0, 6)$ (d) $(6, 0)$

34. If $(3, 0)$ is a point on the graph of $y = f(x)$, which of the following points must be on the graph of $y = \dfrac{1}{2}f(x)$?
(a) $(3, 0)$ (b) $\left(\dfrac{3}{2}, 0\right)$
(c) $\left(0, \dfrac{3}{2}\right)$ (d) $\left(\dfrac{1}{2}, 0\right)$

In Problems 35–64, graph each function using the techniques of shifting, compressing, stretching, and/or reflecting. Start with the graph of the basic function (for example, $y = x^2$) and show all stages.

35. $f(x) = x^2 - 1$

36. $f(x) = x^2 + 4$

37. $g(x) = x^3 + 1$

38. $g(x) = x^3 - 1$

39. $h(x) = \sqrt{x - 2}$

40. $h(x) = \sqrt{x + 1}$

41. $f(x) = (x - 1)^3 + 2$

42. $f(x) = (x + 2)^3 - 3$

43. $g(x) = 4\sqrt{x}$

44. $g(x) = \dfrac{1}{2}\sqrt{x}$

45. $h(x) = \dfrac{1}{2x}$

46. $h(x) = 3\sqrt[3]{x}$

47. $f(x) = -\sqrt[3]{x}$

48. $f(x) = -\sqrt{x}$

49. $g(x) = \sqrt{-x}$

50. $g(x) = \sqrt[3]{-x}$

51. $h(x) = -x^3 + 2$

52. $h(x) = \dfrac{1}{-x} + 2$

53. $f(x) = 2(x + 1)^2 - 3$

54. $f(x) = 3(x - 2)^2 + 1$

55. $g(x) = \sqrt{x - 2} + 1$

56. $g(x) = |x + 1| - 3$

57. $h(x) = \sqrt{-x} - 2$

58. $h(x) = \dfrac{4}{x} + 2$

59. $f(x) = -(x + 1)^3 - 1$

60. $f(x) = -4\sqrt{x - 1}$

61. $g(x) = 2|1 - x|$

62. $g(x) = 4\sqrt{2 - x}$

63. $h(x) = 2 \operatorname{int}(x - 1)$

64. $h(x) = \operatorname{int}(-x)$

In Problems 65–68, the graph of a function f is illustrated. Use the graph of f as the first step toward graphing each of the following functions:

(a) $F(x) = f(x) + 3$ (b) $G(x) = f(x + 2)$ (c) $P(x) = -f(x)$ (d) $H(x) = f(x + 1) - 2$

(e) $Q(x) = \dfrac{1}{2}f(x)$ (f) $g(x) = f(-x)$ (g) $h(x) = f(2x)$

65.

66.

67.

68.

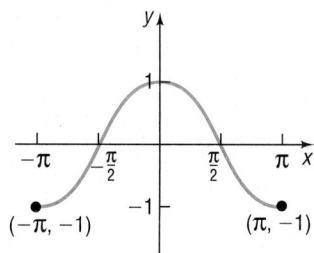

Applications and Extensions

69. Suppose that the x-intercepts of the graph of $y = f(x)$ are −5 and 3.
(a) What are the x-intercepts of the graph of $y = f(x + 2)$?
(b) What are the x-intercepts of the graph of $y = f(x − 2)$?
(c) What are the x-intercepts of the graph of $y = 4f(x)$?
(d) What are the x-intercepts of the graph of $y = f(−x)$?

71. Suppose that the function $y = f(x)$ is increasing on the interval $(−1, 5)$.
(a) Over what interval is the graph of $y = f(x + 2)$ increasing?
(b) Over what interval is the graph of $y = f(x − 5)$ increasing?
(c) What can be said about the graph of $y = −f(x)$?
(d) What can be said about the graph of $y = f(−x)$?

73. The graph of a function f is illustrated in the figure.
(a) Draw the graph of $y = |f(x)|$.
(b) Draw the graph of $y = f(|x|)$.

70. Suppose that the x-intercepts of the graph of $y = f(x)$ are −8 and 1.
(a) What are the x-intercepts of the graph of $y = f(x + 4)$?
(b) What are the x-intercepts of the graph of $y = f(x − 3)$?
(c) What are the x-intercepts of the graph of $y = 2f(x)$?
(d) What are the x-intercepts of the graph of $y = f(−x)$?

72. Suppose that the function $y = f(x)$ is decreasing on the interval $(−2, 7)$.
(a) Over what interval is the graph of $y = f(x + 2)$ decreasing?
(b) Over what interval is the graph of $y = f(x − 5)$ decreasing?
(c) What can be said about the graph of $y = −f(x)$?
(d) What can be said about the graph of $y = f(−x)$?

74. The graph of a function f is illustrated in the figure.
(a) Draw the graph of $y = |f(x)|$.
(b) Draw the graph of $y = f(|x|)$.

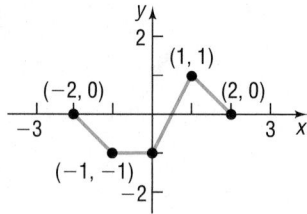

In Problems 75–84, complete the square of each quadratic expression. Then graph each function using the technique of shifting. (If necessary, refer to the Appendix, Section A.4, to review completing the square.)

75. $f(x) = x^2 + 2x$

76. $f(x) = x^2 − 6x$

77. $f(x) = x^2 − 8x + 1$

78. $f(x) = x^2 + 4x + 2$

79. $f(x) = x^2 + x + 1$

80. $f(x) = x^2 − x + 1$

81. $f(x) = 2x^2 − 12x + 19$

82. $f(x) = 3x^2 + 6x + 1$

83. $f(x) = −3x^2 − 12x − 17$

84. $f(x) = −2x^2 − 12x − 13$

85. The equation $y = (x - c)^2$ defines a *family of parabolas*, one parabola for each value of c. On one set of coordinate axes, graph the members of the family for $c = 0$, $c = 3$, and $c = -2$.

86. Repeat Problem 85 for the family of parabolas $y = x^2 + c$.

87. Temperature Measurements The relationship between the Celsius (°C) and Fahrenheit (°F) scales for measuring temperature is given by the equation

$$F = \frac{9}{5}C + 32$$

The relationship between the Celsius (°C) and Kelvin (K) scales is $K = C + 273$. Graph the equation $F = \frac{9}{5}C + 32$ using degrees Fahrenheit on the y-axis and degrees Celsius on the x-axis. Use the techniques introduced in this section to obtain the graph showing the relationship between Kelvin and Fahrenheit temperatures.

88. Period of a Pendulum The period T (in seconds) of a simple pendulum is a function of its length l (in feet) defined by the equation

$$T = 2\pi\sqrt{\frac{l}{g}}$$

where $g \approx 32.2$ feet per second per second is the acceleration of gravity.
(a) Use a graphing utility to graph the function $T = T(l)$.
(b) Now graph the functions $T = T(l + 1)$, $T = T(l + 2)$, and $T = T(l + 3)$.
(c) Discuss how adding to the length l changes the period T.

(d) Now graph the functions $T = T(2l)$, $T = T(3l)$, and $T = T(4l)$.
(e) Discuss how multiplying the length l by factors of 2, 3, and 4 changes the period T.

89. Cigar Company Profits The daily profits of a cigar company from selling x cigars are given by

$$p(x) = -0.05x^2 + 100x - 2000$$

The government wishes to impose a tax on cigars (sometimes called a *sin tax*) that gives the company the option of either paying a flat tax of $10,000 per day or a tax of 10% on profits. As chief financial officer (CFO) of the company, you need to decide which tax is the better option for the company.
(a) On the same screen, graph $Y_1 = p(x) - 10,000$ and $Y_2 = (1 - 0.10)p(x)$.
(b) Based on the graph, which option would you select? Why?
(c) Using the terminology learned in this section, describe each graph in terms of the graph of $p(x)$.
(d) Suppose that the government offered the options of a flat tax of $4800 or a tax of 10% on profits. Which would you select? Why?

Discussion and Writing

90. Suppose that the graph of a function f is known. Explain how the graph of $y = 4f(x)$ differs from the graph of $y = f(4x)$.

1.6 Mathematical Models: Constructing Functions

OBJECTIVE 1 Construct and Analyze Functions

1 Construct and Analyze Functions

Real-world problems often result in mathematical models that involve functions. These functions need to be constructed or built based on the information given. In constructing functions, we must be able to translate the verbal description into the language of mathematics. We do this by assigning symbols to represent the independent and dependent variables and then finding the function or rule that relates these variables.

EXAMPLE 1 **Area of a Rectangle with Fixed Perimeter**

The perimeter of a rectangle is 50 feet. Express its area A as a function of the length l of a side.

Solution

Consult Figure 58. If the length of the rectangle is l and if w is its width, then the sum of the lengths of the sides is the perimeter, 50.

Figure 58

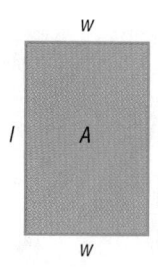

$$l + w + l + w = 50$$
$$2l + 2w = 50$$
$$l + w = 25$$
$$w = 25 - l$$

The area A is length times width, so

$$A = lw = l(25 - l)$$

The area A as a function of l is

$$A(l) = l(25 - l)$$ ∎

Note that we use the symbol A as the dependent variable and also as the name of the function that relates the length l to the area A. This double usage is common in applications and should cause no difficulties.

EXAMPLE 2 | **Economics: Demand Equations**

In economics, revenue R, in dollars, is defined as the amount of money received from the sale of a product and is equal to the unit selling price p, in dollars, of the product times the number x of units actually sold. That is,

$$R = xp$$

In economics, the Law of Demand states that p and x are related: As one increases, the other decreases. Suppose that p and x are related by the following **demand equation**:

$$p = -\frac{1}{10}x + 20, \qquad 0 \leq x \leq 200$$

Express the revenue R as a function of the number x of units sold.

Solution

Since $R = xp$ and $p = -\frac{1}{10}x + 20$, it follows that

$$R(x) = xp = x\left(-\frac{1}{10}x + 20\right) = -\frac{1}{10}x^2 + 20x$$ ∎

 NOW WORK PROBLEM 3.

EXAMPLE 3 | **Finding the Distance from the Origin to a Point on a Graph**

Let $P = (x, y)$ be a point on the graph of $y = x^2 - 1$.

(a) Express the distance d from P to the origin O as a function of x.
(b) What is d if $x = 0$?
(c) What is d if $x = 1$?
(d) What is d if $x = \dfrac{\sqrt{2}}{2}$?
(e) Use a graphing utility to graph the function $d = d(x)$, $x \geq 0$. Rounded to two decimal places, find the value(s) of x at which d has a local minimum. [This gives the point(s) on the graph of $y = x^2 - 1$ closest to the origin.]

Figure 59

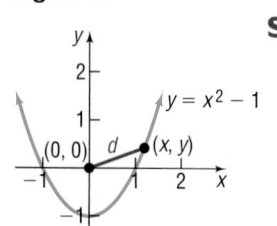

Solution

(a) Figure 59 illustrates the graph of $y = x^2 - 1$. The distance d from P to O is

$$d = \sqrt{(x - 0)^2 + (y - 0)^2} = \sqrt{x^2 + y^2}$$

Since P is a point on the graph of $y = x^2 - 1$, we substitute $x^2 - 1$ for y. Then

$$d(x) = \sqrt{x^2 + (x^2 - 1)^2} = \sqrt{x^4 - x^2 + 1}$$

We have expressed the distance d as a function of x.

(b) If $x = 0$, the distance d is
$$d(0) = \sqrt{1} = 1$$

(c) If $x = 1$, the distance d is
$$d(1) = \sqrt{1 - 1 + 1} = 1$$

(d) If $x = \dfrac{\sqrt{2}}{2}$, the distance d is
$$d\left(\frac{\sqrt{2}}{2}\right) = \sqrt{\left(\frac{\sqrt{2}}{2}\right)^4 - \left(\frac{\sqrt{2}}{2}\right)^2 + 1} = \sqrt{\frac{1}{4} - \frac{1}{2} + 1} = \frac{\sqrt{3}}{2}$$

Figure 60

 (e) Figure 60 shows the graph of $Y_1 = \sqrt{x^4 - x^2 + 1}$. Using the MINIMUM feature on a graphing utility, we find that when $x \approx 0.71$ the value of d is smallest. The local minimum is $d \approx 0.87$ rounded to two decimal places. [Since $d(x)$ is even, by symmetry, it follows that when $x \approx -0.71$ the value of d is also a local minimum.] ■

NOW WORK PROBLEM **9.**

EXAMPLE 4 **Filling a Swimming Pool**

A rectangular swimming pool 20 meters long and 10 meters wide is 4 meters deep at one end and 1 meter deep at the other. Figure 61 illustrates a cross-sectional view of the pool. Water is being pumped into the pool to a height of 3 meters at the deep end.

Figure 61

(a) Find a function that expresses the volume V of water in the pool as a function of the height x of the water at the deep end.

(b) Find the volume when the height is 1 meter.

(c) Find the volume when the height is 2 meters.

(d) At what height is the volume 20 cubic meters? 100 cubic meters?

Solution (a) Let L denote the distance (in meters) measured at water level from the deep end to the short end. Notice that L and x form the sides of a triangle that is similar to the triangle whose sides are 20 meters by 3 meters. This means L and x are related by the equation
$$\frac{L}{x} = \frac{20}{3} \quad \text{or} \quad L = \frac{20x}{3} \qquad 0 \le x \le 3$$

The volume V of water in the pool at any time is
$$V = \left(\begin{array}{c}\text{cross-sectional} \\ \text{triangular area}\end{array}\right)(\text{width}) = \left(\frac{1}{2}Lx\right)(10) \quad \text{cubic meters}$$

Since $L = \dfrac{20x}{3}$, we have
$$V(x) = \left(\frac{1}{2}\cdot\frac{20x}{3}\cdot x\right)(10) = \frac{100}{3}x^2 \quad \text{cubic meters}$$

(b) When the height x of the water is 1 meter, the volume $V = V(x)$ is

$$V(1) = \frac{100}{3} \cdot 1^2 = 33\frac{1}{3} \quad \text{cubic meters}$$

(c) When the height x of the water is 2 meters, the volume $V = V(x)$ is

$$V(2) = \frac{100}{3} \cdot 2^2 = \frac{400}{3} = 133\frac{1}{3} \quad \text{cubic meters}$$

(d) By solving the equation $\dfrac{100}{3}x^2 = 20$, we find that when $x \approx 0.77$ meter, the volume is 20 cubic meters. By solving the equation $\dfrac{100}{3}x^2 = 100$, we find that when $x \approx 1.73$ meters, the volume is 100 cubic meters. ■

EXAMPLE 5 **Area of a Rectangle**

Figure 62

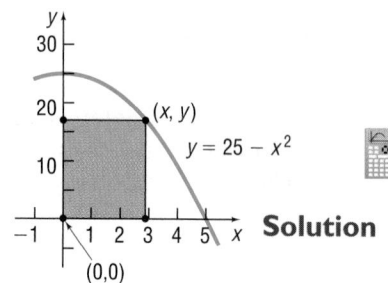

A rectangle has one corner on the graph of $y = 25 - x^2$, another at the origin, a third on the positive y-axis, and the fourth on the positive x-axis. See Figure 62.

(a) Express the area A of the rectangle as a function of x.
(b) What is the domain of A?
(c) Graph $A = A(x)$.
(d) For what value of x is the area largest?

Solution

(a) The area A of the rectangle is $A = xy$, where $y = 25 - x^2$. Substituting this expression for y, we obtain $A(x) = x(25 - x^2) = 25x - x^3$.

(b) Since x represents a side of the rectangle, we have $x > 0$. In addition, the area must be positive, so $y = 25 - x^2 > 0$, which implies that $x^2 < 25$, so $-5 < x < 5$. Combining these restrictions, we have the domain of A as $\{x \mid 0 < x < 5\}$ or $(0, 5)$ using interval notation.

(c) See Figure 63 for the graph of $A = A(x)$.

(d) Using MAXIMUM, we find that the maximum area is 48.11 at $x = 2.89$, each rounded to two decimal places. See Figure 64.

Figure 63

Figure 64

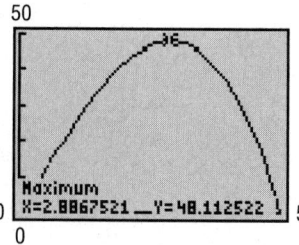

NOW WORK PROBLEM 15.

EXAMPLE 6 **Making a Playpen**[*]

A manufacturer of children's playpens makes a square model that can be opened at one corner and attached at right angles to a wall or, perhaps, the side of a house. If each side is 3 feet in length, the open configuration doubles the available area in which the child can play from 9 square feet to 18 square feet. See Figure 65.

[*]Adapted from Proceedings, *Summer Conference for College Teachers on Applied Mathematics* (University of Missouri, Rolla), 1971.

Now suppose that we place hinges at the outer corners to allow for a configuration like the one shown in Figure 66.

Figure 65

Figure 66

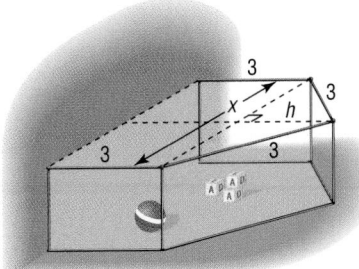

(a) Express the area A of the configuration shown in Figure 66 as a function of the distance x between the two parallel sides.

(b) Find the domain of A.

(c) Find A if $x = 5$.

(d) Graph $A = A(x)$. For what value of x is the area largest? What is the maximum area?

Solution

(a) Refer to Figure 66. The area A that we seek consists of the area of a rectangle (with width 3 and length x) and the area of an isosceles triangle (with base x and two equal sides of length 3). The height h of the triangle may be found using the Pythagorean Theorem.

$$h^2 + \left(\frac{x}{2}\right)^2 = 3^2$$

$$h^2 = 3^2 - \left(\frac{x}{2}\right)^2 = 9 - \frac{x^2}{4} = \frac{36 - x^2}{4}$$

$$h = \frac{1}{2}\sqrt{36 - x^2}$$

The area A enclosed by the playpen is

$$A = \text{area of rectangle} + \text{area of triangle} = 3x + \frac{1}{2}x\left(\frac{1}{2}\sqrt{36 - x^2}\right)$$

$$A(x) = 3x + \frac{x\sqrt{36 - x^2}}{4}$$

Now the area A is expressed as a function of x.

(b) To find the domain of A, we note first that $x > 0$, since x is a length. Also, the expression under the square root must be positive, so

$$36 - x^2 > 0$$
$$x^2 < 36$$
$$-6 < x < 6$$

Combining these restrictions, we find that the domain of A is $0 < x < 6$, or $(0, 6)$ using interval notation.

(c) If $x = 5$, the area is

$$A(5) = 3(5) + \frac{5}{4}\sqrt{36 - (5)^2} \approx 19.15 \text{ square feet}$$

If the length of the playpen is 5 feet, its area is 19.15 square feet.

(d) See Figure 67. The maximum area is about 19.82 square feet, obtained when x is about 5.58 feet.

■

Figure 67

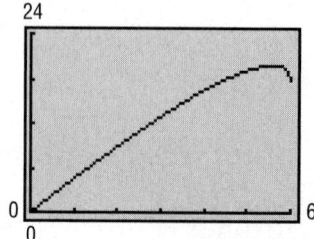

1.6 Assess Your Understanding

Applications and Extensions

1. **Volume of a Cylinder** The volume V of a right circular cylinder of height h and radius r is $V = \pi r^2 h$. If the height is twice the radius, express the volume V as a function of r.

2. **Volume of a Cone** The volume V of a right circular cone is $V = \dfrac{1}{3}\pi r^2 h$. If the height is twice the radius, express the volume V as a function of r.

3. **Demand Equation** The price p, in dollars, and the quantity x sold of a certain product obey the demand equation

$$p = -\frac{1}{6}x + 100, \qquad 0 \le x \le 600$$

 (a) Express the revenue R as a function of x. (Remember, $R = xp$.)
 (b) What is the revenue if 200 units are sold?
 (c) Graph the revenue function using a graphing utility.
 (d) What quantity x maximizes revenue? What is the maximum revenue?
 (e) What price should the company charge to maximize revenue?

4. **Demand Equation** The price p, in dollars, and the quantity x sold of a certain product obey the demand equation

$$p = -\frac{1}{3}x + 100, \qquad 0 \le x \le 300$$

 (a) Express the revenue R as a function of x.
 (b) What is the revenue if 100 units are sold?
 (c) Graph the revenue function using a graphing utility.
 (d) What quantity x maximizes revenue? What is the maximum revenue?
 (e) What price should the company charge to maximize revenue?

5. **Demand Equation** The price p, in dollars, and the quantity x sold of a certain product obey the demand equation

$$x = -5p + 100, \qquad 0 \le p \le 20$$

 (a) Express the revenue R as a function of x.

 (b) What is the revenue if 15 units are sold?
 (c) Graph the revenue function using a graphing utility.
 (d) What quantity x maximizes revenue? What is the maximum revenue?
 (e) What price should the company charge to maximize revenue?

6. **Demand Equation** The price p, in dollars, and the quantity x sold of a certain product obey the demand equation

$$x = -20p + 500, \qquad 0 \le p \le 25$$

 (a) Express the revenue R as a function of x.
 (b) What is the revenue if 20 units are sold?
 (c) Graph the revenue function using a graphing utility.
 (d) What quantity x maximizes revenue? What is the maximum revenue?
 (e) What price should the company charge to maximize revenue?

7. **Enclosing a Rectangular Field** David has available 400 yards of fencing and wishes to enclose a rectangular area.
 (a) Express the area A of the rectangle as a function of the width x of the rectangle.
 (b) What is the domain of A?
 (c) Graph $A = A(x)$ using a graphing utility. For what value of x is the area largest?

8. **Enclosing a Rectangular Field along a River** Beth has 3000 feet of fencing available to enclose a rectangular field. One side of the field lies along a river, so only three sides require fencing.
 (a) Express the area A of the rectangle as a function of x, where x is the length of the side parallel to the river.
 (b) Graph $A = A(x)$ using a graphing utility. For what value of x is the area largest?

9. Let $P = (x, y)$ be a point on the graph of $y = x^2 - 8$.
 (a) Express the distance d from P to the origin as a function of x.
 (b) What is d if $x = 0$?
 (c) What is d if $x = 1$?
 (d) Use a graphing utility to graph $d = d(x)$.
 (e) For what values of x is d smallest?

10. Let $P = (x, y)$ be a point on the graph of $y = x^2 - 8$.
 (a) Express the distance d from P to the point $(0, -1)$ as a function of x.
 (b) What is d if $x = 0$?
 (c) What is d if $x = -1$?
 (d) Use a graphing utility to graph $d = d(x)$.
 (e) For what values of x is d smallest?

11. Let $P = (x, y)$ be a point on the graph of $y = \sqrt{x}$.
 (a) Express the distance d from P to the point $(1, 0)$ as a function of x.
 (b) Use a graphing utility to graph $d = d(x)$.
 (c) For what values of x is d smallest?

12. Let $P = (x, y)$ be a point on the graph of $y = \dfrac{1}{x}$.
 (a) Express the distance d from P to the origin as a function of x.
 (b) Use a graphing utility to graph $d = d(x)$.
 (c) For what values of x is d smallest?

13. A right triangle has one vertex on the graph of $y = x^3$, $x > 0$, at (x, y), another at the origin, and the third on the positive y-axis at $(0, y)$, as shown in the figure. Express the area A of the triangle as a function of x.

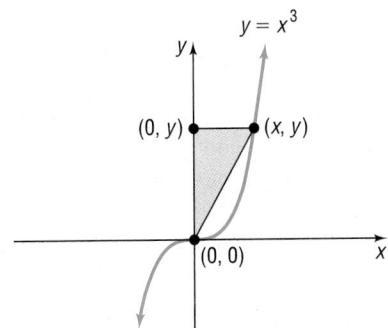

14. A right triangle has one vertex on the graph of $y = 9 - x^2$, $x > 0$, at (x, y), another at the origin, and the third on the positive x-axis at $(x, 0)$. Express the area A of the triangle as a function of x.

15. A rectangle has one corner on the graph of $y = 16 - x^2$, another at the origin, a third on the positive y-axis, and the fourth on the positive x-axis (see the figure).

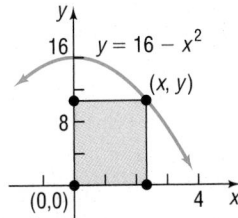

 (a) Express the area A of the rectangle as a function of x.
 (b) What is the domain of A?
 (c) Graph $A = A(x)$. For what value of x is A largest?

16. A rectangle is inscribed in a semicircle of radius 2 (see the figure). Let $P = (x, y)$ be the point in quadrant I that is a vertex of the rectangle and is on the circle.

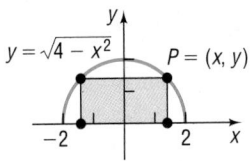

 (a) Express the area A of the rectangle as a function of x.
 (b) Express the perimeter p of the rectangle as a function of x.
 (c) Graph $A = A(x)$. For what value of x is A largest?
 (d) Graph $p = p(x)$. For what value of x is p largest?

17. A rectangle is inscribed in a circle of radius 2 (see the figure). Let $P = (x, y)$ be the point in quadrant I that is a vertex of the rectangle and is on the circle.

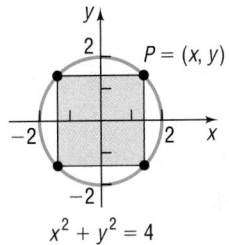

 (a) Express the area A of the rectangle as a function of x.
 (b) Express the perimeter p of the rectangle as a function of x.
 (c) Graph $A = A(x)$. For what value of x is A largest?
 (d) Graph $p = p(x)$. For what value of x is p largest?

18. A circle of radius r is inscribed in a square (see the figure).

 (a) Express the area A of the square as a function of the radius r of the circle.
 (b) Express the perimeter p of the square as a function of r.

19. A wire 10 meters long is to be cut into two pieces. One piece will be shaped as a square, and the other piece will be shaped as a circle (see the figure).

 (a) Express the total area A enclosed by the pieces of wire as a function of the length x of a side of the square.
 (b) What is the domain of A?
 (c) Graph $A = A(x)$. For what value of x is A smallest?

20. A wire 10 meters long is to be cut into two pieces. One piece will be shaped as an equilateral triangle, and the other piece will be shaped as a circle.
 (a) Express the total area A enclosed by the pieces of wire as a function of the length x of a side of the equilateral triangle.
 (b) What is the domain of A?
 (c) Graph $A = A(x)$. For what value of x is A smallest?

21. A wire of length x is bent into the shape of a circle.
 (a) Express the circumference of the circle as a function of x.
 (b) Express the area of the circle as a function of x.

22. A wire of length x is bent into the shape of a square.
 (a) Express the perimeter of the square as a function of x.
 (b) Express the area of the square as a function of x.

23. A semicircle of radius r is inscribed in a rectangle so that the diameter of the semicircle is the length of the rectangle (see the figure).

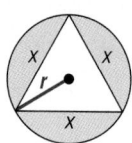

 (a) Express the area A of the rectangle as a function of the radius r of the semicircle.
 (b) Express the perimeter p of the rectangle as a function of r.

24. An equilateral triangle is inscribed in a circle of radius r. See the figure. Express the circumference C of the circle as a function of the length x of a side of the triangle.

 [**Hint:** First show that $r^2 = \dfrac{x^2}{3}$.]

25. An equilateral triangle is inscribed in a circle of radius r. See the figure in Problem 24. Express the area A within the circle, but outside the triangle, as a function of the length x of a side of the triangle.

26. Two cars leave an intersection at the same time. One is headed south at a constant speed of 30 miles per hour, and the other is headed west at a constant speed of 40 miles per hour (see the figure). Express the distance d between the cars as a function of the time t.

 [**Hint:** At $t = 0$, the cars leave the intersection.]

27. Two cars are approaching an intersection. One is 2 miles south of the intersection and is moving at a constant speed of 30 miles per hour. At the same time, the other car is 3 miles east of the intersection and is moving at a constant speed of 40 miles per hour.
 (a) Express the distance d between the cars as a function of time t.
 [**Hint:** At $t = 0$, the cars are 2 miles south and 3 miles east of the intersection, respectively.]
 (b) Use a graphing utility to graph $d = d(t)$. For what value of t is d smallest?

28. **Inscribing a Cylinder in a Sphere** Inscribe a right circular cylinder of height h and radius r in a sphere of fixed radius R. See the illustration. Express the volume V of the cylinder as a function of h.
 [**Hint:** $V = \pi r^2 h$. Note also the right triangle.]

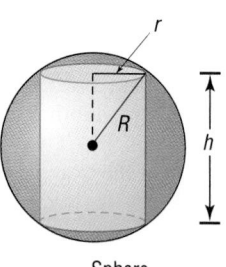

Sphere

29. **Inscribing a Cylinder in a Cone** Inscribe a right circular cylinder of height h and radius r in a cone of fixed radius R and fixed height H. See the illustration. Express the volume V of the cylinder as a function of r.
 [**Hint:** $V = \pi r^2 h$. Note also the similar triangles.]

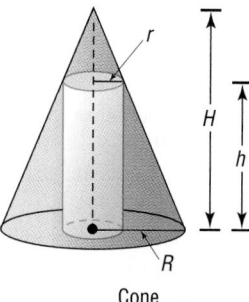

Cone

30. **Installing Cable TV** MetroMedia Cable is asked to provide service to a customer whose house is located 2 miles from the road along which the cable is buried. The nearest connection box for the cable is located 5 miles down the road (see the figure).

(a) If the installation cost is \$10 per mile along the road and \$14 per mile off the road, express the total cost C of installation as a function of the distance x (in miles) from the connection box to the point where the cable installation turns off the road. Give the domain.

(b) Compute the cost if $x = 1$ mile.

(c) Compute the cost if $x = 3$ miles.

(d) Graph the function $C = C(x)$. Use TRACE to see how the cost C varies as x changes from 0 to 5.

(e) What value of x results in the least cost?

31. Time Required to Go from an Island to a Town An island is 2 miles from the nearest point P on a straight shoreline. A town is 12 miles down the shore from P. See the illustration.

(a) If a person can row a boat at an average speed of 3 miles per hour and the same person can walk 5 miles per hour, express the time T that it takes to go from the island to town as a function of the distance x from P to where the person lands the boat.

(b) What is the domain of T?

(c) How long will it take to travel from the island to town if the person lands the boat 4 miles from P?

(d) How long will it take if the person lands the boat 8 miles from P?

32. Water is poured into a container in the shape of a right circular cone with radius 4 feet and height 16 feet (see the figure). Express the volume V of the water in the cone as a function of the height h of the water.

[**Hint:** The volume V of a cone of radius r and height h is $V = \frac{1}{3}\pi r^2 h$.]

Chapter Review

Library of Functions

Linear function (p. 77)

$f(x) = mx + b$

Graph is a line with slope m and y-intercept b.

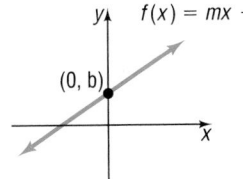

Constant function (p. 77)

$f(x) = b$

Graph is a horizontal line with y-intercept b.

Identity function (p. 78)

$f(x) = x$

Graph is a line with slope 1 and y-intercept 0.

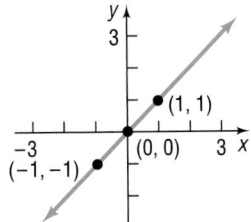

Square function (p. 78)

$f(x) = x^2$

Graph is a parabola with intercept at $(0, 0)$.

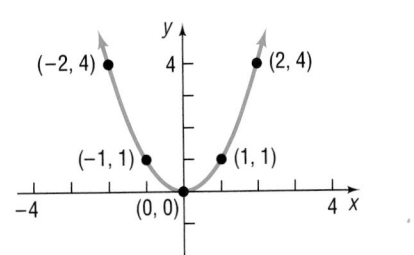

Cube function (p. 78)

$f(x) = x^3$

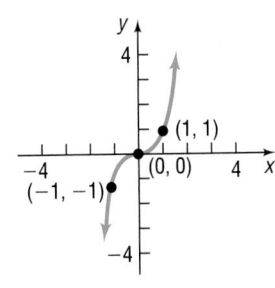

Square root function (p. 78)

$f(x) = \sqrt{x}$

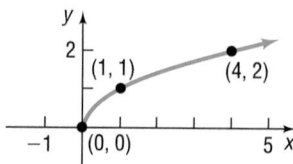

Cube root function (p. 78)

$f(x) = \sqrt[3]{x}$

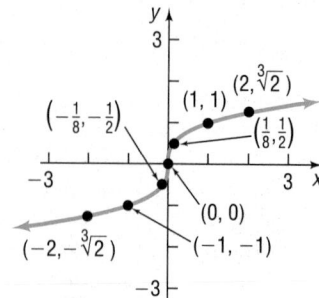

Reciprocal function (p. 79)

$f(x) = \dfrac{1}{x}$

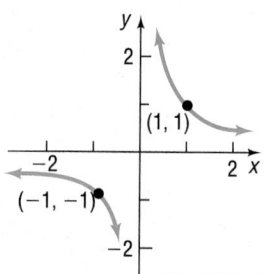

Absolute value function (p. 79)

$f(x) = |x|$

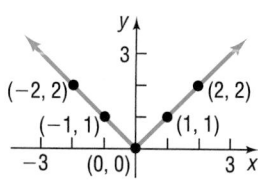

Greatest integer function (pp. 79–80)

$f(x) = \text{int}(x)$

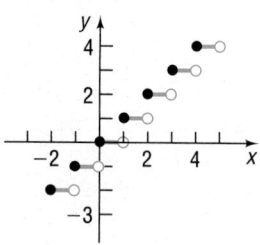

Things to Know

Function (pp. 40–45)	A relation between two sets of real numbers so that each number x in the first set, the domain, has corresponding to it exactly one number y in the second set. The range is the set of y values of the function for the x values in the domain.
	x is the independent variable; y is the dependent variable.
	A function can also be characterized as a set of ordered pairs (x, y) in which no first element is paired with two different second elements.
Function notation (pp. 44–48)	$y = f(x)$
	f is a symbol for the function.
	x is the argument, or independent variable.
	y is the dependent variable.
	The numbers for which $f(x) = 0$, if any, are the zeros of f.
	$f(x)$ is the value of the function at x, or the image of x.
	A function f may be defined implicitly by an equation involving x and y or explicitly by writing $y = f(x)$.
Difference quotient of f (p. 47 and p. 74)	$\dfrac{f(x + h) - f(x)}{h} \quad h \neq 0$
Domain (p. 41 and p. 48)	If unspecified, the domain of a function f is the largest set of real numbers for which $f(x)$ is a real number.
Vertical-line test (p. 55)	A set of points in the plane is the graph of a function if and only if every vertical line intersects the graph in at most one point.
Even function f (p. 63)	$f(-x) = f(x)$ for every x in the domain ($-x$ must also be in the domain).
Odd function f (p. 63)	$f(-x) = -f(x)$ for every x in the domain ($-x$ must also be in the domain).
Increasing function (p. 65)	A function f is increasing on an open interval I if, for any choice of x_1 and x_2 in I, with $x_1 < x_2$, we have $f(x_1) < f(x_2)$.
Decreasing function (p. 65)	A function f is decreasing on an open interval I if, for any choice of x_1 and x_2 in I, with $x_1 < x_2$, we have $f(x_1) > f(x_2)$.
Constant function (p. 65)	A function f is constant on an open interval I if, for all choices of x in I, the values of $f(x)$ are equal.

Local maximum (p. 66)	A function f has a local maximum at c if there is an open interval I containing c so that, for all x in I, $f(x) \leq f(c)$.
Local minimum (p. 66)	A function f has a local minimum at c if there is an open interval I containing c so that, for all x in I, $f(x) \geq f(c)$.
Average rate of change of a function (p. 68)	The average rate of change of f from c to x is $$\frac{\Delta y}{\Delta x} = \frac{f(x) - f(c)}{x - c} \qquad x \neq c$$

Objectives

Section		You should be able to . . .	Review Exercises
1.1	1	Determine whether a relation represents a function (p. 40)	1, 2
	2	Find the value of a function (p. 44)	3–8, 23, 24, 71, 72
	3	Find the domain of a function (p. 48)	9–16
	4	Form the sum, difference, product, and quotient of two functions (p. 49)	17–22
1.2	1	Identify the graph of a function (p. 55)	47–50
	2	Obtain information from or about the graph of a function (p. 56)	25(a)–(e), 26(a)–(e), 27(a), 27(d) 27(f), 27(g), 28(a), 28(d), 28(f), 28(g)
1.3	1	Determine even and odd functions from a graph (p. 63)	27(e), 28(e)
	2	Identify even and odd functions from the equation (p. 64)	29–36
	3	Use a graph to determine where a function is increasing, decreasing, or constant (p. 65)	27(b), 28(b)
	4	Use a graph to locate local maxima and local minima (p. 66)	27(c), 28(c)
	5	Use a graphing utility to approximate local maxima and local minima and to determine where a function is increasing or decreasing (p. 67)	37–40, 75(e), 76(b), 77–80, 81(d), 82(d)
	6	Find the average rate of change of a function (p. 68)	41–46
1.4	1	Graph the functions listed in the library of functions (p. 77)	51–54
	2	Graph piecewise-defined functions (p. 80)	67–70
1.5	1	Graph functions using vertical and horizontal shifts (p. 85)	25(f), 26(f), 26(g) 55, 56, 59–66
	2	Graph functions using compressions and stretches (p. 88)	25(g), 26(h), 57, 58, 65, 66
	3	Graph functions using reflections about the x-axis and the y-axis (p. 91)	25(h), 57, 61, 62, 66
1.6	1	Construct and analyze functions (p. 97)	74–76, 81(a)–(c), 82(a)–(c)

Review Exercises

In Problems 1 and 2, determine whether each relation represents a function. For each function, state the domain and range.

1. $\{(-1, 0), (2, 3), (4, 0)\}$

2. $\{(4, -1), (2, 1), (4, 2)\}$

In Problems 3–8, find the following for each function:

(a) $f(2)$ (b) $f(-2)$ (c) $f(-x)$ (d) $-f(x)$ (e) $f(x - 2)$ (f) $f(2x)$

3. $f(x) = \dfrac{3x}{x^2 - 1}$

4. $f(x) = \dfrac{x^2}{x + 1}$

5. $f(x) = \sqrt{x^2 - 4}$

6. $f(x) = |x^2 - 4|$

7. $f(x) = \dfrac{x^2 - 4}{x^2}$

8. $f(x) = \dfrac{x^3}{x^2 - 9}$

In Problems 9–16, find the domain of each function.

9. $f(x) = \dfrac{x}{x^2 - 9}$

10. $f(x) = \dfrac{3x^2}{x - 2}$

11. $f(x) = \sqrt{2 - x}$

12. $f(x) = \sqrt{x + 2}$

13. $h(x) = \dfrac{\sqrt{x}}{|x|}$

14. $g(x) = \dfrac{|x|}{x}$

15. $f(x) = \dfrac{x}{x^2 + 2x - 3}$

16. $F(x) = \dfrac{1}{x^2 - 3x - 4}$

In Problems 17–22, find $f + g, f - g, f \cdot g,$ and $\dfrac{f}{g}$ for each pair of functions. State the domain of each.

17. $f(x) = 2 - x; \quad g(x) = 3x + 1$

18. $f(x) = 2x - 1; \quad g(x) = 2x + 1$

19. $f(x) = 3x^2 + x + 1; \quad g(x) = 3x$

20. $f(x) = 3x; \quad g(x) = 1 + x + x^2$

21. $f(x) = \dfrac{x + 1}{x - 1}; \quad g(x) = \dfrac{1}{x}$

22. $f(x) = \dfrac{1}{x - 3}; \quad g(x) = \dfrac{3}{x}$

In Problems 23 and 24, find the difference quotient of each function f; that is, find

$$\frac{f(x + h) - f(x)}{h} \qquad h \neq 0$$

23. $f(x) = -2x^2 + x + 1$

24. $f(x) = 3x^2 - 2x + 4$

25. Using the graph of the function f shown:

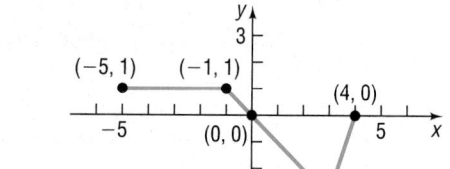

(a) Find the domain and the range of f.

(b) List the intercepts.

(c) Find $f(-2)$.

(d) For what values of x does $f(x) = -3$?

(e) Solve $f(x) > 0$.

(f) Graph $y = f(x - 3)$.

(g) Graph $y = f\left(\dfrac{1}{2}x\right)$.

(h) Graph $y = -f(x)$.

26. Using the graph of the function g shown:

(a) Find the domain and the range of g.

(b) Find $g(-1)$.

(c) List the intercepts.

(d) For what value of x does $g(x) = -3$?

(e) Solve $g(x) > 0$.

(f) Graph $y = g(x - 2)$.

(g) Graph $y = g(x) + 1$.

(h) Graph $y = 2g(x)$.

In Problems 27 and 28, use the graph of the function f to find:

(a) *The domain and the range of f*

(b) *The intervals on which f is increasing, decreasing, or constant*

(c) *The local minima and local maxima*

(d) *Whether the graph is symmetric with respect to the x-axis, the y-axis, or the origin*

(e) *Whether the function is even, odd, or neither*

(f) *The intercepts, if any*

(g) *The zeros of f, if any*

27.

28.

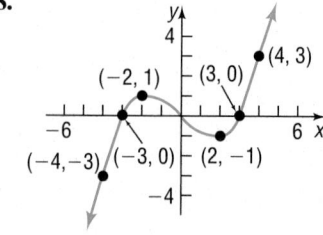

In Problems 29–36, determine (algebraically) whether the given function is even, odd, or neither.

29. $f(x) = x^3 - 4x$

30. $g(x) = \dfrac{4 + x^2}{1 + x^4}$

31. $h(x) = \dfrac{1}{x^4} + \dfrac{1}{x^2} + 1$

32. $F(x) = \sqrt{1 - x^3}$

33. $G(x) = 1 - x + x^3$

34. $H(x) = 1 + x + x^2$

35. $f(x) = \dfrac{x}{1 + x^2}$

36. $g(x) = \dfrac{1 + x^2}{x^3}$

In Problems 37–40, use a graphing utility to graph each function over the indicated interval. Approximate any local maxima and local minima. Determine where the function is increasing and where it is decreasing.

37. $f(x) = 2x^3 - 5x + 1$ $(-3, 3)$

38. $f(x) = -x^3 + 3x - 5$ $(-3, 3)$

39. $f(x) = 2x^4 - 5x^3 + 2x + 1$ $(-2, 3)$

40. $f(x) = -x^4 + 3x^3 - 4x + 3$ $(-2, 3)$

In Problems 41 and 42, find the average rate of change of f:

(a) *From 1 to 2* (b) *From 0 to 1* (c) *From 2 to 4*

41. $f(x) = 8x^2 - x$

42. $f(x) = 2x^3 + x$

In Problems 43–46, find the average rate of change from 2 to x for each function f. Be sure to simplify.

43. $f(x) = 2 - 5x$

44. $f(x) = 2x^2 + 7$

45. $f(x) = 3x - 4x^2$

46. $f(x) = x^2 - 3x + 2$

In Problems 47–50, is the graph shown the graph of a function?

47.

48.

49.

50.
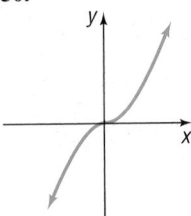

In Problems 51–54, sketch the graph of each function. Be sure to label at least three points.

51. $f(x) = |x|$

52. $f(x) = \sqrt[3]{x}$

53. $f(x) = \sqrt{x}$

54. $f(x) = \dfrac{1}{x}$

In Problems 55–66, graph each function using the techniques of shifting, compressing or stretching, and reflections. Identify any intercepts on the graph. State the domain and, based on the graph, find the range.

55. $F(x) = |x| - 4$

56. $f(x) = |x| + 4$

57. $g(x) = -2|x|$

58. $g(x) = \dfrac{1}{2}|x|$

59. $h(x) = \sqrt{x - 1}$

60. $h(x) = \sqrt{x} - 1$

61. $f(x) = \sqrt{1 - x}$

62. $f(x) = -\sqrt{x + 3}$

63. $h(x) = (x - 1)^2 + 2$

64. $h(x) = (x + 2)^2 - 3$

65. $g(x) = 3(x - 1)^3 + 1$

66. $g(x) = -2(x + 2)^3 - 8$

In Problems 67–70,

 (a) *Find the domain of each function.*

 (c) *Graph each function.*

 (b) *Locate any intercepts.*

 (d) *Based on the graph, find the range.*

67. $f(x) = \begin{cases} 3x & \text{if } -2 < x \le 1 \\ x + 1 & \text{if } x > 1 \end{cases}$

68. $f(x) = \begin{cases} x - 1 & \text{if } -3 < x < 0 \\ 3x - 1 & \text{if } x \ge 0 \end{cases}$

69. $f(x) = \begin{cases} x & \text{if } -4 \le x < 0 \\ 1 & \text{if } x = 0 \\ 3x & \text{if } x > 0 \end{cases}$

70. $f(x) = \begin{cases} x^2 & \text{if } -2 \le x \le 2 \\ 2x - 1 & \text{if } x > 2 \end{cases}$

71. A function f is defined by

$$f(x) = \frac{Ax + 5}{6x - 2}$$

If $f(1) = 4$, find A.

72. A function g is defined by

$$g(x) = \frac{A}{x} + \frac{8}{x^2}$$

If $g(-1) = 0$, find A.

73. Spheres The volume V of a sphere of radius r is $V = \frac{4}{3}\pi r^3$; the surface area S of this sphere is $S = 4\pi r^2$. If the radius doubles, how does the volume change? How does the surface area change?

74. Page Design A page with dimensions of $8\frac{1}{2}$ inches by 11 inches has a border of uniform width x surrounding the printed matter of the page, as shown in the figure.

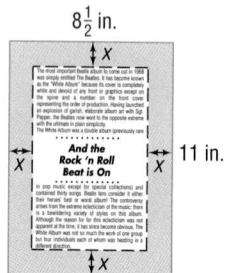

$8\frac{1}{2}$ in.

$\uparrow X$

And the
Rock 'n Roll
Beat is On

X \quad X 11 in.

$\uparrow X$

(a) Write a formula for the area A of the printed part of the page as a function of the width x of the border.
(b) Give the domain and the range of A.
(c) Find the area of the printed the page for borders of widths 1 inch, 1.2 inches, and 1.5 inches.
(d) Graph the function $A = A(x)$.
(e) Use TRACE to determine what margin should be used to obtain an area of 70 square inches and of 50 square inches.

75. Material Needed to Make a Drum A steel drum in the shape of a right circular cylinder is required to have a volume of 100 cubic feet.
(a) Express the amount A of material required to make the drum as a function of the radius r of the cylinder.
(b) How much material is required if the drum is of radius 3 feet?
(c) How much material is required if the drum is of radius 4 feet?
(d) How much material is required if the drum is of radius 5 feet?
(e) Graph $A = A(r)$. For what value of r is A smallest?

76. A rectangle has one vertex on the line $y = 10 - x$, $x > 0$, another at the origin, one on the positive x-axis, and one on the positive y-axis.
(a) Express the area A of the rectangle as a function of x.
(b) Find the largest area A that can be enclosed by the rectangle.

77. Minimizing Marginal Cost The **marginal cost** of a product can be thought of as the cost of producing one additional unit of output. For example, if the marginal cost of producing the 50th product is $6.20, then it cost $6.20 to increase production from 49 to 50 units of output. Callaway Golf Company has determined that the marginal cost C of manufacturing x Big Bertha golf clubs may be expressed by the quadratic function

$$C(x) = 4.9x^2 - 617.4x + 19,600$$

(a) How many clubs should be manufactured to minimize the marginal cost?
(b) At this level of production, what is the marginal cost?

78. Find the point on the line $y = x$ that is closest to the point $(3, 1)$.

79. Find the point on the line $y = x + 1$ that is closest to the point $(4, 1)$.

80. A rectangle has one vertex on the graph of $y = 10 - x^2$, $x > 0$, another at the origin, one on the positive x-axis, and one on the positive y-axis. Find the largest area A that can be enclosed by the rectangle.

81. Constructing a Closed Box A closed box with a square base is required to have a volume of 10 cubic feet.
(a) Express the amount A of material used to make such a box as a function of the length x of a side of the square base.
(b) How much material is required for a base 1 foot by 1 foot?
(c) How much material is required for a base 2 feet by 2 feet?
(d) Graph $A = A(x)$. For what value of x is A smallest?

82. Cost of a Drum A drum in the shape of a right circular cylinder is required to have a volume of 500 cubic centimeters. The top and bottom are made of material that costs 6¢ per square centimeter; the sides are made of material that costs 4¢ per square centimeter.

500 cc

(a) Express the total cost C of the material as a function of the radius r of the cylinder.
(b) What is the cost if the radius is 4 cm?
(c) What is the cost if the radius is 8 cm?
(d) Graph $C = C(r)$. For what value of r is the cost C least?

Chapter Test

1. Determine whether each relation represents a function. For each function, state the domain and range.

 (a) $\{(2, 5), (4, 6), (6, 7), (8, 8)\}$

 (b) $\{(1, 3), (4, -2), (-3, 5), (1, 7)\}$

 (c)

 (d)

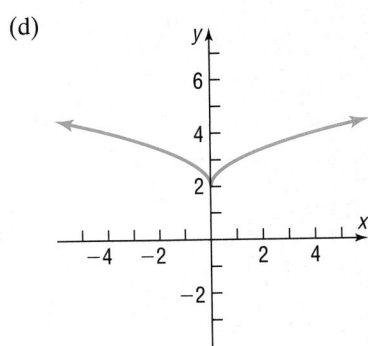

In Problems 2–4, find the domain of each function and evaluate each function at x = −1.

2. $f(x) = \sqrt{4 - 5x}$

3. $g(x) = \dfrac{x + 2}{|x + 2|}$

4. $h(x) = \dfrac{x - 4}{x^2 + 5x - 36}$

5. Using the graph of the function f below:

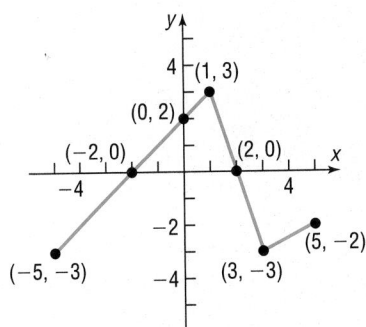

 (a) Find the domain and the range of f.

 (b) List the intercepts.

 (c) Find $f(1)$.

 (d) For what value(s) of x does $f(x) = -3$?

 (e) Solve $f(x) < 0$

 (f) List the zeros of f.

6. Use a graphing utility to graph the function $f(x) = -x^4 + 2x^3 + 4x^2 - 2$ on the interval $(-5, 5)$. Approximate any local maxima and local minima rounded to two decimal places. Determine where the function is increasing and where it is decreasing.

7. Consider the function $g(x) = \begin{cases} 2x + 1 & \text{if } x < -1 \\ x - 4 & \text{if } x \geq -1 \end{cases}$

 (a) Graph the function.

 (b) List the intercepts.

 (c) Find $g(-5)$.

 (d) Find $g(2)$.

8. For the function $f(x) = 3x^2 - 2x + 4$, find the average rate of change of f from 3 to x.

9. For the functions $f(x) = 2x^2 + 1$ and $g(x) = 3x - 2$, find the following and simplify:

 (a) $f - g$

 (b) $f \cdot g$

 (c) $f(x + h) - f(x)$

10. Graph each function using the techniques of shifting, compressing or stretching, and reflections. Start with the graph of the basic function and show all stages.

 (a) $h(x) = -2(x + 1)^3 + 3$

 (b) $g(x) = |x + 4| + 2$

11. The variable interest rate on a student loan changes each July 1 based on the bank prime loan rate. For the years 1992–2004, this rate can be approximated by the model $r(x) = -0.115x^2 + 1.183x + 5.623$, where x is the number of years since 1992 and r is the interest rate as a percent.

 (SOURCE: U.S. Federal Reserve)

 (a) Use a graphing utility to estimate the highest rate during this time period. During which year was the interest rate the highest?

 (b) Use the model to estimate the rate in 2010. Does this value seem reasonable?

12. A community skating rink is in the shape of a rectangle with semicircles attached at the ends. The length of the rectangle is 20 feet less than twice the width. The thickness of the ice is 0.75 inch.

 (a) Write the ice volume, V, as a function of the width, x.

 (b) How much ice is in the rink if the width is 90 feet?

Chapter Projects

1. **Cell Phone Service** This month when you were paying your bills, you noticed that your original service contract for your cell phone service had expired. Since your cell number is portable to another company now, you decided to investigate different companies as well. Noting that you travel outside your local calling area on a regular basis, you decide to stick with only those plans that included roaming charges. Here is what you found:

	Anytime Minutes Included	Charge for Each Extra Minute	Mobile-to-Mobile Minutes	National Long Distance	Nights and Weekends
Company A:					
Plan A1: $49.99	600	$0.45	Unlimited	Included	Unlimited (after 9 pm)
Plan A2: $59.99	900	$0.35	Unlimited	Included	Unlimited (after 7 pm)
Company B:					
Plan B1: $39.99	450 with rollover	$0.45	Unlimited	Included	3000 min (after 9 pm)
Plan B2: $49.99	600 with rollover	$0.40	Unlimited	Included	unlimited (after 9 pm)
Company C:					
Plan C1: $45.00	300	$0.40	Unlimited	Included	Unlimited (after 9 pm)
Plan C2: $60.00	700	$0.40	Unlimited	Included	Unlimited (after 9 pm)

Each plan requires a two-year contract.

(a) Determine the total cost of each plan for the life of the contract, assuming that you stay within the allotted anytime minutes provided by each contract.

(b) If you expect to use 600 anytime minutes and 2500 night and weekend minutes per month, which plan provides the best deal? If you expect to use 600 anytime minutes and 3500 night and weekend minutes, which plan provides the best deal?

(c) Ignoring any night and weekend usage, if you expect to use 425 anytime minutes each month, which option provides the best deal? What if you use 750 anytime minutes per month?

(d) Each monthly charge includes a specific number of peak time minutes in the monthly fee. Write a function for each option, where C is the monthly cost and x is the number of anytime minutes used.

(e) Graph each function from part (d).

(f) For each of the companies A, B, and C, determine the average price per minute for each plan, based on no extra minutes used. For each company, which plan is better?

(g) Now, looking at the three plans that you found to be the best for Companies A, B, and C, in part (f), which of those three seems to be the best deal?

(h) Based on your own cell phone usage, which plan would be the best for you?

SOURCE: Based on rates from the web sites of the companies ATTWireless, Cingular, and Sprint PCS, for the zip code 76201 on October 4, 2004. (*www.attwireless.com*, *www.cingular.com*, *www.sprint.com*)

The following projects are available on the Instructor's Resource Center (IRC):

2. **Project at Motorola** *Pricing Wireless Service*

3. **Cost of Cable**

4. **Oil Spill**

2 Linear and Quadratic Functions

A LOOK BACK Up to now, our discussion has focused on graphs of equations and the idea of a function. We have learned how to graph equations using the point-plotting method, intercepts, and the tests for symmetry. In addition, we have learned what a function is and how to identify whether a relation represents a function. We also introduced properties of functions, such as domain/range, increasing/decreasing, even/odd, and zeros.

A LOOK AHEAD Going forward, we will look at classes of functions. In this chapter, we focus on linear and quadratic functions and their properties and applications.

The Use of Statistics in Making Investment Decisions: Using Beta

One of the most common statistics used by investors is beta, which measures a security or portfolio's movement relative to the S&P 500. When determining beta graphically, returns are plotted on a graph, with the horizontal axis representing the S&P returns and the vertical axis representing the security or portfolio returns; the points that are plotted represent the return of the security versus the return of the S&P 500 at the same time period. A regression line is then drawn, which is a straight line that most closely approximates the points in the graph. Beta represents the slope of that line; a steep slope (45 degrees or greater) indicates that when the stock market moved up or down the security or portfolio moved up or down on average to the same degree or greater, and a flatter slope indicates that when the stock market moved up or down the security or portfolio moved to a lesser degree. Since beta measures variability, it is used as a measure of risk; the greater the beta is, the greater the risk.

SOURCE: *Paul E. Hoffman, Journal of the American Association of Individual Investors, September, 1991;* www.aaii.com

— See Chapter Project 1.

2.1 Properties of Linear Functions

PREPARING FOR THIS SECTION *Before getting started, review the following:*

- Lines (Foundations, Section F.3, pp. 18–26)
- Graphing Equations in Two Variables (Foundations, Section F.2, pp. 8–11)
- Linear Equations (Appendix A, Section A.8, pp. A60–A63)

- Functions (Section 1.1, pp. 40–51)
- The Graph of a Function (Section 1.2, pp. 54–58)
- Properties of Functions (Section 1.3, pp. 63–70)

Now work the 'Are You Prepared?' problems on page 121.

OBJECTIVES 1 Graph Linear Functions
2 Use Average Rate of Change to Identify Linear Functions
3 Determine Whether a Linear Function is Increasing, Decreasing, or Constant
4 Find the Zero of a Linear Function
5 Work with Applications of Linear Functions

1 Graph Linear Functions

In Foundations, Section F.3, we reviewed lines. In particular, we discussed the slope–intercept form of the equation of a line $y = mx + b$. When we write the slope–intercept form of a line using function notation, we have a *linear function*.

> A **linear function** is a function of the form
>
> $$f(x) = mx + b$$
>
> The graph of a linear function is a line with slope m and y-intercept b.

EXAMPLE 1 **Graphing a Linear Function**

Figure 1

Graph the linear function: $f(x) = -3x + 7$

Solution This is a linear function with slope $m = -3$ and y-intercept 7. To graph this function, we start by plotting the point $(0, 7)$, the y-intercept, and use the slope to find an additional point by moving right 1 unit and down 3 units. See Figure 1. ∎

Alternatively, we could have found an additional point by evaluating the function at some $x \neq 0$. For $x = 1$, we find $f(1) = -3(1) + 7 = 4$ and obtain the point $(1, 4)$ on the graph.

NOW WORK PROBLEMS 13(a) AND (b).

2 Use Average Rate of Change to Identify Linear Functions

Look at Table 1, which shows certain values of the independent variable x and corresponding values of the dependent variable y for the function $f(x) = -3x + 7$. Notice that as the value of the independent variable, x, increases by 1 the value of the dependent variable y decreases by 3. That is, the average rate of change of y with respect to x is a constant, -3.

Table 1

x	y = f(x) = -3x + 7	Average Rate of Change = $\dfrac{\Delta y}{\Delta x}$
-2	13	
		$\dfrac{10 - 13}{-1 - (-2)} = \dfrac{-3}{1} = -3$
-1	10	
		$\dfrac{7 - 10}{0 - (-1)} = \dfrac{-3}{1} = -3$
0	7	
		-3
1	4	
		-3
2	1	
		-3
3	-2	

It is not a coincidence that the average rate of change of the linear function $f(x) = -3x + 7$ is the slope of the linear function. That is, $\dfrac{\Delta y}{\Delta x} = m = -3$. The following theorem states this fact.

Theorem

Average Rate of Change of a Linear Function

Linear functions have a constant average rate of change. That is, the average rate of change of a linear function $f(x) = mx + b$ is

$$\frac{\Delta y}{\Delta x} = m$$

Proof The average rate of change of $f(x) = mx + b$ from c to x is

$$\frac{\Delta y}{\Delta x} = \frac{f(x) - f(c)}{x - c} = \frac{(mx + b) - (mc + b)}{x - c} = \frac{mx - mc}{x - c} = \frac{m(x - c)}{x - c} = m \quad \blacksquare$$

Based on the theorem just proved, the average rate of change of the function $g(x) = -\dfrac{2}{5}x + 5$ is $-\dfrac{2}{5}$.

NOW WORK PROBLEM 13(C).

Because the average rate of change of a linear function is constant, we can use the average rate of change to determine whether a particular function is linear or not. Functions that are not linear are said to be **nonlinear**.

| EXAMPLE 2 | Using the Average Rate of Change to Identify Linear Functions |

(a) A strain of E-coli Beu 397-recA441 is placed into a Petri dish at 30° Celsius and allowed to grow. The data shown in Table 2 are collected. The population is measured in grams and the time in hours. Plot the ordered pairs (x, y) in the Cartesian plane and use the average rate of change to determine whether the function is linear.

(b) The data in Table 3 represent the maximum number of heartbeats that a healthy individual should have during a 15-second interval of time while exercising for different ages. Plot the ordered pairs (x, y) in the Cartesian plane and use the average rate of change to determine whether the function is linear.

Table 2

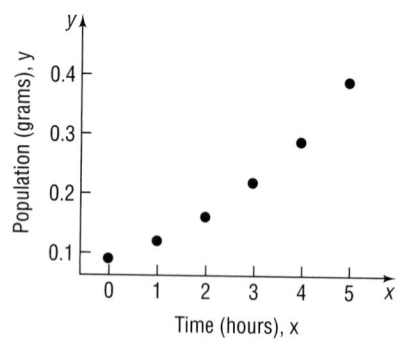

Time (hours), x	Population (grams), y	(x, y)
0	0.09	(0, 0.09)
1	0.12	(1, 0.12)
2	0.16	(2, 0.16)
3	0.22	(3, 0.22)
4	0.29	(4, 0.29)
5	0.39	(5, 0.39)

Table 3

Age, x	Maximum Number of Heart Beats, y	(x, y)
20	50	(20, 50)
30	47.5	(30, 47.5)
40	45	(40, 45)
50	42.5	(50, 42.5)
60	40	(60, 40)
70	37.5	(70, 37.5)

SOURCE: American Heart Association

Solution We compute the average rate of change of each function. If the average rate of change is constant, the function is linear. If the average rate of change is not constant, the function is nonlinear.

(a) Figure 2 shows the points listed in Table 2 plotted in the Cartesian plane. Notice that it is impossible to draw a straight line that contains all the points. Table 4 displays the average rate of change of population.

Figure 2

Population (grams), y — Time (hours), x

Table 4

Time (hours), x	Population (grams), y	Average Rate of Change $= \dfrac{\Delta y}{\Delta x}$
0	0.09	
		$\dfrac{0.12 - 0.09}{1 - 0} = 0.03$
1	0.12	
		0.04
2	0.16	
		0.06
3	0.22	
		0.07
4	0.29	
		0.10
5	0.39	

Because the average rate of change is not constant, we know that the function is not linear. In fact, because the average rate of change is increasing as the value of the independent variable increases, we say that the function is increasing at an increasing rate. So not only is the population increasing over time, but it is also growing more rapidly as time passes.

(b) Figure 3 shows the points listed in Table 3 plotted in the Cartesian plane. We can see that the data in Figure 3 lie on a straight line. Table 5 contains the average rate of change of the maximum number of heartbeats. The average rate of change of the heartbeat data is constant, −0.25 beats per year, so the function is linear.

Figure 3

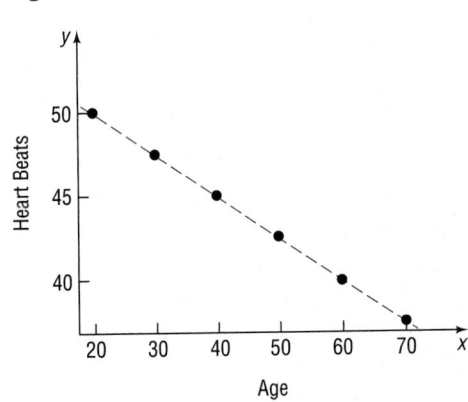

Table 5

Age, x	Maximum Number of Heartbeats, y	Average Rate of Change $= \dfrac{\Delta y}{\Delta x}$
20	50	
		$\dfrac{47.5 - 50}{30 - 20} = -0.25$
30	47.5	
		−0.25
40	45	
		−0.25
50	42.5	
		−0.25
60	40	
		−0.25
70	37.5	

NOW WORK PROBLEM 27.

3 Determine Whether a Linear Function Is Increasing, Decreasing or Constant

From the Seeing the Concept on page 20 in Section F. 3, we know that when the slope m of a linear function is positive $(m > 0)$ the line slants upward from left to right. When the slope m of a linear function is negative $(m < 0)$, the line slants downward from left to right. When the slope m of a linear function is zero $(m = 0)$, the line is horizontal.

Based on these results, we have the following theorem.

Theorem

Increasing, Decreasing, and Constant Linear Functions

A linear function $f(x) = mx + b$ is increasing over its domain if its slope, m, is positive. It is decreasing over its domain if its slope, m, is negative. It is constant over its domain if its slope, m, is zero.

EXAMPLE 3

Determining Whether a Linear Function is Increasing, Decreasing, or Constant

Determine whether the following linear functions are increasing, decreasing, or constant.

(a) $f(x) = 5x - 2$ (b) $g(x) = -2x + 8$ (c) $s(t) = \dfrac{3}{4}t - 4$ (d) $h(z) = 7$

Solution

(a) For the linear function $f(x) = 5x - 2$, the slope is 5, which is positive. The function f is increasing on the interval $(-\infty, \infty)$.

(b) For the linear function $g(x) = -2x + 8$, the slope is -2, which is negative. The function g is decreasing on the interval $(-\infty, \infty)$.

(c) For the linear function $s(t) = \dfrac{3}{4}t - 4$, the slope is $\dfrac{3}{4}$, which is positive. The function s is increasing on the interval $(-\infty, \infty)$.

(d) We can write the linear function h as $h(z) = 0z - 7$. Because the slope is 0, the function is constant on the interval $(-\infty, \infty)$. ∎

NOW WORK PROBLEM 13(d).

4 Find the Zero of a Linear Function

Recall from Section 1.3 that a real number r is a zero of a function f if $f(r) = 0$. So, to find the zeros of a function $y = f(x)$, we solve the equation $f(x) = 0$. The solution(s) represent the zero(s) of f. Any linear function $f(x) = mx + b$ has exactly one zero provided that $m \neq 0$, since, if $m \neq 0$, the equation $mx + b = 0$ has exactly one solution. In other words, a linear function has exactly one zero if it is increasing or decreasing.

EXAMPLE 4	**Finding the Zero of a Linear Function**

(a) Does $f(x) = -\dfrac{1}{2}x + 2$ have a zero?

(b) If it does, find the zero.

(c) Use the zero along with the y-intercept to graph f.

(d) Solve $f(x) > 0$.

Solution

(a) The linear function f is decreasing (do you see why?), so it has one zero.

(b) We solve the equation $f(x) = 0$ to find the zero.

$$f(x) = 0$$

$$-\frac{1}{2}x + 2 = 0 \qquad f(x) = -\frac{1}{2}x + 2$$

$$-\frac{1}{2}x = -2 \qquad \text{Subtract 2 from each side of the equation.}$$

$$x = 4 \qquad \text{Multiply both sides of the equation by } -2.$$

✔ **CHECK:** $f(4) = -\dfrac{1}{2}(4) + 2 = 0.$

The zero of f is 4.

(c) The y-intercept is 2, so the point $(0, 2)$ is on the graph of the function. The zero of f is 4, so the x-intercept is 4 and the point $(4, 0)$ is on the graph. See Figure 4 for the graph of f.

(d) From Figure 4, we see $f(x) > 0$ (above the x-axis) when $x < 4$. Alternatively, we can solve the inequality.

$$-\frac{1}{2}x + 2 > 0$$

$$-\frac{1}{2}x > -2 \qquad \text{Add } -2 \text{ to both sides.}$$

$$-2\left(-\frac{1}{2}x\right) < -2 \cdot (-2) \qquad \text{Multiply both sides by } -2. \text{ Change the direction of the inequality.}$$

$$x < 4 \qquad \text{Simplify.}$$

∎

Figure 4

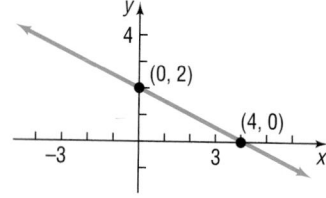

NOW WORK PROBLEM 21.

5 Work with Applications of Linear Functions

There are many applications of linear functions. Let's look at one from accounting.

| EXAMPLE 5 | **Straight-line Depreciation** |

Book value is the value of an asset that a company uses to create its balance sheet. Some companies depreciate their assets using straight-line depreciation so that the value of the asset declines by a fixed amount each year. The amount of the decline depends on the useful life that the company places on the asset. Suppose that a company just purchased a fleet of new cars for its sales force at a cost of $28,000 per car. The company chooses to depreciate each vehicle using the straight-line method over 7 years. This means that each car will depreciate by $\dfrac{\$28,000}{7} = \4000 per year.

(a) Write a linear function that expresses the book value V of each car as a function of its age, x.

(b) Graph the linear function.

(c) What is the book value of each car after 3 years?

(d) Interpret the slope.

(e) When will the book value of each car be $8000?

[**Hint:** Solve the equation $V(x) = 8000$.]

Solution

(a) If we let $V(x)$ represent the value of each car after x years, then $V(0)$ represents the original value of each car, so $V(0) = \$28,000$. The y-intercept of the linear function is $28,000. Because each car depreciates by $4000 per year, the slope of the linear function is $-\$4000$. The linear function that represents the book value V of each car after x years is

$$V(x) = -4000x + 28,000$$

(b) Figure 5 shows the graph of V.

(c) The book value of each car after 3 years is

$$V(3) = -4000(3) + 28,000$$
$$= \$16,000$$

Figure 5

(d) Since the slope of $V(x) = -4000x + 28,000$ is -4000, the average rate of change of book value is $-\$4000/\text{year}$. So for each additional year that passes the book value of the car decreases by $4000.

(e) To find where the book value is $8000, we solve the equation

$$V(x) = 8000$$
$$-4000x + 28,000 = 8000$$
$$-4000x = -20,000 \qquad \text{Subtract \$28,000 from each side}$$
$$x = \frac{-20,000}{-4000} = 5 \qquad \text{Divide by } -4000.$$

The car will have a book value of $8000 when it is 5 years old. ■

NOW WORK PROBLEM 53.

Next we look at an application from economics.

| EXAMPLE 6 | **Supply and Demand** |

The **quantity supplied** of a good is the amount of a product that a company is willing to make available for sale at a given price. The **quantity demanded** of a good is the amount of a product that consumers are willing to purchase at a given price.

Suppose that the quantity supplied, S, and quantity demanded, D, of cellular telephones each month are given by the following functions:

$$S(p) = 60p - 900$$
$$D(p) = -15p + 2850$$

where p is the price (in dollars) of the telephone.

(a) The **equilibrium price** of a product is defined as the price at which quantity supplied equals quantity demanded. That is, the equilibrium price is the price at which $S(p) = D(p)$. Find the equilibrium price of cellular telephones. What is the **equilibrium quantity**, the amount demanded (or supplied) at the equilibrium price?

(b) Determine the prices for which quantity supplied is greater than quantity demanded. That is, solve the inequality $S(p) > D(p)$.

(c) Graph $S = S(p)$, $D = D(p)$ and label the equilibrium price.

Solution (a) To find the equilibrium price, we solve the equation $S(p) = D(p)$.

$$60p - 900 = -15p + 2850 \qquad \text{\small $S(p) = 60p - 900$;}$$
$$\text{\small $D(p) = -15p + 2850$}$$
$$60p = -15p + 3750 \qquad \text{\small Add 900 to each side.}$$
$$75p = 3750 \qquad \text{\small Add 15p to each side.}$$
$$p = \$50 \qquad \text{\small Divide each side by 75.}$$

The equilibrium price is \$50 per cellular phone. To find the equilibrium quantity, we evaluate either $S(p)$ or $D(p)$ at $p = 50$.

$$S(50) = 60(50) - 900 = 2100$$

The equilibrium quantity is 2100 cellular phones. At a price of \$50 per phone, the company will produce and sell 2100 phones each month and have no shortages or excess inventory.

(b) We solve the inequality $S(p) > D(p)$.

$$60p - 900 > -15p + 2850 \qquad \text{\small $S(p) > D(p)$}$$
$$60p > -15p + 3750 \qquad \text{\small Add 900 to each side.}$$
$$75p > 3750 \qquad \text{\small Add 15p to each side.}$$
$$p > \$50 \qquad \text{\small Divide each side by 75.}$$

If the company charges more than \$50 per phone, quantity supplied will exceed quantity demanded. In this case the company will have excess phones in inventory.

(c) Figure 6 shows the graphs of $S = S(p)$ and $D = D(p)$ with the equilibrium point labeled.

Figure 6

NOW WORK PROBLEM 47.

2.1 Assess Your Understanding

'Are You Prepared?'

Answers are given at the end of these exercises. If you get a wrong answer, read the pages listed in red.

1. Graph $y = 2x - 3$. (p. 24)
2. Find the slope of the line joining the points (2, 5) and (−1, 3). (pp. 18–20)
3. List the intercepts of the equation $x^2 + 2y = 4$. (pp. 54–58)
4. Solve: $60x - 900 = -15x + 2850$. (pp. A61–A63)
5. If $f(x) = x^2 - 4$, find $f(-2)$. (pp. 44–46)
6. *True* or *False*: The graph of the function $f(x) = x^2$ is increasing on the interval $(0, \infty)$. (p. 78)

Concepts and Vocabulary

7. For the graph of the linear function $f(x) = mx + b$, m is the _____ and b is the _____.
8. For the graph of the linear function $H(z) = -4z + 3$, the slope is _____ and the y-intercept is _____.
9. If the slope m of the graph of a linear function is _____, the function is increasing over its domain.
10. *True or False* The slope of a line is the average rate of change of the linear function.
11. *True or False* If the average rate of change of a linear function is $\frac{2}{3}$, then if y increases by 3, x will increase by 2.
12. *True or False* The zero of $f(x) = 2x + 8$ is 8.

Skill Building

In Problems 13–20, a linear function is given.
 (a) *Determine the slope and y-intercept of each function.*
 (b) *Use the slope and y-intercept to graph the linear function.*
 (c) *Determine the average rate of change of each function.*
 (d) *Determine whether the linear function is increasing, decreasing, or constant.*

13. $f(x) = 2x + 3$
14. $g(x) = 5x - 4$
15. $h(x) = -3x + 4$
16. $p(x) = -x + 6$

17. $f(x) = \frac{1}{4}x - 3$
18. $h(x) = -\frac{2}{3}x + 4$
19. $F(x) = 4$
20. $G(x) = -2$

In Problems 21–26, (a) find the zero of each linear function and (b) graph each function using the zero and y-intercept.

21. $g(x) = 2x - 8$
22. $g(x) = 3x + 12$
23. $f(x) = -5x + 10$

24. $f(x) = -6x + 12$
25. $H(x) = -\frac{1}{2}x + 4$
26. $G(x) = \frac{1}{3}x - 4$

In Problems 27–34, determine whether the given function is linear or nonlinear. If it is linear, determine the slope.

27.

x	y = f(x)
−2	4
−1	1
0	−2
1	−5
2	−8

28.

x	y = f(x)
−2	1/4
−1	1/2
0	1
1	2
2	4

29.

x	y = f(x)
−2	−8
−1	−3
0	0
1	1
2	0

30.

x	y = f(x)
−2	−4
−1	0
0	4
1	8
2	12

31.

x	y = f(x)
−2	−26
−1	−4
0	2
1	−2
2	−10

32.

x	y = f(x)
−2	−4
−1	−3.5
0	−3
1	−2.5
2	−2

33.

x	y = f(x)
−2	8
−1	8
0	8
1	8
2	8

34.

x	y = f(x)
−2	0
−1	1
0	4
1	9
2	16

35. Suppose that $f(x) = 4x - 1$ and $g(x) = -2x + 5$.
(a) Solve $f(x) = 0$. (b) Solve $f(x) > 0$.
(c) Solve $f(x) = g(x)$. (d) Solve $f(x) \leq g(x)$.
(e) Graph $y = f(x)$ and $y = g(x)$ and label the point that represents the solution to the equation $f(x) = g(x)$.

37. In parts (a)–(f), use the following figure.

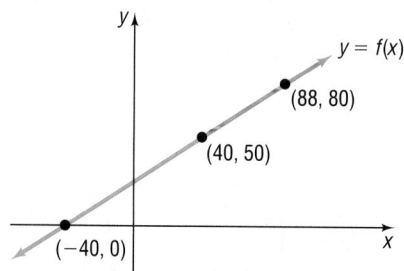

(a) Solve $f(x) = 50$. (b) Solve $f(x) = 80$.
(c) Solve $f(x) = 0$. (d) Solve $f(x) > 50$.
(e) Solve $f(x) \leq 80$. (f) Solve $0 < f(x) < 80$.

39. In parts (a) and (b) use the following figure.

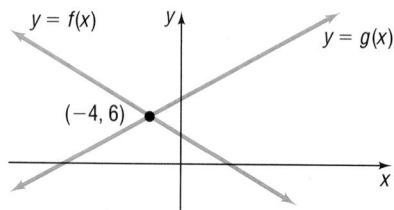

(a) Solve the equation: $f(x) = g(x)$.
(b) Solve the inequality: $f(x) > g(x)$.

41. In parts (a) and (b), use the following figure.

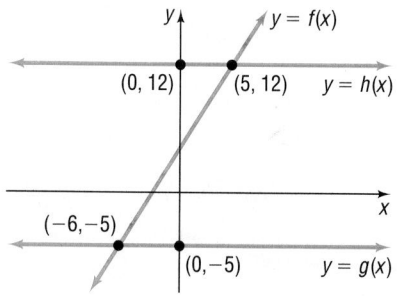

(a) Solve the equation: $f(x) = g(x)$.
(b) Solve the inequality: $g(x) \leq f(x) < h(x)$.

36. Suppose that $f(x) = 3x + 5$ and $g(x) = -2x + 15$.
(a) Solve $f(x) = 0$. (b) Solve $f(x) < 0$.
(c) Solve $f(x) = g(x)$. (d) Solve $f(x) \geq g(x)$.
(e) Graph $y = f(x)$ and $y = g(x)$ and label the point that represents the solution to the equation $f(x) = g(x)$.

38. In parts (a)–(f), use the following figure.

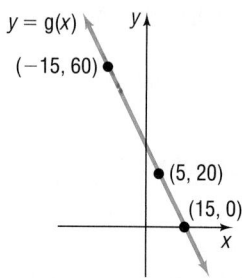

(a) Solve $g(x) = 20$. (b) Solve $g(x) = 60$.
(c) Solve $g(x) = 0$. (d) Solve $g(x) > 20$.
(e) Solve $g(x) \leq 60$. (f) Solve $0 < g(x) < 60$.

40. In parts (a) and (b), use the following figure.

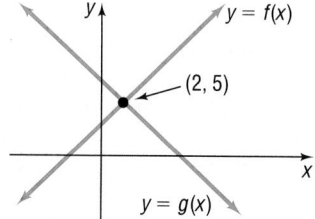

(a) Solve the equation: $f(x) = g(x)$.
(b) Solve the inequality: $f(x) \leq g(x)$.

42. In parts (a) and (b), use the following figure.

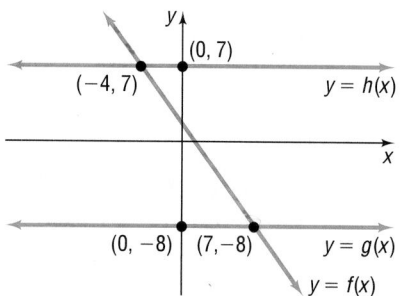

(a) Solve the equation: $f(x) = g(x)$.
(b) Solve the inequality: $g(x) < f(x) \leq h(x)$.

Applications and Extensions

43. Car Rentals The cost C, in dollars, of renting a moving truck for a day is given by the function $C(x) = 0.25x + 35$, where x is the number of miles driven.
(a) What is the cost if you drive $x = 40$ miles?
(b) If the cost of renting the moving truck is $80, how many miles did you drive?
(c) Suppose that you want to keep the cost below $100. What is the maximum number of miles that you can drive?

44. Phone Charges The monthly cost C, in dollars, of a certain cellular phone plan is given by the function $C(x) = 0.38x + 5$, where x is the number of minutes used.
(a) What is the cost if you talk on the phone for $x = 50$ minutes?
(b) Suppose that your monthly bill is $29.32. How many minutes did you use the phone?
(c) Suppose that you budget yourself $60 per month for the phone. What is the maximum number of minutes that you can talk?

45. Disability Benefits The average monthly benefit B, in dollars, for individuals on disability is given by the function

$$B(t) = 19.25t + 585.72$$

where t is the number of years since 1990.
(a) What was the average monthly benefit in 2000 ($t = 10$)?
(b) In what year will the average monthly benefit be $893.72?
(c) In what year will the average monthly benefit exceed $1000?

46. Health Expenditures The total private health expenditures H, in billions of dollars, is given by the function

$$H(t) = 26t + 411$$

 where t is the number of years since 1990.
(a) What was the total private health expenditures in 2000 ($t = 10$)?
(b) In what year will total private health expenditures be $879 billion?
(c) In what year will total private health expenditures exceed $1 trillion ($1,000 billion)?

47. Supply and Demand Suppose that the quantity supplied S and quantity demanded D of T-shirts at a concert are given by the following functions:

$$S(p) = -200 + 50p$$
$$D(p) = 1000 - 25p$$

where p is the price.
(a) Find the equilibrium price for T-shirts at this concert. What is the equilibrium quantity?
(b) Determine the prices for which quantity demanded is greater than quantity supplied.
(c) What do you think will eventually happen to the price of T-shirts if quantity demanded is greater than quantity supplied?

48. Supply and Demand Suppose that the quantity supplied S and quantity demanded D of hot dogs at a baseball game are given by the following functions:

$$S(p) = -2000 + 3000p$$
$$D(p) = 10,000 - 1000p$$

where p is the price.
(a) Find the equilibrium price for hot dogs at the baseball game. What is the equilibrium quantity?
(b) Determine the prices for which quantity demanded is less than quantity supplied.
(c) What do you think will eventually happen to the price of hot dogs if quantity demanded is less than quantity supplied?

49. Taxes The function $T(x) = 0.15(x - 7300) + 730$ represents the tax bill T of a single person whose adjusted gross income is x dollars for income between $7300 and $29,700, inclusive in 2005.
SOURCE: Internal Revenue Service
(a) What is the domain of this linear function?
(b) What is a single filer's tax bill if adjusted gross income is $18,000?

(c) Which variable is independent and which is dependent?
(d) Graph the linear function over the domain specified in part (a).
(e) What is a single filer's adjusted gross income if the tax bill is $2860?

50. Luxury Tax In 2002, major league baseball signed a labor agreement with the players. In this agreement, any team whose payroll exceeds $128 million in 2005 will have to pay a luxury tax of 22.5% (for first-time offenses). The linear function $T(p) = 0.225(p - 128)$ describes the luxury tax T of a team whose payroll is p (in millions of dollars).
(a) What is the implied domain of this linear function?
(b) What is the luxury tax for a team whose payroll is $160 million?
(c) Graph the linear function.
(d) What is the payroll of a team that pays a luxury tax of $11.7 million?

The point at which a company's profits equal zero is called the company's **break-even point.** *For Problems 51 and 52, let R represent a company's revenue, let C represent the company's costs, and let x represent the number of units produced and sold each day.*
(a) *Find the firm's break-even point; that is, find x so that* $R = C$.
(b) *Find the values of x such that* $R(x) > C(x)$. *This represents the number of units that the company must sell to earn a profit.*

51. $R(x) = 8x$
$\quad\ C(x) = 4.5x + 17,500$

52. $R(x) = 12x$
$\quad\ C(x) = 10x + 15,000$

53. Straight-line Depreciation Suppose that a company has just purchased a new computer for $3000. The company chooses to depreciate the computer using the straight-line method over 3 years.
(a) Write a linear function that expresses the book value of the computer as a function of its age.
(b) Graph the linear function.
(c) What is the book value of the computer after 2 years?
(d) When will the computer be worth $2000?

54. Straight-line Depreciation Suppose that a company has just purchased a new machine for its manufacturing facility for $120,000. The company chooses to depreciate the machine using the straight-line method over 10 years.
(a) Write a linear function that expresses the book value of the machine as a function of its age.
(b) Graph the linear function.
(c) What is the book value of the machine after 4 years?
(d) When will the machine be worth $72,000?

55. Cost Function The simplest cost function is the linear cost function, $C(x) = mx + b$, where the y-intercept b represents the fixed costs of operating a business and the slope m represents the cost of each item produced. Suppose that a small bicycle manufacturer has daily fixed costs of $1800 and each bicycle costs $90 to manufacture.
(a) Write a linear function that expresses the cost C of manufacturing x bicycles in a day.
(b) Graph the linear function.
(c) What is the cost of manufacturing 14 bicycles in a day?
(d) How many bicycles could be manufactured for $3780?

56. Cost Function Refer to Problem 55. Suppose that the landlord of the building increases the bicycle manufacturer's rent by $100 per month.
(a) Assuming that the manufacturer is open for business 20 days per month, what are the new daily fixed costs?
(b) Write a linear function that expresses the cost C of manufacturing x bicycles in a day with the higher rent.
(c) Graph the linear function.
(d) What is the cost of manufacturing 14 bicycles in a day?
(e) How many bicycles could be manufactured for $3780?

57. Truck Rentals A truck rental company rents a truck for one day by charging $29 plus $0.07 per mile.
(a) Write a linear function that relates the cost C, in dollars, of renting the truck to the number x of miles driven.
(b) What is the cost of renting the truck if the truck is driven 110 miles? 230 miles?

58. Long Distance A phone company offers a long distance package by charging $5 plus $0.05 per minute.
(a) Write a linear function that relates the cost C, in dollars of talking x minutes.
(b) What is the cost of talking 105 minutes? 180 minutes?

Discussion and Writing

59. Which of the following functions might have the graph shown? (More than one answer is possible.)
(a) $f(x) = 2x - 7$ (b) $g(x) = -3x + 4$
(c) $H(x) = 5$ (d) $F(x) = 3x + 4$
(e) $G(x) = \dfrac{1}{2}x + 2$

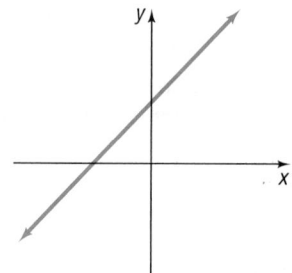

60. Which of the following functions might have the graph shown? (More than one answer is possible.)

(a) $f(x) = 3x + 1$ (b) $g(x) = -2x + 3$
(c) $H(x) = 3$ (d) $F(x) = -4x - 1$
(e) $G(x) = -\dfrac{2}{3}x + 3$

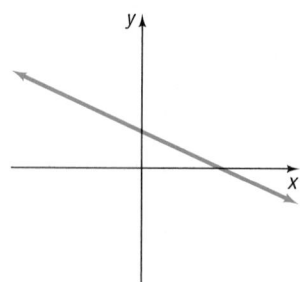

61. Under what circumstances is a linear function $f(x) = mx + b$ odd? Can a linear function ever be even?

62. The accepted symbol used to denote the slope of a line is the letter m. Investigate the origin of this symbolism. Begin by consulting a French dictionary and looking up the French word *monter*. Write a brief essay on your findings.

63. The term grade is used to describe the inclination of a road. How does this term relate to the notion of slope of a line? Is a 4% grade very steep? Investigate the grades of some mountainous roads and determine their slopes. Write a brief essay on your findings.

64. Carpentry Carpenters use the term pitch to describe the steepness of staircases and roofs. How does pitch relate to slope? Investigate typical pitches used for stairs and for roofs. Write a brief essay on your findings.

'Are You Prepared?' Answers

1.

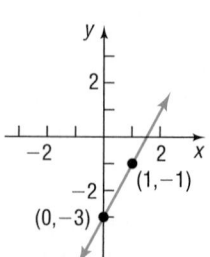

2. $\dfrac{2}{3}$ **3.** $(-2,0), (2,0), (0,2)$ **4.** $\{50\}$ **5.** 0 **6.** True

2.2 Building Linear Functions from Data; Direct Variation

PREPARING FOR THIS SECTION *Before getting started, review the following:*
- Functions (Section 1.1, pp. 40–51)

 Now Work the 'Are You Prepared?' problems on page 130.

OBJECTIVES
1. Draw and Interpret Scatter Diagrams
2. Distinguish between Linear and Nonlinear Relations
3. Use a Graphing Utility to Find the Line of Best Fit
4. Construct a Linear Model Using Direct Variation

1 Draw and Interpret Scatter Diagrams

In Section 2.1, we built linear functions from verbal descriptions. Linear functions can also be constructed by fitting a linear function to data. The first step in finding this relation is to plot the ordered pairs using rectangular coordinates. The resulting graph is called a **scatter diagram**.

EXAMPLE 1 **Drawing and Interpreting a Scatter Diagram**

In baseball, the on-base percentage for a team represents the percentage of time that the players safely reach base. The data given in Table 6 represent the number of runs scored y and the on-base percentage x for teams in the National League during the 2004 baseball season.

Table 6

Team	On-Base Percentage, x	Runs Scored, y	(x, y)
Atlanta	34.3	803	(34.3, 803)
St. Louis	34.4	855	(34.4, 855)
Colorado	34.5	833	(34.5, 833)
Houston	34.2	803	(34.2, 803)
Philadelphia	35.7	840	(35.7, 840)
San Francisco	35.7	850	(35.7, 850)
Pittsburgh	32.1	680	(32.1, 680)
Florida	32.9	718	(32.9, 718)
Chicago Cubs	32.8	789	(32.8, 789)
Arizona	31.0	615	(31.0, 615)
Milwaukee	32.1	634	(32.1, 634)
Montreal	31.3	635	(31.3, 635)
Cincinnati	33.1	750	(33.1, 750)
San Diego	34.2	768	(34.2, 768)
NY Mets	31.7	684	(31.7, 684)
Los Angeles	33.2	761	(33.2, 761)

SOURCE: espn.com

(a) Draw a scatter diagram of the data, treating on-base percentage as the independent variable.

(b) Use a graphing utility to draw a scatter diagram.

(c) Describe what happens as the on-base percentage increases.

Solution (a) To draw a scatter diagram, we plot the ordered pairs listed in Table 6, with the on-base percentage as the x-coordinate and the runs scored as the y-coordinate. See Figure 7(a). Notice that the points in the scatter diagram are not connected.

(b) Figure 7(b) shows a scatter diagram using a TI-84 Plus graphing calculator.

(c) We see from the scatter diagrams that, as the on-base percentage increases, the number of runs scored also increases.

Figure 7

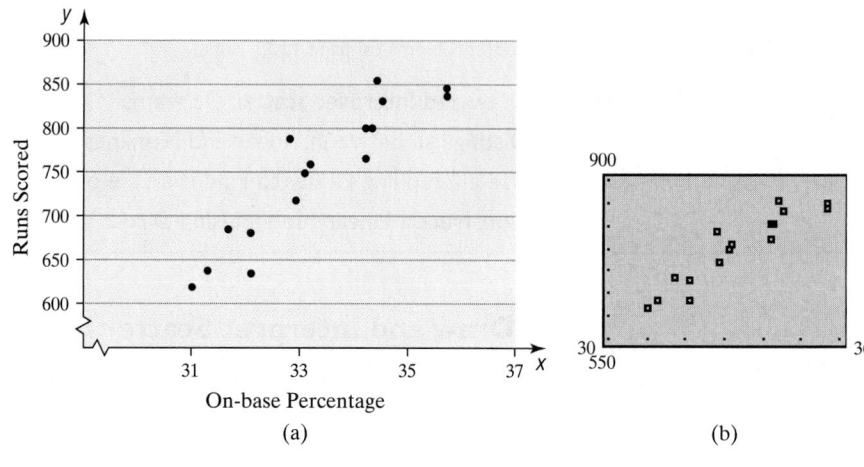

(a)

(b)

NOW WORK PROBLEM 13(a).

2 Distinguish between Linear and Nonlinear Relations

Scatter diagrams are used to help us to see the type of relation that exists between two variables. In this text, we will discuss a variety of different relations that may exist between two variables. For now, we concentrate on distinguishing between linear and nonlinear relations. See Figure 8.

Figure 8

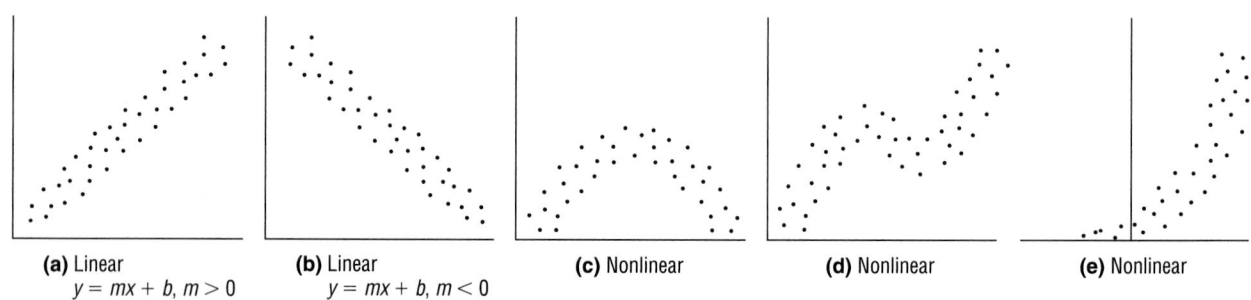

(a) Linear
$y = mx + b, m > 0$

(b) Linear
$y = mx + b, m < 0$

(c) Nonlinear

(d) Nonlinear

(e) Nonlinear

| EXAMPLE 2 | Distinguishing between Linear and Nonlinear Relations |

Determine whether the relation between the two variables in Figure 9 is linear or nonlinear.

Figure 9

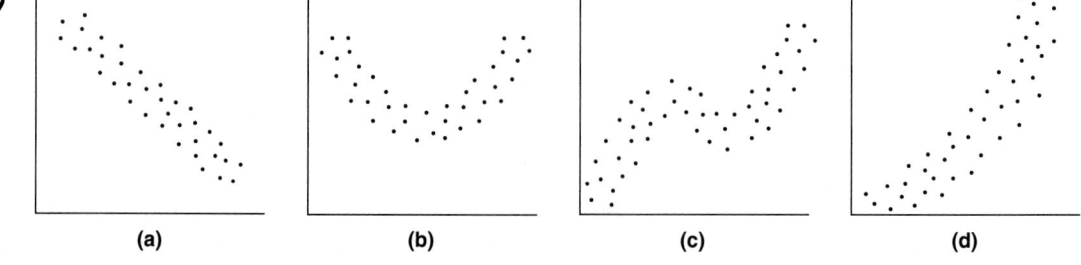

(a)

(b)

(c)

(d)

Solution (a) Linear (b) Nonlinear (c) Nonlinear (d) Nonlinear ■

NOW WORK PROBLEM **7.**

In this section we will study data whose scatter diagrams imply that a linear relation exists between the two variables. Nonlinear data will be discussed in later chapters.

Suppose that the scatter diagram of a set of data appears to be linearly related as in Figure 8(a) or (b). We might wish to find an equation of a line that relates the two variables. One way to obtain an equation for such data is to draw a line through two points on the scatter diagram and determine the equation of the line.

| EXAMPLE 3 | **Finding an Equation for Linearly Related Data** |

Using the data in Table 6 from Example 1,
(a) Select two points and find an equation of the line containing the points.
(b) Graph the line on the scatter diagram obtained in Example 1(b).

Solution (a) Select two points, say $(32.1, 634)$ and $(35.7, 850)$. The slope of the line joining the points $(32.1, 634)$ and $(35.7, 850)$ is

$$m = \frac{850 - 634}{35.7 - 32.1} = \frac{216}{3.6} = 60$$

The equation of the line with slope 60 and passing through $(32.1, 634)$ is found using the point–slope form with $m = 60$, $x_1 = 32.1$, and $y_1 = 634$.

$$y - y_1 = m(x - x_1) \qquad \text{Point-slope form of a line}$$
$$y - 634 = 60(x - 32.1) \qquad x_1 = 32.1, \; y_1 = 634, \; m = 60$$
$$y - 634 = 60x - 1926$$
$$y = 60x - 1292$$

(b) Figure 10 shows the scatter diagram with the graph of the line found in part (a).

Figure 10

On-base Percentage

■

Select two other points and complete the solution. Add the line obtained to Figure 10.

NOW WORK PROBLEMS **13(b)** AND **(c).**

 3 Use a Graphing Utility to Find the Line of Best Fit

The line obtained in Example 3 depends on the selection of points, which will vary from person to person. So the line that we found might be different from the line you found. Although the line we found in Example 3 appears to fit the data well, there may be a line that "fits it better." Do you think your line fits the data better? Is there a *line of best fit*? As it turns out, there is a method for finding the line that best fits linearly related data (called the **line of best fit**).*

| EXAMPLE 4 | **Finding the Line of Best Fit** |

Using the data in Table 6 from Example 1,
(a) Find the line of best fit using a graphing utility.
(b) Graph the line of best fit on the scatter diagram obtained in Example 1(b).
(c) Interpret the slope.
(d) Use the line of best fit to predict the number of runs a team will score if their on-base percentage is 34.1.

Solution
(a) Graphing utilities contain built-in programs that find the line of best fit for a collection of points in a scatter diagram. Upon executing the LINear REGression program, we obtain the results shown in Figure 11. The output that the utility provides shows us the equation $y = ax + b$, where a is the slope of the line and b is the y-intercept. The line of best fit that relates on-base percentage to runs scored may be expressed as the line $y = 51.58x - 967.82$.

(b) Figure 12 shows the graph of the line of best fit, along with the scatter diagram.

Figure 11

Figure 12

(c) The slope of the line of best fit is 51.58, which means that, for every 1 percent increase in the on-base percentage, runs scored increase 51.58, on average.

(d) Letting $x = 34.1$ in the equation of the line of best fit, we obtain $y = 51.58(34.1) - 967.82 \approx 791$ runs. ∎

NOW WORK PROBLEMS 13(d) AND (e).

Does the line of best fit appear to be a good fit? In other words, does the line appear to accurately describe the relation between on-base percentage and runs scored?

And just how "good" is this line of best fit? Look again at Figure 11. The last line of output is $r = 0.928$. This number, called the **correlation coefficient**, r, $-1 \le r \le 1$, is a measure of the strength of the linear relation that exists between two variables. The closer that $|r|$ is to 1, the more perfect the linear relationship is. If r is close to 0, there is little or no linear relationship between the variables. A negative value of r, $r < 0$, indicates that as x increases y decreases; a positive value of r, $r > 0$, indicates that as x increases y does also. The data given in Table 6, having a correlation coefficient of 0.928, are indicative of a strong linear relationship with positive slope.

*We shall not discuss the underlying mathematics of lines of best fit in this book.

4 Construct a Linear Model Using Direct Variation

When a mathematical model is developed for a real-world problem, it often involves relationships between quantities that are expressed in terms of proportionality.

Force is proportional to acceleration.

For an ideal gas held at a constant temperature, pressure and volume are inversely proportional.

The force of attraction between two heavenly bodies is inversely proportional to the square of the distance between them.

Revenue is directly proportional to sales.

Each statement illustrates the idea of **variation**, or how one quantity varies in relation to another quantity. Quantities may vary directly, inversely, or jointly. We discuss direct variation here.

Let x and y denote two quantities. Then y **varies directly** with x, or y is **directly proportional to** x, if there is a nonzero number k such that

$$y = kx$$

The number k is called the **constant of proportionality**.

If y varies directly with x, then y is a linear function of x. The graph in Figure 13 illustrates the relationship between y and x if y varies directly with x and $k > 0, x \geq 0$. Note that the constant of proportionality is, in fact, the slope of the line.

Figure 13
$y = kx, k > 0, x \geq 0$

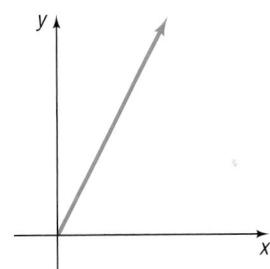

If we know that two quantities vary directly, then knowing the value of each quantity in one instance enables us to write a formula that is true in all cases.

EXAMPLE 5 **Mortgage Payments**

The monthly payment p on a mortgage varies directly with the amount borrowed B. If the monthly payment on a 30-year mortgage is $6.65 for every $1000 borrowed, find a function $p = p(B)$ that relates the monthly payment p to the amount borrowed B for a mortgage with the same terms. Then find the monthly payment p when the amount borrowed B is $120,000. Graph $p = p(B)$.

Solution Because p varies directly with B, we know that

$$p = kB$$

for some constant k. Because $p = 6.65$ when $B = 1000$, it follows that

$$6.65 = k(1000)$$

$$k = 0.00665$$

So we have

$$p = 0.00665B$$

We see that p is a linear function of B: $p(B) = 0.00665B$. In particular, when $B = \$120,000$, we find that

$$p(120,000) = 0.00665(\$120,000) = \$798$$

Figure 14 illustrates the relationship between the monthly payment p and the amount borrowed B.

Figure 14

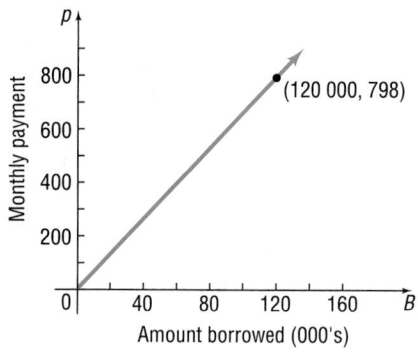

NOW WORK PROBLEM **19.**

2.2 Assess Your Understanding

'Are You Prepared?'

Answers are given at the end of these exercises. If you get a wrong answer, read the pages listed in red.

1. Is the relation $\{(1, 5), (2, 6), (3, 9), (1, 12)\}$ a function? (p. 43)

2. Is the relation defined by the equation $y = 2x + 1$ a function? (p. 44)

Concepts and Vocabulary

3. A _____ _____ is used to help us to see the type of relation, if any, that may exist between two variables.

4. If x and y are two quantities, then y is directly proportional to x if there is a nonzero number k such that _____.

5. *True or False*: The correlation coefficient is a measure of the strength of a linear relation between two variables and must lie between -1 and 1, inclusive.

6. *True or False*: If y varies directly with x, then y is a linear function of x.

Skill Building

In Problems 7–12, examine the scatter diagram and determine whether the type of relation, if any, that may exist is linear or nonlinear.

7.

8.

9.

10.

11.

12.

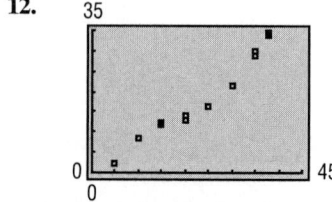

In Problems 13–18,
 (a) *Draw a scatter diagram.*
 (b) *Select two points from the scatter diagram and find the equation of the line containing the points selected.*
 (c) *Graph the line found in part (b) on the scatter diagram.*
 (d) *Use a graphing utility to find the line of best fit.*
 (e) *Use a graphing utility to draw the scatter diagram and graph the line of best fit on it.*

13.

x	3	4	5	6	7	8	9
y	4	6	7	10	12	14	16

14.

x	3	5	7	9	11	13
y	0	2	3	6	9	11

15.

x	−2	−1	0	1	2
y	−4	0	1	4	5

16.

x	−2	−1	0	1	2
y	7	6	3	2	0

17.

x	−20	−17	−15	−14	−10
y	100	120	118	130	140

18.

x	−30	−27	−25	−20	−14
y	10	12	13	13	18

Applications and Extensions

19. Mortgage Payments The monthly payment p on a mortgage varies directly with the amount borrowed B. If the monthly payment on a 30-year mortgage is $6.49 for every $1000 borrowed, find a linear function that relates the monthly payment p to the amount borrowed B for a mortgage with the same terms. Then find the monthly payment p when the amount borrowed B is $145,000.

20. Mortgage Payments The monthly payment p on a mortgage varies directly with the amount borrowed B. If the monthly payment on a 15-year mortgage is $8.99 for every $1000 borrowed, find a linear function that relates the monthly payment p to the amount borrowed B for a mortgage with the same terms. Then find the monthly payment p when the amount borrowed B is $175,000.

21. Revenue Function At the corner Shell station, the revenue R varies directly with the number g of gallons of gasoline sold. If the revenue is $23.40 when the number of gallons sold is 12, find a linear function that relates revenue R to the number g of gallons of gasoline. Then find the revenue R when the number of gallons of gasoline sold is 10.5.

22. Cost Function The cost C of chocolate-covered almonds varies directly with the number A of pounds of almonds purchased. If the cost is $23.75 when the number of pounds of chocolate-covered almonds purchased is 5, find a linear function that relates the cost C to the number A of pounds of almonds purchased. Then find the cost C when the number of pounds of almonds purchased is 3.5.

23. Sand Castles Researchers at Bournemouth University discovered that the amount of water W varies directly with the amount of sand S that is needed to make the perfect sand castle. If 15 gallons of sand is mixed with 1.875 gallons of water, a perfect sand castle can be built. How many gallons of water should be mixed with 40 gallons of sand to create a perfect sand castle?

24. Physics: Falling Objects The velocity v of a falling object is directly proportional to the time t of the fall. If, after 2 seconds, the velocity of the object is 64 feet per second, what will its velocity be after 3 seconds?

25. Candy The following data represent the weight (in grams) of various candy bars and the corresponding number of calories.

Candy Bar	Weight, x	Calories, y
Hershey's Milk Chocolate®	44.28	230
Nestle's Crunch®	44.84	230
Butterfinger®	61.30	270
Baby Ruth®	66.45	280
Almond Joy®	47.33	220
Twix® (with Caramel)	58.00	280
Snickers®	61.12	280
Heath®	39.52	210

SOURCE: Megan Pocius, Student at Joliet Junior College

(a) Draw a scatter diagram of the data treating weight as the independent variable.
(b) What type of relation appears to exist between the weight of a candy bar and the number of calories?
(c) Select two points and find an equation of the line containing the points.
(d) Graph the line on the scatter diagram drawn in part (a).
(e) Predict the number of calories in a candy bar that weighs 62.3 grams.
(f) Interpret the slope of the line found in part (c).

26. Raisins The following data represent the weight (in grams) of a box of raisins and the number of raisins in the box.

Weight (in grams), w	Number of Raisins, N
42.3	87
42.7	91
42.8	93
42.4	87
42.6	89
42.4	90
42.3	82
42.5	86
42.7	86
42.5	86

SOURCE: Jennifer Maxwell, Student at Joliet Junior College

(a) Does the relation defined by the set of ordered pairs (w, N) represent a function?

(b) Draw a scatter diagram of the data treating weight as the independent variable.

(c) Select two points and find the equation of the line containing the points.

(d) Graph the line on the scatter diagram drawn in part (b).

(e) Express the relationship found in part (c) using function notation.

(f) Predict the number of raisins in a box that weighs 42.5 grams.

(g) Interpret the slope of the line found in part (c).

27. **Height versus Head Circumference** A pediatrician wanted to estimate a linear function that relates a child's height, H, to their head circumference, C. She randomly selects nine children from her practice, measures their height and head circumference, and obtains the data shown below. Let H represent the independent variable and C the dependent variable.

(a) Use a graphing utility to draw a scatter diagram.

(b) Use a graphing utility to find the line of best fit to the data. Express the solution using function notation.

(c) Interpret the slope.

(d) Predict the head circumference of a child that is 26 inches tall.

(e) What is the height of a child whose head circumference is 17.4 inches?

Height, H (inches)	Head Circumference, C (inches)
25.25	16.4
25.75	16.9
25	16.9
27.75	17.6
26.5	17.3
27	17.5
26.75	17.3
26.75	17.5
27.5	17.5

Source: Denise Slucki, Student at Joliet Junior College

28. **Gestation Period versus Life Expectancy** A researcher would like to estimate the linear function relating the gestation period of an animal, G, and its life expectancy, L. She collects the following data.

Animal	Gestation (or incubation) Period, G (days)	Life Expectancy, L (years)
Cat	63	11
Chicken	22	7.5
Dog	63	11
Duck	28	10
Goat	151	12
Lion	108	10
Parakeet	18	8
Pig	115	10
Rabbit	31	7
Squirrel	44	9

Source: Time Almanac 2000

Let G represent the independent variable and L the dependent variable.

(a) Use a graphing utility to draw a scatter diagram.

(b) Use a graphing utility to find the line of best fit to the data. Express the solution using function notation.

(c) Interpret the slope.

(d) Predict the life expectancy of an animal whose gestation period is 89 days.

29. **Demand for Jeans** The marketing manager at Levi-Strauss wishes to find a function that relates the demand D for men's jeans and p, the price of the jeans. The following data were obtained based on a price history of the jeans.

Price ($/Pair), p	Demand (Pairs of Jeans Sold per Day), D
20	60
22	57
23	56
23	53
27	52
29	49
30	44

(a) Does the relation defined by the set of ordered pairs (p, D) represent a function?

(b) Draw a scatter diagram of the data.

(c) Using a graphing utility, find the line of best fit relating price and quantity demanded.

(d) Interpret the slope.

(e) Express the relationship found in part (c) using function notation.

(f) What is the domain of the function?

(g) How many jeans will be demanded if the price is $28 a pair?

30. **Advertising and Sales Revenue** A marketing firm wishes to find a function that relates the sales S of a product and A, the amount spent on advertising the product. The data are obtained from past experience. Advertising and sales are measured in thousands of dollars.

Advertising Expenditures, A	Sales, S
20	335
22	339
22.5	338
24	343
24	341
27	350
28.3	351

(a) Does the relation defined by the set of ordered pairs (A, S) represent a function?

(b) Draw a scatter diagram of the data.

(c) Using a graphing utility, find the line of best fit relating advertising expenditures and sales.

(d) Interpret the slope.

(e) Express the relationship found in part (c) using function notation.

(f) What is the domain of the function?

(g) Predict sales if advertising expenditures are $25,000.

Discussion and Writing

31. Maternal Age versus Down Syndrome A biologist would like to know how the age of the mother affects the incidence rate of Down syndrome. The following data represent the age of the mother and the incidence rate of Down syndrome per 1000 pregnancies.

Age of Mother, x	Incidence of Down Syndrome, y
33	2.4
34	3.1
35	4
36	5
37	6.7
38	8.3
39	10
40	13.3
41	16.7
42	22.2
43	28.6
44	33.3
45	50

SOURCE: Hook, E.B., *Journal of the American Medical Association,* 249, 2034-2038, 1983

Draw a scatter diagram treating age of the mother as the independent variable. Explain why it would not make sense to find the line of best fit for these data.

32. Harley-Davidson, Inc., Stock The following data represent the closing year-end price of Harley-Davidson, Inc.,

stock for the years 1990 to 2003. We let $x = 1$ represent 1990, $x = 2$ represent 1991, and so on.

Year, x	Closing Price, y
1990 ($x = 1$)	1.10
1991 ($x = 2$)	2.56
1992 ($x = 3$)	4.30
1993 ($x = 4$)	5.07
1994 ($x = 5$)	6.49
1995 ($x = 6$)	6.76
1996 ($x = 7$)	11.16
1997 ($x = 8$)	13.10
1998 ($x = 9$)	22.98
1999 ($x = 10$)	31.27
2000 ($x = 11$)	38.92
2001 ($x = 12$)	53.27
2002 ($x = 13$)	45.44
2003 ($x = 14$)	46.95
2004 ($x = 15$)	60.43

SOURCE: Yahoo! Finance

Draw a scatter diagram of the data treating the year, x, as the independent variable. Explain why it would not make sense to find the line of best fit for these data.

33. Find the line of best fit for the ordered pairs $(1, 5)$ and $(3, 8)$. What is the correlation coefficient for these data? Why is this result reasonable?

34. What does a correlation coefficient of 0 imply?

35. What would it mean if a teacher states that a student's grade in the class is directly proportional to the amount of time that the student studies?

'Are You Prepared?' Answers

1. No **2.** Yes

2.3 Quadratic Functions and Their Zeros

PREPARING FOR THIS SECTION *Before getting started, review the following:*
- Solving Equations (Appendix A, Section A.8, pp. A60–A64)
- Factoring Polynomials (Appendix A, Section A.4, pp. A29–A35)
- Completing the Square (Appendix A, Section A.4, pp. A35–A36)
- The Graph of a Function (Section 1.2, pp. 54–58)

Now work the 'Are You Prepared?' problems on page 142.

OBJECTIVES 1 Find the Zeros of a Quadratic Function by Factoring
2 Find the Zeros of a Quadratic Function Using the Square Root Method
3 Find the Zeros of a Quadratic Function by Completing the Square
4 Find the Zeros of a Quadratic Function Using the Quadratic Formula
5 Find the Point of Intersection of Two Functions
6 Solve Equations That Are Quadratic in Form

Here are some examples of quadratic functions.

$$F(x) = 3x^2 - 5x + 1, \qquad g(x) = -6x^2 + 1, \qquad H(x) = \frac{1}{2}x^2 + \frac{2}{3}x$$

A **quadratic function** is a function of the form

$$f(x) = ax^2 + bx + c$$

where a, b, and c are real numbers and $a \neq 0$. The domain of a quadratic function consists of all real numbers.

We may also say that a quadratic function is a function defined by a second-degree polynomial in one variable.

Remember, solving the equation $f(x) = 0$ is equivalent to finding the zeros of f and is also equivalent to finding the x-intercepts of the graph of $y = f(x)$. So, to find the zeros, or x-intercepts of the graph, of a quadratic function $f(x) = ax^2 + bx + c$, we must be able to solve the equation $ax^2 + bx + c = 0$. These types of equations are called quadratic equations.

A **quadratic equation** is an equation equivalent to one of the form

$$ax^2 + bx + c = 0 \qquad\qquad \textbf{(1)}$$

where a, b, and c are real numbers and $a \neq 0$.

Examples of quadratic equations are

$$3x^2 - 5x + 1 = 0, \qquad -6x^2 + 1 = 0, \qquad \frac{1}{2}x^2 + \frac{2}{3}x = 0$$

A quadratic equation written in the form $ax^2 + bx + c = 0$ is said to be in **standard form**.

Sometimes a quadratic equation is called a **second-degree equation**, because the left side is a polynomial of degree 2. We shall discuss four ways of solving quadratic equations: by factoring, by the square root method, by completing the square, and by using the quadratic formula.

1 Find the Zeros of a Quadratic Function by Factoring

When a quadratic equation is written in standard form, $ax^2 + bx + c = 0$, it may be possible to factor the expression on the left side as the product of two first-degree polynomials. Then, by setting each factor equal to 0 and solving the resulting linear equations, we obtain the exact solutions of the quadratic equation.

Let's look at an example.

EXAMPLE 1 **Finding the Zeros of a Quadratic Function by Factoring**

Find the zeros of $f(x) = x^2 + x - 12$. List any x-intercepts of the graph of f.

Solution To find the zeros of a function f, we solve the equation $f(x) = 0$.

$$f(x) = 0$$
$$x^2 + x - 12 = 0 \qquad f(x) = x^2 + x - 12$$
$$(x + 4)(x - 3) = 0 \qquad \text{Factor}$$
$$x + 4 = 0 \quad \text{or} \quad x - 3 = 0 \qquad \text{Zero-Product Property}$$
$$x = -4 \quad \text{or} \quad x = 3$$

✔ CHECK: $f(x) = x^2 + x - 12$

$$x = -4: f(-4) = (-4)^2 + (-4) - 12 \qquad x = 3: f(3) = (3)^2 + (3) - 12$$
$$= 16 - 4 - 12 \qquad\qquad\qquad\qquad = 9 + 3 - 12$$
$$= 0 \qquad\qquad\qquad\qquad\qquad\qquad = 0$$

The zeros of $f(x) = x^2 + x - 12$ are -4 and 3. The x-intercepts of the graph of $f(x) = x^2 + x - 12$ are -4 and 3. ∎

NOW WORK PROBLEM 11.

Sometimes, the quadratic expression factors into two linear equations with the same solution. When this happens, the quadratic equation is said to have a **repeated solution**. We also call this solution a **root of multiplicity 2**, or a **double root**.

EXAMPLE 2	**Finding the Zeros of a Quadratic Function by Factoring**

Find the zeros of $f(x) = x^2 - 6x + 9$. List any x-intercepts of the graph of f.

Solution To find the zeros of a function f, we solve the equation $f(x) = 0$.

$$f(x) = 0$$
$$x^2 - 6x + 9 = 0 \qquad f(x) = x^2 - 6x + 9$$
$$(x - 3)(x - 3) = 0 \qquad \text{Factor}$$
$$x - 3 = 0 \quad \text{or} \quad x - 3 = 0 \qquad \text{Zero-Product Property}$$
$$x = 3 \quad \text{or} \quad x = 3$$

We leave the check to you. The only zero of $f(x) = x^2 - 6x + 9$ is 3. The only x-intercept of the graph of $f(x) = x^2 - 6x + 9$ is 3. ∎

NOW WORK PROBLEM 23.

2 Find the Zeros of a Quadratic Function Using the Square Root Method

Suppose that we wish to solve the quadratic equation

$$x^2 = p \tag{2}$$

where $p \geq 0$ is a nonnegative number. We proceed as in the earlier examples.

$$x^2 - p = 0 \qquad \text{Put in standard form.}$$
$$\left(x - \sqrt{p}\right)\left(x + \sqrt{p}\right) = 0 \qquad \text{Factor (over the real numbers).}$$
$$x = \sqrt{p} \quad \text{or} \quad x = -\sqrt{p} \qquad \text{Solve.}$$

We have the following result:

If $x^2 = p$ and $p \geq 0$, then $x = \sqrt{p}$ or $x = -\sqrt{p}$. \qquad (3)

When statement (3) is used, it is called the **Square Root Method**. In statement (3), note that if $p > 0$ the equation $x^2 = p$ has two solutions, $x = \sqrt{p}$ and $x = -\sqrt{p}$. We usually abbreviate these solutions as $x = \pm\sqrt{p}$, read as "x equals plus or minus the square root of p." For example, the two solutions of the equation

$$x^2 = 4$$

are

$$x = \pm\sqrt{4}$$

and, since $\sqrt{4} = 2$, we have

$$x = \pm 2$$

the

| EXAMPLE 3 | **Finding the Zeros of a Quadratic Function Using the Square Root Method** |

Find the zeros of each function.

(a) $f(x) = x^2 - 5$ (b) $g(x) = (x - 2)^2 - 16$

List any x-intercepts of the graph.

Solution (a) To find the zeros of a function f, we solve the equation $f(x) = 0$.

$$f(x) = 0$$
$$x^2 - 5 = 0 \qquad \small f(x) = x^2 - 5$$
$$x^2 = 5 \qquad \small \text{Add 5 to both sides.}$$
$$x = \pm\sqrt{5} \qquad \small \text{Use the Square Root Method.}$$
$$x = \sqrt{5} \quad \text{or} \quad x = -\sqrt{5}$$

The zeros of $f(x) = x^2 - 5$ are $-\sqrt{5}$ and $\sqrt{5}$. The x-intercepts of the graph of $f(x) = x^2 - 5$ are $-\sqrt{5}$ and $\sqrt{5}$.

(b) To find the zeros of a function g, we solve the equation $g(x) = 0$.

$$g(x) = 0$$
$$(x - 2)^2 - 16 = 0 \qquad \small g(x) = (x - 2)^2 - 16$$
$$(x - 2)^2 = 16 \qquad \small \text{Add 16 to both sides.}$$
$$x - 2 = \pm\sqrt{16} \qquad \small \text{Use the Square Root Method.}$$
$$x - 2 = \sqrt{16} \quad \text{or} \quad x - 2 = -\sqrt{16}$$
$$x - 2 = 4 \quad \text{or} \quad x - 2 = -4$$
$$x = 6 \quad \text{or} \quad x = -2$$

The zeros of $g(x) = (x - 2)^2 - 16$ are -2 and 6. The x-intercepts of the graph of $g(x) = (x - 2)^2 - 16$ are -2 and 6. ∎

NOW WORK PROBLEM 27.

3 Find the Zeros of a Quadratic Function by Completing the Square

The next example illustrates how the procedure of completing the square can be used to find the zeros of a quadratic function.

| EXAMPLE 4 | **Finding the Zeros of a Quadratic Function by Completing the Square** |

Find the zeros of $f(x) = x^2 + 5x + 4$ by completing the square. List any x-intercepts of the graph of f.

Solution To find the zeros of a function f, we solve the equation $f(x) = 0$.

$$f(x) = 0$$
$$x^2 + 5x + 4 = 0 \qquad \small f(x) = x^2 + 5x + 4$$

We always begin this procedure by rearranging the equation so that the constant is on the right side.

$$x^2 + 5x + 4 = 0$$
$$x^2 + 5x = -4$$

Since the coefficient of x^2 is 1, we can complete the square on the left side by adding $\left(\dfrac{1}{2} \cdot 5\right)^2 = \dfrac{25}{4}$. Of course, in an equation, whatever we add to the left side must also be added to the right side. We add $\dfrac{25}{4}$ to both sides.

$$x^2 + 5x + \frac{25}{4} = -4 + \frac{25}{4} \qquad \text{Add } \frac{25}{4} \text{ to both sides.}$$

$$\left(x + \frac{5}{2}\right)^2 = \frac{9}{4} \qquad \text{Factor; simplify.}$$

$$x + \frac{5}{2} = \pm\sqrt{\frac{9}{4}} \qquad \text{Use the Square Root Method.}$$

$$x + \frac{5}{2} = \pm\frac{3}{2}$$

$$x = -\frac{5}{2} \pm \frac{3}{2}$$

$$x = -\frac{5}{2} + \frac{3}{2} = -1 \quad \text{or} \quad x = -\frac{5}{2} - \frac{3}{2} = -4$$

We leave the check to you. The zeros of $f(x) = x^2 + 5x + 4$ are -4 and -1. The x-intercepts of the graph of $f(x) = x^2 + 5x + 4$ are -4 and -1. ∎

The zeros of the function in Example 4 can also be obtained by factoring. Rework Example 4 using factoring.

NOW WORK PROBLEM 31.

4 Find the Zeros of a Quadratic Function Using the Quadratic Formula

We can use the method of completing the square to obtain a general formula for solving the quadratic equation

$$ax^2 + bx + c = 0, \qquad a \neq 0$$

As in Example 4, we begin by rearranging the terms as

$$ax^2 + bx = -c, \qquad a > 0$$

NOTE: There is no loss in generality to assume that $a > 0$, since if $a < 0$ we can multiply both sides by -1 to obtain an equivalent equation with a positive leading coefficient. ∎

Since $a > 0$, we can divide both sides by a to get

$$x^2 + \frac{b}{a}x = -\frac{c}{a}$$

Now the coefficient of x^2 is 1. To complete the square on the left side, add the square of $\dfrac{1}{2}$ the coefficient of x: that is, add

$$\left(\frac{1}{2} \cdot \frac{b}{a}\right)^2 = \frac{b^2}{4a^2}$$

to each side. Then

$$x^2 + \frac{b}{a}x + \frac{b^2}{4a^2} = \frac{b^2}{4a^2} - \frac{c}{a}$$

$$\left(x + \frac{b}{2a}\right)^2 = \frac{b^2 - 4ac}{4a^2} \qquad \frac{b^2}{4a^2} - \frac{c}{a} = \frac{b^2}{4a^2} - \frac{4ac}{4a^2} = \frac{b^2 - 4ac}{4a^2} \qquad \textbf{(4)}$$

Provided that $b^2 - 4ac \geq 0$, we now can use the Square Root Method to get

$$x + \frac{b}{2a} = \pm\sqrt{\frac{b^2 - 4ac}{4a^2}}$$

$$x + \frac{b}{2a} = \frac{\pm\sqrt{b^2 - 4ac}}{2a}$$

The square root of a quotient equals the quotient of the square roots. Also, $\sqrt{4a^2} = 2a$ since $a > 0$.

$$x = -\frac{b}{2a} \pm \frac{\sqrt{b^2 - 4ac}}{2a}$$

Add $-\dfrac{b}{2a}$ to both sides.

$$x = \frac{-b \pm \sqrt{b^2 - 4ac}}{2a}$$

Combine the quotients on the right.

What if $b^2 - 4ac$ is negative? Then equation (4) states that the left expression (a real number squared) equals the right expression (a negative number). Since this occurrence is impossible for real numbers, we conclude that if $b^2 - 4ac < 0$ the quadratic equation has no real solution.*

We now state the quadratic formula.

Theorem

Quadratic Formula

Consider the quadratic equation

$$ax^2 + bx + c = 0, \qquad a \neq 0$$

If $b^2 - 4ac < 0$, this equation has no real solution.

If $b^2 - 4ac \geq 0$, the real solution(s) of this equation is (are) given by the **quadratic formula**:

$$x = \frac{-b \pm \sqrt{b^2 - 4ac}}{2a}$$

The quantity $b^2 - 4ac$ is called the **discriminant** of the quadratic equation because its value tells us whether the equation has real solutions, and how many solutions to expect. The discriminant tells us how many real zeros, or x-intercepts, the quadratic function $f(x) = ax^2 + bx + c$ will have as well.

Discriminant of a Quadratic Equation

For a quadratic equation $ax^2 + bx + c = 0$:

1. If $b^2 - 4ac > 0$, there are two unequal real solutions. Therefore, the graph of $f(x) = ax^2 + bx + c$ will have two distinct x-intercepts.
2. If $b^2 - 4ac = 0$, there is a repeated real solution, a root of multiplicity 2. Therefore, the graph of $f(x) = ax^2 + bx + c$ will have one x-intercept.
3. If $b^2 - 4ac < 0$, there is no real solution. Therefore, the graph of $f(x) = ax^2 + bx + c$ will have no x-intercept.

When asked to find the zeros of a quadratic function, always evaluate the discriminant first to see how many real zeros there are.

*If $b^2 - 4ac < 0$, the quadratic function will have complex zeros. We discuss the complex zeros of a quadratic function is Section 2.7.

EXAMPLE 5	**Finding the Zeros of a Quadratic Function Using the Quadratic Formula**

Find the real zeros, if any, of the function $f(x) = 3x^2 - 5x + 1$. List any x-intercepts of the graph of f.

Solution To find the real zeros, we solve the equation $f(x) = 0$.

$$f(x) = 0$$

$$3x^2 - 5x + 1 = 0 \qquad f(x) = 3x^2 - 5x + 1$$

We compare $3x^2 - 5x + 1 = 0$ to $ax^2 + bx + c = 0$ to find a, b, and c.

$$3x^2 - 5x + 1 = 0$$

$$ax^2 + bx + c = 0, \qquad a = 3, b = -5, c = 1$$

With $a = 3, b = -5$, and $c = 1$, we evaluate the discriminant $b^2 - 4ac$.

$$b^2 - 4ac = (-5)^2 - 4(3)(1) = 25 - 12 = 13$$

Since $b^2 - 4ac > 0$, there are two unequal real solutions.
 We use the quadratic formula with $a = 3, b = -5, c = 1$, and $b^2 - 4ac = 13$.

$$x = \frac{-b \pm \sqrt{b^2 - 4ac}}{2a} = \frac{-(-5) \pm \sqrt{13}}{2(3)} = \frac{5 \pm \sqrt{13}}{6}$$

The real zeros of $f(x) = 3x^2 - 5x + 1$ are $\dfrac{5 - \sqrt{13}}{6}$ and $\dfrac{5 + \sqrt{13}}{6}$. The x-intercepts

of the graph of $f(x) = 3x^2 - 5x + 1$ are $\dfrac{5 - \sqrt{13}}{6}$ and $\dfrac{5 + \sqrt{13}}{6}$. ∎

EXAMPLE 6	**Finding the Zeros of a Quadratic Function Using the Quadratic Formula**

Find the real zeros, if any, of the function $f(x) = 3x^2 - 4x + 2$. List any x-intercepts of the graph.

Solution To find the real zeros, we solve the equation $f(x) = 0$.

$$f(x) = 0$$

$$3x^2 - 4x + 2 = 0 \qquad f(x) = 3x^2 - 4x + 2$$

$$ax^2 + bx + c = 0 \qquad \text{Compare to standard form.}$$

With $a = 3, b = -4$, and $c = 2$, we find that

$$b^2 - 4ac = (-4)^2 - 4(3)(2) = 16 - 24 = -8$$

Since $b^2 - 4ac < 0$, the equation has no real solution. Therefore, the function $f(x) = 3x^2 - 5x + 1$ has no real zeros. The graph of $f(x) = 3x^2 - 5x + 1$ has no x-intercept. ∎

NOW WORK PROBLEMS **41** AND **43**.

Summary:

Finding the Zeros, or x-intercepts of the Graph, of a Quadratic Function

STEP 1: Given that $f(x) = ax^2 + bx + c$, write the equation $ax^2 + bx + c = 0$.

STEP 2: Identify a, b, and c.

STEP 3: Evaluate the discriminant, $b^2 - 4ac$.

STEP 4: (a) If the discriminant is negative, the equation has no real solution, so the function has no real zeros. The graph of the function has no x-intercepts.

(b) If the discriminant is nonnegative (greater than or equal to zero), determine whether the left side can be factored. If you can easily spot factors, use the factoring method to solve the equation. Otherwise, use the quadratic formula or the method of completing the square. The solution(s) to the equation is(are) the real zero(s) of the function. The solution(s) also represent(s) the x-intercept(s) of the graph of the function.

5 Find the Point of Intersection of Two Functions

We are often interested in knowing where two functions are equal to each other. For example, if $R = R(x)$ represents the revenue for selling x units of a good and $C = C(x)$ represents the cost of manufacturing x units of a good, we might want to know the number of goods for which revenue equals cost. This requires us to solve $R(x) = C(x)$.

If f and g are two functions, the solution to the equation $f(x) = g(x)$ has a geometric interpretation—it represents the x-coordinate at which the graphs of $y = f(x)$ and $y = g(x)$ intersect.

| EXAMPLE 7 | **Finding the Point of Intersection of the Graph of Two Functions** |

If $f(x) = x^2 + 5x - 3$ and $g(x) = 2x + 1$, solve $f(x) = g(x)$. At what point(s) do the graphs of the two functions intersect?

Solution

$$f(x) = g(x)$$
$$x^2 + 5x - 3 = 2x + 1 \qquad f(x) = x^2 + 5x - 3; g(x) = 2x + 1$$
$$x^2 + 3x - 4 = 0 \qquad \text{Put in standard form.}$$
$$(x + 4)(x - 1) = 0 \qquad \text{Factor.}$$
$$x + 4 = 0 \quad \text{or} \quad x - 1 = 0 \qquad \text{Zero-Product Property}$$
$$x = -4 \quad \text{or} \quad x = 1$$

The x-coordinates of the points of intersection are -4 and 1. To find the y-coordinates of the points of intersection, we evaluate either f or g at $x = -4$ and $x = 1$. It is easier to evaluate g.

$$g(x) = 2x + 1 \qquad\qquad g(x) = 2x + 1$$
$$x = -4: \quad g(-4) = 2(-4) + 1 \qquad x = 1: \quad g(1) = 2(1) + 1$$
$$= -7 \qquad\qquad\qquad = 3$$

The graphs of the two functions intersect at the points $(-4, -7)$ and $(1, 3)$. ∎

NOW WORK PROBLEM 63.

6 Solve Equations That Are Quadratic in Form

So far, this section has been dedicated to finding zeros of a quadratic function $f(x) = ax^2 + bx + c$ by solving the equation $ax^2 + bx + c = 0$. Often, rather than being asked to find the zeros of a quadratic function, we might be asked to find the zeros of a function that can be transformed into a quadratic function with an appropriate substitution.

For example, the function $f(x) = x^4 + x^2 - 12$ is not a quadratic function, but if we let $u = x^2$, we get $f(u) = u^2 + u - 12$, which is a quadratic function. We can find the zeros of this function by solving the equation $u^2 + u - 12 = 0$. Then we can find x by solving $u = x^2$.

In general, if an appropriate substitution u transforms an equation into one of the form

$$au^2 + bu + c = 0, \quad a \neq 0$$

the original equation is called an **equation of the quadratic type** or an **equation quadratic in form**.

The difficulty of solving such an equation lies in the determination that the equation is, in fact, quadratic in form. After you are told an equation is quadratic in form, it is easy enough to see it, but some practice is needed to enable you to recognize them on your own.

EXAMPLE 8 **Finding the Real Zeros of a Function**

Find the real zeros of the function $f(x) = (x + 2)^2 + 11(x + 2) - 12$. Find any x-intercepts of the graph of f.

Solution To find the real zeros, we solve the equation $f(x) = 0$.

$$f(x) = 0$$

$$(x + 2)^2 + 11(x + 2) - 12 = 0 \qquad f(x) = (x + 2)^2 + 11(x + 2) - 12$$

For this equation, let $u = x + 2$. Then $u^2 = (x + 2)^2$, and the original equation

$$(x + 2)^2 + 11(x + 2) - 12 = 0$$

becomes

$$u^2 + 11u - 12 = 0 \qquad \text{Let } u = x + 2.$$

$$(u + 12)(u - 1) = 0 \qquad \text{Factor.}$$

$$u = -12 \quad \text{or} \quad u = 1 \qquad \text{Solve.}$$

But we want to solve for x. Because $u = x + 2$, we have

$$x + 2 = -12 \quad \text{or} \quad x + 2 = 1$$

$$x = -14 \qquad\qquad x = -1$$

The real zeros of f are -14 and -1. The x-intercepts of the graph of f are -14 and -1. ■

EXAMPLE 9 **Finding the Real Zeros of a Function**

Find the real zeros of $g(x) = x + 2\sqrt{x} - 3$. Find any x-intercepts of the graph of g.

Solution To find the real zeros, we solve the equation $g(x) = 0$.

$$g(x) = 0$$

$$x + 2\sqrt{x} - 3 = 0 \qquad g(x) = x + 2\sqrt{x} - 3$$

For the equation $x + 2\sqrt{x} - 3 = 0$, let $u = \sqrt{x}$. Then $u^2 = x$, and the original equation

$$x + 2\sqrt{x} - 3 = 0$$

becomes

$$u^2 + 2u - 3 = 0 \qquad \text{Let } u = \sqrt{x}.$$

$$(u + 3)(u - 1) = 0 \qquad \text{Factor.}$$

$$u = -3 \quad \text{or} \quad u = 1 \qquad \text{Solve.}$$

Since $u = \sqrt{x}$, we have $\sqrt{x} = -3$ or $\sqrt{x} = 1$. The first of these, $\sqrt{x} = -3$, has no real solution, since the square root of a real number is never negative. The second one, $\sqrt{x} = 1$, has the solution $x = 1$.

✔ **CHECK:** $g(x) = x + 2\sqrt{x} - 3$

$$g(1) = 1 + 2\sqrt{1} - 3 = 1 + 2 - 3 = 0$$ ■

The only real zero of the function is 1. The only x-intercept of the graph of the function is 1.

NOW WORK PROBLEM 79.

2.3 Assess Your Understanding

'Are You Prepared?'

Answers are given at the end of these exercises. If you get a wrong answer, read the pages listed in red.

1. Factor: $x^2 - 5x - 6$ (pp. A31–A33)

2. Factor: $2x^2 - x - 3$ (pp. A34–A35)

3. Solve: $2x - 3 = 11$ (pp. A61–A62)

4. Solve: $(x - 3)(3x + 5) = 0$ (p. A64)

5. To complete the square of: $x^2 + 6x$ _____ (add/subtract) _____. (pp. A35–A36)

6. If $f(4) = 10$, what point is on the graph of f? (pp. 57–58)

Concepts and Vocabulary

7. When a quadratic equation has a repeated solution, it is called a(n) _____ root or a root of _____ _____.

8. The quantity $b^2 - 4ac$ is called the _____ of a quadratic equation. If it is _____, the equation has no real solution.

9. *True* or *False*: Quadratic functions always have two real zeros.

10. *True* or *False*: A quadratic equation that is in the form $ax^2 + bx + c = 0$ is said to be in standard form.

Skill Building

In Problems 11–24, find the zeros of each quadratic function by factoring. What are the x-intercepts of the graph of the function?

11. $f(x) = x^2 - 9x$
12. $f(x) = x^2 + 4x$
13. $g(x) = x^2 - 25$
14. $G(x) = x^2 - 9$

15. $F(x) = x^2 + x - 6$
16. $H(x) = x^2 + 7x + 6$
17. $g(x) = 2x^2 - 5x - 3$
18. $f(x) = 3x^2 + 5x + 2$

19. $P(x) = 3x^2 - 48$
20. $H(x) = 2x^2 - 50$
21. $g(x) = x(x + 8) + 12$
22. $f(x) = x(x - 4) - 12$

23. $G(x) = 4x^2 + 9 - 12x$
24. $F(x) = 25x^2 + 16 - 40x$

In Problems 25–30, find the zeros of each quadratic function using the Square Root Method. What are the x-intercepts of the graph of the function?

25. $f(x) = x^2 - 25$
26. $g(x) = x^2 - 36$
27. $g(x) = (x - 1)^2 - 4$

28. $G(x) = (x + 2)^2 - 1$
29. $F(x) = (2x + 3)^2 - 9$
30. $g(x) = (3x - 2)^2 - 4$

In Problems 31–36, find the zeros of each quadratic function by completing the square. What are the x-intercepts of the graph of the function?

31. $f(x) = x^2 + 4x - 21$
32. $f(x) = x^2 - 6x - 13$
33. $g(x) = x^2 - \dfrac{1}{2}x - \dfrac{3}{16}$

34. $g(x) = x^2 + \dfrac{2}{3}x - \dfrac{1}{3}$
35. $F(x) = 3x^2 + x - \dfrac{1}{2}$
36. $G(x) = 2x^2 - 3x - 1$

In Problems 37–48, find the real zeros, if any, of each quadratic function using the quadratic formula. What are the x-intercepts, if any, of the graph of the function?

37. $f(x) = x^2 - 4x + 2$
38. $f(x) = x^2 + 4x + 2$
39. $g(x) = x^2 - 4x - 1$
40. $g(x) = x^2 + 6x + 1$

41. $F(x) = 2x^2 - 5x + 3$
42. $g(x) = 2x^2 + 5x + 3$
43. $P(x) = 4x^2 - x + 2$
44. $H(x) = 4x^2 + x + 1$

45. $f(x) = 4x^2 - 1 + 2x$
46. $f(x) = 2x^2 - 1 + 2x$
47. $G(x) = 4x^2 - 9x - 2$
48. $F(x) = 5x - 4x^2 + 5$

In Problems 49–60, find the real zeros of each quadratic function using any method you wish. What are the x-intercepts, if any, of the graph of the function?

49. $f(x) = x^2 - 5$

50. $f(x) = x^2 - 6$

51. $g(x) = 16x^2 - 8x + 1$

52. $F(x) = 4x^2 - 12x + 9$

53. $G(x) = 10x^2 - 19x - 15$

54. $f(x) = 6x^2 + 7x - 20$

55. $P(x) = 6x^2 - x - 2$

56. $H(x) = 6x^2 + x - 2$

57. $G(x) = x^2 + \sqrt{2}x - \dfrac{1}{2}$

58. $F(x) = \dfrac{1}{2}x^2 - \sqrt{2}x - 1$

59. $f(x) = x^2 + x - 4$

60. $g(x) = x^2 + x - 1$

In Problems 61–66, solve f(x) = g(x). What are the points of intersection of the graphs of the two functions?

61. $f(x) = x^2 + 6x + 3$
$g(x) = 3$

62. $f(x) = x^2 - 4x + 3$
$g(x) = 3$

63. $f(x) = -2x^2 + 1$
$g(x) = 3x + 2$

64. $f(x) = 3x^2 - 7$
$g(x) = 10x + 1$

65. $f(x) = x^2 - x + 1$
$g(x) = 2x^2 - 3x - 14$

66. $f(x) = x^2 + 5x - 3$
$g(x) = 2x^2 + 7x - 27$

In Problems 67–84, find the real zeros of each function. What are the x-intercepts of the graph of the function?

67. $P(x) = x^4 - 6x^2 - 16$

68. $H(x) = x^4 - 3x^2 - 4$

69. $f(x) = x^4 - 5x^2 + 4$

70. $f(x) = x^4 - 10x^2 + 25$

71. $G(x) = 3x^4 - 2x^2 - 1$

72. $F(x) = 2x^4 - 5x^2 - 12$

73. $g(x) = x^6 + 7x^3 - 8$

74. $g(x) = x^6 - 7x^3 - 8$

75. $G(x) = (x + 2)^2 + 7(x + 2) + 12$

76. $f(x) = (2x + 5)^2 - (2x + 5) - 6$

77. $f(x) = (3x + 4)^2 - 6(3x + 4) + 9$

78. $H(x) = (2 - x)^2 + (2 - x) - 20$

79. $P(x) = 2(x + 1)^2 - 5(x + 1) - 3$

80. $H(x) = 3(1 - x)^2 + 5(1 - x) + 2$

81. $G(x) = x - 4\sqrt{x}$

82. $f(x) = x + 8\sqrt{x}$

83. $g(x) = x + \sqrt{x} - 20$

84. $f(x) = x + \sqrt{x} - 2$

Applications and Extensions

85. Dimensions of a Window The function $A(x) = x(x + 2)$ describes the area A of the opening of a rectangular window in which the length is 2 feet more than the width, x. Find the dimensions of the window if the area of the opening is to be 143 square feet by solving $A(x) = 143$.

86. Dimensions of a Window The function $A(x) = x(x + 1)$ describes the area A of the opening of a rectangular window in which the length is 1 centimeter more than the width, x. Find the dimensions of the window if the area of the opening is to be 306 square centimeters by solving $A(x) = 306$.

87. Constructing a Box An open box is to be constructed from a square sheet of sheet metal with dimensions x feet by x feet by removing a square of side 1 foot from each corner and turning up the edges. The volume V of the box is $V(x) = (x - 2)^2$. Find the dimensions of the sheet metal needed to make a box that will hold 4 cubic feet by solving the equation $V(x) = 4$.

88. Constructing a Box Rework Problem 87 if the box needs to hold 16 cubic feet.

89. Physics A ball is thrown vertically upward from the top of a building 96 feet tall with an initial velocity of 80 feet per second. The distance s (in feet) of the ball from the ground after t seconds is $s(t) = 96 + 80t - 16t^2$.
(a) After how many seconds does the ball strike the ground?
(b) After how many seconds will the ball pass the top of the building on its way down?

90. Physics An object is propelled vertically upward with an initial velocity of 20 meters per second. The distance s (in meters) of the object from the ground after t seconds is $s(t) = -4.9t^2 + 20t$.
(a) When will the object be 15 meters above the ground?

(b) When will it strike the ground?
(c) Will the object reach a height of 100 meters?

91. Consecutive Integers The sum S of the consecutive integers $1, 2, 3, \ldots, n$ is given by the function $S(n) = \dfrac{1}{2}n(n + 1)$. Determine the number of consecutive integers, starting with 1, that must be added to get a sum of 210.

92. Geometry The number of diagonals D of a polygon with n sides is given by the function $D(n) = \dfrac{1}{2}n(n - 3)$. Determine the number of sides a polygon with 65 diagonals will have. Is there a polygon with 80 diagonals?

93. Show that the sum of the roots of a quadratic equation is $-\dfrac{b}{a}$.

94. Show that the product of the roots of a quadratic equation is $\dfrac{c}{a}$.

95. Find k such that the function $f(x) = kx^2 + x + k = 0$ has a repeated real zero.

96. Find k such that the function $f(x) = x^2 - kx + 4 = 0$ has a repeated real zero.

97. Show that the real zeros of the function $f(x) = ax^2 + bx + c$ are the negatives of the real zeros of the function $g(x) = ax^2 - bx + c$. Assume that $b^2 - 4ac \geq 0$.

98. Show that the real zeros of the function $f(x) = ax^2 + bx + c$ are the reciprocals of the real zeros of the function $g(x) = cx^2 + bx + a$. Assume that $b^2 - 4ac \geq 0$.

Discussion and Writing

99. Which of the following pairs of equations are equivalent? Explain.
(a) $x^2 = 9$; $x = 3$
(b) $x = \sqrt{9}$; $x = 3$
(c) $(x - 1)(x - 2) = (x - 1)^2$; $x - 2 = x - 1$

100. Describe four methods you might use to find the zeros of a quadratic function. State your preferred method. Explain why you chose it.

101. Explain the benefits of evaluating the discriminant of a quadratic equation before attempting to solve it.

102. Create three quadratic functions, one having two distinct zeros, one having no real zeros, and one having exactly one real zero.

103. The word quadratic seems to imply four (*quad*), yet a quadratic equation is an equation that involves a polynomial of degree 2. Investigate the origin of the term quadratic as it is used in the expression quadratic equation. Write a brief essay on your findings.

104. How many times can the graphs of two different quadratic functions intersect?

'Are You Prepared?' Answers

1. $(x - 6)(x + 1)$ **2.** $(2x - 3)(x + 1)$ **3.** $\{7\}$ **4.** $\left\{-\dfrac{5}{3}, 3\right\}$ **5.** add; 9 **6.** $(4, 10)$

2.4 Properties of Quadratic Functions

PREPARING FOR THIS SECTION *Before getting started, review the following:*

- Intercepts (Foundations, Section F.2, pp. 10–11)
- Graphing Techniques: Transformations (Section 1.5, pp. 85–94)
- Completing the Square (Section A.4, pp. A35–A36)

 Now work the 'Are You Prepared?' problems on page 152.

OBJECTIVES 1 Graph a Quadratic Function Using Transformations
2 Identify the Vertex and Axis of Symmetry of a Quadratic Function
3 Graph a Quadratic Function Using Its Vertex, Axis, and Intercepts
4 Find the Maximum or Minimum Value of a Quadratic Function

Many applications require a knowledge of quadratic functions. For example, suppose that Texas Instruments collects the data shown in Table 7, which relate the number of calculators sold at the price p (in dollars) per calculator. Since the price of a product determines the quantity that will be purchased, we treat price as the independent variable. The relationship between the number x of calculators sold and the price p per calculator may be approximated by the linear equation

$$x = 21{,}000 - 150p$$

Table 7

Price per Calculator, p (Dollars)	Number of Calculators, x
60	11,100
65	10,115
70	9,652
75	8,731
80	8,087
85	7,205
90	6,439

Then the revenue R derived from selling x calculators at the price p per calculator is equal to the unit selling price p of the product times the number x of units actually sold. That is,

$$R = xp$$

$$R(p) = (21,000 - 150p)p$$

$$= -150p^2 + 21,000p$$

So the revenue R is a quadratic function of the price p. Figure 15 illustrates the graph of this revenue function, whose domain is $0 \le p \le 140$, since both x and p must be nonnegative. In Section 2.6 we shall determine the price p that maximizes revenue.

Figure 15

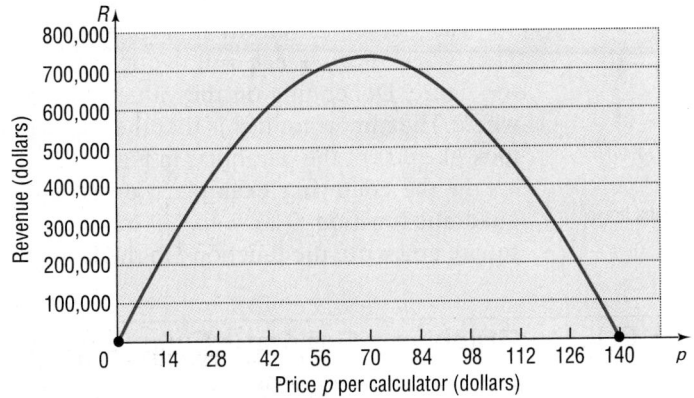

Figure 16
Path of a cannonball

A second situation in which a quadratic function appears involves the motion of a projectile. Based on Newton's second law of motion (force equals mass times acceleration, $F = ma$), it can be shown that, ignoring air resistance, the path of a projectile propelled upward at an inclination to the horizontal is the graph of a quadratic function. See Figure 16 for an illustration. In Section 2.6, we shall analyze the path of a projectile.

1 Graph a Quadratic Function Using Transformations

We know how to graph the quadratic function $f(x) = x^2$. Figure 17 shows the graph of three functions of the form $f(x) = ax^2, a > 0$, for $a = 1$, $a = \dfrac{1}{2}$, and $a = 3$.

Notice that the larger the value of a, the "narrower" the graph is, and the smaller the value of a, the "wider" the graph is.

Figure 18 shows the graphs of $f(x) = ax^2$ for $a < 0$. Notice that these graphs are reflections about the x-axis of the graphs in Figure 17. Based on the results of

Figure 17

Figure 18

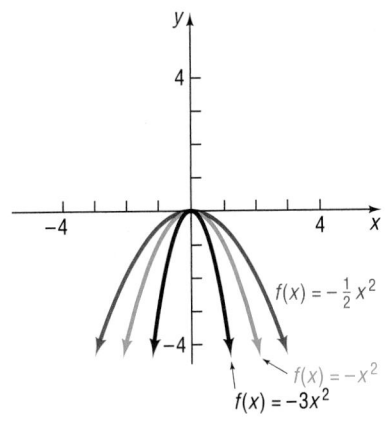

these two figures, we can draw some general conclusions about the graph of $f(x) = ax^2$. First, as $|a|$ increases, the graph becomes "taller" (a vertical stretch), and as $|a|$ gets closer to zero, the graph gets "shorter" (a vertical compression). Second, if a is positive, the graph opens "up," and if a is negative, the graph opens "down."

The graphs in Figures 17 and 18 are typical of the graphs of all quadratic functions, which we call **parabolas**.* Refer to Figure 19, where two parabolas are pictured. The one on the left **opens up** and has a lowest point; the one on the right **opens down** and has a highest point. The lowest or highest point of a parabola is called the **vertex**. The vertical line passing through the vertex in each parabola in Figure 19 is called the **axis of symmetry** (usually abbreviated to **axis**) of the parabola. Because the parabola is symmetric about its axis, the axis of symmetry of a parabola can be used to find additional points on the parabola.

The parabolas shown in Figure 19 are the graphs of a quadratic function $f(x) = ax^2 + bx + c, a \neq 0$. Notice that the coordinate axes are not included in the figure. Depending on the values of a, b, and c, the axes could be placed anywhere. The important fact is that the shape of the graph of a quadratic function will look like one of the parabolas in Figure 19.

In the following example, we use techniques from Section 1.5 to graph a quadratic function $f(x) = ax^2 + bx + c, a \neq 0$. In so doing, we shall complete the square and write the function f in the form $f(x) = a(x - h)^2 + k$.

Figure 19

Graphs of a quadratic function,
$f(x) = ax^2 + bx + c, a \neq 0$

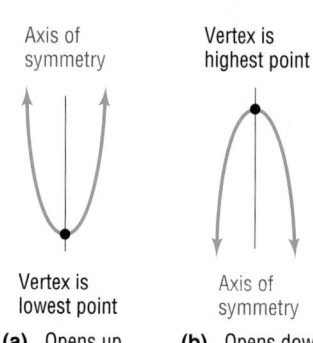

Axis of symmetry

Vertex is highest point

Vertex is lowest point

Axis of symmetry

(a) Opens up
$a > 0$

(b) Opens down
$a < 0$

| EXAMPLE 1 | **Graphing a Quadratic Function Using Transformations** |

Graph the function $f(x) = 2x^2 + 8x + 5$. Find the vertex and axis of symmetry.

Solution We begin by completing the square on the right side.

$$f(x) = 2x^2 + 8x + 5$$
$$= 2(x^2 + 4x) + 5 \qquad \text{Factor out the 2 from } 2x^2 + 8x.$$
$$= 2(x^2 + 4x + 4) + 5 - 8 \qquad \text{Complete the square of } x^2 + 4x. \text{ Notice}$$
$$\qquad\qquad\qquad\qquad\qquad\qquad \text{that the factor of 2 requires that 8 be}$$
$$= 2(x + 2)^2 - 3 \qquad\qquad\qquad \text{added and subtracted.}$$

The graph of f can be obtained in three stages, as shown in Figure 20. Now compare this graph to the graph in Figure 19(a). The graph of $f(x) = 2x^2 + 8x + 5$ is a parabola that opens up and has its vertex (lowest point) at $(-2, -3)$. Its axis of symmetry is the line $x = -2$.

Figure 20

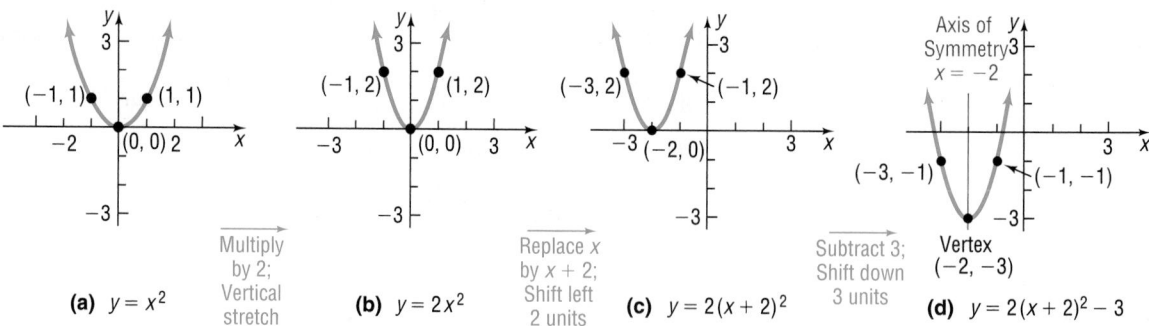

(a) $y = x^2$

Multiply by 2; Vertical stretch

(b) $y = 2x^2$

Replace x by $x + 2$; Shift left 2 units

(c) $y = 2(x + 2)^2$

Subtract 3; Shift down 3 units

(d) $y = 2(x + 2)^2 - 3$

NOW WORK PROBLEM 27.

*We shall study parabolas using a geometric definition later in this book.

The method used in Example 1 can be used to graph any quadratic function $f(x) = ax^2 + bx + c, a \neq 0$, as follows:

$$f(x) = ax^2 + bx + c$$

$$= a\left(x^2 + \frac{b}{a}x\right) + c \qquad \text{Factor out } a \text{ from } ax^2 + bx.$$

$$= a\left(x^2 + \frac{b}{a}x + \frac{b^2}{4a^2}\right) + c - a\left(\frac{b^2}{4a^2}\right) \qquad \text{Complete the square by adding and subtracting } a\left(\frac{b^2}{4a^2}\right). \text{ Look closely at this step!}$$

$$= a\left(x + \frac{b}{2a}\right)^2 + c - \frac{b^2}{4a} \qquad \text{Factor.}$$

$$= a\left(x + \frac{b}{2a}\right)^2 + \frac{4ac - b^2}{4a} \qquad c - \frac{b^2}{4a} = c \cdot \frac{4a}{4a} - \frac{b^2}{4a} = \frac{4ac - b^2}{4a}$$

Based on these results, we conclude the following:

If $h = -\dfrac{b}{2a}$ and $k = \dfrac{4ac - b^2}{4a}$, then

$$f(x) = ax^2 + bx + c = a(x - h)^2 + k \qquad \textbf{(2)}$$

The graph of $f(x) = a(x - h)^2 + k$ is the parabola $y = ax^2$ shifted horizontally h units (replace x by $x - h$) and vertically k units (add k). As a result, the vertex is at (h, k), and the graph opens up if $a > 0$ and down if $a < 0$. The axis of symmetry is the vertical line $x = h$.

For example, compare equation (2) with equation (1) of Example 1.

$$f(x) = 2(x + 2)^2 - 3$$
$$= 2(x - (-2))^2 - 3$$
$$= a(x - h)^2 + k$$

We conclude that $a = 2$, so the graph opens up. Also, we find that $h = -2$ and $k = -3$, so its vertex is at $(-2, -3)$.

2 Identify the Vertex and Axis of Symmetry of a Quadratic Function

We do not need to complete the square to obtain the vertex. In almost every case, it is easier to obtain the vertex of a quadratic function f by remembering that its x-coordinate is $h = -\dfrac{b}{2a}$. The y-coordinate can then be found by evaluating f at $-\dfrac{b}{2a}$ to find $k = f\left(-\dfrac{b}{2a}\right)$.

We summarize these remarks as follows:

Properties of the Graph of a Quadratic Function

$$f(x) = ax^2 + bx + c, \qquad a \neq 0$$

$$\text{Vertex} = \left(-\frac{b}{2a}, f\left(-\frac{b}{2a}\right)\right) \qquad \text{Axis of symmetry: the line } x = -\frac{b}{2a} \qquad \textbf{(1)}$$

Parabola opens up if $a > 0$; the vertex is a minimum point.
Parabola opens down if $a < 0$; the vertex is a maximum point.

EXAMPLE 2	Locating the Vertex without Graphing

Without graphing, locate the vertex and axis of symmetry of the parabola defined by $f(x) = -3x^2 + 6x + 1$. Does it open up or down?

Solution For this quadratic function, $a = -3$, $b = 6$, and $c = 1$. The x-coordinate of the vertex is

$$h = -\frac{b}{2a} = -\frac{6}{-6} = 1$$

The y-coordinate of the vertex is

$$k = f\left(-\frac{b}{2a}\right) = f(1) = -3 + 6 + 1 = 4$$

The vertex is located at the point $(1, 4)$. The axis of symmetry is the line $x = 1$. Because $a = -3 < 0$, the parabola opens down.

3 Graph a Quadratic Function Using Its Vertex, Axis, and Intercepts

The information we gathered in Example 2, together with the location of the intercepts, usually provides enough information to graph $f(x) = ax^2 + bx + c, a \neq 0$.

The y-intercept is the value of f at $x = 0$; that is, the y-intercept is $f(0) = c$.

The x-intercepts, if there are any, are found by solving the quadratic equation

$$ax^2 + bx + c = 0$$

This equation has two, one, or no real solutions, depending on whether the discriminant $b^2 - 4ac$ is positive, 0, or negative. Figure 21 illustrates these possibilities for parabolas that open up.

Figure 21
$f(x) = ax^2 + bx + c, a > 0$

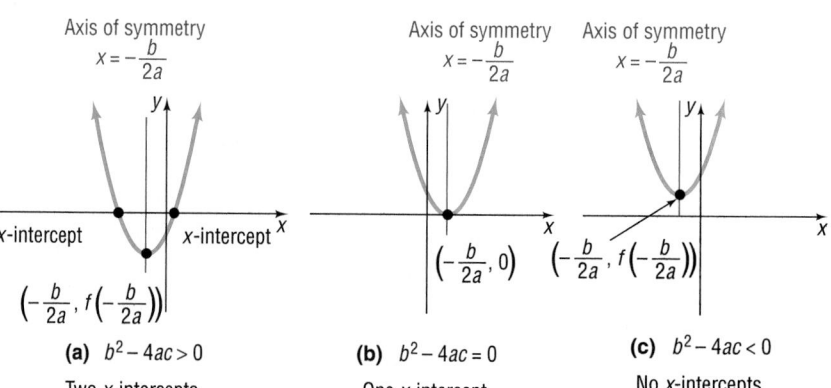

(a) $b^2 - 4ac > 0$
Two x-intercepts

(b) $b^2 - 4ac = 0$
One x-intercept

(c) $b^2 - 4ac < 0$
No x-intercepts

EXAMPLE 3	Graphing a Quadratic Function Using Its Vertex, Axis, and Intercepts

(a) Use the information from Example 2 and the locations of the intercepts to graph $f(x) = -3x^2 + 6x + 1$.

(b) Determine the domain and the range of f.

(c) Determine where f is increasing and where it is decreasing.

Solution (a) In Example 2, we found the vertex to be at $(1, 4)$ and the axis of symmetry to be $x = 1$. The y-intercept is found by letting $x = 0$. The y-intercept is $f(0) = 1$. The x-intercepts are found by solving the equation $f(x) = 0$. This results in the equation

$$-3x^2 + 6x + 1 = 0 \qquad a = -3, b = 6, c = 1$$

The discriminant $b^2 - 4ac = (6)^2 - 4(-3)(1) = 36 + 12 = 48 > 0$, so the equation has two real solutions and the graph has two x-intercepts. Using the quadratic formula, we find that

$$x = \frac{-b + \sqrt{b^2 - 4ac}}{2a} = \frac{-6 + \sqrt{48}}{-6} = \frac{-6 + 4\sqrt{3}}{-6} \approx -0.15$$

and

$$x = \frac{-b - \sqrt{b^2 - 4ac}}{2a} = \frac{-6 - \sqrt{48}}{-6} = \frac{-6 - 4\sqrt{3}}{-6} \approx 2.15$$

The x-intercepts are approximately -0.15 and 2.15.

The graph is illustrated in Figure 22. Notice how we used the y-intercept and the axis of symmetry, $x = 1$, to obtain the additional point $(2, 1)$ on the graph.

(b) The domain of f is the set of all real numbers. Based on the graph, the range of f is the interval $(-\infty, 4]$.

(c) The function f is increasing on the interval $(-\infty, 1)$ and decreasing on the interval $(1, \infty)$. ∎

Figure 22

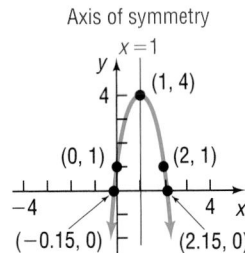

Axis of symmetry
$x = 1$

$(1, 4)$
$(0, 1)$ $(2, 1)$
$(-0.15, 0)$ $(2.15, 0)$

NOW WORK PROBLEM 35.

If the graph of a quadratic function has only one x-intercept or no x-intercepts, it is usually necessary to plot an additional point to obtain the graph.

EXAMPLE 4 **Graphing a Quadratic Function Using Its Vertex, Axis, and Intercepts**

(a) Graph $f(x) = x^2 - 6x + 9$ by determining whether the graph opens up or down and by finding its vertex, axis of symmetry, y-intercept, and x-intercepts, if any.

(b) Determine the domain and the range of f.

(c) Determine where f is increasing and where it is decreasing.

Solution (a) For $f(x) = x^2 - 6x + 9$, we have $a = 1$, $b = -6$, and $c = 9$. Since $a = 1 > 0$, the parabola opens up. The x-coordinate of the vertex is

$$h = -\frac{b}{2a} = -\frac{-6}{2(1)} = 3$$

The y-coordinate of the vertex is

$$k = f(3) = (3)^2 - 6(3) + 9 = 0$$

So the vertex is at $(3, 0)$. The axis of symmetry is the line $x = 3$. The y-intercept is $f(0) = 9$. Since the vertex $(3, 0)$ lies on the x-axis, the graph touches the x-axis at the x-intercept. By using the axis of symmetry and the y-intercept at $(0, 9)$, we can locate the additional point $(6, 9)$ on the graph. See Figure 23.

(b) The domain of f is the set of all real numbers. Based on the graph, the range of f is the interval $[0, \infty)$.

(c) The function f is decreasing on the interval $(-\infty, 3)$ and increasing on the interval $(3, \infty)$. ∎

Figure 23

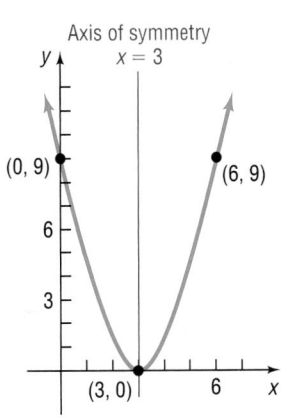

Axis of symmetry
$x = 3$

$(0, 9)$ $(6, 9)$

$(3, 0)$

Graph the function in Example 4 by completing the square and using transformations. Which method do you prefer?

NOW WORK PROBLEM 43.

| **EXAMPLE 5** | **Graphing a Quadratic Function Using Its Vertex, Axis, and Intercepts** |

(a) Graph $f(x) = 2x^2 + x + 1$ by determining whether the graph opens up or down and by finding its vertex, axis of symmetry, y-intercept, and x-intercepts, if any.

(b) Determine the domain and the range of f.

(c) Determine where f is increasing and where it is decreasing.

Solution

(a) For $f(x) = 2x^2 + x + 1$, we have $a = 2$, $b = 1$, and $c = 1$. Since $a = 2 > 0$, the parabola opens up. The x-coordinate of the vertex is

$$h = -\frac{b}{2a} = -\frac{1}{4}$$

NOTE:
In Example 5, since the vertex is above the x-axis and the parabola opens up, we can conclude that the graph of the quadratic function will have no x-intercepts. ◼

The y-coordinate of the vertex is

$$k = f\left(-\frac{1}{4}\right) = 2\left(\frac{1}{16}\right) + \left(-\frac{1}{4}\right) + 1 = \frac{7}{8}$$

So the vertex is at $\left(-\frac{1}{4}, \frac{7}{8}\right)$. The axis of symmetry is the line $x = -\frac{1}{4}$. The y-intercept is $f(0) = 1$. The x-intercept(s), if any, obey the equation $2x^2 + x + 1 = 0$. Since the discriminant $b^2 - 4ac = (1)^2 - 4(2)(1) = -7 < 0$, this equation has no real solutions, and therefore the graph has no x-intercepts. We use the point $(0, 1)$ and the axis of symmetry $x = -\frac{1}{4}$ to locate the additional point $\left(-\frac{1}{2}, 1\right)$ on the graph. See Figure 24.

Figure 24

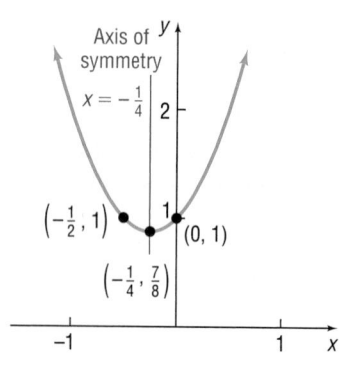

(b) The domain of f is the set of all real numbers. Based on the graph, the range of f is the interval $\left[\frac{7}{8}, \infty\right)$.

(c) The function f is decreasing on the interval $\left(-\infty, -\frac{1}{4}\right)$ and is increasing on the interval $\left(-\frac{1}{4}, \infty\right)$.

✏️ **NOW WORK PROBLEM 47.**

Given the vertex (h, k) and one additional point on the graph of a quadratic function $f(x) = ax^2 + bx + c$, $a \neq 0$, we can use

$$f(x) = a(x - h)^2 + k \qquad\qquad \textbf{(4)}$$

to obtain the quadratic function.

| **EXAMPLE 6** | **Finding the Quadratic Function Given Its Vertex and One Other Point** |

Determine the quadratic function whose vertex is $(1, -5)$ and whose y-intercept is -3. The graph of the parabola is shown in Figure 25.

Solution

The vertex is $(1, -5)$, so $h = 1$ and $k = -5$. Substitute these values into equation (4).

$$f(x) = a(x - h)^2 + k \qquad \text{Equation (4)}$$

$$f(x) = a(x - 1)^2 - 5 \qquad h = 1, k = -5$$

Figure 25

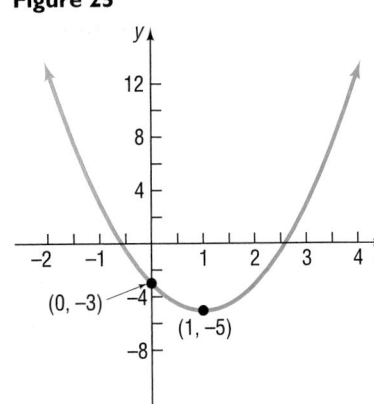

To determine the value of a, we use the fact that $f(0) = -3$ (the y-intercept).

$$f(x) = a(x - 1)^2 - 5$$
$$-3 = a(0 - 1)^2 - 5 \qquad x = 0, y = f(0) = -3$$
$$-3 = a - 5$$
$$a = 2$$

The quadratic function whose graph is shown in Figure 25 is

$$f(x) = a(x - h)^2 + k = 2(x - 1)^2 - 5 = 2x^2 - 4x - 3$$

∎

NOW WORK PROBLEM 53.

Summary

Steps for Graphing a Quadratic Function $f(x) = ax^2 + bx + c, \ a \neq 0$

Option 1

STEP 1: Complete the square in x to write the quadratic function in the form $f(x) = a(x - h)^2 + k$.

STEP 2: Graph the function in stages using transformations.

Option 2

STEP 1: Determine the vertex $\left(-\dfrac{b}{2a}, f\left(-\dfrac{b}{2a} \right) \right)$.

STEP 2: Determine the axis of symmetry, $x = -\dfrac{b}{2a}$.

STEP 3: Determine the y-intercept, $f(0)$.
 (a) If $b^2 - 4ac > 0$, the graph of the quadratic function has two x-intercepts, which are found by solving the equation $ax^2 + bx + c = 0$.
 (b) If $b^2 - 4ac = 0$, the vertex is the x-intercept.
 (c) If $b^2 - 4ac < 0$, there are no x-intercepts.

STEP 5: Determine an additional point by using the y-intercept and the axis of symmetry.

STEP 6: Plot the points and draw the graph.

4 **Find the Maximum or Minimum Value of a Quadratic Function**

The graph of a quadratic function

$$f(x) = ax^2 + bx + c, \qquad a \neq 0$$

is a parabola with vertex at $\left(-\dfrac{b}{2a}, f\left(-\dfrac{b}{2a} \right) \right)$. This vertex is the highest point on the graph if $a < 0$ and the lowest point on the graph if $a > 0$. If the vertex is the highest point $(a < 0)$, then $f\left(-\dfrac{b}{2a} \right)$ is the **maximum value** of f. If the vertex is the lowest point $(a > 0)$, then $f\left(-\dfrac{b}{2a} \right)$ is the **minimum value** of f.

EXAMPLE 7 **Finding the Maximum or Minimum Value of a Quadratic Function**

Determine whether the quadratic function

$$f(x) = x^2 - 4x - 5$$

has a maximum or minimum value. Then find the maximum or minimum value.

Solution We compare $f(x) = x^2 - 4x - 5$ to $f(x) = ax^2 + bx + c$. We conclude that $a = 1, b = -4$, and $c = -5$. Since $a > 0$, the graph of f opens up, so the vertex is a minimum point. The minimum value occurs at

$$x = -\frac{b}{2a} = -\frac{-4}{2(1)} = \frac{4}{2} = 2$$

$a = 1, b = -4$

The minimum value is

$$f\left(-\frac{b}{2a}\right) = f(2) = 2^2 - 4(2) - 5 = 4 - 8 - 5 = -9 \qquad \blacksquare$$

NOW WORK PROBLEM 61.

2.4 Assess Your Understanding

'Are You Prepared?'

Answers are given at the end of these exercises. If you get a wrong answer, read the pages listed in red.

1. List the intercepts of the equation $y = x^2 - 9$. (pp. 10–11)

2. List the intercepts of the equation $y = 2x^2 + 7x - 4$. (pp. 10–11)

3. To complete the square of $x^2 - 5x$, you add the number _____. (pp. A35–A36)

4. To graph $y = (x - 4)^2$, you shift the graph of $y = x^2$ to the _____ a distance of _____ units. (pp. 86–88)

Concepts and Vocabulary

5. The graph of a quadratic function is called a(n) _____.

6. The vertical line passing through the vertex of a parabola is called the _____.

7. The x-coordinate of the vertex of $f(x) = ax^2 + bx + c$, $a \neq 0$, is _____.

8. *True or False*: The graph of $f(x) = 2x^2 + 3x - 4$ opens up.

9. *True or False*: The y-coordinate of the vertex of $f(x) = -x^2 + 4x + 5$ is $f(2)$.

10. *True or False*: If the discriminant $b^2 - 4ac = 0$, the graph of $f(x) = ax^2 + bx + c, a \neq 0$, will touch the x-axis at its vertex.

Skill Building

In Problems 11–18, match each graph to one the following functions.

11. $f(x) = x^2 - 1$

12. $f(x) = -x^2 - 1$

13. $f(x) = x^2 - 2x + 1$

14. $f(x) = x^2 + 2x + 1$

15. $f(x) = x^2 - 2x + 2$

16. $f(x) = x^2 + 2x$

17. $f(x) = x^2 - 2x$

18. $f(x) = x^2 + 2x + 2$

A.

B.

C.

D.

E.

F.

G.

H.

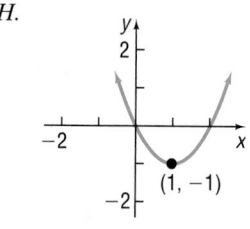

In Problems 19–34, graph the function f by starting with the graph of y = x² and using transformations (shifting, compressing, stretching, and/or reflection).

[Hint: If necessary, write f in the form $f(x) = a(x - h)^2 + k$.]

19. $f(x) = \dfrac{1}{4}x^2$

20. $f(x) = 2x^2$

21. $f(x) = \dfrac{1}{4}x^2 - 2$

22. $f(x) = 2x^2 - 3$

23. $f(x) = \dfrac{1}{4}x^2 + 2$

24. $f(x) = 2x^2 + 4$

25. $f(x) = \dfrac{1}{4}x^2 + 1$

26. $f(x) = -2x^2 - 2$

27. $f(x) = x^2 + 4x + 2$

28. $f(x) = x^2 - 6x - 1$

29. $f(x) = 2x^2 - 4x + 1$

30. $f(x) = 3x^2 + 6x$

31. $f(x) = -x^2 - 2x$

32. $f(x) = -2x^2 + 6x + 2$

33. $f(x) = \dfrac{1}{2}x^2 + x - 1$

34. $f(x) = \dfrac{2}{3}x^2 + \dfrac{4}{3}x - 1$

In Problems 35–52, (a) graph each quadratic function by determining whether its graph opens up or down and by finding its vertex, axis of symmetry, y-intercept, and x-intercepts, if any. (b) Determine the domain and the range of the function. (c) Determine where the function is increasing and where it is decreasing.

35. $f(x) = x^2 + 2x$

36. $f(x) = x^2 - 4x$

37. $f(x) = -x^2 - 6x$

38. $f(x) = -x^2 + 4x$

39. $f(x) = 2x^2 - 8x$

40. $f(x) = 3x^2 + 18x$

41. $f(x) = x^2 + 2x - 8$

42. $f(x) = x^2 - 2x - 3$

43. $f(x) = x^2 + 2x + 1$

44. $f(x) = x^2 + 6x + 9$

45. $f(x) = 2x^2 - x + 2$

46. $f(x) = 4x^2 - 2x + 1$

47. $f(x) = -2x^2 + 2x - 3$

48. $f(x) = -3x^2 + 3x - 2$

49. $f(x) = 3x^2 + 6x + 2$

50. $f(x) = 2x^2 + 5x + 3$

51. $f(x) = -4x^2 - 6x + 2$

52. $f(x) = 3x^2 - 8x + 2$

In Problems 53–58, determine the quadratic function whose graph is given.

53.

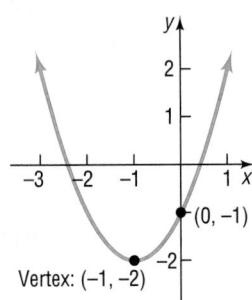

Vertex: (−1, −2)
(0, −1)

54.

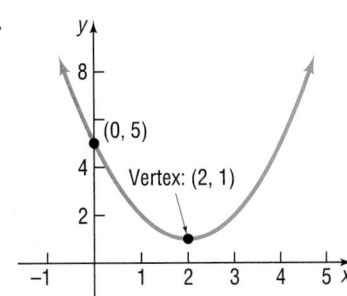

(0, 5)
Vertex: (2, 1)

55.

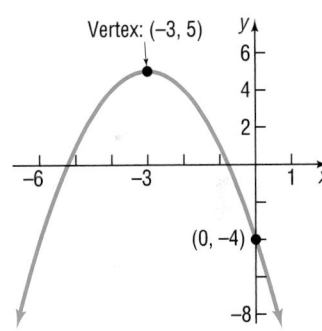

Vertex: (−3, 5)
(0, −4)

56.

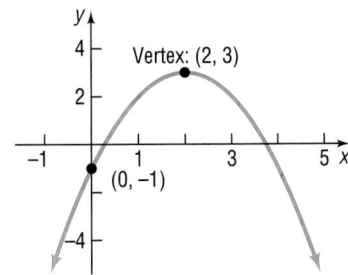

Vertex: (2, 3)
(0, −1)

57.

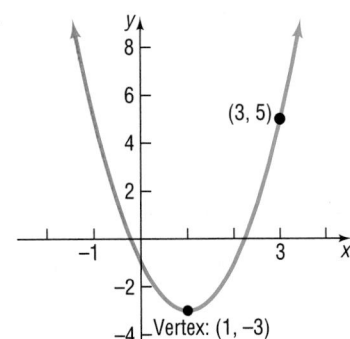

(3, 5)
Vertex: (1, −3)

58.

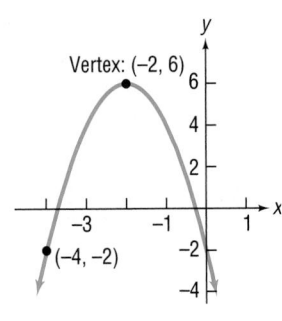

Vertex: (−2, 6)
(−4, −2)

In Problems 59–66, determine, without graphing, whether the given quadratic function has a maximum value or a minimum value and then find the value.

59. $f(x) = 2x^2 + 12x$

60. $f(x) = -2x^2 + 12x$

61. $f(x) = 2x^2 + 12x - 3$

62. $f(x) = 4x^2 - 8x + 3$

63. $f(x) = -x^2 + 10x - 4$

64. $f(x) = -2x^2 + 8x + 3$

65. $f(x) = -3x^2 + 12x + 1$

66. $f(x) = 4x^2 - 4x$

Applications and Extensions

67. The graph of the function $f(x) = ax^2 + bx + c$ has vertex at $(0, 2)$ and passes through the point $(1, 8)$. Find $a, b,$ and c.

68. The graph of the function $f(x) = ax^2 + bx + c$ has vertex at $(1, 4)$ and passes through the point $(-1, -8)$. Find $a, b,$ and c.

In Problems 69–74, for the given functions f and g,

 (a) *Graph f and g on the same Cartesian plane.*
 (b) *Solve $f(x) = g(x)$.*
 (c) *Use the result of part (b) to label the points of intersection of the graphs of f and g.*
 (d) *Shade the region for which $f(x) > g(x)$, that is, the region below f and above g.*

69. $f(x) = 2x - 1$; $g(x) = x^2 - 4$

70. $f(x) = -2x - 1$; $g(x) = x^2 - 9$

71. $f(x) = -x^2 + 4$; $g(x) = -2x + 1$

72. $f(x) = -x^2 + 9$; $g(x) = 2x + 1$

73. $f(x) = -x^2 + 5x$; $g(x) = x^2 + 3x - 4$

74. $f(x) = -x^2 + 7x - 6$; $g(x) = x^2 + x - 6$

Answer Problems 75 and 76 using the following: A quadratic function of the form $f(x) = ax^2 + bx + c$ with $b^2 - 4ac > 0$ may also be written in the form $f(x) = a(x - r_1)(x - r_2)$, where r_1 and r_2 are the x-intercepts of the graph of the quadratic function.

75. (a) Find a quadratic function whose x-intercepts are -3 and 1 with $a = 1; a = 2; a = -2; a = 5$.
 (b) How does the value of a affect the intercepts?
 (c) How does the value of a affect the axis of symmetry?
 (d) How does the value of a affect the vertex?
 (e) Compare the x-coordinate of the vertex with the midpoint of the x-intercepts. What might you conclude?

76. (a) Find a quadratic function whose x-intercepts are -5 and 3 with $a = 1; a = 2; a = -2; a = 5$.
 (b) How does the value of a affect the intercepts?
 (c) How does the value of a affect the axis of symmetry?
 (d) How does the value of a affect the vertex?
 (e) Compare the x-coordinate of the vertex with the midpoint of the x-intercepts. What might you conclude?

77. Suppose that $f(x) = x^2 + 4x - 21$
 (a) What is the vertex of f?
 (b) Find the zeros of f. What are the x-intercepts of the graph of f?
 (c) Solve $f(x) = -21$ for x. What points are on the graph of f?
 (d) Use the information obtained in parts (a)–(c) to graph $f(x) = x^2 + 4x - 21$.

78. Suppose that $f(x) = x^2 + 2x - 8$
 (a) What is the vertex of f?
 (b) Find the zeros of f. What are the x-intercepts of the graph of f?
 (c) Solve $f(x) = -8$ for x. What points are on the graph of f?
 (d) Use the information obtained in parts (a)–(c) to graph $f(x) = x^2 + 2x - 8$.

Discussion and Writing

79. Make up a quadratic function that opens down and has only one x-intercept. Compare yours with others in the class. What are the similarities? What are the differences?

80. On one set of coordinate axes, graph the family of parabolas $f(x) = x^2 + 2x + c$ for $c = -3, c = 0$, and $c = 1$. Describe the characteristics of a member of this family.

81. On one set of coordinate axes, graph the family of parabolas $f(x) = x^2 + bx + 1$ for $b = -4, b = 0$, and $b = 4$. Describe the general characteristics of this family.

82. State the circumstances that cause the graph of a quadratic function $f(x) = ax^2 + bx + c$ to have no x-intercepts.

83. Why does the graph of a quadratic function open up if $a > 0$ and down if $a < 0$?

84. Can a quadratic function have a range of $(-\infty, \infty)$? Justify your answer.

'Are You Prepared?' Answers

1. $(0, -9), (-3, 0), (3, 0)$ **2.** $(-4, 0), \left(\frac{1}{2}, 0\right), (0, -4)$ **3.** $\dfrac{25}{4}$ **4.** right; 4

2.5 Inequalities Involving Quadratic Functions

PREPARING FOR THIS SECTION *Before getting started, review the following:*

- Solving Inequalities (Appendix A, Section A.9, pp. A69–A71)
- Interval Notation (Appendix A, Section A.9, pp. A66–A67)

 Now work the 'Are You Prepared?' problems on page 156.

OBJECTIVE 1 Solve Inequalities Involving a Quadratic Function

1 Solve Inequalities Involving a Quadratic Function

In this section we solve inequalities that involve quadratic functions. We will accomplish this by using their graphs. For example, to solve the inequality

$$ax^2 + bx + c > 0, \qquad a \neq 0$$

we graph the function $f(x) = ax^2 + bx + c$ and, from the graph, determine where it is above the x-axis, that is, where $f(x) > 0$. To solve the inequality $ax^2 + bx + c < 0, a \neq 0$, we graph the function $f(x) = ax^2 + bx + c$ and determine where the graph is below the x-axis. If the inequality is not strict, we include the x-intercepts in the solution.

Let's look at an example.

| EXAMPLE 1 | **Solving an Inequality** |

Solve the inequality $x^2 - 4x - 12 \leq 0$ and graph the solution set.

Figure 26

Solution We graph the function $f(x) = x^2 - 4x - 12$. The intercepts are

y-intercept: $\qquad\qquad\qquad\qquad f(0) = -12$ Evaluate f at 0

x-intercepts (if any):

$$x^2 - 4x - 12 = 0 \qquad \text{Solve } f(x) = 0.$$
$$(x - 6)(x + 2) = 0 \qquad \text{Factor.}$$
$$x - 6 = 0 \quad \text{or} \quad x + 2 = 0 \qquad \text{Apply the Zero- Product Property.}$$
$$x = 6 \quad \text{or} \qquad x = -2$$

The y-intercept is -12; the x-intercepts are -2 and 6.

The vertex is at $x = -\dfrac{b}{2a} = -\dfrac{-4}{2} = 2$. Since $f(2) = -16$, the vertex is $(2, -16)$.

See Figure 26 for the graph.

The graph is below the x-axis for $-2 < x < 6$. Since the original inequality is not strict, we include the x-intercepts. The solution set is $\{x | -2 \leq x \leq 6\}$ or, using interval notation, $[-2, 6]$. See Figure 27 for the graph of the solution set. ∎

Figure 27

NOW WORK PROBLEMS 5 AND 9.

| EXAMPLE 2 | **Solving an Inequality** |

Solve the inequality $2x^2 < x + 10$ and graph the solution set.

Solution *Method 1* We rearrange the inequality so that 0 is on the right side.

$$2x^2 < x + 10$$
$$2x^2 - x - 10 < 0 \qquad \text{Subtract x + 10 from both sides}$$

This inequality is equivalent to the one that we wish to solve.

Next we graph the function $f(x) = 2x^2 - x - 10$. The intercepts are

y-intercept: $\qquad\qquad f(0) = -10$ Evaluate f at 0.

x-intercepts (if any):

$$2x^2 - x - 10 = 0 \qquad \text{Solve } f(x) = 0.$$
$$(2x - 5)(x + 2) = 0 \qquad \text{Factor.}$$
$$2x - 5 = 0 \quad \text{or} \quad x + 2 = 0 \qquad \text{Apply the Zero-Product Property.}$$
$$x = \frac{5}{2} \quad \text{or} \qquad x = -2$$

The y-intercept is -10; the x-intercepts are -2 and $\dfrac{5}{2}$.

Figure 28

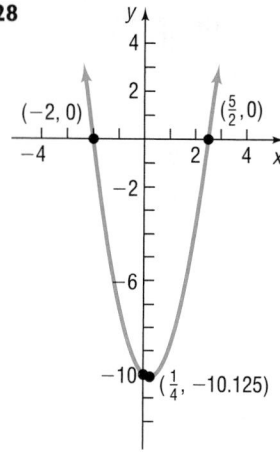

The vertex is at $x = -\dfrac{b}{2a} = -\dfrac{-1}{4} = \dfrac{1}{4}$. Since $f\left(\dfrac{1}{4}\right) = -10.125$, the vertex is $\left(\dfrac{1}{4}, -10.125\right)$. See Figure 28 for the graph.

The graph is below the x-axis between $x = -2$ and $x = \dfrac{5}{2}$. Since the inequality is strict, the solution set is $\left\{x \middle| -2 < x < \dfrac{5}{2}\right\}$ or, using interval notation, $\left(-2, \dfrac{5}{2}\right)$.

Method 2 If $f(x) = 2x^2$ and $g(x) = x + 10$, the inequality that we wish to solve is $f(x) < g(x)$. We graph the functions $f(x) = 2x^2$ and $g(x) = x + 10$. See Figure 29. The graphs intersect where $f(x) = g(x)$. Then

$$2x^2 = x + 10 \qquad\qquad f(x) = g(x)$$
$$2x^2 - x - 10 = 0$$
$$(2x - 5)(x + 2) = 0 \qquad\qquad \text{Factor.}$$
$$2x - 5 = 0 \quad \text{or} \quad x + 2 = 0 \qquad \text{Apply the Zero-Product Property.}$$
$$x = \dfrac{5}{2} \quad \text{or} \qquad x = -2$$

Figure 29

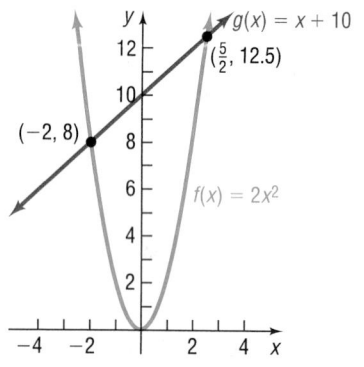

The graphs intersect at the points $(-2, 8)$ and $\left(\dfrac{5}{2}, 12.5\right)$. The inequality that we wish to solve is $f(x) < g(x)$, so we seek where the graph of f is below that of g. This happens between the points of intersection. Since the inequality is strict, the solution set is $\left\{x \middle| -2 < x < \dfrac{5}{2}\right\}$ or, using interval notation, $\left(-2, \dfrac{5}{2}\right)$. See Figure 30 for the graph of the solution set.

Figure 30

$$\begin{array}{ccccc} & & & & \\ -4 & -2 & 0 & 2 & 4 \end{array}$$

NOW WORK PROBLEM 13.

2.5 Assess Your Understanding

'Are You Prepared?'

Answers are given at the end of these exercises. If you get a wrong answer, read the pages listed in red.

1. Solve the inequality $-3x - 2 < 7$ (pp. A69–A70)

2. Write $(-2, 7]$ using inequality notation. (pp. A66–A67)

Skill Building

In Problems 3–6, use the figure to solve each inequality.

3.

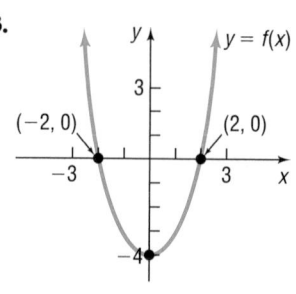

(a) $f(x) > 0$
(b) $f(x) \leq 0$

4.

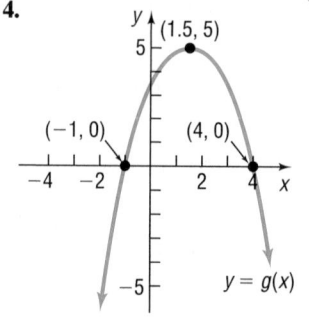

(a) $g(x) < 0$
(b) $g(x) \geq 0$

5.

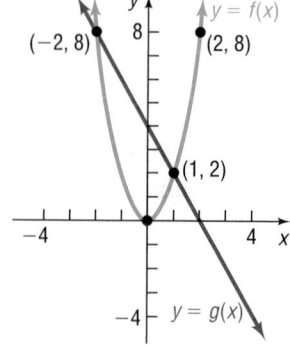

(a) $g(x) \geq f(x)$
(b) $f(x) > g(x)$

6.

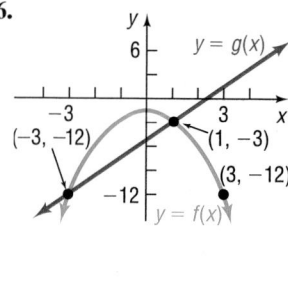

(a) $f(x) < g(x)$
(b) $f(x) \geq g(x)$

In Problems 7–22, solve each inequality.

7. $x^2 - 3x - 10 < 0$ **8.** $x^2 + 3x - 10 > 0$ **9.** $x^2 - 4x > 0$ **10.** $x^2 + 8x > 0$

11. $x^2 - 9 < 0$ **12.** $x^2 - 1 < 0$ **13.** $x^2 + x > 12$ **14.** $x^2 + 7x < -12$

15. $2x^2 < 5x + 3$ **16.** $6x^2 < 6 + 5x$ **17.** $x(x - 7) > 8$ **18.** $x(x + 1) > 20$

19. $4x^2 + 9 < 6x$ **20.** $25x^2 + 16 < 40x$ **21.** $6(x^2 - 1) > 5x$ **22.** $2(2x^2 - 3x) > -9$

23. What is the domain of the function $f(x) = \sqrt{x^2 - 16}$? **24.** What is the domain of the function $f(x) = \sqrt{x - 3x^2}$?

In Problems 25–32, use the functions f and g given below.

(a) Solve $f(x) = 0$. (e) Solve $g(x) \leq 0$.

(b) Solve $g(x) = 0$. (f) Solve $f(x) > g(x)$.

(c) Solve $f(x) = g(x)$. (g) Solve $f(x) \geq 1$.

(d) Solve $f(x) > 0$.

25. $f(x) = x^2 - 1$ **26.** $f(x) = x^2 - 1$ **27.** $f(x) = -x^2 + 1$ **28.** $f(x) = -x^2 + 4$

 $g(x) = 3x + 3$ $g(x) = -3x + 3$ $g(x) = 4x + 1$ $g(x) = -x - 2$

29. $f(x) = x^2 - 4$ **30.** $f(x) = x^2 - 2x + 1$ **31.** $f(x) = x^2 - x - 2$ **32.** $f(x) = -x^2 - x + 1$

 $g(x) = -x^2 + 4$ $g(x) = -x^2 + 1$ $g(x) = x^2 + x - 2$ $g(x) = -x^2 + x + 6$

Applications and Extensions

33. Physics A ball is thrown vertically upward with an initial velocity of 80 feet per second. The distance s (in feet) of the ball from the ground after t seconds is $s(t) = 80t - 16t^2$.

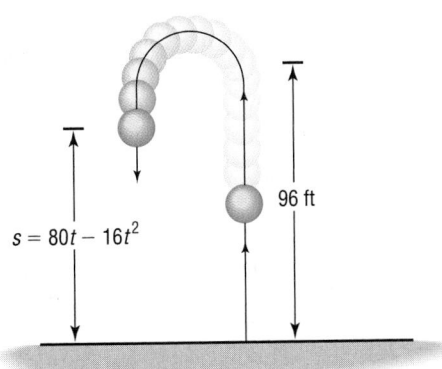

(a) At what time t will the ball strike the ground?

(b) For what time t is the ball more than 96 feet above the ground?

34. Physics A ball is thrown vertically upward with an initial velocity of 96 feet per second. The distance s (in feet) of the ball from the ground after t seconds is $s(t) = 96t - 16t^2$.

(a) At what time t will the ball strike the ground?

(b) For what time t is the ball more than 128 feet above the ground?

35. Revenue Suppose that the manufacturer of a gas clothes dryer has found that, when the unit price is p dollars, the revenue R (in dollars) is

$$R(p) = -4p^2 + 4000p$$

(a) At what price p is revenue zero?

(b) For what range of prices will revenue exceed $800,000?

36. Revenue The John Deere company has found that the revenue from sales of heavy-duty tractors is a function of the unit price p that it charges. If the revenue R is

$$R(p) = -\frac{1}{2}p^2 + 1900p$$

(a) At what prices p is revenue zero?

(b) For what range of prices will revenue exceed $1,200,000?

Discussion and Writing

37. Show that the inequality $(x - 4)^2 \leq 0$ has exactly one solution.

38. Show that the inequality $(x - 2)^2 > 0$ has one real number that is not a solution.

39. Explain why the inequality $x^2 + x + 1 > 0$ has all real numbers as the solution set.

40. Explain why the inequality $x^2 - x + 1 < 0$ has the empty set as solution set.

'Are You Prepared?' Answers

 1. $\{x \mid x > -3\}$ or $(-3, \infty)$ **2.** $-2 < x \leq 7$

2.6 Quadratic Models

PREPARING FOR THIS SECTION *Before getting started, review the following:*

- Problem Solving: Applications of Linear Equations (Appendix A, Section A.10, pp. A74–A79)
- Building Linear Functions from Data (Section 2.2, pp. 125–129)

 Now Work the 'Are You Prepared?' problems on page 162.

OBJECTIVES 1 Solve Applied Problems by Building Quadratic Functions
2 Use a Graphing Utility to Find the Quadratic Function of Best Fit to Data

In this section we will first discuss models that lead to a quadratic function when a verbal description of the problem is given. We end the section by fitting a quadratic function to data, which is another form of modeling.

When a mathematical model leads to a quadratic function, the properties of the graph of the quadratic function can provide important information about the model. In particular, we can use the quadratic function to determine the maximum or minimum value of the function. The fact that the graph of a quadratic function has a maximum or minimum value enables us to answer questions involving **optimization**, that is, finding the maximum or minimum values in models involving quadratic functions.

1 Solve Applied Problems by Building Quadratic Functions

EXAMPLE 1 **Maximizing Revenue**

The marketing department at Texas Instruments has found that, when certain calculators are sold at a price of p dollars per unit, the revenue R (in dollars) as a function of the price p is

$$R(p) = -150p^2 + 21,000p$$

What unit price should be established to maximize revenue? If this price is charged, what is the maximum revenue?

Solution The revenue R is

$$R(p) = -150p^2 + 21,000p \qquad R(p) = ap^2 + bp + c$$

The function R is a quadratic function with $a = -150$, $b = 21,000$, and $c = 0$. Because $a < 0$, the vertex is the highest point on the parabola. The revenue R is therefore a maximum when the price p is

$$p = -\frac{b}{2a} = -\frac{21,000}{2(-150)} = -\frac{21,000}{-300} = \$70.00$$

$$a = -150, b = 21,000$$

The maximum revenue R is

$$R(70) = -150(70)^2 + 21,000(70) = \$735,000$$

See Figure 31 for the graph of R.

Figure 31

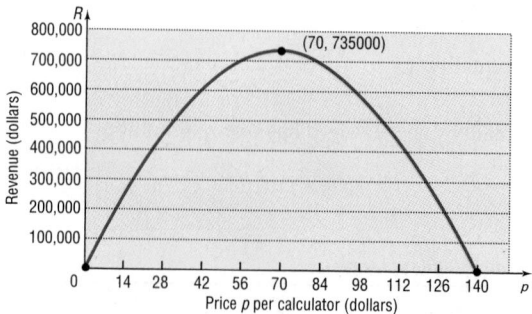

Price p per calculator (dollars)

NOW WORK PROBLEM 3.

| EXAMPLE 2 | **Maximizing the Area Enclosed by a Fence** |

A farmer has 2000 yards of fence to enclose a rectangular field. What are the dimensions of the rectangle that encloses the most area?

Solution

Figure 32

Figure 32 illustrates the situation. The available fence represents the perimeter of the rectangle. If x is the length and w is the width, then

$$2x + 2w = 2000 \qquad \textbf{(1)}$$

The area A of the rectangle is

$$A = xw$$

To express A in terms of a single variable, we solve equation (1) for w and substitute the result in $A = xw$. Then A involves only the variable x. [You could also solve equation (1) for x and express A in terms of w alone. Try it!]

$$2x + 2w = 2000 \qquad \text{Equation (1)}$$
$$2w = 2000 - 2x \qquad \text{Solve for } w.$$
$$w = \frac{2000 - 2x}{2} = 1000 - x$$

Then the area A is

$$A = xw = x(1000 - x) = -x^2 + 1000x$$

Now, A is a quadratic function of x.

$$A(x) = -x^2 + 1000x \qquad a = -1, b = 1000, c = 0$$

Figure 33 shows a graph of $A(x) = -x^2 + 1000x$. Since $a < 0$, the vertex is a maximum point on the graph of A. The maximum value occurs at

Figure 33

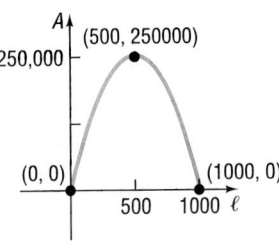

$$x = -\frac{b}{2a} = -\frac{1000}{2(-1)} = 500$$

The maximum value of A is

$$A\left(-\frac{b}{2a}\right) = A(500) = -500^2 + 1000(500) = -250{,}000 + 500{,}000 = 250{,}000$$

The largest rectangle that can be enclosed by 2000 yards of fence has an area of 250,000 square yards. Its dimensions are 500 yards by 500 yards.

NOW WORK PROBLEM 19.

| EXAMPLE 3 | **Analyzing the Motion of a Projectile** |

A projectile is fired from a cliff 500 feet above the water at an inclination of 45° to the horizontal, with a muzzle velocity of 400 feet per second. In physics, it is established that the height h of the projectile above the water is given by

$$h(x) = \frac{-32x^2}{(400)^2} + x + 500$$

where x is the horizontal distance of the projectile from the base of the cliff. See Figure 34.

Figure 34

(a) Find the maximum height of the projectile.

(b) How far from the base of the cliff will the projectile strike the water?

Solution

(a) The height of the projectile is given by a quadratic function.

$$h(x) = \frac{-32x^2}{(400)^2} + x + 500 = \frac{-1}{5000}x^2 + x + 500$$

We are looking for the maximum value of h. Since $a < 0$, the maximum value is obtained at the vertex. We compute

$$x = -\frac{b}{2a} = -\frac{1}{2\left(-\frac{1}{5000}\right)} = \frac{5000}{2} = 2500$$

The maximum height of the projectile is

$$h(2500) = \frac{-1}{5000}(2500)^2 + 2500 + 500 = -1250 + 2500 + 500 = 1750 \text{ ft}$$

(b) The projectile will strike the water when the height is zero. To find the distance x traveled, we need to solve the equation

$$h(x) = \frac{-1}{5000}x^2 + x + 500 = 0$$

─ **Seeing the Concept** ─

Graph

$$h(x) = \frac{-1}{5000}x^2 + x + 500,$$

$$0 \leq x \leq 5500$$

Use MAXIMUM to find the maximum height of the projectile, and use ROOT or ZERO to find the distance from the base of the cliff to where it strikes the water. Compare your results with those obtained in Example 3.

We find the discriminant first.

$$b^2 - 4ac = 1^2 - 4\left(\frac{-1}{5000}\right)(500) = 1.4$$

Then

$$x = \frac{-b \pm \sqrt{b^2 - 4ac}}{2a} = \frac{-1 \pm \sqrt{1.4}}{2\left(-\frac{1}{5000}\right)} \approx \begin{cases} -458 \\ 5458 \end{cases}$$

We discard the negative solution and find that the projectile will strike the water at a distance of about 5458 feet from the base of the cliff. ∎

NOW WORK PROBLEM 7.

| **EXAMPLE 4** | **The Golden Gate Bridge** |

The Golden Gate Bridge, a suspension bridge, spans the entrance to San Francisco Bay. Its 746-foot-tall towers are 4200 feet apart. The bridge is suspended from two huge cables more than 3 feet in diameter; the 90-foot-wide roadway is 220 feet above the water. The cables are parabolic in shape* and touch the road surface at the center of the bridge. Find the height of the cable at a distance of 1000 feet from the center.

Solution

See Figure 35. We begin by choosing the placement of the coordinate axes so that the x-axis coincides with the road surface and the origin coincides with the center of

Figure 35

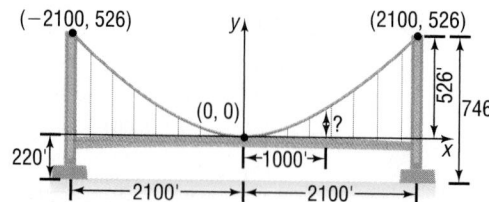

*A cable suspended from two towers is in the shape of a **catenary**, but when a horizontal roadway is suspended from the cable, the cable takes the shape of a parabola.

the bridge. As a result, the twin towers will be vertical (height $746 - 220 = 526$ feet above the road) and located 2100 feet from the center. Also, the cable, which has the shape of a parabola, will extend from the towers, open up, and have its vertex at $(0, 0)$. The choice of placement of the axes enables us to identify the equation of the parabola as $y = ax^2$, $a > 0$. We can also see that the points $(-2100, 526)$ and $(2100, 526)$ are on the graph.

Based on these facts, we can find the value of a in $y = ax^2$.

$$y = ax^2$$
$$526 = a(2100)^2 \qquad x = 2100, y = 526$$
$$a = \frac{526}{(2100)^2}$$

The equation of the parabola is therefore

$$y = \frac{526}{(2100)^2}x^2$$

The height of the cable when $x = 1000$ is

$$y = \frac{526}{(2100)^2}(1000)^2 \approx 119.3 \text{ feet}$$

The cable is 119.3 feet high at a distance of 1000 feet from the center of the bridge. ∎

NOW WORK PROBLEM 23.

2 Use a Graphing Utility to Find the Quadratic Function of Best Fit to Data

In Section 2.2, we found the line of best fit for data that appeared to be linearly related. It was noted that data may also follow a nonlinear relation. Figures 36(a) and (b) show scatter diagrams of data that follow a quadratic relation.

Figure 36

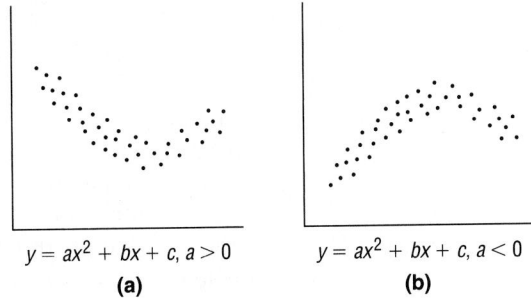

$y = ax^2 + bx + c, a > 0$
(a)

$y = ax^2 + bx + c, a < 0$
(b)

EXAMPLE 5

Fitting a Quadratic Function to Data

A farmer collected the data given in Table 8 on page 162, which shows crop yields Y for various amounts x of fertilizer used.

(a) Draw a scatter diagram of the data. Comment on the type of relation that may exist between the two variables.

(b) Use a graphing utility to find the quadratic function of best fit to these data.

(c) Use the function found in part (b) to determine the optimal amount of fertilizer to apply.

(d) Use the function found in part (b) to predict crop yield when the optimal amount of fertilizer is applied.

(e) Draw the quadratic function of best fit on the scatter diagram.

Table 8

Plot	Fertilizer, x (Pounds/100 ft²)	Yield (Bushels)
1	0	4
2	0	6
3	5	10
4	5	7
5	10	12
6	10	10
7	15	15
8	15	17
9	20	18
10	20	21
11	25	20
12	25	21
13	30	21
14	30	22
15	35	21
16	35	20
17	40	19
18	40	19

Solution

(a) Figure 37 shows the scatter diagram, from which it appears that the data follow a quadratic relation, with $a < 0$.

(b) Upon executing the QUADratic REGression program, we obtain the results shown in Figure 38. The output that the utility provides shows us the equation $y = ax^2 + bx + c$. The quadratic function of best fit is

$$Y(x) = -0.0171x^2 + 1.0765x + 3.8939$$

where x represents the amount of fertilizer used and Y represents crop yield.

(c) Based on the quadratic function of best fit, the optimal amount of fertilizer to apply is

$$x = -\frac{b}{2a} = -\frac{1.0765}{2(-0.0171)} \approx 31.5 \text{ pounds of fertilizer per 100 square feet}$$

(d) We evaluate the function $Y(x)$ for $x = 31.5$.

$$Y(31.5) = -0.0171(31.5)^2 + 1.0765(31.5) + 3.8939 \approx 20.8 \text{ bushels}$$

If we apply 31.5 pounds of fertilizer per 100 square feet, the crop yield will be 20.8 bushels according to the quadratic function of best fit.

(e) Figure 39 shows the graph of the quadratic function found in part (b) drawn on the scatter diagram.

Figure 37

Figure 38

Figure 39

Look again at Figure 38. Notice that the output given by the graphing calculator does not include r, the correlation coefficient. Recall that the correlation coefficient is a measure of the strength of a linear relation that exists between two variables. The graphing calculator does not provide an indication of how well the function fits the data in terms of r since a quadratic function cannot be expressed as a linear function.

NOW WORK PROBLEM 37.

2.6 Assess Your Understanding

'Are You Prepared?'

Answers are given at the end of these exercises. If you get a wrong answer, read the pages listed in red.

1. Translate the following sentence to a mathematical equation: The total revenue R from selling x hot dogs is $3 times the number of hot dogs sold. (pp. A74–A75)

2. Use a graphing utility to find the line of best fit for the following data: (pp. 128–129)

x	3	5	5	6	7	8
y	10	13	12	15	16	19

Applications and Extensions

3. **Maximizing Revenue** Suppose that the manufacturer of a gas clothes dryer has found that, when the unit price is p dollars, the revenue R (in dollars) is

$$R(p) = -4p^2 + 4000p$$

What unit price should be established for the dryer to maximize revenue? What is the maximum revenue?

4. **Maximizing Revenue** The John Deere company has found that the revenue from sales of heavy-duty tractors is a function of the unit price p that it charges. If the revenue R is

$$R(p) = -\frac{1}{2}p^2 + 1900p$$

what unit price p should be charged to maximize revenue? What is the maximum revenue?

5. **Minimizing Marginal Cost** The **marginal cost** of a product can be thought of as the cost of producing one additional unit of output. For example, if the marginal cost of producing the 50th product is $6.20, it cost $6.20 to increase production from 49 to 50 units of output. The marginal cost C (in dollars) to produce x portable 12-inch TVs is given by $C(x) = x^2 - 80x + 2000$. How many TVs should be produced to minimize the marginal cost? What is the minimum marginal cost?

6. **Minimizing Marginal Cost** (See Problem 5.) The marginal cost C (in dollars) of manufacturing x cell phones is given by $C(x) = 5x^2 - 200x + 4000$. How many cell phones should be manufactured to minimize the marginal cost? What is the minimum marginal cost?

7. **Analyzing the Motion of a Projectile** A projectile is fired from a cliff 200 feet above the water at an inclination of 45° to the horizontal, with a muzzle velocity of 50 feet per second. The height h of the projectile above the water is given by

$$h(x) = \frac{-32x^2}{(50)^2} + x + 200$$

where x is the horizontal distance of the projectile from the base of the cliff.
 (a) How far from the base of the cliff is the height of the projectile a maximum?
 (b) Find the maximum height of the projectile.
 (c) How far from the base of the cliff will the projectile strike the water?
 (d) Using a graphing utility, graph the function h, $0 \le x \le 200$.
 (e) Use a graphing utility to verify the solutions found in parts (b) and (c).
 (f) When the height of the projectile is 100 feet above the water, how far is it from the cliff?

8. **Analyzing the Motion of a Projectile** A projectile is fired at an inclination of 45° to the horizontal, with a muzzle velocity of 100 feet per second. The height h of the projectile is given by

$$h(x) = \frac{-32x^2}{(100)^2} + x$$

where x is the horizontal distance of the projectile from the firing point.
 (a) How far from the firing point is the height of the projectile a maximum?
 (b) Find the maximum height of the projectile.
 (c) How far from the firing point will the projectile strike the ground?

 (d) Using a graphing utility, graph the function h, $0 \le x \le 350$.
 (e) Use a graphing utility to verify the results obtained in parts (b) and (c).
 (f) When the height of the projectile is 50 feet above the ground, how far has it traveled horizontally?

9. **Hunting** The function $H(x) = -3.24x^2 + 242.1x - 738.4$ models the number of individuals who engage in hunting activities whose age is x (based on data obtained from the National Sporting Goods Association).
 (a) What is the age at which there are the most hunters? Approximately how many hunters are this age?
 (b) Are the number of hunters increasing or decreasing for individuals who are between 40 and 45 years of age?

10. **Advanced Degrees** The function $P(x) = -0.008x^2 + 0.815x - 9.983$ models the percentage of the U.S. population whose age is x that have earned an advanced degree (more than a bachelor's degree) in March 2003 (based on data obtained from the U.S. Census Bureau).
 (a) What is the age for which the highest percentage of Americans have earned an advanced degree? What is the highest percentage?
 (b) Is the percentage of Americans who have earned an advanced degree increasing or decreasing for individuals between the ages of 40 and 50?

11. **Male Murder Victims** The function $M(x) = 1.00x^2 - 136.74x + 4764.89$ models the number of male murder victims who are x years of age ($20 \le x < 90$) (based on data obtained from the Federal Bureau of Investigation).
 (a) Use the model to approximate the number of male murder victims who are $x = 23$ years of age.
 (b) At what age is the number of male murder victims 1456?
 (c) Describe what happens to the number of male murder victims as age increases from 20 to 65.

12. **Health Care Expenditures** The function $H(x) = 0.004x^2 - 0.197x + 5.406$ models the percentage of total income that an individual who is x years of age spends on health care (based on data obtained from the Bureau of Labor Statistics.)
 (a) Use the model to approximate the percentage of total income an individual who is $x = 45$ years of age spends on health care.
 (b) At what age is the percentage of income spent on health care 10%?
 (c) Using a graphing utility, graph $H = H(x)$.
 (d) Based on the graph drawn in part (c), describe what happens to the percentage of income spent on health care as individuals age.

13. **Business** The monthly revenue R achieved by selling x wristwatches is figured to be $R(x) = 75x - 0.2x^2$. The monthly cost C of selling x wristwatches is $C(x) = 32x + 1750$.
 (a) How many wristwatches must the firm sell to maximize revenue? What is the maximum revenue?
 (b) Profit is given as $P(x) = R(x) - C(x)$. What is the profit function?
 (c) How many wristwatches must the firm sell to maximize profit? What is the maximum profit?
 (d) Provide a reasonable explanation as to why the answers found in parts (a) and (c) differ. Explain why a quadratic function is a reasonable model for revenue.

14. Business The daily revenue R achieved by selling x boxes of candy is figured to be $R(x) = 9.5x - 0.04x^2$. The daily cost C of selling x boxes of candy is $C(x) = 1.25x + 250$.
(a) How many boxes of candy must the firm sell to maximize revenue? What is the maximum revenue?
(b) Profit is given as $P(x) = R(x) - C(x)$. What is the profit function?
(c) How many boxes of candy must the firm sell to maximize profit? What is the maximum profit?
(d) Provide a reasonable explanation as to why the answers found in parts (a) and (c) differ. Explain why a quadratic function is a reasonable model for revenue.

15. Demand Equation The price p (in dollars) and the quantity x sold of a certain product obey the demand equation

$$p = -\frac{1}{6}x + 100, \qquad 0 \le x \le 600$$

(a) Express the revenue R as a function of x. (Remember, $R = xp$.)
(b) What is the revenue if 200 units are sold?
(c) What quantity x maximizes revenue? What is the maximum revenue?
(d) What price should the company charge to maximize revenue?

16. Demand Equation The price p (in dollars) and the quantity x sold of a certain product obey the demand equation

$$p = -\frac{1}{3}x + 100, \qquad 0 \le x \le 300$$

(a) Express the revenue R as a function of x.
(b) What is the revenue if 100 units are sold?
(c) What quantity x maximizes revenue? What is the maximum revenue?
(d) What price should the company charge to maximize revenue?

17. Demand Equation The price p (in dollars) and the quantity x sold of a certain product obey the demand equation
$$x = -5p + 100, \qquad 0 \le p \le 20$$
(a) Express the revenue R as a function of x.
(b) What is the revenue if 15 units are sold?
(c) What quantity x maximizes revenue? What is the maximum revenue?
(d) What price should the company charge to maximize revenue?

18. Demand Equation The price p (in dollars) and the quantity x sold of a certain product obey the demand equation
$$x = -20p + 500, \qquad 0 \le p \le 25$$
(a) Express the revenue R as a function of x.
(b) What is the revenue if 20 units are sold?
(c) What quantity x maximizes revenue? What is the maximum revenue?
(d) What price should the company charge to maximize revenue?

19. Enclosing a Rectangular Field David has available 400 yards of fencing and wishes to enclose a rectangular area.
(a) Express the area A of the rectangle as a function of the width w of the rectangle.
(b) For what value of w is the area largest?
(c) What is the maximum area?

20. Enclosing a Rectangular Field Beth has 3000 feet of fencing available to enclose a rectangular field.
(a) Express the area A of the rectangle as a function of x where x is the length of the rectangle.
(b) For what value of x is the area largest?
(c) What is the maximum area?

21. Enclosing the Most Area with a Fence A farmer with 4000 meters of fencing wants to enclose a rectangular plot that borders on a river. If the farmer does not fence the side along the river, what is the largest area that can be enclosed? (See the figure.)

$4000 - 2x$

22. Enclosing the Most Area with a Fence A farmer with 2000 meters of fencing wants to enclose a rectangular plot that borders on a straight highway. If the farmer does not fence the side along the highway, what is the largest area that can be enclosed?

23. Suspension Bridge A suspension bridge with weight uniformly distributed along its length has twin towers that extend 75 meters above the road surface and are 400 meters apart. The cables are parabolic in shape and are suspended from the tops of the towers. The cables touch the road surface at the center of the bridge. Find the height of the cables at a point 100 meters from the center. (Assume that the road is level.)

24. Architecture A parabolic arch has a span of 120 feet and a maximum height of 25 feet. Choose suitable rectangular coordinate axes and find the equation of the parabola. Then calculate the height of the arch at points 10 feet, 20 feet, and 40 feet from the center.

25. Constructing Rain Gutters A rain gutter is to be made of aluminum sheets that are 12 inches wide by turning up the edges 90°. See the illustration. What depth will provide maximum cross-sectional area and hence allow the most water to flow?

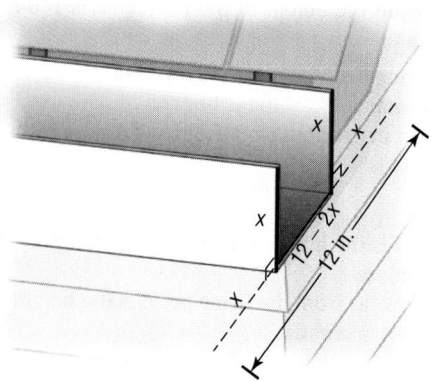

26. Norman Windows A **Norman window** has the shape of a rectangle surmounted by a semicircle of diameter equal to the width of the rectangle (see the figure). If the perimeter of the window is 20 feet, what dimensions will admit the most light (maximize the area)?

[**Hint:** Circumference of a circle $= 2\pi r$; area of a circle $= \pi r^2$, where r is the radius of the circle.]

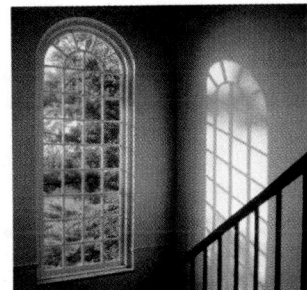

27. Constructing a Stadium A track and field playing area is in the shape of a rectangle with semicircles at each end (see the figure). The inside perimeter of the track is to be 1500 meters. What should the dimensions of the rectangle be so that the area of the rectangle is a maximum?

28. Architecture A special window has the shape of a rectangle surmounted by an equilateral triangle (see the figure). If the perimeter of the window is 16 feet, what dimensions will admit the most light?

[**Hint:** Area of an equilateral triangle $= \left(\dfrac{\sqrt{3}}{4}\right)x^2$, where x is the length of a side of the triangle.]

29. Chemical Reactions A self-catalytic chemical reaction results in the formation of a compound that causes the formation ratio to increase. If the reaction rate V is given by

$$V(x) = kx(a - x), \qquad 0 \le x \le a$$

where k is a positive constant, a is the initial amount of the compound, and x is the variable amount of the compound, for what value of x is the reaction rate a maximum?

30. Calculus: Simpson's Rule The figure shows the graph of $y = ax^2 + bx + c$. Suppose that the points $(-h, y_0)$, $(0, y_1)$, and (h, y_2) are on the graph. It can be shown that the area enclosed by the parabola, the x-axis, and the lines $x = -h$ and $x = h$ is

$$\text{Area} = \frac{h}{3}(2ah^2 + 6c)$$

Show that this area may also be given by

$$\text{Area} = \frac{h}{3}(y_0 + 4y_1 + y_2)$$

31. Use the result obtained in Problem 30 to find the area enclosed by $f(x) = -5x^2 + 8$, the x-axis, and the lines $x = -1$ and $x = 1$.

32. Use the result obtained in Problem 30 to find the area enclosed by $f(x) = 2x^2 + 8$, the x-axis, and the lines $x = -2$ and $x = 2$.

33. Use the result obtained in Problem 30 to find the area enclosed by $f(x) = x^2 + 3x + 5$, the x-axis, and the lines $x = -4$ and $x = 4$.

34. Use the result obtained in Problem 30 to find the area enclosed by $f(x) = -x^2 + x + 4$, the x-axis, and the lines $x = -1$ and $x = 1$.

35. A rectangle has one vertex on the line $y = 10 - x$, $x > 0$, another at the origin, one on the positive x-axis, and one on the positive y-axis. Find the largest area A that can be enclosed by the rectangle.

36. Let $f(x) = ax^2 + bx + c$, where a, b, and c are odd integers. If x is an integer, show that $f(x)$ must be an odd integer.

[**Hint:** x is either an even integer or an odd integer.]

37. Life Cycle Hypothesis An individual's income varies with his or her age. The following table shows the median income I of individuals of different age groups within the United States for 2001. For each age group, let the class midpoint represent the independent variable, x. For the class "65 years and older," we will assume that the class midpoint is 69.5.

Age	Class Midpoint, x	Median Income, I
15–24 years	19.5	$9,301
25–34 years	29.5	$30,510
35–44 years	39.5	$38,340
45–54 years	49.5	$41,104
55–64 years	59.5	$35,637
65 years and older	69.5	$19,688

SOURCE: U.S. Census Bureau

(a) Draw a scatter diagram of the data. Comment on the type of relation that may exist between the two variables.

(b) The quadratic function of best fit to these data is

$$I(x) = -42.7x^2 + 3998x - 51{,}873$$

Use this function to determine the age at which an individual can expect to earn the most income.

(c) Use the function to predict the peak income earned.

(d) Use a graphing utility to verify that the function given in part (b) is the quadratic function of best fit.

(e) With a graphing utility, draw a scatter diagram of the data and then graph the quadratic function of best fit on the scatter diagram.

38. Life Cycle Hypothesis An individual's income varies with his or her age. The following table shows the median income I of individuals of different age groups within the United States for 2002. For each age group, the class midpoint represents the independent variable, x. For the age group "65 years and older," we will assume that the class midpoint is 69.5.

Age	Class Midpoint, x	Median Income, I
15–24 years	19.5	$9,642
25–34 years	29.5	$30,677
35–44 years	39.5	$37,892
45–54 years	49.5	$40,969
55–64 years	59.5	$36,277
65 years and older	69.5	$19,436

Source: U.S. Census Bureau

(a) Draw a scatter diagram of the data. Comment on the type of relation that may exist between the two variables.
(b) The quadratic function of best fit to these data is

$$I(x) = -42.3x^2 + 3963x - 51070$$

Use this function to determine the age at which an individual can expect to earn the most income.
(c) Use the function to predict the peak income earned.
(d) Use a graphing utility to verify that the function given in part (b) is the quadratic function of best fit.
(e) With a graphing utility, draw a scatter diagram of the data and then graph the quadratic function of best fit on the scatter diagram.

39. Height of a Ball A shot-putter throws a ball at an inclination of 45° to the horizontal. The following data represent the height h of the ball at the instant that it has traveled x feet horizontally.

Distance, x	Height, h
20	25
40	40
60	55
80	65
100	71
120	77
140	77
160	75
180	71
200	64

(a) Draw a scatter diagram of the data. Comment on the type of relation that may exist between the two variables.
(b) The quadratic function of best fit to these data is

$$h(x) = -0.0037x^2 + 1.03x + 5.7$$

Use this function to determine how far the ball will travel before it reaches its maximum height.
(c) Use the function to find the maximum height of the ball.
(d) Use a graphing utility to verify that the function given in part (b) is the quadratic function of best fit.
(e) With a graphing utility, draw a scatter diagram of the data and then graph the quadratic function of best fit on the scatter diagram.

40. Miles per Gallon An engineer collects data showing the speed s of a Ford Taurus and its average miles per gallon, M. See the table.

Speed, s	Miles per Gallon, M
30	18
35	20
40	23
40	25
45	25
50	28
55	30
60	29
65	26
65	25
70	25

(a) Draw a scatter diagram of the data. Comment on the type of relation that may exist between the two variables.
(b) The quadratic function of best fit to these data is

$$M(s) = -0.018s^2 + 1.93s - 25.34$$

Use this function to determine the speed that maximizes miles per gallon.
(c) Use the function to predict miles per gallon for a speed of 63 miles per hour.
(d) Use a graphing utility to verify that the function given in part (b) is the quadratic function of best fit.
(e) With a graphing utility, draw a scatter diagram of the data and then graph the quadratic function of best fit on the scatter diagram.

Discussion and Writing

41. Refer to Example 1 on page 158. Notice that if the price charged for the calculators is $0 or $140 the revenue is $0. It is easy to explain why revenue would be $0 if the price charged is $0, but how can revenue be $0 if the price charged is $140?

'Are You Prepared?' Answers

1. $R = 3x$ **2.** $y = 1.7826x + 4.0652$

2.7 Complex Zeros of a Quadratic Function

PREPARING FOR THIS SECTION *Before getting started, review the following:*
- Real numbers (Appendix A, Section A.1, pp. A2–A4)
- Complex numbers (Appendix A, Section A.11, pp. A82–A87)

 Now work the 'Are You Prepared?' problems on page 170.

OBJECTIVE 1 Find the Complex Zeros of a Quadratic Function

1 Find the Complex Zeros of a Quadratic Function

In the complex number system, we can find zeros of quadratic functions when the discriminant, $b^2 - 4ac$, is negative. It is important to remember that finding the complex zeros of a function does not mean that the zeros are not real. Remember, a complex number is a number of the form $a + bi$, where a and b are real numbers. So, if $b = 0$, the complex number is a real number.

| **EXAMPLE 1** | **Finding Complex Zeros of a Quadratic Function** |

Find the complex zeros for each of the following quadratic functions. Graph each function and label the intercepts.
(a) $f(x) = x^2 - 4$ (b) $g(x) = x^2 + 9$

Solution (a) To find the complex zeros of $f(x) = x^2 - 4$, we solve the equation $f(x) = 0$.

$$f(x) = 0$$
$$x^2 - 4 = 0 \qquad f(x) = x^2 - 4$$
$$x^2 = 4 \qquad \text{Add 4 to both sides.}$$
$$x = \pm\sqrt{4} \qquad \text{Use the Square Root Method.}$$
$$x = \pm 2$$

Figure 40

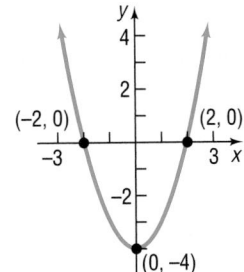

The complex zeros of f are -2 and 2. More specifically, the real zeros of f are -2 and 2. Because the zeros are real, we can also say the x-intercepts of the graph of f are -2 and 2. Figure 40 shows the graph of $f(x) = x^2 - 4$ with the intercepts labeled.

(b) To find the complex zeros of $g(x) = x^2 + 9$, we solve the equation $g(x) = 0$.

$$g(x) = 0$$
$$x^2 + 9 = 0 \qquad g(x) = x^2 + 9$$
$$x^2 = -9 \qquad \text{Subtract 9 from both sides.}$$
$$x = \pm\sqrt{-9} \qquad \text{Use the Square Root Method.}$$
$$x = \pm 3i$$

Figure 41

The complex zeros of g are $-3i$ and $3i$. Because neither of the zeros are real, there are no x-intercepts for the graph of g. Figure 41 shows the graph of $g(x) = x^2 + 9$.

NOW WORK PROBLEM 9.

In Appendix A, Section A.11, we defined the square root of a negative number. We can now restate the quadratic formula without restriction.

Theorem

> In the complex number system, the solutions of the quadratic equation $ax^2 + bx + c = 0$, where $a, b,$ and c are real numbers and $a \neq 0$, are given by the formula
>
> $$x = \frac{-b \pm \sqrt{b^2 - 4ac}}{2a}$$
>
> **(1)**

EXAMPLE 2 **Finding Complex Zeros of a Quadratic Function**

Find the complex zeros for each of the following quadratic functions. Graph each function and label the intercepts.

(a) $f(x) = x^2 + 6x + 2$ (b) $g(x) = x^2 - 4x + 8$

Solution (a) To find the complex zeros of $f(x) = x^2 + 6x + 2$, we solve the equation $f(x) = 0$.

$$f(x) = 0$$
$$x^2 + 6x + 2 = 0 \qquad f(x) = x^2 + 6x + 2$$

Figure 42

Here $a = 1, b = 6, c = 2$, and $b^2 - 4ac = 6^2 - 4(1)(2) = 28$. Using equation (1), we find that

$$x = \frac{-6 \pm \sqrt{28}}{2} = \frac{-6 \pm 2\sqrt{7}}{2} = -3 \pm \sqrt{7}$$

The complex zeros of f are $-3 - \sqrt{7} \approx -5.65$ and $-3 + \sqrt{7} \approx -0.35$. More specifically, the real zeros of f are $-3 - \sqrt{7} \approx -5.65$ and $-3 + \sqrt{7} \approx -0.35$. Because the zeros are real, we can also say the x-intercepts of the graph of f are $-3 - \sqrt{7} \approx -5.65$ and $-3 + \sqrt{7} \approx -0.35$. Figure 42 shows the graph of $f(x) = x^2 + 6x + 2$ with the x-intercepts labeled.

(b) To find the complex zeros of $g(x) = x^2 - 4x + 8$, we solve the equation $g(x) = 0$.

$$g(x) = 0$$
$$x^2 - 4x + 8 = 0 \qquad g(x) = x^2 - 4x + 8$$

Here $a = 1, b = -4, c = 8$, and $b^2 - 4ac = (-4)^2 - 4(1)(8) = -16$. Using equation (1), we find that

$$x = \frac{-(-4) \pm \sqrt{-16}}{2(1)} = \frac{4 \pm 4i}{2} = 2 \pm 2i$$

✔ **CHECK:**

$2 + 2i$: $g(2 + 2i) = (2 + 2i)^2 - 4(2 + 2i) + 8 = 4 + 8i + 4i^2 - 8 - 8i + 8$
$$= 4 - 4 = 0$$

$2 - 2i$: $g(2 - 2i) = (2 - 2i)^2 - 4(2 - 2i) + 8 = 4 - 8i + 4i^2 - 8 + 8i + 8$
$$= 4 - 4 = 0$$

Figure 43

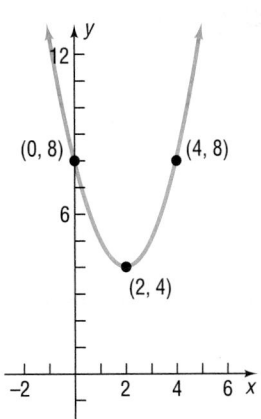

The complex zeros of f are $2 - 2i$ and $2 + 2i$. Because neither of the zeros are real, the graph of g has no x-intercepts. Figure 43 shows the graph of $g(x) = x^2 - 4x + 8$.

NOW WORK PROBLEM 15.

The discriminant $b^2 - 4ac$ of a quadratic equation still serves as a way to determine the character of the solutions.

Character of the Solutions of a Quadratic Equation

In the complex number system, consider a quadratic equation $ax^2 + bx + c = 0$ with real coefficients.

1. If $b^2 - 4ac > 0$, the equation has two unequal real solutions.
2. If $b^2 - 4ac = 0$, the equation has a repeated real solution, a double root.
3. If $b^2 - 4ac < 0$, the equation has two complex solutions that are not real. The complex solutions are conjugates of each other.

The third conclusion in the display is a consequence of the fact that if $b^2 - 4ac = -N < 0$, then, by the quadratic formula, the solutions are

$$x = \frac{-b + \sqrt{b^2 - 4ac}}{2a} = \frac{-b + \sqrt{-N}}{2a} = \frac{-b + \sqrt{N}i}{2a} = \frac{-b}{2a} + \frac{\sqrt{N}}{2a}i$$

and

$$x = \frac{-b - \sqrt{b^2 - 4ac}}{2a} = \frac{-b - \sqrt{-N}}{2a} = \frac{-b - \sqrt{N}i}{2a} = \frac{-b}{2a} - \frac{\sqrt{N}}{2a}i$$

which are conjugates of each other.

Therefore, when $b^2 - 4ac < 0$, the zeros of the quadratic function $f(x) = ax^2 + bx + c$ will be conjugates of each other.

EXAMPLE 3 **Determining the Character of the Solutions of a Quadratic Equation**

Without solving, determine the character of the solutions of each equation.
(a) $3x^2 + 4x + 5 = 0$
(b) $2x^2 + 4x + 1 = 0$
(c) $9x^2 - 6x + 1 = 0$

Solution (a) Here $a = 3, b = 4$, and $c = 5$, so $b^2 - 4ac = 16 - 4(3)(5) = -44$. The solutions are two complex numbers that are not real and are conjugates of each other.
(b) Here $a = 2, b = 4$, and $c = 1$, so $b^2 - 4ac = 16 - 8 = 8$. The solutions are two unequal real numbers.
(c) Here $a = 9, b = -6$, and $c = 1$, so $b^2 - 4ac = 36 - 4(9)(1) = 0$. The solution is a repeated real number, that is, a double root.

NOW WORK PROBLEM 25.

2.7 Assess Your Understanding

'Are You Prepared?'

Answers are given at the end of these exercises. If you get a wrong answer, read the pages listed in red.

1. Name the integers and the rational numbers in the set $\left(-3, 0, \sqrt{2}, \dfrac{6}{5}, \pi\right)$. (p. A2)

2. *True or False*: Rational numbers and irrational numbers are in the set of real numbers. (p. A2)

3. $(2 + i)(3 - 4i) = $ _____ (pp. A83–A84)

4. *True or False*: The conjugate of $2 + 5i$ is $2 - 5i$. (pp. A82–A87)

5. *True or False*: All real numbers are complex numbers. (pp. A82–A84)

6. $\sqrt{-81} = $ _____ . (pp. A82–A84)

Concepts and Vocabulary

7. *True or False*: If $2 - 3i$ is a zero of a quadratic function with real coefficients, then $-2 + 3i$ is also a zero.

8. *True or False*: Consider the quadratic function $f(x) = ax^2 + bx + c$. If $b^2 - 4ac > 0$, the graph of f will have two unequal x-intercepts.

Skill Building

In Problems 9–24, find the complex zeros of each quadratic function. Graph each function and label the intercepts.

9. $f(x) = x^2 + 4$
10. $f(x) = x^2 - 9$
11. $f(x) = x^2 - 16$
12. $f(x) = x^2 + 25$
13. $f(x) = x^2 - 6x + 13$
14. $f(x) = x^2 + 4x + 8$
15. $f(x) = x^2 - 6x + 10$
16. $f(x) = x^2 - 2x + 5$
17. $f(x) = x^2 - 4x + 1$
18. $f(x) = x^2 + 6x + 1$
19. $f(x) = 2x^2 + 2x + 1$
20. $f(x) = 3x^2 + 6x + 4$
21. $f(x) = x^2 + x + 1$
22. $f(x) = x^2 - x + 1$
23. $f(x) = -2x^2 + 8x + 1$
24. $f(x) = -3x^2 + 6x + 1$

In Problems 25–30, without solving, determine the character of the solutions of each equation in the complex number system.

25. $3x^2 - 3x + 4 = 0$
26. $2x^2 - 4x + 1 = 0$
27. $2x^2 + 3x - 4 = 0$
28. $x^2 + 2x + 6 = 0$
29. $9x^2 - 12x + 4 = 0$
30. $4x^2 + 12x + 9 = 0$

Applications and Extensions

31. $2 + 3i$ is a zero of a quadratic function with real coefficients. Find the other zero.

32. $4 - i$ is a zero of a quadratic function with real coefficients. Find the other zero.

'Are You Prepared?' Answers

1. Integers: $\{-3, 0\}$; rational numbers: $\left\{-3, 0, \dfrac{6}{5}\right\}$
2. True
3. $10 - 5i$
4. True
5. True
6. 9i

2.8 Equations and Inequalities Involving the Absolute Value Function

PREPARING FOR THIS SECTION *Before getting started, review the following:*

- Solving Equations (Appendix A, Section A.8, pp. A60–A64)
- Interval Notation; Solving Inequalities (Appendix A, Section A.9, pp. A66–A71)

- Absolute Value Function (Section 1.4, pp. 76–77)
- Graphing Techniques: Transformations (Section 1.5, pp. 85–94)

Now work the 'Are You Prepared?' problems on page 173.

OBJECTIVES 1 Solve Absolute Value Equations

2 Solve Absolute Value Inequalities

1 Solve Absolute Value Equations

Recall that, on the real number line, the absolute value of a equals the distance from the origin to the point whose coordinate is a. For example, there are two points whose distance from the origin is 5 units, -5 and 5. Thus the equation $|x| = 5$ will have the solution set $\{-5, 5\}$. This leads to the following result:

> ### Equations Involving Absolute Value
>
> If a is a positive real number and if u is any algebraic expression, then
>
> $$|u| = a \quad \text{is equivalent to} \quad u = a \quad \text{or} \quad u = -a \qquad \textbf{(1)}$$

| **EXAMPLE 1** | **Solving an Equation Involving Absolute Value** |

Solve the equation: $|x + 2| = 5$

Solution

Figure 44

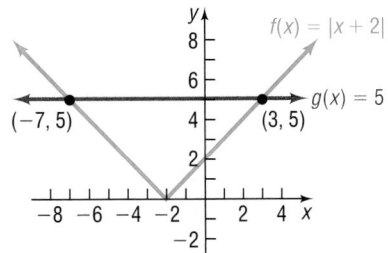

The equation follows the form of equation (1), where $u = x + 2$. There are two possibilities.

$$x + 2 = 5 \quad \text{or} \quad x + 2 = -5$$
$$x = 3 \qquad\qquad x = -7$$

The solution set is $\{-7, 3\}$. ∎

We can visualize the solution by graphing $f(x) = |x + 2|$ and $g(x) = 5$. Figure 44 shows the graphs of $f(x) = |x + 2|$ and $g(x) = 5$. The x-coordinates of the points of intersection are -7 and 3, the solutions of the equation $f(x) = g(x)$.

Figure 45

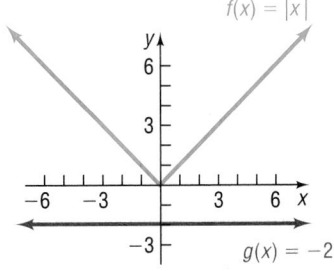

NOW WORK PROBLEMS 13(a) AND 19.

In equation (1), we required that a be a positive number. If $a = 0$, equation (1) becomes $|u| = 0$, which is equivalent to $u = 0$. If a is less than zero, the equation has no real solution. To see why, consider the equation $|x| = -2$. Figure 45 shows the graph of $f(x) = |x|$ and $g(x) = -2$. Notice that the graphs do not intersect, which implies that the equation $|x| = -2$ has no solution. The solution set is the empty set, \varnothing or $\{\ \ \}$.

2 Solve Absolute Value Inequalities

Let's look at an inequality involving absolute value. Once again we will use the graph of the absolute value function to help us visualize the results of the algebra.

| **EXAMPLE 2** | **Solving an Inequality Involving Absolute Value** |

Solve the inequality: $|x| < 4$

Solution

Figure 46

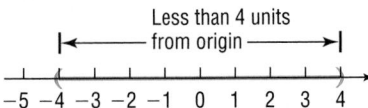

We are looking for all points whose coordinate x is a distance less than 4 units from the origin. See Figure 46 for an illustration.

Because any x between -4 and 4 satisfies the condition $|x| < 4$, the solution set consists of all numbers x for which $-4 < x < 4$, that is, all x in the interval $(-4, 4)$. ∎

Figure 47

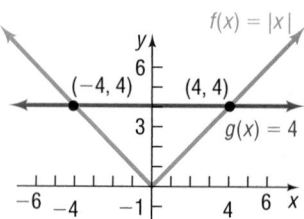

To visualize these results, we graph $f(x) = |x|$ and $g(x) = 4$. See Figure 47. Because we are solving $f(x) < g(x)$, we look for all x-coordinates such that the graph of $f(x)$ is below the graph of $g(x)$. From the graph, we can see that the graph of $f(x) = |x|$ is below the graph of $g(x) = 4$ for all x between -4 and 4. So the solution set consists of all x for which $-4 < x < 4$, or all x in the interval $(-4, 4)$.

The results of Example 2 lead us to the following result.

Inequalities Involving Absolute Value

If a is any positive number and if u is any algebraic expression, then

$	u	< a$ is equivalent to $-a < u < a$		**(2)**
$	u	\leq a$ is equivalent to $-a \leq u \leq a$		**(3)**

In other words, $|u| < a$ is equivalent to $-a < u$ and $u < a$.

EXAMPLE 3 **Solving an Inequality Involving Absolute Value**

Solve the inequality $|2x + 4| \leq 3$, and graph the solution set.

Solution

$$|2x + 4| \leq 3$$ This follows the form of statement (3); the expression $u = 2x + 4$ is inside the absolute value bars.

$$-3 \leq \quad 2x + 4 \quad \leq 3$$ Apply statement (3).

$$-3 - 4 \leq \quad 2x + 4 - 4 \leq 3 - 4$$ Subtract 4 from each part.

$$-7 \leq \quad 2x \quad \leq -1$$ Simplify.

$$\frac{-7}{2} \leq \quad \frac{2x}{2} \quad \leq \frac{-1}{2}$$ Divide each part by 2

$$-\frac{7}{2} \leq \quad x \quad \leq -\frac{1}{2}$$ Simplify.

Figure 48

The solution set is $\left\{ x \left| -\dfrac{7}{2} \leq x \leq -\dfrac{1}{2} \right. \right\}$, that is, all x in the interval $\left[-\dfrac{7}{2}, -\dfrac{1}{2} \right]$. See Figure 48 for a graph of the solution set. ∎

NOW WORK PROBLEMS 13(b) AND 47.

EXAMPLE 4 **Solving an Inequality Involving Absolute Value**

Solve the inequality: $|x| > 3$

Solution

Figure 49

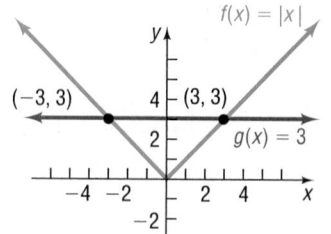

We are looking for all points whose coordinate x is a distance greater than 3 units from the origin. Figure 49 illustrates the situation.

We conclude that any x less than -3 or greater than 3 satisfies the condition $|x| > 3$. Consequently, the solution set consists of all numbers x for which $x < -3$ or $x > 3$, that is, all x in the intervals $(-\infty, -3)$ or $(3, \infty)$. ∎

Figure 50

To visualize these results, we graph $f(x) = |x|$ and $g(x) = 3$. See Figure 50. Because we are solving $f(x) > g(x)$, we look for all x-coordinates such that the graph of $f(x)$ is above the graph of $g(x)$. From the graph, we can see that the graph of $f(x) = |x|$ is above the graph of $g(x) = 3$ for all x between less than -3 or all x greater than 3. So the solution set consists of all x for which $x < -3$ or $x > 3$, that is, all x in the intervals $(-\infty, -3)$ or $(3, \infty)$.

The results of Example 4 lead us to the following result.

WARNING A common error to be avoided is to attempt to write the solution $x < 1$ or $x > 4$ as $1 > x > 4$, which is incorrect, since there are no numbers x for which $1 > x$ and $x > 4$. Another common error is to "mix" the symbols and write $1 < x > 4$, which makes no sense. ■

Inequalities Involving Absolute Value

If a is any positive number and u is any algebraic expression, then

$	u	> a$ is equivalent to $u < -a$ or $u > a$	**(4)**
$	u	\geq a$ is equivalent to $u \leq -a$ or $u \geq a$	**(5)**

EXAMPLE 5 **Solving an Inequality Involving Absolute Value**

Solve the inequality $|2x - 5| > 3$, and graph the solution set.

Solution $|2x - 5| > 3$ This follows the form of statement (4); the expression
 $u = 2x - 5$ is inside the absolute value bars.

$$2x - 5 < -3 \quad \text{or} \quad 2x - 5 > 3 \qquad \text{Apply statement (4).}$$

$$2x - 5 + 5 < -3 + 5 \quad \text{or} \quad 2x - 5 + 5 > 3 + 5 \qquad \text{Add 5 to each part.}$$

$$2x < 2 \quad \text{or} \quad 2x > 8 \qquad \text{Simplify.}$$

$$\frac{2x}{2} < \frac{2}{2} \quad \text{or} \quad \frac{2x}{2} > \frac{8}{2} \qquad \text{Divide each part by 2.}$$

$$x < 1 \quad \text{or} \quad x > 4 \qquad \text{Simplify.}$$

Figure 51 The solution set is $\{x \mid x < 1 \text{ or } x > 4\}$, that is, all x in the intervals $(-\infty, 1)$ or $(4, \infty)$.

-2 -1 0 1 2 3 4 5 6 7

See Figure 51 for a graph of the solution set. ■

NOW WORK PROBLEMS 13(c) AND 51.

2.8 Assess Your Understanding

'Are You Prepared?'

Answers are given at the end of these exercises. If you get a wrong answer, read the pages listed in red.

1. Graph the inequality $x \geq -2$. (p. A5)
2. *True or False*: The absolute value of any real number is positive. (pp. A5–A6)
3. Solve $4x - 3 = 9$. (pp. A61–A62)
4. Solve $3x - 2 > 7$. (pp. A69–A70)
5. Solve $-1 < 2x + 5 < 13$. (p. A71)
6. Use transformations to graph the function $f(x) = |x - 3|$. (pp. 86–88)

Concepts and Vocabulary

7. $|u| = a$ is equivalent to $u = $ _____ or $u = $ _____.
8. $|u| < a$ is equivalent to _____.
9. $|u| < a$ will have no solution if a _____ 0.
10. *True or False*: $|x| = -4$ has no real solution.
11. *True or False*: $|x| > -2$ has no real solution.
12. *True or False*: $|u| > a$ is equivalent to $-a < u < a$.

Skill Building

In Problems 13–16, use the graphs of the functions given to solve each problem.

13. $f(x) = |x + 3|$; $g(x) = 6$
 (a) $f(x) = g(x)$
 (b) $f(x) \leq g(x)$
 (c) $f(x) > g(x)$

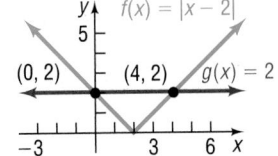

14. $f(x) = |x - 2|$; $g(x) = 2$
 (a) $f(x) = g(x)$
 (b) $f(x) \leq g(x)$
 (c) $f(x) > g(x)$

15. $f(x) = |2x - 1|$; $g(x) = 5$
(a) $f(x) = g(x)$
(b) $f(x) \geq g(x)$
(c) $f(x) < g(x)$

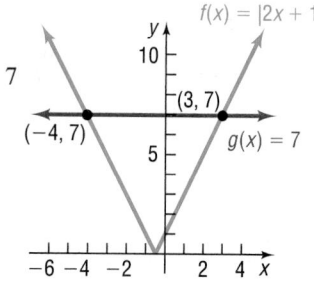

16. $f(x) = |2x + 1|$; $g(x) = 7$
(a) $f(x) = g(x)$
(b) $f(x) \geq g(x)$
(c) $f(x) < g(x)$

In Problems 17–38, solve each equation.

17. $|x| = 6$

18. $|x| = 12$

19. $|2x + 3| = 5$

20. $|3x - 1| = 2$

21. $|1 - 4t| + 8 = 13$

22. $|1 - 2z| + 6 = 9$

23. $|-2x| = 8$

24. $|-x| = 1$

25. $4 - |2x| = 3$

26. $5 - \left|\frac{1}{2}x\right| = 3$

27. $\frac{2}{3}|x| = 9$

28. $\frac{3}{4}|x| = 9$

29. $\left|\frac{x}{3} + \frac{2}{5}\right| = 2$

30. $\left|\frac{x}{2} - \frac{1}{3}\right| = 1$

31. $|u - 2| = -\frac{1}{2}$

32. $|2 - v| = -1$

33. $|x^2 - 9| = 0$

34. $|x^2 - 16| = 0$

35. $|x^2 - 2x| = 3$

36. $|x^2 + x| = 12$

37. $|x^2 + x - 1| = 1$

38. $|x^2 + 3x - 2| = 2$

In Problems 39–58, solve each absolute value inequality. Express your answer using set-builder or interval notation. Graph the solution set.

39. $|x| < 6$

40. $|x| < 9$

41. $|x| > 4$

42. $|x| > 1$

43. $|2x| < 8$

44. $|3x| < 15$

45. $|3x| > 12$

46. $|2x| > 6$

47. $|x - 2| + 2 < 3$

48. $|x + 4| + 3 < 5$

49. $|3t - 2| \leq 4$

50. $|2u + 5| \leq 7$

51. $|x - 3| \geq 2$

52. $|x + 4| \geq 2$

53. $|1 - 4x| - 7 < -2$

54. $|1 - 2x| - 4 < -1$

55. $|1 - 2x| > |-3|$

56. $|2 - 3x| > |-1|$

57. $|2x + 1| < -1$

58. $|3x - 4| \geq 0$

Applications and Extensions

59. Express the fact that x differs from 10 by less than 2 as an inequality involving absolute value. Solve for x.

60. Express the fact that x differs from -6 by less than 3 as an inequality involving absolute value. Solve for x.

61. Express the fact that twice x differs from -1 by more than 5 as an inequality involving absolute value. Solve for x.

62. Express the fact that twice x differs from 3 by more than 1 as an inequality involving absolute value. Solve for x.

63. **Tolerance** A certain rod in an internal combustion engine is supposed to be 5.7 inches long. The tolerance on the rod is 0.0005 inch. If x represents the length of the rod, the acceptable lengths can be expressed as $|x - 5.7| \leq 0.0005$. Determine the acceptable lengths of the rod.

 [SOURCE: WiseCo Piston]

64. **Tolerance** A certain rod in an internal combustion engine is supposed to be 6.125 inches long. The tolerance on the rod is 0.0005 inch. If x represents the length of the rod, the acceptable lengths can be expressed as $|x - 6.125| \leq 0.0005$. Determine the acceptable lengths of the rod.

65. **IQ Scores** According to the Stanford–Binet IQ test, a normal IQ score is 100. It can be shown that anyone with an IQ x that satisfies the inequality $\left|\dfrac{x - 100}{15}\right| > 1.96$ has an unusual IQ score. Determine the IQ scores that would be considered unusual.

66. **Gestation Period** The length of human pregnancy is about 266 days. It can be shown that a mother whose gestation period x satisfies the inequality $\left|\dfrac{x - 266}{16}\right| > 1.96$ has an unusual length of pregnancy. Determine the length of pregnancy that would be considered unusual.

Discussion and Writing

67. The equation $|5x + 1| + 7 = 5$ has no solution. Explain why.

68. The inequality $|x| > -1$ has all real numbers as the solution set. Explain why.

69. The inequality $|2x - 1| \leq 0$ has $\left\{\dfrac{1}{2}\right\}$ as the solution set. Explain why.

'Are You Prepared?' Answers

1.

2. False 3. $\{3\}$ 4. $x > 3$ 5. $-3 < x < 4$ 6.

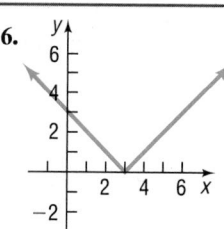

Chapter Review

Things to Know

Linear function (p. 114)

$$f(x) = mx + b$$

Average rate of change $= m$

Graph is a line with slope m and y-intercept b.

Quadratic function (pp. 134, 147–150)

$$f(x) = ax^2 + bx + c, \ a \neq 0$$

Graph is a parabola that opens up if $a > 0$ and opens down if $a < 0$.

Vertex: $\left(-\dfrac{b}{2a}, \ f\left(-\dfrac{b}{2a} \right) \right)$

Axis of symmetry: $x = -\dfrac{b}{2a}$

y-intercept: $f(0)$

x-intercept(s): If any, found by finding the real solutions of the equation $ax^2 + bx + c = 0$

Quadratic equation and quadratic formula

The real solutions of the equation $ax^2 + bx + c = 0, a \neq 0$, are given by $x = \dfrac{-b \pm \sqrt{b^2 - 4ac}}{2a}$, provided $b^2 - 4ac \geq 0$. If $b^2 - 4ac < 0$, there are no real solutions. (p. 138)

In the complex number system, the solutions of the equation $ax^2 + bx + c = 0, a \neq 0$, are given by $x = \dfrac{-b \pm \sqrt{b^2 - 4ac}}{2a}$. (p. 168)

Discriminant (pp. 138 and 169)

If $b^2 - 4ac > 0$, there are two distinct real solutions.

If $b^2 - 4ac = 0$, there is one repeated real solution.

If $b^2 - 4ac < 0$, there are no real solutions, but there are two distinct complex solutions that are not real; the complex solutions are conjugates of each other.

Absolute value

If $|u| = a, a > 0$, then $u = -a$ or $u = a$. (p. 171)

If $|u| \leq a, a > 0$, then $-a \leq u \leq a$. (p. 172)

If $|u| \geq a, a > 0$, then $u \leq -a$ or $u \geq a$. (p. 173)

Objectives

Section		You should be able to . . .	Review Exercises
2.1	1	Graph linear functions (p. 114)	1(a)–6(a), 1(b)–6(b), 7, 8
	2	Use average rate of change to identify linear functions (p. 114)	9, 10
	3	Determine whether a linear function is increasing, decreasing, or constant (p. 117)	1(c)–6(c)
	4	Find the zero of a linear function (p. 118)	7, 8
	5	Work with applications of linear functions (p. 119)	79, 80
2.2	1	Draw and interpret scatter diagrams (p. 125)	93(a), 94(a)
	2	Distinguish between linear and nonlinear relations (p. 126)	93(b) 94(a)
	3	Use a graphing utility to find the line of best fit (p. 128)	93(c)
	4	Construct a linear model using direct variation (p. 129)	81, 82
2.3	1	Find the zeros of a quadratic function by factoring (p. 134)	11–14, 17, 18
	2	Find the zeros of a quadratic function using the Square Root Method (p. 135)	15, 16
	3	Find the zeros of a quadratic function by completing the square (p. 136)	11–22
	4	Find the zeros of a quadratic function using the quadratic formula (p. 137)	11–22
	5	Find the point of intersection of two functions (p. 140)	23–26
	6	Solve equations that are quadratic in form (p. 140)	27–32

2.4	1	Graph a quadratic function using transformations (p. 145)	33–38
	2	Identify the vertex and axis of symmetry of a quadratic function (p. 147)	39–48
	3	Graph a quadratic function using its vertex, axis, and intercepts (p. 148)	39–48
	4	Find the maximum or minimum value of a quadratic function (p. 151)	49–54, 87, 88
2.5	1	Solve inequalities involving a quadratic function (p. 155)	55–58
2.6	1	Solve applied problems by building quadratic functions (p. 158)	83–92
	2	Use a graphing utility to find the quadratic function of best fit to data (p. 161)	94
2.7	1	Find the complex zeros of a quadratic function (p. 167)	59–66
2.8	1	Solve absolute value equations (p. 171)	67–70
	2	Solve absolute value inequalities (p. 171)	71–78

Review Exercises

In Problems 1–6:
 (a) *Determine the slope and y-intercept of each linear function.*
 (b) *Graph each function. Label the intercepts.*
 (c) *Determine whether the function is increasing, decreasing, or constant.*

1. $f(x) = 2x - 5$ **2.** $g(x) = -4x + 7$ **3.** $h(x) = \dfrac{4}{5}x - 6$

4. $F(x) = -\dfrac{1}{3}x + 1$ **5.** $G(x) = 4$ **6.** $H(x) = -3$

7. Find the zero and y-intercept of $f(x) = 2x + 14$. Use the zero and y-intercept to graph f.

8. Find the zero and y-intercept of $f(x) = -\dfrac{1}{4}x + 2$. Use the zero and y-intercept to graph f.

In Problems 9 and 10, determine whether the function is linear or nonlinear. If the function is linear, state its slope.

9.

x	y = f(x)
−1	−2
0	3
1	8
2	13
3	18

10.

x	y = g(x)
−1	−3
0	4
1	7
2	6
3	1

In Problems 11–22, find the zeros of each quadratic function. What are the x-intercepts of the graph of the function?

11. $f(x) = x^2 + x - 72$ **12.** $G(x) = x^2 - 3x - 40$ **13.** $P(x) = 6x^2 - 13x - 5$ **14.** $H(x) = 3x^2 - 17x + 20$

15. $g(x) = (x - 3)^2 - 4$ **16.** $F(x) = (x + 5)^2 - 9$ **17.** $h(x) = 9x^2 + 6x + 1$ **18.** $f(x) = 4x^2 - 4x + 1$

19. $G(x) = 2x^2 - 4x - 1$ **20.** $P(x) = 3x^2 + 5x + 1$ **21.** $f(x) = -2x^2 + x + 1$ **22.** $H(x) = -x^2 - 2x + 5$

In Problems 23–26, solve $f(x) = g(x)$. Graph each function and label the intersection points.

23. $f(x) = (x - 3)^2$; $g(x) = 16$ **24.** $f(x) = x^2 + 6x$; $g(x) = 16$

25. $f(x) = x^2 + 4x - 5$; $g(x) = 4x - 1$ **26.** $f(x) = -2x^2 + 3x + 20$; $g(x) = 3x + 12$

In Problems 27–32, find the real zeros of each function. What are the x-intercepts of the graphs of the function?

27. $f(x) = x^4 - 5x^2 + 4$ **28.** $g(x) = 3x^4 + 4x^2 + 1$ **29.** $F(x) = (x - 3)^2 - 2(x - 3) - 48$

30. $G(x) = 2(x + 4)^2 + 3(x + 4)^2 - 14$ **31.** $h(x) = 3x - 13\sqrt{x} - 10$ **32.** $f(x) = \left(\dfrac{1}{x}\right)^2 - 4\left(\dfrac{1}{x}\right) - 12$

In Problems 33–38, graph each function using transformations (shifting, compressing, stretching, and reflection).

33. $f(x) = (x - 2)^2 + 2$ **34.** $f(x) = (x + 1)^2 - 4$ **35.** $f(x) = -(x - 4)^2$

36. $f(x) = (x - 1)^2 - 3$ **37.** $f(x) = 2(x + 1)^2 + 4$ **38.** $f(x) = -3(x + 2)^2 + 1$

In Problems 39–48, (a) graph each quadratic function by determining whether its graph opens up or down and by finding its vertex, axis of symmetry, y-intercept, and x-intercepts, if any. (b) Determine the domain and the range of the function. (c) Determine where the function is increasing and where it is decreasing.

39. $f(x) = (x - 2)^2 + 2$

40. $f(x) = (x + 1)^2 - 4$

41. $f(x) = \frac{1}{4}x^2 - 16$

42. $f(x) = -\frac{1}{2}x^2 + 2$

43. $f(x) = -4x^2 + 4x$

44. $f(x) = 9x^2 - 6x + 3$

45. $f(x) = \frac{9}{2}x^2 + 3x + 1$

46. $f(x) = -x^2 + x + \frac{1}{2}$

47. $f(x) = 3x^2 + 4x - 1$

48. $f(x) = -2x^2 - x + 4$

In Problems 49–54, determine whether the given quadratic function has a maximum value or a minimum value, and then find the value.

49. $f(x) = 3x^2 - 6x + 4$

50. $f(x) = 2x^2 + 8x + 5$

51. $f(x) = -x^2 + 8x - 4$

52. $f(x) = -x^2 - 10x - 3$

53. $f(x) = -3x^2 + 12x + 4$

54. $f(x) = -2x^2 + 4$

In Problems 55–58, solve each quadratic inequality.

55. $x^2 + 6x - 16 < 0$

56. $3x^2 - 2x - 1 \geq 0$

57. $3x^2 \geq 14x + 5$

58. $4x^2 < 13x - 3$

In Problems 59–66, find the complex zeros of each quadratic function. Graph each function and label the intercepts.

59. $f(x) = x^2 + 8$

60. $f(x) = x^2 - 8$

61. $g(x) = x^2 + 2x - 4$

62. $h(x) = x^2 + 4x + 5$

63. $p(x) = -2x^2 + 4x - 3$

64. $F(x) = -3x^2 + 6x - 5$

65. $f(x) = 4x^2 + 4x + 3$

66. $G(x) = 5x^2 + 5x + 2$

In Problems 67–70, solve each absolute value equation.

67. $|2x + 3| = 7$

68. $|3x - 1| = 5$

69. $|2 - 3x| + 2 = 9$

70. $|1 - 2x| + 1 = 4$

In Problems 71–78, solve each absolute value inequality. Express your answer using set-builder notation or interval notation. Graph the solution set.

71. $|3x + 4| < \frac{1}{2}$

72. $|1 - 2x| < \frac{1}{3}$

73. $|2x - 5| \geq 9$

74. $|3x + 1| \geq 10$

75. $2 + |2 - 3x| \leq 4$

76. $\frac{1}{2} + \left|\frac{2x - 1}{3}\right| \leq 1$

77. $1 - |2 - 3x| < -4$

78. $1 - \left|\frac{2x - 1}{3}\right| < -2$

79. Comparing Phone Companies Marissa must decide between one of two companies as her long-distance phone provider. Company A charges a monthly fee of $7.00 plus $0.06 per minute, while Company B does not have a monthly fee, but charges $0.08 per minute.
(a) Find a linear function that relates cost, C, to total minutes on the phone, x, for each company.
(b) Determine the number of minutes x for which the bill from Company A will equal the bill from Company B.
(c) For how many minutes x will the bill from Company B be less than the bill from Company A?

80. Sales Commissions Bill was just offered a sales position for a computer company. His salary would be $15,000 per year plus 1% of his total annual sales.
(a) Find a linear function that relates Bill's annual salary, S, to his total annual sales, x.
(b) In 2005, Bill had total annual sales of $1,000,000. What was Bill's salary?
(c) What would Bill have to sell to earn $100,000?
(d) Determine the sales required of Bill for his salary to exceed $150,000.

81. Mortgage Payments The monthly payment p on a mortgage varies directly with the amount borrowed B. If the monthly payment on a 30-year mortgage is $854.00 when $130,000 is borrowed, find a linear function that relates the monthly payment p to the amount borrowed B for a mortgage with the same terms. Then find the monthly payment when the amount borrowed is $165,000.

82. Revenue Function At the corner Esso station, the revenue R varies directly with the number g of gallons of gasoline sold. If the revenue is $32.13 when the number of gallons sold is 13.5, find a linear function that relates revenue R to the number g of gallons of gasoline. Then find the revenue when the number of gallons of gasoline sold is 11.2.

83. Business: Determining the Cost of a Charter A group of 20 senior citizens can charter a bus for a one-day excursion trip for $15 per person. The charter company agrees to reduce the price of each ticket by 10¢ for each additional passenger in excess of 20 who goes on the trip, up to a maximum of 44 passengers (the capacity of the bus). If the final bill from the charter company was $482.40, how many seniors went on the trip, and how much did each pay?

84. Pleasing Proportions One formula stating the relationship between the length l and width w of a rectangle of "pleasing proportion" is $l^2 = w(l + w)$. How should a 4-foot by 8-foot sheet of plasterboard be cut so that the result is a rectangle of "pleasing proportion" with a width of 4 feet?

85. Geometry The hypotenuse of a right triangle measures 13 centimeters. Find the lengths of the legs if their sum is 17 centimeters.

86. Geometry The diagonal of a rectangle measures 10 inches. If the length is 2 inches more than the width, find the dimensions of the rectangle.

87. **Minimizing Marginal Cost** Callaway Golf Company has determined that the marginal cost C of manufacturing x Big Bertha golf clubs may be expressed by the quadratic function

$$C(x) = 4.9x^2 - 617.4x + 19{,}600$$

(a) How many clubs should be manufactured to minimize the marginal cost?
(b) At this level of production, what is the marginal cost?

88. **Violent Crimes** The function

$$V(t) = -10.0t^2 + 39.2t + 1862.6$$

models the number V (in thousands) of violent crimes committed in the United States t years after 1990, based on data obtained from the Federal Bureau of Investigation. So $t = 0$ represents 1990. $t = 1$ represents 1991, and so on.
(a) Determine the year in which the most violent crimes were committed.
(b) Approximately how many violent crimes were committed during this year?
(c) Using a graphing utility, graph $V = V(t)$. Were the number of violent crimes increasing or decreasing during the years 1994 to 1998?

89. **Landscaping** A landscape engineer has 200 feet of border to enclose a rectangular pond. What dimensions will result in the largest pond?

90. **Geometry** Find the length and width of a rectangle whose perimeter is 20 feet and whose area is 16 square feet.

91. A rectangle has one vertex on the line $y = 10 - x, x > 0$, another at the origin, one on the positive x-axis, and one on the positive y-axis. Find the largest area A that can be enclosed by the rectangle.

92. **Parabolic Arch Bridge** A horizontal bridge is in the shape of a parabolic arch. Given the information shown in the figure, what is the height h of the arch 2 feet from shore?

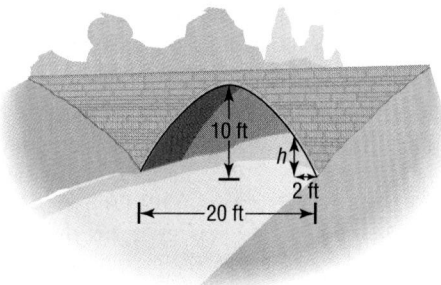

93. **Bone Length** Research performed at NASA, led by Dr. Emily R. Morey-Holton, measured the lengths of the right humerus and right tibia in 11 rats that were sent to space on Spacelab Life Sciences 2. The following data were collected.

(a) Draw a scatter diagram of the data treating length of the right humerus as the independent variable.
(b) Based on the scatter diagram, do you think that there is a linear relation between the length of the right humerus and the length of the right tibia?

Right Humerus (mm), x	Right Tibia (mm), y
24.80	36.05
24.59	35.57
24.59	35.57
24.29	34.58
23.81	34.20
24.87	34.73
25.90	37.38
26.11	37.96
26.63	37.46
26.31	37.75
26.84	38.50

SOURCE: NASA Life Sciences Data Archive

(c) If the two variables appear to be linearly related, use a graphing utility to find the line of best fit relating length of the right humerus and length of the right tibia.
(d) Predict the length of the right tibia on a rat whose right humerus is 26.5 mm.

94. **Advertising** A small manufacturing firm collected the following data on advertising expenditures A (in thousands of dollars) and total revenue R (in thousands of dollars).

Advertising	Total Revenue
20	$6101
22	$6222
25	$6350
25	$6378
27	$6453
28	$6423
29	$6360
31	$6231

(a) Draw a scatter diagram of the data. Comment on the type of relation that may exist between the two variables.
(b) The quadratic function of best fit to these data is

$$R(A) = -7.76A^2 + 411.88A + 942.72$$

Use this function to determine the optimal level of advertising.
(c) Use the function to predict the total revenue when the optimal level of advertising is spent.
(d) Use a graphing utility to verify that the function given in part (b) is the quadratic function of best fit.
(e) Use a graphing utility to draw a scatter diagram of the data and then graph the quadratic function of best fit on the scatter diagram.

Chapter Test

1. For the linear function $f(x) = -4x + 3$,
 (a) Find the zero of f.
 (b) Determine whether f increasing, decreasing, or constant.
 (c) Graph f.

In Problems 2 and 3, find the zeros of each quadratic function.

2. $f(x) = 3x^2 - 2x - 8$
3. $G(x) = -2x^2 + 4x + 1$
4. Given that $f(x) = x^2 + 3x$ and $g(x) = 5x + 3$, solve $f(x) = g(x)$. Graph each function and label the points of intersection.
5. Find the real zeros of $f(x) = (x - 1)^2 + 5(x - 1) + 4$.
6. Graph $f(x) = (x - 3)^2 - 2$ using transformations.
7. For the quadratic function $f(x) = 3x^2 - 12x + 4$,
 (a) Determine whether the graph opens up or down.
 (b) Determine the vertex.
 (c) Determine the axis of symmetry.
 (d) Determine the intercepts.
 (e) Use the information from parts (a)–(d) to graph f.
8. Determine whether $f(x) = -2x^2 + 12x + 3$ has a maximum or minimum value. Then find the maximum or minimum value.
9. Solve $x^2 - 10x + 24 \geq 0$.
10. Find the complex zeros of $f(x) = 2x^2 + 4x + 5$.
11. Solve $|3x + 1| = 8$.

In Problems 12 and 13, solve each absolute value inequality. Express your answer using set-builder notation or interval notation. Graph the solution set.

12. $\left|\dfrac{x + 3}{4}\right| < 2$
13. $|2x + 3| - 4 \geq 3$

14. **RV Rental** The weekly rental cost of a 20-foot recreational vehicle is \$129.50 plus \$0.15 per mile.

 Source: westernrv.com
 (a) Find a linear function that expresses the cost C as a function of miles driven m.
 (b) What is the rental cost if 860 miles are driven?
 (c) How many miles were driven if the rental cost is \$213.80?

15. The variable interest rate on a student loan changes each July 1 based on the bank prime loan rate. For the years 1992–2004, this rate can be approximated by the model

 $$r(x) = -0.115x^2 + 1.183x + 5.623,$$

 where x is the number of years since 1992 and r is the interest rate as a percent.

 Source: U.S. Federal Reserve
 (a) During which year was the interest rate the highest? Determine the highest rate during this time period.
 (b) Use the model to estimate the rate in 2010. Does this value seem reasonable?

Chapter Projects

1. **Analyzing a Stock** The beta, β, of a stock represents the relative risk of a stock compared with a market basket of stocks, such as Standard and Poor's 500 Index of stocks. Beta is computed by finding the slope of the line of best fit between the rate of return of the stock and rate of return of the S&P 500. The rates of return are computed on a weekly basis.

(a) Find the weekly closing price of your favorite stock and the S&P 500 for 20 weeks. One good source is on the Internet at */finance.yahoo.com*.
(b) Compute the rate of return by computing the weekly percentage change in the closing price of your stock and the weekly percentage change in the S&P 500 using the following formula:

$$\text{Weekly \% change} = \frac{P_2 - P_1}{P_1}$$

where P_1 is last week's price and P_2 is this week's price.
(c) Using a graphing utility, find the line of best fit, treating the weekly rate of return of the S&P 500 as the independent variable and the weekly rate of return of your stock as the dependent variable.
(d) What is the beta of your stock?
(e) Compare your result with that of the Value Line Investment Survey found in your library. What might account for any differences?

The following projects are available on the Instructor's Resource Center (IRC):

2. **First and Second Differences**
3. **Cannons**
4. **CBL Experiment**

Cumulative Review

1. Find the distance between the points $P = (-1, 3)$ and $Q = (4, -2)$. Find the midpoint of the line segment P to Q.

2. Which of the following points are on the graph of $y = x^3 - 3x + 1$?
 (a) $(-2, -1)$
 (b) $(2, 3)$
 (c) $(3, 1)$

3. Solve the inequality $5x + 3 \geq 0$ and graph the solution set.

4. Find the equation of the line containing the points $(-1, 4)$ and $(2, -2)$. Express your answer in slope–intercept form and graph the line.

5. Find the equation of the line perpendicular to the line $y = 2x + 1$ and containing the point $(3, 5)$. Express your answer in slope–intercept form and graph the line.

6. Graph the equation $x^2 + y^2 - 4x + 8y - 5 = 0$.

7. Determine whether the following relation represents a function: $\{(-3, 8), (1, 3), (2, 5), (3, 8)\}$.

8. For the function f defined by $f(x) = x^2 - 4x + 1$, evaluate:
 (a) $f(2)$
 (b) $f(x) + f(2)$
 (c) $f(-x)$
 (d) $-f(x)$
 (e) $f(x + 2)$
 (f) $\dfrac{f(x + h) - f(x)}{h}$, $\quad h \neq 0$

9. Find the domain of $h(z) = \dfrac{3z - 1}{6z - 7}$.

10. Determine whether the following graph is the graph of a function.

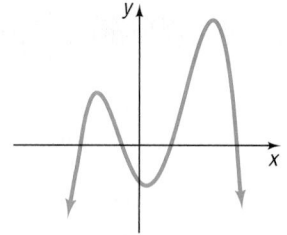

11. Consider the function $f(x) = \dfrac{x}{x + 4}$.
 (a) Is the point $\left(1, \dfrac{1}{4}\right)$ on the graph of f?
 (b) If $x = -2$, what is $f(x)$? What point is on the graph of f?
 (c) If $f(x) = 2$, what is x? What point is on the graph of f?

12. Is the function $f(x) = \dfrac{x^2}{2x + 1}$ even, odd, or neither?

13. Approximate the local maxima and local minima of $f(x) = x^3 - 5x + 1$ on $(-4, 4)$. Determine where the function is increasing and where it is decreasing.

14. If $f(x) = 3x + 5$ and $g(x) = 2x + 1$,
 (a) Solve $f(x) = g(x)$.
 (b) Solve $f(x) > g(x)$.

15. For the graph of the function f,

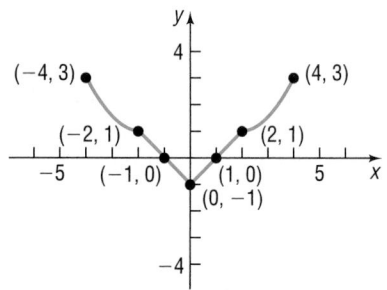

 (a) Find the domain and the range of f.
 (b) Find the intercepts.
 (c) Is the graph of f symmetric with respect to the x-axis, the y-axis, or the origin?
 (d) Find $f(2)$.
 (e) For what value(s) of x is $f(x) = 3$?
 (f) Solve $f(x) < 0$.
 (g) Graph $y = f(x) + 2$
 (h) Graph $y = f(-x)$.
 (i) Graph $y = 2f(x)$.
 (j) Is f even, odd, or neither?
 (k) Find the interval(s) on which f is increasing.

3

Polynomial and Rational Functions

A LOOK BACK In Chapter 1, we began our discussion of functions. We defined domain and range and independent and dependent variables; we found the value of a function and graphed functions. We continued our study of functions by listing the properties that a function might have, like being even or odd, and we created a library of functions, naming key functions and listing their properties, including their graphs.

In Chapter 2, we discussed linear functions and quadratic functions, which belong to the class of *polynomial functions*.

A LOOK AHEAD In this chapter, we look at two general classes of functions: polynomial functions and rational functions, and examine their properties. Polynomial functions are arguably the simplest expressions in algebra. For this reason, they are often used to approximate other, more complicated, functions. Rational functions are ratios of polynomial functions.

We also introduce here methods that can be used to determine the zeros of polynomial functions that are not factored completely. We end the chapter with a discussion of the complex zeros of a polynomial function.

Ragweed Plus Bad Air Will Equal Fall Misery

A hot, dry, smoggy summer is adding up to an insufferable fall for allergy sufferers.

Experts say people with allergies to ragweed, who normally sneeze and wheeze through September, will be dabbing their eyes and nose well into October, even though pollen counts are actually no worse than usual.

The difference this year is that a summer of heavy smog has made the pollen in the air feel much thicker, making it more irritating.

"Smog plays an extremely important role in allergy suffering," says Frances Coates, owner of Ottawa-based Aerobiology Research, which measures pollen levels. "It holds particles in the air a lot longer."

Low rainfall delayed the ragweed season until the end of August. It usually hits early in the month.

SOURCE: Adapted by Louise Surette from *Toronto Star*, September 8, 2001.

—See Chapter Project 1.

3.1 Polynomial Functions and Models

PREPARING FOR THIS SECTION *Before getting started, review the following:*
• Polynomials (Appendix A, Section A.3, pp. A19–A27)
• Using a Graphing Utility to Approximate Local Maxima and Local Minima (Section 1.3, pp. 67–68)
• Intercepts of a Function (Section 1.2, pp. 56–58)
• Graphing Techniques: Transformations (Section 1.5, pp. 85–94)
• Intercepts (Foundations, Section F.2, pp. 10–11)

 Now work the 'Are You Prepared?' problems on page 199.

OBJECTIVES 1 Identify Polynomial Functions and Their Degree
2 Graph Polynomial Functions Using Transformations
3 Identify the Zeros of a Polynomial Function and Their Multiplicity
4 Analyze the Graph of a Polynomial Function
5 Find the Cubic Function of Best Fit to Data

1 Identify Polynomial Functions and Their Degree

In Chapter 2 we studied the linear function $f(x) = mx + b$, which we can write as

$$f(x) = a_1x + a_0$$

and we studied the quadratic function $f(x) = ax^2 + bx + c$, $a \neq 0$, which we can write as

$$f(x) = a_2x^2 + a_1x + a_0, \; a_2 \neq 0$$

Each of these functions is an example of a *polynomial function*.

> A **polynomial function** is a function of the form
>
> $$f(x) = a_nx^n + a_{n-1}x^{n-1} + \cdots + a_1x + a_0 \qquad \textbf{(1)}$$
>
> where $a_n, a_{n-1}, \ldots, a_1, a_0$ are real numbers and n is a nonnegative integer. The domain of a polynomial function is the set of all real numbers.

A polynomial function is a function whose rule is given by a polynomial in one variable. The **degree** of a polynomial function is the degree of the polynomial in one variable, that is, the largest power of x that appears.

Polynomial functions are among the simplest expressions in algebra. They are easy to evaluate: only addition and repeated multiplication are required. Because of this, they are often used to approximate other, more complicated functions. In this section, we investigate properties of this important class of functions.

EXAMPLE 1 **Identifying Polynomial Functions**

Determine which of the following are polynomial functions. For those that are, state the degree; for those that are not, tell why not.

(a) $f(x) = 2 - 3x^4$ (b) $g(x) = \sqrt{x}$ (c) $h(x) = \dfrac{x^2 - 2}{x^3 - 1}$

(d) $F(x) = 0$ (e) $G(x) = 8$ (f) $H(x) = -2x^3(x - 1)^2$

Solution (a) f is a polynomial function of degree 4.

(b) g is not a polynomial function because $g(x) = \sqrt{x} = x^{\frac{1}{2}}$, so the variable x is raised to the $\frac{1}{2}$ power, which is not a nonnegative integer.

(c) h is not a polynomial function. It is the ratio of two polynomials, and the polynomial in the denominator is of positive degree.

(d) F is the zero polynomial function; it is not assigned a degree.

(e) G is a nonzero constant function. It is a polynomial function of degree 0 since $G(x) = 8 = 8x^0$.

(f) $H(x) = -2x^3(x - 1)^2 = -2x^3(x^2 - 2x + 1) = -2x^5 + 4x^4 - 2x^3$. So H is a polynomial function of degree 5. Do you see a way to find the degree of H without multiplying out? ■

═══ NOW WORK PROBLEMS 11 AND 15.

We have already discussed in detail polynomial functions of degrees 0, 1, and 2. See Table 1 for a summary of the properties of the graphs of these polynomial functions.

Table 1

Degree	Form	Name	Graph
No degree	$f(x) = 0$	Zero function	The x-axis
0	$f(x) = a_0,\quad a_0 \neq 0$	Constant function	Horizontal line with y-intercept a_0
1	$f(x) = a_1 x + a_0,\quad a_1 \neq 0$	Linear function	Nonvertical, nonhorizontal line with slope a_1 and y-intercept a_0
2	$f(x) = a_2 x^2 + a_1 x + a_0,\quad a_2 \neq 0$	Quadratic function	Parabola: Graph opens up if $a_2 > 0$; graph opens down if $a_2 < 0$

One objective of this section is to analyze the graph of a polynomial function. If you take a course in calculus, you will learn that the graph of every polynomial function is both smooth and continuous. By **smooth**, we mean that the graph contains no sharp corners or cusps; by **continuous**, we mean that the graph has no gaps or holes and can be drawn without lifting pencil from paper. See Figures 1(a) and (b).

Figure 1

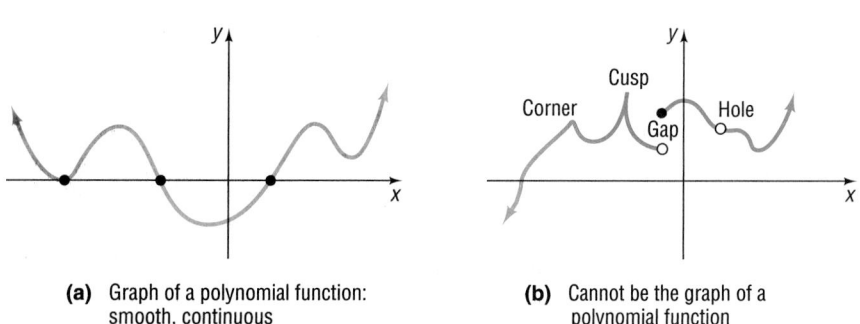

(a) Graph of a polynomial function: smooth, continuous

(b) Cannot be the graph of a polynomial function

Power Functions

We begin the analysis of the graph of a polynomial function by discussing *power functions*, a special kind of polynomial function.

In Words
A power function is a function that is defined by a single monomial.

> A **power function of degree** n is a function of the form
>
> $$f(x) = ax^n \qquad (2)$$
>
> where a is a real number, $a \neq 0$, and $n > 0$ is an integer.

Examples of power functions are

$$f(x) = 3x \qquad f(x) = -5x^2 \qquad f(x) = 8x^3 \qquad f(x) = -5x^4$$

degree 1 degree 2 degree 3 degree 4

The graph of a power function of degree 1, $f(x) = ax$, is a straight line, with slope a, that passes through the origin. The graph of a power function of degree 2, $f(x) = ax^2$, is a parabola, with vertex at the origin, that opens up if $a > 0$ and down if $a < 0$.

If we know how to graph a power function of the form $f(x) = x^n$, a compression or stretch and, perhaps, a reflection about the x-axis will enable us to obtain the graph of $g(x) = ax^n$. Consequently, we shall concentrate on graphing power functions of the form $f(x) = x^n$.

We begin with power functions of even degree of the form $f(x) = x^n$, $n \geq 2$ and n even. The domain of f is the set of all real numbers, and the range is the set of nonnegative real numbers. Such a power function is an even function (do you see why?), so its graph is symmetric with respect to the y-axis. Its graph always contains the origin and the points $(-1, 1)$ and $(1, 1)$.

If $n = 2$, the graph is the familiar parabola $y = x^2$ that opens up, with vertex at the origin. If $n \geq 4$, the graph of $f(x) = x^n$, n even, will be closer to the x-axis than the parabola $y = x^2$ if $-1 < x < 1$ and farther from the x-axis than the parabola $y = x^2$ if $x < -1$ or if $x > 1$. Figure 2(a) illustrates this conclusion. Figure 2(b) shows the graphs of $y = x^4$ and $y = x^8$ for comparison.

Figure 2

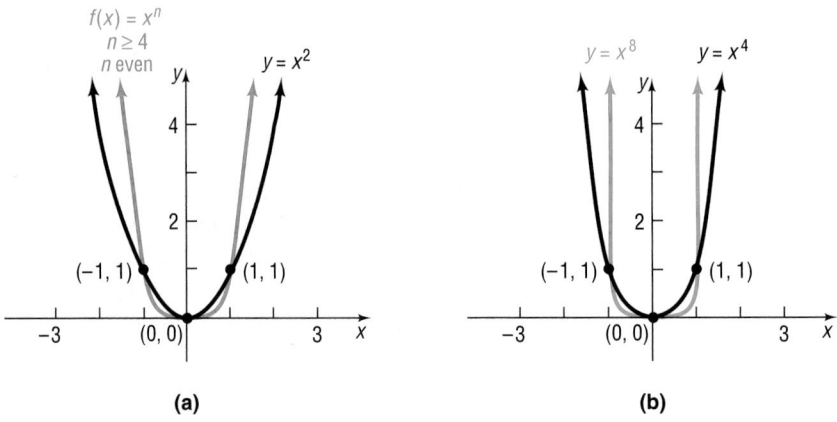

(a) (b)

From Figure 2, we can see that as n increases the graph of $f(x) = x^n$, $n \geq 2$ and n even, tends to flatten out near the origin and to increase very rapidly when x is far from 0. For large n, it may appear that the graph coincides with the x-axis near the origin, but it does not; the graph actually touches the x-axis only at the origin (see Table 2). Also, for large n, it may appear that for $x < -1$ or for $x > 1$ the graph is vertical, but it is not; it is only increasing very rapidly in these intervals. If the graphs were enlarged many times, these distinctions would be clear.

Table 2

	$x = 0.1$	$x = 0.3$	$x = 0.5$
$f(x) = x^8$	10^{-8}	0.0000656	0.0039063
$f(x) = x^{20}$	10^{-20}	$3.487 \cdot 10^{-11}$	0.000001
$f(x) = x^{40}$	10^{-40}	$1.216 \cdot 10^{-21}$	$9.095 \cdot 10^{-13}$

--- **Seeing the Concept** --

Graph $Y_1 = x^4$, $Y_2 = x^8$, and $Y_3 = x^{12}$ using the viewing rectangle $-2 \leq x \leq 2$, $-4 \leq y \leq 16$. Then graph each again using the viewing rectangle $-1 \leq x \leq 1$, $0 \leq y \leq 1$. See Figure 3. TRACE along one of the graphs to confirm that for x close to 0 the graph is above the x-axis and that for $x > 0$ the graph is increasing.

Figure 3

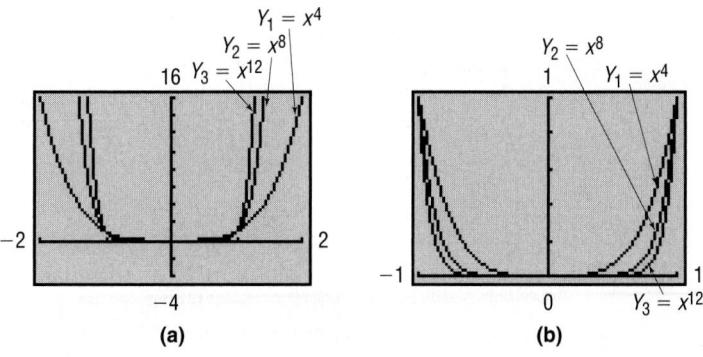

(a) (b)

Properties of Power Functions, $f(x) = x^n$, n Is an Even Integer

1. f is an even function so its graph is symmetric with respect to the y-axis.
2. The domain is the set of all real numbers. The range is the set of nonnegative real numbers.
3. The graph always contains the points $(0, 0)$, $(1, 1)$, and $(-1, 1)$.
4. As the exponent n increases in magnitude, the graph becomes more vertical when $x < -1$ or $x > 1$; but for x near the origin, the graph tends to flatten out and lie closer to the x-axis.

Now we consider power functions of odd degree of the form $f(x) = x^n$, $n \geq 3$ and n odd. The domain and the range of f is the set of real numbers. Such a power function is an odd function (do you see why?), so its graph is symmetric with respect to the origin. Its graph always contains the origin and the points $(-1, -1)$ and $(1, 1)$.

The graph of $f(x) = x^n$ when $n = 3$ has been shown several times and is repeated in Figure 4. If $n \geq 5$, the graph of $f(x) = x^n$, n odd, will be closer to the x-axis than that of $y = x^3$ if $-1 < x < 1$ and farther from the x-axis than that of $y = x^3$ if $x < -1$ or if $x > 1$. Figure 4 also illustrates this conclusion.

Figure 5 shows the graph of $y = x^5$ and the graph of $y = x^9$ for further comparison.

Figure 4 **Figure 5**

 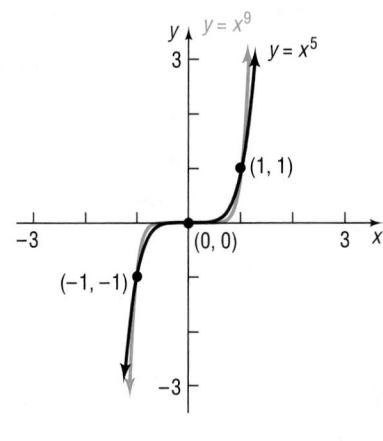

It appears that each graph coincides with the x-axis near the origin, but it does not; each graph actually crosses the x-axis at the origin. Also, it appears that as x increases the graph becomes vertical, but it does not; each graph is increasing very rapidly.

─── **Seeing the Concept** ───

Graph $Y_1 = x^3$, $Y_2 = x^7$, and $Y_3 = x^{11}$ using the viewing rectangle $-2 \le x \le 2$, $-16 \le y \le 16$. Then graph each again using the viewing rectangle $-1 \le x \le 1$, $-1 \le y \le 1$. See Figure 6. TRACE along one of the graphs to confirm that the graph is increasing and crosses the x-axis at the origin.

Figure 6

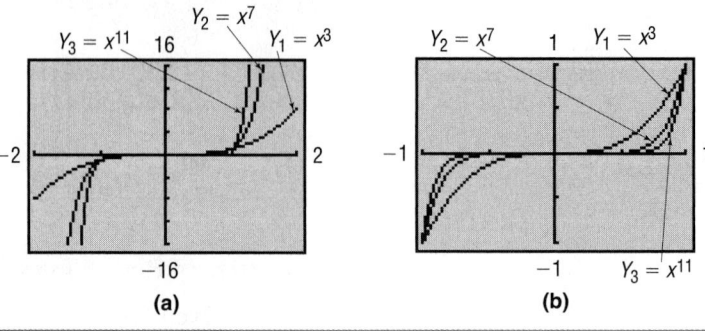

(a)

(b)

To summarize:

> ### Properties of Power Functions, $f(x) = x^n$, n Is an Odd Integer
>
> 1. f is an odd function so its graph is symmetric with respect to the origin.
> 2. The domain and the range are the set of all real numbers.
> 3. The graph always contains the points $(0, 0)$, $(1, 1)$, and $(-1, -1)$.
> 4. As the exponent n increases in magnitude, the graph becomes more vertical when $x < -1$ or $x > 1$; but for x near the origin, the graph tends to flatten out and lie closer to the x-axis.

2 Graph Polynomial Functions Using Transformations

The methods of shifting, compression, stretching, and reflection studied in Section 1.5, when used with the facts just presented, will enable us to graph polynomial functions that are transformations of power functions.

EXAMPLE 2

Graphing Polynomial Functions Using Transformations

Graph: $f(x) = 1 - x^5$

Solution Figure 7 shows the required stages.

Figure 7

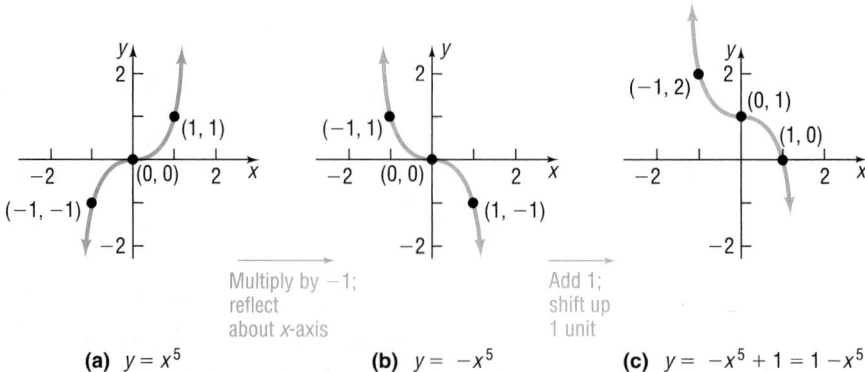

(a) $y = x^5$

(b) $y = -x^5$

(c) $y = -x^5 + 1 = 1 - x^5$

| EXAMPLE 3 | **Graphing Polynomial Functions Using Transformations** |

Graph: $f(x) = \dfrac{1}{2}(x - 1)^4$

Solution Figure 8 shows the required stages.

Figure 8

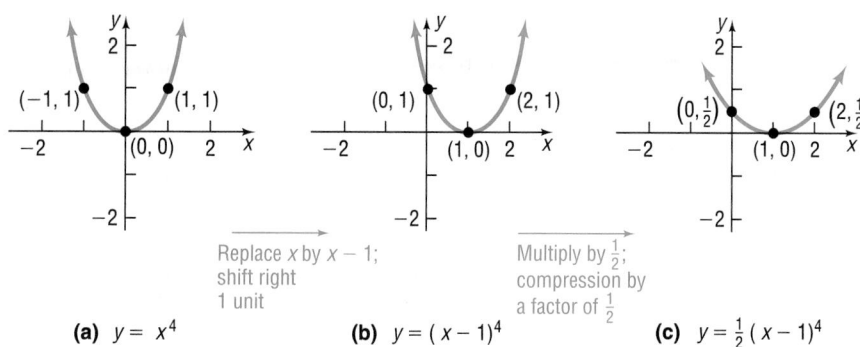

(a) $y = x^4$ (b) $y = (x - 1)^4$ (c) $y = \frac{1}{2}(x - 1)^4$

Replace x by $x - 1$; shift right 1 unit

Multiply by $\frac{1}{2}$; compression by a factor of $\frac{1}{2}$

■

NOW WORK PROBLEMS 23 AND 27.

3 Identify the Zeros of a Polynomial Function and Their Multiplicity

Figure 9 shows the graph of a polynomial function with four x-intercepts. Notice that at the x-intercepts the graph must either cross the x-axis or touch the x-axis. Consequently, between consecutive x-intercepts the graph is either above the x-axis or below the x-axis. We will make use of this property of the graph of a polynomial function shortly.

Figure 9

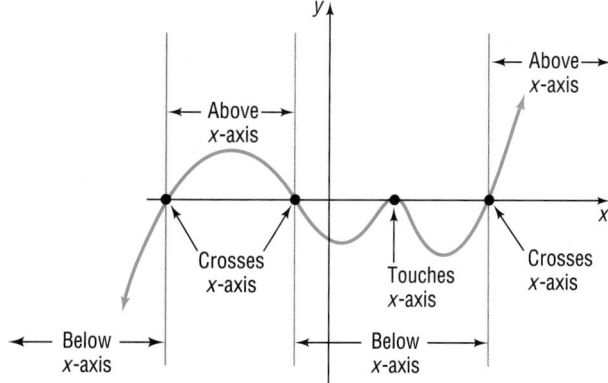

If a polynomial function f is factored completely, it is easy to solve the equation $f(x) = 0$ using the Zero-Product Property and locate the x-intercepts of the graph. For example, if $f(x) = (x - 1)^2(x + 3)$, then the solutions of the equation

$$f(x) = (x - 1)^2(x + 3) = 0$$

are identified as 1 and -3. Based on this result, we make the following observations:

> If f is a polynomial function and r is a real number for which $f(r) = 0$, that is, if r is a (real) zero of f then
>
> (a) r is an x-intercept of the graph of f.
>
> (b) $(x - r)$ is a factor of f.

So the real zeros of a function are the x-intercepts of its graph and they are found by solving the equation $f(x) = 0$.

EXAMPLE 4 | **Finding a Polynomial from Its Zeros**

(a) Find a polynomial of degree 3 whose zeros are $-3, 2,$ and 5.

 (b) Use a graphing utility to graph the polynomial found in part (a) to verify your result.

Solution
(a) If r is a zero of a polynomial f, then $x - r$ is a factor of f. This means that $x - (-3) = x + 3$, $x - 2$, and $x - 5$ are factors of f. As a result, any polynomial of the form

$$f(x) = a(x + 3)(x - 2)(x - 5)$$

where a is any nonzero real number, qualifies.

Figure 10

(b) The value of a causes a stretch, compression, or reflection, but does not affect the x-intercepts. We choose to graph f with $a = 1$. Then

$$f(x) = (x + 3)(x - 2)(x - 5) = x^3 - 4x^2 - 11x + 30$$

Figure 10 shows the graph of f. Notice that the x-intercepts are -3, 2, and 5. ■

— **Seeing the Concept** —

Graph the function found in Example 4 for $a = 2$ and $a = -1$. Does the value of a affect the zeros of f? How does the value of a affect the graph of f?

NOW WORK PROBLEM 37.

If the same factor $x - r$ occurs more than once, r is called a **repeated,** or **multiple, zero of f.** More precisely, we have the following definition.

> If $(x - r)^m$ is a factor of a polynomial f and $(x - r)^{m+1}$ is not a factor of f, then r is called a **zero of multiplicity m of f.***

EXAMPLE 5 | **Identifying Zeros and Their Multiplicities**

For the polynomial

$$f(x) = 5(x - 2)(x + 3)^2\left(x - \frac{1}{2}\right)^4$$

2 is a zero of multiplicity 1 because the exponent on the factor $x - 2$ is 1.
-3 is a zero of multiplicity 2 because the exponent on the factor $x + 3$ is 2.
$\dfrac{1}{2}$ is a zero of multiplicity 4 because the exponent on the factor $x - \dfrac{1}{2}$ is 4. ■

 NOW WORK PROBLEM 45(a).

In Example 5 notice that, if you add the multiplicities $(1 + 2 + 4 = 7)$, you obtain the degree of the polynomial.

Suppose that it is possible to factor completely a polynomial function and, as a result, locate all the x-intercepts of its graph (the real zeros of the function). As mentioned earlier, these x-intercepts then divide the x-axis into open intervals and, on each such interval, the graph of the polynomial will be either above or below the x-axis. Let's look at an example.

*Some books use the terms **multiple root** and **root of multiplicity** m.

| **EXAMPLE 6** | **Graphing a Polynomial Using Its x-Intercepts** |

For the polynomial: $f(x) = x^2(x - 2)$

(a) Find the x- and y-intercepts of the graph of f.

(b) Use the x-intercepts to find the intervals on which the graph of f is above the x-axis and the intervals on which the graph of f is below the x-axis.

(c) Locate other points on the graph and connect all the points plotted with a smooth, continuous curve.

Solution

(a) The y-intercept is $f(0) = 0^2(0 - 2) = 0$. The x-intercepts satisfy the equation

$$f(x) = x^2(x - 2) = 0$$

from which we find

$$x^2 = 0 \quad \text{or} \quad x - 2 = 0$$
$$x = 0 \quad \text{or} \quad x = 2$$

The x-intercepts are 0 and 2.

(b) The two x-intercepts divide the x-axis into three intervals:

$$(-\infty, 0) \qquad (0, 2) \qquad (2, \infty)$$

Since the graph of f crosses or touches the x-axis only at $x = 0$ and $x = 2$, it follows that the graph of f is either above the x-axis $[f(x) > 0]$ or below the x-axis $[f(x) < 0]$ on each of these three intervals. To see where the graph lies, we only need to pick a number in each interval, evaluate f there, and see whether the value is positive (above the x-axis) or negative (below the x-axis). See Table 3.

(c) In constructing Table 3, we obtained three additional points on the graph: $(-1, -3)$, $(1, -1)$, and $(3, 9)$. Figure 11 illustrates these points, the intercepts, and a smooth, continuous curve (the graph of f) connecting them.

Table 3

	0	2	
Interval	$(-\infty, 0)$	$(0, 2)$	$(2, \infty)$
Number Chosen	-1	1	3
Value of f	$f(-1) = -3$	$f(1) = -1$	$f(3) = 9$
Location of Graph	Below x-axis	Below x-axis	Above x-axis
Point on Graph	$(-1, -3)$	$(1, -1)$	$(3, 9)$

Figure 11

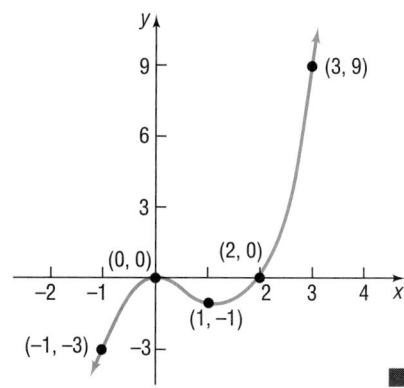

Look again at Table 3. Since the graph of f is below the x-axis on both sides of 0, the graph of f *touches* the x-axis at $x = 0$, a *zero of multiplicity 2*. Since the graph of f is below the x-axis for $x < 2$ and above the x-axis for $x > 2$, the graph of f *crosses* the x-axis at $x = 2$, a *zero of multiplicity 1*.

This suggests the following results:

If r Is a Zero of Even Multiplicity

| Sign of $f(x)$ does not change from one side to the other side of r. | Graph **touches** the x-axis at r. |

If r Is a Zero of Odd Multiplicity

Sign of $f(x)$ changes from one side to the other side of r.

Graph **crosses** the x-axis at r.

NOW WORK PROBLEM 45(b).

Behavior Near a Zero

We have just learned how the multiplicity of a zero can be used to determine whether the graph of a function touches or crosses the x-axis at the zero. However, we can learn more about the behavior of the graph near its zeros than just whether the graph crosses or touches the x-axis. Consider the function $f(x) = x^2(x - 2)$ whose graph is drawn in Figure 11. The zeros of f are 0 and 2. Table 4 shows the value of $f(x) = x^2(x - 2)$ and $y = -2x^2$ for x near 0. Figure 12 shows the points $(-0.1, -0.021), (-0.05, -0.0051)$, and so on, that are on the graph of $f(x) = x^2(x - 2)$ along with the graph of $y = -2x^2$ on the same Cartesian plane. From the table and graph, we can see that the points on the graph of $f(x) = x^2(x - 2)$ and the points on the graph of $y = -2x^2$ are indistinguishable near $x = 0$. So $y = -2x^2$ describes the behavior of the graph of $f(x) = x^2(x - 2)$ near $x = 0$.

Table 4

x	$f(x) = x^2(x - 2)$	$y = -2x^2$
-0.1	-0.021	-0.02
-0.05	-0.0051	-0.005
-0.03	-0.001827	-0.0018
-0.01	-0.0000201	-0.0002
0	0	0
0.01	-0.000199	-0.0002
0.03	-0.001773	-0.0018
0.05	-0.004875	-0.005
0.1	-0.019	-0.02

Figure 12

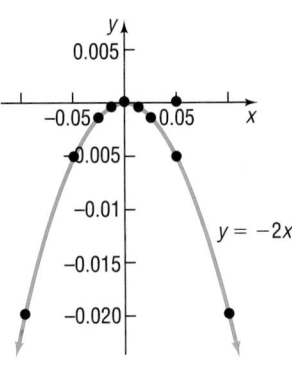

But how did we know that the function $f(x) = x^2(x - 2)$ behaves like $y = -2x^2$ when x is close to 0? In other words, where did $y = -2x^2$ come from? Because the zero of 0 comes from the factor x^2, we evaluate all factors in the function f at 0 with the exception of x^2.

$$f(x) = x^2(x - 2)$$
$$\approx x^2(0 - 2)$$
$$= -2x^2$$

The factor x^2 gives rise to the zero, so we let $x = 0$ in the remaining factors to find the behavior near 0.

This tells us that the graph of $f(x) = x^2(x - 2)$ will behave like the graph of $y = -2x^2$ near $x = 0$.

Now let's discuss the behavior of $f(x) = x^2(x - 2)$ near $x = 2$, the other zero. Because the zero 2 comes from the factor $x - 2$, we evaluate all factors of the function f at 2 with the exception of $x - 2$.

$$f(x) = x^2(x - 2)$$
$$\approx 2^2(x - 2)$$
$$= 4(x - 2)$$

The factor $x - 2$ gives rise to the zero, so we let $x = 2$ in the remaining factors to find the behavior near 2.

So the graph of $f(x) = x^2(x - 2)$ will behave like the graph of $y = 4(x - 2)$ near $x = 2$. Table 5 verifies that $f(x) = x^2(x - 2)$ and $y = 4(x - 2)$ have similar values for x near 2. Figure 13 shows the points $(1.9, -0.361)$, $(1.99, -0.0396)$, and so on, that are on the graph of $f(x) = x^2(x - 2)$ along with the graph of $y = 4(x - 2)$ on the same Cartesian plane. We can see that the points on the graph of $f(x) = x^2(x - 2)$ and the points on the graph of $y = 4(x - 2)$ are indistinguishable near $x = 2$. So $y = 4(x - 2)$ describes the behavior of the graph of $f(x) = x^2(x - 2)$ near $x = 2$.

Table 5

x	$f(x) = x^2(x - 2)$	$y = 4(x - 2)$
1.9	−0.361	−0.4
1.99	−0.0396	−0.04
1.999	−0.003996	−0.004
2	0	0
2.001	0.004004	0.004
2.01	0.0404	0.04
2.1	0.441	0.4

Figure 13

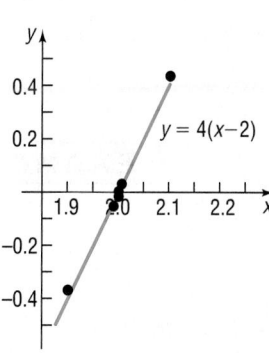

Figure 14 illustrates how we would use this information to begin to graph $f(x) = x^2(x - 2)$.

Figure 14

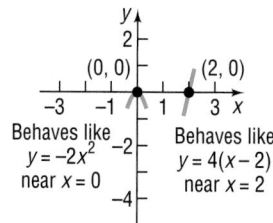

By determining the multiplicity of a zero, we determine whether the graph crosses or touches the x-axis at the zero. By determining the behavior of the graph near the zero, we determine how the graph touches or crosses the x-axis.

✏ **NOW WORK PROBLEM 45(C).**

Turning Points

 Look again at Figure 11. We cannot be sure just how low the graph actually goes between $x = 0$ and $x = 2$. But we do know that somewhere in the interval $(0, 2)$ the graph of f must change direction (from decreasing to increasing). The points at which a graph changes direction are called **turning points.** In calculus, such points are called **local maxima** or **local minima,** and techniques for locating them are given. So we shall not ask for the location of turning points in our graphs. Instead, we will use the following result from calculus, which tells us the maximum number of turning points that the graph of a polynomial function can have.

Theorem

Turning Points

If f is a polynomial function of degree n, then f has at most $n - 1$ turning points.

If the graph of a polynomial function f has $n - 1$ turning points, the degree of f is at least n.

For example, the graph of $f(x) = x^2(x - 2)$ shown in Figure 11 is the graph of a polynomial of degree 3 and has $3 - 1 = 2$ turning points: one at $(0, 0)$ and the other somewhere between $x = 0$ and $x = 2$.

—— **Exploration** ——————————————

A graphing utility can be used to locate the turning points of a graph. Graph $Y_1 = x^2(x - 2)$. Use MINIMUM to find the location of the turning point for $0 < x < 2$. See Figure 15.

Figure 15

✏ **NOW WORK PROBLEM 45(d).**

EXAMPLE 7	**Identifying the Graph of a Polynomial Function**

Which of the graphs in Figure 16 could be the graph of a polynomial function? For those that could, list the zeros and state the least degree the polynomial can have. For those that could not, say why not.

Figure 16

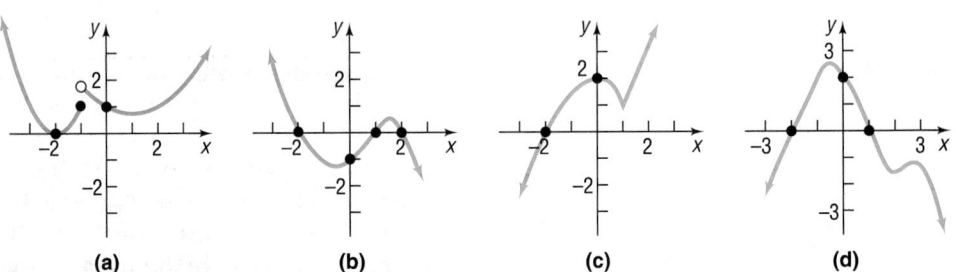

(a) (b) (c) (d)

Solution

(a) The graph in Figure 16(a) cannot be the graph of a polynomial function because of the gap that occurs at $x = -1$. Remember, the graph of a polynomial function is continuous—no gaps or holes.

(b) The graph in Figure 16(b) could be the graph of a polynomial function. It has three zeros, at -2, at 1, and at 2. Since the graph has two turning points, the degree of the polynomial function must be at least 3.

(c) The graph in Figure 16(c) cannot be the graph of a polynomial function because of the cusp at $x = 1$. Remember, the graph of a polynomial function is smooth.

(d) The graph in Figure 16(d) could be the graph of a polynomial function. It has two zeros, at -2 and at 1. Since the graph has three turning points, the degree of the polynomial function is at least 4. ∎

✏ **NOW WORK PROBLEM 57.**

End Behavior

One last remark about Figure 11. For very large values of x, either positive or negative, the graph of $f(x) = x^2(x - 2)$ looks like the graph of $y = x^3$. To see why, we write f in the form

$$f(x) = x^2(x - 2) = x^3 - 2x^2 = x^3\left(1 - \frac{2}{x}\right)$$

Now, for large values of x, either positive or negative, the term $\frac{2}{x}$ is close to 0, so for large values of x

$$f(x) = x^3 - 2x^2 = x^3\left(1 - \frac{2}{x}\right) \approx x^3$$

The behavior of the graph of a function for large values of x, either positive or negative, is referred to as its **end behavior**.

Theorem

End Behavior

For large values of x, either positive or negative, the graph of the polynomial

$$f(x) = a_n x^n + a_{n-1} x^{n-1} + \cdots + a_1 x + a_0$$

resembles the graph of the power function

$$y = a_n x^n$$

Look back at Figures 2 and 4. Based on the above theorem and the previous discussion on power functions, the end behavior of a polynomial can only be of four types. See Figure 17.

Figure 17
End Behavior of
$f(x) = a_n x^n + a_{n-1} x^{n-1} + \cdots + a_1 x + a_0$

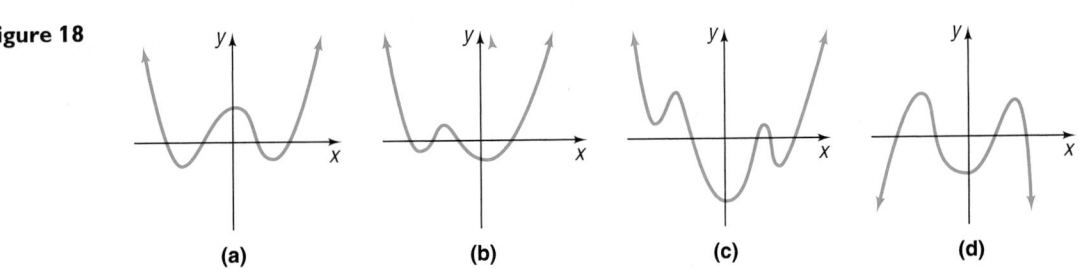

(a)
$n \geq 2$ even; $a_n > 0$

(b)
$n \geq 2$ even; $a_n < 0$

(c)
$n \geq 3$ odd; $a_n > 0$

(d)
$n \geq 3$ odd; $a_n < 0$

For example, if $f(x) = -2x^4 + x^3 + 4x^2 - 7x + 1$, the graph of f will resemble the graph of the power function $y = -2x^4$ for large $|x|$. The graph of f will look like Figure 17(b) for large $|x|$.

NOW WORK PROBLEM 45(e).

EXAMPLE 8

Identifying the Graph of a Polynomial Function

Which of the graphs in Figure 18 could be the graph of

$$f(x) = x^4 + 5x^3 + 5x^2 - 5x - 6?$$

Figure 18

(a)

(b)

(c)

(d)

Solution The y-intercept of f is $f(0) = -6$. We can eliminate the graph in Figure 18(a), whose y-intercept is positive.

We don't have any methods for finding the x-intercepts of f, so we move on to investigate the turning points of each graph. Since f is of degree 4 the graph of f has at most 3 turning points. We eliminate the graph in Figure 18(c) since that graph has 5 turning points.

Now we look at end behavior. For large values of x, the graph of f will behave like the graph of $y = x^4$. This eliminates the graph in Figure 18(d), whose end behavior is like the graph of $y = -x^4$.

Only the graph in Figure 18(b) could be (and, in fact, is) the graph of $f(x) = x^4 + 5x^3 + x^2 - 5x - 6$. ∎

NOW WORK PROBLEM 61.

Summary

Graph of a Polynomial Function $f(x) = a_n x^n + a_{n-1} x^{n-1} + \cdots + a_1 x + a_0, \quad a_n \neq 0$

Degree of the polynomial f: n

Maximum number of turning points: $n - 1$

At a zero of even multiplicity: The graph of f touches the x-axis.

At a zero of odd multiplicity: The graph of f crosses the x-axis.

Between zeros, the graph of f is either above or below the x-axis.

End behavior: For large $|x|$, the graph of f behaves like the graph of $y = a_n x^n$.

4 Analyze the Graph of a Polynomial Function

EXAMPLE 9 **Analyzing the Graph of a Polynomial Function**

For the polynomial $f(x) = x^3 + x^2 - 12x$:

(a) Find the x- and y-intercepts of the graph of f.

(b) Determine whether the graph crosses or touches the x-axis at each x-intercept.

(c) End behavior: find the power function that the graph of f resembles for large values of x.

(d) Determine the maximum number of turning points on the graph of f.

(e) Determine the behavior of the graph of f near each x-intercept.

(f) Put all the information together to obtain the graph of f.

Solution (a) The y-intercept is $f(0) = 0$.
To find the x-intercepts, if any, we first factor f.

$$f(x) = x^3 + x^2 - 12x = x(x^2 + x - 12) = x(x + 4)(x - 3)$$

We find the x-intercepts by solving the equation

$$f(x) = x(x + 4)(x - 3) = 0$$

$x = 0$ or $x + 4 = 0$ or $x - 3 = 0$ Zero-Product Property

$x = 0$ or $x = -4$ or $x = 3$ Solve

The x-intercepts, or zeros of f, are $-4, 0$, and 3.

(b) Since each zero of f is of multiplicity 1, the graph of f will cross the x-axis at each x-intercept.

(c) End behavior: the graph of f resembles that of the power function $y = x^3$ for large values of $|x|$.

(d) The graph of f will contain at most two turning points.

(e) The three x-intercepts are -4, 0, and 3.

Near -4: $f(x) = x(x + 4)(x - 3) \approx -4(x + 4)(-4 - 3) = 28(x + 4)$ A line with slope 28

Near 0: $f(x) = x(x + 4)(x - 3) \approx x(0 + 4)(0 - 3) = -12x$ A line with slope -12

Near 3: $f(x) = x(x + 4)(x - 3) \approx 3(3 + 4)(x - 3) = 21(x - 3)$ A line with slope 21

(f) Figure 19(a) illustrates the information obtained from parts (a), (b), (c), and (e). The graph of f is given in Figure 19(b). Notice that we evaluated f at -5, -2, 1, and 4 to help establish the scale on the y-axis.

Figure 19

(a)

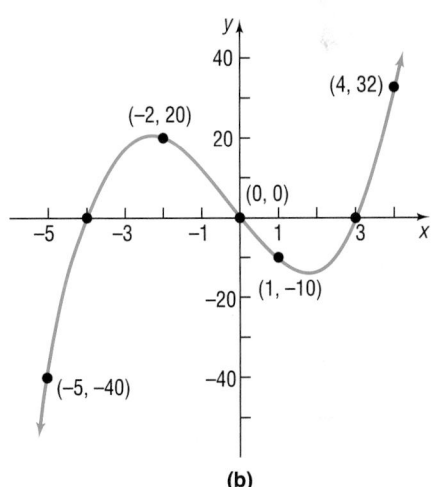

(b)

Exploration

Graph $Y_1 = x^3 + x^2 - 12x$. Compare what you see with Figure 19(b). Use MAXIMUM/MINIMUM to locate the two turning points.

NOW WORK PROBLEM **67**.

EXAMPLE 10 **Analyzing the Graph of a Polynomial Function**

Follow the instructions of Example 9 for the following polynomial:

$$f(x) = x^2(x - 4)(x + 1)$$

Solution (a) The y-intercept is $f(0) = 0$. The x-intercepts satisfy the equation

$$f(x) = x^2(x - 4)(x + 1) = 0$$

So

$$x^2 = 0 \quad \text{or} \quad x - 4 = 0 \quad \text{or} \quad x + 1 = 0$$
$$x = 0 \quad \text{or} \qquad x = 4 \quad \text{or} \qquad x = -1$$

The x-intercepts are $-1, 0$, and 4.

(b) The intercept 0 is a zero of multiplicity 2, so the graph of f will touch the x-axis at 0; 4 and -1 are zeros of multiplicity 1, so the graph of f will cross the x-axis at 4 and -1.

(c) End behavior: the graph of f resembles that of the power function $y = x^4$ for large values of $|x|$.

(d) The graph of f will contain at most three turning points.

(e) The three x-intercepts are $-1, 0$, and 4.

Near -1: $f(x) = x^2(x - 4)(x + 1) \approx (-1)^2(-1 - 4)(x + 1) = -5(x + 1)$ A line with slope -5

Near 0: $f(x) = x^2(x - 4)(x + 1) \approx x^2(0 - 4)(0 + 1) = -4x^2$ A parabola opening down

Near 4: $f(x) = x^2(x - 4)(x + 1) \approx 4^2(x - 4)(4 + 1) = 80(x - 4)$ A line with slope 80

(f) Figure 20(a) illustrates the information obtained from parts (a), (b), (c), and (e). The graph of f is given in Figure 20(b). Notice that we evaluated f at -2, $-\dfrac{1}{2}$, 2, and 5 to help establish the scale on the y-axis.

Figure 20

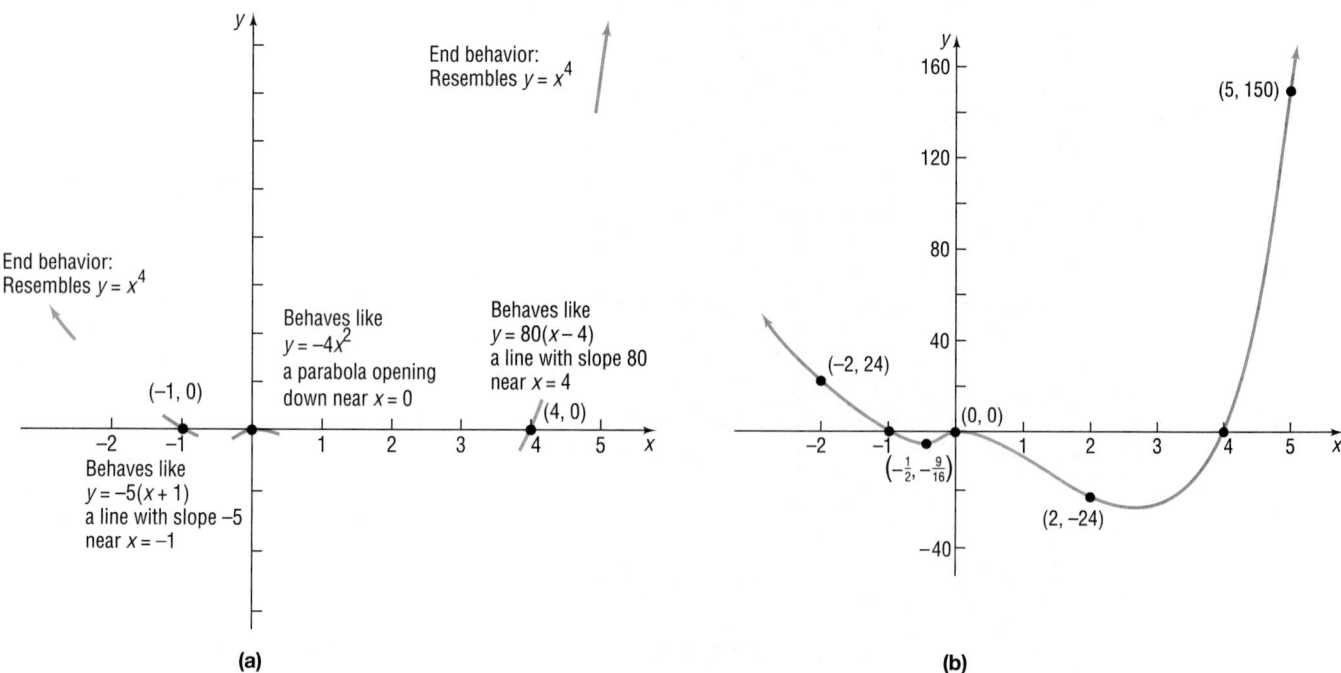

(a)

(b)

--- **Exploration** ---

Graph $Y_1 = x^2(x - 4)(x + 1)$. Compare what you see with Figure 20(b). Use MAXIMUM/MINIMUM to locate the two turning points besides $(0, 0)$.

Summary Steps for Analyzing the Graph of a Polynomial

To analyze the graph of a polynomial function $y = f(x)$, follow these steps:

STEP 1: (a) Find the x-intercepts, if any, by solving the equation $f(x) = 0$.
 (b) Find the y-intercept by letting $x = 0$ and finding the value of $f(0)$.

STEP 2: Determine whether the graph of f crosses or touches the x-axis at each x-intercept.

STEP 3: End behavior: find the power function that the graph of f resembles for large values of $|x|$.

STEP 4: Determine the maximum number of turning points on the graph of f.

STEP 5: Determine the behavior of the graph of f near each x-intercept.

STEP 6: Put all the information together to obtain the graph of f. This usually requires finding additional points on the graph.

NOW WORK PROBLEM **77.**

For polynomial functions that have noninteger coefficients and for polynomials that are not easily factored, we utilize the graphing utility early in the analysis of the graph. This is because the amount of information that can be obtained from algebraic analysis is limited.

EXAMPLE 11

Using a Graphing Utility to Analyze the Graph of a Polynomial Function

For the polynomial $f(x) = x^3 + 2.48x^2 - 4.3155x + 2.484406$:

(a) Find the degree of the polynomial. Determine the end behavior; that is, find the power function that the graph of f resembles for large values of $|x|$.

(b) Graph f using a graphing utility.

(c) Find the x- and y-intercepts of the graph.

(d) Use a TABLE to find points on the graph around each x-intercept. Determine on which intervals the graph is above and below the x-axis.

(e) Determine the local maxima and local minima, if any exist, rounded to two decimal places. That is, locate any turning points.

(f) Use the information obtained in parts (a)–(e) to draw a complete graph of f by hand. Be sure to label the intercepts, turning points, and the points obtained in part (d).

(g) Find the domain of f. Use the graph to find the range of f.

(h) Use the graph to determine where f is increasing and decreasing.

Figure 21

Table 6

Solution

(a) The degree of the polynomial is 3. End behavior: the graph of f resembles that of the power function $y = x^3$ for large values of $|x|$.

(b) See Figure 21 for the graph of f.

(c) The y-intercept is $f(0) = 2.484406$. In Example 9 we could easily factor $f(x)$ to find the x-intercepts. However, it is not readily apparent how $f(x)$ factors in this example. Therefore, we use a graphing utility's ZERO (or ROOT) feature and find the lone x-intercept to be -3.79, rounded to two decimal places.

(d) Table 6 shows values of x around the x-intercept. The points $(-4, -4.57)$ and $(-2, 13.04)$ are on the graph. The graph is below the x-axis on the interval $(-\infty, -3.79)$ and above the x-axis on the interval $(-3.79, \infty)$.

(e) From the graph we see that it has two turning points: one between -3 and -2, the other between 0 and 1. Rounded to two decimal places, the local maximum is 13.36 and occurs at $x = -2.28$; the local minimum is 1 and occurs at $x = 0.63$. The turning points are $(-2.28, 13.36)$ and $(0.63, 1)$.

(f) Figure 22 shows a graph of f drawn by hand using the information obtained in parts (a) to (e).

Figure 22

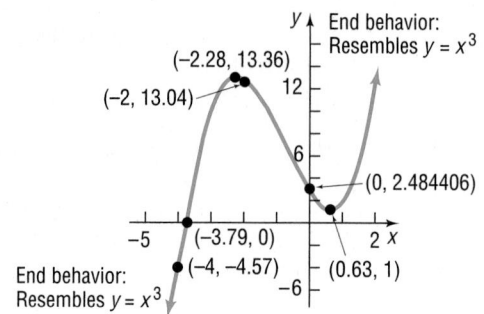

(g) The domain and the range of f are the set of all real numbers.

(h) Based on the graph, f is decreasing on the interval $(-2.28, 0.63)$ and is increasing on the intervals $(-\infty, -2.28)$ and $(0.63, \infty)$. ∎

NOW WORK PROBLEM 89.

5 Find the Cubic Function of Best Fit to Data

In Section 2.2 we found the line of best fit from data and in Section 2.5 we found the quadratic function of best fit. It is also possible to find polynomial functions of best fit. Since most statisticians do not recommend finding polynomials of best fit of degree higher than 3, we restrict our discussion to cubic functions.

Data that follow a cubic relation should look like Figure 23(a) or (b).

Figure 23

 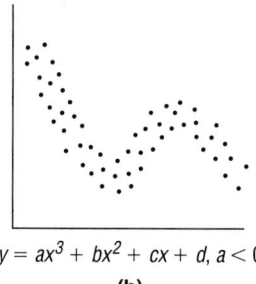

$$y = ax^3 + bx^2 + cx + d, a > 0$$
(a)

$$y = ax^3 + bx^2 + cx + d, a < 0$$
(b)

EXAMPLE 12 | **A Cubic Function of Best Fit**

Table 7

Year, x	Average Miles per Gallon, *M*
1991, 1	17.0
1992, 2	17.3
1993, 3	17.4
1994, 4	17.3
1995, 5	17.3
1996, 6	17.2
1997, 7	17.2
1998, 8	17.2
1999, 9	17.0
2000, 10	17.4
2001, 11	17.6

SOURCE: U. S. Federal Highway Administration

The data in Table 7 represent the average miles per gallon of vans, pickups, and sport utility vehicles (SUVs) in the United States for 1991–2001, where 1 represents 1991, 2 represents 1992, and so on.

(a) Draw a scatter diagram of the data using the year x as the independent variable and average miles per gallon M as the dependent variable. Comment on the type of relation that may exist between the two variables x and M.

(b) Using a graphing utility, find the cubic function of best fit $M = M(x)$ to these data.

(c) Graph the cubic function of best fit on your scatter diagram.

(d) Use the function found in part (b) to predict the average miles per gallon in 2002 ($x = 12$).

Solution

(a) Figure 24 shows the scatter diagram. A cubic relation may exist between the two variables.

(b) Upon executing the CUBIC REGression program, we obtain the results shown in Figure 25. The output the utility provides shows us the equation $y = ax^3 + bx^2 + cx + d$. The cubic function of best fit to the data is $M(x) = 0.0058x^3 - 0.1007x^2 + 0.4950x + 16.6258$.

(c) Figure 26 shows the graph of the cubic function of best fit on the scatter diagram. The function fits the data reasonably well.

Figure 24

Figure 25

Figure 26

(d) We evaluate the function $M(x)$ at $x = 12$.

$$M(12) = 0.0058(12)^3 - 0.1007(12)^2 + 0.4950(12) + 16.6258 \approx 18.1$$

The model predicts that the average miles per gallon of vans, pickups, and SUVs will be 18.1 miles per gallon in 2002. The actual miles per gallon of vans, pickups and SUVs was 17.6 in 2002. So our model overstates the results slightly. ∎

3.1 Assess Your Understanding

'Are You Prepared?'

Answers are given at the end of these exercises. If you get a wrong answer, read the pages listed in red.

1. The intercepts of the equation $9x^2 + 4y = 36$ are _____. (pp. 10–11)

2. *True or False:* The expression $4x^3 - 3.6x^2 - \sqrt{2}$ is a polynomial. (pp. A19–A27)

3. To graph $y = x^2 - 4$, you would shift the graph of $y = x^2$ _____ a distance of _____ units. (pp. 85–86)

4. *True or False:* The x-intercepts of the graph of a function $y = f(x)$ are the real solutions of the equation $f(x) = 0$. (p. 56)

Concepts and Vocabulary

5. The graph of every polynomial function is both _____ and _____.

6. A number r for which $f(r) = 0$ is called a(n) _____ of the function f.

7. If r is a zero of even multiplicity of a function f, the graph of f _____ the x-axis at r.

8. *True or False:* The graph of $f(x) = x^2(x - 3)(x + 4)$ has exactly three x-intercepts.

9. *True or False:* The x-intercepts of the graph of a polynomial function are called turning points.

10. *True or False:* End behavior: the graph of the function $f(x) = 3x^4 - 6x^2 + 2x + 5$ resembles $y = x^4$ for large values of $|x|$.

Skill Building

In Problems 11–22, determine which functions are polynomial functions. For those that are, state the degree. For those that are not, tell why not.

11. $f(x) = 4x + x^3$

12. $f(x) = 5x^2 + 4x^4$

13. $g(x) = \dfrac{1 - x^2}{2}$

14. $h(x) = 3 - \dfrac{1}{2}x$

15. $f(x) = 1 - \dfrac{1}{x}$

16. $f(x) = x(x - 1)$

17. $g(x) = x^{3/2} - x^2 + 2$

18. $h(x) = \sqrt{x}(\sqrt{x} - 1)$

19. $F(x) = 5x^4 - \pi x^3 + \dfrac{1}{2}$

20. $F(x) = \dfrac{x^2 - 5}{x^3}$

21. $G(x) = 2(x - 1)^2(x^2 + 1)$

22. $G(x) = -3x^2(x + 2)^3$

In Problems 23–36, use transformations of the graph of $y = x^4$ or $y = x^5$ to graph each function.

23. $f(x) = (x + 1)^4$

24. $f(x) = (x - 2)^5$

25. $f(x) = x^5 - 3$

26. $f(x) = x^4 + 2$

27. $f(x) = \dfrac{1}{2}x^4$

28. $f(x) = 3x^5$

29. $f(x) = -x^5$

30. $f(x) = -x^4$

31. $f(x) = (x - 1)^5 + 2$

32. $f(x) = (x + 2)^4 - 3$

33. $f(x) = 2(x + 1)^4 + 1$

34. $f(x) = \dfrac{1}{2}(x - 1)^5 - 2$

35. $f(x) = 4 - (x - 2)^5$

36. $f(x) = 3 - (x + 2)^4$

In Problems 37–44, form a polynomial whose zeros and degree are given.

37. Zeros: $-1, 1, 3$; degree 3

38. Zeros: $-2, 2, 3$; degree 3

39. Zeros: $-3, 0, 4$; degree 3

40. Zeros: $-4, 0, 2$; degree 3

41. Zeros: $-4, -1, 2, 3$; degree 4

42. Zeros: $-3, -1, 2, 5$; degree 4

43. Zeros: -1, multiplicity 1; 3, multiplicity 2; degree 3

44. Zeros: -2, multiplicity 2; 4, multiplicity 1; degree 3

In Problems 45–56, for each polynomial function:

(a) List each real zero and its multiplicity.

(b) Determine whether the graph crosses or touches the x-axis at each x-intercept.

(c) Determine the behavior of the graph near each x-intercept (zero).

(d) Determine the maximum number of turning points on the graph.

(e) Determine the end behavior; that is, find the power function that the graph of f resembles for large values of $|x|$.

45. $f(x) = 3(x - 7)(x + 3)^2$

46. $f(x) = 4(x + 4)(x + 3)^3$

47. $f(x) = 4(x^2 + 1)(x - 2)^3$

48. $f(x) = 2(x - 3)(x + 4)^3$

49. $f(x) = -2\left(x + \dfrac{1}{2}\right)^2(x^2 + 4)^2$

50. $f(x) = \left(x - \dfrac{1}{3}\right)^2(x - 1)^3$

51. $f(x) = (x - 5)^3(x + 4)^2$

52. $f(x) = (x + \sqrt{3})^2(x - 2)^4$

53. $f(x) = 3(x^2 + 8)(x^2 + 9)^2$

54. $f(x) = -2(x^2 + 3)^3$

55. $f(x) = -2x^2(x^2 - 2)$

56. $f(x) = 4x(x^2 - 3)$

In Problems 57–60, identify which of the graphs could be the graph of a polynomial function. For those that could be, list the zeros and state the least degree the polynomial can have. For those that could not be, say why not.

57.

58.

59.

60.
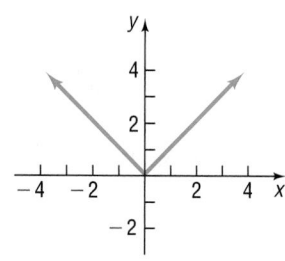

In Problems 61–64, decide which of the polynomial functions in the list might have the given graph. (More than one answer may be possible.)

61.
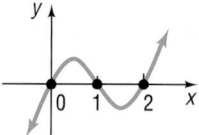

(a) $y = -4x(x - 1)(x - 2)$
(b) $y = x^2(x - 1)^2(x - 2)$
(c) $y = 3x(x - 1)(x - 2)$
(d) $y = x(x - 1)^2(x - 2)^2$
(e) $y = x^3(x - 1)(x - 2)$
(f) $y = -x(1 - x)(x - 2)$

62.
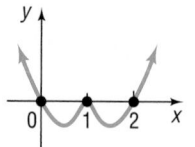

(a) $y = 2x^3(x - 1)(x - 2)^2$
(b) $y = x^2(x - 1)(x - 2)$
(c) $y = x^3(x - 1)^2(x - 2)$
(d) $y = x^2(x - 1)^2(x - 2)^2$
(e) $y = 5x(x - 1)^2(x - 2)$
(f) $y = -2x(x - 1)^2(2 - x)$

63.

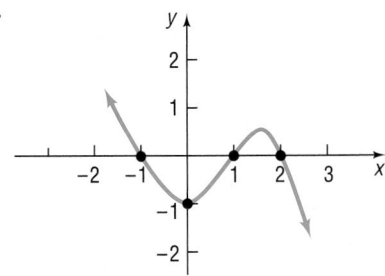

(a) $y = \dfrac{1}{2}(x^2 - 1)(x - 2)$

(b) $y = -\dfrac{1}{2}(x^2 + 1)(x - 2)$

(c) $y = (x^2 - 1)\left(1 - \dfrac{x}{2}\right)$

(d) $y = -\dfrac{1}{2}(x^2 - 1)^2(x - 2)$

(e) $y = \left(x^2 + \dfrac{1}{2}\right)(x^2 - 1)(2 - x)$

(f) $y = -(x - 1)(x - 2)(x + 1)$

64.

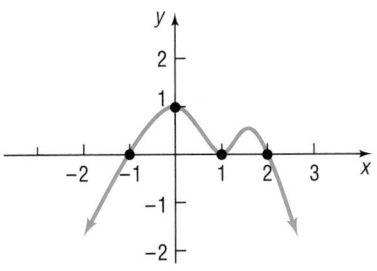

(a) $y = -\dfrac{1}{2}(x^2 - 1)(x - 2)(x + 1)$

(b) $y = -\dfrac{1}{2}(x^2 + 1)(x - 2)(x + 1)$

(c) $y = -\dfrac{1}{2}(x + 1)^2(x - 1)(x - 2)$

(d) $y = (x - 1)^2(x + 1)\left(1 - \dfrac{x}{2}\right)$

(e) $y = -(x - 1)^2(x - 2)(x + 1)$

(f) $y = -\left(x^2 + \dfrac{1}{2}\right)(x - 1)^2(x + 1)(x - 2)$

In Problems 65–88:
 (a) *Find the x- and y-intercepts of each polynomial function* f.
 (b) *Determine whether the graph of f crosses or touches the x-axis at each x-intercept.*
 (c) *End behavior: find the power function that the graph of f resembles for large values of* $|x|$.
 (d) *Determine the maximum number of turning points on the graph of f.*
 (e) *Determine the behavior of the graph of f near each x-intercept.*
 (f) *Put all the information together to obtain the graph of f. (You may need to locate additional points on the graph.)*

65. $f(x) = (x - 1)^2$
66. $f(x) = (x - 2)^3$
67. $f(x) = x^2(x - 3)$
68. $f(x) = x(x + 2)^2$
69. $f(x) = 6x^3(x + 4)$
70. $f(x) = 5x(x - 1)^3$
71. $f(x) = -4x^2(x + 2)$
72. $f(x) = -\dfrac{1}{2}x^3(x + 4)$
73. $f(x) = (x - 1)(x - 2)(x + 4)$
74. $f(x) = (x + 1)(x + 4)(x - 3)$
75. $f(x) = 4x - x^3$
76. $f(x) = x - x^3$
77. $f(x) = x^2(x - 2)(x + 2)$
78. $f(x) = x^2(x - 3)(x + 4)$
79. $f(x) = (x + 2)^2(x - 2)^2$
80. $f(x) = (x + 1)^3(x - 3)$
81. $f(x) = (x - 1)^2(x - 3)(x + 1)$
82. $f(x) = (x + 1)^2(x - 3)(x - 1)$
83. $f(x) = (x + 2)^2(x - 4)^2$
84. $f(x) = (x - 2)^2(x + 2)(x + 4)$
85. $f(x) = x^2(x - 2)(x^2 + 3)$
86. $f(x) = x^2(x^2 + 1)(x + 4)$
87. $f(x) = -x^2(x^2 - 1)(x + 1)$
88. $f(x) = -x^2(x^2 - 4)(x - 5)$

In Problems 89–98, for each polynomial function f:
 (a) *Find the degree of the polynomial. Determine the end behavior: that is, find the power function that the graph of f resembles for large values of* $|x|$.
 (b) *Graph f using a graphing utility.*
 (c) *Find the x- and y-intercepts of the graph, rounded to two decimal places.*
 (d) *Use TABLE to find points on the graph around each x-intercept. Determine on which intervals the graph is above and below the x-axis.*
 (e) *Determine the local maxima and local minima, if any exist, rounded to two decimal places. That is, locate any turning points.*
 (f) *Use the information obtained in parts (a)–(e) to draw a complete graph of f by hand. Be sure to label the intercepts, turning points, and the points obtained in part (d).*
 (g) *Find the domain of f. Use the graph to find the range of f.*
 (h) *Use the graph to determine where f is increasing and where f is decreasing.*

89. $f(x) = x^3 + 0.2x^2 - 1.5876x - 0.31752$
90. $f(x) = x^3 - 0.8x^2 - 4.6656x + 3.73248$
91. $f(x) = x^3 + 2.56x^2 - 3.31x + 0.89$
92. $f(x) = x^3 - 2.91x^2 - 7.668x - 3.8151$
93. $f(x) = x^4 - 2.5x^2 + 0.5625$
94. $f(x) = x^4 - 18.5x^2 + 50.2619$
95. $f(x) = 2x^4 - \pi x^3 + \sqrt{5}x - 4$
96. $f(x) = -1.2x^4 + 0.5x^2 - \sqrt{3}x + 2$
97. $f(x) = -2x^5 - \sqrt{2}x^2 - x - \sqrt{2}$
98. $f(x) = \pi x^5 + \pi x^4 + \sqrt{3}x + 1$

Applications and Extensions

99. Cost of Manufacturing The following data represent the cost C (in thousands of dollars) of manufacturing Chevy Cobalts and the number x of Cobalts produced.

Number of Cobalts Produced, x	Cost, C
0	10
1	23
2	31
3	38
4	43
5	50
6	59
7	70
8	85
9	105
10	135

(a) Draw a scatter diagram of the data using x as the independent variable and C as the dependent variable. Comment on the type of relation that may exist between the two variables C and x.

(b) Find the average rate of change of cost from four to five Cobalts.

(c) What is the average rate of change in cost from eight to nine Cobalts?

(d) Use a graphing utility to find the cubic function of best fit $C = C(x)$.

(e) Graph the cubic function of best fit on the scatter diagram.

(f) Use the function found in part (d) to predict the cost of manufacturing 11 Cobalts.

(g) Interpret the y-intercept.

100. Cost of Printing The following data represent the weekly cost C (in thousands of dollars) of printing textbooks and the number x (in thousands of units) of texts printed.

Number of Text Books, x	Cost, C
0	100
5	128.1
10	144
13	153.5
17	161.2
18	162.6
20	166.3
23	178.9
25	190.2
27	221.8

(a) Draw a scatter diagram of the data using x as the independent variable and C as the dependent variable.

Comment on the type of relation that may exist between the two variables C and x.

(b) Find the average rate of change in cost from 10,000 to 13,000 textbooks.

(c) What is the average rate of change in the cost of producing from 18,000 to 20,000 textbooks?

(d) Use a graphing utility to find the cubic function of best fit $C = C(x)$.

(e) Graph the cubic function of best fit on the scatter diagram.

(f) Use the function found in part (d) to predict the cost of printing 22,000 texts per week.

(g) Interpret the y-intercept.

101. Motor Vehicle Thefts The following data represent the number T (in thousands) of motor vehicle thefts in the United States for the years 1989–2001, where $x = 1$ represents 1989, $x = 2$ represents 1990, and so on.

Year, x	Motor Vehicle Thefts, T
1989, 1	1565
1990, 2	1636
1991, 3	1662
1992, 4	1611
1993, 5	1563
1994, 6	1539
1995, 7	1472
1996, 8	1394
1997, 9	1354
1998, 10	1243
1999, 11	1152
2000, 12	1160
2001, 13	1226

SOURCE: U.S. Federal Bureau of Investigation

(a) Draw a scatter diagram of the data using x as the independent variable and T as the dependent variable. Comment on the type of relation that may exist between the two variables T and x.

(b) Use a graphing utility to find the cubic function of best fit $T = T(x)$.

(c) Graph the cubic function of best fit on the scatter diagram.

(d) Use the function found in part (b) to predict the number of motor vehicle thefts in 2005.

(e) Do you think the function found in part (b) will be useful in predicting motor vehicle thefts in 2010? Why?

102. Average Annual Miles The following data represent the average number of miles driven (in thousands) annually by vans, pickups, and sport utility vehicles (SUVs) for the years 1993–2001, where $x = 1$ represents 1993, $x = 2$ represents 1994, and so on.

Year, x	Average Miles Driven, M
1993, 1	12.4
1994, 2	12.2
1995, 3	12.0
1996, 4	11.8
1997, 5	12.1
1998, 6	12.2
1999, 7	12.0
2000, 8	11.7
2001, 9	11.1

SOURCE: U. S. Federal Highway Administration

(a) Draw a scatter diagram of the data using x as the independent variable and M as the dependent variable. Comment on the type of relation that may exist between the two variables M and x.
(b) Use a graphing utility to find the cubic function of best fit $M = M(x)$.
(c) Graph the cubic function of best fit on the scatter diagram.
(d) Use the function found in part (b) to predict the average miles driven by vans, pickups, and SUVs in 2005.
(e) Do you think the function found in part (b) will be useful in predicting the average miles driven by vans, pickups, and SUVs in 2010? Why?

Discussion and Writing

103. Can the graph of a polynomial function have no y-intercept? Can it have no x-intercepts? Explain.

104. Write a few paragraphs that provide a general strategy for graphing a polynomial function. Be sure to mention the following: degree, intercepts, end behavior, and turning points.

105. Make up a polynomial that has the following characteristics: crosses the x-axis at -1 and 4, touches the x-axis at 0 and 2, and is above the x-axis between 0 and 2. Give your polynomial to a fellow classmate and ask for a written critique of your polynomial.

106. Make up two polynomials, not of the same degree, with the following characteristics: crosses the x-axis at -2, touches the x-axis at 1, and is above the x-axis between -2 and 1. Give your polynomials to a fellow classmate and ask for a written critique of your polynomials.

107. The graph of a polynomial function is always smooth and continuous. Name a function studied earlier that is smooth and not continuous. Name one that is continuous, but not smooth.

108. Which of the following statements are true regarding the graph of the cubic polynomial $f(x) = x^3 + bx^2 + cx + d$? (Give reasons for your conclusions.)
(a) It intersects the y-axis in one and only one point.
(b) It intersects the x-axis in at most three points.
(c) It intersects the x-axis at least once.
(d) For $|x|$ very large, it behaves like the graph of $y = x^3$.
(e) It is symmetric with respect to the origin.
(f) It passes through the origin.

Are You Prepared?' Answers

1. $(-2, 0), (2, 0), (0, 9)$ 2. True 3. down; 4 4. True

3.2 Properties of Rational Functions

PREPARING FOR THIS SECTION *Before getting started, review the following:*

- Rational Expressions (Appendix A, Section A.6, pp. A41–A49)
- Polynomial Division (Appendix A, Section A.4, pp. A25–A27)
- Graph of $f(x) = \dfrac{1}{x}$ (Foundations, Section F.2, Example 12, pp. 14–15)
- Graphing Techniques: Transformations (Section 1.5, pp. 85–94)

Now work the 'Are You Prepared?' problems on page 211.

OBJECTIVES 1 Find the Domain of a Rational Function
2 Find the Vertical Asymptotes of a Rational Function
3 Find the Horizontal or Oblique Asymptotes of a Rational Function

Ratios of integers are called *rational numbers*. Similarly, ratios of polynomial functions are called *rational functions*. Examples of rational functions are

$$R(x) = \frac{x^2 - 4}{x^2 + x + 1} \quad F(x) = \frac{x^3}{x^2 - 4} \quad G(x) = \frac{3x^2}{x^4 - 1}$$

A **rational function** is a function of the form

$$R(x) = \frac{p(x)}{q(x)}$$

where p and q are polynomial functions and q is not the zero polynomial. The domain of a rational function is the set of all real numbers except those for which the denominator q is 0.

1 Find the Domain of a Rational Function

EXAMPLE 1 **Finding the Domain of a Rational Function**

(a) The domain of $R(x) = \dfrac{2x^2 - 4}{x + 5}$ is the set of all real numbers x except -5; that is, the domain is $\{x \mid x \neq -5\}$.

(b) The domain of $R(x) = \dfrac{1}{x^2 - 4}$ is the set of all real numbers x except -2 and 2; that is, the domain is $\{x \mid x \neq -2, x \neq 2\}$.

(c) The domain of $R(x) = \dfrac{x^3}{x^2 + 1}$ is the set of all real numbers.

(d) The domain of $R(x) = \dfrac{3}{x^2 + 2}$ is the set of all real numbers.

(e) The domain of $R(x) = \dfrac{x^2 - 1}{x - 1}$ is the set of all real numbers x except 1; that is, the domain is $\{x \mid x \neq 1\}$. ■

It is important to observe that the functions

$$R(x) = \frac{x^2 - 1}{x - 1} \quad \text{and} \quad f(x) = x + 1$$

are not equal, since the domain of R is $\{x \mid x \neq 1\}$ and the domain of f is the set of all real numbers.

NOW WORK PROBLEM 13.

If $R(x) = \dfrac{p(x)}{q(x)}$ is a rational function and if p and q have no common factors, then the rational function R is said to be in **lowest terms**. For a rational function $R(x) = \dfrac{p(x)}{q(x)}$ in lowest terms, the zeros, if any, of the numerator are the x-intercepts of the graph of R and so will play a major role in the graph of R. The zeros of the denominator of R [that is, the numbers x, if any, for which $q(x) = 0$], although not in the domain of R, also play a major role in the graph of R. We will discuss this role shortly.

We have already discussed the properties of the rational function $f(x) = \dfrac{1}{x}$. (Refer to Example 12, pages 14–15.) The next rational function that we take up is $H(x) = \dfrac{1}{x^2}$.

EXAMPLE 2

Graphing $y = \dfrac{1}{x^2}$

Analyze the graph of: $H(x) = \dfrac{1}{x^2}$

Solution

The domain of $H(x) = \dfrac{1}{x^2}$ is the set of all real numbers x except 0. The graph has no y-intercept, because x can never equal 0. The graph has no x-intercept because the equation $H(x) = 0$ has no solution. Therefore, the graph of H will not cross either of the coordinate axes. Because

$$H(-x) = \frac{1}{(-x)^2} = \frac{1}{x^2} = H(x)$$

H is an even function, so its graph is symmetric with respect to the y-axis.

Table 8

x	$H(x) = \dfrac{1}{x^2}$
$\dfrac{1}{2}$	4
$\dfrac{1}{10}$	100
$\dfrac{1}{100}$	10,000
1	1
2	$\dfrac{1}{4}$
10	$\dfrac{1}{100}$
100	$\dfrac{1}{10,000}$

Table 8 shows the behavior of $H(x) = \dfrac{1}{x^2}$ for selected positive numbers x (we will use symmetry to obtain the graph of H when $x < 0$). From the first three rows of Table 8, we see that, as the values of x approach (get closer to) 0, the values of $H(x)$ become larger and larger positive numbers. When this happens, we say that H is **unbounded in the positive direction.** We symbolize this by writing $H \rightarrow \infty$ (read as "H **approaches infinity**"). In calculus, **limits** are used to convey these ideas. There we use the symbolism $\lim\limits_{x \to 0} H(x) = \infty$, read "the limit of $H(x)$ as x approaches zero equals infinity," to mean that $H(x) \rightarrow \infty$ as $x \rightarrow 0$.

Look at the last four rows of Table 8. As $x \rightarrow \infty$, the values of $H(x)$ approach 0 (the end behavior of the graph). In calculus, this is symbolized by writing $\lim\limits_{x \to \infty} H(x) = 0$. Figure 27 shows the graph. Notice the use of red dashed lines to convey the ideas discussed above.

Figure 27

$H(x) = \dfrac{1}{x^2}$

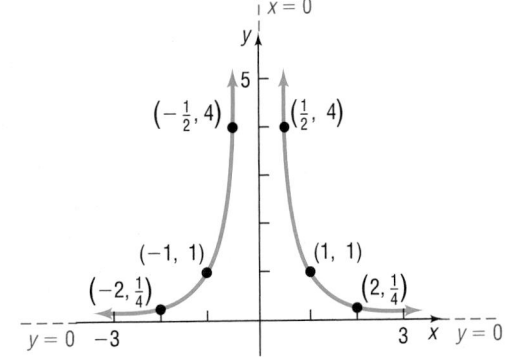

Sometimes transformations (shifting, compressing, stretching, and reflection) can be used to graph a rational function.

EXAMPLE 3

Using Transformations to Graph a Rational Function

Graph the rational function: $R(x) = \dfrac{1}{(x-2)^2} + 1$

Solution

First, we take note of the fact that the domain of R is the set of all real numbers except $x = 2$. To graph R, we start with the graph of $y = \dfrac{1}{x^2}$. See Figure 28 on page 206 for the steps.

Figure 28

Replace x by $x - 2$;
shift right
2 units

Add 1;
shift up
1 unit

(a) $y = \dfrac{1}{x^2}$

(b) $y = \dfrac{1}{(x - 2)^2}$

(c) $y = \dfrac{1}{(x - 2)^2} + 1$

━━━ **NOW WORK PROBLEM 31.**

Asymptotes

In Figure 28(c), notice that as the values of x become more negative, that is, as x becomes **unbounded in the negative direction** ($x \to -\infty$, read as "x **approaches negative infinity**"), the values $R(x)$ approach 1. In fact, we can conclude the following from Figure 28(c):

1. As $x \to -\infty$, the values $R(x)$ approach 1. $[\lim\limits_{x \to -\infty} R(x) = 1]$
2. As x approaches 2, the values $R(x) \to \infty$. $[\lim\limits_{x \to 2} R(x) = \infty]$
3. As $x \to \infty$, the values $R(x)$ approach 1. $[\lim\limits_{x \to \infty} R(x) = 1]$

This behavior of the graph is depicted by the vertical line $x = 2$ and the horizontal line $y = 1$. These lines are called *asymptotes* of the graph, which we define as follows:

> Let R denote a function.
>
> If, as $x \to -\infty$ or as $x \to \infty$, the values of $R(x)$ approach some fixed number, L, then the line $y = L$ is a **horizontal asymptote** of the graph of R.
>
> If, as x approaches some number c, the values $|R(x)| \to \infty$, then the line $x = c$ is a **vertical asymptote** of the graph of R. The graph of R never intersects a vertical asymptote.

Even though the asymptotes of a function are not part of the graph of the function, they provide information about how the graph looks. Figure 29 illustrates some of the possibilities.

Figure 29

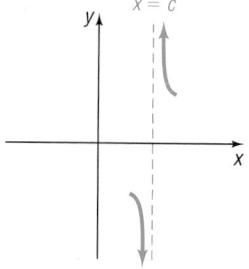

(a) End behavior:
As $x \to \infty$, the values of $R(x)$ approach L $[\lim\limits_{x \to \infty} R(x) = L]$. That is, the points on the graph of R are getting closer to the line $y = L$; $y = L$ is a horizontal asymptote.

(b) End behavior:
As $x \to -\infty$, the values of $R(x)$ approach L $[\lim\limits_{x \to -\infty} R(x) = L]$. That is, the points on the graph of R are getting closer to the line $y = L$; $y = L$ is a horizontal asymptote.

(c) As x approaches c, the values of $|R(x)| \to \infty$ $[\lim\limits_{x \to c^-} R(x) = \infty$; $\lim\limits_{x \to c^+} R(x) = \infty]$. That is, the points on the graph of R are getting closer to the line $x = c$; $x = c$ is a vertical asymptote.

(d) As x approaches c, the values of $|R(x)| \to \infty$ $[\lim\limits_{x \to c^-} R(x) = -\infty$; $\lim\limits_{x \to c^+} R(x) = \infty]$. That is, the points on the graph of R are getting closer to the line $x = c$; $x = c$ is a vertical asymptote.

A horizontal asymptote, when it occurs, describes a certain behavior of the graph as $x \to \infty$ or as $x \to -\infty$, that is, its end behavior. **The graph of a function may intersect a horizontal asymptote**.

A vertical asymptote, when it occurs, describes a certain behavior of the graph when x is close to some number c. **The graph of a function will never intersect a vertical asymptote.**

If an asymptote is neither horizontal nor vertical, it is called **oblique**. Figure 30 shows an oblique asymptote. An oblique asymptote, when it occurs, describes the end behavior of the graph. **The graph of a function may intersect an oblique asymptote.**

Figure 30
Oblique asymptote

 NOW WORK PROBLEM 23.

2 Find the Vertical Asymptotes of a Rational Function

The vertical asymptotes, if any, of a rational function $R(x) = \dfrac{p(x)}{q(x)}$, in lowest terms, are located at the zeros of the denominator of $q(x)$. Suppose that r is a zero of q, so $x - r$ is a factor of q. Now, as x approaches r, symbolized as $x \to r$, the values of $x - r$ approach 0, causing the ratio to become unbounded, that is, causing $|R(x)| \to \infty$. Based on the definition, we conclude that the line $x = r$ is a vertical asymptote.

Theorem

WARNING
If a rational function is not in lowest terms, an application of this theorem will result in an incorrect listing of vertical asymptotes. ∎

Locating Vertical Asymptotes

A rational function $R(x) = \dfrac{p(x)}{q(x)}$, *in lowest terms*, will have a vertical asymptote $x = r$ if r is a real zero of the *denominator q*. That is, if $x - r$ is a factor of the denominator q of a rational function $R(x) = \dfrac{p(x)}{q(x)}$, in lowest terms, R will have the vertical asymptote $x = r$.

EXAMPLE 4

Finding Vertical Asymptotes

Find the vertical asymptotes, if any, of the graph of each rational function.

(a) $R(x) = \dfrac{x}{x^2 - 4}$ (b) $F(x) = \dfrac{x + 3}{x - 1}$

(c) $H(x) = \dfrac{x^2}{x^2 + 1}$ (d) $G(x) = \dfrac{x^2 - 9}{x^2 + 4x - 21}$

Solution

WARNING
In Example 4(a), the vertical asymptotes are x = −2 and x = 2. Do not say that the vertical asymptotes are −2 and 2. ∎

(a) R is in lowest terms and the zeros of the denominator $x^2 - 4$ are -2 and 2. Hence, the lines $x = -2$ and $x = 2$ are the vertical asymptotes of the graph of R.

(b) F is in lowest terms and the only zero of the denominator is 1. The line $x = 1$ is the only vertical asymptote of the graph of F.

(c) H is in lowest terms and the denominator has no real zeros because the equation $x^2 + 1$ has no real solutions. The graph of H has no vertical asymptotes.

(d) Factor $G(x)$ to determine if it is in lowest terms.

$$G(x) = \frac{x^2 - 9}{x^2 + 4x - 21} = \frac{(x + 3)(x - 3)}{(x + 7)(x - 3)} = \frac{x + 3}{x + 7} \qquad x \neq 3$$

The only zero of the denominator of $G(x)$ in lowest terms is -7. Hence, the line $x = -7$ is the only vertical asymptote of the graph of G. ∎

As Example 4 points out, rational functions can have no vertical asymptotes, one vertical asymptote, or more than one vertical asymptote. However, the graph of a rational function will never intersect any of its vertical asymptotes. (Do you know why?)

Exploration

Graph each of the following rational functions:

$$R(x) = \frac{1}{x-1} \qquad R(x) = \frac{1}{(x-1)^2} \qquad R(x) = \frac{1}{(x-1)^3} \qquad R(x) = \frac{1}{(x-1)^4}$$

Each has the vertical asymptote $x = 1$. What happens to the value of $R(x)$ as x approaches 1 from the right side of the vertical asymptote; that is, what is $\lim\limits_{x \to 1^+} R(x)$? What happens to the value of $R(x)$ as x approaches 1 from the left side of the vertical asymptote; that is, what is $\lim\limits_{x \to 1^-} R(x)$? How does the multiplicity of the zero in the denominator affect the graph of R?

NOW WORK PROBLEM 45. (Find the vertical asymptotes, if any.)

3 Find the Horizontal or Oblique Asymptotes of a Rational Function

The procedure for finding horizontal and oblique asymptotes is somewhat more involved. To find such asymptotes, we need to know how the values of a function behave as $x \to -\infty$ or as $x \to \infty$.

If a rational function $R(x)$ is **proper**, that is, if the degree of the numerator is less than the degree of the denominator, then as $x \to -\infty$ or as $x \to \infty$ the value of $R(x)$ approaches 0. Consequently, the line $y = 0$ (the x-axis) is a horizontal asymptote of the graph.

Theorem

> If a rational function is proper, the line $y = 0$ is a horizontal asymptote of its graph.

EXAMPLE 5 **Finding Horizontal Asymptotes**

Find the horizontal asymptotes, if any, of the graph of

$$R(x) = \frac{x - 12}{4x^2 + x + 1}$$

Solution Since the degree of the numerator, 1, is less than the degree of the denominator, 2, the rational function R is proper. We conclude that the line $y = 0$ is a horizontal asymptote of the graph of R. ∎

To see why $y = 0$ is a horizontal asymptote of the function R in Example 5, we need to investigate the behavior of R as $x \to -\infty$ and $x \to \infty$. When $|x|$ is unbounded, the numerator of R, which is $x - 12$, can be approximated by the power function $y = x$, while the denominator of R, which is $4x^2 + x + 1$, can be approximated by the power function $y = 4x^2$. Applying these ideas to $R(x)$, we find

$$R(x) = \frac{x - 12}{4x^2 + x + 1} \underset{\uparrow}{\approx} \frac{x}{4x^2} = \frac{1}{4x} \underset{\uparrow}{\to} 0$$

$$\text{\small For } |x| \text{ unbounded} \qquad\qquad \text{\small As } x \to -\infty \text{ or } x \to \infty$$

This shows that the line $y = 0$ is a horizontal asymptote of the graph of R.

If a rational function $R(x) = \dfrac{p(x)}{q(x)}$ is **improper**, that is, if the degree of the numerator is greater than or equal to the degree of the denominator, we must use long division to write the rational function as the sum of a polynomial $f(x)$ plus a proper rational function $\dfrac{r(x)}{q(x)}$. That is, we write

$$R(x) = \frac{p(x)}{q(x)} = f(x) + \frac{r(x)}{q(x)}$$

where $f(x)$ is a polynomial and $\dfrac{r(x)}{q(x)}$ is a proper rational function. Since $\dfrac{r(x)}{q(x)}$ is

proper, $\dfrac{r(x)}{q(x)} \to 0$ as $x \to -\infty$ or as $x \to \infty$. As a result,

$$R(x) = \frac{p(x)}{q(x)} \to f(x) \qquad \text{as } x \to -\infty \text{ or as } x \to \infty$$

The possibilities are listed next.

1. If $f(x) = b$, a constant, the line $y = b$ is a horizontal asymptote of the graph of R.

2. If $f(x) = ax + b, a \neq 0$, the line $y = ax + b$ is an oblique asymptote of the graph of R.

3. In all other cases, the graph of R approaches the graph of f, and there are no horizontal or oblique asymptotes.

The following examples demonstrate these conclusions.

| **EXAMPLE 6** | **Finding Horizontal or Oblique Asymptotes** |

Find the horizontal or oblique asymptotes, if any, of the graph of

$$H(x) = \frac{3x^4 - x^2}{x^3 - x^2 + 1}$$

Solution Since the degree of the numerator, 4, is larger than the degree of the denominator, 3, the rational function H is improper. To find any horizontal or oblique asymptotes, we use long division.

$$
\begin{array}{r}
3x + 3 \\
x^3 - x^2 + 1\overline{)3x^4 \qquad\quad - x^2 \qquad\quad} \\
\underline{3x^4 - 3x^3 \qquad\quad + 3x} \\
3x^3 - x^2 - 3x \\
\underline{3x^3 - 3x^2 \qquad + 3} \\
2x^2 - 3x - 3
\end{array}
$$

As a result,

$$H(x) = \frac{3x^4 - x^2}{x^3 - x^2 + 1} = 3x + 3 + \frac{2x^2 - 3x - 3}{x^3 - x^2 + 1}$$

As $x \to -\infty$ or as $x \to \infty$,

$$\frac{2x^2 - 3x - 3}{x^3 - x^2 + 1} \approx \frac{2x^2}{x^3} = \frac{2}{x} \to 0$$

As $x \to -\infty$ or as $x \to \infty$, we have $H(x) \to 3x + 3$. We conclude that the graph of the rational function H has an oblique asymptote $y = 3x + 3$. ∎

| **EXAMPLE 7** | **Finding Horizontal or Oblique Asymptotes** |

Find the horizontal or oblique asymptotes, if any, of the graph of

$$R(x) = \frac{8x^2 - x + 2}{4x^2 - 1}$$

Solution Since the degree of the numerator, 2, equals the degree of the denominator, 2, the rational function R is improper. To find any horizontal or oblique asymptotes, we use long division.

$$4x^2 - 1 \overline{\smash{)}\begin{array}{r} 2 \\ 8x^2 - x + 2 \\ \underline{8x^2 - 2} \\ - x + 4 \end{array}}$$

As a result,

$$R(x) = \frac{8x^2 - x + 2}{4x^2 - 1} = 2 + \frac{-x + 4}{4x^2 - 1}$$

Then, as $x \to -\infty$ or as $x \to \infty$,

$$\frac{-x + 4}{4x^2 - 1} \approx \frac{-x}{4x^2} = \frac{-1}{4x} \to 0$$

As $x \to -\infty$ or as $x \to \infty$, we have $R(x) \to 2$. We conclude that $y = 2$ is a horizontal asymptote of the graph. ■

In Example 7, we note that the quotient 2 obtained by long division is the quotient of the leading coefficients of the numerator polynomial and the denominator polynomial $\left(\dfrac{8}{4}\right)$. This means that we can avoid the long division process for rational functions whose numerator and denominator *are of the same degree* and conclude that the quotient of the leading coefficients will give us the horizontal asymptote.

NOW WORK PROBLEM 41.

EXAMPLE 8 **Finding Horizontal or Oblique Asymptotes**

Find the horizontal or oblique asymptotes, if any, of the graph of

$$G(x) = \frac{2x^5 - x^3 + 2}{x^3 - 1}$$

Solution Since the degree of the numerator, 5, is larger than the degree of the denominator, 3, the rational function G is improper. To find any horizontal or oblique asymptotes, we use long division.

$$x^3 - 1 \overline{\smash{)}\begin{array}{r} 2x^2 - 1 \\ 2x^5 - x^3 + 2 \\ \underline{2x^5 - 2x^2} \\ -x^3 + 2x^2 + 2 \\ \underline{-x^3 + 1} \\ 2x^2 + 1 \end{array}}$$

As a result,

$$G(x) = \frac{2x^5 - x^3 + 2}{x^3 - 1} = 2x^2 - 1 + \frac{2x^2 + 1}{x^3 - 1}$$

Then, as $x \to -\infty$ or as $x \to \infty$,

$$\frac{2x^2 + 1}{x^3 - 1} \approx \frac{2x^2}{x^3} = \frac{2}{x} \to 0$$

As $x \to -\infty$ or as $x \to \infty$, we have $G(x) \to 2x^2 - 1$. We conclude that, for large values of $|x|$, the graph of G approaches the graph of $y = 2x^2 - 1$. That is, the graph of G will look like the graph of $y = 2x^2 - 1$ as $x \to -\infty$ or $x \to \infty$. Since $y = 2x^2 - 1$ is not a linear function, G has no horizontal or oblique asymptotes. ■

We now summarize the procedure for finding horizontal and oblique asymptotes.

Summary

Finding Horizontal and Oblique Asymptotes of a Rational Function R

Consider the rational function

$$R(x) = \frac{p(x)}{q(x)} = \frac{a_n x^n + a_{n-1} x^{n-1} + \cdots + a_1 x + a_0}{b_m x^m + b_{m-1} x^{m-1} + \cdots + b_1 x + b_0}$$

in which the degree of the numerator is n and the degree of the denominator is m.

1. If the degree of the numerator is less than the degree of the denominator, R is a proper rational function, and the graph of R will have the horizontal asymptote $y = 0$ (the x-axis).

2. If the degree of the numerator is greater than or equal to the degree of the denominator, then R is improper. Here long division is used.

 (a) If the degree of the numerator equals the degree of the denominator, the quotient obtained will be a number $\dfrac{a_n}{b_m}$, and the line $y = \dfrac{a_n}{b_m}$ is a horizontal asymptote.

 (b) If the degree of the numerator is one more than the degree of the denominator, the quotient obtained is of the form $ax + b$ (a polynomial of degree 1), and the line $y = ax + b$ is an oblique asymptote.

 (c) If the degree of the numerator is two or more than the degree of the denominator, the quotient obtained is a polynomial of degree 2 or higher, and R has neither a horizontal nor an oblique asymptote. In this case, for $|x|$ unbounded, the graph of R will behave like the graph of the quotient.

NOTE The graph of a rational function either has one horizontal or one oblique asymptote or else has no horizontal and no oblique asymptote. ∎

3.2 Assess Your Understanding

'Are You Prepared?'

Answers are given at the end of these exercises. If you get a wrong answer, read the pages listed in red.

1. *True or False:* The quotient of two polynomial expressions is a rational expression. (p. A42)

2. What is the quotient and remainder when $3x^4 - x^2$ is divided by $x^3 - x^2 + 1$. (pp. A25–A27)

3. Graph $y = \dfrac{1}{x}$. (pp. 14–15)

4. Graph $y = 2(x + 1)^2 - 3$ using transformations. (pp. 85–88)

Concepts and Vocabulary

5. The line _____ is a horizontal asymptote of $R(x) = \dfrac{x^3 - 1}{x^3 + 1}$.

6. The line _____ is a vertical asymptote of $R(x) = \dfrac{x - 1}{x + 1}$.

7. For a rational function R, if the degree of the numerator is less than the degree of the denominator, then R is _____.

8. *True or False:* The domain of every rational function is the set of all real numbers.

9. *True or False:* If an asymptote is neither horizontal nor vertical, it is called oblique.

10. *True or False:* If the degree of the numerator of a rational function equals the degree of the denominator, then the ratio of the leading coefficients gives rise to the horizontal asymptote.

Skill Building

In Problems 11–22, find the domain of each rational function.

11. $R(x) = \dfrac{4x}{x - 3}$

12. $R(x) = \dfrac{5x^2}{3 + x}$

13. $H(x) = \dfrac{-4x^2}{(x - 2)(x + 4)}$

14. $G(x) = \dfrac{6}{(x + 3)(4 - x)}$

15. $F(x) = \dfrac{3x(x - 1)}{2x^2 - 5x - 3}$

16. $Q(x) = \dfrac{-x(1 - x)}{3x^2 + 5x - 2}$

17. $R(x) = \dfrac{x}{x^3 - 8}$

18. $R(x) = \dfrac{x}{x^4 - 1}$

19. $H(x) = \dfrac{3x^2 + x}{x^2 + 4}$

20. $G(x) = \dfrac{x - 3}{x^4 + 1}$

21. $R(x) = \dfrac{3(x^2 - x - 6)}{4(x^2 - 9)}$

22. $F(x) = \dfrac{-2(x^2 - 4)}{3(x^2 + 4x + 4)}$

In Problems 23–28, use the graph shown to find:
(a) *The domain and range of each function* (b) *The intercepts, if any* (c) *Horizontal asymptotes, if any*
(d) *Vertical asymptotes, if any* (e) *Oblique asymptotes, if any*

23.

24.

25.

26.

27.

28.

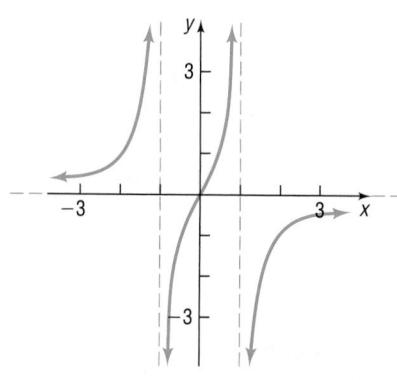

In Problems 29–40, graph each rational function using transformations.

29. $F(x) = 2 + \dfrac{1}{x}$

30. $Q(x) = 3 + \dfrac{1}{x^2}$

31. $R(x) = \dfrac{1}{(x-1)^2}$

32. $R(x) = \dfrac{3}{x}$

33. $H(x) = \dfrac{-2}{x+1}$

34. $G(x) = \dfrac{2}{(x+2)^2}$

35. $R(x) = \dfrac{-1}{x^2+4x+4}$

36. $R(x) = \dfrac{1}{x-1} + 1$

37. $G(x) = 1 + \dfrac{2}{(x-3)^2}$

38. $F(x) = 2 - \dfrac{1}{x+1}$

39. $R(x) = \dfrac{x^2-4}{x^2}$

40. $R(x) = \dfrac{x-4}{x}$

In Problems 41–52, find the vertical, horizontal, and oblique asymptotes, if any, of each rational function.

41. $R(x) = \dfrac{3x}{x+4}$

42. $R(x) = \dfrac{3x+5}{x-6}$

43. $H(x) = \dfrac{x^3-8}{x^2-5x+6}$

44. $G(x) = \dfrac{-x^2+1}{x^2-5x+6}$

45. $T(x) = \dfrac{x^3}{x^4-1}$

46. $P(x) = \dfrac{4x^5}{x^3-1}$

47. $Q(x) = \dfrac{5-x^2}{3x^4}$

48. $F(x) = \dfrac{-2x^2+1}{2x^3+4x^2}$

49. $R(x) = \dfrac{3x^4+4}{x^3+3x}$

50. $R(x) = \dfrac{6x^2+x+12}{3x^2-5x-2}$

51. $G(x) = \dfrac{x^3-1}{x-x^2}$

52. $F(x) = \dfrac{x-1}{x-x^3}$

Applications and Extensions

53. Gravity In physics, it is established that the acceleration due to gravity, g (in meters/sec^2), at a height h meters above sea level is given by

$$g(h) = \frac{3.99 \times 10^{14}}{(6.374 \times 10^6 + h)^2}$$

where 6.374×10^6 is the radius of Earth in meters.
(a) What is the acceleration due to gravity at sea level?
(b) The Sears Tower in Chicago, Illinois, is 443 meters tall. What is the acceleration due to gravity at the top of the Sears Tower?

(c) The peak of Mount Everest is 8848 meters above sea level. What is the acceleration due to gravity on the peak of Mount Everest?

(d) Find the horizontal asymptote of $g(h)$.

(e) Solve $g(h) = 0$. How do you interpret your answer?

54. Population Model A rare species of insect was discovered in the Amazon Rain Forest. To protect the species, environmentalists declare the insect endangered and transplant the insect into a protected area. The population P of the insect t months after being transplanted is

$$P(t) = \frac{50(1 + 0.5t)}{(2 + 0.01t)}$$

(a) How many insects were discovered? In other words, what was the population when $t = 0$?

(b) What will the population be after 5 years?

(c) Determine the horizontal asymptote of $P(t)$. What is the largest population that the protected area can sustain?

Discussion and Writing

55. If the graph of a rational function R has the vertical asymptote $x = 4$, the factor $x - 4$ must be present in the denominator of R. Explain why.

56. If the graph of a rational function R has the horizontal asymptote $y = 2$, the degree of the numerator of R equals the degree of the denominator of R. Explain why.

57. Can the graph of a rational function have both a horizontal and an oblique asymptote? Explain.

58. Make up a rational function that has $y = 2x + 1$ as an oblique asymptote. Explain the methodology that you used.

'Are You Prepared?' Answers

1. True **2.** Quotient: $3x + 3$; Remainder: $2x^2 - 3x - 3$ **3.**

4.

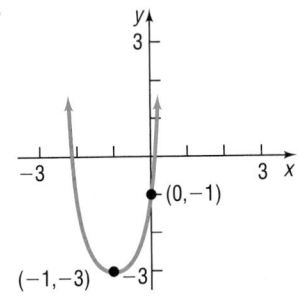

3.3 The Graph of a Rational Function; Inverse and Joint Variation

PREPARING FOR THIS SECTION *Before getting started, review the following:*

• Intercepts (Foundations, Section F.2, pp. 10–11)

Now work the 'Are You Prepared?' problems on page 227.

OBJECTIVES 1 Analyze the Graph of a Rational Function

2 Solve Applied Problems Involving Rational Functions

3 Construct a Model Using Inverse Variation

4 Construct a Model Using Joint or Combined Variation

1 Analyze the Graph of a Rational Function

We commented earlier that calculus provides the tools required to graph a polynomial function accurately. The same holds true for rational functions. However, we can gather together quite a bit of information about their graphs to get an idea of the general shape and position of the graph.

In the examples that follow, we will analyze the graph of a rational function by applying the following steps:

> **Analyzing the Graph of a Rational Function R**
>
> **STEP 1:** Factor the numerator and denominator of R and find its domain. If 0 is in the domain, find the y-intercept, $R(0)$, and plot it.
>
> **STEP 2:** Write R in lowest terms as $\dfrac{p(x)}{q(x)}$ and find the zeros of the numerator; that is, find the real solutions of the equation $p(x) = 0$, if any. These are the x-intercepts of the graph. Determine the behavior of the graph of R near each x-intercept, using the same procedure as for polynomial functions. Plot each x-intercept and indicate the behavior of the graph near it.
>
> **STEP 3:** With R written in lowest terms as $\dfrac{p(x)}{q(x)}$, find the zeros of the denominator; that is, find the real solutions of the equation $q(x) = 0$, if any. These determine the vertical asymptotes of the graph. Graph each vertical asymptote using a dashed line.
>
> **STEP 4:** Locate any horizontal or oblique asymptotes using the procedure given in the previous section. Graph the asymptotes using a dashed line. Determine the points, if any, at which the graph of R intersects these asymptotes. Plot any such points.
>
> **STEP 5:** Using the zeros of the numerator and the denominator of the given equation for R, divide the x-axis into intervals and determine where the graph is above the x-axis and where it is below the x-axis by choosing a number in each interval and evaluating R there. Plot the points found.
>
> **STEP 6:** Analyze the behavior of the graph of R near each asymptote and indicate this behavior on the graph.
>
> **STEP 7:** Put all the information together to obtain the graph of R.

EXAMPLE 1 **Analyzing the Graph of a Rational Function**

Analyze the graph of the rational function: $R(x) = \dfrac{x-1}{x^2-4}$

Solution **STEP 1:** We factor the numerator and denominator of R, obtaining

$$R(x) = \frac{x-1}{(x+2)(x-2)}$$

The domain of R is $\{x \mid x \neq -2, x \neq 2\}$. The y-intercept is

$$R(0) = \frac{-1}{-4} = \frac{1}{4}$$

Plot the point $\left(0, \dfrac{1}{4}\right)$.

STEP 2: R is in lowest terms. The zero of the numerator satisfies the equation $x - 1 = 0$. The only x-intercept is 1.

Near 1: $R(x) = \dfrac{x-1}{(x+2)(x-2)} \approx \dfrac{x-1}{(1+2)(1-2)} = -\dfrac{1}{3}(x-1)$

Plot the point $(1, 0)$ and indicate a line with negative slope there.

STEP 3: R is in lowest terms. The zeros of the denominator are the real solutions of the equation $(x+2)(x-2) = 0$, that is, -2 and 2. The graph of R has two

vertical asymptotes: the lines $x = -2$ and $x = 2$. Graph each of these asymptotes using dashed lines.

STEP 4: The degree of the numerator is less than the degree of the denominator, so R is proper and the line $y = 0$ (the x-axis) is a horizontal asymptote of the graph. Indicate this line by graphing $y = 0$ using a dashed line.

To determine if the graph of R intersects the horizontal asymptote, we solve the equation $R(x) = 0$.

$$\frac{x - 1}{x^2 - 4} = 0$$

$$x - 1 = 0 \qquad \text{Multiply both sides by } x^2 - 4$$

$$x = 1 \qquad \text{Add 1 to both sides}$$

The only solution is $x = 1$, so the graph of R intersects the horizontal asymptote at $(1, 0)$. We have already plotted this point.

STEP 5: The zero of the numerator, 1, and the zeros of the denominator, -2 and 2, divide the x-axis into four intervals:

$$(-\infty, -2) \qquad (-2, 1) \qquad (1, 2) \qquad (2, \infty)$$

Now construct Table 9.

Table 9

Interval	$(-\infty, -2)$	$(-2, 1)$	$(1, 2)$	$(2, \infty)$
Number Chosen	-3	0	$\dfrac{3}{2}$	3
Value of R	$R(-3) = -0.8$	$R(0) = \dfrac{1}{4}$	$R\left(\dfrac{3}{2}\right) = -\dfrac{2}{7}$	$R(3) = 0.4$
Location of Graph	Below x-axis	Above x-axis	Below x-axis	Above x-axis
Point on Graph	$(-3, -0.8)$	$\left(0, \dfrac{1}{4}\right)$	$\left(\dfrac{3}{2}, -\dfrac{2}{7}\right)$	$(3, 0.4)$

Plot the points from Table 9. You should now have Figure 31(a).

STEP 6: Next, we determine the behavior of the graph near the asymptotes. Since the x-axis is a horizontal asymptote and the graph lies below the x-axis for $x < -2$, we can sketch a portion of the graph by placing a small arrow to the far left and under the x-axis. Since the line $x = -2$ is a vertical asymptote and the graph lies below the x-axis for $x < -2$, we continue by placing an arrow well below the x-axis and approaching the line $x = -2$ on the left. Since the graph is above the x-axis for $-2 < x < 1$ and $x = -2$ is a vertical asymptote, the graph will continue on the right of $x = -2$ at the top. Similar explanations account for the other arrows shown. See Figure 31(b).

STEP 7: Figure 31(c) shows the complete graph.

Figure 31

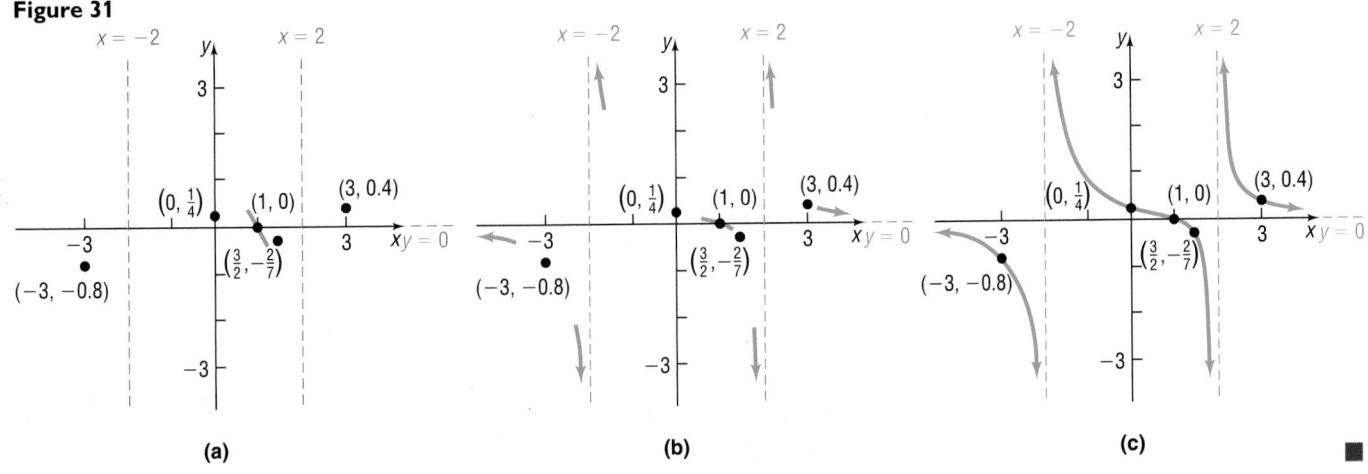

(a) (b) (c)

━━━━━━ **Exploration** ━━━━━━━━━━━━━━━━━━━━━━━━━

Graph $R(x) = \dfrac{x-1}{x^2-4}$

Result The analysis just completed in Example 1 helps us to set the viewing rectangle to obtain a complete graph. Figure 32(a) shows the graph of $R(x) = \dfrac{x-1}{x^2-4}$ in connected mode, and Figure 32(b) shows it in dot mode. Notice in Figure 32(a) that the graph has vertical lines at $x = -2$ and $x = 2$. This is due to the fact that, when the graphing utility is in connected mode, it will connect the dots between consecutive pixels. We know that the graph of R does not cross the lines $x = -2$ and $x = 2$, since R is not defined at $x = -2$ or $x = 2$. So, when graphing rational functions, dot mode should be used to avoid extraneous vertical lines that are not part of the graph.

Figure 32

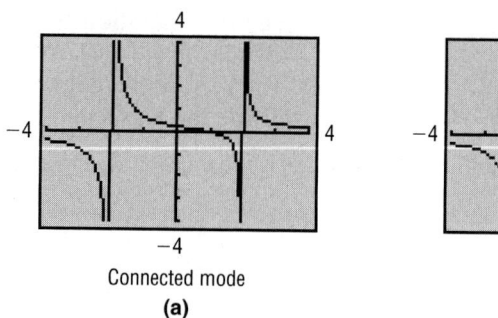

Connected mode
(a)

Dot mode
(b)

━━━━━━━━━━━━━━ **NOW WORK PROBLEM 7.**

| **EXAMPLE 2** | **Analyzing the Graph of a Rational Function** |

Analyze the graph of the rational function: $R(x) = \dfrac{x^2-1}{x}$

Solution **STEP 1:** The domain of R is $\{x \mid x \neq 0\}$. Because x cannot equal 0, there is no y-intercept.

$$R(x) = \frac{(x+1)(x-1)}{x}$$

R is in lowest terms.

STEP 2: The graph has two x-intercepts: -1 and 1.

Near -1: $R(x) = \dfrac{(x+1)(x-1)}{x} \approx \dfrac{(x+1)(-1-1)}{-1} = 2(x+1)$

Near 1: $R(x) = \dfrac{(x+1)(x-1)}{x} \approx \dfrac{(1+1)(x-1)}{1} = 2(x-1)$

Plot the point $(-1, 0)$ and indicate a line with positive slope there. Plot the point $(1, 0)$ and indicate a line with positive slope there.

STEP 3: R is in lowest terms, so the graph of R has the line $x = 0$ (the y-axis) as a vertical asymptote. Graph $x = 0$ using a dashed line.

STEP 4: The rational function R is improper, since the degree of the numerator, 2, is larger than the degree of the denominator, 1. To find any horizontal or oblique asymptotes, we use long division.

$$\begin{array}{r} x\phantom{{}-1} \\ x\overline{)x^2-1} \\ \underline{x^2\phantom{{}-1}} \\ -1 \end{array}$$

The quotient is x, so the line $y = x$ is an oblique asymptote of the graph. Graph $y = x$ using a dashed line.

To determine whether the graph of R intersects the asymptote $y = x$, we solve the equation $R(x) = x$.

$$R(x) = \frac{x^2 - 1}{x} = x$$

$$x^2 - 1 = x^2$$

$$-1 = 0 \qquad \text{Impossible}$$

We conclude that the equation $\dfrac{x^2 - 1}{x} = x$ has no solution, so the graph of R does not intersect the line $y = x$.

STEP 5: The zeros of the numerator are -1 and 1; the zero of the denominator is 0. We divide the x-axis into four intervals:

$$(-\infty, -1) \qquad (-1, 0) \qquad (0, 1) \qquad (1, \infty)$$

Now we construct Table 10.

Table 10

Interval	$(-\infty, -1)$	$(-1, 0)$	$(0, 1)$	$(1, \infty)$
Number chosen	-2	$-\dfrac{1}{2}$	$\dfrac{1}{2}$	2
Value of R	$R(-2) = -\dfrac{3}{2}$	$R\left(-\dfrac{1}{2}\right) = \dfrac{3}{2}$	$R\left(\dfrac{1}{2}\right) = -\dfrac{3}{2}$	$R(2) = \dfrac{3}{2}$
Location of graph	Below x-axis	Above x-axis	Below x-axis	Above x-axis
Point on graph	$\left(-2, -\dfrac{3}{2}\right)$	$\left(-\dfrac{1}{2}, \dfrac{3}{2}\right)$	$\left(\dfrac{1}{2}, -\dfrac{3}{2}\right)$	$\left(2, \dfrac{3}{2}\right)$

Plot the points from Table 10. You should now have Figure 33(a).

STEP 6: Since the graph of R is below the x-axis for $x < -1$ and is above the x-axis for $x > 1$, and since the graph of R does not intersect the oblique asymptote $y = x$, the graph of R will approach the line $y = x$ as shown in Figure 33(b).

Since the graph of R is above the x-axis for $-1 < x < 0$, the graph of R will approach the vertical asymptote $x = 0$ at the top to the left of $x = 0$; since the graph of R is below the x-axis for $0 < x < 1$, the graph of R will approach the vertical asymptote $x = 0$ at the bottom to the right of $x = 0$. See Figure 33(b).

STEP 7: The complete graph is given in Figure 33(c).

Figure 33

(a) (b) (c)

 —— **Seeing the Concept** ——————————————

Graph $R(x) = \dfrac{x^2 - 1}{x}$ and compare what you see with Figure 33(c). Could you have predicted from the graph that $y = x$ is an oblique asymptote? Graph $y = x$ and ZOOM-OUT. What do you observe?

———————————————————————————————

NOW WORK PROBLEM 15.

EXAMPLE 3	**Analyzing the Graph of a Rational Function**

Analyze the graph of the rational function: $R(x) = \dfrac{x^4 + 1}{x^2}$

Solution

STEP 1: R is completely factored. The domain of R is $\{x \mid x \neq 0\}$. There is no y-intercept.

STEP 2: R is in lowest terms. Since $x^4 + 1 = 0$ has no real solutions, there are no x-intercepts.

STEP 3: R is in lowest terms, so $x = 0$ (the y-axis) is a vertical asymptote of R. Graph the line $x = 0$ using dashes.

STEP 4: The rational function R is improper. To find any horizontal or oblique asymptotes, we use long division.

Figure 34

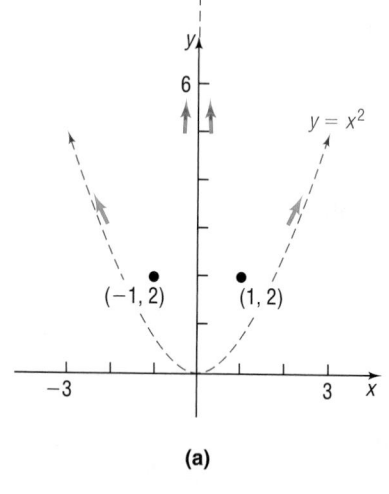

(a)

$$\begin{array}{r} x^2 \\ x^2\overline{)x^4 + 1} \\ \underline{x^4} \\ 1 \end{array}$$

The quotient is x^2, so the graph has no horizontal or oblique asymptotes. However, the graph of R will approach the graph of $y = x^2$ as $x \to -\infty$ and as $x \to \infty$. The graph of R does not intersect $y = x^2$. Graph $y = x^2$ using dashes.

STEP 5: The numerator has no zeros, and the denominator has one zero at 0. We divide the x-axis into the two intervals

$$(-\infty, 0) \qquad (0, \infty)$$

and construct Table 11.

Table 11

Interval	$(-\infty, 0)$	$(0, \infty)$
Number chosen	-1	1
Value of R	$R(-1) = 2$	$R(1) = 2$
Location of graph	Above x-axis	Above x-axis
Point on graph	$(-1, 2)$	$(1, 2)$

Plot the points $(-1, 2)$ and $(1, 2)$.

STEP 6: Since the graph of R is above the x-axis and does not intersect $y = x^2$, we place arrows above $y = x^2$ as shown in Figure 34(a). Also, since the graph of R is above the x-axis, it will approach the vertical asymptote $x = 0$ at the top to the left of $x = 0$ and at the top to the right of $x = 0$. See Figure 34(a).

STEP 7: Figure 34(b) shows the complete graph. ∎

—— Seeing the Concept ——

Graph $R(x) = \dfrac{x^4 + 1}{x^2}$ and compare what you see with Figure 34(b). Use MINIMUM to find the

two turning points. Enter $Y_2 = x^2$ and ZOOM-OUT. What do you see?

━━━━━━━━━▶ N O W W O R K P R O B L E M **1 3 .**

| EXAMPLE 4 | **Analyzing the Graph of a Rational Function** |

Analyze the graph of the rational function: $R(x) = \dfrac{3x^2 - 3x}{x^2 + x - 12}$

Solution **STEP 1:** We factor R to get

$$R(x) = \frac{3x(x - 1)}{(x + 4)(x - 3)}$$

The domain of R is $\{x \mid x \neq -4, x \neq 3\}$. The y-intercept is $R(0) = 0$. Plot the point $(0, 0)$.

STEP 2: R is in lowest terms. Since the real solutions of the equation $3x(x - 1) = 0$ are $x = 0$ and $x = 1$, the graph has two x-intercepts, 0 and 1. We determine the behavior of the graph of R near each x-intercept.

Near 0: $R(x) = \dfrac{3x(x - 1)}{(x + 4)(x - 3)} \approx \dfrac{3x(0 - 1)}{(0 + 4)(0 - 3)} = \dfrac{1}{4}x$

Near 1: $R(x) = \dfrac{3x(x - 1)}{(x + 4)(x - 3)} \approx \dfrac{3(1)(x - 1)}{(1 + 4)(1 - 3)} = -\dfrac{3}{10}(x - 1)$

Plot the point $(0, 0)$ and show a line with positive slope there.
Plot the point $(1, 0)$ and show a line with negative slope there.

STEP 3: R is in lowest terms. Since the real solutions of the equation $(x + 4)(x - 3) = 0$ are $x = -4$ and $x = 3$, the graph of R has two vertical asymptotes, the lines $x = -4$ and $x = 3$. Plot these lines using dashes.

STEP 4: Since the degree of the numerator equals the degree of the denominator, the graph has a horizontal asymptote. To find it, we either use long division or form the quotient of the leading coefficient of the numerator, 3, and the leading coefficient of the denominator, 1. The graph of R has the horizontal asymptote $y = 3$.

To find out whether the graph of R intersects the asymptote, we solve the equation $R(x) = 3$.

$$R(x) = \frac{3x^2 - 3x}{x^2 + x - 12} = 3$$

$$3x^2 - 3x = 3x^2 + 3x - 36$$

$$-6x = -36$$

$$x = 6$$

The graph intersects the line $y = 3$ only at $x = 6$, and $(6, 3)$ is a point on the graph of R. Plot the point $(6, 3)$ and the line $y = 3$ using dashes.

STEP 5: The zeros of the numerator, 0 and 1, and the zeros of the denominator, -4 and 3, divide the x-axis into five intervals:

$$(-\infty, -4) \quad (-4, 0) \quad (0, 1) \quad (1, 3) \quad (3, \infty)$$

Construct Table 12.

Table 12

	$(-\infty, -4)$	$(-4, 0)$	$(0, 1)$	$(1, 3)$	$(3, \infty)$
Interval	$(-\infty, -4)$	$(-4, 0)$	$(0, 1)$	$(1, 3)$	$(3, \infty)$
Number chosen	-5	-2	$\dfrac{1}{2}$	2	4
Value of R	$R(-5) = 11.25$	$R(-2) = -1.8$	$R\left(\dfrac{1}{2}\right) = \dfrac{1}{15}$	$R(2) = -1$	$R(4) = 4.5$
Location of graph	Above x-axis	Below x-axis	Above x-axis	Below x-axis	Above x-axis
Point on graph	$(-5, 11.25)$	$(-2, -1.8)$	$\left(\dfrac{1}{2}, \dfrac{1}{15}\right)$	$(2, -1)$	$(4, 4.5)$

Plot the points from Table 12. Figure 35(a) shows the graph we have so far.

STEP 6: Since the graph of R is above the x-axis for $x < -4$ and only crosses the line $y = 3$ at $(6, 3)$, as x approaches $-\infty$ the graph of R will approach the horizontal asymptote $y = 3$ from above. The graph of R will approach the vertical asymptote $x = -4$ at the top to the left of $x = -4$ and at the bottom to the right of $x = -4$. The graph of R will approach the vertical asymptote $x = 3$ at the bottom to the left of $x = 3$ and at the top to the right of $x = 3$.

We do not know whether the graph of R crosses or touches the line $y = 3$ at $(6, 3)$. To see whether the graph, in fact, crosses or touches the line $y = 3$, we plot an additional point to the right of $(6, 3)$. We use $x = 7$ to find $R(7) = \dfrac{63}{22} < 3$. The graph crosses $y = 3$ at $x = 6$.

Because $(6, 3)$ is the only point where the graph of R intersects the asymptote $y = 3$, the graph must approach the line $y = 3$ from below as $x \to \infty$. See Figure 35(b).

STEP 7: The complete graph is shown in Figure 35(c).

Figure 35

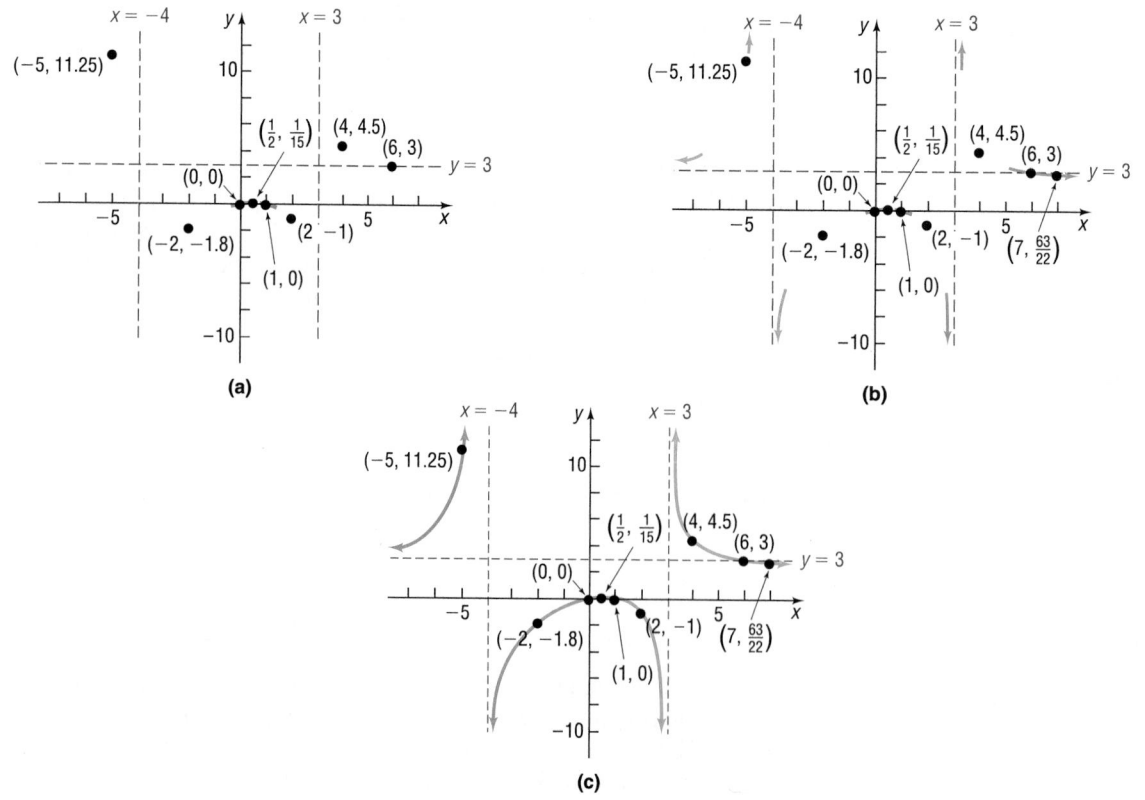

(a)

(b)

(c)

Exploration

Graph $R(x) = \dfrac{3x^2 - 3x}{x^2 + x - 12}$

RESULT Figure 36 shows the graph in connected mode, and Figure 37(a) shows it in dot mode. Neither graph displays clearly the behavior between the two x-intercepts, 0 and 1. Nor do they clearly display the fact that the graph crosses the horizontal asymptote at (6, 3). To see these parts better, we graph R for $-1 \le x \le 2$ [Figure 37(b)] and for $4 \le x \le 60$ [Figure 38(b)].

Figure 36

Connected mode

Figure 37

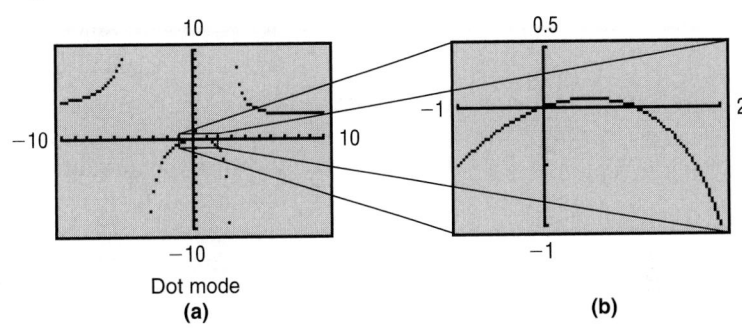

Dot mode
(a)

(b)

Figure 38

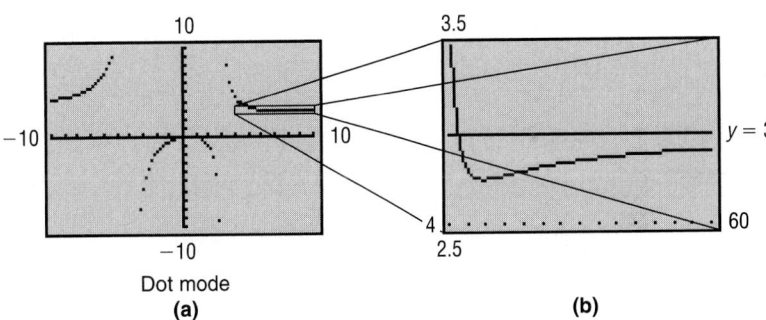

Dot mode
(a)

(b)

The new graphs reflect the behavior produced by the analysis. Furthermore, we observe two turning points, one between 0 and 1 and the other to the right of 4. Rounded to two decimal places, these turning points are (0.52, 0.07) and (11.48, 2.75).

| EXAMPLE 5 | **Analyzing the Graph of a Rational Function with a Hole** |

Analyze the graph of the rational function: $R(x) = \dfrac{2x^2 - 5x + 2}{x^2 - 4}$

Solution **STEP 1:** We factor R and obtain

$$R(x) = \frac{(2x - 1)(x - 2)}{(x + 2)(x - 2)}$$

The domain of R is $\{x \mid x \ne -2, x \ne 2\}$. The y-intercept is $R(0) = -\dfrac{1}{2}$. Plot the point $\left(0, -\dfrac{1}{2}\right)$.

STEP 2: In lowest terms,

$$R(x) = \frac{2x - 1}{x + 2} \qquad x \neq -2$$

The graph has one x-intercept: $\frac{1}{2}$.

Near $\frac{1}{2}$: $\quad R(x) = \frac{2x - 1}{x + 2} \approx \frac{2x - 1}{\frac{1}{2} + 2} = \frac{2}{5}(2x - 1)$

Plot the point $\left(\frac{1}{2}, 0\right)$ showing a line with positive slope.

STEP 3: Look at R in lowest terms. The graph has one vertical asymptote, $x = -2$, since $x + 2$ is the only factor of the denominator of $R(x)$ *in lowest terms*. Remember though, the rational function is undefined at both $x = 2$ and $x = -2$. Graph the line $x = -2$ using dashes.

STEP 4: Since the degree of the numerator equals the degree of the denominator, the graph has a horizontal asymptote. To find it, we either use long division or form the quotient of the leading coefficient of the numerator, 2, and the leading coefficient of the denominator, 1. The graph of R has the horizontal asymptote $y = 2$. Graph the line $y = 2$ using dashes.

To find out whether the graph of R intersects the horizontal asymptote $y = 2$, we solve the equation $R(x) = 2$.

$$R(x) = \frac{2x - 1}{x + 2} = 2$$
$$2x - 1 = 2(x + 2)$$
$$2x - 1 = 2x + 4$$
$$-1 = 4 \qquad \textit{Impossible}$$

The graph does not intersect the line $y = 2$.

STEP 5: Look at the given expression for R. The zeros of the numerator and denominator, $-2, \frac{1}{2}$, and 2, divide the x-axis into four intervals:

$$(-\infty, -2) \qquad \left(-2, \frac{1}{2}\right) \qquad \left(\frac{1}{2}, 2\right) \qquad (2, \infty)$$

Construct Table 13.

Table 13

	-2	$1/2$	2	$\rightarrow x$
Interval	$(-\infty, -2)$	$\left(-2, \dfrac{1}{2}\right)$	$\left(\dfrac{1}{2}, 2\right)$	$(2, \infty)$
Number chosen	-3	-1	1	3
Value of R	$R(-3) = 7$	$R(-1) = -3$	$R(1) = \dfrac{1}{3}$	$R(3) = 1$
Location of graph	Above x-axis	Below x-axis	Above x-axis	Above x-axis
Point on graph	$(-3, 7)$	$(-1, -3)$	$\left(1, \dfrac{1}{3}\right)$	$(3, 1)$

Plot these points.

STEP 6: From Table 13 we know the graph of R is above the x-axis for $x < -2$. From Step 4 we know the graph of R does not intersect the asymptote $y = 2$. Therefore, the graph of R will approach $y = 2$ from above as $x \rightarrow -\infty$ and will approach the vertical asymptote $x = -2$ at the top from the left.

Since the graph of R is below the x-axis for $-2 < x < \dfrac{1}{2}$, the graph of R will approach $x = -2$ at the bottom from the right. Lastly, since the graph of R is above the x-axis for $x > \dfrac{1}{2}$ and does not intersect the horizontal asymptote $y = 2$, the graph of R will approach $y = 2$ from below as $x \to \infty$. See Figure 39(a).

STEP 7: See Figure 39(b) for the complete graph. Since R is not defined at 2, there is a hole at the point $\left(2, \dfrac{3}{4}\right)$.

Figure 39

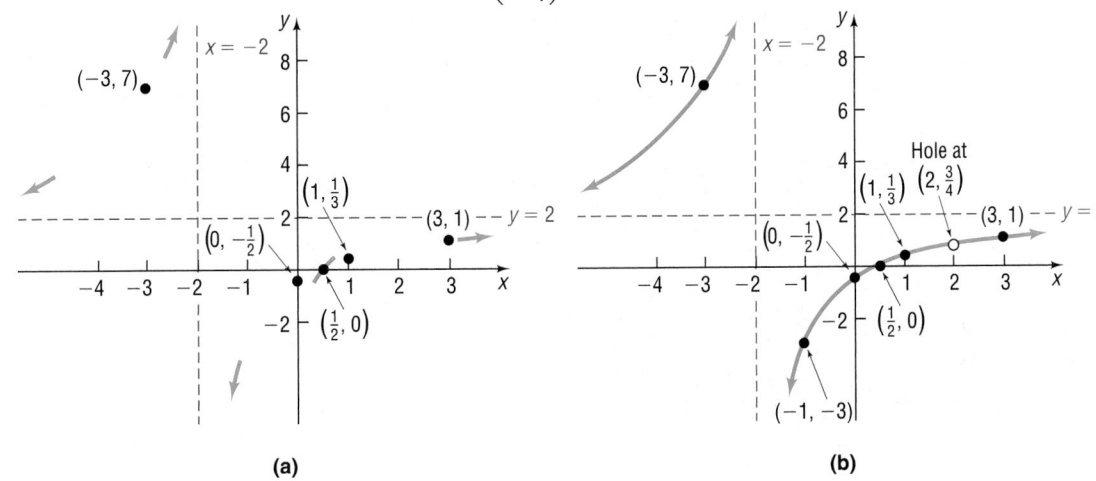

(a) (b)

Exploration

Graph $R(x) = \dfrac{2x^2 - 5x + 2}{x^2 - 4}$. Do you see the hole at $\left(2, \dfrac{3}{4}\right)$? TRACE along the graph. Did you obtain an ERROR at $x = 2$? Are you convinced that an algebraic analysis of a rational function is required in order to accurately interpret the graph obtained with a graphing utility?

NOTE: The coordinates of the hole were obtained by evaluating R in lowest terms at 2. R in lowest terms is $\dfrac{2x - 1}{x + 2}$, which, at $x = 2$, is $\dfrac{2(2) - 1}{2 + 2} = \dfrac{3}{4}$. ▪

As Example 5 shows, **the zeros of the denominator of a rational function give rise to either vertical asymptotes or holes on the graph.**

✏️— **NOW WORK PROBLEM 33.**

We now discuss the problem of finding a rational function from its graph.

| EXAMPLE 6 | **Constructing a Rational Function from Its Graph** |

Find a rational function that might have the graph shown in Figure 40.

Figure 40

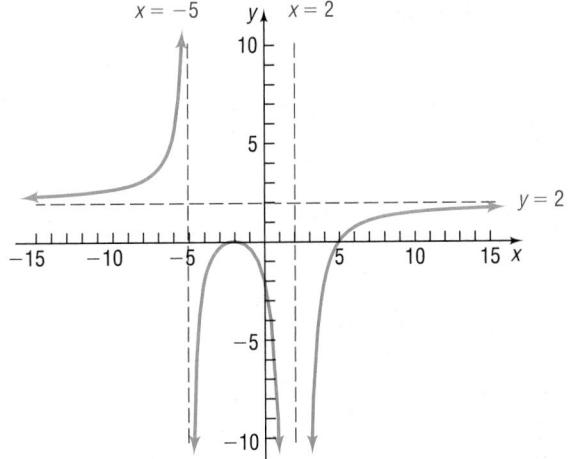

Solution The numerator of a rational function $R(x) = \dfrac{p(x)}{q(x)}$ in lowest terms determines the x-intercepts of its graph. The graph shown in Figure 40 has x-intercepts -2 (even multiplicity; graph touches the x-axis) and 5 (odd multiplicity; graph crosses the x-axis). So one possibility for the numerator is $p(x) = (x + 2)^2(x - 5)$.

The denominator of a rational function in lowest terms determines the vertical asymptotes of its graph. The vertical asymptotes of the graph are $x = -5$ and $x = 2$. Since $R(x)$ approaches ∞ to the left of $x = -5$ and $R(x)$ approaches $-\infty$ to the right of $x = -5$, we know that $(x + 5)$ is a factor of odd multiplicity in $q(x)$. Also, $R(x)$ approaches $-\infty$ on both sides of $x = 2$, so $(x - 2)$ is a factor of even multiplicity in $q(x)$. A possibility for the denominator is $q(x) = (x + 5)(x - 2)^2$. So far we have $R(x) = \dfrac{(x + 2)^2(x - 5)}{(x + 5)(x - 2)^2}$.

However, the horizontal asymptote of the graph given in Figure 40, is $y = 2$, so we know that the degree of the numerator must equal the degree of the denominator and the quotient of leading coefficients must be $\dfrac{2}{1}$. This leads to

$$R(x) = \frac{2(x + 2)^2(x - 5)}{(x + 5)(x - 2)^2}$$

Figure 41

 ✔ CHECK: Figure 41 shows the graph of R on a graphing utility. Since Figure 41 looks similar to Figure 40, we have found a rational function R for the graph in Figure 40. ■

NOW WORK PROBLEM 45.

2 Solve Applied Problems Involving Rational Functions

EXAMPLE 7 **Finding the Least Cost of a Can**

Reynolds Metal Company manufactures aluminum cans in the shape of a cylinder with a capacity of 500 cubic centimeters $\left(\dfrac{1}{2} \text{ liter}\right)$. The top and bottom of the can are made of a special aluminum alloy that costs 0.05¢ per square centimeter. The sides of the can are made of material that costs 0.02¢ per square centimeter.

(a) Express the cost of material for the can as a function of the radius r of the can.

(b) Use a graphing utility to graph the function $C = C(r)$.

(c) What value of r will result in the least cost?

(d) What is this least cost?

Solution (a) Figure 42 illustrates the components of a can in the shape of a right circular cylinder. Notice that the material required to produce a cylindrical can of height h and radius r consists of a rectangle of area $2\pi rh$ and two circles, each of area πr^2. The total cost C (in cents) of manufacturing the can is therefore

Figure 42

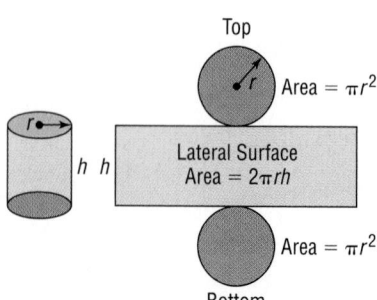

$C = $ Cost of the top and bottom $+$ Cost of the side

$= \underbrace{2(\pi r^2)}_{\substack{\text{Total area} \\ \text{of top and} \\ \text{bottom}}} \underbrace{(0.05)}_{\substack{\text{Cost/unit} \\ \text{area}}} + \underbrace{(2\pi rh)}_{\substack{\text{Total} \\ \text{area of} \\ \text{side}}} \underbrace{(0.02)}_{\substack{\text{Cost/unit} \\ \text{area}}}$

$= 0.10\pi r^2 + 0.04\pi rh$

But we have the additional restriction that the height h and radius r must be chosen so that the volume V of the can is 500 cubic centimeters. Since $V = \pi r^2 h$, we have

$$500 = \pi r^2 h \quad \text{so} \quad h = \frac{500}{\pi r^2}$$

Figure 43

Substituting this expression for h, the cost C, in cents, as a function of the radius r is

$$C(r) = 0.10\pi r^2 + 0.04\pi r \frac{500}{\pi r^2} = 0.10\pi r^2 + \frac{20}{r} = \frac{0.10\pi r^3 + 20}{r}$$

(b) See Figure 43 for the graph of $C = C(r)$.

(c) Using the MINIMUM command, the cost is least for a radius of about 3.17 centimeters.

(d) The least cost is $C(3.17) \approx 9.47¢$. ■

━━━ **NOW WORK PROBLEM 55.**

3 Construct a Model Using Inverse Variation

Figure 44

$y = \dfrac{k}{x}, k > 0, x > 0$

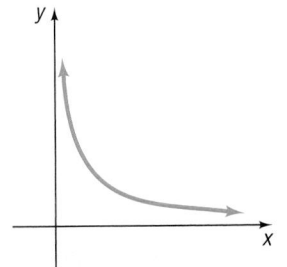

In Section 2.2, we discussed direct variation. We now study **inverse variation**.

Let x and y denote two quantities. Then y **varies inversely** with x, or y is **inversely proportional to** x, if there is a nonzero constant k such that

$$y = \frac{k}{x}$$

The graph in Figure 44 illustrates the relationship between y and x if y varies inversely with x and $k > 0$, $x > 0$.

━━

| EXAMPLE 8 | **Maximum Weight That Can Be Supported by a Piece of Pine** |

Figure 45

See Figure 45. The maximum weight W that can be safely supported by a 2-inch by 4-inch piece of pine varies inversely with its length l. Experiments indicate that the maximum weight a 10-foot-long 2-by-4 pine can support is 500 pounds. Write a general formula relating the maximum weight W (in pounds) to the length l (in feet). Find the maximum weight W that can be safely supported by a length of 25 feet.

Solution Because W varies inversely with l, we know that

$$W = \frac{k}{l}$$

for some constant k. Because $W = 500$ when $l = 10$, we have

$$500 = \frac{k}{10}$$

$$k = 5000$$

So in all cases,

$$W(l) = \frac{5000}{l}$$

Figure 46

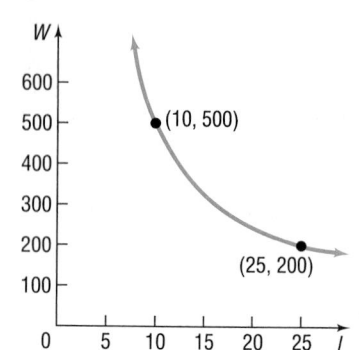

In particular, the maximum weight W that can be safely supported by a piece of pine 25 feet in length is

$$W(25) = \frac{5000}{25} = 200 \text{ pounds}$$

Figure 46 illustrates the relationship between the weight W and the length l. ∎

 NOW WORK PROBLEM 57.

4 Construct a Model Using Joint or Combined Variation

When a variable quantity Q is proportional to the product of two or more other variables, we say that Q **varies jointly** with these quantities. Finally, combinations of direct and/or inverse variation may occur. This is usually referred to as **combined variation**.

Let's look at an example.

EXAMPLE 9

Loss of Heat Through a Wall

The loss of heat through a wall varies jointly with the area of the wall and the difference between the inside and outside temperatures and varies inversely with the thickness of the wall. Write an equation that relates these quantities.

Solution

We begin by assigning symbols to represent the quantities:

L = Heat loss T = Temperature difference

A = Area of wall d = Thickness of wall

Then

$$L = k\frac{AT}{d}$$

where k is the constant of proportionality. ∎

In direct or inverse variation, the quantities that vary may be raised to powers. For example, in the early seventeenth century, Johannes Kepler (1571–1630) discovered that the square of the period of revolution T of a planet around the Sun varies directly with the cube of its mean distance a from the Sun. That is, $T^2 = ka^3$, where k is the constant of proportionality.

EXAMPLE 10

Force of the Wind on a Window

Figure 47

See Figure 47. The force F of the wind on a flat surface positioned at a right angle to the direction of the wind varies jointly with the area A of the surface and the square of the speed v of the wind. A wind of 30 miles per hour blowing on a window measuring 4 feet by 5 feet has a force of 150 pounds. What is the force on a window measuring 3 feet by 4 feet caused by a wind of 50 miles per hour?

Solution

Since F varies jointly with A and v^2, we have

$$F = kAv^2$$

where k is the constant of proportionality. We are told that $F = 150$ when $A = 4 \cdot 5 = 20$ and $v = 30$. Then we have

$$150 = k(20)(900) \qquad F = kAv^2, F = 150, A = 20, v = 30$$

$$k = \frac{1}{120}$$

The general formula is therefore

$$F = \frac{1}{120} A v^2$$

For a wind of 50 miles per hour blowing on a window whose area is $A = 3 \cdot 4 = 12$ square feet, the force F is

$$F = \frac{1}{120}(12)(2500) = 250 \text{ pounds}$$

■

NOW WORK PROBLEM 63.

3.3 Assess Your Understanding

'Are You Prepared?'

The answer is given at the end of these exercises. If you get a wrong answer, read the pages listed in red.

1. The intercepts of the equation $y = \dfrac{x - 1}{x^2 - 4}$ are _____. (pp. 10–11)

Concepts and Vocabulary

2. If the numerator and the denominator of a rational function have no common factors, the rational function is _____.

3. *True or False:* The graph of a polynomial function sometimes has a hole.

4. *True or False:* The graph of a rational function never intersects a horizontal asymptote.

5. *True or False:* The graph of a rational function sometimes intersects a vertical asymptote.

6. *True or False:* The graph of a rational function sometimes has a hole.

Skill Building

In Problems 7–44, follow Steps 1 through 7 on page 214 to analyze the graph of each function.

7. $R(x) = \dfrac{x + 1}{x(x + 4)}$

8. $R(x) = \dfrac{x}{(x - 1)(x + 2)}$

9. $R(x) = \dfrac{3x + 3}{2x + 4}$

10. $R(x) = \dfrac{2x + 4}{x - 1}$

11. $R(x) = \dfrac{3}{x^2 - 4}$

12. $R(x) = \dfrac{6}{x^2 - x - 6}$

13. $P(x) = \dfrac{x^4 + x^2 + 1}{x^2 - 1}$

14. $Q(x) = \dfrac{x^4 - 1}{x^2 - 4}$

15. $H(x) = \dfrac{x^3 - 1}{x^2 - 9}$

16. $G(x) = \dfrac{x^3 + 1}{x^2 + 2x}$

17. $R(x) = \dfrac{x^2}{x^2 + x - 6}$

18. $R(x) = \dfrac{x^2 + x - 12}{x^2 - 4}$

19. $G(x) = \dfrac{x}{x^2 - 4}$

20. $G(x) = \dfrac{3x}{x^2 - 1}$

21. $R(x) = \dfrac{3}{(x - 1)(x^2 - 4)}$

22. $R(x) = \dfrac{-4}{(x + 1)(x^2 - 9)}$

23. $H(x) = \dfrac{x^2 - 1}{x^4 - 16}$

24. $H(x) = \dfrac{x^2 + 4}{x^4 - 1}$

25. $F(x) = \dfrac{x^2 - 3x - 4}{x + 2}$

26. $F(x) = \dfrac{x^2 + 3x + 2}{x - 1}$

27. $R(x) = \dfrac{x^2 + x - 12}{x - 4}$

28. $R(x) = \dfrac{x^2 - x - 12}{x + 5}$

29. $F(x) = \dfrac{x^2 + x - 12}{x + 2}$

30. $G(x) = \dfrac{x^2 - x - 12}{x + 1}$

31. $R(x) = \dfrac{x(x - 1)^2}{(x + 3)^3}$

32. $R(x) = \dfrac{(x - 1)(x + 2)(x - 3)}{x(x - 4)^2}$

33. $R(x) = \dfrac{x^2 + x - 12}{x^2 - x - 6}$

34. $R(x) = \dfrac{x^2 + 3x - 10}{x^2 + 8x + 15}$

35. $R(x) = \dfrac{6x^2 - 7x - 3}{2x^2 - 7x + 6}$

36. $R(x) = \dfrac{8x^2 + 26x + 15}{2x^2 - x - 15}$

37. $R(x) = \dfrac{x^2 + 5x + 6}{x + 3}$

38. $R(x) = \dfrac{x^2 + x - 30}{x + 6}$

39. $f(x) = x + \dfrac{1}{x}$

40. $f(x) = 2x + \dfrac{9}{x}$

41. $f(x) = x^2 + \dfrac{1}{x}$

42. $f(x) = 2x^2 + \dfrac{16}{x}$

43. $f(x) = x + \dfrac{1}{x^3}$

44. $f(x) = 2x + \dfrac{9}{x^3}$

In Problems 45–48, find a rational function that might have the given graph. (More than one answer might be possible.)

45.

46.

47.

48.

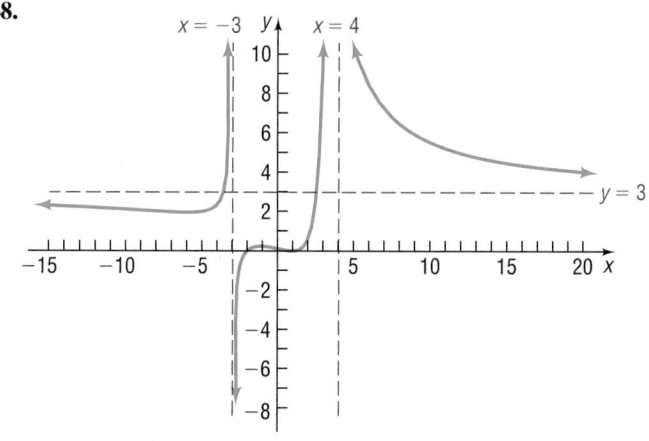

Applications and Extensions

49. Drug Concentration The concentration C of a certain drug in a patient's bloodstream t hours after injection is given by

$$C(t) = \frac{t}{2t^2 + 1}$$

(a) Find the horizontal asymptote of $C(t)$. What happens to the concentration of the drug as t increases?

(b) Using your graphing utility, graph $C(t)$.

(c) Determine the time at which the concentration is highest.

50. Drug Concentration The concentration C of a certain drug in a patient's bloodstream t minutes after injection is given by

$$C(t) = \frac{50t}{t^2 + 25}$$

(a) Find the horizontal asymptote of $C(t)$. What happens to the concentration of the drug as t increases?

(b) Using your graphing utility, graph $C(t)$.

(c) Determine the time at which the concentration is highest.

51. Average Cost In Problem 99, Exercise 3.1, the cost function C (in thousands of dollars) for manufacturing x Chevy Cobalts was found to be

$$C(x) = 0.2x^3 - 2.3x^2 + 14.3x + 10.2$$

Economists define the **average cost function** as

$$\overline{C}(x) = \frac{C(x)}{x}$$

(a) Find the average cost function.

(b) What is the average cost of producing six Cobalts?

(c) What is the average cost of producing nine Cobalts?

(d) Using your graphing utility, graph the average cost function.

(e) Using your graphing utility, find the number of Cobalts that should be produced to minimize the average cost.

(f) What is the minimum average cost?

52. Average Cost In Problem 100, Exercise 3.1, the cost function C (in thousands of dollars) for printing x textbooks (in thousands of units) is found to be

$$C(x) = 0.015x^3 - 0.595x^2 + 9.15x + 98.43$$

(a) Find the average cost function (refer to Problem 51).
(b) What is the average cost of printing 13,000 textbooks per week?
(c) What is the average cost of printing 25,000 textbooks per week?
(d) Using your graphing utility, graph the average cost function.
(e) Using your graphing utility, find the number of textbooks that should be printed to minimize the average cost.
(f) What is the minimum average cost?

53. Minimizing Surface Area United Parcel Service has contracted you to design a closed box with a square base that has a volume of 10,000 cubic inches. See the illustration.

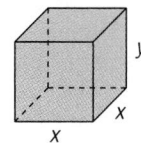

(a) Find a function for the surface area of the box.
(b) Using a graphing utility, graph the function found in part (a).
(c) What is the minimum amount of cardboard that can be used to construct the box?
(d) What are the dimensions of the box that minimize the surface area?
(e) Why might UPS be interested in designing a box that minimizes the surface area?

54. Minimizing Surface Area United Parcel Service has contracted you to design a closed box with a square base that has a volume of 5000 cubic inches. See the illustration.

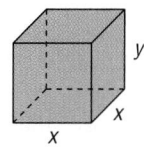

(a) Find a function for the surface area of the box.
(b) Using a graphing utility, graph the function found in part (a).
(c) What is the minimum amount of cardboard that can be used to construct the box?
(d) What are the dimensions of the box that minimize the surface area?
(e) Why might UPS be interested in designing a box that minimizes the surface area?

55. Cost of a Can A can in the shape of a right circular cylinder is required to have a volume of 500 cubic centimeters. The top and bottom are made of material that costs 6¢ per square centimeter, while the sides are made of material that costs 4¢ per square centimeter.
(a) Express the total cost C of the material as a function of the radius r of the cylinder. (Refer to Figure 42.)
(b) Graph $C = C(r)$. For what value of r is the cost C a minimum?

56. Material Needed to Make a Drum A steel drum in the shape of a right circular cylinder is required to have a volume of 100 cubic feet.

(a) Express the amount A of material required to make the drum as a function of the radius r of the cylinder.
(b) How much material is required if the drum's radius is 3 feet?
(c) How much material is required if the drum's radius is 4 feet?
(d) How much material is required if the drum's radius is 5 feet?
(e) Graph $A = A(r)$. For what value of r is A smallest?

57. Demand Suppose that the demand D for candy at the movie theater is inversely related to the price p.
(a) When the price of candy is \$2.75 per bag, the theater sells 156 bags of candy. Express the demand for candy as a function of its price.
(b) Determine the number of bags of candy that will be sold if the price is raised to \$3 a bag.

58. Driving to School The time t that it takes to get to school varies inversely with your average speed s.
(a) Suppose that it takes you 40 minutes to get to school when your average speed is 30 miles per hour. Express the driving time to school as a function of average speed.
(b) Suppose that your average speed to school is 40 miles per hour. How long will it take you to get to school?

59. Pressure The volume of a gas V held at a constant temperature in a closed container varies inversely with its pressure P. If the volume of a gas is 600 cubic centimeters (cm^3) when the pressure is 150 millimeters of mercury (mm Hg), find the volume when the pressure is 200 mm Hg.

60. Resistance The current i in a circuit is inversely proportional to its resistance R measured in ohms. Suppose that when the current in a circuit is 30 amperes the resistance is 8 ohms. Find the current in the same circuit when the resistance is 10 ohms.

61. Weight The weight of an object above the surface of Earth varies inversely with the square of the distance from the center of Earth. If Maria weighs 125 pounds when she is on the surface of Earth (3960 miles from the center), determine Maria's weight if she is at the top of Mount McKinley (3.8 miles from the surface of Earth).

62. Intensity of Light The intensity I of light (measured in foot-candles) varies inversely with the square of the distance from the bulb. Suppose that the intensity of a 100-watt light bulb at a distance of 2 meters is 0.075 foot-candle. Determine the intensity of the bulb at a distance of 5 meters.

63. **Geometry** The volume V of a right circular cylinder varies jointly with the square of its radius r and its height h. The constant of proportionality is π. (See the figure.) Write an equation for V.

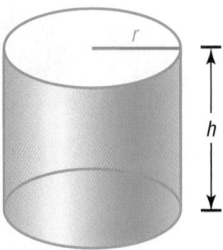

64. **Geometry** The volume V of a right circular cone varies jointly with the square of its radius r and its height h. The constant of proportionality is $\dfrac{\pi}{3}$. (See the figure.) Write an equation for V.

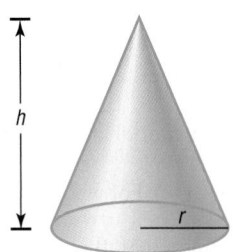

65. **Horsepower** The horsepower (hp) that a shaft can safely transmit varies jointly with its speed (in revolutions per minute, rpm) and the cube of its diameter. If a shaft of a certain material 2 inches in diameter can transmit 36 hp at 75 rpm, what diameter must the shaft have to transmit 45 hp at 125 rpm?

66. **Chemistry: Gas Laws** The volume V of an ideal gas varies directly with the temperature T and inversely with the pres-sure P. Write an equation relating V, T, and P using k as the constant of proportionality. If a cylinder contains oxygen at a temperature of 300 K and a pressure of 15 atmospheres in a volume of 100 liters, what is the constant of proportionali-ty k? If a piston is lowered into the cylinder, decreasing the volume occupied by the gas to 80 liters and raising the tem-perature to 310 K, what is the gas pressure?

67. **Physics: Kinetic Energy** The kinetic energy K of a moving object varies jointly with its mass m and the square of its ve-locity v. If an object weighing 25 kilograms and moving with a velocity of 10 meters per second has a kinetic energy of 1250 Joules, find its kinetic energy when the velocity is 15 meters per second.

68. **Electrical Resistance of a Wire** The electrical resistance of a wire varies directly with the length of the wire and inversely with the square of the diameter of the wire. If a wire 432 feet long and 4 millimeters in diameter has a resistance of 1.24 ohms, find the length of a wire of the same material whose resistance is 1.44 ohms and whose diameter is 3 millimeters.

69. **Measuring the Stress of Materials** The stress in the materi-al of a pipe subject to internal pressure varies jointly with the internal pressure and the internal diameter of the pipe and inversely with the thickness of the pipe. The stress is 100 pounds per square inch when the diameter is 5 inches, the thickness is 0.75 inch, and the internal pressure is 25 pounds per square inch. Find the stress when the internal pressure is 40 pounds per square inch if the diameter is 8 inches and the thickness is 0.50 inch.

70. **Safe Load for a Beam** The maximum safe load for a hori-zontal rectangular beam varies jointly with the width of the beam and the square of the thickness of the beam and in-versely with its length. If an 8-foot beam will support up to 750 pounds when the beam is 4 inches wide and 2 inches thick, what is the maximum safe load in a similar beam 10 feet long, 6 inches wide, and 2 inches thick?

Discussion and Writing

71. Graph each of the following functions:

$$y = \frac{x^2 - 1}{x - 1} \qquad y = \frac{x^3 - 1}{x - 1}$$

$$y = \frac{x^4 - 1}{x - 1} \qquad y = \frac{x^5 - 1}{x - 1}$$

Is $x = 1$ a vertical asymptote? Why not? What is happening for $x = 1$? What do you conjecture about $y = \dfrac{x^n - 1}{x - 1}$, $n \geq 1$ an integer, for $x = 1$?

72. Graph each of the following functions:

$$y = \frac{x^2}{x - 1} \qquad y = \frac{x^4}{x - 1} \qquad y = \frac{x^6}{x - 1} \qquad y = \frac{x^8}{x - 1}$$

What similarities do you see? What differences?

73. Write a few paragraphs that provide a general strategy for graphing a rational function. Be sure to mention the follow-ing: proper, improper, intercepts, and asymptotes.

74. Create a rational function that has the following character-istics: crosses the x-axis at 2; touches the x-axis at -1; one vertical asymptote at $x = -5$ and another at $x = 6$; and one horizontal asymptote, $y = 3$. Compare yours to a fellow classmate's. How do they differ? What are their similarities?

75. Create a rational function that has the following character-istics: crosses the x-axis at 3; touches the x-axis at -2; one vertical asymptote, $x = 1$; and one horizontal asymptote, $y = 2$. Give your rational function to a fellow classmate and ask for a written critique of your rational function.

76. The formula on page 226 attributed to Johannes Kepler is one of the famous three Keplerian Laws of Planetary Mo-tion. Go to the library and research these laws. Write a brief paper about these laws and Kepler's place in history.

77. Using a situation that has not been discussed in the text, write a real-world problem that you think involves two variables that vary directly. Exchange your problem with another student's to solve and critique.

78. Using a situation that has not been discussed in the text, write a real-world problem that you think involves two vari-ables that vary inversely. Exchange your problem with an-other student's to solve and critique.

79. Using a situation that has not been discussed in the text, write a real-world problem that you think involves three variables that vary jointly. Exchange your problem with another student's to solve and critique.

'Are You Prepared?' Answers

1. $\left(0, \dfrac{1}{4}\right), (1, 0)$

3.4 Polynomial and Rational Inequalities

PREPARING FOR THIS SECTION *Before getting started, review the following:*

- Solving Inequalities (Appendix A, Section A.9, pp. A67–A71)

Now work the 'Are You Prepared?' problem on page 235.

OBJECTIVES 1 Solve Polynomial Inequalities
2 Solve Rational Inequalities

In this section we solve inequalities that involve polynomials of degree 3 and higher, as well as some that involve rational expressions. To solve such inequalities, we use the information obtained in the previous three sections about the graph of polynomial and rational functions. The general idea follows:

Suppose that the polynomial or rational inequality is in one of the forms

$$f(x) < 0 \qquad f(x) > 0 \qquad f(x) \le 0 \qquad f(x) \ge 0$$

Locate the zeros of f if f is a polynomial function, and locate the zeros of the numerator and the denominator if f is a rational function. If we use these zeros to divide the real number line into intervals, we know that on each interval the graph of f is either above the x-axis $[f(x) > 0]$ or below the x-axis $[f(x) < 0]$. In other words, we have found the solution of the inequality.

The following steps provide more detail.

Steps for Solving Polynomial and Rational Inequalities

STEP 1: Write the inequality so that a polynomial or rational expression f is on the left side and zero is on the right side in one of the following forms:

$$f(x) > 0 \qquad f(x) \ge 0 \qquad f(x) < 0 \qquad f(x) \le 0$$

For rational expressions, be sure that the left side is written as a single quotient.

STEP 2: Determine the numbers at which the expression f on the left side equals zero and, if the expression is rational, the numbers at which the expression f on the left side is undefined.

STEP 3: Use the numbers found in Step 2 to separate the real number line into intervals.

STEP 4: Select a number in each interval and evaluate f at the number.

(a) If the value of f is positive, then $f(x) > 0$ for all numbers x in the interval.

(b) If the value of f is negative, then $f(x) < 0$ for all numbers x in the interval.

If the inequality is not strict, include the solutions of $f(x) = 0$ in the solution set.

1 Solve Polynomial Inequalities

EXAMPLE 1

Solving a Polynomial Inequality

Solve the inequality $x^4 \le 4x^2$, and graph the solution set.

Solution

STEP 1: Rearrange the inequality so that 0 is on the right side.

$$x^4 \le 4x^2$$
$$x^4 - 4x^2 \le 0 \qquad \text{Subtract } 4x^2 \text{ from both sides of the inequality.}$$

This inequality is equivalent to the one that we wish to solve.

STEP 2: Find the zeros of $f(x) = x^4 - 4x^2$ by solving the equation $x^4 - 4x^2 = 0$.

$$x^4 - 4x^2 = 0$$
$$x^2(x^2 - 4) = 0 \qquad \text{Factor.}$$
$$x^2(x + 2)(x - 2) = 0 \qquad \text{Factor.}$$
$$x = 0 \quad \text{or} \quad x = -2 \quad \text{or} \quad x = 2$$

STEP 3: We use the zeros of f to separate the real number line into three intervals:

$$(-\infty, -2) \qquad (-2, 0) \qquad (2, \infty)$$

STEP 4: Select a number in each interval and evaluate $f(x) = x^4 - 4x^2$ to determine if $f(x)$ is positive or negative. See Table 14.

Table 14

Interval	$(-\infty, -2)$	$(-2, 0)$	$(0, 2)$	$(2, \infty)$
Number chosen	-3	-1	1	3
Value of f	$f(-3) = 45$	$f(-1) = -3$	$f(1) = -3$	$f(3) = 45$
Conclusion	Positive	Negative	Negative	Positive

Based on Table 14, we know that $f(x) < 0$ for all x in the intervals $(-2, 0)$ and $(0, 2)$, that is, for all x such that $-2 < x < 0$ and $0 < x < 2$. However, because the original inequality is not strict, numbers x that satisfy the equation $f(x) = x^4 - 4x^2 = 0$ are also solutions of the inequality $x^4 \le 4x^2$. Thus, we include $-2, 0,$ and -2. The solution set of the given inequality is $\{x|-2 \le x \le 2\}$ or, using interval notation, $[-2, 2]$.

Figure 48

Figure 48 shows the graph of the solution set. ■

NOW WORK PROBLEMS 5 AND 13.

EXAMPLE 2

Solving a Polynomial Inequality

Solve the inequality $x^4 > x$, and graph the solution set.

Solution

STEP 1: Rearrange the inequality so that 0 is on the right side.

$$x^4 > x$$
$$x^4 - x > 0 \qquad \text{Subtract } x \text{ from both sides of the inequality.}$$

This inequality is equivalent to the one that we wish to solve.

STEP 2: Find the zeros of $f(x) = x^4 - x$ by solving $x^4 - x = 0$.

$$x^4 - x = 0$$
$$x(x^3 - 1) = 0 \qquad \text{\small Factor out x.}$$
$$x(x - 1)(x^2 + x + 1) = 0 \qquad \text{\small Factor the difference of two cubes.}$$
$$x = 0 \quad \text{or} \quad x - 1 = 0 \quad \text{or} \quad x^2 + x + 1 = 0 \qquad \text{\small Set each factor equal to zero}$$
$$\text{\small and solve. The equation}$$
$$x = 0 \quad \text{or} \qquad x = 1 \qquad\qquad\qquad \text{\small } x^2 + x + 1 = 0 \text{ has no real}$$
$$\text{\small solutions. (Do you see why?)}$$

STEP 3: We use the zeros 0 and 1 to separate the real number line into three intervals:

$$(-\infty, 0) \qquad (0, 1) \qquad (1, \infty)$$

STEP 4: We choose a number in each interval and evaluate $f(x) = x^4 - x$ to determine if $f(x)$ is positive or negative. See Table 15.

Table 15

	$(-\infty, 0)$	$(0, 1)$	$(1, \infty)$
Interval	$(-\infty, 0)$	$(0, 1)$	$(1, \infty)$
Number chosen	-1	$\frac{1}{2}$	2
Value of f	$f(-1) = 2$	$f\left(\frac{1}{2}\right) = -\frac{7}{16}$	$f(2) = 14$
Conclusion	Positive	Negative	Positive

Based on Table 15, we know that $f(x) > 0$ for all x in the intervals $(-\infty, 0)$ or $(1, \infty)$, that is, for all numbers x for which $x < 0$ or $x > 1$. Because the original inequality is strict, the solution set of the given inequality $\{x \mid x < 0 \text{ or } x > 1\}$ or, using interval notation $(-\infty, 0)$ or $(1, \infty)$. Figure 49 shows the graph of the solution set. ∎

Figure 49

NOW WORK PROBLEM **17.**

2 Solve Rational Inequalities

EXAMPLE 3	**Solving a Rational Inequality**

Solve the inequality $\dfrac{(x + 3)(2 - x)}{(x - 1)^2} > 0$, and graph the solution set.

Solution

STEP 1: The domain of the variable x is $\{x \mid x \neq 1\}$. The inequality is already in a form with 0 on the right side.

STEP 2: Let $f(x) = \dfrac{(x + 3)(2 - x)}{(x - 1)^2}$. The zeros of the numerator of f are -3 and 2; the zero of the denominator is 1.

STEP 3: We use the zeros found in Step 2, -3, 1, and 2, to separate the real number line into four intervals:

$$(-\infty, -3) \qquad (-3, 1) \qquad (1, 2) \qquad (2, \infty)$$

STEP 4: Select a number in each interval and evaluate $f(x) = \dfrac{(x + 3)(2 - x)}{(x - 1)^2}$ to determine if $f(x)$ is positive or negative. See Table 16.

Table 16

Interval	$(-\infty, -3)$	$(-3, 1)$	$(1, 2)$	$(2, \infty)$
Number chosen	-4	0	$\dfrac{3}{2}$	3
Value of f	$f(-4) = -\dfrac{6}{25}$	$f(0) = 6$	$f\left(\dfrac{3}{2}\right) = 9$	$f(3) = -\dfrac{3}{2}$
Conclusion	Negative	Positive	Positive	Negative

Figure 50

Based on Table 16, we know that $f(x) > 0$ for all x in the intervals $(-3, 1)$ or $(1, 2)$, that is, for all x such that $-3 < x < 1$ or $1 < x < 2$. Because the original inequality is strict, the solution set of the given inequality is $\{x | -3 < x < 2, x \neq 1\}$ or, using interval notation, $(-3, 1)$ or $(1, 2)$. Figure 50 shows the graph of the solution set. Notice the hole at $x = 1$ to indicate that 1 is to be excluded. ∎

NOW WORK PROBLEM 21.

EXAMPLE 4 **Solving a Rational Inequality**

Solve the inequality $\dfrac{4x + 5}{x + 2} \geq 3$, and graph the solution set.

Solution **STEP 1:** The domain of the variable x is $\{x | x \neq -2\}$. We rearrange terms so that 0 is on the right side.

$$\frac{4x + 5}{x + 2} - 3 \geq 0 \qquad \text{Subtract 3 from both sides of the inequality.}$$

STEP 2: Let $f(x) = \dfrac{4x + 5}{x + 2} - 3$. To find the zeros of the numerator and the denominator, we express f as a single quotient.

$$f(x) = \frac{4x + 5}{x + 2} - 3$$
$$= \frac{4x + 5}{x + 2} - 3\left(\frac{x + 2}{x + 2}\right) \qquad \text{Multiply } -3 \text{ by } \frac{x + 2}{x + 2}.$$
$$= \frac{4x + 5 - 3x - 6}{x + 2} \qquad \text{Write as a single quotient.}$$
$$= \frac{x - 1}{x + 2} \qquad \text{Combine like terms.}$$

The zero of the numerator of f is 1, and the zero of the denominator is -2.

STEP 3: We use the zeros found in Step 2, -2 and 1, to separate the real number line into three intervals:

$$(-\infty, -2) \qquad (-2, 1) \qquad (1, \infty)$$

STEP 4: Select a number in each interval and evaluate $f(x) = \dfrac{4x + 5}{x + 2} - 3$ to determine if $f(x)$ is positive or negative. See Table 17.

Table 17

Interval	$(-\infty, -2)$	$(-2, 1)$	$(1, \infty)$
Number chosen	-3	0	2
Value of f	$f(-3) = 4$	$f(0) = -\dfrac{1}{2}$	$f(2) = \dfrac{1}{4}$
Conclusion	Positive	Negative	Positive

Based on Table 17, we know that $f(x) > 0$ for all x in the intervals $(-\infty, -2)$ or $(1, \infty)$, that is, for all x such that $x < -2$ or $x > 1$. Because the original inequality is not strict, numbers x that satisfy the equation $f(x) = \dfrac{x-1}{x+2} = 0$ are also solutions of the inequality. Since $\dfrac{x-1}{x+2} = 0$ only if $x = 1$, we conclude that the solution set is $\{x \mid x < -2 \text{ or } x \ge 1\}$ or, using interval notation, $(-\infty, -2)$ or $[1, \infty)$.

Figure 51

Figure 51 shows the graph of the solution set.

NOW WORK PROBLEM 29.

3.4 Assess Your Understanding

'Are You Prepared?'

The answer is given at the end of these exercises. If you get a wrong answer, read the pages listed in red.

1. Solve the inequality: $3 - 4x > 5$. Graph the solution set. (pp. A57–A61)

Concepts and Vocabulary

2. *True or False:* The first step in solving the inequality $x^2 + 4x \ge -4$ is to factor the expression $x^2 + 4x$.

Skill Building

In Problems 3–40, solve each inequality.

3. $(x - 5)^2(x + 2) < 0$

4. $(x - 5)(x + 2)^2 > 0$

5. $x^3 - 4x^2 > 0$

6. $x^3 + 8x^2 < 0$

7. $x^3 - 9x \le 0$

8. $x^3 - x \ge 0$

9. $2x^3 > -8x^2$

10. $3x^3 < -15x^2$

11. $(x - 1)(x^2 + x + 4) \ge 0$

12. $(x + 2)(x^2 - x + 1) \ge 0$

13. $(x - 1)(x - 2)(x - 3) \le 0$

14. $(x + 1)(x + 2)(x + 3) \le 0$

15. $x^3 - 2x^2 - 3x > 0$

16. $x^3 + 2x^2 - 3x > 0$

17. $x^4 > x^2$

18. $x^4 < 9x^2$

19. $x^4 > 1$

20. $x^3 > 1$

21. $\dfrac{x + 1}{x - 1} > 0$

22. $\dfrac{x - 3}{x + 1} > 0$

23. $\dfrac{(x - 1)(x + 1)}{x} \le 0$

24. $\dfrac{(x - 3)(x + 2)}{x - 1} \le 0$

25. $\dfrac{(x - 2)^2}{x^2 - 1} \ge 0$

26. $\dfrac{(x + 5)^2}{x^2 - 4} \ge 0$

27. $6x - 5 < \dfrac{6}{x}$

28. $x + \dfrac{12}{x} < 7$

29. $\dfrac{x + 4}{x - 2} \le 1$

30. $\dfrac{x + 2}{x - 4} \ge 1$

31. $\dfrac{3x - 5}{x + 2} \le 2$

32. $\dfrac{x - 4}{2x + 4} \ge 1$

33. $\dfrac{1}{x - 2} < \dfrac{2}{3x - 9}$

34. $\dfrac{5}{x - 3} > \dfrac{3}{x + 1}$

35. $\dfrac{2x + 5}{x + 1} > \dfrac{x + 1}{x - 1}$

36. $\dfrac{1}{x + 2} > \dfrac{3}{x + 1}$

37. $\dfrac{x^2(3 + x)(x + 4)}{(x + 5)(x - 1)} \ge 0$

38. $\dfrac{x(x^2 + 1)(x - 2)}{(x - 1)(x + 1)} \ge 0$

39. $\dfrac{(3 - x)^3(2x + 1)}{x^3 - 1} < 0$

40. $\dfrac{(2 - x)^3(3x - 2)}{x^3 + 1} < 0$

Application and Extensions

41. For what positive numbers will the cube of a number exceed four times its square?

42. For what positive numbers will the cube of a number be less than the number?

43. What is the domain of the function $f(x) = \sqrt{x^4 - 16}$?

44. What is the domain of the function $f(x) = \sqrt{x^3 - 3x^2}$?

45. What is the domain of the function $f(x) = \sqrt{\dfrac{x - 2}{x + 4}}$?

46. What is the domain of the function $f(x) = \sqrt{\dfrac{x - 1}{x + 4}}$?

In Problems 47–50, determine where the graph of f is below the graph of g by solving the inequality $f(x) \leq g(x)$. Graph f and g together.

47. $f(x) = x^4 - 1$
$g(x) = -2x^2 + 2$

48. $f(x) = x^4 - 1$
$g(x) = x - 1$

49. $f(x) = x^4 - 4$
$g(x) = 3x^2$

50. $f(x) = x^4$
$g(x) = 2 - x^2$

51. Average Cost Suppose that the daily cost C of manufacturing bicycles is given by $C(x) = 80x + 5000$. Then the average daily cost \overline{C} is given by $\overline{C}(x) = \dfrac{80x + 5000}{x}$. How many bicycles must be produced each day in order for the average cost to be no more than \$100?

52. Average Cost See Problem 51. Suppose that the government imposes a \$1000 per day tax on the bicycle manufacturer so that the daily cost C of manufacturing x bicycles is now given by $C(x) = 80x + 6000$. Now the average daily cost \overline{C} is given by $\overline{C}(x) = \dfrac{80x + 6000}{x}$. How many bicycles must be produced each day in order for the average cost to be no more than \$100?

Discussion and Writing

53. Make up an inequality that has no solution. Make up one that has exactly one solution.

54. The inequality $x^4 + 1 < -5$ has no solution. Explain why.

55. A student attempted to solve the inequality $\dfrac{x + 4}{x - 3} \leq 0$ by multiplying both sides of the inequality by $x - 3$ to get

$x + 4 \leq 0$. This led to a solution of $\{x \mid x \leq -4\}$. Is the student correct? Explain.

56. Write a rational inequality whose solution set is $\{x \mid -3 < x \leq 5\}$.

'Are You Prepared?' Answer

1. $\{x \mid x < -\dfrac{1}{2}\}$

(number line from -2 to 1 with points marked at -2, -1, $-\frac{1}{2}$, 0, 1; shaded to the left with open circle at $-\frac{1}{2}$)

3.5 The Real Zeros of a Polynomial Function

PREPARING FOR THIS SECTION *Before getting started, review the following:*

- Evaluating Functions (Section 1.1, pp. 44–47)
- Factoring Polynomials (Appendix A, Section A.4, pp. A29–A35)
- Synthetic Division (Appendix A, Section A.5, pp. A38–A40)
- Polynomial Division (Appendix A, Section A.3, pp. A25–A27)
- Quadratic Formula (Section 2.3, pp. 137–139)

Now work the 'Are You Prepared?' problems on page 248.

OBJECTIVES
1. Use the Remainder and Factor Theorems
2. Use Descartes' Rule of Signs
3. Use the Rational Zeros Theorem
4. Find the Real Zeros of a Polynomial Function
5. Solve Polynomial Equations
6. Use the Theorem for Bounds on Zeros
7. Use the Intermediate Value Theorem

In this section, we discuss techniques that can be used to find the real zeros of a polynomial function. Recall that if r is a real zero of a polynomial function f then $f(r) = 0$, r is an x-intercept of the graph of f, and r is a solution of the equation $f(x) = 0$.

For polynomial and rational functions, we have seen the importance of the zeros for graphing. In most cases, however, the zeros of a polynomial function are

$f(a)$ and $f(m_1)$ that are of opposite sign. Now evaluate f at the midpoint m_2 of a and m_1. Repeat this process until the desired degree of accuracy is obtained. Note that each iteration places the zero in an interval whose length is half that of the previous interval. Use the bisection method to solve Problems 95–102.

[**Hint:** The process ends when both endpoints agree to the desired number of decimal places.]

Discussion and Writing

116. Is $\frac{1}{3}$ a zero of $f(x) = 2x^3 + 3x^2 - 6x + 7$? Explain.

117. Is $\frac{1}{3}$ a zero of $f(x) = 4x^3 - 5x^2 - 3x + 1$? Explain.

118. Is $\frac{3}{5}$ a zero of $f(x) = 2x^6 - 5x^4 + x^3 - x + 1$? Explain.

119. Is $\frac{2}{3}$ a zero of $f(x) = x^7 + 6x^5 - x^4 + x + 2$? Explain.

'Are You Prepared?' Answers

1. 3
2. $(3x + 2)(2x - 1)$
3. Quotient: $3x^3 + 4x^2 + 12x + 43$; Remainder: 125
4. $\left\{ \dfrac{-1 - \sqrt{13}}{2}, \dfrac{-1 + \sqrt{13}}{2} \right\}$

3.6 Complex Zeros; Fundamental Theorem of Algebra

PREPARING FOR THIS SECTION *Before getting started, review the following:*

- Complex Numbers (Appendix A, Section A.11, pp. A82–A87)
- Complex Zeros of a Quadratic Function (Section 2.7, pp. 167–169)

Now work the 'Are You Prepared?' problems on page 255.

> **OBJECTIVES** 1 Use the Conjugate Pairs Theorem
> 2 Find a Polynomial Function with Specified Zeros
> 3 Find the Complex Zeros of a Polynomial

In Section 2.3, we found the real zeros of a quadratic function. That is, we found the real zeros of a polynomial function of degree 2. Then, in Section 2.7 we found the complex zeros of a quadratic function. That is, we found the complex zeros of a polynomial function of degree 2. Recall that finding the complex zeros of a quadratic function required finding all zeros of the form $a + bi$. The zeros will be real if $b = 0$.

In Section 3.5 we found the real zeros of a polynomial function of degree 3 or higher. In this section we will find the complex zeros of a polynomial function of degree 3 or higher.

> A variable in the complex number system is referred to as a **complex variable**. A **complex polynomial function** f of degree n is a function of the form
>
> $$f(x) = a_n x^n + a_{n-1} x^{n-1} + \cdots + a_1 x + a_0 \qquad (1)$$
>
> where $a_n, a_{n-1}, \ldots, a_1, a_0$ are complex numbers, $a_n \neq 0$, n is a nonnegative integer, and x is a complex variable. As before, a_n is called the **leading coefficient** of f. A complex number r is called a **(complex) zero** of f if $f(r) = 0$.

We have learned that some quadratic equations have no real solutions, but that in the complex number system every quadratic equation has a solution, either real or complex. The next result, proved by Karl Friedrich Gauss (1777–1855) when he was 22 years old,[*] gives an extension to complex polynomials. In fact, this result is so important and useful that it has become known as the **Fundamental Theorem of Algebra**.

[*] In all, Gauss gave four different proofs of this theorem, the first one in 1799 being the subject of his doctoral dissertation.

63. $x^4 + 4x^3 + 2x^2 - x + 6 = 0$

64. $x^4 - 2x^3 + 10x^2 - 18x + 9 = 0$

65. $x^3 - \frac{2}{3}x^2 + \frac{8}{3}x + 1 = 0$

66. $x^3 + \frac{3}{2}x^2 + 3x - 2 = 0$

67. $2x^4 - 19x^3 + 57x^2 - 64x + 20 = 0$

68. $2x^4 + x^3 - 24x^2 + 20x + 16 = 0$

In Problems 69–80, find the intercepts of each polynomial function $f(x)$. Graph each polynomial function. [**Hint:** Use the factored form of f (see Problems 45–56).]

69. $f(x) = x^3 + 2x^2 - 5x - 6$

70. $f(x) = x^3 + 8x^2 + 11x - 20$

71. $f(x) = 2x^3 - x^2 + 2x - 1$

72. $f(x) = 2x^3 + x^2 + 2x + 1$

73. $f(x) = x^4 + x^2 - 2$

74. $f(x) = x^4 - 3x^2 - 4$

75. $f(x) = 4x^4 + 7x^2 - 2$

76. $f(x) = 4x^4 + 15x^2 - 4$

77. $f(x) = x^4 + x^3 - 3x^2 - x + 2$

78. $f(x) = x^4 - x^3 - 6x^2 + 4x + 8$

79. $f(x) = 4x^5 - 8x^4 - x + 2$

80. $f(x) = 4x^5 + 12x^4 - x - 3$

In Problems 81–88, find a bound on the real zeros of each polynomial function.

81. $f(x) = x^4 - 3x^2 - 4$

82. $f(x) = x^4 - 5x^2 - 36$

83. $f(x) = x^4 + x^3 - x - 1$

84. $f(x) = x^4 - x^3 + x - 1$

85. $f(x) = 3x^4 + 3x^3 - x^2 - 12x - 12$

86. $f(x) = 3x^4 - 3x^3 - 5x^2 + 27x - 36$

87. $f(x) = 4x^5 - x^4 + 2x^3 - 2x^2 + x - 1$

88. $f(x) = 4x^5 + x^4 + x^3 + x^2 - 2x - 2$

In Problems 89–94, use the Intermediate Value Theorem to show that each polynomial function has a zero in the given interval.

89. $f(x) = 8x^4 - 2x^2 + 5x - 1;$ $[0, 1]$

90. $f(x) = x^4 + 8x^3 - x^2 + 2;$ $[-1, 0]$

91. $f(x) = 2x^3 + 6x^2 - 8x + 2;$ $[-5, -4]$

92. $f(x) = 3x^3 - 10x + 9;$ $[-3, -2]$

93. $f(x) = x^5 - x^4 + 7x^3 - 7x^2 - 18x + 18;$ $[1.4, 1.5]$

94. $f(x) = x^5 - 3x^4 - 2x^3 + 6x^2 + x + 2;$ $[1.7, 1.8]$

In Problems 95–98, each equation has a solution r in the interval indicated. Use the method of Example 10 to approximate this solution correct to two decimal places.

95. $8x^4 - 2x^2 + 5x - 1 = 0;$ $0 \le r \le 1$

96. $x^4 + 8x^3 - x^2 + 2 = 0;$ $-1 \le r \le 0$

97. $2x^3 + 6x^2 - 8x + 2 = 0;$ $-5 \le r \le -4$

98. $3x^3 - 10x + 9 = 0;$ $-3 \le r \le -2$

In Problems 99–102, each polynomial function has exactly one positive zero. Use the method of Example 10 to approximate the zero correct to two decimal places.

99. $f(x) = x^3 + x^2 + x - 4$

100. $f(x) = 2x^4 + x^2 - 1$

101. $f(x) = 2x^4 - 3x^3 - 4x^2 - 8$

102. $f(x) = 3x^3 - 2x^2 - 20$

Applications and Extensions

103. Find k such that $f(x) = x^3 - kx^2 + kx + 2$ has the factor $x - 2$.

104. Find k such that $f(x) = x^4 - kx^3 + kx^2 + 1$ has the factor $x + 2$.

105. What is the remainder when $f(x) = 2x^{20} - 8x^{10} + x - 2$ is divided by $x - 1$?

106. What is the remainder when $f(x) = -3x^{17} + x^9 - x^5 + 2x$ is divided by $x + 1$?

107. Use the Factor Theorem to prove that $x - c$ is a factor of $x^n - c^n$ for any positive integer n.

108. Use the Factor Theorem to prove that $x + c$ is a factor of $x^n + c^n$ if $n \ge 1$ is an odd integer.

109. One solution of the equation $x^3 - 8x^2 + 16x - 3 = 0$ is 3. Find the sum of the remaining solutions.

110. One solution of the equation $x^3 + 5x^2 + 5x - 2 = 0$ is -2. Find the sum of the remaining solutions.

111. What is the length of the edge of a cube if, after a slice 1 inch thick is cut from one side, the volume remaining is 294 cubic inches?

112. What is the length of the edge of a cube if its volume could be doubled by an increase of 6 centimeters in one edge, an increase of 12 centimeters in a second edge, and a decrease of 4 centimeters in the third edge?

113. Let $f(x)$ be a polynomial function whose coefficients are integers. Suppose that r is a real zero of f and that the leading coefficient of f is 1. Use the Rational Zeros Theorem to show that r is either an integer or an irrational number.

114. Prove the Rational Zeros Theorem.

[**Hint:** Let $\frac{p}{q}$, where p and q have no common factors except 1 and -1, be a zero of the polynomial function

$$f(x) = a_n x^n + a_{n-1} x^{n-1} + \cdots + a_1 x + a_0,$$

whose coefficients are all integers. Show that

$$a_n p^n + a_{n-1} p^{n-1} q + \cdots + a_1 p q^{n-1} + a_0 q^n = 0.$$

Now, because p is a factor of the first n terms of this equation, p must also be a factor of the term $a_0 q^n$. Since p is not a factor of q (why?), p must be a factor of a_0. Similarly, q must be a factor of a_n.]

115. Bisection Method for Approximating Zeros of a Function f We begin with two consecutive integers, a and $a + 1$, such that $f(a)$ and $f(a + 1)$ are of opposite sign. Evaluate f at the midpoint m_1 of a and $a + 1$. If $f(m_1) = 0$, then m_1 is the zero of f, and we are finished. Otherwise, $f(m_1)$ is of opposite sign to either $f(a)$ or $f(a + 1)$. Suppose that it is

3.5 Assess Your Understanding

'Are You Prepared?'

Answers are given at the end of these exercises. If you get a wrong answer, read the pages listed in red.

1. Find $f(-1)$ if $f(x) = 2x^2 - x$. (pp. 44–47)

2. Factor the expression $6x^2 + x - 2$. (pp. A29–A35)

3. Find the quotient and remainder if $3x^4 - 5x^3 + 7x - 4$ is divided by $x - 3$. (pp. A25–A27 or A38–A40)

4. Solve the equation $x^2 + x - 3 = 0$. (pp. 137–139)

Concepts and Vocabulary

5. In the process of polynomial division, (Divisor)(Quotient) + _____ = _____.

6. When a polynomial function f is divided by $x - c$, the remainder is _____.

7. If a function f, whose domain is all real numbers, is even and if 4 is a zero of f, then _____ is also a zero.

8. *True or False:* Every polynomial function of degree 3 with real coefficients has exactly three real zeros.

9. *True or False:* The only potential rational zeros of $f(x) = 2x^5 - x^3 + x^2 - x + 1$ are $\pm 1, \pm 2$.

10. *True or False:* If f is a polynomial function of degree 4 and if $f(2) = 5$, then

$$\frac{f(x)}{x - 2} = p(x) + \frac{5}{x - 2}$$

where $p(x)$ is a polynomial of degree 3.

Skill Building

In Problems 11–20, use the Reminder Theorem to find the remainder when $f(x)$ is divided by $x - c$. Then use the Factor Theorem to determine whether $x - c$ is a factor of $f(x)$.

11. $f(x) = 4x^3 - 3x^2 - 8x + 4$; $x - 2$

12. $f(x) = -4x^3 + 5x^2 + 8$; $x + 3$

13. $f(x) = 3x^4 - 6x^3 - 5x + 10$; $x - 2$

14. $f(x) = 4x^4 - 15x^2 - 4$; $x - 2$

15. $f(x) = 3x^6 + 82x^3 + 27$; $x + 3$

16. $f(x) = 2x^6 - 18x^4 + x^2 - 9$; $x + 3$

17. $f(x) = 4x^6 - 64x^4 + x^2 - 15$; $x + 4$

18. $f(x) = x^6 - 16x^4 + x^2 - 16$; $x + 4$

19. $f(x) = 2x^4 - x^3 + 2x - 1$; $x - \dfrac{1}{2}$

20. $f(x) = 3x^4 + x^3 - 3x + 1$; $x + \dfrac{1}{3}$

In Problems 21–32, tell the maximum number of zeros that each polynomial function may have. Then use Descartes' Rule of Signs to determine how many positive and how many negative zeros each polynomial function may have. Do not attempt to find the zeros.

21. $f(x) = -4x^7 + x^3 - x^2 + 2$

22. $f(x) = 5x^4 + 2x^2 - 6x - 5$

23. $f(x) = 2x^6 - 3x^2 - x + 1$

24. $f(x) = -3x^5 + 4x^4 + 2$

25. $f(x) = 3x^3 - 2x^2 + x + 2$

26. $f(x) = -x^3 - x^2 + x + 1$

27. $f(x) = -x^4 + x^2 - 1$

28. $f(x) = x^4 + 5x^3 - 2$

29. $f(x) = x^5 + x^4 + x^2 + x + 1$

30. $f(x) = x^5 - x^4 + x^3 - x^2 + x - 1$ **31.** $f(x) = x^6 - 1$

32. $f(x) = x^6 + 1$

In Problems 33–44, list the potential rational zeros of each polynomial function. Do not attempt to find the zeros.

33. $f(x) = 3x^4 - 3x^3 + x^2 - x + 1$

34. $f(x) = x^5 - x^4 + 2x^2 + 3$

35. $f(x) = x^5 - 6x^2 + 9x - 3$

36. $f(x) = 2x^5 - x^4 - x^2 + 1$

37. $f(x) = -4x^3 - x^2 + x + 2$

38. $f(x) = 6x^4 - x^2 + 2$

39. $f(x) = 6x^4 - x^2 + 9$

40. $f(x) = -4x^3 + x^2 + x + 6$

41. $f(x) = 2x^5 - x^3 + 2x^2 + 12$

42. $f(x) = 3x^5 - x^2 + 2x + 18$

43. $f(x) = 6x^4 + 2x^3 - x^2 + 20$

44. $f(x) = -6x^3 - x^2 + x + 10$

In Problems 45–56, use Descartes' Rule of Signs and the Rational Zeros Theorem to find all the real zeros of each polynomial function. Use the zeros to factor f over the real numbers.

45. $f(x) = x^3 + 2x^2 - 5x - 6$

46. $f(x) = x^3 + 8x^2 + 11x - 20$

47. $f(x) = 2x^3 - x^2 + 2x - 1$

48. $f(x) = 2x^3 + x^2 + 2x + 1$

49. $f(x) = x^4 + x^2 - 2$

50. $f(x) = x^4 - 3x^2 - 4$

51. $f(x) = 4x^4 + 7x^2 - 2$

52. $f(x) = 4x^4 + 15x^2 - 4$

53. $f(x) = x^4 + x^3 - 3x^2 - x + 2$

54. $f(x) = x^4 - x^3 - 6x^2 + 4x + 8$

55. $f(x) = 4x^5 - 8x^4 - x + 2$

56. $f(x) = 4x^5 + 12x^4 - x - 3$

In Problems 57–68, solve each equation in the real number system.

57. $x^4 - x^3 + 2x^2 - 4x - 8 = 0$

58. $2x^3 + 3x^2 + 2x + 3 = 0$

59. $3x^3 + 4x^2 - 7x + 2 = 0$

60. $2x^3 - 3x^2 - 3x - 5 = 0$

61. $3x^3 - x^2 - 15x + 5 = 0$

62. $2x^3 - 11x^2 + 10x + 8 = 0$

$$f(1.20) = -0.23968 \qquad f(1.23) \approx -0.0455613$$
$$f(1.21) \approx -0.1778185 \quad f(1.24) \approx 0.025001$$
$$f(1.22) \approx -0.1131398$$

We conclude that the zero lies between 1.23 and 1.24, and so, correct to two decimal places, the zero is 1.23. ∎

Figure 53

Exploration

We examine the polynomial f given in Example 10. The Theorem on Bounds of Zeros tells us that every zero is between -2 and 2. If we graph f using $-2 \le x \le 2$, (see Figure 53) we see that f has exactly one x-intercept Using ZERO or ROOT, we find this zero to be 1.24 rounded to two decimal places. Correct to two decimal places, the zero is 1.23.

NOW WORK PROBLEM 101.

There are many other numerical techniques for approximating the zeros of a polynomial. The one outlined in Example 10 (a variation of the *bisection method*) has the advantages that it will always work, that it can be programmed rather easily on a computer, and each time it is used another decimal place of accuracy is achieved. See Problem 115 for the bisection method, which places the zero in a succession of intervals, with each new interval being half the length of the preceding one.

HISTORICAL FEATURE

Formulas for the solution of third- and fourth-degree polynomial equations exist, and, while not very practical, they do have an interesting history.

In the 1500s in Italy, mathematical contests were a popular pastime, and persons possessing methods for solving problems kept them secret. (Solutions that were published were already common knowledge.) Niccolo of Brescia (1499–1557), commonly referred to as Tartaglia ("the stammerer"), had the secret for solving cubic (third-degree) equations, which gave him a decided advantage in the contests. Girolamo Cardano (1501–1576) found out that Tartaglia had the secret, and, being interested in cubics, he requested it from Tartaglia. The reluctant Tartaglia hesitated for some time, but finally, swearing Cardano to secrecy with midnight oaths by candlelight, told him the secret. Cardano then published the solution in his book *Ars Magna* (1545), giving Tartaglia the credit but rather compromising the secrecy. Tartaglia exploded into bitter recriminations, and each wrote pamphlets that reflected on the other's mathematics, moral character, and ancestry.

The quartic (fourth-degree) equation was solved by Cardano's student Lodovico Ferrari, and this solution also was included, with credit and this time with permission, in the *Ars Magna*.

Attempts were made to solve the fifth-degree equation in similar ways, all of which failed. In the early 1800s, P. Ruffini, Niels Abel, and Evariste Galois all found ways to show that it is not possible to solve fifth-degree equations by formula, but the proofs required the introduction of new methods. Galois's methods eventually developed into a large part of modern algebra.

Historical Problems

Problems 1–8 develop the Tartaglia–Cardano solution of the cubic equation and show why it is not altogether practical.

1. Show that the general cubic equation $y^3 + by^2 + cy + d = 0$ can be transformed into an equation of the form $x^3 + px + q = 0$ by using the substitution $y = x - \dfrac{b}{3}$.

2. In the equation $x^3 + px + q = 0$, replace x by $H + K$. Let $3HK = -p$, and show that $H^3 + K^3 = -q$.

3. Based on Problem 2, we have the two equations

$$3HK = -p \quad \text{and} \quad H^3 + K^3 = -q$$

Solve for K in $3HK = -p$ and substitute into $H^3 + K^3 = -q$. Then show that

$$H = \sqrt[3]{\dfrac{-q}{2} + \sqrt{\dfrac{q^2}{4} + \dfrac{p^3}{27}}}$$

[**Hint:** Look for an equation that is quadratic in form.]

4. Use the solution for H from Problem 3 and the equation $H^3 + K^3 = -q$ to show that

$$K = \sqrt[3]{\dfrac{-q}{2} - \sqrt{\dfrac{q^2}{4} + \dfrac{p^3}{27}}}$$

5. Use the results from Problems 2–4 to show that the solution of $x^3 + px + q = 0$ is

$$x = \sqrt[3]{\dfrac{-q}{2} + \sqrt{\dfrac{q^2}{4} + \dfrac{p^3}{27}}} + \sqrt[3]{\dfrac{-q}{2} - \sqrt{\dfrac{q^2}{4} + \dfrac{p^3}{27}}}$$

6. Use the result of Problem 5 to solve the equation $x^3 - 6x - 9 = 0$.

7. Use a calculator and the result of Problem 5 to solve the equation $x^3 + 3x - 14 = 0$.

8. Use the methods of this chapter to solve the equation $x^3 + 3x - 14 = 0$.

Although the proof of this result requires advanced methods in calculus, it is easy to "see" why the result is true. Look at Figure 52.

Figure 52
If $f(a) < 0$ and $f(b) > 0$, there is a zero between a and b.

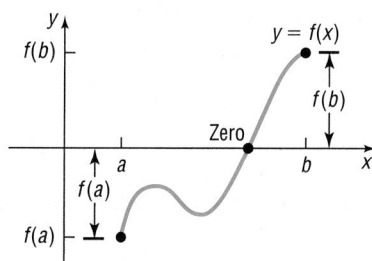

| **EXAMPLE 9** | **Using the Intermediate Value Theorem to Locate Zeros** |

Show that $f(x) = x^5 - x^3 - 1$ has a zero between 1 and 2.

Solution We evaluate f at 1 and at 2.

$$f(1) = -1 \quad \text{and} \quad f(2) = 23$$

Because $f(1) < 0$ and $f(2) > 0$, it follows from the Intermediate Value Theorem that f has a zero between 1 and 2. ∎

NOW WORK PROBLEM 89.

Let's look at the polynomial f of Example 9 more closely. Based on Descartes' Rule of Signs, f has exactly one positive real zero. Based on the Rational Zeros Theorem, 1 is the only potential positive rational zero. Since $f(1) \neq 0$, we conclude that the zero between 1 and 2 is irrational. We can use the Intermediate Value Theorem to approximate it. The steps to use follow:

> **Approximating the Zeros of a Polynomial Function**
>
> **STEP 1:** Find two consecutive integers a and $a + 1$ such that f has a zero between them.
> **STEP 2:** Divide the interval $[a, a + 1]$ into 10 equal subintervals.
> **STEP 3:** Evaluate f at each endpoint of the subintervals until the Intermediate Value Theorem applies; this interval then contains a zero.
> **STEP 4:** Repeat the process starting at Step 2 until the desired accuracy is achieved.

| **EXAMPLE 10** | **Approximating the Zeros of a Polynomial Function** |

Find the positive zero of $f(x) = x^5 - x^3 - 1$ correct to two decimal places.

Solution From Example 9 we know that the positive zero is between 1 and 2. We divide the interval $[1, 2]$ into 10 equal subintervals: $[1, 1.1]$, $[1.1, 1.2]$, $[1.2, 1.3]$, $[1.3, 1.4]$, $[1.4, 1.5]$, $[1.5, 1.6]$, $[1.6, 1.7]$, $[1.7, 1.8]$, $[1.8, 1.9]$, $[1.9, 2]$. Now we find the value of f at each endpoint until the Intermediate Value Theorem applies.

$$f(x) = x^5 - x^3 - 1$$
$$f(1.0) = -1 \qquad\qquad f(1.2) = -0.23968$$
$$f(1.1) = -0.72049 \qquad f(1.3) = 0.51593$$

We can stop here and conclude that the zero is between 1.2 and 1.3. Now we divide the interval $[1.2, 1.3]$ into 10 equal subintervals and proceed to evaluate f at each endpoint.

An example will help to make the theorem clear.

| EXAMPLE 8 | **Using the Theorem for Finding Bounds on Zeros** |

Find a bound to the zeros of each polynomial.

(a) $f(x) = x^5 + 3x^3 - 9x^2 + 5$ (b) $g(x) = 4x^5 - 2x^3 + 2x^2 + 1$

Solution (a) The leading coefficient of f is 1.

$$f(x) = x^5 + 3x^3 - 9x^2 + 5 \quad a_4 = 0, a_3 = 3, a_2 = -9, a_1 = 0, a_0 = 5$$

We evaluate the two expressions in (4).

$$\mathrm{Max}\{1, |a_0| + |a_1| + \cdots + |a_{n-1}|\} = \mathrm{Max}\{1, |5| + |0| + |-9| + |3| + |0|\}$$
$$= \mathrm{Max}\{1, 17\} = 17$$
$$1 + \mathrm{Max}\{|a_0|, |a_1|, \dots, |a_{n-1}|\} = 1 + \mathrm{Max}\{|5|, |0|, |-9|, |3|, |0|\}$$
$$= 1 + 9 = 10$$

The smaller of the two numbers, 10, is the bound. Every zero of f lies between -10 and 10.

(b) First we write g so that it is the product of a constant times a polynomial whose leading coefficient is 1.

$$g(x) = 4x^5 - 2x^3 + 2x^2 + 1 = 4\left(x^5 - \frac{1}{2}x^3 + \frac{1}{2}x^2 + \frac{1}{4}\right)$$

Next we evaluate the two expressions in (4) with $a_4 = 0$, $a_3 = -\frac{1}{2}$, $a_2 = \frac{1}{2}$, $a_1 = 0$, and $a_0 = \frac{1}{4}$.

$$\mathrm{Max}\{1, |a_0| + |a_1| + \cdots + |a_{n-1}|\} = \mathrm{Max}\left\{1, \left|\frac{1}{4}\right| + |0| + \left|\frac{1}{2}\right| + \left|-\frac{1}{2}\right| + |0|\right\}$$
$$= \mathrm{Max}\left\{1, \frac{5}{4}\right\} = \frac{5}{4}$$
$$1 + \mathrm{Max}\{|a_0|, |a_1|, \dots, |a_{n-1}|\} = 1 + \mathrm{Max}\left\{\left|\frac{1}{4}\right|, |0|, \left|\frac{1}{2}\right|, \left|-\frac{1}{2}\right|, |0|\right\}$$
$$= 1 + \frac{1}{2} = \frac{3}{2}$$

The smaller of the two numbers, $\frac{5}{4}$, is the bound. Every zero of g lies between $-\frac{5}{4}$ and $\frac{5}{4}$. ∎

COMMENT: The bounds on the zeros of a polynomial provide good choices for setting Xmin and Xmax of the viewing rectangle. With these choices, all the x-intercepts of the graph can be seen. ∎

— NOW WORK PROBLEM 81.

7 Use the Intermediate Value Theorem

The next result, called the **Intermediate Value Theorem,** is based on the fact that the graph of a polynomial function is continuous; that is, it contains no "holes" or "gaps."

Theorem **Intermediate Value Theorem**

Let f denote a polynomial function. If $a < b$ and if $f(a)$ and $f(b)$ are of opposite sign, there is at least one zero of f between a and b.

5 Solve Polynomial Equations

EXAMPLE 7	**Solving a Polynomial Equation**

Solve the equation: $x^5 - 5x^4 + 12x^3 - 24x^2 + 32x - 16 = 0$

Solution The solutions of this equation are the zeros of the polynomial function

$$f(x) = x^5 - 5x^4 + 12x^3 - 24x^2 + 32x - 16$$

Using the result of Example 6, the real zeros of f are 1 and 2. These are the real solutions of the equation

$$x^5 - 5x^4 + 12x^3 - 24x^2 + 32x - 16 = 0 \qquad \blacksquare$$

NOW WORK PROBLEM 57.

In Example 6, the quadratic factor $x^2 + 4$ that appears in the factored form of f is called *irreducible*, because the polynomial $x^2 + 4$ cannot be factored over the real numbers. In general, we say that a quadratic factor $ax^2 + bx + c$ is **irreducible** if it cannot be factored over the real numbers, that is, if it is prime over the real numbers.

Refer to Examples 5 and 6. The polynomial function of Example 5 has three real zeros, and its factored form contains three linear factors. The polynomial function of Example 6 has two distinct real zeros, and its factored form contains two distinct linear factors and one irreducible quadratic factor.

Theorem Every polynomial function (with real coefficients) can be uniquely factored into a product of linear factors and/or irreducible quadratic factors.

We shall prove this result in Section 3.6, and, in fact, we shall draw several additional conclusions about the zeros of a polynomial function. One conclusion is worth noting now. If a polynomial (with real coefficients) is of odd degree, it must contain at least one linear factor. (Do you see why?) This means that it must have at least one real zero.

COROLLARY A polynomial function (with real coefficients) of odd degree has at least one real zero.

6 Use the Theorem for Bounds on Zeros

The search for the real zeros of a polynomial function can be reduced somewhat if *bounds* on the zeros are found. A number M is a **bound** on the zeros of a polynomial if every zero lies between $-M$ and M, inclusive. That is, M is a bound to the zeros of a polynomial f if

$$-M \le \text{any zero of } f \le M$$

Theorem **Bounds on Zeros**

Let f denote a polynomial function whose leading coefficient is 1.

$$f(x) = x^n + a_{n-1}x^{n-1} + \cdots + a_1 x + a_0$$

A bound M on the zeros of f is the smaller of the two numbers

$$\text{Max}\{1, |a_0| + |a_1| + \cdots + |a_{n-1}|\}, \quad 1 + \text{Max}\{|a_0|, |a_1|, \ldots, |a_{n-1}|\} \qquad (4)$$

where Max{ } means "choose the largest entry in { }."

STEP 3: Because the leading coefficient $a_5 = 1$ and there are no negative zeros, the potential rational zeros are the integers $1, 2, 4, 8,$ and 16, the positive factors of the constant term, 16. We test the potential rational zero 1 first, using synthetic division.

$$
\begin{array}{r|rrrrr}
1) & 1 & -5 & 12 & -24 & 32 & -16 \\
 & & 1 & -4 & 8 & -16 & 16 \\
\hline
 & 1 & -4 & 8 & -16 & 16 & 0
\end{array}
$$

The remainder is $f(1) = 0$, so 1 is a zero and $x - 1$ is a factor of f. Using the entries in the bottom row of the synthetic division, we can begin to factor f.

$$f(x) = x^5 - 5x^4 + 12x^3 - 24x^2 + 32x - 16$$

$$= (x - 1)(x^4 - 4x^3 + 8x^2 - 16x + 16)$$

We now work with the first depressed equation:

$$q_1(x) = x^4 - 4x^3 + 8x^2 - 16x + 16 = 0$$

REPEAT STEP 3: The potential rational zeros of q_1 are still $1, 2, 4, 8,$ and 16. We test 1 first, since it may be a repeated zero.

$$
\begin{array}{r|rrrr}
1) & 1 & -4 & 8 & -16 & 16 \\
 & & 1 & -3 & 5 & -11 \\
\hline
 & 1 & -3 & 5 & -11 & 5
\end{array}
$$

Since the remainder is 5, 1 is not a repeated zero. We try 2 next.

$$
\begin{array}{r|rrrr}
2) & 1 & -4 & 8 & -16 & 16 \\
 & & 2 & -4 & 8 & -16 \\
\hline
 & 1 & -2 & 4 & -8 & 0
\end{array}
$$

The remainder is $f(2) = 0$, so 2 is a zero and $x - 2$ is a factor of f. Again using the bottom row, we find

$$f(x) = x^5 - 5x^4 + 12x^3 - 24x^2 + 32x - 16$$

$$= (x - 1)(x - 2)(x^3 - 2x^2 + 4x - 8)$$

The remaining zeros satisfy the new depressed equation

$$q_2(x) = x^3 - 2x^2 + 4x - 8 = 0$$

Notice that $q_2(x)$ can be factored using grouping. (Alternatively, you could repeat Step 3 and check the potential rational zero 2.) Then

$$x^3 - 2x^2 + 4x - 8 = 0$$
$$x^2(x - 2) + 4(x - 2) = 0$$
$$(x^2 + 4)(x - 2) = 0$$
$$x^2 + 4 = 0 \quad \text{or} \quad x - 2 = 0$$
$$x = 2$$

Since $x^2 + 4 = 0$ has no real solutions, the real zeros of f are 1 and 2, the latter being a zero of multiplicity 2. The factored form of f is

$$f(x) = x^5 - 5x^4 + 12x^3 - 24x^2 + 32x - 16$$
$$= (x - 1)(x - 2)^2(x^2 + 4)$$ ■

NOW WORK PROBLEM **45.**

STEP 4: The depressed equation $2x^2 + 13x + 6 = 0$ is a quadratic equation with discriminant $b^2 - 4ac = 169 - 48 = 121 > 0$. The equation has two real solutions, which can be found by factoring.

$$2x^2 + 13x + 6 = (2x + 1)(x + 6) = 0$$

$$2x + 1 = 0 \quad \text{or} \quad x + 6 = 0$$

$$x = -\frac{1}{2} \qquad x = -6$$

The zeros of f are -6, $-\dfrac{1}{2}$, and 1.

We know from Step 3 that $f(x) = 2x^3 + 11x^2 - 7x - 6$ can be written as $f(x) = (x - 1)(2x^2 + 13x + 6)$. Because $2x^2 + 13x + 6 = (2x + 1)(x + 6)$, we write f in factored form as

$$f(x) = (x - 1)(2x + 1)(x + 6) \qquad \blacksquare$$

Notice that the three zeros of f found in this example are among those given in the list of potential rational zeros in Example 4.

4 Find the Real Zeros of a Polynomial Function

To obtain information about the real zeros of a polynomial function, follow these steps:

> **Steps for Finding the Real Zeros of a Polynomial Function**
>
> **STEP 1:** Use the degree of the polynomial to determine the maximum number of zeros.
>
> **STEP 2:** Use Descartes' Rule of Signs to determine the possible number of positive zeros and negative zeros.
>
> **STEP 3:** (a) If the polynomial has integer coefficients, use the Rational Zeros Theorem to identify those rational numbers that potentially could be zeros.
>
> (b) Use substitution, synthetic division, or long division to test each potential rational zero.
>
> (c) Each time that a zero (and thus a factor) is found, repeat Step 3 on the depressed equation.
>
> **STEP 4:** In attempting to find the zeros, remember to use (if possible) the factoring techniques that you already know (special products, factoring by grouping, and so on).

EXAMPLE 6	**Finding the Real Zeros of a Polynomial Function**

Find the real zeros of $f(x) = x^5 - 5x^4 + 12x^3 - 24x^2 + 32x - 16$. Write f in factored form.

Solution We gather all the information that we can about the zeros.

STEP 1: There are at most five real zeros.

STEP 2: By Descartes' Rule of Signs, there are five, three, or one positive zero. There are no negative zeros because

$$f(-x) = -x^5 - 5x^4 - 12x^3 - 24x^2 - 32x - 16$$

has no variation in sign.

Solution Because f has integer coefficients, we may use the Rational Zeros Theorem. First, we list all the integers p that are factors of the constant term $a_0 = -6$ and all the integers q that are factors of the leading coefficient $a_3 = 2$.

$$p: \quad \pm 1, \pm 2, \pm 3, \pm 6 \qquad \text{Factors of } -6$$
$$q: \quad \pm 1, \pm 2 \qquad \text{Factors of } 2$$

Now we form all possible ratios $\dfrac{p}{q}$.

$$\frac{p}{q}: \quad \pm 1, \pm 2, \pm 3, \pm 6, \pm\frac{1}{2}, \pm\frac{3}{2}$$

If f has a rational zero, it will be found in this list, which contains 12 possibilities. ∎

NOW WORK PROBLEM 33.

In Words

For the polynomial function $f(x) = 2x^3 + 11x^2 - 7x - 6$, we know 5 is not a zero because 5 is not in the list of potential rational zeros. However, -1 may or may not be a zero.

Be sure that you understand what the Rational Zeros Theorem says: For a polynomial with integer coefficients, *if* there is a rational zero, it is one of those listed. It may be the case that the function does not have any rational zeros.

Long division, synthetic division, or substitution can be used to test each potential rational zero to determine whether it is indeed a zero. To make the work easier, integers are usually tested first. Let's continue this example.

EXAMPLE 5 **Finding the Rational Zeros of a Polynomial Function**

Continue working with Example 4 to find the rational zeros of

$$f(x) = 2x^3 + 11x^2 - 7x - 6$$

Write f in factored form.

Solution We gather all the information that we can about the zeros.

STEP 1: There are at most three real zeros.

STEP 2: By Descartes' Rule of Signs, there is one positive real zero. Also, because

$$f(-x) = -2x^3 + 11x^2 + 7x - 6$$

there are two negative zeros or no negative zeros.

STEP 3: Now we use the list of potential rational zeros obtained in Example 4:

$$\pm 1, \pm 2, \pm 3, \pm 6, \pm\frac{1}{2}, \pm\frac{3}{2}.$$

We choose to test the potential rational zero 1 using substitution.

$$f(1) = 2(1)^3 + 11(1)^2 - 7(1) - 6 = 2 + 11 - 7 - 6 = 0$$

Since $f(1) = 0$, 1 is a zero and $x - 1$ is a factor of f. We can use long division or synthetic division to factor f.

$$f(x) = 2x^3 + 11x^2 - 7x - 6$$
$$= (x - 1)(2x^2 + 13x + 6)$$

Now any solution of the equation $2x^2 + 13x + 6 = 0$ will be a zero of f. Because of this, we call the equation $2x^2 + 13x + 6 = 0$ a **depressed equation** of f. Since the degree of the depressed equation of f is less than that of the original polynomial, we work with the depressed equation to find the zeros of f.

Notice that we ignored the zero coefficients in $0x^6$, $0x^5$ and $0x^3$ in counting the number of variations in the sign of $f(x)$. Replacing x by $-x$, we get

$$f(-x) = -3(-x)^7 + 4(-x)^4 + 3(-x)^2 - 2(-x) - 1$$
$$= 3x^7 + 4x^4 + 3x^2 + 2x - 1$$

$+$ to $-$

which has one variation in sign.

Theorem

Descartes' Rule of Signs

Let f denote a polynomial function written in standard form.

The number of positive real zeros of f either equals the number of variations in the sign of the nonzero coefficients of $f(x)$ or else equals that number less an even integer.

The number of negative real zeros of f either equals the number of variations in the sign of the nonzero coefficients of $f(-x)$ or else equals that number less an even integer.

We shall not prove Descartes' Rule of Signs. Let's see how it is used.

EXAMPLE 3

Using the Number of Real Zeros Theorem and Descartes' Rule of Signs

Discuss the real zeros of $f(x) = 3x^6 - 4x^4 + 3x^3 + 2x^2 - x - 3$.

Solution

Because the polynomial is of degree 6, by the Number of Real Zeros Theorem there are at most six real zeros. Since there are three variations in the sign of the nonzero coefficients of $f(x)$, by Descartes' Rule of Signs we expect either three or one positive real zero. To continue, we look at $f(-x)$.

$$f(-x) = 3x^6 - 4x^4 - 3x^3 + 2x^2 + x - 3$$

There are three variations in sign, so we expect either three (or one) negative real zeros. Equivalently, we now know that the graph of f has either three or one positive x-intercept and three or one negative x-intercept. ∎

 NOW WORK PROBLEM 21.

3 Use the Rational Zeros Theorem

The next result, called the **Rational Zeros Theorem,** provides information about the rational zeros of a polynomial *with integer coefficients.*

Theorem

Rational Zeros Theorem

Let f be a polynomial function of degree 1 or higher of the form

$$f(x) = a_n x^n + a_{n-1}x^{n-1} + \cdots + a_1 x + a_0, \qquad a_n \neq 0, \quad a_0 \neq 0$$

where each coefficient is an integer. If $\dfrac{p}{q}$, in lowest terms, is a rational zero of f, then p must be a factor of a_0, and q must be a factor of a_n.

EXAMPLE 4

Listing Potential Rational Zeros

List the potential rational zeros of

$$f(x) = 2x^3 + 11x^2 - 7x - 6$$

has the factor

(a) $x - 1$ (b) $x + 3$

Solution The Factor Theorem states that if $f(c) = 0$ then $x - c$ is a factor.

(a) Because $x - 1$ is of the form $x - c$ with $c = 1$, we find the value of $f(1)$. We choose to use substitution.

$$f(1) = 2(1)^3 - (1)^2 + 2(1) - 3 = 2 - 1 + 2 - 3 = 0$$

By the Factor Theorem, $x - 1$ is a factor of $f(x)$.

(b) To test the factor $x + 3$, we first need to write it in the form $x - c$. Since $x + 3 = x - (-3)$, we find the value of $f(-3)$. We choose to use synthetic division.

$$
\begin{array}{r|rrr}
-3 & 2 & -1 & 2 & -3 \\
 & & -6 & 21 & -69 \\
\hline
 & 2 & -7 & 23 & -72
\end{array}
$$

Because $f(-3) = -72 \neq 0$, we conclude from the Factor Theorem that $x - (-3) = x + 3$ is not a factor of $f(x)$. ■

════════► **NOW WORK PROBLEM 11.**

In Example 2(a), we found that $x - 1$ was a factor of f. To write f in factored form, we use long division or synthetic division. Using synthetic division, we find that

$$
\begin{array}{r|rrr}
1 & 2 & -1 & 2 & -3 \\
 & & 2 & 1 & 3 \\
\hline
 & 2 & 1 & 3 & 0
\end{array}
$$

The quotient is $q(x) = 2x^2 + x + 3$ with a remainder of 0, as expected. We can write f in factored form as

$$f(x) = 2x^3 - x^2 + 2x - 3 = (x - 1)(2x^2 + x + 3)$$

The next theorem concerns the number of real zeros that a polynomial function may have. In counting the zeros of a polynomial, we count each zero as many times as its multiplicity.

Theorem

> **Number of Real Zeros**
>
> A polynomial function cannot have more real zeros than its degree.

Proof The proof is based on the Factor Theorem. If r is a zero of a polynomial function f, then $f(r) = 0$ and, hence, $x - r$ is a factor of $f(x)$. Each zero corresponds to a factor of degree 1. Because f cannot have more first-degree factors than its degree, the result follows. ■

2 Use Descartes' Rule of Signs

Descartes' Rule of Signs provides information about the number and location of the real zeros of a polynomial function written in standard form (descending powers of x). It requires that we count the number of variations in the sign of the coefficients of $f(x)$ and $f(-x)$.

For example, the following polynomial function has two variations in the signs of the coefficients.

$$f(x) = -3x^7 + 4x^4 + 3x^2 - 2x - 1$$
$$= -3x^7 + 0x^6 + 0x^5 + 4x^4 + 0x^3 + 3x^2 - 2x - 1$$

$- \text{ to } +$ $+ \text{ to } -$

Fundamental Theorem of Algebra

Every complex polynomial function $f(x)$ of degree $n \geq 1$ has at least one complex zero.

We shall not prove this result, as the proof is beyond the scope of this book. However, using the Fundamental Theorem of Algebra and the Factor Theorem, we can prove the following result:

Theorem

Every complex polynomial function $f(x)$ of degree $n \geq 1$ can be factored into n linear factors (not necessarily distinct) of the form

$$f(x) = a_n(x - r_1)(x - r_2) \cdot \ldots \cdot (x - r_n) \qquad (2)$$

where $a_n, r_1, r_2, \ldots, r_n$ are complex numbers. That is, every complex polynomial function of degree $n \geq 1$ has exactly n (not necessarily distinct) zeros.

Proof Let

$$f(x) = a_n x^n + a_{n-1} x^{n-1} + \cdots + a_1 x + a_0$$

By the Fundamental Theorem of Algebra, f has at least one zero, say r_1. Then, by the Factor Theorem, $x - r_1$ is a factor, and

$$f(x) = (x - r_1)q_1(x)$$

where $q_1(x)$ is a complex polynomial of degree $n - 1$ whose leading coefficient is a_n. Again by the Fundamental Theorem of Algebra, the complex polynomial $q_1(x)$ has at least one zero, say r_2. By the Factor Theorem, $q_1(x)$ has the factor $x - r_2$, so

$$q_1(x) = (x - r_2)q_2(x)$$

where $q_2(x)$ is a complex polynomial of degree $n - 2$ whose leading coefficient is a_n. Consequently,

$$f(x) = (x - r_1)(x - r_2)q_2(x)$$

Repeating this argument n times, we arrive at

$$f(x) = (x - r_1)(x - r_2) \cdot \ldots \cdot (x - r_n)q_n(x)$$

where $q_n(x)$ is a complex polynomial of degree $n - n = 0$ whose leading coefficient is a_n. Thus, $q_n(x) = a_n x^0 = a_n$, and so

$$f(x) = a_n(x - r_1)(x - r_2) \cdot \ldots \cdot (x - r_n)$$

We conclude that every complex polynomial function $f(x)$ of degree $n \geq 1$ has exactly n (not necessarily distinct) zeros. ∎

1 Use the Conjugate Pairs Theorem

We can use the Fundamental Theorem of Algebra to obtain valuable information about the complex zeros of polynomials whose coefficients are real numbers.

Conjugate Pairs Theorem

Let $f(x)$ be a polynomial whose coefficients are real numbers. If $r = a + bi$ is a zero of f, the complex conjugate $\bar{r} = a - bi$ is also a zero of f.

In other words, for polynomials whose coefficients are real numbers, the zeros occur in conjugate pairs. This result should not be all that surprising since the complex zeros of a quadratic function occurred in conjugate pairs.

Proof Let

$$f(x) = a_n x^n + a_{n-1} x^{n-1} + \cdots + a_1 x + a_0$$

where $a_n, a_{n-1}, \ldots, a_1, a_0$ are real numbers and $a_n \neq 0$. If $r = a + bi$ is a zero of f, then $f(r) = f(a + bi) = 0$, so

$$a_n r^n + a_{n-1} r^{n-1} + \cdots + a_1 r + a_0 = 0$$

We take the conjugate of both sides to get

$$\overline{a_n r^n + a_{n-1} r^{n-1} + \cdots + a_1 r + a_0} = \overline{0}$$

$$\overline{a_n r^n} + \overline{a_{n-1} r^{n-1}} + \cdots + \overline{a_1 r} + \overline{a_0} = \overline{0} \qquad \text{The conjugate of a sum equals the sum of the conjugates (see Appendix A, Section A.11).}$$

$$\overline{a_n}(\overline{r})^n + \overline{a_{n-1}}(\overline{r})^{n-1} + \cdots + \overline{a_1}\overline{r} + \overline{a_0} = \overline{0} \qquad \text{The conjugate of a product equals the product of the conjugates.}$$

$$a_n(\overline{r})^n + a_{n-1}(\overline{r})^{n-1} + \cdots + a_1\overline{r} + a_0 = 0 \qquad \text{The conjugate of a real number equals the real number.}$$

This last equation states that $f(\overline{r}) = 0$; that is, $\overline{r} = a - bi$ is a zero of f. ■

The importance of this result should be clear. Once we know that, say, $3 + 4i$ is a zero of a polynomial with real coefficients, then we know that $3 - 4i$ is also a zero. This result has an important corollary.

COROLLARY

A polynomial f of odd degree with real coefficients has at least one real zero.

Proof Because complex zeros occur as conjugate pairs in a polynomial with real coefficients, there will always be an even number of zeros that are not real numbers. Consequently, since f is of odd degree, one of its zeros has to be a real number. ■

For example, the polynomial $f(x) = x^5 - 3x^4 + 4x^3 - 5$ has at least one zero that is a real number, since f is of degree 5 (odd) and has real coefficients.

EXAMPLE 1 **Using the Conjugate Pairs Theorem**

A polynomial f of degree 5 whose coefficients are real numbers has the zeros $1, 5i$, and $1 + i$. Find the remaining two zeros.

Solution Since f has coefficients that are real numbers, complex zeros appear as conjugate pairs. It follows that $-5i$, the conjugate of $5i$, and $1 - i$, the conjugate of $1 + i$, are the two remaining zeros. ■

NOW WORK PROBLEM 7.

2 Find a Polynomial Function with Specified Zeros

EXAMPLE 2 **Finding a Polynomial Function Whose Zeros Are Given**

Find a polynomial f of degree 4 whose coefficients are real numbers that has the zeros $1, 1$, and $-4 + i$.

Solution Since $-4 + i$ is a zero, by the Conjugate Pairs Theorem, $-4 - i$ must also be a zero of f. Because of the Factor Theorem, if $f(c) = 0$, then $x - c$ is a factor of $f(x)$. So we can now write f as

$$f(x) = a(x - 1)(x - 1)[x - (-4 + i)][x - (-4 - i)]$$

where a is any real number. Then

$$\begin{aligned}
f(x) &= a(x - 1)(x - 1)[x - (-4 + i)][x - (-4 - i)] \\
&= a(x^2 - 2x + 1)[x^2 - (-4 + i)x - (-4 - i)x + (-4 + i)(-4 - i)] \\
&= a(x^2 - 2x + 1)(x^2 + 4x - ix + 4x + ix + 16 + 4i - 4i - i^2) \\
&= a(x^2 - 2x + 1)(x^2 + 8x + 17) \\
&= a(x^4 + 8x^3 + 17x^2 - 2x^3 - 16x^2 - 34x + x^2 + 8x + 17) \\
&= a(x^4 + 6x^3 + 2x^2 - 26x + 17)
\end{aligned}$$

■

Figure 54

───── **Exploration** ─────

Graph the function f found in Example 2 for $a = 1$. Does the value of a affect the zeros of f? How does the value of a affect the graph of f?

RESULT A quick analysis of the polynomial f tells us what to expect:

 At most three turning points.

 For large $|x|$, the graph will behave like $y = x^4$.

 A repeated real zero at 1 so that the graph will touch the x-axis at 1.

 The only x-intercept is at 1; the y-intercept is 17.

Figure 54 shows the complete graph. (Do you see why? The graph has exactly three turning points.) The value of a causes a stretch or compression; a reflection also occurs if $a < 0$. The zeros are not affected.

Now we can prove the theorem we conjectured in Section 3.5.

Theorem Every polynomial function with real coefficients can be uniquely factored over the real numbers into a product of linear factors and/or irreducible quadratic factors.

Proof Every complex polynomial f of degree n has exactly n zeros and can be factored into a product of n linear factors. If its coefficients are real, those zeros that are complex numbers will always occur as conjugate pairs. As a result, if $r = a + bi$ is a complex zero, then so is $\bar{r} = a - bi$. Consequently, when the linear factors $x - r$ and $x - \bar{r}$ of f are multiplied, we have

$$(x - r)(x - \bar{r}) = x^2 - (r + \bar{r})x + r\bar{r} = x^2 - 2ax + a^2 + b^2$$

This second-degree polynomial has real coefficients and is irreducible (over the real numbers). Thus, the factors of f are either linear or irreducible quadratic factors. ■

───── **NOW WORK PROBLEM 17.**

3 Find the Complex Zeros of a Polynomial

<u>**EXAMPLE 3**</u> **Finding the Complex Zeros of a Polynomial**

Find the complex zeros of the polynomial function

$$f(x) = 3x^4 + 5x^3 + 25x^2 + 45x - 18$$

Write f in factored form.

Solution **STEP 1:** The degree of f is 4. So f will have four complex zeros.

STEP 2: Descartes' Rule of Signs provides information about the real zeros. For this polynomial, there is one positive real zero. There are three or one negative real zeros because

$$f(-x) = 3x^4 - 5x^3 + 25x^2 - 45x - 18$$

has three variations in sign.

STEP 3: The Rational Zeros Theorem provides information about the potential rational zeros of polynomials with integer coefficients. For this polynomial (which has integer coefficients), the potential rational zeros are

$$\pm\frac{1}{3}, \quad \pm\frac{2}{3}, \quad \pm 1, \quad \pm 2, \quad \pm 3, \quad \pm 6, \quad \pm 9, \quad \pm 18$$

We test 1 first:
$$1{\overline{)\begin{array}{ccccc} 3 & 5 & 25 & 45 & -18 \\ & 3 & 8 & 33 & 78 \end{array}}}$$
$$\begin{array}{ccccc} 3 & 8 & 33 & 78 & 60 \end{array}$$ 1 is not a zero.

We test -1:
$$-1{\overline{)\begin{array}{ccccc} 3 & 5 & 25 & 45 & -18 \\ & -3 & -2 & -23 & -22 \end{array}}}$$
$$\begin{array}{ccccc} 3 & 2 & 23 & 22 & -40 \end{array}$$ -1 is not a zero.

We test 2:
$$2{\overline{)\begin{array}{ccccc} 3 & 5 & 25 & 45 & -18 \\ & 6 & 22 & 94 & 278 \end{array}}}$$
$$\begin{array}{ccccc} 3 & 11 & 47 & 139 & 260 \end{array}$$ 2 is not a zero.

We test -2:
$$-2{\overline{)\begin{array}{ccccc} 3 & 5 & 25 & 45 & -18 \\ & -6 & 2 & -54 & 18 \end{array}}}$$
$$\begin{array}{ccccc} 3 & -1 & 27 & -9 & 0 \end{array}$$

Since $f(-2) = 0$, then -2 is a zero and $x + 2$ is a factor of f. The depressed equation is

$$3x^3 - x^2 + 27x - 9 = 0$$

STEP 4: We factor by grouping.

$$3x^3 - x^2 + 27x - 9 = 0$$
$$x^2(3x - 1) + 9(3x - 1) = 0 \qquad \text{Factor } x^2 \text{ from } 3x^3 - x^2 \text{ and 9 from } 27x - 9.$$
$$(x^2 + 9)(3x - 1) = 0 \qquad \text{Factor out the common factor } 3x - 1.$$
$$x^2 + 9 = 0 \qquad \text{or} \qquad 3x - 1 = 0 \qquad \text{Apply the Zero-Product Property.}$$
$$x^2 = -9 \qquad \text{or} \qquad x = \frac{1}{3}$$
$$x = -3i, \quad x = 3i \qquad \text{or} \qquad x = \frac{1}{3}$$

The four complex zeros of f are $\left\{-3i, 3i, -2, \dfrac{1}{3}\right\}$.

The factored form of f is

$$f(x) = 3x^4 + 5x^3 + 25x^2 + 45x - 18$$
$$= 3(x + 3i)(x - 3i)(x + 2)\left(x - \frac{1}{3}\right)$$

∎

NOW WORK PROBLEM 33.

3.6 Assess Your Understanding

'Are You Prepared?'

Answers are given at the end of these exercises. If you get the wrong answer, read the pages listed in red.

1. Find the sum and the product of the complex numbers $3 - 2i$ and $-3 + 5i$. (pp. A82–A87)

2. Find the complex zeros of $f(x) = x^2 + 2x + 2$. (pp. 167–169)

Concepts and Vocabulary

3. Every polynomial function of odd degree with real coefficients will have at least _____ real zero(s).

4. If $3 + 4i$ is a zero of a polynomial function of degree 5 with real coefficients, then so is _____.

5. *True or False:* A polynomial function of degree n with real coefficients has exactly n complex zeros. At most n of them are real zeros.

6. *True or False:* A polynomial function of degree 4 with real coefficients could have $-3, 2 + i, 2 - i$, and $-3 + 5i$ as its zeros.

Skill Building

In Problems 7–16, information is given about a polynomial $f(x)$ whose coefficients are real numbers. Find the remaining zeros of f.

7. Degree 3; zeros: $3, 4 - i$

8. Degree 3; zeros: $4, 3 + i$

9. Degree 4; zeros: $i, 1 + i$

10. Degree 4; zeros: $1, 2, 2 + i$

11. Degree 5; zeros: $1, i, 2i$

12. Degree 5; zeros: $0, 1, 2, i$

13. Degree 4; zeros: $i, 2, -2$

14. Degree 4; zeros: $2 - i, -i$

15. Degree 6; zeros: $2, 2 + i, -3 - i, 0$

16. Degree 6; zeros: $i, 3 - 2i, -2 + i$

In Problems 17–22, form a polynomial $f(x)$ with real coefficients having the given degree and zeros.

17. Degree 4; zeros: $3 + 2i; 4$, multiplicity 2

18. Degree 4; zeros: $i, 1 + 2i$

19. Degree 5; zeros: $2; -i; 1 + i$

20. Degree 6; zeros: $i, 4 - i; 2 + i$

21. Degree 4; zeros: 3, multiplicity 2; $-i$

22. Degree 5; zeros: 1, multiplicity 3; $1 + i$

In Problems 23–30, use the given zero to find the remaining zeros of each function.

23. $f(x) = x^3 - 4x^2 + 4x - 16$; zero: $2i$

24. $g(x) = x^3 + 3x^2 + 25x + 75$; zero: $-5i$

25. $f(x) = 2x^4 + 5x^3 + 5x^2 + 20x - 12$; zero: $-2i$

26. $h(x) = 3x^4 + 5x^3 + 25x^2 + 45x - 18$; zero: $3i$

27. $h(x) = x^4 - 9x^3 + 21x^2 + 21x - 130$; zero: $3 - 2i$

28. $f(x) = x^4 - 7x^3 + 14x^2 - 38x - 60$; zero: $1 + 3i$

29. $h(x) = 3x^5 + 2x^4 + 15x^3 + 10x^2 - 528x - 352$; zero: $-4i$

30. $g(x) = 2x^5 - 3x^4 - 5x^3 - 15x^2 - 207x + 108$; zero: $3i$

In Problems 31–40, find the complex zeros of each polynomial function. Write f in factored form.

31. $f(x) = x^3 - 1$

32. $f(x) = x^4 - 1$

33. $f(x) = x^3 - 8x^2 + 25x - 26$

34. $f(x) = x^3 + 13x^2 + 57x + 85$

35. $f(x) = x^4 + 5x^2 + 4$

36. $f(x) = x^4 + 13x^2 + 36$

37. $f(x) = x^4 + 2x^3 + 22x^2 + 50x - 75$

38. $f(x) = x^4 + 3x^3 - 19x^2 + 27x - 252$

39. $f(x) = 3x^4 - x^3 - 9x^2 + 159x - 52$

40. $f(x) = 2x^4 + x^3 - 35x^2 - 113x + 65$

Discussion and Writing

In Problems 41 and 42, explain why the facts given are contradictory.

41. $f(x)$ is a polynomial of degree 3 whose coefficients are real numbers; its zeros are $4 + i, 4 - i$, and $2 + i$.

42. $f(x)$ is a polynomial of degree 3 whose coefficients are real numbers; its zeros are $2, i$, and $3 + i$.

43. $f(x)$ is a polynomial of degree 4 whose coefficients are real numbers; three of its zeros are $2, 1 + 2i$, and $1 - 2i$. Explain why the remaining zero must be a real number.

44. $f(x)$ is a polynomial of degree 4 whose coefficients are real numbers; two of its zeros are -3 and $4 - i$. Explain why one of the remaining zeros must be a real number. Write down one of the missing zeros.

'Are You Prepared?' Answers

1. Sum: $3i$; product: $1 + 21i$ 2. $-1 - i, -1 + i$

Chapter Review

Things to Know

Power function (pp. 183–186)

$f(x) = x^n$, $n \geq 2$ even

Domain: all real numbers Range: nonnegative real numbers

Passes through $(-1, 1)$, $(0, 0)$, $(1, 1)$

Even function

Decreasing on $(-\infty, 0)$, increasing on $(0, \infty)$

$f(x) = x^n$, $n \geq 3$ odd

Domain: all real numbers Range: all real numbers

Passes through $(-1, -1)$, $(0, 0)$, $(1, 1)$

Odd function

Increasing on $(-\infty, \infty)$

Polynomial function (pp. 182, 186–194)

$f(x) = a_n x^n + a_{n-1} x^{n-1}$ $+ \cdots + a_1 x + a_0$, $a_n \neq 0$

Domain: all real numbers

At most $n - 1$ turning points

End behavior: Behaves like $y = a_n x^n$ for large $|x|$

Zeros of a polynomial f (p. 187)

Numbers for which $f(x) = 0$; the real zeros of f are the x-intercepts of the graph of f.

Rational function (p. 204)

$R(x) = \dfrac{p(x)}{q(x)}$

p, q are polynomial functions. Domain: $\{x | q(x) \neq 0\}$

Vertical asymptotes: With $R(x)$ in lowest terms, if $q(r) = 0$, then $x = r$ is a vertical asymptote.

Horizontal or oblique asymptotes: See the summary on page 211.

Inverse Variation (p. 225)

Let x and y denote two quantities. Then y varies inversely with x, or y is inversely proportional to x, if there is a nonzero constant k such that $y = \dfrac{k}{x}$.

Remainder Theorem (p. 237)

If a polynomial $f(x)$ is divided by $x - c$, then the remainder is $f(c)$.

Factor Theorem (p. 238)

$x - c$ is a factor of a polynomial $f(x)$ if and only if $f(c) = 0$.

Descartes' Rule of Signs (p. 240)

Let f denote a polynomial function. The number of positive zeros of f either equals the number of variations in sign of the nonzero coefficients of $f(x)$ or else equals that number less some even integer. The number of negative zeros of f either equals the number of variations in sign of the nonzero coefficients of $f(-x)$ or else equals that number less some even integer.

Rational Zeros Theorem (p. 240)

Let f be a polynomial function of degree 1 or higher of the form

$$f(x) = a_n x^n + a_{n-1} x^{n-1} + \cdots + a_1 x + a_0, \quad a_n \neq 0, a_0 \neq 0$$

where each coefficient is an integer. If $\dfrac{p}{q}$, in lowest terms, is a rational zero of f, then p must be a factor of a_0 and q must be a factor of a_n.

Intermediate Value Theorem (p. 245)

Let f be a polynomial function. If $a < b$ and $f(a)$ and $f(b)$ are of opposite sign, then there is at least one real zero of f between a and b.

Fundamental Theorem of Algebra (p. 251)

Every complex polynomial function $f(x)$ of degree $n \geq 1$ has at least one complex zero.

Conjugate Pairs Theorem (p. 251)

Let $f(x)$ be a polynomial whose coefficients are real numbers. If $r = a + bi$ is a zero of f, then its complex conjugate $\bar{r} = a - bi$ is also a zero of f.

Objectives

Section		You should be able to . . .	Review Exercises
3.1	1	Identify polynomial functions and their degree (p. 182)	1–4
	2	Graph polynomial functions using transformations (p. 186)	5–10

Review Exercises

In Problems 1–4, determine which functions are polynomial functions. For those that are, state the degree. For those that are not, tell why not.

1. $f(x) = 4x^5 - 3x^2 + 5x - 2$ **2.** $f(x) = \dfrac{3x^5}{2x+1}$ **3.** $f(x) = 3x^2 + 5x^{1/2} - 1$ **4.** $f(x) = 3$

In Problems 5–10, graph each function using transformations (shifting, compressing, stretching, and reflection). Show all the stages.

5. $f(x) = (x+2)^3$

6. $f(x) = -x^3 + 3$

7. $f(x) = -(x-1)^4$

8. $f(x) = (x-1)^4 - 2$

9. $f(x) = (x-1)^4 + 2$

10. $f(x) = (1-x)^3$

In Problems 11–18:
 (a) *Find the x- and y-intercepts of each polynomial function f.*
 (b) *Determine whether the graph of f touches or crosses the x-axis at each x-intercept.*
 (c) *End behavior: find the power function that the graph of f resembles for large values of $|x|$.*
 (d) *Determine the maximum number of turning points of the graph of f.*
 (e) *Determine the behavior of the graph of f near each x-intercept.*
 (f) *Put all the information together to obtain the graph of f. (You may need to locate additional points on the graph).*

11. $f(x) = x(x+2)(x+4)$ **12.** $f(x) = x(x-2)(x-4)$ **13.** $f(x) = (x-2)^2(x+4)$

14. $f(x) = (x-2)(x+4)^2$ **15.** $f(x) = -2x^3 + 4x^2$ **16.** $f(x) = -4x^3 + 4x$

17. $f(x) = (x-1)^2(x+3)(x+1)$ **18.** $f(x) = (x-4)(x+2)^2(x-2)$

In Problems 19–22, find the domain of each rational function. Find any horizontal, vertical, or oblique asymptotes.

19. $R(x) = \dfrac{x+2}{x^2-9}$ **20.** $R(x) = \dfrac{x^2+4}{x-2}$ **21.** $R(x) = \dfrac{x^2+3x+2}{(x+2)^2}$ **22.** $R(x) = \dfrac{x^3}{x^3-1}$

In Problems 23–34, discuss each rational function following the seven steps given on page 214.

23. $R(x) = \dfrac{2x-6}{x}$ **24.** $R(x) = \dfrac{4-x}{x}$ **25.** $H(x) = \dfrac{x+2}{x(x-2)}$ **26.** $H(x) = \dfrac{x}{x^2-1}$

27. $R(x) = \dfrac{x^2+x-6}{x^2-x-6}$ **28.** $R(x) = \dfrac{x^2-6x+9}{x^2}$ **29.** $F(x) = \dfrac{x^3}{x^2-4}$ **30.** $F(x) = \dfrac{3x^3}{(x-1)^2}$

31. $R(x) = \dfrac{2x^4}{(x-1)^2}$ **32.** $R(x) = \dfrac{x^4}{x^2-9}$ **33.** $G(x) = \dfrac{x^2-4}{x^2-x-2}$ **34.** $F(x) = \dfrac{(x-1)^2}{x^2-1}$

In Problems 35–44, solve each inequality.

35. $x^3 + x^2 < 4x + 4$ **36.** $x^3 + 4x^2 \geq x + 4$ **37.** $\dfrac{6}{x+3} \geq 1$ **38.** $\dfrac{-2}{1-3x} < 1$

39. $\dfrac{2x-6}{1-x} < 2$ **40.** $\dfrac{3-2x}{2x+5} \geq 2$ **41.** $\dfrac{(x-2)(x-1)}{x-3} \geq 0$ **42.** $\dfrac{x+1}{x(x-5)} \leq 0$

43. $\dfrac{x^2-8x+12}{x^2-16} > 0$ **44.** $\dfrac{x(x^2+x-2)}{x^2+9x+20} \leq 0$

In Problems 45–48, find the remainder R when $f(x)$ is divided by $g(x)$. Is g a factor of f?

45. $f(x) = 8x^3 - 3x^2 + x + 4$; $g(x) = x - 1$ **46.** $f(x) = 2x^3 + 8x^2 - 5x + 5$; $g(x) = x - 2$

47. $f(x) = x^4 - 2x^3 + 15x - 2$; $g(x) = x + 2$ **48.** $f(x) = x^4 - x^2 + 2x + 2$; $g(x) = x + 1$

49. Find the value of $f(x) = 12x^6 - 8x^4 + 1$ at $x = 4$.

50. Find the value of $f(x) = -16x^3 + 18x^2 - x + 2$ at $x = -2$.

In Problems 51 and 52, use Descartes' Rule of Signs to determine how many positive and negative zeros each polynomial function may have. Do not attempt to find the zeros.

51. $f(x) = 12x^8 - x^7 + 8x^4 - 2x^3 + x + 3$ **52.** $f(x) = -6x^5 + x^4 + 5x^3 + x + 1$

53. List all the potential rational zeros of $f(x) = 12x^8 - x^7 + 6x^4 - x^3 + x - 3$.

54. List all the potential rational zeros of $f(x) = -6x^5 + x^4 + 2x^3 - x + 1$.

In Problems 55–60, use Descartes' Rule of Signs and the Rational Zeros Theorem to find all the real zeros of each polynomial function. Use the zeros to factor f over the real numbers.

55. $f(x) = x^3 - 3x^2 - 6x + 8$ **56.** $f(x) = x^3 - x^2 - 10x - 8$

57. $f(x) = 4x^3 + 4x^2 - 7x + 2$ **58.** $f(x) = 4x^3 - 4x^2 - 7x - 2$

59. $f(x) = x^4 - 4x^3 + 9x^2 - 20x + 20$ **60.** $f(x) = x^4 + 6x^3 + 11x^2 + 12x + 18$

In Problems 61–64, solve each equation in the real number system.

61. $2x^4 + 2x^3 - 11x^2 + x - 6 = 0$ **62.** $3x^4 + 3x^3 - 17x^2 + x - 6 = 0$

63. $2x^4 + 7x^3 + x^2 - 7x - 3 = 0$ **64.** $2x^4 + 7x^3 - 5x^2 - 28x - 12 = 0$

In Problems 65–68, find a bound to the zeros of each polynomial function.

65. $f(x) = x^3 - x^2 - 4x + 2$ **66.** $f(x) = x^3 + x^2 - 10x - 5$

67. $f(x) = 2x^3 - 7x^2 - 10x + 35$ **68.** $f(x) = 3x^3 - 7x^2 - 6x + 14$

In Problems 69–72, use the Intermediate Value Theorem to show that each polynomial has a zero in the given interval.

69. $f(x) = 3x^3 - x - 1$; $[0, 1]$ **70.** $f(x) = 2x^3 - x^2 - 3$; $[1, 2]$

71. $f(x) = 8x^4 - 4x^3 - 2x - 1$; $[0, 1]$ **72.** $f(x) = 3x^4 + 4x^3 - 8x - 2$; $[1, 2]$

In Problems 73–76, each polynomial has exactly one positive zero. Approximate the zero correct to two decimal places.

73. $f(x) = x^3 - x - 2$ **74.** $f(x) = 2x^3 - x^2 - 3$

75. $f(x) = 8x^4 - 4x^3 - 2x - 1$ **76.** $f(x) = 3x^4 + 4x^3 - 8x - 2$

In Problems 77–80, information is given about a complex polynomial $f(x)$ whose coefficients are real numbers. Find the remaining zeros of f.

77. Degree 3; zeros: $4 + i, 6$ **78.** Degree 3; zeros: $3 + 4i, 5$

79. Degree 4; zeros: $i, 1 + i$ **80.** Degree 4; zeros: $1, 2, 1 + i$

In Problems 81–88, find the complex zeros of each polynomial function $f(x)$. Write f in factored form.

81. $f(x) = x^3 - 3x^2 - 6x + 8$ **82.** $f(x) = x^3 - x^2 - 10x - 8$

83. $f(x) = 4x^3 + 4x^2 - 7x + 2$ **84.** $f(x) = 4x^3 - 4x^2 - 7x - 2$

85. $f(x) = x^4 - 4x^3 + 9x^2 - 20x + 20$ **86.** $f(x) = x^4 + 6x^3 + 11x^2 + 12x + 18$

87. $f(x) = 2x^4 + 2x^3 - 11x^2 + x - 6$ **88.** $f(x) = 3x^4 + 3x^3 - 17x^2 + x - 6$

89. Weight of a Body The weight of a body varies inversely with the square of its distance from the center of Earth. Assuming that the radius of Earth is 3960 miles, how much would a man weigh at an altitude of 1 mile above Earth's surface if he weighs 200 pounds on Earth's surface?

90. Resistance due to a Conductor The resistance (in ohms) of a circular conductor varies directly with the length of the conductor and inversely with the square of the radius of the conductor. If 50 feet of wire with a radius of 6×10^{-3} inch has a resistance of 10 ohms, what would be the resistance of 100 feet of the same wire if the radius is increased to 7×10^{-3} inch?

91. Making a Can A can in the shape of a right circular cylinder is required to have a volume of 250 cubic centimeters.
(a) Express the amount A of material to make the can as a function of the radius r of the cylinder.
(b) How much material is required if the can is of radius 3 centimeters?
(c) How much material is required if the can is of radius 5 centimeters?
(d) Graph $A = A(r)$. For what value of r is A smallest?

92. AIDS Cases in the United States The following data represent the cumulative number of reported AIDS cases in the United States for 1990–1997.

Year, t	Number of AIDS Cases, A
1990, 1	193,878
1991, 2	251,638
1992, 3	326,648
1993, 4	399,613
1994, 5	457,280
1995, 6	528,215
1996, 7	594,760
1997, 8	653,253

SOURCE: U.S. Center for Disease Control and Prevention

(a) With a graphing utility, draw a scatter diagram of the data.

(b) Find cubic function of best fit to these data. Use this function to predict the cumulative number of AIDS cases reported in the United States in 2000.
(c) Graph the cubic function of best fit on the scatter diagram drawn in part (a).
(d) Do you think the function given in part (b) will be useful in predicting the number of AIDS cases in 2005?

93. Design a polynomial function with the following characteristics: degree 6; four real zeros, one of multiplicity 3; y-intercept 3; behaves like $y = -5x^6$ for large values of $|x|$. Is this polynomial unique? Compare your polynomial with those of other students. What terms will be the same as everyone else's? Add some more characteristics, such as symmetry or naming the real zeros. How does this modify the polynomial?

94. Design a rational function with the following characteristics: three real zeros, one of multiplicity 2; y-intercept 1; vertical asymptotes $x = -2$ and $x = 3$; oblique asymptote $y = 2x + 1$. Is this rational function unique? Compare yours with those of other students. What will be the same as everyone else's? Add some more characteristics, such as symmetry or naming the real zeros. How does this modify the rational function?

95. The illustration shows the graph of a polynomial function.
(a) Is the degree of the polynomial even or odd?
(b) Is the leading coefficient positive or negative?
(c) Is the function even, odd, or neither?
(d) Why is x^2 necessarily a factor of the polynomial?
(e) What is the minimum degree of the polynomial?
(f) Formulate five different polynomials whose graphs could look like the one shown. Compare yours to those of other students. What similarities do you see? What differences?

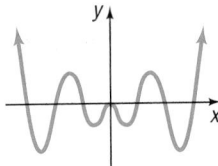

Chapter Test

1. Graph $f(x) = (x - 3)^4 - 2$ using transformations.

2. For the polynomial function $g(x) = 2x^3 + 5x^2 - 28x - 15$,
(a) Determine the maximum number of real zeros that the function may have.
(b) Find bounds to the zeros of the function.
(c) List the potential rational zeros.
(d) Determine the real zeros of g. Factor g over the reals.
(e) Find the x- and y-intercepts of the graph of g.
(f) Determine whether the graph crosses or touches the x-axis at each x-intercept.
(g) Find the power function that the graph of g resembles for large values of $|x|$.
(h) Determine the behavior of the graph of g near each x-intercept.

(i) Put all the information together to obtain the graph of g.

3. Find the complex zeros of $f(x) = x^3 - 4x^2 + 25x - 100$.

4. Solve $3x^3 + 2x - 1 = 8x^2 - 4$ in the complex number system.

In Problems 5 and 6, find the domain of each function. Find any horizontal, vertical, or oblique asymptotes.

5. $g(x) = \dfrac{2x^2 - 14x + 24}{x^2 + 6x - 40}$

6. $r(x) = \dfrac{x^2 + 2x - 3}{x + 1}$

7. Sketch the graph of the function in Problem 6. Label all intercepts, vertical asymptotes, horizontal asymptotes, and oblique asymptotes.

In Problems 8 and 9, write a function that meets the given conditions.

8. Fourth-degree polynomial with real coefficients; zeros: $-2, 0, 3 + i$

9. Rational function; asymptotes: $y = 2, x = 4$; domain: $\{x \mid x \neq 4, x \neq 9\}$

10. Use the Intermediate Value Theorem to show that the function $f(x) = -2x^2 - 3x + 8$ has at least one real zero on the interval $[0, 4]$.

11. Solve: $\dfrac{x + 2}{x - 3} < 2$

Chapter Projects

1. Weed Pollen The table to the left is the Weed pollen count for Lexington, Kentucky, from August 4, 2004, to October 6, 2004, the prime season for weed pollen.

(a) Let the independent variable D represent the date, where $D = 1$ on 8/4, $D = 6$ on 8/9, $D = 9$ on 8/12 ... and $D = 64$ on 10/6. Let the dependent variable P represent the number of grains per cubic meter of weed pollen. Draw a scatter diagram of the data using your graphing utility, and by hand on graph paper.

(b) In the hand-drawn scatter diagram, sketch a smooth curve that fits the data. Try to have the curve pass through as many of the points as closely as possible. How many turning points are there? What does this tell you about the degree of the polynomial that could be fit to the data? If the ends of the graph on the left and on the right are extended downward, how many real zeros are indicated by the graph? How many complex? Why?

(c) Use a graphing utility to determine the quartic function of the best fit. Graph it on your graphing utility. Does it look like what you expected from your hand-drawn graph? Explain.

(d) Use a graphing utility to determine the cubic function of best fit. Graph it on your graphing utility. Does it look like what you expected from your hand-drawn graph? Explain.

(e) Use a graphing utility to determine the quadratic function of best fit. Graph it on your graphing utility. Does it look like what you expected from your hand-drawn graph? Explain.

(f) Which of the three functions seems to fit the best? Explain your reasoning.

(g) Investigate weed pollen counts at other times of the year. (Go to the National Allergy Bureau at www.aaaai.org.) Would the polynomial you found in part (c) work for other 2-month periods? Why or why not? What about for the same period of time in other years?

Date	Grains per cubic meter	Date	Grains per cubic meter
8/4	18	9/8	45
8/9	14	9/10	93
8/12	12	9/13	14
8/13	16	9/15	10
8/17	37	9/17	11
8/18	30	9/20	5
8/23	53	9/22	3
8/24	22	9/27	4
8/25	47	9/29	6
8/26	7	10/1	4
8/31	214	10/4	4
9/2	31	10/5	7
9/3	44	10/6	7
9/6	50		

SOURCE: National Allergy Bureau at the American Academy of Allergy, Asthma and Immunology web site (*www.aaaai.org*)

The following projects are available at the Instructor's Resource Center (IRC):

2. Project at Motorola *How Many Cellphones Can I Make?*

3. Maclaurin Series

4. Theory of Equations

Cumulative Review

1. Find the distance between the points $P = (1, 3)$ and $Q = (-4, 2)$.

2. Solve the inequality $x^2 \geq x$ and graph the solution set.

3. Solve the inequality $x^2 - 3x < 4$ and graph the solution set.

4. Find a linear function with slope -3 that contains the point $(-1, 4)$. Graph the function.

5. Find the equation of the line parallel to the line $y = 2x + 1$ and containing the point $(3, 5)$. Express your answer in slope–intercept form and graph the line.

6. Graph the equation $y = x^3$.

7. Does the relation $\{(3, 6), (1, 3), (2, 5), (3, 8)\}$ represent a function? Why or why not?

8. Solve the equation $x^3 - 6x^2 + 8x = 0$.

9. Solve the inequality $3x + 2 \leq 5x - 1$ and graph the solution set.

10. Find the center and radius of the circle $x^2 + 4x + y^2 - 2y - 4 = 0$. Graph the circle.

11. For the equation $y = x^3 - 9x$, determine the intercepts and test for symmetry.

12. Find an equation of the line perpendicular to $3x - 2y = 7$ that contains the point $(1, 5)$.

13. Is the following graph the graph of a function? Why or why not?

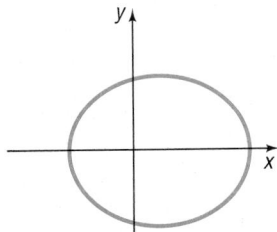

14. For the function $f(x) = x^2 + 5x - 2$, find
 (a) $f(3)$
 (b) $f(-x)$
 (c) $-f(x)$
 (d) $f(3x)$
 (e) $\dfrac{f(x + h) - f(x)}{h}, h \neq 0$

15. Answer the following questions regarding the function
 $$f(x) = \frac{x + 5}{x - 1}$$
 (a) What is the domain of f?
 (b) Is the point $(2, 6)$ on the graph of f?
 (c) If $x = 3$, what is $f(x)$? What point is on the graph of f?
 (d) If $f(x) = 9$, what is x? What point is on the graph of f?

16. Graph the function $f(x) = -3x + 7$.

17. Graph $f(x) = 2x^2 - 4x + 1$ by determining whether its graph opens up or down and by finding its vertex, axis of symmetry, y-intercept, and x-intercepts, if any.

18. Find the average rate of change of $f(x) = x^2 + 3x + 1$ from 1 to x. Use this result to find the slope of the secant line containing $(1, f(1))$ and $(2, f(2))$.

19. In parts (a) to (f) use the following graph.

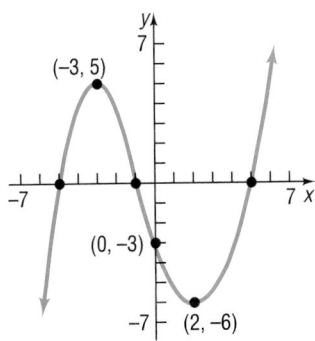

(a) Determine the intercepts.
(b) Based on the graph, tell whether the graph is symmetric with respect to the x-axis, the y-axis, and/or the origin.
(c) Based on the graph, tell whether the function is even, odd, or neither.
(d) List the intervals on which f is increasing. List the intervals on which f is decreasing.
(e) List the numbers, if any, at which f has a local maximum. What are these local maxima?
(f) List the numbers, if any, at which f has a local minimum. What are these local minima?

20. Determine algebraically whether the function
 $$f(x) = \frac{5x}{x^2 - 9}$$
 is even, odd, or neither.

21. For the function $f(x) = \begin{cases} 2x + 1 & \text{if } -3 < x < 2 \\ -3x + 4 & \text{if } x \geq 2 \end{cases}$
 (a) Find the domain of f.
 (b) Locate any intercepts.
 (c) Graph the function.
 (d) Based on the graph, find the range.

22. Graph the function $f(x) = -3(x + 1)^2 + 5$ using transformations.

23. Suppose that $f(x) = x^2 - 5x + 1$ and $g(x) = -4x - 7$.
 (a) Find $f + g$ and state its domain.
 (b) Find $\dfrac{f}{g}$ and state its domain.

24. **Demand Equation** The price p (in dollars) and the quantity x sold of a certain product obey the demand equation
 $$p = -\frac{1}{10}x + 150, \quad 0 \leq x \leq 1500$$
 (a) Express the revenue R as a function of x.
 (b) What is the revenue if 100 units are sold?
 (c) What quantity x maximizes revenue? What is the maximum revenue?
 (d) What price should the company charge to maximize revenue?

4 Exponential and Logarithmic Functions

A LOOK BACK Until now, our study of functions has concentrated on polynomial and rational functions. These functions belong to the class of **algebraic functions**, that is, functions that can be expressed in terms of sums, differences, products, quotients, powers, or roots of polynomials. Functions that are not algebraic are termed **transcendental** (they transcend, or go beyond, algebraic functions).

A LOOK AHEAD In this chapter, we study two transcendental functions: the exponential function and the logarithmic function. These functions occur frequently in a wide variety of applications, such as biology, chemistry, economics, and psychology.

The chapter begins with a discussion of composite, one-to-one, and inverse functions, concepts needed to see the relationship between exponential and logarithmic functions.

The McDonald's Scalding Coffee Case

April 3, 1996

There is a lot of hype about the McDonald's scalding coffee case. No one is in favor of frivolous cases or outlandish results; however, it is important to understand some points that were not reported in most of the stories about the case. McDonald's coffee was not only hot, it was scalding, capable of almost instantaneous destruction of skin, flesh, and muscle.

Plaintiff's expert, a scholar in thermodynamics applied to human skin burns, testified that liquids at 180 degrees will cause a full thickness burn to human skin in two to seven seconds. Other testimony showed that, as the temperature decreases toward 155 degrees, the extent of the burn relative to that temperature decreases exponentially. Thus, if (the) spill had involved coffee at 155 degrees, the liquid would have cooled and given her time to avoid a serious burn.

SOURCE: *ATLA fact sheet © 1995, 1996 Consumer Attorneys of CA. Used with permission of the Association of Trial Lawyers of America.*

—See Chapter Project 1.

4.1 Composite Functions

PREPARING FOR THIS SECTION *Before getting started, review the following:*

• Finding Values of a Function (Section 1.1, pp. 44–47) • Domain of a Function (Section 1.1, pp. 41–44 and 48–49)

Now work the 'Are You Prepared?' problems on page 268.

OBJECTIVES 1 Form a Composite Function
 2 Find the Domain of a Composite Function

1 Form a Composite Function

Figure 1

Suppose that an oil tanker is leaking oil and you want to determine the area of the circular oil patch around the ship. See Figure 1. It is determined that the oil is leaking from the tanker in such a way that the radius of the circular patch of oil around the ship is increasing at a rate of 3 feet per minute. Therefore, the radius r of the oil patch at any time t, in minutes, is given by $r(t) = 3t$. So after 20 minutes, the radius of the oil patch is $r(20) = 3(20) = 60$ feet. The area A of a circle as a function of the radius r is given by $A(r) = \pi r^2$. The area of the circular patch of oil after 20 minutes is $A(60) = \pi(60)^2 = 3600\pi$ square feet.

Notice that $60 = r(20)$, so $A(60) = A(r(20))$. The argument of the function A is itself a function! In general, we can find the area of the oil patch as a function of time t by evaluating $A(r(t))$. The function $A(r(t))$ is a special type of function called a *composite function*.

As another example, consider the function $y = (2x + 3)^2$. If we write $y = f(u) = u^2$ and $u = g(x) = 2x + 3$, then, by a substitution process, we can obtain the original function: $y = f(u) = f(g(x)) = (2x + 3)^2$.

In general, suppose that f and g are two functions and that x is a number in the domain of g. By evaluating g at x, we get $g(x)$. If $g(x)$ is in the domain of f, then we may evaluate f at $g(x)$ and thereby obtain the expression $f(g(x))$. The correspondence from x to $f(g(x))$ is called a *composite function $f \circ g$*.

> Given two functions f and g, the **composite function**, denoted by $f \circ g$ (read as "f composed with g"), is defined by
>
> $$(f \circ g)(x) = f(g(x))$$
>
> The domain of $f \circ g$ is the set of all numbers x in the domain of g such that $g(x)$ is in the domain of f.

Look carefully at Figure 2. Only those x's in the domain of g for which $g(x)$ is in the domain of f can be in the domain of $f \circ g$. The reason is that if $g(x)$ is not in the domain of f then $f(g(x))$ is not defined. Because of this, the domain of $f \circ g$ is a subset of the domain of g; the range of $f \circ g$ is a subset of the range of f.

Figure 2

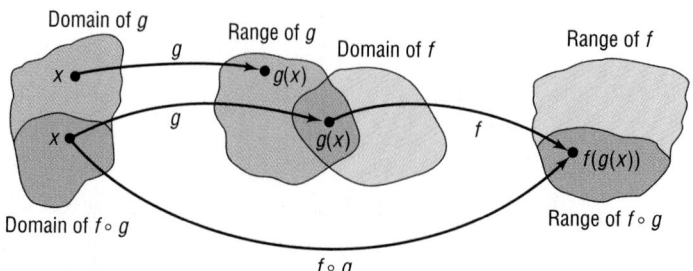

Figure 3 provides a second illustration of the definition. Here, x is the input to the function g, yielding $g(x)$. Then $g(x)$ is the input to the function f, yielding $f(g(x))$. Notice that the "inside" function g in $f(g(x))$ is done first.

Figure 3

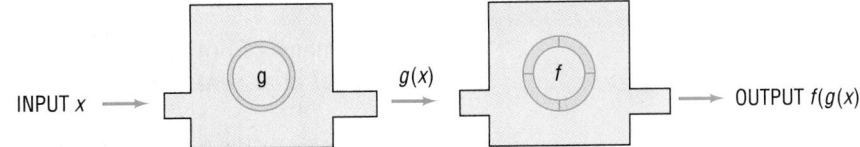

Let's look at some examples.

| EXAMPLE 1 | **Evaluating a Composite Function** |

Suppose that $f(x) = 2x^2 - 3$ and $g(x) = 4x$. Find:

(a) $(f \circ g)(1)$　　(b) $(g \circ f)(1)$　　(c) $(f \circ f)(-2)$　　(d) $(g \circ g)(-1)$

Solution　(a) $(f \circ g)(1) = f(g(1)) = f(4) = 2 \cdot 4^2 - 3 = 29$

$$g(x) = 4x \qquad f(x) = 2x^2 - 3$$
$$g(1) = 4$$

(b) $(g \circ f)(1) = g(f(1)) = g(-1) = 4 \cdot (-1) = -4$

$$f(x) = 2x^2 - 3 \quad g(x) = 4x$$
$$f(1) = -1$$

(c) $(f \circ f)(-2) = f(f(-2)) = f(5) = 2 \cdot 5^2 - 3 = 47$

$$f(-2) = 2(-2)^2 - 3 = 5$$

Figure 4

(d) $(g \circ g)(-1) = g(g(-1)) = g(-4) = 4 \cdot (-4) = -16$

$$g(-1) = -4$$
∎

COMMENT: Graphing calculators can be used to evaluate composite functions.* Let $Y_1 = f(x) = 2x^2 - 3$ and $Y_2 = g(x) = 4x$. Then, using a TI-84 Plus graphing calculator, $(f \circ g)(1)$ is found as shown in Figure 4. Notice that this is the result obtained in Example 1(a). ∎

✏ **NOW WORK PROBLEM 11.**

2 Find the Domain of a Composite Function

| EXAMPLE 2 | **Finding a Composite Function and Its Domain** |

Suppose that $f(x) = x^2 + 3x - 1$ and $g(x) = 2x + 3$.

Find:　(a) $f \circ g$　　　　(b) $g \circ f$

Then find the domain of each composite function.

Solution　The domain of f and the domain of g is the set of all real numbers.

(a) $(f \circ g)(x) = f(g(x)) = f(2x + 3) = (2x + 3)^2 + 3(2x + 3) - 1$

$$f(x) = x^2 + 3x - 1$$

$$= 4x^2 + 12x + 9 + 6x + 9 - 1 = 4x^2 + 18x + 17$$

Since the domains of both f and g is the set of all real numbers, the domain of $f \circ g$ is the set of all real numbers.

*Consult your owner's manual for the appropriate keystrokes.

(b) $(g \circ f)(x) = g(f(x)) = g(x^2 + 3x - 1) = 2(x^2 + 3x - 1) + 3$

$$g(x) = 2x + 3$$

$$= 2x^2 + 6x - 2 + 3 = 2x^2 + 6x + 1$$

Since the domains of both f and g is the set of all real numbers, the domain of $f \circ g$ is the set of all real numbers. ∎

Look back at Figure 2. In determining the domain of the composite function $(f \circ g)(x) = f(g(x))$, keep the following two thoughts in mind about the input x.

1. $g(x)$ must be defined so any x not in the domain of g must be excluded.
2. $f(g(x))$ must be defined so any x for which $g(x)$ is not in the domain of f must be excluded.

EXAMPLE 3 **Finding the Domain of $f \circ g$**

Find the domain of $(f \circ g)(x)$ if $f(x) = \dfrac{1}{x + 2}$ and $g(x) = \dfrac{4}{x - 1}$

Solution For $(f \circ g)(x) = f(g(x))$, we first note that the domain of g is $\{x | x \neq 1\}$, so we exclude 1 from the domain of $f \circ g$. Next, we note that the domain of f is $\{x | x \neq -2\}$, which means that $g(x)$ cannot equal -2. We solve the equation $g(x) = -2$ to determine what values of x to exclude.

$$\frac{4}{x - 1} = -2 \qquad g(x) = -2$$

$$4 = -2(x - 1)$$

$$4 = -2x + 2$$

$$2x = -2$$

$$x = -1$$

We also exclude -1 from the domain of $f \circ g$. The domain of $f \circ g$ is $\{x | x \neq -1, x \neq 1\}$.

✔ **CHECK:** For $x = 1$, $g(x) = \dfrac{4}{x - 1}$ is not defined, so $(f \circ g)(x) = f(g(x))$ is not defined.

For $x = -1$, $g(-1) = \dfrac{4}{-2} = -2$, and $(f \circ g)(-1) = f(g(-1)) = f(-2)$ is not defined. ∎

◂━━━ **NOW WORK PROBLEM 21.**

EXAMPLE 4 **Finding a Composite Function and Its Domain**

Suppose that $f(x) = \dfrac{1}{x + 2}$ and $g(x) = \dfrac{4}{x - 1}$

Find: (a) $f \circ g$ (b) $f \circ f$

Then find the domain of each composite function.

Solution The domain of f is $\{x | x \neq -2\}$ and the domain of g is $\{x | x \neq 1\}$.

(a) $(f \circ g)(x) = f(g(x)) = f\left(\dfrac{4}{x - 1}\right) = \dfrac{1}{\dfrac{4}{x - 1} + 2} = \dfrac{x - 1}{4 + 2(x - 1)} = \dfrac{x - 1}{2x + 2} = \dfrac{x - 1}{2(x + 1)}$

$$f(x) = \frac{1}{x + 2} \qquad \text{Multiply by } \frac{x - 1}{x - 1}.$$

In Example 3, we found the domain of $f \circ g$ to be $\{x | x \neq -1, x \neq 1\}$.

We could also find the domain of $f \circ g$ by first looking at the domain of g: $\{x | x \neq 1\}$. We exclude 1 from the domain of $f \circ g$ as a result. Then we look at $f \circ g$ and notice that x cannot equal -1, since $x = -1$ results in division by 0. So we also exclude -1 from the domain of $f \circ g$. Therefore, the domain of $f \circ g$ is $\{x | x \neq -1, x \neq 1\}$.

(b) $(f \circ f)(x) = f(f(x)) = f\left(\dfrac{1}{x+2}\right) = \dfrac{1}{\dfrac{1}{x+2}+2} = \dfrac{x+2}{1+2(x+2)} = \dfrac{x+2}{2x+5}$

$\qquad\qquad\qquad\qquad\qquad\uparrow\qquad\qquad\quad\uparrow$

$\qquad\qquad\qquad f(x) = \dfrac{1}{x+2} \qquad\qquad \text{Multiply by } \dfrac{x+2}{x+2}.$

The domain of $f \circ f$ consists of those x in the domain of f, $\{x | x \neq -2\}$, for which

$$f(x) = \dfrac{1}{x+2} \neq -2 \qquad \begin{array}{l} \dfrac{1}{x+2} = -2 \\[4pt] 1 = -2(x+2) \\ 1 = -2x - 4 \\ 2x = -5 \\ x = -\dfrac{5}{2} \end{array}$$

or, equivalently,

$$x \neq -\dfrac{5}{2}$$

The domain of $f \circ f$ is $\left\{ x \middle| x \neq -\dfrac{5}{2}, x \neq -2 \right\}$.

We could also find the domain of $f \circ f$ by recognizing that -2 is not in the domain of f and so should be excluded from the domain of $f \circ f$. Then, looking at $f \circ f$, we see that x cannot equal $-\dfrac{5}{2}$. Do you see why? Therefore, the domain of $f \circ f$ is $\left\{ x \middle| x \neq -\dfrac{5}{2}, x \neq -2 \right\}$. ∎

NOW WORK PROBLEMS 33 AND 35.

Look back at Example 2, which illustrates that, in general, $f \circ g \neq g \circ f$. Sometimes $f \circ g$ does equal $g \circ f$, as shown in the next example.

EXAMPLE 5

Showing That Two Composite Functions Are Equal

If $f(x) = 3x - 4$ and $g(x) = \dfrac{1}{3}(x+4)$, show that

$$(f \circ g)(x) = (g \circ f)(x) = x$$

for every x in the domain of $f \circ g$ and $g \circ f$.

Solution

$(f \circ g)(x) = f(g(x))$

$\qquad = f\left(\dfrac{x+4}{3}\right) \qquad g(x) = \dfrac{1}{3}(x+4) = \dfrac{x+4}{3}$

$\qquad = 3\left(\dfrac{x+4}{3}\right) - 4 \qquad$ Substitute $g(x)$ into the rule for f, $f(x) = 3x - 4$.

$\qquad = x + 4 - 4 = x$

$(g \circ f)(x) = g(f(x))$

$\qquad = g(3x - 4) \qquad f(x) = 3x - 4$

$\qquad = \dfrac{1}{3}[(3x - 4) + 4] \qquad$ Substitute $f(x)$ into the rule for g, $g(x) = \dfrac{1}{3}(x+4)$.

$\qquad = \dfrac{1}{3}(3x) = x$

We conclude that $(f \circ g)(x) = (g \circ f)(x) = x$. ∎

— Seeing the Concept —

Using a graphing calculator, let $Y_1 = f(x) = 3x - 4$, $Y_2 = g(x) = \dfrac{1}{3}(x+4)$, $Y_3 = f \circ g$, and $Y_4 = g \circ f$. Using the viewing window $-3 \le x \le 3, -2 \le y \le 2$, graph only Y_3 and Y_4. What do you see? TRACE to verify that $Y_3 = Y_4$.

In Section 4.2, we shall see that there is an important relationship between functions f and g for which $(f \circ g)(x) = (g \circ f)(x) = x$.

 NOW WORK PROBLEM 45.

Calculus Application

Some techniques in calculus require that we be able to determine the components of a composite function. For example, the function $H(x) = \sqrt{x + 1}$ is the composition of the functions f and g, where $f(x) = \sqrt{x}$ and $g(x) = x + 1$, because $H(x) = (f \circ g)(x) = f(g(x)) = f(x + 1) = \sqrt{x + 1}$.

| EXAMPLE 6 | Finding the Components of a Composite Function |

Find functions f and g such that $f \circ g = H$ if $H(x) = (x^2 + 1)^{50}$.

Solution The function H takes $x^2 + 1$ and raises it to the power 50. A natural way to decompose H is to raise the function $g(x) = x^2 + 1$ to the power 50. If we let $f(x) = x^{50}$ and $g(x) = x^2 + 1$, then

$$(f \circ g)(x) = f(g(x))$$
$$= f(x^2 + 1)$$
$$= (x^2 + 1)^{50} = H(x)$$

Figure 5

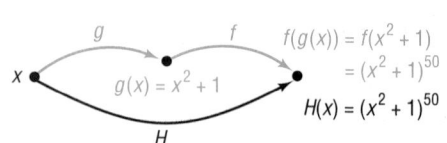

See Figure 5. ∎

Other functions f and g may be found for which $f \circ g = H$ in Example 6. For example, if $f(x) = x^2$ and $g(x) = (x^2 + 1)^{25}$, then

$$(f \circ g)(x) = f(g(x)) = f((x^2 + 1)^{25}) = [(x^2 + 1)^{25}]^2 = (x^2 + 1)^{50}$$

Although the functions f and g found as a solution to Example 6 are not unique, there is usually a "natural" selection for f and g that comes to mind first.

| EXAMPLE 7 | Finding the Components of a Composite Function |

Find functions f and g such that $f \circ g = H$ if $H(x) = \dfrac{1}{x + 1}$.

Solution Here H is the reciprocal of $g(x) = x + 1$. If we let $f(x) = \dfrac{1}{x}$ and $g(x) = x + 1$, we find that

$$(f \circ g)(x) = f(g(x)) = f(x + 1) = \frac{1}{x + 1} = H(x)$$ ∎

 NOW WORK PROBLEM 53.

4.1 Assess Your Understanding

'Are You Prepared?'

Answers are given at the end of these exercises. If you get a wrong answer, read the pages listed in red.

1. Find $f(3)$ if $f(x) = -4x^2 + 5x$. (pp. 45–46)
2. Find $f(3x)$ if $f(x) = 4 - 2x^2$. (pp. 45–46)
3. Find the domain of the function $f(x) = \dfrac{x^2 - 1}{x^2 - 25}$. (p. 48)

Concepts and Vocabulary

4. If $f(x) = x + 1$ and $g(x) = x^3$, then _____ $= (x + 1)^3$.
5. *True or False:* $f(g(x)) = f(x) \cdot g(x)$.
6. *True or False:* The domain of the composite function $(f \circ g)(x)$ is the same as the domain of $g(x)$.

Skill Building

In Problems 7 and 8, evaluate each expression using the values given in the table.

7.
x	−3	−2	−1	0	1	2	3
f(x)	−7	−5	−3	−1	3	5	5
g(x)	8	3	0	−1	0	3	8

 (a) $(f \circ g)(1)$ (b) $(f \circ g)(-1)$ (c) $(g \circ f)(-1)$ (d) $(g \circ f)(0)$ (e) $(g \circ g)(-2)$ (f) $(f \circ f)(-1)$

8.
x	−3	−2	−1	0	1	2	3
f(x)	11	9	7	5	3	1	−1
g(x)	−8	−3	0	1	0	−3	−8

 (a) $(f \circ g)(1)$ (b) $(f \circ g)(2)$ (c) $(g \circ f)(2)$ (d) $(g \circ f)(3)$ (e) $(g \circ g)(1)$ (f) $(f \circ f)(3)$

In Problems 9 and 10, evaluate each expression using the graphs of $y = f(x)$ and $y = g(x)$ shown below.

9. (a) $g(f(-1))$ (b) $g(f(0))$
 (c) $f(g(-1))$ (d) $f(g(4))$

10. (a) $g(f(1))$ (b) $g(f(5))$
 (c) $f(g(0))$ (d) $f(g(2))$

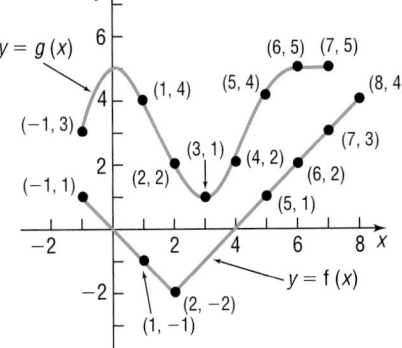

In Problems 11–20, for the given functions f and g, find:

 (a) $(f \circ g)(4)$ (b) $(g \circ f)(2)$ (c) $(f \circ f)(1)$ (d) $(g \circ g)(0)$

11. $f(x) = 2x; \quad g(x) = 3x^2 + 1$

12. $f(x) = 3x + 2; \quad g(x) = 2x^2 - 1$

13. $f(x) = 4x^2 - 3; \quad g(x) = 3 - \dfrac{1}{2}x^2$

14. $f(x) = 2x^2; \quad g(x) = 1 - 3x^2$

15. $f(x) = \sqrt{x}; \quad g(x) = 2x$

16. $f(x) = \sqrt{x + 1}; \quad g(x) = 3x$

17. $f(x) = |x|; \quad g(x) = \dfrac{1}{x^2 + 1}$

18. $f(x) = |x - 2|; \quad g(x) = \dfrac{3}{x^2 + 2}$

19. $f(x) = \dfrac{3}{x + 1}; \quad g(x) = \sqrt[3]{x}$

20. $f(x) = x^{3/2}; \quad g(x) = \dfrac{2}{x + 1}$

In Problems 21–28, find the domain of the composite function $f \circ g$.

21. $f(x) = \dfrac{3}{x - 1}; \quad g(x) = \dfrac{2}{x}$

22. $f(x) = \dfrac{1}{x + 3}; \quad g(x) = \dfrac{-2}{x}$

23. $f(x) = \dfrac{x}{x - 1}; \quad g(x) = \dfrac{-4}{x}$

24. $f(x) = \dfrac{x}{x + 3}; \quad g(x) = \dfrac{2}{x}$

25. $f(x) = \sqrt{x}; \quad g(x) = 2x + 3$

26. $f(x) = x - 2; \quad g(x) = \sqrt{1 - x}$

27. $f(x) = x^2 + 1; \quad g(x) = \sqrt{x - 1}$

28. $f(x) = x^2 + 4; \quad g(x) = \sqrt{x - 2}$

In Problems 29–44, for the given functions f and g, find:

(a) $f \circ g$ (b) $g \circ f$ (c) $f \circ f$ (d) $g \circ g$

State the domain of each composite function.

29. $f(x) = 2x + 3$; $g(x) = 3x$

30. $f(x) = -x$; $g(x) = 2x - 4$

31. $f(x) = 3x + 1$; $g(x) = x^2$

32. $f(x) = x + 1$; $g(x) = x^2 + 4$

33. $f(x) = x^2$; $g(x) = x^2 + 4$

34. $f(x) = x^2 + 1$; $g(x) = 2x^2 + 3$

35. $f(x) = \dfrac{3}{x - 1}$; $g(x) = \dfrac{2}{x}$

36. $f(x) = \dfrac{1}{x + 3}$; $g(x) = \dfrac{-2}{x}$

37. $f(x) = \dfrac{x}{x - 1}$; $g(x) = \dfrac{-4}{x}$

38. $f(x) = \dfrac{x}{x + 3}$; $g(x) = \dfrac{2}{x}$

39. $f(x) = \sqrt{x}$; $g(x) = 2x + 3$

40. $f(x) = \sqrt{x - 2}$; $g(x) = 1 - 2x$

41. $f(x) = x^2 + 1$; $g(x) = \sqrt{x - 1}$

42. $f(x) = x^2 + 4$; $g(x) = \sqrt{x - 2}$

43. $f(x) = ax + b$; $g(x) = cx + d$

44. $f(x) = \dfrac{ax + b}{cx + d}$; $g(x) = mx$

In Problems 45–52, show that $(f \circ g)(x) = (g \circ f)(x) = x$.

45. $f(x) = 2x$; $g(x) = \dfrac{1}{2}x$

46. $f(x) = 4x$; $g(x) = \dfrac{1}{4}x$

47. $f(x) = x^3$; $g(x) = \sqrt[3]{x}$

48. $f(x) = x + 5$; $g(x) = x - 5$

49. $f(x) = 2x - 6$; $g(x) = \dfrac{1}{2}(x + 6)$

50. $f(x) = 4 - 3x$; $g(x) = \dfrac{1}{3}(4 - x)$

51. $f(x) = ax + b$; $g(x) = \dfrac{1}{a}(x - b)$, $a \neq 0$

52. $f(x) = \dfrac{1}{x}$; $g(x) = \dfrac{1}{x}$

In Problems 53–58, find functions f and g so that $f \circ g = H$.

53. $H(x) = (2x + 3)^4$

54. $H(x) = (1 + x^2)^3$

55. $H(x) = \sqrt{x^2 + 1}$

56. $H(x) = \sqrt{1 - x^2}$

57. $H(x) = |2x + 1|$

58. $H(x) = |2x^2 + 3|$

Applications and Extensions

59. If $f(x) = 2x^3 - 3x^2 + 4x - 1$ and $g(x) = 2$, find $(f \circ g)(x)$ and $(g \circ f)(x)$.

60. If $f(x) = \dfrac{x + 1}{x - 1}$, find $(f \circ f)(x)$.

61. If $f(x) = 2x^2 + 5$ and $g(x) = 3x + a$, find a so that the graph of $f \circ g$ crosses the y-axis at 23.

62. If $f(x) = 3x^2 - 7$ and $g(x) = 2x + a$, find a so that the graph of $f \circ g$ crosses the y-axis at 68.

63. **Surface Area of a Balloon** The surface area S (in square meters) of a hot-air balloon is given by

$$S(r) = 4\pi r^2$$

where r is the radius of the balloon (in meters). If the radius r is increasing with time t (in seconds) according to the formula $r(t) = \dfrac{2}{3}t^3, t \geq 0$, find the surface area S of the balloon as a function of the time t.

64. **Volume of a Balloon** The volume V (in cubic meters) of the hot-air balloon described in Problem 63 is given by $V(r) = \dfrac{4}{3}\pi r^3$. If the radius r is the same function of t as in Problem 63, find the volume V as a function of the time t.

65. **Automobile Production** The number N of cars produced at a certain factory in 1 day after t hours of operation is given by $N(t) = 100t - 5t^2, 0 \leq t \leq 10$. If the cost C (in dollars) of producing N cars is $C(N) = 15{,}000 + 8000N$, find the cost C as a function of the time t of operation of the factory.

66. **Environmental Concerns** The spread of oil leaking from a tanker is in the shape of a circle. If the radius r (in feet) of the spread after t hours is $r(t) = 200\sqrt{t}$, find the area A of the oil slick as a function of the time t.

67. **Production Cost** The price p, in dollars, of a certain product and the quantity x sold obey the demand equation

$$p = -\dfrac{1}{4}x + 100, \qquad 0 \leq x \leq 400$$

Suppose that the cost C, in dollars, of producing x units is

$$C = \dfrac{\sqrt{x}}{25} + 600$$

Assuming that all items produced are sold, find the cost C as a function of the price p.

[**Hint:** Solve for x in the demand equation and then form the composite.]

68. **Cost of a Commodity** The price p, in dollars, of a certain commodity and the quantity x sold obey the demand equation

$$p = -\dfrac{1}{5}x + 200, \qquad 0 \leq x \leq 1000$$

Suppose that the cost C, in dollars, of producing x units is

$$C = \dfrac{\sqrt{x}}{10} + 400$$

Assuming that all items produced are sold, find the cost C as a function of the price p.

69. Volume of a Cylinder The volume V of a right circular cylinder of height h and radius r is $V = \pi r^2 h$. If the height is twice the radius, express the volume V as a function of r.

70. Volume of a Cone The volume V of a right circular cone is $V = \dfrac{1}{3}\pi r^2 h$. If the height is twice the radius, express the volume V as a function of r.

71. Foreign Exchange Traders often buy foreign currency in hope of making money when the currency's value changes. For example, on October 5, 2005, one U.S. dollar could purchase 0.8282 Euros and one Euro could purchase 139.5866 yen. Let $f(x)$ represent the number of Euros you can buy with x dollars and let $g(x)$ represent the number of yen you can buy with x Euros.
(a) Find a function that relates dollars to Euros.
(b) Find a function that relates Euros to yen.

(c) Use the results of parts (a) and (b) to find a function that relates dollars to yen. That is, find $g(f(x))$.
(d) What is $g(f(1000))$?

72. Temperature Conversion The function $C(F) = \dfrac{5}{9}(F - 32)$ converts a temperature in degrees Fahrenheit, F, to a temperature in degrees Celsius, C. The function $K(C) = C + 273$, converts a temperature in degrees Celsius to a temperature in kelvins, K.
(a) Find a function that converts a temperature in degrees Fahrenheit to a temperature in kelvins.
(b) Determine 80 degrees Fahrenheit in kelvin.

73. If f and g are odd functions, show that the composite function $f \circ g$ is also odd.

74. If f is an odd function and g is an even function, show that the composite functions $f \circ g$ and $g \circ f$ are both even.

'Are You Prepared?' Answers

1. -21 **2.** $4 - 18x^2$ **3.** $\{x \mid x \neq -5, x \neq 5\}$

4.2 One-to-One Functions; Inverse Functions

PREPARING FOR THIS SECTION *Before getting started, review the following:*

- Functions (Section 1.1, pp. 40–49)
- Increasing/Decreasing Functions (Section 1.3, pp. 65–66)

Now work the 'Are You Prepared?' problems on page 280.

OBJECTIVES
1 Determine Whether a Function Is One-to-One
2 Determine the Inverse of a Function Defined by a Map or an Ordered Pair
3 Obtain the Graph of the Inverse Function from the Graph of the Function
4 Find the Inverse of a Function Defined by an Equation

1 Determine Whether a Function Is One-to-One

In Section 1.1, we presented four different ways to represent a function: as (1) a map, (2) a set of ordered pairs, (3) a graph, and (4) an equation. For example, Figures 6 and 7 illustrate two different functions represented as mappings. The function in Figure 6 shows the correspondence between states and their population (in millions). The function in Figure 7 shows a correspondence between animals and life expectancy.

Figure 6

Figure 7

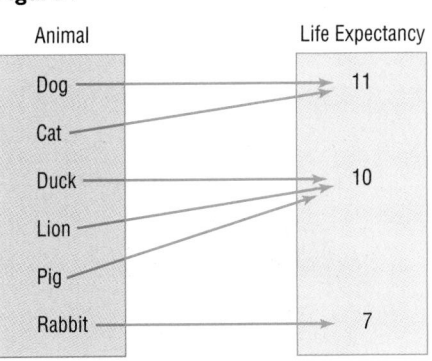

Suppose we asked a group of people to name the state that has a population of 0.8 million based on the function in Figure 6. Everyone in the group would respond South Dakota. Now, if we asked the same group of people to name the animal whose life expectancy is 11 years based on the function in Figure 7, some would respond dog, while others would respond cat. What is the difference between the functions in Figures 6 and 7? In Figure 6, we can see that each element in the domain corresponds to exactly one element in the range. In Figure 7, this is not the case: two different elements in the domain correspond to the same element in the range. We give functions such as the one in Figure 6 a special name.

In Words

A function is not one-to-one if two different inputs correspond to the same output.

A function is **one-to-one** if any two different inputs in the domain correspond to two different outputs in the range. That is, if x_1 and x_2 are two different inputs of a function f, then $f(x_1) \neq f(x_2)$.

Put another way, a function f is one-to-one if no y in the range is the image of more than one x in the domain. A function is not one-to-one if two different elements in the domain correspond to the same element in the range. So the function in Figure 7 is not one-to-one because two different elements in the domain, dog and cat, both correspond to 11. Figure 8 illustrates the distinction of one-to-one functions, non–one-to-one functions, and nonfunctions.

Figure 8

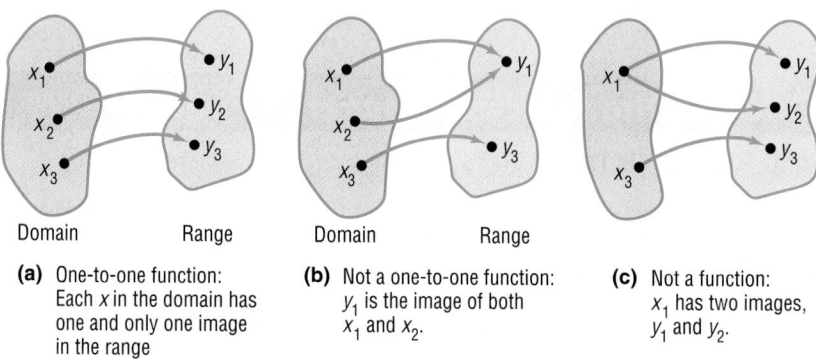

(a) One-to-one function: Each *x* in the domain has one and only one image in the range

(b) Not a one-to-one function: y_1 is the image of both x_1 and x_2.

(c) Not a function: x_1 has two images, y_1 and y_2.

| EXAMPLE 1 | Determining Whether a Function Is One-to-One |

Determine whether the following functions are one-to-one.

(a) For the following function, the domain represents the age of five males and the range represents their HDL (good) cholesterol (mg/dL).

(b) $\{(-2, 6), (-1, 3), (0, 2), (1, 5), (2, 8)\}$

Solution

(a) The function is not one-to-one because there are two different inputs, 55 and 61, that correspond to the same output, 38.

(b) The function is one-to-one because there are no two distinct inputs that correspond to the same output. ∎

- **NOW WORK PROBLEMS 9 AND 13.**

Figure 9

$f(x_1) = f(x_2) = h$ and
$x_1 \neq x_2$; f is not a
one-to-one function.

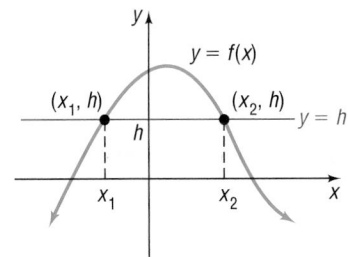

If the graph of a function f is known, there is a simple test, called the **horizontal-line test**, to determine whether f is one-to-one.

Theorem

Horizontal-line Test

If every horizontal line intersects the graph of a function f in at most one point, then f is one-to-one.

The reason that this test works can be seen in Figure 9, where the horizontal line $y = h$ intersects the graph at two distinct points, (x_1, h) and (x_2, h). Since h is the image of both x_1 and x_2, $x_1 \neq x_2$, f is not one-to-one. Based on Figure 9, we can state the horizontal-line test in another way: If the graph of any horizontal line intersects the graph of a function f at more than one point, then f is not one-to-one.

EXAMPLE 2

Using the Horizontal-line Test

For each function, use the graph to determine whether the function is one-to-one.

(a) $f(x) = x^2$ (b) $g(x) = x^3$

Solution

(a) Figure 10(a) illustrates the horizontal-line test for $f(x) = x^2$. The horizontal line $y = 1$ intersects the graph of f twice, at $(1, 1)$ and at $(-1, 1)$, so f is not one-to-one.

(b) Figure 10(b) illustrates the horizontal-line test for $g(x) = x^3$. Because every horizontal line will intersect the graph of g exactly once, it follows that g is one-to-one.

Figure 10

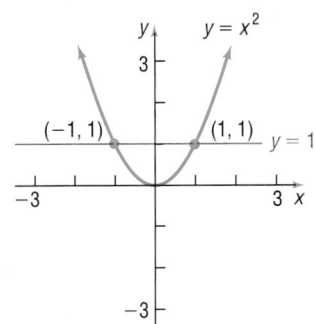

(a) A horizontal line intersects the graph twice; f is not one-to-one

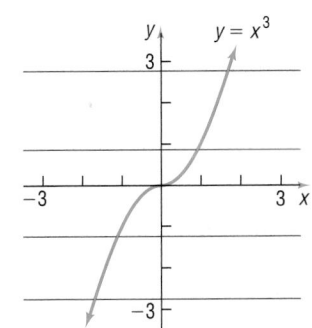

(b) Every horizontal line intersects the graph exactly once; g is one-to-one

NOW WORK PROBLEM 17.

Let's look more closely at the one-to-one function $g(x) = x^3$. This function is an increasing function. Because an increasing (or decreasing) function will always have different y-values for unequal x-values, it follows that a function that is increasing (or decreasing) over its domain is also a one-to-one function.

Theorem

A function that is increasing on an interval I is a one-to-one function on I.

A function that is decreasing on an interval I is a one-to-one function on I.

2 Determine the Inverse of a Function Defined by a Map or an Ordered Pair

In Section 1.1, we said that a function f can be thought of as a machine that receives an input, say x, from the domain, manipulates it, and outputs the value $f(x)$. The **inverse function of f** receives as input $f(x)$, manipulates it, and outputs x. For example, if $f(x) = 2x + 5$, then $f(3) = 11$. Here, the input is 3 and the output is 11.

The inverse of f would "undo" what f does and receive as input 11 and provide as output 3. So, if a function takes inputs and multiplies them by 2, the inverse would take inputs and divide them by 2. For the inverse of a function f to itself be a function, f must be one-to-one.

Recall that we have a variety of ways of representing functions. We will discuss how to find inverses for all four representations of functions: (1) maps, (2) sets of ordered pairs (3) graphs, and (4) equations. We begin with finding inverses of functions represented by maps or sets of ordered pairs.

EXAMPLE 3	**Finding the Inverse of a Function Defined by a Map**

Find the inverse of the following function. Let the domain of the function represent certain states, and let the range represent the state's population (in millions). State the domain and the range of the inverse function.

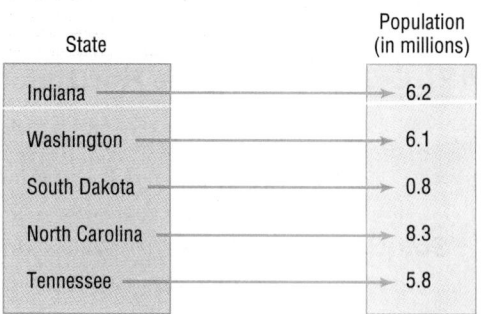

Solution The function is one-to-one, so the inverse will be a function. To find the inverse function, we interchange the elements in the domain with the elements in the range. For example, the function receives as input Indiana and outputs 6.2 million. So the inverse receives as input 6.2 million and outputs Indiana. The inverse function is shown next.

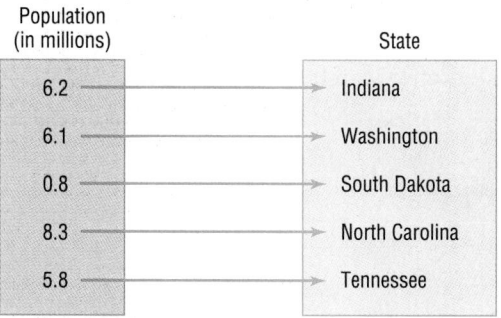

The domain of the inverse function is {6.2, 6.1, 0.8, 8.3, 5.8}. The range of the inverse function is {Indiana, Washington, South Dakota, North Carolina, Tennessee}. ∎

If the function f is a set of ordered pairs (x, y), then the inverse of f is the set of ordered pairs (y, x).

EXAMPLE 4	**Finding the Inverse of a Function Defined by a Set of Ordered Pairs**

Find the inverse of the following one-to-one function:

$$\{(-3, -27), (-2, -8), (-1, -1), (0, 0), (1, 1), (2, 8), (3, 27)\}$$

Solution The inverse of the given function is found by interchanging the entries in each ordered pair and so is given by

$$\{(-27, -3), (-8, -2), (-1, -1), (0, 0), (1, 1), (8, 2), (27, 3)\}$$ ∎

✏️ **NOW WORK PROBLEMS 23 AND 27.**

Figure 11

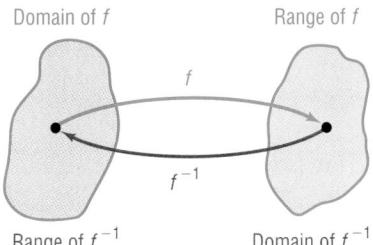

Remember, if f is a one-to-one function, its inverse is a function. Then, to each x in the domain of f, there is exactly one y in the range (because f is a function); and to each y in the range of f, there is exactly one x in the domain (because f is one-to-one). The correspondence from the range of f back to the domain of f is called the inverse function of f and is denoted by the symbol f^{-1}. Figure 11 illustrates this definition.

Based on Figure 11, two facts are now apparent about a function f and its inverse f^{-1}.

$$\text{Domain of } f = \text{Range of } f^{-1} \qquad \text{Range of } f = \text{Domain of } f^{-1}$$

WARNING
Be careful! f^{-1} is a symbol for the inverse function of f. The -1 used in f^{-1} is not an exponent. That is, f^{-1} does not mean the reciprocal of f; $f^{-1}(x)$ is not equal to $\dfrac{1}{f(x)}$. ∎

Look again at Figure 11 to visualize the relationship. If we start with x, apply f, and then apply f^{-1}, we get x back again. If we start with x, apply f^{-1}, and then apply f, we get the number x back again. To put it simply, what f does, f^{-1} undoes, and vice versa. See the illustration that follows.

$$\boxed{\text{Input } x} \xrightarrow{\text{Apply } f} \boxed{f(x)} \xrightarrow{\text{Apply } f^{-1}} \boxed{f^{-1}(f(x)) = x}$$

$$\boxed{\text{Input } x} \xrightarrow{\text{Apply } f^{-1}} \boxed{f^{-1}(x)} \xrightarrow{\text{Apply } f} \boxed{f(f^{-1}(x)) = x}$$

In other words,

$$f^{-1}(f(x)) = x \quad \text{where } x \text{ is in the domain of } f$$
$$f(f^{-1}(x)) = x \quad \text{where } x \text{ is in the domain of } f^{-1}$$

Consider the function $f(x) = 2x$, which multiplies the argument x by 2. The inverse function f^{-1} undoes whatever f does. So the inverse function of f is $f^{-1}(x) = \dfrac{1}{2}x$, which divides the argument by 2. For example, $f(3) = 2(3) = 6$ and $f^{-1}(6) = \dfrac{1}{2}(6) = 3$, so f^{-1} undoes what f did. We can verify this by showing that

Figure 12

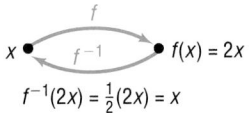

$f^{-1}(2x) = \frac{1}{2}(2x) = x$

$$f^{-1}(f(x)) = f^{-1}(2x) = \frac{1}{2}(2x) = x \quad \text{and} \quad f(f^{-1}(x)) = f\left(\frac{1}{2}x\right) = 2\left(\frac{1}{2}x\right) = x$$

See Figure 12.

| **EXAMPLE 5** | **Verifying Inverse Functions** |

(a) We verify that the inverse of $g(x) = x^3$ is $g^{-1}(x) = \sqrt[3]{x}$ by showing that
$$g^{-1}(g(x)) = g^{-1}(x^3) = \sqrt[3]{x^3} = x \quad \text{for all } x \text{ in the domain of } g$$
$$g(g^{-1}(x)) = g(\sqrt[3]{x}) = (\sqrt[3]{x})^3 = x \quad \text{for all } x \text{ in the domain of } g^{-1}.$$

(b) We verify that the inverse of $f(x) = 2x + 3$ is $f^{-1}(x) = \dfrac{1}{2}(x - 3)$ by showing that

$$f^{-1}(f(x)) = f^{-1}(2x + 3) = \frac{1}{2}[(2x + 3) - 3] = \frac{1}{2}(2x) = x \quad \begin{array}{l}\text{for all } x \text{ in the} \\ \text{domain of } f\end{array}$$

$$f(f^{-1}(x)) = f\left(\frac{1}{2}(x - 3)\right) = 2\left[\frac{1}{2}(x - 3)\right] + 3 = (x - 3) + 3 = x \quad \text{for all } x \text{ in the}$$

domain of f^{-1}. ∎

| EXAMPLE 6 | Verifying Inverse Functions |

Verify that the inverse of $f(x) = \dfrac{1}{x-1}$ is $f^{-1}(x) = \dfrac{1}{x} + 1$. For what values of x is $f^{-1}(f(x)) = x$? For what values of x is $f(f^{-1}(x)) = x$?

Solution The domain of f is $\{x \mid x \neq 1\}$ and the domain of f^{-1} is $\{x \mid x \neq 0\}$ Now,

$$f^{-1}(f(x)) = f^{-1}\left(\frac{1}{x-1}\right) = \frac{1}{\dfrac{1}{x-1}} + 1 = x - 1 + 1 = x \qquad \text{provided } x \neq 1.$$

$$f(f^{-1}(x)) = f\left(\frac{1}{x}+1\right) = \frac{1}{\dfrac{1}{x}+1-1} = \frac{1}{\dfrac{1}{x}} = x \qquad \text{provided } x \neq 0$$

∎

━━━ **NOW WORK PROBLEM 37.**

Figure 13

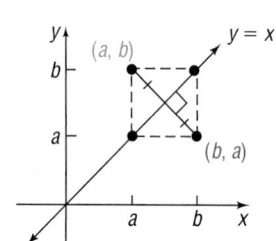

3 Obtain the Graph of the Inverse Function from the Graph of the Function

Suppose that (a, b) is a point on the graph of a one-to-one function f defined by $y = f(x)$. Then $b = f(a)$. This means that $a = f^{-1}(b)$, so (b, a) is a point on the graph of the inverse function f^{-1}. The relationship between the point (a, b) on f and the point (b, a) on f^{-1} is shown in Figure 13. The line segment containing (a, b) and (b, a) is perpendicular to the line $y = x$ and is bisected by the line $y = x$. (Do you see why?) It follows that the point (b, a) on f^{-1} is the reflection about the line $y = x$ of the point (a, b) on f.

Theorem

The graph of a function f and the graph of its inverse f^{-1} are symmetric with respect to the line $y = x$.

Figure 14 illustrates this result. Notice that, once the graph of f is known, the graph of f^{-1} may be obtained by reflecting the graph of f about the line $y = x$.

Figure 14

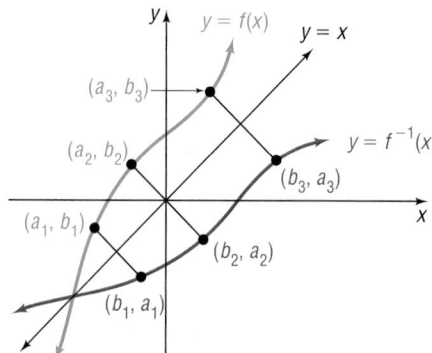

━━━ **Seeing the Concept** ━━━

 Simultaneously graph $Y_1 = x$, $Y_2 = x^3$, and $Y_3 = \sqrt[3]{x}$ on a square screen with $-3 \leq x \leq 3$. What do you observe about the graphs of $Y_2 = x^3$, its inverse $Y_3 = \sqrt[3]{x}$, and the line $Y_1 = x$? Repeat this experiment by simultaneously graphing $Y_1 = x$, $Y_2 = 2x + 3$, and $Y_3 = \dfrac{1}{2}(x - 3)$ on a square screen with $-6 \leq x \leq 3$. Do you see the symmetry of the graph of Y_2 and its inverse Y_3 with respect to the line $Y_1 = x$?

| EXAMPLE 7 | Graphing the Inverse Function |

The graph in Figure 15(a) is that of a one-to-one function $y = f(x)$. Draw the graph of its inverse.

Solution We begin by adding the graph of $y = x$ to Figure 15(a). Since the points $(-2, -1), (-1, 0)$, and $(2, 1)$ are on the graph of f, we know that the points $(-1, -2), (0, -1)$, and $(1, 2)$ must be on the graph of f^{-1}. Keeping in mind that the graph of f^{-1} is the reflection about the line $y = x$ of the graph of f, we can draw f^{-1}. See Figure 15(b).

Figure 15

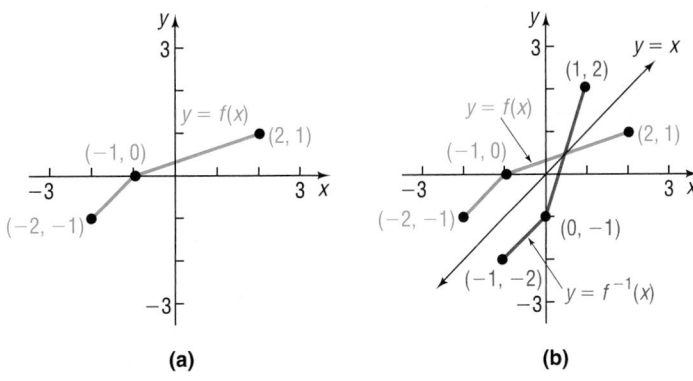

(a) (b)

━━━━━ NOW WORK PROBLEM 31.

4 Find the Inverse of a Function Defined by an Equation

The fact that the graphs of a one-to-one function f and its inverse function f^{-1} are symmetric with respect to the line $y = x$ tells us more. It says that we can obtain f^{-1} by interchanging the roles of x and y in f. Look again at Figure 14. If f is defined by the equation

$$y = f(x)$$

then f^{-1} is defined by the equation

$$x = f(y)$$

The equation $x = f(y)$ defines f^{-1} *implicitly*. If we can solve this equation for y, we will have the *explicit* form of f^{-1}, that is,

$$y = f^{-1}(x)$$

Let's use this procedure to find the inverse of $f(x) = 2x + 3$. (Since f is a linear function and is increasing, we know that f is one-to-one and so has an inverse function.)

EXAMPLE 8 **Finding the Inverse Function**

Find the inverse of $f(x) = 2x + 3$. Also find the domain and range of f and f^{-1}. Graph f and f^{-1} on the same coordinate axes.

Solution In the equation $y = 2x + 3$, interchange the variables x and y. The result,

$$x = 2y + 3$$

is an equation that defines the inverse f^{-1} implicitly. To find the explicit form, we solve for y.

$$2y + 3 = x$$
$$2y = x - 3$$
$$y = \frac{1}{2}(x - 3)$$

The explicit form of the inverse f^{-1} is therefore

$$f^{-1}(x) = \frac{1}{2}(x - 3)$$

which we verified in Example 5(c).

Figure 16

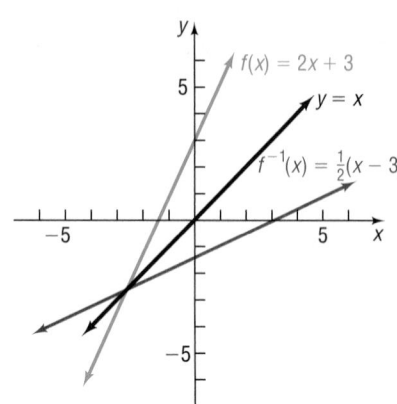

Next we find

$$\text{Domain of } f = \text{Range of } f^{-1} = (-\infty, \infty)$$
$$\text{Range of } f = \text{Domain of } f^{-1} = (-\infty, \infty)$$

The graphs of $f(x) = 2x + 3$ and its inverse $f^{-1}(x) = \dfrac{1}{2}(x - 3)$ are shown in Figure 16. Note the symmetry of the graphs with respect to the line $y = x$. ■

We now outline the steps to follow for finding the inverse of a one-to-one function.

Procedure for Finding the Inverse of a One-to-One Function

STEP 1: In $y = f(x)$, interchange the variables x and y to obtain

$$x = f(y)$$

This equation defines the inverse function f^{-1} implicitly.

STEP 2: If possible, solve the implicit equation for y in terms of x to obtain the explicit form of f^{-1}:

$$y = f^{-1}(x)$$

STEP 3: Check the result by showing that

$$f^{-1}(f(x)) = x \quad \text{and} \quad f(f^{-1}(x)) = x$$

EXAMPLE 9 **Finding the Inverse Function**

The function

$$f(x) = \frac{2x + 1}{x - 1} \qquad x \neq 1$$

is one-to-one. Find its inverse and check the result.

Solution **STEP 1:** Interchange the variables x and y in

$$y = \frac{2x + 1}{x - 1}$$

to obtain

$$x = \frac{2y + 1}{y - 1}$$

STEP 2: Solve for y.

$$x = \frac{2y + 1}{y - 1}$$

$$x(y - 1) = 2y + 1 \qquad \text{Multiply both sides by } y - 1.$$

$$xy - x = 2y + 1 \qquad \text{Apply the Distributive Property.}$$

$$xy - 2y = x + 1 \qquad \text{Subtract } 2y \text{ from both sides; add } x \text{ to both sides.}$$

$$(x - 2)y = x + 1 \qquad \text{Factor.}$$

$$y = \frac{x + 1}{x - 2} \qquad \text{Divide by } x - 2.$$

The inverse is

$$f^{-1}(x) = \frac{x+1}{x-2} \qquad x \neq 2 \qquad \text{Replace } y \text{ by } f^{-1}(x).$$

STEP 3: ✔ CHECK:

$$f^{-1}(f(x)) = f^{-1}\left(\frac{2x+1}{x-1}\right) = \frac{\dfrac{2x+1}{x-1}+1}{\dfrac{2x+1}{x-1}-2} = \frac{2x+1+x-1}{2x+1-2(x-1)} = \frac{3x}{3} = x, \quad x \neq 1$$

$$f(f^{-1}(x)) = f\left(\frac{x+1}{x-2}\right) = \frac{2\left(\dfrac{x+1}{x-2}\right)+1}{\dfrac{x+1}{x-2}-1} = \frac{2(x+1)+x-2}{x+1-(x-2)} = \frac{3x}{3} = x, \quad x \neq 2$$

■

—— **Exploration** ——

In Example 9, we found that, if $f(x) = \dfrac{2x+1}{x-1}$, then $f^{-1}(x) = \dfrac{x+1}{x-2}$. Compare the vertical and horizontal asymptotes of f and f^{-1}. What did you find? Are you surprised?

Result You should have found that the vertical asymptote of f is $x = 1$ and the horizontal asymptote is $y = 2$. The vertical asymptote of f^{-1} is $x = 2$, and the horizontal asymptote is $y = 1$.

━━━ NOW WORK PROBLEM **49**.

We said in Chapter 1 that finding the range of a function f is not easy. However, if f is one-to-one, we can find its range by finding the domain of the inverse function f^{-1}.

EXAMPLE 10 **Finding the Range of a Function**

Find the domain and range of

$$f(x) = \frac{2x+1}{x-1}$$

Solution The domain of f is $\{x \mid x \neq 1\}$. To find the range of f, we first find the inverse f^{-1}. Based on Example 9, we have

$$f^{-1}(x) = \frac{x+1}{x-2}$$

The domain of f^{-1} is $\{x \mid x \neq 2\}$, so the range of f is $\{y \mid y \neq 2\}$. ■

━━━ NOW WORK PROBLEM **63**.

If a function is not one-to-one, its inverse is not a function. Sometimes, though, an appropriate restriction on the domain of such a function will yield a new function that is one-to-one. Let's look at an example of this common practice.

EXAMPLE 11 **Finding the Inverse of a Domain-restricted Function**

Find the inverse of $y = f(x) = x^2$ if $x \geq 0$.

Solution The function $y = x^2$ is not one-to-one. [Refer to Example 2(a).] However, if we restrict the domain of this function to $x \geq 0$, as indicated, we have a new function that is increasing and therefore is one-to-one. As a result, the function defined by $y = f(x) = x^2$, $x \geq 0$, has an inverse function, f^{-1}.

We follow the steps given previously to find f^{-1}.

STEP 1: In the equation $y = x^2$, $x \geq 0$, interchange the variables x and y. The result is

$$x = y^2, \quad y \geq 0$$

This equation defines (implicitly) the inverse function.

STEP 2: We solve for y to get the explicit form of the inverse. Since $y \geq 0$, only one solution for y is obtained: $y = \sqrt{x}$. So $f^{-1}(x) = \sqrt{x}$.

STEP 3: ✔ **CHECK:** $f^{-1}(f(x)) = f^{-1}(x^2) = \sqrt{x^2} = |x| = x$, since $x \geq 0$
$f(f^{-1}(x)) = f(\sqrt{x}) = (\sqrt{x})^2 = x$. ■

Figure 17 illustrates the graphs of $f(x) = x^2$, $x \geq 0$, and $f^{-1}(x) = \sqrt{x}$.

Figure 17

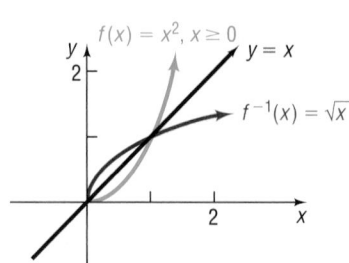

Summary

1. If a function f is one-to-one, then it has an inverse function f^{-1}.
2. Domain of f = Range of f^{-1}; Range of f = Domain of f^{-1}.
3. To verify that f^{-1} is the inverse of f, show that $f^{-1}(f(x)) = x$ for every x in the domain of f and $f(f^{-1}(x)) = x$ for every x in the domain of f^{-1}.
4. The graphs of f and f^{-1} are symmetric with respect to the line $y = x$.
5. To find the range of a one-to-one function f, find the domain of the inverse function f^{-1}.

4.2 Assess Your Understanding

'Are You Prepared?'

Answers are given at the end of these exercises. If you get a wrong answer, read the pages listed in red.

1. Is the set of ordered pairs $\{(1,3),(2,3),(-1,2)\}$ a function? Why or why not? (p. 43)
2. Where is the function $f(x) = x^2$ increasing? Where is it decreasing? (pp. 65–66, 78)
3. What is the domain of $f(x) = \dfrac{x+5}{x^2 + 3x - 18}$? (p. 48)

Concepts and Vocabulary

4. If every horizontal line intersects the graph of a function f at no more than one point, f is a(n) _____ function.
5. If f^{-1} denotes the inverse of a function f, then the graphs of f and f^{-1} are symmetric with respect to the line _____.
6. If the domain of a one-to-one function f is $[4, \infty)$, the range of its inverse, f^{-1}, is _____.
7. *True or False:* If f and g are inverse functions, the domain of f is the same as the domain of g.
8. *True or False:* If f and g are inverse functions, their graphs are symmetric with respect to the line $y = x$.

Skill Building

In Problems 9–16, determine whether the function is one-to-one.

9.

10.

11.

12.

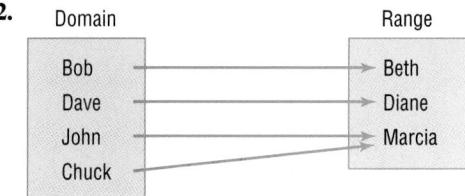

13. $\{(2, 6), (-3, 6), (4, 9), (1, 10)\}$

14. $\{(-2, 5), (-1, 3), (3, 7), (4, 12)\}$

15. $\{(0, 0), (1, 1), (2, 16), (3, 81)\}$

16. $\{(1, 2), (2, 8), (3, 18), (4, 32)\}$

In Problems 17–22, the graph of a function f is given. Use the horizontal-line test to determine whether f is one-to-one.

17.

18.

19.

20.

21.

22.

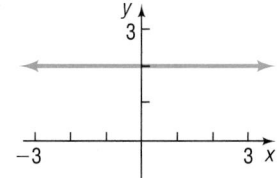

In Problems 23–30, find the inverse of each one-to-one function. State the domain and the range of each inverse function.

23.

Location	Annual Rainfall (inches)
Mt Waialeale, Hawaii	460.00
Monrovia, Liberia	202.01
Pago Pago, American Samoa	196.46
Moulmein, Burma	191.02
Lae, Papua New Guinea	182.87

SOURCE: *Information Please Almanac*

24.

Title	Domestic Gross (in millions)
Star Wars	$461
Star Wars: Episode One – The Phantom Menace	$431
E.T. the Extra Terrestrial	$400
Jurassic Park	$357
Forrest Gump	$330

SOURCE: *Information Please Almanac*

25.

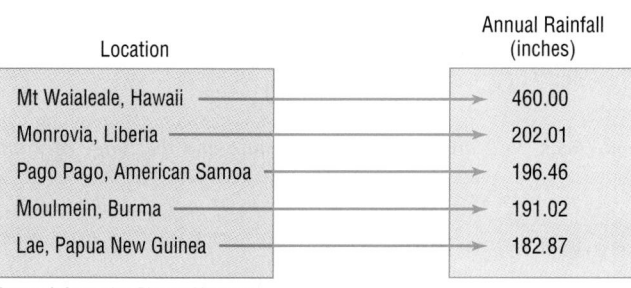

Age	Monthly Cost of Life Insurance
30	$7.09
40	$8.40
45	$11.29

SOURCE: *eterm.com*

26.

State	Unemployment Rate
Virginia	11%
Nevada	5.5%
Tennessee	5.1%
Texas	6.3%

SOURCE: *United States Statistical Abstract*

27. $\{(-3, 5), (-2, 9), (-1, 2), (0, 11), (1, -5)\}$

28. $\{(-2, 2), (-1, 6), (0, 8), (1, -3), (2, 9)\}$

29. $\{(-2, 1), (-3, 2), (-10, 0), (1, 9), (2, 4)\}$

30. $\{(-2, -8), (-1, -1), (0, 0), (1, 1), (2, 8)\}$

In Problems 31–36, the graph of a one-to-one function f is given. Draw the graph of the inverse function f^{-1}. For convenience (and as a hint), the graph of $y = x$ is also given.

31.

32.

33.

34.

35.

36.

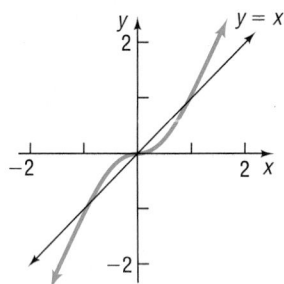

In Problems 37–46, verify that the functions f and g are inverses of each other by showing that $f(g(x)) = x$ and $g(f(x)) = x$. Give any values of x that need to be excluded.

37. $f(x) = 3x + 4;\quad g(x) = \frac{1}{3}(x - 4)$

38. $f(x) = 3 - 2x;\quad g(x) = -\frac{1}{2}(x - 3)$

39. $f(x) = 4x - 8;\quad g(x) = \frac{x}{4} + 2$

40. $f(x) = 2x + 6;\quad g(x) = \frac{1}{2}x - 3$

41. $f(x) = x^3 - 8;\quad g(x) = \sqrt[3]{x + 8}$

42. $f(x) = (x - 2)^2,\ x \geq 2;\quad g(x) = \sqrt{x} + 2$

43. $f(x) = \frac{1}{x};\quad g(x) = \frac{1}{x}$

44. $f(x) = x;\quad g(x) = x$

45. $f(x) = \frac{2x + 3}{x + 4};\quad g(x) = \frac{4x - 3}{2 - x}$

46. $f(x) = \frac{x - 5}{2x + 3};\quad g(x) = \frac{3x + 5}{1 - 2x}$

In Problems 47–58, the function f is one-to-one. Find its inverse and check your answer. State the domain and the range of f and f^{-1}. Graph f, f^{-1}, and $y = x$ on the same coordinate axes.

47. $f(x) = 3x$

48. $f(x) = -4x$

49. $f(x) = 4x + 2$

50. $f(x) = 1 - 3x$

51. $f(x) = x^3 - 1$

52. $f(x) = x^3 + 1$

53. $f(x) = x^2 + 4,\quad x \geq 0$

54. $f(x) = x^2 + 9,\quad x \geq 0$

55. $f(x) = \frac{4}{x}$

56. $f(x) = -\frac{3}{x}$

57. $f(x) = \frac{1}{x - 2}$

58. $f(x) = \frac{4}{x + 2}$

In Problems 59–70, the function f is one-to-one. Find its inverse and check your answer. State the domain of f and find its range using f^{-1}.

59. $f(x) = \frac{2}{3 + x}$

60. $f(x) = \frac{4}{2 - x}$

61. $f(x) = \frac{3x}{x + 2}$

62. $f(x) = \frac{-2x}{x - 1}$

63. $f(x) = \frac{2x}{3x - 1}$

64. $f(x) = \frac{3x + 1}{-x}$

65. $f(x) = \frac{3x + 4}{2x - 3}$

66. $f(x) = \frac{2x - 3}{x + 4}$

67. $f(x) = \frac{2x + 3}{x + 2}$

68. $f(x) = \dfrac{-3x - 4}{x - 2}$

69. $f(x) = \dfrac{x^2 - 4}{2x^2}$, $x > 0$

70. $f(x) = \dfrac{x^2 + 3}{3x^2}$, $x > 0$

Applications and Extensions

71. Use the graph of $y = f(x)$ given in Problem 31 to evaluate the following:
 (a) $f(-1)$ (b) $f(1)$
 (c) $f^{-1}(1)$ (d) $f^{-1}(2)$

72. Use the graph of $y = f(x)$ given in Problem 32 to evaluate the following:
 (a) $f(2)$ (b) $f(1)$
 (c) $f^{-1}(0)$ (d) $f^{-1}(-1)$

73. If $f(7) = 13$ and f is one-to-one, what is $f^{-1}(13)$?

74. If $g(-5) = 3$ and g is one-to-one, what is $g^{-1}(3)$?

75. The domain of a one-to-one function f is $[5, \infty)$, and its range is $[-2, \infty)$. State the domain and the range of f^{-1}.

76. The domain of a one-to-one function f is $[0, \infty)$, and its range is $[5, \infty)$. State the domain and the range of f^{-1}.

77. The domain of a one-to-one function g is the set of all real numbers, and its range is $[0, \infty)$. State the domain and the range of g^{-1}.

78. The domain of a one-to-one function g is $[0, 15]$, and its range is $(0, 8)$. State the domain and the range of g^{-1}.

79. A function $y = f(x)$ is increasing on the interval $(0, 5)$. What conclusions can you draw about the graph of $y = f^{-1}(x)$?

80. A function $y = f(x)$ is decreasing on the interval $(0, 5)$. What conclusions can you draw about the graph of $y = f^{-1}(x)$?

81. Find the inverse of the linear function

$$f(x) = mx + b, \quad m \neq 0$$

82. Find the inverse of the function

$$f(x) = \sqrt{r^2 - x^2}, \quad 0 \leq x \leq r$$

83. A function f has an inverse function. If the graph of f lies in quadrant I, in which quadrant does the graph of f^{-1} lie?

84. A function f has an inverse function. If the graph of f lies in quadrant II, in which quadrant does the graph of f^{-1} lie?

85. The function $f(x) = |x|$ is not one-to-one. Find a suitable restriction on the domain of f so that the new function that results is one-to-one. Then find the inverse of f.

86. The function $f(x) = x^4$ is not one-to-one. Find a suitable restriction on the domain of f so that the new function that results is one-to-one. Then find the inverse of f.

In applications, the symbols used for the independent and dependent variables are based on common usage. So, rather than using $y = f(x)$ to represent a function, an applied problem might use $C = C(q)$ to represent the cost C of manufacturing q units of a good since, in economics, q is used for output. Because of this, the inverse notation f^{-1} used in a pure mathematics problem is not used when finding inverses of applied problems. Rather, the inverse of a function such as $C = C(q)$, will be $q = q(C)$. So, $C = C(q)$ is a function that represents the cost C as a function of the output q, while $q = q(C)$ is a function that represents the output q as a function of the cost C. Problems 87–90 illustrate this idea.

87. Vehicle Stopping Distance Taking into account reaction time, the distance d (in feet) that a car requires to come to a complete stop while traveling r miles per hour is given by the function

$$d(r) = 6.97r - 90.39$$

 (a) Express the speed r at which the car is traveling as a function of the distance d required to come to a complete stop.
 (b) Verify that $r = r(d)$ is the inverse of $d = d(r)$ by showing that $r(d(r)) = r$ and $d(r(d)) = d$.
 (c) Predict the speed that a car was traveling if the distance required to stop is 300 feet.

88. Height and Head Circumference The head circumference C of a child is related to the height H of the child (both in inches) through the function

$$H(C) = 2.15C - 10.53$$

 (a) Express the head circumference C as a function of height H.
 (b) Verify that $C = C(H)$ is the inverse of $H = H(C)$ by showing that $H(C(H)) = H$ and $C(H(C)) = C$.
 (c) Predict the head circumference of a child who is 26 inches tall.

89. Ideal Body Weight The ideal body weight W for men (in kilograms) as a function of height h (in inches) is given by the function

$$W(h) = 50 + 2.3(h - 60)$$

 (a) What is the ideal weight of a 6-foot male?
 (b) Express the height h as a function of weight W.
 (c) Verify that $h = h(W)$ is the inverse of $W = W(h)$ by showing that $h(W(h)) = h$ and $W(h(W)) = W$.
 (d) What is the height of a male who is at his ideal weight of 80 kilograms?

NOTE: The ideal body weight W for women (in kilograms) as a function of height h (in inches) is given by $W(h) = 45.5 + 2.3(h - 60)$.

90. Temperature Conversion The function $F(C) = \dfrac{9}{5}C + 32$ converts a temperature from C degrees Celsius to F degrees Fahrenheit.

 (a) Express the temperature in degrees Celsius C as a function of the temperature in degrees Fahrenheit F.
 (b) Verify that $C = C(F)$ is the inverse of $F = F(C)$ by showing that $C(F(C)) = C$ and $F(C(F)) = F$.
 (c) What is the temperature in degrees Celsius if it is 70 degrees Fahrenheit?

91. Income Taxes The function

$$T(g) = 4090 + 0.28(g - 29,700)$$

represents the 2005 federal income tax T (in dollars) due for a "single" filer whose adjusted gross income is g dollars, where $29,700 \leq g \leq 71,950$.

(a) What is the domain of the function T?
(b) Given that the tax due T is an increasing linear function of adjusted gross income g, find the range of the function T.
(c) Find adjusted gross income g as a function of federal income tax T. What is the domain and the range of this function?

92. Income Taxes The function

$$T(g) = 1460 + 0.15(g - 14,600)$$

represents the 2005 federal income tax T (in dollars) due for a "married filing jointly" filer whose adjusted gross income is g dollars, where $14,600 \leq g \leq 59,400$.

(a) What is the domain of the function T?
(b) Given that the tax due T is an increasing linear function of adjusted gross income g, find the range of the function T.
(c) Find adjusted gross income g as a function of federal income tax T. What is the domain and the range of this function?

93. Gravity on Earth If a rock falls from a height of 100 meters on Earth, the height H (in meters) after t seconds is approximately

$$H(t) = 100 - 4.9t^2$$

(a) In general, quadratic functions are not one-to-one. However, the function $H(t)$ is one-to-one. Why?
(b) Find the inverse of H and verify your result.
(c) How long will it take a rock to fall 80 meters?

94. Period of a Pendulum The period T (in seconds) of a simple pendulum as a function of its length l (in feet) is given by

$$T(l) = 2\pi\sqrt{\frac{l}{32.2}}.$$

(a) Express the length l as a function of the period T.
(b) How long is a pendulum whose period is 3 seconds?

95. Given

$$f(x) = \frac{ax + b}{cx + d}$$

find $f^{-1}(x)$. If $c \neq 0$, under what conditions on $a, b, c,$ and d is $f = f^{-1}$?

Discussion and Writing

96. Can a one-to-one function and its inverse be equal? What must be true about the graph of f for this to happen? Give some examples to support your conclusion.

97. Draw the graph of a one-to-one function that contains the points $(-2, -3)$, $(0, 0)$, and $(1, 5)$. Now draw the graph of its inverse. Compare your graph to those of other students. Discuss any similarities. What differences do you see?

98. Give an example of a function whose domain is the set of real numbers and that is neither increasing nor decreasing on its domain, but is one-to-one.

[**Hint:** Use a piecewise-defined function.]

99. If a function f is even, can it be one-to-one? Explain.

100. Is every odd function one-to-one? Explain.

101. If the graph of a function and its inverse intersect, where must this necessarily occur? Can they intersect anywhere else? Must they intersect?

102. Suppose $C(g)$ represents the cost C of manufacturing g cars. Explain what $C^{-1}(800,000)$ represents.

'Are You Prepared?' Answers

1. Yes; for each input x there is one output y. **2.** Increasing on $(0, \infty)$; decreasing on $(-\infty, 0)$ **3.** $\{x \mid x \neq -6, x \neq 3\}$

4.3 Exponential Functions

PREPARING FOR THIS SECTION *Before getting started, review the following:*

• Exponents (Appendix A, Section A.1, pp. A6–A8, and Section A.7, pp. A52–A57)
• Graphing Techniques: Transformations (Section 1.5, pp. 85–94)

• Average Rate of Change (Section 1.3, pp. 68–70)
• Solving Equations (Appendix A, Section A.8, pp. A60–A64)
• Horizontal Asymptotes (Section 3.2, pp. 208–211)

Now work the 'Are You Prepared?' problems on page 295.

OBJECTIVES 1 Evaluate Exponential Functions
 2 Graph Exponential Functions
 3 Define the Number e
 4 Solve Exponential Equations

1 Evaluate Exponential Functions

In the Appendix, Section A.7, we give a definition for raising a real number a to a rational power. Based on that discussion, we gave meaning to expressions of the form

$$a^r$$

where the base a is a positive real number and the exponent r is a rational number.

 But what is the meaning of a^x, where the base a is a positive real number and the exponent x is an irrational number? Although a rigorous definition requires methods discussed in calculus, the basis for the definition is easy to follow: Select a rational number r that is formed by truncating (removing) all but a finite number of digits from the irrational number x. Then it is reasonable to expect that

$$a^x \approx a^r$$

For example, take the irrational number $\pi = 3.14159\ldots$. Then an approximation to a^π is

$$a^\pi \approx a^{3.14}$$

where the digits after the hundredths position have been removed from the value for π. A better approximation would be

$$a^\pi \approx a^{3.14159}$$

where the digits after the hundred-thousandths position have been removed. Continuing in this way, we can obtain approximations to a^π to any desired degree of accuracy.

Most calculators have an $\boxed{x^y}$ key or a caret key $\boxed{\wedge}$ for working with exponents. To evaluate expressions of the form a^x, enter the base a, then press the $\boxed{x^y}$ key (or the $\boxed{\wedge}$ key), enter the exponent x, and press $\boxed{=}$ (or $\boxed{\text{enter}}$).

EXAMPLE 1 **Using a Calculator to Evaluate Powers of 2**

Using a calculator, evaluate:

(a) $2^{1.4}$ (b) $2^{1.41}$ (c) $2^{1.414}$ (d) $2^{1.4142}$ (e) $2^{\sqrt{2}}$

Solution (a) $2^{1.4} \approx 2.639015822$ (b) $2^{1.41} \approx 2.657371628$

(c) $2^{1.414} \approx 2.66474965$ (d) $2^{1.4142} \approx 2.665119089$

(e) $2^{\sqrt{2}} \approx 2.665144143$ ∎

NOW WORK PROBLEM 11.

It can be shown that the familiar laws for rational exponents hold for real exponents.

Theorem **Laws of Exponents**

If s, t, a, and b are real numbers with $a > 0$ and $b > 0$, then

$$a^s \cdot a^t = a^{s+t} \qquad (a^s)^t = a^{st} \qquad (ab)^s = a^s \cdot b^s$$

$$1^s = 1 \qquad a^{-s} = \frac{1}{a^s} = \left(\frac{1}{a}\right)^s \qquad a^0 = 1 \qquad \text{(1)}$$

We are now ready for the following definition:

An **exponential function** is a function of the form

$$f(x) = a^x$$

where a is a positive real number ($a > 0$) and $a \neq 1$. The domain of f is the set of all real numbers.

WARNING

It is important to distinguish a power function, $g(x) = x^n$, $n \geq 2$, an integer, from an exponential function, $f(x) = a^x$, $a \neq 1$, a real. In a power function, the base is a variable and the exponent is a constant. In an exponential function, the base is a constant and the exponent is a variable. ∎

We exclude the base $a = 1$ because this function is simply the constant function $f(x) = 1^x = 1$. We also need to exclude bases that are negative; otherwise, we would have to exclude many values of x from the domain, such as $x = \dfrac{1}{2}$ and $x = \dfrac{3}{4}$.

[Recall that $(-2)^{1/2} = \sqrt{-2}$, $(-3)^{3/4} = \sqrt[4]{(-3)^3} = \sqrt[4]{-27}$, and so on, are not defined in the set of real numbers.]

Some examples of exponential functions are

$$f(x) = 2^x, \qquad F(x) = \left(\frac{1}{3}\right)^x$$

Notice that in each example the base is a constant and the exponent is a variable.

You may wonder what role the base a plays in the exponential function $f(x) = a^x$. We use the following Exploration to find out.

Exploration

(a) Evaluate $f(x) = 2^x$ at $x = -2, -1, 0, 1, 2,$ and 3.
(b) Evaluate $g(x) = 3x + 2$ at $x = -2, -1, 0, 1, 2,$ and 3.
(c) Comment on the pattern that exists in the values of f and g.

Result (a) Table 1 shows the values of $f(x) = 2^x$ for $x = -2, -1, 0, 1, 2,$ and 3.
(b) Table 2 shows the values of $g(x) = 3x + 2$ for $x = -2, -1, 0, 1, 2,$ and 3.

Table 1

x	$f(x) = 2^x$
-2	$f(-2) = 2^{-2} = \dfrac{1}{2^2} = \dfrac{1}{4}$
-1	$\dfrac{1}{2}$
0	1
1	2
2	4
3	8

Table 2

x	$g(x) = 3x + 2$
-2	$g(-2) = 3(-2) + 2 = -4$
-1	-1
0	2
1	5
2	8
3	11

(c) In Table 1 we notice that each value of the exponential function $f(x) = a^x = 2^x$ could be found by multiplying the previous value of the function by the base, $a = 2$. For example,

$$f(-1) = 2 \cdot f(-2) = 2 \cdot \frac{1}{4} = \frac{1}{2}, \quad f(0) = 2 \cdot f(-1) = 2 \cdot \frac{1}{2} = 1, \quad f(1) = 2 \cdot f(0) = 2 \cdot 1 = 2$$

and so on.

Put another way, we see that the ratio of consecutive outputs is constant for unit increases in the input. The constant equals the value of the base a of the exponential function. For example, for the function $f(x) = 2^x$, we notice that

$$\frac{f(-1)}{f(-2)} = \frac{\frac{1}{2}}{\frac{1}{4}} = 2, \qquad \frac{f(1)}{f(0)} = \frac{2}{1} = 2, \qquad \frac{f(x+1)}{f(x)} = \frac{2^{x+1}}{2^x} = 2$$

and so on.

From Table 2 we see that the function $g(x) = 3x + 2$ does not have the ratio of consecutive outputs that are constant because it is not exponential. For example,

$$\frac{g(-1)}{g(-2)} = \frac{-1}{-4} = \frac{1}{4} \neq \frac{g(1)}{g(0)} = \frac{5}{2}$$

Instead, because $g(x) = 3x + 2$ is a linear function, for unit increases in the input, the outputs increase by a fixed amount equal to the value of the slope, 3.

The results of the Exploration lead to the following result.

Theorem

For an exponential function $f(x) = a^x, a > 0, a \neq 1$, if x is any real number, then

$$\frac{f(x+1)}{f(x)} = a$$

In Words
For an exponential function $f(x) = a^x$, for 1-unit changes in the input x, the ratio of consecutive outputs is the constant a.

Proof

$$\frac{f(x+1)}{f(x)} = \frac{a^{x+1}}{a^x} = a^{x+1-x} = a^1 = a \qquad\blacksquare$$

NOW WORK PROBLEM **21**.

2 Graph Exponential Functions

First, we graph the exponential function $f(x) = 2^x$.

EXAMPLE 2

Graphing an Exponential Function

Graph the exponential function: $f(x) = 2^x$

Solution

The domain of $f(x) = 2^x$ consists of all real numbers. We begin by locating some points on the graph of $f(x) = 2^x$, as listed in Table 3 on page 288.

Since $2^x > 0$ for all x, the range of f is the interval $(0, \infty)$. From this, we conclude that the graph has no x-intercepts, and, in fact, the graph will lie above the x-axis. As Table 3 indicates, the y-intercept is 1. Table 3 also indicates that as $x \to -\infty$ the value of $f(x) = 2^x$ gets closer and closer to 0. We conclude that the x-axis is a horizontal asymptote to the graph as $x \to -\infty$. This gives us the end behavior of the graph for x large and negative.

To determine the end behavior for x large and positive, look again at Table 3. As $x \to \infty$, $f(x) = 2^x$ grows very quickly, causing the graph of $f(x) = 2^x$ to rise very rapidly. It is apparent that f is an increasing function and hence is one-to-one.

Using all this information, we plot some of the points from Table 3 and connect them with a smooth, continuous curve, as shown in Figure 18.

Table 3	x	$f(x) = 2^x$
	-10	$2^{-10} \approx 0.00098$
	-3	$2^{-3} = \dfrac{1}{8}$
	-2	$2^{-2} = \dfrac{1}{4}$
	-1	$2^{-1} = \dfrac{1}{2}$
	0	$2^0 = 1$
	1	$2^1 = 2$
	2	$2^2 = 4$
	3	$2^3 = 8$
	10	$2^{10} = 1024$

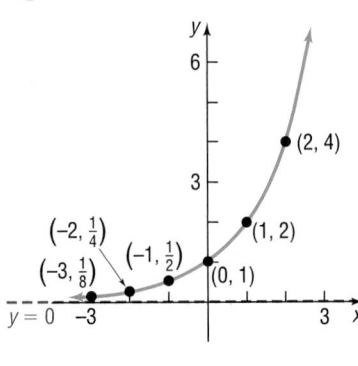

Figure 18

As we shall see, graphs that look like the one in Figure 18 occur very frequently in a variety of situations. For example, look at the graph in Figure 19, which illustrates the number of cellular telephone subscribers at the end of each year from 1985 to 2004. We might conclude from this graph that the number of cellular telephone subscribers is growing *exponentially*.

Figure 19

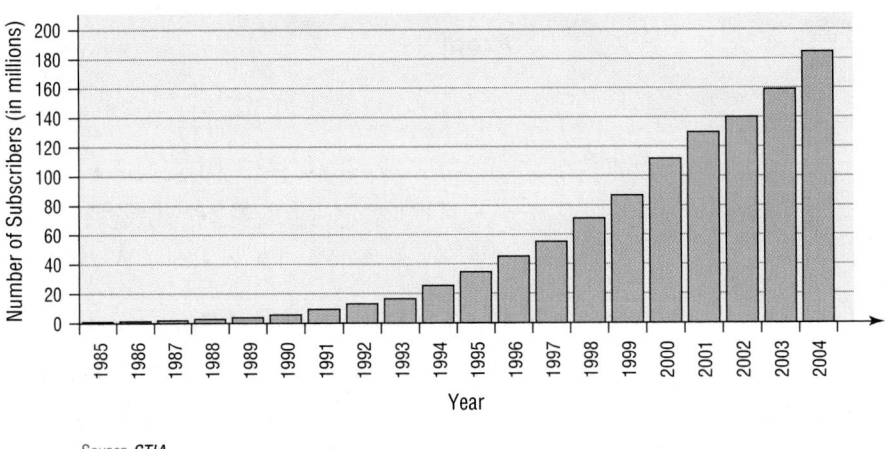

Number of Cellular Phone Subscribers at Year End

SOURCE: *CTIA*

Figure 20

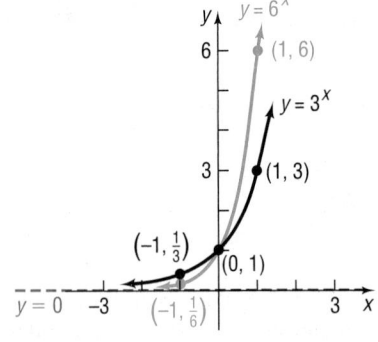

We shall have more to say about situations that lead to exponential growth later in this chapter. For now, we continue to seek properties of the exponential functions.

The graph of $f(x) = 2^x$ in Figure 18 is typical of all exponential functions that have a base larger than 1. Such functions are increasing functions and hence are one-to-one. Their graphs lie above the x-axis, pass through the point $(0, 1)$, and thereafter rise rapidly as $x \to \infty$. As $x \to -\infty$, the x-axis ($y = 0$) is a horizontal asymptote. There are no vertical asymptotes. Finally, the graphs are smooth and continuous, with no corners or gaps.

Figure 20 illustrates the graphs of two more exponential functions whose bases are larger than 1. Notice that for the larger base the graph is steeper when $x > 0$ and is closer to the x-axis when $x < 0$.

───── **Seeing the Concept** ─────

Graph $Y_1 = 2^x$ and compare what you see to Figure 18. Clear the screen and graph $Y_1 = 3^x$ and $Y_2 = 6^x$ and compare what you see to Figure 20. Clear the screen and graph $Y_1 = 10^x$ and $Y_2 = 100^x$. What viewing rectangle seems to work best?

The following display summarizes the information that we have about $f(x) = a^x, a > 1$.

Figure 21
$f(x) = a^x, a > 1$

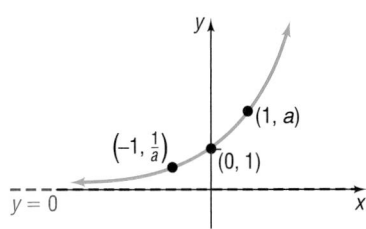

Properties of the Exponential Function $f(x) = a^x, a > 1$

1. The domain is the set of all real numbers; the range is the set of positive real numbers.
2. There are no x-intercepts; the y-intercept is 1.
3. The x-axis ($y = 0$) is a horizontal asymptote as $x \to -\infty$.
4. $f(x) = a^x, a > 1$, is an increasing function and is one-to-one.
5. The graph of f contains the points $(0, 1)$, $(1, a)$, and $\left(-1, \dfrac{1}{a}\right)$.
6. The graph of f is smooth and continuous, with no corners or gaps. See Figure 21.

Now we consider $f(x) = a^x$ when $0 < a < 1$.

EXAMPLE 3

Graphing an Exponential Function

Graph the exponential function: $f(x) = \left(\dfrac{1}{2}\right)^x$

Table 4

x	$f(x) = \left(\dfrac{1}{2}\right)^x$
-10	$\left(\dfrac{1}{2}\right)^{-10} = 1024$
-3	$\left(\dfrac{1}{2}\right)^{-3} = 8$
-2	$\left(\dfrac{1}{2}\right)^{-2} = 4$
-1	$\left(\dfrac{1}{2}\right)^{-1} = 2$
0	$\left(\dfrac{1}{2}\right)^{0} = 1$
1	$\left(\dfrac{1}{2}\right)^{1} = \dfrac{1}{2}$
2	$\left(\dfrac{1}{2}\right)^{2} = \dfrac{1}{4}$
3	$\left(\dfrac{1}{2}\right)^{3} = \dfrac{1}{8}$
10	$\left(\dfrac{1}{2}\right)^{10} \approx 0.00098$

Solution The domain of $f(x) = \left(\dfrac{1}{2}\right)^x$ consists of all real numbers. As before, we locate some points on the graph by creating Table 4. Since $\left(\dfrac{1}{2}\right)^x > 0$ for all x, the range of f is the interval $(0, \infty)$. The graph lies above the x-axis and so has no x-intercepts. The y-intercept is 1. As $x \to -\infty$, $f(x) = \left(\dfrac{1}{2}\right)^x$ grows very quickly. As $x \to \infty$, the values of $f(x)$ approach 0. The x-axis ($y = 0$) is a horizontal asymptote as $x \to \infty$. It is apparent that f is a decreasing function and hence is one-to-one. Figure 22 illustrates the graph.

Figure 22

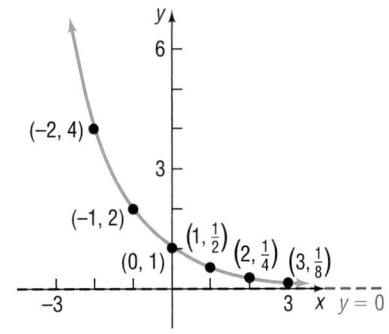

We could have obtained the graph of $y = \left(\dfrac{1}{2}\right)^x$ from the graph of $y = 2^x$ using transformations. The graph of $y = \left(\dfrac{1}{2}\right)^x = 2^{-x}$ is a reflection about the y-axis of the graph of $y = 2^x$ (Replace x by $-x$). See Figures 23(a) and (b).

Figure 23

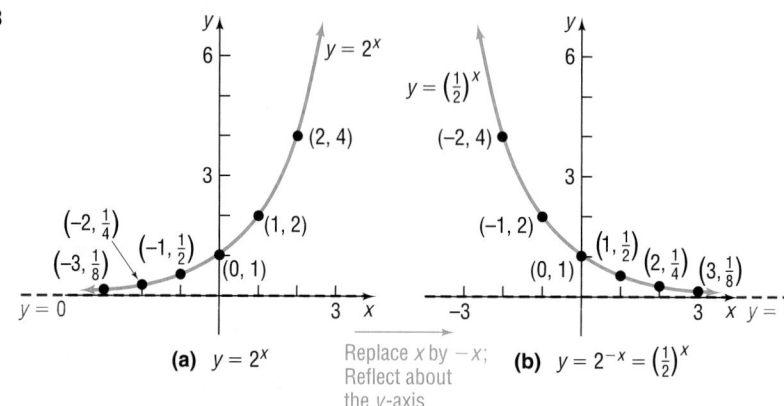

(a) $y = 2^x$ Replace x by $-x$; Reflect about the y-axis (b) $y = 2^{-x} = \left(\dfrac{1}{2}\right)^x$

Figure 24

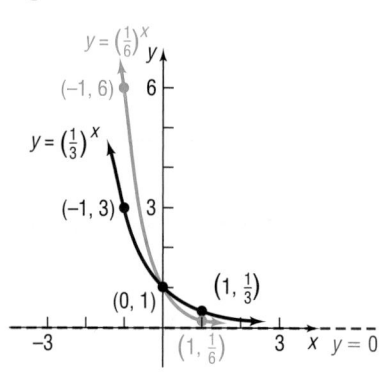

The graph of $f(x) = \left(\dfrac{1}{2}\right)^x$ in Figure 22 is typical of all exponential functions that have a base between 0 and 1. Such functions are decreasing and one-to-one. Their graphs lie above the x-axis and pass through the point $(0, 1)$. The graphs rise rapidly as $x \to -\infty$. As $x \to \infty$, the x-axis ($y = 0$) is a horizontal asymptote. There are no vertical asymptotes. Finally, the graphs are smooth and continuous, with no corners or gaps.

Figure 24 illustrates the graphs of two more exponential functions whose bases are between 0 and 1. Notice that the smaller base results in a graph that is steeper when $x < 0$. When $x > 0$, the graph of the equation with the smaller base is closer to the x-axis.

The following display summarizes the information that we have about the function $f(x) = a^x$, $0 < a < 1$.

Properties of the Exponential Function $f(x) = a^x$, $0 < a < 1$

1. The domain is the set of all real numbers; the range is the set of positive real numbers.

2. There are no x-intercepts; the y-intercept is 1.

3. The x-axis ($y = 0$) is a horizontal asymptote as $x \to \infty$.

4. $f(x) = a^x$, $0 < a < 1$, is a decreasing function and is one-to-one.

5. The graph of f contains the points $(0, 1)$, $(1, a)$, and $\left(-1, \dfrac{1}{a}\right)$.

6. The graph of f is smooth and continuous, with no corners or gaps. See Figure 25.

Figure 25
$f(x) = a^x$, $0 < a < 1$

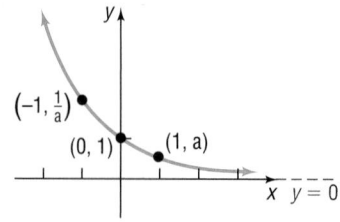

| EXAMPLE 4 | **Graphing Exponential Functions Using Transformations** |

Graph $f(x) = 2^{-x} - 3$ and determine the domain, range, and horizontal asymptote of f.

Solution We begin with the graph of $y = 2^x$. Figure 26 shows the stages.

Figure 26

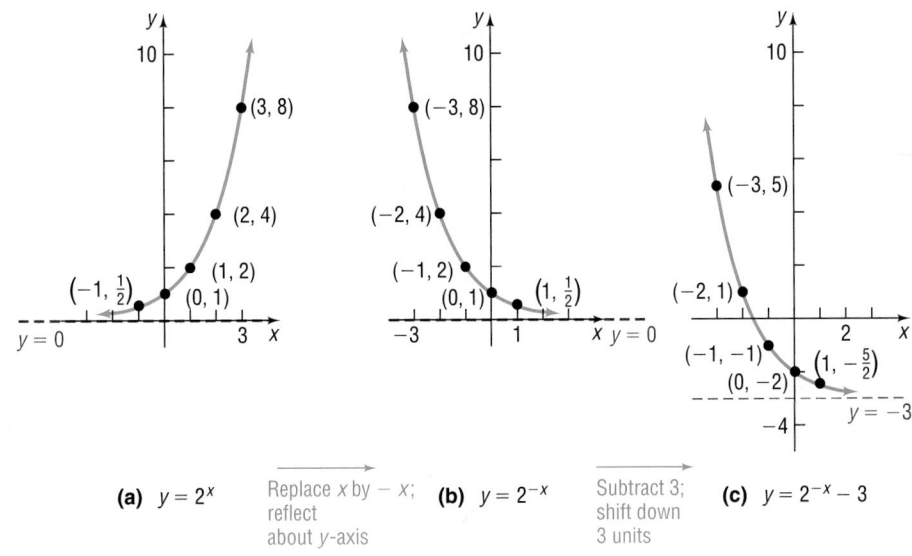

(a) $y = 2^x$ Replace x by $-x$;
reflect
about y-axis **(b)** $y = 2^{-x}$ Subtract 3;
shift down
3 units **(c)** $y = 2^{-x} - 3$

As Figure 26(c) illustrates, the domain of $f(x) = 2^{-x} - 3$ is the interval $(-\infty, \infty)$ and the range is the interval $(-3, \infty)$. The horizontal asymptote of f is the line $y = -3$. ■

 NOW WORK PROBLEM 37.

3 Define the Number e

As we shall see shortly, many problems that occur in nature require the use of an exponential function whose base is a certain irrational number, symbolized by the letter e.

Let's look at one way of arriving at this important number e.

The **number e** is defined as the number that the expression

$$\left(1 + \frac{1}{n}\right)^n \tag{2}$$

approaches as $n \to \infty$. In calculus, this is expressed using limit notation as

$$e = \lim_{n \to \infty} \left(1 + \frac{1}{n}\right)^n$$

Table 5 illustrates what happens to the defining expression (2) as n takes on increasingly large values. The last number in the right column in the table is correct to nine decimal places and is the same as the entry given for e on your calculator (if expressed correctly to nine decimal places).

Table 5

n	$\dfrac{1}{n}$	$1 + \dfrac{1}{n}$	$\left(1 + \dfrac{1}{n}\right)^n$
1	1	2	2
2	0.5	1.5	2.25
5	0.2	1.2	2.48832
10	0.1	1.1	2.59374246
100	0.01	1.01	2.704813829
1,000	0.001	1.001	2.716923932
10,000	0.0001	1.0001	2.718145927
100,000	0.00001	1.00001	2.718268237
1,000,000	0.000001	1.000001	2.718280469
1,000,000,000	10^{-9}	$1 + 10^{-9}$	2.718281827

The exponential function $f(x) = e^x$, whose base is the number e, occurs with such frequency in applications that it is usually referred to as *the* exponential function. Indeed, most calculators have the key $\boxed{e^x}$ or $\boxed{\exp(x),}$ which may be used to evaluate the exponential function for a given value of x.*

Now use your calculator to approximate e^x for $x = -2$, $x = -1$, $x = 0$, $x = 1$, and $x = 2$, as we have done to create Table 6. The graph of the exponential function $f(x) = e^x$ is given in Figure 27. Since $2 < e < 3$, the graph of $y = e^x$ lies between the graphs of $y = 2^x$ and $y = 3^x$. Do you see why? (Refer to Figures 18 and 20.)

— **Seeing the Concept** —

Graph $Y_1 = e^x$ and compare what you see to Figure 27. Use eVALUEate or TABLE to verify the points on the graph shown in Figure 27. Now graph $Y_2 = 2^x$ and $Y_3 = 3^x$ on the same screen as $Y_1 = e^x$. Notice that the graph of $Y_1 = e^x$ lies between these two graphs.

Table 6

x	e^x
-2	0.14
-1	0.37
0	1
1	2.72
2	7.39

Figure 27
$y = e^x$

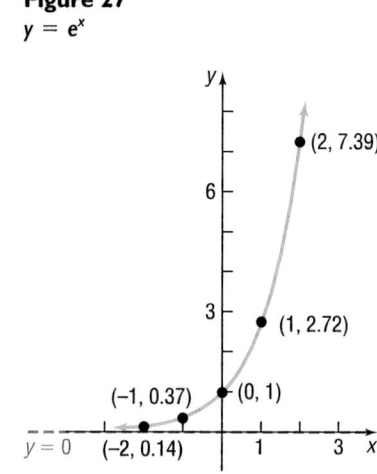

| EXAMPLE 5 | **Graphing Exponential Functions Using Transformations** |

Graph $f(x) = -e^{x-3}$ and determine the domain, the range, and horizontal asymptote of f.

Solution We begin with the graph of $y = e^x$. Figure 28 shows the stages.

*If your calculator does not have this key, refer to your Owner's Manual.

Figure 28

(a) $y = e^x$ *Multiply by −1; Reflect about x-axis.* **(b)** $y = -e^x$ *Replace x by x − 3; Shift right 3 units.* **(c)** $y = -e^{x-3}$

As Figure 28(c) illustrates, the domain of $f(x) = -e^{x-3}$ is the interval $(-\infty, \infty)$, and the range is the interval $(-\infty, 0)$. The horizontal asymptote is the line $y = 0$.

■

NOW WORK PROBLEM **49**.

4 Solve Exponential Equations

Equations that involve terms of the form a^x, $a > 0$, $a \neq 1$, are often referred to as **exponential equations**. Such equations can sometimes be solved by appropriately applying the Laws of Exponents and property (3) below.

In Words

When two exponential expressions with the same base are equal, then their exponents are equal.

$$\text{If} \quad a^u = a^v, \quad \text{then} \quad u = v \tag{3}$$

Property (3) is a consequence of the fact that exponential functions are one-to-one. To use property (3), each side of the equality must be written with the same base.

EXAMPLE 6 **Solving an Exponential Equation**

Solve: $3^{x+1} = 81$

Solution Since $81 = 3^4$, we can write the equation as

$$3^{x+1} = 81 = 3^4$$

Now we have the same base, 3, on each side, so we can set the exponents equal to each other to obtain

$$x + 1 = 4$$
$$x = 3$$

The solution set is $\{3\}$.

■

NOW WORK PROBLEM **59**.

EXAMPLE 7 **Solving an Exponential Equation**

Solve: $e^{-x^2} = (e^x)^2 \cdot \dfrac{1}{e^3}$

Solution We use the Laws of Exponents first to get the base e on the right side.

$$(e^x)^2 \cdot \frac{1}{e^3} = e^{2x} \cdot e^{-3} = e^{2x-3}$$

As a result,

$$e^{-x^2} = e^{2x-3}$$
$$-x^2 = 2x - 3 \qquad \text{Apply Property (3).}$$
$$x^2 + 2x - 3 = 0 \qquad \text{Place the quadratic equation in standard form.}$$
$$(x + 3)(x - 1) = 0 \qquad \text{Factor.}$$
$$x = -3 \quad \text{or} \quad x = 1 \qquad \text{Use the Zero-Product Property.}$$

The solution set is $\{-3, 1\}$. ∎

Many applications involve the exponential functions. Let's look at one.

EXAMPLE 8 **Exponential Probability**

Between 9:00 PM and 10:00 PM cars arrive at Burger King's drive-thru at the rate of 12 cars per hour (0.2 car per minute). The following formula from statistics can be used to determine the probability that a car will arrive within t minutes of 9:00 PM.

$$F(t) = 1 - e^{-0.2t}$$

(a) Determine the probability that a car will arrive within 5 minutes of 9 PM (that is, before 9:05 PM).

(b) Determine the probability that a car will arrive within 30 minutes of 9 PM (before 9:30 PM).

(c) What value does F approach as t becomes unbounded in the positive direction?

(d) Graph F using your graphing utility.

Solution (a) The probability that a car will arrive within 5 minutes is found by evaluating $F(t)$ at $t = 5$.

$$F(5) = 1 - e^{-0.2(5)} \approx 0.63212$$
$$\uparrow$$
$$\text{Use a calculator}$$

We conclude that there is a 63% probability that a car will arrive within 5 minutes.

(b) The probability that a car will arrive within 30 minutes is found by evaluating $F(t)$ at $t = 30$.

$$F(30) = 1 - e^{-0.2(30)} \approx 0.9975$$
$$\uparrow$$
$$\text{Use a calculator}$$

There is a 99.75% probability that a car will arrive within 30 minutes.

(c) As time passes, the probability that a car will arrive increases. The value that F approaches can be found by letting $t \to \infty$. Since $e^{-0.2t} = \dfrac{1}{e^{0.2t}}$, it follows that $e^{-0.2t} \to 0$ as $t \to \infty$. We conclude that F approaches 1 as t gets large.

(d) See Figure 29 for the graph of F. ∎

Figure 29

 NOW WORK PROBLEM 101.

Summary

Properties of the Exponential Function

$f(x) = a^x, \quad a > 1$ Domain: the interval $(-\infty, \infty)$; range: the interval $(0, \infty)$
x-intercepts: none; y-intercept: 1
Horizontal asymptote: x-axis $(y = 0)$ as $x \to -\infty$
Increasing; one-to-one; smooth; continuous
See Figure 21 for a typical graph.

$f(x) = a^x, \quad 0 < a < 1$ Domain: the interval $(-\infty, \infty)$; Range: the interval $(0, \infty)$.
x-intercepts: none; y-intercept: 1
Horizontal asymptote: x-axis $(y = 0)$ as $x \to \infty$
Decreasing; one-to-one; smooth; continuous
See Figure 25 for a typical graph.

If $a^u = a^v$, then $u = v$.

4.3 Assess Your Understanding

'Are You Prepared?'

Answers are given at the end of these exercises. If you get a wrong answer, read the pages listed in red.

1. $4^3 =$ _____; $8^{2/3} =$ _____; $3^{-2} =$ _____. (pp. A6–A7 and pp. A55–A56)

2. Solve: $5x - 2 = 3$ (pp. A60–A63)

3. *True or False:* To graph $y = (x - 2)^3$, shift the graph of $y = x^3$ to the left 2 units. (pp. 86–87)

4. Find the average rate of change of $f(x) = 3x - 5$ from $x = 0$ to $x = c$. (pp. 68–70)

5. *True or False:* The function $f(x) = \dfrac{2x}{x - 3}$ has $y = 2$ as a horizontal asymptote. (pp. 208–211)

Concepts and Vocabulary

6. The graph of every exponential function $f(x) = a^x, a > 0, a \ne 1$, passes through three points: _____, _____, and _____.

7. If the graph of the exponential function $f(x) = a^x, a > 0, a \ne 1$, is decreasing, then a must be less than _____.

8. If $3^x = 3^4$, then $x =$ _____.

9. *True or False:* The graphs of $y = 3^x$ and $y = \left(\dfrac{1}{3}\right)^x$ are identical.

10. *True or False:* The range of the exponential function $f(x) = a^x, a > 0, a \ne 1$, is the set of all real numbers.

Skill Building

In Problems 11–20, approximate each number using a calculator. Express your answer rounded to three decimal places.

11. (a) $3^{2.2}$ (b) $3^{2.23}$ (c) $3^{2.236}$ (d) $3^{\sqrt{5}}$
12. (a) $5^{1.7}$ (b) $5^{1.73}$ (c) $5^{1.732}$ (d) $5^{\sqrt{3}}$
13. (a) $2^{3.14}$ (b) $2^{3.141}$ (c) $2^{3.1415}$ (d) 2^{π}
14. (a) $2^{2.7}$ (b) $2^{2.71}$ (c) $2^{2.718}$ (d) 2^e
15. (a) $3.1^{2.7}$ (b) $3.14^{2.71}$ (c) $3.141^{2.718}$ (d) π^e
16. (a) $2.7^{3.1}$ (b) $2.71^{3.14}$ (c) $2.718^{3.141}$ (d) e^{π}
17. $e^{1.2}$
18. $e^{-1.3}$
19. $e^{-0.85}$
20. $e^{2.1}$

In Problems 21–28, determine whether the given function is exponential or not. For those that are exponential functions, identify the value of the base. [**Hint:** *Look at the ratio of consecutive values.*]

21.
x	$f(x)$
−1	3
0	6
1	12
2	18
3	30

22.
x	$g(x)$
−1	2
0	5
1	8
2	11
3	14

23.
x	$H(x)$
−1	$\frac{1}{4}$
0	1
1	4
2	16
3	64

24.
x	$F(x)$
−1	$\frac{2}{3}$
0	1
1	$\frac{3}{2}$
2	$\frac{9}{4}$
3	$\frac{27}{8}$

25.

x	f(x)
−1	$\frac{3}{2}$
0	3
1	6
2	12
3	24

26.

x	g(x)
−1	6
0	1
1	0
2	3
3	10

27.

x	H(x)
−1	2
0	4
1	6
2	8
3	10

28.

x	F(x)
−1	$\frac{1}{2}$
0	$\frac{1}{4}$
1	$\frac{1}{8}$
2	$\frac{1}{16}$
3	$\frac{1}{32}$

In Problems 29–36, the graph of an exponential function is given. Match each graph to one of the following functions.

A. $y = 3^x$ B. $y = 3^{-x}$ C. $y = -3^x$ D. $y = -3^{-x}$
E. $y = 3^x - 1$ F. $y = 3^{x-1}$ G. $y = 3^{1-x}$ H. $y = 1 - 3^x$

29.

30.

31.

32.

33.

34.

35.

36.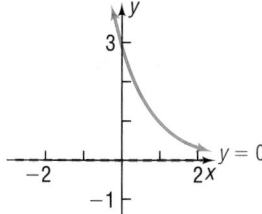

In Problems 37–44, use transformations to graph each function. Determine the domain, range, and horizontal asymptote of each function.

37. $f(x) = 2^x + 1$ **38.** $f(x) = 3^x - 2$ **39.** $f(x) = 3^{x-1}$ **40.** $f(x) = 2^{x+2}$

41. $f(x) = 3 \cdot \left(\frac{1}{2}\right)^x$ **42.** $f(x) = 4 \cdot \left(\frac{1}{3}\right)^x$ **43.** $f(x) = 3^{-x} - 2$ **44.** $f(x) = -3^x + 1$

45. $f(x) = 2 + 3(4^x)$ **46.** $f(x) = 1 - 3(2^x)$ **47.** $f(x) = 2 + 3^{x/2}$ **48.** $f(x) = 1 - 2^{-x/3}$

In Problems 49–56, begin with the graph of $y = e^x$ [Figure 27] and use transformations to graph each function. Determine the domain, range, and horizontal asymptote of each function.

49. $f(x) = e^{-x}$ **50.** $f(x) = -e^x$ **51.** $f(x) = e^{x+2}$ **52.** $f(x) = e^x - 1$

53. $f(x) = 5 - e^{-x}$ **54.** $f(x) = 9 - 3e^{-x}$ **55.** $f(x) = 2 - e^{-x/2}$ **56.** $f(x) = 7 - 3e^{2x}$

In Problems 57–76, solve each equation.

57. $7^x = 7^3$ **58.** $5^x = 5^{-6}$ **59.** $2^{-x} = 16$ **60.** $3^{-x} = 81$

61. $\left(\frac{1}{5}\right)^x = \frac{1}{25}$ **62.** $\left(\frac{1}{4}\right)^x = \frac{1}{64}$ **63.** $2^{2x-1} = 4$ **64.** $5^{x+3} = \frac{1}{5}$

65. $3^{x^3} = 9^x$ **66.** $4^{x^2} = 2^x$ **67.** $8^{-x+14} = 16^x$ **68.** $9^{-x+15} = 27^x$

69. $3^{x^2-7} = 27^{2x}$ **70.** $5^{x^2+8} = 125^{2x}$ **71.** $4^x \cdot 2^{x^2} = 16^2$ **72.** $9^{2x} \cdot 27^{x^2} = 3^{-1}$

73. $e^x = e^{3x+8}$ **74.** $e^{3x} = e^{2-x}$ **75.** $e^{x^2} = e^{3x} \cdot \frac{1}{e^2}$ **76.** $(e^4)^x \cdot e^{x^2} = e^{12}$

77. Suppose that $f(x) = 2^x$.
 (a) What is $f(4)$? What point is on the graph of f?
 (b) If $f(x) = \dfrac{1}{16}$, what is x? What point is on the graph of f?

78. Suppose that $f(x) = 3^x$.
 (a) What is $f(4)$? What point is on the graph of f?
 (b) If $f(x) = \dfrac{1}{9}$, what is x? What point is on the graph of f?

79. Suppose that $g(x) = 4^x + 2$.
 (a) What is $g(-1)$? What point is on the graph of g?
 (b) If $g(x) = 66$, what is x? What point is on the graph of g?

80. Suppose that $g(x) = 5^x - 3$.
 (a) What is $g(-1)$? What point is on the graph of g?
 (b) If $g(x) = 122$, what is x? What point is on the graph of g?

81. Suppose that $H(x) = 3 \cdot \left(\dfrac{1}{2}\right)^x - 2$.
 (a) What is $H(-2)$? What point is on the graph of H?
 (b) If $H(x) = -\dfrac{13}{8}$, what is x? What point is on the graph of H?

82. Suppose that $F(x) = -2 \cdot \left(\dfrac{1}{3}\right)^x + 1$.
 (a) What is $F(-1)$? What point is on the graph of F?
 (b) If $F(x) = -53$, what is x? What point is on the graph of F?

83. If $4^x = 7$, what does 4^{-2x} equal?

84. If $2^x = 3$, what does 4^{-x} equal?

85. If $3^{-x} = 2$, what does 3^{2x} equal?

86. If $5^{-x} = 3$, what does 5^{3x} equal?

In Problems 87–90, determine the exponential function whose graph is given.

87.

88.

89.

90.

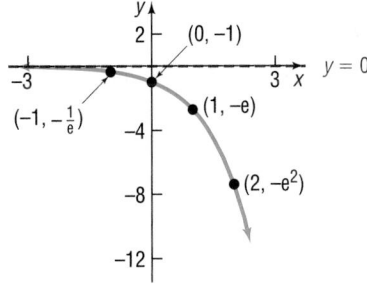

In Problems 91–94, graph each function. Based on the graph, state the domain and the range and find any intercepts.

91. $f(x) = \begin{cases} e^{-x} & \text{if } x < 0 \\ e^x & \text{if } x \geq 0 \end{cases}$

92. $f(x) = \begin{cases} e^x & \text{if } x < 0 \\ e^{-x} & \text{if } x \geq 0 \end{cases}$

93. $f(x) = \begin{cases} -e^x & \text{if } x < 0 \\ -e^{-x} & \text{if } x \geq 0 \end{cases}$

94. $f(x) = \begin{cases} -e^{-x} & \text{if } x < 0 \\ -e^x & \text{if } x \geq 0 \end{cases}$

Applications and Extensions

95. Optics If a single pane of glass obliterates 3% of the light passing through it, the percent p of light that passes through n successive panes is given approximately by the function

$$p(n) = 100(0.97)^n$$

 (a) What percent of light will pass through 10 panes?
 (b) What percent of light will pass through 25 panes?

96. Atmospheric Pressure The atmospheric pressure p on a balloon or plane decreases with increasing height. This pressure, measured in millimeters of mercury, is related to the height h (in kilometers) above sea level by the function

$$p(h) = 760e^{-0.145h}$$

 (a) Find the atmospheric pressure at a height of 2 kilometers (over a mile).
 (b) What is it at a height of 10 kilometers (over 30,000 feet)?

97. Depreciation The price p of a Honda Civic DX Sedan that is x years old is given by

$$p(x) = 16{,}630(0.90)^x$$

(a) How much does a 3-year-old Civic DX Sedan cost?
(b) How much does a 9-year-old Civic DX Sedan cost?

98. Healing of Wounds The normal healing of wounds can be modeled by an exponential function. If A_0 represents the original area of the wound and if A equals the area of the wound, then the function

$$A(n) = A_0 e^{-0.35n}$$

describes the area of a wound after n days following an injury when no infection is present to retard the healing. Suppose that a wound initially had an area of 100 square millimeters.
(a) If healing is taking place, how large will the area of the wound be after 3 days?
(b) How large will it be after 10 days?

99. Drug Medication The function

$$D(h) = 5e^{-0.4h}$$

can be used to find the number of milligrams D of a certain drug that is in a patient's bloodstream h hours after the drug has been administered. How many milligrams will be present after 1 hour? After 6 hours?

100. Spreading of Rumors A model for the number N of people in a college community who have heard a certain rumor is

$$N = P(1 - e^{-0.15d})$$

where P is the total population of the community and d is the number of days that have elapsed since the rumor began. In a community of 1000 students, how many students will have heard the rumor after 3 days?

101. Exponential Probability Between 12:00 PM and 1:00 PM, cars arrive at Citibank's drive-thru at the rate of 6 cars per hour (0.1 car per minute). The following formula from probability can be used to determine the probability that a car will arrive within t minutes of 12:00 PM:

$$F(t) = 1 - e^{-0.1t}$$

(a) Determine the probability that a car will arrive within 10 minutes of 12:00 PM (that is, before 12:10 PM).
(b) Determine the probability that a car will arrive within 40 minutes of 12:00 PM (before 12:40 PM).
(c) What value does F approach as t becomes unbounded in the positive direction?
(d) Graph F using a graphing utility.
(e) Using TRACE, determine how many minutes are needed for the probability to reach 50%.

102. Exponential Probability Between 5:00 PM and 6:00 PM, cars arrive at Jiffy Lube at the rate of 9 cars per hour (0.15 car per minute). The following formula from probability can be used to determine the probability that a car will arrive within t minutes of 5:00 PM:

$$F(t) = 1 - e^{-0.15t}$$

(a) Determine the probability that a car will arrive within 15 minutes of 5:00 PM (that is, before 5:15 PM).
(b) Determine the probability that a car will arrive within 30 minutes of 5:00 PM (before 5:30 PM).
(c) What value does F approach as t becomes unbounded in the positive direction?

(d) Graph F using a graphing utility.
(e) Using TRACE, determine how many minutes are needed for the probability to reach 60%.

103. Poisson Probability Between 5:00 PM and 6:00 PM, cars arrive at McDonald's drive-thru at the rate of 20 cars per hour. The following formula from probability can be used to determine the probability that x cars will arrive between 5:00 PM and 6:00 PM.

$$P(x) = \frac{20^x e^{-20}}{x!}$$

where

$$x! = x \cdot (x - 1) \cdot (x - 2) \cdots\cdots 3 \cdot 2 \cdot 1$$

(a) Determine the probability that $x = 15$ cars will arrive between 5:00 PM and 6:00 PM.
(b) Determine the probability that $x = 20$ cars will arrive between 5:00 PM and 6:00 PM.

104. Poisson Probability People enter a line for the *Demon Roller Coaster* at the rate of 4 per minute. The following formula from probability can be used to determine the probability that x people will arrive within the next minute.

$$P(x) = \frac{4^x e^{-4}}{x!}$$

where

$$x! = x \cdot (x - 1) \cdot (x - 2) \cdots\cdots 3 \cdot 2 \cdot 1$$

(a) Determine the probability that $x = 5$ people will arrive within the next minute.
(b) Determine the probability that $x = 8$ people will arrive within the next minute.

105. Relative Humidity The relative humidity is the ratio (expressed as a percent) of the amount of water vapor in the air to the maximum amount that it can hold at a specific temperature. The relative humidity, R, is found using the following formula:

$$R = 10^{\left(\frac{4221}{T+459.4} - \frac{4221}{D+459.4} + 2\right)}$$

where T is the air temperature (in °F) and D is the dew point temperature (in °F).
(a) Determine the relative humidity if the air temperature is 50° Fahrenheit and the dew point temperature is 41° Fahrenheit.
(b) Determine the relative humidity if the air temperature is 68° Fahrenheit and the dew point temperature is 59° Fahrenheit.
(c) What is the relative humidity if the air temperature and the dew point temperature are the same?

106. Learning Curve Suppose that a student has 500 vocabulary words to learn. If the student learns 15 words after 5 minutes, the function

$$L(t) = 500(1 - e^{-0.0061t})$$

approximates the number of words L that the student will learn after t minutes.
(a) How many words will the student learn after 30 minutes?
(b) How many words will the student learn after 60 minutes?

107. Current in a *RL* Circuit The equation governing the amount of current I (in amperes) after time t (in seconds) in a single *RL* circuit consisting of a resistance R (in ohms), an inductance L (in henrys), and an electromotive force E (in volts) is

$$I = \frac{E}{R}\left[1 - e^{-(R/L)t}\right]$$

(a) If $E = 120$ volts, $R = 10$ ohms, and $L = 5$ henrys, how much current I_1 is flowing after 0.3 second? After 0.5 second? After 1 second?
(b) What is the maximum current?
(c) Graph this function $I = I_1(t)$, measuring I along the y-axis and t along the x-axis.
(d) If $E = 120$ volts, $R = 5$ ohms, and $L = 10$ henrys, how much current I_2 is flowing after 0.3 second? After 0.5 second? After 1 second?
(e) What is the maximum current?
(f) Graph this function $I = I_2(t)$ on the same coordinate axes as $I_1(t)$.

108. Current in a *RC* Circuit The equation governing the amount of current I (in amperes) after time t (in microseconds) in a single *RC* circuit consisting of a resistance R (in ohms), a capacitance C (in microfarads), and an electromotive force E (in volts) is

$$I = \frac{E}{R}e^{-t/(RC)}$$

(a) If $E = 120$ volts, $R = 2000$ ohms, and $C = 1.0$ microfarad, how much current I_1 is flowing initially $(t = 0)$? After 1000 microseconds? After 3000 microseconds?
(b) What is the maximum current?
(c) Graph this function $I = I_1(t)$, measuring I along the y-axis and t along the x-axis.
(d) If $E = 120$ volts, $R = 1000$ ohms, and $C = 2.0$ microfarads, how much current I_2 is flowing initially? After 1000 microseconds? After 3000 microseconds?
(e) What is the maximum current?
(f) Graph this function $I = I_2(t)$ on the same coordinate axes as $I_1(t)$.

109. Another Formula for e Use a calculator to compute the values of

$$2 + \frac{1}{2!} + \frac{1}{3!} + \cdots + \frac{1}{n!}$$

for $n = 4, 6, 8,$ and 10. Compare each result with e.
[**Hint:** $1! = 1$, $2! = 2 \cdot 1$, $3! = 3 \cdot 2 \cdot 1$, $n! = n(n - 1) \cdots \cdots (3)(2)(1)$.]

110. Another Formula for e Use a calculator to compute the various values of the expression. Compare the values to e.

$$2 + \cfrac{1}{1 + \cfrac{1}{2 + \cfrac{2}{3 + \cfrac{3}{4 + \cfrac{4}{\text{etc.}}}}}}$$

111. Difference Quotient If $f(x) = a^x$, show that

$$\frac{f(x + h) - f(x)}{h} = a^x \cdot \frac{a^h - 1}{h}$$

112. If $f(x) = a^x$, show that $f(A + B) = f(A) \cdot f(B)$.

113. If $f(x) = a^x$, show that $f(-x) = \dfrac{1}{f(x)}$.

114. If $f(x) = a^x$, show that $f(\alpha x) = [f(x)]^\alpha$.

Problems 115 and 116 provide definitions for two other transcendental functions.

115. The **hyperbolic sine function**, designated by sinh x, is defined as

$$\sinh x = \frac{1}{2}(e^x - e^{-x})$$

(a) Show that $f(x) = \sinh x$ is an odd function.
(b) Graph $f(x) = \sinh x$ using a graphing utility.

116. The **hyperbolic cosine function**, designated by cosh x, is defined as

$$\cosh x = \frac{1}{2}(e^x + e^{-x})$$

(a) Show that $f(x) = \cosh x$ is an even function.
(b) Graph $f(x) = \cosh x$ using a graphing utility.
(c) Refer to Problem 115. Show that, for every x,

$$(\cosh x)^2 - (\sinh x)^2 = 1$$

117. Historical Problem Pierre de Fermat (1601–1665) conjectured that the function

$$f(x) = 2^{(2^x)} + 1$$

for $x = 1, 2, 3, \ldots$, would always have a value equal to a prime number. But Leonhard Euler (1707–1783) showed that this formula fails for $x = 5$. Use a calculator to determine the prime numbers produced by f for $x = 1, 2, 3, 4$. Then show that $f(5) = 641 \times 6{,}700{,}417$, which is not prime.

Discussion and Writing

118. The bacteria in a 4-liter container double every minute. After 60 minutes the container is full. How long did it take to fill half the container?

119. Explain in your own words what the number e is. Provide at least two applications that use this number.

120. Do you think that there is a power function that increases more rapidly than an exponential function whose base is greater than 1? Explain.

121. As the base a of an exponential function $f(x) = a^x, a > 1$, increases, what happens to the behavior of its graph for $x > 0$? What happens to the behavior of its graph for $x < 0$?

122. The graphs of $y = a^{-x}$ and $y = \left(\dfrac{1}{a}\right)^x$ are identical. Why?

'Are You Prepared?' Answers

1. $64; 4; \dfrac{1}{9}$ **2.** $\{1\}$ **3.** False **4.** 3 **5.** True

4.4 Logarithmic Functions

PREPARING FOR THIS SECTION *Before getting started, review the following:*

- Solving Inequalities (Appendix A, Section A.9, pp. A69–A71)
- Quadratic Inequalities (Section 2.5, pp. 154–156)
- Polynomial and Rational Inequalities (Section 3.4, pp. 231–235)

Now work the 'Are You Prepared?' problems on page 308.

OBJECTIVES
1. Change Exponential Expressions to Logarithmic Expressions and Logarithmic Expressions to Exponential Expressions
2. Evaluate Logarithmic Expressions
3. Determine the Domain of a Logarithmic Function
4. Graph Logarithmic Functions
5. Solve Logarithmic Equations

Recall that a one-to-one function $y = f(x)$ has an inverse function that is defined (implicitly) by the equation $x = f(y)$. In particular, the exponential function $y = f(x) = a^x, a > 0, a \neq 1$, is one-to-one and hence has an inverse function that is defined implicitly by the equation

$$x = a^y, \quad a > 0, \quad a \neq 1$$

This inverse function is so important that it is given a name, the *logarithmic function*.

> The **logarithmic function to the base a**, where $a > 0$ and $a \neq 1$, is denoted by $y = \log_a x$ (read as "y is the logarithm to the base a of x") and is defined by
>
> $$y = \log_a x \quad \text{if and only if} \quad x = a^y$$
>
> The domain of the logarithmic function $y = \log_a x$ is $x > 0$.

As the above definition illustrates, a *logarithm* is a name for a certain exponent.

EXAMPLE 1	Relating Logarithms to Exponents

(a) If $y = \log_3 x$, then $x = 3^y$. For example, $4 = \log_3 81$ is equivalent to $81 = 3^4$.

(b) If $y = \log_5 x$, then $x = 5^y$. For example, $-1 = \log_5\left(\dfrac{1}{5}\right)$ is equivalent to $\dfrac{1}{5} = 5^{-1}$.

■

1 Change Exponential Expressions to Logarithmic Expressions and Logarithmic Expressions to Exponential Expressions

We can use the definition of a logarithm to convert from exponential form to logarithmic form, and vice versa, as the following two examples illustrate.

EXAMPLE 2 | **Changing Exponential Expressions to Logarithmic Expressions**

Change each exponential expression to an equivalent expression involving a logarithm.

(a) $1.2^3 = m$ (b) $e^b = 9$ (c) $a^4 = 24$

Solution We use the fact that $y = \log_a x$ and $x = a^y$, $a > 0$, $a \neq 1$, are equivalent.

(a) If $1.2^3 = m$, then $3 = \log_{1.2} m$. (b) If $e^b = 9$, then $b = \log_e 9$.

(c) If $a^4 = 24$, then $4 = \log_a 24$. ∎

➤ **NOW WORK PROBLEM 9.**

EXAMPLE 3 | **Changing Logarithmic Expressions to Exponential Expressions**

Change each logarithmic expression to an equivalent expression involving an exponent.

(a) $\log_a 4 = 5$ (b) $\log_e b = -3$ (c) $\log_3 5 = c$

Solution (a) If $\log_a 4 = 5$, then $a^5 = 4$. (b) If $\log_e b = -3$, then $e^{-3} = b$.

(c) If $\log_3 5 = c$, then $3^c = 5$. ∎

➤ **NOW WORK PROBLEM 17.**

2 Evaluate Logarithmic Expressions

To find the exact value of a logarithm, we write the logarithm in exponential notation and use the fact that if $a^u = a^v$ then $u = v$.

EXAMPLE 4 | **Finding the Exact Value of a Logarithmic Expression**

Find the exact value of:

(a) $\log_2 16$ (b) $\log_3 \dfrac{1}{27}$

Solution (a) $y = \log_2 16$

$2^y = 16$ Change to exponential form.
$2^y = 2^4$ $16 = 2^4$
$y = 4$ Equate exponents.

Therefore, $\log_2 16 = 4$.

(b) $y = \log_3 \dfrac{1}{27}$

$3^y = \dfrac{1}{27}$ Change to exponential form.

$3^y = 3^{-3}$ $\dfrac{1}{27} = \dfrac{1}{3^3} = 3^{-3}$

$y = -3$ Equate exponents.

Therefore, $\log_3 \dfrac{1}{27} = -3$. ∎

➤ **NOW WORK PROBLEM 25.**

3 Determine the Domain of a Logarithmic Function

The logarithmic function $y = \log_a x$ has been defined as the inverse of the exponential function $y = a^x$. That is, if $f(x) = a^x$, then $f^{-1}(x) = \log_a x$. Based on the discussion given in Section 4.2 on inverse functions, for a function f and its inverse f^{-1}, we have

$$\text{Domain of } f^{-1} = \text{Range of } f \quad \text{and} \quad \text{Range of } f^{-1} = \text{Domain of } f$$

Consequently, it follows that

> Domain of the logarithmic function = Range of the exponential function = $(0, \infty)$
>
> Range of the logarithmic function = Domain of the exponential function = $(-\infty, \infty)$

In the next box, we summarize some properties of the logarithmic function:

> $y = \log_a x$ (defining equation: $x = a^y$)
>
> Domain: $0 < x < \infty$ Range: $-\infty < y < \infty$

The domain of a logarithmic function consists of the *positive* real numbers, so the argument of a logarithmic function must be greater than zero.

EXAMPLE 5 **Finding the Domain of a Logarithmic Function**

Find the domain of each logarithmic function.

(a) $F(x) = \log_2(x + 3)$ (b) $g(x) = \log_5\left(\dfrac{1 + x}{1 - x}\right)$ (c) $h(x) = \log_{1/2}|x|$

Solution
(a) The domain of F consists of all x for which $x + 3 > 0$, that is, $x > -3$. Using interval notation, the domain of f is $(-3, \infty)$.

(b) The domain of g is restricted to

$$\frac{1 + x}{1 - x} > 0$$

Solving this inequality, we find that the domain of g consists of all x between -1 and 1, that is, $-1 < x < 1$ or, using interval notation, $(-1, 1)$.

(c) Since $|x| > 0$, provided that $x \neq 0$, the domain of h consists of all real numbers except zero or, using interval notation, $(-\infty, 0)$ or $(0, \infty)$. ∎

═══ **NOW WORK PROBLEMS 39 AND 45.**

4 Graph Logarithmic Functions

Since exponential functions and logarithmic functions are inverses of each other, the graph of the logarithmic function $y = \log_a x$ is the reflection about the line $y = x$ of the graph of the exponential function $y = a^x$, as shown in Figure 30.

Figure 30

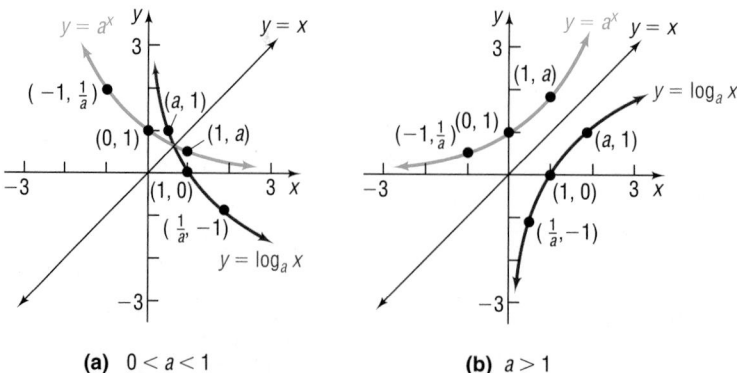

(a) $0 < a < 1$ (b) $a > 1$

For example, to graph $y = \log_2 x$, graph $y = 2^x$ and reflect it about the line $y = x$. See Figure 31. To graph $y = \log_{1/3} x$, graph $y = \left(\dfrac{1}{3}\right)^x$ and reflect it about the line $y = x$. See Figure 32.

Figure 31

Figure 32

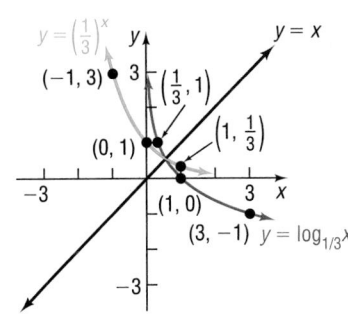

NOW WORK PROBLEM 59.

Properties of the Logarithmic Function $f(x) = \log_a x$

1. The domain is the set of positive real numbers; the range is the set of all real numbers.

2. The x-intercept of the graph is 1. There is no y-intercept.

3. The y-axis ($x = 0$) is a vertical asymptote of the graph.

4. A logarithmic function is decreasing if $0 < a < 1$ and increasing if $a > 1$.

5. The graph of f contains the points $(1, 0)$, $(a, 1)$, and $\left(\dfrac{1}{a}, -1\right)$.

6. The graph is smooth and continuous, with no corners or gaps.

If the base of a logarithmic function is the number e, then we have the **natural logarithm function**. This function occurs so frequently in applications that it is given a special symbol, **ln** (from the Latin, *logarithmus naturalis*). That is,

$$y = \ln x \quad \text{if and only if} \quad x = e^y \qquad\qquad \textbf{(1)}$$

Since $y = \ln x$ and the exponential function $y = e^x$ are inverse functions, we can obtain the graph of $y = \ln x$ by reflecting the graph of $y = e^x$ about the line $y = x$. See Figure 33.

Using a calculator with an ⌐ln⌐ key, we can obtain other points on the graph of $f(x) = \ln x$. See Table 7.

Figure 33

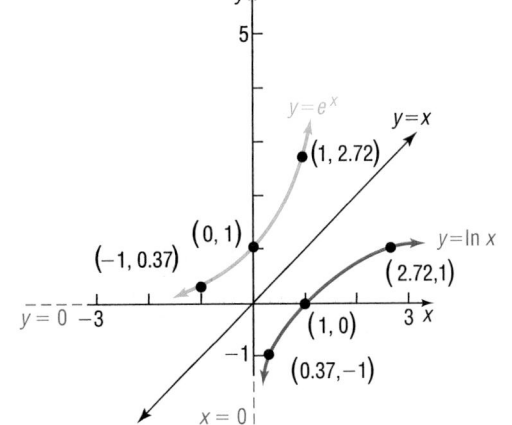

Table 7

x	ln x
$\dfrac{1}{2}$	-0.69
2	0.69
3	1.10

—— **Seeing the Concept** ——

Graph $Y_1 = e^x$ and $Y_2 = \ln x$ on the same square screen. Use eVALUEate to verify the points on the graph given in Figure 33. Do you see the symmetry of the two graphs with respect to the line $y = x$?

EXAMPLE 6

Graphing Logarithmic Functions Using Transformations

Graph $f(x) = -\ln(x + 2)$ by starting with the graph of $y = \ln x$. Determine the domain, range, and vertical asymptote.

Solution

The domain consists of all x for which

$$x + 2 > 0 \quad \text{or} \quad x > -2$$

To obtain the graph of $y = -\ln(x + 2)$, we use the steps illustrated in Figure 34.

Figure 34

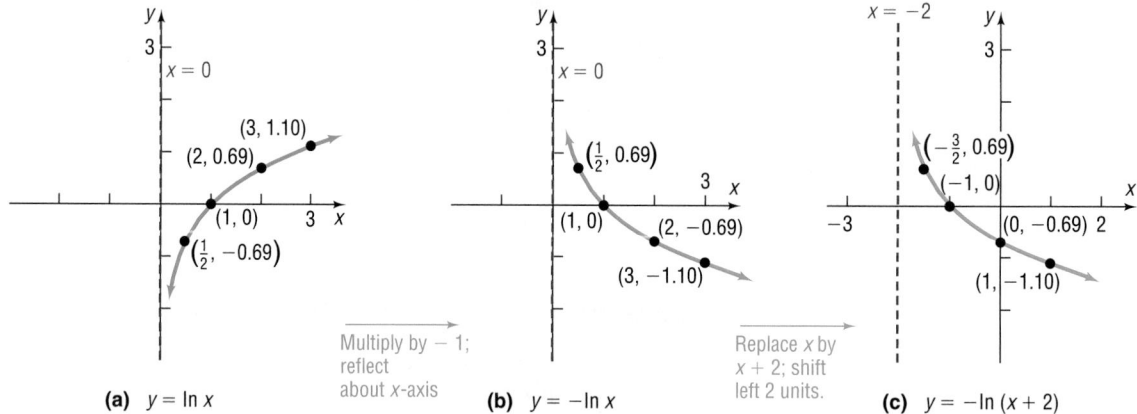

(a) $y = \ln x$ Multiply by -1; reflect about x-axis (b) $y = -\ln x$ Replace x by $x + 2$; shift left 2 units. (c) $y = -\ln(x + 2)$

The range of $f(x) = -\ln(x + 2)$ is the interval $(-\infty, \infty)$, and the vertical asymptote is $x = -2$. [Do you see why? The original asymptote $(x = 0)$ is shifted to the left 2 units.] ∎

NOW WORK PROBLEM 71.

If the base of a logarithmic function is the number 10, then we have the **common logarithm function**. If the base a of the logarithmic function is not indicated, it is understood to be 10. That is,

$$y = \log x \quad \text{if and only if} \quad x = 10^y$$

Since $y = \log x$ and the exponential function $y = 10^x$ are inverse functions, we can obtain the graph of $y = \log x$ by reflecting the graph of $y = 10^x$ about the line $y = x$. See Figure 35.

Figure 35

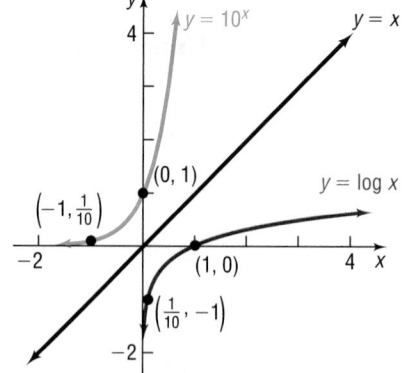

EXAMPLE 7

Graphing Logarithmic Functions Using Transformations

Graph $f(x) = 3 \log(x - 1)$. Determine the domain, range, and vertical asymptote of f.

Solution The domain consists of all x for which

$$x - 1 > 0 \quad \text{or} \quad x > 1$$

To obtain the graph of $y = 3 \log(x - 1)$, we use the steps illustrated in Figure 36.

Figure 36

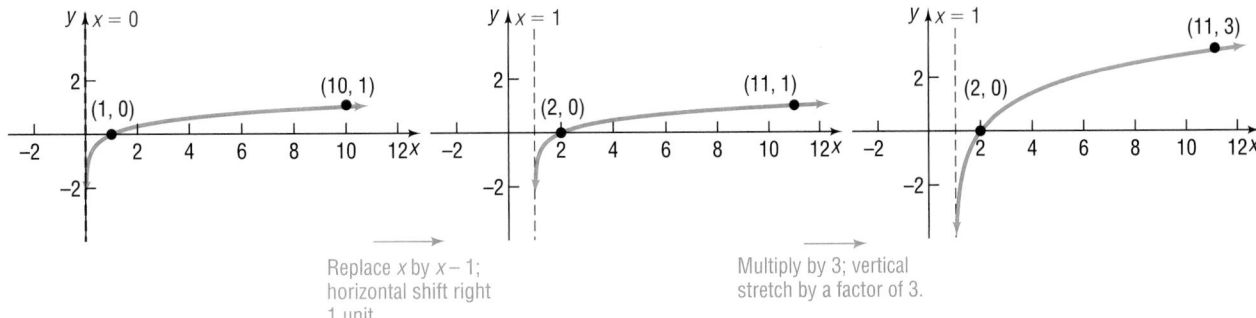

Replace x by $x - 1$;
horizontal shift right
1 unit

Multiply by 3; vertical
stretch by a factor of 3.

(a) $y = \log x$ **(b)** $y = \log(x - 1)$ **(c)** $y = 3 \log(x - 1)$

The range of $f(x) = 3 \log(x - 1)$ is the interval $(-\infty, \infty)$, and the vertical asymptote is $x = 1$. ∎

 NOW WORK PROBLEM 83.

5 Solve Logarithmic Equations

Equations that contain logarithms are called **logarithmic equations**. Care must be taken when solving logarithmic equations algebraically. Be sure to check each apparent solution in the original equation and discard any that are extraneous. In the expression $\log_a M$, remember that a and M are positive and $a \neq 1$.

Some logarithmic equations can be solved by changing from a logarithmic expression to an exponential expression.

EXAMPLE 8	**Solving a Logarithmic Equation**

Solve: (a) $\log_3(4x - 7) = 2$ (b) $\log_x 64 = 2$

Solution (a) We can obtain an exact solution by changing the logarithm to exponential form.

$$\log_3(4x - 7) = 2$$
$$4x - 7 = 3^2 \quad \textit{Change to exponential form.}$$
$$4x - 7 = 9$$
$$4x = 16$$
$$x = 4$$

✔ **CHECK:** $\log_3(4x - 7) = \log_3(16 - 7) = \log_3 9 = 2 \quad 3^2 = 9$

The solution set is {4}.

(b) We can obtain an exact solution by changing the logarithm to exponential form.

$$\log_x 64 = 2$$
$$x^2 = 64 \qquad \textit{Change to exponential form.}$$
$$x = \pm\sqrt{64} = \pm 8 \quad \textit{Square Root Method}$$

The base of a logarithm is always positive. As a result, we discard -8. We check the solution 8.

✔ **CHECK:** $\log_8 64 = 2 \quad 8^2 = 64$

The solution set is {8}. ∎

| EXAMPLE 9 | **Using Logarithms to Solve Exponential Equations** |

Solve: $e^{2x} = 5$

Solution We can obtain an exact solution by changing the exponential equation to logarithmic form.

$$e^{2x} = 5$$

$$\ln 5 = 2x \qquad \text{\textit{Change to a logarithmic expression using the}}$$
$$\qquad\qquad \text{\textit{fact that if } } e^y = x, \text{ \textit{then } } y = \ln x.$$

$$x = \frac{\ln 5}{2} \qquad \text{\textit{Exact solution}}$$

$$\approx 0.805 \qquad \text{\textit{Approximate solution}}$$

The solution set is $\left\{\dfrac{\ln 5}{2}\right\}$. ∎

───── **NOW WORK PROBLEMS 87 AND 99.**

| EXAMPLE 10 | **Alcohol and Driving** |

The blood alcohol concentration (BAC) is the amount of alcohol in a person's bloodstream. A BAC of 0.04% means that a person has 4 parts alcohol per 10,000 parts blood in the body. Relative risk is defined as the likelihood of one event occurring divided by the likelihood of a second event occurring. For example, if an individual with a BAC of 0.02% is 1.4 times as likely to have a car accident as an individual that has not been drinking, the relative risk of an accident with a BAC of 0.02% is 1.4. Recent medical research suggests that the relative risk R of having an accident while driving a car can be modeled by the equation

$$R = e^{kx}$$

where x is the percent of concentration of alcohol in the bloodstream and k is a constant.

(a) Medical research indicates that the relative risk of a person having an accident with a BAC of 0.02% is 1.4. Find the constant k in the equation.
(b) Using this value of k, what is the relative risk if the concentration is 0.17%?
(c) Using this same value of k, what BAC corresponds to a relative risk of 100?
(d) If the law asserts that anyone with a relative risk of 5 or more should not have driving privileges, at what concentration of alcohol in the bloodstream should a driver be arrested and charged with a DUI (driving under the influence)?

Solution (a) For a concentration of alcohol in the blood of 0.02% and a relative risk of 1.4, we let $x = 0.02$ and $R = 1.4$ in the equation and solve for k.

$$R = e^{kx}$$
$$1.4 = e^{k(0.02)} \qquad \text{\textit{R = 1.4; x = 0.02}}$$
$$0.02k = \ln 1.4 = 0.336472 \qquad \text{\textit{Change to a logarithmic expression.}}$$
$$k = 16.82 \qquad \text{\textit{Solve for k.}}$$

(b) For a concentration of 0.17%, we have $x = 0.17$. Using $k = 16.82$ in the equation, we find the relative risk R to be

$$R = e^{kx} = e^{(16.82)(0.17)} = 17.5$$

For a concentration of alcohol in the blood of 0.17% the relative risk of an accident is about 17.5. That is, a person with a BAC of 0.17% is 17.5 times as likely to have a car accident as a person with no alcohol in the bloodstream.

(c) For a relative risk of 100, we have $R = 100$. Using $k = 16.82$ in the equation $R = e^{kx}$, we find the concentration x of alcohol in the blood obeys

$$100 = e^{16.82x} \qquad R = e^{kx}, R = 100; k = 16.82$$

$$16.82x = \ln 100 = 4.60517 \qquad \text{Change to a logarithmic expression.}$$

$$x = 0.27 \qquad \text{Solve for } x.$$

NOTE
A BAC of 0.30 results in a loss of consciousness in most people. ■

For a concentration of alcohol in the blood of 0.27% the relative risk of an accident is 100.

(d) For a relative risk of 5, we have $R = 5$. Using $k = 16.82$ in the equation $R = e^{kx}$, we find the concentration x of alcohol in the bloodstream obeys

$$5 = e^{16.82x}$$

$$16.82x = \ln 5 = 1.60944$$

$$x = 0.096$$

NOTE
Most states use 0.08 or 0.10 as the blood alcohol content at which a DUI citation is given. ■

A driver with a BAC of 0.096% or more should be arrested and charged with DUI. ■

Summary

Properties of the Logarithmic Function

$f(x) = \log_a x, \quad a > 1$
($y = \log_a x$ means $x = a^y$)

Domain: the interval $(0, \infty)$; range: the interval $(-\infty, \infty)$
x-intercept: 1; y-intercept: none; vertical asymptote: $x = 0$ (y-axis); increasing; one-to-one
See Figure 37(a) for a typical graph.

$f(x) = \log_a x, \quad 0 < a < 1$
($y = \log_a x$ means $x = a^y$)

Domain: the interval $(0, \infty)$; range: the interval $(-\infty, \infty)$
x-intercept: 1; y-intercept: none; vertical asymptote: $x = 0$ (y-axis); decreasing; one-to-one
See Figure 37(b) for a typical graph.

Figure 37

(a) $a > 1$ (b) $0 < a < 1$

4.4 Assess Your Understanding

'Are You Prepared?'

Answers are given at the end of these exercises. If you get a wrong answer, read the pages listed in red.

1. Solve the inequality: $3x - 7 \leq 8 - 2x$ (pp. A69–A70)

2. Solve the inequality: $x^2 - x - 6 > 0$ (pp.154–156)

3. Solve the inequality: $\dfrac{x - 1}{x + 4} > 0$ (pp. 233–235)

Concepts and Vocabulary

4. The domain of the logarithmic function $f(x) = \log_a x$ is _____.

5. The graph of every logarithmic function $f(x) = \log_a x$, $a > 0, a \neq 1$, passes through three points: _____, _____, and _____.

6. If the graph of a logarithmic function $f(x) = \log_a x$, $a > 0, a \neq 1$, is increasing, then its base must be larger than _____.

7. *True or False:* If $y = \log_a x$, then $y = a^x$.

8. *True or False:* The graph of $f(x) = \log_a x, a > 0, a \neq 1$, has an x-intercept equal to 1 and no y-intercept .

Skill Building

In Problems 9–20, change each exponential expression to an equivalent expression involving a logarithm.

9. $9 = 3^2$

10. $16 = 4^2$

11. $a^2 = 1.6$

12. $a^3 = 2.1$

13. $2^x = 7.2$

14. $3^x = 4.6$

15. $e^x = 8$

16. $e^{2.2} = M$

In Problems 17–24, change each logarithmic expression to an equivalent expression involving an exponent.

17. $\log_2 8 = 3$

18. $\log_3\left(\dfrac{1}{9}\right) = -2$

19. $\log_a 3 = 6$

20. $\log_b 4 = 2$

21. $\log_3 2 = x$

22. $\log_2 6 = x$

23. $\ln 4 = x$

24. $\ln x = 4$

In Problems 25–36, find the exact value of each logarithm without using a calculator.

25. $\log_2 1$

26. $\log_8 8$

27. $\log_5 25$

28. $\log_3\left(\dfrac{1}{9}\right)$

29. $\log_{1/2} 16$

30. $\log_{1/3} 9$

31. $\log_{10}\sqrt{10}$

32. $\log_5 \sqrt[3]{25}$

33. $\log_{\sqrt{2}} 4$

34. $\log_{\sqrt{3}} 9$

35. $\ln\sqrt{e}$

36. $\ln e^3$

In Problems 37–48, find the domain of each function.

37. $f(x) = \ln(x - 3)$

38. $g(x) = \ln(x - 1)$

39. $F(x) = \log_2 x^2$

40. $H(x) = \log_5 x^3$

41. $f(x) = 3 - 2\log_4\left[\dfrac{x}{2} - 5\right]$

42. $g(x) = 8 + 5\ln(2x + 3)$

43. $f(x) = \ln\left(\dfrac{1}{x + 1}\right)$

44. $g(x) = \ln\left(\dfrac{1}{x - 5}\right)$

45. $g(x) = \log_5\left(\dfrac{x + 1}{x}\right)$

46. $h(x) = \log_3\left(\dfrac{x}{x - 1}\right)$

47. $f(x) = \sqrt{\ln x}$

48. $g(x) = \dfrac{1}{\ln x}$

In Problems 49–56, use a calculator to evaluate each expression. Round your answer to three decimal places.

49. $\ln\dfrac{5}{3}$

50. $\dfrac{\ln 5}{3}$

51. $\dfrac{\ln\dfrac{10}{3}}{0.04}$

52. $\dfrac{\ln\dfrac{2}{3}}{-0.1}$

53. $\dfrac{\ln 4 + \ln 2}{\log 4 + \log 2}$

54. $\dfrac{\log 15 + \log 20}{\ln 15 + \ln 20}$

55. $\dfrac{2\ln 5 + \log 50}{\log 4 - \ln 2}$

56. $\dfrac{3\log 80 - \ln 5}{\log 5 + \ln 20}$

57. Find a so that the graph of $f(x) = \log_a x$ contains the point $(2, 2)$.

58. Find a so that the graph of $f(x) = \log_a x$ contains the point $\left(\dfrac{1}{2}, -4\right)$.

In Problems 59–62, graph each function and its inverse on the same Cartesian plane.

59. $f(x) = 3^x; f^{-1}(x) = \log_3 x$

60. $f(x) = 4^x; f^{-1}(x) = \log_4 x$

61. $f(x) = \left(\dfrac{1}{2}\right)^x; f^{-1}(x) = \log_{\frac{1}{2}} x$

62. $f(x) = \left(\dfrac{1}{3}\right)^x; f^{-1}(x) = \log_{\frac{1}{3}} x$

In Problems 63–70, the graph of a logarithmic function is given. Match each graph to one of the following functions:

A. $y = \log_3 x$ B. $y = \log_3(-x)$ C. $y = -\log_3 x$ D. $y = -\log_3(-x)$

E. $y = \log_3 x - 1$ F. $y = \log_3(x - 1)$ G. $y = \log_3(1 - x)$ H. $y = 1 - \log_3 x$

63. **64.** **65.** **66.**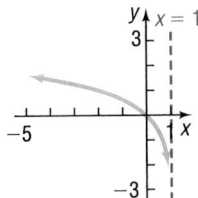

67. **68.** **69.** **70.**

In Problems 71–86, use transformations to graph each function. Determine the domain, the range, and the vertical asymptote of each function.

71. $f(x) = \ln(x + 4)$

72. $f(x) = \ln(x - 3)$

73. $f(x) = 2 + \ln x$

74. $f(x) = -\ln(-x)$

75. $g(x) = \ln(2x)$

76. $h(x) = \ln\left(\dfrac{1}{2}x\right)$

77. $f(x) = 3 \ln x$

78. $f(x) = -2 \ln x$

79. $f(x) = \log(x - 4)$

80. $f(x) = \log(x + 5)$

81. $h(x) = 4 \log_2 x$

82. $g(x) = -3 \log_2 x$

83. $F(x) = \log(2x)$

84. $G(x) = \log(5x)$

85. $h(x) = 3 + \log_3(x + 2)$

86. $g(x) = 2 - \log_3(x + 1)$

In Problems 87–106, solve each equation.

87. $\log_3 x = 2$

88. $\log_5 x = 3$

89. $\log_2(2x + 1) = 3$

90. $\log_3(3x - 2) = 2$

91. $\log_x 4 = 2$

92. $\log_x\left(\dfrac{1}{8}\right) = 3$

93. $\ln e^x = 5$

94. $\ln e^{-2x} = 8$

95. $\log_4 64 = x$

96. $\log_5 625 = x$

97. $\log_3 243 = 2x + 1$

98. $\log_6 36 = 5x + 3$

99. $e^{3x} = 10$

100. $e^{-2x} = \dfrac{1}{3}$

101. $e^{2x+5} = 8$

102. $e^{-2x+1} = 13$

103. $\log_3(x^2 + 1) = 2$

104. $\log_5(x^2 + x + 4) = 2$

105. $\log_2 8^x = -3$

106. $\log_3 3^x = -1$

107. Suppose that $G(x) = \log_3(2x + 1)$.
 (a) What is the domain of G?
 (b) What is $G(40)$? What point is on the graph of G?
 (c) If $G(x) = 2$, what is x? What point is on the graph of G?

108. Suppose that $F(x) = \log_2(x + 1) - 3$.
 (a) What is the domain of F?
 (b) What is $F(7)$? What point is on the graph of f?
 (c) If $F(x) = -1$, what is x? What point is on the graph of F?

In Problems 109–112, graph each function. Based on the graph, state the domain and the range and find any intercepts.

109. $f(x) = \begin{cases} \ln(-x) & \text{if } x < 0 \\ \ln x & \text{if } x > 0 \end{cases}$

110. $f(x) = \begin{cases} \ln(-x) & \text{if } x \le -1 \\ -\ln(-x) & \text{if } -1 < x < 0 \end{cases}$

111. $f(x) = \begin{cases} -\ln x & \text{if } 0 < x < 1 \\ \ln x & \text{if } x \ge 1 \end{cases}$

112. $f(x) = \begin{cases} \ln x & \text{if } 0 < x < 1 \\ -\ln x & \text{if } x \ge 1 \end{cases}$

In Problems 113–116, find the inverse of each function.

113. $f(x) = e^{2x+3}$ **114.** $f(x) = 3^{x-5} + 4$ **115.** $f(x) = \log_2(5x) - 3$ **116.** $f(x) = \log_5(x + 1) + 4$

Applications and Extensions

In Problems 117–120, (a) graph each function and state its domain, range, and asymptote; (b) determine the inverse function; (c) use the graph obtained in (a) to graph the inverse and state its domain, range, and asymptote.

117. $f(x) = 4^x$ **118.** $f(x) = 5^x$ **119.** $f(x) = 2^{x+3}$ **120.** $f(x) = 3^x - 2$

121. Chemistry The pH of a chemical solution is given by the formula

$$\text{pH} = -\log_{10}[H^+]$$

where $[H^+]$ is the concentration of hydrogen ions in moles per liter. Values of pH range from 0 (acidic) to 14 (alkaline).
(a) What is the pH of a solution for which $[H^+]$ is 0.1?
(b) What is the pH of a solution for which $[H^+]$ is 0.01?
(c) What is the pH of a solution for which $[H^+]$ is 0.001?
(d) What happens to pH as the hydrogen ion concentration decreases?
(e) Determine the hydrogen ion concentration of an orange (pH = 3.5).
(f) Determine the hydrogen ion concentration of human blood (pH = 7.4).

122. Diversity Index Shannon's diversity index is a measure of the diversity of a population. The diversity index is given by the formula

$$H = -(p_1 \log p_1 + p_2 \log p_2 + \cdots + p_n \log p_n)$$

where p_1 is the proportion of the population that is species 1, p_2 is the proportion of the population that is species 2, and so on.
(a) According to the U.S. Census Bureau, the distribution of race in the United States in 2000 was as follows:

Race	Proportion
American Indian or Native Alaskan	0.014
Asian	0.041
Black or African American	0.128
Hispanic	0.124
Native Hawaiian or Pacific Islander	0.003
White	0.690

SOURCE: U.S. Census Bureau

Compute the diversity index of the United States in 2000.
(b) The largest value of the diversity index is given by $H_{max} = \log(S)$, where S is the number of categories of race. Compute H_{max}.
(c) The **evenness ratio** is given by $E_H = \dfrac{H}{H_{max}}$, where $0 \le E_H \le 1$. If $E_H = 1$, there is complete evenness. Compute the evenness ratio for the United States.
(d) Obtain the distribution of race for the United States in 1990 from the Census Bureau. Compute Shannon's diversity index. Is the United States becoming more diverse? Why?

123. Atmospheric Pressure The atmospheric pressure p on a balloon or an aircraft decreases with increasing height. This pressure, measured in millimeters of mercury, is related to the height h (in kilometers) above sea level by the formula

$$p = 760e^{-0.145h}$$

(a) Find the height of an aircraft if the atmospheric pressure is 320 millimeters of mercury.
(b) Find the height of a mountain if the atmospheric pressure is 667 millimeters of mercury.

124. Healing of Wounds The normal healing of wounds can be modeled by an exponential function. If A_0 represents the original area of the wound and if A equals the area of the wound, then the formula

$$A = A_0 e^{-0.35n}$$

describes the area of a wound after n days following an injury when no infection is present to retard the healing. Suppose that a wound initially had an area of 100 square millimeters.
(a) If healing is taking place, after how many days will the wound be one-half its original size?
(b) How long before the wound is 10% of its original size?

125. Exponential Probability Between 12:00 PM and 1:00 PM, cars arrive at Citibank's drive-thru at the rate of 6 cars per hour (0.1 car per minute). The following formula from statistics can be used to determine the probability that a car will arrive within t minutes of 12:00 PM.

$$F(t) = 1 - e^{-0.1t}$$

(a) Determine how many minutes are needed for the probability to reach 50%.
(b) Determine how many minutes are needed for the probability to reach 80%.
(c) Is it possible for the probability to equal 100%? Explain.

126. Exponential Probability Between 5:00 PM and 6:00 PM, cars arrive at Jiffy Lube at the rate of 9 cars per hour (0.15 car per minute). The following formula from statistics can be used to determine the probability that a car will arrive within t minutes of 5:00 PM.

$$F(t) = 1 - e^{-0.15t}$$

(a) Determine how many minutes are needed for the probability to reach 50%.
(b) Determine how many minutes are needed for the probability to reach 80%.

127. Drug Medication The formula

$$D = 5e^{-0.4h}$$

can be used to find the number of milligrams D of a certain drug that is in a patient's bloodstream h hours after the drug has been administered. When the number of milligrams reaches 2, the drug is to be administered again. What is the time between injections?

128. Spreading of Rumors A model for the number N of people in a college community who have heard a certain rumor is

$$N = P(1 - e^{-0.15d})$$

where P is the total population of the community and d is the number of days that have elapsed since the rumor began. In a community of 1000 students, how many days will elapse before 450 students have heard the rumor?

129. Current in a *RL* Circuit The equation governing the amount of current I (in amperes) after time t (in seconds) in a simple *RL* circuit consisting of a resistance R (in ohms), an inductance L (in henrys), and an electromotive force E (in volts) is

$$I = \frac{E}{R}[1 - e^{-(R/L)t}]$$

If $E = 12$ volts, $R = 10$ ohms, and $L = 5$ henrys, how long does it take to obtain a current of 0.5 ampere? Of 1.0 ampere? Graph the equation.

130. Learning Curve Psychologists sometimes use the function

$$L(t) = A(1 - e^{-kt})$$

to measure the amount L learned at time t. The number A represents the amount to be learned, and the number k measures the rate of learning. Suppose that a student has an amount A of 200 vocabulary words to learn. A psychologist determines that the student learned 20 vocabulary words after 5 minutes.
(a) Determine the rate of learning k.
(b) Approximately how many words will the student have learned after 10 minutes?
(c) After 15 minutes?
(d) How long does it take for the student to learn 180 words?

Loudness of Sound *Problems 131–134 use the following discussion: The **loudness** $L(x)$, measured in decibels, of a sound of intensity x, measured in watts per square meter, is defined as $L(x) = 10 \log \dfrac{x}{I_0}$, where $I_0 = 10^{-12}$ watt per square meter is the least intense sound that a human ear can detect. Determine the loudness, in decibels, of each of the following sounds.*

131. Normal conversation: intensity of $x = 10^{-7}$ watt per square meter.

132. Heavy city traffic: intensity of $x = 10^{-3}$ watt per square meter.

133. Amplified rock music: intensity of 10^{-1} watt per square meter.

134. Diesel truck traveling 40 miles per hour 50 feet away: intensity 10 times that of a passenger car traveling 50 miles per hour 50 feet away whose loudness is 70 decibels.

*Problems 135 and 136 use the following discussion: The **Richter scale** is one way of converting seismographic readings into numbers that provide an easy reference for measuring the magnitude M of an earthquake. All earthquakes are compared to a **zero-level earthquake** whose seismographic reading measures 0.001 millimeter at a distance of 100 kilometers from the epicenter. An earthquake whose seismographic reading measures x millimeters has **magnitude** $M(x)$, given by*

$$M(x) = \log\left(\frac{x}{x_0}\right)$$

where $x_0 = 10^{-3}$ is the reading of a zero-level earthquake the same distance from its epicenter. Determine the magnitude of the following earthquakes.

135. Magnitude of an Earthquake Mexico City in 1985: seismographic reading of 125,892 millimeters 100 kilometers from the center.

136. Magnitude of an Earthquake San Francisco in 1906: seismographic reading of 7943 millimeters 100 kilometers from the center.

137. Alcohol and Driving The concentration of alcohol in a person's bloodstream is measurable. Suppose that the relative risk R of having an accident while driving a car can be modeled by the equation

$$R = e^{kx}$$

where x is the percent of concentration of alcohol in the bloodstream and k is a constant.

(a) Suppose that a concentration of alcohol in the bloodstream of 0.03 percent results in a relative risk of an accident of 1.4. Find the constant k in the equation.
(b) Using this value of k, what is the relative risk if the concentration is 0.17 percent?
(c) Using the same value of k, what concentration of alcohol corresponds to a relative risk of 100?
(d) If the law asserts that anyone with a relative risk of having an accident of 5 or more should not have driving privileges, at what concentration of alcohol in the bloodstream should a driver be arrested and charged with a DUI?
(e) Compare this situation with that of Example 10. If you were a lawmaker, which situation would you support? Give your reasons.

Discussion and Writing

138. Is there any function of the form $y = x^\alpha, 0 < \alpha < 1$, that increases more slowly than a logarithmic function whose base is greater than 1? Explain.

139. In the definition of the logarithmic function, the base a is not allowed to equal 1. Why?

140. Critical Thinking In buying a new car, one consideration might be how well the price of the car holds up over time. Different makes of cars have different depreciation rates. One way to compute a depreciation rate for a car is given here. Suppose that the current prices of a certain Mercedes automobile are as follows:

Age in Years					
New	**1**	**2**	**3**	**4**	**5**
$38,000	$36,600	$32,400	$28,750	$25,400	$21,200

Use the formula New = Old(e^{Rt}) to find R, the annual depreciation rate, for a specific time t. When might be the best time to trade in the car? Consult the NADA ("blue") book and compare two like models that you are interested in. Which has the better depreciation rate?

'Are You Prepared?' Answers

1. $x \leq 3$ **2.** $x < -2$ or $x > 3$ **3.** $x < -4$ or $x > 1$

4.5 Properties of Logarithms

OBJECTIVES 1 Work with the Properties of Logarithms
2 Write a Logarithmic Expression as a Sum or Difference of Logarithms
3 Write a Logarithmic Expression as a Single Logarithm
4 Evaluate Logarithms Whose Base Is Neither 10 nor e

1 Work with the Properties of Logarithms

Logarithms have some very useful properties that can be derived directly from the definition and the laws of exponents.

EXAMPLE 1 **Establishing Properties of Logarithms**

(a) Show that $\log_a 1 = 0$. (b) Show that $\log_a a = 1$.

Solution (a) This fact was established when we graphed $y = \log_a x$ (see Figure 30). To show the result algebraically, let $y = \log_a 1$. Then

$$y = \log_a 1$$
$$a^y = 1 \qquad \text{Change to an exponent.}$$
$$a^y = a^0 \qquad a^0 = 1$$
$$y = 0 \qquad \text{Solve for y.}$$
$$\log_a 1 = 0 \qquad y = \log_a 1$$

(b) Let $y = \log_a a$. Then

$$y = \log_a a$$
$$a^y = a \qquad \text{Change to an exponent.}$$
$$a^y = a^1 \qquad a = a^1$$
$$y = 1 \qquad \text{Solve for y.}$$
$$\log_a a = 1 \qquad y = \log_a a$$ ∎

To summarize:

$$\log_a 1 = 0 \qquad \log_a a = 1$$

Theorem **Properties of Logarithms**

In the properties given next, M and a are positive real numbers, $a \neq 1$, and r is any real number.

(continued)

The number $\log_a M$ is the exponent to which a must be raised to obtain M. That is,

$$a^{\log_a M} = M \qquad (1)$$

The logarithm to the base a of a raised to a power equals that power. That is,

$$\log_a a^r = r \qquad (2)$$

The proof uses the fact that $y = a^x$ and $y = \log_a x$ are inverses.

Proof of Property (1) For inverse functions,

$$f(f^{-1}(x)) = x \quad \text{for all } x \text{ in the domain of } f^{-1}$$

Using $f(x) = a^x$ and $f^{-1}(x) = \log_a x$, we find

$$f(f^{-1}(x)) = a^{\log_a x} = x \quad \text{for} \quad x > 0$$

Now let $x = M$ to obtain $a^{\log_a M} = M$, where $M > 0$. ■

Proof of Property (2) For inverse functions,

$$f^{-1}(f(x)) = x \quad \text{for all } x \text{ in the domain of } f.$$

Using $f(x) = a^x$ and $f^{-1}(x) = \log_a x$, we find

$$f^{-1}(f(x)) = \log_a a^x = x \quad \text{for all real numbers } x.$$

Now let $x = r$ to obtain $\log_a a^r = r$, where r is any real number. ■

EXAMPLE 2 **Using Properties (1) and (2)**

(a) $2^{\log_2 \pi} = \pi$ (b) $\log_{0.2} 0.2^{-\sqrt{2}} = -\sqrt{2}$ (c) $\ln e^{kt} = kt$ ■

NOW WORK PROBLEM 9.

Other useful properties of logarithms are given next.

Theorem **Properties of Logarithms**

In the following properties, M, N, and a are positive real numbers, $a \neq 1$, and r is any real number.

The Log of a Product Equals the Sum of the Logs

$$\log_a(MN) = \log_a M + \log_a N \qquad (3)$$

The Log of a Quotient Equals the Difference of the Logs

$$\log_a\left(\frac{M}{N}\right) = \log_a M - \log_a N \qquad (4)$$

The Log of a Power Equals the Product of the Power and the Log

$$\log_a M^r = r \log_a M \qquad (5)$$

We shall derive properties (3) and (5) and leave the derivation of property (4) as an exercise (see Problem 103).

Proof of Property (3) Let $A = \log_a M$ and let $B = \log_a N$. These expressions are equivalent to the exponential expressions

$$a^A = M \quad \text{and} \quad a^B = N$$

Now

$$\log_a(MN) = \log_a(a^A a^B) = \log_a a^{A+B} \qquad \text{Law of Exponents}$$

$$= A + B \qquad \text{Property (2) of logarithms}$$

$$= \log_a M + \log_a N$$

∎

Proof of Property (5) Let $A = \log_a M$. This expression is equivalent to

$$a^A = M$$

Now

$$\log_a M^r = \log_a(a^A)^r = \log_a a^{rA} \qquad \text{Law of Exponents}$$

$$= rA \qquad \text{Property (2) of logarithms}$$

$$= r \log_a M$$

∎

➤ **NOW WORK PROBLEM 13.**

2 Write a Logarithmic Expression as a Sum or Difference of Logarithms

 Logarithms can be used to transform products into sums, quotients into differences, and powers into factors. Such transformations prove useful in certain types of calculus problems.

EXAMPLE 3 **Writing a Logarithmic Expression as a Sum of Logarithms**

Write $\log_a\left(x\sqrt{x^2 + 1}\right)$, $x > 0$, as a sum of logarithms. Express all powers as factors.

Solution $\log_a\left(x\sqrt{x^2 + 1}\right) = \log_a x + \log_a \sqrt{x^2 + 1} \qquad \log_a(M \cdot N) = \log_a M + \log_a N$

$$= \log_a x + \log_a(x^2 + 1)^{1/2}$$

$$= \log_a x + \frac{1}{2}\log_a(x^2 + 1) \qquad \log_a M^r = r \log_a M$$

∎

EXAMPLE 4 **Writing a Logarithmic Expression as a Difference of Logarithms**

Write

$$\ln \frac{x^2}{(x - 1)^3}, \qquad x > 1$$

as a difference of logarithms. Express all powers as factors.

Solution $\ln \dfrac{x^2}{(x - 1)^3} = \ln x^2 - \ln(x - 1)^3 = 2 \ln x - 3 \ln(x - 1)$

\uparrow $\qquad\qquad\qquad\qquad\qquad \uparrow$

$\log_a\left(\dfrac{M}{N}\right) = \log_a M - \log_a N \qquad \log_a M^r = r \log_a M$

∎

EXAMPLE 5 **Writing a Logarithmic Expression as a Sum and Difference of Logarithms**

Write

$$\log_a \frac{\sqrt{x^2 + 1}}{x^3(x + 1)^4}, \qquad x > 0$$

as a sum and difference of logarithms. Express all powers as factors.

WARNING
In using properties (3) through (5), be careful about the values that the variable may assume. For example, the domain of the variable for $\log_a x$ is $x > 0$ and for $\log_a(x - 1)$ it is $x > 1$. If we add these functions, the domain is $x > 1$. That is, the equality

$$\log_a x + \log_a(x - 1) = \log_a[x(x - 1)]$$

is true only for $x > 1$. ∎

Solution

$$\log_a \frac{\sqrt{x^2 + 1}}{x^3(x + 1)^4} = \log_a \sqrt{x^2 + 1} - \log_a[x^3(x + 1)^4] \qquad \text{Property (4)}$$

$$= \log_a \sqrt{x^2 + 1} - [\log_a x^3 + \log_a(x + 1)^4] \qquad \text{Property (3)}$$

$$= \log_a(x^2 + 1)^{1/2} - \log_a x^3 - \log_a(x + 1)^4$$

$$= \frac{1}{2}\log_a(x^2 + 1) - 3\log_a x - 4\log_a(x + 1) \qquad \text{Property (5)}$$

∎

━━━ **NOW WORK PROBLEM 45.**

3 Write a Logarithmic Expression as a Single Logarithm

Another use of properties (3) through (5) is to write sums and/or differences of logarithms with the same base as a single logarithm. This skill will be needed to solve certain logarithmic equations discussed in the next section.

| **EXAMPLE 6** | **Writing Expressions as a Single Logarithm** |

Write each of the following as a single logarithm.

(a) $\log_a 7 + 4\log_a 3$ (b) $\frac{2}{3}\ln 8 - \ln(3^4 - 8)$

(c) $\log_a x + \log_a 9 + \log_a(x^2 + 1) - \log_a 5$

Solution

(a) $\log_a 7 + 4\log_a 3 = \log_a 7 + \log_a 3^4 \qquad r\log_a M = \log_a M^r$

$$= \log_a 7 + \log_a 81$$

$$= \log_a(7 \cdot 81) \qquad \log_a M + \log_a N = \log_a(M \cdot N)$$

$$= \log_a 567$$

(b) $\frac{2}{3}\ln 8 - \ln(3^4 - 8) = \ln 8^{2/3} - \ln(81 - 8) \qquad r\log_a M = \log_a M^r$

$$= \ln 4 - \ln 73$$

$$= \ln\left(\frac{4}{73}\right) \qquad \log_a M - \log_a N = \log_a\left(\frac{M}{N}\right)$$

(c) $\log_a x + \log_a 9 + \log_a(x^2 + 1) - \log_a 5 = \log_a(9x) + \log_a(x^2 + 1) - \log_a 5$

$$= \log_a[9x(x^2 + 1)] - \log_a 5$$

$$= \log_a\left[\frac{9x(x^2 + 1)}{5}\right]$$

∎

WARNING A common error made by some students is to express the logarithm of a sum as the sum of logarithms.

$$\log_a(M + N) \quad \text{is not equal to} \quad \log_a M + \log_a N$$

Correct statement $\log_a(MN) = \log_a M + \log_a N$ Property (3)

Another common error is to express the difference of logarithms as the quotient of logarithms.

$$\log_a M - \log_a N \quad \text{is not equal to} \quad \frac{\log_a M}{\log_a N}$$

Correct statement $\log_a M - \log_a N = \log_a\left(\frac{M}{N}\right)$ Property (4)

A third common error is to express a logarithm raised to a power as the product of the power times the logarithm.

$$(\log_a M)^r \quad \text{is not equal to} \quad r\log_a M$$

Correct statement $\log_a M^r = r\log_a M$ Property (5)

∎

━━━ **NOW WORK PROBLEM 51.**

Two other properties of logarithms that we need to know are consequences of the fact that the logarithmic function $y = \log_a x$ is one-to-one.

Theorem

> **Properties of Logarithms**
>
> In the following properties, M, N, and a are positive real numbers, $a \neq 1$.
>
> | If $M = N$, then $\log_a M = \log_a N$. | (6) |
> | If $\log_a M = \log_a N$, then $M = N$. | (7) |

When property (6) is used, we start with the equation $M = N$ and say "take the logarithm of both sides" to obtain $\log_a M = \log_a N$.

Properties (6) and (7) are useful for solving *exponential and logarithmic equations*, a topic discussed in the next section.

4 Evaluate Logarithms Whose Base Is Neither 10 nor e

Logarithms to the base 10, common logarithms, were used to facilitate arithmetic computations before the widespread use of calculators. (See the Historical Feature at the end of this section.) Natural logarithms, that is, logarithms whose base is the number e, remain very important because they arise frequently in the study of natural phenomena.

Common logarithms are usually abbreviated by writing **log**, with the base understood to be 10, just as natural logarithms are abbreviated by **ln**, with the base understood to be e.

Most calculators have both $\boxed{\log}$ and $\boxed{\ln}$ keys to calculate the common logarithm and natural logarithm of a number. Let's look at an example to see how to approximate logarithms having a base other than 10 or e.

EXAMPLE 7 **Approximating Logarithms Whose Base Is Neither 10 nor e**

Approximate $\log_2 7$. Round the answer to four decimal places.

Solution Let $y = \log_2 7$. Then $2^y = 7$, so

$$2^y = 7$$
$$\ln 2^y = \ln 7 \qquad \text{Property (6)}$$
$$y \ln 2 = \ln 7 \qquad \text{Property (5)}$$
$$y = \frac{\ln 7}{\ln 2} \qquad \text{Exact value}$$
$$y \approx 2.8074 \qquad \text{Approximate value rounded to four decimal places} \quad \blacksquare$$

Example 7 shows how to approximate a logarithm whose base is 2 by changing to logarithms involving the base e. In general, we use the **Change-of-Base Formula**.

Theorem

> **Change-of-Base Formula**
>
> If $a \neq 1$, $b \neq 1$, and M are positive real numbers, then
>
> $$\log_a M = \frac{\log_b M}{\log_b a} \qquad (8)$$

Proof We derive this formula as follows: Let $y = \log_a M$. Then

$$a^y = M$$

$$\log_b a^y = \log_b M \qquad \text{Property (6)}$$

$$y \log_b a = \log_b M \qquad \text{Property (5)}$$

$$y = \frac{\log_b M}{\log_b a} \qquad \text{Solve for } y.$$

$$\log_a M = \frac{\log_b M}{\log_b a} \qquad y = \log_a M \qquad \blacksquare$$

Since calculators have keys only for $\boxed{\log}$ and $\boxed{\ln}$, in practice, the Change-of-Base Formula uses either $b = 10$ or $b = e$. That is,

$$\log_a M = \frac{\log M}{\log a} \quad \text{and} \quad \log_a M = \frac{\ln M}{\ln a} \qquad \textbf{(9)}$$

EXAMPLE 8	**Using the Change-of-Base Formula**

Approximate: (a) $\log_5 89$

(b) $\log_{\sqrt{2}} \sqrt{5}$

Round answers to four decimal places.

Solution (a) $\log_5 89 = \dfrac{\log 89}{\log 5} \approx \dfrac{1.949390007}{0.6989700043} \approx 2.7889$

or

$$\log_5 89 = \frac{\ln 89}{\ln 5} \approx \frac{4.48863637}{1.609437912} \approx 2.7889$$

(b) $\log_{\sqrt{2}} \sqrt{5} = \dfrac{\log \sqrt{5}}{\log \sqrt{2}} = \dfrac{\frac{1}{2} \log 5}{\frac{1}{2} \log 2} \approx 2.3219$

or

$$\log_{\sqrt{2}} \sqrt{5} = \frac{\ln \sqrt{5}}{\ln \sqrt{2}} = \frac{\frac{1}{2} \ln 5}{\frac{1}{2} \ln 2} \approx 2.3219 \qquad \blacksquare$$

━━━ NOW WORK PROBLEMS **17** AND **65**.

COMMENT: To graph logarithmic functions when the base is different from e or 10 requires the Change-of-Base Formula. For example, to graph $y = \log_2 x$, we would instead graph $y = \dfrac{\ln x}{\ln 2}$. Try it. ∎

━━━ NOW WORK PROBLEM **73**.

Summary

Properties of Logarithms

In the list that follows, $a > 0$, $a \neq 1$, $b > 0$, and $b \neq 1$; also, $M > 0$ and $N > 0$.

Definition $\qquad\qquad\qquad\qquad y = \log_a x$ means $x = a^y$

Properties of logarithms $\qquad\quad \log_a 1 = 0; \quad \log_a a = 1 \qquad\qquad\qquad \log_a M^r = r \log_a M$

$\qquad\qquad\qquad\qquad\qquad\qquad\quad a^{\log_a M} = M; \quad \log_a a^r = r \qquad\qquad$ If $M = N$, then $\log_a M = \log_a N$.

$\qquad\qquad\qquad\qquad\qquad\qquad\quad \log_a(MN) = \log_a M + \log_a N \qquad$ If $\log_a M = \log_a N$, then $M = N$.

$$\log_a\left(\frac{M}{N}\right) = \log_a M - \log_a N$$

Change-of-Base Formula $\qquad\qquad \log_a M = \dfrac{\log_b M}{\log_b a}$

HISTORICAL FEATURE

John Napier
(1550–1617)

Logarithms were invented about 1590 by John Napier (1550–1617) and Joost Bürgi (1552–1632), working independently. Napier, whose work had the greater influence, was a Scottish lord, a secretive man whose neighbors were inclined to believe him to be in league with the devil. His approach to logarithms was very different from ours; it was based on the relationship between arithmetic and geometric sequences, discussed in a later chapter, and not on the inverse function relationship of logarithms to exponential functions (described in Section 4.4). Napier's tables, pub-

lished in 1614, listed what would now be called *natural logarithms* of sines and were rather difficult to use. A London professor, Henry Briggs, became interested in the tables and visited Napier. In their conversations, they developed the idea of common logarithms, which were published in 1617. Their importance for calculation was immediately recognized, and by 1650 they were being printed as far away as China. They remained an important calculation tool until the advent of the inexpensive handheld calculator about 1972, which has decreased their calculational, but not their theoretical, importance.

A side effect of the invention of logarithms was the popularization of the decimal system of notation for real numbers.

4.5 Assess Your Understanding

Concepts and Vocabulary

1. The logarithm of a product equals the _____ of the logarithms.

2. If $\log_8 M = \dfrac{\log_5 7}{\log_5 8}$, then $M = $ _____.

3. $\log_a M^r = $ _____.

4. *True or False:* $\ln(x + 3) - \ln(2x) = \dfrac{\ln(x + 3)}{\ln(2x)}$

5. *True or False:* $\log_2(3x^4) = 4 \log_2(3x)$

6. *True or False:* $\log_2 16 = \dfrac{\ln 16}{\ln 2}$

Skill Building

In Problems 7–22, use properties of logarithms to find the exact value of each expression. Do not use a calculator.

7. $\log_3 3^{71}$ **8.** $\log_2 2^{-13}$ **9.** $\ln e^{-4}$ **10.** $\ln e^{\sqrt{2}}$

11. $2^{\log_2 7}$ **12.** $e^{\ln 8}$ **13.** $\log_8 2 + \log_8 4$ **14.** $\log_6 9 + \log_6 4$

15. $\log_6 18 - \log_6 3$ **16.** $\log_8 16 - \log_8 2$ **17.** $\log_2 6 \cdot \log_6 4$ **18.** $\log_3 8 \cdot \log_8 9$

19. $3^{\log_3 5 - \log_3 4}$ **20.** $5^{\log_5 6 + \log_5 7}$ **21.** $e^{\log_{e^2} 16}$ **22.** $e^{\log_{e^2} 9}$

In Problems 23–30, suppose that $\ln 2 = a$ and $\ln 3 = b$. Use properties of logarithms to write each logarithm in terms of a and b.

23. $\ln 6$ **24.** $\ln \dfrac{2}{3}$ **25.** $\ln 1.5$ **26.** $\ln 0.5$

27. $\ln 8$ **28.** $\ln 27$ **29.** $\ln \sqrt[5]{6}$ **30.** $\ln \sqrt[4]{\dfrac{2}{3}}$

In Problems 31–50, write each expression as a sum and/or difference of logarithms. Express powers as factors.

31. $\log_5(25x)$ **32.** $\log_3 \dfrac{x}{9}$ **33.** $\log_2 z^3$ **34.** $\log_7(x^5)$

35. $\ln(ex)$　　　　　　**36.** $\ln \dfrac{e}{x}$　　　　　　**37.** $\ln(xe^x)$　　　　　　**38.** $\ln \dfrac{x}{e^x}$

39. $\log_a(u^2 v^3)$, $u > 0, v > 0$　**40.** $\log_2\left(\dfrac{a}{b^2}\right)$, $a > 0, b > 0$　**41.** $\ln\left(x^2\sqrt{1-x}\right)$, $0 < x < 1$　**42.** $\ln\left(x\sqrt{1+x^2}\right)$, $x > 0$

43. $\log_2\left(\dfrac{x^3}{x-3}\right)$, $x > 3$　　**44.** $\log_5\left(\dfrac{\sqrt[3]{x^2+1}}{x^2-1}\right)$, $x > 1$　　**45.** $\log\left[\dfrac{x(x+2)}{(x+3)^2}\right]$, $x > 0$　**46.** $\log\left[\dfrac{x^3\sqrt{x+1}}{(x-2)^2}\right]$, $x > 2$

47. $\ln\left[\dfrac{x^2-x-2}{(x+4)^2}\right]^{1/3}$, $x > 2$　　　　　　　　　　　**48.** $\ln\left[\dfrac{(x-4)^2}{x^2-1}\right]^{2/3}$, $x > 4$

49. $\ln\dfrac{5x\sqrt{1+3x}}{(x-4)^3}$, $x > 4$　　　　　　　　　　　　**50.** $\ln\left[\dfrac{5x^2\sqrt[3]{1-x}}{4(x+1)^2}\right]$, $0 < x < 1$

In Problems 51–64, write each expression as a single logarithm.

51. $3\log_5 u + 4\log_5 v$　　　　　　**52.** $2\log_3 u - \log_3 v$　　　　　　**53.** $\log_3 \sqrt{x} - \log_3 x^3$

54. $\log_2\left(\dfrac{1}{x}\right) + \log_2\left(\dfrac{1}{x^2}\right)$　　**55.** $\log_4(x^2-1) - 5\log_4(x+1)$　　**56.** $\log(x^2+3x+2) - 2\log(x+1)$

57. $\ln\left(\dfrac{x}{x-1}\right) + \ln\left(\dfrac{x+1}{x}\right) - \ln(x^2-1)$　**58.** $\log\left(\dfrac{x^2+2x-3}{x^2-4}\right) - \log\left(\dfrac{x^2+7x+6}{x+2}\right)$　**59.** $8\log_2\sqrt{3x-2} - \log_2\left(\dfrac{4}{x}\right) + \log_2 4$

60. $21\log_3\sqrt[3]{x} + \log_3(9x^2) - \log_3 9$　　**61.** $2\log_a(5x^3) - \dfrac{1}{2}\log_a(2x+3)$　　**62.** $\dfrac{1}{3}\log(x^3+1) + \dfrac{1}{2}\log(x^2+1)$

63. $2\log_2(x+1) - \log_2(x+3) - \log_2(x-1)$　　　　　**64.** $3\log_5(3x+1) - 2\log_5(2x-1) - \log_5 x$

In Problems 65–72, use the Change-of-Base Formula and a calculator to evaluate each logarithm. Round your answer to three decimal places.

65. $\log_3 21$　　　　　**66.** $\log_5 18$　　　　　**67.** $\log_{1/3} 71$　　　　　**68.** $\log_{1/2} 15$

69. $\log_{\sqrt{2}} 7$　　　　　**70.** $\log_{\sqrt{5}} 8$　　　　　**71.** $\log_\pi e$　　　　　**72.** $\log_\pi \sqrt{2}$

In Problems 73–78, graph each function using a graphing utility and the Change-of-Base Formula.

73. $y = \log_4 x$　　　　　**74.** $y = \log_5 x$　　　　　**75.** $y = \log_2(x+2)$　　　　**76.** $y = \log_4(x-3)$

77. $y = \log_{x-1}(x+1)$　　　　　　　　　　　**78.** $y = \log_{x+2}(x-2)$

Applications and Extensions

79. If $f(x) = \ln x$, $g(x) = e^x$, and $h(x) = x^2$, find:
(a) $(f \circ g)(x)$. What is the domain of $f \circ g$?
(b) $(g \circ f)(x)$. What is the domain of $g \circ f$?
(c) $(f \circ g)(5)$
(d) $(f \circ h)(x)$
(e) $(f \circ h)(e)$

80. If $f(x) = \log_2 x$, $g(x) = 2^x$, and $h(x) = 4x$, find:
(a) $(f \circ g)(x)$. What is the domain of $f \circ g$?
(b) $(g \circ f)(x)$. What is the domain of $g \circ f$?
(c) $(f \circ g)(3)$
(d) $(f \circ h)(x)$
(e) $(f \circ h)(8)$

In Problems 81–90, express y as a function of x. The constant C is a positive number.

81. $\ln y = \ln x + \ln C$　　　　　　　　　**82.** $\ln y = \ln(x + C)$

83. $\ln y = \ln x + \ln(x+1) + \ln C$　　　　**84.** $\ln y = 2\ln x - \ln(x+1) + \ln C$

85. $\ln y = 3x + \ln C$　　　　　　　　　　**86.** $\ln y = -2x + \ln C$

87. $\ln(y-3) = -4x + \ln C$　　　　　　　**88.** $\ln(y+4) = 5x + \ln C$

89. $3\ln y = \dfrac{1}{2}\ln(2x+1) - \dfrac{1}{3}\ln(x+4) + \ln C$　　**90.** $2\ln y = -\dfrac{1}{2}\ln x + \dfrac{1}{3}\ln(x^2+1) + \ln C$

91. Find the value of $\log_2 3 \cdot \log_3 4 \cdot \log_4 5 \cdot \log_5 6 \cdot \log_6 7 \cdot \log_7 8$.

92. Find the value of $\log_2 4 \cdot \log_4 6 \cdot \log_6 8$.

93. Find the value of $\log_2 3 \cdot \log_3 4 \cdot \cdots \cdot \log_n(n+1) \cdot \log_{n+1} 2$.

94. Find the value of $\log_2 2 \cdot \log_2 4 \cdot \cdots \cdot \log_2 2^n$.

95. Show that $\log_a\left(x + \sqrt{x^2-1}\right) + \log_a\left(x - \sqrt{x^2-1}\right) = 0$.

96. Show that $\log_a\left(\sqrt{x} + \sqrt{x-1}\right) + \log_a\left(\sqrt{x} - \sqrt{x-1}\right) = 0$.

97. Show that $\ln(1 + e^{2x}) = 2x + \ln(1 + e^{-2x})$.

98. Difference Quotient If $f(x) = \log_a x$, show that $\dfrac{f(x+h) - f(x)}{h} = \log_a\left(1 + \dfrac{h}{x}\right)^{1/h}$, $h \neq 0$.

99. If $f(x) = \log_a x$, show that $-f(x) = \log_{1/a} x$.

100. If $f(x) = \log_a x$, show that $f(AB) = f(A) + f(B)$.

101. If $f(x) = \log_a x$, show that $f\left(\dfrac{1}{x}\right) = -f(x)$.

102. If $f(x) = \log_a x$, show that $f(x^\alpha) = \alpha f(x)$.

103. Show that $\log_a\left(\dfrac{M}{N}\right) = \log_a M - \log_a N$, where a, M, and N are positive real numbers, and $a \neq 1$.

104. Show that $\log_a\left(\dfrac{1}{N}\right) = -\log_a N$, where a and N are positive real numbers, and $a \neq 1$.

Discussion and Writing

105. Graph $Y_1 = \log(x^2)$ and $Y_2 = 2\log(x)$ using a graphing utility. Are they equivalent? What might account for any differences in the two functions?

106. Write an example that illustrates why $(\log_a x)^r \neq r\log_a x$.

107. Write an example that illustrates why $\log_2(x + y) \neq \log_2 x + \log_2 y$.

108. Does $3^{\log_3(-5)} = -5$? Why or why not?

4.6 Logarithmic and Exponential Equations

PREPARING FOR THIS SECTION *Before getting started, review the following:*

- Solving Equations Using a Graphing Utility (Appendix B, Section B.4, pp. B9–B11)
- Solving Quadratic Equations (Section 2.3, pp. 133–142)

Now work the 'Are You Prepared?' problems on page 323.

OBJECTIVES
1. Solve Logarithmic Equations Using the Properties of Logarithms
2. Solve Exponential Equations
3. Solve Logarithmic and Exponential Equations Using a Graphing Utility

1 Solve Logarithmic Equations Using the Properties of Logarithms

In Section 4.4 we solved logarithmic equations by changing a logarithm to exponential form. Often, however, some manipulation of the equation (usually using the properties of logarithms) is required before we can change to exponential form.

Our practice will be to solve equations, whenever possible, by finding exact solutions using algebraic methods. When algebraic methods cannot be used, approximate solutions will be obtained using a graphing utility. The reader is encouraged to pay particular attention to the form of equations for which exact solutions are possible.

EXAMPLE 1 | **Solving a Logarithmic Equation**

Solve: $2\log_5 x = \log_5 9$

Solution Because each logarithm is to the same base, 5, we can obtain an exact solution as follows:

$$2\log_5 x = \log_5 9$$
$$\log_5 x^2 = \log_5 9 \qquad \log_a M^r = r\log_a M$$
$$x^2 = 9 \qquad \text{If } \log_a M = \log_a N, \text{ then } M = N.$$
$$x = 3 \quad \text{or} \quad \cancel{x = -3} \qquad \text{Recall that logarithms of negative numbers are not defined, so, in the expression } 2\log_5 x, x \text{ must be positive. Therefore, } -3 \text{ is extraneous and we discard it.}$$

✔ **CHECK:** $2\log_5 3 \overset{?}{=} \log_5 9$
$$\log_5 3^2 \overset{?}{=} \log_5 9 \qquad r\log_a M = \log_a M^r$$
$$\log_5 9 = \log_5 9$$

The solution set is $\{3\}$.

◀────── NOW WORK PROBLEM 9.

EXAMPLE 2 **Solving a Logarithmic Equation**

Solve: $\log_4(x + 3) + \log_4(2 - x) = 1$

Solution To obtain an exact solution, we need to express the left side as a single logarithm. Then we will change the expression to exponential form.

$$\log_4(x + 3) + \log_4(2 - x) = 1$$
$$\log_4[(x + 3)(2 - x)] = 1 \qquad \log_a M + \log_a N = \log_a(MN)$$
$$(x + 3)(2 - x) = 4^1 = 4 \qquad \text{Change to an exponential expression.}$$
$$-x^2 - x + 6 = 4 \qquad \text{Simplify.}$$
$$x^2 + x - 2 = 0 \qquad \text{Place the quadratic equation in standard form.}$$
$$(x + 2)(x - 1) = 0 \qquad \text{Factor.}$$
$$x = -2 \quad \text{or} \quad x = 1 \qquad \text{Zero-Product Property}$$

WARNING:
A negative solution is not automatically extraneous. You must determine whether the potential solution causes the argument of any logarithmic expression in the equation to be negative. ■

Since the arguments of each logarithmic expression in the equation are positive for both $x = -2$ and $x = 1$, neither is extraneous. We leave the check to you. The solution set is $\{-2, 1\}$. ■

> **NOW WORK PROBLEM 13.**

2 Solve Exponential Equations

In Sections 4.3 and 4.4, we solved certain exponential equations algebraically by expressing each side of the equation with the same base. However, many exponential equations cannot be rewritten so that each side has the same base. In such cases, algebraic techniques can sometimes be used to obtain a solution.

EXAMPLE 3 **Solving an Exponential Equation**

Solve: $4^x - 2^x - 12 = 0$

Solution We note that $4^x = (2^2)^x = 2^{2x} = (2^x)^2$, so the equation is actually quadratic in form, and we can rewrite it as

$$(2^x)^2 - 2^x - 12 = 0 \qquad \text{Let } u = 2^x; \text{ then } u^2 - u - 12 = 0.$$

Now we can factor as usual.

$$(2^x - 4)(2^x + 3) = 0 \qquad (u - 4)(u + 3) = 0$$
$$2^x - 4 = 0 \quad \text{or} \quad 2^x + 3 = 0 \qquad u - 4 = 0 \text{ or } u + 3 = 0$$
$$2^x = 4 \qquad\qquad 2^x = -3 \qquad u = 2^x = 4 \qquad u = 2^x = -3$$

The equation on the left has the solution $x = 2$, since $2^x = 4 = 2^2$; the equation on the right has no solution, since $2^x > 0$ for all x. The only solution is 2. The solution set is $\{2\}$. ■

In the solution of the previous example, we were able to write the exponential expression using the same base after utilizing some algebra, obtaining an exact solution to the equation. When this is not possible, logarithms can sometimes be used to obtain the solution.

EXAMPLE 4	**Solving an Exponential Equation**

Solve: $2^x = 5$

Solution Since 5 cannot be written as an integral power of 2, we write the exponential equation as the equivalent logarithmic equation.

$$2^x = 5$$

$$x = \log_2 5 = \underset{\uparrow}{\frac{\ln 5}{\ln 2}}$$

 Change-of-Base Formula (9), Section 4.5

Alternatively, we can solve the equation $2^x = 5$ by taking the natural logarithm (or common logarithm) of each side. Taking the natural logarithm,

$$2^x = 5$$
$$\ln 2^x = \ln 5 \qquad \text{If } M = N, \text{ then } \ln M = \ln N.$$
$$x \ln 2 = \ln 5 \qquad \ln M^r = r \ln M$$
$$x = \frac{\ln 5}{\ln 2} \qquad \text{Exact solution}$$
$$\approx 2.322 \qquad \text{Approximate solution}$$

The solution set is $\left\{ \dfrac{\ln 5}{\ln 2} \right\}$. ∎

➤ **NOW WORK PROBLEM 23.**

EXAMPLE 5	**Solving an Exponential Equation**

Solve: $8 \cdot 3^x = 5$

Solution

$$8 \cdot 3^x = 5$$

$$3^x = \frac{5}{8} \qquad\qquad \text{Solve for } 3^x.$$

$$x = \log_3\left(\frac{5}{8}\right) = \frac{\ln\left(\frac{5}{8}\right)}{\ln 3} \qquad \text{Exact solution}$$

$$\approx -0.428 \qquad \text{Approximate solution}$$

The solution set is $\left\{ \dfrac{\ln\left(\frac{5}{8}\right)}{\ln 3} \right\}$. ∎

EXAMPLE 6	**Solving an Exponential Equation**

Solve: $5^{x-2} = 3^{3x+2}$

Solution Because the bases are different, we first apply Property (6), Section 4.5 (take the natural logarithm of each side), and then use appropriate properties of logarithms. The result is an equation in x that we can solve.

$$5^{x-2} = 3^{3x+2}$$
$$\ln 5^{x-2} = \ln 3^{3x+2} \qquad\qquad \text{If } M = N, \ln M = \ln N.$$
$$(x - 2)\ln 5 = (3x + 2)\ln 3 \qquad\qquad \ln M^r = r \ln M$$
$$(\ln 5)x - 2\ln 5 = (3\ln 3)x + 2\ln 3 \qquad\qquad \text{Distribute.}$$
$$(\ln 5)x - (3\ln 3)x = 2\ln 3 + 2\ln 5 \qquad\qquad \text{Place terms involving } x \text{ on the left.}$$
$$(\ln 5 - 3\ln 3)x = 2(\ln 3 + \ln 5) \qquad\qquad \text{Factor.}$$
$$x = \frac{2(\ln 3 + \ln 5)}{\ln 5 - 3\ln 3} \qquad\qquad \text{Exact solution}$$
$$\approx -3.212 \qquad\qquad \text{Approximate solution}$$

The solution set is $\left\{\dfrac{2(\ln 3 + \ln 5)}{\ln 5 - 3\ln 3}\right\}$. ∎

NOW WORK PROBLEM 31.

 3 Solve Logarithmic and Exponential Equations Using a Graphing Utility

The algebraic techniques introduced in this section to obtain exact solutions apply only to certain types of logarithmic and exponential equations. Solutions for other types are usually studied in calculus, using numerical methods. For such types, we can use a graphing utility to approximate the solution.

EXAMPLE 7	**Solving Equations Using a Graphing Utility**

Figure 38

Solve: $x + e^x = 2$
Express the solution(s) rounded to two decimal places.

Solution The solution is found by graphing $Y_1 = x + e^x$ and $Y_2 = 2$. Y_1 is an increasing function (do you know why?), and so there is only one point of intersection for Y_1 and Y_2. Figure 38 shows the graphs of Y_1 and Y_2. Using the INTERSECT command, the solution is 0.44 rounded to two decimal places. ∎

NOW WORK PROBLEM 73.

4.6 Assess Your Understanding

'Are You Prepared?'

Answers are given at the end of these exercises. If you get a wrong answer, read the pages listed in red.

1. Solve $x^2 - 7x - 30 = 0$. (pp. 134–135)

2. Solve $(x + 3)^2 - 4(x + 3) + 3 = 0$. (pp. 140–142)

3. Approximate the solution(s) to $x^3 = x^2 - 5$ using a graphing utility. (p. B10)

4. Approximate the solution(s) to $x^3 - 2x + 2 = 0$ using a graphing utility. (pp. B9–B10)

Skill Building

In Problems 5–60, solve each equation.

5. $\log_4(x + 2) = \log_4 8$

6. $\log_5(2x + 3) = \log_5 3$

7. $\dfrac{1}{2}\log_3 x = 2\log_3 2$

8. $-2\log_4 x = \log_4 9$

9. $2\log_5 x = 3\log_5 4$

10. $3\log_2 x = -\log_2 27$

11. $3\log_2(x - 1) + \log_2 4 = 5$

12. $2\log_3(x + 4) - \log_3 9 = 2$

13. $\log x + \log(x + 15) = 2$

14. $\log_4 x + \log_4(x - 3) = 1$

15. $\ln x + \ln(x + 2) = 4$

16. $\ln(x + 1) - \ln x = 2$

17. $\log_3(x + 12) + \log_3(x + 4) = 2$

18. $\log_2(x + 11) + \log_2(x + 7) = 5$

19. $2^{2x} + 2^x - 12 = 0$

20. $3^{2x} + 3^x - 2 = 0$

21. $3^{2x} + 3^{x+1} - 4 = 0$

22. $2^{2x} + 2^{x+2} - 12 = 0$

23. $2^x = 10$

24. $3^x = 14$

25. $8^{-x} = 1.2$

26. $2^{-x} = 1.5$

27. $3^{1-2x} = 4^x$

28. $2^{x+1} = 5^{1-2x}$

29. $\left(\dfrac{3}{5}\right)^x = 7^{1-x}$

30. $\left(\dfrac{4}{3}\right)^{1-x} = 5^x$

31. $1.2^x = (0.5)^{-x}$

32. $0.3^{1+x} = 1.7^{2x-1}$

33. $\pi^{1-x} = e^x$

34. $e^{x+3} = \pi^x$

35. $5(2^{3x}) = 8$

36. $0.3(4^{0.2x}) = 0.2$

37. $16^x + 4^{x+1} - 3 = 0$

38. $9^x - 3^{x+1} + 1 = 0$

39. $25^x - 8 \cdot 5^x = -16$

40. $36^x - 6 \cdot 6^x = -9$

41. $3 \cdot 4^x + 4 \cdot 2^x + 8 = 0$

42. $2 \cdot 49^x + 11 \cdot 7^x + 5 = 0$

43. $4^x - 10 \cdot 4^{-x} = 3$

44. $3^x - 14 \cdot 3^{-x} = 5$

45. $\log_a(x - 1) - \log_a(x + 6) = \log_a(x - 2) - \log_a(x + 3)$

46. $\log_a x + \log_a(x - 2) = \log_a(x + 4)$

47. $\log_{1/3}(x^2 + x) - \log_{1/3}(x^2 - x) = -1$

48. $\log_4(x^2 - 9) - \log_4(x + 3) = 3$

49. $\log_2(x + 1) - \log_4 x = 1$

[**Hint:** Change $\log_4 x$ to base 2.]

50. $\log_2(3x + 2) - \log_4 x = 3$

51. $\log_{16} x + \log_4 x + \log_2 x = 7$

52. $\log_9 x + 3 \log_3 x = 14$

53. $\left(\sqrt[3]{2}\right)^{2-x} = 2^{x^2}$

54. $\log_2 x^{\log_2 x} = 4$

55. $\dfrac{e^x + e^{-x}}{2} = 1$

[**Hint:** Multiply each side by e^x.]

56. $\dfrac{e^x + e^{-x}}{2} = 3$

57. $\dfrac{e^x - e^{-x}}{2} = 2$

58. $\dfrac{e^x - e^{-x}}{2} = -2$

59. $\log_5 x + \log_3 x = 1$

[**Hint:** Use the Change-of-Base Formula and factor $\log x$.]

60. $\log_2 x + \log_6 x = 3$

61. $f(x) = \log_2(x + 3)$ and $g(x) = \log_2(3x + 1)$.
(a) Solve $f(x) = 3$. What point is on the graph of f?
(b) Solve $g(x) = 4$. What point is on the graph of g?
(c) Solve $f(x) = g(x)$. Do the graphs of f and g intersect? If so, where?
(d) Solve $f(x) + g(x) = 7$.
(e) Solve $f(x) - g(x) = 2$.

62. $f(x) = \log_3(x + 5)$ and $g(x) = \log_3(x - 1)$.
(a) Solve $f(x) = 2$. What point is on the graph of f?
(b) Solve $g(x) = 3$. What point is on the graph of g?
(c) Solve $f(x) = g(x)$. Do the graphs of f and g intersect? If so, where?
(d) Solve $f(x) + g(x) = 3$.
(e) Solve $f(x) - g(x) = 2$.

63. (a) If $f(x) = 3^{x+1}$ and $g(x) = 2^{x+2}$, graph f and g on the same Cartesian plane.
(b) Find the point(s) of intersection of the graphs of f and g by solving $f(x) = g(x)$. Round answers to three decimal places. Label this point on the graph drawn in part (a).
(c) Based on the graph, solve $f(x) > g(x)$.

64. (a) If $f(x) = 5^{x-1}$ and $g(x) = 2^{x+1}$, graph f and g on the same Cartesian plane.
(b) Find the point(s) of intersection of the graphs of f and g by solving $f(x) = g(x)$. Label this point on the graph drawn in part (a).
(c) Based on the graph, solve $f(x) > g(x)$.

65. (a) Graph $f(x) = 3^x$ and $g(x) = 10$ on the same Cartesian plane.
(b) Shade the region bounded by the y-axis, $f(x) = 3^x$, and $g(x) = 10$ on the graph drawn in part (a).
(c) Solve $f(x) = g(x)$ and label the point of intersection on the graph drawn in part (b).

66. (a) Graph $f(x) = 2^x$ and $g(x) = 12$ on the same Cartesian plane.
(b) Shade the region bounded by the y-axis, $f(x) = 2^x$, and $g(x) = 12$ on the graph drawn in part (a).
(c) Solve $f(x) = g(x)$ and label the point of intersection on the graph drawn in part (b).

67. (a) Graph $f(x) = 2^{x+1}$ and $g(x) = 2^{-x+2}$ on the same Cartesian plane.
(b) Shade the region bounded by the y-axis, $f(x) = 2^{x+1}$, and $g(x) = 2^{-x+2}$ on the graph draw in part (a).
(c) Solve $f(x) = g(x)$ and label the point of intersection on the graph drawn in part (b).

68. (a) Graph $f(x) = 3^{-x+1}$ and $g(x) = 3^{x-2}$ on the same Cartesian plane.
(b) Shade the region bounded by the y-axis, $f(x) = 3^{-x+1}$, and $g(x) = 3^{x-2}$ on the graph draw in part (a).
(c) Solve $f(x) = g(x)$ and label the point of intersection on the graph drawn in part (b).

69. (a) Graph $f(x) = 2^x - 4$.
(b) Find the zero of f.
(c) Based on the graph, solve $f(x) < 0$.

70. (a) Graph $g(x) = 3^x - 9$.
(b) Find the zero of g.
(c) Based on the graph, solve $g(x) > 0$.

In Problems 71–84, use a graphing utility to solve each equation. Express your answer rounded to two decimal places.

71. $\log_5(x + 1) - \log_4(x - 2) = 1$

72. $\log_2(x - 1) - \log_6(x + 2) = 2$

73. $e^x = -x$

74. $e^{2x} = x + 2$

75. $e^x = x^2$

76. $e^x = x^3$

77. $\ln x = -x$

78. $\ln(2x) = -x + 2$

79. $\ln x = x^3 - 1$

80. $\ln x = -x^2$

81. $e^x + \ln x = 4$

82. $e^x - \ln x = 4$

83. $e^{-x} = \ln x$

84. $e^{-x} = -\ln x$

Applications and Extensions

85. A Population Model The population of the United States in 2000 was 282 million people. In addition, the population of the United States was growing at a rate of 1.1% per year. Assuming that this growth rate continues, the model $P(t) = 282(1.011)^{t-2000}$ represents the population P (in millions of people) in year t.
(a) According to this model, when will the population of the United States be 303 million people?
(b) According to this model, when will the population of the United States be 355 million people?

SOURCE: *Statistical Abstract of the United States*

86. A Population Model The population of the world in 2000 was 6.08 billion people. In addition, the population of the world was growing at a rate of 1.26% per year. Assuming that this growth rate continues, the model $P(t) = 6.08(1.0126)^{t-2000}$ represents the population P (in millions of people) in year t.

(a) According to this model, when will the population of the world be 9.17 billion people?

(b) According to this model, when will the population of the world be 11.55 billion people?

SOURCE: *Statistical Abstract of the United States*

87. Depreciation The value V of a Chevy Cavalier Coupe that is t years old can be modeled by $V(t) = 14{,}512(0.82)^t$.

(a) According to the model, when will the car be worth $9000?

(b) According to the model, when will the car be worth $4000?

(c) According to the model, when will the car be worth $2000?

SOURCE: *Kelley Blue Book*

88. Depreciation The value V of a Dodge Stratus that is t years old can be modeled by $V(t) = 19{,}282(0.84)^t$.

(a) According to the model, when will the car be worth $15,000?

(b) According to the model, when will the car be worth $8000?

(c) According to the model, when will the car be worth $2000?

SOURCE: *Kelley Blue Book*

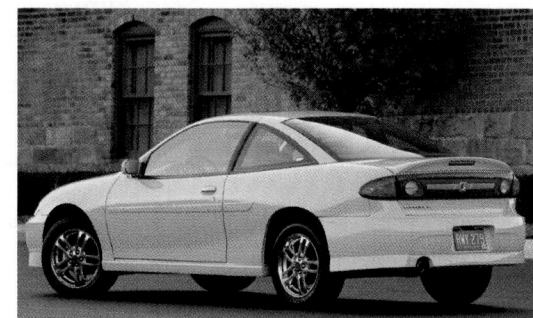

Discussion and Writing

89. Fill in reasons for each step in the following two solutions.

Solve: $\log_3(x-1)^2 = 2$

Solution A	**Solution B**
$\log_3(x-1)^2 = 2$	$\log_3(x-1)^2 = 2$
$(x-1)^2 = 3^2 = 9$ _____	$2\log_3(x-1) = 2$ _____
$(x-1) = \pm 3$ _____	$\log_3(x-1) = 1$ _____
$x - 1 = -3$ or $x - 1 = 3$ _____	$x - 1 = 3^1 = 3$ _____
$x = -2$ or $x = 4$ _____	$x = 4$ _____

Both solutions given in Solution A check. Explain what caused the solution $x = -2$ to be lost in Solution B.

'Are You Prepared?' Answers

1. $\{-3, 10\}$ **2.** $\{-2, 0\}$ **3.** $\{-1.43\}$ **4.** $\{-1.77\}$

4.7 Compound Interest

PREPARING FOR THIS SECTION *Before getting started, review the following:*

• Simple Interest (Appendix A, Section A.10, pp. A75–A76)

Now work the 'Are You Prepared?' problems on page 332.

OBJECTIVES 1 Determine the Future Value of a Lump Sum of Money
2 Calculate Effective Rates of Return
3 Determine the Present Value of a Lump Sum of Money
4 Determine the Time Required to Double or Triple a Lump Sum of Money

1 Determine the Future Value of a Lump Sum of Money

Interest is money paid for the use of money. The total amount borrowed (whether by an individual from a bank in the form of a loan or by a bank from an individual in the form of a savings account) is called the **principal**. The **rate of interest**, expressed as a percent, is the amount charged for the use of the principal for a given period of time, usually on a yearly (that is, per annum) basis.

> **Simple Interest Formula**
>
> If a principal of P dollars is borrowed for a period of t years at a per annum interest rate r, expressed as a decimal, the interest I charged is
>
> $$I = Prt \qquad\qquad (1)$$
>
> Interest charged according to formula (1) is called **simple interest**.

In working with problems involving interest, we define the term **payment period** as follows:

Annually:	Once per year	**Monthly:**	12 times per year
Semiannually:	Twice per year	**Daily:**	365 times per year*
Quarterly:	Four times per year		

When the interest due at the end of a payment period is added to the principal so that the interest computed at the end of the next payment period is based on this new principal amount (old principal + interest), the interest is said to have been **compounded**. **Compound interest** is interest paid on principal and previously earned interest.

EXAMPLE 1 | **Computing Compound Interest**

A credit union pays interest of 8% per annum compounded quarterly on a certain savings plan. If $1000 is deposited in such a plan and the interest is left to accumulate, how much is in the account after 1 year?

Solution We use the simple interest formula, $I = Prt$. The principal P is $1000 and the rate of interest is 8% = 0.08. After the first quarter of a year, the time t is $\frac{1}{4}$ year, so the interest earned is

$$I = Prt = (\$1000)(0.08)\left(\frac{1}{4}\right) = \$20$$

The new principal is $P + I = \$1000 + \$20 = \$1020$. At the end of the second quarter, the interest on this principal is

$$I = (\$1020)(0.08)\left(\frac{1}{4}\right) = \$20.40$$

At the end of the third quarter, the interest on the new principal of $1020 + $20.40 = $1040.40 is

$$I = (\$1040.40)(0.08)\left(\frac{1}{4}\right) = \$20.81$$

Finally, after the fourth quarter, the interest is

$$I = (\$1061.21)(0.08)\left(\frac{1}{4}\right) = \$21.22$$

After 1 year the account contains $1061.21 + $21.22 = $1082.43. ∎

*Most banks use a 360-day "year." Why do you think they do?

The pattern of the calculations performed in Example 1 leads to a general formula for compound interest. To fix our ideas, let P represent the principal to be invested at a per annum interest rate r that is compounded n times per year, so the time of each compounding period is $\dfrac{1}{n}$ years. (For computing purposes, r is expressed as a decimal.) The interest earned after each compounding period is given by formula (1).

$$\text{Interest} = \text{principal} \times \text{rate} \times \text{time} = P \cdot r \cdot \frac{1}{n} = P \cdot \left(\frac{r}{n}\right)$$

The amount A after one compounding period is

$$A = P + P \cdot \left(\frac{r}{n}\right) = P \cdot \left(1 + \frac{r}{n}\right)$$

After two compounding periods, the amount A, based on the new principal $P \cdot \left(1 + \dfrac{r}{n}\right)$, is

$$A = \underbrace{P \cdot \left(1 + \frac{r}{n}\right)}_{\substack{\text{New} \\ \text{principal}}} + \underbrace{P \cdot \left(1 + \frac{r}{n}\right)\left(\frac{r}{n}\right)}_{\substack{\text{Interest on} \\ \text{new principal}}} = P \cdot \left(1 + \frac{r}{n}\right)\left(1 + \frac{r}{n}\right) = P \cdot \left(1 + \frac{r}{n}\right)^2$$

After three compounding periods, the amount A is

$$A = P \cdot \left(1 + \frac{r}{n}\right)^2 + P \cdot \left(1 + \frac{r}{n}\right)^2\left(\frac{r}{n}\right) = P \cdot \left(1 + \frac{r}{n}\right)^2 \cdot \left(1 + \frac{r}{n}\right) = P \cdot \left(1 + \frac{r}{n}\right)^3$$

Continuing this way, after n compounding periods (1 year), the amount A is

$$A = P \cdot \left(1 + \frac{r}{n}\right)^n$$

Because t years will contain $n \cdot t$ compounding periods, after t years we have

$$A = P \cdot \left(1 + \frac{r}{n}\right)^{nt}$$

Theorem

Compound Interest Formula

The amount A after t years due to a principal P invested at an annual interest rate r compounded n times per year is

$$A = P \cdot \left(1 + \frac{r}{n}\right)^{nt} \tag{2}$$

— **Exploration** —————

To see the effects of compounding interest monthly on an initial deposit of \$1, graph $Y_1 = \left(1 + \dfrac{r}{12}\right)^{12x}$ with $r = 0.06$ and $r = 0.12$ for $0 \leq x \leq 30$. What is the future value of \$1 in 30 years when the interest rate per annum is $r = 0.06$ (6%)? What is the future value of \$1 in 30 years when the interest rate per annum is $r = 0.12$ (12%)? Does doubling the interest rate double the future value?

For example, to rework Example 1, we would use $P = \$1000$, $r = 0.08$, $n = 4$ (quarterly compounding), and $t = 1$ year to obtain

$$A = P \cdot \left(1 + \frac{r}{n}\right)^{nt} = 1000\left(1 + \frac{0.08}{4}\right)^{4 \cdot 1} = \$1082.43$$

In equation (2), the amount A is typically referred to as the **future value** of the account, while P is called the **present value**.

━━━━━ **NOW WORK PROBLEM 3.**

EXAMPLE 2

Comparing Investments Using Different Compounding Periods

Investing $1000 at an annual rate of 10% compounded annually, semiannually, quarterly, monthly, and daily will yield the following amounts after 1 year:

Annual compounding ($n = 1$):

$$A = P \cdot (1 + r)$$
$$= (\$1000)(1 + 0.10) = \$1100.00$$

Semiannual compounding ($n = 2$):

$$A = P \cdot \left(1 + \frac{r}{2}\right)^2$$
$$= (\$1000)(1 + 0.05)^2 = \$1102.50$$

Quarterly compounding ($n = 4$):

$$A = P \cdot \left(1 + \frac{r}{4}\right)^4$$
$$= (\$1000)(1 + 0.025)^4 = \$1103.81$$

Monthly compounding ($n = 12$):

$$A = P \cdot \left(1 + \frac{r}{12}\right)^{12}$$
$$= (\$1000)(1 + 0.00833)^{12} = \$1104.71$$

Daily compounding ($n = 365$):

$$A = P \cdot \left(1 + \frac{r}{365}\right)^{365}$$
$$= (\$1000)(1 + 0.000274)^{365} = \$1105.16 \quad \blacksquare$$

From Example 2, we can see that the effect of compounding more frequently is that the amount after 1 year is higher: $1000 compounded 4 times a year at 10% results in $1103.81, $1000 compounded 12 times a year at 10% results in $1104.71, and $1000 compounded 365 times a year at 10% results in $1105.16. This leads to the following question: What would happen to the amount after 1 year if the number of times that the interest is compounded were increased without bound?

Let's find the answer. Suppose that P is the principal, r is the per annum interest rate, and n is the number of times that the interest is compounded each year. The amount after 1 year is

$$A = P \cdot \left(1 + \frac{r}{n}\right)^n$$

Rewrite this expression as follows:

$$A = P \cdot \left(1 + \frac{r}{n}\right)^n = P \cdot \left(1 + \frac{1}{\frac{n}{r}}\right)^n = P \cdot \left[\left(1 + \frac{1}{\frac{n}{r}}\right)^{n/r}\right]^r = P \cdot \left[\left(1 + \frac{1}{h}\right)^h\right]^r \qquad \textbf{(3)}$$
$$\underset{h = \frac{n}{r}}{\uparrow}$$

Now suppose that the number n of times that the interest is compounded per year gets larger and larger; that is, suppose that $n \to \infty$. Then $h = \dfrac{n}{r} \to \infty$, and the expression in brackets equals e. [Refer to equation (2) on page 291.] That is, $A \to Pe^r$.

Table 8 compares $\left(1 + \dfrac{r}{n}\right)^n$, for large values of n, to e^r for $r = 0.05$, $r = 0.10$, $r = 0.15$, and $r = 1$. The larger that n gets, the closer $\left(1 + \dfrac{r}{n}\right)^n$ gets to e^r.

No matter how frequent the compounding, the amount after 1 year has the definite ceiling Pe^r.

Table 8

	$\left(1 + \frac{r}{n}\right)^n$			
	$n = 100$	$n = 1000$	$n = 10,000$	e^r
$r = 0.05$	1.0512580	1.0512698	1.051271	1.0512711
$r = 0.10$	1.1051157	1.1051654	1.1051704	1.1051709
$r = 0.15$	1.1617037	1.1618212	1.1618329	1.1618342
$r = 1$	2.7048138	2.7169239	2.7181459	2.7182818

When interest is compounded so that the amount after 1 year is Pe^r, we say the interest is **compounded continuously**.

Theorem

> **Continuous Compounding**
>
> The amount A after t years due to a principal P invested at an annual interest rate r compounded continuously is
>
> $$A = Pe^{rt} \qquad (4)$$

EXAMPLE 3

Using Continuous Compounding

The amount A that results from investing a principal P of $1000 at an annual rate r of 10% compounded continuously for a time t of 1 year is

$$A = \$1000e^{0.10} = (\$1000)(1.10517) = \$1105.17 \qquad \blacksquare$$

NOW WORK PROBLEM **11.**

2 Calculate Effective Rates of Return

The **effective rate of interest** is the equivalent annual simple rate of interest that would yield the same amount as compounding after 1 year. For example, based on Example 3, a principal of $1000 will result in $1105.17 at a rate of 10% compounded continuously. To get this same amount using a simple rate of interest would require that interest of $1105.17 − $1000.00 = $105.17 be earned on the principal. Since $105.17 is 10.517% of $1000, a simple rate of interest of 10.517% is needed to equal 10% compounded continuously. The effective rate of interest of 10% compounded continuously is 10.517%.

Based on the results of Examples 2 and 3, we find the following comparisons:

	Annual Rate	Effective Rate
Annual compounding	10%	10%
Semiannual compounding	10%	10.25%
Quarterly compounding	10%	10.381%
Monthly compounding	10%	10.471%
Daily compounding	10%	10.516%
Continuous compounding	10%	10.517%

EXAMPLE 4

Computing the Effective Rate of Interest

On January 2, 2004, $2000 is placed in an Individual Retirement Account (IRA) that will pay interest of 7% per annum compounded continuously.

(a) What will the IRA be worth on January 1, 2024?

(b) What is the effective annual rate of interest?

Solution

(a) The amount A after 20 years is

$$A = Pe^{rt} = \$2000e^{(0.07)(20)} = \$8110.40$$

(b) First, we compute the interest earned on $2000 at $r = 7\%$ compounded continuously for 1 year.

$$A = \$2000e^{0.07(1)} = \$2145.02$$

So the interest earned is $2145.02 − $2000.00 = $145.02. Use the simple interest formula $I = Prt$, with $I = 145.02, $P = 2000, and $t = 1$, and solve for r, the effective annual rate of interest.

$$\$145.02 = \$2000 \cdot r \cdot 1 \quad \text{I = Prt}$$

$$r = \frac{\$145.02}{\$2000} = 0.07251$$

The effective annual rate of interest is 7.251%. ∎

— Exploration —————

For the IRA described in Example 4, how long will it be until $A = \$4000$? 6000?

[**Hint:** Graph $Y_1 = 2000e^{0.07x}$ and $Y_2 = 4000$. Use INTERSECT to find x.]

NOW WORK PROBLEM 23.

3 **Determine the Present Value of a Lump Sum of Money**

When people engaged in finance speak of the "time value of money," they are usually referring to the *present value* of money. The **present value** of A dollars to be received at a future date is the principal that you would need to invest now so that it will grow to A dollars in the specified time period. The present value of money to be received at a future date is always less than the amount to be received, since the amount to be received will equal the present value (money invested now) *plus* the interest accrued over the time period.

We use the compound interest formula (2) to get a formula for present value. If P is the present value of A dollars to be received after t years at a per annum interest rate r compounded n times per year, then, by formula (2),

$$A = P \cdot \left(1 + \frac{r}{n}\right)^{nt}$$

To solve for P, we divide both sides by $\left(1 + \frac{r}{n}\right)^{nt}$. The result is

$$\frac{A}{\left(1 + \frac{r}{n}\right)^{nt}} = P \quad \text{or} \quad P = A \cdot \left(1 + \frac{r}{n}\right)^{-nt}$$

Theorem

Present Value Formulas

The present value P of A dollars to be received after t years, assuming a per annum interest rate r compounded n times per year, is

$$P = A \cdot \left(1 + \frac{r}{n}\right)^{-nt} \tag{5}$$

If the interest is compounded continuously,

$$P = Ae^{-rt} \tag{6}$$

To prove (6), solve formula (4) for P.

EXAMPLE 5

Computing the Value of a Zero-Coupon Bond

A zero-coupon (noninterest-bearing) bond can be redeemed in 10 years for $1000. How much should you be willing to pay for it now if you want a return of

(a) 8% compounded monthly?

(b) 7% compounded continuously?

Solution (a) We are seeking the present value of $1000. We use formula (5) with $A = \$1000$, $n = 12$, $r = 0.08$, and $t = 10$.

$$P = A \cdot \left(1 + \frac{r}{n}\right)^{-nt} = \$1000\left(1 + \frac{0.08}{12}\right)^{-12(10)} = \$450.52$$

For a return of 8% compounded monthly, you should pay $450.52 for the bond.

(b) Here we use formula (6) with $A = \$1000$, $r = 0.07$, and $t = 10$.

$$P = Ae^{-rt} = \$1000e^{-(0.07)(10)} = \$496.59$$

For a return of 7% compounded continuously, you should pay $496.59 for the bond. ∎

 NOW WORK PROBLEM 13.

4 Determine the Time Required to Double or Triple a Lump Sum of Money

EXAMPLE 6	**Rate of Interest Required to Double an Investment**

What annual rate of interest compounded annually should you seek if you want to double your investment in 5 years?

Solution If P is the principal and we want P to double, the amount A will be $2P$. We use the compound interest formula with $n = 1$ and $t = 5$ to find r.

$$A = P \cdot \left(1 + \frac{r}{n}\right)^{nt}$$

$$2P = P \cdot (1 + r)^5 \qquad\qquad A = 2P, n = 1, t = 5$$

$$2 = (1 + r)^5 \qquad\qquad \text{Cancel the } P\text{'s.}$$

$$1 + r = \sqrt[5]{2} \qquad\qquad \text{Take the fifth root of each side.}$$

$$r = \sqrt[5]{2} - 1 \approx 1.148698 - 1 = 0.148698$$

The annual rate of interest needed to double the principal in 5 years is 14.87%. ∎

 NOW WORK PROBLEM 25.

EXAMPLE 7	**Doubling and Tripling Time for an Investment**

(a) How long will it take for an investment to double in value if it earns 5% compounded continuously?

(b) How long will it take to triple at this rate?

Solution (a) If P is the initial investment and we want P to double, the amount A will be $2P$. We use formula (4) for continuously compounded interest with $r = 0.05$. Then

$$A = Pe^{rt}$$

$$2P = Pe^{0.05t} \qquad\qquad A = 2P, r = 0.05$$

$$2 = e^{0.05t} \qquad\qquad \text{Cancel the } P\text{'s.}$$

$$0.05t = \ln 2 \qquad\qquad \text{Rewrite as a logarithm.}$$

$$t = \frac{\ln 2}{0.05} \approx 13.86 \qquad \text{Solve for } t.$$

It will take about 14 years to double the investment.

(b) To triple the investment, we set $A = 3P$ in formula (4).

$$A = Pe^{rt}$$
$$3P = Pe^{0.05t} \qquad \textit{A = 3P, r = 0.05}$$
$$3 = e^{0.05t} \qquad \textit{Cancel the P's.}$$
$$0.05t = \ln 3 \qquad \textit{Rewrite as a logarithm.}$$
$$t = \frac{\ln 3}{0.05} \approx 21.97 \qquad \textit{Solve for t.}$$

It will take about 22 years to triple the investment. ■

━━━━━ **NOW WORK PROBLEM 31.**

4.7 Assess Your Understanding

'Are You Prepared?'

Answers are given at the end of these exercises. If you get a wrong answer, read the pages listed in red.

1. What is the interest due if $500 is borrowed for 6 months at a simple interest rate of 6% per annum? (pp. A75–A76)

2. If you borrow $5000 and, after 9 months, pay off the loan in the amount of $5500, what per annum rate of interest was charged? (pp. A75–A76)

Skill Building

In Problems 3–12, find the amount that results from each investment.

3. $100 invested at 4% compounded quarterly after a period of 2 years

4. $50 invested at 6% compounded monthly after a period of 3 years

5. $500 invested at 8% compounded quarterly after a period of $2\frac{1}{2}$ years

6. $300 invested at 12% compounded monthly after a period of $1\frac{1}{2}$ years

7. $600 invested at 5% compounded daily after a period of 3 years

8. $700 invested at 6% compounded daily after a period of 2 years

9. $10 invested at 11% compounded continuously after a period of 2 years

10. $40 invested at 7% compounded continuously after a period of 3 years

11. $100 invested at 10% compounded continuously after a period of $2\frac{1}{4}$ years

12. $100 invested at 12% compounded continuously after a period of $3\frac{3}{4}$ years

In Problems 13–22, find the principal needed now to get each amount; that is, find the present value.

13. To get $100 after 2 years at 6% compounded monthly

14. To get $75 after 3 years at 8% compounded quarterly

15. To get $1000 after $2\frac{1}{2}$ years at 6% compounded daily

16. To get $800 after $3\frac{1}{2}$ years at 7% compounded monthly

17. To get $600 after 2 years at 4% compounded quarterly

18. To get $300 after 4 years at 3% compounded daily

19. To get $80 after $3\frac{1}{4}$ years at 9% compounded continuously

20. To get $800 after $2\frac{1}{2}$ years at 8% compounded continuously

21. To get $400 after 1 year at 10% compounded continuously

22. To get $1000 after 1 year at 12% compounded continuously

23. Find the effective rate of interest for $5\frac{1}{4}$% compounded quarterly.

24. What interest rate compounded quarterly will give an effective interest rate of 7%?

25. What rate of interest compounded annually is required to double an investment in 3 years?

26. What rate of interest compounded annually is required to double an investment in 10 years?

In Problems 27–30, which of the two rates would yield the larger amount in 1 year?

[**Hint:** *Start with a principal of $10,000 in each instance.*]

27. 6% compounded quarterly or $6\frac{1}{4}$% compounded annually

28. 9% compounded quarterly or $9\frac{1}{4}$% compounded annually

29. 9% compounded monthly or 8.8% compounded daily

30. 8% compounded semiannually or 7.9% compounded daily

31. How long does it take for an investment to double in value if it is invested at 8% per annum compounded monthly? How long does it take if the interest is compounded continuously?

32. How long does it take for an investment to double in value if it is invested at 10% per annum compounded monthly? How long does it take if the interest is compounded continuously?

Applications and Extensions

33. If Tanisha has $100 to invest at 8% per annum compounded monthly, how long will it be before she has $150? If the compounding is continuous, how long will it be?

34. If Angela has $100 to invest at 10% per annum compounded monthly, how long will it be before she has $175? If the compounding is continuous, how long will it be?

35. How many years will it take for an initial investment of $10,000 to grow to $25,000? Assume a rate of interest of 6% compounded continuously.

36. How many years will it take for an initial investment of $25,000 to grow to $80,000? Assume a rate of interest of 7% compounded continuously.

37. What will a $90,000 house cost 5 years from now if the inflation rate over that period averages 3% compounded annually?

38. Sears charges 1.25% per month on the unpaid balance for customers with charge accounts (interest is compounded monthly). A customer charges $200 and does not pay her bill for 6 months. What is the bill at that time?

39. Jerome will be buying a used car for $15,000 in 3 years. How much money should he ask his parents for now so that, if he invests it at 5% compounded continuously, he will have enough to buy the car?

40. John will require $3000 in 6 months to pay off a loan that has no prepayment privileges. If he has the $3000 now, how much of it should he save in an account paying 3% compounded monthly so that in 6 months he will have exactly $3000?

41. George is contemplating the purchase of 100 shares of a stock selling for $15 per share. The stock pays no dividends. The history of the stock indicates that it should grow at an annual rate of 15% per year. How much will the 100 shares of stock be worth in 5 years?

42. Tracy is contemplating the purchase of 100 shares of a stock selling for $15 per share. The stock pays no dividends. Her broker says that the stock will be worth $20 per share in 2 years. What is the annual rate of return on this investment?

43. A business purchased for $650,000 in 2001 is sold in 2004 for $850,000. What is the annual rate of return for this investment?

44. Tanya has just inherited a diamond ring appraised at $5000. If diamonds have appreciated in value at an annual rate of 8%, what was the value of the ring 10 years ago when the ring was purchased?

45. Jim places $1000 in a bank account that pays 5.6% compounded continuously. After 1 year, will he have enough money to buy a computer system that costs $1060? If another bank will pay Jim 5.9% compounded monthly, is this a better deal?

46. On January 1, Kim places $1000 in a certificate of deposit that pays 6.8% compounded continuously and matures in 3 months. Then Kim places the $1000 and the interest in a passbook account that pays 5.25% compounded monthly. How much does Kim have in the passbook account on May 1?

47. Will invests $2000 in a bond trust that pays 9% interest compounded semiannually. His friend Henry invests $2000 in a certificate of deposit that pays $8\frac{1}{2}$% compounded continuously. Who has more money after 20 years, Will or Henry?

48. Suppose that April has access to an investment that will pay 10% interest compounded continuously. Which is better: To be given $1000 now so that she can take advantage of this investment opportunity or to be given $1325 after 3 years?

49. Colleen and Bill have just purchased a house for $150,000, with the seller holding a second mortgage of $50,000. They promise to pay the seller $50,000 plus all accrued interest 5 years from now. The seller offers them three interest options on the second mortgage:
(a) Simple interest at 12% per annum
(b) $11\frac{1}{2}$% interest compounded monthly
(c) $11\frac{1}{4}$% interest compounded continuously

Which option is best; that is, which results in the least interest on the loan?

50. The First National Bank advertises that it pays interest on savings accounts at the rate of 4.25% compounded daily. Find the effective rate if the bank uses 360 days in determining the daily rate. What is it if the bank uses 365 days?

*Problems 51–54 involve zero-coupon bonds. A **zero-coupon bond** is a bond that is sold now at a discount and will pay its face value at the time when it matures; no interest payments are made.*

51. A zero-coupon bond can be redeemed in 20 years for $10,000. How much should you be willing to pay for it now if you want a return of:
(a) 10% compounded monthly?
(b) 10% compounded continuously?

52. A child's grandparents are considering buying a $40,000 face value zero-coupon bond at birth so that she will have enough money for her college education 17 years later. If they want a rate of return of 8% compounded annually, what should they pay for the bond?

53. How much should a $10,000 face value zero-coupon bond, maturing in 10 years, be sold for now if its rate of return is to be 8% compounded annually?

54. If Pat pays $12,485.52 for a $25,000 face value zero-coupon bond that matures in 8 years, what is his annual rate of return?

55. Time to Double or Triple an Investment The formula

$$t = \frac{\ln m}{n \ln\left(1 + \frac{r}{n}\right)}$$

can be used to find the number of years t required to multiply an investment m times when r is the per annum interest rate compounded n times a year.
(a) How many years will it take to double the value of an IRA that compounds annually at the rate of 12%?

(b) How many years will it take to triple the value of a savings account that compounds quarterly at an annual rate of 6%?

(c) Give a derivation of this formula.

56. Time to Reach an Investment Goal The formula

$$t = \frac{\ln A - \ln P}{r}$$

can be used to find the number of years t required for an investment P to grow to a value A when compounded continuously at an annual rate r.

(a) How long will it take to increase an initial investment of $1000 to $8000 at an annual rate of 10%?

(b) What annual rate is required to increase the value of a $2000 IRA to $30,000 in 35 years?

(c) Give a derivation of this formula.

Discussion and Writing

57. Explain in your own words what the term *compound interest* means. What does *continuous compounding* mean?

58. Explain in your own words the meaning of *present value*.

59. Critical Thinking You have just contracted to buy a house and will seek financing in the amount of $100,000. You go to several banks. Bank 1 will lend you $100,000 at the rate of 8.75% amortized over 30 years with a loan origination fee of 1.75%. Bank 2 will lend you $100,000 at the rate of 8.375% amortized over 15 years with a loan origination fee of 1.5%. Bank 3 will lend you $100,000 at the rate of 9.125% amortized over 30 years with no loan origination fee. Bank 4 will lend you $100,000 at the rate of 8.625% amortized over 15 years with no loan origination fee. Which loan would you take? Why? Be sure to have sound reasons for your choice.

Use the information in the table to assist you. If the amount of the monthly payment does not matter to you, which loan would you take? Again, have sound reasons for your choice. Compare your final decision with others in the class. Discuss.

	Monthly Payment	Loan Origination Fee
Bank 1	$786.70	$1,750.00
Bank 2	$977.42	$1,500.00
Bank 3	$813.63	$0.00
Bank 4	$990.68	$0.00

'Are You Prepared?' Answers

1. $15 **2.** 13.33%

4.8 Exponential Growth and Decay; Newton's Law; Logistic Growth and Decay

OBJECTIVES 1 Find Equations of Populations That Obey the Law of Uninhibited Growth
2 Find Equations of Populations That Obey the Law of Decay
3 Use Newton's Law of Cooling
4 Use Logistic Models

1 Find Equations of Populations That Obey the Law of Uninhibited Growth

Many natural phenomena have been found to follow the law that an amount A varies with time t according to the function

$$A(t) = A_0 e^{kt} \tag{1}$$

Here A_0 is the original amount $(t = 0)$ and $k \neq 0$ is a constant.

If $k > 0$, then equation (1) states that the amount A is increasing over time; if $k < 0$, the amount A is decreasing over time. In either case, when an amount A varies over time according to equation (1), it is said to follow the **exponential law** or the **law of uninhibited growth** $(k > 0)$ **or decay** $(k < 0)$. See Figure 39.

For example, we saw in Section 4.7 that continuously compounded interest follows the law of uninhibited growth. In this section we shall look at three additional phenomena that follow the exponential law.

Figure 39

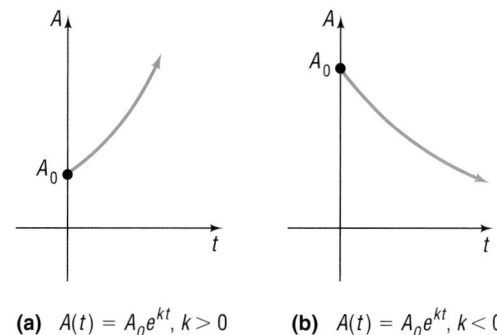

(a) $A(t) = A_0 e^{kt}, k > 0$ **(b)** $A(t) = A_0 e^{kt}, k < 0$

Cell division is the growth process of many living organisms, such as amoebas, plants, and human skin cells. Based on an ideal situation in which no cells die and no by-products are produced, the number of cells present at a given time follows the law of uninhibited growth. Actually, however, after enough time has passed, growth at an exponential rate will cease due to the influence of factors such as lack of living space and dwindling food supply. The law of uninhibited growth accurately reflects only the early stages of the cell division process.

The cell division process begins with a culture containing N_0 cells. Each cell in the culture grows for a certain period of time and then divides into two identical cells. We assume that the time needed for each cell to divide in two is constant and does not change as the number of cells increases. These new cells then grow, and eventually each divides in two, and so on.

Uninhibited Growth of Cells

A model that gives the number N of cells in a culture after a time t has passed (in the early stages of growth) is

$$N(t) = N_0 e^{kt}, \qquad k > 0 \qquad \text{(2)}$$

where N_0 is the initial number of cells and k is a positive constant that represents the growth rate of the cells.

In using formula (2) to model the growth of cells, we are using a function that yields positive real numbers, even though we are counting the number of cells, which must be an integer. This is a common practice in many applications.

EXAMPLE 1

Bacterial Growth

A colony of bacteria grows according to the law of uninhibited growth according to the function $N(t) = 100e^{0.045t}$, where N is measured in grams and t is measured in days.

(a) Determine the initial amount of bacteria.

(b) What is the growth rate of the bacteria?

(c) What is the population after 5 days?

(d) How long will it take for the population to reach 140 grams?

(e) What is the doubling time for the population?

Solution (a) The initial amount of bacteria, N_0, is obtained when $t = 0$, so

$$N_0 = N(0) = 100e^{0.045(0)} = 100 \text{ grams}$$

(b) Compare $N(t) = 100e^{0.045t}$ to $N(t) = 100e^{kt}$. The value of k, 0.045, indicates a growth rate of 4.5%.

(c) The population after 5 days is $N(5) = 100e^{0.045(5)} = 125.2$ grams.

(d) To find how long it takes for the population to reach 140 grams, we solve the equation $N(t) = 140$.

$$100e^{0.045t} = 140$$
$$e^{0.045t} = 1.4 \qquad \text{\textit{Divide both sides of the equation by 100.}}$$
$$0.045t = \ln 1.4 \qquad \text{\textit{Rewrite as a logarithm.}}$$
$$t = \frac{\ln 1.4}{0.045} \qquad \text{\textit{Divide both sides of the equation by 0.045.}}$$
$$\approx 7.5 \text{ days}$$

(e) The population doubles when $N(t) = 200$ grams, so we find the doubling time by solving the equation $200 = 100e^{0.045t}$ for t.

$$200 = 100e^{0.045t}$$
$$2 = e^{0.045t} \qquad \text{\textit{Divide both sides of the equation by 100.}}$$
$$\ln 2 = 0.045t \qquad \text{\textit{Rewrite as a logarithm.}}$$
$$t = \frac{\ln 2}{0.045} \qquad \text{\textit{Divide both sides of the equation by 0.045.}}$$
$$\approx 15.4 \text{ days}$$

The population doubles approximately every 15.4 days.

NOW WORK PROBLEM 1.

EXAMPLE 2 **Bacterial Growth**

A colony of bacteria increases according to the law of uninhibited growth.

(a) If the number of bacteria doubles in 3 hours, find the function that gives the number of cells in the culture.

(b) How long will it take for the size of the colony to triple?

(c) How long will it take for the population to double a second time (that is, increase four times)?

Solution (a) Using formula (2), the number N of cells at a time t is

$$N(t) = N_0 e^{kt}$$

where N_0 is the initial number of bacteria present and k is a positive number. We first seek the number k. The number of cells doubles in 3 hours, so we have

$$N(3) = 2N_0$$

But $N(3) = N_0 e^{k(3)}$, so

$$N_0 e^{k(3)} = 2N_0$$
$$e^{3k} = 2 \qquad \text{\textit{Divide both sides by N_0.}}$$
$$3k = \ln 2 \qquad \text{\textit{Write the exponential equation as a logarithm.}}$$
$$k = \frac{1}{3}\ln 2$$

Formula (2) for this growth process is therefore

$$N(t) = N_0 e^{\left(\frac{1}{3}\ln 2\right)t}$$

(b) The time t needed for the size of the colony to triple requires that $N = 3N_0$. We substitute $3N_0$ for N to get

$$3N_0 = N_0 e^{\left(\frac{1}{3}\ln 2\right)t}$$

$$3 = e^{\left(\frac{1}{3}\ln 2\right)t}$$

$$\left(\frac{1}{3}\ln 2\right)t = \ln 3$$

$$t = \frac{3\ln 3}{\ln 2} \approx 4.755 \text{ hours}$$

It will take about 4.755 hours or 4 hours, 45 minutes for the size of the colony to triple.

(c) If a population doubles in 3 hours, it will double a second time in 3 more hours, for a total time of 6 hours. ■

2 Find Equations of Populations That Obey the Law of Decay

Radioactive materials follow the law of uninhibited decay.

Uninhibited Radioactive Decay

The amount A of a radioactive material present at time t is given by

$$A(t) = A_0 e^{kt}, \qquad k < 0 \tag{3}$$

where A_0 is the original amount of radioactive material and k is a negative number that represents the rate of decay.

All radioactive substances have a specific **half-life**, which is the time required for half of the radioactive substance to decay. In **carbon dating**, we use the fact that all living organisms contain two kinds of carbon, carbon 12 (a stable carbon) and carbon 14 (a radioactive carbon with a half-life of 5600 years). While an organism is living, the ratio of carbon 12 to carbon 14 is constant. But when an organism dies, the original amount of carbon 12 present remains unchanged, whereas the amount of carbon 14 begins to decrease. This change in the amount of carbon 14 present relative to the amount of carbon 12 present makes it possible to calculate when an organism died.

EXAMPLE 3

Estimating the Age of Ancient Tools

Traces of burned wood along with ancient stone tools in an archeological dig in Chile were found to contain approximately 1.67% of the original amount of carbon 14. If the half-life of carbon 14 is 5600 years, approximately when was the tree cut and burned?

Solution Using formula (3), the amount A of carbon 14 present at time t is

$$A(t) = A_0 e^{kt}$$

where A_0 is the original amount of carbon 14 present and k is a negative number. We first seek the number k. To find it, we use the fact that after 5600 years half of the original amount of carbon 14 remains, so $A(5600) = \frac{1}{2}A_0$. Then

$$\frac{1}{2}A_0 = A_0 e^{k(5600)}$$

$$\frac{1}{2} = e^{5600k} \qquad \text{Divide both sides of the equation by } A_0.$$

$$5600k = \ln\frac{1}{2} \qquad \text{Rewrite as a logarithm.}$$

$$k = \frac{1}{5600}\ln\frac{1}{2} \approx -0.000124$$

Formula (3) therefore becomes

$$A(t) = A_0 e^{\frac{\ln\frac{1}{2}}{5600}t}$$

If the amount A of carbon 14 now present is 1.67% of the original amount, it follows that

$$0.0167 A_0 = A_0 e^{\frac{\ln\frac{1}{2}}{5600}t}$$

$$0.0167 = e^{\frac{\ln\frac{1}{2}}{5600}t} \qquad \textit{Divide both sides of the equation by } A_0.$$

$$\frac{\ln\frac{1}{2}}{5600}t = \ln 0.0167 \qquad \textit{Rewrite as a logarithm.}$$

$$t = \frac{5600}{\ln\frac{1}{2}} \ln 0.0167 \approx 33{,}062 \text{ years}$$

The tree was cut and burned about 33,062 years ago. Some archeologists use this conclusion to argue that humans lived in the Americas 33,000 years ago, much earlier than is generally accepted. ∎

━━━━━━━ **NOW WORK PROBLEM 3.**

3 Use Newton's Law of Cooling

Newton's Law of Cooling[*] states that the temperature of a heated object decreases exponentially over time toward the temperature of the surrounding medium.

> **Newton's Law of Cooling**
>
> The temperature u of a heated object at a given time t can be modeled by the following function:
>
> $$u(t) = T + (u_0 - T)e^{kt}, \qquad k < 0 \qquad (4)$$
>
> where T is the constant temperature of the surrounding medium, u_0 is the initial temperature of the heated object, and k is a negative constant.

EXAMPLE 4

Using Newton's Law of Cooling

An object is heated to 100°C (degrees Celsius) and is then allowed to cool in a room whose air temperature is 30°C.

(a) If the temperature of the object is 80°C after 5 minutes, when will its temperature be 50°C?

(b) Determine the elapsed time before the temperature of the object is 35°C.

(c) What do you notice about the temperature as time passes?

Solution

(a) Using formula (4) with $T = 30$ and $u_0 = 100$, the temperature (in degrees Celsius) of the object at time t (in minutes) is

$$u(t) = 30 + (100 - 30)e^{kt} = 30 + 70e^{kt}$$

[*]Named after Sir Isaac Newton (1642–1727), one of the cofounders of calculus.

where k is a negative constant. To find k, we use the fact that $u = 80$ when $t = 5$. Then

$$80 = 30 + 70e^{k(5)}$$

$$50 = 70e^{5k}$$

$$e^{5k} = \frac{50}{70}$$

$$5k = \ln \frac{5}{7}$$

$$k = \frac{1}{5} \ln \frac{5}{7} \approx -0.0673$$

Formula (4) therefore becomes

$$u(t) = 30 + 70e^{\frac{\ln \frac{5}{7}}{5}t} \tag{5}$$

We want to find t when $u = 50°C$, so

$$50 = 30 + 70e^{\frac{\ln \frac{5}{7}}{5}t}$$

$$20 = 70e^{\frac{\ln \frac{5}{7}}{5}t}$$

$$e^{\frac{\ln \frac{5}{7}}{5}t} = \frac{20}{70}$$

$$\frac{\ln \frac{5}{7}}{5}t = \ln \frac{2}{7}$$

$$t = \frac{5}{\ln \frac{5}{7}} \ln \frac{2}{7} \approx 18.6 \text{ minutes}$$

The temperature of the object will be 50°C after about 18.6 minutes or 18 minutes, 37 seconds.

(b) If $u = 35°C$, then, based on equation (5), we have

$$35 = 30 + 70e^{\frac{\ln \frac{5}{7}}{5}t}$$

$$5 = 70e^{\frac{\ln \frac{5}{7}}{5}t}$$

$$e^{\frac{\ln \frac{5}{7}}{5}t} = \frac{5}{70}$$

$$\frac{\ln \frac{5}{7}}{5}t = \ln \frac{5}{70}$$

$$t = \frac{\ln \frac{5}{70}}{\frac{\ln \frac{5}{7}}{5}} \approx 39.2 \text{ minutes}$$

The object will reach a temperature of 35°C after about 39.2 minutes.

(c) Look at equation (5). As t increases, the value of $e^{\frac{\ln\frac{5}{7}}{5}t}$ approaches zero, so the value of u, the temperature of the object, approaches 30°C, the air temperature of the room. ∎

NOW WORK PROBLEM 13.

4 Use Logistic Models

The exponential growth model $A(t) = A_0 e^{kt}$, $k > 0$, assumes uninhibited growth, meaning that the value of the function grows without limit. Recall that we stated that cell division could be modeled using this function, assuming that no cells die and no by-products are produced. However, cell division would eventually be limited by factors such as living space and food supply. The **logistic model**, given next, can describe situations where the growth or decay of the dependent variable is limited.

Logistic Model

In a logistic growth model, the population P after time t is given by the function

$$P(t) = \frac{c}{1 + ae^{-bt}} \tag{6}$$

where a, b, and c are constants with $c > 0$. The model is a growth model if $b > 0$; the model is a decay model if $b < 0$.

The number c is called the **carrying capacity** (for growth models) because the value $P(t)$ approaches c as t approaches infinity; that is, $\lim_{t\to\infty} P(t) = c$. The number $|b|$ is the growth rate for $b > 0$ and the decay rate for $b < 0$. Figure 40(a) shows the graph of a typical logistic growth function, and Figure 40(b) shows the graph of a typical logistic decay function.

Figure 40

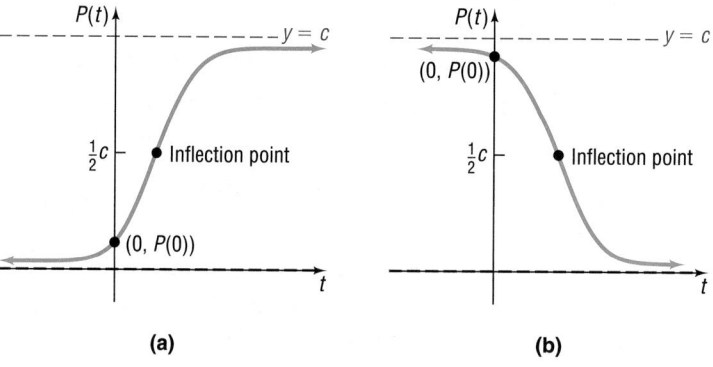

(a) (b)

Based on the figures, we have the following properties of logistic growth functions.

Properties of the Logistic Growth Function, Equation (6)

1. The domain is the set of all real numbers. The range is the interval $(0, c)$, where c is the carrying capacity.
2. There are no x-intercepts; the y-intercept is $P(0)$.
3. There are two horizontal asymptotes: $y = 0$ and $y = c$.
4. $P(t)$ is an increasing function if $b > 0$ and a decreasing function if $b < 0$.

5. There is an **inflection point** where $P(t)$ equals $\dfrac{1}{2}$ of the carrying capacity. The inflection point is the point on the graph where the graph changes from being curved upward to curved downward for growth functions and the point where the graph changes from being curved downward to curved upward for decay functions.

6. The graph is smooth and continuous, with no corners or gaps.

EXAMPLE 5	**Fruit Fly Population**

Fruit flies are placed in a half-pint milk bottle with a banana (for food) and yeast plants (for food and to provide a stimulus to lay eggs). Suppose that the fruit fly population after t days is given by

$$P(t) = \frac{230}{1 + 56.5e^{-0.37t}}$$

(a) State the carrying capacity and the growth rate.

(b) Determine the initial population.

(c) What is the population after 5 days?

(d) How long does it take for the population to reach 180?

 (e) Use a graphing utility to determine how long it takes for the population to reach one-half of the carrying capacity by graphing $Y_1 = P(t)$, and $Y_2 = 115$ and using INTERSECT.

Solution

(a) As $t \to \infty$, $e^{-0.37t} \to 0$ and $P(t) \to 230/1$. The carrying capacity of the half-pint bottle is 230 fruit flies. The growth rate is $|b| = |0.37| = 37\%$.

(b) To find the initial number of fruit flies in the half-pint bottle, we evaluate $P(0)$.

$$P(0) = \frac{230}{1 + 56.5e^{-0.37(0)}}$$
$$= \frac{230}{1 + 56.5}$$
$$= 4$$

So initially there were 4 fruit flies in the half-pint bottle.

(c) To find the number of fruit flies in the half-pint bottle after 5 days, we evaluate $P(5)$.

$$P(5) = \frac{230}{1 + 56.5e^{-0.37(5)}} \approx 23 \text{ fruit flies}$$

After 5 days, there are approximately 23 fruit flies in the bottle.

(d) To determine when the population of fruit flies will be 180, we solve the equation $P(t) = 180$.

$$\frac{230}{1 + 56.5e^{-0.37t}} = 180$$

$$230 = 180(1 + 56.5e^{-0.37t})$$ Divide both sides by 180.

$$1.2778 = 1 + 56.5e^{-0.37t}$$ Subtract 1 from both sides.

$$0.2778 = 56.5e^{-0.37t}$$ Divide both sides by 56.5.

$$0.0049 = e^{-0.37t}$$ Rewrite as a logarithmic expression.

$$\ln(0.0049) = -0.37t$$ Divide both sides by -0.37.

$$t \approx 14.4 \text{ days}$$

It will take approximately 14.4 days (14 days, 9 hours) for the population to reach 180 fruit flies.

(e) One-half of the carrying capacity is 115 fruit flies. We solve $P(t) = 115$ by graphing $Y_1 = \dfrac{230}{1 + 56.5e^{-0.37t}}$ and $Y_2 = 115$ and using INTERSECT. See Figure 41. The population will reach one-half of the carrying capacity in about 10.9 days (10 days, 22 hours).

Figure 41

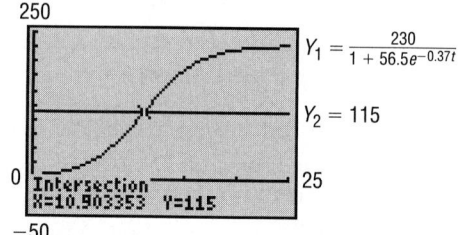

Look back at Figure 41. Notice the point where the graph reaches 115 fruit flies (one-half of the carrying capacity): the graph changes from being curved upward to being curved downward. Using the language of calculus, we say the graph changes from increasing at an increasing rate to increasing at a decreasing rate. For any logistic growth function, when the population reaches one-half the carrying capacity, the population growth starts to slow down.

— Exploration —

On the same viewing rectangle, graph

$Y_1 = \dfrac{500}{1 + 24e^{-0.03t}}$ and $Y_2 = \dfrac{500}{1 + 24e^{-0.08t}}$

What effect does the growth rate $|b|$ have on the logistic growth function?

 NOW WORK PROBLEM 21.

EXAMPLE 6 | **Wood Products**

The EFISCEN wood product model classifies wood products according to their life-span. There are four classifications: short (1 year), medium short (4 years), medium long (16 years), and long (50 years). Based on data obtained from the European Forest Institute, the percentage of remaining wood products after t years for wood products with long life-spans (such as those used in the building industry) is given by

$$P(t) = \frac{100.3952}{1 + 0.0316e^{0.0581t}}$$

(a) What is the decay rate?

(b) What is the percentage of remaining wood products after 10 years?

(c) How long does it take for the percentage of remaining wood products to reach 50%?

(d) Explain why the numerator given in the model is reasonable.

Solution

(a) The decay rate is $|b| = |-0.0581| = 5.81\%$.

(b) Evaluate $P(10)$.

$$P(10) = \frac{100.3952}{1 + 0.0316e^{0.0581(10)}} \approx 95.0$$

So 95% of long life-span wood products remain after 10 years.

(c) Solve the equation $P(t) = 50$.

$$\frac{100.3952}{1 + 0.0316e^{0.0581t}} = 50$$

$$100.3952 = 50(1 + 0.0316e^{0.0581t})$$

$$2.0079 = 1 + 0.0316e^{0.0581t} \quad \text{Divide both sides by 50.}$$

$$1.0079 = 0.0316e^{0.0581t} \quad \text{Subtract 1 from both sides.}$$

$$31.8956 = e^{0.0581t} \quad \text{Divide both sides by 0.0316.}$$

$$\ln(31.8956) = 0.0581t \quad \text{Rewrite as a logarithmic expression.}$$

$$t \approx 59.6 \text{ years} \quad \text{Divide both sides by 0.0581.}$$

It will take approximately 59.6 years for the percentage of long life-span wood products remaining to reach 50%.

(d) The numerator of 100.3952 is reasonable because the maximum percentage of wood products remaining that is possible is 100%. ∎

4.8 Assess Your Understanding

Applications and Extensions

1. **Growth of an Insect Population** The size P of a certain insect population at time t (in days) obeys the function $P(t) = 500e^{0.02t}$.
 (a) Determine the number of insects at $t = 0$ days.
 (b) What is the growth rate of the insect population?
 (c) What is the population after 10 days?
 (d) When will the insect population reach 800?
 (e) When will the insect population double?

2. **Growth of Bacteria** The number N of bacteria present in a culture at time t (in hours) obeys the law of uninhibited growth $N(t) = 1000e^{0.01t}$.
 (a) Determine the number of bacteria at $t = 0$ hours.
 (b) What is the growth rate of the bacteria?
 (c) What is the population after 4 hours?
 (d) When will the number of bacteria reach 1700?
 (e) When will the number of bacteria double?

3. **Radioactive Decay** Strontium 90 is a radioactive material that decays according to the function $A(t) = A_0 e^{-0.0244t}$, where A_0 is the initial amount present and A is the amount present at time t (in years). Assume that a scientist has a sample of 500 grams of strontium 90.
 (a) What is the decay rate of strontium 90?
 (b) How much strontium 90 is left after 10 years?
 (c) When will 400 grams of strontium 90 be left?
 (d) What is the half-life of strontium 90?

4. **Radioactive Decay** Iodine 131 is a radioactive material that decays according to the function $A(t) = A_0 e^{-0.087t}$, where A_0 is the initial amount present and A is the amount present at time t (in days). Assume that a scientist has a sample of 100 grams of iodine 131.
 (a) What is the decay rate of iodine 131?
 (b) How much iodine 131 is left after 9 days?
 (c) When will 70 grams of iodine 131 be left?
 (d) What is the half-life of iodine 131?

5. **Growth of a Colony of Mosquitoes** The population of a colony of mosquitoes obeys the law of uninhibited growth.

If there are 1000 mosquitoes initially and there are 1800 after 1 day, what is the size of the colony after 3 days? How long is it until there are 10,000 mosquitoes?

6. **Bacterial Growth** A culture of bacteria obeys the law of uninhibited growth. If 500 bacteria are present initially and there are 800 after 1 hour, how many will be present in the culture after 5 hours? How long is it until there are 20,000 bacteria?

7. **Population Growth** The population of a southern city follows the exponential law. If the population doubled in size over an 18-month period and the current population is 10,000, what will the population be 2 years from now?

8. **Population Decline** The population of a midwestern city follows the exponential law. If the population decreased from 900,000 to 800,000 from 2003 to 2005, what will the population be in 2007?

9. **Radioactive Decay** The half-life of radium is 1690 years. If 10 grams are present now, how much will be present in 50 years?

10. **Radioactive Decay** The half-life of radioactive potassium is 1.3 billion years. If 10 grams are present now, how much will be present in 100 years? In 1000 years?

11. **Estimating the Age of a Tree** A piece of charcoal is found to contain 30% of the carbon 14 that it originally had. When did the tree from which the charcoal came die? Use 5600 years as the half-life of carbon 14.

12. **Estimating the Age of a Fossil** A fossilized leaf contains 70% of its normal amount of carbon 14. How old is the fossil?

13. **Cooling Time of a Pizza** A pizza baked at 450°F is removed from the oven at 5:00 PM into a room that is a constant 70°F. After 5 minutes, the pizza is at 300°F.
 (a) At what time can you begin eating the pizza if you want its temperature to be 135°F?
 (b) Determine the time that needs to elapse before the pizza is at 160°F.

(c) What do you notice about the temperature as time passes?

14. **Newton's Law of Cooling** A thermometer reading 72°F is placed in a refrigerator where the temperature is a constant 38°F.
 (a) If the thermometer reads 60°F after 2 minutes, what will it read after 7 minutes?
 (b) How long will it take before the thermometer reads 39°F?
 (d) Determine the time needed to elapse before the thermometer reads 45°F.
 (e) What do you notice about the temperature as time passes?

15. **Newton's Law of Heating** A thermometer reading 8°C is brought into a room with a constant temperature of 35°C. If the thermometer reads 15°C after 3 minutes, what will it read after being in the room for 5 minutes? For 10 minutes? [**Hint:** You need to construct a formula similar to equation (4).]

16. **Thawing Time of a Steak** A frozen steak has a temperature of 28°F. It is placed in a room with a constant temperature of 70°F. After 10 minutes, the temperature of the steak has risen to 35°F. What will the temperature of the steak be after 30 minutes? How long will it take the steak to thaw to a temperature of 45°F? (See the hint given for Problem 15.)

17. **Decomposition of Salt in Water** Salt (NaCl) decomposes in water into sodium (NA^+) and chloride (Cl^-) ions according to the law of uninhibited decay. If the initial amount of salt is 25 kilograms and, after 10 hours, 15 kilograms of salt is left, how much salt is left after 1 day? How long does it take until $\frac{1}{2}$ kilogram of salt is left?

18. **Voltage of a Conductor** The voltage of a certain conductor decreases over time according to the law of uninhibited decay. If the initial voltage is 40 volts, and 2 seconds later it is 10 volts, what is the voltage after 5 seconds?

19. **Radioactivity from Chernobyl** After the release of radioactive material into the atmosphere from a nuclear power plant at Chernobyl (Ukraine) in 1986, the hay in Austria was contaminated by iodine 131 (half-life 8 days). If it is all right to feed the hay to cows when 10% of the iodine 131 remains, how long did the farmers need to wait to use this hay?

20. **Pig Roasts** The hotel Bora-Bora is having a pig roast. At noon, the chef put the pig in a large earthen oven. The pig's original temperature was 75°F. At 2:00 PM the chef checked the pig's temperature and was upset because it had reached only 100°F. If the oven's temperature remains a constant 325°F, at what time may the hotel serve its guests, assuming that pork is done when it reaches 175°F?

21. **Proportion of the Population That Owns a DVD Player** The logistic growth model

$$P(t) = \frac{0.9}{1 + 6e^{-0.32t}}$$

relates the proportion of U.S. households that own a DVD player to the year. Let $t = 0$ represent 2000, $t = 1$ represent 2001, and so on.
 (a) Determine the maximum proportion of households that will own a DVD player.
 (b) What proportion of households owned a DVD player in 2000 ($t = 0$)?
 (c) What proportion of households owned a DVD player in 2005 ($t = 5$)?
 (d) When will 0.8 (80%) of U.S. households own a DVD player?
 (e) How long will it be before 0.45 (45%) of the population own a DVD player?

22. **Market Penetration of Intel's Coprocessor** The logistic growth model

$$P(t) = \frac{0.90}{1 + 3.5e^{-0.339t}}$$

relates the proportion of new personal computers (PCs) sold at Best Buy that have Intel's latest coprocessor t months after it has been introduced.
 (a) Determine the maximum proportion of PCs sold at Best Buy that will have Intel's latest coprocessor.
 (b) What proportion of computers sold at Best Buy will have Intel's latest coprocessor when it is first introduced ($t = 0$)?
 (c) What proportion of PCs sold will have Intel's latest coprocessor $t = 4$ months after it is introduced?
 (d) When will 0.75 (75%) of PCs sold at Best Buy have Intel's latest coprocessor?
 (e) How long will it be before 0.45 (45%) of the PCs sold by Best Buy have Intel's latest coprocessor?

23. **Population of a Bacteria Culture** The logistic growth model

$$P(t) = \frac{1000}{1 + 32.33e^{-0.439t}}$$

represents the population (in grams) of a bacterium after t hours.
 (a) Determine the carrying capacity of the environment.
 (b) What is the growth rate of the bacteria?
 (c) Determine the initial population size.
 (d) What is the population after 9 hours?
 (e) When will the population be 700 grams?
 (f) How long does it take for the population to reach one-half of the carrying capacity?

24. **Population of an Endangered Species** Often environmentalists capture an endangered species and transport the species to a controlled environment where the species can produce offspring and regenerate its population. Suppose that six American bald eagles are captured, transported to Montana,

and set free. Based on experience, the environmentalists expect the population to grow according to the model

$$P(t) = \frac{500}{1 + 83.33e^{-0.162t}}$$

where t is measured in years.

(a) Determine the carrying capacity of the environment.

(b) What is the growth rate of the bald eagle?
(c) What is the population after 3 years?
(d) When will the population be 300 eagles?
(e) How long does it take for the population to reach one-half of the carrying capacity?

25. The *Challenger* Disaster After the *Challenger* disaster in 1986, a study was made of the 23 launches that preceded the fatal flight. A mathematical model was developed involving the relationship between the Fahrenheit temperature x around the O-rings and the number y of eroded or leaky primary O-rings. The model stated that

$$y = \frac{6}{1 + e^{-(5.085 - 0.1156x)}}$$

where the number 6 indicates the 6 primary O-rings on the spacecraft.
(a) What is the predicted number of eroded or leaky primary O-rings at a temperature of 100°F?
(b) What is the predicted number of eroded or leaky primary O-rings at a temperature of 60°F?
(c) What is the predicted number of eroded or leaky primary O-rings at a temperature of 30°F?
(d) Graph the equation using a graphing utility. At what temperature is the predicted number of eroded or leaky O-rings 1? 3? 5?

SOURCE: Linda Tappin, "Analyzing Data Relating to the *Challenger* Disaster," *Mathematics Teacher*, Vol. 87, No. 6, September 1994, pp. 423–426.

4.9 Building Exponential, Logarithmic, and Logistic Models from Data

PREPARING FOR THIS SECTION *Before getting started, review the following:*

• Scatter Diagrams; Linear Curve Fitting (Section 2.2, pp. 125–129) • Quadratic Functions of Best Fit (Section 2.6, pp. 161–162)

OBJECTIVES 1 Use a Graphing Utility to Fit an Exponential Function to Data
2 Use a Graphing Utility to Fit a Logarithmic Function to Data
3 Use a Graphing Utility to Fit a Logistic Function to Data

In Section 2.2, we discussed how to find the linear function of best fit ($y = ax + b$), and in Section 2.6, we discussed how to find the quadratic function of best fit ($y = ax^2 + bx + c$).

In this section we will discuss how to use a graphing utility to find equations of best fit that describe the relation between two variables when the relation is thought to be exponential ($y = ab^x$), logarithmic ($y = a + b \ln x$), or logistic $\left(y = \dfrac{c}{1 + ae^{-bx}} \right)$. As before, we draw a scatter diagram of the data to help to determine the appropriate model to use.

Figure 42 shows scatter diagrams that will typically be observed for the three models. Below each scatter diagram are any restrictions on the values of the parameters.

Figure 42

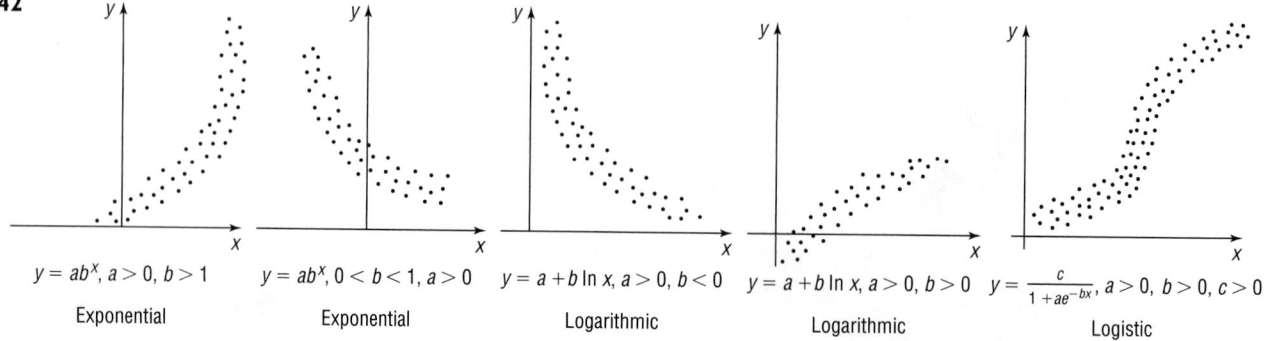

$y = ab^x, a > 0, b > 1$ $y = ab^x, 0 < b < 1, a > 0$ $y = a + b \ln x, a > 0, b < 0$ $y = a + b \ln x, a > 0, b > 0$ $y = \dfrac{c}{1 + ae^{-bx}}, a > 0, b > 0, c > 0$

Exponential Exponential Logarithmic Logarithmic Logistic

Most graphing utilities have REGression options that fit data to a specific type of curve. Once the data have been entered and a scatter diagram obtained, the type of curve that you want to fit to the data is selected. Then that REGression option is used to obtain the curve of *best fit* of the type selected.

The correlation coefficient r will appear only if the model can be written as a linear expression. As it turns out, r will appear for the linear, power, exponential, and logarithmic models, since these models can be written as a linear expression. Remember, the closer $|r|$ is to 1, the better the fit.

Let's look at some examples.

 1 Use a Graphing Utility to Fit an Exponential Function to Data

We saw in Section 4.7 that the future value of money behaves exponentially, and we saw in Section 4.8 that growth and decay models also behave exponentially. The next example shows how data can lead to an exponential model.

| EXAMPLE 1 | **Fitting an Exponential Function to Data** |

Beth is interested in finding a function that explains the closing price of Harley Davidson stock at the end of each year. She obtains the data shown in Table 9.

(a) Using a graphing utility, draw a scatter diagram with year as the independent variable.

(b) Using a graphing utility, fit an exponential function to the data.

(c) Express the function found in part (b) in the form $A = A_0 e^{kt}$.

(d) Graph the exponential function found in part (b) or (c) on the scatter diagram.

(e) Using the solution to part (b) or (c), predict the closing price of Harley David-son stock at the end of 2004.

(f) Interpret the value of k found in part (c).

Solution

(a) Enter the data into the graphing utility, letting 1 represent 1987, 2 represent 1988, and so on. We obtain the scatter diagram shown in Figure 43.

(b) A graphing utility fits the data in Figure 43 to an exponential function of the form $y = ab^x$ by using the EXPonential REGression option. From Figure 44 we find $y = ab^x = 0.47547(1.3582)^x$. Notice that $|r|$ is close to 1 indicating a good fit.

Table 9

Year, x	Closing Price, y
1987 (x = 1)	0.392
1988 (x = 2)	0.7652
1989 (x = 3)	1.1835
1990 (x = 4)	1.1609
1991 (x = 5)	2.6988
1992 (x = 6)	4.5381
1993 (x = 7)	5.3379
1994 (x = 8)	6.8032
1995 (x = 9)	7.0328
1996 (x = 10)	11.5585
1997 (x = 11)	13.4799
1998 (x = 12)	23.5424
1999 (x = 13)	31.9342
2000 (x = 14)	39.7277
2001 (x = 15)	54.31
2002 (x = 16)	46.20
2003 (x = 17)	47.53

Source: finance.yahoo.com

Figure 43

Figure 44

```
ExpReg
y=a*b^x
a=.4754667252
b=1.358201358
r²=.9660173956
r=.9828618395
```

(c) To express $y = ab^x$ in the form $A = A_0e^{kt}$, where $x = t$ and $y = A$, we proceed as follows:

$$ab^x = A_0e^{kt}, \qquad x = t$$

When $x = t = 0$, we find $a = A_0$. This leads to

$$a = A_0, \qquad b^x = e^{kt}$$
$$b^x = (e^k)^t$$
$$b = e^k \qquad x = t$$

Since $y = ab^x = 0.47547(1.3582)^x$, we find that $a = 0.47547$ and $b = 1.3582$:

$$a = A_0 = 0.47547 \quad \text{and} \quad b = 1.3582 = e^k$$

We want to find k, so we rewrite $1.3582 = e^k$ as a logarithm and obtain

$$k = \ln(1.3582) \approx 0.3062$$

As a result, $A = A_0e^{kt} = 0.47547e^{0.3062t}$.

(d) See Figure 45 for the graph of the exponential function of best fit.

(e) Let $t = 18$ (end of 2004) in the function found in part (c). The predicted closing price of Harley Davidson stock at the end of 2004 is

$$A = 0.47547e^{0.3062(18)} = \$117.70$$

(f) The value of k represents the annual interest rate compounded continuously.

$$A = A_0e^{kt} = 0.47547e^{0.3062t}$$
$$= Pe^{rt} \qquad \text{Equation (4), Section 4.7}$$

The price of Harley Davidson stock has grown at an annual rate of 30.62% (compounded continuously) between 1987 and 2003. ∎

Figure 45

NOW WORK PROBLEM **1**.

 2 Use a Graphing Utility to Fit a Logarithmic Function to Data

Many relations between variables do not follow an exponential model; instead, the independent variable is related to the dependent variable using a logarithmic model.

EXAMPLE 2

Table 10

Atmospheric Pressure, p	Height, h
760	0
740	0.184
725	0.328
700	0.565
650	1.079
630	1.291
600	1.634
580	1.862
550	2.235

Fitting a Logarithmic Function to Data

Jodi, a meteorologist, is interested in finding a function that explains the relation between the height of a weather balloon (in kilometers) and the atmospheric pressure (measured in millimeters of mercury) on the balloon. She collects the data shown in Table 10.

(a) Using a graphing utility, draw a scatter diagram of the data with atmospheric pressure as the independent variable.

(b) It is known that the relation between atmospheric pressure and height follows a logarithmic model. Using a graphing utility, fit a logarithmic function to the data.

(c) Draw the logarithmic function found in part (b) on the scatter diagram.

(d) Use the function found in part (b) to predict the height of the weather balloon if the atmospheric pressure is 560 millimeters of mercury.

Solution (a) After entering the data into the graphing utility, we obtain the scatter diagram shown in Figure 46.

(b) A graphing utility fits the data in Figure 46 to a logarithmic function of the form $y = a + b \ln x$ by using the LOGarithm REGression option. See Figure 47. The logarithmic function of best fit to the data is

$$h(p) = 45.7863 - 6.9025 \ln p$$

where h is the height of the weather balloon and p is the atmospheric pressure. Notice that $|r|$ is close to 1, indicating a good fit.

(c) Figure 48 shows the graph of $h(p) = 45.7863 - 6.9025 \ln p$ on the scatter diagram.

Figure 46

Figure 47

Figure 48

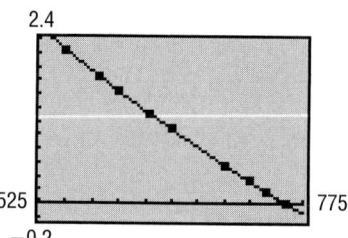

(d) Using the function found in part (b), Jodi predicts the height of the weather balloon when the atmospheric pressure is 560 to be

$$h(560) = 45.7863 - 6.9025 \ln 560$$
$$\approx 2.108 \text{ kilometers}$$

 NOW WORK PROBLEM 7.

 3 **Use a Graphing Utility to Fit a Logistic Function to Data**

Logistic growth models can be used to model situations for which the value of the dependent variable is limited. Many real-world situations conform to this scenario. For example, the population of the human race is limited by the availability of natural resources such as food and shelter. When the value of the dependent variable is limited, a logistic growth model is often appropriate.

EXAMPLE 3 **Fitting a Logistic Function to Data**

The data in Table 11 represent the amount of yeast biomass in a culture after t hours.

Table 11

Time (in hours)	Yeast Biomass	Time (in hours)	Yeast Biomass
0	9.6	10	513.3
1	18.3	11	559.7
2	29.0	12	594.8
3	47.2	13	629.4
4	71.1	14	640.8
5	119.1	15	651.1
6	174.6	16	655.9
7	257.3	17	659.6
8	350.7	18	661.8
9	441.0		

SOURCE: Tor Carlson (Über Geschwindigkeit und Grösse der Hefevermehrung in Würze, Biochemische Zeitschrift, Bd. 57, pp. 313–334, 1913)

(a) Using a graphing utility, draw a scatter diagram of the data with time as the independent variable.

(b) Using a graphing utility, fit a logistic function to the data.

(c) Using a graphing utility, graph the function found in part (b) on the scatter diagram.

(d) What is the predicted carrying capacity of the culture?

(e) Use the function found in part (b) to predict the population of the culture at $t = 19$ hours.

Solution

(a) See Figure 49 for a scatter diagram of the data.

(b) A graphing utility fits a logistic growth model of the form $y = \dfrac{c}{1 + ae^{-bx}}$ by using the LOGISTIC regression option. See Figure 50. The logistic function of best fit to the data is

$$y = \frac{663.0}{1 + 71.6e^{-0.5470x}}$$

where y is the amount of yeast biomass in the culture and x is the time.

Figure 49

Figure 50

Figure 51

(c) See Figure 51 for the graph of the logistic function of best fit.

(d) Based on the logistic growth function found in part (b), the carrying capacity of the culture is 663.

(e) Using the logistic growth function found in part (b), the predicted amount of yeast biomass at $t = 19$ hours is

$$y = \frac{663.0}{1 + 71.6e^{-0.5470(19)}} = 661.5$$

NOW WORK PROBLEM **9.**

4.9 Assess Your Understanding

Applications and Extensions

1. **Biology** A strain of E-coli Beu 397-recA441 is placed into a nutrient broth at 30°Celsius and allowed to grow. The following data are collected. Theory states that the number of bacteria in the petri dish will initially grow according to the law of uninhibited growth. The population is measured using an optical device in which the amount of light that passes through the petri dish is measured.

Time (hours), x	Population, y
0	0.09
2.5	0.18
3.5	0.26
4.5	0.35
6	0.50

SOURCE: Dr. Polly Lavery, Joliet Junior College

(a) Draw a scatter diagram treating time as the independent variable.

(b) Using a graphing utility, fit an exponential function to the data.

(c) Express the function found in part (b) in the form $N(t) = N_0 e^{kt}$.

(d) Graph the exponential function found in part (b) or (c) on the scatter diagram.

(e) Use the exponential function from part (b) or (c) to predict the population at $x = 7$ hours.

(f) Use the exponential function from part (b) or (c) to predict when the population will reach 0.75.

2. Biology A strain of E-coli SC18del-recA718 is placed into a nutrient broth at 30°Celsius and allowed to grow. The following data are collected. Theory states that the number of bacteria in the petri dish will initially grow according to the law of uninhibited growth. The population is measured using an optical device in which the amount of light that passes through the petri dish is measured.

Time (hours), x	Population, y
2.5	0.175
3.5	0.38
4.5	0.63
4.75	0.76
5.25	1.20

Source: Dr. Polly Lavery, Joliet Junior College

(a) Draw a scatter diagram treating time as the independent variable.

(b) Using a graphing utility, fit an exponential function to the data.

(c) Express the function found in part (b) in the form $N(t) = N_0 e^{kt}$.

(d) Graph the exponential function found in part (b) or (c) on the scatter diagram.

(e) Use the exponential function from part (b) or (c) to predict the population at $x = 6$ hours.

(f) Use the exponential function from part (b) or (c) to predict when the population will reach 2.1.

3. Chemistry A chemist has a 100-gram sample of a radioactive material. He records the amount of radioactive material every week for 6 weeks and obtains the following data:

Week	Weight (in Grams)
0	100.0
1	88.3
2	75.9
3	69.4
4	59.1
5	51.8
6	45.5

(a) Using a graphing utility, draw a scatter diagram with week as the independent variable.

(b) Using a graphing utility, fit an exponential function to the data.

(c) Express the function found in part (b) in the form $A(t) = A_0 e^{kt}$.

(d) Graph the exponential function found in part (b) or (c) on the scatter diagram.

(e) From the result found in part (b), determine the half-life of the radioactive material.

(f) How much radioactive material will be left after 50 weeks?

(g) When will there be 20 grams of radioactive material?

4. Chemistry A chemist has a 1000-gram sample of a radioactive material. She records the amount of radioactive material remaining in the sample every day for a week and obtains the following data:

Day	Weight (in Grams)
0	1000.0
1	897.1
2	802.5
3	719.8
4	651.1
5	583.4
6	521.7
7	468.3

(a) Using a graphing utility, draw a scatter diagram with day as the independent variable.

(b) Using a graphing utility, fit an exponential function to the data.

(c) Express the function found in part (b) in the form $A(t) = A_0 e^{kt}$.

(d) Graph the exponential function found in part (b) or (c) on the scatter diagram.

(e) From the result found in part (b), find the half-life of the radioactive material.

(f) How much radioactive material will be left after 20 days?

(g) When will there be 200 grams of radioactive material?

5. Finance The following data represent the amount of money an investor has in an investment account each year for 10 years. She wishes to determine the average annual rate of return on her investment.

Year	Value of Account
1994	$10,000
1995	$10,573
1996	$11,260
1997	$11,733
1998	$12,424
1999	$13,269
2000	$13,968
2001	$14,823
2002	$15,297
2003	$16,539

(a) Using a graphing utility, draw a scatter diagram with time as the independent variable and the value of the account as the dependent variable.

(b) Using a graphing utility, fit an exponential function to the data.

(c) Based on the answer in part (b), what was the average annual rate of return from this account over the past 10 years?

(d) If the investor plans on retiring in 2021, what will the predicted value of this account be?

(e) When will the account be worth $50,000?

6. Finance The following data show the amount of money an investor has in an investment account each year for 7 years. He wishes to determine the average annual rate of return on his investment.

Year	Value of Account
1997	$20,000
1998	$21,516
1999	$23,355
2000	$24,885
2001	$27,434
2002	$30,053
2003	$32,622

(a) Using a graphing utility, draw a scatter diagram with time as the independent variable and the value of the account as the dependent variable.

(b) Using a graphing utility, fit an exponential function to the data.

(c) Based on the answer to part (b), what was the average annual rate of return from this account over the past 7 years?

(d) If the investor plans on retiring in 2020, what will the predicted value of his account be?

(e) When will the account be worth $80,000?

7. Economics and Marketing The following data represent the price and quantity demanded in 2005 for IBM personal computers.

Price ($/Computer)	Quantity Demanded
2300	152
2000	159
1700	164
1500	171
1300	176
1200	180
1000	189

(a) Using a graphing utility, draw a scatter diagram of the data with price as the dependent variable.

(b) Using a graphing utility, fit a logarithmic function to the data.

(c) Using a graphing utility, draw the logarithmic function found in part (b) on the scatter diagram.

(d) Use the function found in part (b) to predict the number of IBM personal computers that would be demanded if the price were $1650.

8. Economics and Marketing The following data represent the price and quantity supplied in 2005 for IBM personal computers.

Price ($/Computer)	Quantity Supplied
2300	180
2000	173
1700	160
1500	150
1300	137
1200	130
1000	113

(a) Using a graphing utility, draw a scatter diagram of the data with price as the dependent variable.

(b) Using a graphing utility, fit a logarithmic function to the data.

(c) Using a graphing utility, draw the logarithmic function found in part (b) on the scatter diagram.

(d) Use the function found in part (b) to predict the number of IBM personal computers that would be supplied if the price were $1650.

9. Population Model The following data represent the population of the United States. An ecologist is interested in finding a function that describes the population of the United States.

Year	Population
1900	76,212,168
1910	92,228,496
1920	106,021,537
1930	123,202,624
1940	132,164,569
1950	151,325,798
1960	179,323,175
1970	203,302,031
1980	226,542,203
1990	248,709,873
2000	281,421,906

SOURCE: U.S. Census Bureau

(a) Using a graphing utility, draw a scatter diagram of the data using the year as the independent variable and population as the dependent variable.

(b) Using a graphing utility, fit a logistic function to the data.
(c) Using a graphing utility, draw the function found in part (b) on the scatter diagram.
(d) Based on the function found in part (b), what is the carrying capacity of the United States?
(e) Use the function found in part (b) to predict the population of the United States in 2004.
(f) When will the United States population be 300,000,000?
(g) Compare actual U.S. Census figures to the prediction found in part (e).

10. Population Model The following data represent the world population. An ecologist is interested in finding a function that describes the world population.

Year	Population (in Billions)
1993	5.531
1994	5.611
1995	5.691
1996	5.769
1997	5.847
1998	5.925
1999	6.003
2000	6.080
2001	6.157

Source: U.S. Census Bureau

(a) Using a graphing utility, draw a scatter diagram of the data using year as the independent variable and population as the dependent variable.
(b) Using a graphing utility, fit a logistic function to the data.
(c) Using a graphing utility, draw the function found in part (b) on the scatter diagram.
(d) Based on the function found in part (b), what is the carrying capacity of the world?
(e) Use the function found in part (b) to predict the population of the world in 2004.
(f) When will world population be 7 billion?
(g) Compare actual U.S. Census figures to the prediction found in part (e).

11. Population Model The following data represent the population of Illinois. An urban economist is interested in finding a model that describes the population of Illinois

(a) Using a graphing utility, draw a scatter diagram of the data using year as the independent variable and population as the dependent variable.
(b) Using a graphing utility, fit a logistic function to the data.
(c) Using a graphing utility, draw the function found in part (b) on the scatter diagram.

Year	Population
1900	4,821,550
1910	5,638,591
1920	6,485,280
1930	7,630,654
1940	7,897,241
1950	8,712,176
1960	10,081,158
1970	11,110,285
1980	11,427,409
1990	11,430,602
2000	12,419,293

Source: U.S. Census Bureau

(d) Based on the function found in part (b), what is the carrying capacity of Illinois?
(e) Use the function found in part (b) to predict the population of Illinois in 2010.

12. Population Model The following data represent the population of Pennsylvania. An urban economist is interested in finding a model that describes the population of Pennsylvania.

(a) Using a graphing utility, draw a scatter diagram of the data using year as the independent variable and population as the dependent variable.
(b) Using a graphing utility, fit a logistic function to the data.
(c) Using a graphing utility, draw the function found in part (b) on the scatter diagram.
(d) Based on the function found in part (b), what is the carrying capacity of Pennsylvania?
(e) Use the function found in part (b) to predict the population of Pennsylvania in 2010.

Year	Population
1900	6,302,115
1910	7,665,111
1920	8,720,017
1930	9,631,350
1940	9,900,180
1950	10,498,012
1960	11,319,366
1970	11,800,766
1980	11,864,720
1990	11,881,643
2000	12,281,054

Source: U.S. Census Bureau

Chapter Review

Things to Know

Composite function (p. 264)	$(f \circ g)(x) = f(g(x))$
One-to-one function f (p. 272)	A function whose inverse is also a function
	For any choice of elements x_1, x_2 in the domain of f, if $x_1 \neq x_2$, then $f(x_1) \neq f(x_2)$.
Horizontal-line test (p. 273)	If every horizontal line intersects the graph of a function f in at most one point, f is one-to-one.
Inverse function f^{-1} of f (pp. 273–276)	Domain of f = range of f^{-1}; range of f = domain of f^{-1}
	$f^{-1}(f(x)) = x$ for all x in the domain of f, and $f(f^{-1}(x)) = x$ for all x in the domain of f^{-1}.
	Graphs of f and f^{-1} are symmetric with respect to the line $y = x$.

Properties of the exponential function (pp. 289 and 290)

$f(x) = a^x, \quad a > 1$
- Domain: the interval $(-\infty, \infty)$
- Range: the interval $(0, \infty)$
- x-intercepts: none; y-intercept: 1
- Horizontal asymptote: x-axis ($y = 0$) as $x \to -\infty$
- Increasing; one-to-one; smooth; continuous
- See Figure 21 for a typical graph.

$f(x) = a^x, \quad 0 < a < 1$
- Domain: the interval $(-\infty, \infty)$
- Range: the interval $(0, \infty)$
- x-intercepts: none; y-intercept: 1
- Horizontal asymptote: x-axis ($y = 0$) as $x \to \infty$
- Decreasing; one-to-one; smooth; continuous
- See Figure 25 for a typical graph.

Number e (p. 291)	Value approached by the expression $\left(1 + \dfrac{1}{n}\right)^n$ as $n \to \infty$; that is, $\displaystyle\lim_{n \to \infty}\left(1 + \dfrac{1}{n}\right)^n = e$
Property of exponents (p. 293)	If $a^u = a^v$, then $u = v$.

Properties of the logarithmic function (pp. 302–303)

$f(x) = \log_a x, \quad a > 1$
($y = \log_a x$ means $x = a^y$)
- Domain: the interval $(0, \infty)$
- Range: the interval $(-\infty, \infty)$
- x-intercept: 1; y-intercept; none
- Vertical asymptote: $x = 0$ (y-axis)
- Increasing; one-to-one; smooth; continuous
- See Figure 30(b) for a typical graph.

$f(x) = \log_a x, \quad 0 < a < 1$
($y = \log_a x$ means $x = a^y$)
- Domain: the interval $(0, \infty)$
- Range: the interval $(-\infty, \infty)$
- x-intercept: 1; y-intercept; none
- Vertical asymptote: $x = 0$ (y-axis)
- Decreasing; one-to-one; smooth; continuous
- See Figure 30(a) for a typical graph.

Natural logarithm (p. 303)	$y = \ln x$ means $x = e^y$.

Properties of logarithms (pp. 312, 313, 316)

$\log_a 1 = 0 \qquad \log_a a = 1 \qquad a^{\log_a M} = M \qquad \log_a a^r = r$

$\log_a(MN) = \log_a M + \log_a N \qquad \log_a\left(\dfrac{M}{N}\right) = \log_a M - \log_a N$

$\log_a M^r = r \log_a M$

If $M = N$, then $\log_a M = \log_a N$.

If $\log_a M = \log_a N$, then $M = N$.

Formulas

Change-of-Base Formula (p. 316)	$\log_a M = \dfrac{\log_b M}{\log_b a}$
Compound Interest Formula (p. 327)	$A = P \cdot \left(1 + \dfrac{r}{n}\right)^{nt}$
Continuous compounding (p. 329)	$A = Pe^{rt}$
Present Value Formulas (p. 330)	$P = A \cdot \left(1 + \dfrac{r}{n}\right)^{-nt}$ or $P = Ae^{-rt}$
Growth and decay (p. 334)	$A(t) = A_0 e^{kt}$
Newton's Law of Cooling (p. 338)	$u(t) = T + (u_0 - T)e^{kt}, \quad k < 0$
Logistic model (p. 340)	$P(t) = \dfrac{c}{1 + ae^{-bt}}$

Objectives

Section		You should be able to . . .	Review Exercises
4.1	1	Form a composite function (p. 264)	1–12
	2	Find the domain of a composite function (p. 265)	7–12
4.2	1	Determine whether a function is one-to-one (p. 271)	13(a), 14(a), 15, 16
	2	Determine the inverse of a function defined by a map on or an ordered pair (p. 273)	13(b), 14(b)
	3	Obtain the graph of the inverse function from the graph of the function (p. 276)	15, 16
	4	Find the inverse of a function defined by an equation (p. 277)	17–22
4.3	1	Evaluate exponential functions (p. 285)	23(a), (c); 24(a), (c), 87(a), 97(a)
	2	Graph exponential functions (p. 287)	55–59, 62, 63
	3	Define the number e (p. 291)	59, 62, 63
	4	Solve exponential equations (p. 293)	65–68, 73, 74, 76–78, 87(b), 97(d)
4.4	1	Change exponential expressions to logarithmic expressions and logarithmic expressions to exponential expressions (p. 301)	25–28
	2	Evaluate logarithmic expressions (p. 301)	23(b), (d), 24(b), (d), 33–34, 83(b), 84(b), 85, 86, 88(a), 89
	3	Determine the domain of a logarithmic function (p. 302)	29–32
	4	Graph logarithmic functions (p. 302)	60, 61, 64, 83(a), 84(a)
	5	Solve logarithmic equations (p. 305)	69, 70, 75, 83(c), 84(c), 88(b)
4.5	1	Work with the properties of logarithms (p. 312)	35–38
	2	Write a logarithmic expression as a sum or difference of logarithms (p. 314)	39–44
	3	Write a logarithmic expression as a single logarithm (p. 315)	45–50
	4	Evaluate logarithms whose base is neither 10 nor e (p. 316)	51, 52
4.6	1	Solve logarithmic equations using the properties of logarithms (p. 320)	79, 80
	2	Solve exponential equations (p. 321)	71, 72, 81–82
	3	Solve logarithmic and exponential equations using a graphing utility (p. 323)	65–82
4.7	1	Determine the future value of a lump sum of money (p. 326)	90, 92
	2	Calculate effective rates of return (p. 329)	90
	3	Determine the present value of a lump sum of money (p. 330)	91
	4	Determine the time required to double or triple a lump sum of money (p. 331)	90
4.8	1	Find equations of populations that obey the law of uninhibited growth (p. 334)	95
	2	Find equations of populations that obey the law of decay (p. 337)	93, 96
	3	Use Newton's Law of Cooling (p. 338)	94
	4	Use logistic models (p. 340)	97

4.9	1	Use a graphing utility to fit an exponential function to data (p. 346)	98
	2	Use a graphing utility to fit a logarithmic function to data (p. 347)	99
	3	Use a graphing utility to fit a logistic function to data (p. 348)	100

Review Exercises

In Problems 1–6, for the given functions f and g find:

(a) $(f \circ g)(2)$ (b) $(g \circ f)(-2)$ (c) $(f \circ f)(4)$ (d) $(g \circ g)(-1)$

1. $f(x) = 3x - 5;$ $g(x) = 1 - 2x^2$ **2.** $f(x) = 4 - x;$ $g(x) = 1 + x^2$

3. $f(x) = \sqrt{x + 2};$ $g(x) = 2x^2 + 1$ **4.** $f(x) = 1 - 3x^2;$ $g(x) = \sqrt{4 - x}$

5. $f(x) = e^x;$ $g(x) = 3x - 2$ **6.** $f(x) = \dfrac{2}{1 + 2x^2};$ $g(x) = 3x$

In Problems 7–12, find $f \circ g$, $g \circ f$, $f \circ f$, and $g \circ g$ for each pair of functions. State the domain of each composite function.

7. $f(x) = 2 - x;$ $g(x) = 3x + 1$ **8.** $f(x) = 2x - 1;$ $g(x) = 2x + 1$

9. $f(x) = 3x^2 + x + 1;$ $g(x) = |3x|$ **10.** $f(x) = \sqrt{3x};$ $g(x) = 1 + x + x^2$

11. $f(x) = \dfrac{x + 1}{x - 1};$ $g(x) = \dfrac{1}{x}$ **12.** $f(x) = \sqrt{x - 3};$ $g(x) = \dfrac{3}{x}$

In Problems 13 and 14, (a) verify that the function is one-to-one, and (b) find the inverse of the given function.

13. $\{(1, 2), (3, 5), (5, 8), (6, 10)\}$ **14.** $\{(-1, 4), (0, 2), (1, 5), (3, 7)\}$

In Problems 15 and 16, state why the graph of the function is one-to-one. Then draw the graph of the inverse function f^{-1}. For convenience (and as a hint), the graph of $y = x$ is also given.

15.

16.

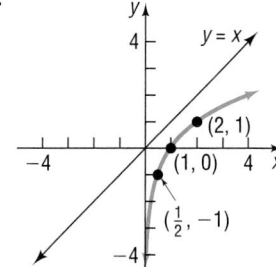

In Problems 17–22, the function f is one-to-one. Find the inverse of each function and check your answer. Find the domain and the range of f and f^{-1}.

17. $f(x) = \dfrac{2x + 3}{5x - 2}$ **18.** $f(x) = \dfrac{2 - x}{3 + x}$ **19.** $f(x) = \dfrac{1}{x - 1}$

20. $f(x) = \sqrt{x - 2}$ **21.** $f(x) = \dfrac{3}{x^{1/3}}$ **22.** $f(x) = x^{1/3} + 1$

In Problems 23 and 24, $f(x) = 3^x$ and $g(x) = \log_3 x$.

23. Evaluate: (a) $f(4)$ (b) $g(9)$ (c) $f(-2)$ (d) $g\left(\dfrac{1}{27}\right)$

24. Evaluate: (a) $f(1)$ (b) $g(81)$ (c) $f(-4)$ (d) $g\left(\dfrac{1}{243}\right)$

In Problems 25 and 26, convert each exponential expression to an equivalent expression involving a logarithm. In Problems 27 and 28, convert each logarithmic expression to an equivalent expression involving an exponent.

25. $5^2 = z$ **26.** $a^5 = m$ **27.** $\log_5 u = 13$ **28.** $\log_a 4 = 3$

In Problems 29–32, find the domain of each logarithmic function.

29. $f(x) = \log(3x - 2)$ **30.** $F(x) = \log_5(2x + 1)$ **31.** $H(x) = \log_2(x^2 - 3x + 2)$ **32.** $F(x) = \ln(x^2 - 9)$

In Problems 33–38, evaluate each expression. Do not use a calculator.

33. $\log_2\left(\dfrac{1}{8}\right)$

34. $\log_3 81$

35. $\ln e^{\sqrt{2}}$

36. $e^{\ln 0.1}$

37. $2^{\log_2 0.4}$

38. $\log_2 2^{\sqrt{3}}$

In Problems 39–44, write each expression as the sum and/or difference of logarithms. Express powers as factors.

39. $\log_3\left(\dfrac{uv^2}{w}\right)$, $u > 0, v > 0, w > 0$ **40.** $\log_2\left(a^2\sqrt{b}\right)^4$, $a > 0, b > 0$ **41.** $\log\left(x^2\sqrt{x^3 + 1}\right)$, $x > 0$

42. $\log_5\left(\dfrac{x^2 + 2x + 1}{x^2}\right)$, $x > 0$ **43.** $\ln\left(\dfrac{x\sqrt[3]{x^2 + 1}}{x - 3}\right)$, $x > 3$ **44.** $\ln\left(\dfrac{2x + 3}{x^2 - 3x + 2}\right)^2$, $x > 2$

In Problems 45–50, write each expression as a single logarithm.

45. $3\log_4 x^2 + \dfrac{1}{2}\log_4 \sqrt{x}$

46. $-2\log_3\left(\dfrac{1}{x}\right) + \dfrac{1}{3}\log_3 \sqrt{x}$

47. $\ln\left(\dfrac{x - 1}{x}\right) + \ln\left(\dfrac{x}{x + 1}\right) - \ln(x^2 - 1)$

48. $\log(x^2 - 9) - \log(x^2 + 7x + 12)$

49. $2\log 2 + 3\log x - \dfrac{1}{2}[\log(x + 3) + \log(x - 2)]$

50. $\dfrac{1}{2}\ln(x^2 + 1) - 4\ln\dfrac{1}{2} - \dfrac{1}{2}[\ln(x - 4) + \ln x]$

In Problems 51 and 52, use the Change-of-Base Formula and a calculator to evaluate each logarithm. Round your answer to three decimal places.

51. $\log_4 19$

52. $\log_2 21$

In Problems 53 and 54, graph each function using a graphing utility and the Change-of-Base Formula.

53. $y = \log_3 x$

54. $y = \log_7 x$

In Problems 55–64, use transformations to graph each function. Determine the domain, range, and any asymptotes.

55. $f(x) = 2^{x-3}$ **56.** $f(x) = -2^x + 3$ **57.** $f(x) = \dfrac{1}{2}(3^{-x})$ **58.** $f(x) = 1 + 3^{2x}$

59. $f(x) = 1 - e^x$ **60.** $f(x) = 3 + \ln x$ **61.** $f(x) = \dfrac{1}{2}\ln(x + 3)$ **62.** $f(x) = 3e^x$

63. $f(x) = 3 - e^{-x}$ **64.** $f(x) = 4 - \ln(-x)$

In Problems 65–82, solve each equation.

65. $4^{1-2x} = 2$ **66.** $8^{6+3x} = 4$ **67.** $3^{x^2+x} = \sqrt{3}$ **68.** $4^{x-x^2} = \dfrac{1}{2}$

69. $\log_x 64 = -3$ **70.** $\log_{\sqrt{2}} x = -6$ **71.** $5^x = 3^{x+2}$ **72.** $5^{x+2} = 7^{x-2}$

73. $9^{2x} = 27^{3x-4}$ **74.** $25^{2x} = 5^{x^2-12}$ **75.** $\log_3 \sqrt{x - 2} = 2$ **76.** $2^{x+1} \cdot 8^{-x} = 4$

77. $8 = 4^{x^2} \cdot 2^{5x}$ **78.** $2^x \cdot 5 = 10^x$ **79.** $\log_6(x + 3) + \log_6(x + 4) = 1$

80. $\log(7x - 12) = 2\log x$ **81.** $e^{1-x} = 5$ **82.** $e^{1-2x} = 4$

83. Suppose that $f(x) = \log_2(x - 2) + 1$.
 (a) Graph f.
 (b) What is $f(6)$? What point is on the graph of f?
 (c) Solve $f(x) = 4$. What point is on the graph of f?
 (d) What is the zero of f?
 (e) Based on the graph drawn in part (a) and the zero found in part (d), solve $f(x) > 0$.
 (f) Find $f^{-1}(x)$. Graph f^{-1} on the same Cartesian plane as f.

84. Suppose that $f(x) = \log_3(x + 1) - 4$.
 (a) Graph f.
 (b) What is $f(8)$? What point is on the graph of f?
 (c) Solve $f(x) = -3$. What point is on the graph of f?
 (d) What is the zero of f?
 (e) Based on the graph drawn in part (a) and the zero found in part (d), solve $f(x) < 0$.
 (f) Find $f^{-1}(x)$. Graph f^{-1} on the same Cartesian plane as f.

In Problems 85 and 86, use the following result: If x is the atmospheric pressure (measured in millimeters of mercury), then the formula for the altitude h(x) (measured in meters above sea level) is

$$h(x) = (30T + 8000) \log\left(\frac{P_0}{x}\right)$$

where T is the temperature (in degrees Celsius) and P_0 is the atmospheric pressure at sea level, which is approximately 760 millimeters of mercury.

85. Finding the Altitude of an Airplane At what height is a Piper Cub whose instruments record an outside temperature of 0°C and a barometric pressure of 300 millimeters of mercury?

86. Finding the Height of a Mountain How high is a mountain if instruments placed on its peak record a temperature of 5°C and a barometric pressure of 500 millimeters of mercury?

87. Amplifying Sound An amplifier's power output P (in watts) is related to its decibel voltage gain d by the formula $P = 25e^{0.1d}$.

(a) Find the power output for a decibel voltage gain of 4 decibels.
(b) For a power output of 50 watts, what is the decibel voltage gain?

88. Limiting Magnitude of a Telescope A telescope is limited in its usefulness by the brightness of the star that it is aimed at and by the diameter of its lens. One measure of a star's brightness is its *magnitude*; the dimmer the star, the larger its magnitude. A formula for the limiting magnitude L of a telescope, that is, the magnitude of the dimmest star that it can be used to view, is given by

$$L = 9 + 5.1 \log d$$

where d is the diameter (in inches) of the lens.
(a) What is the limiting magnitude of a 3.5-inch telescope?
(b) What diameter is required to view a star of magnitude 14?

89. Salvage Value The number of years n for a piece of machinery to depreciate to a known salvage value can be found using the formula

$$n = \frac{\log s - \log i}{\log(1 - d)}$$

where s is the salvage value of the machinery, i is its initial value, and d is the annual rate of depreciation.
(a) How many years will it take for a piece of machinery to decline in value from $90,000 to $10,000 if the annual rate of depreciation is 0.20 (20%)?
(b) How many years will it take for a piece of machinery to lose half of its value if the annual rate of depreciation is 15%?

90. Funding a College Education A child's grandparents purchase a $10,000 bond fund that matures in 18 years to be used for her college education. The bond fund pays 4% interest compounded semiannually. How much will the bond fund be worth at maturity? What is the effective rate of in-

terest? How long would it take the bond to double in value under these terms?

91. Funding a College Education A child's grandparents wish to purchase a bond that matures in 18 years to be used for her college education. The bond pays 4% interest compounded semiannually. How much should they pay so that the bond will be worth $85,000 at maturity?

92. Funding an IRA First Colonial Bankshares Corporation advertised the following IRA investment plans.

Target IRA Plans	
For each $5000 Maturity Value Desired	
Deposit:	At a Term of:
$620.17	20 Years
$1045.02	15 Years
$1760.92	10 Years
$2967.26	5 Years

(a) Assuming continuous compounding, what annual rate of interest did they offer?
(b) First Colonial Bankshares claims that $4000 invested today will have a value of over $32,000 in 20 years. Use the answer found in part (a) to find the actual value of $4000 in 20 years. Assume continuous compounding.

93. Estimating the Date That a Prehistoric Man Died The bones of a prehistoric man found in the desert of New Mexico contain approximately 5% of the original amount of carbon 14. If the half-life of carbon 14 is 5600 years, approximately how long ago did the man die?

94. Temperature of a Skillet A skillet is removed from an oven whose temperature is 450°F and placed in a room whose temperature is 70°F. After 5 minutes, the temperature of the skillet is 400°F. How long will it be until its temperature is 150°F?

95. World Population The growth rate of the world's population in 2003 was k = 1.16% = 0.0116. The population of the world in 2003 was 6,302,486,693. Letting t = 0 represent 2003, use the uninhibited growth model to predict the world's population in the year 2010.
Source: U.S. Census Bureau

96. Radioactive Decay The half-life of radioactive cobalt is 5.27 years. If 100 grams of radioactive cobalt is present now, how much will be present in 20 years? In 40 years?

97. Logistic Growth The logistic growth model

$$P(t) = \frac{0.8}{1 + 1.67e^{-0.16t}}$$

represents the proportion of new cars with a Global Positioning System (GPS). Let $t = 0$ represent 2003, $t = 1$ represent 2004, and so on.
(a) What proportion of new cars in 2003 had a GPS?
(b) Determine the maximum proportion of new cars that have a GPS.
(c) Using a graphing utility, graph $P(t)$.
(d) When will 75% of new cars have a GPS?

98. CBL Experiment The following data were collected by placing a temperature probe in a portable heater, removing the probe, and then recording temperature over time.

Time (sec.)	Temperature (°F)
0	165.07
1	164.77
2	163.99
3	163.22
4	162.82
5	161.96
6	161.20
7	160.45
8	159.35
9	158.61
10	157.89
11	156.83
12	156.11
13	155.08
14	154.40
15	153.72

According to Newton's Law of Cooling, these data should follow an exponential model.
(a) Using a graphing utility, draw a scatter diagram for the data.
(b) Using a graphing utility, fit an exponential function to the data.
(c) Graph the exponential function found in part (b) on the scatter diagram.
(d) Predict how long it will take for the probe to reach a temperature of 110°F.

99. Wind Chill Factor The following data represent the wind speed (mph) and wind chill factor at an air temperature of 15°F.

Wind Speed (mph)	Wind Chill Factor (°F)
5	7
10	3
15	0
20	−2
25	−4
30	−5
35	−7

SOURCE: U.S. National Weather Service

(a) Using a graphing utility, draw a scatter diagram with wind speed as the independent variable.
(b) Using a graphing utility, fit a logarithmic function to the data.
(c) Using a graphing utility, draw the logarithmic function found in part (b) on the scatter diagram.
(d) Use the function found in part (b) to predict the wind chill factor if the air temperature is 15°F and the wind speed is 23 mph.

100. Spreading of a Disease Jack and Diane live in a small town of 50 people. Unfortunately, Jack and Diane both have a cold. Those who come in contact with someone who has this cold will themselves catch the cold. The following data represent the number of people in the small town who have caught the cold after t days.

Days, t	Number of People with Cold, C
0	2
1	4
2	8
3	14
4	22
5	30
6	37
7	42
8	44

(a) Using a graphing utility, draw a scatter diagram of the data. Comment on the type of relation that appears to exist between the days and number of people with a cold.
(b) Using a graphing utility, fit a logistic function to the data.
(c) Graph the function found in part (b) on the scatter diagram.
(d) According to the function found in part (b), what is the maximum number of people who will catch the cold? In reality, what is the maximum number of people who could catch the cold?
(e) Sometime between the second and third day, 10 people in the town had a cold. According to the model found in part (b), when did 10 people have a cold?
(f) How long will it take for 46 people to catch the cold?

Chapter Test

1. Given $f(x) = \dfrac{x + 2}{x - 2}$ and $g(x) = 2x + 5$, find:

 (a) $f \circ g$ and state its domain (b) $(g \circ f)(-2)$

 (c) $(f \circ g)(-2)$

2. Determine whether the function is one-to-one.

 (a) $y = 4x^2 + 3$ (b) $y = \sqrt{x + 3} - 5$

3. Find the inverse of $f(x) = \dfrac{2}{3x - 5}$ and check your answer. State the domain and the range of f and f^{-1}.

4. If the point $(3, -5)$ is on the graph of a one-to-one function f, what point must be on the graph of f^{-1}?

In Problems 5–7, find the unknown value without using a calculator.

5. $3^x = 243$ **6.** $\log_b 16 = 2$ **7.** $\log_5 x = 4$

In Problems 8–11, use a calculator to evaluate each expression. Round your answer to three decimal places.

8. $e^3 + 2$ **9.** $\log 20$

10. $\log_3 21$ **11.** $\ln 133$

In Problems 12 and 13, use transformations to graph each function. Determine the domain, range, and any asymptotes.

12. $f(x) = 4^{x+1} - 2$ **13.** $g(x) = 1 - \log_5(x - 2)$

In Problems 14–19, solve each equation.

14. $5^{x+2} = 125$ **15.** $\log(x + 9) = 2$

16. $8 - 2e^{-x} = 4$ **17.** $\log(x^2 + 3) = \log(x + 6)$

18. $7^{x+3} = e^x$ **19.** $\log_2(x - 4) + \log_2(x + 4) = 3$

20. Write $\log_2\left(\dfrac{4x^3}{x^2 - 3x - 18}\right)$ as the sum and/or difference of logarithms. Express powers as factors.

21. A 50-mg sample of a radioactive substance decays to 34 mg after 30 days. How long will it take for there to be 2 mg remaining?

22. The average cost of college at 4-year private colleges was $19,710 in 2003–2004. This was a 6% increase from the previous year.

 (a) If the cost of college increases by 6% each year, what will be the average cost of college at 4-year private colleges in 2013–2014?

 (b) College savings plans allow individuals to put money aside now to help pay for college later. If one such plan offers a rate of 5% compounded continuously, how much should be put in a college savings plan in 2003 to pay for 1 year of the cost of college at a 4-year private college in 2013?

23. The decibel level, D, of sound is given by the equation $D = 10 \log\left(\dfrac{I}{I_0}\right)$, where I is the intensity of the sound and $I_0 = 10^{-12}$ watt per square meter.

 (a) If the shout of a single person measures 80 decibels, how loud will the sound be if two people shout at the same time? That is, how loud would the sound be if the intensity doubled?

 (b) The pain threshold for sound is 125 decibels. If the Athens Olympic Stadium 2004 (Olympiako Stadio Athinas 'Spyros Louis') can seat 74,400 people, how many people in the crowd need to shout at the same time for the resulting sound level to meet or exceed the pain threshold? (Ignore any possible sound dampening.)

24. The table shows the estimated number of U.S. cell phone subscribers (in millions), y, from 1985 to 2003. Use a graphing utility to make a scatter diagram of the data. Fit a logistic model to the data and use the model to predict the number of U.S. cell phone subscribers in 2007. Let $x =$ the number of years since 1985.

 SOURCE: Cellular Telecommunications & Internet Association

x	0	2	4	6	8	10	12	14	16	18
y	0.34	1.23	3.51	7.56	16.01	33.76	55.31	86.05	128.37	158.72

Chapter Projects

1. Hot Coffee A fast-food restaurant wants a special container to hold coffee. The restaurant wishes the container to quickly cool the coffee from 200° to 130°F and keep the liquid between 110° and 130°F as long as possible. The restaurant has three containers to select from.

 1. The CentiKeeper Company has a container that reduces the temperature of a liquid from 200° to 100°F in 30 minutes by maintaining a constant temperature of 70°F.

 2. The TempControl Company has a container that reduces the temperature of a liquid from 200° to 110°F in 25 minutes by maintaining a constant temperature of 60°F.

 3. The Hot'n'Cold Company has a container that reduces the temperature of a liquid from 200° to 120°F in 20 minutes by maintaining a constant temperature of 65°F.

 You need to recommend which container the restaurant should purchase.

(a) Use Newton's Law of Cooling to find a function relating the temperature of the liquid over time for each container.

(b) How long does it take each container to lower the coffee temperature from 200° to 130°F?

(c) How long will the coffee temperature remain between 110° and 130°F? This temperature is considered the optimal drinking temperature.

(d) Graph each function using a graphing utility.

(e) Which company would you recommend to the restaurant? Why?

(f) How might the cost of the container affect your decision?

The following projects are available on the Instructor's Resource Center (IRC):

2. **Project at Motorola** *Thermal Fatigue of Solder Connections*

3. **Depreciation of a New Car**

4. **CBL Experiment**

Cumulative Review

1. Is the following graph the graph of a function? If it is, is the function one-to-one?

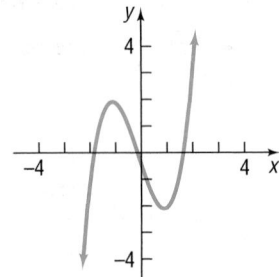

2. For the function $f(x) = 2x^2 - 3x + 1$, find the following:
 (a) $f(3)$ (b) $f(-x)$ (c) $f(x + h)$

3. Determine which of the following points are on the graph of $x^2 + y^2 = 1$.
 (a) $\left(\dfrac{1}{2}, \dfrac{1}{2}\right)$ (b) $\left(\dfrac{1}{2}, \dfrac{\sqrt{3}}{2}\right)$

4. Solve the equation $3(x - 2) = 4(x + 5)$.

5. Graph the line $2x - 4y = 16$.

6. (a) Graph the quadratic function $f(x) = -x^2 + 2x - 3$ by determining whether its graph opens up or down and by finding its vertex, axis of symmetry, y-intercept, and x-intercept(s), if any.
 (b) Solve $f(x) \le 0$.

7. Determine the quadratic function whose graph is given in the figure.

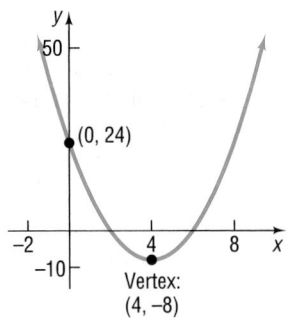

8. Graph $f(x) = 3(x + 1)^3 - 2$ using transformations.

9. Given that $f(x) = x^2 + 2$ and $g(x) = \dfrac{2}{x - 3}$, find $f(g(x))$ and state its domain. What is $f(g(5))$?

10. For the polynomial function $f(x) = 4x^3 + 9x^2 - 30x - 8$:
 (a) Find the real zeros of f.
 (b) Determine the intercepts of the graph of f.
 (c) Use a graphing utility to approximate the local maxima and local minima.
 (d) Draw a complete graph of f. Be sure to label the intercepts and turning points.

11. For the function $g(x) = 3^x + 2$:
 (a) Graph g using transformations. State the domain, range, and horizontal asymptote of g.
 (b) Determine the inverse of g. State the domain, range, and vertical asymptote of g^{-1}.
 (c) On the same graph as g, graph g^{-1}.

12. Solve the equation $4^{x-3} = 8^{2x}$.

13. Solve the equation: $\log_3(x + 1) + \log_3(2x - 3) = \log_9 9$.

14. Suppose that $f(x) = \log_3(x + 2)$. Solve:
 (a) $f(x) = 0$.
 (b) $f(x) > 0$.
 (c) $f(x) = 3$.

15. **Data Analysis** The following data represent the percent of all drivers by age that have been stopped by the police for any reason within the past year. The median age represents the midpoint of the upper and lower limit for the age range.

Age Range	Median Age, x	Percentage Stopped, y
16–19	17.5	18.2
20–29	24.5	16.8
30–39	34.5	11.3
40–49	44.5	9.4
50–59	54.5	7.7
≥ 60	69.5	3.8

(a) Using your graphing utility, draw a scatter diagram of the data treating median age, x, as the independent variable.

(b) Determine a model that you feel best describes the relation between median age and percentage stopped. You may choose from among linear, quadratic, cubic, power, exponential, logarithmic, or logistic models.

(c) Provide a justification for the model that you selected in part (b).

5 Conics

A LOOK BACK In the Foundations chapter, we introduced rectangular coordinates and showed how geometry problems can be solved algebraically. We defined a circle geometrically and then used the distance formula and rectangular coordinates to obtain an equation for a circle.

A LOOK AHEAD In this chapter we give geometric definitions for the conics and use the distance formula and rectangular coordinates to obtain their equations.

Historically, Apollonius (200 BC) was among the first to study *conics* and discover some of their interesting properties. Today, conics are still studied because of their many uses. *Paraboloids of revolution* (parabolas rotated about their axes of symmetry) are used as signal collectors (the satellite dishes used with radar and cable TV, for example), as solar energy collectors, and as reflectors (telescopes, light projection, and so on). The planets circle the Sun in approximately *elliptical* orbits. Elliptical surfaces can be used to reflect signals such as light and sound from one place to another. And *hyperbolas* can be used to determine the location of lightning strikes.

The Greeks used the methods of Euclidean geometry to study conics. However, we shall use the more powerful methods of analytic geometry, bringing to bear both algebra and geometry, for our study of conics.

Pluto's Unusual Orbit

Pluto is about 39 times as far from the Sun as Earth is. Its average distance from the Sun is about 3,647,240,000 miles (5,869,660,000 kilometers). Pluto travels around the Sun in an elliptical (oval-shaped) orbit. At some point in its orbit, it comes closer to the Sun than Neptune, usually the second farthest planet. It stays inside Neptune's orbit for about 20 Earth-years. This event occurs every 248 Earth-years, which is about the same number of Earth-years it takes Pluto to travel once around the Sun. Pluto entered Neptune's orbit on January 23, 1979, and remained there until February 11, 1999. Pluto will remain the outermost planet until the year 2227.

SOURCE: Reprinted with permission of The Associated Press.

—See Chapter Project 1.

5.1 Conics

OBJECTIVE 1 Know the Names of the Conics

1 Know the Names of the Conics

The word *conic* derives from the word *cone*, which is a geometric figure that can be constructed in the following way: Let *a* and *g* be two distinct lines that intersect at a point *V*. Keep the line *a* fixed. Now rotate the line *g* about *a* while maintaining the same angle between *a* and *g*. The collection of points swept out (generated) by the line *g* is called a **(right circular) cone**. See Figure 1. The fixed line *a* is called the **axis** of the cone; the point *V* is its **vertex**; the lines that pass through *V* and make the same angle with *a* as *g* are **generators** of the cone. Each generator is a line that lies entirely on the cone. The cone consists of two parts, called **nappes**, that intersect at the vertex.

Figure 1

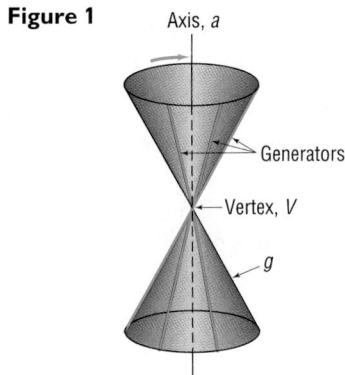

Axis, *a*

Generators

Vertex, *V*

g

Conics, an abbreviation for **conic sections**, are curves that result from the intersection of a (right circular) cone and a plane. The conics we shall study arise when the plane does not contain the vertex, as shown in Figure 2. These conics are **circles** when the plane is perpendicular to the axis of the cone and intersects each generator; **ellipses** when the plane is tilted slightly so that it intersects each generator, but intersects only one nappe of the cone; **parabolas** when the plane is tilted farther so that it is parallel to one (and only one) generator and intersects only one nappe of the cone; and **hyperbolas** when the plane intersects both nappes.

If the plane does contain the vertex, the intersection of the plane and the cone is a point, a line, or a pair of intersecting lines. These are usually called **degenerate conics**.

Figure 2

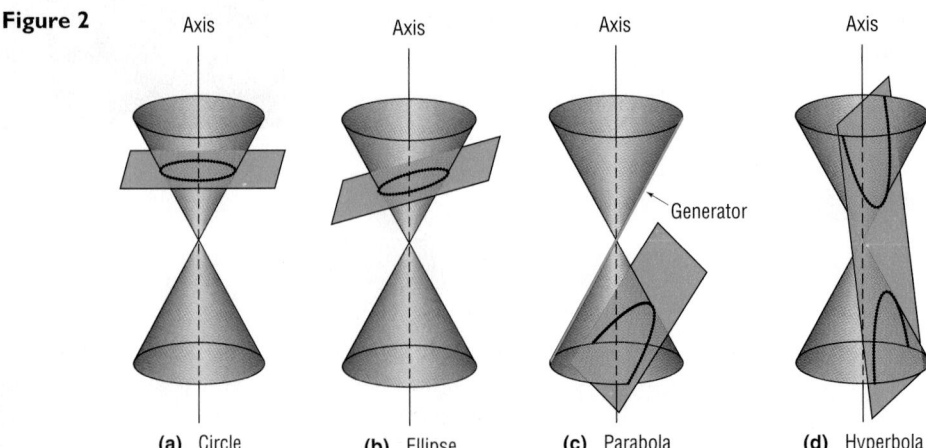

Axis Axis Axis Axis

Generator

(a) Circle **(b)** Ellipse **(c)** Parabola **(d)** Hyperbola

5.2 The Parabola

PREPARING FOR THIS SECTION *Before getting started, review the following:*

- Distance Formula (Fundamentals, Section F.1, p. 3)
- Symmetry (Fundamentals, Section F.2, pp. 11–13)
- Square Root Method (Section 2.3, p. 135–136)

- Completing the Square (Appendix A, Section A.4, pp. A35–A36)
- Graphing Techniques: Transformations (Section 1.5, pp. 85–94)

Now work the 'Are You Prepared?' problems on page 369.

OBJECTIVES
1 Analyze Parabolas with Vertex at the Origin
2 Analyze Parabolas with Vertex at (h, k)
3 Solve Applied Problems Involving Parabolas

We stated earlier (Section 2.4) that the graph of a quadratic function is a parabola. In this section, we begin with a geometric definition of a parabola and use it to obtain an equation.

A **parabola** is the collection of all points P in the plane that are the same distance from a fixed point F as they are from a fixed line D. The point F is called the **focus** of the parabola, and the line D is its **directrix**. As a result, a parabola is the set of points P for which

$$d(F, P) = d(P, D) \qquad \textbf{(1)}$$

Figure 3 shows a parabola (in blue). The line through the focus F and perpendicular to the directrix D is called the **axis of symmetry** of the parabola. The point of intersection of the parabola with its axis of symmetry is called the **vertex** V.

Figure 3

Because the vertex V lies on the parabola, it must satisfy equation (1): $d(F, V) = d(V, D)$. The vertex is midway between the focus and the directrix. We shall let a equal the distance $d(F, V)$ from F to V. Now we are ready to derive an equation for a parabola. To do this, we use a rectangular system of coordinates positioned so that the vertex V, focus F, and directrix D of the parabola are conveniently located.

1 Analyze Parabolas with Vertex at the Origin

If we choose to locate the vertex V at the origin $(0, 0)$, we can conveniently position the focus F on either the x-axis or the y-axis. First, we consider the case where the focus F is on the positive x-axis, as shown in Figure 4. Because the distance from F to V is a, the coordinates of F will be $(a, 0)$ with $a > 0$. Similarly, because the distance from V to the directrix D is also a and, because D must be perpendicular to the x-axis (since the x-axis is the axis of symmetry), the equation of the directrix D must be $x = -a$.

Now, if $P = (x, y)$ is any point on the parabola, P must obey equation (1):

$$d(F, P) = d(P, D)$$

So we have

$$\sqrt{(x - a)^2 + y^2} = |x + a| \qquad \text{Use the Distance Formula.}$$
$$(x - a)^2 + y^2 = (x + a)^2 \qquad \text{Square both sides.}$$
$$x^2 - 2ax + a^2 + y^2 = x^2 + 2ax + a^2 \qquad \text{Remove parentheses.}$$
$$y^2 = 4ax \qquad \text{Simplify.}$$

Figure 4
$y^2 = 4ax$

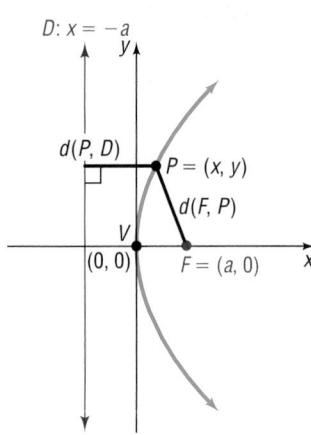

Theorem

Equation of a Parabola; Vertex at $(0, 0)$, Focus at $(a, 0)$, $a > 0$

The equation of a parabola with vertex at $(0, 0)$, focus at $(a, 0)$, and directrix $x = -a$, $a > 0$, is

$$y^2 = 4ax \qquad\qquad (2)$$

EXAMPLE 1 | **Finding the Equation of a Parabola and Graphing It**

Find an equation of the parabola with vertex at $(0, 0)$ and focus at $(3, 0)$. Graph the equation.

Solution

The distance from the vertex $(0, 0)$ to the focus $(3, 0)$ is $a = 3$. Based on equation (2), the equation of this parabola is

$$y^2 = 4ax$$
$$y^2 = 12x \qquad a = 3$$

To graph this parabola, it is helpful to plot the two points on the graph above and below the focus. To locate them, we let $x = 3$. Then

$$y^2 = 12x = 12(3) = 36$$
$$y = \pm 6 \qquad \text{Solve for } y.$$

The points on the parabola above and below the focus are $(3, 6)$ and $(3, -6)$. These points help in graphing the parabola because they determine the "opening." See Figure 5.

Figure 5

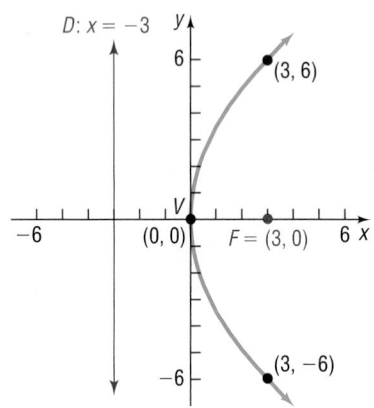

In general, the points on a parabola $y^2 = 4ax$ that lie above and below the focus $(a, 0)$ are each at a distance $2a$ from the focus. This follows from the fact that if $x = a$ then $y^2 = 4ax = 4a^2$, so $y = \pm 2a$. The line segment joining these two points is called the **latus rectum**; its length is $4a$.

COMMENT: To graph the parabola $y^2 = 12x$ discussed in Example 1, we need to graph the two functions $Y_1 = \sqrt{12x}$ and $Y_2 = -\sqrt{12x}$. Do this and compare what you see with Figure 5. ∎

NOW WORK PROBLEM 19.

By reversing the steps we used to obtain equation (2), it follows that the graph of an equation of the form of equation (2), $y^2 = 4ax$, is a parabola; its vertex is at $(0, 0)$, its focus is at $(a, 0)$, its directrix is the line $x = -a$, and its axis of symmetry is the x-axis.

For the remainder of this section, the direction "**Analyze the equation**" will mean to find the vertex, focus, and directrix of the parabola and graph it.

EXAMPLE 2 **Analyzing the Equation of a Parabola**

Analyze the equation: $y^2 = 8x$

Solution The equation $y^2 = 8x$ is of the form $y^2 = 4ax$, where $4a = 8$ so that $a = 2$. Consequently, the graph of the equation is a parabola with vertex at $(0, 0)$ and focus on the positive x-axis at $(a, 0) = (2, 0)$. The directrix is the vertical line $x = -2$. The two points that determine the latus rectum are obtained by letting $x = 2$. Then $y^2 = 16$, so $y = \pm 4$. The points $(2, -4)$ and $(2, 4)$ determine the latus rectum. See Figure 6 for the graph. ∎

Recall that we arrived at equation (2) after placing the focus on the positive x-axis. If the focus is placed on the negative x-axis, positive y-axis, or negative y-axis, a different form of the equation for the parabola results. The four forms of the equation of a parabola with vertex at $(0, 0)$ and focus on a coordinate axis a distance a from $(0, 0)$ are given in Table 1, and their graphs are given in Figure 7. Notice that each graph is symmetric with respect to its axis of symmetry.

Figure 6

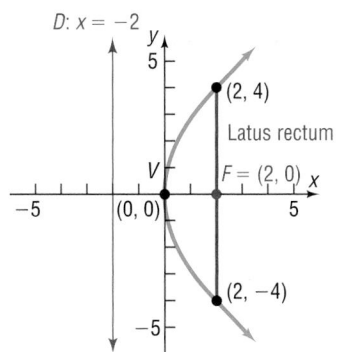

Table 1

EQUATIONS OF A PARABOLA VERTEX AT (0, 0); FOCUS ON AN AXIS; $a > 0$				
Vertex	**Focus**	**Directrix**	**Equation**	**Description**
$(0,0)$	$(a, 0)$	$x = -a$	$y^2 = 4ax$	Parabola, axis of symmetry is the x-axis, opens right
$(0,0)$	$(-a, 0)$	$x = a$	$y^2 = -4ax$	Parabola, axis of symmetry is the x-axis, opens left
$(0,0)$	$(0, a)$	$y = -a$	$x^2 = 4ay$	Parabola, axis of symmetry is the y-axis, opens up
$(0,0)$	$(0, -a)$	$y = a$	$x^2 = -4ay$	Parabola, axis of symmetry is the y-axis, opens down

Figure 7

(a) $y^2 = 4ax$

(b) $y^2 = -4ax$

(c) $x^2 = 4ay$

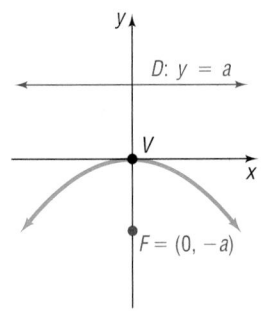

(d) $x^2 = -4ay$

Figure 8

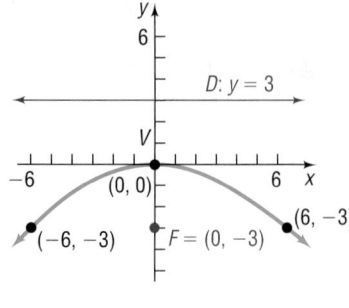

| EXAMPLE 3 | Analyzing the Equation of a Parabola |

Analyze the equation: $x^2 = -12y$

Solution The equation $x^2 = -12y$ is of the form $x^2 = -4ay$, with $a = 3$. Consequently, the graph of the equation is a parabola with vertex at $(0, 0)$, focus at $(0, -3)$ and directrix the line $y = 3$. The parabola opens down, and its axis of symmetry is the y-axis. To obtain the points defining the latus rectum, let $y = -3$. Then $x^2 = 36$, so $x = \pm 6$. The points $(-6, -3)$ and $(6, -3)$ determine the latus rectum. See Figure 8 for the graph. ∎

NOW WORK PROBLEM **39**.

| EXAMPLE 4 | Finding the Equation of a Parabola |

Find the equation of the parabola with focus at $(0, 4)$ and directrix the line $y = -4$. Graph the equation.

Figure 9

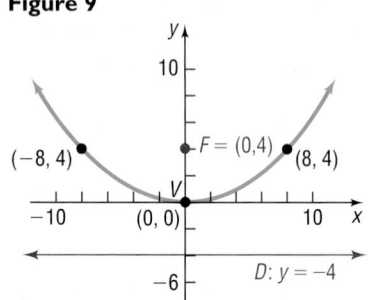

Solution A parabola whose focus is at $(0, 4)$ and whose directrix is the horizontal line $y = -4$ will have its vertex at $(0, 0)$. (Do you see why? The vertex is midway between the focus and the directrix.) Since the focus is on the positive y-axis at $(0, 4)$, the equation of this parabola is of the form $x^2 = 4ay$, with $a = 4$; that is,

$$x^2 = 4ay = 4(4)y = 16y$$
$$\uparrow$$
$$a = 4$$

The points $(8, 4)$ and $(-8, 4)$ determine the latus rectum. Figure 9 shows the graph of $x^2 = 16y$. ∎

| EXAMPLE 5 | Finding the Equation of a Parabola |

Find the equation of a parabola with vertex at $(0, 0)$ if its axis of symmetry is the x-axis and its graph contains the point $\left(-\frac{1}{2}, 2\right)$. Find its focus and directrix, and graph the equation.

Solution

The vertex is at the origin, the axis of symmetry is the x-axis, and the graph contains a point in the second quadrant, so the parabola opens to the left. We see from Table 1 that the form of the equation is

$$y^2 = -4ax$$

Because the point $\left(-\frac{1}{2}, 2\right)$ is on the parabola, the coordinates $x = -\frac{1}{2}, y = 2$ must satisfy the equation. Substituting $x = -\frac{1}{2}$ and $y = 2$ into the equation, we find that

Figure 10

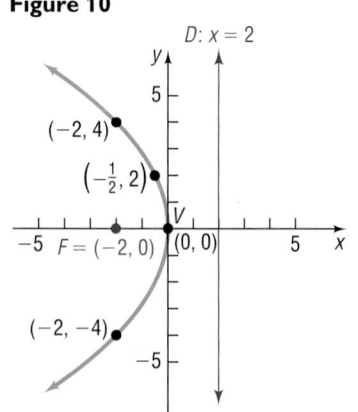

$$4 = -4a\left(-\frac{1}{2}\right) \qquad y^2 = -4ax; x = -\frac{1}{2}, y = 2$$
$$a = 2$$

The equation of the parabola is

$$y^2 = -4(2)x = -8x$$

The focus is at $(-2, 0)$ and the directrix is the line $x = 2$. Letting $x = -2$, we find $y^2 = 16$, so $y = \pm 4$. The points $(-2, 4)$ and $(-2, -4)$ determine the latus rectum. See Figure 10. ∎

NOW WORK PROBLEM **27**.

2 Analyze Parabolas with Vertex at (h, k)

If a parabola with vertex at the origin and axis of symmetry along a coordinate axis is shifted horizontally h units and then vertically k units, the result is a parabola with vertex at (h, k) and axis of symmetry parallel to a coordinate axis. The equations of such parabolas have the same forms as those in Table 1, but with x replaced by $x - h$ (the horizontal shift) and y replaced by $y - k$ (the vertical shift). Table 2 gives the forms of the equations of such parabolas. Figures 11(a)–(d) illustrate the graphs for $h > 0, k > 0$.

Table 2

PARABOLAS WITH VERTEX AT (h, k); AXIS OF SYMMETRY PARALLEL TO A COORDINATE AXIS; a > 0				
Vertex	**Focus**	**Directrix**	**Equation**	**Description**
(h, k)	$(h + a, k)$	$x = h - a$	$(y - k)^2 = 4a(x - h)$	Parabola, axis of symmetry parallel to the x-axis, opens right
(h, k)	$(h - a, k)$	$x = h + a$	$(y - k)^2 = -4a(x - h)$	Parabola, axis of symmetry parallel to the x-axis, opens left
(h, k)	$(h, k + a)$	$y = k - a$	$(x - h)^2 = 4a(y - k)$	Parabola, axis of symmetry parallel to the y-axis, opens up
(h, k)	$(h, k - a)$	$y = k + a$	$(x - h)^2 = -4a(y - k)$	Parabola, axis of symmetry parallel to the y-axis, opens down

Figure 11

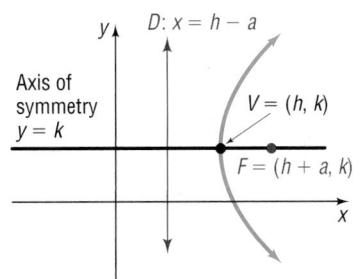

(a) $(y - k)^2 = 4a(x - h)$

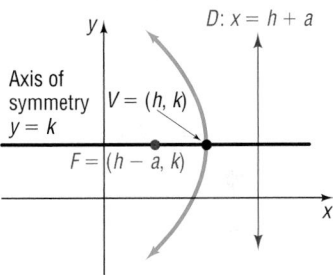

(b) $(y - k)^2 = -4a(x - h)$

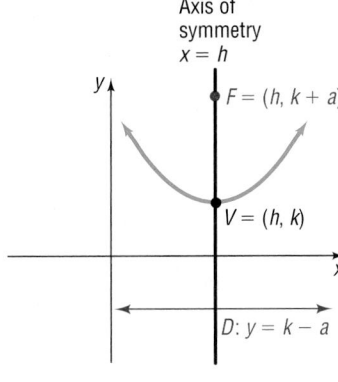

(c) $(x - h)^2 = 4a(y - k)$

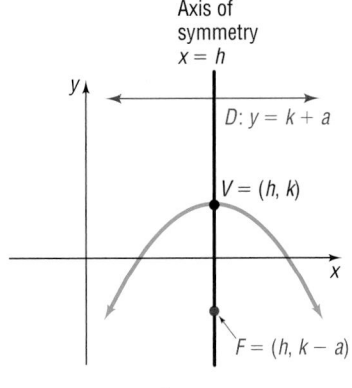

(d) $(x - h)^2 = -4a(y - k)$

EXAMPLE 6 | **Finding the Equation of a Parabola, Vertex Not at Origin**

Find an equation of the parabola with vertex at $(-2, 3)$ and focus at $(0, 3)$. Graph the equation.

Figure 12

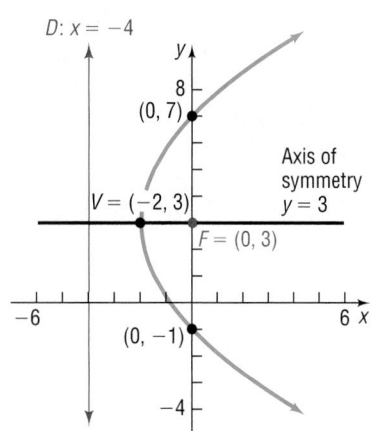

Solution The vertex $(-2, 3)$ and focus $(0, 3)$ both lie on the horizontal line $y = 3$ (the axis of symmetry). The distance a from the vertex $(-2, 3)$ to the focus $(0, 3)$ is $a = 2$. Also, because the focus lies to the right of the vertex, we know that the parabola opens to the right. Consequently, the form of the equation is

$$(y - k)^2 = 4a(x - h)$$

where $(h, k) = (-2, 3)$ and $a = 2$. Therefore, the equation is

$$(y - 3)^2 = 4 \cdot 2[x - (-2)]$$
$$(y - 3)^2 = 8(x + 2)$$

If $x = 0$, then $(y - 3)^2 = 16$. Then $y - 3 = \pm 4$, so $y = -1$ or $y = 7$. The points $(0, -1)$ and $(0, 7)$ determine the latus rectum; the line $x = -4$ is the directrix. See Figure 12. ∎

> **NOW WORK PROBLEM 29.**

Polynomial equations define parabolas whenever they involve two variables that are quadratic in one variable and linear in the other. To analyze this type of equation, we first complete the square of the variable that is quadratic.

EXAMPLE 7	**Analyzing the Equation of a Parabola**

Analyze the equation: $x^2 + 4x - 4y = 0$

Figure 13

Solution To analyze the equation $x^2 + 4x - 4y = 0$, we complete the square involving the variable x.

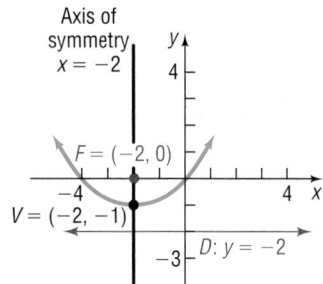

$$x^2 + 4x - 4y = 0$$
$$x^2 + 4x = 4y \qquad \text{Isolate the terms involving x on the left side.}$$
$$x^2 + 4x + 4 = 4y + 4 \qquad \text{Complete the square on the left side.}$$
$$(x + 2)^2 = 4(y + 1) \qquad \text{Factor.}$$

This equation is of the form $(x - h)^2 = 4a(y - k)$, with $h = -2, k = -1$, and $a = 1$. The graph is a parabola with vertex at $(h, k) = (-2, -1)$ that opens up. The focus is at $(-2, 0)$, and the directrix is the line $y = -2$. See Figure 13. ∎

> **NOW WORK PROBLEM 47.**

3 Solve Applied Problems Involving Parabolas

Parabolas find their way into many applications. For example, as we discussed in Section 2.6, suspension bridges have cables in the shape of a parabola. Another property of parabolas that is used in applications is their reflecting property.

Suppose that a mirror is shaped like a **paraboloid of revolution**, a surface formed by rotating a parabola about its axis of symmetry. If a light (or any other emitting source) is placed at the focus of the parabola, all the rays emanating from the light will reflect off the mirror in lines parallel to the axis of symmetry. This principle is used in the design of searchlights, flashlights, certain automobile headlights, and other such devices. See Figure 14.

Conversely, suppose that rays of light (or other signals) emanate from a distant source so that they are essentially parallel. When these rays strike the surface of a parabolic mirror whose axis of symmetry is parallel to these rays, they are reflected to a single point at the focus. This principle is used in the design of some solar energy devices, satellite dishes, and the mirrors used in some types of telescopes. See Figure 15.

Figure 14
Searchlight

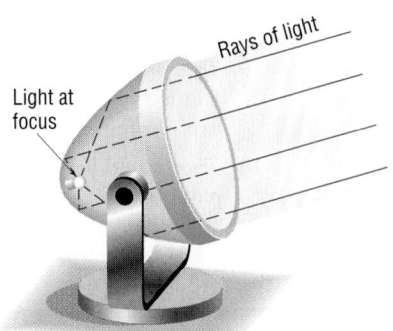

Rays of light

Light at focus

Figure 15
Telescope

| EXAMPLE 8 | **Satellite Dish** |

A satellite dish is shaped like a paraboloid of revolution. The signals that emanate from a satellite strike the surface of the dish and are reflected to a single point, where the receiver is located. If the dish is 8 feet across at its opening and 3 feet deep at its center, at what position should the receiver be placed?

Figure 16

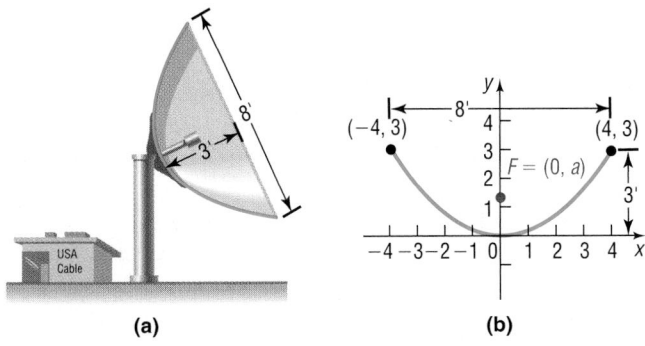

(a) (b)

Solution Figure 16(a) shows the satellite dish. We draw the parabola used to form the dish on a rectangular coordinate system so that the vertex of the parabola is at the origin and its focus is on the positive y-axis. See Figure 16(b).

The form of the equation of the parabola is

$$x^2 = 4ay$$

and its focus is at $(0, a)$. Since $(4, 3)$ is a point on the graph, we have

$$4^2 = 4a(3) \qquad x = 4, y = 3.$$

$$a = \frac{4}{3} \qquad \text{Solve for } a.$$

The receiver should be located $1\frac{1}{3}$ feet from the base of the dish, along its axis of symmetry. ∎

 NOW WORK PROBLEM 63.

5.2 Assess Your Understanding

'Are You Prepared?'

Answers are given at the end of these exercises. If you get a wrong answer, read the pages listed in red.

1. The formula for the distance d from $P_1 = (x_1, y_1)$ to $P_2 = (x_2, y_2)$ is $d =$ _____. (p. 3)

2. To complete the square of $x^2 - 4x$, add _____. (pp. A35–A36)

3. Use the Square Root Method to find the real solutions of $(x + 4)^2 = 9$. (pp. 135–136)

4. The point that is symmetric with respect to the x-axis to the point $(-2, 5)$ is _____. (pp. 11–13)

5. To graph $y = (x - 3)^2 + 1$, shift the graph of $y = x^2$ to the right _____ units and then _____ 1 unit. (pp. 85–94)

Concepts and Vocabulary

6. A(n) _____ is the collection of all points in the plane such that the distance from each point to a fixed point equals its distance to a fixed line.

7. The surface formed by rotating a parabola about its axis of symmetry is called a _____ _____ _____.

8. *True or False:* The vertex of a parabola is a point on the parabola that also is on its axis of symmetry.

9. *True or False:* If a light is placed at the focus of a parabola, all the rays reflected off the parabola will be parallel to the axis of symmetry.

10. *True or False:* The graph of a quadratic function is a parabola.

Skill Building

In Problems 11–18, the graph of a parabola is given. Match each graph to its equation.

A. $y^2 = 4x$

B. $x^2 = 4y$

C. $y^2 = -4x$

D. $x^2 = -4y$

E. $(y - 1)^2 = 4(x - 1)$

F. $(x + 1)^2 = 4(y + 1)$

G. $(y - 1)^2 = -4(x - 1)$

H. $(x + 1)^2 = -4(y + 1)$

11.

12.

13.

14.

15.

16.

17.

18.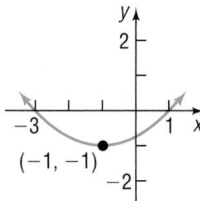

In Problems 19–36, find the equation of the parabola described. Find the two points that define the latus rectum, and graph the equation.

 19. Focus at $(4, 0)$; vertex at $(0, 0)$

20. Focus at $(0, 2)$; vertex at $(0, 0)$

21. Focus at $(0, -3)$; vertex at $(0, 0)$

22. Focus at $(-4, 0)$; vertex at $(0, 0)$

23. Focus at $(-2, 0)$; directrix the line $x = 2$

24. Focus at $(0, -1)$; directrix the line $y = 1$

25. Directrix the line $y = -\frac{1}{2}$; vertex at $(0, 0)$

26. Directrix the line $x = -\frac{1}{2}$; vertex at $(0, 0)$

27. Vertex at $(0, 0)$; axis of symmetry the y-axis; containing the point $(2, 3)$

28. Vertex at $(0, 0)$; axis of symmetry the x-axis; containing the point $(2, 3)$

29. Vertex at $(2, -3)$; focus at $(2, -5)$

30. Vertex at $(4, -2)$; focus at $(6, -2)$

31. Vertex at $(-1, -2)$; focus at $(0, -2)$

32. Vertex at $(3, 0)$; focus at $(3, -2)$

33. Focus at $(-3, 4)$; directrix the line $y = 2$

34. Focus at $(2, 4)$; directrix the line $x = -4$

35. Focus at $(-3, -2)$; directrix the line $x = 1$

36. Focus at $(-4, 4)$; directrix the line $y = -2$

In Problems 37–54, find the vertex, focus, and directrix of each parabola. Graph the equation.

37. $x^2 = 4y$

38. $y^2 = 8x$

39. $y^2 = -16x$

40. $x^2 = -4y$

41. $(y - 2)^2 = 8(x + 1)$

42. $(x + 4)^2 = 16(y + 2)$

43. $(x - 3)^2 = -(y + 1)$

44. $(y + 1)^2 = -4(x - 2)$

45. $(y + 3)^2 = 8(x - 2)$

46. $(x - 2)^2 = 4(y - 3)$

47. $y^2 - 4y + 4x + 4 = 0$

48. $x^2 + 6x - 4y + 1 = 0$

49. $x^2 + 8x = 4y - 8$

50. $y^2 - 2y = 8x - 1$

51. $y^2 + 2y - x = 0$

52. $x^2 - 4x = 2y$

53. $x^2 - 4x = y + 4$

54. $y^2 + 12y = -x + 1$

In Problems 55–62, write an equation for each parabola.

55.

56.

57.

58.

59.

60.

61.

62.
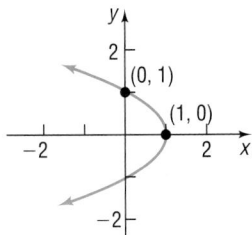

Applications and Extensions

63. **Satellite Dish** A satellite dish is shaped like a paraboloid of revolution. The signals that emanate from a satellite strike the surface of the dish and are reflected to a single point, where the receiver is located. If the dish is 10 feet across at its opening and 4 feet deep at its center, at what position should the receiver be placed?

64. **Constructing a TV Dish** A cable TV receiving dish is in the shape of a paraboloid of revolution. Find the location of the receiver, which is placed at the focus, if the dish is 6 feet across at its opening and 2 feet deep.

65. **Constructing a Flashlight** The reflector of a flashlight is in the shape of a paraboloid of revolution. Its diameter is 4 inches and its depth is 1 inch. How far from the vertex should the light bulb be placed so that the rays will be reflected parallel to the axis?

66. **Constructing a Headlight** A sealed-beam headlight is in the shape of a paraboloid of revolution. The bulb, which is placed at the focus, is 1 inch from the vertex. If the depth is to be 2 inches, what is the diameter of the headlight at its opening?

67. **Suspension Bridge** The cables of a suspension bridge are in the shape of a parabola, as shown in the figure. The towers supporting the cable are 600 feet apart and 80 feet high. If the cables touch the road surface midway between the towers, what is the height of the cable at a point 150 feet from the center of the bridge?

68. **Suspension Bridge** The cables of a suspension bridge are in the shape of a parabola. The towers supporting the cable are 400 feet apart and 100 feet high. If the cables are at a height of 10 feet midway between the towers, what is the height of the cable at a point 50 feet from the center of the bridge?

69. **Searchlight** A searchlight is shaped like a paraboloid of revolution. If the light source is located 2 feet from the base along the axis of symmetry and the opening is 5 feet across, how deep should the searchlight be?

70. **Searchlight** A searchlight is shaped like a paraboloid of revolution. If the light source is located 2 feet from the base along the axis of symmetry and the depth of the searchlight is 4 feet, what should the width of the opening be?

71. **Solar Heat** A mirror is shaped like a paraboloid of revolution and will be used to concentrate the rays of the sun at its focus, creating a heat source. (See the figure.) If the mirror is 20 feet across at its opening and is 6 feet deep, where will the heat source be concentrated?

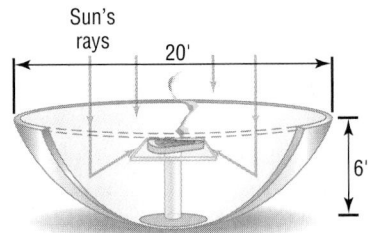

72. **Reflecting Telescope** A reflecting telescope contains a mirror shaped like a paraboloid of revolution. If the mirror is 4 inches across at its opening and is 3 feet deep, where will the collected light be concentrated?

73. Parabolic Arch Bridge A bridge is built in the shape of a parabolic arch. The bridge has a span of 120 feet and a maximum height of 25 feet. See the illustration. Choose a suitable rectangular coordinate system and find the height of the arch at distances of 10, 30, and 50 feet from the center.

74. Parabolic Arch Bridge A bridge is to be built in the shape of a parabolic arch and is to have a span of 100 feet. The height of the arch a distance of 40 feet from the center is to be 10 feet. Find the height of the arch at its center.

75. Show that an equation of the form

$$Ax^2 + Ey = 0, \qquad A \neq 0, E \neq 0$$

is the equation of a parabola with vertex at $(0, 0)$ and axis of symmetry the y-axis. Find its focus and directrix.

76. Show that an equation of the form

$$Cy^2 + Dx = 0, \qquad C \neq 0, D \neq 0$$

is the equation of a parabola with vertex at $(0, 0)$ and axis of symmetry the x-axis. Find its focus and directrix.

77. Show that the graph of an equation of the form

$$Ax^2 + Dx + Ey + F = 0, \qquad A \neq 0$$

(a) Is a parabola if $E \neq 0$.
(b) Is a vertical line if $E = 0$ and $D^2 - 4AF = 0$.
(c) Is two vertical lines if $E = 0$ and $D^2 - 4AF > 0$.
(d) Contains no points if $E = 0$ and $D^2 - 4AF < 0$.

78. Show that the graph of an equation of the form

$$Cy^2 + Dx + Ey + F = 0, \qquad C \neq 0$$

(a) Is a parabola if $D \neq 0$.
(b) Is a horizontal line if $D = 0$ and $E^2 - 4CF = 0$.
(c) Is two horizontal lines if $D = 0$ and $E^2 - 4CF > 0$.
(d) Contains no points if $D = 0$ and $E^2 - 4CF < 0$.

'Are You Prepared?' Answers

1. $d = \sqrt{(x_2 - x_1)^2 + (y_2 - y_1)^2}$ **2.** 4 **3.** $x + 4 = \pm 3; \{-7, -1\}$ **4.** $(-2, -5)$ **5.** 3; up

5.3 The Ellipse

PREPARING FOR THIS SECTION *Before getting started, review the following:*

• Distance Formula (Foundations, Section F.1, p. 3)
• Completing the Square (Appendix A, Section A.4, pp. A35–A36)
• Intercepts (Foundations, Section F.2, pp. 10–11)
• Symmetry (Foundations, Section F.2, pp. 11–13)
• Circles (Foundations, Section F.4, pp. 33–36)
• Graphing Techniques: Transformations (Section 1.5, pp. 85–94)

Now work the 'Are You Prepared?' problems on page 379.

OBJECTIVES 1 Analyze Ellipses with Center at the Origin
2 Analyze Ellipses with Center at (h, k)
3 Solve Applied Problems Involving Ellipses

Figure 17

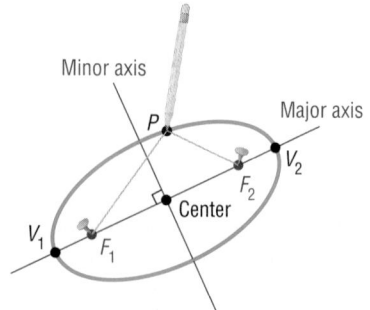

An **ellipse** is the collection of all points in the plane the sum of whose distances from two fixed points, called the **foci**, is a constant.

The definition actually contains within it a physical means for drawing an ellipse. Find a piece of string (the length of this string is the constant referred to in the definition). Then take two thumbtacks (the foci) and stick them on a piece of cardboard so that the distance between them is less than the length of the string. Now attach the ends of the string to the thumbtacks and, using the point of a pencil, pull the string taut. See Figure 17. Keeping the string taut, rotate the pencil around the two thumbtacks. The pencil traces out an ellipse, as shown in Figure 17.

In Figure 17, the foci are labeled F_1 and F_2. The line containing the foci is called the **major axis**. The midpoint of the line segment joining the foci is the **center** of the ellipse. The line through the center and perpendicular to the major axis is the **minor axis**.

The two points of intersection of the ellipse and the major axis are the **vertices**, V_1 and V_2, of the ellipse. The distance from one vertex to the other is the **length of the major axis**. The ellipse is symmetric with respect to its major axis, with respect to its minor axis, and with respect to its center.

1 Analyze Ellipses with Center at the Origin

Figure 18

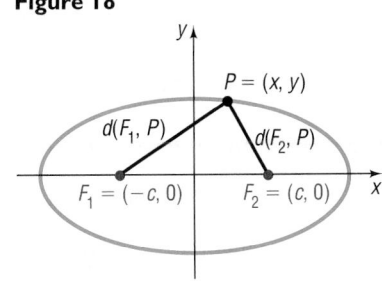

With these ideas in mind, we are now ready to find the equation of an ellipse in a rectangular coordinate system. First, we place the center of the ellipse at the origin. Second, we position the ellipse so that its major axis coincides with a coordinate axis. Suppose that the major axis coincides with the x-axis, as shown in Figure 18. If c is the distance from the center to a focus, one focus will be at $F_1 = (-c, 0)$ and the other at $F_2 = (c, 0)$. As we shall see, it is convenient to let $2a$ denote the constant distance referred to in the definition. Then, if $P = (x, y)$ is any point on the ellipse, we have

$$d(F_1, P) + d(F_2, P) = 2a \qquad \text{Sum of the distances from } P \text{ to the foci equals a constant, } 2a.$$

$$\sqrt{(x + c)^2 + y^2} + \sqrt{(x - c)^2 + y^2} = 2a \qquad \text{Use the Distance Formula.}$$

$$\sqrt{(x + c)^2 + y^2} = 2a - \sqrt{(x - c)^2 + y^2} \qquad \text{Isolate one radical.}$$

$$(x + c)^2 + y^2 = 4a^2 - 4a\sqrt{(x - c)^2 + y^2} \qquad \text{Square both sides.}$$
$$+ (x - c)^2 + y^2$$

$$x^2 + 2cx + c^2 + y^2 = 4a^2 - 4a\sqrt{(x - c)^2 + y^2} \qquad \text{Remove parentheses.}$$
$$+ x^2 - 2cx + c^2 + y^2$$

$$4cx - 4a^2 = -4a\sqrt{(x - c)^2 + y^2} \qquad \text{Simplify; isolate the radical.}$$

$$cx - a^2 = -a\sqrt{(x - c)^2 + y^2} \qquad \text{Divide each side by 4.}$$

$$(cx - a^2)^2 = a^2[(x - c)^2 + y^2] \qquad \text{Square both sides again.}$$

$$c^2x^2 - 2a^2cx + a^4 = a^2(x^2 - 2cx + c^2 + y^2) \qquad \text{Remove parentheses.}$$

$$(c^2 - a^2)x^2 - a^2y^2 = a^2c^2 - a^4 \qquad \text{Rearrange the terms.}$$

$$(a^2 - c^2)x^2 + a^2y^2 = a^2(a^2 - c^2) \qquad \text{Multiply each side by } -1; \text{ factor } a^2 \text{ on the right side.} \qquad \textbf{(1)}$$

To obtain points on the ellipse off the x-axis, it must be that $a > c$. To see why, look again at Figure 18.

$$d(F_1, P) + d(F_2, P) > d(F_1, F_2) \qquad \text{The sum of the lengths of two sides of a triangle is greater than the length of the third side.}$$

$$2a > 2c \qquad d(F_1, P) + d(F_2, P) = 2a; \ d(F_1, F_2) = 2c$$

$$a > c$$

Since $a > c$, we also have $a^2 > c^2$, so $a^2 - c^2 > 0$. Let $b^2 = a^2 - c^2$, $b > 0$. Then $a > b$ and equation (1) can be written as

$$b^2x^2 + a^2y^2 = a^2b^2$$

$$\frac{x^2}{a^2} + \frac{y^2}{b^2} = 1 \qquad \text{Divide each side by } a^2b^2.$$

Theorem | **Equation of an Ellipse; Center at (0, 0); Major Axis along the *x*-Axis**

An equation of the ellipse with center at $(0, 0)$, foci at $(-c, 0)$ and $(c, 0)$, and vertices at $(-a, 0)$ and $(a, 0)$ is

$$\frac{x^2}{a^2} + \frac{y^2}{b^2} = 1, \qquad \text{where } a > b > 0 \text{ and } b^2 = a^2 - c^2 \qquad (2)$$

The major axis is the *x*-axis.

As you can verify, the ellipse defined by equation (2) is symmetric with respect to the *x*-axis, *y*-axis, and origin.

Because the major axis is the *x*-axis, we find the vertices of the ellipse defined by equation (2) by letting $y = 0$. The vertices satisfy the equation $\frac{x^2}{a^2} = 1$, the solutions of which are $x = \pm a$. Consequently, the vertices of the ellipse given by equation (2) are $V_1 = (-a, 0)$ and $V_2 = (a, 0)$. The *y*-intercepts of the ellipse, found by letting $x = 0$, have coordinates $(0, -b)$ and $(0, b)$. These four intercepts, $(a, 0)$, $(-a, 0)$, $(0, b)$, and $(0, -b)$, are used to graph the ellipse. See Figure 19.

Notice in Figure 19 the right triangle formed with the points $(0, 0)$, $(c, 0)$, and $(0, b)$. Because $b^2 = a^2 - c^2$ (or $b^2 + c^2 = a^2$), the distance from the focus at $(c, 0)$ to the point $(0, b)$ is a.

Figure 19

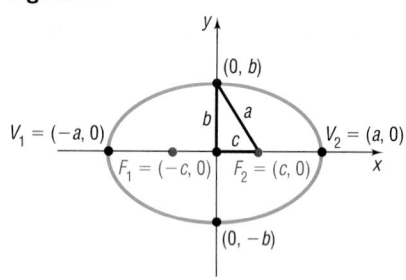

EXAMPLE 1 | **Finding an Equation of an Ellipse**

Find an equation of the ellipse with center at the origin, one focus at $(3, 0)$, and a vertex at $(-4, 0)$. Graph the equation.

Figure 20

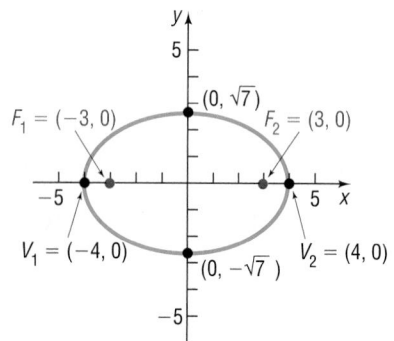

Solution

The ellipse has its center at the origin and, since the given focus and vertex lie on the *x*-axis, the major axis is the *x*-axis. The distance from the center, $(0, 0)$, to one of the foci, $(3, 0)$, is $c = 3$. The distance from the center, $(0, 0)$, to one of the vertices, $(-4, 0)$, is $a = 4$. From equation (2), it follows that

$$b^2 = a^2 - c^2 = 16 - 9 = 7$$

so an equation of the ellipse is

$$\frac{x^2}{16} + \frac{y^2}{7} = 1$$

Figure 20 shows the graph. ∎

Notice in Figure 20 how we used the intercepts of the equation to graph the ellipse. Following this practice will make it easier for you to obtain an accurate graph of an ellipse when graphing by hand.

 COMMENT: The intercepts of the ellipse also provide information about how to set the viewing rectangle for graphing an ellipse. To graph the ellipse

$$\frac{x^2}{16} + \frac{y^2}{7} = 1$$

discussed in Example 1, we set the viewing rectangle using a square screen that includes the intercepts, perhaps $-4.5 \le x \le 4.5$, $-3 \le y \le 3$. Then we proceed to solve the equation for y:

$$\frac{x^2}{16} + \frac{y^2}{7} = 1$$

$$\frac{y^2}{7} = 1 - \frac{x^2}{16} \qquad \text{Subtract } \frac{x^2}{16} \text{ from each side.}$$

$$y^2 = 7\left(1 - \frac{x^2}{16}\right) \qquad \text{Multiply both sides by 7.}$$

$$y = \pm\sqrt{7\left(1 - \frac{x^2}{16}\right)} \qquad \text{Take the square root of each side.}$$

Figure 21

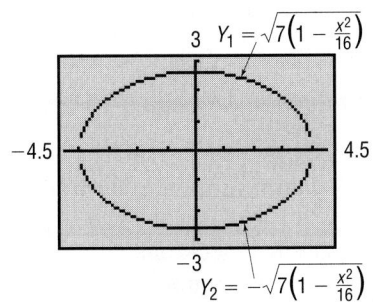

Now graph the two functions

$$Y_1 = \sqrt{7\left(1 - \frac{x^2}{16}\right)} \quad \text{and} \quad Y_2 = -\sqrt{7\left(1 - \frac{x^2}{16}\right)}.$$

Figure 21 shows the result. ■

✏ ━━━ **NOW WORK PROBLEM 27.**

An equation of the form of equation (2), with $a > b$, is the equation of an ellipse with center at the origin, foci on the x-axis at $(-c, 0)$ and $(c, 0)$, where $c^2 = a^2 - b^2$, and major axis along the x-axis.

For the remainder of this section, the direction "**Analyze the equation**" will mean to find the center, major axis, foci, and vertices of the ellipse and graph it.

EXAMPLE 2	**Analyzing the Equation of an Ellipse**

Analyze the equation: $\dfrac{x^2}{25} + \dfrac{y^2}{9} = 1$

Solution The given equation is of the form of equation (2), with $a^2 = 25$ and $b^2 = 9$. The equation is that of an ellipse with center $(0, 0)$ and major axis along the x-axis. The vertices are at $(\pm a, 0) = (\pm 5, 0)$. Because $b^2 = a^2 - c^2$, we find that

$$c^2 = a^2 - b^2 = 25 - 9 = 16$$

The foci are at $(\pm c, 0) = (\pm 4, 0)$. Figure 22 shows the graph.

Figure 22

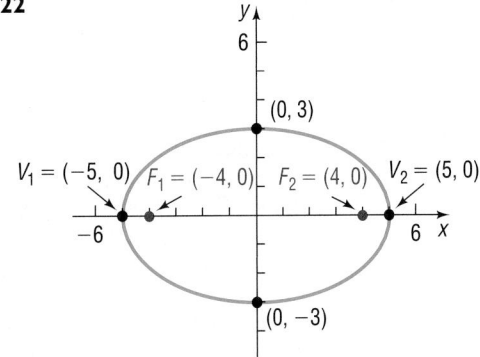

✏ ━━━ **NOW WORK PROBLEM 17.**

If the major axis of an ellipse with center at $(0, 0)$ lies on the y-axis, the foci are at $(0, -c)$ and $(0, c)$. Using the same steps as before, the definition of an ellipse leads to the following result:

Theorem

Equation of an Ellipse; Center at (0, 0); Major Axis along the y-Axis

An equation of the ellipse with center at $(0, 0)$, foci at $(0, -c)$ and $(0, c)$, and vertices at $(0, -a)$ and $(0, a)$ is

$$\frac{x^2}{b^2} + \frac{y^2}{a^2} = 1, \qquad \text{where } a > b > 0 \text{ and } b^2 = a^2 - c^2 \qquad \textbf{(3)}$$

The major axis is the y-axis.

Figure 23

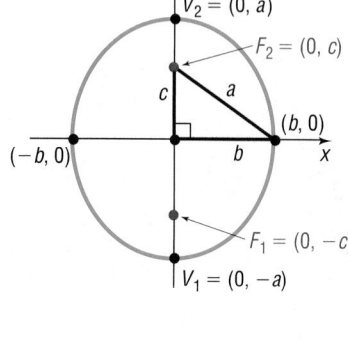

Figure 23 illustrates the graph of such an ellipse. Again, notice the right triangle with the points at $(0, 0)$, $(b, 0)$, and $(0, c)$.

Look closely at equations (2) and (3). Although they may look alike, there is a difference! In equation (2), the larger number, a^2, is in the denominator of the x^2-term, so the major axis of the ellipse is along the x-axis. In equation (3), the larger number, a^2, is in the denominator of the y^2-term, so the major axis is along the y-axis.

EXAMPLE 3	**Analyzing the Equation of an Ellipse**

Analyze the equation: $9x^2 + y^2 = 9$

Figure 24

Solution

To put the equation in proper form, we divide each side by 9.

$$x^2 + \frac{y^2}{9} = 1$$

The larger number, 9, is in the denominator of the y^2-term so, based on equation (3), this is the equation of an ellipse with center at the origin and major axis along the y-axis. Also, we conclude that $a^2 = 9$, $b^2 = 1$, and $c^2 = a^2 - b^2 = 9 - 1 = 8$. The vertices are at $(0, \pm a) = (0, \pm 3)$, and the foci are at $(0, \pm c) = (0, \pm 2\sqrt{2})$. Figure 24 shows the graph.

NOW WORK PROBLEM 21.

EXAMPLE 4	**Finding an Equation of an Ellipse**

Find an equation of the ellipse having one focus at $(0, 2)$ and vertices at $(0, -3)$ and $(0, 3)$. Graph the equation.

Figure 25

Solution

Because the vertices are at $(0, -3)$ and $(0, 3)$, the center of this ellipse is at their midpoint, the origin. Also, its major axis lies on the y-axis. The distance from the center, $(0, 0)$, to one of the foci, $(0, 2)$, is $c = 2$. The distance from the center, $(0, 0)$, to one of the vertices, $(0, 3)$, is $a = 3$. So $b^2 = a^2 - c^2 = 9 - 4 = 5$. The form of the equation of this ellipse is given by equation (3).

$$\frac{x^2}{b^2} + \frac{y^2}{a^2} = 1$$

$$\frac{x^2}{5} + \frac{y^2}{9} = 1$$

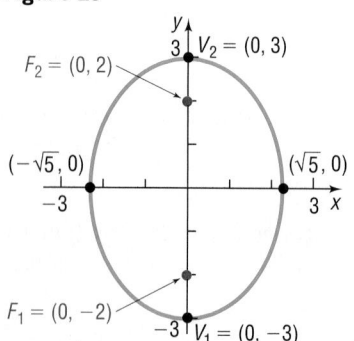

Figure 25 shows the graph.

NOW WORK PROBLEM 29.

The circle may be considered a special kind of ellipse. To see why, let $a = b$ in equation (2) or (3). Then

$$\frac{x^2}{a^2} + \frac{y^2}{a^2} = 1$$
$$x^2 + y^2 = a^2$$

This is the equation of a circle with center at the origin and radius a. The value of c is

$$c^2 = a^2 - b^2 = 0$$

We conclude that the closer the two foci of an ellipse are to the center, the more the ellipse will look like a circle.

2 Analyze Ellipses with Center at (h, k)

If an ellipse with center at the origin and major axis coinciding with a coordinate axis is shifted horizontally h units and then vertically k units, the result is an ellipse with center at (h, k) and major axis parallel to a coordinate axis. The equations of such ellipses have the same forms as those given in equations (2) and (3), except that x is replaced by $x - h$ (the horizontal shift) and y is replaced by $y - k$ (the vertical shift). Table 3 gives the forms of the equations of such ellipses and Figure 26 shows their graphs.

Table 3

ELLIPSES WITH CENTER AT (h, k) AND MAJOR AXIS PARALLEL TO A COORDINATE AXIS

Center	Major Axis	Foci	Vertices	Equation
(h, k)	Parallel to the x-axis	$(h + c, k)$ $(h - c, k)$	$(h + a, k)$ $(h - a, k)$	$\dfrac{(x - h)^2}{a^2} + \dfrac{(y - k)^2}{b^2} = 1$, $a > b$ and $b^2 = a^2 - c^2$
(h, k)	Parallel to the y-axis	$(h, k + c)$ $(h, k - c)$	$(h, k + a)$ $(h, k - a)$	$\dfrac{(x - h)^2}{b^2} + \dfrac{(y - k)^2}{a^2} = 1$, $a > b$ and $b^2 = a^2 - c^2$

Figure 26

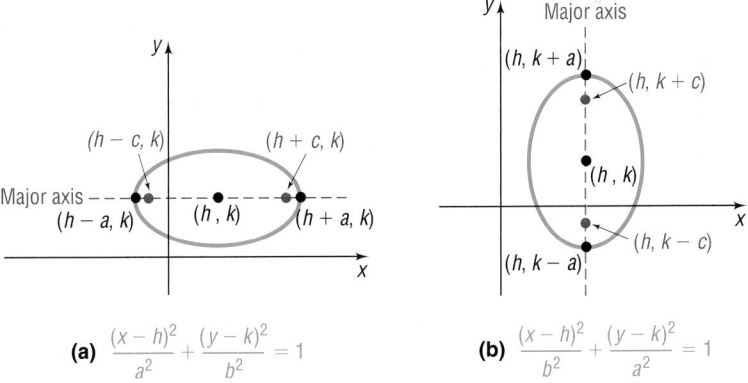

(a) $\dfrac{(x - h)^2}{a^2} + \dfrac{(y - k)^2}{b^2} = 1$ (b) $\dfrac{(x - h)^2}{b^2} + \dfrac{(y - k)^2}{a^2} = 1$

EXAMPLE 5

Finding an Equation of an Ellipse, Center Not at the Origin

Find an equation for the ellipse with center at $(2, -3)$, one focus at $(3, -3)$, and one vertex at $(5, -3)$. Graph the equation.

Solution

The center is at $(h, k) = (2, -3)$, so $h = 2$ and $k = -3$. Since the center, focus, and vertex all lie on the line $y = -3$, the major axis is parallel to the x-axis. The distance from the center $(2, -3)$ to a focus $(3, -3)$ is $c = 1$; the distance from the center $(2, -3)$ to a vertex $(5, -3)$ is $a = 3$. Then $b^2 = a^2 - c^2 = 9 - 1 = 8$. The form of the equation is

$$\frac{(x - h)^2}{a^2} + \frac{(y - k)^2}{b^2} = 1, \quad \text{where } h = 2, k = -3, a = 3, b = 2\sqrt{2}$$
$$\frac{(x - 2)^2}{9} + \frac{(y + 3)^2}{8} = 1$$

Figure 27

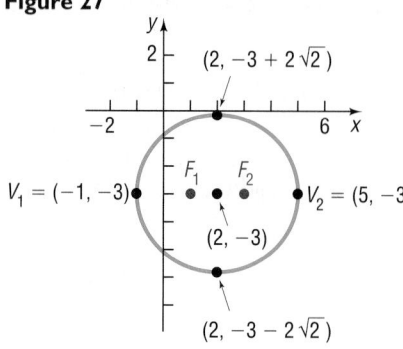

To graph the equation, we use the center $(h, k) = (2, -3)$ to locate the vertices. The major axis is parallel to the x-axis, so the vertices are $a = 3$ units left and right of the center $(2, -3)$. Therefore, the vertices are

$$V_1 = (2 - 3, -3) = (-1, -3) \quad \text{and} \quad V_2 = (2 + 3, -3) = (5, -3)$$

Since $c = 1$ and the major axis is parallel to the x-axis, the foci are 1 unit left and right of the center. Therefore, the foci are

$$F_1 = (2 - 1, -3) = (1, -3) \quad \text{and} \quad F_2 = (2 + 1, -3) = (3, -3)$$

Finally, we use the value of $b = 2\sqrt{2}$ to find the two points above and below the center.

$$\left(2, -3 - 2\sqrt{2}\right) \quad \text{and} \quad \left(2, -3 + 2\sqrt{2}\right)$$

Figure 27 shows the graph. ■

NOW WORK PROBLEM 55.

EXAMPLE 6

Analyzing the Equation of an Ellipse

Analyze the equation: $4x^2 + y^2 - 8x + 4y + 4 = 0$

Solution

We proceed to complete the squares in x and in y.

Figure 28

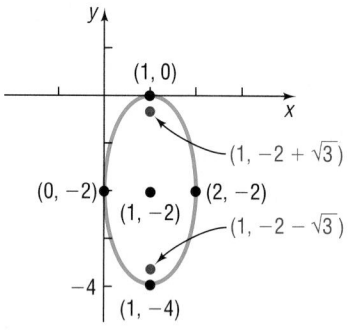

$$4x^2 + y^2 - 8x + 4y + 4 = 0$$
$$4x^2 - 8x + y^2 + 4y = -4 \qquad \text{\textit{Group like variables; place the constant on the right side.}}$$
$$4(x^2 - 2x) + (y^2 + 4y) = -4 \qquad \text{\textit{Factor out 4 from the first two terms.}}$$
$$4(x^2 - 2x + 1) + (y^2 + 4y + 4) = -4 + 4 + 4 \qquad \text{\textit{Complete each square.}}$$
$$4(x - 1)^2 + (y + 2)^2 = 4 \qquad \text{\textit{Factor.}}$$
$$(x - 1)^2 + \frac{(y + 2)^2}{4} = 1 \qquad \text{\textit{Divide each side by 4.}}$$

This is the equation of an ellipse with center at $(1, -2)$ and major axis parallel to the y-axis. Since $a^2 = 4$ and $b^2 = 1$, we have $c^2 = a^2 - b^2 = 4 - 1 = 3$. The vertices are at $(h, k \pm a) = (1, -2 \pm 2)$ or $(1, 0)$ and $(1, -4)$. The foci are at $(h, k \pm c) = (1, -2 \pm \sqrt{3})$ or $(1, -2 - \sqrt{3})$ and $(1, -2 + \sqrt{3})$. Figure 28 shows the graph. ■

NOW WORK PROBLEM 47.

3 Solve Applied Problems Involving Ellipses

Ellipses are found in many applications in science and engineering. For example, the orbits of the planets around the Sun are elliptical, with the Sun's position at a focus. See Figure 29.

Figure 29

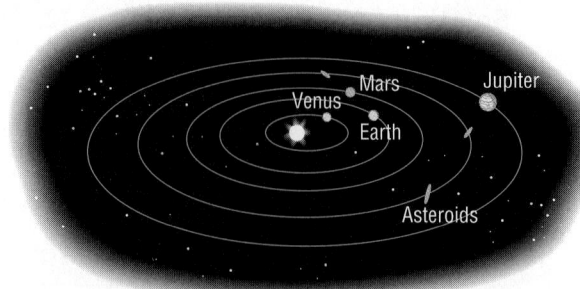

Stone and concrete bridges are often shaped as semielliptical arches. Elliptical gears are used in machinery when a variable rate of motion is required.

Ellipses also have an interesting reflection property. If a source of light (or sound) is placed at one focus, the waves transmitted by the source will reflect off the ellipse and concentrate at the other focus. This is the principle behind *whispering galleries*, which are rooms designed with elliptical ceilings. A person standing at one focus of the ellipse can whisper and be heard by a person standing at the other focus, because all the sound waves that reach the ceiling are reflected to the other person.

| EXAMPLE 7 | A Whispering Gallery |

The whispering gallery in the Museum of Science and Industry in Chicago is 47.3 feet long. The distance from the center of the room to the foci is 20.3 feet. Find an equation that describes the shape of the room. How high is the room at its center?

SOURCE: Chicago Museum of Science and Industry Web site

Solution We set up a rectangular coordinate system so that the center of the ellipse is at the origin and the major axis is along the x-axis. The equation of the ellipse is

$$\frac{x^2}{a^2} + \frac{y^2}{b^2} = 1$$

Since the length of the room is 47.3 feet, the distance from the center of the room to each vertex (the end of the room) will be $\frac{47.3}{2} = 23.65$ feet; so $a = 23.65$ feet. The distance from the center of the room to each focus is $c = 20.3$ feet. See Figure 30. Since $b^2 = a^2 - c^2$, we find $b^2 = 23.65^2 - 20.3^2 = 147.2325$. An equation that describes the shape of the room is given by

$$\frac{x^2}{23.65^2} + \frac{y^2}{147.2325} = 1$$

The height of the room at its center is $b = \sqrt{147.2325} \approx 12.1$ feet. ■

Figure 30

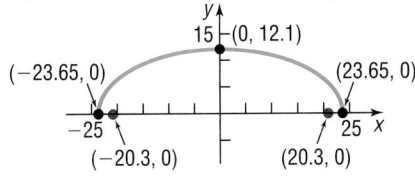

5.3 Assess Your Understanding

'Are You Prepared?'

Answers are given at the end of these exercises. If you get a wrong answer, read the pages listed in red.

1. The distance d from $P_1 = (2, -5)$ to $P_2 = (4, -2)$ is $d =$ _____. (p. 3)

2. To complete the square of $x^2 - 3x$, add _____. (pp. A35–A36)

3. Find the intercepts of the equation $y^2 = 16 - 4x^2$. (pp. 10–11)

4. The point that is symmetric with respect to the y-axis to the point $(-2, 5)$ is _____. (pp. 11–13)

5. To graph $y = (x + 1)^2 - 4$, shift the graph of $y = x^2$ to the (left/right) _____ unit(s) and then (up/down) _____ unit(s). (pp. 85–94)

6. The standard equation of a circle with center at $(2, -3)$ and radius 1 is _____. (pp. 33–36)

Concepts and Vocabulary

7. A(n) _____ is the collection of all points in the plane the sum of whose distances from two fixed points is a constant.

8. For an ellipse, the foci lie on a line called the _____ axis.

9. For the ellipse $\frac{x^2}{4} + \frac{y^2}{25} = 1$, the vertices are the points _____ and _____.

10. *True or False:* The foci, vertices, and center of an ellipse lie on a line called the axis of symmetry.

11. *True or False:* If the center of an ellipse is at the origin and the foci lie on the y-axis, the ellipse is symmetric with respect to the x-axis, the y-axis, and the origin.

12. *True or False:* A circle is a certain type of ellipse.

Skill Building

In Problems 13–16, the graph of an ellipse is given. Match each graph to its equation.

A. $\dfrac{x^2}{4} + y^2 = 1$　　　B. $x^2 + \dfrac{y^2}{4} = 1$　　　C. $\dfrac{x^2}{16} + \dfrac{y^2}{4} = 1$　　　D. $\dfrac{x^2}{4} + \dfrac{y^2}{16} = 1$

13. 　　**14.** 　　**15.** 　　**16.**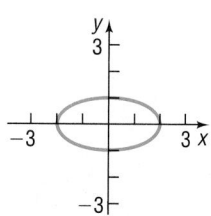

In Problems 17–26, find the vertices and foci of each ellipse. Graph each equation.

17. $\dfrac{x^2}{25} + \dfrac{y^2}{4} = 1$　　**18.** $\dfrac{x^2}{9} + \dfrac{y^2}{4} = 1$　　**19.** $\dfrac{x^2}{9} + \dfrac{y^2}{25} = 1$　　**20.** $x^2 + \dfrac{y^2}{16} = 1$

21. $4x^2 + y^2 = 16$　　**22.** $x^2 + 9y^2 = 18$　　**23.** $4y^2 + x^2 = 8$　　**24.** $4y^2 + 9x^2 = 36$

25. $x^2 + y^2 = 16$　　**26.** $x^2 + y^2 = 4$

In Problems 27–38, find an equation for each ellipse. Graph the equation.

27. Center at $(0, 0)$;　focus at $(3, 0)$;　vertex at $(5, 0)$

28. Center at $(0, 0)$;　focus at $(-1, 0)$;　vertex at $(3, 0)$

29. Center at $(0, 0)$;　focus at $(0, -4)$;　vertex at $(0, 5)$

30. Center at $(0, 0)$;　focus at $(0, 1)$;　vertex at $(0, -2)$

31. Foci at $(\pm 2, 0)$;　length of the major axis is 6

32. Foci at $(0, \pm 2)$;　length of the major axis is 8

33. Focus at $(-4, 0)$;　vertices at $(\pm 5, 0)$

34. Focus at $(0, -4)$;　vertices at $(0, \pm 8)$

35. Foci at $(0, \pm 3)$;　x-intercepts are ± 2

36. Vertices at $(\pm 4, 0)$;　y-intercepts are ± 1

37. Center at $(0, 0)$;　vertex at $(0, 4)$;　$b = 1$

38. Vertices at $(\pm 5, 0)$;　$c = 2$

In Problems 39–42, write an equation for each ellipse.

39. 　　**40.** 　　**41.** 　　**42.**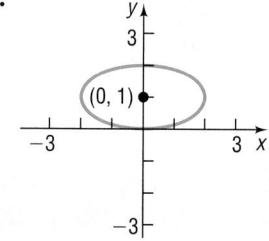

In Problems 43–54, analyze each equation; that is, find the center, foci, and vertices of each ellipse. Graph each equation.

43. $\dfrac{(x-3)^2}{4} + \dfrac{(y+1)^2}{9} = 1$　　**44.** $\dfrac{(x+4)^2}{9} + \dfrac{(y+2)^2}{4} = 1$　　**45.** $(x+5)^2 + 4(y-4)^2 = 16$

46. $9(x-3)^2 + (y+2)^2 = 18$　　**47.** $x^2 + 4x + 4y^2 - 8y + 4 = 0$　　**48.** $x^2 + 3y^2 - 12y + 9 = 0$

49. $2x^2 + 3y^2 - 8x + 6y + 5 = 0$　　**50.** $4x^2 + 3y^2 + 8x - 6y = 5$　　**51.** $9x^2 + 4y^2 - 18x + 16y - 11 = 0$

52. $x^2 + 9y^2 + 6x - 18y + 9 = 0$　　**53.** $4x^2 + y^2 + 4y = 0$　　**54.** $9x^2 + y^2 - 18x = 0$

In Problems 55–64, find an equation for each ellipse. Graph the equation.

55. Center at $(2, -2)$; vertex at $(7, -2)$; focus at $(4, -2)$

56. Center at $(-3, 1)$; vertex at $(-3, 3)$; focus at $(-3, 0)$

57. Vertices at $(4, 3)$ and $(4, 9)$; focus at $(4, 8)$

58. Foci at $(1, 2)$ and $(-3, 2)$; vertex at $(-4, 2)$

59. Foci at $(5, 1)$ and $(-1, 1)$; length of the major axis is 8

60. Vertices at $(2, 5)$ and $(2, -1)$;　$c = 2$

61. Center at $(1, 2)$; focus at $(4, 2)$; contains the point $(1, 3)$

62. Center at $(1, 2)$; focus at $(1, 4)$; contains the point $(2, 2)$

63. Center at $(1, 2)$; vertex at $(4, 2)$; contains the point $(1, 3)$

64. Center at $(1, 2)$; vertex at $(1, 4)$; contains the point $(2, 2)$

In Problems 65–68, graph each function. Be sure to label all the intercepts.

[**Hint:** *Notice that each function is half an ellipse.*]

65. $f(x) = \sqrt{16 - 4x^2}$ **66.** $f(x) = \sqrt{9 - 9x^2}$ **67.** $f(x) = -\sqrt{64 - 16x^2}$ **68.** $f(x) = -\sqrt{4 - 4x^2}$

Applications and Extensions

69. Semielliptical Arch Bridge An arch in the shape of the upper half of an ellipse is used to support a bridge that is to span a river 20 meters wide. The center of the arch is 6 meters above the center of the river (see the figure). Write an equation for the ellipse in which the x-axis coincides with the water level and the y-axis passes through the center of the arch.

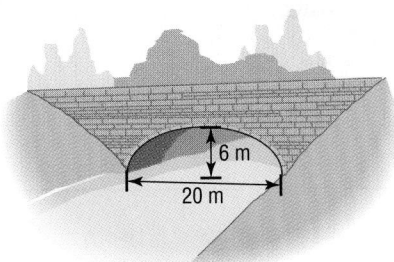

70. Semielliptical Arch Bridge The arch of a bridge is a semi-ellipse with a horizontal major axis. The span is 30 feet, and the top of the arch is 10 feet above the major axis. The roadway is horizontal and is 2 feet above the top of the arch. Find the vertical distance from the roadway to the arch at 5-foot intervals along the roadway.

71. Whispering Gallery A hall 100 feet in length is to be designed as a whispering gallery. If the foci are located 25 feet from the center, how high will the ceiling be at the center?

72. Whispering Gallery Jim, standing at one focus of a whispering gallery, is 6 feet from the nearest wall. His friend is standing at the other focus, 100 feet away. What is the length of this whispering gallery? How high is its elliptical ceiling at the center?

73. Semielliptical Arch Bridge A bridge is built in the shape of a semielliptical arch. The bridge has a span of 120 feet and a maximum height of 25 feet. Choose a suitable rectangular coordinate system and find the height of the arch at distances of 10, 30, and 50 feet from the center.

74. Semielliptical Arch Bridge A bridge is to be built in the shape of a semielliptical arch and is to have a span of 100 feet. The height of the arch, at a distance of 40 feet from the center, is to be 10 feet. Find the height of the arch at its center.

75. Semielliptical Arch An arch in the form of half an ellipse is 40 feet wide and 15 feet high at the center. Find the height of the arch at intervals of 10 feet along its width.

76. Semielliptical Arch Bridge An arch for a bridge over a highway is in the form of half an ellipse. The top of the arch is 20 feet above the ground level (the major axis). The highway has four lanes, each 12 feet wide; a center safety strip 8 feet wide; and two side strips, each 4 feet wide. What should the span of the bridge be (the length of its major axis) if the height 28 feet from the center is to be 13 feet?

*In Problems 77–80, use the fact that the orbit of a planet about the Sun is an ellipse, with the Sun at one focus. The **aphelion** of a planet is its greatest distance from the Sun, and the **perihelion** is its shortest distance. The **mean distance** of a planet from the Sun is the length of the semimajor axis of the elliptical orbit. See the illustration.*

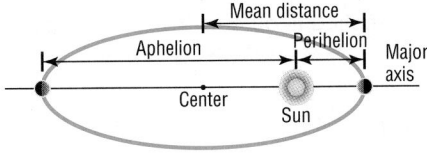

77. Earth The mean distance of Earth from the Sun is 93 million miles. If the aphelion of Earth is 94.5 million miles, what is the perihelion? Write an equation for the orbit of Earth around the Sun.

78. Mars The mean distance of Mars from the Sun is 142 million miles. If the perihelion of Mars is 128.5 million miles, what is the aphelion? Write an equation for the orbit of Mars about the Sun.

79. Jupiter The aphelion of Jupiter is 507 million miles. If the distance from the Sun to the center of its elliptical orbit is 23.2 million miles, what is the perihelion? What is the mean distance? Write an equation for the orbit of Jupiter around the Sun.

80. Pluto The perihelion of Pluto is 4551 million miles, and the distance of the Sun from the center of its elliptical orbit is 897.5 million miles. Find the aphelion of Pluto. What is the mean distance of Pluto from the Sun? Write an equation for the orbit of Pluto about the Sun.

81. Racetrack Design Consult the figure. A racetrack is in the shape of an ellipse, 100 feet long and 50 feet wide. What is the width 10 feet from a vertex?

82. Racetrack Design A racetrack is in the shape of an ellipse 80 feet long and 40 feet wide. What is the width 10 feet from a vertex?

83. Show that an equation of the form

$$Ax^2 + Cy^2 + F = 0, \qquad A \neq 0, C \neq 0, F \neq 0$$

where A and C are of the same sign and F is of opposite sign,

(a) Is the equation of an ellipse with center at $(0, 0)$ if $A \neq C$.

(b) Is the equation of a circle with center $(0, 0)$ if $A = C$.

84. Show that the graph of an equation of the form

$$Ax^2 + Cy^2 + Dx + Ey + F = 0, \qquad A \neq 0, C \neq 0$$

where A and C are of the same sign,

(a) Is an ellipse if $\dfrac{D^2}{4A} + \dfrac{E^2}{4C} - F$ is the same sign as A.

(b) Is a point if $\dfrac{D^2}{4A} + \dfrac{E^2}{4C} - F = 0$.

(c) Contains no points if $\dfrac{D^2}{4A} + \dfrac{E^2}{4C} - F$ is of opposite sign to A.

Discussion and Writing

85. The **eccentricity** e of an ellipse is defined as the number $\dfrac{c}{a}$, where a and c are the numbers given in equation (2). Because $a > c$, it follows that $e < 1$. Write a brief paragraph about the general shape of each of the following ellipses. Be sure to justify your conclusions.

(a) Eccentricity close to 0
(b) Eccentricity = 0.5
(c) Eccentricity close to 1

'Are You Prepared?' Answers

1. $\sqrt{13}$ **2.** $\dfrac{9}{4}$ **3.** $(-2, 0), (2, 0), (0, -4), (0, 4)$ **4.** $(2, 5)$ **5.** left: 1; down: 4 **6.** $(x - 2)^2 + (y + 3)^2 = 1$

5.4 The Hyperbola

PREPARING FOR THIS SECTION *Before getting started, review the following:*

- Distance Formula (Foundations, Section F.1, p. 3)
- Completing the Square (Appendix A, Section A.4, pp. A35–A36)
- Intercepts (Foundations, Section F.2, pp. 10–11)
- Symmetry (Foundations, Section F.2, pp. 11–13)
- Asymptotes (Section 3.2, pp. 206–211)
- Graphing Techniques: Transformations (Section 1.5, pp. 85–94)
- Square Root Method (Section 2.3, pp. 135–136)

Now work the 'Are You Prepared?' problems on page 392.

OBJECTIVES
1 Analyze Hyperbolas with Center at the Origin
2 Find the Asymptotes of a Hyperbola
3 Analyze Hyperbolas with Center at (h, k)
4 Solve Applied Problems Involving Hyperbolas

Figure 31

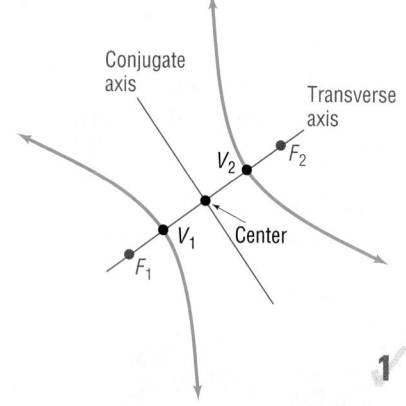

A **hyperbola** is the collection of all points in the plane the difference of whose distances from two fixed points, called the **foci**, is a constant.

Figure 31 illustrates a hyperbola with foci F_1 and F_2. The line containing the foci is called the **transverse axis**. The midpoint of the line segment joining the foci is the **center** of the hyperbola. The line through the center and perpendicular to the transverse axis is the **conjugate axis**. The hyperbola consists of two separate curves, called **branches**, that are symmetric with respect to the transverse axis, conjugate axis, and center. The two points of intersection of the hyperbola and the transverse axis are the **vertices**, V_1 and V_2, of the hyperbola.

1 Analyze Hyperbolas with Center at the Origin

With these ideas in mind, we are now ready to find the equation of a hyperbola in the rectangular coordinate system. First, we place the center at the origin. Next, we position the hyperbola so that its transverse axis coincides with a coordinate axis. Suppose that the transverse axis coincides with the x-axis, as shown in Figure 32.

Figure 32
$d(F_1, P) - d(F_2, P) = \pm 2a$

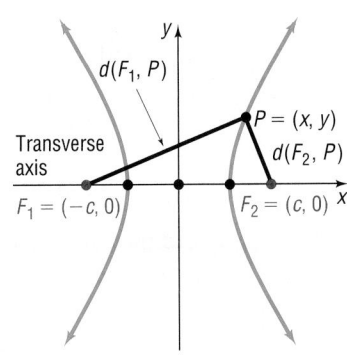

If c is the distance from the center to a focus, one focus will be at $F_1 = (-c, 0)$ and the other at $F_2 = (c, 0)$. Now we let the constant difference of the distances from any point $P = (x, y)$ on the hyperbola to the foci F_1 and F_2 be denoted by $\pm 2a$. (If P is on the right branch, the $+$ sign is used; if P is on the left branch, the $-$ sign is used.) The coordinates of P must satisfy the equation

$$d(F_1, P) - d(F_2, P) = \pm 2a$$
Difference of the distances from P to the foci equals $\pm 2a$.

$$\sqrt{(x + c)^2 + y^2} - \sqrt{(x - c)^2 + y^2} = \pm 2a$$
Use the Distance Formula.

$$\sqrt{(x + c)^2 + y^2} = \pm 2a + \sqrt{(x - c)^2 + y^2}$$
Isolate one radical.

$$(x + c)^2 + y^2 = 4a^2 \pm 4a\sqrt{(x - c)^2 + y^2} + (x - c)^2 + y^2$$
Square both sides.

Next we remove the parentheses.

$$x^2 + 2cx + c^2 + y^2 = 4a^2 \pm 4a\sqrt{(x - c)^2 + y^2} + x^2 - 2cx + c^2 + y^2$$

$$4cx - 4a^2 = \pm 4a\sqrt{(x - c)^2 + y^2}$$
Simplify; isolate the radical.

$$cx - a^2 = \pm a\sqrt{(x - c)^2 + y^2}$$
Divide each side by 4.

$$(cx - a^2)^2 = a^2[(x - c)^2 + y^2]$$
Square both sides.

$$c^2x^2 - 2ca^2x + a^4 = a^2(x^2 - 2cx + c^2 + y^2)$$
Simplify.

$$c^2x^2 + a^4 = a^2x^2 + a^2c^2 + a^2y^2$$
Remove parentheses and simplify.

$$(c^2 - a^2)x^2 - a^2y^2 = a^2c^2 - a^4$$
Rearrange terms.

$$(c^2 - a^2)x^2 - a^2y^2 = a^2(c^2 - a^2)$$
Factor a^2 on the right side. **(1)**

To obtain points on the hyperbola off the x-axis, it must be that $a < c$. To see why, look again at Figure 32.

$$d(F_1, P) < d(F_2, P) + d(F_1, F_2)$$
Use triangle F_1PF_2.

$$d(F_1, P) - d(F_2, P) < d(F_1, F_2)$$

$$2a < 2c$$
P is on the right branch, so $d(F_1, P) - d(F_2, P) = 2a$; $d(F_1, F_2) = 2c$.

$$a < c$$

Since $a < c$, we also have $a^2 < c^2$, so $c^2 - a^2 > 0$. Let $b^2 = c^2 - a^2, b > 0$. Then equation (1) can be written as

$$b^2x^2 - a^2y^2 = a^2b^2$$

$$\frac{x^2}{a^2} - \frac{y^2}{b^2} = 1$$
Divide each side by a^2b^2.

To find the vertices of the hyperbola defined by this equation, let $y = 0$. The vertices satisfy the equation $\dfrac{x^2}{a^2} = 1$, the solutions of which are $x = \pm a$. Consequently, the vertices of the hyperbola are $V_1 = (-a, 0)$ and $V_2 = (a, 0)$. Notice that the distance from the center $(0, 0)$ to either vertex is a.

Theorem

Equation of a Hyperbola; Center at (0, 0); Transverse Axis along the x-Axis

An equation of the hyperbola with center at $(0, 0)$, foci at $(-c, 0)$ and $(c, 0)$, and vertices at $(-a, 0)$ and $(a, 0)$ is

$$\frac{x^2}{a^2} - \frac{y^2}{b^2} = 1, \quad \text{where } b^2 = c^2 - a^2 \qquad (2)$$

The transverse axis is the x-axis.

Figure 33

$\dfrac{x^2}{a^2} - \dfrac{y^2}{b^2} = 1, \quad b^2 = c^2 - a^2$

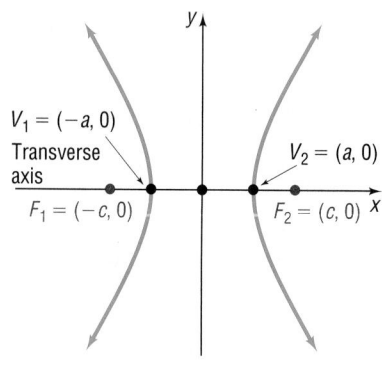

See Figure 33. As you can verify, the hyperbola defined by equation (2) is symmetric with respect to the x-axis, y-axis, and origin. To find the y-intercepts, if any, let $x = 0$ in equation (2). This results in the equation $\dfrac{y^2}{b^2} = -1$, which has no real solution. We conclude that the hyperbola defined by equation (2) has no y-intercepts. In fact, since $\dfrac{x^2}{a^2} - 1 = \dfrac{y^2}{b^2} \geq 0$, it follows that $\dfrac{x^2}{a^2} \geq 1$. There are no points on the graph for $-a < x < a$.

EXAMPLE 1

Finding and Graphing an Equation of a Hyperbola

Find an equation of the hyperbola with center at the origin, one focus at $(3, 0)$, and one vertex at $(-2, 0)$. Graph the equation.

Solution

The hyperbola has its center at the origin, and the transverse axis coincides with the x-axis. One focus is at $(c, 0) = (3, 0)$, so $c = 3$. One vertex is at $(-a, 0) = (-2, 0)$, so $a = 2$. From equation (2), it follows that $b^2 = c^2 - a^2 = 9 - 4 = 5$, so an equation of the hyperbola is

$$\frac{x^2}{4} - \frac{y^2}{5} = 1$$

To graph a hyperbola, it is helpful to locate and plot other points on the graph. For example, to find the points above and below the foci, we let $x = \pm 3$. Then

$$\frac{x^2}{4} - \frac{y^2}{5} = 1$$

$$\frac{(\pm 3)^2}{4} - \frac{y^2}{5} = 1 \qquad x = \pm 3$$

$$\frac{9}{4} - \frac{y^2}{5} = 1$$

$$\frac{y^2}{5} = \frac{5}{4}$$

$$y^2 = \frac{25}{4}$$

$$y = \pm \frac{5}{2}$$

Figure 34

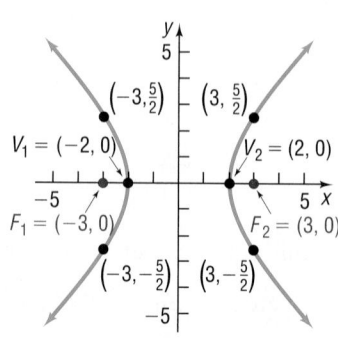

The points above and below the foci are $\left(\pm 3, \dfrac{5}{2}\right)$ and $\left(\pm 3, -\dfrac{5}{2}\right)$. These points determine the "opening" of the hyperbola. See Figure 34. ∎

COMMENT: To graph the hyperbola $\dfrac{x^2}{4} - \dfrac{y^2}{5} = 1$ discussed in Example 1, we need to graph the two functions $Y_1 = \sqrt{5}\sqrt{\dfrac{x^2}{4} - 1}$ and $Y_2 = -\sqrt{5}\sqrt{\dfrac{x^2}{4} - 1}$. Do this and compare what you see with Figure 34. ∎

— **NOW WORK PROBLEM 17.**

An equation of the form of equation (2) is the equation of a hyperbola with center at the origin, foci on the x-axis at $(-c, 0)$ and $(c, 0)$, where $c^2 = a^2 + b^2$, and transverse axis along the x-axis.

For the remainder of this section, the direction "**Analyze the equation**" will mean to find the center, transverse axis, vertices, and foci of the hyperbola and graph it.

EXAMPLE 2

Analyzing the Equation of a Hyperbola

Analyze the equation: $\dfrac{x^2}{16} - \dfrac{y^2}{4} = 1$

Solution The given equation is of the form of equation (2), with $a^2 = 16$ and $b^2 = 4$. The graph of the equation is a hyperbola with center at $(0, 0)$ and transverse axis along the x-axis. Also, we know that $c^2 = a^2 + b^2 = 16 + 4 = 20$. The vertices are at $(\pm a, 0) = (\pm 4, 0)$, and the foci are at $(\pm c, 0) = (\pm 2\sqrt{5}, 0)$.

To locate the points on the graph above and below the foci, we let $x = \pm 2\sqrt{5}$. Then

$$\frac{x^2}{16} - \frac{y^2}{4} = 1$$

$$\frac{\left(\pm 2\sqrt{5}\right)^2}{16} - \frac{y^2}{4} = 1 \qquad x = \pm 2\sqrt{5}$$

$$\frac{20}{16} - \frac{y^2}{4} = 1$$

$$\frac{5}{4} - \frac{y^2}{4} = 1$$

$$\frac{y^2}{4} = \frac{1}{4}$$

$$y = \pm 1$$

The points above and below the foci are $(\pm 2\sqrt{5}, 1)$ and $(\pm 2\sqrt{5}, -1)$. See Figure 35.

Figure 35

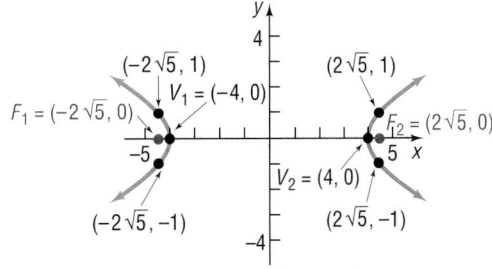

The next result gives the form of the equation of a hyperbola with center at the origin and transverse axis along the y-axis.

Theorem

Equation of a Hyperbola; Center at (0, 0); Transverse Axis along the y-Axis

An equation of the hyperbola with center at $(0,0)$, foci at $(0,-c)$ and $(0,c)$, and vertices at $(0,-a)$ and $(0,a)$ is

$$\frac{y^2}{a^2} - \frac{x^2}{b^2} = 1, \qquad \text{where } b^2 = c^2 - a^2 \qquad (3)$$

The transverse axis is the y-axis.

Figure 36

$\frac{y^2}{a^2} - \frac{x^2}{b^2} = 1, b^2 = c^2 - a^2$

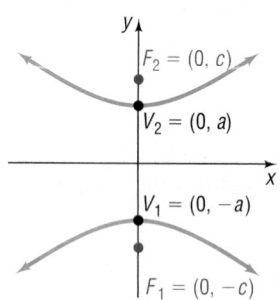

Figure 36 shows the graph of a typical hyperbola defined by equation (3).

An equation of the form of equation (2), $\frac{x^2}{a^2} - \frac{y^2}{b^2} = 1$, is the equation of a hyperbola with center at the origin, foci on the x-axis at $(-c,0)$ and $(c,0)$, where $c^2 = a^2 + b^2$, and transverse axis along the x-axis.

An equation of the form of equation (3), $\frac{y^2}{a^2} - \frac{x^2}{b^2} = 1$, is the equation of a hyperbola with center at the origin, foci on the y-axis at $(0,-c)$ and $(0,c)$, where $c^2 = a^2 + b^2$, and transverse axis along the y-axis.

Notice the difference in the forms of equations (2) and (3). When the y^2-term is subtracted from the x^2-term, the transverse axis is along the x-axis. When the x^2-term is subtracted from the y^2-term, the transverse axis is along the y-axis.

EXAMPLE 3

Analyzing the Equation of a Hyperbola

Analyze the equation: $y^2 - 4x^2 = 4$

Solution

To put the equation in proper form, we divide each side by 4:

$$\frac{y^2}{4} - x^2 = 1$$

Since the x^2-term is subtracted from the y^2-term, the equation is that of a hyperbola with center at the origin and transverse axis along the y-axis. Also, comparing the above equation to equation (3), we find $a^2 = 4, b^2 = 1$, and $c^2 = a^2 + b^2 = 5$. The vertices are at $(0, \pm a) = (0, \pm 2)$, and the foci are at $(0, \pm c) = (0, \pm\sqrt{5})$.

To locate other points on the graph, we let $x = \pm 2$. Then

Figure 37

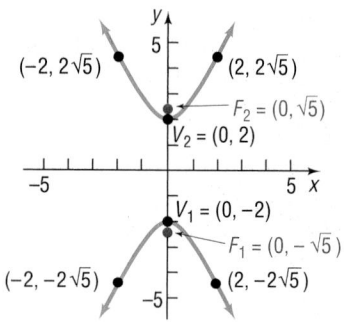

$$y^2 - 4x^2 = 4$$
$$y^2 - 4(\pm 2)^2 = 4 \qquad x = \pm 2$$
$$y^2 - 16 = 4$$
$$y^2 = 20$$
$$y = \pm 2\sqrt{5}$$

Four other points on the graph are $(\pm 2, 2\sqrt{5})$ and $(\pm 2, -2\sqrt{5})$. See Figure 37. ∎

EXAMPLE 4

Finding an Equation of a Hyperbola

Find an equation of the hyperbola having one vertex at $(0,2)$ and foci at $(0,-3)$ and $(0,3)$. Graph the equation.

Solution

Since the foci are at $(0,-3)$ and $(0,3)$, the center of the hyperbola is at their midpoint, the origin. Also, the transverse axis is along the y-axis. The given informa-

Figure 38

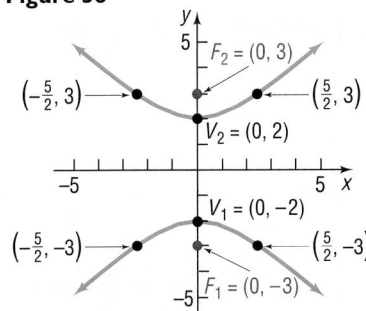

tion also reveals that $c = 3$, $a = 2$, and $b^2 = c^2 - a^2 = 9 - 4 = 5$. The form of the equation of the hyperbola is given by equation (3):

$$\frac{y^2}{a^2} - \frac{x^2}{b^2} = 1$$

$$\frac{y^2}{4} - \frac{x^2}{5} = 1$$

Let $y = \pm 3$ to obtain points on the graph across from the foci. See Figure 38. ■

NOW WORK PROBLEM 19.

Look at the equations of the hyperbolas in Examples 2 and 4. For the hyperbola in Example 2, $a^2 = 16$ and $b^2 = 4$, so $a > b$; for the hyperbola in Example 4, $a^2 = 4$ and $b^2 = 5$, so $a < b$. We conclude that, for hyperbolas, there are no requirements involving the relative sizes of a and b. Contrast this situation to the case of an ellipse, in which the relative sizes of a and b dictate which axis is the major axis. Hyperbolas have another feature to distinguish them from ellipses and parabolas: Hyperbolas have asymptotes.

2 Find the Asymptotes of a Hyperbola

Recall from Section 3.2 that a horizontal or oblique asymptote of a graph is a line with the property that the distance from the line to points on the graph approaches 0 as $x \rightarrow -\infty$ or as $x \rightarrow \infty$. The asymptotes provide information about the end behavior of the graph of a hyperbola.

Theorem

Asymptotes of a Hyperbola

The hyperbola $\dfrac{x^2}{a^2} - \dfrac{y^2}{b^2} = 1$ has the two oblique asymptotes

$$y = \frac{b}{a}x \quad \text{and} \quad y = -\frac{b}{a}x \qquad \qquad (4)$$

Proof We begin by solving for y in the equation of the hyperbola.

$$\frac{x^2}{a^2} - \frac{y^2}{b^2} = 1$$

$$\frac{y^2}{b^2} = \frac{x^2}{a^2} - 1$$

$$y^2 = b^2\left(\frac{x^2}{a^2} - 1\right)$$

Since $x \neq 0$, we can rearrange the right side in the form

$$y^2 = \frac{b^2 x^2}{a^2}\left(1 - \frac{a^2}{x^2}\right)$$

$$y = \pm \frac{bx}{a}\sqrt{1 - \frac{a^2}{x^2}}$$

Now, as $x \rightarrow -\infty$ or as $x \rightarrow \infty$, the term $\dfrac{a^2}{x^2}$ approaches 0, so the expression under the radical approaches 1. Thus, as $x \rightarrow -\infty$ or as $x \rightarrow \infty$, the value of y approaches $\pm \dfrac{bx}{a}$; that is, the graph of the hyperbola approaches the lines

$$y = -\frac{b}{a}x \quad \text{and} \quad y = \frac{b}{a}x$$

These lines are oblique asymptotes of the hyperbola. ■

Figure 39

$$\frac{x^2}{a^2} - \frac{y^2}{b^2} = 1$$

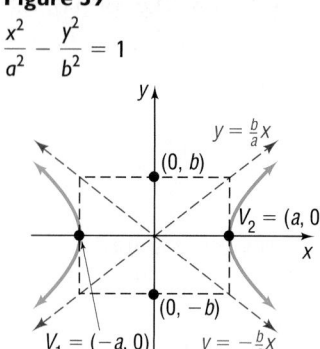

The asymptotes of a hyperbola are not part of the hyperbola, but they do serve as a guide for graphing a hyperbola. For example, suppose that we want to graph the equation

$$\frac{x^2}{a^2} - \frac{y^2}{b^2} = 1$$

We begin by plotting the vertices $(-a, 0)$ and $(a, 0)$. Then we plot the points $(0, -b)$ and $(0, b)$ and use these four points to construct a rectangle, as shown in Figure 39. The diagonals of this rectangle have slopes $\frac{b}{a}$ and $-\frac{b}{a}$, and their extensions are the asymptotes $y = \frac{b}{a}x$ and $y = -\frac{b}{a}x$ of the hyperbola. If we graph the asymptotes, we can use them to establish the "opening" of the hyperbola and avoid plotting other points.

Theorem

Asymptotes of a Hyperbola

The hyperbola $\dfrac{y^2}{a^2} - \dfrac{x^2}{b^2} = 1$ has the two oblique asymptotes

$$y = \frac{a}{b}x \quad \text{and} \quad y = -\frac{a}{b}x \tag{5}$$

You are asked to prove this result in Problem 72.

For the remainder of this section, the direction "**Analyze the equation**" will mean to find the center, transverse axis, vertices, foci, and asymptotes of the hyperbola and graph it.

EXAMPLE 5 | **Analyzing the Equation of a Hyperbola**

Analyze the equation: $\dfrac{y^2}{4} - x^2 = 1$

Figure 40

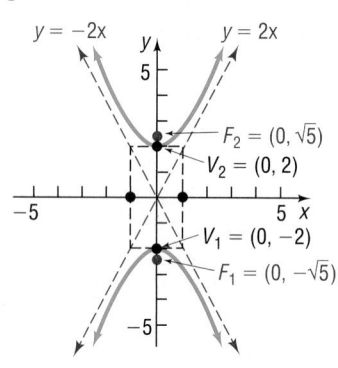

Solution Since the x^2-term is subtracted from the y^2-term, the equation is of the form of equation (3) and is a hyperbola with center at the origin and transverse axis along the y-axis. Also, comparing this equation to equation (3), we find that $a^2 = 4, b^2 = 1$, and $c^2 = a^2 + b^2 = 5$. The vertices are at $(0, \pm a) = (0, \pm 2)$, and the foci are at $(0, \pm c) = (0, \pm\sqrt{5})$. Using equation (5), the asymptotes are the lines $y = \frac{a}{b}x = 2x$ and $y = -\frac{a}{b}x = -2x$. Form the rectangle containing the points $(0, \pm a) = (0, \pm 2)$ and $(\pm b, 0) = (\pm 1, 0)$. The extensions of the diagonals of this rectangle are the asymptotes. Now graph the rectangle, the asymptotes, and the hyperbola. See Figure 40. ∎

EXAMPLE 6 | **Analyzing the Equation of a Hyperbola**

Analyze the equation: $9x^2 - 4y^2 = 36$

Solution Divide each side of the equation by 36 to put the equation in proper form.

$$\frac{x^2}{4} - \frac{y^2}{9} = 1$$

We now proceed to analyze the equation. The center of the hyperbola is the origin. Since the x^2-term is first in the equation, we know that the transverse axis is along the x-axis and the vertices and foci will lie on the x-axis. Using equation (2), we find

Figure 41

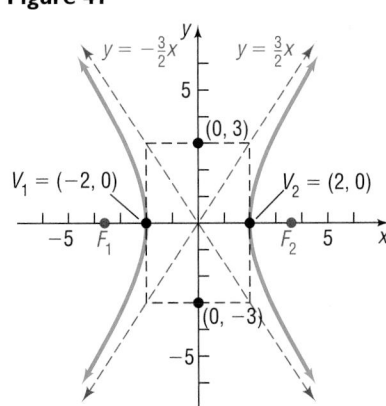

$a^2 = 4$, $b^2 = 9$, and $c^2 = a^2 + b^2 = 13$. The vertices are $a = 2$ units left and right of the center at $(\pm a, 0) = (\pm 2, 0)$, the foci are $c = \sqrt{13}$ units left and right of center at $(\pm c, 0) = (\pm\sqrt{13}, 0)$, and the asymptotes have the equations

$$y = \frac{b}{a}x = \frac{3}{2}x \quad \text{and} \quad y = -\frac{b}{a}x = -\frac{3}{2}x$$

To graph the hyperbola, form the rectangle containing the points $(\pm a, 0)$ and $(0, \pm b)$, that is, $(-2, 0)$, $(2, 0)$, $(0, -3)$, and $(0, 3)$. The extensions of the diagonals of this rectangle are the asymptotes. See Figure 41 for the graph. ∎

NOW WORK PROBLEM 29.

3 Analyze Hyperbolas with Center at (*h*, *k*)

If a hyperbola with center at the origin and transverse axis coinciding with a coordinate axis is shifted horizontally h units and then vertically k units, the result is a hyperbola with center at (h, k) and transverse axis parallel to a coordinate axis. The equations of such hyperbolas have the same forms as those given in equations (2) and (3), except that x is replaced by $x - h$ (the horizontal shift) and y is replaced by $y - k$ (the vertical shift). Table 4 gives the forms of the equations of such hyperbolas. See Figure 42 for typical graphs.

Table 4

HYPERBOLAS WITH CENTER AT (*h*, *k*) AND TRANSVERSE AXIS PARALLEL TO A COORDINATE AXIS					
Center	**Transverse Axis**	**Foci**	**Vertices**	**Equation**	**Asymptotes**
(h, k)	Parallel to the x-axis	$(h \pm c, k)$	$(h \pm a, k)$	$\dfrac{(x - h)^2}{a^2} - \dfrac{(y - k)^2}{b^2} = 1$, $b^2 = c^2 - a^2$	$y - k = \pm\dfrac{b}{a}(x - h)$
(h, k)	Parallel to the y-axis	$(h, k \pm c)$	$(h, k \pm a)$	$\dfrac{(y - k)^2}{a^2} - \dfrac{(x - h)^2}{b^2} = 1$, $b^2 = c^2 - a^2$	$y - k = \pm\dfrac{a}{b}(x - h)$

Figure 42

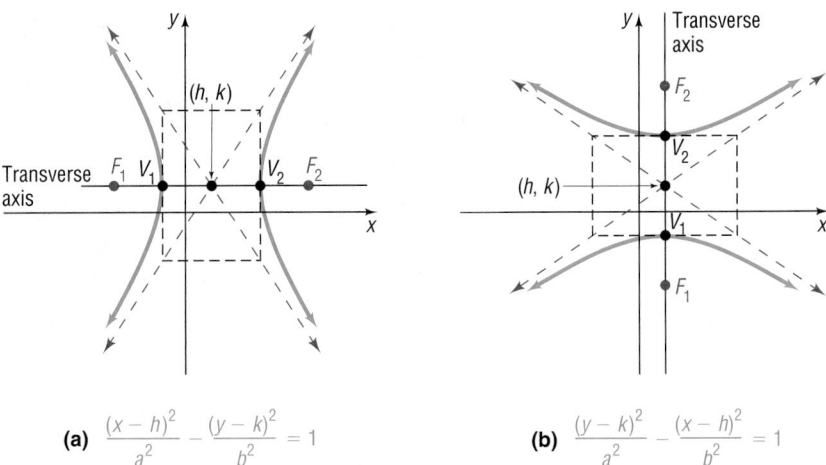

(a) $\dfrac{(x - h)^2}{a^2} - \dfrac{(y - k)^2}{b^2} = 1$ (b) $\dfrac{(y - k)^2}{a^2} - \dfrac{(x - h)^2}{b^2} = 1$

EXAMPLE 7

Finding an Equation of a Hyperbola, Center Not at the Origin

Find an equation for the hyperbola with center at $(1, -2)$, one focus at $(4, -2)$, and one vertex at $(3, -2)$. Graph the equation.

Solution The center is at $(h, k) = (1, -2)$, so $h = 1$ and $k = -2$. Since the center, focus, and vertex all lie on the line $y = -2$, the transverse axis is parallel to the x-axis. The

Figure 43

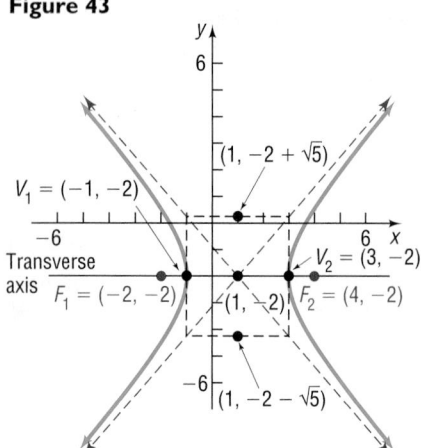

distance from the center $(1, -2)$ to the focus $(4, -2)$ is $c = 3$; the distance from the center $(1, -2)$ to the vertex $(3, -2)$ is $a = 2$. Thus, $b^2 = c^2 - a^2 = 9 - 4 = 5$. The equation is

$$\frac{(x - h)^2}{a^2} - \frac{(y - k)^2}{b^2} = 1$$

$$\frac{(x - 1)^2}{4} - \frac{(y + 2)^2}{5} = 1$$

See Figure 43. ∎

NOW WORK PROBLEM 39.

EXAMPLE 8

Analyzing the Equation of a Hyperbola

Analyze the equation: $-x^2 + 4y^2 - 2x - 16y + 11 = 0$

Solution

We complete the squares in x and in y.

$$-x^2 + 4y^2 - 2x - 16y + 11 = 0$$

$$-(x^2 + 2x) + 4(y^2 - 4y) = -11 \qquad \text{Group terms.}$$

$$-(x^2 + 2x + 1) + 4(y^2 - 4y + 4) = -11 - 1 + 16 \qquad \text{Complete each square.}$$

$$-(x + 1)^2 + 4(y - 2)^2 = 4$$

$$(y - 2)^2 - \frac{(x + 1)^2}{4} = 1 \qquad \text{Divide each side by 4.}$$

Figure 44

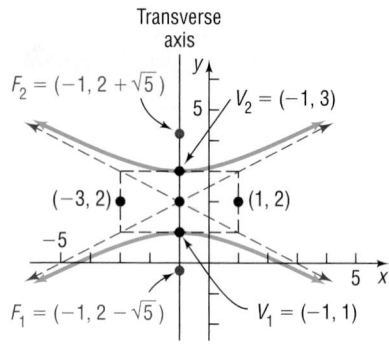

This is the equation of a hyperbola with center at $(-1, 2)$ and transverse axis parallel to the y-axis. Also, $a^2 = 1$ and $b^2 = 4$, so $c^2 = a^2 + b^2 = 5$. Since the transverse axis is parallel to the y-axis, the vertices and foci are located a and c units above and below the center, respectively. The vertices are at $(h, k \pm a) = (-1, 2 \pm 1)$, or $(-1, 1)$ and $(-1, 3)$. The foci are at $(h, k \pm c) = \left(-1, 2 \pm \sqrt{5}\right)$. The asymptotes are $y - 2 = \frac{1}{2}(x + 1)$ and $y - 2 = -\frac{1}{2}(x + 1)$. Figure 44 shows the graph. ∎

NOW WORK PROBLEM 53.

4 Solve Applied Problems Involving Hyperbolas

Figure 45

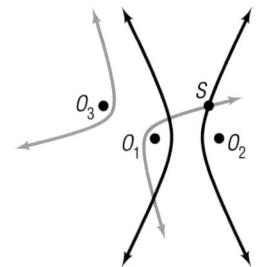

Look at Figure 45. Suppose that three microphones are located at points O_1, O_2, and O_3 (the foci of the two hyperbolas). In addition, suppose that a gun is fired at S and the microphone at O_1 records the gun shot 1 second after the microphone at O_2. Because sound travels at about 1100 feet per second, we conclude that the microphone at O_1 is 1100 feet farther from the gunshot than O_2. We can model this situation by saying that S lies on a branch of a hyperbola with foci at O_1 and O_2. (Do you see why? The difference of the distances from S to O_1 and from S to O_2 is the constant 1100.) If the third microphone at O_3 records the gunshot 2 seconds after O_1, then S will lie on a branch of a second hyperbola with foci at O_1 and O_3. In this case, the constant difference will be 2200. The intersection of the two hyperbolas will identify the location of S.

EXAMPLE 9

Lightning Strikes

Suppose that two people standing 1 mile apart both see a flash of lightning. After a period of time, the person standing at point A hears the thunder. One second later, the person standing at point B hears the thunder. If the person at B is due west of the person at A and the lightning strike is known to occur due north of the person standing at point A, where did the lightning strike?

Solution See Figure 46 in which the ordered pair (x, y) represents the location of the lightning strike. We know that sound travels at 1100 feet per second, so the person at point A is 1100 feet closer to the lightning strike than the person at point B. Since the difference of the distance from (x, y) to A and the distance from (x, y) to B is the constant 1100, the point (x, y) lies on a hyperbola whose foci are at A and B.

Figure 46

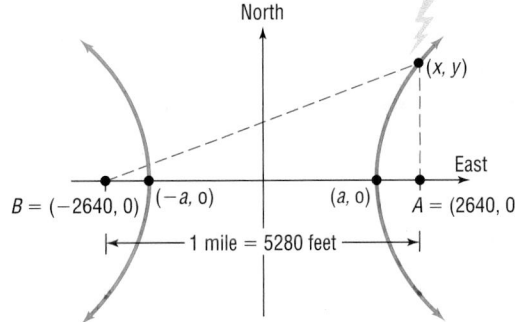

An equation of the hyperbola is

$$\frac{x^2}{a^2} - \frac{y^2}{b^2} = 1$$

where $2a = 1100$ so $a = 550$.

Because the distance between the two people is 1 mile (5280 feet) and each person is at a focus of the hyperbola, we have

$$2c = 5280$$
$$c = \frac{5280}{2} = 2640$$

Since $b^2 = c^2 - a^2 = 2640^2 - 550^2 = 6{,}667{,}100$, the equation of the hyperbola that describes the location of the lightning strike is

$$\frac{x^2}{550^2} - \frac{y^2}{6{,}667{,}100} = 1$$

Refer again to Figure 46. Since the lightning strikes due north of the individual at the point $A = (2640, 0)$, we let $x = 2640$ and solve the resulting equation.

$$\frac{2640^2}{550^2} - \frac{y^2}{6{,}667{,}100} = 1$$

$$-\frac{y^2}{6{,}667{,}100} = -22.04 \qquad \text{Subtract } \frac{2640^2}{550^2} \text{ from both sides.}$$

$$y^2 = 146{,}942{,}884 \qquad \text{Multiply both sides by } -6{,}667{,}100.$$

$$y = 12{,}122 \qquad y > 0 \text{ since the lightning strike}$$
$$\text{occured in quadrant I}$$

✔ **CHECK:** The difference between the distance from $(2640, 12\,122)$ to the person at the point $B = (-2640, 0)$ and the distance from $(2640, 12\,122)$ to the person at the point $A = (2640, 0)$, should be 1100. Using the distance formula, we find the difference in the distances is

$$\sqrt{(2640 - (-2640))^2 + (12{,}122 - 0)^2} - \sqrt{(2640 - 2640)^2 + (12{,}122 - 0)^2} = 1100$$

as required.

The lightning strike is 12,122 feet north of the person standing at point A. ∎

━ NOW WORK PROBLEM 65.

5.4 Assess Your Understanding

'Are You Prepared?'

Answers are given at the end of these exercises. If you get a wrong answer, read the pages listed in red.

1. The distance d from $P_1 = (3, -4)$ to $P_2 = (-2, 1)$ is $d =$ _____. (p. 3)

2. To complete the square of $x^2 + 5x$, add _____. (pp. A35–A36)

3. Find the intercepts of the equation $y^2 = 9 + 4x^2$. (pp. 10–11)

4. *True or False:* The equation $y^2 = 9 + x^2$ is symmetric with respect to the x-axis, the y-axis, and the origin. (pp. 11–13)

5. To graph $y = (x - 5)^3 - 4$, shift the graph of $y = x^3$ to the (left/right) _____ unit(s) and then (up/down) _____ unit(s). (pp. 85–94)

6. Find the vertical asymptotes, if any, and the horizontal or oblique asymptotes, if any, of $y = \dfrac{x^2 - 9}{x^2 - 4}$. (pp. 206–211)

Concepts and Vocabulary

7. A(n) _____ is the collection of points in the plane the difference of whose distances from two fixed points is a constant.

8. For a hyperbola, the foci lie on a line called the _____ _____.

9. The asymptotes of the hyperbola $\dfrac{x^2}{4} - \dfrac{y^2}{9} = 1$ are _____ and _____.

10. *True or False:* The foci of a hyperbola lie on a line called the axis of symmetry.

11. *True or False:* Hyperbolas always have asymptotes.

12. *True or False:* A hyperbola will never intersect its transverse axis.

Skill Building

In Problems 13–16, the graph of a hyperbola is given. Match each graph to its equation.

A. $\dfrac{x^2}{4} - y^2 = 1$

B. $x^2 - \dfrac{y^2}{4} = 1$

C. $\dfrac{y^2}{4} - x^2 = 1$

D. $y^2 - \dfrac{x^2}{4} = 1$

13.

14.

15.

16.
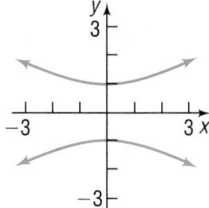

In Problems 17–26, find an equation for the hyperbola described. Graph the equation.

17. Center at $(0, 0)$; focus at $(3, 0)$; vertex at $(1, 0)$

18. Center at $(0, 0)$; focus at $(0, 5)$; vertex at $(0, 3)$

19. Center at $(0, 0)$; focus at $(0, -6)$; vertex at $(0, 4)$

20. Center at $(0, 0)$; focus at $(-3, 0)$; vertex at $(2, 0)$

21. Foci at $(-5, 0)$ and $(5, 0)$; vertex at $(3, 0)$

22. Focus at $(0, 6)$; vertices at $(0, -2)$ and $(0, 2)$

23. Vertices at $(0, -6)$ and $(0, 6)$; asymptote the line $y = 2x$

24. Vertices at $(-4, 0)$ and $(4, 0)$; asymptote the line $y = 2x$

25. Foci at $(-4, 0)$ and $(4, 0)$; asymptote the line $y = -x$

26. Foci at $(0, -2)$ and $(0, 2)$; asymptote the line $y = -x$

In Problems 27–34, find the center, transverse axis, vertices, foci, and asymptotes. Graph each equation.

27. $\dfrac{x^2}{25} - \dfrac{y^2}{9} = 1$

28. $\dfrac{y^2}{16} - \dfrac{x^2}{4} = 1$

29. $4x^2 - y^2 = 16$

30. $4y^2 - x^2 = 16$

31. $y^2 - 9x^2 = 9$

32. $x^2 - y^2 = 4$

33. $y^2 - x^2 = 25$

34. $2x^2 - y^2 = 4$

In Problems 35–38, write an equation for each hyperbola.

35.

36.

37.

38.
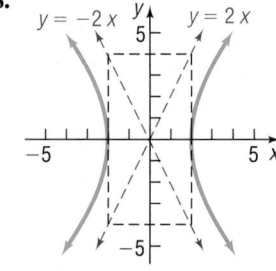

In Problems 39–46, find an equation for the hyperbola described. Graph the equation.

39. Center at $(4, -1)$; focus at $(7, -1)$; vertex at $(6, -1)$

40. Center at $(-3, 1)$; focus at $(-3, 6)$; vertex at $(-3, 4)$

41. Center at $(-3, -4)$; focus at $(-3, -8)$; vertex at $(-3, -2)$

42. Center at $(1, 4)$; focus at $(-2, 4)$; vertex at $(0, 4)$

43. Foci at $(3, 7)$ and $(7, 7)$; vertex at $(6, 7)$

44. Focus at $(-4, 0)$ vertices at $(-4, 4)$ and $(-4, 2)$

45. Vertices at $(-1, -1)$ and $(3, -1)$;
asymptote the line $y + 1 = \dfrac{3}{2}(x - 1)$

46. Vertices at $(1, -3)$ and $(1, 1)$;
asymptote the line $y + 1 = \dfrac{3}{2}(x - 1)$

In Problems 47–60, find the center, transverse axis, vertices, foci, and asymptotes. Graph each equation.

47. $\dfrac{(x - 2)^2}{4} - \dfrac{(y + 3)^2}{9} = 1$

48. $\dfrac{(y + 3)^2}{4} - \dfrac{(x - 2)^2}{9} = 1$

49. $(y - 2)^2 - 4(x + 2)^2 = 4$

50. $(x + 4)^2 - 9(y - 3)^2 = 9$

51. $(x + 1)^2 - (y + 2)^2 = 4$

52. $(y - 3)^2 - (x + 2)^2 = 4$

53. $x^2 - y^2 - 2x - 2y - 1 = 0$

54. $y^2 - x^2 - 4y + 4x - 1 = 0$

55. $y^2 - 4x^2 - 4y - 8x - 4 = 0$

56. $2x^2 - y^2 + 4x + 4y - 4 = 0$

57. $4x^2 - y^2 - 24x - 4y + 16 = 0$

58. $2y^2 - x^2 + 2x + 8y + 3 = 0$

59. $y^2 - 4x^2 - 16x - 2y - 19 = 0$

60. $x^2 - 3y^2 + 8x - 6y + 4 = 0$

In Problems 61–64, graph each function. Be sure to label any intercepts.

[**Hint:** *Notice that each function is half a hyperbola.*]

61. $f(x) = \sqrt{16 + 4x^2}$

62. $f(x) = -\sqrt{9 + 9x^2}$

63. $f(x) = -\sqrt{-25 + x^2}$

64. $f(x) = \sqrt{-1 + x^2}$

Applications and Extensions

65. Fireworks Display Suppose that two people standing 2 miles apart both see the burst from a fireworks display. After a period of time, the first person standing at point A hears the burst. One second later, the second person standing at point B hears the burst. If the person at point B is due west of the person at point A and if the display is known to occur due north of the person at point A, where did the fireworks display occur?

66. Lightning Strikes Suppose that two people standing 1 mile apart both see a flash of lightning. After a period of time, the first person standing at point A hears the thunder. Two seconds later, the second person standing at point B hears the thunder. If the person at point B is due west of the person at point A and if the lightning strike is known to occur due north of the person standing at point A, where did the lightning strike?

67. Rutherford's Experiment In May 1911, Ernest Rutherford published a paper in *Philosophical Magazine*. In this article, he described the motion of alpha particles as they are shot at a piece of gold foil 0.00004 cm thick. Before conducting this experiment, Rutherford expected that the alpha particles would shoot through the foil just as a bullet would shoot through snow. Instead, a small fraction of the alpha particles bounced off the foil. This led to the conclusion that the nucleus of an atom is dense, while the remainder of the atom is sparse. Only the density of the nucleus could cause the alpha particles to deviate from their path. The figure shows a diagram from Rutherford's paper that indicates that the deflected alpha particles follow the path of one branch of a hyperbola.

(a) Find an equation of the asymptotes under this scenario.

(b) If the vertex of the path of the alpha particles is 10 cm from the center of the hyperbola, find an equation that describes the path of the particle.

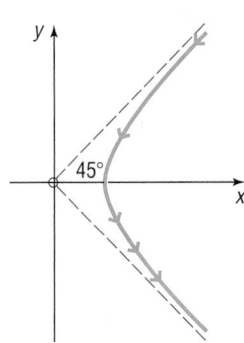

68. An Explosion Two recording devices are set 2400 feet apart, with the device at point A to the west of the device at point B. At a point between the devices, 300 feet from point B, a small amount of explosive is detonated. The recording devices record the time until the sound reaches each. How far directly north of point B should a second explosion be done so that the measured time difference recorded by the devices is the same as that for the first detonation?

69. The **eccentricity** e of a hyperbola is defined as the number $\dfrac{c}{a}$, where a and c are the numbers given in equation (2). Because $c > a$, it follows that $e > 1$. Describe the general shape of a hyperbola whose eccentricity is close to 1. What is the shape if e is very large?

70. A hyperbola for which $a = b$ is called an **equilateral hyperbola**. Find the eccentricity e of an equilateral hyperbola.

[**NOTE:** The eccentricity of a hyperbola is defined in Problem 69.]

71. Two hyperbolas that have the same set of asymptotes are called **conjugate**. Show that the hyperbolas

$$\frac{x^2}{4} - y^2 = 1 \quad \text{and} \quad y^2 - \frac{x^2}{4} = 1$$

are conjugate. Graph each hyperbola on the same set of coordinate axes.

72. Prove that the hyperbola

$$\frac{y^2}{a^2} - \frac{x^2}{b^2} = 1$$

has the two oblique asymptotes

$$y = \frac{a}{b}x \quad \text{and} \quad y = -\frac{a}{b}x$$

73. Show that the graph of an equation of the form

$$Ax^2 + Cy^2 + F = 0, \qquad A \neq 0, C \neq 0, F \neq 0$$

where A and C are of opposite sign, is a hyperbola with center at $(0, 0)$.

74. Show that the graph of an equation of the form

$$Ax^2 + Cy^2 + Dx + Ey + F = 0, \qquad A \neq 0, C \neq 0$$

where A and C are of opposite sign,

(a) Is a hyperbola if $\dfrac{D^2}{4A} + \dfrac{E^2}{4C} - F \neq 0$.

(b) Is two intersecting lines if

$$\frac{D^2}{4A} + \frac{E^2}{4C} - F = 0$$

'Are You Prepared?' Answers

1. $5\sqrt{2}$ **2.** $\dfrac{25}{4}$ **3.** $(0, -3), (0, 3)$ **4.** True **5.** right, 5; down, 4 **6.** Vertical: $x = -2, x = 2$; Horizontal: $y = 1$

Chapter Review

Things to Know

Equations

Parabola	See Tables 1 and 2 (pp. 365 and 367).
Ellipse	See Table 3 (p. 377).
Hyperbola	See Table 4 (p. 389).

Definitions

Parabola (p. 363)	Set of points P in the plane for which $d(F, P) = d(P, D)$, where F is the focus and D is the directrix
Ellipse (p. 372)	Set of points P in the plane, the sum of whose distances from two fixed points (the foci) is a constant
Hyperbola (p. 382)	Set of points P in the plane, the difference of whose distances from two fixed points (the foci) is a constant

Objectives

Section		You should be able to . . .	Review Exercises
5.1	1	Know the names of the conics (p. 362)	1–32
5.2	1	Analyze parabolas with vertex at the origin (p. 364)	1, 2, 21, 24
	2	Analyze parabolas with vertex at (h, k) (p. 367)	7, 11, 12, 17, 18, 27, 30
	3	Solve applied problems involving parabolas (p. 368)	41, 42
5.3	1	Analyze ellipses with center at the origin (p. 373)	5, 6, 10, 22, 25
	2	Analyze ellipses with center at (h, k) (p. 377)	14–16, 19, 28, 31
	3	Solve applied problems involving ellipses (p. 378)	43, 44
5.4	1	Analyze hyperbolas with center at the origin (p. 382)	3, 4, 8, 9, 23, 26
	2	Find the asymptotes of a hyperbola (p. 387)	3, 4, 8, 9
	3	Analyze hyperbolas with center at (h, k) (p. 389)	13, 20, 29, 32–36
	4	Solve applied problems involving hyperbolas (p. 390)	45

Review Exercises

In Problems 1–20, identify each equation. If it is a parabola, gives its vertex, focus, and directrix; if it is an ellipse, give its center, vertices, and foci; if it is a hyperbola, give its center, vertices, foci, and asymptotes.

1. $y^2 = -16x$

2. $16x^2 = y$

3. $\dfrac{x^2}{25} - y^2 = 1$

4. $\dfrac{y^2}{25} - x^2 = 1$

5. $\dfrac{y^2}{25} + \dfrac{x^2}{16} = 1$

6. $\dfrac{x^2}{9} + \dfrac{y^2}{16} = 1$

7. $x^2 + 4y = 4$

8. $3y^2 - x^2 = 9$

9. $4x^2 - y^2 = 8$

10. $9x^2 + 4y^2 = 36$

11. $x^2 - 4x = 2y$

12. $2y^2 - 4y = x - 2$

13. $y^2 - 4y - 4x^2 + 8x = 4$

14. $4x^2 + y^2 + 8x - 4y + 4 = 0$

15. $4x^2 + 9y^2 - 16x - 18y = 11$

16. $4x^2 + 9y^2 - 16x + 18y = 11$

17. $4x^2 - 16x + 16y + 32 = 0$

18. $4y^2 + 3x - 16y + 19 = 0$

19. $9x^2 + 4y^2 - 18x + 8y = 23$

20. $x^2 - y^2 - 2x - 2y = 1$

In Problems 21–36, obtain an equation of the conic described. Graph the equation.

21. Parabola; focus at $(-2, 0)$; directrix the line $x = 2$

22. Ellipse; center at $(0, 0)$; focus at $(0, 3)$; vertex at $(0, 5)$

23. Hyperbola; center at $(0,0)$; focus at $(0,4)$; vertex at $(0,-2)$

24. Parabola; vertex at $(0, 0)$; directrix the line $y = -3$

25. Ellipse; foci at $(-3, 0)$ and $(3, 0)$; vertex at $(4, 0)$

26. Hyperbola; vertices at $(-2, 0)$ and $(2, 0)$; focus at $(4, 0)$

27. Parabola; vertex at $(2, -3)$; focus at $(2, -4)$

28. Ellipse; center at $(-1, 2)$; focus at $(0, 2)$; vertex at $(2, 2)$

29. Hyperbola; center at $(-2, -3)$; focus at $(-4, -3)$; vertex at $(-3, -3)$

30. Parabola; focus at $(3, 6)$; directrix the line $y = 8$

31. Ellipse; foci at $(-4, 2)$ and $(-4, 8)$; vertex at $(-4, 10)$

32. Hyperbola; vertices at $(-3, 3)$ and $(5, 3)$; focus at $(7, 3)$

33. Center at $(-1, 2)$; $a = 3$; $c = 4$; transverse axis parallel to the x-axis

34. Center at $(4, -2)$; $a = 1$; $c = 4$; transverse axis parallel to the y-axis

35. Vertices at $(0, 1)$ and $(6, 1)$; asymptote the line $3y + 2x = 9$

36. Vertices at $(4, 0)$ and $(4, 4)$; asymptote the line $y + 2x = 10$

37. Find an equation of the hyperbola whose foci are the vertices of the ellipse $4x^2 + 9y^2 = 36$ and whose vertices are the foci of this ellipse.

38. Find an equation of the ellipse whose foci are the vertices of the hyperbola $x^2 - 4y^2 = 16$ and whose vertices are the foci of this hyperbola.

39. Describe the collection of points in a plane so that the distance from each point to the point $(3, 0)$ is three-fourths of its distance from the line $x = \dfrac{16}{3}$.

40. Describe the collection of points in a plane so that the distance from each point to the point $(5, 0)$ is five-fourths of its distance from the line $x = \dfrac{16}{5}$.

41. Mirrors A mirror is shaped like a paraboloid of revolution. If a light source is located 1 foot from the base along the axis of symmetry and the opening is 2 feet across, how deep should the mirror be?

42. Parabolic Arch Bridge A bridge is built in the shape of a parabolic arch. The bridge has a span of 60 feet and a maximum height of 20 feet. Find the height of the arch at distances of 5, 10, and 20 feet from the center.

43. Semi-elliptical Arch Bridge A bridge is built in the shape of a semi-elliptical arch. The bridge has a span of 60 feet and a maximum height of 20 feet. Find the height of the arch at distances of 5, 10, and 20 feet from the center.

44. Whispering Galleries The figure shows the specifications for an elliptical ceiling in a hall designed to be a whispering gallery. Where are the foci located in the hall?

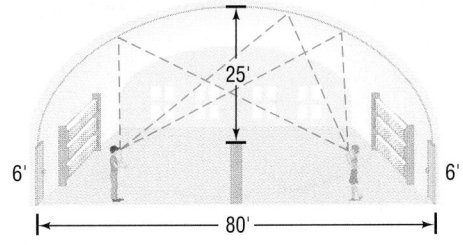

45. Calibrating Instruments In a test of their recording devices, a team of seismologists positioned two of the devices 2000 feet apart, with the device at point A to the west of the device at point B. At a point between the devices and 200 feet from point B, a small amount of explosive was detonated and a note made of the time at which the sound reached each device. A second explosion is to be carried out at a point directly north of point B. How far north should the site of the second explosion be chosen so that the measured time difference recorded by the devices for the second detonation is the same as that recorded for the first detonation?

Chapter Test

In Problems 1–3, identify each equation. If it is a parabola, give its vertex, focus, and directrix; if an ellipse, give its center, vertices, and foci; if a hyperbola, give its center, vertices, foci, and asymptotes.

1. $\dfrac{(x+1)^2}{4} - \dfrac{y^2}{9} = 1$

2. $8y = (x-1)^2 - 4$

3. $2x^2 + 3y^2 + 4x - 6y = 13$

In Problems 4–6, obtain an equation of the conic described; graph the equation.

4. parabola: focus $(-1, 4.5)$, vertex $(-1, 3)$

5. ellipse: center $(0, 0)$, vertex $(0, -4)$, focus $(0, 3)$

6. hyperbola: center $(2, 2)$, vertex $(2, 4)$, contains the point $\left(2 + \sqrt{10}, 5\right)$

7. A parabolic reflector (paraboloid of revolution) is used by TV crews at football games to pick up the referee's announcements, quarterback signals, and so on. A microphone is placed at the focus of the parabola. If a certain reflector is 4 feet wide and 1.5 feet deep, where should the microphone be placed?

Chapter Projects

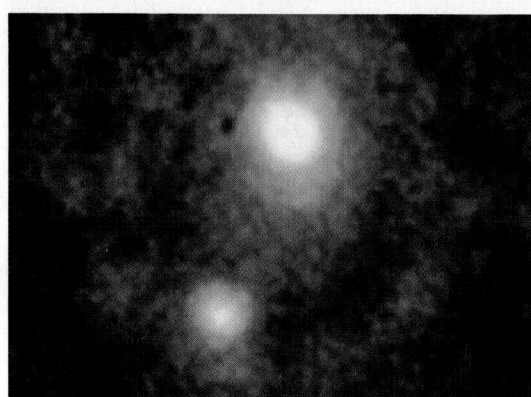

1. The Orbits of Neptune and Pluto The orbit of a planet about the Sun is an ellipse, with the Sun at one focus. The **aphelion** of a planet is its greatest distance from the Sun and the **perihelion** is its shortest distance. The **mean distance** of a planet from the Sun is the length of the semimajor axis of the elliptical orbit. See the illustration.

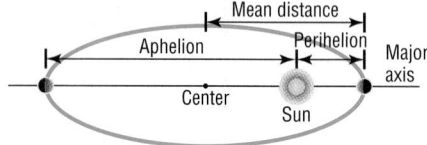

(a) The aphelion of Neptune is 4532.2×10^6 km and its perihelion is 4458.0×10^6 km. Find the equation for the orbit of Neptune around the Sun.

(b) The aphelion of Pluto is 7381.2×10^6 km and its perihelion is 4445.8×10^6 km. Find the equation for the orbit of Pluto around the Sun.

(c) Graph the orbits of Pluto and Neptune on a graphing utility. Knowing that the orbits of the planets intersect, what is wrong with the graphs you obtained?

(d) The graphs of the orbits drawn in part (c) have the same center, but their foci lie in different locations. To see an accurate representation, the location of the Sun (a focus) needs to be the same for both graphs. This can be accomplished by shifting Pluto's orbit to the left. The shift amount is equal to Pluto's distance from the center [in the graph in part (c)] to the Sun minus Neptune's distance from the center to the Sun. Find the new equation representing the orbit of Pluto.

(e) Graph the equation for the orbit of Pluto found in part (d) along with the equation of the orbit of Neptune. Do you see that Pluto's orbit is sometimes inside Neptune's?

(f) Find the point(s) of intersection of the two orbits.

(g) Do you think two planets will ever collide?

The following projects can be found on the Instructor's Resource Center (IRC):

2. **Project at Motorola** *Distorted Deployable Space Reflector Antennas*

3. **Constructing a Bridge Over the East River**

Cumulative Review

1. For $f(x) = -3x^2 + 5x - 2$, find

$$\frac{f(x + h) - f(x)}{h} \qquad h \neq 0$$

2. (a) Find the domain and range of $y = 3^x + 2$.

(b) Find the inverse of $y = 3^x + 2$ and state its domain and range.

3. Solve the equation $9x^4 + 33x^3 - 71x^2 - 57x - 10 = 0$.

4. For what numbers x is $6 - x \geq x^2$?

5. Find an equation for each of the following graphs:

(a) Line:

(b) Circle:

(c) Ellipse:

(d) Parabola:

(e) Hyperbola:

(f) Exponential:

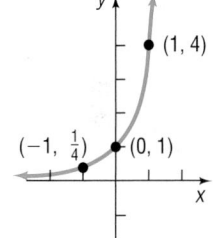

6. If $f(x) = \log_4(x - 2)$:

(a) Solve $f(x) = 2$.

(b) Solve $f(x) \leq 2$.

6 Systems of Equations and Inequalities

A LOOK BACK In Appendix A, Sections A.8 and A.9, and in Chapters 2, 3, and 4, we solved equations and inequalities involving a single variable.

A LOOK AHEAD In this chapter we take up the problem of solving equations and inequalities containing two or more variables. There are various ways to solve such problems:

The *method of substitution* for solving equations in several unknowns goes back to ancient times.

The *method of elimination*, although it had existed for centuries, was put into systematic order by Karl Friedrich Gauss (1777–1855) and by Camille Jordan (1838–1922).

The theory of *matrices* was developed in 1857 by Arthur Cayley (1821–1895), although only later were matrices used as we use them in this chapter. Matrices have become a very flexible instrument, useful in almost all areas of mathematics.

The method of *determinants* was invented by Takakazu Seki Kôwa (1642–1708) in 1683 in Japan and by Gottfried Wilhelm von Leibniz (1646–1716) in 1693 in Germany. *Cramer's Rule* is named after Gabriel Cramer (1704–1752) of Switzerland, who popularized the use of determinants for solving linear systems.

Section 6.5, on *partial fraction decomposition*, provides an application of systems of equations. This particular application is one that is used in integral calculus.

Section 6.8 introduces *linear programming*, a modern application of linear inequalities. This topic is particularly useful for students interested in operations research.

Economic Outcomes

Annual Earnings of Young Adults For both males and females, earnings increase with education: full-time workers with at least a bachelor's degree have higher median earnings than those with less education. For example, in 2002, male college graduates earned 65 percent more than male high school completers. Females with a bachelor's or higher degree earned 71 percent more than female high school completers. Males and females who dropped out of high school earned 23 and 27 percent less, respectively, than male and female high school completers.

The median earnings of young adults who have at least a bachelor's degree declined in the 1970s relative to their counterparts who were high school completers, before increasing between 1980 and 2002. Males with a bachelor's degree or higher had earnings 19 percent higher than male high school completers in 1980 and had earnings 65 percent higher in 2002. Among females, those with at least a bachelor's degree had earnings 34 percent higher than female high school completers in 1980, compared with earnings 71 percent higher in 2002.

—See Chapter Project 1.

399

6.1 Systems of Linear Equations: Substitution and Elimination

PREPARING FOR THIS SECTION *Before getting started, review the following:*

• Linear Equations (Appendix A, Section A.8, pp. A61–A63) • Lines (Foundations, Section F.3, pp. 18–28)

Now work the 'Are You Prepared?' problems on page 411.

OBJECTIVES 1 Solve Systems of Equations by Substitution
2 Solve Systems of Equations by Elimination
3 Identify Inconsistent Systems of Equations Containing Two Variables
4 Express the Solution of a System of Dependent Equations Containing Two Variables
5 Solve Systems of Three Equations Containing Three Variables
6 Identify Inconsistent Systems of Equations Containing Three Variables
7 Express the Solution of a System of Dependent Equations Containing Three Variables

We begin with an example.

EXAMPLE 1 **Movie Theater Ticket Sales**

A movie theater sells tickets for $8.00 each, with seniors receiving a discount of $2.00. One evening the theater took in $3580 in revenue. If x represents the number of tickets sold at $8.00 and y the number of tickets sold at the discounted price of $6.00, write an equation that relates these variables.

Solution Each nondiscounted ticket brings in $8.00, so x tickets will bring in $8x$ dollars. Similarly, y discounted tickets bring in $6y$ dollars. Since the total brought in is $3580, we must have

$$8x + 6y = 3580$$ ∎

In Example 1, suppose that we also know that 525 tickets were sold that evening. Then we have another equation relating the variables x and y:

$$x + y = 525$$

The two equations

$$8x + 6y = 3580$$
$$x + y = 525$$

form a *system* of equations.

In general, a **system of equations** is a collection of two or more equations, each containing one or more variables. Example 2 gives some illustrations of systems of equations.

EXAMPLE 2 **Examples of Systems of Equations**

(a) $\begin{cases} 2x + y = 5 & (1) \\ -4x + 6y = -2 & (2) \end{cases}$ Two equations containing two variables, x and y

(b) $\begin{cases} x + y^2 = 5 & (1) \quad \text{Two equations containing two variables, } x \text{ and } y \\ 2x + y = 4 & (2) \end{cases}$

(c) $\begin{cases} x + y + z = 6 & (1) \quad \text{Three equations containing three variables, } x, y, \text{ and } z \\ 3x - 2y + 4z = 9 & (2) \\ x - y - z = 0 & (3) \end{cases}$

(d) $\begin{cases} x + y + z = 5 & (1) \quad \text{Two equations containing three variables, } x, y, \text{ and } z \\ x - y = 2 & (2) \end{cases}$

(e) $\begin{cases} x + y + z = 6 & (1) \quad \text{Four equations containing three variables, } x, y, \text{ and } z \\ 2x + 2z = 4 & (2) \\ y + z = 2 & (3) \\ x = 4 & (4) \end{cases}$ ∎

We use a brace, as shown, to remind us that we are dealing with a system of equations. We also will find it convenient to number each equation in the system.

A **solution** of a system of equations consists of values for the variables that are solutions of each equation of the system. To **solve** a system of equations means to find all solutions of the system.

For example, $x = 2$, $y = 1$ is a solution of the system in Example 2(a), because

$$\begin{cases} 2x + y = 5 & (1) \\ -4x + 6y = -2 & (2) \end{cases} \qquad \begin{cases} 2(2) + 1 = 4 + 1 = 5 \\ -4(2) + 6(1) = -8 + 6 = -2 \end{cases}$$

A solution of the system in Example 2(b) is $x = 1$, $y = 2$, because

$$\begin{cases} x + y^2 = 5 & (1) \\ 2x + y = 4 & (2) \end{cases} \qquad \begin{cases} 1 + 2^2 = 1 + 4 = 5 \\ 2(1) + 2 = 2 + 2 = 4 \end{cases}$$

Another solution of the system in Example 2(b) is $x = \dfrac{11}{4}$, $y = -\dfrac{3}{2}$, which you can check for yourself.

A solution of the system in Example 2(c) is $x = 3$, $y = 2$, $z = 1$, because

$$\begin{cases} x + y + z = 6 & (1) \\ 3x - 2y + 4z = 9 & (2) \\ x - y - z = 0 & (3) \end{cases} \qquad \begin{cases} 3 + 2 + 1 = 6 & (1) \\ 3(3) - 2(2) + 4(1) = 9 - 4 + 4 = 9 & (2) \\ 3 - 2 - 1 = 0 & (3) \end{cases}$$

Note that $x = 3$, $y = 3$, $z = 0$ is not a solution of the system in Example 2(c).

$$\begin{cases} x + y + z = 6 & (1) \\ 3x - 2y + 4z = 9 & (2) \\ x - y - z = 0 & (3) \end{cases} \qquad \begin{cases} 3 + 3 + 0 = 6 & (1) \\ 3(3) - 2(3) + 4(0) = 3 \neq 9 & (2) \\ 3 - 3 - 0 = 0 & (3) \end{cases}$$

Although these values satisfy equations (1) and (3), they do not satisfy equation (2). Any solution of the system must satisfy *each* equation of the system.

━━━━ NOW WORK PROBLEM 9.

When a system of equations has at least one solution, it is said to be **consistent**; otherwise, it is called **inconsistent**.

An equation in n variables is said to be **linear** if it is equivalent to an equation of the form

$$a_1x_1 + a_2x_2 + \cdots + a_nx_n = b$$

where x_1, x_2, \ldots, x_n are n distinct variables, a_1, a_2, \ldots, a_n, b are constants, and at least one of the a's is not 0.

Some examples of linear equations are

$$2x + 3y = 2 \qquad 5x - 2y + 3z = 10 \qquad 8x + 8y - 2z + 5w = 0$$

If each equation in a system of equations is linear, we have a **system of linear equations**. The systems in Examples 2(a), (c), (d), and (e) are linear, whereas the system in Example 2(b) is nonlinear. In this chapter we shall solve linear systems in Sections 6.1 to 6.3. We discuss nonlinear systems in Section 6.6.

We begin by discussing a system of two linear equations containing two variables. We can view the problem of solving such a system as a geometry problem. The graph of each equation in such a system is a line. So a system of two equations containing two variables represents a pair of lines. The lines either (1) intersect or (2) are parallel or (3) are **coincident** (that is, identical).

1. If the lines intersect, the system of equations has one solution, given by the point of intersection. The system is **consistent** and the equations are **independent**. See Figure 1(a).

2. If the lines are parallel, the system of equations has no solution, because the lines never intersect. The system is **inconsistent**. See Figure 1(b).

3. If the lines are coincident, the system of equations has infinitely many solutions, represented by the totality of points on the line. The system is **consistent** and the equations are **dependent**. See Figure 1(c).

Figure 1

 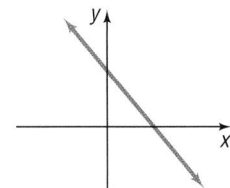

(a) Intersecting lines; system has one solution

(b) Parallel lines; system has no solution

(c) Coincident lines; system has infinitely many solutions

EXAMPLE 3 **Graphing a System of Linear Equations**

Graph the system:
$$\begin{cases} 2x + y = 5 & (1) \\ -4x + 6y = 12 & (2) \end{cases}$$

Figure 2

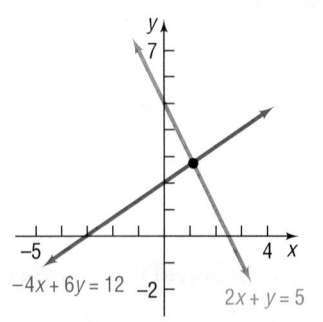

Solution Equation (1) in slope–intercept form is $y = -2x + 5$, which has slope -2 and y-intercept 5. Equation (2) in slope–intercept form is $y = \frac{2}{3}x + 2$, which has slope $\frac{2}{3}$ and y-intercept 2. Figure 2 shows their graphs. ∎

From the graph in Figure 2, we see that the lines intersect, so the system given in Example 3 is consistent. We can also use the graph as a means of approximating the solution. For this system, the solution appears to be close to the point $(1, 3)$. The actual solution, which you should verify, is $\left(\dfrac{9}{8}, \dfrac{11}{4}\right)$.

—— **Seeing the Concept** ——

Graph the lines $2x + y = 5$ ($Y_1 = -2x + 5$) and $-4x + 6y = 12$ $\left(Y_2 = \dfrac{2}{3}x + 2\right)$ and compare what you see with Figure 2. Use INTERSECT to verify that the point of intersection is (1.125, 2.75).

1 Solve Systems of Equations by Substitution

To obtain exact solutions, we use algebraic methods. A number of methods are available to us for solving systems of linear equations algebraically. In this section, we introduce two methods: *substitution* and *elimination*. We illustrate the **method of substitution** by solving the system given in Example 3.

EXAMPLE 4 **Solving a System of Linear Equations by Substitution**

Solve: $\begin{cases} 2x + y = 5 & (1) \\ -4x + 6y = 12 & (2) \end{cases}$

Solution We solve the first equation for y, obtaining

$$2x + y = 5 \qquad (1)$$
$$y = -2x + 5 \qquad \text{Subtract 2x from each side of (1).}$$

We substitute this result for y into the second equation. The result is an equation containing just the variable x, which we can then solve.

$$-4x + 6y = 12 \qquad (2)$$
$$-4x + 6(-2x + 5) = 12 \qquad \text{Substitute } y = -2x + 5 \text{ in (2).}$$
$$-4x - 12x + 30 = 12 \qquad \text{Remove parentheses.}$$
$$-16x = -18 \qquad \begin{array}{l} \text{Combine like terms and subtract 30} \\ \text{from both sides.} \end{array}$$
$$x = \frac{-18}{-16} = \frac{9}{8} \qquad \text{Divide each side by } -16.$$

Once we know that $x = \dfrac{9}{8}$, we can easily find the value of y by **back-substitution**, that is, by substituting $\dfrac{9}{8}$ for x in one of the original equations. We will use the first equation.

$$2x + y = 5 \qquad (1)$$
$$y = -2x + 5 \qquad \text{Subtract 2x from each side.}$$
$$y = -2\left(\frac{9}{8}\right) + 5 \qquad \text{Substitute } x = \frac{9}{8} \text{ in (1).}$$
$$= \frac{-9}{4} + \frac{20}{4} = \frac{11}{4}$$

The solution of the system is $x = \dfrac{9}{8} = 1.125$, $y = \dfrac{11}{4} = 2.75$.

✔ **CHECK:** $\begin{cases} 2x + y = 5: & 2\left(\dfrac{9}{8}\right) + \dfrac{11}{4} = \dfrac{9}{4} + \dfrac{11}{4} = \dfrac{20}{4} = 5 \\ \\ -4x + 6y = 12: & -4\left(\dfrac{9}{8}\right) + 6\left(\dfrac{11}{4}\right) = -\dfrac{9}{2} + \dfrac{33}{2} = \dfrac{24}{2} = 12 \end{cases}$ ∎

The method used to solve the system in Example 4 is called **substitution**. The steps to be used are outlined next.

Steps for Solving by Substitution

STEP 1: Pick one of the equations and solve for one of the variables in terms of the remaining variables.

STEP 2: Substitute the result into the remaining equations.

STEP 3: If one equation in one variable results, solve this equation. Otherwise repeat Steps 1 and 2 until a single equation with one variable remains.

STEP 4: Find the values of the remaining variables by back-substitution.

STEP 5: Check the solution found.

NOW USE SUBSTITUTION TO WORK PROBLEM 19.

2 Solve Systems of Equations by Elimination

A second method for solving a system of linear equations is the *method of elimination*. This method is usually preferred over substitution if substitution leads to fractions or if the system contains more than two variables. Elimination also provides the necessary motivation for solving systems using matrices (the subject of Section 6.2).

The idea behind the method of elimination is to replace the original system of equations by an equivalent system so that adding two of the equations eliminates a variable. The rules for obtaining equivalent equations are the same as those studied earlier. However, we may also interchange any two equations of the system and/or replace any equation in the system by the sum (or difference) of that equation and a nonzero multiple of any other equation in the system.

In Words

When using elimination, we want to get the coefficients of one of the variables to be negatives of one another.

Rules for Obtaining an Equivalent System of Equations

1. Interchange any two equations of the system.
2. Multiply (or divide) each side of an equation by the same nonzero constant.
3. Replace any equation in the system by the sum (or difference) of that equation and a nonzero multiple of any other equation in the system.

An example will give you the idea. As you work through the example, pay particular attention to the pattern being followed.

EXAMPLE 5

Solving a System of Linear Equations by Elimination

Solve: $\begin{cases} 2x + 3y = 1 & (1) \\ -x + y = -3 & (2) \end{cases}$

Solution

We multiply each side of equation (2) by 2 so that the coefficients of x in the two equations are negatives of one another. The result is the equivalent system

$$\begin{cases} 2x + 3y = 1 & (1) \\ -2x + 2y = -6 & (2) \end{cases}$$

If we add equations (1) and (2) we obtain an equation containing just the variable y, which we can solve.

$$\begin{cases} 2x + 3y = 1 & (1) \\ \underline{-2x + 2y = -6} & (2) \\ 5y = -5 & \text{Add (1) and (2).} \\ y = -1 & \text{Solve for y.} \end{cases}$$

We back-substitute this value for y in equation (1) and simplify.

$$2x + 3y = 1 \qquad \text{(1)}$$
$$2x + 3(-1) = 1 \qquad \text{Substitute } y = -1 \text{ in (1).}$$
$$2x = 4 \qquad \text{Simplify.}$$
$$x = 2 \qquad \text{Solve for } x.$$

The solution of the original system is $x = 2$, $y = -1$. We leave it to you to check the solution. ∎

The procedure used in Example 5 is called the **method of elimination**. Notice the pattern of the solution. First, we eliminated the variable x from the second equation. Then we back-substituted; that is, we substituted the value found for y back into the first equation to find x.

➤ **NOW USE ELIMINATION TO WORK PROBLEM 19.**

Let's return to the movie theater example Example 1.

| EXAMPLE 6 | **Movie Theater Ticket Sales** |

A movie theater sells tickets for $8.00 each, with seniors receiving a discount of $2.00. One evening the theater sold 525 tickets and took in $3580 in revenue. How many of each type of ticket were sold?

Solution If x represents the number of tickets sold at $8.00 and y the number of tickets sold at the discounted price of $6.00, the given information results in the system of equations

$$\begin{cases} 8x + 6y = 3580 & \text{(1)} \\ x + y = 525 & \text{(2)} \end{cases}$$

We use the method of elimination. First, multiply the second equation by -6, and then add the equations.

$$\begin{cases} 8x + 6y = 3580 & \text{(1)} \\ -6x - 6y = -3150 & \text{(2)} \end{cases}$$
$$\overline{ 2x = 430} \qquad \text{Add the equations}$$
$$x = 21$$

Since $x + y = 525$, then $y = 525 - x = 525 - 215 = 310$. So 215 nondiscounted tickets and 310 senior discount tickets were sold. ∎

3 Identify Inconsistent Systems of Equations Containing Two Variables

The previous examples dealt with consistent systems of equations that had a single solution. The next two examples deal with two other possibilities that may occur, the first being a system that has no solution.

| EXAMPLE 7 | **An Inconsistent System of Linear Equations** |

Solve: $\begin{cases} 2x + y = 5 & \text{(1)} \\ 4x + 2y = 8 & \text{(2)} \end{cases}$

Solution We choose to use the method of substitution and solve equation (1) for y.

$$2x + y = 5 \qquad \text{(1)}$$
$$y = -2x + 5 \qquad \text{Subtract } 2x \text{ from each side.}$$

Now substitute $y = -2x + 5$ for y in equation (2) and solve for x.

$$4x + 2y = 8 \quad \text{(2)}$$
$$4x + 2(-2x + 5) = 8 \quad \text{Substitute } y = -2x + 5 \text{ in (2)}.$$
$$4x - 4x + 10 = 8 \quad \text{Remove parentheses.}$$
$$0 = -2 \quad \text{Subtract 10 from both sides.}$$

Figure 3

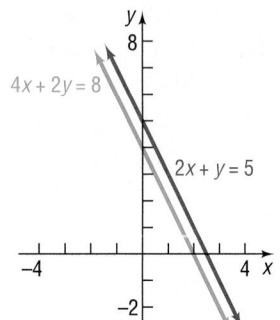

This equation is false. We conclude that the system has no solution and is therefore inconsistent. ∎

Figure 3 illustrates the pair of lines whose equations form the system in Example 7. Notice that the graphs of the two equations are lines, each with slope -2; one has a y-intercept of 5, the other a y-intercept of 4. The lines are parallel and have no point of intersection. This geometric statement is equivalent to the algebraic statement that the system has no solution.

—— **Seeing the Concept** ——

Graph the lines $2x + y = 5$ ($Y_1 = -2x + 5$) and $4x + 2y = 8$ ($Y_2 = -2x + 4$) and compare what you see with Figure 3. How can you be sure that the lines are parallel?

4 Express the Solution of a System of Dependent Equations Containing Two Variables

EXAMPLE 8 **Solving a System of Dependent Equations**

Solve: $\begin{cases} 2x + y = 4 & \text{(1)} \\ -6x - 3y = -12 & \text{(2)} \end{cases}$

Solution We choose to use the method of elimination.

$$\begin{cases} 2x + y = 4 & \text{(1)} \\ -6x - 3y = -12 & \text{(2)} \end{cases}$$

$$\begin{cases} 6x + 3y = 12 & \text{(1)} \quad \text{Multiply each side of equation (1) by 3.} \\ -6x - 3y = -12 & \text{(2)} \end{cases}$$

$$\begin{cases} 6x + 3y = 12 & \text{(1)} \\ 0 = 0 & \text{(2)} \quad \text{Replace equation (2) by the sum of equations (1) and (2).} \end{cases}$$

The original system is equivalent to a system containing one equation, so the equations are dependent. This means that any values of x and y for which $6x + 3y = 12$ or, equivalently, $2x + y = 4$ are solutions. For example, $x = 2, y = 0; x = 0, y = 4; x = -2, y = 8; x = 4, y = -4$; and so on, are solutions. There are, in fact, infinitely many values of x and y for which $2x + y = 4$, so the original system has infinitely many solutions. We will write the solution of the original system either as

$$y = -2x + 4$$

where x can be any real number, or as

$$x = -\frac{1}{2}y + 2$$

where y can be any real number. ∎

Figure 4

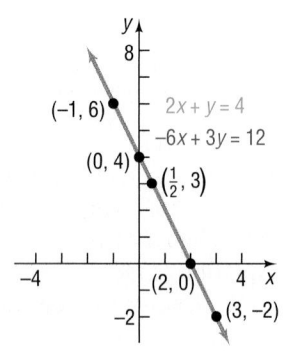

Figure 4 illustrates the situation presented in Example 8. Notice that the graphs of the two equations are lines, each with slope -2 and each with y-intercept 4. The

lines are coincident. Notice also that equation (2) in the original system is -3 times equation (1), indicating that the two equations are dependent.

For the system in Example 8, we can write down some of the infinite number of solutions by assigning values to x and then finding $y = -2x + 4$.

If $x = -2$, then $y = -2(-2) + 4 = 8$.

If $x = 0$, then $y = 4$.

If $x = 2$, then $y = 0$.

The ordered pairs $(-2, 8)$, $(0, 4)$, and $(2, 0)$ are three of the points on the line in Figure 4.

 ────── **Seeing the Concept** ──────────────────────────

Graph the lines $2x + y = 4$ ($Y_1 = -2x + 4$) and $-6x - 3y = -12$ ($Y_2 = -2x + 4$) and compare what you see with Figure 4. How can you be sure that the lines are coincident?

 NOW WORK PROBLEMS 25 AND 29.

5 Solve Systems of Three Equations Containing Three Variables

Just as with a system of two linear equations containing two variables, a system of three linear equations containing three variables also has either (1) exactly one solution (a consistent system with independent equations), or (2) no solution (an inconsistent system), or (3) infinitely many solutions (a consistent system with dependent equations).

We can view the problem of solving a system of three linear equations containing three variables as a geometry problem. The graph of each equation in such a system is a plane in space. A system of three linear equations containing three variables represents three planes in space. Figure 5 illustrates some of the possibilities.

Figure 5

(a) Consistent system; one solution

(b) Consistent system; infinite number of solutions

(c) Inconsistent system; no solution

Recall that a **solution** to a system of equations consists of values for the variables that are solutions of each equation of the system. For example, $x = 3$, $y = -1$, $z = -5$ is a solution to the system of equations

$$\begin{cases} x + y + z = -3 & (1) \\ 2x - 3y + 6z = -21 & (2) \\ -3x + 5y = -14 & (3) \end{cases}$$

$3 + (-1) + (-5) = -3$
$2(3) - 3(-1) + 6(-5) = 6 + 3 - 30 = -21$
$-3(3) + 5(-1) = -9 - 5 = -14$

because these values of the variables are solutions of each equation.

Typically, when solving a system of three linear equations containing three variables, we use the method of elimination. Recall that the idea behind the method of elimination is to form equivalent equations so that adding two of the equations eliminates a variable.

Let's see how elimination works on a system of three equations containing three variables.

EXAMPLE 9 | **Solving a System of Three Linear Equations with Three Variables**

Use the method of elimination to solve the system of equations.

$$\begin{cases} x + y - z = -1 & (1) \\ 4x - 3y + 2z = 16 & (2) \\ 2x - 2y - 3z = 5 & (3) \end{cases}$$

Solution For a system of three equations, we attempt to eliminate one variable at a time, using pairs of equations until an equation with a single variable remains. Our strategy for solving this system will be to use equation (1) to eliminate the variable x from equations (2) and (3).

We begin by multiplying each side of equation (1) by -4 and adding the result to equation (2). (Do you see why? The coefficients of x are now negatives of one another.) We also multiply equation (1) by -2 and add the result to equation (3). Notice that these two procedures result in the removal of the variable x from equations (2) and (3).

$$\begin{array}{l} x + y - z = -1 \;\;(1) \quad \text{Multiply by } -4 \\ 4x - 3y + 2z = 16 \;\;(2) \end{array} \qquad \begin{array}{l} -4x - 4y + 4z = 4 \;\;(1) \\ \underline{4x - 3y + 2z = 16} \;\;(2) \\ -7y + 6z = 20 \quad \text{Add.} \end{array}$$

$$\begin{array}{l} x + y - z = -1 \;\;(1) \quad \text{Multiply by } -2 \\ 2x - 2y - 3z = 5 \;\;(3) \end{array} \qquad \begin{array}{l} -2x - 2y + 2z = 2 \;\;(1) \\ \underline{2x - 2y - 3z = 5} \;\;(3) \\ -4y - z = 7 \quad \text{Add.} \end{array}$$

$$\begin{cases} x + y - z = -1 & (1) \\ -7y + 6z = 20 & (2) \\ -4y - z = 7 & (3) \end{cases}$$

We now concentrate on equations (2) and (3), treating them as a system of two equations containing two variables. It is easiest to eliminate z. We multiply each side of equation (3) by 6 and add equations (2) and (3). The result is the new equation (3).

$$\begin{array}{l} -7y + 6z = 20 \;\;(2) \\ -4y - z = 7 \;\;(3) \quad \text{Multiply by 6} \end{array} \qquad \begin{array}{l} -7y + 6z = 20 \;\;(2) \\ \underline{-24y - 6z = 42} \;\;(3) \\ -31y = 62 \quad \text{Add.} \end{array} \qquad \longrightarrow \qquad \begin{cases} x + y - z = -1 & (1) \\ -7y + 6z = 20 & (2) \\ -31y = 62 & (3) \end{cases}$$

We now solve equation (3) for y by dividing both sides of the equation by -31.

$$\begin{cases} x + y - z = -1 & (1) \\ -7y + 6z = 20 & (2) \\ y = -2 & (3) \end{cases}$$

Back-substitute $y = -2$ in equation (2) and solve for z.

$$\begin{array}{ll} -7y + 6z = 20 & (2) \\ -7(-2) + 6z = 20 & \text{Substitute } y = -2 \text{ in (2).} \\ 6z = 6 & \text{Subtract 14 from both sides of the equation.} \\ z = 1 & \text{Divide both sides of the equation by 6.} \end{array}$$

Finally, we back-substitute $y = -2$ and $z = 1$ in equation (1) and solve for x.

$$\begin{array}{ll} x + y - z = -1 & (1) \\ x + (-2) - 1 = -1 & \text{Substitute } y = -2 \text{ and } z = 1 \text{ in (1).} \\ x - 3 = -1 & \text{Simplify.} \\ x = 2 & \text{Add 3 to both sides.} \end{array}$$

The solution of the original system is $x = 2$, $y = -2$, $z = 1$. You should verify this solution. ∎

Look back over the solution given in Example 9. Note the pattern of removing one of the variables from two of the equations, followed by solving this system of two equations and two unknowns. Although which variables to remove is your choice, the methodology remains the same for all systems.

NOW WORK PROBLEM 43.

6 Identify Inconsistent Systems of Equations Containing Three Variables

| EXAMPLE 10 | An Inconsistent System of Linear Equations |

Solve: $\begin{cases} 2x + y - z = -2 & (1) \\ x + 2y - z = -9 & (2) \\ x - 4y + z = 1 & (3) \end{cases}$

Solution Our strategy is the same as in Example 9. However, in this system, it seems easiest to eliminate the variable z first. Do you see why?

Multiply each side of equation (1) by -1 and add the result to equation (2). Also, add equations (2) and (3).

$$\begin{array}{l} -2x - y + z = 2 \quad \text{(1) Multiply by } -1. \\ \underline{x + 2y - z = -9} \quad \text{(2)} \\ -x + y \qquad = -7 \quad \text{Add.} \end{array}$$

$$\begin{array}{l} x + 2y - z = -9 \quad \text{(2)} \\ \underline{x - 4y + z = 1} \quad \text{(3)} \\ 2x - 2y \qquad = -8 \quad \text{Add.} \end{array}$$

$\begin{cases} 2x + y - z = -2 & (1) \\ -x + y = -7 & (2) \\ 2x - 2y = -8 & (3) \end{cases}$

We now concentrate on equations (2) and (3), treating them as a system of two equations containing two variables. Multiply each side of equation (2) by 2 and add the result to equation (3).

$$\begin{array}{ll} -x + y = -7 & \text{(2) Multiply by 2.} \\ 2x - 2y = -8 & \text{(3)} \end{array} \qquad \begin{array}{l} -2x + 2y = -14 \quad \text{(2)} \\ \underline{2x - 2y = -8} \quad \text{(3)} \\ 0 = -22 \quad \text{Add.} \end{array} \longrightarrow \begin{cases} 2x + y - z = -2 & (1) \\ -x + y = -7 & (2) \\ 0 = -22 & (3) \end{cases}$$

Equation (3) has no solution and the system is inconsistent. ∎

7 Express the Solution of a System of Dependent Equations Containing Three Variables

| EXAMPLE 11 | Solving a System of Dependent Equations |

Solve: $\begin{cases} x - 2y - z = 8 & (1) \\ 2x - 3y + z = 23 & (2) \\ 4x - 5y + 5z = 53 & (3) \end{cases}$

Solution Our plan is to eliminate x from equations (2) and (3). Multiply each side of equation (1) by -2 and add the result to equation (2). Also, multiply each side of equation (1) by -4 and add the result to equation (3).

$$\begin{array}{ll} x - 2y - z = 8 & \text{(1) Multiply by } -2. \\ 2x - 3y + z = 23 & \text{(2)} \end{array} \qquad \begin{array}{l} -2x + 4y + 2z = -16 \quad \text{(1)} \\ \underline{2x - 3y + z = 23} \quad \text{(2)} \\ y + 3z = 7 \quad \text{Add.} \end{array}$$

$$\begin{array}{ll} x - 2y - z = 8 & \text{(1) Multiply by } -4. \\ 4x - 5y + 5z = 53 & \text{(3)} \end{array} \qquad \begin{array}{l} -4x + 8y + 4z = -32 \quad \text{(1)} \\ \underline{4x - 5y + 5z = 53} \quad \text{(2)} \\ 3y + 9z = 21 \quad \text{Add.} \end{array} \longrightarrow \begin{cases} x - 2y - z = 8 & (1) \\ y + 3z = 7 & (2) \\ 3y + 9z = 21 & (3) \end{cases}$$

Treat equations (2) and (3) as a system of two equations containing two variables, and eliminate the variable y by multiplying both sides of equation (2) by -3 and adding the result to equation (3).

$$y + 3z = 7 \quad \text{Multiply by } -3. \qquad \begin{array}{r} -3y - 9z = -21 \\ 3y + 9z = 21 \quad \text{Add.} \\ \hline 0 = 0 \end{array} \longrightarrow \begin{cases} x - 2y - z = 8 \ (1) \\ y + 3z = 7 \ (2) \\ 0 = 0 \ (3) \end{cases}$$

The original system is equivalent to a system containing two equations, so the equations are dependent and the system has infinitely many solutions. If we solve equation (2) for y, we can express y in terms of z as $y = -3z + 7$. Substitute this expression into equation (1) to determine x in terms of z.

$$
\begin{aligned}
x - 2y - z &= 8 & &\text{(1)} \\
x - 2(-3z + 7) - z &= 8 & &\text{Substitute } y = -3z + 7 \text{ in (1).} \\
x + 6z - 14 - z &= 8 & &\text{Remove parentheses.} \\
x + 5z &= 22 & &\text{Combine like terms.} \\
x &= -5z + 22 & &\text{Solve for } x.
\end{aligned}
$$

We will write the solution to the system as

$$\begin{cases} x = -5z + 22 \\ y = -3z + 7 \end{cases}$$

where z can be any real number.

This way of writing the solution makes it easier to find specific solutions of the system. To find specific solutions, choose any value of z and use the equations $x = -5z + 22$ and $y = -3z + 7$ to determine x and y. For example, if $z = 0$, then $x = 22$ and $y = 7$, and if $z = 1$, then $x = 17$ and $y = 4$. ∎

 NOW WORK PROBLEM 45.

Two points in the Cartesian plane determine a unique line. Given three noncollinear points, we can find the (unique) quadratic function whose graph contains these three points.

EXAMPLE 12 | **Curve Fitting**

Find real numbers a, b, and c so that the graph of the quadratic function $y = ax^2 + bx + c$ contains the points $(-1, -4)$, $(1, 6)$, and $(3, 0)$.

Solution

We require that the three points satisfy the equation $y = ax^2 + bx + c$.

For the point $(-1, -4)$ we have: $-4 = a(-1)^2 + b(-1) + c \quad -4 = a - b + c$
For the point $(1, 6)$ we have: $6 = a(1)^2 + b(1) + c \quad 6 = a + b + c$
For the point $(3, 0)$ we have: $0 = a(3)^2 + b(3) + c \quad 0 = 9a + 3b + c$

We wish to determine a, b, and c so that each equation is satisfied. That is, we want to solve the following system of three equations containing three variables:

$$\begin{cases} a - b + c = -4 & \text{(1)} \\ a + b + c = 6 & \text{(2)} \\ 9a + 3b + c = 0 & \text{(3)} \end{cases}$$

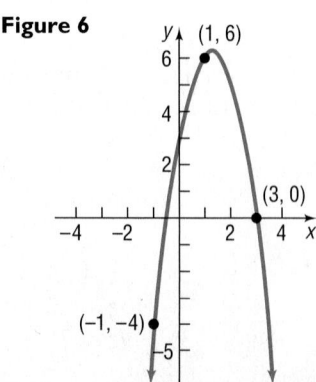

Figure 6

Solving this system of equations, we obtain $a = -2$, $b = 5$, and $c = 3$. So the quadratic function whose graph contains the points $(-1, -4)$, $(1, 6)$, and $(3, 0)$ is

$$y = -2x^2 + 5x + 3 \qquad y = ax^2 + bx + c, \quad a = -2, b = 5, c = 3$$

Figure 6 shows the graph of the function along with the three points. ∎

6.1 Assess Your Understanding

'Are You Prepared?'

Answers are given at the end of these exercises. If you get a wrong answer, read the pages listed in red.

1. Solve the equation: $3x + 4 = 8 - x$. (pp. A61–A63)

2. (a) Graph the line: $3x + 4y = 12$. (p. 25)
 (b) What is the slope of a line parallel to this line? (pp. 24–27)

Concepts and Vocabulary

3. If a system of equations has no solution, it is said to be _____.

4. If a system of equations has one or more solutions, the system is said to be _____.

5. *True or False:* A system of two linear equations containing two variables always has at least one solution.

6. *True or False:* A solution of a system of equations consists of values for the variables that are solutions of each equation of the system.

Skill Building

In Problems 7–16, verify that the values of the variables listed are solutions of the system of equations.

7. $\begin{cases} 2x - y = 5 \\ 5x + 2y = 8 \end{cases}$

$x = 2, y = -1$

8. $\begin{cases} 3x + 2y = 2 \\ x - 7y = -30 \end{cases}$

$x = -2, y = 4$

9. $\begin{cases} 3x - 4y = 4 \\ \frac{1}{2}x - 3y = -\frac{1}{2} \end{cases}$

$x = 2, y = \frac{1}{2}$

10. $\begin{cases} 2x + \frac{1}{2}y = 0 \\ 3x - 4y = -\frac{19}{2} \end{cases}$

$x = -\frac{1}{2}, y = 2$

11. $\begin{cases} x - y = 3 \\ \frac{1}{2}x + y = 3 \end{cases}$

$x = 4, y = 1$

12. $\begin{cases} x - y = 3 \\ -3x + y = 1 \end{cases}$

$x = -2, y = -5$

13. $\begin{cases} 3x + 3y + 2z = 4 \\ x - y - z = 0 \\ 2y - 3z = -8 \end{cases}$

$x = 1, y = -1, z = 2$

14. $\begin{cases} 4x - z = 7 \\ 8x + 5y - z = 0 \\ -x - y + 5z = 6 \end{cases}$

$x = 2, y = -3, z = 1$

15. $\begin{cases} 3x + 3y + 2z = 4 \\ x - 3y + z = 10 \\ 5x - 2y - 3z = 8 \end{cases}$

$x = 2, y = -2, z = 2$

16. $\begin{cases} 4x - 5z = 6 \\ 5y - z = -17 \\ -x - 6y + 5z = 24 \end{cases}$

$x = 4, y = -3, z = 2$

In Problems 17–54, solve each system of equations. If the system has no solution, say that it is inconsistent.

17. $\begin{cases} x + y = 8 \\ x - y = 4 \end{cases}$

18. $\begin{cases} x + 2y = 5 \\ x + y = 3 \end{cases}$

19. $\begin{cases} 5x - y = 13 \\ 2x + 3y = 12 \end{cases}$

20. $\begin{cases} x + 3y = 5 \\ 2x - 3y = -8 \end{cases}$

21. $\begin{cases} 3x = 24 \\ x + 2y = 0 \end{cases}$

22. $\begin{cases} 4x + 5y = -3 \\ -2y = -4 \end{cases}$

23. $\begin{cases} 3x - 6y = 2 \\ 5x + 4y = 1 \end{cases}$

24. $\begin{cases} 2x + 4y = \frac{2}{3} \\ 3x - 5y = -10 \end{cases}$

25. $\begin{cases} 2x + y = 1 \\ 4x + 2y = 3 \end{cases}$

26. $\begin{cases} x - y = 5 \\ -3x + 3y = 2 \end{cases}$

27. $\begin{cases} 2x - y = 0 \\ 3x + 2y = 7 \end{cases}$

28. $\begin{cases} 3x + 3y = -1 \\ 4x + y = \frac{8}{3} \end{cases}$

29. $\begin{cases} x + 2y = 4 \\ 2x + 4y = 8 \end{cases}$

30. $\begin{cases} 3x - y = 7 \\ 9x - 3y = 21 \end{cases}$

31. $\begin{cases} 2x - 3y = -1 \\ 10x + y = 11 \end{cases}$

32. $\begin{cases} 3x - 2y = 0 \\ 5x + 10y = 4 \end{cases}$

33. $\begin{cases} 2x + 3y = 6 \\ x - y = \frac{1}{2} \end{cases}$

34. $\begin{cases} \frac{1}{2}x + y = -2 \\ x - 2y = 8 \end{cases}$

35. $\begin{cases} \frac{1}{2}x + \frac{1}{3}y = 3 \\ \frac{1}{4}x - \frac{2}{3}y = -1 \end{cases}$

36. $\begin{cases} \frac{1}{3}x - \frac{3}{2}y = -5 \\ \frac{3}{4}x + \frac{1}{3}y = 11 \end{cases}$

37. $\begin{cases} 3x - 5y = 3 \\ 15x + 5y = 21 \end{cases}$

38. $\begin{cases} 2x - y = -1 \\ x + \dfrac{1}{2}y = \dfrac{3}{2} \end{cases}$

39. $\begin{cases} \dfrac{1}{x} + \dfrac{1}{y} = 8 \\ \dfrac{3}{x} - \dfrac{5}{y} = 0 \end{cases}$

40. $\begin{cases} \dfrac{4}{x} - \dfrac{3}{y} = 0 \\ \dfrac{6}{x} + \dfrac{3}{2y} = 2 \end{cases}$

[**Hint:** Let $u = \dfrac{1}{x}$ and $v = \dfrac{1}{y}$, and solve for u and v.

Then $x = \dfrac{1}{u}$ and $y = \dfrac{1}{v}$.]

41. $\begin{cases} x - y = 6 \\ 2x - 3z = 16 \\ 2y + z = 4 \end{cases}$

42. $\begin{cases} 2x + y = -4 \\ -2y + 4z = 0 \\ 3x - 2z = -11 \end{cases}$

43. $\begin{cases} x - 2y + 3z = 7 \\ 2x + y + z = 4 \\ -3x + 2y - 2z = -10 \end{cases}$

44. $\begin{cases} 2x + y - 3z = 0 \\ -2x + 2y + z = -7 \\ 3x - 4y - 3z = 7 \end{cases}$

45. $\begin{cases} x - y - z = 1 \\ 2x + 3y + z = 2 \\ 3x + 2y = 0 \end{cases}$

46. $\begin{cases} 2x - 3y - z = 0 \\ -x + 2y + z = 5 \\ 3x - 4y - z = 1 \end{cases}$

47. $\begin{cases} x - y - z = 1 \\ -x + 2y - 3z = -4 \\ 3x - 2y - 7z = 0 \end{cases}$

48. $\begin{cases} 2x - 3y - z = 0 \\ 3x + 2y + 2z = 2 \\ x + 5y + 3z = 2 \end{cases}$

49. $\begin{cases} 2x - 2y + 3z = 6 \\ 4x - 3y + 2z = 0 \\ -2x + 3y - 7z = 1 \end{cases}$

50. $\begin{cases} 3x - 2y + 2z = 6 \\ 7x - 3y + 2z = -1 \\ 2x - 3y + 4z = 0 \end{cases}$

51. $\begin{cases} x + y - z = 6 \\ 3x - 2y + z = -5 \\ x + 3y - 2z = 14 \end{cases}$

52. $\begin{cases} x - y + z = -4 \\ 2x - 3y + 4z = -15 \\ 5x + y - 2z = 12 \end{cases}$

53. $\begin{cases} x + 2y - z = -3 \\ 2x - 4y + z = -7 \\ -2x + 2y - 3z = 4 \end{cases}$

54. $\begin{cases} x + 4y - 3z = -8 \\ 3x - y + 3z = 12 \\ x + y + 6z = 1 \end{cases}$

Applications and Extensions

55. The perimeter of a rectangular floor is 90 feet. Find the dimensions of the floor if the length is twice the width.

56. The length of fence required to enclose a rectangular field is 3000 meters. What are the dimensions of the field if it is known that the difference between its length and width is 50 meters?

57. Cost of Fast Food Four large cheeseburgers and two chocolate shakes cost a total of $7.90. Two shakes cost 15¢ more than one cheeseburger. What is the cost of a cheeseburger? What is the cost of a shake?

58. Movie Theater Tickets A movie theater charges $9.00 for adults and $7.00 for senior citizens. On a day when 325 people paid an admission, the total receipts were $2495. How many who paid were adults? How many were seniors?

59. Mixing Nuts A store sells cashews for $5.00 per pound and peanuts for $1.50 per pound. The manager decides to mix 30 pounds of peanuts with some cashews and sell the mixture for $3.00 per pound. How many pounds of cashews should be mixed with the peanuts so that the mixture will produce the same revenue as would selling the nuts separately?

60. Financial Planning A recently retired couple need $12,000 per year to supplement their Social Security. They have $150,000 to invest to obtain this income. They have decided on two investment options: AA bonds yielding 10% per annum and a Bank Certificate yielding 5%.
(a) How much should be invested in each to realize exactly $12,000?
(b) If, after 2 years, the couple requires $14,000 per year in income, how should they reallocate their investment to achieve the new amount?

61. Computing Wind Speed With a tail wind, a small Piper aircraft can fly 600 miles in 3 hours. Against this same wind, the Piper can fly the same distance in 4 hours. Find the average wind speed and the average airspeed of the Piper.

3 hours 4 hours

600 mi.

62. Computing Wind Speed The average airspeed of a single-engine aircraft is 150 miles per hour. If the aircraft flew the same distance in 2 hours with the wind as it flew in 3 hours against the wind, what was the wind speed?

63. Restaurant Management A restaurant manager wants to purchase 200 sets of dishes. One design costs $25 per set, while another costs $45 per set. If she only has $7400 to spend, how many of each design should be ordered?

64. Cost of Fast Food One group of people purchased 10 hot dogs and 5 soft drinks at a cost of $12.50. A second bought 7 hot dogs and 4 soft drinks at a cost of $9.00. What is the cost of a single hot dog? A single soft drink?

We paid $12.50.
How much is one hot dog?
How much is one cola?

We paid $9.00.
How much is one hot dog?
How much is one cola?

65. **Computing a Refund** The grocery store we use does not mark prices on its goods. My wife went to this store, bought three 1-pound packages of bacon and two cartons of eggs, and paid a total of $7.45. Not knowing that she went to the store, I also went to the same store, purchased two 1-pound packages of bacon and three cartons of eggs, and paid a total of $6.45. Now we want to return two 1-pound packages of bacon and two cartons of eggs. How much will be refunded?

66. **Finding the Current of a Stream** Pamela requires 3 hours to swim 15 miles downstream on the Illinois River. The return trip upstream takes 5 hours. Find Pamela's average speed in still water. How fast is the current? (Assume that Pamela's speed is the same in each direction.)

67. **Pharmacy** A doctor's prescription calls for a daily intake containing 40 mg of vitamin C and 30 mg of vitamin D. Your pharmacy stocks two liquids that can be used: one contains 20% vitamin C and 30% vitamin D, the other 40% vitamin C and 20% vitamin D. How many milligrams of each compound should be mixed to fill the prescription?

68. **Pharmacy** A doctor's prescription calls for the creation of pills that contain 12 units of vitamin B_{12} and 12 units of vitamin E. Your pharmacy stocks two powders that can be used to make these pills: one contains 20% vitamin B_{12} and 30% vitamin E, the other 40% vitamin B_{12} and 20% vitamin E. How many units of each powder should be mixed in each pill?

69. **Curve Fitting** Find real numbers a, b, and c so that the graph of the function $y = ax^2 + bx + c$ contains the points $(-1, 4)$, $(2, 3)$, and $(0, 1)$.

70. **Curve Fitting** Find real numbers a, b, and c so that the graph of the function $y = ax^2 + bx + c$ contains the points $(-1, -2)$, $(1, -4)$, and $(2, 4)$.

71. **IS-LM Model in Economics** In economics, the IS curve is a linear equation that represents all combinations of income Y and interest rates r that maintain an equilibrium in the market for goods in the economy. The LM curve is a linear equation that represents all combinations of income Y and interest rates r that maintain an equilibrium in the market for money in the economy. In an economy, suppose the equilibrium level of income (in millions of dollars) and interest rates satisfy the system of equations

$$\begin{cases} 0.06Y - 5000r = 240 \\ 0.06Y + 6000r = 900 \end{cases}$$

Find the equilibrium level of income and interest rates.

72. **IS-LM Model in Economics** In economics, the IS curve is a linear equation that represents all combinations of income Y and interest rates r that maintain an equilibrium in the market for goods in the economy. The LM curve is a linear equation that represents all combinations of income Y and interest rates r that maintain an equilibrium in the market for money in the economy. In an economy, suppose the equilibrium level of income (in millions of dollars) and interest rates satisfy the system of equations

$$\begin{cases} 0.05Y - 1000r = 10 \\ 0.05Y + 800r = 100 \end{cases}$$

Find the equilibrium level of income and interest rates.

73. **Electricity: Kirchhoff's Rules** An application of Kirchhoff's Rules to the circuit shown results in the following system of equations:

$$\begin{cases} I_2 = I_1 + I_3 \\ 5 - 3I_1 - 5I_2 = 0 \\ 10 - 5I_2 - 7I_3 = 0 \end{cases}$$

Find the currents I_1, I_2, and I_3.

SOURCE: *Physics for Scientists & Engineers* 3/E by Serway. © 1990. Reprinted with permission of Brooks/Cole, a division of Thomson Learning.

74. **Electricity: Kirchhoff's Rules** An application of Kirchhoff's Rules to the circuit shown results in the following system of equations:

$$\begin{cases} I_3 = I_1 + I_2 \\ 8 = 4I_3 + 6I_2 \\ 8I_1 = 4 + 6I_2 \end{cases}$$

Find the currents I_1, I_2, and I_3.

SOURCE: *Physics for Scientists & Engineers* 3/E by Serway. © 1990. Reprinted with permission of Brooks/Cole, a division of Thomson Learning.

75. **Theater Revenues** A Broadway theater has 500 seats, divided into orchestra, main, and balcony seating. Orchestra seats sell for $50, main seats for $35, and balcony seats for $25. If all the seats are sold, the gross revenue to the theater is $17,100. If all the main and balcony seats are sold, but only half the orchestra seats are sold, the gross revenue is $14,600. How many are there of each kind of seat?

76. **Theater Revenues** A movie theater charges $8.00 for adults, $4.50 for children, and $6.00 for senior citizens. One day the theater sold 405 tickets and collected $2320 in receipts. There were twice as many children's tickets sold as adult tickets. How many adults, children, and senior citizens went to the theater that day?

77. Nutrition A dietitian wishes a patient to have a meal that has 66 grams of protein, 94.5 grams of carbohydrates, and 910 milligrams of calcium. The hospital food service tells the dietitian that the dinner for today is chicken, corn, and 2% milk. Each serving of chicken has 30 grams of protein, 35 grams of carbohydrates, and 200 milligrams of calcium. Each serving of corn has 3 grams of protein, 16 grams of carbohydrates, and 10 milligrams of calcium. Each glass of 2% milk has 9 grams of protein, 13 grams of carbohydrates, and 300 milligrams of calcium. How many servings of each food should the dietitian provide for the patient?

78. Investments Kelly has $20,000 to invest. As her financial planner, you recommend that she diversify into three investments: Treasury bills that yield 5% simple interest, Treasury bonds that yield 7% simple interest, and corporate bonds that yield 10% simple interest. Kelly wishes to earn $1390 per year in income. Also, Kelly wants her investment in Treasury bills to be $3000 more than her investment in corporate bonds. How much money should Kelly place in each investment?

79. Prices of Fast Food One group of customers bought 8 deluxe hamburgers, 6 orders of large fries, and 6 large colas for $26.10. A second group ordered 10 deluxe hamburgers, 6 large fries, and 8 large colas and paid $31.60. Is there sufficient information to determine the price of each food item? If not, construct a table showing the various possibilities. Assume that the hamburgers cost between $1.75 and $2.25, the fries between $0.75 and $1.00, and the colas between $0.60 and $0.90.

80. Prices of Fast Food Use the information given in Problem 79. Suppose that a third group purchased 3 deluxe hamburgers, 2 large fries, and 4 large colas for $10.95. Now is there sufficient information to determine the price of each food item? If so, determine each price.

81. Painting a House Three painters, Beth, Bill, and Edie, working together, can paint the exterior of a home in 10 hours. Bill and Edie together have painted a similar house in 15 hours. One day, all three worked on this same kind of house for 4 hours, after which Edie left. Beth and Bill required 8 more hours to finish. Assuming no gain or loss in efficiency, how long should it take each person to complete such a job alone?

Discussion and Writing

82. Make up a system of three linear equations containing three variables that has:
(a) No solution
(b) Exactly one solution
(c) Infinitely many solutions
Give the three systems to a friend to solve and critique.

83. Write a brief paragraph outlining your strategy for solving a system of two linear equations containing two variables.

84. Do you prefer the method of substitution or the method of elimination for solving a system of two linear equations containing two variables? Give reasons.

'Are You Prepared?' Answers

1. $\{1\}$ **2.** (a)

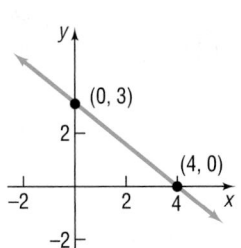

(b) $-\dfrac{3}{4}$

6.2 Systems of Linear Equations: Matrices

OBJECTIVES
1 Write the Augmented Matrix of a System of Linear Equations
2 Write the System from the Augmented Matrix
3 Perform Row Operations on a Matrix
4 Solve a System of Linear Equations Using Matrices

The systematic approach of the method of elimination for solving a system of linear equations provides another method of solution that involves a simplified notation.



Consider the following system of linear equations:

$$\begin{cases} x + 4y = 14 \\ 3x - 2y = 0 \end{cases}$$

If we choose not to write the symbols used for the variables, we can represent this system as

$$\left[\begin{array}{cc|c} 1 & 4 & 14 \\ 3 & -2 & 0 \end{array}\right]$$

where it is understood that the first column represents the coefficients of the variable x, the second column the coefficients of y, and the third column the constants on the right side of the equal signs. The vertical line serves as a reminder of the equal signs. The large square brackets are used to denote a *matrix* in algebra.

A **matrix** is defined as a rectangular array of numbers,

$$\begin{array}{c} \\ \text{Row 1} \\ \text{Row 2} \\ \vdots \\ \text{Row } i \\ \vdots \\ \text{Row } m \end{array} \begin{array}{ccccc} \text{Column 1} & \text{Column 2} & \text{Column } j & \text{Column } n \\ \left[\begin{array}{ccccccc} a_{11} & a_{12} & \cdots & a_{1j} & \cdots & a_{1n} \\ a_{21} & a_{22} & \cdots & a_{2j} & \cdots & a_{2n} \\ \vdots & \vdots & & \vdots & & \vdots \\ a_{i1} & a_{i2} & \cdots & a_{ij} & \cdots & a_{in} \\ \vdots & \vdots & & \vdots & & \vdots \\ a_{m1} & a_{m2} & \cdots & a_{mj} & \cdots & a_{mn} \end{array}\right] \end{array} \quad (1)$$

Each number a_{ij} of the matrix has two indexes: the **row index** i and the **column index** j. The matrix shown in display (1) has m rows and n columns. The numbers a_{ij} are usually referred to as the **entries** of the matrix. For example, a_{23} refers to the entry in the second row, third column.

1 Write the Augmented Matrix of a System of Linear Equations

Now we will use matrix notation to represent a system of linear equations. The matrix used to represent a system of linear equations is called an **augmented matrix**. In writing the augmented matrix of a system, the variables of each equation must be on the left side of the equal sign and the constants on the right side. A variable that does not appear in an equation has a coefficient of 0.

EXAMPLE 1 **Writing the Augmented Matrix of a System of Linear Equations**

Write the augmented matrix of each system of equations.

(a) $\begin{cases} 3x - 4y = -6 & (1) \\ 2x - 3y = -5 & (2) \end{cases}$

(b) $\begin{cases} 2x - y + z = 0 & (1) \\ x + z - 1 = 0 & (2) \\ x + 2y - 8 = 0 & (3) \end{cases}$

Solution (a) The augmented matrix is

$$\left[\begin{array}{cc|c} 3 & -4 & -6 \\ 2 & -3 & -5 \end{array}\right]$$

(b) Care must be taken that the system be written so that the coefficients of all variables are present (if any variable is missing, its coefficient is 0). Also, all

constants must be to the right of the equal sign. We need to rearrange the given system as follows:

$$\begin{cases} 2x - y + z = 0 & (1) \\ x + z - 1 = 0 & (2) \\ x + 2y - 8 = 0 & (3) \end{cases}$$

$$\begin{cases} 2x - y + z = 0 & (1) \\ x + 0 \cdot y + z = 1 & (2) \\ x + 2y + 0 \cdot z = 8 & (3) \end{cases}$$

The augmented matrix is

$$\begin{bmatrix} 2 & -1 & 1 & | & 0 \\ 1 & 0 & 1 & | & 1 \\ 1 & 2 & 0 & | & 8 \end{bmatrix}$$ ∎

If we do not include the constants to the right of the equal sign, that is, to the right of the vertical bar in the augmented matrix of a system of equations, the resulting matrix is called the **coefficient matrix** of the system. For the systems discussed in Example 1, the coefficient matrices are

$$\begin{bmatrix} 3 & -4 \\ 2 & -3 \end{bmatrix} \text{ and } \begin{bmatrix} 2 & -1 & 1 \\ 1 & 0 & 1 \\ 1 & 2 & 0 \end{bmatrix}$$

NOW WORK PROBLEM 7.

2 Write the System from the Augmented Matrix

EXAMPLE 2

Writing the System of Linear Equations from the Augmented Matrix

Write the system of linear equations corresponding to each augmented matrix.

(a) $\begin{bmatrix} 5 & 2 & | & 13 \\ -3 & 1 & | & -10 \end{bmatrix}$

(b) $\begin{bmatrix} 3 & -1 & -1 & | & 7 \\ 2 & 0 & 2 & | & 8 \\ 0 & 1 & 1 & | & 0 \end{bmatrix}$

Solution

(a) The matrix has two rows and so represents a system of two equations. The two columns to the left of the vertical bar indicate that the system has two variables. If x and y are used to denote these variables, the system of equations is

$$\begin{cases} 5x + 2y = 13 & (1) \\ -3x + y = -10 & (2) \end{cases}$$

(b) Since the augmented matrix has three rows, it represents a system of three equations. Since there are three columns to the left of the vertical bar, the system contains three variables. If $x, y,$ and z are the three variables, the system of equations is

$$\begin{cases} 3x - y - z = 7 & (1) \\ 2x + 2z = 8 & (2) \\ y + z = 0 & (3) \end{cases}$$ ∎

3 Perform Row Operations on a Matrix

Row operations on a matrix are used to solve systems of equations when the system is written as an augmented matrix. There are three basic row operations.

Row Operations

1. Interchange any two rows.

2. Replace a row by a nonzero multiple of that row.

3. Replace a row by the sum of that row and a constant nonzero multiple of some other row.

These three row operations correspond to the three rules given earlier for obtaining an equivalent system of equations. When a row operation is performed on a matrix, the resulting matrix represents a system of equations equivalent to the system represented by the original matrix.

For example, consider the augmented matrix

$$\begin{bmatrix} 1 & 2 & | & 3 \\ 4 & -1 & | & 2 \end{bmatrix}$$

Suppose that we want to apply a row operation to this matrix that results in a matrix whose entry in row 2, column 1 is a 0. The row operation to use is

Multiply each entry in row 1 by -4 and add the result to the corresponding entries in row 2. **(2)**

If we use R_2 to represent the new entries in row 2 and we use r_1 and r_2 to represent the original entries in rows 1 and 2, respectively, we can represent the row operation in statement (2) by

$$R_2 = -4r_1 + r_2$$

Then

$$\begin{bmatrix} 1 & 2 & | & 3 \\ 4 & -1 & | & 2 \end{bmatrix} \xrightarrow[R_2 = -4r_1 + r_2]{} \begin{bmatrix} 1 & 2 & | & 3 \\ -4(1)+4 & -4(2)+(-1) & | & -4(3)+2 \end{bmatrix} = \begin{bmatrix} 1 & 2 & | & 3 \\ 0 & -9 & | & -10 \end{bmatrix}$$

As desired, we now have the entry 0 in row 2, column 1.

EXAMPLE 3 Applying a Row Operation to an Augmented Matrix

Apply the row operation $R_2 = -3r_1 + r_2$ to the augmented matrix

$$\begin{bmatrix} 1 & -2 & | & 2 \\ 3 & -5 & | & 9 \end{bmatrix}$$

Solution The row operation $R_2 = -3r_1 + r_2$ tells us that the entries in row 2 are to be replaced by the entries obtained after multiplying each entry in row 1 by -3 and adding the result to the corresponding entries in row 2.

$$\begin{bmatrix} 1 & -2 & | & 2 \\ 3 & -5 & | & 9 \end{bmatrix} \xrightarrow[R_2 = -3r_1 + r_2]{} \begin{bmatrix} 1 & -2 & | & 2 \\ -3(1)+3 & (-3)(-2)+(-5) & | & -3(2)+9 \end{bmatrix} = \begin{bmatrix} 1 & -2 & | & 2 \\ 0 & 1 & | & 3 \end{bmatrix}$$

■

NOW WORK PROBLEM **17**.

| **EXAMPLE 4** | **Finding a Particular Row Operation** |

Find a row operation that will result in the augmented matrix

$$\begin{bmatrix} 1 & -2 & | & 2 \\ 0 & 1 & | & 3 \end{bmatrix}$$

having a 0 in row 1, column 2.

Solution We want a 0 in row 1, column 2. This result can be accomplished by multiplying row 2 by 2 and adding the result to row 1. That is, we apply the row operation $R_1 = 2r_2 + r_1$.

$$\begin{bmatrix} 1 & -2 & | & 2 \\ 0 & 1 & | & 3 \end{bmatrix} \rightarrow \begin{bmatrix} 2(0)+1 & 2(1)+(-2) & | & 2(3)+2 \\ 0 & 1 & | & 3 \end{bmatrix} = \begin{bmatrix} 1 & 0 & | & 8 \\ 0 & 1 & | & 3 \end{bmatrix}$$

$$R_1 = 2r_2 + r_1$$

A word about the notation that we have introduced. A row operation such as $R_1 = 2r_2 + r_1$ changes the entries in row 1. Note also that for this type of row operation we change the entries in a given row by multiplying the entries in some other row by an appropriate nonzero number and adding the results to the original entries of the row to be changed.

4 Solve a System of Linear Equations Using Matrices

To solve a system of linear equations using matrices, we use row operations on the augmented matrix of the system to obtain a matrix that is in *row echelon form*.

A matrix is in **row echelon form** when

1. The entry in row 1, column 1 is a 1, and 0's appear below it.

2. The first nonzero entry in each row after the first row is a 1, 0's appear below it, and it appears to the right of the first nonzero entry in any row above.

3. Any rows that contain all 0's to the left of the vertical bar appear at the bottom.

For example, for a system of three equations containing three variables with a unique solution, the augmented matrix is in row echelon form if it is of the form

$$\begin{bmatrix} 1 & a & b & | & d \\ 0 & 1 & c & | & e \\ 0 & 0 & 1 & | & f \end{bmatrix}$$

where $a, b, c, d, e,$ and f are real numbers. The last row of this augmented matrix states that $z = f$. We can then determine the value of y using back-substitution with $z = f$, since row 2 represents the equation $y + cz = e$. Finally, x is determined using back-substitution again.

Two advantages of solving a system of equations by writing the augmented matrix in row echelon form are the following:

1. The process is algorithmic; that is, it consists of repetitive steps that can be programmed on a computer.

2. The process works on any system of linear equations, no matter how many equations or variables are present.

The next example shows how to write a matrix in row echelon form.

| EXAMPLE 5 | **Solving a System of Linear Equations Using Matrices (Row Echelon Form)** |

$$\text{Solve:} \begin{cases} 2x + 2y \quad\;\; = 6 & (1) \\ x + \;\; y + z = 1 & (2) \\ 3x + 4y - z = 13 & (3) \end{cases}$$

Solution First, we write the augmented matrix that represents this system.

$$\begin{bmatrix} 2 & 2 & 0 & | & 6 \\ 1 & 1 & 1 & | & 1 \\ 3 & 4 & -1 & | & 13 \end{bmatrix}$$

The first step requires getting the entry 1 in row 1, column 1. An interchange of rows 1 and 2 is the easiest way to do this. [Note that this is equivalent to interchanging equations (1) and (2) of the system.]

$$\begin{bmatrix} 1 & 1 & 1 & | & 1 \\ 2 & 2 & 0 & | & 6 \\ 3 & 4 & -1 & | & 13 \end{bmatrix}$$

Next, we want a 0 in row 2, column 1 and a 0 in row 3, column 1. We use the row operations $R_2 = -2r_1 + r_2$ and $R_3 = -3r_1 + r_3$ to accomplish this. Notice that row 1 is unchanged using these row operations. Also, do you see that performing these row operations simultaneously is the same as doing one followed by the other?

$$\begin{bmatrix} 1 & 1 & 1 & | & 1 \\ 2 & 2 & 0 & | & 6 \\ 3 & 4 & -1 & | & 13 \end{bmatrix} \rightarrow \begin{bmatrix} 1 & 1 & 1 & | & 1 \\ 0 & 0 & -2 & | & 4 \\ 0 & 1 & -4 & | & 10 \end{bmatrix}$$

$$R_2 = -2r_1 + r_2$$
$$R_3 = -3r_1 + r_3$$

Now we want the entry 1 in row 2, column 2. Interchanging rows 2 and 3 will accomplish this.

$$\begin{bmatrix} 1 & 1 & 1 & | & 1 \\ 0 & 0 & -2 & | & 4 \\ 0 & 1 & -4 & | & 10 \end{bmatrix} \rightarrow \begin{bmatrix} 1 & 1 & 1 & | & 1 \\ 0 & 1 & -4 & | & 10 \\ 0 & 0 & -2 & | & 4 \end{bmatrix}$$

Finally, we want a 1 in row 3, column 3. To obtain it, we use the row operation $R_3 = -\dfrac{1}{2}r_3$. The result is

$$\begin{bmatrix} 1 & 1 & 1 & | & 1 \\ 0 & 1 & -4 & | & 10 \\ 0 & 0 & -2 & | & 4 \end{bmatrix} \rightarrow \begin{bmatrix} 1 & 1 & 1 & | & 1 \\ 0 & 1 & -4 & | & 10 \\ 0 & 0 & 1 & | & -2 \end{bmatrix}$$

$$R_3 = -\frac{1}{2}r_3$$

This matrix is the row echelon form of the augmented matrix. The third row of this matrix represents the equation $z = -2$. Using $z = -2$, we back-substitute into the equation $y - 4z = 10$ (from the second row) and obtain

$$y - 4z = 10$$
$$y - 4(-2) = 10 \qquad z = -2$$
$$y = 2 \qquad \text{Solve for } y.$$

Finally, we back-substitute $y = 2$ and $z = -2$ into the equation $x + y + z = 1$ (from the first row) and obtain

$$x + y + z = 1$$
$$x + 2 + (-2) = 1 \qquad y = 2, z = -2$$
$$x = 1 \qquad \text{Solve for x.}$$

The solution of the system is $x = 1, y = 2, z = -2$. ∎

The steps that we used to solve the system of linear equations in Example 5 can be summarized as follows:

Matrix Method for Solving a System of Linear Equations (Row Echelon Form)

STEP 1: Write the augmented matrix that represents the system.

STEP 2: Perform row operations that place the entry 1 in row 1, column 1.

STEP 3: Perform row operations that leave the entry 1 in row 1, column 1 unchanged, while causing 0's to appear below it in column 1.

STEP 4: Perform row operations that place the entry 1 in row 2, column 2, but leave the entries in columns to the left unchanged. If it is impossible to place a 1 in row 2, column 2, proceed to place a 1 in row 2, column 3. Once a 1 is in place, perform row operations to place 0's below it. [Place any rows that contain only 0's on the left side of the vertical bar, at the bottom of the matrix.]

STEP 5: Now repeat Step 4, placing a 1 in the next row, but one column to the right. Continue until the bottom row or the vertical bar is reached.

STEP 6: The matrix that results is the row echelon form of the augmented matrix. Analyze the system of equations corresponding to it to solve the original system.

In Words
To obtain an augmented matrix in row echelon form
• Add rows, exchange rows, or multiply the row by a nonzero constant
• Work from top to bottom, and left to right
• Get ones in the main diagonal with zeros below the ones.

In the next example, we solve a system of linear equations using these steps.

EXAMPLE 6 | **Solving a System of Linear Equations Using Matrices (Row Echelon Form)**

Solve:
$$\begin{cases} x - y + z = 8 & (1) \\ 2x + 3y - z = -2 & (2) \\ 3x - 2y - 9z = 9 & (3) \end{cases}$$

Solution **STEP 1:** The augmented matrix of the system is

$$\begin{bmatrix} 1 & -1 & 1 & \vert & 8 \\ 2 & 3 & -1 & \vert & -2 \\ 3 & -2 & -9 & \vert & 9 \end{bmatrix}$$

STEP 2: Because the entry 1 is already present in row 1, column 1, we can go to step 3.

STEP 3: Perform the row operations $R_2 = -2r_1 + r_2$ and $R_3 = -3r_1 + r_3$. Each of these leaves the entry 1 in row 1, column 1 unchanged, while causing 0's to appear under it.

$$\begin{bmatrix} 1 & -1 & 1 & \vert & 8 \\ 2 & 3 & -1 & \vert & -2 \\ 3 & -2 & -9 & \vert & 9 \end{bmatrix} \rightarrow \begin{bmatrix} 1 & -1 & 1 & \vert & 8 \\ 0 & 5 & -3 & \vert & -18 \\ 0 & 1 & -12 & \vert & -15 \end{bmatrix}$$
$$R_2 = -2r_1 + r_2$$
$$R_3 = -3r_1 + r_3$$

In Words
Notice we find multiples of the entry in row 1, column 1 to obtain zeros in the entries below the one.

STEP 4: The easiest way to obtain the entry 1 in row 2, column 2 without altering column 1 is to interchange rows 2 and 3. (Another way would be to multiply row 2 by $\frac{1}{5}$, but this introduces fractions).

$$\begin{bmatrix} 1 & -1 & 1 & | & 8 \\ 0 & 1 & -12 & | & -15 \\ 0 & 5 & -3 & | & -18 \end{bmatrix}$$

To get a 0 under the 1 in row 2, column 2, perform the row operation $R_3 = -5r_2 + r_3$.

$$\begin{bmatrix} 1 & -1 & 1 & | & 8 \\ 0 & 1 & -12 & | & -15 \\ 0 & 5 & -3 & | & -18 \end{bmatrix} \rightarrow \begin{bmatrix} 1 & -1 & 1 & | & 8 \\ 0 & 1 & -12 & | & -15 \\ 0 & 0 & 57 & | & 57 \end{bmatrix}$$
$$\uparrow$$
$$R_3 = -5r_2 + r_3$$

STEP 5: Continuing, we obtain a 1 in row 3, column 3 by using $R_3 = \frac{1}{57}r_3$.

$$\begin{bmatrix} 1 & -1 & 1 & | & 8 \\ 0 & 1 & -12 & | & -15 \\ 0 & 0 & 57 & | & 57 \end{bmatrix} \rightarrow \begin{bmatrix} 1 & -1 & 1 & | & 8 \\ 0 & 1 & -12 & | & -15 \\ 0 & 0 & 1 & | & 1 \end{bmatrix}$$
$$\uparrow$$
$$R_3 = \frac{1}{57}r_3$$

STEP 6: The matrix on the right is the row echelon form of the augmented matrix. The system of equations represented by the matrix in row echelon form is

$$\begin{cases} x - y + z = 8 & (1) \\ \quad\ \ y - 12z = -15 & (2) \\ \qquad\qquad z = 1 & (3) \end{cases}$$

Using $z = 1$, we back-substitute to get

$$\begin{cases} x - y + 1 = 8 & (1) \\ \quad\ \ y - 12(1) = -15 & (2) \end{cases} \longrightarrow \begin{cases} x - y = 7 & (1) \\ \quad\ \ y = -3 & (2) \end{cases}$$
$$\text{Simplify.}$$

We get $y = -3$, and back-substituting into $x - y = 7$, we find that $x = 4$. The solution of the system is $x = 4$, $y = -3$, $z = 1$. ∎

Sometimes it is advantageous to write a matrix in **reduced row echelon form**. In this form, row operations are used to obtain entries that are 0 above (as well as below) the leading 1 in a row. For example, the row echelon form obtained in the solution to Example 6 is

$$\begin{bmatrix} 1 & -1 & 1 & | & 8 \\ 0 & 1 & -12 & | & -15 \\ 0 & 0 & 1 & | & 1 \end{bmatrix}$$

To write this matrix in reduced row echelon form, we proceed as follows:

$$\begin{bmatrix} 1 & -1 & 1 & | & 8 \\ 0 & 1 & -12 & | & -15 \\ 0 & 0 & 1 & | & 1 \end{bmatrix} \rightarrow \begin{bmatrix} 1 & 0 & -11 & | & -7 \\ 0 & 1 & -12 & | & -15 \\ 0 & 0 & 1 & | & 1 \end{bmatrix} \rightarrow \begin{bmatrix} 1 & 0 & 0 & | & 4 \\ 0 & 1 & 0 & | & -3 \\ 0 & 0 & 1 & | & 1 \end{bmatrix}$$
$$\qquad\qquad\qquad\quad \uparrow \qquad\qquad\qquad\qquad\qquad\quad \uparrow$$
$$\qquad\qquad\qquad R_1 = r_2 + r_1 \qquad\qquad\qquad R_1 = 11r_3 + r_1$$
$$\qquad\qquad\qquad\qquad\qquad\qquad\qquad\qquad\qquad\quad R_2 = 12r_3 + r_2$$

The matrix is now written in reduced row echelon form. The advantage of writing the matrix in this form is that the solution to the system, $x = 4$, $y = -3$, $z = 1$, is readily found, without the need to back-substitute. Another advantage will be seen in Section 6.4, where the inverse of a matrix is discussed. The methodology used to write a matrix in reduced row echelon form is called the **Gauss–Jordan method**.

NOW WORK PROBLEMS 37 AND 47.

The matrix method for solving a system of linear equations also identifies systems that have infinitely many solutions and systems that are inconsistent. Let's see how.

| **EXAMPLE 7** | **Solving a Dependent System of Linear Equations Using Matrices** |

Solve: $\begin{cases} 6x - y - z = 4 & (1) \\ -12x + 2y + 2z = -8 & (2) \\ 5x + y - z = 3 & (3) \end{cases}$

Solution We start with the augmented matrix of the system and proceed to obtain a 1 in row 1, column 1 with zeros below.

$$\begin{bmatrix} 6 & -1 & -1 & | & 4 \\ -12 & 2 & 2 & | & -8 \\ 5 & 1 & -1 & | & 3 \end{bmatrix} \rightarrow \begin{bmatrix} 1 & -2 & 0 & | & 1 \\ -12 & 2 & 2 & | & -8 \\ 5 & 1 & -1 & | & 3 \end{bmatrix} \rightarrow \begin{bmatrix} 1 & -2 & 0 & | & 1 \\ 0 & -22 & 2 & | & 4 \\ 0 & 11 & -1 & | & -2 \end{bmatrix}$$

$$\uparrow \qquad\qquad\qquad \uparrow$$
$$R_1 = -1r_3 + r_1 \qquad\qquad R_2 = 12r_1 + r_2$$
$$R_3 = -5r_1 + r_3$$

Obtaining a 1 in row 2, column 2 without altering column 1 can be accomplished by $R_2 = -\dfrac{1}{22}r_2$ or by $R_3 = \dfrac{1}{11}r_3$ and interchanging rows 2 and 3 or by $R_2 = \dfrac{23}{11}r_3 + r_2$. We shall use the first of these.

$$\begin{bmatrix} 1 & -2 & 0 & | & 1 \\ 0 & -22 & 2 & | & 4 \\ 0 & 11 & -1 & | & -2 \end{bmatrix} \rightarrow \begin{bmatrix} 1 & -2 & 0 & | & 1 \\ 0 & 1 & -\dfrac{1}{11} & | & -\dfrac{2}{11} \\ 0 & 11 & -1 & | & -2 \end{bmatrix} \rightarrow \begin{bmatrix} 1 & -2 & 0 & | & 1 \\ 0 & 1 & -\dfrac{1}{11} & | & -\dfrac{2}{11} \\ 0 & 0 & 0 & | & 0 \end{bmatrix}$$

$$\uparrow \qquad\qquad\qquad\qquad \uparrow$$
$$R_2 = -\dfrac{1}{22}r_2 \qquad\qquad R_3 = -11r_2 + r_3$$

This matrix is in row echelon form. Because the bottom row consists entirely of 0's, the system actually consists of only two equations.

$$\begin{cases} x - 2y = 1 & (1) \\ y - \dfrac{1}{11}z = -\dfrac{2}{11} & (2) \end{cases}$$

To make it easier to write down some of the solutions, we express both x and y in terms of z.

From the second equation, $y = \dfrac{1}{11}z - \dfrac{2}{11}$. Now back-substitute this solution for y into the first equation to get

$$x = 2y + 1 = 2\left(\dfrac{1}{11}z - \dfrac{2}{11}\right) + 1 = \dfrac{2}{11}z + \dfrac{7}{11}$$

The original system is equivalent to the system

$$\begin{cases} x = \dfrac{2}{11}z + \dfrac{7}{11} & (1) \\[2mm] y = \dfrac{1}{11}z - \dfrac{2}{11} & (2) \end{cases}$$

where z can be any real number.

Let's look at the situation. The original system of three equations is equivalent to a system containing two equations. This means that any values of x, y, z that satisfy both

$$x = \frac{2}{11}z + \frac{7}{11} \quad \text{and} \quad y = \frac{1}{11}z - \frac{2}{11}$$

will be solutions. For example, $z = 0, x = \dfrac{7}{11}, y = -\dfrac{2}{11}; z = 1, x = \dfrac{9}{11}, y = -\dfrac{1}{11};$ and $z = -1, x = \dfrac{5}{11}, y = -\dfrac{3}{11}$ are some of the solutions of the original system.

There are, in fact, infinitely many values of $x, y,$ and z for which the two equations are satisfied. That is, the original system has infinitely many solutions. We will write the solution of the original system as

$$\begin{cases} x = \dfrac{2}{11}z + \dfrac{7}{11} \\[2mm] y = \dfrac{1}{11}z - \dfrac{2}{11} \end{cases}$$

where z can be any real number. ■

We can also find the solution by writing the augmented matrix in reduced row echelon form. Starting with the row echelon form, we have

$$\begin{bmatrix} 1 & -2 & 0 & \bigm| & 1 \\ 0 & 1 & -\frac{1}{11} & \bigm| & -\frac{2}{11} \\ 0 & 0 & 0 & \bigm| & 0 \end{bmatrix} \xrightarrow[R_1 = 2r_2 + r_1]{} \begin{bmatrix} 1 & 0 & -\frac{2}{11} & \bigm| & \frac{7}{11} \\ 0 & 1 & -\frac{1}{11} & \bigm| & -\frac{2}{11} \\ 0 & 0 & 0 & \bigm| & 0 \end{bmatrix}$$

The matrix on the right is in reduced row echelon form. The corresponding system of equations is

$$\begin{cases} x - \dfrac{2}{11}z = \dfrac{7}{11} & (1) \\[2mm] y - \dfrac{1}{11}z = -\dfrac{2}{11} & (2) \end{cases}$$

or, equivalently,

$$\begin{cases} x = \dfrac{2}{11}z + \dfrac{7}{11} & (1) \\[2mm] y = \dfrac{1}{11}z - \dfrac{2}{11} & (2) \end{cases}$$

where z can be any real number.

NOW WORK PROBLEM 53.

EXAMPLE 8 | **Solving an Inconsistent System of Linear Equations Using Matrices**

Solve: $\begin{cases} x + y + z = 6 \\ 2x - y - z = 3 \\ x + 2y + 2z = 0 \end{cases}$

Solution We proceed as follows, beginning with the augmented matrix.

$$\begin{bmatrix} 1 & 1 & 1 & | & 6 \\ 2 & -1 & -1 & | & 3 \\ 1 & 2 & 2 & | & 0 \end{bmatrix} \rightarrow \begin{bmatrix} 1 & 1 & 1 & | & 6 \\ 0 & -3 & -3 & | & -9 \\ 0 & 1 & 1 & | & -6 \end{bmatrix} \rightarrow \begin{bmatrix} 1 & 1 & 1 & | & 6 \\ 0 & 1 & 1 & | & -6 \\ 0 & -3 & -3 & | & -9 \end{bmatrix} \rightarrow \begin{bmatrix} 1 & 1 & 1 & | & 6 \\ 0 & 1 & 1 & | & -6 \\ 0 & 0 & 0 & | & -27 \end{bmatrix}$$

$R_2 = -2r_1 + r_2$
$R_3 = -1r_1 + r_3$ Interchange rows 2 and 3. $R_3 = 3r_2 + r_3$

This matrix is in row echelon form. The bottom row is equivalent to the equation

$$0x + 0y + 0z = -27$$

which has no solution. Hence, the original system is inconsistent. ∎

NOW WORK PROBLEM 27.

The matrix method is especially effective for systems of equations for which the number of equations and the number of variables are unequal. Here, too, such a system is either inconsistent or consistent. If it is consistent, it will have either exactly one solution or infinitely many solutions.

Let's look at a system of four equations containing three variables.

EXAMPLE 9 | **Solving a System of Linear Equations Using Matrices**

Solve: $\begin{cases} x - 2y + z = 0 & (1) \\ 2x + 2y - 3z = -3 & (2) \\ y - z = -1 & (3) \\ -x + 4y + 2z = 13 & (4) \end{cases}$

Solution We proceed as follows, beginning with the augmented matrix.

$$\begin{bmatrix} 1 & -2 & 1 & | & 0 \\ 2 & 2 & -3 & | & -3 \\ 0 & 1 & -1 & | & -1 \\ -1 & 4 & 2 & | & 13 \end{bmatrix} \rightarrow \begin{bmatrix} 1 & -2 & 1 & | & 0 \\ 0 & 6 & -5 & | & -3 \\ 0 & 1 & -1 & | & -1 \\ 0 & 2 & 3 & | & 13 \end{bmatrix} \rightarrow \begin{bmatrix} 1 & -2 & 1 & | & 0 \\ 0 & 1 & -1 & | & -1 \\ 0 & 6 & -5 & | & -3 \\ 0 & 2 & 3 & | & 13 \end{bmatrix}$$

$R_2 = -2r_1 + r_2$
$R_4 = r_1 + r_4$ Interchange rows 2 and 3.

$$\rightarrow \begin{bmatrix} 1 & -2 & 1 & | & 0 \\ 0 & 1 & -1 & | & -1 \\ 0 & 0 & 1 & | & 3 \\ 0 & 0 & 5 & | & 15 \end{bmatrix} \rightarrow \begin{bmatrix} 1 & -2 & 1 & | & 0 \\ 0 & 1 & -1 & | & -1 \\ 0 & 0 & 1 & | & 3 \\ 0 & 0 & 0 & | & 0 \end{bmatrix}$$

$R_3 = -6r_2 + r_3$
$R_4 = -2r_2 + r_4$ $R_4 = -5r_3 + r_4$

We could stop here, since the matrix is in row echelon form, and back-substitute $z = 3$ to find x and y. Or we can continue to obtain the reduced row echelon form.

$$\rightarrow \begin{bmatrix} 1 & 0 & -1 & | & -2 \\ 0 & 1 & -1 & | & -1 \\ 0 & 0 & 1 & | & 3 \\ 0 & 0 & 0 & | & 0 \end{bmatrix} \rightarrow \begin{bmatrix} 1 & 0 & 0 & | & 1 \\ 0 & 1 & 0 & | & 2 \\ 0 & 0 & 1 & | & 3 \\ 0 & 0 & 0 & | & 0 \end{bmatrix}$$

$R_1 = 2r_2 + r_1$ $R_1 = r_3 + r_1$
$R_2 = r_3 + r_2$

The matrix is now in reduced row echelon form, and we can see that the solution is $x = 1, y = 2, z = 3$. ■

― **NOW WORK PROBLEM 69.**

EXAMPLE 10 **Nutrition**

A dietitian at Cook County Hospital wants a patient to have a meal that has 65 grams of protein, 95 grams of carbohydrates, and 905 milligrams of calcium. The hospital food service tells the dietitian that the dinner for today is chicken a la king, baked potatoes, and 2% milk. Each serving of chicken a la king has 30 grams of protein, 35 grams of carbohydrates, and 200 milligrams of calcium. Each serving of baked potatoes contains 4 grams of protein, 33 grams of carbohydrates, and 10 milligrams of calcium. Each glass of 2% milk contains 9 grams of protein, 13 grams of carbohydrates, and 300 milligrams of calcium. How many servings of each food should the dietitian provide for the patient?

Solution Let c, p, and m represent the number of servings of chicken a la king, baked potatoes, and milk, respectively. The dietitian wants the patient to have 65 grams of protein. Each serving of chicken a la king has 30 grams of protein, so c servings will have $30c$ grams of protein. Each serving of baked potatoes contains 4 grams of protein, so p potatoes will have $4p$ grams of protein. Finally, each glass of milk has 9 grams of protein, so m glasses of milk will have $9m$ grams of protein. The same logic will result in equations for carbohydrates and calcium, and we have the following system of equations:

$$\begin{cases} 30c + 4p + 9m = 65 & \text{Protein equation} \\ 35c + 33p + 13m = 95 & \text{Carbohydrate equation} \\ 200c + 10p + 300m = 905 & \text{Calcium equation} \end{cases}$$

We begin with the augmented matrix and proceed as follows:

$$\begin{bmatrix} 30 & 4 & 9 & | & 65 \\ 35 & 33 & 13 & | & 95 \\ 200 & 10 & 300 & | & 905 \end{bmatrix} \rightarrow \begin{bmatrix} 1 & \frac{2}{15} & \frac{3}{10} & | & \frac{13}{6} \\ 35 & 33 & 13 & | & 95 \\ 200 & 10 & 300 & | & 905 \end{bmatrix} \rightarrow \begin{bmatrix} 1 & \frac{2}{15} & \frac{3}{10} & | & \frac{13}{6} \\ 0 & \frac{85}{3} & \frac{5}{2} & | & \frac{115}{6} \\ 0 & -\frac{50}{3} & 240 & | & \frac{1415}{3} \end{bmatrix}$$

$$R_1 = \left(\frac{1}{30}\right)r_1 \qquad\qquad R_2 = -35r_1 + r_2 \\ R_3 = -200r_1 + r_3$$

$$\rightarrow \begin{bmatrix} 1 & \frac{2}{15} & \frac{3}{10} & | & \frac{13}{6} \\ 0 & 1 & \frac{3}{34} & | & \frac{23}{34} \\ 0 & -\frac{50}{3} & 240 & | & \frac{1415}{3} \end{bmatrix} \rightarrow \begin{bmatrix} 1 & \frac{2}{15} & \frac{3}{10} & | & \frac{13}{6} \\ 0 & 1 & \frac{3}{34} & | & \frac{23}{34} \\ 0 & 0 & \frac{4105}{17} & | & \frac{8210}{17} \end{bmatrix} \rightarrow \begin{bmatrix} 1 & \frac{2}{15} & \frac{3}{10} & | & \frac{13}{6} \\ 0 & 1 & \frac{3}{34} & | & \frac{23}{34} \\ 0 & 0 & 1 & | & 2 \end{bmatrix}$$

$$R_2 = \left(\frac{3}{85}\right)r_2 \qquad R_3 = \left(\frac{50}{3}\right)r_2 + r_3 \qquad R_3 = \left(\frac{17}{4105}\right)r_3$$

The matrix is now in echelon form. The final matrix represents the system

$$\begin{cases} c + \dfrac{2}{15}p + \dfrac{3}{10}m = \dfrac{13}{6} & \text{(1)} \\[2mm] p + \dfrac{3}{34}m = \dfrac{23}{34} & \text{(2)} \\[2mm] m = 2 & \text{(3)} \end{cases}$$

From (3), we determine that 2 glasses of milk should be served. Back-substitute $m = 2$ into equation (2) to find that $p = \dfrac{1}{2}$, so $\dfrac{1}{2}$ of a baked potato should be served. Back-substitute these values into equation (1) and find that $c = 1.5$, so 1.5 servings of chicken *a la king* should be given to the patient to meet the dietary requirements. ∎

COMMENT: Most graphing utilities have the capability to put an augmented matrix into row echelon form (ref) and also reduced row echelon form (rref). See Appendix B, Section B.7, for a discussion. ∎

6.2 Assess Your Understanding

Concepts and Vocabulary

1. An m by n rectangular array of numbers is called a(n) _____.

2. The matrix used to represent a system of linear equations is called a(n) _____ matrix.

3. *True or False*: The augmented matrix of a system of two equations containing three variables has two rows and four columns.

4. *True or False*: The matrix $\begin{bmatrix} 1 & 3 & | & -2 \\ 0 & 1 & | & 5 \\ 0 & 0 & | & 0 \end{bmatrix}$ is in row echelon form.

Skill Building

In Problems 5–16, write the augmented matrix of the given system of equations.

5. $\begin{cases} x - 5y = 5 \\ 4x + 3y = 6 \end{cases}$

6. $\begin{cases} 3x + 4y = 7 \\ 4x - 2y = 5 \end{cases}$

7. $\begin{cases} 2x + 3y - 6 = 0 \\ 4x - 6y + 2 = 0 \end{cases}$

8. $\begin{cases} 9x - y = 0 \\ 3x - y - 4 = 0 \end{cases}$

9. $\begin{cases} 0.01x - 0.03y = 0.06 \\ 0.13x + 0.10y = 0.20 \end{cases}$

10. $\begin{cases} \dfrac{4}{3}x - \dfrac{3}{2}y = \dfrac{3}{4} \\[2mm] -\dfrac{1}{4}x + \dfrac{1}{3}y = \dfrac{2}{3} \end{cases}$

11. $\begin{cases} x - y + z = 10 \\ 3x + 3y = 5 \\ x + y + 2z = 2 \end{cases}$

12. $\begin{cases} 5x - y - z = 0 \\ x + y = 5 \\ 2x - 3z = 2 \end{cases}$

13. $\begin{cases} x + y - z = 2 \\ 3x - 2y = 2 \\ 5x + 3y - z = 1 \end{cases}$

14. $\begin{cases} 2x + 3y - 4z = 0 \\ x - 5z + 2 = 0 \\ x + 2y - 3z = -2 \end{cases}$

15. $\begin{cases} x - y - z = 10 \\ 2x + y + 2z = -1 \\ -3x + 4y = 5 \\ 4x - 5y + z = 0 \end{cases}$

16. $\begin{cases} x - y + 2z - w = 5 \\ x + 3y - 4z + 2w = 2 \\ 3x - y - 5z - w = -1 \end{cases}$

In Problems 17–24, perform each row operation on the given augmented matrix.

17. $\begin{bmatrix} 1 & -3 & | & -2 \\ 2 & -5 & | & 5 \end{bmatrix}$ $R_2 = -2r_1 + r_2$

18. $\begin{bmatrix} 1 & -3 & | & -3 \\ 2 & -5 & | & -4 \end{bmatrix}$ $R_2 = -2r_1 + r_2$

19. $\begin{bmatrix} 1 & -3 & 4 & | & 3 \\ 2 & -5 & 6 & | & 6 \\ -3 & 3 & 4 & | & 6 \end{bmatrix}$
(a) $R_2 = -2r_1 + r_2$
(b) $R_3 = 3r_1 + r_3$

20. $\begin{bmatrix} 1 & -3 & 3 & | & -5 \\ 2 & -5 & -3 & | & -5 \\ -3 & -2 & 4 & | & 6 \end{bmatrix}$
(a) $R_2 = -2r_1 + r_2$
(b) $R_3 = 3r_1 + r_3$

21. $\begin{bmatrix} 1 & -3 & 2 & | & -6 \\ 2 & -5 & 3 & | & -4 \\ -3 & -6 & 4 & | & 6 \end{bmatrix}$
(a) $R_2 = -2r_1 + r_2$
(b) $R_3 = 3r_1 + r_3$

22. $\begin{bmatrix} 1 & -3 & -4 & | & -6 \\ 2 & -5 & 6 & | & -6 \\ -3 & 1 & 4 & | & 6 \end{bmatrix}$
(a) $R_2 = -2r_1 + r_2$
(b) $R_3 = 3r_1 + r_3$

23. $\begin{bmatrix} 1 & -3 & 1 & | & -2 \\ 2 & -5 & 6 & | & -2 \\ -3 & 1 & 4 & | & 6 \end{bmatrix}$ (a) $R_2 = -2r_1 + r_2$ 24. $\begin{bmatrix} 1 & -3 & -1 & | & 2 \\ 2 & -5 & 2 & | & 6 \\ -3 & -6 & 4 & | & 6 \end{bmatrix}$ (a) $R_2 = -2r_1 + r_2$
 (b) $R_3 = 3r_1 + r_3$ (b) $R_3 = 3r_1 + r_3$

In Problems 25–36, the reduced row echelon form of a system of linear equations is given. Write the system of equations corresponding to the given matrix. Use x, y; or x, y, z; or x_1, x_2, x_3, x_4 as variables. Determine whether the system is consistent or inconsistent. If it is consistent, give the solution.

25. $\begin{bmatrix} 1 & 0 & | & 5 \\ 0 & 1 & | & -1 \end{bmatrix}$ 26. $\begin{bmatrix} 1 & 0 & | & -4 \\ 0 & 1 & | & 0 \end{bmatrix}$ 27. $\begin{bmatrix} 1 & 0 & 0 & | & 1 \\ 0 & 1 & 0 & | & 2 \\ 0 & 0 & 0 & | & 3 \end{bmatrix}$

28. $\begin{bmatrix} 1 & 0 & 0 & | & 0 \\ 0 & 1 & 0 & | & 0 \\ 0 & 0 & 0 & | & 2 \end{bmatrix}$ 29. $\begin{bmatrix} 1 & 0 & 2 & | & -1 \\ 0 & 1 & -4 & | & -2 \\ 0 & 0 & 0 & | & 0 \end{bmatrix}$ 30. $\begin{bmatrix} 1 & 0 & 4 & | & 4 \\ 0 & 1 & 3 & | & 2 \\ 0 & 0 & 0 & | & 0 \end{bmatrix}$

31. $\begin{bmatrix} 1 & 0 & 0 & 0 & | & 1 \\ 0 & 1 & 0 & 1 & | & 2 \\ 0 & 0 & 1 & 2 & | & 3 \end{bmatrix}$ 32. $\begin{bmatrix} 1 & 0 & 0 & 0 & | & 1 \\ 0 & 1 & 0 & 2 & | & 2 \\ 0 & 0 & 1 & 3 & | & 0 \end{bmatrix}$ 33. $\begin{bmatrix} 1 & 0 & 0 & 4 & | & 2 \\ 0 & 1 & 1 & 3 & | & 3 \\ 0 & 0 & 0 & 0 & | & 0 \end{bmatrix}$

34. $\begin{bmatrix} 1 & 0 & 0 & 0 & | & 1 \\ 0 & 1 & 0 & 0 & | & 2 \\ 0 & 0 & 1 & 2 & | & 3 \end{bmatrix}$ 35. $\begin{bmatrix} 1 & 0 & 0 & 1 & | & -2 \\ 0 & 1 & 0 & 2 & | & 2 \\ 0 & 0 & 1 & -1 & | & 0 \\ 0 & 0 & 0 & 0 & | & 0 \end{bmatrix}$ 36. $\begin{bmatrix} 1 & 0 & 0 & 0 & | & 1 \\ 0 & 1 & 0 & 0 & | & 2 \\ 0 & 0 & 1 & 0 & | & 3 \\ 0 & 0 & 0 & 1 & | & 0 \end{bmatrix}$

In Problems 37–72, solve each system of equations using matrices (row operations). If the system has no solution, say that it is inconsistent.

37. $\begin{cases} x + y = 8 \\ x - y = 4 \end{cases}$ 38. $\begin{cases} x + 2y = 5 \\ x + y = 3 \end{cases}$ 39. $\begin{cases} 2x - 4y = -2 \\ 3x + 2y = 3 \end{cases}$

40. $\begin{cases} 3x + 3y = 3 \\ 4x + 2y = \dfrac{8}{3} \end{cases}$ 41. $\begin{cases} x + 2y = 4 \\ 2x + 4y = 8 \end{cases}$ 42. $\begin{cases} 3x - y = 7 \\ 9x - 3y = 21 \end{cases}$

43. $\begin{cases} 2x + 3y = 6 \\ x - y = \dfrac{1}{2} \end{cases}$ 44. $\begin{cases} \dfrac{1}{2}x + y = -2 \\ x - 2y = 8 \end{cases}$ 45. $\begin{cases} 3x - 5y = 3 \\ 15x + 5y = 21 \end{cases}$

46. $\begin{cases} 2x - y = -1 \\ x + \dfrac{1}{2}y = \dfrac{3}{2} \end{cases}$ 47. $\begin{cases} x - y = 6 \\ 2x - 3z = 16 \\ 2y + z = 4 \end{cases}$ 48. $\begin{cases} 2x + y = -4 \\ -2y + 4z = 0 \\ 3x - 2z = -11 \end{cases}$

49. $\begin{cases} x - 2y + 3z = 7 \\ 2x + y + z = 4 \\ -3x + 2y - 2z = -10 \end{cases}$ 50. $\begin{cases} 2x + y - 3z = 0 \\ -2x + 2y + z = -7 \\ 3x - 4y - 3z = 7 \end{cases}$ 51. $\begin{cases} 2x - 2y - 2z = 2 \\ 2x + 3y + z = 2 \\ 3x + 2y = 0 \end{cases}$

52. $\begin{cases} 2x - 3y - z = 0 \\ -x + 2y + z = 5 \\ 3x - 4y - z = 1 \end{cases}$ 53. $\begin{cases} -x + y + z = -1 \\ -x + 2y - 3z = -4 \\ 3x - 2y - 7z = 0 \end{cases}$ 54. $\begin{cases} 2x - 3y - z = 0 \\ 3x + 2y + 2z = 2 \\ x + 5y + 3z = 2 \end{cases}$

55. $\begin{cases} 2x - 2y + 3z = 6 \\ 4x - 3y + 2z = 0 \\ -2x + 3y - 7z = 1 \end{cases}$ 56. $\begin{cases} 3x - 2y + 2z = 6 \\ 7x - 3y + 2z = -1 \\ 2x - 3y + 4z = 0 \end{cases}$ 57. $\begin{cases} x + y - z = 6 \\ 3x - 2y + z = -5 \\ x + 3y - 2z = 14 \end{cases}$

58. $\begin{cases} x - y + z = -4 \\ 2x - 3y + 4z = -15 \\ 5x + y - 2z = 12 \end{cases}$ 59. $\begin{cases} x + 2y - z = -3 \\ 2x - 4y + z = -7 \\ -2x + 2y - 3z = 4 \end{cases}$ 60. $\begin{cases} x + 4y - 3z = -8 \\ 3x - y + 3z = 12 \\ x + y + 6z = 1 \end{cases}$

61. $\begin{cases} 3x + y - z = \dfrac{2}{3} \\ 2x - y + z = 1 \\ 4x + 2y = \dfrac{8}{3} \end{cases}$

62. $\begin{cases} x + y = 1 \\ 2x - y + z = 1 \\ x + 2y + z = \dfrac{8}{3} \end{cases}$

63. $\begin{cases} x + y + z + w = 4 \\ 2x - y + z = 0 \\ 3x + 2y + z - w = 6 \\ x - 2y - 2z + 2w = -1 \end{cases}$

64. $\begin{cases} x + y + z + w = 4 \\ -x + 2y + z = 0 \\ 2x + 3y + z - w = 6 \\ -2x + y - 2z + 2w = -1 \end{cases}$

65. $\begin{cases} x + 2y + z = 1 \\ 2x - y + 2z = 2 \\ 3x + y + 3z = 3 \end{cases}$

66. $\begin{cases} x + 2y - z = 3 \\ 2x - y + 2z = 6 \\ x - 3y + 3z = 4 \end{cases}$

67. $\begin{cases} x - y + z = 5 \\ 3x + 2y - 2z = 0 \end{cases}$

68. $\begin{cases} 2x + y - z = 4 \\ -x + y + 3z = 1 \end{cases}$

69. $\begin{cases} 2x + 3y - z = 3 \\ x - y - z = 0 \\ -x + y + z = 0 \\ x + y + 3z = 5 \end{cases}$

70. $\begin{cases} x - 3y + z = 1 \\ 2x - y - 4z = 0 \\ x - 3y + 2z = 1 \\ x - 2y = 5 \end{cases}$

71. $\begin{cases} 4x + y + z - w = 4 \\ x - y + 2z + 3w = 3 \end{cases}$

72. $\begin{cases} -4x + y = 5 \\ 2x - y + z - w = 5 \\ z + w = 4 \end{cases}$

Applications and Extensions

73. Curve Fitting Find the function $y = ax^2 + bx + c$ whose graph contains the points $(1, 2)$, $(-2, -7)$, and $(2, -3)$.

74. Curve Fitting Find the function $y = ax^2 + bx + c$ whose graph contains the points $(1, -1)$, $(3, -1)$, and $(-2, 14)$.

75. Curve Fitting Find the function $f(x) = ax^3 + bx^2 + cx + d$ for which $f(-3) = -112, f(-1) = -2, f(1) = 4,$ and $f(2) = 13$.

76. Curve Fitting Find the function $f(x) = ax^3 + bx^2 + cx + d$ for which $f(-2) = -10, f(-1) = 3, f(1) = 5,$ and $f(3) = 15$.

77. Nutrition A dietitian at Palos Community Hospital wants a patient to have a meal that has 78 grams of protein, 59 grams of carbohydrates, and 75 milligrams of vitamin A. The hospital food service tells the dietitian that the dinner for today is salmon steak, baked eggs, and acorn squash. Each serving of salmon steak has 30 grams of protein, 20 grams of carbohydrates, and 2 milligrams of vitamin A. Each serving of baked eggs contains 15 grams of protein, 2 grams of carbohydrates, and 20 milligrams of vitamin A. Each serving of acorn squash contains 3 grams of protein, 25 grams of carbohydrates, and 32 milligrams of vitamin A. How many servings of each food should the dietitian provide for the patient?

78. Nutrition A dietitian at General Hospital wants a patient to have a meal that has 47 grams of protein, 58 grams of carbohydrates, and 630 milligrams of calcium. The hospital food service tells the dietitian that the dinner for today is pork chops, corn on the cob, and 2% milk. Each serving of pork chops has 23 grams of protein, 0 grams of carbohydrates, and 10 milligrams of calcium. Each serving of corn on the cob contains 3 grams of protein, 16 grams of carbohydrates, and 10 milligrams of calcium. Each glass of 2% milk contains 9 grams of protein, 13 grams of carbohydrates, and 300 milligrams of calcium. How many servings of each food should the dietitian provide for the patient?

79. Financial Planning Carletta has $10,000 to invest. As her financial consultant, you recommend that she invest in Treasury bills that yield 6%, Treasury bonds that yield 7%, and corporate bonds that yield 8%. Carletta wants to have an annual income of $680, and the amount invested in corporate bonds must be half that invested in Treasury bills. Find the amount in each investment.

80. Financial Planning John has $20,000 to invest. As his financial consultant, you recommend that he invest in Treasury bills that yield 5%, Treasury bonds that yield 7%, and corporate bonds that yield 9%. John wants to have an annual income of $1280, and the amount invested in Treasury bills must be two times the amount invested in corporate bonds. Find the amount in each investment.

81. Production To manufacture an automobile requires painting, drying, and polishing. Epsilon Motor Company produces three types of cars: the Delta, the Beta, and the Sigma. Each Delta requires 10 hours for painting, 3 hours for drying, and 2 hours for polishing. A Beta requires 16 hours for painting, 5 hours for drying, and 3 hours for polishing, while a Sigma requires 8 hours for painting, 2 hours for drying, and 1 hour for polishing. If the company has 240 hours for painting, 69 hours for drying, and 41 hours for polishing per month, how many of each type of car are produced?

82. Production A Florida juice company completes the preparation of its products by sterilizing, filling, and labeling bottles. Each case of orange juice requires 9 minutes for sterilizing, 6 minutes for filling, and 1 minute for labeling. Each case of grapefruit juice requires 10 minutes for sterilizing, 4 minutes for filling, and 2 minutes for labeling. Each case of tomato juice requires 12 minutes for sterilizing, 4 minutes for filling, and 1 minute for labeling. If the company runs the sterilizing machine for 398 minutes, the filling machine for 164 minutes, and the labeling machine for 58 minutes, how many cases of each type of juice are prepared?

83. Electricity: Kirchhoff's Rules An application of Kirchhoff's Rules to the circuit shown results in the following system of equations:

$$\begin{cases} -4 + 8 - 2I_2 = 0 \\ 8 = 5I_4 + I_1 \\ 4 = 3I_3 + I_1 \\ I_3 + I_4 = I_1 \end{cases}$$

Find the currents I_1, I_2, I_3, and I_4.

SOURCE: Based on Raymond Serway, *Physics*, 3rd ed. (Philadelphia: Saunders, 1990), Prob. 34, p. 790.

84. Electricity: Kirchhoff's Rules An application of Kirchhoff's Rules to the circuit shown results in the following system of equations:

$$\begin{cases} I_1 = I_3 + I_2 \\ 24 - 6I_1 - 3I_3 = 0 \\ 12 + 24 - 6I_1 - 6I_2 = 0 \end{cases}$$

Find the currents I_1, I_2, and I_3.

SOURCE: Ibid., Prob. 38, p. 791.

85. Financial Planning Three retired couples each require an additional annual income of $2000 per year. As their financial consultant, you recommend that they invest some money in Treasury bills that yield 7%, some money in cor-

porate bonds that yield 9%, and some money in junk bonds that yield 11%. Prepare a table for each couple showing the various ways that their goals can be achieved:
(a) If the first couple has $20,000 to invest.
(b) If the second couple has $25,000 to invest.
(c) If the third couple has $30,000 to invest.
(d) What advice would you give each couple regarding the amount to invest and the choices available?

[**Hint:** Higher yields generally carry more risk.]

86. Financial Planning A young couple has $25,000 to invest. As their financial consultant, you recommend that they invest some money in Treasury bills that yield 7%, some money in corporate bonds that yield 9%, and some money in junk bonds that yield 11%. Prepare a table showing the various ways that this couple can achieve the following goals:
(a) The couple wants $1500 per year in income.
(b) The couple wants $2000 per year in income.
(c) The couple wants $2500 per year in income.
(d) What advice would you give this couple regarding the income that they require and the choices available?

[**Hint:** Higher yields generally carry more risk.]

87. Pharmacy A doctor's prescription calls for a daily intake of a supplement containing 40 mg of vitamin C and 30 mg of vitamin D. Your pharmacy stocks three supplements that can be used: one contains 20% vitamin C and 30% vitamin D; a second, 40% vitamin C and 20% vitamin D; and a third, 30% vitamin C and 50% vitamin D. Create a table showing the possible combinations that could be used to fill the prescription.

88. Pharmacy A doctor's prescription calls for the creation of pills that contain 12 units of vitamin B_{12} and 12 units of vitamin E. Your pharmacy stocks three powders that can be used to make these pills: one contains 20% vitamin B_{12} and 30% vitamin E; a second, 40% vitamin B_{12} and 20% vitamin E; and a third, 30% vitamin B_{12} and 40% vitamin E. Create a table showing the possible combinations of each powder that could be mixed in each pill.

Discussion and Writing

89. Write a brief paragraph or two that outline your strategy for solving a system of linear equations using matrices.

90. When solving a system of linear equations using matrices, do you prefer to place the augmented matrix in row echelon form or in reduced row echelon form? Give reasons for your choice.

91. Make up a system of three linear equations containing three variables that has:
(a) No solution
(b) Exactly one solution
(c) Infinitely many solutions
Give the three systems to a friend to solve and critique.

6.3 Systems of Linear Equations: Determinants

OBJECTIVES
1 Evaluate 2 by 2 Determinants
2 Use Cramer's Rule to Solve a System of Two Equations Containing Two Variables
3 Evaluate 3 by 3 Determinants
4 Use Cramer's Rule to Solve a System of Three Equations Containing Three Variables
5 Know Properties of Determinants

In the preceding section, we described a method of using matrices to solve a system of linear equations. This section deals with yet another method for solving systems of linear equations; however, it can be used only when the number of equations equals the number of variables. Although the method will work for any system (provided that the number of equations equals the number of variables), it is most often used for systems of two equations containing two variables or three equations containing three variables. This method, called *Cramer's Rule*, is based on the concept of a *determinant*.

1 Evaluate 2 by 2 Determinants

If a, b, c, and d are four real numbers, the symbol

$$D = \begin{vmatrix} a & b \\ c & d \end{vmatrix}$$

is called a **2 by 2 determinant**. Its value is the number $ad - bc$; that is,

$$D = \begin{vmatrix} a & b \\ c & d \end{vmatrix} = ad - bc \qquad \textbf{(1)}$$

The following device may be helpful for remembering the value of a 2 by 2 determinant:

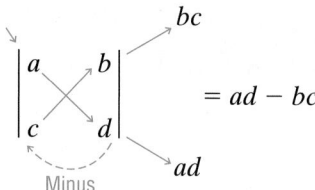

$$\begin{vmatrix} a & b \\ c & d \end{vmatrix} = ad - bc$$

EXAMPLE 1 **Evaluating a 2 × 2 Determinant**

Evaluate: $\begin{vmatrix} 3 & -2 \\ 6 & 1 \end{vmatrix}$

Solution
$$\begin{vmatrix} 3 & -2 \\ 6 & 1 \end{vmatrix} = (3)(1) - (6)(-2)$$
$$= 3 - (-12)$$
$$= 15$$

━━━ **NOW WORK PROBLEM 7.**

2 Use Cramer's Rule to Solve a System of Two Equations Containing Two Variables

Let's now see the role that a 2 by 2 determinant plays in the solution of a system of two equations containing two variables. Consider the system

$$\begin{cases} ax + by = s & (1) \\ cx + dy = t & (2) \end{cases} \qquad \textbf{(2)}$$

We shall use the method of elimination to solve this system.

Provided $d \neq 0$ and $b \neq 0$, this system is equivalent to the system

$$\begin{cases} adx + bdy = sd & \text{(1)} \quad \text{Multiply by } d. \\ bcx + bdy = tb & \text{(2)} \quad \text{Multiply by } b. \end{cases}$$

On subtracting the second equation from the first equation, we get

$$(ad - bc)x + 0 \cdot y = sd - tb$$

which can be rewritten using determinant notation as

$$\begin{vmatrix} a & b \\ c & d \end{vmatrix} x = \begin{vmatrix} s & b \\ t & d \end{vmatrix}$$

If $D = \begin{vmatrix} a & b \\ c & d \end{vmatrix} = ad - bc \neq 0$, we can solve for x to get

$$x = \frac{\begin{vmatrix} s & b \\ t & d \end{vmatrix}}{\begin{vmatrix} a & b \\ c & d \end{vmatrix}} = \frac{\begin{vmatrix} s & b \\ t & d \end{vmatrix}}{D} \qquad \textbf{(3)}$$

Return now to the original system (2). Provided that $a \neq 0$ and $c \neq 0$, the system is equivalent to

$$\begin{cases} acx + bcy = cs & \text{(1)} \quad \text{Multiply by } c. \\ acx + ady = at & \text{(2)} \quad \text{Multiply by } a. \end{cases}$$

On subtracting the first equation from the second equation, we get

$$0 \cdot x + (ad - bc)y = at - cs$$

which can be rewritten using determinant notation as

$$\begin{vmatrix} a & b \\ c & d \end{vmatrix} y = \begin{vmatrix} a & s \\ c & t \end{vmatrix}$$

If $D = \begin{vmatrix} a & b \\ c & d \end{vmatrix} = ad - bc \neq 0$, we can solve for y to get

$$y = \frac{\begin{vmatrix} a & s \\ c & t \end{vmatrix}}{\begin{vmatrix} a & b \\ c & d \end{vmatrix}} = \frac{\begin{vmatrix} a & s \\ c & t \end{vmatrix}}{D} \qquad \textbf{(4)}$$

Equations (3) and (4) lead us to the following result, called **Cramer's Rule**.

Theorem

Cramer's Rule for Two Equations Containing Two Variables

The solution to the system of equations

$$\begin{cases} ax + by = s & (1) \\ cx + dy = t & (2) \end{cases} \tag{5}$$

is given by

$$x = \frac{\begin{vmatrix} s & b \\ t & d \end{vmatrix}}{\begin{vmatrix} a & b \\ c & d \end{vmatrix}}, \qquad y = \frac{\begin{vmatrix} a & s \\ c & t \end{vmatrix}}{\begin{vmatrix} a & b \\ c & d \end{vmatrix}} \tag{6}$$

provided that

$$D = \begin{vmatrix} a & b \\ c & d \end{vmatrix} = ad - bc \neq 0$$

In the derivation given for Cramer's Rule above, we assumed that none of the numbers $a, b, c,$ and d was 0. In Problem 60 you will be asked to complete the proof under the less stringent condition that $D = ad - bc \neq 0$.

Now look carefully at the pattern in Cramer's Rule. The denominator in the solution (6) is the determinant of the coefficients of the variables.

$$\begin{cases} ax + by = s \\ cx + dy = t \end{cases} \qquad D = \begin{vmatrix} a & b \\ c & d \end{vmatrix}$$

In the solution for x, the numerator is the determinant, denoted by D_x, formed by replacing the entries in the first column (the coefficients of x) of D by the constants on the right side of the equal sign.

$$D_x = \begin{vmatrix} s & b \\ t & d \end{vmatrix}$$

In the solution for y, the numerator is the determinant, denoted by D_y, formed by replacing the entries in the second column (the coefficients in y) of D by the constants on the right side of the equal sign.

$$D_y = \begin{vmatrix} a & s \\ c & t \end{vmatrix}$$

Cramer's Rule then states that, if $D \neq 0$,

$$x = \frac{D_x}{D}, \qquad y = \frac{D_y}{D} \tag{7}$$

EXAMPLE 2

Solving a System of Linear Equations Using Determinants

Use Cramer's Rule, if applicable, to solve the system

$$\begin{cases} 3x - 2y = 4 & (1) \\ 6x + y = 13 & (2) \end{cases}$$

Solution The determinant D of the coefficients of the variables is

$$D = \begin{vmatrix} 3 & -2 \\ 6 & 1 \end{vmatrix} = (3)(1) - (6)(-2) = 15$$

Because $D \neq 0$, Cramer's Rule (7) can be used.

$$x = \frac{D_x}{D} = \frac{\begin{vmatrix} 4 & -2 \\ 13 & 1 \end{vmatrix}}{15} \qquad y = \frac{D_y}{D} = \frac{\begin{vmatrix} 3 & 4 \\ 6 & 13 \end{vmatrix}}{15}$$

$$= \frac{(4)(1) - (13)(-2)}{15} \qquad = \frac{(3)(13) - (6)(4)}{15}$$

$$= \frac{30}{15} \qquad = \frac{15}{15}$$

$$= 2 \qquad = 1$$

The solution is $x = 2$, $y = 1$. ■

In attempting to use Cramer's Rule, if the determinant D of the coefficients of the variables is found to equal 0 (so that Cramer's Rule is not applicable), the system either is inconsistent or has infinitely many solutions.

NOW WORK PROBLEM 15.

3 Evaluate 3 by 3 Determinants

To use Cramer's Rule to solve a system of three equations containing three variables, we need to define a 3 by 3 determinant.

A **3 by 3 determinant** is symbolized by

$$\begin{vmatrix} a_{11} & a_{12} & a_{13} \\ a_{21} & a_{22} & a_{23} \\ a_{31} & a_{32} & a_{33} \end{vmatrix} \tag{8}$$

in which a_{11}, a_{12}, \ldots, are real numbers.

As with matrices, we use a double subscript to identify an entry by indicating its row and column numbers. For example, the entry a_{23} is in row 2, column 3.

The value of a 3 by 3 determinant may be defined in terms of 2 by 2 determinants by the following formula:

$$\begin{vmatrix} a_{11} & a_{12} & a_{13} \\ a_{21} & a_{22} & a_{23} \\ a_{31} & a_{32} & a_{33} \end{vmatrix} = a_{11}\begin{vmatrix} a_{22} & a_{23} \\ a_{32} & a_{33} \end{vmatrix} \overset{\text{Minus}}{-} a_{12}\begin{vmatrix} a_{21} & a_{23} \\ a_{31} & a_{33} \end{vmatrix} + a_{13}\begin{vmatrix} a_{21} & a_{22} \\ a_{31} & a_{32} \end{vmatrix} \tag{9}$$

2 by 2 determinant left after removing the row and column containing a_{11} / 2 by 2 determinant left after removing the row and column containing a_{12} / 2 by 2 determinant left after removing the row and column containing a_{13}

The 2 by 2 determinants shown in formula (9) are called **minors** of the 3 by 3 determinant. For an n by n determinant, the **minor** M_{ij} of entry a_{ij} is the determinant resulting from removing the ith row and jth column.

| EXAMPLE 3 | **Finding Minors of a 3 by 3 Determinant** |

For the determinant $A = \begin{vmatrix} 2 & -1 & 3 \\ -2 & 5 & 1 \\ 0 & 6 & -9 \end{vmatrix}$, find: (a) M_{12} (b) M_{23}

Solution (a) M_{12} is the determinant that results from removing the first row and second column from A.

$$A = \begin{vmatrix} 2 & -1 & 3 \\ -2 & 5 & 1 \\ 0 & 6 & -9 \end{vmatrix} \qquad M_{12} = \begin{vmatrix} -2 & 1 \\ 0 & -9 \end{vmatrix} = (-2)(-9) - (0)(1) = 18$$

(b) M_{23} is the determinant that results from removing the second row and third column from A.

$$A = \begin{vmatrix} 2 & -1 & 3 \\ -2 & 5 & 1 \\ 0 & 6 & -9 \end{vmatrix} \qquad M_{23} = \begin{vmatrix} 2 & -1 \\ 0 & 6 \end{vmatrix} = (2)(6) - (0)(-1) = 12$$

 ∎

Referring back to formula (9), we see that each element a_{ij} is multiplied by its minor, but sometimes this term is added and other times, subtracted. To determine whether to add or subtract a term, we must consider the *cofactor*.

> For an n by n determinant A, the **cofactor** of entry a_{ij}, denoted by A_{ij}, is given by
>
> $$A_{ij} = (-1)^{i+j} M_{ij}$$
>
> where M_{ij} is the minor of entry a_{ij}.

The exponent of $(-1)^{i+j}$ is the sum of the row and column of the entry a_{ij}, so if $i + j$ is even, $(-1)^{i+j}$ will equal 1, and if $i + j$ is odd, $(-1)^{i+j}$ will equal -1.

To find the value of a determinant, multiply each entry in any row or column by its cofactor and sum the results. This process is referred to as **expanding across a row or column**. For example, the value of the 3 by 3 determinant in formula (9) was found by expanding across row 1.

If we choose to expand down column 2, we obtain

$$\begin{vmatrix} a_{11} & a_{12} & a_{13} \\ a_{21} & a_{22} & a_{23} \\ a_{31} & a_{32} & a_{33} \end{vmatrix} = (-1)^{1+2} a_{12} \begin{vmatrix} a_{21} & a_{23} \\ a_{31} & a_{33} \end{vmatrix} + (-1)^{2+2} a_{22} \begin{vmatrix} a_{11} & a_{13} \\ a_{31} & a_{33} \end{vmatrix} + (-1)^{3+2} a_{32} \begin{vmatrix} a_{11} & a_{13} \\ a_{21} & a_{23} \end{vmatrix}$$

↑ Expand down column 2.

If we choose to expand across row 3, we obtain

$$\begin{vmatrix} a_{11} & a_{12} & a_{13} \\ a_{21} & a_{22} & a_{23} \\ a_{31} & a_{32} & a_{33} \end{vmatrix} = (-1)^{3+1} a_{31} \begin{vmatrix} a_{12} & a_{13} \\ a_{22} & a_{23} \end{vmatrix} + (-1)^{3+2} a_{32} \begin{vmatrix} a_{11} & a_{13} \\ a_{21} & a_{23} \end{vmatrix} + (-1)^{3+3} a_{33} \begin{vmatrix} a_{11} & a_{12} \\ a_{21} & a_{22} \end{vmatrix}$$

↑ Expand across row 3.

It can be shown that the value of a determinant does not depend on the choice of the row or column used in the expansion. However, expanding across a row or column that has an element equal to 0 reduces the amount of work needed to compute the value of the determinant.

EXAMPLE 4 **Evaluating a 3 × 3 Determinant**

Find the value of the 3 by 3 determinant: $\begin{vmatrix} 3 & 4 & -1 \\ 4 & 6 & 2 \\ 8 & -2 & 3 \end{vmatrix}$

Solution We choose to expand across row 1.

$$\begin{vmatrix} 3 & 4 & -1 \\ 4 & 6 & 2 \\ 8 & -2 & 3 \end{vmatrix} = (-1)^{1+1}3 \begin{vmatrix} 6 & 2 \\ -2 & 3 \end{vmatrix} + (-1)^{1+2}4 \begin{vmatrix} 4 & 2 \\ 8 & 3 \end{vmatrix} + (-1)^{1+3}(-1) \begin{vmatrix} 4 & 6 \\ 8 & -2 \end{vmatrix}$$

$$= 3(18 + 4) - 4(12 - 16) + (-1)(-8 - 48)$$
$$= 3(22) - 4(-4) + (-1)(-56)$$
$$= 66 + 16 + 56 = 138$$

We could also find the value of the 3 by 3 determinant in Example 4 by expanding down column 3.

$$\begin{vmatrix} 3 & 4 & -1 \\ 4 & 6 & 2 \\ 8 & -2 & 3 \end{vmatrix} = (-1)^{1+3}(-1) \begin{vmatrix} 4 & 6 \\ 8 & -2 \end{vmatrix} + (-1)^{2+3}2 \begin{vmatrix} 3 & 4 \\ 8 & -2 \end{vmatrix} + (-1)^{3+3}3 \begin{vmatrix} 3 & 4 \\ 4 & 6 \end{vmatrix}$$

$$= -1(-8 - 48) - 2(-6 - 32) + 3(18 - 16)$$
$$= 56 + 76 + 6 = 138$$

━━━ NOW WORK PROBLEM 11.

4 **Use Cramer's Rule to Solve a System of Three Equations Containing Three Variables**

Consider the following system of three equations containing three variables.

$$\begin{cases} a_{11}x + a_{12}y + a_{13}z = c_1 \\ a_{21}x + a_{22}y + a_{23}z = c_2 \\ a_{31}x + a_{32}y + a_{33}z = c_3 \end{cases} \tag{10}$$

It can be shown that if the determinant D of the coefficients of the variables is not 0, that is, if

$$D = \begin{vmatrix} a_{11} & a_{12} & a_{13} \\ a_{21} & a_{22} & a_{23} \\ a_{31} & a_{32} & a_{33} \end{vmatrix} \neq 0$$

the unique solution of system (10) is given by

Cramer's Rule for Three Equations Containing Three Variables

$$x = \frac{D_x}{D}, \qquad y = \frac{D_y}{D}, \qquad z = \frac{D_z}{D}$$

where

$$D_x = \begin{vmatrix} c_1 & a_{12} & a_{13} \\ c_2 & a_{22} & a_{23} \\ c_3 & a_{32} & a_{33} \end{vmatrix}, \qquad D_y = \begin{vmatrix} a_{11} & c_1 & a_{13} \\ a_{21} & c_2 & a_{23} \\ a_{31} & c_3 & a_{33} \end{vmatrix}, \qquad D_z = \begin{vmatrix} a_{11} & a_{12} & c_1 \\ a_{21} & a_{22} & c_2 \\ a_{31} & a_{32} & c_3 \end{vmatrix}$$

Do you see the similarity of this pattern and the pattern observed earlier for a system of two equations containing two variables?

| EXAMPLE 5 | **Using Cramer's Rule** |

Use Cramer's Rule, if applicable, to solve the following system:

$$\begin{cases} 2x + y - z = 3 & (1) \\ -x + 2y + 4z = -3 & (2) \\ x - 2y - 3z = 4 & (3) \end{cases}$$

Solution The value of the determinant D of the coefficients of the variables is

$$D = \begin{vmatrix} 2 & 1 & -1 \\ -1 & 2 & 4 \\ 1 & -2 & -3 \end{vmatrix} = (-1)^{1+1}2\begin{vmatrix} 2 & 4 \\ -2 & -3 \end{vmatrix} + (-1)^{1+2}1\begin{vmatrix} -1 & 4 \\ 1 & -3 \end{vmatrix} + (-1)^{1+3}(-1)\begin{vmatrix} -1 & 2 \\ 1 & -2 \end{vmatrix}$$

$$= 2(2) - 1(-1) + (-1)(0)$$

$$= 4 + 1 = 5$$

Because $D \neq 0$, we proceed to find the values of D_x, D_y, and D_z.

$$D_x = \begin{vmatrix} 3 & 1 & -1 \\ -3 & 2 & 4 \\ 4 & -2 & -3 \end{vmatrix} = (-1)^{1+1}3\begin{vmatrix} 2 & 4 \\ -2 & -3 \end{vmatrix} + (-1)^{1+2}1\begin{vmatrix} -3 & 4 \\ 4 & -3 \end{vmatrix} + (-1)^{1+3}(-1)\begin{vmatrix} -3 & 2 \\ 4 & -2 \end{vmatrix}$$

$$= 3(2) - 1(-7) + (-1)(-2) = 15$$

$$D_y = \begin{vmatrix} 2 & 3 & -1 \\ -1 & -3 & 4 \\ 1 & 4 & -3 \end{vmatrix} = (-1)^{1+1}2\begin{vmatrix} -3 & 4 \\ 4 & -3 \end{vmatrix} + (-1)^{1+2}3\begin{vmatrix} -1 & 4 \\ 1 & -3 \end{vmatrix} + (-1)^{1+3}(-1)\begin{vmatrix} -1 & -3 \\ 1 & 4 \end{vmatrix}$$

$$= 2(-7) - 3(-1) + (-1)(-1)$$

$$= -14 + 3 + 1 = -10$$

$$D_z = \begin{vmatrix} 2 & 1 & 3 \\ -1 & 2 & -3 \\ 1 & -2 & 4 \end{vmatrix} = (-1)^{1+1}2\begin{vmatrix} 2 & -3 \\ -2 & 4 \end{vmatrix} + (-1)^{1+2}1\begin{vmatrix} -1 & -3 \\ 1 & 4 \end{vmatrix} + (-1)^{1+3}3\begin{vmatrix} -1 & 2 \\ 1 & -2 \end{vmatrix}$$

$$= 2(2) - 1(-1) + 3(0) = 5$$

As a result,

$$x = \frac{D_x}{D} = \frac{15}{5} = 3, \qquad y = \frac{D_y}{D} = \frac{-10}{5} = -2, \qquad z = \frac{D_z}{D} = \frac{5}{5} = 1$$

The solution is $x = 3$, $y = -2$, $z = 1$. ∎

If the determinant of the coefficients of the variables of a system of three linear equations containing three variables is 0, Cramer's Rule is not applicable. In such a case, the system either is inconsistent or has infinitely many solutions. To solve such a system, use elimination or row reduction on the augmented matrix.

 NOW WORK PROBLEM 33.

5 Know Properties of Determinants

Determinants have several properties that are sometimes helpful for obtaining their value. We list some of them here.

Theorem The value of a determinant changes sign if any two rows (or any two columns) are interchanged. **(11)**

Proof for 2 by 2 Determinants

$$\begin{vmatrix} a & b \\ c & d \end{vmatrix} = ad - bc \quad \text{and} \quad \begin{vmatrix} c & d \\ a & b \end{vmatrix} = bc - ad = -(ad - bc) \qquad ∎$$

EXAMPLE 6 **Demonstrating Theorem (11)**

$$\begin{vmatrix} 3 & 4 \\ 1 & 2 \end{vmatrix} = 6 - 4 = 2 \qquad \begin{vmatrix} 1 & 2 \\ 3 & 4 \end{vmatrix} = 4 - 6 = -2$$ ∎

Theorem If all the entries in any row (or any column) equal 0, the value of the determinant is 0. **(12)**

Proof Expand across the row (or down the column) containing the 0's. ∎

Theorem If any two rows (or any two columns) of a determinant have corresponding entries that are equal, the value of the determinant is 0. **(13)**

You are asked to prove this result for a 3 by 3 determinant in which the entries in column 1 equal the entries in column 3 in Problem 63.

EXAMPLE 7 **Demonstrating Theorem (13)**

$$\begin{vmatrix} 1 & 2 & 3 \\ 1 & 2 & 3 \\ 4 & 5 & 6 \end{vmatrix} = (-1)^{1+1}1\begin{vmatrix} 2 & 3 \\ 5 & 6 \end{vmatrix} + (-1)^{1+2}2\begin{vmatrix} 1 & 3 \\ 4 & 6 \end{vmatrix} + (-1)^{1+3}3\begin{vmatrix} 1 & 2 \\ 4 & 5 \end{vmatrix}$$

$$= 1(-3) - 2(-6) + 3(-3) = -3 + 12 - 9 = 0$$ ∎

Theorem If any row (or any column) of a determinant is multiplied by a nonzero number k, the value of the determinant is also changed by a factor of k. **(14)**

You are asked to prove this result for a 3 by 3 determinant using row 2 in Problem 62.

EXAMPLE 8 **Demonstrating Theorem (14)**

$$\begin{vmatrix} 1 & 2 \\ 4 & 6 \end{vmatrix} = 6 - 8 = -2$$

$$\begin{vmatrix} k & 2k \\ 4 & 6 \end{vmatrix} = 6k - 8k = -2k = k(-2) = k\begin{vmatrix} 1 & 2 \\ 4 & 6 \end{vmatrix}$$ ∎

Theorem If the entries of any row (or any column) of a determinant are multiplied by a nonzero number k and the result is added to the corresponding entries of another row (or column), the value of the determinant remains unchanged. **(15)**

In Problem 64, you are asked to prove this result for a 3 by 3 determinant using rows 1 and 2.

EXAMPLE 9 **Demonstrating Theorem (15)**

$$\begin{vmatrix} 3 & 4 \\ 5 & 2 \end{vmatrix} = -14 \qquad \begin{vmatrix} 3 & 4 \\ 5 & 2 \end{vmatrix} \rightarrow \begin{vmatrix} -7 & 0 \\ 5 & 2 \end{vmatrix} = -14$$

Multiply row 2 by -2 and add to row 1. ∎

6.3 Assess Your Understanding

Concepts and Vocabulary

1. Cramer's Rule uses _____ to solve a system of linear equations.

2. $D = \begin{vmatrix} a & b \\ c & d \end{vmatrix} = $ _____

3. *True or False:* A 3 by 3 determinant can never equal 0.

4. *True or False:* The value of a determinant remains unchanged if any two rows or any two columns are interchanged.

Skill Building

In Problems 5–14, find the value of each determinant.

5. $\begin{vmatrix} 3 & 1 \\ 4 & 2 \end{vmatrix}$

6. $\begin{vmatrix} 6 & 1 \\ 5 & 2 \end{vmatrix}$

7. $\begin{vmatrix} 6 & 4 \\ -1 & 3 \end{vmatrix}$

8. $\begin{vmatrix} 8 & -3 \\ 4 & 2 \end{vmatrix}$

9. $\begin{vmatrix} -3 & -1 \\ 4 & 2 \end{vmatrix}$

10. $\begin{vmatrix} -4 & 2 \\ -5 & 3 \end{vmatrix}$

11. $\begin{vmatrix} 3 & 4 & 2 \\ 1 & -1 & 5 \\ 1 & 2 & -2 \end{vmatrix}$

12. $\begin{vmatrix} 1 & 3 & -2 \\ 6 & 1 & -5 \\ 8 & 2 & 3 \end{vmatrix}$

13. $\begin{vmatrix} 4 & -1 & 2 \\ 6 & -1 & 0 \\ 1 & -3 & 4 \end{vmatrix}$

14. $\begin{vmatrix} 3 & -9 & 4 \\ 1 & 4 & 0 \\ 8 & -3 & 1 \end{vmatrix}$

In Problems 15–42, solve each system of equations using Cramer's Rule if it is applicable. If Cramer's Rule is not applicable, say so.

15. $\begin{cases} x + y = 8 \\ x - y = 4 \end{cases}$

16. $\begin{cases} x + 2y = 5 \\ x - y = 3 \end{cases}$

17. $\begin{cases} 5x - y = 13 \\ 2x + 3y = 12 \end{cases}$

18. $\begin{cases} x + 3y = 5 \\ 2x - 3y = -8 \end{cases}$

19. $\begin{cases} 3x = 24 \\ x + 2y = 0 \end{cases}$

20. $\begin{cases} 4x + 5y = -3 \\ -2y = -4 \end{cases}$

21. $\begin{cases} 3x - 6y = 24 \\ 5x + 4y = 12 \end{cases}$

22. $\begin{cases} 2x + 4y = 16 \\ 3x - 5y = -9 \end{cases}$

23. $\begin{cases} 3x - 2y = 4 \\ 6x - 4y = 0 \end{cases}$

24. $\begin{cases} -x + 2y = 5 \\ 4x - 8y = 6 \end{cases}$

25. $\begin{cases} 2x - 4y = -2 \\ 3x + 2y = 3 \end{cases}$

26. $\begin{cases} 3x + 3y = 3 \\ 4x + 2y = \dfrac{8}{3} \end{cases}$

27. $\begin{cases} 2x - 3y = -1 \\ 10x + 10y = 5 \end{cases}$

28. $\begin{cases} 3x - 2y = 0 \\ 5x + 10y = 4 \end{cases}$

29. $\begin{cases} 2x + 3y = 6 \\ x - y = \dfrac{1}{2} \end{cases}$

30. $\begin{cases} \dfrac{1}{2}x + y = -2 \\ x - 2y = 8 \end{cases}$

31. $\begin{cases} 3x - 5y = 3 \\ 15x + 5y = 21 \end{cases}$

32. $\begin{cases} 2x - y = -1 \\ x + \dfrac{1}{2}y = \dfrac{3}{2} \end{cases}$

33. $\begin{cases} x + y - z = 6 \\ 3x - 2y + z = -5 \\ x + 3y - 2z = 14 \end{cases}$

34. $\begin{cases} x - y + z = -4 \\ 2x - 3y + 4z = -15 \\ 5x + y - 2z = 12 \end{cases}$

35. $\begin{cases} x + 2y - z = -3 \\ 2x - 4y + z = -7 \\ -2x + 2y - 3z = 4 \end{cases}$

36. $\begin{cases} x + 4y - 3z = -8 \\ 3x - y + 3z = 12 \\ x + y + 6z = 1 \end{cases}$

37. $\begin{cases} x - 2y + 3z = 1 \\ 3x + y - 2z = 0 \\ 2x - 4y + 6z = 2 \end{cases}$

38. $\begin{cases} x - y + 2z = 5 \\ 3x + 2y = 4 \\ -2x + 2y - 4z = -10 \end{cases}$

39. $\begin{cases} x + 2y - z = 0 \\ 2x - 4y + z = 0 \\ -2x + 2y - 3z = 0 \end{cases}$

40. $\begin{cases} x + 4y - 3z = 0 \\ 3x - y + 3z = 0 \\ x + y + 6z = 0 \end{cases}$

41. $\begin{cases} x - 2y + 3z = 0 \\ 3x + y - 2z = 0 \\ 2x - 4y + 6z = 0 \end{cases}$

42. $\begin{cases} x - y + 2z = 0 \\ 3x + 2y = 0 \\ -2x + 2y - 4z = 0 \end{cases}$

In Problems 43–48, solve for x.

43. $\begin{vmatrix} x & x \\ 4 & 3 \end{vmatrix} = 5$

44. $\begin{vmatrix} x & 1 \\ 3 & x \end{vmatrix} = -2$

45. $\begin{vmatrix} x & 1 & 1 \\ 4 & 3 & 2 \\ -1 & 2 & 5 \end{vmatrix} = 2$

46. $\begin{vmatrix} 3 & 2 & 4 \\ 1 & x & 5 \\ 0 & 1 & -2 \end{vmatrix} = 0$

47. $\begin{vmatrix} x & 2 & 3 \\ 1 & x & 0 \\ 6 & 1 & -2 \end{vmatrix} = 7$

48. $\begin{vmatrix} x & 1 & 2 \\ 1 & x & 3 \\ 0 & 1 & 2 \end{vmatrix} = -4x$

In Problems 49–56, use properties of determinants to find the value of each determinant if it is known that

$$\begin{vmatrix} x & y & z \\ u & v & w \\ 1 & 2 & 3 \end{vmatrix} = 4$$

49. $\begin{vmatrix} 1 & 2 & 3 \\ u & v & w \\ x & y & z \end{vmatrix}$

50. $\begin{vmatrix} x & y & z \\ u & v & w \\ 2 & 4 & 6 \end{vmatrix}$

51. $\begin{vmatrix} x & y & z \\ -3 & -6 & -9 \\ u & v & w \end{vmatrix}$

52. $\begin{vmatrix} 1 & 2 & 3 \\ x - u & y - v & z - w \\ u & v & w \end{vmatrix}$

53. $\begin{vmatrix} 1 & 2 & 3 \\ x-3 & y-6 & z-9 \\ 2u & 2v & 2w \end{vmatrix}$ **54.** $\begin{vmatrix} x & y & z-x \\ u & v & w-u \\ 1 & 2 & 2 \end{vmatrix}$ **55.** $\begin{vmatrix} 1 & 2 & 3 \\ 2x & 2y & 2z \\ u-1 & v-2 & w-3 \end{vmatrix}$ **56.** $\begin{vmatrix} x+3 & y+6 & z+9 \\ 3u-1 & 3v-2 & 3w-3 \\ 1 & 2 & 3 \end{vmatrix}$

Applications and Extensions

57. Geometry: Equation of a Line An equation of the line containing the two points (x_1, y_1) and (x_2, y_2) may be expressed as the determinant

$$\begin{vmatrix} x & y & 1 \\ x_1 & y_1 & 1 \\ x_2 & y_2 & 1 \end{vmatrix} = 0$$

Prove this result by expanding the determinant and comparing the result to the 2-point form of the equation of a line.

58. Geometry: Collinear Points Using the result obtained in Problem 57, show that three distinct points (x_1, y_1), (x_2, y_2), and (x_3, y_3) are collinear (lie on the same line) if and only if

$$\begin{vmatrix} x_1 & y_1 & 1 \\ x_2 & y_2 & 1 \\ x_3 & y_3 & 1 \end{vmatrix} = 0$$

59. Show that $\begin{vmatrix} x^2 & x & 1 \\ y^2 & y & 1 \\ z^2 & z & 1 \end{vmatrix} = (y-z)(x-y)(x-z)$.

60. Complete the proof of Cramer's Rule for two equations containing two variables.

[**Hint:** In system (5), page 432, if $a = 0$, then $b \neq 0$ and $c \neq 0$, since $D = -bc \neq 0$. Now show that equation (6) provides a solution of the system when $a = 0$. There are then three remaining cases: $b = 0$, $c = 0$, and $d = 0$.]

61. Interchange columns 1 and 3 of a 3 by 3 determinant. Show that the value of the new determinant is -1 times the value of the original determinant.

62. Multiply each entry in row 2 of a 3 by 3 determinant by the number k, $k \neq 0$. Show that the value of the new determinant is k times the value of the original determinant.

63. Prove that a 3 by 3 determinant in which the entries in column 1 equal those in column 3 has the value 0.

64. Prove that, if row 2 of a 3 by 3 determinant is multiplied by k, $k \neq 0$, and the result is added to the entries in row 1, there is no change in the value of the determinant.

6.4 Matrix Algebra

OBJECTIVES 1 Find the Sum and Difference of Two Matrices
2 Find Scalar Multiples of a Matrix
3 Find the Product of Two Matrices
4 Find the Inverse of a Matrix
5 Solve a System of Linear Equations Using Inverse Matrices

In Section 6.2, we defined a matrix as a rectangular array of real numbers and used an augmented matrix to represent a system of linear equations. There is, however, a branch of mathematics, called **linear algebra**, that deals with matrices in such a way that an algebra of matrices is permitted. In this section, we provide a survey of how this **matrix algebra** is developed.

Before getting started, we restate the definition of a matrix.

A **matrix** is defined as a rectangular array of numbers:

$$\begin{array}{c} \\ \text{Row 1} \\ \text{Row 2} \\ \vdots \\ \text{Row } i \\ \vdots \\ \text{Row } m \end{array} \begin{bmatrix} \overset{\text{Column 1}}{a_{11}} & \overset{\text{Column 2}}{a_{12}} & \cdots & \overset{\text{Column } j}{a_{1j}} & \cdots & \overset{\text{Column } n}{a_{1n}} \\ a_{21} & a_{22} & \cdots & a_{2j} & \cdots & a_{2n} \\ \vdots & \vdots & & \vdots & & \vdots \\ a_{i1} & a_{i2} & \cdots & a_{ij} & \cdots & a_{in} \\ \vdots & \vdots & & \vdots & & \vdots \\ a_{m1} & a_{m2} & \cdots & a_{mj} & \cdots & a_{mn} \end{bmatrix}$$

Each number a_{ij} of the matrix has two indexes: the **row index** i and the **column index** j. The matrix shown above has m rows and n columns. The numbers a_{ij} are usually referred to as the **entries** of the matrix. For example, a_{23} refers to the entry in the second row, third column.

Let's begin with an example that illustrates how matrices can be used to conveniently represent an array of information.

EXAMPLE 1 **Arranging Data in a Matrix**

In a survey of 900 people, the following information was obtained:

200 males	Thought federal defense spending was too high
150 males	Thought federal defense spending was too low
45 males	Had no opinion
315 females	Thought federal defense spending was too high
125 females	Thought federal defense spending was too low
65 females	Had no opinion

We can arrange these data in a rectangular array as follows:

	Too High	Too Low	No Opinion
Male	200	150	45
Female	315	125	65

or as the matrix

$$\begin{bmatrix} 200 & 150 & 45 \\ 315 & 125 & 65 \end{bmatrix}$$

This matrix has two rows (representing male and female) and three columns (representing "too high," "too low," and "no opinion"). ∎

The matrix we developed in Example 1 has 2 rows and 3 columns. In general, a matrix with m rows and n columns is called an ***m* by *n* matrix**. The matrix we developed in Example 1 is a 2 by 3 matrix and contains $2 \cdot 3 = 6$ entries. An m by n matrix will contain $m \cdot n$ entries.

If an m by n matrix has the same number of rows as columns, that is, if $m = n$, then the matrix is referred to as a **square matrix**.

EXAMPLE 2 **Examples of Matrices**

(a) $\begin{bmatrix} 5 & 0 \\ -6 & 1 \end{bmatrix}$ A 2 by 2 square matrix (b) $\begin{bmatrix} 1 & 0 & 3 \end{bmatrix}$ A 1 by 3 matrix

(c) $\begin{bmatrix} 6 & -2 & 4 \\ 4 & 3 & 5 \\ 8 & 0 & 1 \end{bmatrix}$ A 3 by 3 square matrix ∎

1 Find the Sum and Difference of Two Matrices

We begin our discussion of matrix algebra by first defining what is meant by two matrices being equal and then defining the operations of addition and subtraction. It is important to note that these definitions require each matrix to have the same number of rows *and* the same number of columns as a prerequisite for equality and for addition and subtraction.

We usually represent matrices by capital letters, such as A, B, and C.

Two m by n matrices A and B are said to be **equal**, written as

$$A = B$$

provided that each entry a_{ij} in A is equal to the corresponding entry b_{ij} in B.

For example,

$$\begin{bmatrix} 2 & 1 \\ 0.5 & -1 \end{bmatrix} = \begin{bmatrix} \sqrt{4} & 1 \\ \frac{1}{2} & -1 \end{bmatrix} \quad \text{and} \quad \begin{bmatrix} 3 & 2 & 1 \\ 0 & 1 & -2 \end{bmatrix} = \begin{bmatrix} \sqrt{9} & \sqrt{4} & 1 \\ 0 & 1 & \sqrt[3]{-8} \end{bmatrix}$$

$$\begin{bmatrix} 4 & 1 \\ 6 & 1 \end{bmatrix} \neq \begin{bmatrix} 4 & 0 \\ 6 & 1 \end{bmatrix} \qquad \text{Because the entries in row 1, column 2 are not equal}$$

$$\begin{bmatrix} 4 & 1 & 2 \\ 6 & 1 & 2 \end{bmatrix} \neq \begin{bmatrix} 4 & 1 & 2 & 3 \\ 6 & 1 & 2 & 4 \end{bmatrix} \qquad \text{Because the matrix on the left is 2 by 3 and the matrix on the right is 2 by 4}$$

Suppose that A and B represent two m by n matrices. We define their **sum** $A + B$ to be the m by n matrix formed by adding the corresponding entries a_{ij} of A and b_{ij} of B. The **difference** $A - B$ is defined as the m by n matrix formed by subtracting the entries b_{ij} in B from the corresponding entries a_{ij} in A. Addition and subtraction of matrices are allowed only for matrices having the same number m of rows and the same number n of columns. For example, a 2 by 3 matrix and a 2 by 4 matrix cannot be added or subtracted.

EXAMPLE 3

Adding and Subtracting Matrices

Suppose that

$$A = \begin{bmatrix} 2 & 4 & 8 & -3 \\ 0 & 1 & 2 & 3 \end{bmatrix} \quad \text{and} \quad B = \begin{bmatrix} -3 & 4 & 0 & 1 \\ 6 & 8 & 2 & 0 \end{bmatrix}$$

Find: (a) $A + B$ (b) $A - B$

Solution (a) $A + B = \begin{bmatrix} 2 & 4 & 8 & -3 \\ 0 & 1 & 2 & 3 \end{bmatrix} + \begin{bmatrix} -3 & 4 & 0 & 1 \\ 6 & 8 & 2 & 0 \end{bmatrix}$

$$= \begin{bmatrix} 2 + (-3) & 4 + 4 & 8 + 0 & -3 + 1 \\ 0 + 6 & 1 + 8 & 2 + 2 & 3 + 0 \end{bmatrix} \qquad \text{Add corresponding entries.}$$

$$= \begin{bmatrix} -1 & 8 & 8 & -2 \\ 6 & 9 & 4 & 3 \end{bmatrix}$$

(b) $A - B = \begin{bmatrix} 2 & 4 & 8 & -3 \\ 0 & 1 & 2 & 3 \end{bmatrix} - \begin{bmatrix} -3 & 4 & 0 & 1 \\ 6 & 8 & 2 & 0 \end{bmatrix}$

$$= \begin{bmatrix} 2 - (-3) & 4 - 4 & 8 - 0 & -3 - 1 \\ 0 - 6 & 1 - 8 & 2 - 2 & 3 - 0 \end{bmatrix} \qquad \text{Subtract corresponding entries.}$$

$$= \begin{bmatrix} 5 & 0 & 8 & -4 \\ -6 & -7 & 0 & 3 \end{bmatrix}$$

■

Figure 7

—— Seeing the Concept ——

Graphing utilities can make the sometimes tedious process of matrix algebra easy. In fact, most graphing calculators can handle matrices as large as 9 by 9, some even larger ones. Enter the matrices into a graphing utility. Name them [A] and [B]. Figure 7 shows the results of adding and subtracting [A] and [B].

NOW WORK PROBLEM **7.**

Many of the algebraic properties of sums of real numbers are also true for sums of matrices. Suppose that A, B, and C are m by n matrices. Then matrix addition is **commutative**. That is,

Commutative Property

$$A + B = B + A$$

Matrix addition is also **associative**. That is,

Associative Property

$$(A + B) + C = A + (B + C)$$

Although we shall not prove these results, the proofs, as the following example illustrates, are based on the commutative and associative properties for real numbers.

EXAMPLE 4 | **Demonstrating the Commutative Property**

$$\begin{bmatrix} 2 & 3 & -1 \\ 4 & 0 & 7 \end{bmatrix} + \begin{bmatrix} -1 & 2 & 1 \\ 5 & -3 & 4 \end{bmatrix} = \begin{bmatrix} 2 + (-1) & 3 + 2 & -1 + 1 \\ 4 + 5 & 0 + (-3) & 7 + 4 \end{bmatrix}$$

$$= \begin{bmatrix} -1 + 2 & 2 + 3 & 1 + (-1) \\ 5 + 4 & -3 + 0 & 4 + 7 \end{bmatrix}$$

$$= \begin{bmatrix} -1 & 2 & 1 \\ 5 & -3 & 4 \end{bmatrix} + \begin{bmatrix} 2 & 3 & -1 \\ 4 & 0 & 7 \end{bmatrix} \quad \blacksquare$$

A matrix whose entries are all equal to 0 is called a **zero matrix**. Each of the following matrices is a zero matrix.

$$\begin{bmatrix} 0 & 0 \\ 0 & 0 \end{bmatrix} \quad \text{2 by 2 square zero matrix} \qquad \begin{bmatrix} 0 & 0 & 0 \\ 0 & 0 & 0 \end{bmatrix} \quad \text{2 by 3 zero matrix} \qquad \begin{bmatrix} 0 & 0 & 0 \end{bmatrix} \quad \text{1 by 3 zero matrix}$$

Zero matrices have properties similar to the real number 0. If A is an m by n matrix and 0 is an m by n zero matrix,

$$A + 0 = A$$

In other words, the zero matrix is the additive identity in matrix algebra.

2 Find Scalar Multiples of a Matrix

We can also multiply a matrix by a real number. If k is a real number and A is an m by n matrix, the matrix kA is the m by n matrix formed by multiplying each entry a_{ij} in A by k. The number k is sometimes referred to as a **scalar**, and the matrix kA is called a **scalar multiple** of A.

EXAMPLE 5 | **Operations Using Matrices**

Suppose that

$$A = \begin{bmatrix} 3 & 1 & 5 \\ -2 & 0 & 6 \end{bmatrix} \qquad B = \begin{bmatrix} 4 & 1 & 0 \\ 8 & 1 & -3 \end{bmatrix} \qquad C = \begin{bmatrix} 9 & 0 \\ -3 & 6 \end{bmatrix}$$

Find: (a) $4A$ (b) $\dfrac{1}{3}C$ (c) $3A - 2B$

Solution (a) $4A = 4\begin{bmatrix} 3 & 1 & 5 \\ -2 & 0 & 6 \end{bmatrix} = \begin{bmatrix} 4 \cdot 3 & 4 \cdot 1 & 4 \cdot 5 \\ 4(-2) & 4 \cdot 0 & 4 \cdot 6 \end{bmatrix} = \begin{bmatrix} 12 & 4 & 20 \\ -8 & 0 & 24 \end{bmatrix}$

(b) $\dfrac{1}{3}C = \dfrac{1}{3}\begin{bmatrix} 9 & 0 \\ -3 & 6 \end{bmatrix} = \begin{bmatrix} \dfrac{1}{3} \cdot 9 & \dfrac{1}{3} \cdot 0 \\ \dfrac{1}{3}(-3) & \dfrac{1}{3} \cdot 6 \end{bmatrix} = \begin{bmatrix} 3 & 0 \\ -1 & 2 \end{bmatrix}$

(c) $3A - 2B = 3\begin{bmatrix} 3 & 1 & 5 \\ -2 & 0 & 6 \end{bmatrix} - 2\begin{bmatrix} 4 & 1 & 0 \\ 8 & 1 & -3 \end{bmatrix}$

$= \begin{bmatrix} 3 \cdot 3 & 3 \cdot 1 & 3 \cdot 5 \\ 3(-2) & 3 \cdot 0 & 3 \cdot 6 \end{bmatrix} - \begin{bmatrix} 2 \cdot 4 & 2 \cdot 1 & 2 \cdot 0 \\ 2 \cdot 8 & 2 \cdot 1 & 2(-3) \end{bmatrix}$

$= \begin{bmatrix} 9 & 3 & 15 \\ -6 & 0 & 18 \end{bmatrix} - \begin{bmatrix} 8 & 2 & 0 \\ 16 & 2 & -6 \end{bmatrix}$

$= \begin{bmatrix} 9 - 8 & 3 - 2 & 15 - 0 \\ -6 - 16 & 0 - 2 & 18 - (-6) \end{bmatrix}$

$= \begin{bmatrix} 1 & 1 & 15 \\ -22 & -2 & 24 \end{bmatrix}$ ∎

✔ CHECK: Enter the matrices [A], [B], and [C] into a graphing utility. Then find $4A$, $\dfrac{1}{3}C$, and $3A - 2B$. ∎

━━━ NOW WORK PROBLEM **11**.

We list next some of the algebraic properties of scalar multiplication. Let h and k be real numbers, and let A and B be m by n matrices. Then

Properties of Scalar Multiplication

$$k(hA) = (kh)A$$
$$(k + h)A = kA + hA$$
$$k(A + B) = kA + kB$$

3 Find the Product of Two Matrices

Unlike the straightforward definition for adding two matrices, the definition for multiplying two matrices is not what we might expect. In preparation for this definition, we need the following definitions:

A **row vector** R is a 1 by n matrix

$$R = \begin{bmatrix} r_1 & r_2 & \cdots & r_n \end{bmatrix}$$

A **column vector** C is an n by 1 matrix

$$C = \begin{bmatrix} c_1 \\ c_2 \\ \vdots \\ c_n \end{bmatrix}$$

The **product** *RC* of *R* times *C* is defined as the number

$$RC = [r_1 \quad r_2 \cdots r_n] \begin{bmatrix} c_1 \\ c_2 \\ \vdots \\ c_n \end{bmatrix} = r_1c_1 + r_2c_2 + \cdots + r_nc_n$$

Notice that a row vector and a column vector can be multiplied only if they contain the same number of entries.

EXAMPLE 6

The Product of a Row Vector by a Column Vector

If $R = [3 \quad -5 \quad 2]$ and $C = \begin{bmatrix} 3 \\ 4 \\ -5 \end{bmatrix}$, then

$$RC = [3 \quad -5 \quad 2] \begin{bmatrix} 3 \\ 4 \\ -5 \end{bmatrix} = 3 \cdot 3 + (-5)4 + 2(-5) = 9 - 20 - 10 = -21 \quad \blacksquare$$

Let's look at an application of the product of a row vector by a column vector.

EXAMPLE 7

Using Matrices to Compute Revenue

A clothing store sells men's shirts for $40, silk ties for $20, and wool suits for $400. Last month, the store had sales consisting of 100 shirts, 200 ties, and 50 suits. What was the total revenue due to these sales?

Solution We set up a row vector *R* to represent the prices of each item and a column vector *C* to represent the corresponding number of items sold. Then

$$\begin{array}{cc} \overset{\text{Prices}}{\underset{\text{Shirts Ties Suits}}{}} & \overset{\text{Number}}{\underset{\text{sold}}{}} \\ R = [40 \quad 20 \quad 400], & C = \begin{bmatrix} 100 \\ 200 \\ 50 \end{bmatrix} \begin{array}{l} \text{Shirts} \\ \text{Ties} \\ \text{Suits} \end{array} \end{array}$$

The total revenue obtained is the product *RC*. That is,

$$RC = [40 \ 20 \ 400] \begin{bmatrix} 100 \\ 200 \\ 50 \end{bmatrix}$$

$$= \underbrace{40 \cdot 100}_{\text{Shirt revenue}} + \underbrace{20 \cdot 200}_{\text{Tie revenue}} + \underbrace{400 \cdot 50}_{\text{Suit revenue}} = \underbrace{\$28,000}_{\text{Total revenue}} \quad \blacksquare$$

The definition for multiplying two matrices is based on the definition of a row vector times a column vector.

Let *A* denote an *m* by *r* matrix, and let *B* denote an *r* by *n* matrix. The **product** *AB* is defined as the *m* by *n* matrix whose entry in row *i*, column *j* is the product of the *i*th row of *A* and the *j*th column of *B*.

The definition of the product AB of two matrices A and B, in this order, requires that the number of columns of A equal the number of rows of B; otherwise, no product is defined.

In Words
To find the product AB, the number of columns in A must equal the number of rows in B.

An example will help to clarify the definition.

EXAMPLE 8 **Multiplying Two Matrices**

Find the product AB if

$$A = \begin{bmatrix} 2 & 4 & -1 \\ 5 & 8 & 0 \end{bmatrix} \quad \text{and} \quad B = \begin{bmatrix} 2 & 5 & 1 & 4 \\ 4 & 8 & 0 & 6 \\ -3 & 1 & -2 & -1 \end{bmatrix}$$

Solution First, we note that A is 2 by 3 and B is 3 by 4, so the product AB is defined and will be a 2 by 4 matrix.

Suppose that we want the entry in row 2, column 3 of AB. To find it, we find the product of the row vector from row 2 of A and the column vector from column 3 of B.

$$\underset{\text{Row 2 of } A}{[5 \quad 8 \quad 0]} \underset{\text{Column 3 of } B}{\begin{bmatrix} 1 \\ 0 \\ -2 \end{bmatrix}} = 5 \cdot 1 + 8 \cdot 0 + 0(-2) = 5$$

So far, we have

$$AB = \begin{bmatrix} \underline{\quad} & \underline{\quad} & \overset{\overset{\text{Column 3}}{\downarrow}}{5} & \underline{\quad} \end{bmatrix} \quad \leftarrow \text{Row 2}$$

Now, to find the entry in row 1, column 4 of AB, we find the product of row 1 of A and column 4 of B.

$$\underset{\text{Row 1 of } A}{[2 \quad 4 \quad -1]} \underset{\text{Column 4 of } B}{\begin{bmatrix} 4 \\ 6 \\ -1 \end{bmatrix}} = 2 \cdot 4 + 4 \cdot 6 + (-1)(-1) = 33$$

Continuing in this fashion, we find AB.

$$AB = \begin{bmatrix} 2 & 4 & -1 \\ 5 & 8 & 0 \end{bmatrix} \begin{bmatrix} 2 & 5 & 1 & 4 \\ 4 & 8 & 0 & 6 \\ -3 & 1 & -2 & -1 \end{bmatrix}$$

$$= \begin{bmatrix} \begin{array}{l}\text{Row 1 of } A \\ \text{times} \\ \text{column 1 of } B\end{array} & \begin{array}{l}\text{Row 1 of } A \\ \text{times} \\ \text{column 2 of } B\end{array} & \begin{array}{l}\text{Row 1 of } A \\ \text{times} \\ \text{column 3 of } B\end{array} & \begin{array}{l}\text{Row 1 of } A \\ \text{times} \\ \text{column 4 of } B\end{array} \\ \begin{array}{l}\text{Row 2 of } A \\ \text{times} \\ \text{column 1 of } B\end{array} & \begin{array}{l}\text{Row 2 of } A \\ \text{times} \\ \text{column 2 of } B\end{array} & \begin{array}{l}\text{Row 2 of } A \\ \text{times} \\ \text{column 3 of } B\end{array} & \begin{array}{l}\text{Row 2 of } A \\ \text{times} \\ \text{column 4 of } B\end{array} \end{bmatrix}$$

$$= \begin{bmatrix} 2 \cdot 2 + 4 \cdot 4 + (-1)(-3) & 2 \cdot 5 + 4 \cdot 8 + (-1)1 & 2 \cdot 1 + 4 \cdot 0 + (-1)(-2) & 33 \text{ (from earlier)} \\ 5 \cdot 2 + 8 \cdot 4 + 0(-3) & 5 \cdot 5 + 8 \cdot 8 + 0 \cdot 1 & 5 \text{ (from earlier)} & 5 \cdot 4 + 8 \cdot 6 + 0(-1) \end{bmatrix}$$

$$= \begin{bmatrix} 23 & 41 & 4 & 33 \\ 42 & 89 & 5 & 68 \end{bmatrix}$$ ∎

NOW WORK PROBLEM 23.

Notice that for the matrices given in Example 8 the product BA is not defined, because B is 3 by 4 and A is 2 by 3.

 ✔ CHECK: Enter the matrices [A] and [B]. Then find AB. (See what happens if you try to find BA.) ∎

Another result that can occur when multiplying two matrices is illustrated in the next example.

EXAMPLE 9 **Multiplying Two Matrices**

If

$$A = \begin{bmatrix} 2 & 1 & 3 \\ 1 & -1 & 0 \end{bmatrix} \quad \text{and} \quad B = \begin{bmatrix} 1 & 0 \\ 2 & 1 \\ 3 & 2 \end{bmatrix}$$

find: (a) AB (b) BA

Solution (a) $AB = \begin{bmatrix} 2 & 1 & 3 \\ 1 & -1 & 0 \end{bmatrix} \begin{bmatrix} 1 & 0 \\ 2 & 1 \\ 3 & 2 \end{bmatrix} = \begin{bmatrix} 13 & 7 \\ -1 & -1 \end{bmatrix}$

 2 by 3 3 by 2 2 by 2

(b) $BA = \begin{bmatrix} 1 & 0 \\ 2 & 1 \\ 3 & 2 \end{bmatrix} \begin{bmatrix} 2 & 1 & 3 \\ 1 & -1 & 0 \end{bmatrix} = \begin{bmatrix} 2 & 1 & 3 \\ 5 & 1 & 6 \\ 8 & 1 & 9 \end{bmatrix}$

 3 by 2 2 by 3 3 by 3 ∎

Notice in Example 9 that AB is 2 by 2 and BA is 3 by 3. It is possible for both AB and BA to be defined, yet be unequal. In fact, even if A and B are both n by n matrices so that AB and BA are each defined and n by n, AB and BA will usually be unequal.

EXAMPLE 10 **Multiplying Two Square Matrices**

If

$$A = \begin{bmatrix} 2 & 1 \\ 0 & 4 \end{bmatrix} \quad \text{and} \quad B = \begin{bmatrix} -3 & 1 \\ 1 & 2 \end{bmatrix}$$

find: (a) AB (b) BA

Solution (a) $AB = \begin{bmatrix} 2 & 1 \\ 0 & 4 \end{bmatrix} \begin{bmatrix} -3 & 1 \\ 1 & 2 \end{bmatrix} = \begin{bmatrix} -5 & 4 \\ 4 & 8 \end{bmatrix}$

(b) $BA = \begin{bmatrix} -3 & 1 \\ 1 & 2 \end{bmatrix} \begin{bmatrix} 2 & 1 \\ 0 & 4 \end{bmatrix} = \begin{bmatrix} -6 & 1 \\ 2 & 9 \end{bmatrix}$ ∎

The preceding examples demonstrate that an important property of real numbers, the commutative property of multiplication, is not shared by matrices. In general:

Theorem Matrix multiplication is not commutative.

NOW WORK PROBLEMS 13 AND 15.

Next we give two of the properties of real numbers that are shared by matrices. Assuming that each product and sum is defined, we have the following:

Associative Property

$$A(BC) = (AB)C$$

Distributive Property

$$A(B + C) = AB + AC$$

For an n by n square matrix, the entries located in row i, column i, $1 \le i \le n$, are called the **diagonal entries** or **the main diagonal**. An n by n square matrix whose diagonal entries are 1's, while all other entries are 0's, is called the **identity matrix I_n**. For example,

$$I_2 = \begin{bmatrix} 1 & 0 \\ 0 & 1 \end{bmatrix}, \qquad I_3 = \begin{bmatrix} 1 & 0 & 0 \\ 0 & 1 & 0 \\ 0 & 0 & 1 \end{bmatrix}$$

and so on.

EXAMPLE 11

Multiplication with an Identity Matrix

Let

$$A = \begin{bmatrix} -1 & 2 & 0 \\ 0 & 1 & 3 \end{bmatrix} \quad \text{and} \quad B = \begin{bmatrix} 3 & 2 \\ 4 & 6 \\ 5 & 2 \end{bmatrix}$$

Find: (a) AI_3 (b) I_2A (c) BI_2

Solution

(a) $AI_3 = \begin{bmatrix} -1 & 2 & 0 \\ 0 & 1 & 3 \end{bmatrix} \begin{bmatrix} 1 & 0 & 0 \\ 0 & 1 & 0 \\ 0 & 0 & 1 \end{bmatrix} = \begin{bmatrix} -1 & 2 & 0 \\ 0 & 1 & 3 \end{bmatrix} = A$

(b) $I_2A = \begin{bmatrix} 1 & 0 \\ 0 & 1 \end{bmatrix} \begin{bmatrix} -1 & 2 & 0 \\ 0 & 1 & 3 \end{bmatrix} = \begin{bmatrix} -1 & 2 & 0 \\ 0 & 1 & 3 \end{bmatrix} = A$

(c) $BI_2 = \begin{bmatrix} 3 & 2 \\ 4 & 6 \\ 5 & 2 \end{bmatrix} \begin{bmatrix} 1 & 0 \\ 0 & 1 \end{bmatrix} = \begin{bmatrix} 3 & 2 \\ 4 & 6 \\ 5 & 2 \end{bmatrix} = B$ ∎

Example 11 demonstrates the following property:

Identity Property

If A is an m by n matrix, then

$$I_mA = A \quad \text{and} \quad AI_n = A$$

If A is an n by n square matrix, $AI_n = I_nA = A$.

An identity matrix has properties similar to those of the real number 1. In other words, the identity matrix is a multiplicative identity in matrix algebra.

4 Find the Inverse of a Matrix

Let A be a square n by n matrix. If there exists an n by n matrix A^{-1}, read "A inverse," for which

$$AA^{-1} = A^{-1}A = I_n$$

then A^{-1} is called the **inverse** of the matrix A.

As we shall soon see, not every square matrix has an inverse. When a matrix A does have an inverse A^{-1}, then A is said to be **nonsingular**. If a matrix A has no inverse, it is called **singular**.[*]

EXAMPLE 12 **Multiplying a Matrix by Its Inverse**

Show that the inverse of

$$A = \begin{bmatrix} 3 & 1 \\ 2 & 1 \end{bmatrix} \quad \text{is} \quad A^{-1} = \begin{bmatrix} 1 & -1 \\ -2 & 3 \end{bmatrix}$$

Solution We need to show that $AA^{-1} = A^{-1}A = I_2$.

$$AA^{-1} = \begin{bmatrix} 3 & 1 \\ 2 & 1 \end{bmatrix}\begin{bmatrix} 1 & -1 \\ -2 & 3 \end{bmatrix} = \begin{bmatrix} 1 & 0 \\ 0 & 1 \end{bmatrix} = I_2$$

$$A^{-1}A = \begin{bmatrix} 1 & -1 \\ -2 & 3 \end{bmatrix}\begin{bmatrix} 3 & 1 \\ 2 & 1 \end{bmatrix} = \begin{bmatrix} 1 & 0 \\ 0 & 1 \end{bmatrix} = I_2$$

∎

We now show one way to find the inverse of

$$A = \begin{bmatrix} 3 & 1 \\ 2 & 1 \end{bmatrix}$$

Suppose that A^{-1} is given by

$$A^{-1} = \begin{bmatrix} x & y \\ z & w \end{bmatrix} \tag{1}$$

where x, y, z, and w are four variables. Based on the definition of an inverse, if A has an inverse,

$$AA^{-1} = I_2$$

$$\begin{bmatrix} 3 & 1 \\ 2 & 1 \end{bmatrix}\begin{bmatrix} x & y \\ z & w \end{bmatrix} = \begin{bmatrix} 1 & 0 \\ 0 & 1 \end{bmatrix}$$

$$\begin{bmatrix} 3x + z & 3y + w \\ 2x + z & 2y + w \end{bmatrix} = \begin{bmatrix} 1 & 0 \\ 0 & 1 \end{bmatrix}$$

Because corresponding entries must be equal, it follows that this matrix equation is equivalent to two systems of linear equations.

$$\begin{cases} 3x + z = 1 \\ 2x + z = 0 \end{cases} \qquad \begin{cases} 3y + w = 0 \\ 2y + w = 1 \end{cases}$$

The augmented matrix of each system is

$$\begin{bmatrix} 3 & 1 & | & 1 \\ 2 & 1 & | & 0 \end{bmatrix} \qquad \begin{bmatrix} 3 & 1 & | & 0 \\ 2 & 1 & | & 1 \end{bmatrix} \tag{2}$$

[*]If the determinant of A is zero, A is singular. (Refer to Section 6.3.)

The usual procedure would be to transform each augmented matrix into reduced row echelon form. Notice, though, that the left sides of the augmented matrices are equal, so the same row operations (see Section 6.2) can be used to reduce each one. We find it more efficient to combine the two augmented matrices (2) into a single matrix, as shown next, and then transform it into reduced row echelon form.

$$\left[\begin{array}{cc|cc} 3 & 1 & 1 & 0 \\ 2 & 1 & 0 & 1 \end{array}\right]$$

We attempt to transform the left side into an identity matrix.

$$\left[\begin{array}{cc|cc} 3 & 1 & 1 & 0 \\ 2 & 1 & 0 & 1 \end{array}\right] \xrightarrow[R_1 = -1r_2 + r_1]{} \left[\begin{array}{cc|cc} 1 & 0 & 1 & -1 \\ 2 & 1 & 0 & 1 \end{array}\right]$$

$$\xrightarrow[R_2 = -2r_1 + r_2]{} \left[\begin{array}{cc|cc} 1 & 0 & 1 & -1 \\ 0 & 1 & -2 & 3 \end{array}\right] \qquad (3)$$

Matrix (3) is in reduced row echelon form.

Now we reverse the earlier step of combining the two augmented matrices in (2) and write the single matrix (3) as two augmented matrices.

$$\left[\begin{array}{cc|c} 1 & 0 & 1 \\ 0 & 1 & -2 \end{array}\right] \text{ and } \left[\begin{array}{cc|c} 1 & 0 & -1 \\ 0 & 1 & 3 \end{array}\right]$$

We conclude from these matrices that $x = 1, z = -2$, and $y = -1, w = 3$. Substituting these values into matrix (1), we find that

$$A^{-1} = \left[\begin{array}{cc} 1 & -1 \\ -2 & 3 \end{array}\right]$$

Notice in display (3) that the 2 by 2 matrix to the right of the vertical bar is, in fact, the inverse of A. Also notice that the identity matrix I_2 is the matrix that appears to the left of the vertical bar. These observations and the procedures followed above will work in general.

> **Procedure for Finding the Inverse of a Nonsingular Matrix**
>
> To find the inverse of an n by n nonsingular matrix A, proceed as follows:
>
> **STEP 1:** Form the matrix $[A|I_n]$.
> **STEP 2:** Transform the matrix $[A|I_n]$ into reduced row echelon form.
> **STEP 3:** The reduced row echelon form of $[A|I_n]$ will contain the identity matrix I_n on the left of the vertical bar; the n by n matrix on the right of the vertical bar is the inverse of A.

In other words, if A is nonsingular, we begin with the matrix $[A|I_n]$ and, after transforming it into reduced row echelon form, we end up with the matrix $[I_n|A^{-1}]$.

Let's look at another example.

EXAMPLE 13 **Finding the Inverse of a Matrix**

The matrix

$$A = \left[\begin{array}{ccc} 1 & 1 & 0 \\ -1 & 3 & 4 \\ 0 & 4 & 3 \end{array}\right]$$

is nonsingular. Find its inverse.

Solution First, we form the matrix

$$[A\,|\,I_3] = \begin{bmatrix} 1 & 1 & 0 & 1 & 0 & 0 \\ -1 & 3 & 4 & 0 & 1 & 0 \\ 0 & 4 & 3 & 0 & 0 & 1 \end{bmatrix}$$

Next, we use row operations to transform $[A\,|\,I_3]$ into reduced row echelon form.

$$\begin{bmatrix} 1 & 1 & 0 & 1 & 0 & 0 \\ -1 & 3 & 4 & 0 & 1 & 0 \\ 0 & 4 & 3 & 0 & 0 & 1 \end{bmatrix} \rightarrow \begin{bmatrix} 1 & 1 & 0 & 1 & 0 & 0 \\ 0 & 4 & 4 & 1 & 1 & 0 \\ 0 & 4 & 3 & 0 & 0 & 1 \end{bmatrix} \rightarrow \begin{bmatrix} 1 & 1 & 0 & 1 & 0 & 0 \\ 0 & 1 & 1 & \dfrac{1}{4} & \dfrac{1}{4} & 0 \\ 0 & 4 & 3 & 0 & 0 & 1 \end{bmatrix}$$

$$\uparrow \qquad\qquad\qquad\qquad\qquad\qquad \uparrow$$
$$R_2 = r_1 + r_2 \qquad\qquad\qquad\qquad\qquad R_2 = \frac{1}{4}r_2$$

$$\rightarrow \begin{bmatrix} 1 & 0 & -1 & \dfrac{3}{4} & -\dfrac{1}{4} & 0 \\ 0 & 1 & 1 & \dfrac{1}{4} & \dfrac{1}{4} & 0 \\ 0 & 0 & -1 & -1 & -1 & 1 \end{bmatrix} \rightarrow \begin{bmatrix} 1 & 0 & -1 & \dfrac{3}{4} & -\dfrac{1}{4} & 0 \\ 0 & 1 & 1 & \dfrac{1}{4} & \dfrac{1}{4} & 0 \\ 0 & 0 & 1 & 1 & 1 & -1 \end{bmatrix}$$

$$\uparrow \qquad\qquad\qquad\qquad\qquad\qquad \uparrow$$
$$R_1 = -1r_2 + r_1 \qquad\qquad\qquad\qquad R_3 = -1r_3$$
$$R_3 = -4r_2 + r_3$$

$$\rightarrow \begin{bmatrix} 1 & 0 & 0 & \dfrac{7}{4} & \dfrac{3}{4} & -1 \\ 0 & 1 & 0 & -\dfrac{3}{4} & -\dfrac{3}{4} & 1 \\ 0 & 0 & 1 & 1 & 1 & -1 \end{bmatrix}$$

$$\uparrow$$
$$R_1 = r_3 + r_1$$
$$R_2 = -1r_3 + r_2$$

The matrix $[A\,|\,I_3]$ is now in reduced row echelon form, and the identity matrix I_3 is on the left of the vertical bar. Hence, the inverse of A is

$$A^{-1} = \begin{bmatrix} \dfrac{7}{4} & \dfrac{3}{4} & -1 \\ -\dfrac{3}{4} & -\dfrac{3}{4} & 1 \\ 1 & 1 & -1 \end{bmatrix}$$

You can (and should) verify that this is the correct inverse by showing that

$$AA^{-1} = A^{-1}A = I_3. \qquad\blacksquare$$

Figure 8

✔ CHECK: Enter the matrix A into a graphing utility. Figure 8 shows A^{-1}.

NOW WORK PROBLEM 31.

If transforming the matrix $[A\,|\,I_n]$ into reduced row echelon form does not result in the identity matrix I_n to the left of the vertical bar, A is singular and has no inverse. The next example demonstrates such a matrix.

EXAMPLE 14 **Showing That a Matrix Has No Inverse**

Show that the matrix $A = \begin{bmatrix} 4 & 6 \\ 2 & 3 \end{bmatrix}$ has no inverse.

Solution Proceeding as in Example 13, we form the matrix

$$[A|I_2] = \begin{bmatrix} 4 & 6 & | & 1 & 0 \\ 2 & 3 & | & 0 & 1 \end{bmatrix}$$

Then we use row operations to transform $[A|I_2]$ into reduced row echelon form.

$$[A|I_2] = \begin{bmatrix} 4 & 6 & | & 1 & 0 \\ 2 & 3 & | & 0 & 1 \end{bmatrix} \rightarrow \begin{bmatrix} 1 & \frac{3}{2} & | & \frac{1}{4} & 0 \\ 2 & 3 & | & 0 & 1 \end{bmatrix} \rightarrow \begin{bmatrix} 1 & \frac{3}{2} & | & \frac{1}{4} & 0 \\ 0 & 0 & | & -\frac{1}{2} & 1 \end{bmatrix}$$

$$\uparrow \qquad\qquad\qquad \uparrow$$
$$R_1 = \frac{1}{4} r_1 \qquad\qquad R_2 = -2r_1 + r_2$$

The matrix $[A|I_2]$ is sufficiently reduced for us to see that the identity matrix cannot appear to the left of the vertical bar. We conclude that A is singular and so has no inverse. ∎

✔ CHECK: Enter the matrix A. Try to find its inverse. What happens? ∎

NOW WORK PROBLEM **59.**

5 Solve a System of Linear Equations Using Inverse Matrices

Inverse matrices can be used to solve systems of equations in which the number of equations is the same as the number of variables.

EXAMPLE 15 **Using the Inverse Matrix to Solve a System of Linear Equations**

Solve the system of equations: $\begin{cases} x + y & = 3 \\ -x + 3y + 4z & = -3 \\ 4y + 3z & = 2 \end{cases}$

Solution If we let

$$A = \begin{bmatrix} 1 & 1 & 0 \\ -1 & 3 & 4 \\ 0 & 4 & 3 \end{bmatrix}, \qquad X = \begin{bmatrix} x \\ y \\ z \end{bmatrix}, \qquad B = \begin{bmatrix} 3 \\ -3 \\ 2 \end{bmatrix}$$

the original system of equations can be written compactly as the matrix equation

$$AX = B \qquad\qquad (4)$$

We know from Example 13 that the matrix A has the inverse A^{-1}, so we multiply each side of equation (4) by A^{-1}.

$$AX = B$$
$$A^{-1}(AX) = A^{-1}B \qquad \text{Multiply both sides by } A^{-1}.$$
$$(A^{-1}A)X = A^{-1}B \qquad \text{Associative Property of multiplication}$$
$$I_3 X = A^{-1}B \qquad \text{Definition of an inverse matrix}$$
$$X = A^{-1}B \qquad \text{Property of the identity matrix} \qquad (5)$$

Now we use (5) to find $X = \begin{bmatrix} x \\ y \\ z \end{bmatrix}$.

Solution $X = \begin{bmatrix} x \\ y \\ z \end{bmatrix} = A^{-1}B = \begin{bmatrix} \frac{7}{4} & \frac{3}{4} & -1 \\ -\frac{3}{4} & -\frac{3}{4} & 1 \\ 1 & 1 & -1 \end{bmatrix} \begin{bmatrix} 3 \\ -3 \\ 2 \end{bmatrix} = \begin{bmatrix} 1 \\ 2 \\ -2 \end{bmatrix}$

\uparrow
Example 13

So $x = 1$, $y = 2$, $z = -2$. ∎

The method used in Example 15 to solve a system of equations is particularly useful when it is necessary to solve several systems of equations in which the constants appearing to the right of the equal signs change, while the coefficients of the variables on the left side remain the same. See Problems 39–58 for some illustrations. Be careful; this method can only be used if the inverse exists. If it does not exist, row reduction must be used since the system is either inconsistent or dependent.

✏️ **NOW WORK PROBLEM 43.**

HISTORICAL FEATURE

**Arthur Cayley
(1821–1895)**

Matrices were invented in 1857 by Arthur Cayley (1821–1895) as a way of efficiently computing the result of substituting one linear system into another (see Historical Problem 2). The resulting system had incredible richness, in the sense that a wide variety of mathematical systems could be mimicked by the matrices. Cayley and his friend James J. Sylvester (1814–1897) spent much of the rest of their lives elaborating the theory. The torch was then passed to Georg Frobenius (1849–1917), whose deep investigations established a central place for matrices in modern mathematics. In 1924, rather to the surprise of physicists, it was found that matrices (with complex numbers in them) were exactly the right tool for describing the behavior of atomic systems. Today, matrices are used in a wide variety of applications.

Historical Problems

1. **Matrices and Complex Numbers** Frobenius emphasized in his research how matrices could be used to mimic other mathematical systems. Here, we mimic the behavior of complex numbers using matrices. Mathematicians call such a relationship an *isomorphism*.

$$\text{Complex number} \longleftrightarrow \text{Matrix}$$

$$a + bi \longleftrightarrow \begin{bmatrix} a & b \\ -b & a \end{bmatrix}$$

Note that the complex number can be read off the top line of the matrix. Thus,

$$2 + 3i \longleftrightarrow \begin{bmatrix} 2 & 3 \\ -3 & 2 \end{bmatrix} \quad \text{and} \quad \begin{bmatrix} 4 & -2 \\ 2 & 4 \end{bmatrix} \longleftrightarrow 4 - 2i$$

 (a) Find the matrices corresponding to $2 - 5i$ and $1 + 3i$.
 (b) Multiply the two matrices.
 (c) Find the corresponding complex number for the matrix found in part (b).
 (d) Multiply $2 - 5i$ and $1 + 3i$. The result should be the same as that found in part (c).

The process also works for addition and subtraction. Try it for yourself.

2. **Cayley's Definition of Matrix Multiplication** Cayley invented matrix multiplication to simplify the following problem:

$$\begin{cases} u = ar + bs \\ v = cr + ds \end{cases} \qquad \begin{cases} x = ku + lv \\ y = mu + nv \end{cases}$$

 (a) Find x and y in terms of r and s by substituting u and v from the first system of equations into the second system of equations.
 (b) Use the result of part (a) to find the 2 by 2 matrix A in

$$\begin{bmatrix} x \\ y \end{bmatrix} = A \begin{bmatrix} r \\ s \end{bmatrix}$$

 (c) Now look at the following way to do it. Write the equations in matrix form.

$$\begin{bmatrix} u \\ v \end{bmatrix} = \begin{bmatrix} a & b \\ c & d \end{bmatrix} \begin{bmatrix} r \\ s \end{bmatrix} \qquad \begin{bmatrix} x \\ y \end{bmatrix} = \begin{bmatrix} k & l \\ m & n \end{bmatrix} \begin{bmatrix} u \\ v \end{bmatrix}$$

 So

$$\begin{bmatrix} x \\ y \end{bmatrix} = \begin{bmatrix} k & l \\ m & n \end{bmatrix} \begin{bmatrix} a & b \\ c & d \end{bmatrix} \begin{bmatrix} r \\ s \end{bmatrix}$$

Do you see how Cayley defined matrix multiplication?

6.4 Assess Your Understanding

Concepts and Vocabulary

1. A matrix B, for which $AB = I_n$, the identity matrix, is called the _____ of A.

2. A matrix that has the same number of rows as columns is called a(n) _____ matrix.

3. In the algebra of matrices, the matrix that has properties similar to the number 1 is called the _____ matrix.

4. *True or False:* Every square matrix has an inverse.

5. *True or False:* Matrix multiplication is commutative.

6. *True or False:* Any pair of matrices can be multiplied.

Skill Building

In Problems 7–22, use the following matrices to compute the given expression.

$$A = \begin{bmatrix} 0 & 3 & -5 \\ 1 & 2 & 6 \end{bmatrix}, \qquad B = \begin{bmatrix} 4 & 1 & 0 \\ -2 & 3 & -2 \end{bmatrix}, \qquad C = \begin{bmatrix} 4 & 1 \\ 6 & 2 \\ -2 & 3 \end{bmatrix}$$

7. $A + B$

8. $A - B$

9. $4A$

10. $-3B$

11. $3A - 2B$

12. $2A + 4B$

13. AC

14. BC

15. CA

16. CB

17. $C(A + B)$

18. $(A + B)C$

19. $AC - 3I_2$

20. $CA + 5I_3$

21. $CA - CB$

22. $AC + BC$

In Problems 23–28, find the product.

23. $\begin{bmatrix} 2 & -2 \\ 1 & 0 \end{bmatrix}\begin{bmatrix} 2 & 1 & 4 & 6 \\ 3 & -1 & 3 & 2 \end{bmatrix}$

24. $\begin{bmatrix} 4 & 1 \\ 2 & 1 \end{bmatrix}\begin{bmatrix} -6 & 6 & 1 & 0 \\ 2 & 5 & 4 & -1 \end{bmatrix}$

25. $\begin{bmatrix} 1 & 2 & 3 \\ 0 & -1 & 4 \end{bmatrix}\begin{bmatrix} 1 & 2 \\ -1 & 0 \\ 2 & 4 \end{bmatrix}$

26. $\begin{bmatrix} 1 & -1 \\ -3 & 2 \\ 0 & 5 \end{bmatrix}\begin{bmatrix} 2 & 8 & -1 \\ 3 & 6 & 0 \end{bmatrix}$

27. $\begin{bmatrix} 1 & 0 & 1 \\ 2 & 4 & 1 \\ 3 & 6 & 1 \end{bmatrix}\begin{bmatrix} 1 & 3 \\ 6 & 2 \\ 8 & -1 \end{bmatrix}$

28. $\begin{bmatrix} 4 & -2 & 3 \\ 0 & 1 & 2 \\ -1 & 0 & 1 \end{bmatrix}\begin{bmatrix} 2 & 6 \\ 1 & -1 \\ 0 & 2 \end{bmatrix}$

In Problems 29–38, each matrix is nonsingular. Find the inverse of each matrix.

29. $\begin{bmatrix} 2 & 1 \\ 1 & 1 \end{bmatrix}$

30. $\begin{bmatrix} 3 & -1 \\ -2 & 1 \end{bmatrix}$

31. $\begin{bmatrix} 6 & 5 \\ 2 & 2 \end{bmatrix}$

32. $\begin{bmatrix} -4 & 1 \\ 6 & -2 \end{bmatrix}$

33. $\begin{bmatrix} 2 & 1 \\ a & a \end{bmatrix}, \quad a \neq 0$

34. $\begin{bmatrix} b & 3 \\ b & 2 \end{bmatrix}, \quad b \neq 0$

35. $\begin{bmatrix} 1 & -1 & 1 \\ 0 & -2 & 1 \\ -2 & -3 & 0 \end{bmatrix}$

36. $\begin{bmatrix} 1 & 0 & 2 \\ -1 & 2 & 3 \\ 1 & -1 & 0 \end{bmatrix}$

37. $\begin{bmatrix} 1 & 1 & 1 \\ 3 & 2 & -1 \\ 3 & 1 & 2 \end{bmatrix}$

38. $\begin{bmatrix} 3 & 3 & 1 \\ 1 & 2 & 1 \\ 2 & -1 & 1 \end{bmatrix}$

In Problems 39–58, use the inverses found in Problems 29–38 to solve each system of equations.

39. $\begin{cases} 2x + y = 8 \\ x + y = 5 \end{cases}$

40. $\begin{cases} 3x - y = 8 \\ -2x + y = 4 \end{cases}$

41. $\begin{cases} 2x + y = 0 \\ x + y = 5 \end{cases}$

42. $\begin{cases} 3x - y = 4 \\ -2x + y = 5 \end{cases}$

43. $\begin{cases} 6x + 5y = 7 \\ 2x + 2y = 2 \end{cases}$

44. $\begin{cases} -4x + y = 0 \\ 6x - 2y = 14 \end{cases}$

45. $\begin{cases} 6x + 5y = 13 \\ 2x + 2y = 5 \end{cases}$

46. $\begin{cases} -4x + y = 5 \\ 6x - 2y = -9 \end{cases}$

47. $\begin{cases} 2x + y = -3 \\ ax + ay = -a \end{cases} \quad a \neq 0$

48. $\begin{cases} bx + 3y = 2b + 3 \\ bx + 2y = 2b + 2 \end{cases} \quad b \neq 0$

49. $\begin{cases} 2x + y = \dfrac{7}{a} \\ ax + ay = 5 \end{cases} \quad a \neq 0$

50. $\begin{cases} bx + 3y = 14 \\ bx + 2y = 10 \end{cases} \quad b \neq 0$

51. $\begin{cases} x - y + z = 0 \\ -2y + z = -1 \\ -2x - 3y = -5 \end{cases}$

52. $\begin{cases} x + 2z = 6 \\ -x + 2y + 3z = -5 \\ x - y = 6 \end{cases}$

53. $\begin{cases} x - y + z = 2 \\ -2y + z = 2 \\ -2x - 3y = \dfrac{1}{2} \end{cases}$

54. $\begin{cases} x + 2z = 2 \\ -x + 2y + 3z = -\dfrac{3}{2} \\ x - y = 2 \end{cases}$

55. $\begin{cases} x + y + z = 9 \\ 3x + 2y - z = 8 \\ 3x + y + 2z = 1 \end{cases}$ **56.** $\begin{cases} 3x + 3y + z = 8 \\ x + 2y + z = 5 \\ 2x - y + z = 4 \end{cases}$ **57.** $\begin{cases} x + y + z = 2 \\ 3x + 2y - z = \dfrac{7}{3} \\ 3x + y + 2z = \dfrac{10}{3} \end{cases}$ **58.** $\begin{cases} 3x + 3y + z = 1 \\ x + 2y + z = 0 \\ 2x - y + z = 4 \end{cases}$

In Problems 59–64, show that each matrix has no inverse.

59. $\begin{bmatrix} 4 & 2 \\ 2 & 1 \end{bmatrix}$

60. $\begin{bmatrix} -3 & \dfrac{1}{2} \\ 6 & -1 \end{bmatrix}$

61. $\begin{bmatrix} 15 & 3 \\ 10 & 2 \end{bmatrix}$

62. $\begin{bmatrix} -3 & 0 \\ 4 & 0 \end{bmatrix}$

63. $\begin{bmatrix} -3 & 1 & -1 \\ 1 & -4 & -7 \\ 1 & 2 & 5 \end{bmatrix}$

64. $\begin{bmatrix} 1 & 1 & -3 \\ 2 & -4 & 1 \\ -5 & 7 & 1 \end{bmatrix}$

In Problems 65–68, use a graphing utility to find the inverse, if it exists, of each matrix. Round answers to two decimal places.

65. $\begin{bmatrix} 25 & 61 & -12 \\ 18 & -2 & 4 \\ 8 & 35 & 21 \end{bmatrix}$ **66.** $\begin{bmatrix} 18 & -3 & 4 \\ 6 & -20 & 14 \\ 10 & 25 & -15 \end{bmatrix}$ **67.** $\begin{bmatrix} 44 & 21 & 18 & 6 \\ -2 & 10 & 15 & 5 \\ 21 & 12 & -12 & 4 \\ -8 & -16 & 4 & 9 \end{bmatrix}$ **68.** $\begin{bmatrix} 16 & 22 & -3 & 5 \\ 21 & -17 & 4 & 8 \\ 2 & 8 & 27 & 20 \\ 5 & 15 & -3 & -10 \end{bmatrix}$

In Problems 69–72, use the idea behind Example 15 with a graphing utility to solve the following systems of equations. Round answers to two decimal places.

69. $\begin{cases} 25x + 61y - 12z = 10 \\ 18x - 12y + 7y = -9 \\ 3x + 4y - z = 12 \end{cases}$ **70.** $\begin{cases} 25x + 61y - 12z = 15 \\ 18x - 12y + 7z = -3 \\ 3x + 4y - z = 12 \end{cases}$ **71.** $\begin{cases} 25x + 61y - 12z = 21 \\ 18x - 12y + 7z = 7 \\ 3x + 4y - z = -2 \end{cases}$ **72.** $\begin{cases} 25x + 61y - 12z = 25 \\ 18x - 12y + 7z = 10 \\ 3x + 4y - z = -4 \end{cases}$

Applications and Extensions

73. Computing the Cost of Production The Acme Steel Company is a producer of stainless steel and aluminum containers. On a certain day, the following stainless steel containers were manufactured: 500 with 10-gallon capacity, 350 with 5-gallon capacity, and 400 with 1-gallon capacity. On the same day, the following aluminum containers were manufactured: 700 with 10-gallon capacity, 500 with 5-gallon capacity, and 850 with 1-gallon capacity.
 (a) Find a 2 by 3 matrix representing the above data. Find a 3 by 2 matrix to represent the same data.
 (b) If the amount of material used in the 10-gallon containers is 15 pounds, the amount used in the 5-gallon containers is 8 pounds, and the amount used in the 1-gallon containers is 3 pounds, find a 3 by 1 matrix representing the amount of material used.
 (c) Multiply the 2 by 3 matrix found in part (a) and the 3 by 1 matrix found in part (b) to get a 2 by 1 matrix showing the day's usage of material.
 (d) If stainless steel costs Acme $0.10 per pound and aluminum costs $0.05 per pound, find a 1 by 2 matrix representing cost.
 (e) Multiply the matrices found in parts (c) and (d) to determine the total cost of the day's production.

74. Computing Profit Rizza Ford has two locations, one in the city and the other in the suburbs. In January, the city loca-

tion sold 400 subcompacts, 250 intermediate-size cars, and 50 SUVs; in February, it sold 350 subcompacts, 100 intermediates, and 30 SUVs. At the suburban location in January, 450 subcompacts, 200 intermediates, and 140 SUVs were sold. In February, the suburban location sold 350 subcompacts, 300 intermediates, and 100 SUVs.
 (a) Find 2 by 3 matrices that summarize the sales data for each location for January and February (one matrix for each month).
 (b) Use matrix addition to obtain total sales for the two-month period.
 (c) The profit on each kind of car is $100 per subcompact, $150 per intermediate, and $200 per SUV. Find a 3 by 1 matrix representing this profit.
 (d) Multiply the matrices found in parts (b) and (c) to get a 2 by 1 matrix showing the profit at each location.

75. Consider the 2 by 2 square matrix

$$A = \begin{bmatrix} a & b \\ c & d \end{bmatrix}$$

If $D = ad - bc \neq 0$, show that A is nonsingular and that

$$A^{-1} = \frac{1}{D}\begin{bmatrix} d & -b \\ -c & a \end{bmatrix}$$

Discussion and Writing

76. Create a situation different from any found in the text that can be represented by a matrix.

6.5 Partial Fraction Decomposition

PREPARING FOR THIS SECTION *Before getting started, review the following:*

• Identity (Appendix A, Section A.8 p. A60)

• Proper and Improper Rational Functions (Section 3.2, p. 208)

• Factoring Polynomials (Appendix A, Section A.4, pp. A29–A35)

• Fundamental Theorem of Algebra (Section 3.6, pp. 251–253)

 Now work the 'Are You Prepared?' problems on page 461.

OBJECTIVES 1 Decompose $\dfrac{P}{Q}$, Where Q Has Only Nonrepeated Linear Factors

2 Decompose $\dfrac{P}{Q}$, Where Q Has Repeated Linear Factors

3 Decompose $\dfrac{P}{Q}$, Where Q Has a Nonrepeated Irreducible Quadratic Factor

4 Decompose $\dfrac{P}{Q}$, Where Q Has a Repeated Irreducible Quadratic Factor

Consider the problem of adding two rational expressions:

$$\frac{3}{x+4} \quad \text{and} \quad \frac{2}{x-3}$$

The result is

$$\frac{3}{x+4} + \frac{2}{x-3} = \frac{3(x-3) + 2(x+4)}{(x+4)(x-3)} = \frac{5x-1}{x^2+x-12}$$

The reverse procedure, of starting with the rational expression $\dfrac{5x-1}{x^2+x-12}$ and writing it as the sum (or difference) of the two simpler fractions $\dfrac{3}{x+4}$ and $\dfrac{2}{x-3}$, is referred to as **partial fraction decomposition**, and the two simpler fractions are called **partial fractions**. Decomposing a rational expression into a sum of partial fractions is important in solving certain types of calculus problems. This section presents a systematic way to decompose rational expressions.

We begin by recalling that a rational expression is the ratio of two polynomials, say, P and $Q \neq 0$. We assume that P and Q have no common factors. Recall also that a rational expression $\dfrac{P}{Q}$ is called **proper** if the degree of the polynomial in the numerator is less than the degree of the polynomial in the denominator. Otherwise, the rational expression is termed **improper**.

Because any improper rational expression can be reduced by long division to a mixed form consisting of the sum of a polynomial and a proper rational expression, we shall restrict the discussion that follows to proper rational expressions.

The partial fraction decomposition of the rational expression $\dfrac{P}{Q}$ depends on the factors of the denominator Q. Recall (from Section 3.7) that any polynomial whose coefficients are real numbers can be factored (over the real numbers) into products of linear and/or irreducible quadratic factors. This means that the denominator Q of the rational expression $\dfrac{P}{Q}$ will contain only factors of one or both of the following types:

1. *Linear factors* of the form $x - a$, where a is a real number.

2. *Irreducible quadratic factors* of the form $ax^2 + bx + c$, where a, b, and c are real numbers, $a \neq 0$, and $b^2 - 4ac < 0$ (which guarantees that $ax^2 + bx + c$ cannot be written as the product of two linear factors with real coefficients).

As it turns out, there are four cases to be examined. We begin with the case for which Q has only nonrepeated linear factors.

1 Decompose $\dfrac{P}{Q}$, Where Q Has Only Nonrepeated Linear Factors

Case 1: Q has only nonrepeated linear factors.

Under the assumption that Q has only nonrepeated linear factors, the polynomial Q has the form

$$Q(x) = (x - a_1)(x - a_2) \cdots (x - a_n)$$

where none of the numbers a_1, a_2, \ldots, a_n is equal. In this case, the partial fraction decomposition of $\dfrac{P}{Q}$ is of the form

$$\frac{P(x)}{Q(x)} = \frac{A_1}{x - a_1} + \frac{A_2}{x - a_2} + \cdots + \frac{A_n}{x - a_n} \qquad \textbf{(1)}$$

where the numbers A_1, A_2, \ldots, A_n are to be determined.

We show how to find these numbers in the example that follows.

EXAMPLE 1 **Nonrepeated Linear Factors**

Write the partial fraction decomposition of $\dfrac{x}{x^2 - 5x + 6}$

Solution First, we factor the denominator,

$$x^2 - 5x + 6 = (x - 2)(x - 3)$$

and conclude that the denominator contains only nonrepeated linear factors. Then we decompose the rational expression according to equation (1):

$$\frac{x}{x^2 - 5x + 6} = \frac{A}{x - 2} + \frac{B}{x - 3} \qquad \textbf{(2)}$$

where A and B are to be determined. To find A and B, we clear the fractions by multiplying each side by $(x - 2)(x - 3) = x^2 - 5x + 6$. The result is

$$x = A(x - 3) + B(x - 2) \qquad \textbf{(3)}$$

or

$$x = (A + B)x + (-3A - 2B)$$

This equation is an identity in x. We equate the coefficients of like powers of x to get

$$\begin{cases} 1 = A + B & \text{Equate coefficients of x: 1x = (A + B)x.} \\ 0 = -3A - 2B & \text{Equate coefficients of } x^0, \text{ the constants: } 0x^0 = (-3A - 2B)x^0. \end{cases}$$

This system of two equations containing two variables, A and B, can be solved using whatever method you wish. Solving it, we get

$$A = -2 \qquad B = 3$$

From equation (2), the partial fraction decomposition is

$$\frac{x}{x^2 - 5x + 6} = \frac{-2}{x - 2} + \frac{3}{x - 3}$$

✔ **CHECK:** The decomposition can be checked by adding the rational expressions.

$$\frac{-2}{x - 2} + \frac{3}{x - 3} = \frac{-2(x - 3) + 3(x - 2)}{(x - 2)(x - 3)} = \frac{x}{(x - 2)(x - 3)}$$

$$= \frac{x}{x^2 - 5x + 6} \qquad ■$$

The numbers to be found in the partial fraction decomposition can sometimes be found more readily by using suitable choices for x (which may include complex numbers) in the identity obtained after fractions have been cleared. In Example 1, the identity after clearing fractions is equation (3):

$$x = A(x - 3) + B(x - 2)$$

If we let $x = 2$ in this expression, the term containing B drops out, leaving $2 = A(-1)$, or $A = -2$. Similarly, if we let $x = 3$, the term containing A drops out, leaving $3 = B$. As before, $A = -2$ and $B = 3$.

 NOW WORK PROBLEM 13.

2 Decompose $\dfrac{P}{Q}$, Where Q Has Repeated Linear Factors

Case 2: Q has repeated linear factors.

If the polynomial Q has a repeated linear factor, say $(x - a)^n$, $n \geq 2$ an integer, then, in the partial fraction decomposition of $\dfrac{P}{Q}$, we allow for the terms

$$\frac{A_1}{x - a} + \frac{A_2}{(x - a)^2} + \cdots + \frac{A_n}{(x - a)^n}$$

where the numbers A_1, A_2, \ldots, A_n are to be determined.

EXAMPLE 2

Repeated Linear Factors

Write the partial fraction decomposition of $\dfrac{x + 2}{x^3 - 2x^2 + x}$

Solution First, we factor the denominator,

$$x^3 - 2x^2 + x = x(x^2 - 2x + 1) = x(x - 1)^2$$

and find that the denominator has the nonrepeated linear factor x and the repeated linear factor $(x - 1)^2$. By Case 1, we must allow for the term $\dfrac{A}{x}$ in the decomposition; and, by Case 2, we must allow for the terms $\dfrac{B}{x - 1} + \dfrac{C}{(x - 1)^2}$ in the decomposition. We write

$$\frac{x + 2}{x^3 - 2x^2 + x} = \frac{A}{x} + \frac{B}{x - 1} + \frac{C}{(x - 1)^2} \tag{4}$$

Again, we clear fractions by multiplying each side by $x^3 - 2x^2 + x = x(x - 1)^2$. The result is the identity

$$x + 2 = A(x - 1)^2 + Bx(x - 1) + Cx \tag{5}$$

If we let $x = 0$ in this expression, the terms containing B and C drop out, leaving $2 = A(-1)^2$, or $A = 2$. Similarly, if we let $x = 1$, the terms containing A and B drop out, leaving $3 = C$. Then equation (5) becomes

$$x + 2 = 2(x - 1)^2 + Bx(x - 1) + 3x$$

Now let $x = 2$ (any choice other than 0 or 1 will work as well). The result is

$$4 = 2(1)^2 + B(2)(1) + 3(2)$$
$$4 = 2 + 2B + 6$$
$$2B = -4$$
$$B = -2$$

We have $A = 2$, $B = -2$, and $C = 3$.
From equation (4), the partial fraction decomposition is

$$\frac{x + 2}{x^3 - 2x^2 + x} = \frac{2}{x} + \frac{-2}{x - 1} + \frac{3}{(x - 1)^2}$$ ∎

EXAMPLE 3 **Repeated Linear Factors**

Write the partial fraction decomposition of $\dfrac{x^3 - 8}{x^2(x - 1)^3}$

Solution The denominator contains the repeated linear factor x^2 and the repeated linear factor $(x - 1)^3$. The partial fraction decomposition takes the form

$$\frac{x^3 - 8}{x^2(x - 1)^3} = \frac{A}{x} + \frac{B}{x^2} + \frac{C}{x - 1} + \frac{D}{(x - 1)^2} + \frac{E}{(x - 1)^3} \tag{6}$$

As before, we clear fractions and obtain the identity
$$x^3 - 8 = Ax(x - 1)^3 + B(x - 1)^3 + Cx^2(x - 1)^2 + Dx^2(x - 1) + Ex^2 \tag{7}$$
Let $x = 0$. (Do you see why this choice was made?) Then
$$-8 = B(-1)$$
$$B = 8$$
Now let $x = 1$ in equation (7). Then
$$-7 = E$$
Use $B = 8$ and $E = -7$ in equation (7) and collect like terms.
$$x^3 - 8 = Ax(x - 1)^3 + 8(x - 1)^3$$
$$+ Cx^2(x - 1)^2 + Dx^2(x - 1) - 7x^2$$
$$x^3 - 8 - 8(x^3 - 3x^2 + 3x - 1) + 7x^2 = Ax(x - 1)^3 + Cx^2(x - 1)^2 + Dx^2(x - 1)$$
$$-7x^3 + 31x^2 - 24x = x(x - 1)[A(x - 1)^2 + Cx(x - 1) + Dx]$$
$$x(x - 1)(-7x + 24) = x(x - 1)[A(x - 1)^2 + Cx(x - 1) + Dx]$$
$$-7x + 24 = A(x - 1)^2 + Cx(x - 1) + Dx \tag{8}$$

We now work with equation (8). Let $x = 0$. Then
$$24 = A$$
Now let $x = 1$ in equation (8). Then
$$17 = D$$
Use $A = 24$ and $D = 17$ in equation (8) and collect like terms.
$$-7x + 24 = 24(x - 1)^2 + Cx(x - 1) + 17x$$
Now let $x = 2$. Then
$$-14 + 24 = 24 + C(2) + 34$$
$$-48 = 2C$$
$$-24 = C$$

We now know all the numbers A, B, C, D, and E, so, from equation (6), we have the decomposition

$$\frac{x^3 - 8}{x^2(x - 1)^3} = \frac{24}{x} + \frac{8}{x^2} + \frac{-24}{x - 1} + \frac{17}{(x - 1)^2} + \frac{-7}{(x - 1)^3}$$ ∎

The method employed in Example 3, although somewhat tedious, is still preferable to solving the system of five equations containing five variables that the expansion of equation (6) leads to.

 NOW WORK PROBLEM 19.

The final two cases involve irreducible quadratic factors. A quadratic factor is irreducible if it cannot be factored into linear factors with real coefficients. A quadratic expression $ax^2 + bx + c$ is irreducible whenever $b^2 - 4ac < 0$. For example, $x^2 + x + 1$ and $x^2 + 4$ are irreducible.

3 **Decompose $\dfrac{P}{Q}$, Where Q Has a Nonrepeated Irreducible Quadratic Factor**

> **Case 3: Q contains a nonrepeated irreducible quadratic factor.**
>
> If Q contains a nonrepeated irreducible quadratic factor of the form $ax^2 + bx + c$, then, in the partial fraction decomposition of $\dfrac{P}{Q}$, allow for the term
>
> $$\frac{Ax + B}{ax^2 + bx + c}$$
>
> where the numbers A and B are to be determined.

EXAMPLE 4 **Nonrepeated Irreducible Quadratic Factor**

Write the partial fraction decomposition of $\dfrac{3x - 5}{x^3 - 1}$

Solution We factor the denominator,

$$x^3 - 1 = (x - 1)(x^2 + x + 1)$$

and find that it has a nonrepeated linear factor $x - 1$ and a nonrepeated irreducible quadratic factor $x^2 + x + 1$. We allow for the term $\dfrac{A}{x - 1}$ by Case 1, and we allow for the term $\dfrac{Bx + C}{x^2 + x + 1}$ by Case 3. We write

$$\frac{3x - 5}{x^3 - 1} = \frac{A}{x - 1} + \frac{Bx + C}{x^2 + x + 1}$$ **(9)**

We clear fractions by multiplying each side of equation (9) by $x^3 - 1 = (x - 1)(x^2 + x + 1)$ to obtain the identity

$$3x - 5 = A(x^2 + x + 1) + (Bx + C)(x - 1)$$ **(10)**

Now let $x = 1$. Then equation (10) gives $-2 = A(3)$, or $A = -\dfrac{2}{3}$. We use this value of A in equation (10) and simplify.

$$3x - 5 = -\frac{2}{3}(x^2 + x + 1) + (Bx + C)(x - 1)$$

$$3(3x - 5) = -2(x^2 + x + 1) + 3(Bx + C)(x - 1) \qquad \text{Multiply each side by 3.}$$

$$9x - 15 = -2x^2 - 2x - 2 + 3(Bx + C)(x - 1)$$

$$2x^2 + 11x - 13 = 3(Bx + C)(x - 1) \qquad \text{Collect terms.}$$

$$(2x + 13)(x - 1) = 3(Bx + C)(x - 1) \qquad \text{Factor the left side.}$$

$$2x + 13 = 3Bx + 3C \qquad \text{Cancel } x - 1 \text{ on each side.}$$

$$2 = 3B \quad \text{and} \quad 13 = 3C \qquad \text{Equate coefficients.}$$

$$B = \frac{2}{3} \qquad C = \frac{13}{3}$$

From equation (9), we see that

$$\frac{3x - 5}{x^3 - 1} = \frac{-\dfrac{2}{3}}{x - 1} + \frac{\dfrac{2}{3}x + \dfrac{13}{3}}{x^2 + x + 1}$$ ∎

NOW WORK PROBLEM 21.

4 Decompose $\dfrac{P}{Q}$, Where Q Has a Repeated Irreducible Quadratic Factor

Case 4: Q contains a repeated irreducible quadratic factor.

If the polynomial Q contains a repeated irreducible quadratic factor $(ax^2 + bx + c)^n$, $n \geq 2$, n an integer, then, in the partial fraction decomposition of $\dfrac{P}{Q}$, allow for the terms

$$\frac{A_1 x + B_1}{ax^2 + bx + c} + \frac{A_2 x + B_2}{(ax^2 + bx + c)^2} + \cdots + \frac{A_n x + B_n}{(ax^2 + bx + c)^n}$$

where the numbers $A_1, B_1, A_2, B_2, \ldots, A_n, B_n$ are to be determined.

EXAMPLE 5 | **Repeated Irreducible Quadratic Factor**

Write the partial fraction decomposition of $\dfrac{x^3 + x^2}{(x^2 + 4)^2}$

Solution The denominator contains the repeated irreducible quadratic factor $(x^2 + 4)^2$, so we write

$$\frac{x^3 + x^2}{(x^2 + 4)^2} = \frac{Ax + B}{x^2 + 4} + \frac{Cx + D}{(x^2 + 4)^2} \qquad \textbf{(11)}$$

We clear fractions to obtain

$$x^3 + x^2 = (Ax + B)(x^2 + 4) + Cx + D$$

Collecting like terms yields

$$x^3 + x^2 = Ax^3 + Bx^2 + (4A + C)x + D + 4B$$

Equating coefficients, we arrive at the system

$$\begin{cases} A = 1 \\ B = 1 \\ 4A + C = 0 \\ D + 4B = 0 \end{cases}$$

The solution is $A = 1, B = 1, C = -4, D = -4$. From equation (11),

$$\frac{x^3 + x^2}{(x^2 + 4)^2} = \frac{x + 1}{x^2 + 4} + \frac{-4x - 4}{(x^2 + 4)^2}$$

∎

NOW WORK PROBLEM 35.

6.5 Assess Your Understanding

'Are You Prepared?'

Answers are given at the end of these exercises. If you get a wrong answer, read the pages listed in red.

1. *True or False:* The equation $(x - 1)^2 - 1 = x(x - 2)$ is an example of an identity. (p. A60)

2. *True or False:* The rational expression $\dfrac{5x^2 - 1}{x^3 + 1}$ is proper. (p. 208)

3. Factor completely: $3x^4 + 6x^3 + 3x^2$. (pp. A29–A35)

4. *True or False:* Every polynomial with real numbers as coefficients can be factored into products of linear and/or irreducible quadratic factors. (p. 253)

Skill Building

In Problems 5–12, tell whether the given rational expression is proper or improper. If improper, rewrite it as the sum of a polynomial and a proper rational expression.

5. $\dfrac{x}{x^2 - 1}$

6. $\dfrac{5x + 2}{x^3 - 1}$

7. $\dfrac{x^2 + 5}{x^2 - 4}$

8. $\dfrac{3x^2 - 2}{x^2 - 1}$

9. $\dfrac{5x^3 + 2x - 1}{x^2 - 4}$

10. $\dfrac{3x^4 + x^2 - 2}{x^3 + 8}$

11. $\dfrac{x(x - 1)}{(x + 4)(x - 3)}$

12. $\dfrac{2x(x^2 + 4)}{x^2 + 1}$

In Problems 13–46, write the partial fraction decomposition of each rational expression.

13. $\dfrac{4}{x(x - 1)}$

14. $\dfrac{3x}{(x + 2)(x - 1)}$

15. $\dfrac{1}{x(x^2 + 1)}$

16. $\dfrac{1}{(x + 1)(x^2 + 4)}$

17. $\dfrac{x}{(x - 1)(x - 2)}$

18. $\dfrac{3x}{(x + 2)(x - 4)}$

19. $\dfrac{x^2}{(x - 1)^2(x + 1)}$

20. $\dfrac{x + 1}{x^2(x - 2)}$

21. $\dfrac{1}{x^3 - 8}$

22. $\dfrac{2x + 4}{x^3 - 1}$

23. $\dfrac{x^2}{(x - 1)^2(x + 1)^2}$

24. $\dfrac{x + 1}{x^2(x - 2)^2}$

25. $\dfrac{x - 3}{(x + 2)(x + 1)^2}$

26. $\dfrac{x^2 + x}{(x + 2)(x - 1)^2}$

27. $\dfrac{x + 4}{x^2(x^2 + 4)}$

28. $\dfrac{10x^2 + 2x}{(x - 1)^2(x^2 + 2)}$

29. $\dfrac{x^2 + 2x + 3}{(x + 1)(x^2 + 2x + 4)}$

30. $\dfrac{x^2 - 11x - 18}{x(x^2 + 3x + 3)}$

31. $\dfrac{x}{(3x - 2)(2x + 1)}$

32. $\dfrac{1}{(2x + 3)(4x - 1)}$

33. $\dfrac{x}{x^2 + 2x - 3}$

34. $\dfrac{x^2 - x - 8}{(x + 1)(x^2 + 5x + 6)}$

35. $\dfrac{x^2 + 2x + 3}{(x^2 + 4)^2}$

36. $\dfrac{x^3 + 1}{(x^2 + 16)^2}$

37. $\dfrac{7x + 3}{x^3 - 2x^2 - 3x}$

38. $\dfrac{x^5 + 1}{x^6 - x^4}$

39. $\dfrac{x^2}{x^3 - 4x^2 + 5x - 2}$

40. $\dfrac{x^2 + 1}{x^3 + x^2 - 5x + 3}$

41. $\dfrac{x^3}{(x^2 + 16)^3}$

42. $\dfrac{x^2}{(x^2 + 4)^3}$

43. $\dfrac{4}{2x^2 - 5x - 3}$

44. $\dfrac{4x}{2x^2 + 3x - 2}$

45. $\dfrac{2x + 3}{x^4 - 9x^2}$

46. $\dfrac{x^2 + 9}{x^4 - 2x^2 - 8}$

'Are You Prepared?' Answers

1. True 2. True 3. $3x^2(x + 1)^2$ 4. True

6.6 Systems of Nonlinear Equations

PREPARING FOR THIS SECTION *Before getting started, review the following:*

- Lines (Foundations, Section F.3, pp. 18–28)
- Circles (Foundations, Section F.4, pp. 33–36)
- Parabolas (Section 5.2, pp. 363–368)

- Ellipses (Section 5.3, pp. 372–378)
- Hyperbolas (Section 5.4, pp. 382–390)

Now work the 'Are You Prepared?' problems on page 467.

OBJECTIVES 1 Solve a System of Nonlinear Equations Using Substitution
2 Solve a System of Nonlinear Equations Using Elimination

1 Solve a System of Nonlinear Equations Using Substitution

In Section 6.1, we observed that the solution to a system of linear equations could be found geometrically by determining the point(s) of intersection (if any) of the equations in the system. Similarly, when solving systems of nonlinear equations, the solution(s) also represents the point(s) of intersection (if any) of the graphs of the equations.

There is no general methodology for solving a system of nonlinear equations. At times substitution is best; other times, elimination is best; and sometimes neither of these methods works. Experience and a certain degree of imagination are your allies here.

Before we begin, two comments are in order.

1. If the system contains two variables and if the equations in the system are easy to graph, then graph them. By graphing each equation in the system, you can get an idea of how many solutions a system has and approximately where they are located.

2. Extraneous solutions can creep in when solving nonlinear systems, so it is imperative that all apparent solutions be checked.

EXAMPLE 1 **Solving a System of Nonlinear Equations**

Solve the following system of equations:

$$\begin{cases} 3x - y = -2 & (1) \quad \text{A line} \\ 2x^2 - y = 0 & (2) \quad \text{A parabola} \end{cases}$$

Solution Using Substitution

Figure 9

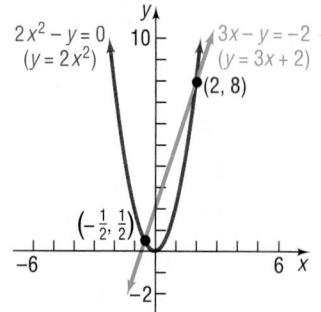

First, we notice that the system contains two variables and that we know how to graph each equation. See Figure 9. The system apparently has two solutions.

We will use substitution to solve the system. Equation (1) is easily solved for y.

$$3x - y = -2 \qquad \text{Equation (1)}$$
$$y = 3x + 2$$

We substitute this expression for y in equation (2). The result is an equation containing just the variable x, which we can then solve.

$$2x^2 - y = 0 \qquad \text{Equation (2)}$$
$$2x^2 - (3x + 2) = 0 \qquad \text{Substitute } 3x + 2 \text{ for } y.$$
$$2x^2 - 3x - 2 = 0 \qquad \text{Remove parentheses.}$$
$$(2x + 1)(x - 2) = 0 \qquad \text{Factor.}$$
$$2x + 1 = 0 \quad \text{or} \quad x - 2 = 0 \qquad \text{Apply the Zero-Product Property.}$$
$$x = -\frac{1}{2} \quad \text{or} \qquad x = 2$$

Using these values for x in $y = 3x + 2$, we find

$$y = 3\left(-\frac{1}{2}\right) + 2 = \frac{1}{2} \quad \text{or} \quad y = 3(2) + 2 = 8$$

The apparent solutions are $x = -\dfrac{1}{2}$, $y = \dfrac{1}{2}$ and $x = 2$, $y = 8$.

✔ CHECK: For $x = -\dfrac{1}{2}$, $y = \dfrac{1}{2}$:

$$\begin{cases} 3\left(-\dfrac{1}{2}\right) - \dfrac{1}{2} = -\dfrac{3}{2} - \dfrac{1}{2} = -2 & (1) \\ 2\left(-\dfrac{1}{2}\right)^2 - \dfrac{1}{2} = 2\left(\dfrac{1}{4}\right) - \dfrac{1}{2} = 0 & (2) \end{cases}$$

For $x = 2$, $y = 8$:

$$\begin{cases} 3(2) - 8 = 6 - 8 = -2 & (1) \\ 2(2)^2 - 8 = 2(4) - 8 = 0 & (2) \end{cases}$$

Each solution checks. Now we know that the graphs in Figure 9 intersect at the points $\left(-\dfrac{1}{2}, \dfrac{1}{2}\right)$ and $(2, 8)$. ∎

 ✔ CHECK: Graph $3x - y = -2$ ($Y_1 = 3x + 2$) and $2x^2 - y = 0$ ($Y_2 = 2x^2$) and compare what you see with Figure 9. Use INTERSECT (twice) to find the two points of intersection. ∎

✏ **NOW WORK PROBLEM 15 USING SUBSTITUTION.**

2 Solve a System of Nonlinear Equations Using Elimination

Our next example illustrates how the method of elimination works for nonlinear systems.

EXAMPLE 2

Solving a System of Nonlinear Equations

Solve: $\begin{cases} x^2 + y^2 = 13 & (1) \quad \text{A circle} \\ x^2 - y = 7 & (2) \quad \text{A parabola} \end{cases}$

Solution Using Elimination

First, we graph each equation, as shown in Figure 10. Based on the graph, we expect four solutions. By subtracting equation (2) from equation (1), the variable x can be eliminated.

Figure 10

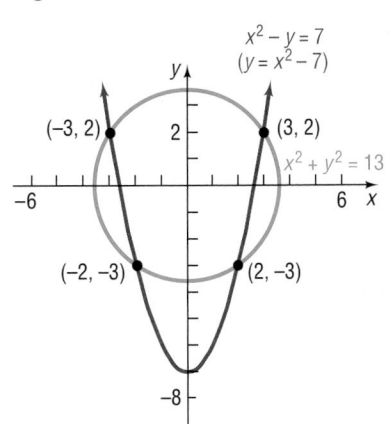

$$\begin{cases} x^2 + y^2 = 13 \\ \underline{x^2 - y = 7} \qquad \text{Subtract} \\ y^2 + y = 6 \end{cases}$$

This quadratic equation in y can be solved by factoring.

$$y^2 + y - 6 = 0$$
$$(y + 3)(y - 2) = 0$$
$$y = -3 \quad \text{or} \quad y = 2$$

We use these values for y in equation (2) to find x.

If $y = 2$, then $x^2 = y + 7 = 9$, so $x = 3$ or -3.
If $y = -3$, then $x^2 = y + 7 = 4$, so $x = 2$ or -2.

We have four solutions: $x = 3$, $y = 2$; $x = -3$, $y = 2$; $x = 2$, $y = -3$; and $x = -2$, $y = -3$.

You should verify that, in fact, these four solutions also satisfy equation (1), so all four are solutions of the system. The four points, $(3, 2)$, $(-3, 2)$, $(2, -3)$, and $(-2, -3)$, are the points of intersection of the graphs. Look again at Figure 10. ∎

✔ CHECK: Graph $x^2 + y^2 = 13$ and $x^2 - y = 7$. (Remember that to graph $x^2 + y^2 = 13$ requires two functions: $Y_1 = \sqrt{13 - x^2}$ and $Y_2 = -\sqrt{13 - x^2}$.) ∎

NOW WORK PROBLEM **13** USING ELIMINATION.

EXAMPLE 3

Solving a System of Nonlinear Equations

Solve: $\begin{cases} x^2 + x + y^2 - 3y + 2 = 0 & (1) \\ x + 1 + \dfrac{y^2 - y}{x} = 0 & (2) \end{cases}$

Solution Using Elimination

Since it is not straightforward how to graph the equations in the system, we proceed directly to use the method of elimination.

First, we multiply equation (2) by x to eliminate the fraction. The result is an equivalent system because x cannot be 0. [Look at equation (2) to see why.]

$\begin{cases} x^2 + x + y^2 - 3y + 2 = 0 & (1) \\ x^2 + x + y^2 - y = 0 & (2) \end{cases}$

Now subtract equation (2) from equation (1) to eliminate x. The result is

$$-2y + 2 = 0$$
$$y = 1 \quad \text{Solve for } y.$$

To find x, we back-substitute $y = 1$ in equation (1):

$$x^2 + x + y^2 - 3y + 2 = 0 \qquad \text{Equation (1)}$$
$$x^2 + x + 1 - 3 + 2 = 0 \qquad \text{Substitute 1 for } y \text{ in (1).}$$
$$x^2 + x = 0 \qquad \text{Simplify.}$$
$$x(x + 1) = 0 \qquad \text{Factor.}$$
$$x = 0 \quad \text{or} \quad x = -1 \qquad \text{Apply the Zero-Product Property.}$$

Because x cannot be 0, the value $x = 0$ is extraneous, and we discard it.

✔ CHECK: We now check $x = -1$, $y = 1$:

$$\begin{cases} (-1)^2 + (-1) + 1^2 - 3(1) + 2 = 1 - 1 + 1 - 3 + 2 = 0 & (1) \\ -1 + 1 + \dfrac{1^2 - 1}{-1} = 0 + \dfrac{0}{-1} = 0 & (2) \end{cases}$$

The solution is $x = -1$, $y = 1$.

NOW WORK PROBLEMS **29** AND **49**.

EXAMPLE 4

Solving a System of Nonlinear Equations

Solve: $\begin{cases} x^2 - y^2 = 4 & (1) \quad \text{A hyperbola} \\ y = x^2 & (2) \quad \text{A parabola} \end{cases}$

Solution Either substitution or elimination can be used here. We use substitution and replace x^2 by y in equation (1). The result is

$$x^2 - y^2 = 4 \qquad \text{Equation (1)}$$
$$y - y^2 = 4 \qquad y = x^2$$
$$y^2 - y + 4 = 0 \qquad \text{Place in standard form}$$

This is a quadratic equation. Its discriminant is $(-1)^2 - 4 \cdot 1 \cdot 4 = 1 - 16 = -15 < 0$. The equation has no real solutions, so the system is inconsistent. The graphs of these two equations do not intersect. See Figure 11. ∎

Figure 11

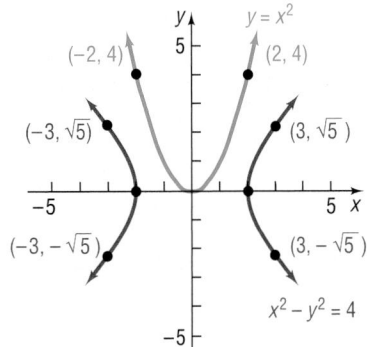

| **EXAMPLE 5** | **Solving a System of Nonlinear Equations** |

Solve: $\begin{cases} 3xy - 2y^2 = -2 & (1) \\ 9x^2 + 4y^2 = 10 & (2) \end{cases}$

Solution We multiply equation (1) by 2 and add the result to equation (2) to eliminate the y^2 terms.

$$\begin{cases} 6xy - 4y^2 = -4 & (1) \\ 9x^2 + 4y^2 = 10 & (2) \end{cases}$$
$$9x^2 + 6xy = 6 \qquad \text{Add.}$$
$$3x^2 + 2xy = 2 \qquad \text{Divide each side by 3.}$$

Since $x \neq 0$ (do you see why?), we can solve for y in this equation to get

$$y = \frac{2 - 3x^2}{2x} \qquad x \neq 0 \qquad (3)$$

Now substitute for y in equation (2) of the system.

$$9x^2 + 4y^2 = 10 \qquad \text{Equation (2)}$$
$$9x^2 + 4\left(\frac{2 - 3x^2}{2x}\right)^2 = 10 \qquad \text{Substitute } y = \frac{2 - 3x^2}{2x} \text{ in (2).}$$
$$9x^2 + \frac{4 - 12x^2 + 9x^4}{x^2} = 10$$
$$9x^4 + 4 - 12x^2 + 9x^4 = 10x^2 \qquad \text{Multiply both sides by } x^2.$$
$$18x^4 - 22x^2 + 4 = 0 \qquad \text{Subtract } 10x^2 \text{ from both sides.}$$
$$9x^4 - 11x^2 + 2 = 0 \qquad \text{Divide both sides by 2.}$$

This quadratic equation (in x^2) can be factored:

$$(9x^2 - 2)(x^2 - 1) = 0$$

$$9x^2 - 2 = 0 \quad \text{or} \quad x^2 - 1 = 0$$

$$x^2 = \frac{2}{9} \quad \text{or} \quad x^2 = 1$$

$$x = \pm\sqrt{\frac{2}{9}} = \pm\frac{\sqrt{2}}{3} \quad \text{or} \quad x = \pm 1$$

To find y, we use equation (3):

$$\text{If } x = \frac{\sqrt{2}}{3}: \quad y = \frac{2 - 3x^2}{2x} = \frac{2 - \frac{2}{3}}{2\left(\frac{\sqrt{2}}{3}\right)} = \frac{4}{2\sqrt{2}} = \sqrt{2}$$

$$\text{If } x = -\frac{\sqrt{2}}{3}: \quad y = \frac{2 - 3x^2}{2x} = \frac{2 - \frac{2}{3}}{2\left(-\frac{\sqrt{2}}{3}\right)} = \frac{4}{-2\sqrt{2}} = -\sqrt{2}$$

$$\text{If } x = 1: \quad y = \frac{2 - 3x^2}{2x} = \frac{2 - 3(1)^2}{2} = -\frac{1}{2}$$

$$\text{If } x = -1: \quad y = \frac{2 - 3x^2}{2x} = \frac{2 - 3(-1)^2}{-2} = \frac{1}{2}$$

The system has four solutions. Check them for yourself. ∎

- NOW WORK PROBLEM 47.

| EXAMPLE 6 | **Running a Long Distance Race** |

In a 50-mile race, the winner crosses the finish line 1 mile ahead of the second-place runner and 4 miles ahead of the third-place runner. Assuming that each runner maintains a constant speed throughout the race, by how many miles does the second-place runner beat the third-place runner?

|←————— 3 miles —————→|←— 1 mile —→|

Solution Let v_1, v_2, and v_3 denote the speeds of the first-, second-, and third-place runners, respectively. Let t_1 and t_2 denote the times (in hours) required for the first-place runner and second-place runner to finish the race. Then we have the system of equations

$$\begin{cases} 50 = v_1 t_1 & \text{(1)} \quad \textit{First-place runner goes 50 miles in } t_1 \textit{ hours.} \\ 49 = v_2 t_1 & \text{(2)} \quad \textit{Second-place runner goes 49 miles in } t_1 \textit{ hours.} \\ 46 = v_3 t_1 & \text{(3)} \quad \textit{Third-place runner goes 46 miles in } t_1 \textit{ hours.} \\ 50 = v_2 t_2 & \text{(4)} \quad \textit{Second-place runner goes 50 miles in } t_2 \textit{ hours.} \end{cases}$$

We seek the distance d of the third-place runner from the finish at time t_2. At time t_2, the third-place runner has gone a distance of $v_3 t_2$ miles, so the distance d remaining is $50 - v_3 t_2$. Now

$$d = 50 - v_3 t_2$$

$$= 50 - v_3 \left(t_1 \cdot \frac{t_2}{t_1} \right)$$

$$= 50 - (v_3 t_1) \cdot \frac{t_2}{t_1}$$

$$= 50 - 46 \cdot \frac{\dfrac{50}{v_2}}{\dfrac{50}{v_1}} \qquad \begin{cases} \text{From (3), } v_3 t_1 = 46 \\[4pt] \text{From (4), } t_2 = \dfrac{50}{v_2} \\[4pt] \text{From (1), } t_1 = \dfrac{50}{v_1} \end{cases}$$

$$= 50 - 46 \cdot \frac{v_1}{v_2}$$

$$= 50 - 46 \cdot \frac{50}{49} \qquad \text{Form the quotient of (1) and (2).}$$

$$\approx 3.06 \text{ miles} \qquad\qquad \blacksquare$$

HISTORICAL FEATURE

In the beginning of this section, we said that imagination and experience are important in solving systems of nonlinear equations. Indeed, these kinds of problems lead into some of the deepest and most difficult parts of modern mathematics. Look again at the graphs in Examples 1 and 2 of this section (Figures 9 and 10). We see that Example 1 has two solutions, and Example 2 has four solutions. We might conjecture that the number of solutions is equal to the product of the degrees of the equations involved.

This conjecture was indeed made by Etienne Bezout (1730–1783), but working out the details took about 150 years. It turns out that, to arrive at the correct number of intersections, we must count not only the complex number intersections, but also those intersections that, in a certain sense, lie at infinity. For example, a parabola and a line lying on the axis of the parabola intersect at the vertex and at infinity. This topic is part of the study of algebraic geometry.

Historical Problem

A papyrus dating back to 1950 BC contains the following problem: "A given surface area of 100 units of area shall be represented as the sum of two squares whose sides are to each other as $1 : \dfrac{3}{4}$."

Solve for the sides by solving the system of equations

$$\begin{cases} x^2 + y^2 = 100 \\[4pt] x = \dfrac{3}{4}y \end{cases}$$

6.6 Assess Your Understanding

'Are You Prepared?'

Answers are given at the end of these exercises. If you get a wrong answer, read the pages listed in red.

1. Graph the equation: $y = 3x + 2$ (pp. 18–28)

2. Graph the equation: $y + 4 = x^2$ (pp. 364–366)

3. Graph the equation: $y^2 = x^2 - 1$ (pp. 382–389)

4. Graph the equation: $x^2 + 4y^2 = 4$ (pp. 373–377)

Skill Building

In Problems 5–24, graph each equation of the system. Then solve the system to find the points of intersection.

5. $\begin{cases} y = x^2 + 1 \\ y = x + 1 \end{cases}$

6. $\begin{cases} y = x^2 + 1 \\ y = 4x + 1 \end{cases}$

7. $\begin{cases} y = \sqrt{36 - x^2} \\ y = 8 - x \end{cases}$

8. $\begin{cases} y = \sqrt{4 - x^2} \\ y = 2x + 4 \end{cases}$

9. $\begin{cases} y = \sqrt{x} \\ y = 2 - x \end{cases}$　　**10.** $\begin{cases} y = \sqrt{x} \\ y = 6 - x \end{cases}$　　**11.** $\begin{cases} x = 2y \\ x = y^2 - 2y \end{cases}$　　**12.** $\begin{cases} y = x - 1 \\ y = x^2 - 6x + 9 \end{cases}$

13. $\begin{cases} x^2 + y^2 = 4 \\ x^2 + 2x + y^2 = 0 \end{cases}$　　**14.** $\begin{cases} x^2 + y^2 = 8 \\ x^2 + y^2 + 4y = 0 \end{cases}$　　**15.** $\begin{cases} y = 3x - 5 \\ x^2 + y^2 = 5 \end{cases}$　　**16.** $\begin{cases} x^2 + y^2 = 10 \\ y = x + 2 \end{cases}$

17. $\begin{cases} x^2 + y^2 = 4 \\ y^2 - x = 4 \end{cases}$　　**18.** $\begin{cases} x^2 + y^2 = 16 \\ x^2 - 2y = 8 \end{cases}$　　**19.** $\begin{cases} xy = 4 \\ x^2 + y^2 = 8 \end{cases}$　　**20.** $\begin{cases} x^2 = y \\ xy = 1 \end{cases}$

21. $\begin{cases} x^2 + y^2 = 4 \\ y = x^2 - 9 \end{cases}$　　**22.** $\begin{cases} xy = 1 \\ y = 2x + 1 \end{cases}$　　**23.** $\begin{cases} y = x^2 - 4 \\ y = 6x - 13 \end{cases}$　　**24.** $\begin{cases} x^2 + y^2 = 10 \\ xy = 3 \end{cases}$

In Problems 25–54, solve each system. Use any method you wish.

25. $\begin{cases} 2x^2 + y^2 = 18 \\ xy = 4 \end{cases}$　　**26.** $\begin{cases} x^2 - y^2 = 21 \\ x + y = 7 \end{cases}$　　**27.** $\begin{cases} y = 2x + 1 \\ 2x^2 + y^2 = 1 \end{cases}$

28. $\begin{cases} x^2 - 4y^2 = 16 \\ 2y - x = 2 \end{cases}$　　**29.** $\begin{cases} x + y + 1 = 0 \\ x^2 + y^2 + 6y - x = -5 \end{cases}$　　**30.** $\begin{cases} 2x^2 - xy + y^2 = 8 \\ xy = 4 \end{cases}$

31. $\begin{cases} 4x^2 - 3xy + 9y^2 = 15 \\ 2x + 3y = 5 \end{cases}$　　**32.** $\begin{cases} 2y^2 - 3xy + 6y + 2x + 4 = 0 \\ 2x - 3y + 4 = 0 \end{cases}$　　**33.** $\begin{cases} x^2 - 4y^2 + 7 = 0 \\ 3x^2 + y^2 = 31 \end{cases}$

34. $\begin{cases} 3x^2 - 2y^2 + 5 = 0 \\ 2x^2 - y^2 + 2 = 0 \end{cases}$　　**35.** $\begin{cases} 7x^2 - 3y^2 + 5 = 0 \\ 3x^2 + 5y^2 = 12 \end{cases}$　　**36.** $\begin{cases} x^2 - 3y^2 + 1 = 0 \\ 2x^2 - 7y^2 + 5 = 0 \end{cases}$

37. $\begin{cases} x^2 + 2xy = 10 \\ 3x^2 - xy = 2 \end{cases}$　　**38.** $\begin{cases} 5xy + 13y^2 + 36 = 0 \\ xy + 7y^2 = 6 \end{cases}$　　**39.** $\begin{cases} 2x^2 + y^2 = 2 \\ x^2 - 2y^2 + 8 = 0 \end{cases}$

40. $\begin{cases} y^2 - x^2 + 4 = 0 \\ 2x^2 + 3y^2 = 6 \end{cases}$　　**41.** $\begin{cases} x^2 + 2y^2 = 16 \\ 4x^2 - y^2 = 24 \end{cases}$　　**42.** $\begin{cases} 4x^2 + 3y^2 = 4 \\ 2x^2 - 6y^2 = -3 \end{cases}$

43. $\begin{cases} \dfrac{5}{x^2} - \dfrac{2}{y^2} + 3 = 0 \\ \dfrac{3}{x^2} + \dfrac{1}{y^2} = 7 \end{cases}$　　**44.** $\begin{cases} \dfrac{2}{x^2} - \dfrac{3}{y^2} + 1 = 0 \\ \dfrac{6}{x^2} - \dfrac{7}{y^2} + 2 = 0 \end{cases}$　　**45.** $\begin{cases} \dfrac{1}{x^4} + \dfrac{6}{y^4} = 6 \\ \dfrac{2}{x^4} - \dfrac{2}{y^4} = 19 \end{cases}$

46. $\begin{cases} \dfrac{1}{x^4} - \dfrac{1}{y^4} = 1 \\ \dfrac{1}{x^4} + \dfrac{1}{y^4} = 4 \end{cases}$　　**47.** $\begin{cases} x^2 - 3xy + 2y^2 = 0 \\ x^2 + xy = 6 \end{cases}$　　**48.** $\begin{cases} x^2 - xy - 2y^2 = 0 \\ xy + x + 6 = 0 \end{cases}$

49. $\begin{cases} y^2 + y + x^2 - x - 2 = 0 \\ y + 1 + \dfrac{x - 2}{y} = 0 \end{cases}$　　**50.** $\begin{cases} x^3 - 2x^2 + y^2 + 3y - 4 = 0 \\ x - 2 + \dfrac{y^2 - y}{x^2} = 0 \end{cases}$　　**51.** $\begin{cases} \log_x y = 3 \\ \log_x(4y) = 5 \end{cases}$

52. $\begin{cases} \log_x(2y) = 3 \\ \log_x(4y) = 2 \end{cases}$　　**53.** $\begin{cases} \ln x = 4 \ln y \\ \log_3 x = 2 + 2 \log_3 y \end{cases}$　　**54.** $\begin{cases} \ln x = 5 \ln y \\ \log_2 x = 3 + 2 \log_2 y \end{cases}$

In Problems 55–60, graph each equation and find the point(s) of intersection, if any.

55. The line $x + 2y = 0$ and
the circle $(x - 1)^2 + (y - 1)^2 = 5$

56. The line $x + 2y + 6 = 0$ and
the circle $(x + 1)^2 + (y + 1)^2 = 5$

57. The circle $(x - 1)^2 + (y + 2)^2 = 4$
and the parabola $y^2 + 4y - x + 1 = 0$

58. The circle $(x + 2)^2 + (y - 1)^2 = 4$
and the parabola $y^2 - 2y - x - 5 = 0$

59. The graph of $y = \dfrac{4}{x - 3}$ and
the circle $x^2 - 6x + y^2 + 1 = 0$

60. The graph of $y = \dfrac{4}{x + 2}$ and
the circle $x^2 + 4x + y^2 - 4 = 0$

In Problems 61–68, use a graphing utility to solve each system of equations. Express the solution(s) rounded to two decimal places.

61. $\begin{cases} y = x^{2/3} \\ y = e^{-x} \end{cases}$

62. $\begin{cases} y = x^{3/2} \\ y = e^{-x} \end{cases}$

63. $\begin{cases} x^2 + y^3 = 2 \\ x^3 y = 4 \end{cases}$

64. $\begin{cases} x^3 + y^2 = 2 \\ x^2 y = 4 \end{cases}$

65. $\begin{cases} x^4 + y^4 = 12 \\ xy^2 = 2 \end{cases}$

66. $\begin{cases} x^4 + y^4 = 6 \\ xy = 1 \end{cases}$

67. $\begin{cases} xy = 2 \\ y = \ln x \end{cases}$

68. $\begin{cases} x^2 + y^2 = 4 \\ y = \ln x \end{cases}$

Applications and Extensions

69. The difference of two numbers is 2 and the sum of their squares is 10. Find the numbers.

70. The sum of two numbers is 7 and the difference of their squares is 21. Find the numbers.

71. The product of two numbers is 4 and the sum of their squares is 8. Find the numbers.

72. The product of two numbers is 10 and the difference of their squares is 21. Find the numbers.

73. The difference of two numbers is the same as their product, and the sum of their reciprocals is 5. Find the numbers.

74. The sum of two numbers is the same as their product, and the difference of their reciprocals is 3. Find the numbers.

75. The ratio of a to b is $\dfrac{2}{3}$. The sum of a and b is 10. What is the ratio of $a + b$ to $b - a$?

76. The ratio of a to b is $4 : 3$. The sum of a and b is 14. What is the ratio of $a - b$ to $a + b$?

77. Geometry The perimeter of a rectangle is 16 inches and its area is 15 square inches. What are its dimensions?

78. Geometry An area of 52 square feet is to be enclosed by two squares whose sides are in the ratio of $2 : 3$. Find the sides of the squares.

79. Geometry Two circles have circumferences that add up to 12π centimeters and areas that add up to 20π square centimeters. Find the radius of each circle.

80. Geometry The altitude of an isosceles triangle drawn to its base is 3 centimeters, and its perimeter is 18 centimeters. Find the length of its base.

81. The Tortoise and the Hare In a 21-meter race between a tortoise and a hare, the tortoise leaves 9 minutes before the hare. The hare, by running at an average speed of 0.5 meter per hour faster than the tortoise, crosses the finish line 3 minutes before the tortoise. What are the average speeds of the tortoise and the hare?

82. Running a Race In a 1-mile race, the winner crosses the finish line 10 feet ahead of the second-place runner and 20 feet ahead of the third-place runner. Assuming that each runner maintains a constant speed throughout the race, by how many feet does the second-place runner beat the third-place runner?

83. Constructing a Box A rectangular piece of cardboard, whose area is 216 square centimeters, is made into an open box by cutting a 2-centimeter square from each corner and turning up the sides. See the figure. If the box is to have a volume of 224 cubic centimeters, what size cardboard should you start with?

84. Constructing a Cylindrical Tube A rectangular piece of cardboard, whose area is 216 square centimeters, is made into a cylindrical tube by joining together two sides of the rectangle. See the figure. If the tube is to have a volume of 224 cubic centimeters, what size cardboard should you start with?

85. Fencing A farmer has 300 feet of fence available to enclose a 4500-square-foot region in the shape of adjoining squares, with sides of length x and y. See the figure. Find x and y.

86. Bending Wire A wire 60 feet long is cut into two pieces. Is it possible to bend one piece into the shape of a square and the other into the shape of a circle so that the total area enclosed by the two pieces is 100 square feet? If this is possible, find the length of the side of the square and the radius of the circle.

87. Geometry Find formulas for the length l and width w of a rectangle in terms of its area A and perimeter P.

88. Geometry Find formulas for the base b and one of the equal sides l of an isosceles triangle in terms of its altitude h and perimeter P.

89. Descartes's Method of Equal Roots Descartes's method for finding tangents depends on the idea that, for many graphs, the tangent line at a given point is the *unique* line that intersects the graph at that point only. We will apply his method to find an equation of the tangent line to the parabola $y = x^2$ at the point $(2, 4)$. See the figure.

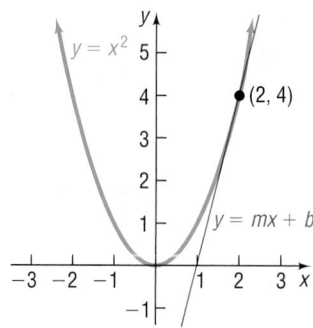

First, we know that the equation of the tangent line must be in the form $y = mx + b$. Using the fact that the point $(2, 4)$ is on the line, we can solve for b in terms of m and get the equation $y = mx + (4 - 2m)$. Now we want $(2, 4)$ to be the *unique* solution to the system

$$\begin{cases} y = x^2 \\ y = mx + 4 - 2m \end{cases}$$

From this system, we get $x^2 - mx + (2m - 4) = 0$. By using the quadratic formula, we get

$$x = \frac{m \pm \sqrt{m^2 - 4(2m - 4)}}{2}$$

To obtain a unique solution for x, the two roots must be equal; in other words, the discriminant $m^2 - 4(2m - 4)$ must be 0. Complete the work to get m, and write an equation of the tangent line.

In Problems 90–96, use Descartes's method from Problem 89 to find the equation of the line tangent to each graph at the given point.

90. $x^2 + y^2 = 10$; at $(1, 3)$
91. $y = x^2 + 2$; at $(1, 3)$
92. $x^2 + y = 5$; at $(-2, 1)$

93. $2x^2 + 3y^2 = 14$; at $(1, 2)$
94. $3x^2 + y^2 = 7$; at $(-1, 2)$
95. $x^2 - y^2 = 3$; at $(2, 1)$

96. $2y^2 - x^2 = 14$; at $(2, 3)$

97. If r_1 and r_2 are two solutions of a quadratic equation $ax^2 + bx + c = 0$, it can be shown that

$$r_1 + r_2 = -\frac{b}{a} \quad \text{and} \quad r_1 r_2 = \frac{c}{a}$$

Solve this system of equations for r_1 and r_2.

Discussion and Writing

98. A circle and a line intersect at most twice. A circle and a parabola intersect at most four times. Deduce that a circle and the graph of a polynomial of degree 3 intersect at most six times. What do you conjecture about a polynomial of degree 4? What about a polynomial of degree n? Can you explain your conclusions using an algebraic argument?

99. Suppose that you are the manager of a sheet metal shop. A customer asks you to manufacture 10,000 boxes, each box being open on top. The boxes are required to have a square base and a 9-cubic-foot capacity. You construct the boxes by cutting out a square from each corner of a square piece of sheet metal and folding along the edges.

(a) What are the dimensions of the square to be cut if the area of the square piece of sheet metal is 100 square feet?
(b) Could you make the box using a smaller piece of sheet metal? Make a list of the dimensions of the box for various pieces of sheet metal.

'Are You Prepared?' Answers

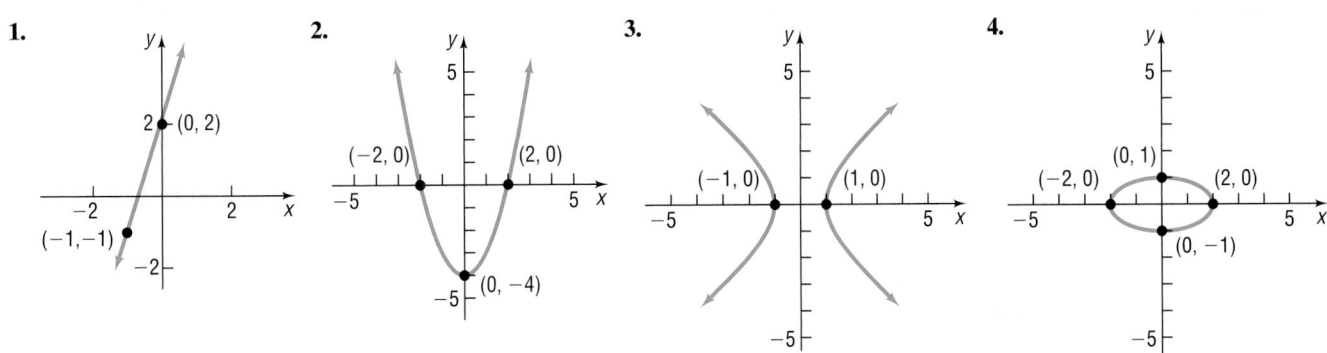

6.7 Systems of Inequalities

PREPARING FOR THIS SECTION *Before getting started, review the following:*

- Solving Linear Inequalities (Appendix A, Section A.9, pp. A69–A70)
- Lines (Foundations, Section F.3, pp. 18–28)
- Circles (Foundations, Section F.4, pp. 33–36)
- Graphing Techniques: Transformation (Section 1.5, pp. 85–94)

Now work the 'Are You Prepared?' problems on page 477.

OBJECTIVES 1 Graph an Inequality
2 Graph a System of Inequalities

In the Appendix, Section A.9, we discuss inequalities in one variable. In this section, we discuss inequalities in two variables.

| EXAMPLE 1 | **Examples of Inequalities in Two Variables** |

(a) $3x + y \leq 6$ (b) $x^2 + y^2 < 4$ (c) $y^2 > x$ ∎

1 Graph an Inequality

An inequality in two variables x and y is **satisfied** by an ordered pair (a, b) if, when x is replaced by a and y by b, a true statement results. The **graph of an inequality in two variables** x and y consists of all points (x, y) whose coordinates satisfy the inequality.

Let's look at an example.

| EXAMPLE 2 | **Graphing an Inequality** |

Graph the linear inequality: $3x + y \leq 6$

Solution We begin by graphing the equation

$$3x + y = 6$$

formed by replacing (for now) the ≤ symbol with an = sign. The graph of the equation is a line. See Figure 12(a). This line is part of the graph of the inequality that we seek because the inequality is nonstrict. (Do you see why? We are seeking points for which $3x + y$ is less than *or equal to* 6.)

Now let's test a few randomly selected points to see whether they belong to the graph of the inequality.

	$3x + y \leq 6$	**Conclusion**
$(4, -1)$	$3(4) + (-1) = 11 > 6$	Does not belong to graph
$(5, 5)$	$3(5) + 5 = 20 > 6$	Does not belong to graph
$(-1, 2)$	$3(-1) + 2 = -1 \leq 6$	Belongs to graph
$(-2, -2)$	$3(-2) + (-2) = -8 \leq 6$	Belongs to graph

Look again at Figure 12(a). Notice that the two points that belong to the graph both lie on the same side of the line, and the two points that do not belong to the graph lie on the opposite side. As it turns out, this is always the case. The graph we seek consists of all points that lie on the same side of the line as $(-1, 2)$ and $(-2, -2)$ and is shown as the shaded region in Figure 12(b).

Figure 12

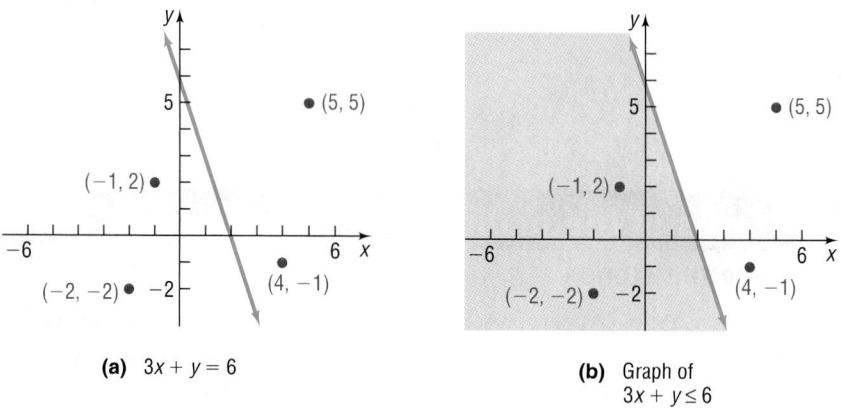

(a) $3x + y = 6$ ‎ ‎ ‎ ‎ ‎ ‎ **(b)** Graph of $3x + y \leq 6$

━ **NOW WORK PROBLEM 15.**

The graph of any inequality in two variables may be obtained in a like way. First, the equation corresponding to the inequality is graphed, using dashes if the inequality is strict and solid marks if it is nonstrict. This graph, in almost every case, will separate the xy-plane into two or more regions. In each region, either all points satisfy the inequality or no points satisfy the inequality. The use of a single test point in each region is all that is required to determine whether the points of that region are part of the graph. The steps to follow are given next.

NOTE
The strict inequalities are $<$ or $>$. The nonstrict inequalities are \leq or \geq. ∎

> **Steps for Graphing an Inequality**
>
> **STEP 1:** Replace the inequality symbol by an equal sign and graph the resulting equation. If the inequality is strict, use dashes; if it is nonstrict, use a solid mark. This graph separates the xy-plane into two or more regions.
>
> **STEP 2:** In each region, select a test point P.
> (a) If the coordinates of P satisfy the inequality, so do all the points in that region. Indicate this by shading the region.
> (b) If the coordinates of P do not satisfy the inequality, none of the points in that region does.

EXAMPLE 3 **Graphing an Inequality**

Graph: $x^2 + y^2 \leq 4$

Solution First, we graph the equation $x^2 + y^2 = 4$, a circle of radius 2, center at the origin. A solid circle will be used because the inequality is not strict. We use two test points, one inside the circle, the other outside.

Inside $(0, 0)$: $x^2 + y^2 = 0^2 + 0^2 = 0 \leq 4$ *Belongs to the graph*
Outside $(4, 0)$: $x^2 + y^2 = 4^2 + 0^2 = 16 > 4$ *Does not belong to the graph*

All the points inside and on the circle satisfy the inequality. See Figure 13.

Figure 13

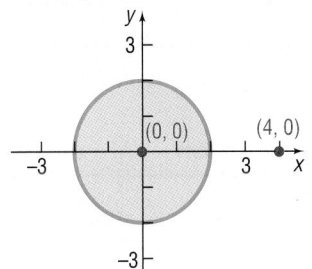

NOW WORK PROBLEM **17**.

Linear Inequalities

Figure 14

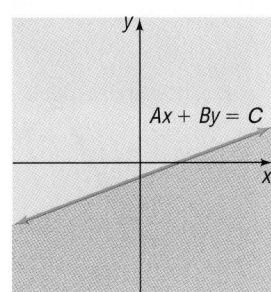

Linear inequalities are inequalities in one of the forms

$$Ax + By < C, \quad Ax + By > C, \quad Ax + By \leq C, \quad Ax + By \geq C$$

where A and B are not both zero.

The graph of the corresponding equation of a linear inequality is a line that separates the xy-plane into two regions, called **half-planes**. See Figure 14.

As shown, $Ax + By = C$ is the equation of the boundary line and it divides the plane into two half-planes: one for which $Ax + By < C$ and the other for which $Ax + By > C$. Because of this, for linear inequalities, only one test point is required.

EXAMPLE 4

Graphing Linear Inequalities

Graph: (a) $y < 2$ (b) $y \geq 2x$

Solution (a) The graph of the equation $y = 2$ is a horizontal line and is not part of the graph of the inequality. Since $(0, 0)$ satisfies the inequality, the graph consists of the half-plane below the line $y = 2$. See Figure 15.

(b) The graph of the equation $y = 2x$ is a line and is part of the graph of the inequality. Using $(3, 0)$ as a test point, we find it does not satisfy the inequality $[0 < 2 \cdot 3]$. Points in the half-plane on the opposite side of $(3, 0)$ satisfy the inequality. See Figure 16.

Figure 15

Figure 16

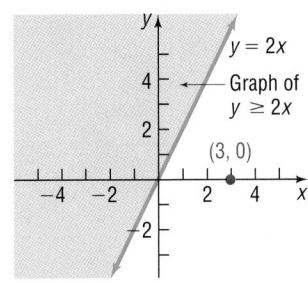

COMMENT: A graphing utility can be used to graph inequalities. To see how, read Section B.6 in Appendix B.

NOW WORK PROBLEM **13**.

2 Graph a System of Inequalities

The **graph of a system of inequalities** in two variables x and y is the set of all points (x, y) that simultaneously satisfy *each* inequality in the system. The graph of a system of inequalities can be obtained by graphing each inequality individually and then determining where, if at all, they intersect.

| EXAMPLE 5 | **Graphing a System of Linear Inequalities** |

Graph the system: $\begin{cases} x + y \geq 2 \\ 2x - y \leq 4 \end{cases}$

Solution We use the test point $(0, 0)$ on each inequality. For example, $(0, 0)$ does not satisfy $x + y \geq 2$, so we shade above the line $x + y = 2$. See Figure 17(a). Also, $(0, 0)$ does satisfy $2x - y \leq 4$, so we shade above the line $2x - y = 4$. See Figure 17(b). The intersection of the shaded regions gives us the result presented in Figure 17(c).

Figure 17

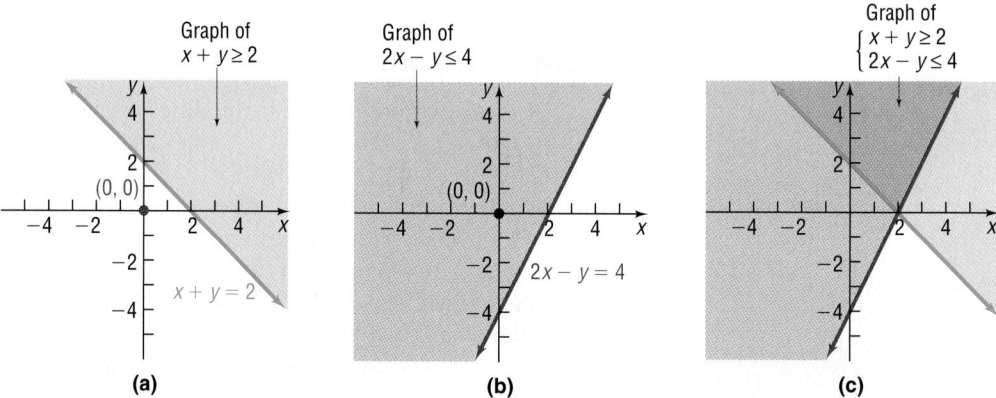

(a) (b) (c)

NOW WORK PROBLEM 23.

| EXAMPLE 6 | **Graphing a System of Linear Inequalities** |

Graph the system: $\begin{cases} x + y \leq 2 \\ x + y \geq 0 \end{cases}$

Solution See Figure 18. The overlapping purple-shaded region between the two boundary lines is the graph of the system.

Figure 18

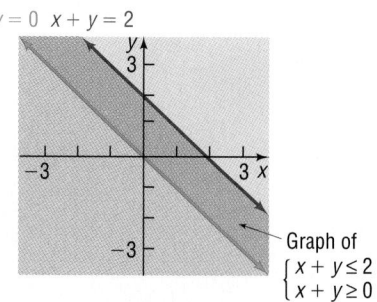

NOW WORK PROBLEM 29.

| EXAMPLE 7 | **Graphing a System of Linear Inequalities** |

Graph the system: $\begin{cases} 2x - y \geq 0 \\ 2x - y \geq 2 \end{cases}$

Solution See Figure 19. The overlapping purple-shaded region is the graph of the system. Note that the graph of the system is identical to the graph of the single inequality $2x - y \geq 2$.

Figure 19

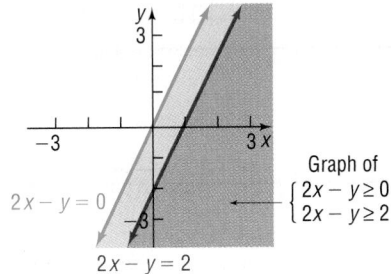

| EXAMPLE 8 | **Graphing a System of Linear Inequalities** |

Graph the system: $\begin{cases} x + 2y \leq 2 \\ x + 2y \geq 6 \end{cases}$

Solution See Figure 20. Because no overlapping region results, there are no points in the xy-plane that simultaneously satisfy each inequality. The system has no solution.

Figure 20

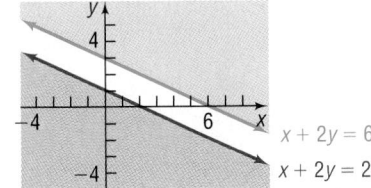

| EXAMPLE 9 | **Graphing a System of Nonlinear Inequalities** |

Graph the region below the graph of $x + y = 2$ and above the graph of $y = x^2 - 4$ by graphing the system:

$$\begin{cases} y \geq x^2 - 4 \\ x + y \leq 2 \end{cases}$$

Label all points of intersection.

Solution Figure 21 shows the graph of the region above the graph of the parabola $y = x^2 - 4$ and below the graph of the line $x + y = 2$. The points of intersection are found by solving the system of equations

$$\begin{cases} y = x^2 - 4 \\ x + y = 2 \end{cases}$$

Figure 21

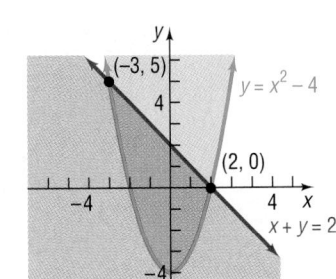

Using substitution, we find

$$x + (x^2 - 4) = 2$$

$$x^2 + x - 6 = 0$$

$$(x + 3)(x - 2) = 0$$

$$x = -3 \text{ or } x = 2$$

The two points of intersection are $(-3, 5)$ and $(2, 0)$. ∎

NOW WORK PROBLEM 37.

EXAMPLE 10	**Graphing a System of Four Linear Inequalities**

Graph the system:
$$\begin{cases} x + y \geq 3 \\ 2x + y \geq 4 \\ x \geq 0 \\ y \geq 0 \end{cases}$$

Solution The two inequalities $x \geq 0$ and $y \geq 0$ require the graph of the system to be in quadrant I. We concentrate on the remaining two inequalities. The intersection of the graphs of these two inequalities and quadrant I, shown in light gray in Figure 22, is the graph of the system.

Figure 22

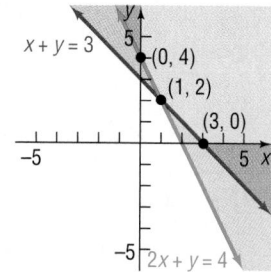

∎

EXAMPLE 11	**Financial Planning**

A retired couple can invest up to $25,000. As their financial adviser, you recommend that they place at least $15,000 in Treasury bills yielding 6% and at most $5000 in corporate bonds yielding 9%.

(a) Using x to denote the amount of money invested in Treasury bills and y the amount invested in corporate bonds, write a system of linear inequalities that describes the possible amounts of each investment. We shall assume that x and y are in thousands of dollars.

(b) Graph the system.

Solution (a) The system of linear inequalities is

$$\begin{cases} x \geq 0 & \text{\textit{x and y are nonnegative variables since they represent}} \\ y \geq 0 & \text{\textit{money invested in thousands of dollars.}} \\ x + y \leq 25 & \text{\textit{The total of the two investments, x + y, cannot}} \\ & \text{\textit{exceed \$25,000.}} \\ x \geq 15 & \text{\textit{At least \$15,000 in Treasury bills}} \\ y \leq 5 & \text{\textit{At most \$5000 in corporate bonds}} \end{cases}$$

(b) See the shaded region in Figure 23. Note that the inequalities $x \geq 0$ and $y \geq 0$ require that the graph of the system be in quadrant I.

Figure 23

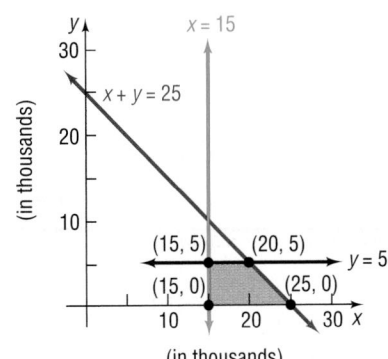

The graph of the system of linear inequalities in Figure 23 is said to be **bounded**, because it can be contained within some circle of sufficiently large radius. A graph that cannot be contained in any circle is said to be **unbounded**. For example, the graph of the system of linear inequalities in Figure 22 is unbounded, since it extends indefinitely in the positive x and positive y directions.

Notice in Figures 22 and 23 that those points belonging to the graph that are also points of intersection of boundary lines have been plotted. Such points are referred to as **vertices** or **corner points** of the graph. The system graphed in Figure 22 has three corner points: $(0, 4)$, $(1, 2)$, and $(3, 0)$. The system graphed in Figure 23 has four corner points: $(15, 0)$, $(25, 0)$, $(20, 5)$, and $(15, 5)$.

These ideas will be used in the next section in developing a method for solving linear programming problems, an important application of linear inequalities.

━━━━━► NOW WORK PROBLEM **45**.

6.7 Assess Your Understanding

'Are You Prepared?'

Answers are given at the end of these exercises. If you get a wrong answer, read the pages listed in red.

1. Solve the inequality: $3x + 4 < 8 - x$. (pp. A69–A70)

2. Graph the equation: $3x - 2y = 6$. (pp. 18–28)

3. Graph the equation: $x^2 + y^2 = 9$. (pp. 33–36)

4. Graph the equation: $y = x^2 + 4$. (pp. 85–86)

5. *True or False:* The lines $2x + y = 4$ and $4x + 2y = 0$ are parallel. (pp. 26–27)

6. The graph of $y = (x - 2)^2$ may be obtained by shifting the graph of _____ to the (left/right) a distance of _____ units. (pp. 86–87)

Concepts and Vocabulary

7. An inequality in two variables x and y is _____ by an ordered pair (a, b) if, when x is replaced by a and y by b, a true statement results.

8. The graph of a linear inequality is called a(n) _____.

9. *True or False:* The graph of a linear inequality is a line.

10. *True or False:* The graph of a system of linear inequalities is sometimes unbounded.

Skill Building

In Problems 11–22, graph each inequality.

11. $x \geq 0$

12. $y \geq 0$

13. $x \geq 4$

14. $y \leq 2$

15. $2x + y \geq 6$

16. $3x + 2y \leq 6$

17. $x^2 + y^2 > 1$

18. $x^2 + y^2 \leq 9$

19. $y \leq x^2 - 1$

20. $y > x^2 + 2$

21. $xy \geq 4$

22. $xy \leq 1$

In Problems 23–34, graph each system of linear inequalities.

23. $\begin{cases} x + y \le 2 \\ 2x + y \ge 4 \end{cases}$

24. $\begin{cases} 3x - y \ge 6 \\ x + 2y \le 2 \end{cases}$

25. $\begin{cases} 2x - y \le 4 \\ 3x + 2y \ge -6 \end{cases}$

26. $\begin{cases} 4x - 5y \le 0 \\ 2x - y \ge 2 \end{cases}$

27. $\begin{cases} 2x - 3y \le 0 \\ 3x + 2y \le 6 \end{cases}$

28. $\begin{cases} 4x - y \ge 2 \\ x + 2y \ge 2 \end{cases}$

29. $\begin{cases} x - 2y \le 6 \\ 2x - 4y \ge 0 \end{cases}$

30. $\begin{cases} x + 4y \le 8 \\ x + 4y \ge 4 \end{cases}$

31. $\begin{cases} 2x + y \ge -2 \\ 2x + y \ge 2 \end{cases}$

32. $\begin{cases} x - 4y \le 4 \\ x - 4y \ge 0 \end{cases}$

33. $\begin{cases} 2x + 3y \ge 6 \\ 2x + 3y \le 0 \end{cases}$

34. $\begin{cases} 2x + y \ge 0 \\ 2x + y \ge 2 \end{cases}$

In Problems 35–42, graph each system of inequalities.

35. $\begin{cases} x^2 + y^2 \le 9 \\ x + y \ge 3 \end{cases}$

36. $\begin{cases} x^2 + y^2 \ge 9 \\ x + y \le 3 \end{cases}$

37. $\begin{cases} y \ge x^2 - 4 \\ y \le x - 2 \end{cases}$

38. $\begin{cases} y^2 \le x \\ y \ge x \end{cases}$

39. $\begin{cases} x^2 + y^2 \le 16 \\ y \ge x^2 - 4 \end{cases}$

40. $\begin{cases} x^2 + y^2 \le 25 \\ y \le x^2 - 5 \end{cases}$

41. $\begin{cases} xy \ge 4 \\ y \ge x^2 + 1 \end{cases}$

42. $\begin{cases} y + x^2 \le 1 \\ y \ge x^2 - 1 \end{cases}$

In Problems 43–52, graph each system of linear inequalities. Tell whether the graph is bounded or unbounded, and label the corner points.

43. $\begin{cases} x \ge 0 \\ y \ge 0 \\ 2x + y \le 6 \\ x + 2y \le 6 \end{cases}$

44. $\begin{cases} x \ge 0 \\ y \ge 0 \\ x + y \ge 4 \\ 2x + 3y \ge 6 \end{cases}$

45. $\begin{cases} x \ge 0 \\ y \ge 0 \\ x + y \ge 2 \\ 2x + y \ge 4 \end{cases}$

46. $\begin{cases} x \ge 0 \\ y \ge 0 \\ 3x + y \le 6 \\ 2x + y \le 2 \end{cases}$

47. $\begin{cases} x \ge 0 \\ y \ge 0 \\ x + y \ge 2 \\ 2x + 3y \le 12 \\ 3x + y \le 12 \end{cases}$

48. $\begin{cases} x \ge 0 \\ y \ge 0 \\ x + y \ge 2 \\ x + y \le 10 \\ 2x + y \le 3 \end{cases}$

49. $\begin{cases} x \ge 0 \\ y \ge 0 \\ x + y \ge 2 \\ x + y \le 8 \\ 2x + y \le 10 \end{cases}$

50. $\begin{cases} x \ge 0 \\ y \ge 0 \\ x + y \ge 2 \\ x + y \le 8 \\ x + 2y \ge 1 \end{cases}$

51. $\begin{cases} x \ge 0 \\ y \ge 0 \\ x + 2y \ge 1 \\ x + 2y \le 10 \end{cases}$

52. $\begin{cases} x \ge 0 \\ y \ge 0 \\ x + 2y \ge 1 \\ x + 2y \le 10 \\ x + y \ge 2 \\ x + y \le 8 \end{cases}$

In Problems 53–56, write a system of linear inequalities that has the given graph.

53.

54.

55.

56.

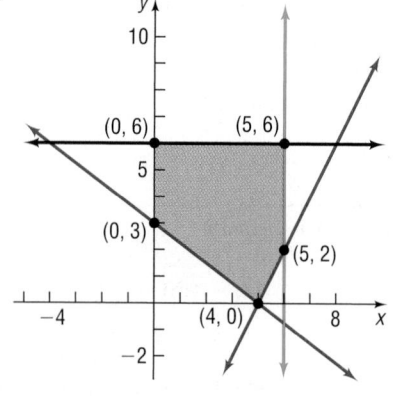

Applications and Extensions

57. Financial Planning A retired couple has up to $50,000 to invest. As their financial adviser, you recommend that they place at least $35,000 in Treasury bills yielding 7% and at most $10,000 in corporate bonds yielding 10%.
(a) Using x to denote the amount of money invested in Treasury bills and y the amount invested in corporate bonds, write a system of linear inequalities that describes the possible amounts of each investment.
(b) Graph the system and label the corner points.

58. Manufacturing Trucks Mike's Toy Truck Company manufacturers two models of toy trucks, a standard model and a deluxe model. Each standard model requires 2 hours for painting and 3 hours for detail work; each deluxe model requires 3 hours for painting and 4 hours for detail work. Two painters and three detail workers are employed by the company, and each works 40 hours per week.
(a) Using x to denote the number of standard model trucks and y to denote the number of deluxe model trucks, write a system of linear inequalities that describes the possible number of each model of truck that can be manufactured in a week.
(b) Graph the system and label the corner points.

59. Blending Coffee Bill's Coffee House, a store that specializes in coffee, has available 75 pounds of A grade coffee and 120 pounds of B grade coffee. These will be blended into 1 pound packages as follows: An economy blend that contains 4 ounces of A grade coffee and 12 ounces of B grade coffee and a superior blend that contains 8 ounces of A grade coffee and 8 ounces of B grade coffee.
(a) Using x to denote the number of packages of the economy blend and y to denote the number of packages of the superior blend, write a system of linear inequalities that describes the possible number of packages of each kind of blend.
(b) Graph the system and label the corner points.

60. Mixed Nuts Nola's Nuts, a store that specializes in selling nuts, has available 90 pounds of cashews and 120 pounds of peanuts. These are to be mixed in 12-ounce packages as follows: a lower-priced package containing 8 ounces of peanuts and 4 ounces of cashews and a quality package containing 6 ounces of peanuts and 6 ounces of cashews.
(a) Use x to denote the number of lower-priced packages and use y to denote the number of quality packages. Write a system of linear inequalities that describes the possible number of each kind of package.
(b) Graph the system and label the corner points.

61. Transporting Goods A small truck can carry no more than 1600 pounds of cargo nor more than 150 cubic feet of cargo. A printer weighs 20 pounds and occupies 3 cubic feet of space. A microwave oven weighs 30 pounds and occupies 2 cubic feet of space.
(a) Using x to represent the number of microwave ovens and y to represent the number of printers, write a system of linear inequalities that describes the number of ovens and printers that can be hauled by the truck.
(b) Graph the system and label the corner points.

'Are You Prepared?' Answers

1. $\{x \mid x < 1\}$
2.
3.
4.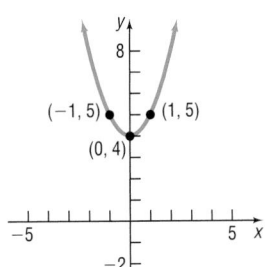
5. True
6. $y = x^2$; right; 2

6.8 Linear Programming

OBJECTIVES 1 Set up a Linear Programming Problem
2 Solve a Linear Programming Problem

Historically, linear programming evolved as a technique for solving problems involving resource allocation of goods and materials for the U.S. Air Force during

World War II. Today, linear programming techniques are used to solve a wide variety of problems, such as optimizing airline scheduling and establishing telephone lines. Although most practical linear programming problems involve systems of several hundred linear inequalities containing several hundred variables, we will limit our discussion to problems containing only two variables, because we can solve such problems using graphing techniques.[*]

1 Set up a Linear Programming Problem

We begin by returning to Example 11 of the previous section.

EXAMPLE 1 | **Financial Planning**

A retired couple has up to $25,000 to invest. As their financial adviser, you recommend that they place at least $15,000 in Treasury bills yielding 6% and at most $5000 in corporate bonds yielding 9%. How much money should be placed in each investment so that income is maximized? ∎

The problem given in Example 1 is typical of a *linear programming problem*. The problem requires that a certain linear expression, the income, be maximized. If I represents income, x the amount invested in Treasury bills at 6%, and y the amount invested in corporate bonds at 9%, then

$$I = 0.06x + 0.09y$$

We shall assume, as before, that I, x, and y are in thousands of dollars.

The linear expression $I = 0.06x + 0.09y$ is called the **objective function**. Furthermore, the problem requires that the maximum income be achieved under certain conditions or **constraints**, each of which is a linear inequality involving the variables. (See Example 11 in Section 6.7.) The linear programming problem given in Example 1 may be restated as

$$\text{Maximize} \qquad I = 0.06x + 0.09y$$

subject to the conditions that

$$\begin{aligned} x \geq 0, \quad y &\geq 0 \\ x + y &\leq 25 \\ x &\geq 15 \\ y &\leq 5 \end{aligned}$$

In general, every linear programming problem has two components:

1. A linear objective function that is to be maximized or minimized
2. A collection of linear inequalities that must be satisfied simultaneously

A **linear programming problem** in two variables x and y consists of maximizing (or minimizing) a linear objective function

$$z = Ax + By, \qquad A \text{ and } B \text{ are real numbers, not both } 0$$

subject to certain conditions, or constraints, expressible as linear inequalities in x and y.

[*]The **simplex method** is a way to solve linear programming problems involving many inequalities and variables. This method was developed by George Dantzig in 1946 and is particularly well suited for computerization. In 1984, Narendra Karmarkar of Bell Laboratories discovered a way of solving large linear programming problems that improves on the simplex method.

2 Solve a Linear Programing Problem

To maximize (or minimize) the quantity $z = Ax + By$, we need to identify points (x, y) that make the expression for z the largest (or smallest) possible. But not all points (x, y) are eligible; only those that also satisfy each linear inequality (constraint) can be used. We refer to each point (x, y) that satisfies the system of linear inequalities (the constraints) as a **feasible point**. In a linear programming problem, we seek the feasible point(s) that maximizes (or minimizes) the objective function.

Let's look again at the linear programming problem in Example 1.

EXAMPLE 2 **Analyzing a Linear Programming Problem**

Consider the linear programming problem

$$\text{Maximize} \quad I = 0.06x + 0.09y$$

subject to the conditions that

$$x \leq 0$$
$$y \geq 0$$
$$x + y \leq 25$$
$$x \geq 15$$
$$y \leq 5$$

Graph the constraints. Then graph the objective function for $I = 0, 0.9, 1.35, 1.65,$ and 1.8.

Solution Figure 24 shows the graph of the constraints. We superimpose on this graph the graph of the objective function for the given values of I.

Figure 24

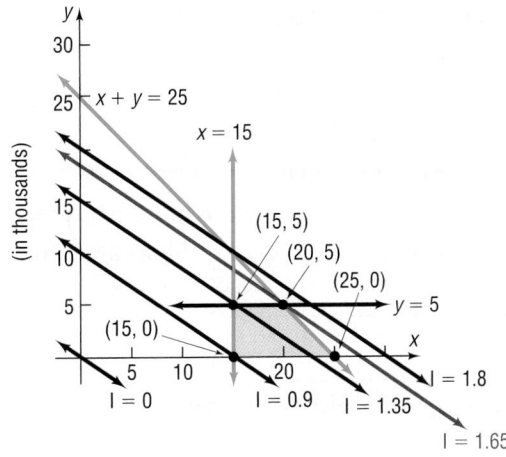

For $I = 0$, the objective function is the line $0 = 0.06x + 0.09y$.

For $I = 0.9$, the objective function is the line $0.9 = 0.06x + 0.09y$.

For $I = 1.35$, the objective function is the line $1.35 = 0.06x + 0.09y$.

For $I = 1.65$, the objective function is the line $1.65 = 0.06x + 0.09y$.

For $I = 1.8$, the objective function is the line $1.8 = 0.06x + 0.09y$. ■

A **solution** to a linear programming problem consists of a feasible point that maximizes (or minimizes) the objective function, together with the corresponding value of the objective function.

One condition for a linear programming problem in two variables to have a solution is that the graph of the feasible points be bounded. (Refer to page 477.)

If none of the feasible points maximizes (or minimizes) the objective function, or if there are no feasible points, the linear programming problem has no solution.

Consider the linear programming problem stated in Example 2, and look again at Figure 24. The feasible points are the points that lie in the shaded region. For example, $(20, 3)$ is a feasible point, as are $(15, 5)$, $(20, 5)$, $(18, 4)$, and so on. To find the solution of the problem requires that we find a feasible point (x, y) that makes $I = 0.06x + 0.09y$ as large as possible. Notice that as I increases in value from $I = 0$ to $I = 0.9$ to $I = 1.35$ to $I = 1.65$ to $I = 1.8$ we obtain a collection of parallel lines. Furthermore, notice that the largest value of I that can be obtained using feasible points is $I = 1.65$, which corresponds to the line $1.65 = 0.06x + 0.09y$. Any larger value of I results in a line that does not pass through any feasible points. Finally, notice that the feasible point that yields $I = 1.65$ is the point $(20, 5)$, a corner point. These observations form the basis of the following result, which we state without proof.

Theorem

> ### Location of the Solution of a Linear Programming Problem
>
> If a linear programming problem has a solution, it is located at a corner point of the graph of the feasible points.
>
> If a linear programming problem has multiple solutions, at least one of them is located at a corner point of the graph of the feasible points.
>
> In either case, the corresponding value of the objective function is unique.

We shall not consider here linear programming problems that have no solution. As a result, we can outline the procedure for solving a linear programming problem as follows:

Procedure for Solving a Linear Programming Problem

STEP 1: Write an expression for the quantity to be maximized (or minimized). This expression is the objective function.

STEP 2: Write all the constraints as a system of linear inequalities and graph the system.

STEP 3: List the corner points of the graph of the feasible points.

STEP 4: List the corresponding values of the objective function at each corner point. The largest (or smallest) of these is the solution.

EXAMPLE 3	**Solving a Minimum Linear Programming Problem**

Minimize the expression

$$z = 2x + 3y$$

subject to the constraints

$$y \leq 5, \qquad x \leq 6 \qquad x + y \geq 2, \qquad x \geq 0, \qquad y \geq 0$$

Solution The objective function is $z = 2x + 3y$. We seek the smallest value of z that can occur if x and y are solutions of the system of linear inequalities

$$\begin{cases} y \leq 5 \\ x \leq 6 \\ x + y \geq 2 \\ x \geq 0 \\ y \geq 0 \end{cases}$$

The graph of this system (the set of feasible points) is shown as the shaded region in Figure 25. We have also plotted the corner points. Table 1 lists the corner points and the corresponding values of the objective function. From the table, we can see that the minimum value of z is 4, and it occurs at the point $(2, 0)$.

Figure 25

Table 1

Corner Point (x, y)	Value of the Objective Function $z = 2x + 3y$
$(0, 2)$	$z = 2(0) + 3(2) = 6$
$(0, 5)$	$z = 2(0) + 3(5) = 15$
$(6, 5)$	$z = 2(6) + 3(5) = 27$
$(6, 0)$	$z = 2(6) + 3(0) = 12$
$(2, 0)$	$z = 2(2) + 3(0) = 4$

NOW WORK PROBLEMS **5** AND **11**.

EXAMPLE 4 **Maximizing Profit**

At the end of every month, after filling orders for its regular customers, a coffee company has some pure Colombian coffee and some special-blend coffee remaining. The practice of the company has been to package a mixture of the two coffees into 1-pound packages as follows: a low-grade mixture containing 4 ounces of Colombian coffee and 12 ounces of special-blend coffee and a high-grade mixture containing 8 ounces of Colombian and 8 ounces of special-blend coffee. A profit of $0.30 per package is made on the low-grade mixture, whereas a profit of $0.40 per package is made on the high-grade mixture. This month, 120 pounds of special-blend coffee and 100 pounds of pure Colombian coffee remain. How many packages of each mixture should be prepared to achieve a maximum profit? Assume that all packages prepared can be sold.

Solution We begin by assigning symbols for the two variables.

$$x = \text{Number of packages of the low-grade mixture}$$

$$y = \text{Number of packages of the high-grade mixture}$$

If P denotes the profit, then

$$P = \$0.30x + \$0.40y$$

This expression is the objective function. We seek to maximize P subject to certain constraints on x and y. Because x and y represent numbers of packages, the only meaningful values for x and y are nonnegative integers. So we have the two constraints

$$x \geq 0, \quad y \geq 0 \quad \text{Nonnegative constraints}$$

We also have only so much of each type of coffee available. For example, the total amount of Colombian coffee used in the two mixtures cannot exceed 100 pounds, or 1600 ounces. Because we use 4 ounces in each low-grade package and 8 ounces in each high-grade package, we are led to the constraint

$$4x + 8y \leq 1600 \quad \text{Colombian coffee constraint}$$

Similarly, the supply of 120 pounds, or 1920 ounces, of special-blend coffee leads to the constraint

$$12x + 8y \leq 1920 \quad \text{Special-blend coffee constraint}$$

The linear programming problem may be stated as

$$\text{Maximize} \quad P = 0.3x + 0.4y$$

subject to the constraints

$$x \geq 0, \quad y \geq 0, \quad 4x + 8y \leq 1600, \quad 12x + 8y \leq 1920$$

The graph of the constraints (the feasible points) is illustrated in Figure 26. We list the corner points and evaluate the objective function at each. In Table 2, we can see that the maximum profit, $84, is achieved with 40 packages of the low-grade mixture and 180 packages of the high-grade mixture.

Figure 26

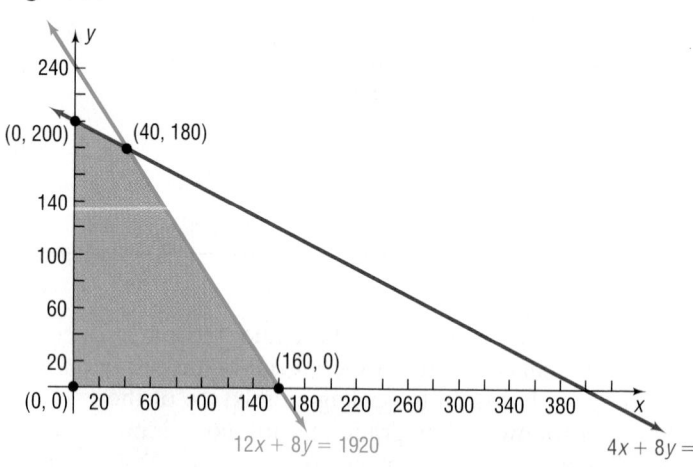

$12x + 8y = 1920$ $4x + 8y = 1600$

Table 2

Corner Point (x, y)	Value of Profit $P = 0.3x + 0.4y$
(0, 0)	$P = 0$
(0, 200)	$P = 0.3(0) + 0.4(200) = \80
(40, 180)	$P = 0.3(40) + 0.4(180) = \84
(160, 0)	$P = 0.3(160) + 0.4(0) = \48

NOW WORK PROBLEM 19.

6.8 Assess Your Understanding

Concepts and Vocabulary

1. A linear programming problem requires that a linear expression, called the _____ _____, be maximized or minimized.

2. *True or False:* If a linear programming problem has a solution, it is located at a corner point of the graph of the feasible points.

Skill Building

In Problems 3–8, find the maximum and minimum value of the given objective function of a linear programming problem. The figure illustrates the graph of the feasible points.

3. $z = x + y$

4. $z = 2x + 3y$

5. $z = x + 10y$

6. $z = 10x + y$

7. $z = 5x + 7y$

8. $z = 7x + 5y$

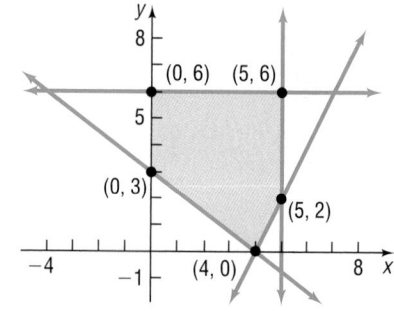

In Problems 9–18, solve each linear programming problem.

9. Maximize $z = 2x + y$ subject to $x \geq 0, \quad y \geq 0, \quad x + y \leq 6, \quad x + y \geq 1$

10. Maximize $z = x + 3y$ subject to $x \geq 0, \quad y \geq 0, \quad x + y \geq 3, \quad x \leq 5, \quad y \leq 7$

11. Minimize $z = 2x + 5y$ subject to $x \geq 0, \quad y \geq 0, \quad x + y \geq 2, \quad x \leq 5, \quad y \leq 3$

12. Minimize $z = 3x + 4y$ subject to $x \geq 0, \quad y \geq 0, \quad 2x + 3y \geq 6, \quad x + y \leq 8$

13. Maximize $z = 3x + 5y$ subject to $x \geq 0, \quad y \geq 0, \quad x + y \geq 2, \quad 2x + 3y \leq 12, \quad 3x + 2y \leq 12$

14. Maximize $z = 5x + 3y$ subject to $x \geq 0$, $y \geq 0$, $x + y \geq 2$, $x + y \leq 8$, $2x + y \leq 10$

15. Minimize $z = 5x + 4y$ subject to $x \geq 0$, $y \geq 0$, $x + y \geq 2$, $2x + 3y \leq 12$, $3x + y \leq 12$

16. Minimize $z = 2x + 3y$ subject to $x \geq 0$, $y \geq 0$, $x + y \geq 3$, $x + y \leq 9$, $x + 3y \geq 6$

17. Maximize $z = 5x + 2y$ subject to $x \geq 0$, $y \geq 0$, $x + y \leq 10$, $2x + y \geq 10$, $x + 2y \geq 10$

18. Maximize $z = 2x + 4y$ subject to $x \geq 0$, $y \geq 0$, $2x + y \geq 4$, $x + y \leq 9$

Applications and Extensions

19. **Maximizing Profit** A manufacturer of skis produces two types: downhill and cross-country. Use the following table to determine how many of each kind of ski should be produced to achieve a maximum profit. What is the maximum profit? What would the maximum profit be if the maximum time available for manufacturing is increased to 48 hours?

	Downhill	Cross-country	Maximum Time Available
Manufacturing time per ski	2 hours	1 hour	40 hours
Finishing time per ski	1 hour	1 hour	32 hours
Profit per ski	$70	$50	

20. **Farm Management** A farmer has 70 acres of land available for planting either soybeans or wheat. The cost of preparing the soil, the workdays required, and the expected profit per acre planted for each type of crop are given in the following table:

	Soybeans	Wheat
Preparation cost per acre	$60	$30
Workdays required per acre	3	4
Profit per acre	$180	$100

The farmer cannot spend more than $1800 in preparation costs nor use more than a total of 120 workdays. How many acres of each crop should be planted to maximize the profit? What is the maximum profit? What is the maximum profit if the farmer is willing to spend no more than $2400 on preparation?

21. **Farm Management** A small farm in Illinois has 100 acres of land available on which to grow corn and soybeans. The following table shows the cultivation cost per acre, the labor cost per acre, and the expected profit per acre. The column on the right shows the amount of money available for each of these expenses. Find the number of acres of each crop that should be planted to maximize profit.

	Soybeans	Corn	Money Available
Cultivation cost per acre	$40	$60	$1800
Labor cost per acre	$60	$60	$2400
Profit per acre	$200	$250	

22. **Dietary Requirements** A certain diet requires at least 60 units of carbohydrates, 45 units of protein, and 30 units of fat each day. Each ounce of Supplement A provides 5 units of carbohydrates, 3 units of protein, and 4 units of fat. Each ounce of Supplement B provides 2 units of carbohydrates, 2 units of protein, and 1 unit of fat. If Supplement A costs $1.50 per ounce and Supplement B costs $1.00 per ounce, how many ounces of each supplement should be taken daily to minimize the cost of the diet?

23. **Production Scheduling** In a factory, machine 1 produces 8-inch pliers at the rate of 60 units per hour and 6-inch pliers at the rate of 70 units per hour. Machine 2 produces 8-inch pliers at the rate of 40 units per hour and 6-inch pliers at the

rate of 20 units per hour. It costs $50 per hour to operate machine 1, and machine 2 costs $30 per hour to operate. The production schedule requires that at least 240 units of 8-inch pliers and at least 140 units of 6-inch pliers be produced during each 10-hour day. Which combination of machines will cost the least money to operate?

24. **Farm Management** An owner of a fruit orchard hires a crew of workers to prune at least 25 of his 50 fruit trees. Each newer tree requires one hour to prune, while each older tree needs one-and-a-half hours. The crew contracts to work for at least 30 hours and charge $15 for each newer tree and $20 for each older tree. To minimize his cost, how many of each kind of tree will the orchard owner have pruned? What will be the cost?

25. **Managing a Meat Market** A meat market combines ground beef and ground pork in a single package for meat loaf. The ground beef is 75% lean (75% beef, 25% fat) and costs the market $0.75 per pound. The ground pork is 60% lean and costs the market $0.45 per pound. The meat loaf must be at least 70% lean. If the market wants to use at least 50 lb of its available pork, but no more than 200 lb of its available ground beef, how much ground beef should be mixed with ground pork so that the cost is minimized?

26. **Return on Investment** An investment broker is instructed by her client to invest up to $20,000, some in a junk bond yielding 9% per annum and some in Treasury bills yielding 7% per annum. The client wants to invest at least $8000 in T-bills and no more than $12,000 in the junk bond.

(a) How much should the broker recommend that the client place in each investment to maximize income if the client insists that the amount invested in T-bills must equal or exceed the amount placed in junk bonds?

(b) How much should the broker recommend that the client place in each investment to maximize income if the client insists that the amount invested in T-bills must not exceed the amount placed in junk bonds?

27. **Maximizing Profit on Ice Skates** A factory manufactures two kinds of ice skates: racing skates and figure skates. The racing skates require 6 work-hours in the fabrication department, whereas the figure skates require 4 work-hours there. The racing skates require 1 work-hour in the finishing department, whereas the figure skates require 2 work-hours there. The fabricating department has available at most 120 work-hours per day, and the finishing department has no more than 40 work-hours per day available. If the profit on each racing skate is $10 and the profit on each figure skate is $12, how many of each should be manufactured each day to maximize profit? (Assume that all skates made are sold.)

28. **Financial Planning** A retired couple has up to $50,000 to place in fixed-income securities. Their financial adviser suggests two securities to them: one is an AAA bond that yields 8% per annum; the other is a certificate of deposit (CD) that yields 4%. After careful consideration of the alternatives, the couple decides to place at most $20,000 in the AAA bond and at least $15,000 in the CD. They also instruct the financial adviser to place at least as much in the CD as in the AAA bond. How should the financial adviser proceed to maximize the return on their investment?

29. **Product Design** An entrepreneur is having a design group produce at least six samples of a new kind of fastener that he wants to market. It costs $9.00 to produce each metal fastener and $4.00 to produce each plastic fastener. He wants to have at least two of each version of the fastener and needs to have all the samples 24 hours from now. It takes 4 hours to produce each metal sample and 2 hours to produce each plastic sample. To minimize the cost of the samples, how many of each kind should the entrepreneur order? What will be the cost of the samples?

30. **Animal Nutrition** Kevin's dog Amadeus likes two kinds of canned dog food. Gourmet Dog costs 40 cents a can and

has 20 units of a vitamin complex; the calorie content is 75 calories. Chow Hound costs 32 cents a can and has 35 units of vitamins and 50 calories. Kevin likes Amadeus to have at least 1175 units of vitamins a month and at least 2375 calories during the same time period. Kevin has space to store only 60 cans of dog food at a time. How much of each kind of dog food should Kevin buy each month to minimize his cost?

31. **Airline Revenue** An airline has two classes of service: first class and coach. Management's experience has been that each aircraft should have at least 8 but no more than 16 first-class seats and at least 80 but not more than 120 coach seats.
(a) If management decides that the ratio of first class to coach seats should never exceed 1:12, with how many of each type of seat should an aircraft be configured to maximize revenue?
(b) If management decides that the ratio of first class to coach seats should never exceed 1:8, with how many of each type of seat should an aircraft be configured to maximize revenue?
(c) If you were management, what would you do?
[**Hint:** Assume that the airline charges $C for a coach seat and $F for a first-class seat; $C > 0$, $F > C$.]

32. **Minimizing Cost** A farm that specializes in raising frying chickens supplements the regular chicken feed with four vitamins. The owner wants the supplemental food to contain at least 50 units of vitamin I, 90 units of vitamin II, 60 units of vitamin III, and 100 units of vitamin IV per 100 ounces of feed. Two supplements are available: Supplement A, which contains 5 units of vitamin I, 25 units of vitamin II, 10 units of vitamin III, and 35 units of vitamin IV per ounce, and Supplement B, which contains 25 units of vitamin I, 10 units of vitamin II, 10 units of vitamin III, and 20 units of vitamin IV per ounce. If Supplement A costs $0.06 per ounce and Supplement B costs $0.08 per ounce, how much of each supplement should the manager of the farm buy to add to each 100 ounces of feed to keep the total cost at a minimum, while still meeting the owner's vitamin specifications?

Discussion and Writing

33. Explain in your own words what a linear programming problem is and how it can be solved.

Chapter Review

Things to Know

Systems of equations (pp. 400–402)

Systems with no solutions are inconsistent. Systems with a solution are consistent.

Consistent systems of linear equations have either a unique solution or an infinite number of solutions.

Determinants and Cramer's Rule (pp. 430, 432, 433 and 435)

Matrix (pp. 415 and 439)	Rectangular array of numbers, called entries
m by n matrix (p. 440)	Matrix with m rows and n columns
Identity matrix I_n (p. 447)	An n by n square matrix whose diagonal entries are 1's, while all other entries are 0's
Inverse of a matrix (p. 448)	A^{-1} is the inverse of A if $AA^{-1} = A^{-1}A = I_n$
Nonsingular matrix (p. 448)	A square matrix that has an inverse

Linear programming (p. 480)

Maximize (or minimize) a linear objective function, $z = Ax + By$, subject to certain conditions, or constraints, expressible as linear inequalities in x and y. A feasible point (x, y) is a point that satisfies the constraints of a linear programming problem.

Location of solution (p. 481)

If a linear programming problem has a solution, it is located at a corner point of the graph of the feasible points. If a linear programming problem has multiple solutions, at least one of them is located at a corner point of the graph of the feasible points. In either case, the corresponding value of the objective function is unique.

Objectives

Section		You should be able to . . .	Review Exercises
6.1	1	Solve systems of equations by substitution (p. 403)	1–14, 101, 102, 105–107
	2	Solve systems of equations by elimination (p. 404)	1–14, 101, 102, 105–107
	3	Identify inconsistent systems of equations containing two variables (p. 405)	9, 10, 13, 98
	4	Express the solution of a system of dependent equations containing two variables (p. 406)	14, 97
	5	Solve systems of three equations containing three variables (p. 407)	15–18, 99, 100, 103
	6	Identify inconsistent systems of equations containing three variables (p. 409)	18
	7	Express the solution of a system of dependent equations containing three variables (p. 409)	17
6.2	1	Write the augmented matrix of a system of linear equations (p. 415)	35–44
	2	Write the system from the augmented matrix (p. 416)	19, 20
	3	Perform row operations on a matrix (p. 416)	35–44
	4	Solve a system of linear equations using matrices (p. 418)	35–44
6.3	1	Evaluate 2 by 2 determinants (p. 430)	45, 46
	2	Use Cramer's Rule to solve a system of two equations containing two variables (p. 430)	51–54
	3	Evaluate 3 by 3 determinants (p. 433)	47–50
	4	Use Cramer's Rule to solve a system of three equations containing three variables (p. 435)	55, 56
	5	Know properties of determinants (p. 436)	57, 58
6.4	1	Find the sum and difference of two matrices (p. 440)	21, 22
	2	Find scalar multiples of a matrix (p. 442)	23, 24
	3	Find the product of two matrices (p. 443)	25–28
	4	Find the inverse of a matrix (p. 448)	29–34
	5	Solve a system of linear equations using inverse matrices (p. 451)	35–37, 39, 40, 43, 44
6.5	1	Decompose $\dfrac{P}{Q}$, where Q has only nonrepeated linear factors (p. 456)	59, 60
	2	Decompose $\dfrac{P}{Q}$, where Q has repeated linear factors (p. 457)	61, 62
	3	Decompose $\dfrac{P}{Q}$, where Q has a nonrepeated irreducible quadratic factor (p. 459)	63, 64, 67, 68
	4	Decompose $\dfrac{P}{Q}$, where Q has a repeated irreducible quadratic factor (p. 460)	65, 66
6.6	1	Solve a system of nonlinear equations using substitution (p. 462)	69–78
	2	Solve a system of nonlinear equations using elimination (p. 463)	69–78
6.7	1	Graph an inequality (p. 471)	79–82
	2	Graph a system of inequalities (p. 474)	83–92, 104
6.8	1	Set up a linear programming problem (p. 480)	108, 109
	2	Solve a linear programming problem (p. 481)	93–96, 108, 109

Review Exercises

In Problems 1–18, solve each system of equations using the method of substitution or the method of elimination. If the system has no solution, say that it is inconsistent.

1. $\begin{cases} 2x - y = 5 \\ 5x + 2y = 8 \end{cases}$

2. $\begin{cases} 2x + 3y = 2 \\ 7x - y = 3 \end{cases}$

3. $\begin{cases} 3x - 4y = 4 \\ x - 3y = \dfrac{1}{2} \end{cases}$

4. $\begin{cases} 2x + y = 0 \\ 5x - 4y = -\dfrac{13}{2} \end{cases}$

5. $\begin{cases} x - 2y - 4 = 0 \\ 3x + 2y - 4 = 0 \end{cases}$ **6.** $\begin{cases} x - 3y + 5 = 0 \\ 2x + 3y - 5 = 0 \end{cases}$ **7.** $\begin{cases} y = 2x - 5 \\ x = 3y + 4 \end{cases}$ **8.** $\begin{cases} x = 5y + 2 \\ y = 5x + 2 \end{cases}$

9. $\begin{cases} x - 3y + 4 = 0 \\ \dfrac{1}{2}x - \dfrac{3}{2}y + \dfrac{4}{3} = 0 \end{cases}$ **10.** $\begin{cases} x + \dfrac{1}{4}y = 2 \\ y + 4x + 2 = 0 \end{cases}$ **11.** $\begin{cases} 2x + 3y - 13 = 0 \\ 3x - 2y = 0 \end{cases}$ **12.** $\begin{cases} 4x + 5y = 21 \\ 5x + 6y = 42 \end{cases}$

13. $\begin{cases} 3x - 2y = 8 \\ x - \dfrac{2}{3}y = 12 \end{cases}$ **14.** $\begin{cases} 2x + 5y = 10 \\ 4x + 10y = 20 \end{cases}$ **15.** $\begin{cases} x + 2y - z = 6 \\ 2x - y + 3z = -13 \\ 3x - 2y + 3z = -16 \end{cases}$

16. $\begin{cases} x + 5y - z = 2 \\ 2x + y + z = 7 \\ x - y + 2z = 11 \end{cases}$ **17.** $\begin{cases} 2x - 4y + z = -15 \\ x + 2y - 4z = 27 \\ 5x - 6y - 2z = -3 \end{cases}$ **18.** $\begin{cases} x - 4y + 3z = 15 \\ -3x + y - 5z = -5 \\ -7x - 5y - 9z = 10 \end{cases}$

In Problems 19 and 20, write the system of equations corresponding to the given augmented matrix.

19. $\begin{bmatrix} 3 & 2 & | & 8 \\ 1 & 4 & | & -1 \end{bmatrix}$

20. $\begin{bmatrix} 1 & 2 & 5 & | & -2 \\ 5 & 0 & -3 & | & 8 \\ 2 & -1 & 0 & | & 0 \end{bmatrix}$

In Problems 21–28, use the following matrices to compute each expression.

$$A = \begin{bmatrix} 1 & 0 \\ 2 & 4 \\ -1 & 2 \end{bmatrix}, \quad B = \begin{bmatrix} 4 & -3 & 0 \\ 1 & 1 & -2 \end{bmatrix}, \quad C = \begin{bmatrix} 3 & -4 \\ 1 & 5 \\ 5 & 2 \end{bmatrix}$$

21. $A + C$ **22.** $A - C$ **23.** $6A$ **24.** $-4B$

25. AB **26.** BA **27.** CB **28.** BC

In Problems 29–34, find the inverse, if there is one, of each matrix. If there is not an inverse, say that the matrix is singular.

29. $\begin{bmatrix} 4 & 6 \\ 1 & 3 \end{bmatrix}$ **30.** $\begin{bmatrix} -3 & 2 \\ 1 & -2 \end{bmatrix}$ **31.** $\begin{bmatrix} 1 & 3 & 3 \\ 1 & 2 & 1 \\ 1 & -1 & 2 \end{bmatrix}$

32. $\begin{bmatrix} 3 & 1 & 2 \\ 3 & 2 & -1 \\ 1 & 1 & 1 \end{bmatrix}$ **33.** $\begin{bmatrix} 4 & -8 \\ -1 & 2 \end{bmatrix}$ **34.** $\begin{bmatrix} -3 & 1 \\ -6 & 2 \end{bmatrix}$

In Problems 35–44, solve each system of equations using matrices. If the system has no solution, say that it is inconsistent.

35. $\begin{cases} 3x - 2y = 1 \\ 10x + 10y = 5 \end{cases}$ **36.** $\begin{cases} 3x + 2y = 6 \\ x - y = -\dfrac{1}{2} \end{cases}$ **37.** $\begin{cases} 5x - 6y - 3z = 6 \\ 4x - 7y - 2z = -3 \\ 3x + y - 7z = 1 \end{cases}$

38. $\begin{cases} 2x + y + z = 5 \\ 4x - y - 3z = 1 \\ 8x + y - z = 5 \end{cases}$ **39.** $\begin{cases} x - 2z = 1 \\ 2x + 3y = -3 \\ 4x - 3y - 4z = 3 \end{cases}$ **40.** $\begin{cases} x + 2y - z = 2 \\ 2x - 2y + z = -1 \\ 6x + 4y + 3z = 5 \end{cases}$

41. $\begin{cases} x - y + z = 0 \\ x - y - 5z - 6 = 0 \\ 2x - 2y + z - 1 = 0 \end{cases}$ **42.** $\begin{cases} 4x - 3y + 5z = 0 \\ 2x + 4y - 3z = 0 \\ 6x + 2y + z = 0 \end{cases}$ **43.** $\begin{cases} x - y - z - t = 1 \\ 2x + y - z + 2t = 3 \\ x - 2y - 2z - 3t = 0 \\ 3x - 4y + z + 5t = -3 \end{cases}$

44. $\begin{cases} x - 3y + 3z - t = 4 \\ x + 2y - z = -3 \\ x + 3z + 2t = 3 \\ x + y + 5z = 6 \end{cases}$

In Problems 45–50, find the value of each determinant.

45. $\begin{vmatrix} 3 & 4 \\ 1 & 3 \end{vmatrix}$ **46.** $\begin{vmatrix} -4 & 0 \\ 1 & 3 \end{vmatrix}$ **47.** $\begin{vmatrix} 1 & 4 & 0 \\ -1 & 2 & 6 \\ 4 & 1 & 3 \end{vmatrix}$

48. $\begin{vmatrix} 2 & 3 & 10 \\ 0 & 1 & 5 \\ -1 & 2 & 3 \end{vmatrix}$

49. $\begin{vmatrix} 2 & 1 & -3 \\ 5 & 0 & 1 \\ 2 & 6 & 0 \end{vmatrix}$

50. $\begin{vmatrix} -2 & 1 & 0 \\ 1 & 2 & 3 \\ -1 & 4 & 2 \end{vmatrix}$

In Problems 51–56, use Cramer's Rule, if applicable, to solve each system.

51. $\begin{cases} x - 2y = 4 \\ 3x + 2y = 4 \end{cases}$

52. $\begin{cases} x - 3y = -5 \\ 2x + 3y = 5 \end{cases}$

53. $\begin{cases} 2x + 3y - 13 = 0 \\ 3x - 2y = 0 \end{cases}$

54. $\begin{cases} 3x - 4y - 12 = 0 \\ 5x + 2y + 6 = 0 \end{cases}$

55. $\begin{cases} x + 2y - z = 6 \\ 2x - y + 3z = -13 \\ 3x - 2y + 3z = -16 \end{cases}$

56. $\begin{cases} x - y + z = 8 \\ 2x + 3y - z = -2 \\ 3x - y - 9z = 9 \end{cases}$

In Problems 57 and 58, use properties of determinants to find the value of each determinant if it is known that

$$\begin{vmatrix} x & y \\ a & b \end{vmatrix} = 8$$

57. $\begin{vmatrix} 2x & y \\ 2a & b \end{vmatrix}$

58. $\begin{vmatrix} y & x \\ b & a \end{vmatrix}$

In Problems 59–68, write the partial fraction decomposition of each rational expression.

59. $\dfrac{6}{x(x-4)}$

60. $\dfrac{x}{(x+2)(x-3)}$

61. $\dfrac{x-4}{x^2(x-1)}$

62. $\dfrac{2x-6}{(x-2)^2(x-1)}$

63. $\dfrac{x}{(x^2+9)(x+1)}$

64. $\dfrac{3x}{(x-2)(x^2+1)}$

65. $\dfrac{x^3}{(x^2+4)^2}$

66. $\dfrac{x^3+1}{(x^2+16)^2}$

67. $\dfrac{x^2}{(x^2+1)(x^2-1)}$

68. $\dfrac{4}{(x^2+4)(x^2-1)}$

In Problems 69–78, solve each system of equations.

69. $\begin{cases} 2x + y + 3 = 0 \\ x^2 + y^2 = 5 \end{cases}$

70. $\begin{cases} x^2 + y^2 = 16 \\ 2x - y^2 = -8 \end{cases}$

71. $\begin{cases} 2xy + y^2 = 10 \\ 3y^2 - xy = 2 \end{cases}$

72. $\begin{cases} 3x^2 - y^2 = 1 \\ 7x^2 - 2y^2 - 5 = 0 \end{cases}$

73. $\begin{cases} x^2 + y^2 = 6y \\ x^2 = 3y \end{cases}$

74. $\begin{cases} 2x^2 + y^2 = 9 \\ x^2 + y^2 = 9 \end{cases}$

75. $\begin{cases} 3x^2 + 4xy + 5y^2 = 8 \\ x^2 + 3xy + 2y^2 = 0 \end{cases}$

76. $\begin{cases} 3x^2 + 2xy - 2y^2 = 6 \\ xy - 2y^2 + 4 = 0 \end{cases}$

77. $\begin{cases} x^2 - 3x + y^2 + y = -2 \\ \dfrac{x^2 - x}{y} + y + 1 = 0 \end{cases}$

78. $\begin{cases} x^2 + x + y^2 = y + 2 \\ x + 1 = \dfrac{2 - y}{x} \end{cases}$

In Problems 79–82 graph each inequality.

79. $3x + 4y \le 12$

80. $2x - 3y \ge 6$

81. $y \le x^2$

82. $x \ge y^2$

In Problems 83–88, graph each system of inequalities. Tell whether the graph is bounded or unbounded, and label the corner points.

83. $\begin{cases} -2x + y \le 2 \\ x + y \ge 2 \end{cases}$

84. $\begin{cases} x - 2y \le 6 \\ 2x + y \ge 2 \end{cases}$

85. $\begin{cases} x \ge 0 \\ y \ge 0 \\ x + y \le 4 \\ 2x + 3y \le 6 \end{cases}$

86. $\begin{cases} x \ge 0 \\ y \ge 0 \\ 3x + y \ge 6 \\ 2x + y \ge 2 \end{cases}$

87. $\begin{cases} x \ge 0 \\ y \ge 0 \\ 2x + y \le 8 \\ x + 2y \ge 2 \end{cases}$

88. $\begin{cases} x \ge 0 \\ y \ge 0 \\ 3x + y \le 9 \\ 2x + 3y \ge 6 \end{cases}$

In Problems 89–92, graph each system of inequalities.

89. $\begin{cases} x^2 + y^2 \le 16 \\ x + y \ge 2 \end{cases}$

90. $\begin{cases} y^2 \le x - 1 \\ x - y \le 3 \end{cases}$

91. $\begin{cases} y \le x^2 \\ xy \le 4 \end{cases}$

92. $\begin{cases} x^2 + y^2 \ge 1 \\ x^2 + y^2 \le 4 \end{cases}$

In Problems 93–96, solve each linear programming problem.

93. Maximize $z = 3x + 4y$ subject to $x \geq 0, y \geq 0, 3x + 2y \geq 6, x + y \leq 8$

94. Maximize $z = 2x + 4y$ subject to $x \geq 0, y \geq 0, x + y \leq 6, x \geq 2$

95. Minimize $z = 3x + 5y$ subject to $x \geq 0, y \geq 0, x + y \geq 1, 3x + 2y \leq 12, x + 3y \leq 12$

96. Minimize $z = 3x + y$ subject to $x \geq 0, y \geq 0, x \leq 8, y \leq 6, 2x + y \geq 4$

97. Find A so that the system of equations has infinitely many solutions.

$$\begin{cases} 2x + 5y = 5 \\ 4x + 10y = A \end{cases}$$

98. Find A so that the system in Problem 97 is inconsistent.

99. Curve Fitting Find the quadratic function $y = ax^2 + bx + c$ that passes through the three points $(0, 1)$, $(1, 0)$, and $(-2, 1)$.

100. Curve Fitting Find the general equation of the circle that passes through the three points $(0, 1)$, $(1, 0)$, and $(-2, 1)$.

[**Hint:** The general equation of a circle is
$$x^2 + y^2 + Dx + Ey + F = 0.]$$

101. Blending Coffee A coffee distributor is blending a new coffee that will cost \$3.90 per pound. It will consist of a blend of \$3.00 per pound coffee and \$6.00 per pound coffee. What amounts of each type of coffee should be mixed to achieve the desired blend?

[**Hint**: Assume that the weight of the blended coffee is 100 pounds.]

$3.00/lb $3.90/lb $6.00/lb

102. Farming A 1000-acre farm in Illinois is used to grow corn and soybeans. The cost per acre for raising corn is \$65, and the cost per acre for soybeans is \$45. If \$54,325 has been budgeted for costs and all the acreage is to be used, how many acres should be allocated for each crop?

103. Cookie Orders A cookie company makes three kinds of cookies, oatmeal raisin, chocolate chip, and shortbread, packaged in small, medium, and large boxes. The small box contains 1 dozen oatmeal raisin and 1 dozen chocolate chip; the medium box has 2 dozen oatmeal raisin, 1 dozen chocolate chip, and 1 dozen shortbread; the large box contains 2 dozen oatmeal raisin, 2 dozen chocolate chip, and 3 dozen shortbread. If you require exactly 15 dozen oatmeal raisin, 10 dozen chocolate chip, and 11 dozen shortbread, how many of each size box should you buy?

104. Mixed Nuts A store that specializes in selling nuts has available 72 pounds of cashews and 120 pounds of peanuts. These are to be mixed in 12-ounce packages as follows: a lower-priced package containing 8 ounces of peanuts and 4 ounces of cashews and a quality package containing 6 ounces of peanuts and 6 ounces of cashews.

(a) Use x to denote the number of lower-priced packages and use y to denote the number of quality packages. Write a system of linear inequalities that describes the possible number of each kind of package.

(b) Graph the system and label the corner points.

105. Determining the Speed of the Current of the Aguarico River On a recent trip to the Cuyabeno Wildlife Reserve in the Amazon region of Ecuador, a 100-kilometer trip by speedboat was taken down the Aguarico River from Chiritza to the Flotel Orellana. As I watched the Amazon unfold, I wondered how fast the speedboat was going and how fast the current of the white-water Aguarico River was. I timed the trip downstream at 2.5 hours and the return trip at 3 hours. What were the two speeds?

106. Finding the Speed of the Jet Stream On a flight between Midway Airport in Chicago and Ft. Lauderdale, Florida, a Boeing 737 jet maintains an airspeed of 475 miles per hour. If the trip from Chicago to Ft. Lauderdale takes 2 hours, 30 minutes and the return flight takes 2 hours, 50 minutes, what is the speed of the jet stream? (Assume that the speed of the jet stream remains constant at the various altitudes of the plane and that the plane flies with the jet stream one way and against it the other way.)

107. Constant Rate Jobs If Bruce and Bryce work together for 1 hour and 20 minutes, they will finish a certain job. If Bryce and Marty work together for 1 hour and 36 minutes, the same job can be finished. If Marty and Bruce work together, they can complete this job in 2 hours and 40 minutes. How long will it take each of them working alone to finish the job?

108. Maximizing Profit on Figurines A factory manufactures two kinds of ceramic figurines: a dancing girl and a mermaid. Each requires three processes: molding, painting, and glazing. The daily labor available for molding is no more than 90 work-hours, labor available for painting does not exceed 120 work-hours, and labor available for glazing is no more than 60 work-hours. The dancing girl requires 3 work-hours for molding, 6 work-hours for painting, and 2 work-hours for glazing. The mermaid requires 3 work-hours for molding, 4 work-hours for painting, and 3 work-hours for glazing. If the profit on each figurine is \$25 for dancing girls and \$30 for mermaids, how many of each should be produced each day to maximize profit? If management decides to produce the number of each figurine that maximizes profit, determine which of these processes has work-hours assigned to it that are not used.

109. Minimizing Production Cost A factory produces gasoline engines and diesel engines. Each week the factory is obligated to deliver at least 20 gasoline engines and at least 15 diesel engines. Due to physical limitations, however, the factory cannot make more than 60 gasoline engines nor more than 40 diesel engines in any given week. Finally, to prevent layoffs, a total of at least 50 engines must be produced. If gasoline engines cost $450 each to produce and diesel engines cost $550 each to produce, how many of each should be produced per week to mini-mize the cost? What is the excess capacity of the factory; that is, how many of each kind of engine is being produced in excess of the number that the factory is obligated to deliver?

110. Describe four ways of solving a system of three linear equations containing three variables. Which method do you prefer? Why?

Chapter Test

In Problems 1–4, solve each system of equations using the method of substitution or the method of elimination. If the system has no solution, say that it is inconsistent.

1. $\begin{cases} -2x + y = -7 \\ 4x + 3y = 9 \end{cases}$

2. $\begin{cases} \frac{1}{3}x - 2y = 1 \\ 5x - 30y = 18 \end{cases}$

3. $\begin{cases} x - y + 2z = 5 \\ 3x + 4y - z = -2 \\ 5x + 2y + 3z = 8 \end{cases}$

4. $\begin{cases} 3x + 2y - 8z = -3 \\ -x - \frac{2}{3}y + z = 1 \\ 6x - 3y + 15z = 8 \end{cases}$

5. Write the augmented matrix corresponding to the system of equations: $\begin{cases} 4x - 5y + z = 0 \\ -2x - y + 6 = -19 \\ x + 5y - 5z = 10 \end{cases}$

6. Write the system of equations corresponding to the augmented matrix: $\begin{bmatrix} 3 & 2 & 4 & -6 \\ 1 & 0 & 8 & 2 \\ -2 & 1 & 3 & -11 \end{bmatrix}$

In Problems 7–10, use the given matrices to compute each expression.

$$A = \begin{bmatrix} 1 & -1 \\ 0 & -4 \\ 3 & 2 \end{bmatrix}, \quad B = \begin{bmatrix} 1 & -2 & 5 \\ 0 & 3 & 1 \end{bmatrix}, \quad C = \begin{bmatrix} 4 & 6 \\ 1 & -3 \\ -1 & 8 \end{bmatrix}$$

7. $2A + C$ **8.** $A - 3C$ **9.** AC **10.** BA

In Problems 11 and 12, find the inverse of each nonsingular matrix.

11. $A = \begin{bmatrix} 3 & 2 \\ 5 & 4 \end{bmatrix}$

12. $B = \begin{bmatrix} 1 & -1 & 1 \\ 2 & 5 & -1 \\ 2 & 3 & 0 \end{bmatrix}$

In Problems 13–16, solve each system of equations using matrices. If the system has no solution, say that it is inconsistent.

13. $\begin{cases} 6x + 3y = 12 \\ 2x - y = -2 \end{cases}$

14. $\begin{cases} x + \frac{1}{4}y = 7 \\ 8x + 2y = 56 \end{cases}$

15. $\begin{cases} x + 2y + 4z = -3 \\ 2x + 7y + 15z = -12 \\ 4x + 7y + 13z = -10 \end{cases}$

16. $\begin{cases} 2x + 2y - 3z = 5 \\ x - y + 2z = 8 \\ 3x + 5y - 8z = -2 \end{cases}$

In Problems 17 and 18, find the value of each determinant.

17. $\begin{vmatrix} -2 & 5 \\ 3 & 7 \end{vmatrix}$

18. $\begin{vmatrix} 2 & -4 & 6 \\ 1 & 4 & 0 \\ -1 & 2 & -4 \end{vmatrix}$

In Problems 19 and 20, use Cramer's Rule, if possible, to solve each system.

19. $\begin{cases} 4x + 3y = -23 \\ 3x - 5y = 19 \end{cases}$

20. $\begin{cases} 4x - 3y + 2z = 15 \\ -2x + y - 3z = -15 \\ 5x - 5y + 2z = 18 \end{cases}$

In Problems 21 and 22, solve each system of equations.

21. $\begin{cases} 3x^2 + y^2 = 12 \\ y^2 = 9x \end{cases}$

22. $\begin{cases} 2y^2 - 3x^2 = 5 \\ y - x = 1 \end{cases}$

23. Graph the system of inequalities: $\begin{cases} x^2 + y^2 \le 100 \\ 4x - 3y \ge 0 \end{cases}$

In Problems 24 and 25, write the partial fraction decomposition of each rational expression.

24. $\dfrac{3x + 7}{(x + 3)^2}$

25. $\dfrac{4x^2 - 3}{x(x^2 + 3)^2}$

26. Graph the system of inequalities. Tell whether the graph is bounded or unbounded, and label all corner points.

$$\begin{cases} x \geq 0 \\ y \geq 0 \\ x + 2y \geq 8 \\ 2x - 3y \geq 2 \end{cases}$$

27. Maximize $z = 5x + 8y$ subject to $x \geq 0$, $2x + y \leq 8$, and $x - 3y \leq -3$.

28. Megan went clothes shopping and bought 2 pairs of flare jeans, 2 camisoles, and 4 T-shirts for $90.00. At the same store, Paige bought one pair of flare jeans and 3 T-shirts for $42.50, while Kara bought 1 pair of flare jeans, 3 camisoles, and 2 T-shirts for $62.00. Determine the price of each clothing item.

Chapter Projects

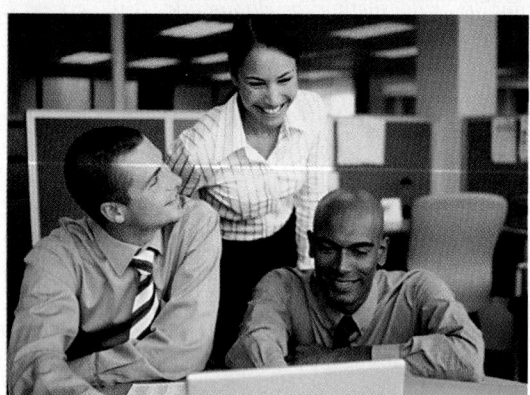

1. Markov Chains
A **Markov chain** (or process) is one in which future outcomes are determined by a current state. Future outcomes are based on probabilities. The probability of moving to a certain state depends only on the state previously occupied and does not vary with time. An example of a Markov chain is the maximum education achieved by children based on the highest education attained by their parents, where the states are (1) earned college degree, (2) high-school diploma only, (3) elementary school only. If p_{ij} is the probability of moving from state i to state j, the **transition matrix** is the $m \times m$ matrix

$$P = \begin{bmatrix} p_{11} & p_{12} & \cdots & p_{1m} \\ \vdots & \vdots & & \vdots \\ p_{m1} & p_{m2} & \cdots & p_{mm} \end{bmatrix}$$

The table represents the probabilities of the highest educational level of children based on the highest educational level of their parents. For example, the table shows that the probability p_{21} is 40% that parents with a high-school education (row 2) will have children with a college education (column 1).

(a) Convert the percentages to decimals.

(b) What is the transition matrix?

(c) Sum across the rows. What do you notice? Why do you think that you obtained this result?

(d) If P is the transition matrix of a Markov chain, the (i, j)th entry of P^n (nth power of P) gives the probability of passing from state i to state j in n stages. What is the probability that a grandchild of a college graduate is a college graduate?

(e) What is the probability that the grandchild of a high school graduate finishes college?

(f) The row vector $v^{(0)} = [0.267 \quad 0.574 \quad 0.159]$ represents the proportion of the U.S. population that has college, high school, and elementary school, respectively, as the highest educational level in 2002.* In a Markov chain the probability distribution $v^{(k)}$ after k stages is $v^{(k)} = v^{(0)}P^k$, where P^k is the kth power of the transition matrix. What will be the distribution of highest educational attainment of the grandchildren of the current population?

(g) Calculate P^3, P^4, P^5, \ldots. Continue until the matrix does not change. This is called the long-run distribution. What is the long-run distribution of highest educational attainment of the population?

*SOURCE: U.S. Census Bureau.

Highest Educational Level of Parents	Maximum Education That Children Achieve		
	College	**High School**	**Elementary**
College	80%	18%	2%
High school	40%	50%	10%
Elementary	20%	60%	20%

The following projects are available at the Instructor's Resource Center (IRC):

2. Project at Motorola *Error Control Codings*

3. Using Matrices to Find the Line of Best Fit

4. CBL Experiment

Cumulative Review

In Problems 1–6, solve each equation.

1. $2x^2 - x = 0$

2. $\sqrt{3x + 1} = 4$

3. $2x^3 - 3x^2 - 8x - 3 = 0$

4. $3^x = 9^{x+1}$

5. $\log_3(x - 1) + \log_3(2x + 1) = 2$

6. $3^x = e$

7. Determine whether the function $g(x) = \dfrac{2x^3}{x^4 + 1}$ is even, odd, or neither. Is the graph of g symmetric with respect to the x-axis, y-axis, or origin?

8. Find the center and radius of the circle $x^2 + y^2 - 2x + 4y - 11 = 0$. Graph the circle.

9. Graph $f(x) = 3^{x-2} + 1$ using transformations. What is the domain, range, and horizontal asymptote of f?

10. The function $f(x) = \dfrac{5}{x + 2}$ is one-to-one. Find f^{-1}. Find the domain and the range of f and the domain and the range of f^{-1}.

11. Graph each equation.
 (a) $y = 3x + 6$
 (b) $x^2 + y^2 = 4$
 (c) $y = x^3$
 (d) $y = \dfrac{1}{x}$
 (e) $y = \sqrt{x}$
 (f) $y = e^x$
 (g) $y = \ln x$

12. $f(x) = x^3 - 3x + 5$:
 (a) Using a graphing utility, graph f and approximate the zero(s) of f.
 (b) Using a graphing utility, approximate the local maxima and local minima.
 (c) Determine the intervals on which f is increasing.

7 Sequences; Induction; the Binomial Theorem

A LOOK BACK, A LOOK AHEAD This chapter may be divided into three independent parts: Sections 7.1–7.3, Section 7.4, and Section 7.5.

In Chapter 1, we defined a function and its domain, which was usually some set of real numbers. In Sections 7.1–7.3, we discuss a sequence, which is a function whose domain is the set of positive integers.

Throughout this text, where it seemed appropriate, we have given proofs of many of the results. In Section 7.4, a technique for proving theorems involving natural numbers is discussed.

In the Appendix, Section A.3, there are formulas for expanding $(x + a)^2$ and $(x + a)^3$. In Section 7.5, we discuss the Binomial Theorem, a formula for the expansion of $(x + a)^n$, where n is any positive integer.

The topics introduced in this chapter are covered in more detail in courses titled *Discrete Mathematics*. Applications of these topics can be found in the fields of computer science, engineering, business and economics, the social sciences, and the physical and biological sciences.

The Future of the World Population

WASHINGTON—World population is projected to increase 46 percent by 2050, with most of the growth occurring in the less industrialized areas of the globe.

Projected growth in population by continent, 2003–2050: North America 41.8%; Latin America and the Caribbean 46.2%; Oceania 55.6%; Europe −8.8%; Asia 39.8%; Africa 118.8%.

Note: The United Nations classifies the countries of Latin America and the Caribbean, Asia, Oceania, and Africa as less industrialized with the exception of Australia, New Zealand, and Japan.

Africa's population could soar by more than 1 billion over the next half-century, further straining food and water supplies and social services in areas already struggling, according to a new report.

The latest edition of the "World Population Data Sheet" estimates the global population will rise 46 percent between now and 2050 to about 9 billion, a level also predicted by the United Nations and other groups.

European nations, more industrialized and prosperous, are expected to lose population because of falling birth rates and low immigration.

The U.S. population is expected to grow 45 percent to 422 million in 2050, paced by a stable birth rate and high levels of immigration.

But most of the world's growth will be in developing nations. India's population is estimated to grow 52 percent to 1.6 billion by 2050, when it will surpass China as the world's largest country.

Africa is predicted to more than double in population to 1.9 billion by midcentury.

<small>SOURCE: *The Houston Chronicle* (Houston, TX), July 23, 2003, p. 12.</small>

—*See Chapter Project 1.*

7.1 Sequences

PREPARING FOR THIS SECTION *Before getting started, review the following concept:*

• Functions (Section 1.1, pp. 40–49)

Now work the 'Are You Prepared?' problems on page 502.

OBJECTIVES
1 Write the First Several Terms of a Sequence
2 Write the Terms of a Sequence Defined by a Recursive Formula
3 Use Summation Notation
4 Find the Sum of a Sequence

A **sequence** is a function whose domain is the set of positive integers.

Because a sequence is a function, it will have a graph. In Figure 1(a), we have the graph of the function $f(x) = \dfrac{1}{x}, x > 0$. If all the points on this graph were removed except those whose x-coordinates are positive integers, that is, if all points were removed except $(1, 1)$, $\left(2, \dfrac{1}{2}\right)$, $\left(3, \dfrac{1}{3}\right)$, and so on, the remaining points would be the graph of the sequence $f(n) = \dfrac{1}{n}$, as shown in Figure 1(b). Notice that we use n to represent the independent variable in a sequence. This serves to remind us that n is a positive integer.

Figure 1

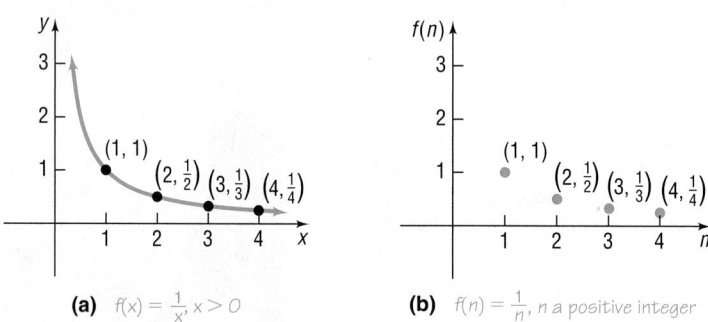

(a) $f(x) = \frac{1}{x}, x > 0$

(b) $f(n) = \frac{1}{n}, n$ a positive integer

1 Write the First Several Terms of a Sequence

A sequence is usually represented by listing its values in order. For example, the sequence whose graph is given in Figure 1(b) might be represented as

$$f(1), f(2), f(3), f(4), \ldots \quad \text{or} \quad 1, \frac{1}{2}, \frac{1}{3}, \frac{1}{4}, \ldots$$

The list never ends, as the ellipsis indicates. The numbers in this ordered list are called the **terms** of the sequence.

In dealing with sequences, we usually use subscripted letters, such as a_1, to represent the first term, a_2 for the second term, a_3 for the third term, and so on.

For the sequence $f(n) = \dfrac{1}{n}$, we write

$$a_1 = f(1) = 1, \quad a_2 = f(2) = \frac{1}{2}, \quad a_3 = f(3) = \frac{1}{3}, \quad a_4 = f(4) = \frac{1}{4}, \ldots, \quad a_n = f(n) = \frac{1}{n}, \ldots$$

In other words, we usually do not use the traditional function notation $f(n)$ for sequences. For this particular sequence, we have a rule for the nth term, which is $a_n = \dfrac{1}{n}$, so it is easy to find any term of the sequence.

When a formula for the nth term (sometimes called the **general term**) of a sequence is known, rather than write out the terms of the sequence, we usually represent the entire sequence by placing braces around the formula for the nth term.

For example, the sequence whose nth term is $b_n = \left(\dfrac{1}{2}\right)^n$ may be represented as

$$\{b_n\} = \left\{\left(\frac{1}{2}\right)^n\right\}$$

or by

$$b_1 = \frac{1}{2}, \quad b_2 = \frac{1}{4}, \quad b_3 = \frac{1}{8}, \ldots, \quad b_n = \left(\frac{1}{2}\right)^n, \ldots$$

EXAMPLE 1	**Writing the First Several Terms of a Sequence**

Write down the first six terms of the following sequence and graph it.

$$\{a_n\} = \left\{\frac{n-1}{n}\right\}$$

Solution The first six terms of the sequence are

$$a_1 = 0, \quad a_2 = \frac{1}{2}, \quad a_3 = \frac{2}{3}, \quad a_4 = \frac{3}{4}, \quad a_5 = \frac{4}{5}, \quad a_6 = \frac{5}{6}$$

See Figure 2 for the graph. ■

Figure 2

 COMMENT: *Graphing utilities can be used to write the terms of a sequence and graph them. Figure 3 shows the sequence given in Example 1 generated on a TI-84 Plus graphing calculator. We can see the first few terms of the sequence on the viewing window. You need to press the right arrow key to scroll right to see the remaining terms of the sequence. Figure 4 shows a graph of the sequence. Notice that the first term of the sequence is not visible since it lies on the x-axis. TRACEing the graph will allow you to see the terms of the sequence. The TABLE feature can also be used to generate the terms of the sequence. See Table 1.*

Figure 3 **Figure 4** **Table 1**

 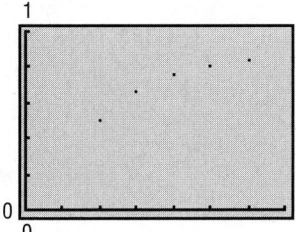

NOW WORK PROBLEM **17.**

EXAMPLE 2	**Writing the First Several Terms of a Sequence**

Figure 5

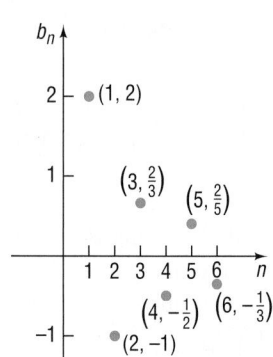

Write down the first six terms of the following sequence and graph it.

$$\{b_n\} = \left\{(-1)^{n+1}\left(\frac{2}{n}\right)\right\}$$

Solution The first six terms of the sequence are

$$b_1 = 2, \quad b_2 = -1, \quad b_3 = \frac{2}{3}, \quad b_4 = -\frac{1}{2}, \quad b_5 = \frac{2}{5}, \quad b_6 = -\frac{1}{3}$$

See Figure 5 for the graph. ∎

Notice in the sequence $\{b_n\}$ in Example 2 that the signs of the terms **alternate**. When this occurs, we use factors such as $(-1)^{n+1}$, which equals 1 if n is odd and -1 if n is even, or $(-1)^n$, which equals -1 if n is odd and 1 if n is even.

EXAMPLE 3	**Writing the First Several Terms of a Sequence**

Figure 6

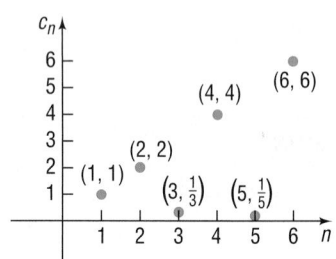

Write down the first six terms of the following sequence and graph it.

$$\{c_n\} = \begin{cases} n & \text{if } n \text{ is even} \\ \dfrac{1}{n} & \text{if } n \text{ is odd} \end{cases}$$

Solution The first six terms of the sequence are

$$c_1 = 1, \quad c_2 = 2, \quad c_3 = \frac{1}{3}, \quad c_4 = 4, \quad c_5 = \frac{1}{5}, \quad c_6 = 6$$

See Figure 6 for the graph. ∎

NOW WORK PROBLEM **19.**

Sometimes a sequence is indicated by an observed pattern in the first few terms that makes it possible to infer the makeup of the nth term. In the example that follows, a sufficient number of terms of the sequence is given so that a natural choice for the nth term is suggested.

EXAMPLE 4	**Determining a Sequence from a Pattern**

(a) $e, \dfrac{e^2}{2}, \dfrac{e^3}{3}, \dfrac{e^4}{4}, \ldots \qquad\qquad a_n = \dfrac{e^n}{n}$

(b) $1, \dfrac{1}{3}, \dfrac{1}{9}, \dfrac{1}{27}, \ldots \qquad\qquad b_n = \dfrac{1}{3^{n-1}}$

(c) $1, 3, 5, 7, \ldots \qquad\qquad c_n = 2n - 1$

(d) $1, 4, 9, 16, 25, \ldots \qquad\qquad d_n = n^2$

(e) $1, -\dfrac{1}{2}, \dfrac{1}{3}, -\dfrac{1}{4}, \dfrac{1}{5}, \ldots \qquad e_n = (-1)^{n+1}\left(\dfrac{1}{n}\right)$ ∎

NOW WORK PROBLEM **27.**

The Factorial Symbol

If $n \geq 0$ is an integer, the **factorial symbol** $n!$ is defined as follows:

$$0! = 1 \qquad 1! = 1$$
$$n! = n(n-1) \cdot \cdots \cdot 3 \cdot 2 \cdot 1 \qquad \text{if } n \geq 2$$

Table 2

n	0	1	2	3	4	5	6
$n!$	1	1	2	6	24	120	720

For example, $2! = 2 \cdot 1 = 2$, $3! = 3 \cdot 2 \cdot 1 = 6$, $4! = 4 \cdot 3 \cdot 2 \cdot 1 = 24$, and so on. Table 2 lists the values of $n!$ for $0 \le n \le 6$.

Because

$$n! = n\underbrace{(n-1)(n-2) \cdot \cdots \cdot 3 \cdot 2 \cdot 1}_{(n-1)!}$$

we can use the formula

$$n! = n(n-1)!$$

to find successive factorials. For example, because $6! = 720$, we have

$$7! = 7 \cdot 6! = 7(720) = 5040$$

and

$$8! = 8 \cdot 7! = 8(5040) = 40{,}320$$

COMMENT
Your calculator has a factorial key. Use it to see how fast factorials increase in value. Find the value of 69!. What happens when you try to find 70!? In fact, 70! is larger than 10^{100} (a googol), the largest number most calculators can display. ■

NOW WORK PROBLEM **11**.

2 Write the Terms of a Sequence Defined by a Recursive Formula

A second way of defining a sequence is to assign a value to the first (or the first few) term(s) and specify the nth term by a formula or equation that involves one or more of the terms preceding it. Sequences defined this way are said to be defined **recursively**, and the rule or formula is called a **recursive formula**.

EXAMPLE 5

Writing the Terms of a Recursively Defined Sequence

Write down the first five terms of the following recursively defined sequence.

$$s_1 = 1, \qquad s_n = ns_{n-1}$$

Solution The first term is given as $s_1 = 1$. To get the second term, we use $n = 2$ in the formula $s_n = ns_{n-1}$ to get $s_2 = 2s_1 = 2 \cdot 1 = 2$. To get the third term, we use $n = 3$ in the formula to get $s_3 = 3s_2 = 3 \cdot 2 = 6$. To get a new term requires that we know the value of the preceding term. The first five terms are

$$s_1 = 1$$
$$s_2 = 2 \cdot 1 = 2$$
$$s_3 = 3 \cdot 2 = 6$$
$$s_4 = 4 \cdot 6 = 24$$
$$s_5 = 5 \cdot 24 = 120$$

Do you recognize this sequence? $s_n = n!$ ■

EXAMPLE 6

Writing the Terms of a Recursively Defined Sequence

Write down the first five terms of the following recursively defined sequence.

$$u_1 = 1, \qquad u_2 = 1, \qquad u_{n+2} = u_n + u_{n+1}$$

Solution We are given the first two terms. To get the third term requires that we know both of the previous two terms. That is,

$$u_1 = 1$$
$$u_2 = 1$$
$$u_3 = u_1 + u_2 = 1 + 1 = 2$$
$$u_4 = u_2 + u_3 = 1 + 2 = 3$$
$$u_5 = u_3 + u_4 = 2 + 3 = 5$$

■

The sequence defined in Example 6 is called a **Fibonacci sequence**, and the terms of this sequence are called **Fibonacci numbers**. These numbers appear in a wide variety of applications (see Problems 85–88).

NOW WORK PROBLEMS 35 AND 43.

3 Use Summation Notation

It is often important to be able to find the sum of the first n terms of a sequence $\{a_n\}$, that is,

$$a_1 + a_2 + a_3 + \cdots + a_n$$

Rather than write down all these terms, we introduce a more concise way to express the sum, called **summation notation**. Using summation notation, we write the sum as

$$a_1 + a_2 + a_3 + \cdots + a_n = \sum_{k=1}^{n} a_k$$

The symbol Σ (the Greek letter sigma, which is an S in our alphabet) is simply an instruction to sum, or add up, the terms. The integer k is called the **index** of the sum; it tells you where to start the sum and where to end it. The expression

$$\sum_{k=1}^{n} a_k$$

is an instruction to add the terms a_k of the sequence $\{a_n\}$ starting with $k = 1$ and ending with $k = n$. We read the expression as "the sum of a_k from $k = 1$ to $k = n$."

EXAMPLE 7

Expanding Summation Notation

Write out each sum.

(a) $\displaystyle\sum_{k=1}^{n} \frac{1}{k}$　　　　　　　　　　　　(b) $\displaystyle\sum_{k=1}^{n} k!$

Solution　(a) $\displaystyle\sum_{k=1}^{n} \frac{1}{k} = 1 + \frac{1}{2} + \frac{1}{3} + \cdots + \frac{1}{n}$　　(b) $\displaystyle\sum_{k=1}^{n} k! = 1! + 2! + \cdots + n!$　■

EXAMPLE 8

Writing a Sum in Summation Notation

Express each sum using summation notation.

(a) $1^2 + 2^2 + 3^2 + \cdots + 9^2$　　　　　(b) $1 + \dfrac{1}{2} + \dfrac{1}{4} + \dfrac{1}{8} + \cdots + \dfrac{1}{2^{n-1}}$

Solution　(a) The sum $1^2 + 2^2 + 3^2 + \cdots + 9^2$ has 9 terms, each of the form k^2, and starts at $k = 1$ and ends at $k = 9$:

$$1^2 + 2^2 + 3^2 + \cdots + 9^2 = \sum_{k=1}^{9} k^2$$

(b) The sum

$$1 + \frac{1}{2} + \frac{1}{4} + \frac{1}{8} + \cdots + \frac{1}{2^{n-1}}$$

has n terms, each of the form $\dfrac{1}{2^{k-1}}$, and starts at $k = 1$ and ends at $k = n$:

$$1 + \frac{1}{2} + \frac{1}{4} + \frac{1}{8} + \cdots + \frac{1}{2^{n-1}} = \sum_{k=1}^{n} \frac{1}{2^{k-1}}$$　■

The index of summation need not always begin at 1 or end at n; for example, we could have expressed the sum in Example 8(b) as

$$\sum_{k=0}^{n-1} \frac{1}{2^k} = 1 + \frac{1}{2} + \frac{1}{4} + \cdots + \frac{1}{2^{n-1}}$$

Letters other than k may be used as the index. For example,

$$\sum_{j=1}^{n} j! \quad \text{and} \quad \sum_{i=1}^{n} i!$$

each represent the same sum as the one given in Example 7(b).

NOW WORK PROBLEMS 51 AND 61.

4 Find the Sum of a Sequence

Next we list some properties of sequences using summation notation. These properties are useful for adding the terms of a sequence.

Theorem

Properties of Sequences

If $\{a_n\}$ and $\{b_n\}$ are two sequences and c is a real number, then:

$$\sum_{k=1}^{n} c = \underbrace{c + c + \cdots + c}_{n \text{ terms}} = cn \tag{1}$$

$$\sum_{k=1}^{n} (ca_k) = ca_1 + ca_2 + \cdots + ca_n = c(a_1 + a_2 + \cdots + a_n) = c\sum_{k=1}^{n} a_k \tag{2}$$

$$\sum_{k=1}^{n} (a_k + b_k) = \sum_{k=1}^{n} a_k + \sum_{k=1}^{n} b_k \tag{3}$$

$$\sum_{k=1}^{n} (a_k - b_k) = \sum_{k=1}^{n} a_k - \sum_{k=1}^{n} b_k \tag{4}$$

$$\sum_{k=1}^{n} a_k = \sum_{k=1}^{j} a_k + \sum_{k=j+1}^{n} a_k, \quad \text{where } 0 < j < n \tag{5}$$

$$\sum_{k=1}^{n} k = 1 + 2 + 3 + \cdots + n = \frac{n(n+1)}{2} \tag{6}$$

$$\sum_{k=1}^{n} k^2 = 1^2 + 2^2 + 3^2 + \cdots + n^2 = \frac{n(n+1)(2n+1)}{6} \tag{7}$$

$$\sum_{k=1}^{n} k^3 = 1^3 + 2^3 + 3^3 + \cdots + n^3 = \left[\frac{n(n+1)}{2}\right]^2 \tag{8}$$

We shall not prove these properties. The proofs of (1) through (5) are based on properties of real numbers; the proofs of (7) and (8) require mathematical induction, which is discussed in Section 7.4. See Problem 91 for a derivation of (6).

EXAMPLE 9 **Finding the Sum of a Sequence**

Find the sum of each sequence.

(a) $\displaystyle\sum_{k=1}^{5} (3k)$ (b) $\displaystyle\sum_{k=1}^{3} (k^3 + 1)$ (c) $\displaystyle\sum_{k=1}^{4} (k^2 - 7k + 2)$

Solution

(a) $\displaystyle\sum_{k=1}^{5} (3k) = 3 \sum_{k=1}^{5} k$ *Property (2)*

$\qquad\qquad = 3\left(\dfrac{5(5+1)}{2}\right)$ *Property (6)*

$\qquad\qquad = 3(15)$

$\qquad\qquad = 45$

(b) $\displaystyle\sum_{k=1}^{3} (k^3 + 1) = \sum_{k=1}^{3} k^3 + \sum_{k=1}^{3} 1$ *Property (3)*

$\qquad\qquad = \left(\dfrac{3(3+1)}{2}\right)^2 + 1(3)$ *Properties (1) and (8)*

$\qquad\qquad = 36 + 3$

$\qquad\qquad = 39$

(c) $\displaystyle\sum_{k=1}^{4} (k^2 - 7k + 2) = \sum_{k=1}^{4} k^2 - \sum_{k=1}^{4} (7k) + \sum_{k=1}^{4} 2$ *Properties (3), (4)*

$\qquad\qquad = \sum_{k=1}^{4} k^2 - 7\sum_{k=1}^{4} k + \sum_{k=1}^{4} 2$ *Property (2)*

$\qquad\qquad = \dfrac{4(4+1)(2\cdot 4+1)}{6} - 7\left(\dfrac{4(4+1)}{2}\right) + 2(4)$ *Properties (1), (6), (7)*

$\qquad\qquad = 30 - 70 + 8$

$\qquad\qquad = -32$ ∎

NOW WORK PROBLEM 73.

7.1 Assess Your Understanding

'Are You Prepared?'

Answers are given at the end of these exercises. If you get a wrong answer, read the pages listed in red.

1. For the function $f(x) = \dfrac{x-1}{x}$, find $f(2)$ and $f(3)$. (pp. 44–47)

2. *True or False:* A function is a relation between two sets D and R so that each element x in the first set D is related to exactly one element y in the second set R. (pp. 40–44)

Concepts and Vocabulary

3. A(n) _____ is a function whose domain is the set of positive integers.

4. For the sequence $\{s_n\} = \{4n - 1\}$, the first term is $s_1 = $ _____ and the fourth term is $s_4 = $ _____.

5. $\displaystyle\sum_{k=1}^{4} (2k) = $ _____.

6. *True or False:* Sequences are sometimes defined recursively.

7. *True or False:* A sequence is a function.

8. *True or False:* $\displaystyle\sum_{k=1}^{2} k = 3$

Skill Building

In Problems 9–14, evaluate each factorial expression.

9. $10!$ **10.** $9!$ **11.** $\dfrac{9!}{6!}$ **12.** $\dfrac{12!}{10!}$ **13.** $\dfrac{3!\,7!}{4!}$ **14.** $\dfrac{5!\,8!}{3!}$

In Problems 15–26, write down the first five terms of each sequence.

15. $\{s_n\} = \{n\}$

16. $\{s_n\} = \{n^2 + 1\}$

17. $\{a_n\} = \left\{\dfrac{n}{n+2}\right\}$

18. $\{b_n\} = \left\{\dfrac{2n+1}{2n}\right\}$

19. $\{c_n\} = \{(-1)^{n+1} n^2\}$

20. $\{d_n\} = \left\{(-1)^{n-1}\left(\dfrac{n}{2n-1}\right)\right\}$

21. $\{s_n\} = \left\{\dfrac{2^n}{3^n + 1}\right\}$

22. $\{s_n\} = \left\{\left(\dfrac{4}{3}\right)^n\right\}$

23. $\{t_n\} = \left\{\dfrac{(-1)^n}{(n+1)(n+2)}\right\}$

24. $\{a_n\} = \left\{\dfrac{3^n}{n}\right\}$

25. $\{b_n\} = \left\{\dfrac{n}{e^n}\right\}$

26. $\{c_n\} = \left\{\dfrac{n^2}{2^n}\right\}$

In Problems 27–34, the given pattern continues. Write down the nth term of a sequence $\{a_n\}$ suggested by the pattern.

27. $\dfrac{1}{2}, \dfrac{2}{3}, \dfrac{3}{4}, \dfrac{4}{5}, \cdots$

28. $\dfrac{1}{1 \cdot 2}, \dfrac{1}{2 \cdot 3}, \dfrac{1}{3 \cdot 4}, \dfrac{1}{4 \cdot 5}, \cdots$

29. $1, \dfrac{1}{2}, \dfrac{1}{4}, \dfrac{1}{8}, \cdots$

30. $\dfrac{2}{3}, \dfrac{4}{9}, \dfrac{8}{27}, \dfrac{16}{81}, \cdots$

31. $1, -1, 1, -1, 1, -1, \ldots$

32. $1, \dfrac{1}{2}, 3, \dfrac{1}{4}, 5, \dfrac{1}{6}, 7, \dfrac{1}{8}, \ldots$

33. $1, -2, 3, -4, 5, -6, \ldots$

34. $2, -4, 6, -8, 10, \ldots$

In Problems 35–48, a sequence is defined recursively. Write down the first five terms.

35. $a_1 = 2;\quad a_n = 3 + a_{n-1}$

36. $a_1 = 3;\quad a_n = 4 - a_{n-1}$

37. $a_1 = -2;\quad a_n = n + a_{n-1}$

38. $a_1 = 1;\quad a_n = n - a_{n-1}$

39. $a_1 = 5;\quad a_n = 2a_{n-1}$

40. $a_1 = 2;\quad a_n = -a_{n-1}$

41. $a_1 = 3;\quad a_n = \dfrac{a_{n-1}}{n}$

42. $a_1 = -2;\quad a_n = n + 3a_{n-1}$

43. $a_1 = 1;\quad a_2 = 2;\quad a_n = a_{n-1} \cdot a_{n-2}$

44. $a_1 = -1;\quad a_2 = 1;\quad a_n = a_{n-2} + na_{n-1}$ **45.** $a_1 = A;\quad a_n = a_{n-1} + d$

46. $a_1 = A;\quad a_n = ra_{n-1},\quad r \neq 0$

47. $a_1 = \sqrt{2};\quad a_n = \sqrt{2 + a_{n-1}}$

48. $a_1 = \sqrt{2};\quad a_n = \sqrt{\dfrac{a_{n-1}}{2}}$

In Problems 49–58, write out each sum.

49. $\displaystyle\sum_{k=1}^{n} (k + 2)$

50. $\displaystyle\sum_{k=1}^{n} (2k + 1)$

51. $\displaystyle\sum_{k=1}^{n} \dfrac{k^2}{2}$

52. $\displaystyle\sum_{k=1}^{n} (k + 1)^2$

53. $\displaystyle\sum_{k=0}^{n} \dfrac{1}{3^k}$

54. $\displaystyle\sum_{k=0}^{n} \left(\dfrac{3}{2}\right)^k$

55. $\displaystyle\sum_{k=0}^{n-1} \dfrac{1}{3^{k+1}}$

56. $\displaystyle\sum_{k=0}^{n-1} (2k + 1)$

57. $\displaystyle\sum_{k=2}^{n} (-1)^k \ln k$

58. $\displaystyle\sum_{k=3}^{n} (-1)^{k+1} 2^k$

In Problems 59–68, express each sum using summation notation.

59. $1 + 2 + 3 + \cdots + 20$

60. $1^3 + 2^3 + 3^3 + \cdots + 8^3$

61. $\dfrac{1}{2} + \dfrac{2}{3} + \dfrac{3}{4} + \cdots + \dfrac{13}{13 + 1}$

62. $1 + 3 + 5 + 7 + \cdots + [2(12) - 1]$

63. $1 - \dfrac{1}{3} + \dfrac{1}{9} - \dfrac{1}{27} + \cdots + (-1)^6\left(\dfrac{1}{3^6}\right)$

64. $\dfrac{2}{3} - \dfrac{4}{9} + \dfrac{8}{27} - \cdots + (-1)^{11+1}\left(\dfrac{2}{3}\right)^{11}$

65. $3 + \dfrac{3^2}{2} + \dfrac{3^3}{3} + \cdots + \dfrac{3^n}{n}$

66. $\dfrac{1}{e} + \dfrac{2}{e^2} + \dfrac{3}{e^3} + \cdots + \dfrac{n}{e^n}$

67. $a + (a + d) + (a + 2d) + \cdots + (a + nd)$

68. $a + ar + ar^2 + \cdots + ar^{n-1}$

In Problems 69–80, find the sum of each sequence.

69. $\displaystyle\sum_{k=1}^{10} 5$

70. $\displaystyle\sum_{k=1}^{20} 8$

71. $\displaystyle\sum_{k=1}^{6} k$

72. $\displaystyle\sum_{k=1}^{4} (-k)$

73. $\displaystyle\sum_{k=1}^{5} (5k + 3)$

74. $\displaystyle\sum_{k=1}^{6} (3k - 7)$

75. $\displaystyle\sum_{k=1}^{3} (k^2 + 4)$

76. $\displaystyle\sum_{k=0}^{4} (k^2 - 4)$

77. $\displaystyle\sum_{k=1}^{6} (-1)^k 2^k$

78. $\displaystyle\sum_{k=1}^{4} (-1)^k 3^k$

79. $\displaystyle\sum_{k=1}^{4} (k^3 - 1)$

80. $\displaystyle\sum_{k=0}^{3} (k^3 + 2)$

Applications and Extensions

81. Credit Card Debt John has a balance of $3000 on his Discover card that charges 1% interest per month on any unpaid balance. John can afford to pay $100 toward the balance each month. His balance each month after making a $100 payment is given by the recursively defined sequence

$$B_0 = \$3000, \qquad B_n = 1.01B_{n-1} - 100$$

Determine John's balance after making the first payment. That is, determine B_1.

82. Trout Population A pond currently has 2000 trout in it. A fish hatchery decides to add an additional 20 trout each month. In addition, it is known that the trout population is growing 3% per month. The size of the population after n months is given by the recursively defined sequence

$$p_0 = 2000, \qquad p_n = 1.03p_{n-1} + 20$$

How many trout are in the pond after two months? That is, what is p_2?

83. Car Loans Phil bought a car by taking out a loan for $18,500 at 0.5% interest per month. Phil's normal monthly payment is $434.47 per month, but he decides that he can afford to pay $100 extra toward the balance each month. His balance each month is given by the recursively defined sequence

$$B_0 = \$18,500, \qquad B_n = 1.005B_{n-1} - 534.47$$

Determine Phil's balance after making the first payment. That is, determine B_1.

84. Environmental Control The Environmental Protection Agency (EPA) determines that Maple Lake has 250 tons of pollutant as a result of industrial waste and that 10% of the pollutant present is neutralized by solar oxidation every year. The EPA imposes new pollution control laws that result in 15 tons of new pollutant entering the lake each year. The amount of pollutant in the lake after n years is given by the recursively defined sequence

$$p_0 = 250, \qquad p_n = 0.9p_{n-1} + 15$$

Determine the amount of pollutant in the lake after 2 years. That is, determine p_2.

85. Growth of a Rabbit Colony A colony of rabbits begins with one pair of mature rabbits, which will produce a pair of offspring (one male, one female) each month. Assume that all rabbits mature in 1 month and produce a pair of offspring (one male, one female) after 2 months. If no rabbits ever die, how many pairs of mature rabbits are there after 7 months?

[**Hint:** A Fibonacci sequence models this colony. Do you see why?]

1 mature pair
1 mature pair
2 mature pairs
3 mature pairs

86. Fibonacci Sequence Let

$$u_n = \frac{\left(1 + \sqrt{5}\right)^n - \left(1 - \sqrt{5}\right)^n}{2^n\sqrt{5}}$$

define the nth term of a sequence.
(a) Show that $u_1 = 1$ and $u_2 = 1$.
(b) Show that $u_{n+2} = u_{n+1} + u_n$.
(c) Draw the conclusion that $\{u_n\}$ is a Fibonacci sequence.

87. Pascal's Triangle Divide the triangular array shown (called Pascal's triangle) using diagonal lines as indicated. Find the sum of the numbers in each of these diagonal rows. Do you recognize this sequence?

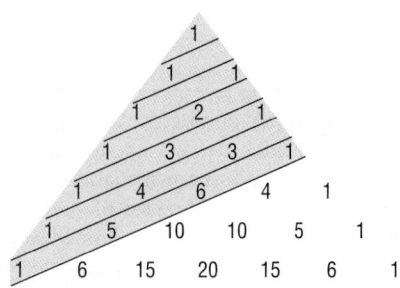

88. Fibonacci Sequence Use the result of Problem 86 to do the following problems:
(a) Write the first 10 terms of the Fibonacci sequence.
(b) Compute the ratio $\dfrac{u_{n+1}}{u_n}$ for the first 10 terms.
(c) As n gets large, what number does the ratio approach? This number is referred to as the **golden ratio**. Rectangles whose sides are in this ratio were considered pleasing to the eye by the Greeks. For example, the facade of the Parthenon was constructed using the golden ratio.
(d) Compute the ratio $\dfrac{u_n}{u_{n+1}}$ for the first 10 terms.
(e) As n gets large, what number does the ratio approach? This number is also referred to as the **golden ratio**. This ratio is believed to have been used in the construction of the Great Pyramid in Egypt. The ratio equals the sum of the areas of the four face triangles divided by the total surface area of the Great Pyramid.

89. Approximating $f(x) = e^x$ In calculus, it can be shown that

$$f(x) = e^x = \sum_{k=0}^{\infty} \frac{x^k}{k!}$$

We can approximate the value of $f(x) = e^x$ for any x using the following sum

$$f(x) = e^x \approx \sum_{k=0}^{n} \frac{x^k}{k!}$$

for some n.
(a) Approximate $f(1.3)$ with $n = 4$.
(b) Approximate with $f(1.3)$ $n = 7$.
(c) Use a calculator to approximate $f(1.3)$.
(d) Using trial and error along with a graphing utility's SEQuence mode, determine the value of n required to approximate $f(1.3)$ correct to eight decimal places.

90. Approximating $f(x) = e^x$ Refer to Problem 89.
(a) Approximate $f(-2.4)$ with $n = 3$.
(b) Approximate $f(-2.4)$ with $n = 6$.
(c) Use a calculator to approximate $f(-2.4)$.
(d) Using trial and error along with a graphing utility's SEQuence mode, determine the value of n required to approximate $f(-2.4)$ correct to eight decimal places.

91. Show that

$$1 + 2 + \cdots + (n - 1) + n = \frac{n(n + 1)}{2}$$

[**Hint:** Let

$$S = 1 + 2 + \cdots + (n - 1) + n$$
$$S = n + (n - 1) + (n - 2) + \cdots + 1$$

Add these equations. Then

$$2S = \underbrace{[1 + n] + [2 + (n - 1)] + \cdots + [n + 1]}_{n \text{ terms in brackets}}$$

Now complete the derivation.]

Discussion and Writing

92. Investigate various applications that lead to a Fibonacci sequence, such as art, architecture, or financial markets. Write an essay on these applications.

'Are You Prepared?' Answers

1. $f(2) = \dfrac{1}{2}; f(3) = \dfrac{2}{3}$ **2.** True

7.2 Arithmetic Sequences

OBJECTIVES **1** Determine If a Sequence Is Arithmetic
 2 Find a Formula for an Arithmetic Sequence
 3 Find the Sum of an Arithmetic Sequence

1 Determine If a Sequence Is Arithmetic

When the difference between successive terms of a sequence is always the same number, the sequence is called **arithmetic**. An **arithmetic sequence**[*] may be defined recursively as $a_1 = a$, $a_n - a_{n-1} = d$, or as

$$a_1 = a, \qquad a_n = a_{n-1} + d \qquad \qquad \textbf{(1)}$$

where $a_1 = a$ and d are real numbers. The number a is the first term, and the number d is called the **common difference**.

The terms of an arithmetic sequence with first term a_1 and common difference d follow the pattern

$$a_1, \quad a_1 + d, \quad a_1 + 2d, \quad a_1 + 3d, \dots$$

EXAMPLE 1 **Determining If a Sequence Is Arithmetic**

The sequence

$$4, \quad 6, \quad 8, \quad 10, \dots$$

is arithmetic since the difference of successive terms is 2. The first term is $a_1 = 4$, and the common difference is 2. ∎

EXAMPLE 2 **Determining If a Sequence Is Arithmetic**

Show that the following sequence is arithmetic. Find the first term and the common difference.

$$\{s_n\} = \{3n + 5\}$$

Solution The first term is $s_1 = 3 \cdot 1 + 5 = 8$. The nth and $(n - 1)$st terms of the sequence $\{s_n\}$ are

$$s_n = 3n + 5 \quad \text{and} \quad s_{n-1} = 3(n - 1) + 5 = 3n + 2$$

Their difference is

$$s_n - s_{n-1} = (3n + 5) - (3n + 2) = 5 - 2 = 3$$

Since the difference of two successive terms is constant, the sequence is arithmetic and the common difference is 3. ∎

[*]Sometimes called an **arithmetic progression**.

EXAMPLE 3	**Determining If a Sequence Is Arithmetic**

Show that the sequence $\{t_n\} = \{4 - n\}$ is arithmetic. Find the first term and the common difference.

Solution The first term is $t_1 = 4 - 1 = 3$. The nth and $(n - 1)$st terms are

$$t_n = 4 - n \quad \text{and} \quad t_{n-1} = 4 - (n - 1) = 5 - n$$

Their difference is

$$t_n - t_{n-1} = (4 - n) - (5 - n) = 4 - 5 = -1$$

Since the difference of two successive terms is constant, $\{t_n\}$ is an arithmetic sequence whose common difference is -1. ∎

 NOW WORK PROBLEM 5.

2 Find a Formula for an Arithmetic Sequence

Suppose that a is the first term of an arithmetic sequence whose common difference is d. We seek a formula for the nth term, a_n. To see the pattern, we write down the first few terms.

$$a_1 = a$$
$$a_2 = a_1 + d = a_1 + 1 \cdot d$$
$$a_3 = a_2 + d = (a_1 + d) + d = a_1 + 2 \cdot d$$
$$a_4 = a_3 + d = (a_1 + 2 \cdot d) + d = a_1 + 3 \cdot d$$
$$a_5 = a_4 + d = (a_1 + 3 \cdot d) + d = a_1 + 4 \cdot d$$
$$\vdots$$
$$a_n = a_{n-1} + d = [a_1 + (n - 2)d] + d = a_1 + (n - 1)d$$

We are led to the following result:

Theorem

> **nth Term of an Arithmetic Sequence**
>
> For an arithmetic sequence $\{a_n\}$ whose first term is a_1 and whose common difference is d, the nth term is determined by the formula
>
> $$a_n = a_1 + (n - 1)d \qquad (2)$$

EXAMPLE 4	**Finding a Particular Term of an Arithmetic Sequence**

Find the thirteenth term of the arithmetic sequence: $2, 6, 10, 14, 18, \ldots$

Solution The first term of this arithmetic sequence is $a_1 = 2$, and the common difference is 4. By formula (2), the nth term is

$$a_n = 2 + (n - 1)4$$

The thirteenth term is

$$a_{13} = 2 + 12 \cdot 4 = 50$$

∎

— Exploration —
Use a graphing utility to find the thirteenth term of the sequence given in Example 4. Use it to find the twentieth and fiftieth terms.

| EXAMPLE 5 | **Finding a Recursive Formula for an Arithmetic Sequence** |

The eighth term of an arithmetic sequence is 75, and the twentieth term is 39.
(a) Find the first term and the common difference.
(b) Give a recursive formula for the sequence.
(c) What is the nth term of the sequence?

Solution

(a) By formula (2), we know that $a_n = a_1 + (n - 1)d$. As a result,

$$\begin{cases} a_8 = a_1 + 7d = 75 \\ a_{20} = a_1 + 19d = 39 \end{cases}$$

This is a system of two linear equations containing two variables, a_1 and d, which we can solve by elimination. Subtracting the second equation from the first equation, we get

$$-12d = 36$$
$$d = -3$$

With $d = -3$, we use $a_1 + 7d = 75$ and find that $a_1 = 75 - 7d = 75 - 7(-3) = 96$. The first term is $a_1 = 96$, and the common difference is $d = -3$.

(b) Using formula (1), a recursive formula for this sequence is

$$a_1 = 96, \qquad a_n = a_{n-1} - 3$$

(c) Using formula (2), a formula for the nth term of the sequence $\{a_n\}$ is

$$a_n = a_1 + (n - 1)d = 96 + (n - 1)(-3) = 99 - 3n \qquad \blacksquare$$

 Exploration
Graph the recursive formula from Example 5, $a_1 = 96$, $a_n = a_{n-1} - 3$, using a graphing utility. Conclude that the graph of the recursive formula behaves like the graph of a linear function. How is d, the common difference, related to m, the slope of a line?

 NOW WORK PROBLEMS 21 AND 27.

3 Find the Sum of an Arithmetic Sequence

The next result gives a formula for finding the sum of the first n terms of an arithmetic sequence.

Theorem

Sum of the First n Terms of an Arithmetic Sequence

Let $\{a_n\}$ be an arithmetic sequence with first term a_1 and common difference d. The sum S_n of the first n terms of $\{a_n\}$ may be found in two ways:

$$S_n = \frac{n}{2}[2a_1 + (n - 1)d] \qquad \textbf{(3)}$$

$$S_n = \frac{n}{2}(a_1 + a_n) \qquad \textbf{(4)}$$

Proof

$$\begin{aligned}
S_n &= a_1 + a_2 + a_3 + \cdots + a_n && \text{Sum of first } n \text{ terms} \\
&= a_1 + (a_1 + d) + (a_1 + 2d) + \cdots + [a_1 + (n - 1)d] && \text{Formula (2)} \\
&= \underbrace{(a_1 + a_1 + \cdots + a)}_{n \text{ terms}} + [d + 2d + \cdots + (n - 1)d] && \text{Rearrange terms} \\
&= na_1 + d[1 + 2 + \cdots + (n - 1)] \\
&= na_1 + d\left[\frac{(n - 1)n}{2}\right] && \text{Property 6, Section 7.1} \\
&= na_1 + \frac{n}{2}(n - 1)d \\
&= \frac{n}{2}[2a_1 + (n - 1)d] && \text{Factor out } \frac{n}{2}; \text{This is Formula (3)} \\
&= \frac{n}{2}[a_1 + a_1 + (n - 1)d] \\
&= \frac{n}{2}(a_1 + a_n) && \text{Use Formula (2); This is Formula (4)} \qquad \blacksquare
\end{aligned}$$

There are two ways to find the sum of the first n terms of an arithmetic sequence. Notice that formula (3) involves the first term and common difference, whereas formula (4) involves the first term and the nth term. Use whichever form is easier.

| **EXAMPLE 6** | **Finding the Sum of n Terms of an Arithmetic Sequence** |

Find the sum S_n of the first n terms of the sequence $\{3n + 5\}$; that is, find

$$8 + 11 + 14 + \cdots + (3n + 5)$$

Solution The sequence $\{3n + 5\}$ is an arithmetic sequence with first term $a_1 = 8$ and the nth term $(3n + 5)$. To find the sum S_n, we use formula (4).

$$S_n = \frac{n}{2}(a_1 + a_n) = \frac{n}{2}[8 + (3n + 5)] = \frac{n}{2}(3n + 13) \qquad \blacksquare$$

NOW WORK PROBLEM **35**.

| **EXAMPLE 7** | **Using a Graphing Utility to Find the Sum of 20 Terms of an Arithmetic Sequence** |

Figure 7

```
sum(seq(9.5n+2.6
,n,1,20,1)
            2047
```

Use a graphing utility to find the sum of the first 20 terms of the sequence $\{9.5n + 2.6\}$.

Solution Figure 7 shows the results obtained using a TI-84 Plus graphing calculator.

The sum of the first 20 terms of the sequence $\{9.5n + 2.6\}$ is 2047. $\qquad \blacksquare$

WORK EXAMPLE **7** USING FORMULA **(3)**.

NOW WORK PROBLEM **43**.

| **EXAMPLE 8** | **Creating a Floor Design** |

A ceramic tile floor is designed in the shape of a trapezoid 20 feet wide at the base and 10 feet wide at the top. See Figure 8. The tiles, 12 inches by 12 inches, are to be placed so that each successive row contains one less tile than the preceding row. How many tiles will be required?

Figure 8

The bottom row requires 20 tiles and the top row, 10 tiles. Since each successive row requires one less tile, the total number of tiles required is

$$S = 20 + 19 + 18 + \cdots + 11 + 10$$

Solution

This is the sum of an arithmetic sequence; the common difference is -1. The number of terms to be added is $n = 11$, with the first term $a_1 = 20$ and the last term $a_{11} = 10$. The sum S is

$$S = \frac{n}{2}(a_1 + a_{11}) = \frac{11}{2}(20 + 10) = 165$$

In all, 165 tiles will be required.

7.2 Assess Your Understanding

Concepts and Vocabulary

1. In a(n) _____ sequence, the difference between successive terms is a constant.

2. *True or False:* In an arithmetic sequence the sum of the first and last terms equals twice the sum of all the terms.

Skill Building

In Problems 3–12, show that each sequence is arithmetic. Find the common difference and write out the first four terms.

3. $\{s_n\} = \{n + 4\}$

4. $\{s_n\} = \{n - 5\}$

5. $\{a_n\} = \{2n - 5\}$

6. $\{b_n\} = \{3n + 1\}$

7. $\{c_n\} = \{6 - 2n\}$

8. $\{d_n\} = \{4 - 2n\}$

9. $\{t_n\} = \left\{\dfrac{1}{2} - \dfrac{1}{3}n\right\}$

10. $\{t_n\} = \left\{\dfrac{2}{3} + \dfrac{n}{4}\right\}$

11. $\{s_n\} = \{\ln 3^n\}$

12. $\{s_n\} = \{e^{\ln n}\}$

In Problems 13–20, find the nth term of the arithmetic sequence $\{a_n\}$ whose initial term a and common difference d are given. What is the fifth term?

13. $a_1 = 2;\quad d = 3$

14. $a_1 = -2;\quad d = 4$

15. $a_1 = 5;\quad d = -3$

16. $a_1 = 6;\quad d = -2$

17. $a_1 = 0;\quad d = \dfrac{1}{2}$

18. $a_1 = 1;\quad d = -\dfrac{1}{3}$

19. $a_1 = \sqrt{2};\quad d = \sqrt{2}$

20. $a_1 = 0;\quad d = \pi$

In Problems 21–26, find the indicated term in each arithmetic sequence.

21. 12th term of $2, 4, 6, \ldots$

22. 8th term of $-1, 1, 3, \ldots$

23. 10th term of $1, -2, -5, \ldots$

24. 9th term of $5, 0, -5, \ldots$

25. 8th term of $a, a + b, a + 2b, \ldots$

26. 7th term of $2\sqrt{5}, 4\sqrt{5}, 6\sqrt{5}, \ldots$

In Problems 27–34, find the first term and the common difference of the arithmetic sequence described. Give a recursive formula for the sequence. Find a formula for the nth term.

27. 8th term is 8; 20th term is 44

28. 4th term is 3; 20th term is 35

29. 9th term is -5; 15th term is 31

30. 8th term is 4; 18th term is -96

31. 15th term is 0; 40th term is -50

32. 5th term is -2; 13th term is 30

33. 14th term is -1; 18th term is -9

34. 12th term is 4; 18th term is 28

In Problems 35–42, find each sum.

35. $1 + 3 + 5 + \cdots + (2n - 1)$

36. $2 + 4 + 6 + \cdots + 2n$

37. $7 + 12 + 17 + \cdots + (2 + 5n)$

38. $-1 + 3 + 7 + \cdots + (4n - 5)$

39. $2 + 4 + 6 + \cdots + 70$

40. $1 + 3 + 5 + \cdots + 59$

41. $5 + 9 + 13 + \cdots + 49$

42. $2 + 5 + 8 + \cdots + 41$

For Problems 43–48, use a graphing utility to find the sum of each sequence.

43. $\{3.45n + 4.12\}, \quad n = 20$

44. $\{2.67n - 1.23\}, \quad n = 25$

45. $2.8 + 5.2 + 7.6 + \cdots + 36.4$

46. $5.4 + 7.3 + 9.2 + \cdots + 32$

47. $4.9 + 7.48 + 10.06 + \cdots + 66.82$

48. $3.71 + 6.9 + 10.09 + \cdots + 80.27$

Applications and Extensions

49. Find x so that $x + 3, 2x + 1$, and $5x + 2$ are consecutive terms of an arithmetic sequence.

50. Find x so that $2x, 3x + 2$, and $5x + 3$ are consecutive terms of an arithmetic sequence.

51. Drury Lane Theater The Drury Lane Theater has 25 seats in the first row and 30 rows in all. Each successive row contains one additional seat. How many seats are in the theater?

52. Football Stadium The corner section of a football stadium has 15 seats in the first row and 40 rows in all. Each successive row contains two additional seats. How many seats are in this section?

53. Creating a Mosaic A mosaic is designed in the shape of an equilateral triangle, 20 feet on each side. Each tile in the mosaic is in the shape of an equilateral triangle, 12 inches to a side. The tiles are to alternate in color as shown in the illustration. How many tiles of each color will be required?

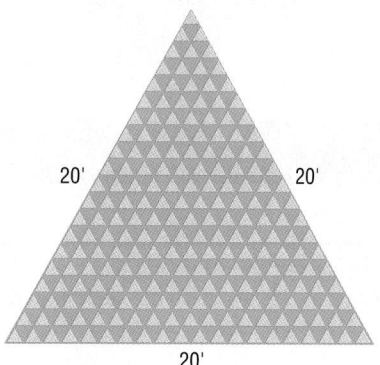

20' 20'

20'

54. Constructing a Brick Staircase A brick staircase has a total of 30 steps. The bottom step requires 100 bricks. Each successive step requires two less bricks than the prior step.
(a) How many bricks are required for the top step?
(b) How many bricks are required to build the staircase?

55. Stadium Construction How many rows are in the corner section of a stadium containing 2040 seats if the first row has 10 seats and each successive row has 4 additional seats?

56. Salary Suppose that you just received a job offer with a starting salary of \$35,000 per year and a guaranteed raise of \$1400 per year. How many years will it take before your aggregate salary is \$280,000?

[**Hint:** Your aggregate salary after 2 years is \$35,000 + (\$35,000 + \$1400).]

Discussion and Writing

57. Make up an arithmetic sequence. Give it to a friend and ask for its twentieth term.

58. Describe the similarities and differences between arithmetic sequences and linear functions.

7.3 Geometric Sequences; Geometric Series

PREPARING FOR THIS SECTION *Before getting started, review the following:*

- Compound Interest (Section 4.7, pp. 325–332)

Now work the 'Are You Prepared?' problems on page 518.

OBJECTIVES 1 Determine If a Sequence Is Geometric
2 Find a Formula for a Geometric Sequence
3 Find the Sum of a Geometric Sequence
4 Find the Sum of a Geometric Series
5 Solve Annuity Problems

1 Determine If a Sequence Is Geometric

When the ratio of successive terms of a sequence is always the same nonzero number, the sequence is called **geometric**. A **geometric sequence**[*] may be defined recursively as $a_1 = a$, $\dfrac{a_n}{a_{n-1}} = r$, or as

$$a_1 = a, \qquad a_n = ra_{n-1} \qquad\qquad \textbf{(1)}$$

where $a_1 = a$ and $r \neq 0$ are real numbers. The number a_1 is the first term, and the nonzero number r is called the **common ratio**.

The terms of a geometric sequence with first term a_1 and common ratio r follow the pattern

$$a_1, \quad a_1 r, \quad a_1 r^2, \quad a_1 r^3, \ldots$$

EXAMPLE 1 **Determining If a Sequence Is Geometric**

The sequence

$$2, \ 6, \ 18, \ 54, \ 162, \ldots$$

[*]Sometimes called a **geometric progression**.

is geometric since the ratio of successive terms is $3\left(\dfrac{6}{2} = \dfrac{18}{6} = \dfrac{54}{18} = \cdots = 3\right)$. The first term is $a_1 = 2$, and the common ratio is 3. ∎

EXAMPLE 2 **Determining If a Sequence Is Geometric**

Show that the following sequence is geometric.
$$\{s_n\} = 2^{-n}$$
Find the first term and the common ratio.

Solution The first term is $s_1 = 2^{-1} = \dfrac{1}{2}$. The nth and $(n-1)$st terms of the sequence $\{s_n\}$ are
$$s_n = 2^{-n} \quad \text{and} \quad s_{n-1} = 2^{-(n-1)}$$

Their ratio is
$$\frac{s_n}{s_{n-1}} = \frac{2^{-n}}{2^{-(n-1)}} = 2^{-n+(n-1)} = 2^{-1} = \frac{1}{2}$$

Because the ratio of successive terms is a nonzero constant, the sequence $\{s_n\}$ is geometric with common ratio $\dfrac{1}{2}$. ∎

EXAMPLE 3 **Determining If a Sequence Is Geometric**

Show that the following sequence is geometric.
$$\{t_n\} = \{4^n\}$$
Find the first term and the common ratio.

Solution The first term is $t_1 = 4^1 = 4$. The nth and $(n-1)$st terms are
$$t_n = 4^n \quad \text{and} \quad t_{n-1} = 4^{n-1}$$

Their ratio is
$$\frac{t_n}{t_{n-1}} = \frac{4^n}{4^{n-1}} = 4^{n-(n-1)} = 4$$

The sequence, $\{t_n\}$, is a geometric sequence with common ratio 4. ∎

NOW WORK PROBLEM **11**.

2 Find a Formula for a Geometric Sequence

Suppose that a_1 is the first term of a geometric sequence with common ratio $r \neq 0$. We seek a formula for the nth term a_n. To see the pattern, we write down the first few terms:
$$a_1 = a_1 \cdot 1 = a_1 r^0$$
$$a_2 = r a_1 = a_1 r^1$$
$$a_3 = r a_2 = r(a_1 r) = a_1 r^2$$
$$a_4 = r a_3 = r(a_1 r^2) = a_1 r^3$$
$$a_5 = r a_4 = r(a_1 r^3) = a_1 r^4$$
$$\vdots$$
$$a_n = r a_{n-1} = r(a_1 r^{n-2}) = a_1 r^{n-1}$$

We are led to the following result:

Theorem

*n*th Term of a Geometric Sequence

For a geometric sequence $\{a_n\}$ whose first term is a_1 and whose common ratio is r, the *n*th term is determined by the formula

$$a_n = a_1 r^{n-1} \qquad r \neq 0 \tag{2}$$

EXAMPLE 4

Finding a Particular Term of a Geometric Sequence

(a) Find the ninth term of the geometric sequence: $\quad 10,\ 9,\ \dfrac{81}{10},\ \dfrac{729}{100} \cdots$

(b) Find a recursive formula for this sequence.

Solution

(a) The first term of this geometric sequence is $a_1 = 10$ and the common ratio is $\dfrac{9}{10}$.

$\left(\text{Use } \dfrac{9}{10} \text{ or } \dfrac{\frac{81}{10}}{9} = \dfrac{9}{10} \text{ or any two successive terms.}\right)$ By formula (2), the *n*th term is

$$a_n = 10\left(\frac{9}{10}\right)^{n-1}$$

The ninth term is

$$a_9 = 10\left(\frac{9}{10}\right)^{9-1} = 10\left(\frac{9}{10}\right)^{8} \approx 4.3046721$$

(b) The first term in the sequence is 10 and the common ratio is $r = \dfrac{9}{10}$. Using formula (1), the recursive formula is $a_1 = 10,\ a_n = \dfrac{9}{10} a_{n-1}$. ∎

— Exploration ——————

Use a graphing utility to find the ninth term of the sequence given in Example 4. Use it to find the twentieth and fiftieth terms. Now use a graphing utility to graph the recursive formula found in Example 4(b). Conclude that the graph of the recursive formula behaves like the graph of an exponential function. How is r, the common ratio, related to a, the base of the exponential function $y = a^x$?

NOW WORK PROBLEMS 33 AND 41.

3 Find the Sum of a Geometric Sequence

The next result gives us a formula for finding the sum of the first n terms of a geometric sequence.

Theorem

Sum of the First *n* Terms of a Geometric Sequence

Let $\{a_n\}$ be a geometric sequence with first term a_1 and common ratio r, where $r \neq 0, r \neq 1$. The sum S_n of the first n terms of $\{a_n\}$ is

$$S_n = a_1 \cdot \frac{1 - r^n}{1 - r} \qquad r \neq 0, 1 \tag{3}$$

Proof The sum S_n of the first n terms of $\{a_n\} = \{a_1 r^{n-1}\}$ is

$$S_n = a_1 + a_1 r + \cdots + a_1 r^{n-1} \tag{4}$$

Multiply each side by r to obtain

$$r S_n = a_1 r + a_1 r^2 + \cdots + a_1 r^n \tag{5}$$

Now, subtract (5) from (4). The result is

$$S_n - r S_n = a_1 - a_1 r^n$$
$$(1 - r) S_n = a_1 (1 - r^n)$$

Since $r \neq 1$, we can solve for S_n.

$$S_n = a_1 \cdot \frac{1 - r^n}{1 - r}$$ ∎

EXAMPLE 5 **Finding the Sum of n Terms of a Geometric Sequence**

Find the sum S_n of the first n terms of the sequence $\left\{ \left(\frac{1}{2} \right)^n \right\}$; that is, find

$$\frac{1}{2} + \frac{1}{4} + \frac{1}{8} + \cdots + \left(\frac{1}{2} \right)^n$$

Solution The sequence $\left\{ \left(\frac{1}{2} \right)^n \right\}$ is a geometric sequence with $a_1 = \frac{1}{2}$ and $r = \frac{1}{2}$. The sum S_n that we seek is the sum of the first n terms of the sequence, so we use formula (3) to get

$$S_n = \sum_{k=1}^{n} \left(\frac{1}{2} \right)^k = \frac{1}{2} + \frac{1}{4} + \frac{1}{8} + \cdots + \left(\frac{1}{2} \right)^n$$

$$= \frac{1}{2} \left[\frac{1 - \left(\frac{1}{2} \right)^n}{1 - \frac{1}{2}} \right] \qquad \text{Formula (3)}$$

$$= \frac{1}{2} \left[\frac{1 - \left(\frac{1}{2} \right)^n}{\frac{1}{2}} \right]$$

$$= 1 - \left(\frac{1}{2} \right)^n$$ ∎

 NOW WORK PROBLEM 47.

EXAMPLE 6 **Using a Graphing Utility to Find the Sum of a Geometric Sequence**

Use a graphing utility to find the sum of the first 15 terms of the sequence $\left\{ \left(\frac{1}{3} \right)^n \right\}$; that is, find

$$\frac{1}{3} + \frac{1}{9} + \frac{1}{27} + \cdots + \left(\frac{1}{3} \right)^{15}$$

Figure 9

```
sum(seq((1/3)^n,
n,1,15,1))
      .4999999652
```

Solution Figure 9 shows the result obtained using a TI-84 Plus graphing calculator. The sum of the first 15 terms of the sequence $\left\{ \left(\frac{1}{3} \right)^n \right\}$ is 0.4999999652. ∎

NOW WORK PROBLEM 53.

4 Find the Sum of a Geometric Series

An infinite sum of the form

$$a_1 + a_1 r + a_1 r^2 + \cdots + a_1 r^{n-1} + \cdots$$

with first term a_1 and common ratio r, is called an **infinite geometric series** and is denoted by

$$\sum_{k=1}^{\infty} a_1 r^{k-1}$$

Based on formula (3), the sum S_n of the first n terms of a geometric series is

$$S_n = a_1 \cdot \frac{1 - r^n}{1 - r} = \frac{a_1}{1 - r} - \frac{a_1 r^n}{1 - r} \qquad (6)$$

If this finite sum S_n approaches a number L as $n \to \infty$, we say the infinite geometric series $\sum_{k=1}^{\infty} a_1 r^{k-1}$ **converges.** We call L the **sum of the infinite geometric series,** and we write

$$L = \sum_{k=1}^{\infty} a_1 r^{k-1}$$

If a series does not converge, it is called a **divergent series**.

Theorem

> **Sum of an Infinite Geometric Series**
>
> If $|r| < 1$, the sum of the infinite geometric series $\sum_{k=1}^{\infty} a_1 r^{k-1}$ is
>
> $$\sum_{k=1}^{\infty} a_1 r^{k-1} = \frac{a_1}{1 - r} \qquad (7)$$

Intuitive Proof Since $|r| < 1$, it follows that $|r^n|$ approaches 0 as $n \to \infty$. Then, based on formula (6), the sum S_n approaches $\dfrac{a_1}{1 - r}$ as $n \to \infty$. ∎

EXAMPLE 7 **Finding the Sum of a Geometric Series**

Find the sum of the geometric series: $2 + \dfrac{4}{3} + \dfrac{8}{9} + \cdots$

Solution The first term is $a_1 = 2$, and the common ratio is

Exploration

Use a graphing utility to graph $U_n = 2\left(\dfrac{2}{3}\right)^{n-1}$ in sequence mode. TRACE the graph for large values of n. What happens to the value of U_n as n increases without bound? What can you conclude about $\sum_{n=1}^{\infty} 2\left(\dfrac{2}{3}\right)^{n-1}$?

$$r = \frac{\dfrac{4}{3}}{2} = \frac{4}{6} = \frac{2}{3}$$

Since $|r| < 1$, we use formula (7) to find that

$$2 + \frac{4}{3} + \frac{8}{9} + \cdots = \frac{2}{1 - \dfrac{2}{3}} = 6 \qquad ■$$

NOW WORK PROBLEM 59.

EXAMPLE 8 **Repeating Decimals**

Show that the repeating decimal $0.999\ldots$ equals 1.

Solution $0.999\ldots = \dfrac{9}{10} + \dfrac{9}{100} + \dfrac{9}{1000} + \cdots$

The decimal $0.999\ldots$ is a geometric series with first term $\dfrac{9}{10}$ and common ratio $\dfrac{1}{10}$. Using formula (7), we find

$$0.999\ldots = \frac{\dfrac{9}{10}}{1 - \dfrac{1}{10}} = \frac{\dfrac{9}{10}}{\dfrac{9}{10}} = 1 \qquad ■$$

| EXAMPLE 9 | **Pendulum Swings** |

Figure 10

Initially, a pendulum swings through an arc of 18 inches. See Figure 10. On each successive swing, the length of the arc is 0.98 of the previous length.

(a) What is the length of the arc of the 10th swing?

(b) On which swing is the length of the arc first less than 12 inches?

(c) After 15 swings, what total distance will the pendulum have swung?

(d) When it stops, what total distance will the pendulum have swung?

Solution

(a) The length of the first swing is 18 inches.

The length of the second swing is $0.98(18)$ inches.

The length of the third swing is $0.98(0.98)(18) = 0.98^2(18)$ inches.

The length of the arc of the 10th swing is
$$(0.98)^9(18) \approx 15.007 \text{ inches}$$

(b) The length of the arc of the nth swing is $(0.98)^{n-1}(18)$. For this to be exactly 12 inches requires that

$$(0.98)^{n-1}(18) = 12$$

$$(0.98)^{n-1} = \frac{12}{18} = \frac{2}{3} \qquad \text{Divide both sides by 18.}$$

$$n - 1 = \log_{0.98}\left(\frac{2}{3}\right) \qquad \text{Express as a logarithm.}$$

$$n = 1 + \frac{\ln\left(\frac{2}{3}\right)}{\ln 0.98} \approx 1 + 20.07 = 21.07 \qquad \begin{array}{l}\text{Solve for } n; \text{ use the Change}\\ \text{of Base Formula.}\end{array}$$

The length of the arc of the pendulum exceeds 12 inches on the 21st swing and is first less than 12 inches on the 22nd swing.

(c) After 15 swings, the pendulum will have swung the following total distance L:

$$L = \underset{\text{1st}}{18} + \underset{\text{2nd}}{0.98(18)} + \underset{\text{3rd}}{(0.98)^2(18)} + \underset{\text{4th}}{(0.98)^3(18)} + \cdots + \underset{\text{15th}}{(0.98)^{14}(18)}$$

This is the sum of a geometric sequence. The common ratio is 0.98; the first term is 18. The sum has 15 terms, so

$$L = 18 \cdot \frac{1 - 0.98^{15}}{1 - 0.98} \approx 18(13.07) \approx 235.3 \text{ inches}$$

The pendulum will have swung through 235.3 inches after 15 swings.

(d) When the pendulum stops, it will have swung the following total distance T:

$$T = 18 + 0.98(18) + (0.98)^2(18) + (0.98)^3(18) + \cdots$$

This is the sum of a geometric series. The common ratio is $r = 0.98$; the first term is $a_1 = 18$. The sum is

$$T = \frac{a_1}{1 - r} = \frac{18}{1 - 0.98} = 900$$

The pendulum will have swung a total of 900 inches when it finally stops. ∎

NOW WORK PROBLEM 73.

5 Solve Annuity Problems

In Section 4.7 we developed the compound interest formula that gives the future value when a fixed amount of money is deposited in an account that pays interest compounded periodically. Often, though, money is invested in small amounts at periodic intervals. An **annuity** is a sequence of equal periodic deposits. The periodic deposits may be made annually, quarterly, monthly, or daily.

When deposits are made at the same time that the interest is credited, the annuity is called **ordinary.** We will only deal with ordinary annuities here. The **amount of an annuity** is the sum of all deposits made plus all interest paid.

Suppose the interest that an account earns is i percent per payment period (expressed as a decimal). For example, if an account pays 12% compounded monthly (12 times a year), then $i = \dfrac{0.12}{12} = 0.01$. If an account pays 8% compounded quarterly (4 times a year), then $i = \dfrac{0.08}{4} = 0.02$.

To develop a formula for the amount of an annuity, suppose that P is deposited each payment period for n payment periods in an account that earns i percent per payment period. When the last deposit is made at the nth payment period, the first deposit of P has earned interest compounded for $n - 1$ payment periods, the second deposit of P has earned interest compounded for $n - 2$ payment periods, and so on. Table 3 shows the value of each deposit after n deposits have been made.

Table 3

Deposit	1	2	3	...	n − 1	n
Amount	$P(1+i)^{n-1}$	$P(1+i)^{n-2}$	$P(1+i)^{n-3}$...	$P(1+i)$	P

The amount A of the annuity is the sum of the amounts shown in Table 3; that is,

$$A = P(1+i)^{n-1} + P(1+i)^{n-2} + \cdots + P(1+i) + P$$
$$= P[1 + (1+i) + \cdots + (1+i)^{n-1}]$$

The expression in brackets is the sum of a geometric sequence with n terms and a common ratio of $(1+i)$. As a result,

$$A = P[1 + (1+i) + \cdots + (1+i)^{n-2} + (1+i)^{n-1}]$$
$$= P\frac{1-(1+i)^n}{1-(1+i)} = P\frac{1-(1+i)^n}{-i} = P\frac{(1+i)^n - 1}{i}$$

We have established the following result:

Theorem

Amount of an Annuity

Suppose P is the deposit in dollars made at each payment period for an annuity paying i percent interest per payment period. The amount A of the annuity after n deposits is

$$A = P\frac{(1+i)^n - 1}{i} \tag{8}$$

Note: In using formula (8), remember that when the nth deposit is made, the first deposit has earned interest for $n - 1$ compounding periods.

EXAMPLE 10 | **Determining the Amount of an Annuity**

To save for retirement, Brett decides to place $2000 into an Individual Retirement Account (IRA) each year for the next 30 years. What will the value of the IRA be when Brett makes his 30th deposit? Assume that the rate of return of the IRA is 10% per annum compounded annually.

| EXAMPLE 1 | **Using Mathematical Induction** |

Show that the following statement is true for all natural numbers n.

$$1 + 3 + 5 + \cdots + (2n - 1) = n^2 \qquad \textbf{(2)}$$

Solution We need to show first that statement (2) holds for $n = 1$. Because $1 = 1^2$, statement (2) is true for $n = 1$. Condition I holds.

Next, we need to show that Condition II holds. Suppose that we know for some k that

$$1 + 3 + \cdots + (2k - 1) = k^2 \qquad \textbf{(3)}$$

We wish to show that, based on equation (3), statement (2) holds for $k + 1$. We look at the sum of the first $k + 1$ positive odd integers to determine whether this sum equals $(k + 1)^2$.

$$1 + 3 + \cdots + (2k - 1) + [2(k + 1) - 1] = \underbrace{[1 + 3 + \cdots + (2k - 1)]}_{= \, k^2 \text{ by equation (3)}} + (2k + 1)$$

$$= k^2 + (2k + 1)$$
$$= k^2 + 2k + 1 = (k + 1)^2$$

Conditions I and II are satisfied; by the Principle of Mathematical Induction, statement (2) is true for all natural numbers n. ∎

| EXAMPLE 2 | **Using Mathematical Induction** |

Show that the following statement is true for all natural numbers n.

$$2^n > n$$

Solution First, we show that the statement $2^n > n$ holds when $n = 1$. Because $2^1 = 2 > 1$, the inequality is true for $n = 1$. Condition I holds.

Next, we assume, for some natural number k, that $2^k > k$. We wish to show that the formula holds for $k + 1$; that is, we wish to show that $2^{k+1} > k + 1$. Now

$$2^{k+1} = 2 \cdot 2^k \underset{\substack{\uparrow \\ \text{We know that} \\ 2^k > k.}}{>} 2 \cdot k = k + k \underset{\substack{\uparrow \\ k \geq 1}}{\geq} k + 1$$

If $2^k > k$, then $2^{k+1} > k + 1$, so Condition II of the Principle of Mathematical Induction is satisfied. The statement $2^n > n$ is true for all natural numbers n. ∎

| EXAMPLE 3 | **Using Mathematical Induction** |

Show that the following formula is true for all natural numbers n.

$$1 + 2 + 3 + \cdots + n = \frac{n(n + 1)}{2} \qquad \textbf{(4)}$$

Solution First, we show that formula (4) is true when $n = 1$. Because

$$\frac{1(1 + 1)}{2} = \frac{1(2)}{2} = 1$$

Condition I of the Principle of Mathematical Induction holds.

Next, we assume that formula (4) holds for some k, and we determine whether the formula then holds for $k + 1$. We assume that

$$1 + 2 + 3 + \cdots + k = \frac{k(k + 1)}{2} \quad \text{for some } k \qquad \textbf{(5)}$$

7.4 Mathematical Induction

OBJECTIVE 1 Prove Statements Using Mathematical Induction

1 Prove Statements Using Mathematical Induction

Mathematical induction is a method for proving that statements involving natural numbers are true for all natural numbers.*

For example, the statement "$2n$ is always an even integer" can be proved true for all natural numbers n by using mathematical induction. Also, the statement "the sum of the first n positive odd integers equals n^2," that is,

$$1 + 3 + 5 + \cdots + (2n - 1) = n^2 \qquad \textbf{(1)}$$

can be proved for all natural numbers n by using mathematical induction.

Before stating the method of mathematical induction, let's try to gain a sense of the power of the method. We shall use the statement in equation (1) for this purpose by restating it for various values of $n = 1, 2, 3, \ldots$.

$n = 1$	The sum of the first positive odd integer is 1^2; $1 = 1^2$.
$n = 2$	The sum of the first 2 positive odd integers is 2^2; $1 + 3 = 4 = 2^2$.
$n = 3$	The sum of the first 3 positive odd integers is 3^2; $1 + 3 + 5 = 9 = 3^2$.
$n = 4$	The sum of the first 4 positive odd integers is 4^2; $1 + 3 + 5 + 7 = 16 = 4^2$.

Although from this pattern we might conjecture that statement (1) is true for any choice of n, can we really be sure that it does not fail for some choice of n? The method of proof by mathematical induction will, in fact, prove that the statement is true for all n.

Theorem

The Principle of Mathematical Induction

Suppose that the following two conditions are satisfied with regard to a statement about natural numbers:

CONDITION I: The statement is true for the natural number 1.

CONDITION II: If the statement is true for some natural number k, it is also true for the next natural number $k + 1$.

Then the statement is true for all natural numbers.

We shall not prove this principle. However, we can provide a physical interpretation that will help us to see why the principle works. Think of a collection of natural numbers obeying a statement as a collection of infinitely many dominoes. See Figure 11.

Figure 11

Now, suppose that we are told two facts:

1. The first domino is pushed over.

2. If one domino falls over, say the kth domino, so will the next one, the $(k + 1)$st domino.

Is it safe to conclude that *all* the dominoes fall over? The answer is yes, because if the first one falls (Condition I), the second one does also (by Condition II); and if the second one falls, so does the third (by Condition II); and so on.

Now let's prove some statements about natural numbers using mathematical induction.

*Recall that the natural numbers are the numbers $1, 2, 3, 4, \ldots$. In other words, the terms *natural numbers* and *positive integers* are synonymous.

82. Look at the figure below. What fraction of the square is eventually shaded if the indicated shading process continues indefinitely?

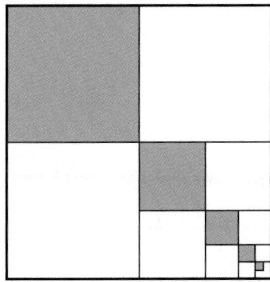

83. Multiplier Suppose that, throughout the U.S. economy, individuals spend 90% of every additional dollar that they earn. Economists would say that an individual's **marginal propensity to consume** is 0.90. For example, if Jane earns an additional dollar, she will spend 0.9(1) = $0.90 of it. The individual that earns $0.90 (from Jane) will spend 90% of it or $0.81. This process of spending continues and results in an infinite geometric series as follows:

$$1, 0.90, 0.90^2, 0.90^3, 0.90^4, \ldots$$

The sum of this infinite geometric series is called the **multiplier.** What is the multiplier if individuals spend 90% of every additional dollar that they earn?

84. Multiplier Refer to Problem 83. Suppose that the marginal propensity to consume throughout the U.S. economy is 0.95. What is the multiplier for the U.S. economy?

85. Stock Price One method of pricing a stock is to discount the stream of future dividends of the stock. Suppose that a stock pays $P per year in dividends and, historically, the dividend has been increased i% per year. If you desire an annual rate of return of r%, this method of pricing a stock states that the price that you should pay is the present value of an infinite stream of payments:

$$\text{Price} = P + P\frac{1+i}{1+r} + P\left(\frac{1+i}{1+r}\right)^2 + P\left(\frac{1+i}{1+r}\right)^3 + \cdots$$

The price of the stock is the sum of an infinite geometric series. Suppose that a stock pays an annual dividend of $4.00 and, historically, the dividend has been increased 3% per year. You desire an annual rate of return of 9%. What is the most you should pay for the stock?

86. Stock Price Refer to Problem 85. Suppose that a stock pays an annual dividend of $2.50 and, historically, the dividend has increased 4% per year. You desire an annual rate of return of 11%. What is the most that you should pay for the stock?

87. A Rich Man's Promise A rich man promises to give you $1000 on September 1, 2001. Each day thereafter he will give you $\frac{9}{10}$ of what he gave you the previous day. What is the first date on which the amount you receive is less than 1¢? How much have you received when this happens?

Discussion and Writing

88. Critical Thinking You are interviewing for a job and receive two offers:

 A: $20,000 to start, with guaranteed annual increases of 6% for the first 5 years

 B: $22,000 to start, with guaranteed annual increases of 3% for the first 5 years

Which offer is best if your goal is to be making as much as possible after 5 years? Which is best if your goal is to make as much money as possible over the contract (5 years)?

89. Critical Thinking Which of the following choices, A or B, results in more money?

 A: To receive $1000 on day 1, $999 on day 2, $998 on day 3, with the process to end after 1000 days

 B: To receive $1 on day 1, $2 on day 2, $4 on day 3, for 19 days

90. Critical Thinking You have just signed a 7-year professional football league contract with a beginning salary of $2,000,000 per year. Management gives you the following options with regard to your salary over the 7 years.

 1. A bonus of $100,000 each year
 2. An annual increase of 4.5% per year beginning after 1 year
 3. An annual increase of $95,000 per year beginning after 1 year

Which option provides the most money over the 7-year period? Which the least? Which would you choose? Why?

91. Can a sequence be both arithmetic and geometric? Give reasons for your answer.

92. Make up a geometric sequence. Give it to a friend and ask for its 20th term.

93. Make up two infinite geometric series, one that has a sum and one that does not. Give them to a friend and ask for the sum of each series.

94. Describe the similarities and differences between geometric sequences and exponential functions.

'Are You Prepared?' Answers

1. $1082.43 **2.** $9513.28

In Problems 59–68, find the sum of each infinite geometric series.

59. $1 + \dfrac{1}{3} + \dfrac{1}{9} + \cdots$

60. $2 + \dfrac{4}{3} + \dfrac{8}{9} + \cdots$

61. $8 + 4 + 2 + \cdots$

62. $6 + 2 + \dfrac{2}{3} + \cdots$

63. $2 - \dfrac{1}{2} + \dfrac{1}{8} - \dfrac{1}{32} + \cdots$

64. $1 - \dfrac{3}{4} + \dfrac{9}{16} - \dfrac{27}{64} + \cdots$

65. $\displaystyle\sum_{k=1}^{\infty} 5\left(\dfrac{1}{4}\right)^{k-1}$

66. $\displaystyle\sum_{k=1}^{\infty} 8\left(\dfrac{1}{3}\right)^{k-1}$

67. $\displaystyle\sum_{k=1}^{\infty} 6\left(-\dfrac{2}{3}\right)^{k-1}$

68. $\displaystyle\sum_{k=1}^{\infty} 4\left(-\dfrac{1}{2}\right)^{k-1}$

Applications and Extensions

69. Find x so that x, $x + 2$, and $x + 3$ are consecutive terms of a geometric sequence.

70. Find x so that $x - 1$, x, and $x + 2$ are consecutive terms of a geometric sequence.

71. **Salary Increases** Suppose that you have just been hired at an annual salary of $18,000 and expect to receive annual increases of 5%. What will your salary be when you begin your fifth year?

72. **Equipment Depreciation** A new piece of equipment cost a company $15,000. Each year, for tax purposes, the company depreciates the value by 15%. What value should the company give the equipment after 5 years?

73. **Pendulum Swings** Initially, a pendulum swings through an arc of 2 feet. On each successive swing, the length of the arc is 0.9 of the previous length.
 (a) What is the length of the arc of the 10th swing?
 (b) On which swing is the length of the arc first less than 1 foot?
 (c) After 15 swings, what total length will the pendulum have swung?
 (d) When it stops, what total length will the pendulum have swung?

74. **Bouncing Balls** A ball is dropped from a height of 30 feet. Each time it strikes the ground, it bounces up to 0.8 of the previous height.

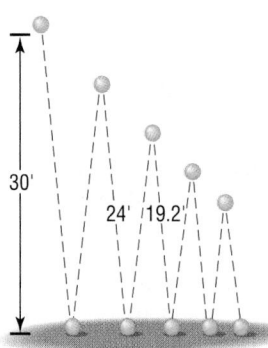

 (a) What height will the ball bounce up to after it strikes the ground for the third time?
 (b) How high will it bounce after it strikes the ground for the nth time?
 (c) How many times does the ball need to strike the ground before its bounce is less than 6 inches?
 (d) What total distance does the ball travel before it stops bouncing?

75. **Retirement** Christine contributes $100 each month to her 401(k). What will be the value of Christine's 401(k) after the 360th deposit (30 years) if the per annum rate of return is assumed to be 12% compounded monthly?

76. **Saving for a Home** Jolene wants to purchase a new home. Suppose that she invests $400 per month into a mutual fund. If the per annum rate of return of the mutual fund is assumed to be 10% compounded monthly, how much will Jolene have for a down payment after the 36th deposit (3 years)?

77. **Tax Sheltered Annuity** Don contributes $500 at the end of each quarter to a tax sheltered annuity (TSA). What will the value of the TSA be after the 80th deposit (20 years) if the per annum rate of return is assumed to be 8% compounded quarterly?

78. **Retirement** Ray contributes $1000 to an Individual Retirement Account (IRA) semiannually. What will the value of the IRA be when Ray makes his 30th deposit (after 15 years) if the per annum rate of return is assumed to be 10% compounded semiannually?

79. **Sinking Fund** Scott and Alice want to purchase a vacation home in 10 years and need $50,000 for a down payment. How much should they place in a savings account each month if the per annum rate of return is assumed to be 6% compounded monthly?

80. **Sinking Fund** For a child born in 1996, a 4-year college education at a public university is projected to $150,000. Assuming an 8% per annum rate of return compounded monthly, how much must be contributed to a college fund every month to have $150,000 in 18 years when the child begins college?

81. **Grains of Wheat on a Chess Board** In an old fable, a commoner who had saved the king's life was told he could ask the king for any just reward. Being a shrewd man, the commoner said, "A simple wish, sire. Place one grain of wheat on the first square of a chessboard, two grains on the second square, four grains on the third square, continuing until you have filled the board. This is all I seek." Compute the total number of grains needed to do this to see why the request, seemingly simple, could not be granted. (A chessboard consists of $8 \times 8 = 64$ squares.)

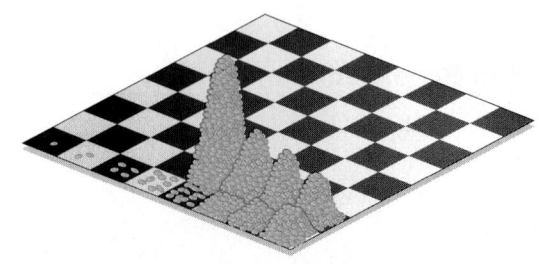

7.3 Assess Your Understanding

'Are You Prepared?'

Answers are given at the end of these exercises. If you get a wrong answer, read the pages listed in red.

1. If $1000 is invested at 4% per annum compounded semiannually, how much is in the account after two years? (pp. 325–332)

2. How much do you need to invest now at 5% per annum compounded monthly so that in 1 year you will have $10,000? (pp. 325–332)

Concepts and Vocabulary

3. In a(n) _____ sequence the ratio of successive terms is a constant.

4. If $|r| < 1$, the sum of the geometric series $\sum_{k=1}^{\infty} ar^{k-1}$ is _____.

5. A sequence of equal periodic deposits is called a(n) _____.

6. *True or False:* A geometric sequence may be defined recursively.

7. *True or False:* In a geometric sequence the common ratio is always a positive number.

8. *True or False:* For a geometric sequence with first term a_1 and common ratio r, where $r \neq 0, r \neq 1$, the sum of the first n terms is $S_n = a_1 \cdot \dfrac{1 - r^n}{1 - r}$.

Skill Building

In Problems 9–18, a geometric sequence is given. Find the common ratio and write out the first four terms.

9. $\{s_n\} = \{3^n\}$

10. $\{s_n\} = \{(-5)^n\}$

11. $\{a_n\} = \left\{-3\left(\dfrac{1}{2}\right)^n\right\}$

12. $\{b_n\} = \left\{\left(\dfrac{5}{2}\right)^n\right\}$

13. $\{c_n\} = \left\{\dfrac{2^{n-1}}{4}\right\}$

14. $\{d_n\} = \left\{\dfrac{3^n}{9}\right\}$

15. $\{e_n\} = \{2^{n/3}\}$

16. $\{f_n\} = \{3^{2n}\}$

17. $\{t_n\} = \left\{\dfrac{3^{n-1}}{2^n}\right\}$

18. $\{u_n\} = \left\{\dfrac{2^n}{3^{n-1}}\right\}$

In Problems 19–32, determine whether the given sequence is arithmetic, geometric, or neither. If the sequence is arithmetic, find the common difference; if it is geometric, find the common ratio.

19. $\{n + 2\}$

20. $\{2n - 5\}$

21. $\{4n^2\}$

22. $\{5n^2 + 1\}$

23. $\left\{3 - \dfrac{2}{3}n\right\}$

24. $\left\{8 - \dfrac{3}{4}n\right\}$

25. $1, 3, 6, 10, \ldots$

26. $2, 4, 6, 8, \ldots$

27. $\left\{\left(\dfrac{2}{3}\right)^n\right\}$

28. $\left\{\left(\dfrac{5}{4}\right)^n\right\}$

29. $-1, -2, -4, -8, \ldots$

30. $1, 1, 2, 3, 5, 8, \ldots$

31. $\{3^{n/2}\}$

32. $\{(-1)^n\}$

In Problems 33–40, find the fifth term and the nth term of the geometric sequence whose initial term a and common ratio r are given.

33. $a = 2; \quad r = 3$

34. $a = -2; \quad r = 4$

35. $a = 5; \quad r = -1$

36. $a = 6; \quad r = -2$

37. $a = 0; \quad r = \dfrac{1}{2}$

38. $a = 1; \quad r = -\dfrac{1}{3}$

39. $a = \sqrt{2}; \quad r = \sqrt{2}$

40. $a = 0; \quad r = \dfrac{1}{\pi}$

In Problems 41–46, find the indicated term of each geometric sequence.

41. 7th term of $1, \dfrac{1}{2}, \dfrac{1}{4}, \ldots$

42. 8th term of $1, 3, 9, \ldots$

43. 9th term of $1, -1, 1, \ldots$

44. 10th term of $-1, 2, -4, \ldots$

45. 8th term of $0.4, 0.04, 0.004, \ldots$

46. 7th term of $0.1, 1.0, 10.0, \ldots$

In Problems 47–52, find each sum.

47. $\dfrac{1}{4} + \dfrac{2}{4} + \dfrac{2^2}{4} + \dfrac{2^3}{4} + \cdots + \dfrac{2^{n-1}}{4}$

48. $\dfrac{3}{9} + \dfrac{3^2}{9} + \dfrac{3^3}{9} + \cdots + \dfrac{3^n}{9}$

49. $\sum_{k=1}^{n} \left(\dfrac{2}{3}\right)^k$

50. $\sum_{k=1}^{n} 4 \cdot 3^{k-1}$

51. $-1 - 2 - 4 - 8 - \cdots - (2^{n-1})$

52. $2 + \dfrac{6}{5} + \dfrac{18}{25} + \cdots + 2\left(\dfrac{3}{5}\right)^{n-1}$

For Problems 53–58, use a graphing utility to find the sum of each geometric sequence.

53. $\dfrac{1}{4} + \dfrac{2}{4} + \dfrac{2^2}{4} + \dfrac{2^3}{4} + \cdots + \dfrac{2^{14}}{4}$

54. $\dfrac{3}{9} + \dfrac{3^2}{9} + \dfrac{3^3}{9} + \cdots + \dfrac{3^{15}}{9}$

55. $\sum_{n=1}^{15} \left(\dfrac{2}{3}\right)^n$

56. $\sum_{n=1}^{15} 4 \cdot 3^{n-1}$

57. $-1 - 2 - 4 - 8 - \cdots - 2^{14}$

58. $2 + \dfrac{6}{5} + \dfrac{18}{25} + \cdots + 2\left(\dfrac{3}{5}\right)^{15}$

Solution This is an ordinary annuity with $n = 30$ annual deposits of $P = \$2000$. The rate of interest per payment period is $i = \dfrac{0.10}{1} = 0.10$. The amount A of the annuity after 30 deposits is

$$A = 2000\left\{\frac{(1 + 0.10)^{30} - 1}{0.10}\right\} = \$2000(164.494023) = \$328,988.05 \quad \blacksquare$$

EXAMPLE 11 **Determining the Amount of an Annuity**

To save for her daughter's college education, Ms. Miranda decides to put $50 aside every month in a credit union account paying 10% interest compounded monthly. She begins this savings program when her daughter is 3 years old. How much will she have saved by the time she makes the 180th deposit? How old is her daughter at this time?

Solution This is an annuity with $P = \$50$, $n = 180$, and $i = \dfrac{0.10}{12}$. The amount A saved is

$$A = 50\left[\frac{\left(1 + \dfrac{0.10}{12}\right)^{180} - 1}{\dfrac{0.10}{12}}\right] = \$50(414.47035) = \$20,723.52$$

Since there are 12 deposits per year, when the 180th deposit is made $\dfrac{180}{12} = 15$ years have passed and Ms. Miranda's daughter is 18 years old. $\quad \blacksquare$

NOW WORK PROBLEM **77.**

HISTORICAL FEATURE

Fibonacci

Sequences are among the oldest objects of mathematical investigation, having been studied for over 3500 years. After the initial steps, however, little progress was made until about 1600.

Arithmetic and geometric sequences appear in the Rhind papyrus, a mathematical text containing 85 problems copied around 1650 BC by the Egyptian scribe Ahmes from an earlier work (see Historical Problem 1). Fibonacci (AD 1220) wrote about problems similar to those found in the Rhind papyrus, leading one to suspect that Fibonacci may have had material available that is now lost. This material would have been in the non-Euclidean Greek tradition of Heron (about AD 75) and Diophantus (about AD

250). One problem, again modified slightly, is still with us in the familiar puzzle rhyme "As I was going to St. Ives ..." (see Historical Problem 2).

The Rhind papyrus indicates that the Egyptians knew how to add up the terms of an arithmetic or geometric sequence, as did the Babylonians. The rule for summing up a geometric sequence is found in Euclid's *Elements* (Book IX, 35, 36), where, like all Euclid's algebra, it is presented in a geometric form.

Investigations of other kinds of sequences began in the 1500s, when algebra became sufficiently developed to handle the more complicated problems. The development of calculus in the 1600s added a powerful new tool, especially for finding the sum of infinite series, and the subject continues to flourish today.

Historical Problems

1. *Arithmetic sequence problem from the Rhind papyrus (statement modified slightly for clarity)* One hundred loaves of bread are to be divided among five people so that the amounts that they receive form an arithmetic sequence. The first two together receive one-seventh of what the last three receive. How many loaves does each receive?

 [*Partial answer:* First person receives $1\frac{2}{3}$ loaves.]

2. The following old English children's rhyme resembles one of the Rhind papyrus problems.

 > As I was going to St. Ives
 > I met a man with seven wives

 > Each wife had seven sacks
 > Each sack had seven cats
 > Each cat had seven kits [kittens]
 > Kits, cats, sacks, wives
 > How many were going to St. Ives?

 (a) Assuming that the speaker and the cat fanciers met by traveling in opposite directions, what is the answer?

 (b) How many kittens are being transported?

 (c) Kits, cats, sacks, wives; how many?

Now we need to show that

$$1 + 2 + 3 + \cdots + k + (k + 1) = \frac{(k + 1)(k + 1 + 1)}{2} = \frac{(k + 1)(k + 2)}{2}$$

We do this as follows:

$$1 + 2 + 3 + \cdots + k + (k + 1) = \underbrace{[1 + 2 + 3 + \cdots + k]}_{= \frac{k(k+1)}{2} \quad \text{by equation (5)}} + (k + 1)$$

$$= \frac{k(k + 1)}{2} + (k + 1)$$

$$= \frac{k^2 + k + 2k + 2}{2}$$

$$= \frac{k^2 + 3k + 2}{2} = \frac{(k + 1)(k + 2)}{2}$$

Condition II also holds. As a result, formula (4) is true for all natural numbers n. ■

─────── **NOW WORK PROBLEM 1.**

EXAMPLE 4	**Using Mathematical Induction**

Show that $3^n - 1$ is divisible by 2 for all natural numbers n.

Solution First, we show that the statement is true when $n = 1$. Because $3^1 - 1 = 3 - 1 = 2$ is divisible by 2, the statement is true when $n = 1$. Condition I is satisfied.

Next, we assume that the statement holds for some k, and we determine whether the statement then holds for $k + 1$. We assume that $3^k - 1$ is divisible by 2 for some k. We need to show that $3^{k+1} - 1$ is divisible by 2. Now

$$3^{k+1} - 1 = 3^{k+1} - 3^k + 3^k - 1 \qquad \text{Subtract and add } 3^k.$$
$$= 3^k(3 - 1) + (3^k - 1) = 3^k \cdot 2 + (3^k - 1)$$

Because $3^k \cdot 2$ is divisible by 2 and $3^k - 1$ is divisible by 2, it follows that $3^k \cdot 2 + (3^k - 1) = 3^{k+1} - 1$ is divisible by 2. Condition II is also satisfied. As a result, the statement "$3^n - 1$ is divisible by 2" is true for all natural numbers n. ■

WARNING The conclusion that a statement involving natural numbers is true for all natural numbers is made only after both Conditions I and II of the Principle of Mathematical Induction have been satisfied. Problem 27 demonstrates a statement for which only Condition I holds, but the statement is not true for all natural numbers. Problem 28 demonstrates a statement for which only Condition II holds, but the statement is not true for any natural number. ■

7.4 Assess Your Understanding

Skill Building

In Problems 1–26, use the Principle of Mathematical Induction to show that the given statement is true for all natural numbers n.

1. $2 + 4 + 6 + \cdots + 2n = n(n + 1)$

2. $1 + 5 + 9 + \cdots + (4n - 3) = n(2n - 1)$

3. $3 + 4 + 5 + \cdots + (n + 2) = \frac{1}{2}n(n + 5)$

4. $3 + 5 + 7 + \cdots + (2n + 1) = n(n + 2)$

5. $2 + 5 + 8 + \cdots + (3n - 1) = \frac{1}{2}n(3n + 1)$

6. $1 + 4 + 7 + \cdots + (3n - 2) = \frac{1}{2}n(3n - 1)$

7. $1 + 2 + 2^2 + \cdots + 2^{n-1} = 2^n - 1$

8. $1 + 3 + 3^2 + \cdots + 3^{n-1} = \frac{1}{2}(3^n - 1)$

9. $1 + 4 + 4^2 + \cdots + 4^{n-1} = \frac{1}{3}(4^n - 1)$

10. $1 + 5 + 5^2 + \cdots + 5^{n-1} = \frac{1}{4}(5^n - 1)$

11. $\dfrac{1}{1\cdot2} + \dfrac{1}{2\cdot3} + \dfrac{1}{3\cdot4} + \cdots + \dfrac{1}{n(n+1)} = \dfrac{n}{n+1}$

12. $\dfrac{1}{1\cdot3} + \dfrac{1}{3\cdot5} + \dfrac{1}{5\cdot7} + \cdots + \dfrac{1}{(2n-1)(2n+1)} = \dfrac{n}{2n+1}$

13. $1^2 + 2^2 + 3^2 + \cdots + n^2 = \dfrac{1}{6}n(n+1)(2n+1)$

14. $1^3 + 2^3 + 3^3 + \cdots + n^3 = \dfrac{1}{4}n^2(n+1)^2$

15. $4 + 3 + 2 + \cdots + (5-n) = \dfrac{1}{2}n(9-n)$

16. $-2 - 3 - 4 - \cdots - (n+1) = -\dfrac{1}{2}n(n+3)$

17. $1\cdot2 + 2\cdot3 + 3\cdot4 + \cdots + n(n+1) = \dfrac{1}{3}n(n+1)(n+2)$

18. $1\cdot2 + 3\cdot4 + 5\cdot6 + \cdots + (2n-1)(2n) = \dfrac{1}{3}n(n+1)(4n-1)$

19. $n^2 + n$ is divisible by 2.

20. $n^3 + 2n$ is divisible by 3.

21. $n^2 - n + 2$ is divisible by 2.

22. $n(n+1)(n+2)$ is divisible by 6.

23. If $x > 1$, then $x^n > 1$.

24. If $0 < x < 1$, then $0 < x^n < 1$.

25. $a - b$ is a factor of $a^n - b^n$.
 [**Hint:** $a^{k+1} - b^{k+1} = a(a^k - b^k) + b^k(a - b)$]

26. $a + b$ is a factor of $a^{2n+1} + b^{2n+1}$.

Applications and Extensions

27. Show that the statement "$n^2 - n + 41$ is a prime number" is true for $n = 1$, but is not true for $n = 41$.

28. Show that the formula
$$2 + 4 + 6 + \cdots + 2n = n^2 + n + 2$$
obeys Condition II of the Principle of Mathematical Induction. That is, show that if the formula is true for some k it is also true for $k + 1$. Then show that the formula is false for $n = 1$ (or for any other choice of n).

29. Use mathematical induction to prove that if $r \neq 1$ then
$$a + ar + ar^2 + \cdots + ar^{n-1} = a\dfrac{1 - r^n}{1 - r}$$

30. Use mathematical induction to prove that
$$a + (a + d) + (a + 2d)$$
$$+ \cdots + [a + (n-1)d] = na + d\dfrac{n(n-1)}{2}$$

31. **Extended Principle of Mathematical Induction** The Extended Principle of Mathematical Induction states that if Conditions I and II hold, that is,

(I) A statement is true for a natural number j.

(II) If the statement is true for some natural number $k \geq j$, then it is also true for the next natural number $k + 1$.

then the statement is true for all natural numbers $\geq j$.

Use the Extended Principle of Mathematical Induction to show that the number of diagonals in a convex polygon of n sides is $\dfrac{1}{2}n(n-3)$.

[**Hint:** Begin by showing that the result is true when $n = 4$ (Condition I).]

32. **Geometry** Use the Extended Principle of Mathematical Induction to show that the sum of the interior angles of a convex polygon of n sides equals $(n-2)\cdot180°$.

Discussion and Writing

33. How would you explain the Principle of Mathematical Induction to a friend?

7.5 The Binomial Theorem

OBJECTIVES 1 Evaluate $\dbinom{n}{j}$

2 Use the Binomial Theorem

Formulas have been given for expanding $(x + a)^n$ for $n = 2$ and $n = 3$. The *Binomial Theorem*[*] is a formula for the expansion of $(x + a)^n$ for any positive integer n. If $n = 1, 2, 3$, and 4, the expansion of $(x + a)^n$ is straightforward.

[*]The name *binomial* is derived from the fact that $x + a$ is a binomial; that is, it contains two terms.

$$(x + a)^1 = x + a$$

$$(x + a)^2 = x^2 + 2ax + a^2$$

$$(x + a)^3 = x^3 + 3ax^2 + 3a^2x + a^3$$

$$(x + a)^4 = x^4 + 4ax^3 + 6a^2x^2 + 4a^3x + a^4$$

Notice that each expansion of $(x + a)^n$ begins with x^n and ends with a^n. As you read from left to right, the powers of x are decreasing by 1, while the powers of a are increasing by 1. Also, the number of terms that appears equals $n + 1$. Notice, too, that the degree of each monomial in the expansion equals n. For example, in the expansion of $(x + a)^3$, each monomial $(x^3, 3ax^2, 3a^2x, a^3)$ is of degree 3. As a result, we might conjecture that the expansion of $(x + a)^n$ would look like this:

$$(x + a)^n = x^n + \underline{\quad}ax^{n-1} + \underline{\quad}a^2x^{n-2} + \cdots + \underline{\quad}a^{n-1}x + a^n$$

where the blanks are numbers to be found. This is, in fact, the case, as we shall see shortly.

Before we can fill in the blanks, we need to introduce the symbol $\binom{n}{j}$.

1 Evaluate $\binom{n}{j}$

We define the symbol $\binom{n}{j}$, read "n taken j at a time," as follows:

> If j and n are integers with $0 \leq j \leq n$, the symbol $\binom{n}{j}$ is defined as
>
> $$\binom{n}{j} = \frac{n!}{j!(n - j)!} \qquad \textbf{(1)}$$

EXAMPLE 1

Evaluating $\binom{n}{j}$

Find:

(a) $\binom{3}{1}$ (b) $\binom{4}{2}$ (c) $\binom{8}{7}$ (d) $\binom{65}{15}$

Solution

(a) $\binom{3}{1} = \dfrac{3!}{1!(3 - 1)!} = \dfrac{3!}{1! \, 2!} = \dfrac{3 \cdot 2 \cdot 1}{1(2 \cdot 1)} = \dfrac{6}{2} = 3$

(b) $\binom{4}{2} = \dfrac{4!}{2!(4 - 2)!} = \dfrac{4!}{2! \, 2!} = \dfrac{4 \cdot 3 \cdot 2 \cdot 1}{(2 \cdot 1)(2 \cdot 1)} = \dfrac{24}{4} = 6$

(c) $\binom{8}{7} = \dfrac{8!}{7!(8 - 7)!} = \dfrac{8!}{7! \, 1!} = \dfrac{8 \cdot 7!}{7! \cdot 1!} = \dfrac{8}{1} = 8$

\uparrow

$8! = 8 \cdot 7!$

Figure 12

```
65 nCr 15
    2.073746998ᴇ14
```

(d) Figure 12 shows the solution using a TI-84 Plus graphing calculator. So,

$$\binom{65}{15} \approx 2.073746998 \times 10^{14}$$

■

 NOW WORK PROBLEM 5.

Four useful formulas involving the symbol $\binom{n}{j}$ are

$$\binom{n}{0} = 1 \qquad \binom{n}{1} = n \qquad \binom{n}{n-1} = n \qquad \binom{n}{n} = 1$$

Proof $\quad \binom{n}{0} = \dfrac{n!}{0!(n-0)!} = \dfrac{\cancel{n!}}{0! \, \cancel{n!}} = \dfrac{1}{1} = 1$

$\qquad \binom{n}{1} = \dfrac{n!}{1!(n-1)!} = \dfrac{n!}{(n-1)!} = \dfrac{n(n-1)!}{(n-1)!} = n$

You are asked to show the remaining two formulas in Problem 45. ∎

Suppose that we arrange the various values of the symbol $\binom{n}{j}$ in a triangular display, as shown next and in Figure 13.

$$\binom{0}{0}$$

$$\binom{1}{0} \quad \binom{1}{1}$$

$$\binom{2}{0} \quad \binom{2}{1} \quad \binom{2}{2}$$

$$\binom{3}{0} \quad \binom{3}{1} \quad \binom{3}{2} \quad \binom{3}{3}$$

$$\binom{4}{0} \quad \binom{4}{1} \quad \binom{4}{2} \quad \binom{4}{3} \quad \binom{4}{4}$$

$$\binom{5}{0} \quad \binom{5}{1} \quad \binom{5}{2} \quad \binom{5}{3} \quad \binom{5}{4} \quad \binom{5}{5}$$

Figure 13
Pascal triangle

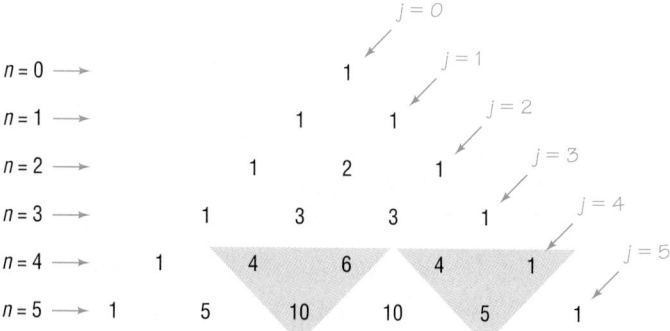

This display is called the **Pascal triangle**, named after Blaise Pascal (1623–1662), a French mathematician.

The Pascal triangle has 1's down the sides. To get any other entry, add the two nearest entries in the row above it. The shaded triangles in Figure 13 illustrate this feature of the Pascal triangle. Based on this feature, the row corresponding to $n = 6$ is found as follows:

$n = 5 \rightarrow \qquad 1 \quad 5 \quad 10 \quad 10 \quad 5 \quad 1$

$n = 6 \rightarrow \quad 1 \quad 6 \quad 15 \quad 20 \quad 15 \quad 6 \quad 1$

Later we shall prove that this addition always works (see the theorem on page 529).

Although the Pascal triangle provides an interesting and organized display of the symbol $\binom{n}{j}$, in practice it is not all that helpful. For example, if you wanted to know the value of $\binom{12}{5}$, you would need to produce 13 rows of the triangle before seeing the answer. It is much faster to use the definition (1).

2 Use the Binomial Theorem

Now we are ready to state the **Binomial Theorem**.

Theorem

Binomial Theorem

Let x and a be real numbers. For any positive integer n, we have

$$(x + a)^n = \binom{n}{0}x^n + \binom{n}{1}ax^{n-1} + \cdots + \binom{n}{j}a^j x^{n-j} + \cdots + \binom{n}{n}a^n$$

$$= \sum_{j=0}^{n} \binom{n}{j}x^{n-j}a^j \qquad\qquad (2)$$

Now you know why we needed to introduce the symbol $\binom{n}{j}$; these symbols are the numerical coefficients that appear in the expansion of $(x + a)^n$. Because of this, the symbol $\binom{n}{j}$ is called a **binomial coefficient**.

EXAMPLE 2

Expanding a Binomial

Use the Binomial Theorem to expand $(x + 2)^5$.

Solution

In the Binomial Theorem, let $a = 2$ and $n = 5$. Then

$$(x + 2)^5 = \binom{5}{0}x^5 + \binom{5}{1}2x^4 + \binom{5}{2}2^2x^3 + \binom{5}{3}2^3x^2 + \binom{5}{4}2^4x + \binom{5}{5}2^5$$

↑
Use equation (2).

$$= 1 \cdot x^5 + 5 \cdot 2x^4 + 10 \cdot 4x^3 + 10 \cdot 8x^2 + 5 \cdot 16x + 1 \cdot 32$$

↑
Use row $n = 5$ of the Pascal triangle or formula (1) for $\binom{n}{j}$.

$$= x^5 + 10x^4 + 40x^3 + 80x^2 + 80x + 32 \qquad \blacksquare$$

EXAMPLE 3

Expanding a Binomial

Expand $(2y - 3)^4$ using the Binomial Theorem.

Solution

First, we rewrite the expression $(2y - 3)^4$ as $[2y + (-3)]^4$. Now we use the Binomial Theorem with $n = 4$, $x = 2y$, and $a = -3$.

$$[2y + (-3)]^4 = \binom{4}{0}(2y)^4 + \binom{4}{1}(-3)(2y)^3 + \binom{4}{2}(-3)^2(2y)^2$$

$$+ \binom{4}{3}(-3)^3(2y) + \binom{4}{4}(-3)^4$$

$$= 1 \cdot 16y^4 + 4(-3)8y^3 + 6 \cdot 9 \cdot 4y^2 + 4(-27)2y + 1 \cdot 81$$

↑
Use row $n = 4$ of the Pascal triangle or formula (1) for $\binom{n}{j}$.

$$= 16y^4 - 96y^3 + 216y^2 - 216y + 81$$

In this expansion, note that the signs alternate due to the fact that $a = -3 < 0$. \blacksquare

NOW WORK PROBLEM **21.**

| EXAMPLE 4 | **Finding a Particular Coefficient in a Binomial Expansion** |

Find the coefficient of y^8 in the expansion of $(2y + 3)^{10}$.

Solution

We write out the expansion using the Binomial Theorem.

$$(2y + 3)^{10} = \binom{10}{0}(2y)^{10} + \binom{10}{1}(2y)^9(3)^1 + \binom{10}{2}(2y)^8(3)^2 + \binom{10}{3}(2y)^7(3)^3$$

$$+ \binom{10}{4}(2y)^6(3)^4 + \cdots + \binom{10}{9}(2y)(3)^9 + \binom{10}{10}(3)^{10}$$

From the third term in the expansion, the coefficient of y^8 is

$$\binom{10}{2}(2)^8(3)^2 = \frac{10!}{2!\,8!} \cdot 2^8 \cdot 9 = \frac{10 \cdot 9 \cdot 8!}{2 \cdot 8!} \cdot 2^8 \cdot 9 = 103{,}680 \qquad ∎$$

As this solution demonstrates, we can use the Binomial Theorem to find a particular term in an expansion without writing the entire expansion.

Based on the expansion of $(x + a)^n$, the term containing x^j is

$$\binom{n}{n - j}a^{n-j}x^j \qquad (3)$$

For example, we can solve Example 4 by using formula (3) with $n = 10$, $a = 3$, $x = 2y$, and $j = 8$. Then the term containing y^8 is

$$\binom{10}{10 - 8}3^{10-8}(2y)^8 = \binom{10}{2} \cdot 3^2 \cdot 2^8 \cdot y^8 = \frac{10!}{2!\,8!} \cdot 9 \cdot 2^8 y^8$$

$$= \frac{10 \cdot 9 \cdot 8!}{2 \cdot 8!} \cdot 9 \cdot 2^8 y^8 = 103{,}680y^8$$

| EXAMPLE 5 | **Finding a Particular Term in a Binomial Expansion** |

Find the sixth term in the expansion of $(x + 2)^9$.

Solution A

We expand using the Binomial Theorem until the sixth term is reached.

$$(x + 2)^9 = \binom{9}{0}x^9 + \binom{9}{1}x^8 \cdot 2 + \binom{9}{2}x^7 \cdot 2^2 + \binom{9}{3}x^6 \cdot 2^3 + \binom{9}{4}x^5 \cdot 2^4$$

$$+ \binom{9}{5}x^4 \cdot 2^5 + \cdots$$

The sixth term is

$$\binom{9}{5}x^4 \cdot 2^5 = \frac{9!}{5!\,4!} \cdot x^4 \cdot 32 = 4032x^4$$

Solution B

The sixth term in the expansion of $(x + 2)^9$, which has 10 terms total, contains x^4. (Do you see why?) By formula (3), the sixth term is

$$\binom{9}{9 - 4}2^{9-4}x^4 = \binom{9}{5}2^5x^4 = \frac{9!}{5!\,4!} \cdot 32x^4 = 4032x^4 \qquad ∎$$

NOW WORK PROBLEMS 29 AND 35.

Next we show that the *triangular addition* feature of the Pascal triangle illustrated in Figure 13 always works.

Theorem If n and j are integers with $1 \leq j \leq n$, then

$$\binom{n}{j-1} + \binom{n}{j} = \binom{n+1}{j} \qquad \text{(4)}$$

Proof

$$\binom{n}{j-1} + \binom{n}{j} = \frac{n!}{(j-1)![n-(j-1)]!} + \frac{n!}{j!(n-j)!}$$

$$= \frac{n!}{(j-1)!(n-j+1)!} + \frac{n!}{j!(n-j)!}$$

$$= \frac{jn!}{j(j-1)!(n-j+1)!} + \frac{(n-j+1)n!}{j!(n-j+1)(n-j)!}$$

Multiply the first term by $\frac{j}{j}$ and the second term by $\frac{n-j+1}{n-j+1}$. Now the denominators are equal.

$$= \frac{jn!}{j!(n-j+1)!} + \frac{(n-j+1)n!}{j!(n-j+1)!}$$

$$= \frac{jn! + (n-j+1)n!}{j!(n-j+1)!}$$

$$= \frac{n!(j+n-j+1)}{j!(n-j+1)!}$$

$$= \frac{n!(n+1)}{j!(n-j+1)!} = \frac{(n+1)!}{j![(n+1)-j]!} = \binom{n+1}{j}$$ ∎

HISTORICAL FEATURE

Omar Khayyám
(1050–1123)

The case $n = 2$ of the Binomial Theorem, $(a + b)^2$, was known to Euclid in 300 BC, but the general law seems to have been discovered by the Persian mathematician and astronomer Omar Khayyám (1050–1123), who is also well known as the author of the *Rubáiyát*, a collection of four-line poems making observations on the human condition. Omar Khayyám did not state the Binomial Theorem explicitly, but he claimed to have a method for extracting third, fourth, fifth roots, and so on. A little study shows that one must know the Binomial Theorem to create such a method.

The heart of the Binomial Theorem is the formula for the numerical coefficients, and, as we saw, they can be written out in a symmetric triangular form. The Pascal triangle appears first in the books of Yang Hui (about 1270) and Chu Shih-chieh (1303). Pascal's name is attached to the triangle because of the many applications he made of it, especially to counting and probability. In establishing these results, he was one of the earliest users of mathematical induction.

Many people worked on the proof of the Binomial Theorem, which was finally completed for all n (including complex numbers) by Niels Abel (1802–1829).

7.5 Assess Your Understanding

Concepts and Vocabulary

1. The _____ _____ is a triangular display of the binomial coefficients.

2. $\binom{6}{2} = $ _____

3. *True or False:* $\binom{n}{j} = \dfrac{j!}{(n-j)! \, n!}$

4. The _____ _____ can be used to expand expressions like $(2x + 3)^6$.

Skill Building

In Problems 5–16, evaluate each expression.

5. $\dbinom{5}{3}$

6. $\dbinom{7}{3}$

7. $\dbinom{7}{5}$

8. $\dbinom{9}{7}$

9. $\dbinom{50}{49}$

10. $\dbinom{100}{98}$

11. $\dbinom{1000}{1000}$

12. $\dbinom{1000}{0}$

13. $\dbinom{55}{23}$

14. $\dbinom{60}{20}$

15. $\dbinom{47}{25}$

16. $\dbinom{37}{19}$

In Problems 17–28, expand each expression using the Binomial Theorem.

17. $(x + 1)^5$

18. $(x - 1)^5$

19. $(x - 2)^6$

20. $(x + 3)^5$

21. $(3x + 1)^4$

22. $(2x + 3)^5$

23. $(x^2 + y^2)^5$

24. $(x^2 - y^2)^6$

25. $\left(\sqrt{x} + \sqrt{2}\right)^6$

26. $\left(\sqrt{x} - \sqrt{3}\right)^4$

27. $(ax + by)^5$

28. $(ax - by)^4$

In Problems 29–42, use the Binomial Theorem to find the indicated coefficient or term.

29. The coefficient of x^6 in the expansion of $(x + 3)^{10}$

30. The coefficient of x^3 in the expansion of $(x - 3)^{10}$

31. The coefficient of x^7 in the expansion of $(2x - 1)^{12}$

32. The coefficient of x^3 in the expansion of $(2x + 1)^{12}$

33. The coefficient of x^7 in the expansion of $(2x + 3)^9$

34. The coefficient of x^2 in the expansion of $(2x - 3)^9$

35. The fifth term in the expansion of $(x + 3)^7$

36. The third term in the expansion of $(x - 3)^7$

37. The third term in the expansion of $(3x - 2)^9$

38. The sixth term in the expansion of $(3x + 2)^8$

39. The coefficient of x^0 in the expansion of $\left(x^2 + \dfrac{1}{x}\right)^{12}$

40. The coefficient of x^0 in the expansion of $\left(x - \dfrac{1}{x^2}\right)^9$

41. The coefficient of x^4 in the expansion of $\left(x - \dfrac{2}{\sqrt{x}}\right)^{10}$

42. The coefficient of x^2 in the expansion of $\left(\sqrt{x} + \dfrac{3}{\sqrt{x}}\right)^8$

Applications and Extensions

43. Use the Binomial Theorem to find the numerical value of $(1.001)^5$ correct to five decimal places.
 [**Hint:** $(1.001)^5 = (1 + 10^{-3})^5$]

44. Use the Binomial Theorem to find the numerical value of $(0.998)^6$ correct to five decimal places.

45. Show that $\dbinom{n}{n-1} = n$ and $\dbinom{n}{n} = 1$.

46. Show that if n and j are integers with $0 \le j \le n$ then
$$\dbinom{n}{j} = \dbinom{n}{n-j}$$

Conclude that the Pascal triangle is symmetric with respect to a vertical line drawn from the topmost entry.

47. If n is a positive integer, show that
$$\dbinom{n}{0} + \dbinom{n}{1} + \cdots + \dbinom{n}{n} = 2^n$$

[**Hint:** $2^n = (1 + 1)^n$; now use the Binomial Theorem.]

48. If n is a positive integer, show that
$$\dbinom{n}{0} - \dbinom{n}{1} + \dbinom{n}{2} - \cdots + (-1)^n\dbinom{n}{n} = 0$$

49. $\dbinom{5}{0}\left(\dfrac{1}{4}\right)^5 + \dbinom{5}{1}\left(\dfrac{1}{4}\right)^4\left(\dfrac{3}{4}\right) + \dbinom{5}{2}\left(\dfrac{1}{4}\right)^3\left(\dfrac{3}{4}\right)^2$

$+ \dbinom{5}{3}\left(\dfrac{1}{4}\right)^2\left(\dfrac{3}{4}\right)^3 + \dbinom{5}{4}\left(\dfrac{1}{4}\right)\left(\dfrac{3}{4}\right)^4 + \dbinom{5}{5}\left(\dfrac{3}{4}\right)^5 = ?$

50. **Stirling's Formula** An approximation for $n!$, when n is large, is given by
$$n! \approx \sqrt{2n\pi}\left(\dfrac{n}{e}\right)^n\left(1 + \dfrac{1}{12n - 1}\right)$$

Calculate $12!$, $20!$, and $25!$ on your calculator. Then use Stirling's formula to approximate $12!$, $20!$, and $25!$.

Chapter Review

Things to Know

Sequence (p. 496)	A function whose domain is the set of positive integers.		
Factorials (p. 498)	$0! = 1, 1! = 1, n! = n(n-1) \cdot \cdots \cdot 3 \cdot 2 \cdot 1$ if $n \geq 2$		
Arithmetic sequence (pp. 505 and 506)	$a_1 = a$, $a_n = a_{n-1} + d$, where $a_1 = a = $ first term, $d = $ common difference		
	$a_n = a_1 + (n-1)d$		
Sum of the first n terms of an arithmetic sequence (p. 507)	$S_n = \dfrac{n}{2}[2a_1 + (n-1)d] = \dfrac{n}{2}(a_1 + a_n)$		
Geometric sequence (pp. 510 and 512)	$a_1 = a$, $a_n = ra_{n-1}$, where $a_1 = a = $ first term, $r = $ common ratio		
	$a_n = a_1 r^{n-1}$, $r \neq 0$		
Sum of the first n terms of a geometric sequence (p. 512)	$S_n = a_1 \dfrac{1 - r^n}{1 - r}$, $r \neq 0, 1$		
Infinite geometric series (p. 513)	$a_1 + a_1 r + \cdots + a_1 r^{n-1} + \cdots = \displaystyle\sum_{k=1}^{\infty} a_1 r^{k-1}$		
Sum of an infinite geometric series (p. 514)	$\displaystyle\sum_{k=1}^{\infty} a_1 r^{k-1} = \dfrac{a_1}{1 - r}$, $	r	< 1$
Principle of Mathematical Induction (p. 521)	Suppose the following two conditions are satisfied.		
	Condition I: The statement is true for the natural number 1.		
	Condition II: If the statement is true for some natural number k, it is also true for $k + 1$.		
	Then the statement is true for all natural numbers n.		
Binomial coefficient (p. 525)	$\dbinom{n}{j} = \dfrac{n!}{j!(n-j)!}$		
Pascal triangle (p. 526)	See Figure 13.		
Binomial Theorem (p. 527)	$(x + a)^n = \dbinom{n}{0}x^n + \dbinom{n}{1}ax^{n-1} + \cdots + \dbinom{n}{j}a^j x^{n-j} + \cdots + \dbinom{n}{n}a^n = \displaystyle\sum_{j=0}^{n}\dbinom{n}{j}x^{n-j}a^j$		

Objectives

Section		You should be able to . . .	Review Exercises
7.1	1	Write the first several terms of a sequence (p. 496)	1–4
	2	Write the terms of a sequence defined by a recursive formula (p. 499)	5–8
	3	Use summation notation (p. 500)	9–12
	4	Find the sum of a sequence (p. 501)	25–30
7.2	1	Determine if a sequence is arithmetic (p. 505)	13–24
	2	Find a formula for an arithmetic sequence (p. 506)	31, 32, 35, 37–40, 63, 64
	3	Find the sum of an arithmetic sequence (p. 507)	13, 14, 19, 20, 63, 64
7.3	1	Determine if a sequence is geometric (p. 510)	13–24
	2	Find a formula for a geometric sequence (p. 511)	17, 18, 21, 22, 33, 34, 36, 68
	3	Find the sum of a geometric sequence (p. 512)	17, 18, 21, 22, 65(a)-(c)
	4	Find the sum of a geometric series (p. 513)	41–46, 65(d)
	5	Solve annuity problems (p. 515)	66, 67
7.4	1	Prove statements using mathematical induction (p. 521)	47–52
7.5	1	Evaluate $\dbinom{n}{j}$ (p. 525)	53–54
	2	Use the Binomial Theorem (p. 527)	55–62

Review Exercises

In Problems 1–8, write down the first five terms of each sequence.

1. $\{a_n\} = \left\{(-1)^n\left(\dfrac{n+3}{n+2}\right)\right\}$ **2.** $\{b_n\} = \{(-1)^{n+1}(2n+3)\}$ **3.** $\{c_n\} = \left\{\dfrac{2^n}{n^2}\right\}$ **4.** $\{d_n\} = \left\{\dfrac{e^n}{n}\right\}$

5. $a_1 = 3; \quad a_n = \dfrac{2}{3}a_{n-1}$ **6.** $a_1 = 4; \quad a_n = -\dfrac{1}{4}a_{n-1}$ **7.** $a_1 = 2; \quad a_n = 2 - a_{n-1}$ **8.** $a_1 = -3; \quad a_n = 4 + a_{n-1}$

In Problems 9 and 10, write out each sum.

9. $\displaystyle\sum_{k=1}^{4}(4k+2)$ **10.** $\displaystyle\sum_{k=1}^{3}(3-k^2)$

In Problems 11 and 12, express each sum using summation notation.

11. $1 - \dfrac{1}{2} + \dfrac{1}{3} - \dfrac{1}{4} + \cdots + \dfrac{1}{13}$ **12.** $2 + \dfrac{2^2}{3} + \dfrac{2^3}{3^2} + \cdots + \dfrac{2^{n+1}}{3^n}$

In Problems 13–24, determine whether the given sequence is arithmetic, geometric, or neither. If the sequence is arithmetic, find the common difference and the sum of the first n terms. If the sequence is geometric, find the common ratio and the sum of the first n terms.

13. $\{a_n\} = \{n+5\}$ **14.** $\{b_n\} = \{4n+3\}$ **15.** $\{c_n\} = \{2n^3\}$ **16.** $\{d_n\} = \{2n^2 - 1\}$

17. $\{s_n\} = \{2^{3n}\}$ **18.** $\{u_n\} = \{3^{2n}\}$ **19.** $0, 4, 8, 12, \ldots$ **20.** $1, -3, -7, -11, \ldots$

21. $3, \dfrac{3}{2}, \dfrac{3}{4}, \dfrac{3}{8}, \dfrac{3}{16}, \ldots$ **22.** $5, -\dfrac{5}{3}, \dfrac{5}{9}, -\dfrac{5}{27}, \dfrac{5}{81}, \ldots$ **23.** $\dfrac{2}{3}, \dfrac{3}{4}, \dfrac{4}{5}, \dfrac{5}{6}, \ldots$ **24.** $\dfrac{3}{2}, \dfrac{5}{4}, \dfrac{7}{6}, \dfrac{9}{8}, \dfrac{11}{10}, \ldots$

In Problems 25–30, evaluate each sum.

25. $\displaystyle\sum_{k=1}^{5}(k^2+12)$ **26.** $\displaystyle\sum_{k=1}^{3}(k+2)^2$ **27.** $\displaystyle\sum_{k=1}^{10}(3k-9)$

28. $\displaystyle\sum_{k=1}^{9}(-2k+8)$ **29.** $\displaystyle\sum_{k=1}^{7}\left(\dfrac{1}{3}\right)^k$ **30.** $\displaystyle\sum_{k=1}^{10}(-2)^k$

In Problems 31–36, find the indicated term in each sequence.

[Hint: Find the general term first.]

31. 9th term of $3, 7, 11, 15, \ldots$ **32.** 8th term of $1, -1, -3, -5, \ldots$ **33.** 11th term of $1, \dfrac{1}{10}, \dfrac{1}{100}, \ldots$

34. 11th term of $1, 2, 4, 8, \ldots$ **35.** 9th term of $\sqrt{2}, 2\sqrt{2}, 3\sqrt{2}, \ldots$ **36.** 9th term of $\sqrt{2}, 2, 2^{3/2}, \ldots$

In Problems 37–40, find a general formula for each arithmetic sequence.

37. 7th term is 31; 20th term is 96 **38.** 8th term is -20; 17th term is -47

39. 10th term is 0; 18th term is 8 **40.** 12th term is 30; 22nd term is 50

In Problems 41–46, find the sum of each infinite geometric series.

41. $3 + 1 + \dfrac{1}{3} + \dfrac{1}{9} + \cdots$ **42.** $2 + 1 + \dfrac{1}{2} + \dfrac{1}{4} + \cdots$ **43.** $2 - 1 + \dfrac{1}{2} - \dfrac{1}{4} + \cdots$

44. $6 - 4 + \dfrac{8}{3} - \dfrac{16}{9} + \cdots$ **45.** $\displaystyle\sum_{k=1}^{\infty}4\left(\dfrac{1}{2}\right)^{k-1}$ **46.** $\displaystyle\sum_{k=1}^{\infty}3\left(-\dfrac{3}{4}\right)^{k-1}$

In Problems 47–52, use the Principle of Mathematical Induction to show that the given statement is true for all natural numbers.

47. $3 + 6 + 9 + \cdots + 3n = \dfrac{3n}{2}(n+1)$ **48.** $2 + 6 + 10 + \cdots + (4n-2) = 2n^2$

49. $2 + 6 + 18 + \cdots + 2\cdot3^{n-1} = 3^n - 1$ **50.** $3 + 6 + 12 + \cdots + 3\cdot2^{n-1} = 3(2^n - 1)$

51. $1^2 + 4^2 + 7^2 + \cdots + (3n-2)^2 = \dfrac{1}{2}n(6n^2 - 3n - 1)$ **52.** $1\cdot3 + 2\cdot4 + 3\cdot5 + \cdots + n(n+2) = \dfrac{n}{6}(n+1)(2n+7)$

In Problems 53 and 54, evaluate each binomial coefficient.

53. $\dbinom{5}{2}$ **54.** $\dbinom{8}{6}$

In Problems 55–58, expand each expression using the Binomial Theorem.

55. $(x + 2)^5$ **56.** $(x - 3)^4$ **57.** $(2x + 3)^5$ **58.** $(3x - 4)^4$

59. Find the coefficient of x^7 in the expansion of $(x + 2)^9$.

60. Find the coefficient of x^3 in the expansion of $(x - 3)^8$.

61. Find the coefficient of x^2 in the expansion of $(2x + 1)^7$.

62. Find the coefficient of x^6 in the expansion of $(2x + 1)^8$.

63. Constructing a Brick Staircase A brick staircase has a total of 25 steps. The bottom step requires 80 bricks. Each successive step requires three less bricks than the prior step.
(a) How many bricks are required for the top step?
(b) How many bricks are required to build the staircase?

64. Creating a Floor Design A mosaic tile floor is designed in the shape of a trapezoid 30 feet wide at the base and 15 feet wide at the top. The tiles, 12 inches by 12 inches, are to be placed so that each successive row contains one less tile than the row below. How many tiles will be required?

[**Hint:** Refer to Figure 8, page 508.]

65. Bouncing Balls A ball is dropped from a height of 20 feet. Each time it strikes the ground, it bounces up to three-quarters of the previous height.
(a) What height will the ball bounce up to after it strikes the ground for the third time?

(b) How high will it bounce after it strikes the ground for the nth time?
(c) How many times does the ball need to strike the ground before its bounce is less than 6 inches?
(d) What total distance does the ball travel before it stops bouncing?

66. Retirement Planning Chris gets paid once a month and contributes $200 each pay period into his 401(k). If Chris plans on retiring in 20 years, what will be the value of his 401(k) if the per annum rate of return of the 401(k) is 10% compounded monthly?

67. Retirement Planning Jacky contributes $500 every quarter to an IRA. If Jacky plans on retiring in 30 years, what will be the value of the IRA if the per annum rate of return of the IRA is 8% compounded quarterly?

68. Salary Increases Your friend has just been hired at an annual salary of $20,000. If she expects to receive annual increases of 4%, what will be her salary as she begins her fifth year?

Chapter Test

In Problems 1 and 2, write down the first five terms of each sequence.

1. $\{s_n\} = \left\{\dfrac{n^2 - 1}{n + 8}\right\}$

2. $a_1 = 4, a_n = 3a_{n-1} + 2$

In Problems 3 and 4, write out each sum. Evaluate each sum.

3. $\displaystyle\sum_{k=1}^{3} (-1)^{k+1}\left(\dfrac{k + 1}{k^2}\right)$

4. $\displaystyle\sum_{k=1}^{4}\left[\left(\dfrac{2}{3}\right)^k - k\right]$

5. Write the following sum using summation notation.

$$-\dfrac{2}{5} + \dfrac{3}{6} - \dfrac{4}{7} + \cdots + \dfrac{11}{14}$$

In Problems 6–11, determine whether the given sequence is arithmetic, geometric, or neither. If the sequence is arithmetic, find the common difference and the sum of the first n terms. If the sequence is geometric, find the common ratio and the sum of the first n terms.

6. $6, 12, 36, 144, \ldots$

7. $\left\{-\dfrac{1}{2} \cdot 4^n\right\}$

8. $-2, -10, -18, -26, \ldots$

9. $\left\{-\dfrac{n}{2} + 7\right\}$

10. $25, 10, 4, \dfrac{8}{5}, \ldots$

11. $\left\{\dfrac{2n - 3}{2n + 1}\right\}$

12. Find the sum of the infinite geometric series
$256 - 64 + 16 - 4 + \cdots$

13. Expand $(3m + 2)^5$ using the Binomial Theorem.

14. Use the Principle of Mathematical Induction to show that the given statement is true for all natural numbers.
$$\left(1 + \dfrac{1}{1}\right)\left(1 + \dfrac{1}{2}\right)\left(1 + \dfrac{1}{3}\right)\cdots\left(1 + \dfrac{1}{n}\right) = n + 1$$

15. A 2004 Dodge Durango sold for $31,000. If the vehicle loses 15% of its value each year, how much will it be worth after 10 years?

16. A weightlifter begins his routine by benching 100 pounds and increases the weight by 30 pounds for each set. If he does 10 repetitions in each set, what is the total weight lifted after 5 sets?

Chapter Projects

1. Population Growth The size of the population of the United States essentially depends on its current population, the birth and death rates of the population, and immigration. Suppose that b represents the birth rate of the U.S. population and d represents its death rate. Then $r = b - d$ represents the growth rate of the population, where r varies from year to year. The U.S. population after n years can be modeled using the recursive function

$$p_n = (1 + r)p_{n-1} + I$$

where I represents net immigration into the United States.

(a) Using data from the National Center for Health Statistics *www.fedstats.gov*, determine the birth and death rates for all races for the most recent year that data are available. Birth rates are given as the number of live births per 1000 population, while death rates are given as the number of deaths per 100,000 population. Each must be computed as the number of births (deaths) per individual. For example, in 1990, the birth rate was 16.7 per 1000 and the death rate was 863.8 per 100,000, so

$$b = \frac{16.7}{1000} = 0.0167, \text{ while } d = \frac{863.8}{100,000} = 0.008638.$$

Next, using data from the Immigration and Naturalization Service *www.fedstats.gov*, determine the net immigration to the United States for the same year used to obtain b and d in part (a).

(b) Determine the value of r, the growth rate of the population.

(c) Find a recursive formula for the population of the United States.

(d) Use the recursive formula to predict the population of the United States in the following year. In other words, if data are available up to the year 2003, predict the U.S. population in 2004.

(e) Compare your prediction to actual data.

(f) Do you think the recursive formula found in part (c) will be useful in predicting future populations? Why or why not?

The following projects are available at the Instructor's Resource Center (IRC):

2. Project at Motorola *Digital Wireless Communication*

3. Economics

4. Standardized Tests

Cumulative Review

1. Find all the solutions, real and complex, of the equation

$$\left| x^2 \right| = 9$$

2. (a) Graph the circle $x^2 + y^2 = 100$ and the parabola $y = 3x^2$.

(b) Solve the system of equations: $\begin{cases} x^2 + y^2 = 100 \\ \quad\quad y = 3x^2 \end{cases}$

(c) Where do the circle and the parabola intersect?

3. Solve the equation $2e^x = 5$.

4. Find an equation of the line with slope 5 and x-intercept 2.

5. Find the general equation of the circle whose center is the point $(-1, 2)$ if $(3, 5)$ is a point on the circle.

6. $f(x) = \dfrac{3x}{x - 2}$, $g(x) = 2x + 1$

Find:

(a) $(f \circ g)(2)$ (b) $(g \circ f)(4)$ (c) $(f \circ g)(x)$

(d) The domain of $(f \circ g)(x)$

(e) $(g \circ f)(x)$

(f) The domain of $(g \circ f)(x)$

(g) The function g^{-1} and its domain

(h) The function f^{-1} and its domain

7. Find the equation of an ellipse with center at the origin, a focus at $(0, 3)$, and a vertex at $(0, 4)$.

8. Find the equation of a parabola with vertex at $(-1, 2)$ and focus at $(-1, 3)$.

8 Counting and Probability

The Two-children Problem

PROBLEM: A woman and a man (unrelated) each have two children. At least one of the woman's children is a boy and the man's older child is a boy. Do the chances that the woman has two boys equal the chances that the man has two boys?

The above problem was posed to Marilyn vos Savant in her column *Ask Marilyn*. Her original answer, based on theoretical probabilities, was that the chances that the woman has two boys are 1 in 3 and the chances that the man has two boys are 1 in 2. This is found by looking at the sample space of two-child families: BB, BG, GB, GG. In the case of the man, we know that his older child is a boy and the sample space reduces to BB and BG. Hence the probability that he has two boys is 1 out of 2. In the case of the woman, since we only know that she has at least one boy, the sample space reduces to BB, BG, and GB. Thus, her chances of having two boys is 1 out of 3.

The answer about the woman's chances created quite a bit of controversy, resulting in many letters that challenged the correctness of her answer (*Parade*, July 27, 1997). Marilyn proposed that readers with exactly two children and at least one boy write in and tell the sex of both their children.

—*See Chapter Project 1.*

8.1 Sets and Counting

OBJECTIVES 1 Find All the Subsets of a Set
2 Find the Intersection and Union of Sets
3 Find the Complement of a Set
4 Count the Number of Elements in a Set

A **set** is a well-defined collection of distinct objects. The objects of a set are called its **elements**. By **well-defined**, we mean that there is a rule that enables us to determine whether a given object is an element of the set. If a set has no elements, it is called the **empty set**, or **null set**, and is denoted by the symbol ∅.

Because the elements of a set are distinct, we never repeat elements. For example, we would never write $\{1, 2, 3, 2\}$; the correct listing is $\{1, 2, 3\}$. Because a set is a collection, the order in which the elements are listed is immaterial. $\{1, 2, 3\}$, $\{1, 3, 2\}$, $\{2, 1, 3\}$, and so on, all represent the same set.

EXAMPLE 1 | **Writing the Elements of a Set**

Write the set consisting of the possible results (outcomes) from tossing a coin twice. Use H for *heads* and T for *tails*.

Solution In tossing a coin twice, we can get heads each time, HH; or heads the first time and tails the second, HT; or tails the first time and heads the second, TH; or tails each time, TT. Because no other possibilities exist, the set of outcomes is

$$\{HH, HT, TH, TT\}$$ ∎

1 Find All the Subsets of a Set

We now look at ways that two sets can be compared, beginning with set equality.

If two sets A and B have precisely the same elements, we say that A and B are **equal** and write $A = B$.

If each element of a set A is also an element of a set B, we say that A is a **subset** of B and write $A \subseteq B$.

If $A \subseteq B$ and $A \neq B$, we say that A is a **proper subset** of B and write $A \subset B$.

If $A \subseteq B$, every element in set A is also in set B, but B may or may not have additional elements. If $A \subset B$, every element in A is also in B, and B has at least one element not found in A.

Finally, we agree that the empty set is a subset of every set; that is,

$$\emptyset \subseteq A \quad \text{for any set } A$$

EXAMPLE 2 | **Finding All the Subsets of a Set**

Write down all the subsets of the set $\{a, b, c\}$.

Solution To organize our work, we write down all the subsets with no elements, then those with one element, then those with two elements, and finally those with three elements. These will give us all the subsets. Do you see why?

0 Elements	1 Element	2 Elements	3 Elements
∅	$\{a\}, \{b\}, \{c\}$	$\{a, b\}, \{b, c\}, \{a, c\}$	$\{a, b, c\}$

∎

NOW WORK PROBLEM 25.

2 Find the Intersection and Union of Sets

> If A and B are sets, the **intersection** of A with B, denoted $A \cap B$, is the set consisting of elements that belong to both A and B. The **union** of A with B, denoted $A \cup B$, is the set consisting of elements that belong to either A or B, or both.

EXAMPLE 3 | **Finding the Intersection and Union of Sets**

Let $A = \{1, 3, 5, 8\}$, $B = \{3, 5, 7\}$, and $C = \{2, 4, 6, 8\}$. Find:

(a) $A \cap B$ (b) $A \cup B$ (c) $B \cap (A \cup C)$

Solution

(a) $A \cap B = \{1, 3, 5, 8\} \cap \{3, 5, 7\} = \{3, 5\}$

(b) $A \cup B = \{1, 3, 5, 8\} \cup \{3, 5, 7\} = \{1, 3, 5, 7, 8\}$

(c) $B \cap (A \cup C) = \{3, 5, 7\} \cap [\{1, 3, 5, 8\} \cup \{2, 4, 6, 8\}]$

$\qquad\qquad\qquad = \{3, 5, 7\} \cap \{1, 2, 3, 4, 5, 6, 8\} = \{3, 5\}$ ∎

 NOW WORK PROBLEM 9.

3 Find the Complement of a Set

Usually, in working with sets, we designate a **universal set** U, the set consisting of all the elements that we wish to consider. Once a universal set has been designated, we can consider elements of the universal set not found in a given set.

> If A is a set, the **complement** of A, denoted \overline{A}, is the set consisting of all the elements in the universal set that are not in A.[*]

EXAMPLE 4 | **Finding the Complement of a Set**

If the universal set is $U = \{1, 2, 3, 4, 5, 6, 7, 8, 9\}$ and if $A = \{1, 3, 5, 7, 9\}$, then $\overline{A} = \{2, 4, 6, 8\}$. ∎

It follows that $A \cup \overline{A} = U$ and $A \cap \overline{A} = \varnothing$. Do you see why?

 NOW WORK PROBLEM 17.

It is often helpful to draw pictures of sets. Such pictures, called **Venn diagrams**, represent sets as circles enclosed in a rectangle, which represents the universal set. Such diagrams often help us to visualize various relationships among sets. See Figure 1.

If we know that $A \subset B$, we might use the Venn diagram in Figure 2(a). If we know that A and B have no elements in common, that is, if $A \cap B = \varnothing$, we might use the Venn diagram in Figure 2(b). The sets A and B in Figure 2(b) are said to be **disjoint**.

Figure 1

Figure 2

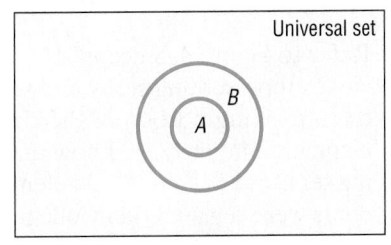

(a) $A \subset B$
proper subset

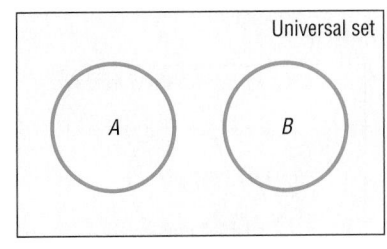

(b) $A \cap B = \varnothing$
disjoint sets

[*]Some books use the notation A' for the complement of A.

Figures 3(a), 3(b), and 3(c) use Venn diagrams to illustrate the definitions of intersection, union, and complement, respectively.

Figure 3

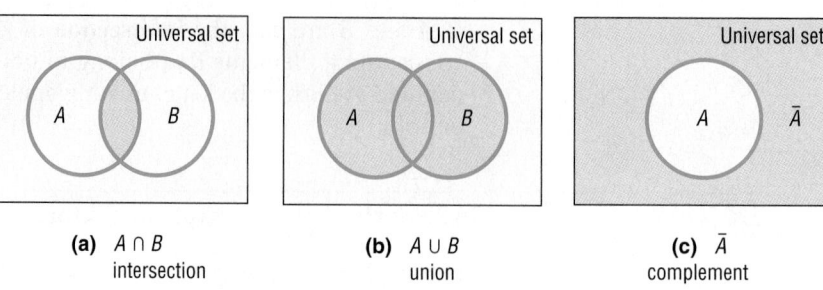

(a) $A \cap B$
intersection

(b) $A \cup B$
union

(c) \bar{A}
complement

4 Count the Number of Elements in a Set

As you count the number of students in a classroom or the number of pennies in your pocket, what you are really doing is matching, on a one-to-one basis, each object to be counted with the set of counting numbers $1, 2, 3, \ldots, n$, for some number n. If a set A matched up in this fashion with the set $\{1, 2, \ldots, 25\}$, you would conclude that there are 25 elements in the set A. We use the notation $n(A) = 25$ to indicate that there are 25 elements in the set A.

Because the empty set has no elements, we write

$$n(\varnothing) = 0$$

If the number of elements in a set is a nonnegative integer, we say that the set is **finite**. Otherwise, it is **infinite**. We shall concern ourselves only with finite sets.

Look again at Example 2. A set with 3 elements has $2^3 = 8$ subsets. This result can be generalized.

If A is a set with n elements, A has 2^n subsets.

For example, the set $\{a, b, c, d, e\}$ has $2^5 = 32$ subsets.

EXAMPLE 5

Analyzing Survey Data

In a survey of 100 college students, 35 were registered in College Algebra, 52 were registered in Computer Science I, and 18 were registered in both courses.

(a) How many students were registered in College Algebra or Computer Science I?

(b) How many were registered in neither course?

Solution

(a) First, let A = set of students in College Algebra

 B = set of students in Computer Science I

Then the given information tells us that

$$n(A) = 35, \quad n(B) = 52, \quad n(A \cap B) = 18,$$

Figure 4

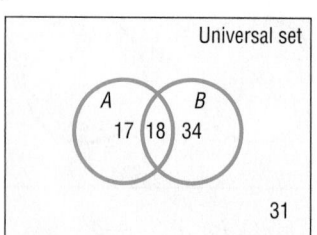

Refer to Figure 4. Since $n(A \cap B) = 18$, we know that the common part of the circles representing set A and set B has 18 elements. In addition, we know that the remaining portion of the circle representing set A will have $35 - 18 = 17$ elements. Similarly, we know that the remaining portion of the circle representing set B has $52 - 18 = 34$ elements. We conclude that $17 + 18 + 34 = 69$ students were registered in College Algebra or Computer Science I.

(b) Since 100 students were surveyed, it follows that $100 - 69 = 31$ were registered in neither course. ■

NOW WORK PROBLEMS 33 AND 39.

The solution to Example 5 contains the basis for a general counting formula. If we count the elements in each of two sets A and B, we necessarily count twice any elements that are in both A and B, that is, those elements in $A \cap B$. To count correctly the elements that are in A or B, that is, to find $n(A \cup B)$, we need to subtract those in $A \cap B$ from $n(A) + n(B)$.

Theorem

Counting Formula

If A and B are finite sets,

$$n(A \cup B) = n(A) + n(B) - n(A \cap B) \qquad (1)$$

Refer back to Example 5. Using (1), we have

$$n(A \cup B) = n(A) + n(B) - n(A \cap B)$$
$$= 35 + 52 - 18$$
$$= 69$$

There are 69 students registered in College Algebra or Computer Science I.

A special case of the counting formula (1) occurs if A and B have no elements in common. In this case, $A \cap B = \emptyset$, so $n(A \cap B) = 0$.

Theorem

Addition Principle of Counting

If two sets A and B have no elements in common, that is,

$$\text{if } A \cap B = \emptyset, \text{ then } n(A \cup B) = n(A) + n(B) \qquad (2)$$

We can generalize formula (2).

Theorem

General Addition Principle of Counting

If, for n sets A_1, A_2, \ldots, A_n, no two have elements in common,

$$n(A_1 \cup A_2 \cup \cdots \cup A_n) = n(A_1) + n(A_2) + \cdots + n(A_n) \qquad (3)$$

EXAMPLE 6 **Counting**

As of June 2002, federal agencies employed 93,445 full-time personnel authorized to make arrests and to carry firearms. Table 1 lists the type of law-enforcement officer and the corresponding number of full-time officers. No officer is classified as more than one type of officer.

Table 1

Type of Officer	Number of Full-time Federal Officers
Criminal (investigation/enforcement)	37,208
Police response and patrol	20,955
Corrections	16,915
Noncriminal (investigation/inspection)	12,801
Court operations	4,090
Security/protection	1,320
Other	156

SOURCE: Bureau of Justice Statistics

(a) How many full-time law-enforcement officers in the U.S. federal government were criminal officers or corrections officers?

(b) How many full-time law-enforcement officers in the U.S. federal government were criminal officers, corrections officers, or noncriminal officers?

Solution Let A represent the set of criminal officers, B represent the set of corrections officers, and C represent the set of noncriminal officers. No two of the sets A, B, and C have elements in common since a single officer cannot be classified as more than one type of officer. Then

$$n(A) = 37{,}208 \qquad n(B) = 16{,}915 \qquad n(C) = 12{,}801$$

(a) Using formula (2), we have

$$n(A \cup B) = n(A) + n(B) = 37{,}208 + 16{,}915 = 54{,}123$$

There were 54,123 officers that were criminal or corrections officers.

(b) Using formula (3), we have

$$n(A \cup B \cup C) = n(A) + n(B) + n(C) = 37{,}208 + 16{,}915 + 12{,}801 = 66{,}924$$

There were 66,924 officers that were criminal, corrections, or noncriminal officers. ∎

✏ **NOW WORK PROBLEM 43.**

8.1 Assess Your Understanding

Concepts and Vocabulary

1. The _____ of A and B consists of all elements in either A or B or both.

2. The _____ of A with B consists of all elements in both A and B.

3. *True or False:* The intersection of two sets is always a subset of their union.

4. *True or False:* If A is a set, the complement of A is the set of all the elements in the universal set that are not in A.

Skill Building

In Problems 5–14, use $A = \{1, 3, 5, 7, 9\}$, $B = \{1, 5, 6, 7\}$, and $C = \{1, 2, 4, 6, 8, 9\}$ to find each set.

5. $A \cup B$ **6.** $A \cup C$ **7.** $A \cap B$ **8.** $A \cap C$ **9.** $(A \cup B) \cap C$

10. $(A \cap C) \cup (B \cap C)$ **11.** $(A \cap B) \cup C$ **12.** $(A \cup B) \cup C$ **13.** $(A \cup C) \cap (B \cup C)$ **14.** $(A \cap B) \cap C$

In Problems 15–24, use U = universal set = $\{0, 1, 2, 3, 4, 5, 6, 7, 8, 9\}$, $A = \{1, 3, 4, 5, 9\}$, $B = \{2, 4, 6, 7, 8\}$, and $C = \{1, 3, 4, 6\}$ to find each set.

15. \overline{A} **16.** \overline{C} **17.** $\overline{A \cap B}$ **18.** $\overline{B \cup C}$ **19.** $\overline{A} \cup \overline{B}$

20. $\overline{B} \cap \overline{C}$ **21.** $\overline{A \cap \overline{C}}$ **22.** $\overline{\overline{B} \cup C}$ **23.** $\overline{A \cup B \cup C}$ **24.** $\overline{A \cap B \cap C}$

25. Write down all the subsets of $\{a, b, c, d\}$.

26. Write down all the subsets of $\{a, b, c, d, e\}$.

27. If $n(A) = 15$, $n(B) = 20$, and $n(A \cap B) = 10$, find $n(A \cup B)$.

28. If $n(A) = 30$, $n(B) = 40$, and $n(A \cup B) = 45$, find $n(A \cap B)$.

29. If $n(A \cup B) = 50$, $n(A \cap B) = 10$, and $n(B) = 20$, find $n(A)$.

30. If $n(A \cup B) = 60$, $n(A \cap B) = 40$, and $n(A) = n(B)$, find $n(A)$.

In Problems 31–38, use the information given in the figure.

31. How many are in set A?

32. How many are in set B?

33. How many are in A or B?

34. How many are in A and B?

35. How many are in A but not C?

36. How many are not in A?

37. How many are in A and B and C?

38. How many are in A or B or C?

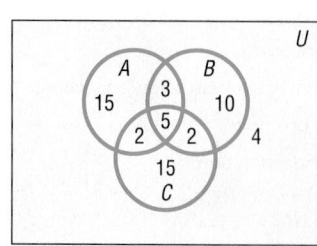

Applications and Extensions

39. Analyzing Survey Data In a consumer survey of 500 people, 200 indicated that they would be buying a major appliance within the next month, 150 indicated that they would buy a car, and 25 said that they would purchase both a major appliance and a car. How many will purchase neither? How many will purchase only a car?

40. Analyzing Survey Data In a student survey, 200 indicated that they would attend Summer Session I and 150 indicated Summer Session II. If 75 students plan to attend both summer sessions and 275 indicated that they would attend neither session, how many students participated in the survey?

41. Analyzing Survey Data In a survey of 100 investors in the stock market,

 50 owned shares in IBM
 40 owned shares in AT&T
 45 owned shares in GE
 20 owned shares in both IBM and GE
 15 owned shares in both AT&T and GE
 20 owned shares in both IBM and AT&T
 5 owned shares in all three

(a) How many of the investors surveyed did not have shares in any of the three companies?
(b) How many owned just IBM shares?
(c) How many owned just GE shares?
(d) How many owned neither IBM nor GE?
(e) How many owned either IBM or AT&T but no GE?

42. Classifying Blood Types Human blood is classified as either Rh+ or Rh−. Blood is also classified by type: A, if it contains an A antigen; B, if it contains a B antigen; AB, if it contains both A and B antigens; and O, if it contains neither antigen. Draw a Venn diagram illustrating the various blood types. Based on this classification, how many different kinds of blood are there?

43. The following data represent the marital status of males 18 years old and older in 2002.

Marital Status	Number (in thousands)
Married	61,212
Widowed	2,632
Divorced	8,659
Never married	28,107

Source: Current Population Survey

(a) Determine the number of males 18 years old and older who are widowed or divorced.
(b) Determine the number of males 18 years old and older who are married, widowed, or divorced.

44. The following data represent the marital status of females 18 years old and older in 2002.

Marital Status	Number (in thousands)
Married	62,037
Widowed	11,404
Divorced	12,236
Never married	23,036

Source: Current Population Survey

(a) Determine the number of females 18 years old and older who are widowed or divorced.
(b) Determine the number of females 18 years old and older who are married, widowed, or divorced.

Discussion and Writing

45. Make up a problem different from any found in the text that requires the addition principle of counting to solve. Give it to a friend to solve and critique.

46. Investigate the notion of counting as it relates to infinite sets. Write an essay on your findings.

8.2 Permutations and Combinations

PREPARING FOR THIS SECTION *Before getting started, review the following:*

• Factorial (Section 7.1, pp. 498–499).

Now work the 'Are You Prepared?' problems on page 549.

OBJECTIVES 1 Solve Counting Problems Using the Multiplication Principle
 2 Solve Counting Problems Using Permutations
 3 Solve Counting Problems Using Combinations
 4 Solve Counting Problems Using Permutations Involving *n* Nondistinct Objects

1 Solve Counting Problems Using the Multiplication Principle

Counting plays a major role in many diverse areas, such as probability, statistics, and computer science; counting techniques are a part of a branch of mathematics called **combinatorics**. In this section we shall look at special types of counting problems and develop general formulas for solving them.

We begin with an example that will demonstrate a general counting principle.

EXAMPLE 1 **Counting the Number of Possible Meals**

The fixed-price dinner at Mabenka Restaurant provides the following choices:

Appetizer: soup or salad
Entree: baked chicken, broiled beef patty, baby beef liver, or roast beef au jus
Dessert: ice cream or cheese cake

How many different meals can be ordered?

Solution Ordering such a meal requires three separate decisions:

| **Choose an Appetizer** | **Choose an Entree** | **Choose a Dessert** |
| 2 choices | 4 choices | 2 choices |

Look at the **tree diagram** in Figure 5. We see that, for each choice of appetizer, there are 4 choices of entrees. And for each of these $2 \cdot 4 = 8$ choices, there are 2 choices for dessert. A total of

$$2 \cdot 4 \cdot 2 = 16$$

different meals can be ordered.

Figure 5

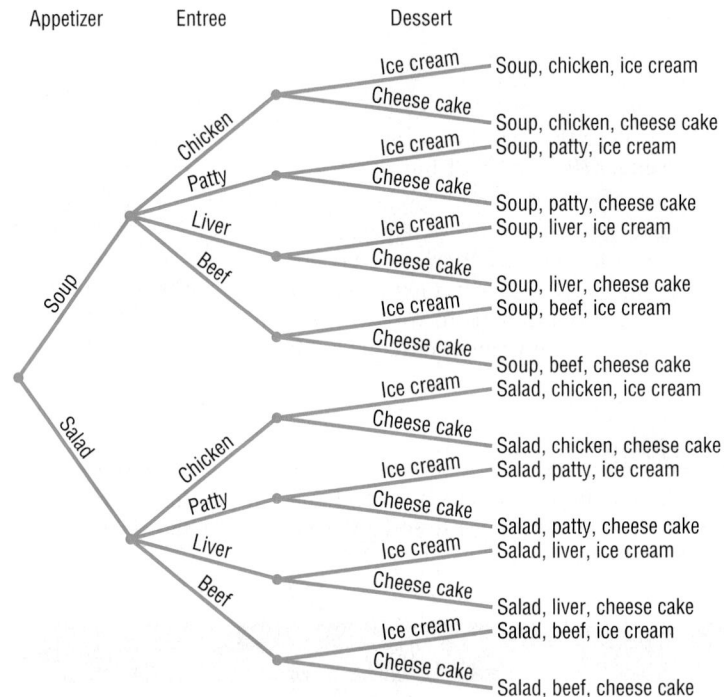

Theorem **Multiplication Principle of Counting**

If a task consists of a sequence of choices in which there are p selections for the first choice, q selections for the second choice, r selections for the third choice, and so on, the task of making these selections can be done in

$$p \cdot q \cdot r \cdot \cdots$$

different ways.

EXAMPLE 2 **Forming Codes**

How many two-symbol codewords can be formed if the first symbol is an uppercase letter and the second symbol is a digit?

Solution It sometimes helps to begin by listing some of the possibilities. The code consists of an uppercase letter followed by a digit, so some possibilities are A1, A2, B3, X0, and so on. The task consists of making two selections: the first selection requires choosing an uppercase letter (26 choices) and the second task requires choosing a digit (10 choices). By the Multiplication Principle, there are

$$26 \cdot 10 = 260$$

different codewords of the type described. ∎

NOW WORK PROBLEM 31.

2 Solve Counting Problems Using Permutations

We begin with a definition.

> A **permutation** is an ordered arrangement of r objects chosen from n objects.

We discuss three types of permutations:

1. The n objects are distinct (different), and repetition is allowed in the selection of r of them. [Distinct, with repetition]
2. The n objects are distinct (different), and repetition is not allowed in the selection of r of them, where $r \leq n$. [Distinct, without repetition]
3. The n objects are not distinct, and we use all of them in the arrangement. [Not distinct]

We take up the first two types here and deal with the third type at the end of this section.

The first type of permutation is handled using the Multiplication Principle.

EXAMPLE 3 **Counting Airport Codes [Permutation: Distinct, with Repetition]**

The International Airline Transportation Association (IATA) assigns three-letter codes to represent airport locations. For example, the airport code for Ft. Lauderdale, Florida, is FLL. Notice that repetition is allowed in forming this code. How many airport codes are possible?

Solution We are choosing 3 letters from 26 letters and arranging them in order. In the ordered arrangement a letter may be repeated. This is an example of a permutation with repetition in which 3 objects are chosen from 26 distinct objects.

The task of counting the number of such arrangements consists of making three selections. Each selection requires choosing a letter of the alphabet (26 choices). By the Multiplication Principle, there are

$$26 \cdot 26 \cdot 26 = 26^3 = 17{,}576$$

different airport codes. ∎

The solution given to Example 3 can be generalized.

Theorem **Permutations: Distinct Objects with Repetition**

The number of ordered arrangements of r objects chosen from n objects, in which the n objects are distinct and repetition is allowed, is n^r.

NOW WORK PROBLEM 35.

We begin the discussion of permutations in which the objects are distinct and repetition is not allowed with an example.

EXAMPLE 4

Forming Codes
[Permutation: Distinct, without Repetition]

Suppose that we wish to establish a three-letter code using any of the 26 uppercase letters of the alphabet, but we require that no letter be used more than once. How many different three-letter codes are there?

Solution

Some of the possibilities are ABC, ABD, ABZ, ACB, CBA, and so on. The task consists of making three selections. The first selection requires choosing from 26 letters. Because no letter can be used more than once, the second selection requires choosing from 25 letters. The third selection requires choosing from 24 letters. (Do you see why?) By the Multiplication Principle, there are

$$26 \cdot 25 \cdot 24 = 15,600$$

different three-letter codes with no letter repeated. ∎

For the second type of permutation, we introduce the following notation.

The notation $P(n, r)$ represents the number of ordered arrangements of r objects chosen from n distinct objects, where $r \leq n$ and repetition is not allowed.

For example, the question posed in Example 4 asks for the number of ways that the 26 letters of the alphabet can be arranged in order using three nonrepeated letters. The answer is

$$P(26, 3) = 26 \cdot 25 \cdot 24 = 15,600$$

EXAMPLE 5

Lining Up People

In how many ways can 5 people be lined up?

Solution

The 5 people are distinct. Once a person is in line, that person will not be repeated elsewhere in the line; and, in lining up people, order is important. We have a permutation of 5 objects taken 5 at a time. We can line up 5 people in

$$P(5, 5) = \underbrace{5 \cdot 4 \cdot 3 \cdot 2 \cdot 1}_{5 \text{ factors}} = 120 \text{ ways}$$ ∎

 NOW WORK PROBLEM 37.

To arrive at a formula for $P(n, r)$, we note that the task of obtaining an ordered arrangement of n objects in which only $r \leq n$ of them are used, without repeating any of them, requires making r selections. For the first selection, there are n choices; for the second selection, there are $n - 1$ choices; for the third selection, there are $n - 2$ choices; . . . ; for the rth selection, there are $n - (r - 1)$ choices. By the Multiplication Principle, we have

$$
\begin{array}{cccc}
\text{1st} & \text{2nd} & \text{3rd} & \text{rth}
\end{array}
$$
$$P(n, r) = n \cdot (n - 1) \cdot (n - 2) \cdot \cdots \cdot [n - (r - 1)]$$
$$= n \cdot (n - 1) \cdot (n - 2) \cdot \cdots \cdot (n - r + 1)$$

This formula for $P(n, r)$ can be compactly written using factorial notation.*

$$P(n, r) = n \cdot (n - 1) \cdot (n - 2) \cdot \cdots \cdot (n - r + 1)$$

$$= n \cdot (n - 1) \cdot (n - 2) \cdot \cdots \cdot (n - r + 1) \cdot \frac{(n - r) \cdot \cdots \cdot 3 \cdot 2 \cdot 1}{(n - r) \cdot \cdots \cdot 3 \cdot 2 \cdot 1} = \frac{n!}{(n - r)!}$$

*Recall that $0! = 1$, $1! = 1$, $2! = 2 \cdot 1, \ldots, n! = n(n - 1) \cdot \cdots \cdot 3 \cdot 2 \cdot 1$.

Theorem

Permutations of *r* Objects Chosen from *n* Distinct Objects without Repetition

The number of arrangements of *n* objects using $r \leq n$ of them, in which

1. the *n* objects are distinct,
2. once an object is used it cannot be repeated, and
3. order is important,

is given by the formula

$$P(n, r) = \frac{n!}{(n - r)!} \qquad (1)$$

EXAMPLE 6

Computing Permutations

Evaluate: (a) $P(7, 3)$ (b) $P(6, 1)$ (c) $P(52, 5)$

Solution We shall work parts (a) and (b) in two ways.

(a) $P(7, 3) = \underbrace{7 \cdot 6 \cdot 5}_{3 \text{ factors}} = 210$

or

$$P(7, 3) = \frac{7!}{(7 - 3)!} = \frac{7!}{4!} = \frac{7 \cdot 6 \cdot 5 \cdot 4!}{4!} = 210$$

(b) $P(6, 1) = \underbrace{6}_{1 \text{ factor}} = 6$

Figure 6

or

$$P(6, 1) = \frac{6!}{(6 - 1)!} = \frac{6!}{5!} = \frac{6 \cdot 5!}{5!} = 6$$

 (c) Figure 6 shows the solution using a TI-84 Plus graphing calculator. So
$$P(52, 5) = 311{,}875{,}200$$ ■

➤ **NOW WORK PROBLEM 7.**

EXAMPLE 7

The Birthday Problem

All we know about Shannon, Patrick, and Ryan is that they have different birthdays. If we listed all the possible ways this could occur, how many would there be? Assume that there are 365 days in a year.

Solution This is an example of a permutation in which 3 birthdays are selected from a possible 365 days, and no birthday may repeat itself. The number of ways that this can occur is

$$P(365, 3) = \frac{365!}{(365 - 3)!} = \frac{365 \cdot 364 \cdot 363 \cdot 362!}{362!} = 365 \cdot 364 \cdot 363 = 48{,}228{,}180$$

There are 48,228,180 ways in a group of three people for each to have a different birthday. ■

➤ **NOW WORK PROBLEM 53.**

3 Solve Counting Problems Using Combinations

In a permutation, order is important. For example, the arrangements *ABC*, *CAB*, *BAC*, ... are considered different arrangements of the letters *A*, *B*, and *C*. In many situations, though, order is unimportant. For example, in the card game of poker, the order in which the cards are received does not matter; it is the *combination* of the cards that matters.

> A **combination** is an arrangement, without regard to order, of *r* objects selected from *n* distinct objects without repetition, where $r \leq n$. The notation $C(n, r)$ represents the number of combinations of *n* distinct objects using *r* of them.

EXAMPLE 8

Listing Combinations

List all the combinations of the 4 objects *a*, *b*, *c*, *d* taken 2 at a time. What is $C(4, 2)$?

Solution One combination of *a*, *b*, *c*, *d* taken 2 at a time is

$$ab$$

We exclude *ba* from the list because order is not important in a combination (this means that we do not distinguish *ab* from *ba*). The list of all such combinations (convince yourself of this) is

$$ab, \quad ac, \quad ad, \quad bc, \quad bd, \quad cd$$

so

$$C(4, 2) = 6 \qquad \blacksquare$$

We can find a formula for $C(n, r)$ by noting that the only difference between a permutation of type 2 (distinct, without repetition) and a combination is that we disregard order in combinations. To determine $C(n, r)$, we need only eliminate from the formula for $P(n, r)$ the number of permutations that were simply rearrangements of a given set of *r* objects. This can be determined from the formula for $P(n, r)$ by calculating $P(r, r) = r!$. So, if we divide $P(n, r)$ by $r!$, we will have the desired formula for $C(n, r)$:

$$C(n, r) = \frac{P(n, r)}{r!} = \frac{\dfrac{n!}{(n-r)!}}{r!} = \frac{n!}{(n-r)!r!}$$

↑ Use formula (1).

We have proved the following result:

Theorem

Number of Combinations of *n* Distinct Objects Taken *r* at a Time

The number of arrangements of *n* objects using $r \leq n$ of them, in which

1. the *n* objects are distinct,
2. once an object is used, it cannot be repeated, and
3. order is not important,

is given by the formula

$$C(n, r) = \frac{n!}{(n-r)!r!} \qquad (2)$$

Based on formula (2), we discover that the symbol $C(n, r)$ and the symbol $\binom{n}{r}$ for the binomial coefficients are, in fact, the same. The Pascal triangle (see Section 7.5) can be used to find the value of $C(n, r)$. However, because it is more practical and convenient, we will use formula (2) instead.

EXAMPLE 9 | **Using Formula (2)**

Use formula (2) to find the value of each expression.

(a) $C(3, 1)$ (b) $C(6, 3)$ (c) $C(n, n)$ (d) $C(n, 0)$ (e) $C(52, 5)$

Solution

(a) $C(3, 1) = \dfrac{3!}{(3 - 1)!1!} = \dfrac{3!}{2!1!} = \dfrac{3 \cdot 2 \cdot 1}{2 \cdot 1 \cdot 1} = 3$

(b) $C(6, 3) = \dfrac{6!}{(6 - 3)!3!} = \dfrac{6 \cdot 5 \cdot 4 \cdot 3!}{3! \cdot 3!} = \dfrac{6 \cdot 5 \cdot 4}{6} = 20$

Figure 7

52 nCr 5
 2598960

(c) $C(n, n) = \dfrac{n!}{(n - n)!n!} = \dfrac{n!}{0!n!} = \dfrac{1}{1} = 1$

(d) $C(n, 0) = \dfrac{n!}{(n - 0)!0!} = \dfrac{n!}{n!0!} = \dfrac{1}{1} = 1$

(e) Figure 7 shows the solution using a TI-84 Plus graphing calculator. So

$$C(52, 5) = 2{,}598{,}960$$

NOW WORK PROBLEM **15**.

EXAMPLE 10 | **Forming Committees**

How many different committees of 3 people can be formed from a pool of 7 people?

Solution

The 7 people are distinct. More important, though, is the observation that the order of being selected for a committee is not significant. The problem asks for the number of combinations of 7 objects taken 3 at a time.

$$C(7, 3) = \frac{7!}{4!3!} = \frac{7 \cdot 6 \cdot 5 \cdot 4!}{4!3!} = \frac{7 \cdot 6 \cdot 5}{6} = 35$$

EXAMPLE 11 | **Forming Committees**

In how many ways can a committee consisting of 2 faculty members and 3 students be formed if 6 faculty members and 10 students are eligible to serve on the committee?

Solution

The problem can be separated into two parts: the number of ways that the faculty members can be chosen, $C(6, 2)$, and the number of ways that the student members can be chosen, $C(10, 3)$. By the Multiplication Principle, the committee can be formed in

$$C(6, 2) \cdot C(10, 3) = \frac{6!}{4!2!} \cdot \frac{10!}{7!3!} = \frac{6 \cdot 5 \cdot 4!}{4!2!} \cdot \frac{10 \cdot 9 \cdot 8 \cdot 7!}{7!3!}$$

$$= \frac{30}{2} \cdot \frac{720}{6} = 1800 \text{ ways}$$

NOW WORK PROBLEM **55**.

4 Solve Counting Problems Using Permutations Involving n Nondistinct Objects

We begin with an example.

EXAMPLE 12

Forming Different Words

How many different words (real or imaginary) can be formed using all the letters in the word REARRANGE?

Solution

Each word formed will have 9 letters: 3 R's, 2 A's, 2 E's, 1 N, and 1 G. To construct each word, we need to fill in 9 positions with the 9 letters:

$$\overline{1} \quad \overline{2} \quad \overline{3} \quad \overline{4} \quad \overline{5} \quad \overline{6} \quad \overline{7} \quad \overline{8} \quad \overline{9}$$

The process of forming a word consists of five tasks:

Task 1: Choose the positions for the 3 R's.

Task 2: Choose the positions for the 2 A's.

Task 3: Choose the positions for the 2 E's.

Task 4: Choose the position for the 1 N.

Task 5: Choose the position for the 1 G.

Task 1 can be done in $C(9, 3)$ ways. There then remain 6 positions to be filled, so Task 2 can be done in $C(6, 2)$ ways. There remain 4 positions to be filled, so Task 3 can be done in $C(4, 2)$ ways. There remain 2 positions to be filled, so Task 4 can be done in $C(2, 1)$ ways. The last position can be filled in $C(1, 1)$ way. Using the Multiplication Principle, the number of possible words that can be formed is

$$C(9, 3) \cdot C(6, 2) \cdot C(4, 2) \cdot C(2, 1) \cdot C(1, 1) = \frac{9!}{3! \cdot 6!} \cdot \frac{6!}{2! \cdot 4!} \cdot \frac{4!}{2! \cdot 2!} \cdot \frac{2!}{1! \cdot 1!} \cdot \frac{1!}{0! \cdot 1!}$$

$$= \frac{9!}{3! \cdot 2! \cdot 2! \cdot 1! \cdot 1!} \qquad \blacksquare$$

The form of the answer to Example 12 is suggestive of a general result. Had the letters in REARRANGE each been different, there would have been $P(9, 9) = 9!$ possible words formed. This is the numerator of the answer. The presence of 3 R's, 2 A's, and 2 E's reduces the number of different words, as the entries in the denominator illustrate. We are led to the following result:

Theorem

Permutations Involving _n_ Objects That Are Not Distinct

The number of permutations of n objects of which n_1 are of one kind, n_2 are of a second kind, ..., and n_k are of a kth kind is given by

$$\boxed{\frac{n!}{n_1! \cdot n_2! \cdot \cdots \cdot n_k!}} \qquad \textbf{(3)}$$

where $n = n_1 + n_2 + \cdots + n_k$.

EXAMPLE 13

Arranging Flags

How many different vertical arrangements are there of 8 flags if 4 are white, 3 are blue, and 1 is red?

Solution

We seek the number of permutations of 8 objects, of which 4 are of one kind, 3 are of a second kind, and 1 is of a third kind. Using formula (3), we find that there are

$$\frac{8!}{4! \cdot 3! \cdot 1!} = \frac{8 \cdot 7 \cdot 6 \cdot 5 \cdot 4!}{4! \cdot 3! \cdot 1!} = 280 \text{ different arrangements} \qquad \blacksquare$$

NOW WORK PROBLEM 57.

8.2 Assess Your Understanding

'Are You Prepared?'

Answers are given at the end of these exercises. If you get a wrong answer, read the pages listed in red.

1. $0! = $ ____; $1! = $ ____. (p. 498)

2. *True or False:* $n! = \dfrac{(n+1)!}{n}$. (pp. 498–499)

Concepts and Vocabulary

3. A(n) ____ is an ordered arrangement of r objects chosen from n objects.

4. A(n) ____ is an arrangement of r objects chosen from n distinct objects, without repetition and without regard to order.

5. *True or False:* In a combination problem, order is not important.

6. *True or False:* In some permutation problems, once an object is used, it cannot be repeated.

Skill Building

In Problems 7–14, find the value of each permutation.

7. $P(6,2)$
8. $P(7,2)$
9. $P(4,4)$
10. $P(8,8)$
11. $P(7,0)$
12. $P(9,0)$
13. $P(8,4)$
14. $P(8,3)$

In Problems 15–22, use formula (2) to find the value of each combination.

15. $C(8,2)$
16. $C(8,6)$
17. $C(7,4)$
18. $C(6,2)$
19. $C(15,15)$
20. $C(18,1)$
21. $C(26,13)$
22. $C(18,9)$

Applications and Extensions

23. List all the ordered arrangements of 5 objects a, b, c, d, and e choosing 3 at a time without repetition. What is $P(5,3)$?

24. List all the ordered arrangements of 5 objects a, b, c, d, and e choosing 2 at a time without repetition. What is $P(5,2)$?

25. List all the ordered arrangements of 4 objects $1, 2, 3$, and 4 choosing 3 at a time without repetition. What is $P(4,3)$?

26. List all the ordered arrangements of 6 objects $1, 2, 3, 4, 5$, and 6 choosing 3 at a time without repetition. What is $P(6,3)$?

27. List all the combinations of 5 objects a, b, c, d, and e taken 3 at a time. What is $C(5,3)$?

28. List all the combinations of 5 objects a, b, c, d, and e taken 2 at a time. What is $C(5,2)$?

29. List all the combinations of 4 objects $1, 2, 3$, and 4 taken 3 at a time. What is $C(4,3)$?

30. List all the combinations of 6 objects $1, 2, 3, 4, 5$, and 6 taken 3 at a time. What is $C(6,3)$?

31. **Shirts and Ties** A man has 5 shirts and 3 ties. How many different shirt and tie arrangements can he wear?

32. **Blouses and Skirts** A woman has 3 blouses and 5 skirts. How many different outfits can she wear?

33. **Forming Codes** How many two-letter codes can be formed using the letters A, B, C, and D? Repeated letters are allowed.

34. **Forming Codes** How many two-letter codes can be formed using the letters A, B, C, D, and E? Repeated letters are allowed.

35. **Forming Numbers** How many three-digit numbers can be formed using the digits 0 and 1? Repeated digits are allowed.

36. **Forming Numbers** How many three-digit numbers can be formed using the digits $0, 1, 2, 3, 4, 5, 6, 7, 8$, and 9? Repeated digits are allowed.

37. **Lining Up People** In how many ways can 4 people be lined up?

38. **Stacking Boxes** In how many ways can 5 different boxes be stacked?

39. **Forming Codes** How many different three-letter codes are there if only the letters A, B, C, D, and E can be used and no letter can be used more than once?

40. **Forming Codes** How many different four-letter codes are there if only the letters A, B, C, D, E, and F can be used and no letter can be used more than once?

41. **Stocks on the NYSE** Companies whose stocks are listed on the New York Stock Exchange (NYSE) have their company name represented by either 1, 2, or 3 letters (repetition of letters is allowed). What is the maximum number of companies that can be listed on the NYSE?

42. **Stocks on the NASDAQ** Companies whose stocks are listed on the NASDAQ stock exchange have their company name represented by either 4 or 5 letters (repetition of letters is allowed). What is the maximum number of companies that can be listed on the NASDAQ?

43. **Establishing Committees** In how many ways can a committee of 4 students be formed from a pool of 7 students?

44. **Establishing Committees** In how many ways can a committee of 3 professors be formed from a department having 8 professors?

45. **Possible Answers on a True/False Test** How many arrangements of answers are possible for a true/false test with 10 questions?

46. Possible Answers on a Multiple-choice Test How many arrangements of answers are possible in a multiple-choice test with 5 questions, each of which has 4 possible answers?

47. Four-digit Numbers How many four-digit numbers can be formed using the digits 0, 1, 2, 3, 4, 5, 6, 7, 8, and 9 if the first digit cannot be 0? Repeated digits are allowed.

48. Five-digit Numbers How many five-digit numbers can be formed using the digits 0, 1, 2, 3, 4, 5, 6, 7, 8, and 9 if the first digit cannot be 0 or 1? Repeated digits are allowed.

49. Arranging Books Five different mathematics books are to be arranged on a student's desk. How many arrangements are possible?

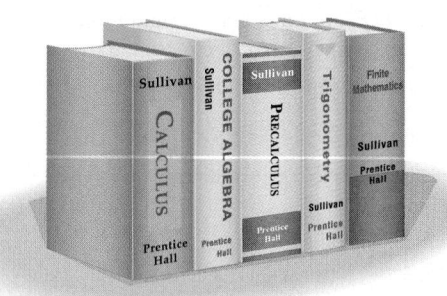

50. Forming License Plate Numbers How many different license plate numbers can be made using 2 letters followed by 4 digits selected from the digits 0 through 9, if
(a) letters and digits may be repeated?
(b) letters may be repeated, but digits may not be repeated?
(c) neither letters nor digits may be repeated?

51. Stock Portfolios As a financial planner, you are asked to select one stock each from the following groups: 8 DOW stocks, 15 NASDAQ stocks, and 4 global stocks. How many different portfolios are possible?

52. Combination Locks A combination lock displays 50 numbers. To open it, you turn to a number, then rotate clockwise to a second number, and then counterclockwise to the third number.
(a) How many different lock combinations are there?
(b) Comment on the description of such a lock as a *combination* lock.

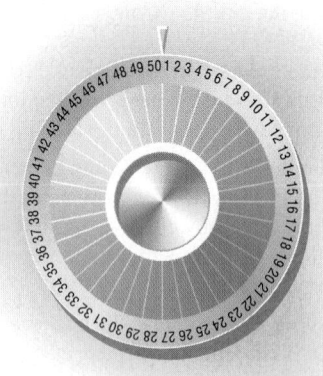

53. Birthday Problem In how many ways can 2 people each have different birthdays? Assume that there are 365 days in a year.

54. Birthday Problem In how many ways can 5 people each have different birthdays? Assume that there are 365 days in a year.

55. Forming a Committee A student dance committee is to be formed consisting of 2 boys and 3 girls. If the membership is to be chosen from 4 boys and 8 girls, how many different committees are possible?

56. Forming a Committee The student relations committee of a college consists of 2 administrators, 3 faculty members, and 5 students. Four administrators, 8 faculty members, and 20 students are eligible to serve. How many different committees are possible?

57. Forming Words How many different 9-letter words (real or imaginary) can be formed from the letters in the word ECONOMICS?

58. Forming Words How many different 11-letter words (real or imaginary) can be formed from the letters in the word MATHEMATICS?

59. Selecting Objects An urn contains 7 white balls and 3 red balls. Three balls are selected. In how many ways can the 3 balls be drawn from the total of 10 balls:
(a) If 2 balls are white and 1 is red?
(b) If all 3 balls are white?
(c) If all 3 balls are red?

60. Selecting Objects An urn contains 15 red balls and 10 white balls. Five balls are selected. In how many ways can the 5 balls be drawn from the total of 25 balls:
(a) If all 5 balls are red?
(b) If 3 balls are red and 2 are white?
(c) If at least 4 are red balls?

61. Senate Committees The U.S. Senate has 100 members. Suppose that it is desired to place each senator on exactly 1 of 7 possible committees. The first committee has 22 members, the second has 13, the third has 10, the fourth has 5, the fifth has 16, and the sixth and seventh have 17 apiece. In how many ways can these committees be formed?

62. Football Teams A defensive football squad consists of 25 players. Of these, 10 are linemen, 10 are linebackers, and 5 are safeties. How many different teams of 5 linemen, 3 linebackers, and 3 safeties can be formed?

63. Baseball In the American Baseball League, a designated hitter may be used. How many batting orders is it possible for a manager to use? (There are 9 regular players on a team.)

64. Baseball In the National Baseball League, the pitcher usually bats ninth. If this is the case, how many batting orders is it possible for a manager to use?

65. Baseball Teams A baseball team has 15 members. Four of the players are pitchers, and the remaining 11 members can play any position. How many different teams of 9 players can be formed?

66. World Series In the World Series the American League team (*A*) and the National League team (*N*) play until one team wins four games. If the sequence of winners is designated by letters (for example, *NAAAA* means that the National League team won the first game and the American League won the next four), how many different sequences are possible?

67. Basketball Teams A basketball team has 6 players who play guard (2 of 5 starting positions). How many different teams are possible, assuming that the remaining 3 positions are filled and it is not possible to distinguish a left guard from a right guard?

68. Basketball Teams On a basketball team of 12 players, 2 only play center, 3 only play guard, and the rest play forward (5 players on a team: 2 forwards, 2 guards, and 1 center). How many different teams are possible, assuming that it is not possible to distinguish left and right guards and left and right forwards?

Discussion and Writing

69. Create a problem different from any found in the text that requires the Multiplication Principle to solve. Give it to a friend to solve and critique.

70. Create a problem different from any found in the text that requires a permutation to solve. Give it to a friend to solve and critique.

71. Create a problem different from any found in the text that requires a combination to solve. Give it to a friend to solve and critique.

72. Explain the difference between a permutation and a combination. Give an example to illustrate your explanation.

'Are You Prepared?' Answers

1. $1; 1$ **2.** False

8.3 Probability

> **OBJECTIVES** 1 Construct Probability Models
> 2 Compute Probabilities of Equally Likely Outcomes
> 3 Use the Addition Rule to Find Probabilities
> 4 Use the Complement Rule to Find Probabilities

Probability is an area of mathematics that deals with experiments that yield random results, yet admit a certain regularity. Such experiments do not always produce the same result or outcome, so the result of any one observation is not predictable. However, the results of the experiment over a long period do produce regular patterns that enable us to predict with remarkable accuracy.

EXAMPLE 1 **Tossing a Fair Coin**

In tossing a fair coin, we know that the outcome is either a head or a tail. On any particular throw, we cannot predict what will happen, but, if we toss the coin many times, we observe that the number of times that a head comes up is approximately equal to the number of times that we get a tail. It seems reasonable, therefore, to assign a probability of $\frac{1}{2}$ that a head comes up and a probability of $\frac{1}{2}$ that a tail comes up. ∎

1 Construct Probability Models

The discussion in Example 1 constitutes the construction of a **probability model** for the experiment of tossing a fair coin once. A probability model has two components: a sample space and an assignment of probabilities. A **sample space** *S* is a set whose elements represent all the possibilities that can occur as a result of the experiment. Each element of *S* is called an **outcome**. To each outcome, we assign a number, called the **probability** of that outcome, which has two properties:

1. The probability assigned to each outcome is nonnegative.

2. The sum of all the probabilities equals 1.

If a probability model has the sample space

$$S = \{e_1, e_2, \ldots, e_n\}$$

where e_1, e_2, \ldots, e_n are the possible outcomes, and if $P(e_1), P(e_2), \ldots, P(e_n)$ denote the respective probabilities of these outcomes, then

$$P(e_1) \geq 0,\ P(e_2) \geq 0, \ldots,\ P(e_n) \geq 0 \qquad \textbf{(1)}$$

$$\sum_{i=1}^{n} P(e_i) = P(e_1) + P(e_2) + \cdots + P(e_n) = 1 \qquad \textbf{(2)}$$

EXAMPLE 2 **Determining Probability Models**

In a bag of M&Ms,™ the candies are colored red, green, blue, brown, yellow, and orange. Suppose that a candy is drawn from the bag and the color is recorded. The sample space of this experiment is {red, green, blue, brown, yellow, orange}. Determine which of the following are probability models.

(a)

Outcome	Probability
red	0.3
green	0.15
blue	0
brown	0.15
yellow	0.2
orange	0.2

(b)

Outcome	Probability
red	0.1
green	0.1
blue	0.1
brown	0.4
yellow	0.2
orange	0.3

(c)

Outcome	Probability
red	0.3
green	−0.3
blue	0.2
brown	0.4
yellow	0.2
orange	0.2

(d)

Outcome	Probability
red	0
green	0
blue	0
brown	0
yellow	1
orange	0

Solution

(a) This model is a probability model since all the outcomes have probabilities that are nonnegative and the sum of the probabilities is 1.

(b) This model is not a probability model because the sum of the probabilities is not 1.

(c) This model is not a probability model because $P(\text{green})$ is less than 0. Recall, all probabilities must be nonnegative.

(d) This model is a probability model because all the outcomes have probabilities that are nonnegative, and the sum of the probabilities is 1. Notice that $P(\text{yellow}) = 1$, meaning that this outcome will occur with 100% certainty each time that the experiment is repeated. This means that the bag of M&Ms™ has only yellow candies. ■

NOW WORK PROBLEM 7.

EXAMPLE 3 **Constructing a Probability Model**

An experiment consists of rolling a fair die once. A die is a cube with each face having either 1, 2, 3, 4, 5, or 6 dots on it. See Figure 8. Construct a probability model for this experiment.

Figure 8

Solution A sample space S consists of all the possibilities that can occur. Because rolling the die will result in one of six faces showing, the sample space S consists of

$$S = \{1, 2, 3, 4, 5, 6\}$$

Because the die is fair, one face is no more likely to occur than another. As a result, our assignment of probabilities is

$$P(1) = \frac{1}{6}, \qquad P(2) = \frac{1}{6},$$

$$P(3) = \frac{1}{6}, \qquad P(4) = \frac{1}{6},$$

$$P(5) = \frac{1}{6}, \qquad P(6) = \frac{1}{6} \qquad \blacksquare$$

Now suppose that a die is loaded (weighted) so that the probability assignments are

$$P(1) = 0, \quad P(2) = 0, \quad P(3) = \frac{1}{3}, \quad P(4) = \frac{2}{3}, \quad P(5) = 0, \quad P(6) = 0$$

This assignment would be made if the die were loaded so that only a 3 or 4 could occur and the 4 is twice as likely as the 3 to occur. This assignment is consistent with the definition, since each assignment is nonnegative and the sum of all the probability assignments equals 1.

NOW WORK PROBLEM 23.

EXAMPLE 4 **Constructing a Probability Model**

An experiment consists of tossing a coin. The coin is weighted so that heads (H) is three times as likely to occur as tails (T). Construct a probability model for this experiment.

Solution The sample space S is $S = \{H, T\}$. If x denotes the probability that a tail occurs,

$$P(T) = x \quad \text{and} \quad P(H) = 3x$$

Since the sum of the probabilities of the possible outcomes must equal 1, we have

$$P(T) + P(H) = x + 3x = 1$$
$$4x = 1$$
$$x = \frac{1}{4}$$

We assign the probabilities

$$P(T) = \frac{1}{4}, \qquad P(H) = \frac{3}{4} \qquad \blacksquare$$

NOW WORK PROBLEM 27.

In working with probability models, the term **event** is used to describe a set of possible outcomes of the experiment. An event E is some subset of the sample space S. The **probability of an event** E, $E \neq \emptyset$, denoted by $P(E)$, is defined as the sum of the probabilities of the outcomes in E. We can also think of the probability of an event E as the likelihood that the event E occurs. If $E = \emptyset$, then $P(E) = 0$; if $E = S$, then $P(E) = P(S) = 1$.

2 Compute Probabilities of Equally Likely Outcomes

When the same probability is assigned to each outcome of the sample space, the experiment is said to have **equally likely outcomes**.

Theorem

Probability for Equally Likely Outcomes

If an experiment has n equally likely outcomes and if the number of ways that an event E can occur is m, then the probability of E is

$$P(E) = \frac{\text{Number of ways that } E \text{ can occur}}{\text{Number of possible outcomes}} = \frac{m}{n} \qquad (3)$$

If S is the sample space of this experiment,

$$P(E) = \frac{n(E)}{n(S)} \qquad (4)$$

EXAMPLE 5

Calculating Probabilities of Events Involving Equally Likely Outcomes

Calculate the probability that in a 3-child family there are 2 boys and 1 girl. Assume equally likely outcomes.

Figure 9

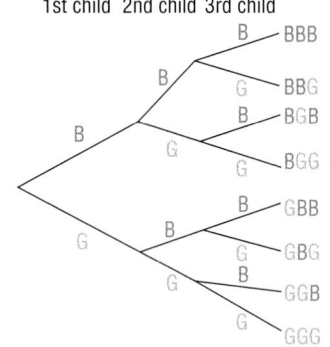

1st child 2nd child 3rd child

Solution

We begin by constructing a tree diagram to help in listing the possible outcomes of the experiment. See Figure 9, where B stands for boy and G for girl. The sample space S of this experiment is

$$S = \{\text{BBB, BBG, BGB, BGG, GBB, GBG, GGB, GGG}\}$$

so $n(S) = 8$.

We wish to know the probability of the event E: "having two boys and one girl." From Figure 9, we conclude that $E = \{\text{BBG, BGB, GBB}\}$, so $n(E) = 3$. Since the outcomes are equally likely, the probability of E is

$$P(E) = \frac{n(E)}{n(S)} = \frac{3}{8} \qquad ∎$$

NOW WORK PROBLEM 37.

So far, we have calculated probabilities of single events. We will now compute probabilities of multiple events, called **compound probabilities**.

EXAMPLE 6

Computing Compound Probabilities

Consider the experiment of rolling a single fair die. Let E represent the event "roll an odd number," and let F represent the event "roll a 1 or 2."

(a) Write the event E and F.

(b) Write the event E or F.

(c) Compute $P(E)$. Compute $P(F)$.

(d) Compute $P(E \cap F)$.

(e) Compute $P(E \cup F)$.

Solution The sample space S of the experiment is $\{1, 2, 3, 4, 5, 6\}$, so $n(S) = 6$. Since the die is fair, the outcomes are equally likely. The event E: "roll an odd number" is $\{1, 3, 5\}$, and the event F: "roll a 1 or 2" is $\{1, 2\}$, so $n(E) = 3$ and $n(F) = 2$.

(a) The word *and* in probability means the intersection of two events. The event E and F is

$$E \cap F = \{1, 3, 5\} \cap \{1, 2\} = \{1\} \qquad n(E \cap F) = 1$$

(b) The word *or* in probability means the union of the two events. The event E or F is

$$E \cup F = \{1, 3, 5\} \cup \{1, 2\} = \{1, 2, 3, 5\} \qquad n(E \cup F) = 4$$

(c) We use formula (4).

$$P(E) = \frac{n(E)}{n(S)} = \frac{3}{6} = \frac{1}{2} \qquad P(F) = \frac{n(F)}{n(S)} = \frac{2}{6} = \frac{1}{3}$$

(d) $P(E \cap F) = \dfrac{n(E \cap F)}{n(S)} = \dfrac{1}{6}$

(e) $P(E \cup F) = \dfrac{n(E \cup F)}{n(S)} = \dfrac{4}{6} = \dfrac{2}{3}$ ∎

3 Use the Addition Rule to Find Probabilities

The **Addition Rule** can be used to find the probability of the union of two events.

Theorem

> **Addition Rule**
>
> For any two events E and F,
>
> $$P(E \cup F) = P(E) + P(F) - P(E \cap F) \tag{5}$$

For example, we can use the Addition Rule to find $P(E \cup F)$ in Example 6(e). Then

$$P(E \cup F) = P(E) + P(F) - P(E \cap F) = \frac{1}{2} + \frac{1}{3} - \frac{1}{6} = \frac{3}{6} + \frac{2}{6} - \frac{1}{6} = \frac{4}{6} = \frac{2}{3}$$

as before.

EXAMPLE 7 **Computing Probabilities of Compound Events Using the Addition Rule**

If $P(E) = 0.2$, $P(F) = 0.3$, and $P(E \cap F) = 0.1$, find the probability of E or F. That is, find $P(E \cup F)$.

Solution We use the Addition Rule, formula (5).

$$\text{Probability of } E \text{ or } F = P(E \cup F) = P(E) + P(F) - P(E \cap F)$$
$$= 0.2 + 0.3 - 0.1 = 0.4 \qquad ∎$$

A Venn diagram can sometimes be used to obtain probabilities. To construct a Venn diagram representing the information in Example 7, we draw two sets E and F. We begin with the fact that $P(E \cap F) = 0.1$. See Figure 10(a) on page 556. Then, since $P(E) = 0.2$ and $P(F) = 0.3$, we fill in E with $0.2 - 0.1 = 0.1$ and F with $0.3 - 0.1 = 0.2$. See Figure 10(b). Since $P(S) = 1$, we complete the diagram by

inserting $1 - (0.1 + 0.1 + 0.2) = 0.6$ outside the circles. See Figure 10(c). Now it is easy to see, for example, that the probability of F, but not E, is 0.2. Also, the probability of neither E nor F is 0.6.

Figure 10

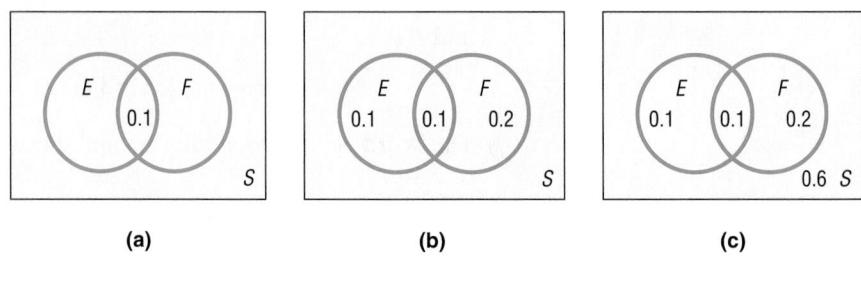

(a) (b) (c)

NOW WORK PROBLEM 45.

If events E and F are disjoint so that $E \cap F = \emptyset$, we say they are **mutually exclusive**. In this case, $P(E \cap F) = 0$, and the Addition Rule takes the following form:

Theorem

Mutually Exclusive Events

If E and F are mutually exclusive events,

$$P(E \cup F) = P(E) + P(F) \qquad (6)$$

EXAMPLE 8 **Computing Compound Probabilities of Mutually Exclusive Events**

If $P(E) = 0.4$ and $P(F) = 0.25$, and E and F are mutually exclusive, find $P(E \cup F)$.

Solution Since E and F are mutually exclusive, we use formula (6).

$$P(E \cup F) = P(E) + P(F) = 0.4 + 0.25 = 0.65 \qquad ■$$

NOW WORK PROBLEM 47.

4 Use the Complement Rule to Find Probabilities

Recall, if A is a set, the complement of A, denoted \overline{A}, is the set of all elements in the universal set U not in A. We similarly define the complement of an event.

Complement of an Event

Let S denote the sample space of an experiment, and let E denote an event. The **complement of E**, denoted \overline{E}, is the set of all outcomes in the sample space S that are not outcomes in the event E.

The complement of an event E, that is, \overline{E}, in a sample space S has the following two properties:

$$E \cap \overline{E} = \emptyset, \qquad E \cup \overline{E} = S$$

Since E and \overline{E} are mutually exclusive, it follows from (6) that

$$P(E \cup \overline{E}) = P(S) = 1, \qquad P(E) + P(\overline{E}) = 1, \qquad P(\overline{E}) = 1 - P(E)$$

We have the following result:

Theorem

> **Computing Probabilities of Complementary Events**
>
> If E represents any event and \overline{E} represents the complement of E,
>
> $$P(\overline{E}) = 1 - P(E) \qquad\qquad (7)$$

EXAMPLE 9	**Computing Probabilities Using Complements**

On the local news the weather reporter stated that the probability of rain tomorrow is 40%. What is the probability that it will not rain?

Solution The complement of the event "rain" is "no rain."

$$P(\text{no rain}) = 1 - P(\text{rain}) = 1 - 0.4 = 0.6$$

There is a 60% chance of no rain tomorrow. ∎

✏️ NOW WORK PROBLEM **51**.

EXAMPLE 10	**Birthday Problem**

What is the probability that in a group of 10 people at least 2 people have the same birthday? Assume that there are 365 days in a year.

Solution We assume that a person is as likely to be born on one day as another, so we have equally likely outcomes.

We first determine the number of outcomes in the sample space S. There are 365 possibilities for each person's birthday. Since there are 10 people in the group, there are 365^{10} possibilities for the birthdays. [For one person in the group, there are 365 days on which his or her birthday can fall; for two people, there are $(365)(365) = 365^2$ pairs of days; and, in general, using the Multiplication Principle, for n people there are 365^n possibilities.] So

$$n(S) = 365^{10}$$

We wish to find the probability of the event E: "at least two people have the same birthday." It is difficult to count the elements in this set; it is much easier to count the elements of the complementary event \overline{E}: "no two people have the same birthday."

We find $n(\overline{E})$ as follows: Choose one person at random. There are 365 possibilities for his or her birthday. Choose a second person. There are 364 possibilities for this birthday, if no two people are to have the same birthday. Choose a third person. There are 363 possibilities left for this birthday. Finally, we arrive at the tenth person. There are 356 possibilities left for this birthday. By the Multiplication Principle, the total number of possibilities is

$$n(\overline{E}) = 365 \cdot 364 \cdot 363 \cdot \,\cdots\, \cdot 356$$

The probability of the event \overline{E} is

$$P(\overline{E}) = \frac{n(\overline{E})}{n(S)} = \frac{365 \cdot 364 \cdot 363 \cdot \,\cdots\, \cdot 356}{365^{10}} \approx 0.883$$

The probability of two or more people in a group of 10 people having the same birthday is then

$$P(E) = 1 - P(\overline{E}) \approx 1 - 0.883 = 0.117 \qquad ∎$$

The birthday problem can be solved for any group size. The following table gives the probabilities for two or more people having the same birthday for various group sizes. Notice that the probability is greater than $\frac{1}{2}$ for any group of 23 or more people.

						Number of People										
	5	10	15	20	21	22	23	24	25	30	40	50	60	70	80	90
Probability That Two or More Have the Same Birthday	0.027	0.117	0.253	0.411	0.444	0.476	0.507	0.538	0.569	0.706	0.891	0.970	0.994	0.99916	0.99991	0.99999

➡ **NOW WORK PROBLEM 69.**

HISTORICAL FEATURE

**Blaise Pascal
(1623–1662)**

Set theory, counting, and probability first took form as a systematic theory in an exchange of letters (1654) between Pierre de Fermat (1601–1665) and Blaise Pascal (1623–1662). They discussed the problem of how to divide the stakes in a game that is interrupted before completion, knowing how many points each player needs to win. Fermat solved the problem by listing all possibilities and counting the favorable ones, whereas Pascal made use of the triangle that now bears his name. As mentioned in the text, the entries in Pascal's triangle are equivalent to $C(n, r)$. This recognition of the role of $C(n, r)$ in counting is the foundation of all further developments.

The first book on probability, the work of Christiaan Huygens (1629–1695), appeared in 1657. In it, the notion of mathematical expectation is explored. This allows the calculation of the profit or loss that a gambler might expect, knowing the probabilities involved in the game (see the Historical Problems that follow).

Although Girolamo Cardano (1501–1576) wrote a treatise on probability, it was not published until 1663 in Cardano's collected works, and this was too late to have any effect on the development of the theory.

In 1713, the posthumously published *Ars Conjectandi* of Jakob Bernoulli (1654–1705) gave the theory the form it would have until 1900. Recently, both combinatorics (counting) and probability have undergone rapid development due to the use of computers.

A final comment about notation. The notations $C(n, r)$ and $P(n, r)$ are variants of a form of notation developed in England after 1830. The notation $\binom{n}{r}$ for $C(n, r)$ goes back to Leonhard Euler (1707–1783), but is now losing ground because it has no clearly related symbolism of the same type for permutations. The set symbols \cup and \cap were introduced by Giuseppe Peano (1858–1932) in 1888 in a slightly different context. The inclusion symbol \subset was introduced by E. Schroeder (1841–1902) about 1890. The treatment of set theory in the text is due to George Boole (1815–1864), who wrote $A + B$ for $A \cup B$ and AB for $A \cap B$ (statisticians still use AB for $A \cap B$).

Historical Problems

1. *The Problem Discussed by Fermat and Pascal* A game between two equally skilled players, A and B, is interrupted when A needs 2 points to win and B needs 3 points. In what proportion would the stakes be divided?

 (a) *Fermat's solution* List all possible outcomes that can occur as a result of four more plays. The probabilities for A to win and B to win then determine how the stakes should be divided.

 (b) *Pascal's solution* Use combinations to determine the number of ways that the 2 points needed for A to win could occur in four plays. Then use combinations to determine the number of ways that the 3 points needed for B to win could occur. This is trickier than it looks, since A can win with 2 points in either two plays, three plays, or four plays. Compute the probabilities and compare with the results in part (a).

2. *Huygen's Mathematical Expectation* In a game with n possible outcomes with probabilities p_1, p_2, \ldots, p_n, suppose that the net winnings are w_1, w_2, \ldots, w_n, respectively. Then the mathematical expectation is

 $$E = p_1 w_1 + p_2 w_2 + \cdots + p_n w_n$$

 The number E represents the profit or loss per game in the long run. The following problems are a modification of those of Huygens.

 (a) A fair die is tossed. A gambler wins \$3 if he throws a 6 and \$6 if he throws a 5. What is his expectation?
 [Hint: $w_1 = w_2 = w_3 = w_4 = 0$**]**

 (b) A gambler plays the same game as in part (a), but now the gambler must pay \$1 to play. This means that $w_5 = \$5, w_6 = \2, and $w_1 = w_2 = w_3 = w_4 = -\1. What is the expectation?

8.3 Assess Your Understanding

Concepts and Vocabulary

1. When the same probability is assigned to each outcome of a sample space, the experiment is said to have _____ _____ outcomes.

2. The _____ of an event E is the set of all outcomes in the sample space S that are not outcomes in the event E.

3. *True or False:* The probability of an event can never equal 0.

4. *True or False:* In a probability model, the sum of all probabilities is 1.

Skill Building

5. In a probability model, which of the following numbers could be the probability of an outcome?

$$0, \quad 0.01, \quad 0.35, \quad -0.4, \quad 1, \quad 1.4$$

6. In a probability model, which of the following numbers could be the probability of an outcome?

$$1.5, \quad \frac{1}{2}, \quad \frac{3}{4}, \quad \frac{2}{3}, \quad 0, \quad -\frac{1}{4}$$

7. Determine whether the following is a probability model.

Outcome	Probability
1	0.2
2	0.3
3	0.1
4	0.4

8. Determine whether the following is a probability model.

Outcome	Probability
Jim	0.4
Bob	0.3
Faye	0.1
Patricia	0.2

9. Determine whether the following is a probability model.

Outcome	Probability
Linda	0.3
Jean	0.2
Grant	0.1
Ron	0.3

10. Determine whether the following is a probability model.

Outcome	Probability
Lanny	0.3
Joanne	0.2
Nelson	0.1
Rich	0.5
Judy	−0.1

In Problems 11–16, construct a probability model for each experiment.

11. Tossing a fair coin twice
12. Tossing two fair coins once
13. Tossing two fair coins, then a fair die
14. Tossing a fair coin, a fair die, and then a fair coin
15. Tossing three fair coins once
16. Tossing one fair coin three times

In Problems 17–22, use the following spinners to construct a probability model for each experiment.

 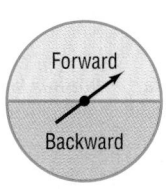

Spinner I Spinner II Spinner III

17. Spin spinner I, then spinner II. What is the probability of getting a 2 or a 4, followed by Red?
18. Spin spinner III, then spinner II. What is the probability of getting Forward, followed by Yellow or Green?
19. Spin spinner I, then II, then III. What is the probability of getting a 1, followed by Red or Green, followed by Backward?
20. Spin spinner II, then I, then III. What is the probability of getting Yellow, followed by a 2 or a 4, followed by Forward?
21. Spin spinner I twice, then spinner II. What is the probability of getting a 2, followed by a 2 or a 4, followed by Red or Green?
22. Spin spinner III, then spinner I twice. What is the probability of getting Forward, followed by a 1 or a 3, followed by a 2 or a 4?

In Problems 23–26, consider the experiment of tossing a coin twice. The table lists six possible assignments of probabilities for this experiment. Using this table, answer the following questions.

	Sample Space			
Assignments	HH	HT	TH	TT
A	$\frac{1}{4}$	$\frac{1}{4}$	$\frac{1}{4}$	$\frac{1}{4}$
B	0	0	0	1
C	$\frac{3}{16}$	$\frac{5}{16}$	$\frac{5}{16}$	$\frac{3}{16}$
D	$\frac{1}{2}$	$\frac{1}{2}$	$-\frac{1}{2}$	$\frac{1}{2}$
E	$\frac{1}{4}$	$\frac{1}{4}$	$\frac{1}{4}$	$\frac{1}{8}$
F	$\frac{1}{9}$	$\frac{2}{9}$	$\frac{2}{9}$	$\frac{4}{9}$

23. Which of the assignments of probabilities are consistent with the definition of a probability model?
24. Which of the assignments of probabilities should be used if the coin is known to be fair?
25. Which of the assignments of probabilities should be used if the coin is known to always come up tails?
26. Which of the assignments of probabilities should be used if tails is twice as likely as heads to occur?
27. Assigning Probabilities A coin is weighted so that heads is four times as likely as tails to occur. What probability should we assign to heads? to tails?
28. Assigning Probabilities A coin is weighted so that tails is twice as likely as heads to occur. What probability should we assign to heads? to tails?

29. **Assigning Probabilities** A die is weighted so that an odd-numbered face is twice as likely to occur as an even-numbered face. What probability should we assign to each face?

30. **Assigning Probabilities** A die is weighted so that a six cannot appear. The other faces occur with the same probability. What probability should we assign to each face?

For Problems 31–34, let the sample space be $S = \{1, 2, 3, 4, 5, 6, 7, 8, 9, 10\}$. *Suppose that the outcomes are equally likely.*

31. Compute the probability of the event $E = \{1, 2, 3\}$.

32. Compute the probability of the event $F = \{3, 5, 9, 10\}$.

33. Compute the probability of the event E: "an even number."

34. Compute the probability of the event F: "an odd number."

For Problems 35 and 36, an urn contains 5 white marbles, 10 green marbles, 8 yellow marbles, and 7 black marbles.

35. If one marble is selected, determine the probability that it is white.

36. If one marble is selected, determine the probability that it is black.

In Problems 37–40, assume equally likely outcomes.

37. Determine the probability of having 3 boys in a 3-child family.

38. Determine the probability of having 3 girls in a 3-child family.

39. Determine the probability of having 1 girl and 3 boys in a 4-child family.

40. Determine the probability of having 2 girls and 2 boys in a 4-child family.

For Problems 41–44, two fair dice are rolled.

41. Determine the probability that the sum of the two dice is 7.

42. Determine the probability that the sum of the two dice is 11.

43. Determine the probability that the sum of the two dice is 3.

44. Determine the probability that the sum of the two dice is 12.

In Problems 45–48, find the probability of the indicated event if $P(A) = 0.25$ *and* $P(B) = 0.45$.

45. $P(A \cup B)$ if $P(A \cap B) = 0.15$

46. $P(A \cap B)$ if $P(A \cup B) = 0.6$

47. $P(A \cup B)$ if A, B are mutually exclusive

48. $P(A \cap B)$ if A, B are mutually exclusive

49. If $P(A) = 0.60$, $P(A \cup B) = 0.85$, and $P(A \cap B) = 0.05$, find $P(B)$.

50. If $P(B) = 0.30$, $P(A \cup B) = 0.65$, and $P(A \cap B) = 0.15$, find $P(A)$.

51. According to the Federal Bureau of Investigation, in 2002 there was a 26.5% probability of theft from a motor vehicle. If a victim of theft is randomly selected, what is the probability that he or she was not a victim of theft from a motor vehicle?

52. According to the Federal Bureau of Investigation, in 2002 there was a 3.9% probability of theft involving a bicycle. If a victim of theft is randomly selected, what is the probability that he or she was not a victim of bicycle theft?

53. In Chicago, there is a 30% probability that Memorial Day will have a high temperature in the 70s. What is the probability that next Memorial Day will not have a high temperature in the 70s in Chicago?

54. In Chicago, there is a 4% probability that Memorial Day will have a low temperature in the 30s. What is the probability that next Memorial Day will not have a low temperature in the 30s in Chicago?

For Problems 55–58, a golf ball is selected at random from a container. If the container has 9 white balls, 8 green balls, and 3 orange balls, find the probability of each event.

55. The golf ball is white or green.

56. The golf ball is white or orange.

57. The golf ball is not white.

58. The golf ball is not green.

59. On the "Price is Right" there is a game in which a bag is filled with 3 strike chips and 5 numbers. Let's say that the numbers in the bag are 0, 1, 3, 6, and 9. What is the probability of selecting a strike chip or the number 1?

60. Another game on the "Price is Right" requires the contestant to spin a wheel with numbers 5, 10, 15, 20, . . . , 100. What is the probability that the contestant spins 100 or 30?

Problems 61–64, are based on a consumer survey of annual incomes in 100 households. The following table gives the data.

Income	$0–9999	$10,000–19,999	$20,000–29,999	$30,000–39,999	$40,000 or more
Number of households	5	35	30	20	10

61. What is the probability that a household has an annual income of $30,000 or more?

62. What is the probability that a household has an annual income between $10,000 and $29,999, inclusive?

63. What is the probability that a household has an annual income of less than $20,000?

64. What is the probability that a household has an annual income of $20,000 or more?

65. Surveys In a survey about the number of TV sets in a house, the following probability table was constructed:

Number of TV sets	0	1	2	3	4 or more
Probability	0.05	0.24	0.33	0.21	0.17

Find the probability of a house having:
(a) 1 or 2 TV sets
(b) 1 or more TV sets
(c) 3 or fewer TV sets
(d) 3 or more TV sets
(e) Fewer than 2 TV sets
(f) Fewer than 1 TV set
(g) 1, 2, or 3 TV sets
(h) 2 or more TV sets

66. Checkout Lines Through observation, it has been determined that the probability for a given number of people waiting in line at the "5 items or less" checkout register of a supermarket is as follows:

Number waiting in line	0	1	2	3	4 or more
Probability	0.10	0.15	0.20	0.24	0.31

Find the probability of:
(a) At most 2 people in line
(b) At least 2 people in line
(c) At least 1 person in line

67. In a certain Algebra and Trigonometry class, there are 18 freshmen and 15 sophomores. Of the 18 freshmen, 10 are male, and of the 15 sophomores, 8 are male. Find the probability that a randomly selected student is:
(a) A freshman or female
(b) A sophomore or male

68. The faculty of the mathematics department at Joliet Junior College is composed of 4 females and 9 males. Of the 4 females, 2 are under the age of 40, and of the males 3 are under age 40. Find the probability that a randomly selected faculty member is:
(a) Female or under age 40
(b) Male or over age 40

69. Birthday Problem What is the probability that at least 2 people have the same birthday in a group of 12 people? Assume that there are 365 days in a year.

70. Birthday Problem What is the probability that at least 2 people have the same birthday in a group of 35 people? Assume that there are 365 days in a year.

71. Winning a Lottery In a certain lottery, there are ten balls, numbered 1, 2, 3, 4, 5, 6, 7, 8, 9, 10. Of these, five are drawn in order. If you pick five numbers that match those drawn in the correct order, you win $1,000,000. What is the probability of winning such a lottery?

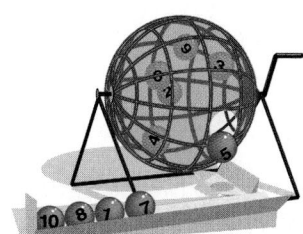

Chapter Review

Things to Know

Set (p. 536)

Well-defined collection of distinct objects, called elements

Empty set or null set (p. 536)	\varnothing	Set that has no elements
Equality (p. 536)	$A = B$	A and B have the same elements.
Subset (p. 536)	$A \subseteq B$	Each element of A is also an element of B.
Intersection (p. 537)	$A \cap B$	Set consisting of elements that belong to both A and B
Union (p. 537)	$A \cup B$	Set consisting of elements that belong to either A or B, or both
Universal set (p. 537)	U	Set consisting of all the elements that we wish to consider
Complement (p. 537)	\overline{A}	Set consisting of elements of the universal set that are not in A
Finite set (p. 538)		The number of elements in the set is a nonnegative integer.
Infinite set (p. 538)		A set that is not finite.

Counting formula (p. 539)

$n(A \cup B) = n(A) + n(B) - n(A \cap B)$

Addition Principle (p. 539)

If $A \cap B = \varnothing$, then $n(A \cup B) = n(A) + n(B)$.

Multiplication Principle (p. 542)

If a task consists of a sequence of choices in which there are p selections for the first choice, q selections for the second choice, and so on, the task of making these selections can be done in $p \cdot q \cdot \cdots$ different ways.

Permutation (p. 543)

An ordered arrangement of r objects chosen from n objects

Permutation: Distinct, with repetition (p. 543)	n^r The n objects are distinct (different), and repetition is allowed in the selection of r of them.
Permutation: Distinct, without repetition (p. 544–545)	$P(n, r) = n(n - 1) \cdot \cdots \cdot [n - (r - 1)] = \dfrac{n!}{(n - r)!}$ The n objects are distinct (different), and repetition is not allowed in the selection of r of them, where $r \le n$.
Combination (p. 546)	$C(n, r) = \dfrac{P(n, r)}{r!} = \dfrac{n!}{(n - r)!r!}$ An arrangement, without regard to order, of r objects selected from n distinct objects, where $r \le n$
Permutation: Not distinct (p. 548)	$\dfrac{n!}{n_1!n_2!\cdots n_k!}$ The number of permutations of n objects of which n_1 are of one kind, n_2 are of a second kind, \ldots, and n_k are of a kth kind, where $n = n_1 + n_2 + \cdots + n_k$
Sample space (p. 551)	Set whose elements represent the possible outcomes that can occur as a result of an experiment
Probability (p. 551)	A nonnegative number assigned to each outcome of a sample space; the sum of all the probabilities of the outcomes equals 1.
Equally likely outcomes (p. 554)	$P(E) = \dfrac{n(E)}{n(S)}$ The same probability is assigned to each outcome.
Addition Rule (p. 555)	$P(E \cup F) = P(E) + P(F) - P(E \cap F)$
Complement of an event (p. 556)	$P(\overline{E}) = 1 - P(E)$

Objectives

Section		You should be able to . . .	Review Exercises
12.1	1	Find all the subsets of a set (p. 536)	1, 2
	2	Find the intersection and union of sets (p. 537)	3–6
	3	Find the complement of a set (p. 537)	7–10
	4	Count the number of elements in a set (p. 538)	11–18
12.2	1	Solve counting problems using the Multiplication Principle (p. 541)	23–26, 32–36
	2	Solve counting problems using permutations (p. 543)	19, 20, 27, 28, 41(a)
	3	Solve counting problems using combinations (p. 546)	21, 22, 29–31, 39, 40
	4	Solve counting problems using permutations involving n nondistinct objects (p. 547)	37, 38
12.3	1	Construct probability models (p. 551)	41(b)
	2	Compute probabilities of equally likely outcomes (p. 554)	41(b), 42(a), 43(a), 44–47
	3	Use the Addition Rule to find probabilities (p. 555)	48
	4	Use the Complement Rule to find probabilities (p. 556)	41(c), 42(b), 43(b), 44

Review Exercises

1. Write down all the subsets of the set {Dave, Joanne, Erica}. **2.** Write down all the subsets of the set {Green, Blue, Red}.

In Problems 3–10, use U = universal set = {1, 2, 3, 4, 5, 6, 7, 8, 9}, A = {1, 3, 5, 7}, B = {3, 5, 6, 7, 8}, and C = {2, 3, 7, 8, 9} to find each set.

3. $A \cup B$ **4.** $B \cup C$ **5.** $A \cap C$ **6.** $A \cap B$

7. $\overline{A} \cup \overline{B}$ **8.** $\overline{B} \cap \overline{C}$ **9.** $\overline{B \cap C}$ **10.** $\overline{A \cup B}$

11. If $n(A) = 8, n(B) = 12$, and $n(A \cap B) = 3$, find $n(A \cup B)$. **12.** If $n(A) = 12, n(A \cup B) = 30$, and $n(A \cap B) = 6$, find $n(B)$.

In Problems 13–18, use the information supplied in the figure.

13. How many are in A?

14. How many are in A or B?

15. How many are in A and C?

16. How many are not in B?

17. How many are in neither A nor C?

18. How many are in B but not in C?

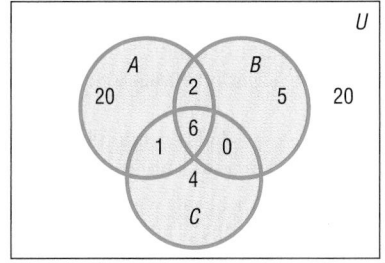

In Problems 19–22, compute the given expression.

19. $P(8, 3)$ **20.** $P(7, 3)$ **21.** $C(8, 3)$ **22.** $C(7, 3)$

23. Stocking a Store A clothing store sells pure wool and polyester–wool suits. Each suit comes in 3 colors and 10 sizes. How many suits are required for a complete assortment?

24. Wiring In connecting a certain electrical device, 5 wires are to be connected to 5 different terminals. How many different wirings are possible if 1 wire is connected to each terminal?

25. Baseball On a given day, the American Baseball League schedules 7 games. How many different outcomes are possible, assuming that each game is played to completion?

26. Baseball On a given day, the National Baseball League schedules 6 games. How many different outcomes are possible, assuming that each game is played to completion?

27. Choosing Seats If 4 people enter a bus having 9 vacant seats, in how many ways can they be seated?

28. Arranging Letters How many different arrangements are there of the letters in the word ROSE?

29. Choosing a Team In how many ways can a squad of 4 relay runners be chosen from a track team of 8 runners?

30. Writing a Test A professor has 10 similar problems to put on a test with 3 problems. How many different tests can she design?

31. Baseball In how many ways can 2 teams from 14 teams in the American League be chosen without regard to which team is at home?

32. Arranging Books on a Shelf There are 5 different French books and 5 different Spanish books. How many ways are there to arrange them on a shelf if:
 (a) Books of the same language must be grouped together, French on the left, Spanish on the right?
 (b) French and Spanish books must alternate in the grouping, beginning with a French book?

33. Telephone Numbers Using the digits $0, 1, 2, \ldots, 9$, how many 7-digit numbers can be formed if the first digit cannot be 0 or 9 and if the last digit is greater than or equal to 2 and less than or equal to 3? Repeated digits are allowed.

34. Home Choices A contractor constructs homes with 5 different choices of exterior finish, 3 different roof arrangements, and 4 different window designs. How many different types of homes can be built?

35. License Plate Possibilities A license plate consists of 1 letter, excluding O and I, followed by a 4-digit number that cannot have a 0 in the lead position. How many different plates are possible?

36. Binary Codes Using the digits 0 and 1, how many different numbers consisting of 8 digits can be formed?

37. Forming Different Words How many different words, real or imaginary, can be formed using all the letters in the word MISSING?

38. Arranging Flags How many different vertical arrangements are there of 10 flags if 4 are white, 3 are blue, 2 are green, and 1 is red?

39. Forming Committees A group of 9 people is going to be formed into committees of 4, 3, and 2 people. How many committees can be formed if:
 (a) A person can serve on any number of committees?
 (b) No person can serve on more than one committee?

40. Forming Committees A group consists of 5 men and 8 women. A committee of 4 is to be formed from this group, and policy dictates that at least 1 woman be on this committee.
 (a) How many committees can be formed that contain exactly 1 man?
 (b) How many committees can be formed that contain exactly 2 women?
 (c) How many committees can be formed that contain at least 1 man?

41. Birthday Problem For this problem, assume that a year has 365 days.
 (a) How many ways can 18 people have different birthdays?
 (b) What is the probability that nobody has the same birthday in a group of 18 people?
 (c) What is the probability in a group of 18 people that at least 2 people have the same birthday?

42. Death Rates According to the U.S. National Center for Health Statistics, 29% of all deaths in 2001 were due to heart disease.
 (a) What is the probability that a randomly selected death in 2001 was due to heart disease?
 (b) What is the probability that a randomly selected death in 2001 was not due to heart disease?

43. Unemployment According to the U.S. Bureau of Labor Statistics, 5.8% of the U.S. labor force was unemployed in 2002.
 (a) What is the probability that a randomly selected member of the U.S. labor force was unemployed in 2002?
 (b) What is the probability that a randomly selected member of the U.S. labor force was not unemployed in 2002?

44. From a box containing three 40-watt bulbs, six 60-watt bulbs, and eleven 75-watt bulbs, a bulb is drawn at random. What is the probability that the bulb is 40 watts? What is the probability that it is not a 75-watt bulb?

45. You have four $1 bills, three $5 bills, and two $10 bills in your wallet. If you pick a bill at random, what is the probability that it will be a $1 bill?

46. Each of the letters in the word ROSE is written on an index card and the cards are then shuffled. When the cards are dealt, what is the probability that they spell the word ROSE?

47. Each of the numbers, $1, 2, \ldots, 100$ is written on an index card and the cards are then shuffled. If a card is selected at random, what is the probability that the number on the card is divisible by 5? What is the probability that the card selected is either a 1 or names a prime number?

48. At the Milex tune-up and brake repair shop, the manager has found that a car will require a tune-up with a probability of 0.6, a brake job with a probability of 0.1, and both with a probability of 0.02.
 (a) What is the probability that a car requires either a tune-up or a brake job?
 (b) What is the probability that a car requires a tune-up but not a brake job?
 (c) What is the probability that a car requires neither a tune-up nor a brake job?

Chapter Test

In Problems 1–6, use U = Universal set = $\{0, 1, 2, 3, 4, 5, 6, 7, 8, 9\}$, $A = \{0, 1, 4, 9\}$, $B = \{2, 4, 6, 8\}$, and $C = \{1, 3, 5, 7, 9\}$ to find each set.

1. $A \cap B$

2. $A \cup C$

3. $(A \cup B) \cap C$

4. $\overline{A \cup B}$

5. \overline{C}

6. $\overline{A \cap \overline{C}}$

In Problems 7–10, a survey of 70 college freshmen asked whether students planned to take biology, chemistry, or physics during their first year. Use the diagram to answer each question.

7. How many of the surveyed students plan to take physics during their first year?

8. How many of the surveyed students do not plan to take biology, chemistry, or physics during their first year?

9. How many of the surveyed students plan to take only biology and chemistry during their first year?

10. How many of the surveyed students plan to take physics or chemistry during their first year?

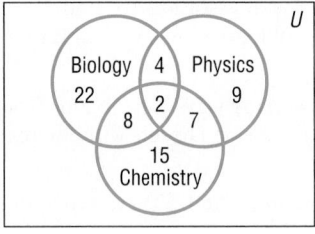

In Problems 11–13, compute the given expression.

11. 7!

12. $P(10, 6)$

13. $C(11, 5)$

14. M&M's® offers customers the opportunity to create their own color mix of candy. There are 21 colors to choose from, and customers are allowed to select up to 6 different colors. How many different color mixes are possible, assuming that no color is selected more than once?

15. How many distinct 8-letter words (real or imaginary) can be formed from the letters in the word REDEEMED?

16. In horse racing, an exacta bet requires the bettor to pick the first two horses in the exact order. If there are 8 horses in a race, in how many ways could you make an exacta bet?

17. On February 20, 2004, the Ohio Bureau of Motor Vehicles unveiled the state's new license plate format. The plate con-

sists of three letters (A–Z) followed by 4 digits (0–9). Assume that all letters and digits may be used except that the third letter cannot be O, I, or Z. If repetitions are allowed, how many different plates are possible?

18. Kiersten applies for admission to the University of Southern California (USC) and Florida State University (FSU). She estimates that she has a 60% chance of being admitted to USC, a 70% chance of being admitted to FSU, and a 35% chance of being admitted to both universities.
 (a) What is the probability that she will be admitted to either USC or FSU?
 (b) What is the probability that she will not be admitted to FSU?

19. A cooler contains 8 bottles of Pepsi, 5 bottles of Coke, 4 bottles of Mountain Dew, and 3 bottles of IBC.
(a) What is the probability that a bottle chosen at random is Coke?
(b) What is the probability that a bottle chosen at random is either Pepsi or IBC?

20. A study on the age distribution of a community college gave the following table:

Age	17 and under	18–20	21–24	25–34	35–64	65 and over
Probability	0.03	???	0.23	0.29	0.25	0.01

What must be the probability that a randomly selected student at the college is between 18 and 20 years old?

21. Powerball is a multistate lottery where 5 white balls from a drum with 53 balls and 1 red ball from a drum with 42 red balls are selected. For a $1 ticket, players get one chance at winning the jackpot by matching all 6 numbers. What is the probability of selecting the winning numbers on a $1 play?

22. If you roll 1 die five times, what is the probability that you obtain exactly 2 fours?

Chapter Projects

1. Simulation In the Winter 1998 edition of *Eightysomething!*, Mike Koehler uses simulation to calculate the following probabilities: "A woman and man (unrelated) each have two children. At least one of the woman's children is a boy, and the man's older child is a boy. Do the chances that the woman has two boys equal the chances that the man has two boys?" Perform a simulation to answer the question.

The following projects are available at the Instructor's Resource Center (IRC):

2. Project at Motorola *Probability of Error Digital Wireless Comunications.*

3. Surveys

4. Law of Large Numbers

Cumulative Review

1. Solve: $3x^2 - 2x = -1$

2. Graph $f(x) = x^2 + 4x - 5$ by determining whether the graph opens up or down and by finding the vertex, axis of symmetry, and intercepts.

3. Graph $f(x) = 2(x + 1)^2 - 4$ using transformations.

4. Solve: $|x - 4| \le 0.01$

5. Find the complex zeros of
$$f(x) = 5x^4 - 9x^3 - 7x^2 - 31x - 6$$

6. Graph $g(x) = 3^{x-1} + 5$ using transformations. Determine the domain, the range, and horizontal asymptote of g.

7. What is the exact value of $\log_3 9$?

8. Solve: $\log_2(3x - 2) + \log_2 x = 4$

9. Solve the system:
$$\begin{cases} x - 2y + z = 15 \\ 3x + y - 3z = -8 \\ -2x + 4y - z = -27 \end{cases}$$

10. What is the 33rd term in the sequence $-3, 1, 5, 9, \ldots$? What is the sum of the first 20 terms?

Review | Appendix A

OUTLINE

A.1 Algebra Essentials

PREPARING FOR THIS BOOK *Before getting started, read "To the Student" at the beginning of this book on page xxiv.*

OBJECTIVES 1 Evaluate Algebraic Expressions

 2 Determine the Domain of a Variable

 3 Graph Inequalities

 4 Find Distance on the Real Number Line

 5 Use the Laws of Exponents

 6 Evaluate Square Roots

Sets

When we want to treat a collection of similar but distinct objects as a whole, we use the idea of a **set**. For example, the set of *digits* consists of the collection of numbers 0, 1, 2, 3, 4, 5, 6, 7, 8, and 9. If we use the symbol D to denote the set of digits, we can write

$$D = \{0, 1, 2, 3, 4, 5, 6, 7, 8, 9\}$$

In this notation, the braces $\{\ \}$ are used to enclose the objects, or **elements**, in the set. This method of denoting a set is called the **roster method**. A second way to denote a set is to use **set-builder notation**, where the set D of digits is written as

$$D = \{\quad x \quad | \quad x \text{ is a digit}\}$$

Read as "D is the set of all x such that x is a digit."

| EXAMPLE 1 | **Using Set-builder Notation and the Roster Method** |

(a) $E = \{x | x \text{ is an even digit}\} = \{0, 2, 4, 6, 8\}$

(b) $O = \{x | x \text{ is an odd digit}\} = \{1, 3, 5, 7, 9\}$ ∎

In listing the elements of a set, we do not list an element more than once because the elements of a set are distinct. Also, the order in which the elements are listed does not matter. For example, $\{2, 3\}$ and $\{3, 2\}$ both represent the same set.

If every element of a set A is also an element of a set B, we say that A is a **subset** of B. If two sets A and B have the same elements, we say that A **equals** B. For example, $\{1, 2, 3\}$ is a subset of $\{1, 2, 3, 4, 5\}$; and $\{1, 2, 3\}$ equals $\{2, 3, 1\}$.

Finally, if a set has no elements, it is called the **empty set**, or the **null set**, and it is denoted by the symbol \emptyset.

Real Numbers

Real numbers are represented by symbols such as

$$25, \quad 0, \quad -3, \quad \frac{1}{2}, \quad -\frac{5}{4}, \quad 0.125, \quad \sqrt{2}, \quad \pi, \quad \sqrt[3]{-2}, \quad 0.666\ldots$$

The set of **counting numbers**, or **natural numbers**, contains the numbers in the set $\{1, 2, 3, 4, \ldots\}$. (The three dots, called an **ellipsis**, indicate that the pattern continues indefinitely.) The set of **integers** contains the numbers in the set $\{\ldots, -3, -2, -1, 0, 1, 2, 3, \ldots\}$. A **rational number** is a number that can be expressed as a *quotient* $\dfrac{a}{b}$ of two integers, where the integer b cannot be 0. Examples of rational numbers are $\dfrac{3}{4}, \dfrac{5}{2}, \dfrac{0}{4}$, and $-\dfrac{2}{3}$. Since $\dfrac{a}{1} = a$ for any integer a, every integer is also a rational number. Real numbers that are not rational are called **irrational**. Examples of irrational numbers are $\sqrt{2}$ and π (the Greek letter pi), which equals the constant ratio of the circumference to the diameter of a circle. See Figure 1.

Real numbers can be represented as **decimals**. Rational real numbers have decimal representations that either **terminate** or are nonterminating with **repeating** blocks of digits. For example, $\dfrac{3}{4} = 0.75$, which terminates; and $\dfrac{2}{3} = 0.666\ldots$, in which the digit 6 repeats indefinitely. Irrational real numbers have decimal representations that neither repeat nor terminate. For example, $\sqrt{2} = 1.414213\ldots$ and $\pi = 3.14159\ldots$. In practice, the decimal representation of an irrational number is given as an approximation. We use the symbol \approx (read as "approximately equal to") to write $\sqrt{2} \approx 1.4142$ and $\pi \approx 3.1416$.

Two properties of real numbers that we shall use often are given next.

Suppose that a, b, and c are real numbers.

Figure 1 $\pi = \dfrac{C}{d}$

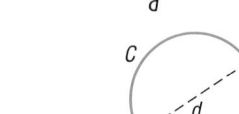

Distributive Property

$$a \cdot (b + c) = ab + ac$$

Zero-Product Property

If $ab = 0$, then either $a = 0$ or $b = 0$ or both equal 0.

The Distributive Property can be used to remove parentheses:

$$2(x + 3) = 2x + 2 \cdot 3 = 2x + 6$$

The Zero-Product Property will be used to solve equations (Section A.8). If $2x = 0$, then $2 = 0$ or $x = 0$. Since $2 \neq 0$, it follows that $x = 0$.

Constants and Variables

In algebra we use letters to represent numbers. If the letter used is to represent *any* number from a given set of numbers, it is called a **variable**. A **constant** is either a fixed number, such as 5 or $\sqrt{3}$, or a letter that represents a fixed (possibly unspecified) number.

Constants and variables are combined using the operations of addition, subtraction, multiplication, and division to form *algebraic expressions*. Examples of algebraic expressions include

$$x + 3 \qquad \frac{3}{1 - t} \qquad 7x - 2y$$

1 Evaluate Algebraic Expressions

To evaluate an algebraic expression, substitute for each variable its numerical value.

| EXAMPLE 2 | **Evaluating an Algebraic Expression** |

Evaluate each expression if $x = 3$ and $y = -1$.

(a) $x + 3y$ (b) $5xy$ (c) $\dfrac{3y}{2 - 2x}$

Solution (a) Substitute 3 for x and -1 for y in the expression $x + 3y$.

$$x + 3y = 3 + 3(-1) = 3 + (-3) = 0$$
$$\uparrow$$
$$x = 3, y = -1$$

(b) If $x = 3$ and $y = -1$, then

$$5xy = 5(3)(-1) = -15$$

(c) If $x = 3$ and $y = -1$, then

$$\frac{3y}{2 - 2x} = \frac{3(-1)}{2 - 2(3)} = \frac{-3}{2 - 6} = \frac{-3}{-4} = \frac{3}{4}$$ ∎

NOW WORK PROBLEM 9.

2 Determine the Domain of a Variable

In working with expressions or formulas involving variables, the variables may be allowed to take on values from only a certain set of numbers. For example, in the formula for the area A of a circle of radius r, $A = \pi r^2$, the variable r is necessarily restricted to the positive real numbers. In the expression $\dfrac{1}{x}$, the variable x cannot take on the value 0, since division by 0 is not defined.

The set of values that a variable in an expression may assume is called the **domain of the variable**.

| EXAMPLE 3 | **Finding the Domain of a Variable** |

The domain of the variable x in the expression

$$\frac{5}{x - 2}$$

is $\{x \mid x \neq 2\}$, since, if $x = 2$, the denominator becomes 0, so the expression is not defined. ∎

| EXAMPLE 4 | **Circumference of a Circle** |

In the formula for the circumference C of a circle of radius r,

$$C = 2\pi r$$

the domain of the variable r, representing the radius of the circle, is the set of positive real numbers. The domain of the variable C, representing the circumference of the circle, is also the set of positive real numbers. ∎

In describing the domain of a variable, we may use either set notation or words, whichever is more convenient.

➤ **NOW WORK PROBLEM 17.**

The Real Number Line

The real numbers can be represented by points on a line called the **real number line**. There is a one-to-one correspondence between real numbers and points on a line. That is, every real number corresponds to a point on the line, and each point on the line has a unique real number associated with it.

Pick a point on the line somewhere in the center, and label it O. This point, called the **origin**, corresponds to the real number 0. See Figure 2. The point 1 unit to the right of O corresponds to the number 1. The distance between 0 and 1 determines the **scale** of the number line. For example, the point associated with the number 2 is twice as far from 0 as 1 is. Notice that an arrowhead on the right end of the line indicates the direction in which the numbers increase. Figure 2 also shows the points associated with the irrational numbers $\sqrt{2}$ and π. Points to the left of the origin correspond to the real numbers -1, -2, and so on.

Figure 2
Real number line

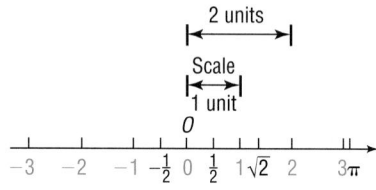

> The real number associated with a point P is called the **coordinate** of P, and the line whose points have been assigned coordinates is called the **real number line**.

➤ **NOW WORK PROBLEM 29.**

The real number line consists of three classes of real numbers, as shown in Figure 3.

Figure 3

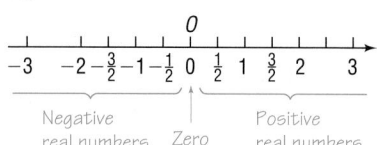

1. The **negative real numbers** are the coordinates of points to the left of the origin O.
2. The real number **zero** is the coordinate of the origin O.
3. The **positive real numbers** are the coordinates of points to the right of the origin O.

3 Graph Inequalities

An important property of the real number line follows from the fact that, given two numbers (points) a and b, either a is to the left of b, a is at the same location as b, or a is to the right of b. See Figure 4.

Figure 4

(a) $a < b$

(b) $a = b$

(c) $a > b$

If a is to the left of b, we say that "a is less than b" and write $a < b$. If a is to the right of b, we say that "a is greater than b" and write $a > b$. If a is at the same location as b, then $a = b$. If a is either less than or equal to b, we write $a \leq b$. Similarly, $a \geq b$ means that a is either greater than or equal to b. Collectively, the symbols $<, >, \leq$, and \geq are called **inequality symbols**.

Note that $a < b$ and $b > a$ mean the same thing. It does not matter whether we write $2 < 3$ or $3 > 2$.

Furthermore, if $a < b$ or if $b > a$, then the difference $b - a$ is positive. Do you see why?

An **inequality** is a statement in which two expressions are related by an inequality symbol. The expressions are referred to as the **sides** of the inequality. Statements of the form $a < b$ or $b > a$ are called **strict inequalities**, while statements of the form $a \leq b$ or $b \geq a$ are called **nonstrict inequalities**.

Based on the discussion thus far, we conclude that

$a > 0$	is equivalent to a is positive
$a < 0$	is equivalent to a is negative

We sometimes read $a > 0$ by saying that "a is positive." If $a \geq 0$, either $a > 0$ or $a = 0$, and we may read this as "a is nonnegative."

 NOW WORK PROBLEMS 33 AND 43.

We shall find it useful in later work to graph inequalities on the real number line.

| **EXAMPLE 5** | **Graphing Inequalities** |

(a) On the real number line, graph all numbers x for which $x > 4$.

(b) On the real number line, graph all numbers x for which $x \leq 5$.

Figure 5 **Solution**

(a) See Figure 5. Notice that we use a left parenthesis to indicate that the number 4 is not part of the graph.

Figure 6

(b) See Figure 6. Notice that we use a right bracket to indicate that the number 5 is part of the graph. ∎

NOW WORK PROBLEM 49.

Figure 7

Absolute Value

The *absolute value* of a number a is the distance from 0 to a on the number line. For example, -4 is 4 units from 0; and 3 is 3 units from 0. See Figure 7. Thus, the absolute value of -4 is 4, and the absolute value of 3 is 3.

A more formal definition of absolute value is given next.

The **absolute value** of a real number a, denoted by the symbol $|a|$, is defined by the rules

$$|a| = a \quad \text{if } a \geq 0 \quad \text{and} \quad |a| = -a \quad \text{if } a < 0$$

For example, since $-4 < 0$, the second rule must be used to get $|-4| = -(-4) = 4$.

EXAMPLE 6 **Computing Absolute Value**

(a) $|8| = 8$ (b) $|0| = 0$ (c) $|-15| = -(-15) = 15$ ■

NOW WORK PROBLEM **51**.

4 Find Distance on the Real Number Line

Look again at Figure 7. The distance from -4 to 3 is 7 units. This distance is the difference $3 - (-4)$, obtained by subtracting the smaller coordinate from the larger. However, since $|3 - (-4)| = |7| = 7$ and $|-4 - 3| = |-7| = 7$, we can use absolute value to calculate the distance between two points without being concerned about which one has a smaller coordinate.

If P and Q are two points on a real number line with coordinates a and b, respectively, the **distance between P and Q**, denoted by $d(P, Q)$, is

$$d(P, Q) = |b - a|$$

Since $|b - a| = |a - b|$, it follows that $d(P, Q) = d(Q, P)$.

EXAMPLE 7 **Finding Distance on a Number Line**

Let P, Q, and R be points on a real number line with coordinates -5, 7, and -3, respectively. Find the distance

(a) between P and Q (b) between Q and R

Solution See Figure 8.

Figure 8

(a) $d(P, Q) = |7 - (-5)| = |12| = 12$

(b) $d(Q, R) = |-3 - 7| = |-10| = 10$ ■

NOW WORK PROBLEM **65**.

5 Use the Laws of Exponents

Integer exponents provide a shorthand device for representing repeated multiplications of a real number.

If a is a real number and n is a positive integer, then the symbol a^n represents the product of n factors of a. That is,

$$a^n = \underbrace{a \cdot a \cdot \ldots \cdot a}_{n \text{ factors}}$$

where it is understood that $a^1 = a$. Then, $a^2 = a \cdot a$, $a^3 = a \cdot a \cdot a$, and so on. In the expression a^n, a is called the **base** and n is called the **exponent**, or **power**. We read a^n as "a raised to the power n" or as "a to the nth power." We usually read a^2 as "a squared" and a^3 as "a cubed."

In working with exponents, the operation of *raising to a power* is performed before any other operation. For example,

$$4 \cdot 3^2 + 5 = 4 \cdot 9 + 5 = 36 + 5 = 41 \qquad -2^4 = -16 \qquad 2^2 + 3^2 = 4 + 9 = 13$$

Parentheses are used to indicate operations to be performed first. For example,

$$(-2)^4 = (-2)(-2)(-2)(-2) = 16 \qquad (2 + 3)^2 = 5^2 = 25$$

If $a \neq 0$, we define

$$a^0 = 1 \qquad \text{if } a \neq 0$$

If $a \neq 0$ and if n is a positive integer, then we define

$$a^{-n} = \frac{1}{a^n} \qquad \text{if } a \neq 0$$

With these definitions, the symbol a^n is defined for any integer n.

The following properties, called the **laws of exponents**, can be proved using the preceding definitions. In the list, a and b are real numbers, and m and n are integers.

Laws of Exponents

$$a^m a^n = a^{m+n} \qquad (a^m)^n = a^{mn} \qquad (ab)^n = a^n b^n$$

$$\frac{a^m}{a^n} = a^{m-n} = \frac{1}{a^{n-m}} \quad \text{if } a \neq 0 \qquad \left(\frac{a}{b}\right)^n = \frac{a^n}{b^n} \quad \text{if } b \neq 0$$

EXAMPLE 8

Using the Laws of Exponents

Write each expression so that all exponents are positive.

(a) $\dfrac{x^5 y^{-2}}{x^3 y}$ $\quad x \neq 0, y \neq 0$ \qquad (b) $\left(\dfrac{x^{-3}}{3y^{-1}}\right)^{-2}$ $\quad x \neq 0, y \neq 0$

Solution

(a) $\dfrac{x^5 y^{-2}}{x^3 y} = \dfrac{x^5}{x^3} \cdot \dfrac{y^{-2}}{y} = x^{5-3} \cdot y^{-2-1} = x^2 y^{-3} = x^2 \cdot \dfrac{1}{y^3} = \dfrac{x^2}{y^3}$

(b) $\left(\dfrac{x^{-3}}{3y^{-1}}\right)^{-2} = \dfrac{(x^{-3})^{-2}}{(3y^{-1})^{-2}} = \dfrac{x^6}{3^{-2}(y^{-1})^{-2}} = \dfrac{x^6}{\dfrac{1}{9}y^2} = \dfrac{9x^6}{y^2}$ \qquad ■

NOW WORK PROBLEMS **71** AND **81.**

6 Evaluate Square Roots

A real number is squared when it is raised to the power 2. The inverse of squaring is finding a **square root**. For example, since $6^2 = 36$ and $(-6)^2 = 36$, the numbers 6 and -6 are square roots of 36.

The symbol $\sqrt{}$, called a **radical sign**, is used to denote the **principal**, or nonnegative, square root. Thus, $\sqrt{36} = 6$.

> In general, if a is a nonnegative real number, the nonnegative number b such that $b^2 = a$ is **the principal square root** of a and is denoted by $b = \sqrt{a}$.

The following comments are noteworthy:

1. Negative numbers do not have square roots (in the real number system), because the square of any real number is *nonnegative*. For example, $\sqrt{-4}$ is not a real number, because there is no real number whose square is -4.
2. The principal square root of 0 is 0, since $0^2 = 0$. That is, $\sqrt{0} = 0$.
3. The principal square root of a positive number is positive.
4. If $c \geq 0$, then $\left(\sqrt{c}\right)^2 = c$. For example, $\left(\sqrt{2}\right)^2 = 2$ and $\left(\sqrt{3}\right)^2 = 3$.

| EXAMPLE 9 | **Evaluating Square Roots** |

(a) $\sqrt{64} = 8$ (b) $\sqrt{\dfrac{1}{16}} = \dfrac{1}{4}$ (c) $\left(\sqrt{1.4}\right)^2 = 1.4$ (d) $\sqrt{(-3)^2} = |-3| = 3$ ∎

Examples 9(a) and (b) are examples of square roots of perfect squares, since $64 = 8^2$ and $\dfrac{1}{16} = \left(\dfrac{1}{4}\right)^2$.

Notice the need for the absolute value in Example 9(d). Since $a^2 \geq 0$, the principal square root of a^2 is defined whether $a > 0$ or $a < 0$. However, since the principal square root is nonnegative, we need the absolute value to ensure the nonnegative result.

In general, we have

$$\sqrt{a^2} = |a| \tag{1}$$

| EXAMPLE 10 | **Using Equation (1)** |

(a) $\sqrt{(2.3)^2} = |2.3| = 2.3$ (b) $\sqrt{(-2.3)^2} = |-2.3| = 2.3$ (c) $\sqrt{x^2} = |x|$ ∎

NOW WORK PROBLEM 77.

Calculators

Calculators are finite machines. As a result, they are incapable of displaying decimals that contain a large number of digits. For example, some calculators are capable of displaying only eight digits. When a number requires more than eight digits,

the calculator either truncates or rounds. To see how your calculator handles decimals, divide 2 by 3. How many digits do you see? Is the last digit a 6 or a 7? If it is a 6, your calculator truncates; if it is a 7, your calculator rounds.

There are different kinds of calculators. An **arithmetic** calculator can only add, subtract, multiply, and divide numbers; therefore, this type is not adequate for this course. **Scientific** calculators have all the capabilities of arithmetic calculators and also contain **function keys** labeled ln, log, sin, cos, tan, x^y, inv, and so on. As you proceed through this text, you will discover how to use many of the function keys. **Graphing** calculators have all the capabilities of scientific calculators and contain a screen on which graphs can be displayed. We use the ▨ symbol whenever a graphing calculator needs to be used. In this book the use of a graphing calculator is optional.

A.1 Assess Your Understanding

Concepts and Vocabulary

1. A _____ is a letter used in algebra to represent any number from a given set of numbers.

2. On the real number line, the real number zero is the coordinate of the _____.

3. An inequality of the form $a > b$ is called a(n) _____ inequality.

4. In the expression 2^4, the number 2 is called the _____ and 4 is called the _____.

5. $\sqrt{(-5)^2} =$ _____.

6. *True or False:* The distance between two points on the real number line is always greater than zero.

7. *True or False:* The absolute value of a real number is always greater than zero.

8. *True or False:* To multiply two expressions having the same base, retain the base and multiply the exponents.

Skill Building

In Problems 9–16, find the value of each expression if $x = -2$ and $y = 3$.

9. $x + 2y$

10. $3x + y$

11. $5xy + 2$

12. $-2x + xy$

13. $\dfrac{2x}{x - y}$

14. $\dfrac{x + y}{x - y}$

15. $\dfrac{3x + 2y}{2 + y}$

16. $\dfrac{2x - 3}{y}$

In Problems 17–24, determine which of the value(s) given below, if any, must be excluded from the domain of the variable in each expression.

(a) $x = 3$

(b) $x = 1$

(c) $x = 0$

(d) $x = -1$

17. $\dfrac{x^2 - 1}{x}$

18. $\dfrac{x^2 + 1}{x}$

19. $\dfrac{x}{x^2 - 9}$

20. $\dfrac{x}{x^2 + 9}$

21. $\dfrac{x^2}{x^2 + 1}$

22. $\dfrac{x^3}{x^2 - 1}$

23. $\dfrac{x^2 + 5x - 10}{x^3 - x}$

24. $\dfrac{-9x^2 - x + 1}{x^3 + x}$

In Problems 25–28, determine the domain of the variable x in each expression.

25. $\dfrac{4}{x - 5}$

26. $\dfrac{-6}{x + 4}$

27. $\dfrac{x}{x + 4}$

28. $\dfrac{x - 2}{x - 6}$

29. On the real number line, label the points with coordinates 0, $1, -1, \dfrac{5}{2}, -2.5, \dfrac{3}{4}$, and 0.25.

30. Repeat Problem 29 for the coordinates 0, $-2, 2, -1.5, \dfrac{3}{2}, \dfrac{1}{3}$, and $\dfrac{2}{3}$.

In Problems 31–40, replace the question mark by $<$, $>$, or $=$, whichever is correct.

31. $\dfrac{1}{2}$? 0 **32.** 5 ? 6 **33.** -1 ? -2 **34.** -3 ? $-\dfrac{5}{2}$ **35.** π ? 3.14

36. $\sqrt{2}$? 1.41 **37.** $\dfrac{1}{2}$? 0.5 **38.** $\dfrac{1}{3}$? 0.33 **39.** $\dfrac{2}{3}$? 0.67 **40.** $\dfrac{1}{4}$? 0.25

In Problems 41–46, write each statement as an inequality.

41. x is positive. **42.** z is negative. **43.** x is less than 2.
44. y is greater than -5. **45.** x is less than or equal to 1. **46.** x is greater than or equal to 2.

In Problems 47–50, graph the numbers x on the real number line.

47. $x \geq -2$ **48.** $x < 4$ **49.** $x > -1$ **50.** $x \leq 7$

In Problems 51–60, find the value of each expression if $x = 3$ and $y = -2$.

51. $|x + y|$ **52.** $|x - y|$ **53.** $|x| + |y|$ **54.** $|x| - |y|$ **55.** $\dfrac{|x|}{x}$

56. $\dfrac{|y|}{y}$ **57.** $|4x - 5y|$ **58.** $|3x + 2y|$ **59.** $||4x| - |5y||$ **60.** $3|x| + 2|y|$

In Problems 61–66, use the real number line below to compute each distance.

$$\begin{array}{ccccccccccc} & A & & B & C & D & & E & & & \\ \hline -4 & -3 & -2 & -1 & 0 & 1 & 2 & 3 & 4 & 5 & 6 \end{array}$$

61. $d(C, D)$ **62.** $d(C, A)$ **63.** $d(D, E)$ **64.** $d(C, E)$ **65.** $d(A, E)$ **66.** $d(D, B)$

In Problems 67–78, simplify each expression.

67. $(-4)^2$ **68.** -4^2 **69.** 4^{-2} **70.** -4^{-2} **71.** $3^{-6} \cdot 3^4$ **72.** $4^{-2} \cdot 4^3$

73. $(3^{-2})^{-1}$ **74.** $(2^{-1})^{-3}$ **75.** $\sqrt{25}$ **76.** $\sqrt{36}$ **77.** $\sqrt{(-4)^2}$ **78.** $\sqrt{(-3)^2}$

In Problems 79–88, simplify each expression. Express the answer so that all exponents are positive. Whenever an exponent is 0 or negative, we assume that the base is not 0.

79. $(8x^3)^{-2}$ **80.** $(-4x^2)^{-1}$ **81.** $(x^2 y^{-1})^2$ **82.** $(x^{-1} y)^3$ **83.** $\dfrac{x^{-2} y^3}{xy^4}$

84. $\dfrac{x^{-2} y}{xy^2}$ **85.** $\dfrac{(-2)^3 x^4 (yz)^2}{3^2 xy^3 z}$ **86.** $\dfrac{4x^{-2}(yz)^{-1}}{2^3 x^4 y}$ **87.** $\left(\dfrac{3x^{-1}}{4y^{-1}}\right)^{-2}$ **88.** $\left(\dfrac{5x^{-2}}{6y^{-2}}\right)^{-3}$

Applications and Extensions

In Problems 89–100, express each statement as an equation involving the indicated variables.

89. Area of a Rectangle The area A of a rectangle is the product of its length l and its width w.

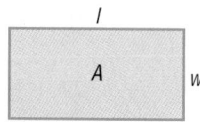

90. Perimeter of a Rectangle The perimeter P of a rectangle is twice the sum of its length l and its width w.

91. Circumference of a Circle The circumference C of a circle is the product of π and its diameter d.

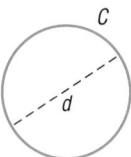

92. Area of a Triangle The area A of a triangle is one-half the product of its base b and its height h.

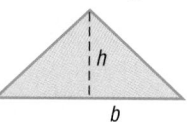

93. Area of an Equilateral Triangle The area A of an equilateral triangle is $\dfrac{\sqrt{3}}{4}$ times the square of the length x of one side.

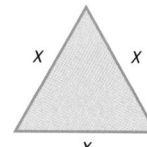

94. Perimeter of an Equilateral Triangle The perimeter P of an equilateral triangle is 3 times the length x of one side.

95. Volume of a Sphere The volume V of a sphere is $\frac{4}{3}$ times π times the cube of the radius r.

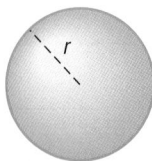

96. Surface Area of a Sphere The surface area S of a sphere is 4 times π times the square of the radius r.

97. Volume of a Cube The volume V of a cube is the cube of the length x of a side.

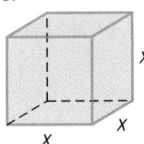

98. Surface Area of a Cube The surface area S of a cube is 6 times the square of the length x of a side.

99. U.S. Voltage In the United States, normal household voltage is 115 volts. It is acceptable for the actual voltage x to differ from normal by at most 5 volts. A formula that describes this is

$$|x - 115| \leq 5$$

(a) Show that a voltage of 113 volts is acceptable.
(b) Show that a voltage of 109 volts is not acceptable.

100. Foreign Voltage In countries other than the United States, normal household voltage is 220 volts. It is acceptable for the actual voltage x to differ from normal by at most 8 volts. A formula that describes this is

$$|x - 220| \leq 8$$

(a) Show that a voltage of 214 volts is acceptable.
(b) Show that a voltage of 209 volts is not acceptable.

101. Making Precision Ball Bearings The FireBall Company manufactures ball bearings for precision equipment. One of their products is a ball bearing with a stated radius of 3 centimeters (cm). Only ball bearings with a radius within 0.01 cm of this stated radius are acceptable. If x is the radius of a ball bearing, a formula describing this situation is

$$|x - 3| \leq 0.01$$

(a) Is a ball bearing of radius $x = 2.999$ acceptable?
(b) Is a ball bearing of radius $x = 2.89$ acceptable?

102. Body Temperature Normal human body temperature is 98.6°F. A temperature x that differs from normal by at least 1.5°F is considered unhealthy. A formula that describes this is

$$|x - 98.6| \geq 1.5$$

(a) Show that a temperature of 97°F is unhealthy.
(b) Show that a temperature of 100°F is not unhealthy.

103. Does $\frac{1}{3}$ equal 0.333? If not, which is larger? By how much?

104. Does $\frac{2}{3}$ equal 0.666? If not, which is larger? By how much?

Discussion and Writing

105. Is there a positive real number "closest" to 0?

106. I'm thinking of a number! It lies between 1 and 10; its square is rational and lies between 1 and 10. The number is larger than π. Correct to two decimal places, name the number. Now think of your own number, describe it, and challenge a fellow student to name it.

107. Write a brief paragraph that illustrates the similarities and differences between "less than" ($<$) and "less than or equal" (\leq).

A.2 Geometry Essentials

OBJECTIVES 1 Use the Pythagorean Theorem and Its Converse
2 Know Geometry Formulas
3 Understand Congruent Triangles and Similar Triangles

In this section we review some topics studied in geometry that we shall need for our study of algebra.

1 Use the Pythagorean Theorem and Its Converse

Figure 9

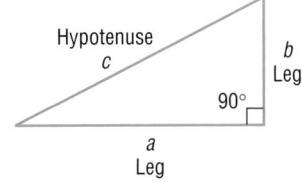

The *Pythagorean Theorem* is a statement about *right triangles*. A **right triangle** is one that contains a **right angle**, that is, an angle of 90°. The side of the triangle opposite the 90° angle is called the **hypotenuse**; the remaining two sides are called **legs**. In Figure 9 we have used c to represent the length of the hypotenuse and a and b to represent the lengths of the legs. Notice the use of the symbol \ulcorner to show the 90° angle. We now state the Pythagorean Theorem.

Pythagorean Theorem

In a right triangle, the square of the length of the hypotenuse is equal to the sum of the squares of the lengths of the legs. That is, in the right triangle shown in Figure 9,

$$c^2 = a^2 + b^2 \qquad (1)$$

| EXAMPLE 1 | Finding the Hypotenuse of a Right Triangle |

In a right triangle, one leg has length 4 and the other has length 3. What is the length of the hypotenuse?

Solution Since the triangle is a right triangle, we use the Pythagorean Theorem with $a = 4$ and $b = 3$ to find the length c of the hypotenuse. From equation (1), we have

$$c^2 = a^2 + b^2$$
$$c^2 = 4^2 + 3^2 = 16 + 9 = 25$$
$$c = \sqrt{25} = 5$$

 NOW WORK PROBLEM 13.

The converse of the Pythagorean Theorem is also true.

Converse of the Pythagorean Theorem

In a triangle, if the square of the length of one side equals the sum of the squares of the lengths of the other two sides, the triangle is a right triangle. The 90° angle is opposite the longest side.

| EXAMPLE 2 | Verifying That a Triangle Is a Right Triangle |

Show that a triangle whose sides are of lengths 5, 12, and 13 is a right triangle. Identify the hypotenuse.

Solution We square the lengths of the sides.

$$5^2 = 25, \qquad 12^2 = 144, \qquad 13^2 = 169$$

Notice that the sum of the first two squares (25 and 144) equals the third square (169). Hence, the triangle is a right triangle. The longest side, 13, is the hypotenuse. See Figure 10.

Figure 10

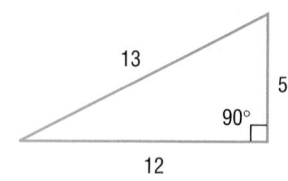

NOW WORK PROBLEM 21.

| EXAMPLE 3 | Applying the Pythagorean Theorem |

The tallest inhabited building in the world is the Sears Tower in Chicago. If the observation tower is 1450 feet above ground level, how far can a person standing in the observation tower see (with the aid of a telescope)? Use 3960 miles for the radius of Earth. See Figure 11.

SOURCE: Council on Tall Buildings and Urban Habitat (1997); Sears Tower No. 1 for tallest roof (1450 ft) and tallest occupied floor (1431 ft).

Figure 11

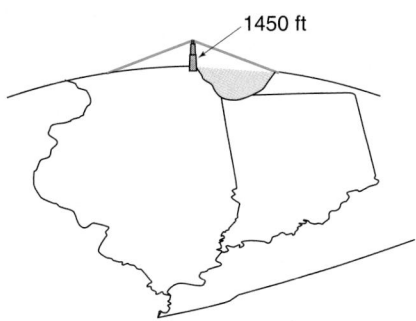

1450 ft

Solution From the center of Earth, draw two radii: one through the Sears Tower and the other to the farthest point a person can see from the tower. See Figure 12. Apply the Pythagorean Theorem to the right triangle.

Since 1 mile $= 5280$ feet, then 1450 feet $= \dfrac{1450}{5280}$ miles. So we have

Figure 12

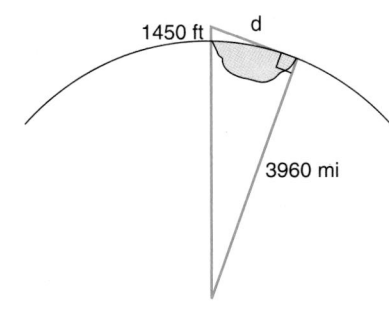

1450 ft d

3960 mi

$$d^2 + (3960)^2 = \left(3960 + \frac{1450}{5280} \right)^2$$

$$d^2 = \left(3960 + \frac{1450}{5280} \right)^2 - (3960)^2 \approx 2175.08$$

$$d \approx 46.64$$

A person can see about 47 miles from the observation tower. ■

➤ **NOW WORK PROBLEM 51.**

2 Know Geometry Formulas

Certain formulas from geometry are useful in solving algebra problems. We list some of these formulas next.

For a rectangle of length l and width w,

w

l

$$\text{Area} = lw \qquad \text{Perimeter} = 2l + 2w$$

For a triangle with base b and altitude h,

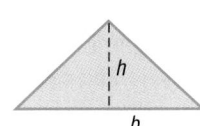

h

b

$$\text{Area} = \frac{1}{2}bh$$

For a circle of radius r (diameter $d = 2r$),

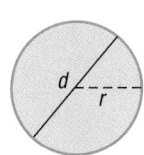

d r

$$\text{Area} = \pi r^2 \qquad \text{Circumference} = 2\pi r = \pi d$$

For a closed rectangular box of length l, width w, and height h,

h

w

l

$$\text{Volume} = lwh \qquad \text{Surface area} = 2lh + 2wh + 2lw$$

For a sphere of radius r,

$$\text{Volume} = \frac{4}{3}\pi r^3 \qquad \text{Surface area} = 4\pi r^2$$

For a right circular cylinder of height h and radius r,

$$\text{Volume} = \pi r^2 h \qquad \text{Surface area} = 2\pi r^2 + 2\pi rh$$

✏️ **NOW WORK PROBLEM 29.**

EXAMPLE 4 **Using Geometry Formulas**

A Christmas tree ornament is in the shape of a semicircle on top of a triangle. How many square centimeters (cm) of copper is required to make the ornament if the height of the triangle is 6 cm and the base is 4 cm?

Solution See Figure 13. The amount of copper required equals the shaded area. This area is the sum of the area of the triangle and the semicircle. The triangle has height $h = 6$ and base $b = 4$. The semicircle has diameter $d = 4$, so its radius is $r = 2$.

Figure 13

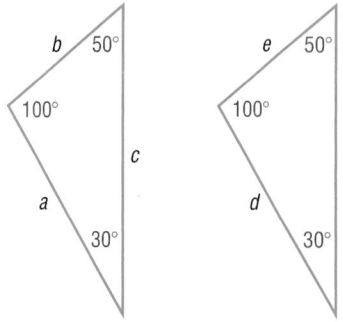

$$\text{Area} = \text{Area of triangle} + \text{Area of semicircle}$$

$$= \frac{1}{2}bh + \frac{1}{2}\pi r^2 = \frac{1}{2}(4)(6) + \frac{1}{2}\pi \cdot 2^2 \qquad b = 4; h = 6; r = 2$$

$$= 12 + 2\pi \approx 18.28 \text{ cm}^2$$

About 18.28 cm² of copper is required. ∎

✏️ **NOW WORK PROBLEM 47.**

3 **Understand Congruent Triangles and Similar Triangles**

In Words

Two triangles are congruent if they have the same size and shape.

Throughout the text we will make reference to triangles. We begin with a discussion of *congruent* triangles. According to dictionary.com, the word **congruent** means coinciding exactly when superimposed. So two angles are congruent if they have the same measure. Two line segments are congruent if they have the same length. Two triangles are congruent if each of the corresponding angles is the same measure and each of the corresponding sides is the same length.

In Figure 14, corresponding angles are equal and, the lengths of the corresponding sides are equal: $a = d$, $b = e$, and $c = f$. We conclude that these triangles are congruent.

FIGURE 14
Congruent triangles

It is not necessary to verify that all three angles and all three sides are the same measure to determine whether two triangles are congruent.

Determining Congruent Triangles

1. **ANGLE-SIDE-ANGLE CASE.** Two triangles are congruent if two of the angles are equal and the lengths of the corresponding sides between the two angles are equal.

 For example, in Figure 15(a), the two triangles are congruent because two angles and the included side are equal.

2. **SIDE-SIDE-SIDE CASE.** Two triangles are congruent if the lengths of the corresponding sides of the triangle are equal.

 For example, in Figure 15(b), the two triangles are congruent because the three corresponding sides are all equal.

3. **SIDE-ANGLE-SIDE CASE.** Two triangles are congruent if the lengths of two corresponding sides are equal and the angles between the two sides are the same.

 For example, in Figure 15(c), the two triangles are congruent because two sides and the included angle are equal.

FIGURE 15

| (a) | (b) | (c) |

We contrast congruent triangles with *similar* triangles.

In Words

Two triangles are similar if they have the same shape, but (possibly) different sizes.

Two triangles are **similar** if the corresponding angles are equal and the lengths of the corresponding sides are proportional.

For example, the triangles in Figure 16 are similar because the corresponding angles are equal. In addition, the lengths of the corresponding sides are proportional because each side in the triangle on the right is twice as long as each corresponding side in the triangle on the left. That is, the ratio of the corresponding sides is a constant: $\dfrac{d}{a} = \dfrac{e}{b} = \dfrac{f}{c} = 2$.

FIGURE 16

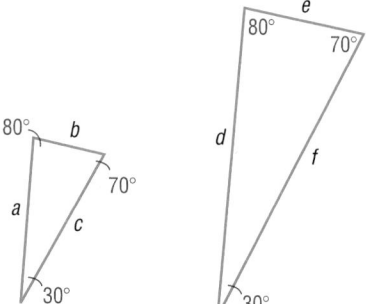

It is not necessary to verify that all three angles are equal and all three sides are proportional to determine whether two triangles are congruent.

Determining Similar Triangles

1. **ANGLE-ANGLE CASE.** Two triangles are similar if two of the corresponding angles are equal.

 For example, in Figure 17(a), the two triangles are similar because two angles are equal.

2. **SIDE-SIDE-SIDE CASE.** Two triangles are similar if the lengths of all three sides of each triangle are proportional.

 For example, in Figure 17(b), the two triangles are similar because

 $$\frac{10}{30} = \frac{5}{15} = \frac{6}{18} = \frac{1}{3}.$$

3. **SIDE-ANGLE-SIDE CASE.** Two triangles are similar if two corresponding sides are proportional and the angle between the two sides are equal.

 For example, in Figure 17(c), the two triangles are similar because $\frac{4}{6} = \frac{12}{18} = \frac{2}{3}$ and the angle between the sides are equal.

FIGURE 17

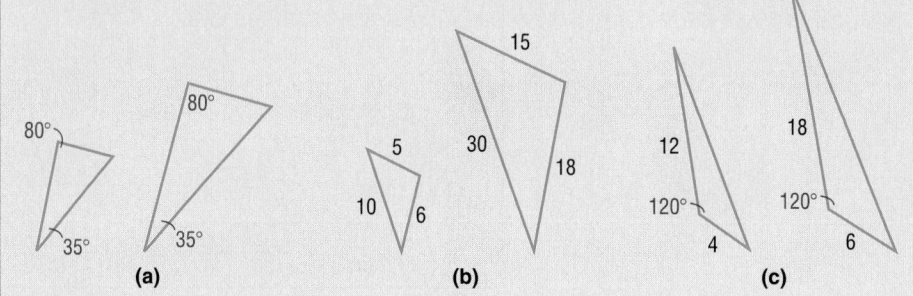

(a) (b) (c)

EXAMPLE 5	**Using Similar Triangles**

Given that the triangles in Figure 18 are similar, find the missing length.

FIGURE 18

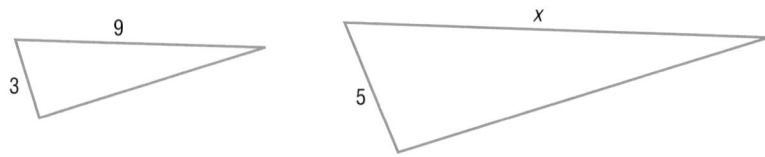

Solution Because the triangles are similar, the corresponding sides are proportional. That is, $\frac{3}{5} = \frac{9}{x}$. We solve this equation for x.

$$\frac{3}{5} = \frac{9}{x}$$

$$5x \cdot \frac{3}{5} = 5x \cdot \frac{9}{x} \qquad \text{Multiply both sides by 5x.}$$

$$3x = 45 \qquad \text{Simplify.}$$

$$x = 15 \qquad \text{Divide both sides by 3.}$$

The missing length is 15 units. ∎

NOW WORK PROBLEM **41.**

A.2 Assess Your Understanding

Concepts and Vocabulary

1. A _____ triangle is one that contains an angle of 90 degrees. The longest side is called the _____.

2. For a triangle with base b and altitude h, a formula for the area A is _____.

3. The formula for the circumference C of a circle of radius r is _____.

4. Two triangles are_____ if corresponding angles are equal and the lengths of the corresponding sides are proportional.

5. *True or False:* In a right triangle, the square of the length of the longest side equals the sum of the squares of the lengths of the other two sides.

6. *True or False:* The triangle with sides of length 6, 8, and 10 is a right triangle.

7. *True or False*: The volume of a sphere of radius r is $\frac{4}{3}\pi r^2$.

8. *True* or *False*: The triangles shown are congruent.

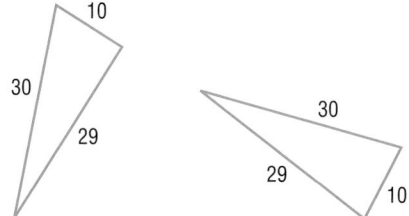

9. *True* or *False*: The triangles shown are similar.

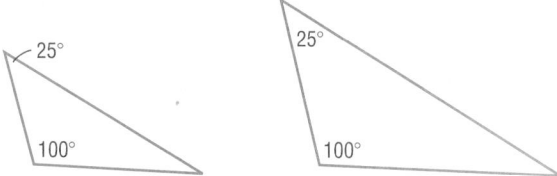

10. *True* or *False*: The triangles shown are similar.

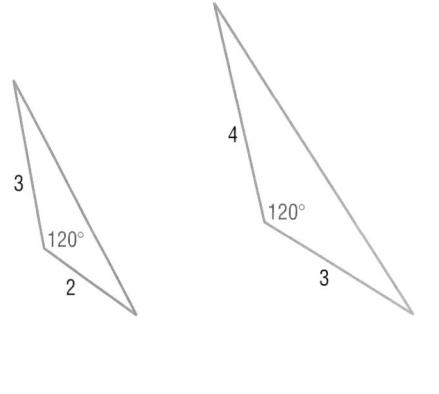

Skill Building

In Problems 11-16, the lengths of the legs of a right triangle are given. Find the hypotenuse.

11. $a = 5$, $b = 12$ 12. $a = 6$, $b = 8$ 13. $a = 10$, $b = 24$
14. $a = 4$, $b = 3$ 15. $a = 7$, $b = 24$ 16. $a = 14$, $b = 48$

In Problems 17–24, the lengths of the sides of a triangle are given. Determine which are right triangles. For those that are, identify the hypotenuse.

17. $3, 4, 5$ 18. $6, 8, 10$ 19. $4, 5, 6$ 20. $2, 2, 3$
21. $7, 24, 25$ 22. $10, 24, 26$ 23. $6, 4, 3$ 24. $5, 4, 7$

25. Find the area A of a rectangle with length 4 inches and width 2 inches.

26. Find the area A of a rectangle with length 9 centimeters and width 4 centimeters.

27. Find the area A of a triangle with height 4 inches and base 2 inches.

28. Find the area A of a triangle with height 9 centimeters and base 4 centimeters.

29. Find the area A and circumference C of a circle of radius 5 meters.

30. Find the area A and circumference C of a circle of radius 2 feet.

31. Find the volume V and surface area S of a rectangular box with length 8 feet, width 4 feet, and height 7 feet.

32. Find the volume V and surface area S of a rectangular box with length 9 inches, width 4 inches, and height 8 inches.

33. Find the volume V and surface area S of a sphere of radius 4 centimeters.

34. Find the volume V and surface area S of a sphere of radius 3 feet.

35. Find the volume V and surface area S of a right circular cylinder with radius 9 inches and height 8 inches.

36. Find the volume V and surface area S of a right circular cylinder with radius 8 inches and height 9 inches.

In Problems 37–40, find the area of the shaded region.

37.

2

2

38.

2

2

39.

2

2

40.

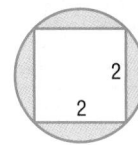

2

2

In Problems 41–44, each pair of triangles is similar. Find the missing length.

41.

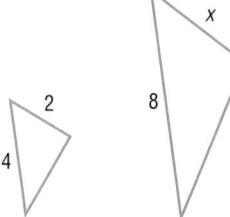

x

8

2

4

42.

16

12

x

6

43.

20

45

30

x

44.

10

50

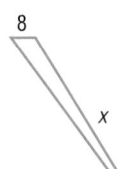

8

x

Applications and Extensions

45. How many feet does a wheel with a diameter of 16 inches travel after four revolutions?

46. How many revolutions will a circular disk with a diameter of 4 feet have completed after it has rolled 20 feet?

47. In the figure shown, *ABCD* is a square, with each side of length 6 feet. The width of the border (shaded portion) between the outer square *EFGH* and *ABCD* is 2 feet. Find the area of the border.

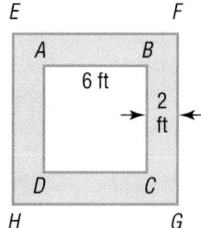

E F
A B
6 ft
2 ft
D C
H G

48. Refer to the figure. Square *ABCD* has an area of 100 square feet; square *BEFG* has an area of 16 square feet. What is the area of the triangle *CGF*?

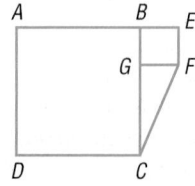

A B E
G F
D C

49. Architecture A **Norman window** consists of a rectangle surmounted by a semicircle. Find the area of the Norman window shown in the illustration. How much wood frame is needed to enclose the window?

6'

4'

50. Construction A circular swimming pool, 20 feet in diameter, is enclosed by a wooden deck that is 3 feet wide. What is the area of the deck? How much fence is required to enclose the deck?

3'

20'

In Problems 51–53, use the facts that the radius of Earth is 3960 miles and 1 mile = 5280 feet.

51. **How Far Can You See?** The conning tower of the U.S.S. *Silversides*, a World War II submarine now permanently stationed in Muskegon, Michigan, is approximately 20 feet above sea level. How far can you see from the conning tower?

52. **How Far Can You See?** A person who is 6 feet tall is standing on the beach in Fort Lauderdale, Florida, and looks out onto the Atlantic Ocean. Suddenly, a ship appears on the horizon. How far is the ship from shore?

53. **How Far Can You See?** The deck of a destroyer is 100 feet above sea level. How far can a person see from the deck? How far can a person see from the bridge, which is 150 feet above sea level?

54. Suppose that m and n are positive integers with $m > n$. If $a = m^2 - n^2$, $b = 2mn$, and $c = m^2 + n^2$, show that a, b, and c are the lengths of the sides of a right triangle. (This formula can be used to find the sides of a right triangle that are integers, such as 3, 4, 5; 5, 12, 13; and so on. Such triplets of integers are called **Pythagorean triples.**)

Discussion and Writing

55. You have 1000 feet of flexible pool siding and wish to construct a swimming pool. Experiment with rectangular-shaped pools with perimeters of 1000 feet. How do their areas vary? What is the shape of the rectangle with the largest area? Now compute the area enclosed by a circular pool with a perimeter (circumference) of 1000 feet. What would be your choice of shape for the pool? If rectangular, what is your preference for dimensions? Justify your choice. If your only consideration is to have a pool that encloses the most area, what shape should you use?

56. **The Gibb's Hill Lighthouse, Southampton, Bermuda,** in operation since 1846, stands 117 feet high on a hill 245 feet high, so its beam of light is 362 feet above sea level. A brochure states that the light itself can be seen on the horizon about 26 miles distant. Verify the correctness of this information. The brochure further states that ships 40 miles away can see the light and planes flying at 10,000 feet can see it 120 miles away. Verify the accuracy of these statements. What assumption did the brochure make about the height of the ship?

A.3 Polynomials

OBJECTIVES 1 Recognize Monomials
2 Recognize Polynomials
3 Add and Subtract Polynomials
4 Multiply Polynomials
5 Know Formulas for Special Products
6 Divide Polynomials Using Long Division
7 Work with Polynomials in Two Variables

We have described algebra as a generalization of arithmetic in which letters are used to represent real numbers. From now on, we shall use the letters at the end of the alphabet, such as x, y, and z, to represent variables and the letters at the beginning of the alphabet, such as a, b, and c, to represent constants. In the expressions $3x + 5$ and $ax + b$, it is understood that x is a variable and that a and b are constants, even though the constants a and b are unspecified. As you will find out, the context usually makes the intended meaning clear.

1 Recognize Monomials

NOTE

The nonnegative integers are the integers 0, 1, 2, 3, ■

A **monomial** in one variable is the product of a constant times a variable raised to a nonnegative integer power. A monomial is of the form

$$ax^k$$

where a is a constant, x is a variable, and $k \geq 0$ is an integer. The constant a is called the **coefficient** of the monomial. If $a \neq 0$, then k is called the **degree** of the monomial.

EXAMPLE 1

Examples of Monomials

Monomial	Coefficient	Degree	
(a) $6x^2$	6	2	
(b) $-\sqrt{2}x^3$	$-\sqrt{2}$	3	
(c) 3	3	0	Since $3 = 3 \cdot 1 = 3x^0, x \neq 0$
(d) $-5x$	-5	1	Since $-5x = -5x^1$
(e) x^4	1	4	Since $x^4 = 1 \cdot x^4$ ■

Now let's look at some expressions that are not monomials.

EXAMPLE 2

Examples of Nonmonomial Expressions

(a) $3x^{1/2}$ is not a monomial, since the exponent of the variable x is $\dfrac{1}{2}$ and $\dfrac{1}{2}$ is not a nonnegative integer.

(b) $4x^{-3}$ is not a monomial, since the exponent of the variable x is -3 and -3 is not a nonnegative integer. ■

 NOW WORK PROBLEM 7.

2 Recognize Polynomials

Two monomials with the same variable raised to the same power are called **like terms**. For example, $2x^4$ and $-5x^4$ are like terms. In contrast, the monomials $2x^3$ and $2x^5$ are not like terms.

We can add or subtract like terms using the Distributive Property. For example,

$$2x^2 + 5x^2 = (2 + 5)x^2 = 7x^2 \quad \text{and} \quad 8x^3 - 5x^3 = (8 - 5)x^3 = 3x^3$$

The sum or difference of two monomials having different degrees is called a **binomial**. The sum or difference of three monomials with three different degrees is called a **trinomial**. For example,

$$x^2 - 2 \text{ is a binomial.}$$
$$x^3 - 3x + 5 \text{ is a trinomial.}$$
$$2x^2 + 5x^2 + 2 = 7x^2 + 2 \text{ is a binomial.}$$

In Words

A polynomial is a sum of monomials.

A **polynomial** in one variable is an algebraic expression of the form

$$a_n x^n + a_{n-1} x^{n-1} + \cdots + a_1 x + a_0 \tag{1}$$

where $a_n, a_{n-1}, \ldots, a_1, a_0$ are constants,* called the **coefficients** of the polynomial, $n \geq 0$ is an integer, and x is a variable. If $a_n \neq 0$, it is called the **leading coefficient**, and n is called the **degree** of the polynomial.

*The notation a_n is read as "a sub n." The number n is called a **subscript** and should not be confused with an exponent. We use subscripts to distinguish one constant from another when a large or undetermined number of constants is required.

The monomials that make up a polynomial are called its **terms**. If all the coefficients are 0, the polynomial is called the **zero polynomial**, which has no degree.

Polynomials are usually written in **standard form**, beginning with the nonzero term of highest degree and continuing with terms in descending order according to degree. If a power of x is missing, it is because its coefficient is zero.

EXAMPLE 3 **Examples of Polynomials**

Polynomial	Coefficients	Degree
$-8x^3 + 4x^2 - 6x + 2$	$-8, 4, -6, 2$	3
$3x^2 - 5 = 3x^2 + 0 \cdot x + (-5)$	$3, 0, -5$	2
$8 - 2x + x^2 = 1 \cdot x^2 + (-2)x + 8$	$1, -2, 8$	2
$5x + \sqrt{2} = 5x^1 + \sqrt{2}$	$5, \sqrt{2}$	1
$3 = 3 \cdot 1 = 3 \cdot x^0$	3	0
0	0	No degree

Although we have been using x to represent the variable, letters such as y or z are also commonly used.

$3x^4 - x^2 + 2$ is a polynomial (in x) of degree 4.

$9y^3 - 2y^2 + y - 3$ is a polynomial (in y) of degree 3.

$z^5 + \pi$ is a polynomial (in z) of degree 5.

Algebraic expressions such as

$$\frac{1}{x} \quad \text{and} \quad \frac{x^2 + 1}{x + 5}$$

are not polynomials. The first is not a polynomial because $\dfrac{1}{x} = x^{-1}$ has an exponent that is not a nonnegative integer. Although the second expression is the quotient of two polynomials, the polynomial in the denominator has degree greater than 0, so the expression cannot be a polynomial.

 NOW WORK PROBLEM 17.

3 Add and Subtract Polynomials

Polynomials are added and subtracted by combining like terms.

EXAMPLE 4 **Adding Polynomials**

Find the sum of the polynomials:

$$8x^3 - 2x^2 + 6x - 2 \quad \text{and} \quad 3x^4 - 2x^3 + x^2 + x$$

Solution We shall find the sum in two ways.

Horizontal Addition: The idea here is to group the like terms and then combine them.

$$(8x^3 - 2x^2 + 6x - 2) + (3x^4 - 2x^3 + x^2 + x)$$
$$= 3x^4 + (8x^3 - 2x^3) + (-2x^2 + x^2) + (6x + x) - 2$$
$$= 3x^4 + 6x^3 - x^2 + 7x - 2$$

Vertical Addition: The idea here is to vertically line up the like terms in each polynomial and then add the coefficients.

$$
\begin{array}{r}
8x^3 - 2x^2 + 6x - 2 \\
(+) \quad 3x^4 - 2x^3 + x^2 + x \\
\hline
3x^4 + 6x^3 - x^2 + 7x - 2
\end{array}
$$

We can subtract two polynomials horizontally or vertically as well.

EXAMPLE 5 **Subtracting Polynomials**

Find the difference: $(3x^4 - 4x^3 + 6x^2 - 1) - (2x^4 - 8x^2 - 6x + 5)$

Solution ***Horizontal Subtraction:***

$(3x^4 - 4x^3 + 6x^2 - 1) - (2x^4 - 8x^2 - 6x + 5)$

$= 3x^4 - 4x^3 + 6x^2 - 1 + (-2x^4 + 8x^2 + 6x - 5)$

Be sure to change the sign of each term in the second polynomial.

$= (3x^4 - 2x^4) + (-4x^3) + (6x^2 + 8x^2) + 6x + (-1 - 5)$

Group like terms.

$= x^4 - 4x^3 + 14x^2 + 6x - 6$

Vertical Subtraction: We line up like terms, change the sign of each coefficient of the second polynomial, and add.

$$
\begin{array}{r}
3x^4 - 4x^3 + 6x^2 - 1 = \\
(-) \quad [2x^4 - 8x^2 - 6x + 5] = (+)
\end{array}
\quad
\begin{array}{r}
3x^4 - 4x^3 + 6x^2 - 1 \\
-2x^4 + 8x^2 + 6x - 5 \\
\hline
x^4 - 4x^3 + 14x^2 + 6x - 6
\end{array}
$$

The choice of which of these methods to use for adding and subtracting polynomials is left to you. To save space, we shall most often use the horizontal format.

COMMENT: Vertical subtraction will be used when we divide polynomials.

NOW WORK PROBLEM 29.

4 **Multiply Polynomials**

Two monomials may be multiplied using the Laws of Exponents and the Commutative and Associative Properties. For example,

$$(2x^3) \cdot (5x^4) = (2 \cdot 5) \cdot (x^3 \cdot x^4) = 10x^{3+4} = 10x^7$$

Products of polynomials are found by repeated use of the Distributive Property and the Laws of Exponents. Again, you have a choice of horizontal or vertical format.

EXAMPLE 6 **Multiplying Polynomials**

Find the product: $(2x + 5)(x^2 - x + 2)$

Solution *Horizontal Multiplication:*

$$(2x + 5)(x^2 - x + 2) = 2x(x^2 - x + 2) + 5(x^2 - x + 2)$$
↑
Distributive Property

$$= (2x \cdot x^2 - 2x \cdot x + 2x \cdot 2) + (5 \cdot x^2 - 5 \cdot x + 5 \cdot 2)$$
↑
Distributive Property

$$= (2x^3 - 2x^2 + 4x) + (5x^2 - 5x + 10)$$
↑
Law of Exponents

$$= 2x^3 + 3x^2 - x + 10$$
↑
Combine like terms

Vertical Multiplication: The idea here is very much like multiplying a two-digit number by a three-digit number.

$$
\begin{array}{r}
x^2 - x + 2 \\
2x + 5 \\
\hline
2x^3 - 2x^2 + 4x \\
(+) \quad 5x^2 - 5x + 10 \\
\hline
2x^3 + 3x^2 - x + 10
\end{array}
$$

This line is $2x(x^2 - x + 2)$. ■
This line is $5(x^2 - x + 2)$.
Sum of the above two lines.

NOW WORK PROBLEM **41**.

5 Know Formulas for Special Products

Certain products, which we call **special products**, occur frequently in algebra. We can calculate them easily using the **FOIL** (*F*irst, *O*uter, *I*nner, *L*ast) method of multiplying two binomials.

```
        ┌──── Outer ────┐
      ┌─ First ─┐
```
$$(ax + b)(cx + d) = ax(cx + d) + b(cx + d)$$
```
      └ Inner ┘
        └──── Last ────┘
```

$$
\begin{aligned}
&= \overbrace{ax \cdot cx}^{\text{First}} + \overbrace{ax \cdot d}^{\text{Outer}} + \overbrace{b \cdot cx}^{\text{Inner}} + \overbrace{b \cdot d}^{\text{Last}} \\
&= acx^2 + adx + bcx + bd \\
&= acx^2 + (ad + bc)x + bd
\end{aligned}
$$

EXAMPLE 7	**Using FOIL**

(a) $(x - 3)(x + 3) = x^2 + 3x - 3x - 9 = x^2 - 9$
 F O I L
(b) $(x + 2)^2 = (x + 2)(x + 2) = x^2 + 2x + 2x + 4 = x^2 + 4x + 4$
(c) $(x - 3)^2 = (x - 3)(x - 3) = x^2 - 3x - 3x + 9 = x^2 - 6x + 9$
(d) $(x + 3)(x + 1) = x^2 + x + 3x + 3 = x^2 + 4x + 3$
(e) $(2x + 1)(3x + 4) = 6x^2 + 8x + 3x + 4 = 6x^2 + 11x + 4$ ■

NOW WORK PROBLEMS **47** AND **55**.

Some products have been given special names because of their form. The following special products are based on Examples 7(a), (b), and (c).

A24 APPENDIX A Review

Difference of Two Squares

$$(x - a)(x + a) = x^2 - a^2 \qquad \textbf{(2)}$$

Squares of Binomials, or Perfect Squares

$$(x + a)^2 = x^2 + 2ax + a^2 \qquad \textbf{(3a)}$$
$$(x - a)^2 = x^2 - 2ax + a^2 \qquad \textbf{(3b)}$$

EXAMPLE 8 **Using Special Product Formulas**

(a) $(x - 5)(x + 5) = x^2 - 5^2 = x^2 - 25$ Difference of two squares

(b) $(x + 7)^2 = x^2 + 2 \cdot 7 \cdot x + 7^2 = x^2 + 14x + 49$ Square of a binomial

(c) $(2x + 1)^2 = (2x)^2 + 2 \cdot 1 \cdot 2x + 1^2 = 4x^2 + 4x + 1$ Notice that we used 2x in place of x in formula (3a).

(d) $(3x - 4)^2 = (3x)^2 - 2 \cdot 4 \cdot 3x + 4^2 = 9x^2 - 24x + 16$ Replace x by 3x in formula (3b). ∎

NOW WORK PROBLEMS **65, 67, AND 69.**

Let's look at some more examples that lead to general formulas.

EXAMPLE 9 **Cubing a Binomial**

(a) $(x + 2)^3 = (x + 2)(x + 2)^2 = (x + 2)(x^2 + 4x + 4)$ Formula (3a)
$$= (x^3 + 4x^2 + 4x) + (2x^2 + 8x + 8)$$
$$= x^3 + 6x^2 + 12x + 8$$

(b) $(x - 1)^3 = (x - 1)(x - 1)^2 = (x - 1)(x^2 - 2x + 1)$ Formula (3b)
$$= (x^3 - 2x^2 + x) - (x^2 - 2x + 1)$$
$$= x^3 - 3x^2 + 3x - 1 \qquad ∎$$

Cubes of Binomials, or Perfect Cubes

$$(x + a)^3 = x^3 + 3ax^2 + 3a^2x + a^3 \qquad \textbf{(4a)}$$
$$(x - a)^3 = x^3 - 3ax^2 + 3a^2x - a^3 \qquad \textbf{(4b)}$$

NOW WORK PROBLEM **85.**

EXAMPLE 10 **Forming the Difference of Two Cubes**

$$(x - 1)(x^2 + x + 1) = x(x^2 + x + 1) - 1(x^2 + x + 1)$$
$$= x^3 + x^2 + x - x^2 - x - 1$$
$$= x^3 - 1 \qquad ∎$$

EXAMPLE 11 **Forming the Sum of Two Cubes**

$$(x + 2)(x^2 - 2x + 4) = x(x^2 - 2x + 4) + 2(x^2 - 2x + 4)$$
$$= x^3 - 2x^2 + 4x + 2x^2 - 4x + 8$$
$$= x^3 + 8 \qquad ∎$$

Examples 10 and 11 lead to two more special products.

Difference of Two Cubes

$$(x - a)(x^2 + ax + a^2) = x^3 - a^3 \tag{5}$$

Sum of Two Cubes

$$(x + a)(x^2 - ax + a^2) = x^3 + a^3 \tag{6}$$

6 Divide Polynomials Using Long Division

The procedure for dividing two polynomials is similar to the procedure for dividing two integers.

EXAMPLE 12 | **Dividing Two Integers**

Divide 842 by 15.

Solution

$$
\begin{array}{r}
56 \quad \leftarrow \text{Quotient} \\
15\overline{)842} \quad \leftarrow \text{Dividend} \\
\underline{75} \quad \leftarrow 5 \cdot 15 \ (\text{subtract}) \\
92 \\
\underline{90} \quad \leftarrow 6 \cdot 15 \ (\text{subtract}) \\
2 \quad \leftarrow \text{Remainder}
\end{array}
$$

Divisor \rightarrow

So, $\dfrac{842}{15} = 56 + \dfrac{2}{15}$. ■

In the long division process detailed in Example 12, the number 15 is called the **divisor**, the number 842 is called the **dividend**, the number 56 is called the **quotient**, and the number 2 is called the **remainder**.

To check the answer obtained in a division problem, multiply the quotient by the divisor and add the remainder. The answer should be the dividend.

$$(\text{Quotient})(\text{Divisor}) + \text{Remainder} = \text{Dividend}$$

For example, we can check the results obtained in Example 12 as follows:

$$(56)(15) + 2 = 840 + 2 = 842$$

To divide two polynomials, we first must write each polynomial in standard form. The process then follows a pattern similar to that of Example 12. The next example illustrates the procedure.

EXAMPLE 13 | **Dividing Two Polynomials**

Find the quotient and the remainder when

$$3x^3 + 4x^2 + x + 7 \quad \text{is divided by} \quad x^2 + 1$$

Solution Each polynomial is in standard form. The dividend is $3x^3 + 4x^2 + x + 7$, and the divisor is $x^2 + 1$.

STEP 1: Divide the leading term of the dividend, $3x^3$, by the leading term of the divisor, x^2. Enter the result, $3x$, over the term $3x^3$, as follows:

$$x^2 + 1 \overline{\smash{)}3x^3 + 4x^2 + x + 7}$$ with $3x$ above.

STEP 2: Multiply $3x$ by $x^2 + 1$ and enter the result below the dividend.

$$\begin{array}{r} 3x \\ x^2 + 1 \overline{\smash{)}3x^3 + 4x^2 + x + 7} \\ \underline{3x^3 + 3x} \end{array}$$

$\leftarrow 3x \cdot (x^2 + 1) = 3x^3 + 3x$

Notice that we align the 3x term under the x to make the next step easier.

STEP 3: Subtract and bring down the remaining terms.

$$\begin{array}{r} 3x \\ x^2 + 1 \overline{\smash{)}3x^3 + 4x^2 + x + 7} \\ \underline{3x^3 + 3x} \\ 4x^2 - 2x + 7 \end{array}$$

\leftarrow Subtract (change the signs and add).
\leftarrow Bring down the $4x^2$ and the 7.

STEP 4: Repeat Steps 1–3 using $4x^2 - 2x + 7$ as the dividend.

$$\begin{array}{r} 3x + 4 \\ x^2 + 1 \overline{\smash{)}3x^3 + 4x^2 + x + 7} \\ \underline{3x^3 + 3x} \\ 4x^2 - 2x + 7 \\ \underline{4x^2 + 4} \\ -2x + 3 \end{array}$$

\leftarrow Divide $4x^2$ by x^2 to get 4.
\leftarrow Multiply $x^2 + 1$ by 4; subtract.

Since x^2 does not divide $-2x$ evenly (that is, the result is not a monomial), the process ends. The quotient is $3x + 4$, and the remainder is $-2x + 3$.

✔ **CHECK:** (Quotient)(Divisor) + Remainder

$$= (3x + 4)(x^2 + 1) + (-2x + 3)$$
$$= 3x^3 + 4x^2 + 3x + 4 + (-2x + 3)$$
$$= 3x^3 + 4x^2 + x + 7 = \text{Dividend}$$

Then

$$\frac{3x^3 + 4x^2 + x + 7}{x^2 + 1} = 3x + 4 + \frac{-2x + 3}{x^2 + 1}$$

■

The next example combines the steps involved in long division.

EXAMPLE 14 | **Dividing Two Polynomials**

Find the quotient and the remainder when

$$x^4 - 3x^3 + 2x - 5 \quad \text{is divided by} \quad x^2 - x + 1$$

Solution In setting up this division problem, it is necessary to leave a space for the missing x^2 term in the dividend.

$$
\begin{array}{r}
x^2 - 2x \ - 3 \qquad \leftarrow \text{Quotient} \\
\text{Divisor} \rightarrow \quad x^2 - x + 1 \overline{)\ x^4 - 3x^3 \qquad\quad + 2x - 5} \quad \leftarrow \text{Dividend} \\
\text{Subtract} \rightarrow \quad \underline{x^4 - \ x^3 + \ x^2} \qquad\qquad \\
-2x^3 - \ x^2 + 2x - 5 \\
\text{Subtract} \rightarrow \quad \underline{-2x^3 + 2x^2 - 2x} \qquad \\
-3x^2 + 4x - 5 \\
\text{Subtract} \rightarrow \quad \underline{-3x^2 + 3x - 3} \\
x - 2 \quad \leftarrow \text{Remainder}
\end{array}
$$

✔ **Check:** (Quotient)(Divisor) + Remainder

$$
\begin{aligned}
&= (x^2 - 2x - 3)(x^2 - x + 1) + x - 2 \\
&= x^4 - x^3 + x^2 - 2x^3 + 2x^2 - 2x - 3x^2 + 3x - 3 + x - 2 \\
&= x^4 - 3x^3 + 2x - 5 = \text{Dividend}
\end{aligned}
$$

As a result,

$$
\frac{x^4 - 3x^3 + 2x - 5}{x^2 - x + 1} = x^2 - 2x - 3 + \frac{x - 2}{x^2 - x + 1}
$$ ∎

The process of dividing two polynomials leads to the following result:

Theorem Let Q be a polynomial of positive degree and let P be a polynomial whose degree is greater than or equal to the degree of Q. The remainder after dividing P by Q is either the zero polynomial or a polynomial whose degree is less than the degree of the divisor Q.

NOW WORK PROBLEM 93.

7 Work with Polynomials in Two Variables

A **monomial in two variables** x and y has the form $ax^n y^m$, where a is a constant, x and y are variables, and n and m are nonnegative integers. The **degree** of a monomial is the sum of the powers of the variables.

For example,

$$
2xy^3, \quad x^2 y^2, \quad \text{and} \quad x^3 y
$$

are monomials, each of which has degree 4.

A **polynomial in two variables** x and y is the sum of one or more monomials in two variables. The **degree of a polynomial** in two variables is the highest degree of all the monomials with nonzero coefficients.

EXAMPLE 15 **Examples of Polynomials in Two Variables**

$$
3x^2 + 2x^3 y + 5 \qquad \pi x^3 - y^2 \qquad x^4 + 4x^3 y - xy^3 + y^4
$$

Two variables, Two variables, Two variables,
degree is 4. degree is 3. degree is 4. ∎

Multiplying polynomials in two variables is handled in the same way as polynomials in one variable.

| **EXAMPLE 16** | **Using a Special Product Formula** |

To multiply $(2x - y)^2$, use the Squares of Binomials (3b) formula with $2x$ instead of x and y instead of a.

$$(2x - y)^2 = (2x)^2 - 2 \cdot y \cdot 2x + y^2 = 4x^2 - 4xy + y^2$$ ∎

➤ **NOW WORK PROBLEM 79.**

A.3 Assess Your Understanding

Concepts and Vocabulary

1. The polynomial $3x^4 - 2x^3 + 13x^2 - 5$ is of degree _____. The leading coefficient is _____.

2. $(x^2 - 4)(x^2 + 4) = $ _____.

3. $(x - 2)(x^2 + 2x + 4) = $ _____.

4. *True or False:* $4x^{-2}$ is a monomial of degree -2.

5. *True or False:* The degree of the product of two nonzero polynomials equals the sum of their degrees.

6. *True or False:* $(x + a)(x^2 + ax + a) = x^3 + a^3$.

Skill Building

In Problems 7–16, tell whether the expression is a monomial. If it is, name the variable(s) and the coefficient and give the degree of the monomial.

7. $2x^3$ 8. $-4x^2$ 9. $\dfrac{8}{x}$ 10. $-2x^{-3}$ 11. $-2xy^2$

12. $5x^2y^3$ 13. $\dfrac{8x}{y}$ 14. $-\dfrac{2x^2}{y^3}$ 15. $x^2 + y^2$ 16. $3x^2 + 4$

In Problems 17–26, tell whether the expression is a polynomial. If it is, give its degree.

17. $3x^2 - 5$ 18. $1 - 4x$ 19. 5 20. $-\pi$ 21. $3x^2 - \dfrac{5}{x}$

22. $\dfrac{3}{x} + 2$ 23. $2y^3 - \sqrt{2}$ 24. $10z^2 + z$ 25. $\dfrac{x^2 + 5}{x^3 - 1}$ 26. $\dfrac{3x^3 + 2x - 1}{x^2 + x + 1}$

In Problems 27–46, add, subtract, or multiply, as indicated. Express your answer as a single polynomial in standard form.

27. $(x^2 + 4x + 5) + (3x - 3)$

28. $(x^3 + 3x^2 + 2) + (x^2 - 4x + 4)$

29. $(x^3 - 2x^2 + 5x + 10) - (2x^2 - 4x + 3)$

30. $(x^2 - 3x - 4) - (x^3 - 3x^2 + x + 5)$

31. $(6x^5 + x^3 + x) + (5x^4 - x^3 + 3x^2)$

32. $(10x^5 - 8x^2) + (3x^3 - 2x^2 + 6)$

33. $(x^2 - 3x + 1) + 2(3x^2 + x - 4)$

34. $-2(x^2 + x + 1) + (-5x^2 - x + 2)$

35. $6(x^3 + x^2 - 3) - 4(2x^3 - 3x^2)$

36. $8(4x^3 - 3x^2 - 1) - 6(4x^3 + 8x - 2)$

37. $(x^2 - x + 2) + (2x^2 - 3x + 5) - (x^2 + 1)$

38. $(x^2 + 1) - (4x^2 + 5) + (x^2 + x - 2)$

39. $9(y^2 - 3y + 4) - 6(1 - y^2)$

40. $8(1 - y^3) + 4(1 + y + y^2 + y^3)$

41. $x(x^2 + x - 4)$

42. $4x^2(x^3 - x + 2)$

43. $-2x^2(4x^3 + 5)$

44. $5x^3(3x - 4)$

45. $(x + 1)(x^2 + 2x - 4)$

46. $(2x - 3)(x^2 + x + 1)$

In Problems 47–64, multiply the polynomials using the FOIL method. Express your answer as a single polynomial in standard form.

47. $(x + 2)(x + 4)$

48. $(x + 3)(x + 5)$

49. $(2x + 5)(x + 2)$

50. $(3x + 1)(2x + 1)$

51. $(x - 4)(x + 2)$

52. $(x + 4)(x - 2)$

53. $(x - 3)(x - 2)$

54. $(x - 5)(x - 1)$

55. $(2x + 3)(x - 2)$

56. $(2x - 4)(3x + 1)$

57. $(-2x + 3)(x - 4)$

58. $(-3x - 1)(x + 1)$

59. $(-x - 2)(-2x - 4)$

60. $(-2x - 3)(3 - x)$

61. $(x - 2y)(x + y)$

62. $(2x + 3y)(x - y)$

63. $(-2x - 3y)(3x + 2y)$

64. $(x - 3y)(-2x + y)$

In Problems 65–88, multiply the polynomials using the special product formulas. Express your answer as a single polynomial in standard form.

65. $(x - 7)(x + 7)$

66. $(x - 1)(x + 1)$

67. $(2x + 3)(2x - 3)$

68. $(3x + 2)(3x - 2)$

69. $(x + 4)^2$

70. $(x + 5)^2$

71. $(x - 4)^2$

72. $(x - 5)^2$

73. $(3x + 4)(3x - 4)$

74. $(5x - 3)(5x + 3)$

75. $(2x - 3)^2$

76. $(3x - 4)^2$

77. $(x + y)(x - y)$

78. $(x + 3y)(x - 3y)$

79. $(3x + y)(3x - y)$

80. $(3x + 4y)(3x - 4y)$

| EXAMPLE 1 | Identifying Common Monomial Factors |

Polynomial	Common Monomial Factor	Remaining Factor	Factored Form
$2x + 4$	2	$x + 2$	$2x + 4 = 2(x + 2)$
$3x - 6$	3	$x - 2$	$3x - 6 = 3(x - 2)$
$2x^2 - 4x + 8$	2	$x^2 - 2x + 4$	$2x^2 - 4x + 8 = 2(x^2 - 2x + 4)$
$8x - 12$	4	$2x - 3$	$8x - 12 = 4(2x - 3)$
$x^2 + x$	x	$x + 1$	$x^2 + x = x(x + 1)$
$x^3 - 3x^2$	x^2	$x - 3$	$x^3 - 3x^2 = x^2(x - 3)$
$6x^2 + 9x$	$3x$	$2x + 3$	$6x^2 + 9x = 3x(2x + 3)$

Notice that, once all common monomial factors have been removed from a polynomial, the remaining factor is either a prime polynomial of degree 1 or a polynomial of degree 2 or higher. (Do you see why?)

NOW WORK PROBLEM 5.

1 Factor the Difference of Two Squares and the Sum and Difference of Two Cubes

When you factor a polynomial, first check for common monomial factors. Then see whether you can use one of the special formulas discussed in the previous section.

Difference of Two Squares	$x^2 - a^2 = (x - a)(x + a)$
Perfect Squares	$x^2 + 2ax + a^2 = (x + a)^2$
	$x^2 - 2ax + a^2 = (x - a)^2$
Sum of Two Cubes	$x^3 + a^3 = (x + a)(x^2 - ax + a^2)$
Difference of Two Cubes	$x^3 - a^3 = (x - a)(x^2 + ax + a^2)$

| EXAMPLE 2 | Factoring the Difference of Two Squares |

Factor completely: $x^2 - 4$

Solution We notice that $x^2 - 4$ is the difference of two squares, x^2 and 2^2.
$$x^2 - 4 = (x - 2)(x + 2)$$

| EXAMPLE 3 | Factoring the Difference of Two Cubes |

Factor completely: $x^3 - 1$

Solution Because $x^3 - 1$ is the difference of two cubes, x^3 and 1^3, we find that
$$x^3 - 1 = (x - 1)(x^2 + x + 1)$$

| EXAMPLE 4 | Factoring the Sum of Two Cubes |

Factor completely: $x^3 + 8$

Solution Because $x^3 + 8$ is the sum of two cubes, x^3 and 2^3, we have
$$x^3 + 8 = (x + 2)(x^2 - 2x + 4)$$

81. $(x + y)^2$ **82.** $(x - y)^2$ **83.** $(x - 2y)^2$ **84.** $(2x + 3y)^2$

85. $(x - 2)^3$ **86.** $(x + 1)^3$ **87.** $(2x + 1)^3$ **88.** $(3x - 2)^3$

In Problems 89–104, find the quotient and the remainder. Check your work by verifying that

$$(Quotient)(Divisor) + Remainder = Dividend$$

89. $4x^3 - 3x^2 + x + 1$ divided by $x + 2$ **90.** $3x^3 - x^2 + x - 2$ divided by $x + 2$

91. $4x^3 - 3x^2 + x + 1$ divided by x^2 **92.** $3x^3 - x^2 + x - 2$ divided by x^2

93. $5x^4 - 3x^2 + x + 1$ divided by $x^2 + 2$ **94.** $5x^4 - x^2 + x - 2$ divided by $x^2 + 2$

95. $4x^5 - 3x^2 + x + 1$ divided by $2x^3 - 1$ **96.** $3x^5 - x^2 + x - 2$ divided by $3x^3 - 1$

97. $2x^4 - 3x^3 + x + 1$ divided by $2x^2 + x + 1$ **98.** $3x^4 - x^3 + x - 2$ divided by $3x^2 + x + 1$

99. $-4x^3 + x^2 - 4$ divided by $x - 1$ **100.** $-3x^4 - 2x - 1$ divided by $x - 1$

101. $1 - x^2 + x^4$ divided by $x^2 + x + 1$ **102.** $1 - x^2 + x^4$ divided by $x^2 - x + 1$

103. $x^3 - a^3$ divided by $x - a$ **104.** $x^5 - a^5$ divided by $x - a$

Discussion and Writing

105. Explain why the degree of the product of two nonzero polynomials equals the sum of their degrees.

106. Explain why the degree of the sum of two polynomials of different degrees equals the larger of their degrees.

107. Give a careful statement about the degree of the sum of two polynomials of the same degree.

108. Do you prefer adding two polynomials using the horizontal method or the vertical method? Write a brief position paper defending your choice.

109. Do you prefer to memorize the rule for the square of a binomial $(x + a)^2$ or to use FOIL to obtain the product? Write a brief position paper defending your choice.

A.4 Factoring Polynomials

> **OBJECTIVES**
> 1 Factor the Difference of Two Squares and the Sum and the Difference of Two Cubes
> 2 Factor Perfect Squares
> 3 Factor a Second-degree Polynomial: $x^2 + Bx + C$
> 4 Factor by Grouping
> 5 Factor a Second-degree Polynomial: $Ax^2 + Bx + C, A \neq 1$
> 6 Complete the Square

Consider the following product:

$$(2x + 3)(x - 4) = 2x^2 - 5x - 12$$

The two polynomials on the left side are called **factors** of the polynomial on the right side. Expressing a given polynomial as a product of other polynomials, that is, finding the factors of a polynomial, is called **factoring**.

We shall restrict our discussion here to factoring polynomials in one variable into products of polynomials in one variable, where all coefficients are integers. We call this **factoring over the integers**.

Any polynomial can be written as the product of 1 times itself or as -1 times its additive inverse. If a polynomial cannot be written as the product of two other polynomials (excluding 1 and -1), then the polynomial is said to be **prime**. When a polynomial has been written as a product consisting only of prime factors, it is said to be **factored completely**. Examples of prime polynomials are

$$2, \quad 3, \quad 5, \quad x, \quad x + 1, \quad x - 1, \quad 3x + 4$$

The first factor to look for in a factoring problem is a common monomial factor present in each term of the polynomial. If one is present, use the Distributive Property to factor it out.

EXAMPLE 5	**Factoring the Difference of Two Squares**

Factor completely: $x^4 - 16$

Solution Because $x^4 - 16$ is the difference of two squares, $x^4 = (x^2)^2$ and $16 = 4^2$, we have

$$x^4 - 16 = (x^2 - 4)(x^2 + 4)$$

But $x^2 - 4$ is also the difference of two squares. Thus,

$$x^4 - 16 = (x^2 - 4)(x^2 + 4) = (x - 2)(x + 2)(x^2 + 4)$$ ∎

NOW WORK PROBLEMS **15** AND **33**.

2 Factor Perfect Squares

When the first term and third term of a trinomial are both positive and are perfect squares, such as $x^2, 9x^2, 1,$ and $4,$ check to see whether the trinomial is a perfect square.

EXAMPLE 6	**Factoring Perfect Squares**

Factor completely: $x^2 + 6x + 9$

Solution The first term, x^2, and the third term, $9 = 3^2$, are perfect squares. Because the middle term $6x$ is twice the product of x and 3, we have a perfect square.

$$x^2 + 6x + 9 = (x + 3)^2$$ ∎

EXAMPLE 7	**Factoring Perfect Squares**

Factor completely: $9x^2 - 6x + 1$

Solution The first term, $9x^2 = (3x)^2$, and the third term, $1 = 1^2$, are perfect squares. Because the middle term $-6x$ is -2 times the product of $3x$ and 1, we have a perfect square.

$$9x^2 - 6x + 1 = (3x - 1)^2$$ ∎

EXAMPLE 8	**Factoring Perfect Squares**

Factor completely: $25x^2 + 30x + 9$

Solution The first term, $25x^2 = (5x)^2$, and the third term, $9 = 3^2$, are perfect squares. Because the middle term $30x$ is twice the product of $5x$ and 3, we have a perfect square.

$$25x^2 + 30x + 9 = (5x + 3)^2$$ ∎

NOW WORK PROBLEMS **25** AND **93**.

If a trinomial is not a perfect square, it may be possible to factor it using the technique discussed next.

3 Factor a Second-degree Polynomial: $x^2 + Bx + C$

The idea behind factoring a second-degree polynomial like $x^2 + Bx + C$ is to see whether it can be made equal to the product of two, possibly equal, first-degree polynomials.

For example, we know that

$$(x + 3)(x + 4) = x^2 + 7x + 12$$

The factors of $x^2 + 7x + 12$ are $x + 3$ and $x + 4$. Notice the following:

$$x^2 + 7x + 12 = (x + 3)(x + 4)$$

 12 is the product of 3 and 4
 7 is the sum of 3 and 4

In general, if $x^2 + Bx + C = (x + a)(x + b)$, then $ab = C$ and $a + b = B$.

To factor a second-degree polynomial $x^2 + Bx + C$, find integers whose product is C and whose sum is B. That is, if there are numbers a, b, where $ab = C$ and $a + b = B$, then

$$x^2 + Bx + C = (x + a)(x + b)$$

EXAMPLE 9 **Factoring Trinomials**

Factor completely: $x^2 + 7x + 10$

Solution First, determine all integers whose product is 10 and then compute their sums.

Integers whose product is 10	1, 10	−1, −10	2, 5	−2, −5
Sum	11	−11	7	−7

The integers 2 and 5 have a product of 10 and add up to 7, the coefficient of the middle term. As a result, we have

$$x^2 + 7x + 10 = (x + 2)(x + 5)$$ ■

EXAMPLE 10 **Factoring Trinomials**

Factor completely: $x^2 - 6x + 8$

Solution First, determine all integers whose product is 8 and then compute each sum.

Integers whose product is 8	1, 8	−1, −8	2, 4	−2, −4
Sum	9	−9	6	−6

Since −6 is the coefficient of the middle term,

$$x^2 - 6x + 8 = (x - 2)(x - 4)$$ ■

EXAMPLE 11 **Factoring Trinomials**

Factor completely: $x^2 - x - 12$

Solution First, determine all integers whose product is −12 and then compute each sum.

Integers whose product is −12	1, −12	−1, 12	2, −6	−2, 6	3, −4	−3, 4
Sum	−11	11	−4	4	−1	1

Since −1 is the coefficient of the middle term,

$$x^2 - x - 12 = (x + 3)(x - 4)$$ ■

EXAMPLE 12 **Factoring Trinomials**

Factor completely: $x^2 + 4x - 12$

Solution The integers -2 and 6 have a product of -12 and have the sum 4. Thus,

$$x^2 + 4x - 12 = (x - 2)(x + 6)$$ ■

To avoid errors in factoring, always check your answer by multiplying it out to see if the result equals the original expression.

When none of the possibilities works, the polynomial is prime.

EXAMPLE 13 **Identifying Prime Polynomials**

Show that $x^2 + 9$ is prime.

Solution First, list the integers whose product is 9 and then compute their sums.

Integers whose product is 9	1, 9	−1, −9	3, 3	−3, −3
Sum	10	−10	6	−6

Since the coefficient of the middle term in $x^2 + 9 = x^2 + 0x + 9$ is 0 and none of the sums equals 0, we conclude that $x^2 + 9$ is prime. ■

Example 13 demonstrates a more general result:

Theorem Any polynomial of the form $x^2 + a^2$, a real, is prime.

NOW WORK PROBLEMS 39 AND 77.

4 Factor by Grouping

Sometimes a common factor does not occur in every term of the polynomial, but in each of several groups of terms that together make up the polynomial. When this happens, the common factor can be factored out of each group by means of the Distributive Property. This technique is called **factoring by grouping**.

EXAMPLE 14 **Factoring by Grouping**

Factor completely by grouping: $(x^2 + 2)x + (x^2 + 2) \cdot 3$

Solution Notice the common factor $x^2 + 2$. By applying the Distributive Property, we have

$$(x^2 + 2)x + (x^2 + 2) \cdot 3 = (x^2 + 2)(x + 3)$$

Since $x^2 + 2$ and $x + 3$ are prime, the factorization is complete. ■

The next example shows a factoring problem that occurs in calculus.

EXAMPLE 15 **Factoring by Grouping**

Factor completely by grouping: $3(x - 1)^2(x + 2)^4 + 4(x - 1)^3(x + 2)^3$

Solution Here, $(x - 1)^2(x + 2)^3$ is a common factor of $3(x - 1)^2(x + 2)^4$ and of $4(x - 1)^3(x + 2)^3$. As a result,

$$3(x - 1)^2(x + 2)^4 + 4(x - 1)^3(x + 2)^3 = (x - 1)^2(x + 2)^3[3(x + 2) + 4(x - 1)]$$
$$= (x - 1)^2(x + 2)^3[3x + 6 + 4x - 4]$$
$$= (x - 1)^2(x + 2)^3(7x + 2)$$ ■

EXAMPLE 16	**Factoring by Grouping**

Factor completely by grouping: $x^3 - 4x^2 + 2x - 8$

Solution To see if factoring by grouping will work, group the first two terms and the last two terms. Then look for a common factor in each group. In this example, we can factor x^2 from $x^3 - 4x^2$ and 2 from $2x - 8$. The remaining factor in each case is the same, $x - 4$. This means that factoring by grouping will work, as follows:

$$\begin{aligned} x^3 - 4x^2 + 2x - 8 &= (x^3 - 4x^2) + (2x - 8) \\ &= x^2(x - 4) + 2(x - 4) \\ &= (x - 4)(x^2 + 2) \end{aligned}$$

Since $x^2 + 2$ and $x - 4$ are prime, the factorization is complete. ■

NOW WORK PROBLEMS **51** AND **127.**

5 Factor a Second-degree Polynomial: $Ax^2 + Bx + C, A \neq 1$

To factor a second-degree polynomial $Ax^2 + Bx + C$, when $A \neq 1$ and $A, B,$ and C have no common factors, follow these steps:

Steps for Factoring $Ax^2 + Bx + C$,
where $A \neq 1$, and $A, B,$ and C Have No Common Factors

STEP 1: Find the value of AC.
STEP 2: Find integers whose product is AC that add up to B. That is, find a and b so that $ab = AC$ and $a + b = B$.
STEP 3: Write $Ax^2 + Bx + C = Ax^2 + ax + bx + C$.
STEP 4: Factor this last expression by grouping.

EXAMPLE 17	**Factoring Trinomials**

Factor completely: $2x^2 + 5x + 3$

Solution Comparing $2x^2 + 5x + 3$ to $Ax^2 + Bx + C$, we find that $A = 2, B = 5,$ and $C = 3$.

STEP 1: The value of AC is $2 \cdot 3 = 6$.

STEP 2: Determine the integers whose product is $AC = 6$ and compute their sums.

Integers whose product is 6	1, 6	−1, −6	2, 3	−2, −3
Sum	7	−7	5	−5

STEP 3: The integers whose product is 6 that add up to $B = 5$ are 2 and 3.

$$2x^2 + 5x + 3 = 2x^2 + 2x + 3x + 3$$

STEP 4: Factor by grouping.

$$\begin{aligned} 2x^2 + 2x + 3x + 3 &= (2x^2 + 2x) + (3x + 3) \\ &= 2x(x + 1) + 3(x + 1) \\ &= (x + 1)(2x + 3) \end{aligned}$$

Thus,

$$2x^2 + 5x + 3 = (x + 1)(2x + 3)$$ ■

EXAMPLE 18 **Factoring Trinomials**

Factor completely: $2x^2 - x - 6$

Solution Comparing $2x^2 - x - 6$ to $Ax^2 + Bx + C$, we find that $A = 2$, $B = -1$, and $C = -6$.

STEP 1: The value of AC is $2 \cdot (-6) = -12$.

STEP 2: Determine the integers whose product is $AC = -12$ and compute their sums.

Integers whose product is -12	$1, -12$	$-1, 12$	$2, -6$	$-2, 6$	$3, -4$	$-3, 4$
Sum	-11	11	-4	4	-1	1

STEP 3: The integers whose product is -12 that add up to $B = -1$ are -4 and 3.

$$2x^2 - x - 6 = 2x^2 - 4x + 3x - 6$$

STEP 4: Factor by grouping.

$$2x^2 - 4x + 3x - 6 = (2x^2 - 4x) + (3x - 6)$$
$$= 2x(x - 2) + 3(x - 2)$$
$$= (x - 2)(2x + 3)$$

As a result,

$$2x^2 - x - 6 = (x - 2)(2x + 3)$$ ∎

NOW WORK PROBLEM 57.

Summary

Type of Polynomial	Method	Example
Any polynomial	Look for common monomial factors. (Always do this first!)	$6x^2 + 9x = 3x(2x + 3)$
Binomials of degree 2 or higher	Check for a special product: Difference of two squares, $x^2 - a^2$	$x^2 - 16 = (x - 4)(x + 4)$
	Difference of two cubes, $x^3 - a^3$	$x^3 - 64 = (x - 4)(x^2 + 4x + 16)$
	Sum of two cubes, $x^3 + a^3$	$x^3 + 27 = (x + 3)(x^2 - 3x + 9)$
Trinomials of degree 2	Check for a perfect square, $(x \pm a)^2$ (p. A31)	$x^2 + 8x + 16 = (x + 4)^2$
		$x^2 - 10x + 25 = (x - 5)^2$
	Factoring $x^2 + Bx + C$ (p. A31)	$x^2 - x - 2 = (x - 2)(x + 1)$
	Factoring $Ax^2 + Bx + C$ (p. A34)	$6x^2 + x - 1 = (2x + 1)(3x - 1)$
Three or more terms	Grouping	$2x^3 - 3x^2 + 4x - 6 = (2x - 3)(x^2 + 2)$

6 Complete the Square

We now introduce the method of completing the square. The idea behind completing the square in one variable is to "adjust" an expression of the form $x^2 + bx$ to make it a perfect square. Perfect squares are trinomials of the form

$$x^2 + 2ax + a^2 = (x + a)^2 \text{ or } x^2 - 2ax + a^2 = (x - a)^2$$

For example, $x^2 + 6x + 9$ is a perfect square because $x^2 + 6x + 9 = (x + 3)^2$. And $p^2 - 12p + 36$ is a perfect square because $p^2 - 12p + 36 = (p - 6)^2$.

So how do we "adjust" $x^2 + bx$ to make it a perfect square? We do it by adding a number. For example, to make $x^2 + 6x$ a perfect square we would add 9. But how do we know to add 9? If we divide the coefficient on the first-degree term, 6, by 2, and then square the result, we obtain 9. This approach works in general.

> **Completing the Square**
>
> Identify the coefficient of the first-degree term. Multiply this coefficient by 1/2 and then square the result. That is, determine the value of b in $x^2 + bx$ and compute $\left(\dfrac{1}{2}b\right)^2$.

EXAMPLE 19 **Completing the Square**

Determine the number that must be added to each expression to complete the square. Then factor the expression.

Start	Add	Result	Factored Form
$y^2 + 8y$	$\left(\dfrac{1}{2} \cdot 8\right)^2 = 16$	$y^2 + 8y + 16$	$(y + 4)^2$
$x^2 + 12x$	$\left(\dfrac{1}{2} \cdot 12\right)^2 = 36$	$x^2 + 12x + 36$	$(x + 6)^2$
$a^2 - 20a$	$\left(\dfrac{1}{2} \cdot (-20)\right)^2 = 100$	$a^2 - 20a + 100$	$(a - 10)^2$
$p^2 - 5p$	$\left(\dfrac{1}{2} \cdot (-5)\right)^2 = \dfrac{25}{4}$	$p^2 - 5p + \dfrac{25}{4}$	$\left(p - \dfrac{5}{2}\right)^2$

Notice that the factored form of a perfect square is either

$$x^2 + bx + \left(\frac{b}{2}\right)^2 = \left(x + \frac{b}{2}\right)^2 \quad \text{or} \quad x^2 - bx + \left(\frac{b}{2}\right)^2 = \left(x - \frac{b}{2}\right)^2$$

NOW WORK PROBLEM 117.

Figure 19

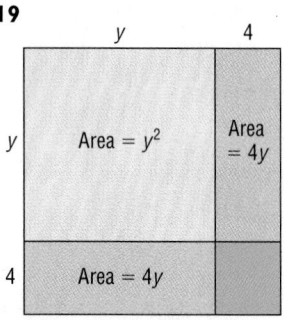

Are you wondering why we call making an expression a perfect square "completing the square"? Look at the square in Figure 19. It's area is $(y + 4)^2$. The yellow area is y^2 and each orange area is $4y$ (for a total area of $8y$). The sum of these areas is $y^2 + 8y$. To complete the square we need to add the area of the green region: $4 \cdot 4 = 16$. As a result, $y^2 + 8y + 16 = (y + 4)^2$.

A.4 Assess Your Understanding

Concepts and Vocabulary

1. If factored completely, $3x^3 - 12x =$ _____.

2. If a polynomial cannot be written as the product of two other polynomials (excluding 1 and -1), then the polynomial is said to be _____.

3. *True or False:* The polynomial $x^2 + 4$ is prime.

4. *True or False:* $3x^3 - 2x^2 - 6x + 4 = (3x - 2)(x^2 + 2)$.

Skill Building

In Problems 5–14, factor each polynomial by removing the common monomial factor.

5. $3x + 6$
6. $7x - 14$
7. $ax^2 + a$
8. $ax - a$
9. $x^3 + x^2 + x$
10. $x^3 - x^2 + x$
11. $2x^2 - 2x$
12. $3x^2 - 3x$
13. $3x^2y - 6xy^2 + 12xy$
14. $60x^2y - 48xy^2 + 72x^3y$

In Problems 15–22, factor the difference of two squares.

15. $x^2 - 1$
16. $x^2 - 4$
17. $4x^2 - 1$
18. $9x^2 - 1$
19. $x^2 - 16$
20. $x^2 - 25$
21. $25x^2 - 4$
22. $36x^2 - 9$

In Problems 23–32, factor the perfect squares.

23. $x^2 + 2x + 1$ **24.** $x^2 - 4x + 4$ **25.** $x^2 + 4x + 4$ **26.** $x^2 - 2x + 1$ **27.** $x^2 - 10x + 25$

28. $x^2 + 10x + 25$ **29.** $4x^2 + 4x + 1$ **30.** $9x^2 + 6x + 1$ **31.** $16x^2 + 8x + 1$ **32.** $25x^2 + 10x + 1$

In Problems 33–38, factor the sum or difference of two cubes.

33. $x^3 - 27$ **34.** $x^3 + 125$ **35.** $x^3 + 27$ **36.** $27 - 8x^3$ **37.** $8x^3 + 27$ **38.** $64 - 27x^3$

In Problems 39–50, factor each polynomial.

39. $x^2 + 5x + 6$ **40.** $x^2 + 6x + 8$ **41.** $x^2 + 7x + 6$ **42.** $x^2 + 9x + 8$

43. $x^2 + 7x + 10$ **44.** $x^2 + 11x + 10$ **45.** $x^2 - 10x + 16$ **46.** $x^2 - 17x + 16$

47. $x^2 - 7x - 8$ **48.** $x^2 - 2x - 8$ **49.** $x^2 + 7x - 8$ **50.** $x^2 + 2x - 8$

In Problems 51–56, factor by grouping.

51. $2x^2 + 4x + 3x + 6$ **52.** $3x^2 - 3x + 2x - 2$ **53.** $2x^2 - 4x + x - 2$

54. $3x^2 + 6x - x - 2$ **55.** $6x^2 + 9x + 4x + 6$ **56.** $9x^2 - 6x + 3x - 2$

In Problems 57–68, factor each polynomial.

57. $3x^2 + 4x + 1$ **58.** $2x^2 + 3x + 1$ **59.** $2z^2 + 5z + 3$ **60.** $6z^2 + 5z + 1$

61. $3x^2 + 2x - 8$ **62.** $3x^2 + 10x + 8$ **63.** $3x^2 - 2x - 8$ **64.** $3x^2 - 10x + 8$

65. $3x^2 + 14x + 8$ **66.** $3x^2 - 14x + 8$ **67.** $3x^2 + 10x - 8$ **68.** $3x^2 - 10x - 8$

In Problems 69–116, factor completely each polynomial. If the polynomial cannot be factored, say it is prime.

69. $x^2 - 36$ **70.** $x^2 - 9$ **71.** $2 - 8x^2$ **72.** $3 - 27x^2$

73. $x^2 + 7x + 10$ **74.** $x^2 + 5x + 4$ **75.** $x^2 - 10x + 21$ **76.** $x^2 - 6x + 8$

77. $4x^2 - 8x + 32$ **78.** $3x^2 - 12x + 15$ **79.** $x^2 + 4x + 16$ **80.** $x^2 + 12x + 36$

81. $15 + 2x - x^2$ **82.** $14 + 6x - x^2$ **83.** $3x^2 - 12x - 36$ **84.** $x^3 + 8x^2 - 20x$

85. $y^4 + 11y^3 + 30y^2$ **86.** $3y^3 - 18y^2 - 48y$ **87.** $4x^2 + 12x + 9$ **88.** $9x^2 - 12x + 4$

89. $6x^2 + 8x + 2$ **90.** $8x^2 + 6x - 2$ **91.** $x^4 - 81$ **92.** $x^4 - 1$

93. $x^6 - 2x^3 + 1$ **94.** $x^6 + 2x^3 + 1$ **95.** $x^7 - x^5$ **96.** $x^8 - x^5$

97. $16x^2 + 24x + 9$ **98.** $9x^2 - 24x + 16$ **99.** $5 + 16x - 16x^2$ **100.** $5 + 11x - 16x^2$

101. $4y^2 - 16y + 15$ **102.** $9y^2 + 9y - 4$ **103.** $1 - 8x^2 - 9x^4$ **104.** $4 - 14x^2 - 8x^4$

105. $x(x + 3) - 6(x + 3)$ **106.** $5(3x - 7) + x(3x - 7)$ **107.** $(x + 2)^2 - 5(x + 2)$ **108.** $(x - 1)^2 - 2(x - 1)$

109. $(3x - 2)^3 - 27$ **110.** $(5x + 1)^3 - 1$ **111.** $3(x^2 + 10x + 25) - 4(x + 5)$

112. $7(x^2 - 6x + 9) + 5(x - 3)$ **113.** $x^3 + 2x^2 - x - 2$ **114.** $x^3 - 3x^2 - x + 3$

115. $x^4 - x^3 + x - 1$ **116.** $x^4 + x^3 + x + 1$

In Problems 117–122, determine the number that should be added to complete the square of each expression. Then factor each expression.

117. $x^2 + 10x$ **118.** $p^2 + 14p$ **119.** $y^2 - 6y$

120. $x^2 - 4x$ **121.** $x^2 - \frac{1}{2}x$ **122.** $x^2 + \frac{1}{3}x$

Applications and Extensions

In Problems 123–132, expressions that occur in calculus are given. Factor completely each expression.

123. $2(3x + 4)^2 + (2x + 3) \cdot 2(3x + 4) \cdot 3$ **124.** $5(2x + 1)^2 + (5x - 6) \cdot 2(2x + 1) \cdot 2$

125. $2x(2x + 5) + x^2 \cdot 2$ **126.** $3x^2(8x - 3) + x^3 \cdot 8$

127. $2(x + 3)(x - 2)^3 + (x + 3)^2 \cdot 3(x - 2)^2$ **128.** $4(x + 5)^3(x - 1)^2 + (x + 5)^4 \cdot 2(x - 1)$

129. $(4x - 3)^2 + x \cdot 2(4x - 3) \cdot 4$ **130.** $3x^2(3x + 4)^2 + x^3 \cdot 2(3x + 4) \cdot 3$

131. $2(3x - 5) \cdot 3(2x + 1)^3 + (3x - 5)^2 \cdot 3(2x + 1)^2 \cdot 2$ **132.** $3(4x + 5)^2 \cdot 4(5x + 1)^2 + (4x + 5)^3 \cdot 2(5x + 1) \cdot 5$

133. Show that $x^2 + 4$ is prime. **134.** Show that $x^2 + x + 1$ is prime.

Discussion and Writing

135. Make up a polynomial that factors into a perfect square.

136. Explain to a fellow student what you look for first when presented with a factoring problem. What do you do next?

A.5 Synthetic Division

OBJECTIVE 1 Divide Polynomials Using Synthetic Division

1 Divide Polynomials Using Synthetic Division

To find the quotient as well as the remainder when a polynomial of degree 1 or higher is divided by $x - c$, a shortened version of long division, called **synthetic division**, makes the task simpler.

To see how synthetic division works, we will use long division to divide the polynomial $2x^3 - x^2 + 3$ by $x - 3$.

$$
\begin{array}{r}
2x^2 + 5x\ + 15 \quad \leftarrow Quotient \\
x - 3\overline{)2x^3 - x^2 + 3} \\
\underline{2x^3 - 6x^2} \\
5x^2 \\
\underline{5x^2 - 15x} \\
15x + 3 \\
\underline{15x - 45} \\
48 \quad \leftarrow Remainder
\end{array}
$$

✔ **CHECK:** (Divisor) · (Quotient) + Remainder

$$
\begin{aligned}
&= (x - 3)(2x^2 + 5x + 15) + 48 \\
&= 2x^3 + 5x^2 + 15x - 6x^2 - 15x - 45 + 48 \\
&= 2x^3 - x^2 + 3
\end{aligned}
$$

The process of synthetic division arises from rewriting the long division in a more compact form, using simpler notation. For example, in the long division above, the terms in blue are not really necessary because they are identical to the terms directly above them. With these terms removed, we have

$$
\begin{array}{r}
2x^2 + 5x\ + 15 \\
x - 3\overline{)2x^3 - x^2 + 3} \\
\underline{- 6x^2} \\
5x^2 \\
\underline{- 15x} \\
15x \\
\underline{- 45} \\
48
\end{array}
$$

Most of the x's that appear in this process can also be removed, provided that we are careful about positioning each coefficient. In this regard, we will need to use 0 as the coefficient of x in the dividend, because that power of x is missing. Now we have

$$
\begin{array}{r}
2x^2 + 5x + 15 \\
x - 3\overline{)2 - 1 0 3} \\
\underline{- 6} \\
5 \\
\underline{- 15} \\
15 \\
\underline{- 45} \\
48
\end{array}
$$

We can make this display more compact by moving the lines up until the numbers in blue align horizontally.

$$
\begin{array}{r}
2x^2 + 5x + 15 \qquad \text{Row 1}\\
x-3\overline{)2 \quad -1 \quad\;\; 0 \quad\;\; 3} \quad \text{Row 2}\\
-6 \;\; -15 \;\; -45 \quad \text{Row 3}\\
\hline
\bigcirc \quad\; 5 \quad\; 15 \quad\; 48 \quad \text{Row 4}
\end{array}
$$

Because the leading coefficient of the divisor is always 1, we know that the leading coefficient of the dividend will also be the leading coefficient of the quotient. So we place the leading coefficient of the quotient, 2, in the circled position. Now, the first three numbers in row 4 are precisely the coefficients of the quotient, and the last number in row 4 is the remainder. Thus, row 1 is not really needed, so we can compress the process to three rows, where the bottom row contains both the coefficients of the quotient and the remainder.

$$
\begin{array}{r}
x-3\overline{)2 \quad -1 \quad\;\; 0 \quad\;\; 3} \quad \text{Row 1}\\
-6 \;\; -15 \;\; -45 \quad \text{Row 2 (subtract)}\\
\hline
2 \quad\; 5 \quad\; 15 \quad\; 48 \quad \text{Row 3}
\end{array}
$$

Recall that the entries in row 3 are obtained by subtracting the entries in row 2 from those in row 1. Rather than subtracting the entries in row 2, we can change the sign of each entry and add. With this modification, our display will look like this:

$$
\begin{array}{r}
x-3\overline{)2 \quad -1 \quad\;\; 0 \quad\;\; 3} \quad \text{Row 1}\\
6 \;\; 15 \;\; 45 \quad \text{Row 2 (add)}\\
\hline
2 \quad\; 5 \quad\; 15 \quad\; 48 \quad \text{Row 3}
\end{array}
$$

Notice that the entries in row 2 are three times the prior entries in row 3. Our last modification to the display replaces the $x-3$ by 3. The entries in row 3 give the quotient and the remainder, as shown next.

$$
\begin{array}{r}
3\overline{)2 \quad -1 \quad\;\; 0 \quad\;\; 3} \quad \text{Row 1}\\
6 \;\; 15 \;\; 45 \quad \text{Row 2 (add)}\\
\hline
2 \quad\; 5 \quad\; 15 \quad\; 48 \quad \text{Row 3}
\end{array}
$$

Quotient Remainder

$$2x^2 + 5x + 15 \qquad 48$$

Let's go through an example step by step.

EXAMPLE 1

Using Synthetic Division to Find the Quotient and Remainder

Use synthetic division to find the quotient and remainder when

$$x^3 - 4x^2 - 5 \quad \text{is divided by} \quad x - 3$$

Solution **STEP 1:** Write the dividend in descending powers of x. Then copy the coefficients, remembering to insert a 0 for any missing powers of x.

$$1 \quad -4 \quad 0 \quad -5 \quad \text{Row 1}$$

STEP 2: Insert the usual division symbol. In synthetic division, the divisor is of the form $x - c$, and c is the number placed to the left of the division symbol. Here, since the divisor is $x - 3$, we insert 3 to the left of the division symbol.

$$3\overline{)1 \quad -4 \quad 0 \quad -5} \quad \text{Row 1}$$

STEP 3: Bring the 1 down two rows, and enter it in row 3.

$$
\begin{array}{r}
3\overline{)1 \quad -4 \quad 0 \quad -5} \quad \text{Row 1}\\
\downarrow \qquad\qquad\qquad \text{Row 2}\\
\hline
1 \qquad\qquad\qquad\;\; \text{Row 3}
\end{array}
$$

A.5 Assess Your Understanding

Concepts and Vocabulary

1. To check division, the expression that is being divided, the dividend, should equal the product of the _____ and the _____ plus the _____.

2. To divide $2x^3 - 5x + 1$ by $x + 3$ using synthetic division, the first step is to write _____)‾‾‾‾

3. *True or False:* In using synthetic division, the divisor is always a polynomial of degree 1, whose leading coefficient is 1.

4. *True or False:* $-2\overline{)\begin{array}{rrrr} 5 & 3 & 2 & 1 \\ & -10 & 14 & -32 \\ \hline 5 & -7 & 16 & -31 \end{array}}$ means $\dfrac{5x^3 + 3x^2 + 2x + 1}{x + 2} = 5x^2 - 7x + 16 + \dfrac{-31}{x + 2}$

Skill Building

In Problems 5–16, use synthetic division to find the quotient and remainder when:

5. $x^3 - x^2 + 2x + 4$ is divided by $x - 2$

6. $x^3 + 2x^2 - 3x + 1$ is divided by $x + 1$

7. $3x^3 + 2x^2 - x + 3$ is divided by $x - 3$

8. $-4x^3 + 2x^2 - x + 1$ is divided by $x + 2$

9. $x^5 - 4x^3 + x$ is divided by $x + 3$

10. $x^4 + x^2 + 2$ is divided by $x - 2$

11. $4x^6 - 3x^4 + x^2 + 5$ is divided by $x - 1$

12. $x^5 + 5x^3 - 10$ is divided by $x + 1$

13. $0.1x^3 + 0.2x$ is divided by $x + 1.1$

14. $0.1x^2 - 0.2$ is divided by $x + 2.1$

15. $x^5 - 1$ is divided by $x - 1$

16. $x^5 + 1$ is divided by $x + 1$

In Problems 17–26, use synthetic division to determine whether $x - c$ is a factor of the given polynomial.

17. $4x^3 - 3x^2 - 8x + 4$; $x - 2$

18. $-4x^3 + 5x^2 + 8$; $x + 3$

19. $3x^4 - 6x^3 - 5x + 10$; $x - 2$

20. $4x^4 - 15x^2 - 4$; $x - 2$

21. $3x^6 + 82x^3 + 27$; $x + 3$

22. $2x^6 - 18x^4 + x^2 - 9$; $x + 3$

23. $4x^6 - 64x^4 + x^2 - 15$; $x + 4$

24. $x^6 - 16x^4 + x^2 - 16$; $x + 4$

25. $2x^4 - x^3 + 2x - 1$; $x - \dfrac{1}{2}$

26. $3x^4 + x^3 - 3x + 1$; $x + \dfrac{1}{3}$

Applications and Extensions

27. Find the sum of a, b, c, and d if

$$\frac{x^3 - 2x^2 + 3x + 5}{x + 2} = ax^2 + bx + c + \frac{d}{x + 2}$$

Discussion and Writing

28. When dividing a polynomial by $x - c$, do you prefer to use long division or synthetic division? Does the value of c make a difference to you in choosing? Give reasons.

A.6 Rational Expressions

OBJECTIVES 1 Reduce a Rational Expression to Lowest Terms

2 Multiply and Divide Rational Expressions

3 Add and Subtract Rational Expressions

4 Use the Least Common Multiple Method

5 Simplify Complex Rational Expressions

1 Reduce a Rational Expression to Lowest Terms

If we form the quotient of two polynomials, the result is called a **rational expression**. Some examples of rational expressions are

(a) $\dfrac{x^3 + 1}{x}$ (b) $\dfrac{3x^2 + x - 2}{x^2 + 5}$ (c) $\dfrac{x}{x^2 - 1}$ (d) $\dfrac{xy^2}{(x - y)^2}$

Expressions (a), (b), and (c) are rational expressions in one variable, x, whereas (d) is a rational expression in two variables, x and y.

Rational expressions are described in the same manner as rational numbers. In expression (a), the polynomial $x^3 + 1$ is called the **numerator**, and x is called the **denominator**. When the numerator and denominator of a rational expression contain no common factors (except 1 and -1), we say that the rational expression is **reduced to lowest terms**, or **simplified**.

The polynomial in the denominator of a rational expression cannot be equal to 0 because division by 0 is not defined. For example, for the expression $\dfrac{x^3 + 1}{x}$, x cannot take on the value 0. The domain of the variable x is $\{x \mid x \neq 0\}$.

A rational expression is reduced to lowest terms by factoring completely the numerator and the denominator and canceling any common factors by using the Cancellation Property:

$$\frac{a\cancel{c}}{b\cancel{c}} = \frac{a}{b} \qquad \text{if } b \neq 0, c \neq 0 \tag{1}$$

| EXAMPLE 1 | **Reducing a Rational Expression to Lowest Terms** |

Reduce to lowest terms: $\dfrac{x^2 + 4x + 4}{x^2 + 3x + 2}$

Solution We begin by factoring the numerator and the denominator.

$$x^2 + 4x + 4 = (x + 2)(x + 2)$$
$$x^2 + 3x + 2 = (x + 2)(x + 1)$$

WARNING
Apply the Cancellation Property only to rational expressions written in factored form. Be sure to cancel only common factors! ∎

Since a common factor, $x + 2$, appears, the original expression is not in lowest terms. To reduce it to lowest terms, we use the Cancellation Property:

$$\frac{x^2 + 4x + 4}{x^2 + 3x + 2} = \frac{\cancel{(x + 2)}(x + 2)}{\cancel{(x + 2)}(x + 1)} = \frac{x + 2}{x + 1} \qquad x \neq -2, -1 \qquad \blacksquare$$

| EXAMPLE 2 | **Reducing Rational Expressions to Lowest Terms** |

Reduce each rational expression to lowest terms.

(a) $\dfrac{x^3 - 8}{x^3 - 2x^2}$

(b) $\dfrac{8 - 2x}{x^2 - x - 12}$

Solution (a) $\dfrac{x^3 - 8}{x^3 - 2x^2} = \dfrac{\cancel{(x - 2)}(x^2 + 2x + 4)}{x^2\cancel{(x - 2)}} = \dfrac{x^2 + 2x + 4}{x^2} \qquad x \neq 0, 2$

(b) $\dfrac{8 - 2x}{x^2 - x - 12} = \dfrac{2(4 - x)}{(x - 4)(x + 3)} = \dfrac{2(-1)\cancel{(x - 4)}}{\cancel{(x - 4)}(x + 3)} = \dfrac{-2}{x + 3} \qquad x \neq -3, 4 \quad \blacksquare$

NOW WORK PROBLEM 5.

2 Multiply and Divide Rational Expressions

The rules for multiplying and dividing rational expressions are the same as the rules for multiplying and dividing rational numbers. If $\dfrac{a}{b}$ and $\dfrac{c}{d}$, $b \neq 0, d \neq 0$, are two rational expressions, then

$$\frac{a}{b} \cdot \frac{c}{d} = \frac{ac}{bd} \qquad \text{if } b \neq 0, d \neq 0 \tag{2}$$

$$\frac{\dfrac{a}{b}}{\dfrac{c}{d}} = \frac{a}{b} \cdot \frac{d}{c} = \frac{ad}{bc} \qquad \text{if } b \neq 0, c \neq 0, d \neq 0 \tag{3}$$

In using equations (2) and (3) with rational expressions, be sure first to factor each polynomial completely so that common factors can be canceled. Leave your answer in factored form.

EXAMPLE 3

Multiplying and Dividing Rational Expressions

Perform the indicated operation and simplify the result. Leave your answer in factored form.

(a) $\dfrac{x^2 - 2x + 1}{x^3 + x} \cdot \dfrac{4x^2 + 4}{x^2 + x - 2}$

(b) $\dfrac{\dfrac{x + 3}{x^2 - 4}}{\dfrac{x^2 - x - 12}{x^3 - 8}}$

Solution

(a)
$$\frac{x^2 - 2x + 1}{x^3 + x} \cdot \frac{4x^2 + 4}{x^2 + x - 2} = \frac{(x - 1)^2}{x(x^2 + 1)} \cdot \frac{4(x^2 + 1)}{(x + 2)(x - 1)}$$

$$= \frac{(x - 1)^2 (4)\cancel{(x^2 + 1)}}{x\cancel{(x^2 + 1)}(x + 2)\cancel{(x - 1)}}$$

$$= \frac{4(x - 1)}{x(x + 2)} \qquad x \neq -2, 0, 1$$

(b)
$$\frac{\dfrac{x + 3}{x^2 - 4}}{\dfrac{x^2 - x - 12}{x^3 - 8}} = \frac{x + 3}{x^2 - 4} \cdot \frac{x^3 - 8}{x^2 - x - 12}$$

$$= \frac{x + 3}{(x - 2)(x + 2)} \cdot \frac{(x - 2)(x^2 + 2x + 4)}{(x - 4)(x + 3)}$$

$$= \frac{\cancel{(x + 3)}\cancel{(x - 2)}(x^2 + 2x + 4)}{\cancel{(x - 2)}(x + 2)(x - 4)\cancel{(x + 3)}}$$

$$= \frac{x^2 + 2x + 4}{(x + 2)(x - 4)} \qquad x \neq -3, -2, 2, 4$$

∎

NOW WORK PROBLEMS 17 AND 25.

3 Add and Subtract Rational Expressions

The rules for adding and subtracting rational expressions are the same as the rules for adding and subtracting rational numbers. Thus, if the denominators of

In Words

To add (or subract) two rational expressions with the same denominator, keep the common denominator and add (or subtract) the numerators.

two rational expressions to be added (or subtracted) are equal, we add (or subtract) the numerators and keep the common denominator.

> If $\dfrac{a}{b}$ and $\dfrac{c}{b}$ are two rational expressions, then
>
> $$\frac{a}{b}+\frac{c}{b}=\frac{a+c}{b} \qquad \frac{a}{b}-\frac{c}{b}=\frac{a-c}{b} \qquad \text{if } b \neq 0 \qquad \textbf{(4)}$$

EXAMPLE 4 **Adding and Subtracting Rational Expressions with Equal Denominators**

Perform the indicated operation and simplify the result. Leave your answer in factored form.

(a) $\dfrac{2x^2-4}{2x+5}+\dfrac{x+3}{2x+5} \qquad x \neq -\dfrac{5}{2}$ (b) $\dfrac{x}{x-3}-\dfrac{3x+2}{x-3} \qquad x \neq 3$

Solution (a) $\dfrac{2x^2-4}{2x+5}+\dfrac{x+3}{2x+5}=\dfrac{(2x^2-4)+(x+3)}{2x+5}$

$$=\frac{2x^2+x-1}{2x+5}=\frac{(2x-1)(x+1)}{2x+5}$$

(b) $\dfrac{x}{x-3}-\dfrac{3x+2}{x-3}=\dfrac{x-(3x+2)}{x-3}=\dfrac{x-3x-2}{x-3}$

$$=\frac{-2x-2}{x-3}=\frac{-2(x+1)}{x-3}$$ ∎

EXAMPLE 5 **Adding Rational Expressions Whose Denominators Are Additive Inverses of Each Other**

Perform the indicated operation and simplify the result. Leave your answer in factored form.

$$\frac{2x}{x-3}+\frac{5}{3-x} \qquad x \neq 3$$

Solution Notice that the denominators of the two rational expressions are different. However, the denominator of the second expression is the additive inverse of the denominator of the first. That is,

$$3-x=-x+3=-1\cdot(x-3)=-(x-3)$$

Then

$$\frac{2x}{x-3}+\frac{5}{3-x}=\frac{2x}{x-3}+\frac{5}{-(x-3)}=\frac{2x}{x-3}+\frac{-5}{x-3}$$

$$\underset{\uparrow}{3-x=-(x-3)} \qquad \underset{\uparrow}{\frac{a}{-b}=\frac{-a}{b}}$$

$$=\frac{2x+(-5)}{x-3}=\frac{2x-5}{x-3}$$ ∎

NOW WORK PROBLEMS 37 AND 43.

If the denominators of two rational expressions to be added or subtracted are not equal, we can use the general formulas for adding and subtracting quotients.

$$\frac{a}{b} + \frac{c}{d} = \frac{a \cdot d}{b \cdot d} + \frac{b \cdot c}{b \cdot d} = \frac{ad + bc}{bd} \qquad \text{if } b \neq 0, d \neq 0 \qquad \textbf{(5a)}$$

$$\frac{a}{b} - \frac{c}{d} = \frac{a \cdot d}{b \cdot d} - \frac{b \cdot c}{b \cdot d} = \frac{ad - bc}{bd} \qquad \text{if } b \neq 0, d \neq 0 \qquad \textbf{(5b)}$$

EXAMPLE 6

Adding and Subtracting Rational Expressions with Unequal Denominators

Perform the indicated operation and simplify the result. Leave your answer in factored form.

(a) $\dfrac{x-3}{x+4} + \dfrac{x}{x-2} \qquad x \neq -4, 2$ (b) $\dfrac{x^2}{x^2-4} - \dfrac{1}{x} \qquad x \neq -2, 0, 2$

Solution (a) $\dfrac{x-3}{x+4} + \dfrac{x}{x-2} \underset{\underset{(5a)}{\uparrow}}{=} \dfrac{x-3}{x+4} \cdot \dfrac{x-2}{x-2} + \dfrac{x+4}{x+4} \cdot \dfrac{x}{x-2}$

$$= \frac{(x-3)(x-2) + (x+4)(x)}{(x+4)(x-2)}$$

$$= \frac{x^2 - 5x + 6 + x^2 + 4x}{(x+4)(x-2)} = \frac{2x^2 - x + 6}{(x+4)(x-2)}$$

(b) $\dfrac{x^2}{x^2-4} - \dfrac{1}{x} \underset{\underset{(5b)}{\uparrow}}{=} \dfrac{x^2}{x^2-4} \cdot \dfrac{x}{x} - \dfrac{x^2-4}{x^2-4} \cdot \dfrac{1}{x} = \dfrac{x^2(x) - (x^2-4)(1)}{(x^2-4)(x)}$

$$= \frac{x^3 - x^2 + 4}{(x-2)(x+2)(x)}$$

∎

NOW WORK PROBLEM 47.

4 Use the Least Common Multiple Method

If the denominators of two rational expressions to be added (or subtracted) have common factors, we usually do not use the general rules given by equations (5a) and (5b). Just as with fractions, we apply the **least common multiple (LCM) method**. The LCM method uses the polynomial of least degree that has each denominator polynomial as a factor.

The LCM Method for Adding or Subtracting Rational Expressions

The Least Common Multiple (LCM) method requires four steps:

STEP 1: Factor completely the polynomial in the denominator of each rational expression.

STEP 2: The LCM of the denominator is the product of each of these factors raised to a power equal to the greatest number of times that the factor occurs in the polynomials.

STEP 3: Write each rational expression using the LCM as the common denominator.

STEP 4: Add or subtract the rational expressions using equation (4).

Let's work an example that only requires Steps 1 and 2.

EXAMPLE 7	**Finding the Least Common Multiple**

Find the least common multiple of the following pair of polynomials:

$$x(x - 1)^2(x + 1) \quad \text{and} \quad 4(x - 1)(x + 1)^3$$

Solution **STEP 1:** The polynomials are already factored completely as

$$x(x - 1)^2(x + 1) \quad \text{and} \quad 4(x - 1)(x + 1)^3$$

STEP 2: Start by writing the factors of the left-hand polynomial. (Or you could start with the one on the right.)

$$x(x - 1)^2(x + 1)$$

Now look at the right-hand polynomial. Its first factor, 4, does not appear in our list, so we insert it.

$$4x(x - 1)^2(x + 1)$$

The next factor, $x - 1$, is already in our list, so no change is necessary. The final factor is $(x + 1)^3$. Since our list has $x + 1$ to the first power only, we replace $x + 1$ in the list by $(x + 1)^3$. The LCM is

$$4x(x - 1)^2(x + 1)^3$$ ■

Notice that the LCM is, in fact, the polynomial of least degree that contains $x(x - 1)^2(x + 1)$ and $4(x - 1)(x + 1)^3$ as factors.

NOW WORK PROBLEM 53.

EXAMPLE 8	**Using the Least Common Multiple to Add Rational Expressions**

Perform the indicated operation and simplify the result. Leave your answer in factored form.

$$\frac{x}{x^2 + 3x + 2} + \frac{2x - 3}{x^2 - 1} \quad x \neq -2, -1, 1$$

Solution **STEP 1:** Factor completely the polynomials in the denominators.

$$x^2 + 3x + 2 = (x + 2)(x + 1)$$
$$x^2 - 1 = (x - 1)(x + 1)$$

STEP 2: The LCM is $(x + 2)(x + 1)(x - 1)$. Do you see why?

STEP 3: Write each rational expression using the LCM as the denominator.

$$\frac{x}{x^2 + 3x + 2} = \frac{x}{(x + 2)(x + 1)} = \frac{x}{(x + 2)(x + 1)} \cdot \frac{x - 1}{x - 1} = \frac{x(x - 1)}{(x + 2)(x + 1)(x - 1)}$$

Multiply numerator and denominator by x − 1 to get the LCM in the denominator.

$$\frac{2x - 3}{x^2 - 1} = \frac{2x - 3}{(x - 1)(x + 1)} = \frac{2x - 3}{(x - 1)(x + 1)} \cdot \frac{x + 2}{x + 2} = \frac{(2x - 3)(x + 2)}{(x - 1)(x + 1)(x + 2)}$$

Multiply numerator and denominator by x + 2 to get the LCM in the denominator.

STEP 4: Now we can add by using equation (4).

$$\frac{x}{x^2+3x+2}+\frac{2x-3}{x^2-1}=\frac{x(x-1)}{(x+2)(x+1)(x-1)}+\frac{(2x-3)(x+2)}{(x+2)(x+1)(x-1)}$$

$$=\frac{(x^2-x)+(2x^2+x-6)}{(x+2)(x+1)(x-1)}$$

$$=\frac{3x^2-6}{(x+2)(x+1)(x-1)}=\frac{3(x^2-2)}{(x+2)(x+1)(x-1)}$$ ∎

EXAMPLE 9 **Using the Least Common Multiple to Subtract Rational Expressions**

Perform the indicated operations and simplify the result. Leave your answer in factored form.

$$\frac{3}{x^2+x}-\frac{x+4}{x^2+2x+1}\qquad x\neq -1,0$$

Solution **STEP 1:** Factor completely the polynomials in the denominators.

$$x^2+x=x(x+1)$$
$$x^2+2x+1=(x+1)^2$$

STEP 2: The LCM is $x(x+1)^2$.

STEP 3: Write each rational expression using the LCM as the denominator.

$$\frac{3}{x^2+x}=\frac{3}{x(x+1)}=\frac{3}{x(x+1)}\cdot\frac{x+1}{x+1}=\frac{3(x+1)}{x(x+1)^2}$$

$$\frac{x+4}{x^2+2x+1}=\frac{x+4}{(x+1)^2}=\frac{x+4}{(x+1)^2}\cdot\frac{x}{x}=\frac{x(x+4)}{x(x+1)^2}$$

STEP 4: Subtract, using equation (4).

$$\frac{3}{x^2+x}-\frac{x+4}{x^2+2x+1}=\frac{3(x+1)}{x(x+1)^2}-\frac{x(x+4)}{x(x+1)^2}$$

$$=\frac{3(x+1)-x(x+4)}{x(x+1)^2}$$

$$=\frac{3x+3-x^2-4x}{x(x+1)^2}$$

$$=\frac{-x^2-x+3}{x(x+1)^2}$$ ∎

NOW WORK PROBLEM **63**.

5 Simplify Complex Rational Expressions

When sums and/or differences of rational expressions appear as the numerator and/or denominator of a quotient, the quotient is called a **complex rational expression.**[*] For example,

$$\frac{1+\dfrac{1}{x}}{1-\dfrac{1}{x}}\quad\text{and}\quad\frac{\dfrac{x^2}{x^2-4}-3}{\dfrac{x-3}{x+2}-1}$$

[*]Some texts use the term **complex fraction**.

are complex rational expressions. To **simplify** a complex rational expression means to write it as a rational expression reduced to lowest terms. This can be accomplished in either of two ways.

Simplifying a Complex Rational Expression

METHOD 1: Treat the numerator and denominator of the complex rational expression separately, performing whatever operations are indicated and simplifying the results. Follow this by simplifying the resulting rational expression.

METHOD 2: Find the LCM of the denominators of all rational expressions that appear in the complex rational expression. Multiply the numerator and denominator of the complex rational expression by the LCM and simplify the result.

We will use both methods in the next example. By carefully studying each method, you can discover situations in which one method may be easier to use than the other.

EXAMPLE 10 | **Simplifying a Complex Rational Expression**

Simplify: $\dfrac{\dfrac{1}{2}+\dfrac{3}{x}}{\dfrac{x+3}{4}}$ $x \neq -3, 0$

Solution *Method 1:* First, we perform the indicated operation in the numerator, and then we divide.

$$\frac{\dfrac{1}{2}+\dfrac{3}{x}}{\dfrac{x+3}{4}} = \frac{\dfrac{1\cdot x + 2\cdot 3}{2\cdot x}}{\dfrac{x+3}{4}} = \frac{\dfrac{x+6}{2x}}{\dfrac{x+3}{4}} = \frac{x+6}{2x}\cdot\frac{4}{x+3}$$

$$\underset{\text{Rule for adding quotients}}{\uparrow}\qquad\qquad\underset{\text{Rule for dividing quotients}}{\uparrow}$$

$$= \frac{(x+6)\cdot 4}{2\cdot x\cdot(x+3)} = \frac{2\cdot 2\cdot(x+6)}{2\cdot x\cdot(x+3)} = \frac{2(x+6)}{x(x+3)}$$

$$\underset{\text{Rule for multiplying quotients}}{\uparrow}$$

Method 2: The rational expressions that appear in the complex rational expression are

$$\frac{1}{2}, \quad \frac{3}{x}, \quad \frac{x+3}{4}$$

The LCM of their denominators is $4x$. We multiply the numerator and denominator of the complex rational expression by $4x$ and then simplify.

$$\frac{\dfrac{1}{2}+\dfrac{3}{x}}{\dfrac{x+3}{4}} = \frac{4x\cdot\left(\dfrac{1}{2}+\dfrac{3}{x}\right)}{4x\cdot\left(\dfrac{x+3}{4}\right)} = \frac{4x\cdot\dfrac{1}{2}+4x\cdot\dfrac{3}{x}}{\dfrac{4x\cdot(x+3)}{4}}$$

$$\underset{\substack{\text{Multiply the}\\\text{numerator and}\\\text{denominator by } 4x.}}{\uparrow}\qquad\qquad\underset{\substack{\text{Distributive Property}\\\text{in numerator}}}{\uparrow}$$

$$= \frac{2\cdot 2x\cdot\dfrac{1}{2}+4x\cdot\dfrac{3}{x}}{\dfrac{4x\cdot(x+3)}{4}} = \frac{2x+12}{x(x+3)} = \frac{2(x+6)}{x(x+3)}$$

$$\underset{\text{Simplify}}{\uparrow}\qquad\underset{\text{Factor}}{\uparrow}$$

∎

EXAMPLE 11 | **Simplifying a Complex Rational Expression**

Simplify: $\dfrac{\dfrac{x^2}{x-4} + 2}{\dfrac{2x-2}{x} - 1}$

Solution We will use Method 1.

$$\frac{\dfrac{x^2}{x-4} + 2}{\dfrac{2x-2}{x} - 1} = \frac{\dfrac{x^2}{x-4} + \dfrac{2(x-4)}{x-4}}{\dfrac{2x-2}{x} - \dfrac{x}{x}} = \frac{\dfrac{x^2 + 2x - 8}{x-4}}{\dfrac{2x-2-x}{x}}$$

$$= \frac{\dfrac{(x+4)(x-2)}{x-4}}{\dfrac{x-2}{x}} = \frac{(x+4)\cancel{(x-2)}}{x-4} \cdot \frac{x}{\cancel{x-2}}$$

$$= \frac{(x+4) \cdot x}{x-4} \qquad \blacksquare$$

 ✎ ➤ **NOW WORK PROBLEM 73.**

Application

EXAMPLE 12 | **Solving an Application in Electricity**

Figure 20

An electrical circuit contains two resistors connected in parallel, as shown in Figure 20. If the resistance of each is R_1 and R_2 ohms, respectively, their combined resistance R is given by the formula

$$R = \frac{1}{\dfrac{1}{R_1} + \dfrac{1}{R_2}}$$

Express R as a rational expression; that is, simplify the right-hand side of this formula. Evaluate the rational expression if $R_1 = 6$ ohms and $R_2 = 10$ ohms.

Solution We will use Method 2. If we consider 1 as the fraction $\dfrac{1}{1}$, then the rational expressions in the complex rational expression are

$$\frac{1}{1}, \quad \frac{1}{R_1}, \quad \frac{1}{R_2}$$

The LCM of the denominators is $R_1 R_2$. We multiply the numerator and denominator of the complex rational expression by $R_1 R_2$ and simplify.

$$\frac{1}{\dfrac{1}{R_1} + \dfrac{1}{R_2}} = \frac{1 \cdot R_1 R_2}{\left(\dfrac{1}{R_1} + \dfrac{1}{R_2}\right) \cdot R_1 R_2} = \frac{R_1 R_2}{\dfrac{1}{R_1} \cdot R_1 R_2 + \dfrac{1}{R_2} \cdot R_1 R_2} = \frac{R_1 R_2}{R_2 + R_1}$$

Thus,

$$R = \frac{R_1 R_2}{R_2 + R_1}$$

If $R_1 = 6$ and $R_2 = 10$, then

$$R = \frac{6 \cdot 10}{10 + 6} = \frac{60}{16} = \frac{15}{4} \quad \text{ohms} \qquad \blacksquare$$

A.6 Assess Your Understanding

Concepts and Vocabulary

1. When the numerator and denominator of a rational expression contain no common factors (except 1 and −1), the rational expression is in _____ _____.

2. LCM is an abbreviation for _____ _____ _____.

3. *True or False:* The rational expression $\dfrac{2x^3 - 4x}{x - 2}$ is reduced to lowest terms.

4. *True or False:* The LCM of $2x^3 + 6x^2$ and $6x^4 + 4x^3$ is $4x^3(x + 1)$.

Skill Building

In Problems 5–16, reduce each rational expression to lowest terms.

5. $\dfrac{3x + 9}{x^2 - 9}$

6. $\dfrac{4x^2 + 8x}{12x + 24}$

7. $\dfrac{x^2 - 2x}{3x - 6}$

8. $\dfrac{15x^2 + 24x}{3x^2}$

9. $\dfrac{24x^2}{12x^2 - 6x}$

10. $\dfrac{x^2 + 4x + 4}{x^2 - 16}$

11. $\dfrac{y^2 - 25}{2y^2 - 8y - 10}$

12. $\dfrac{3y^2 - y - 2}{3y^2 + 5y + 2}$

13. $\dfrac{x^2 + 4x - 5}{x^2 - 2x + 1}$

14. $\dfrac{x - x^2}{x^2 + x - 2}$

15. $\dfrac{x^2 + 5x - 14}{2 - x}$

16. $\dfrac{2x^2 + 5x - 3}{1 - 2x}$

In Problems 17–34, perform the indicated operation and simplify the result. Leave your answer in factored form.

17. $\dfrac{3x + 6}{5x^2} \cdot \dfrac{x}{x^2 - 4}$

18. $\dfrac{3}{2x} \cdot \dfrac{x^2}{6x + 10}$

19. $\dfrac{4x^2}{x^2 - 16} \cdot \dfrac{x - 4}{2x}$

20. $\dfrac{12}{x^2 - x} \cdot \dfrac{x^2 - 1}{4x - 2}$

21. $\dfrac{4x - 8}{-3x} \cdot \dfrac{12}{12 - 6x}$

22. $\dfrac{6x - 27}{5x} \cdot \dfrac{2}{4x - 18}$

23. $\dfrac{x^2 - 3x - 10}{x^2 + 2x - 35} \cdot \dfrac{x^2 + 4x - 21}{x^2 + 9x + 14}$

24. $\dfrac{x^2 + x - 6}{x^2 + 4x - 5} \cdot \dfrac{x^2 - 25}{x^2 + 2x - 15}$

25. $\dfrac{\dfrac{6x}{x^2 - 4}}{\dfrac{3x - 9}{2x + 4}}$

26. $\dfrac{\dfrac{12x}{5x + 20}}{\dfrac{4x^2}{x^2 - 16}}$

27. $\dfrac{\dfrac{8x}{x^2 - 1}}{\dfrac{10x}{x + 1}}$

28. $\dfrac{\dfrac{x - 2}{4x}}{\dfrac{x^2 - 4x + 4}{12x}}$

29. $\dfrac{\dfrac{4 - x}{4 + x}}{\dfrac{4x}{x^2 - 16}}$

30. $\dfrac{\dfrac{3 + x}{3 - x}}{\dfrac{x^2 - 9}{9x^3}}$

31. $\dfrac{\dfrac{x^2 + 7x + 12}{x^2 - 7x + 12}}{\dfrac{x^2 + x - 12}{x^2 - x - 12}}$

32. $\dfrac{\dfrac{x^2 + 7x + 6}{x^2 + x - 6}}{\dfrac{x^2 + 5x - 6}{x^2 + 5x + 6}}$

33. $\dfrac{\dfrac{2x^2 - x - 28}{3x^2 - x - 2}}{\dfrac{4x^2 + 16x + 7}{3x^2 + 11x + 6}}$

34. $\dfrac{\dfrac{9x^2 + 3x - 2}{12x^2 + 5x - 2}}{\dfrac{9x^2 - 6x + 1}{8x^2 - 10x - 3}}$

In Problems 35–52, perform the indicated operations and simplify the result. Leave your answer in factored form.

35. $\dfrac{x}{2} + \dfrac{5}{2}$

36. $\dfrac{3}{x} - \dfrac{6}{x}$

37. $\dfrac{x^2}{2x - 3} - \dfrac{4}{2x - 3}$

38. $\dfrac{3x^2}{2x - 1} - \dfrac{9}{2x - 1}$

39. $\dfrac{x + 1}{x - 3} + \dfrac{2x - 3}{x - 3}$

40. $\dfrac{2x - 5}{3x + 2} + \dfrac{x + 4}{3x + 2}$

41. $\dfrac{3x + 5}{2x - 1} - \dfrac{2x - 4}{2x - 1}$

42. $\dfrac{5x - 4}{3x + 4} - \dfrac{x + 1}{3x + 4}$

43. $\dfrac{4}{x - 2} + \dfrac{x}{2 - x}$

44. $\dfrac{6}{x - 1} - \dfrac{x}{1 - x}$

45. $\dfrac{4}{x - 1} - \dfrac{2}{x + 2}$

46. $\dfrac{2}{x + 5} - \dfrac{5}{x - 5}$

47. $\dfrac{x}{x + 1} + \dfrac{2x - 3}{x - 1}$

48. $\dfrac{3x}{x - 4} + \dfrac{2x}{x + 3}$

49. $\dfrac{x - 3}{x + 2} - \dfrac{x + 4}{x - 2}$

50. $\dfrac{2x - 3}{x - 1} - \dfrac{2x + 1}{x + 1}$

51. $\dfrac{x}{x^2 - 4} + \dfrac{1}{x}$

52. $\dfrac{x - 1}{x^3} + \dfrac{x}{x^2 + 1}$

In Problems 53–60, find the LCM of the given polynomials.

53. $x^2 - 4, \quad x^2 - x - 2$

54. $x^2 - x - 12, \quad x^2 - 8x + 16$

55. $x^3 - x, \quad x^2 - x$

56. $3x^2 - 27, \quad 2x^2 - x - 15$

57. $4x^3 - 4x^2 + x, \quad 2x^3 - x^2, \quad x^3$

58. $x - 3, \quad x^2 + 3x, \quad x^3 - 9x$

59. $x^3 - x, \quad x^3 - 2x^2 + x, \quad x^3 - 1$

60. $x^2 + 4x + 4, \quad x^3 + 2x^2, \quad (x + 2)^3$

In Problems 61–72, perform the indicated operations and simplify the result. Leave your answer in factored form.

61. $\dfrac{x}{x^2 - 7x + 6} - \dfrac{x}{x^2 - 2x - 24}$

62. $\dfrac{x}{x - 3} - \dfrac{x + 1}{x^2 + 5x - 24}$

63. $\dfrac{4x}{x^2 - 4} - \dfrac{2}{x^2 + x - 6}$

64. $\dfrac{3x}{x - 1} - \dfrac{x - 4}{x^2 - 2x + 1}$

65. $\dfrac{3}{(x - 1)^2(x + 1)} + \dfrac{2}{(x - 1)(x + 1)^2}$

66. $\dfrac{2}{(x + 2)^2(x - 1)} - \dfrac{6}{(x + 2)(x - 1)^2}$

67. $\dfrac{x + 4}{x^2 - x - 2} - \dfrac{2x + 3}{x^2 + 2x - 8}$

68. $\dfrac{2x - 3}{x^2 + 8x + 7} - \dfrac{x - 2}{(x + 1)^2}$

69. $\dfrac{1}{x} - \dfrac{2}{x^2 + x} + \dfrac{3}{x^3 - x^2}$

70. $\dfrac{x}{(x - 1)^2} + \dfrac{2}{x} - \dfrac{x + 1}{x^3 - x^2}$

71. $\dfrac{1}{h}\left(\dfrac{1}{x + h} - \dfrac{1}{x}\right)$

72. $\dfrac{1}{h}\left[\dfrac{1}{(x + h)^2} - \dfrac{1}{x^2}\right]$

In Problems 73–82, perform the indicated operations and simplify the result. Leave your answer in factored form.

73. $\dfrac{1 + \dfrac{1}{x}}{1 - \dfrac{1}{x}}$

74. $\dfrac{4 + \dfrac{1}{x^2}}{3 - \dfrac{1}{x^2}}$

75. $\dfrac{2 - \dfrac{x + 1}{x}}{3 + \dfrac{x - 1}{x + 1}}$

76. $\dfrac{1 - \dfrac{x}{x + 1}}{2 - \dfrac{x - 1}{x}}$

77. $\dfrac{\dfrac{x + 4}{x - 2} - \dfrac{x - 3}{x + 1}}{x + 1}$

78. $\dfrac{\dfrac{x - 2}{x + 1} - \dfrac{x}{x - 2}}{x + 3}$

79. $\dfrac{\dfrac{x - 2}{x + 2} + \dfrac{x - 1}{x + 1}}{\dfrac{x}{x + 1} - \dfrac{2x - 3}{x}}$

80. $\dfrac{\dfrac{2x + 5}{x} - \dfrac{x}{x - 3}}{\dfrac{x^2}{x - 3} - \dfrac{(x + 1)^2}{x + 3}}$

81. $1 - \dfrac{1}{1 - \dfrac{1}{x}}$

82. $1 - \dfrac{1}{1 - \dfrac{1}{1 - x}}$

Applications and Extensions

In Problems 83–90, expressions that occur in calculus are given. Reduce each expression to lowest terms.

83. $\dfrac{(2x + 3) \cdot 3 - (3x - 5) \cdot 2}{(3x - 5)^2}$

84. $\dfrac{(4x + 1) \cdot 5 - (5x - 2) \cdot 4}{(5x - 2)^2}$

85. $\dfrac{x \cdot 2x - (x^2 + 1) \cdot 1}{(x^2 + 1)^2}$

86. $\dfrac{x \cdot 2x - (x^2 - 4) \cdot 1}{(x^2 - 4)^2}$

87. $\dfrac{(3x + 1) \cdot 2x - x^2 \cdot 3}{(3x + 1)^2}$

88. $\dfrac{(2x - 5) \cdot 3x^2 - x^3 \cdot 2}{(2x - 5)^2}$

89. $\dfrac{(x^2 + 1) \cdot 3 - (3x + 4) \cdot 2x}{(x^2 + 1)^2}$

90. $\dfrac{(x^2 + 9) \cdot 2 - (2x - 5) \cdot 2x}{(x^2 + 9)^2}$

91. The Lensmaker's Equation The focal length f of a lens with index of refraction n is

$$\dfrac{1}{f} = (n - 1)\left[\dfrac{1}{R_1} + \dfrac{1}{R_2}\right]$$

where R_1 and R_2 are the radii of curvature of the front and back surfaces of the lens. Express f as a rational expression. Evaluate the rational expression for $n = 1.5$, $R_1 = 0.1$ meter, and $R_2 = 0.2$ meter.

92. Electrical Circuits An electrical circuit contains three resistors connected in parallel. If the resistance of each is $R_1, R_2,$ and R_3 ohms, respectively, their combined resistance R is given by the formula

$$\dfrac{1}{R} = \dfrac{1}{R_1} + \dfrac{1}{R_2} + \dfrac{1}{R_3}$$

Express R as a rational expression. Evaluate R for $R_1 = 5$ ohms, $R_2 = 4$ ohms, and $R_3 = 10$ ohms.

Discussion and Writing

93. The following expressions are called **continued fractions**:

$$1 + \dfrac{1}{x}, \quad 1 + \dfrac{1}{1 + \dfrac{1}{x}}, \quad 1 + \dfrac{1}{1 + \dfrac{1}{1 + \dfrac{1}{x}}}, \quad 1 + \dfrac{1}{1 + \dfrac{1}{1 + \dfrac{1}{1 + \dfrac{1}{x}}}}, \quad \cdots$$

Each simplifies to an expression of the form

$$\dfrac{ax + b}{bx + c}$$

Trace the successive values of $a, b,$ and c as you "continue" the fraction. Can you discover the patterns that these values follow? Go to the library and research Fibonacci numbers. Write a report on your findings.

94. Explain to a fellow student when you would use the LCM method to add two rational expressions. Give two examples of adding two rational expressions, one in which you use the LCM and the other in which you do not.

95. Which of the two methods given in the text for simplifying complex rational expressions do you prefer? Write a brief paragraph stating the reasons for your choice.

A.7 nth Roots; Rational Exponents

PREPARING FOR THIS SECTION *Before getting started, review the following:*

• Exponents, Square Roots (Appendix A, Section A.1, pp. A6–A8)

 Now Work the 'Are You Prepared?' problems on page A57

OBJECTIVES 1 Work with nth Roots
2 Simplify Radicals
3 Rationalize Denominators
4 Simplify Expressions with Rational Exponents

1 Work with nth Roots

The **principal nth root of a real number** a, $n \geq 2$ an integer, symbolized by $\sqrt[n]{a}$, is defined as follows:

$$\sqrt[n]{a} = b \quad \text{means} \quad a = b^n$$

where $a \geq 0$ and $b \geq 0$ if n is even and a, b are any real numbers if n is odd.

Notice that if a is negative and n is even then $\sqrt[n]{a}$ is not defined. When it is defined, the principal nth root of a number is unique.

The symbol $\sqrt[n]{a}$ for the principal nth root of a is called a **radical**; the integer n is called the **index**, and a is called the **radicand**. If the index of a radical is 2, we call $\sqrt[2]{a}$ the **square root** of a and omit the index 2 by simply writing \sqrt{a}. If the index is 3, we call $\sqrt[3]{a}$ the **cube root** of a.

EXAMPLE 1 | **Simplifying Principal nth Roots**

(a) $\sqrt[3]{8} = \sqrt[3]{2^3} = 2$ (b) $\sqrt[3]{-64} = \sqrt[3]{(-4)^3} = -4$

(c) $\sqrt[4]{\dfrac{1}{16}} = \sqrt[4]{\left(\dfrac{1}{2}\right)^4} = \dfrac{1}{2}$ (d) $\sqrt[6]{(-2)^6} = |-2| = 2$ ∎

These are examples of **perfect roots**, since each simplifies to a rational number. Notice the absolute value in Example 1(d). If n is even, the principal nth root must be nonnegative.

In general, if $n \geq 2$ is an integer and a is a real number, we have

$$\sqrt[n]{a^n} = a \qquad \text{if } n \geq 3 \text{ is odd} \qquad \textbf{(1a)}$$
$$\sqrt[n]{a^n} = |a| \qquad \text{if } n \geq 2 \text{ is even} \qquad \textbf{(1b)}$$

NOW WORK PROBLEM 7.

Radicals provide a way of representing many irrational real numbers. For example, there is no rational number whose square is 2. Using radicals, we can say that $\sqrt{2}$ is the positive number whose square is 2.

| EXAMPLE 2 | **Using a Calculator to Approximate Roots** |

Use a calculator to approximate $\sqrt[5]{16}$.

Figure 24 **Solution** Figure 21 shows the result using a TI-84 plus graphing calculator. ∎

```
5*√16
        1.741101127
```

✏ **N O W W O R K P R O B L E M 101.**

2 Simplify Radicals

Let $n \geq 2$ and $m \geq 2$ denote positive integers, and let a and b represent real numbers. Assuming that all radicals are defined, we have the following properties:

> **Properties of Radicals**
>
> $$\sqrt[n]{ab} = \sqrt[n]{a}\,\sqrt[n]{b} \tag{2a}$$
>
> $$\sqrt[n]{\dfrac{a}{b}} = \dfrac{\sqrt[n]{a}}{\sqrt[n]{b}} \tag{2b}$$
>
> $$\sqrt[n]{a^m} = (\sqrt[n]{a})^m \tag{2c}$$

When used in reference to radicals, the direction to "simplify" will mean to remove from the radicals any perfect roots that occur as factors. Let's look at some examples of how the preceding rules are applied to simplify radicals.

| EXAMPLE 3 | **Simplifying Radicals** |

(a) $\sqrt{32} = \sqrt{16 \cdot 2} = \sqrt{16} \cdot \sqrt{2} = 4\sqrt{2}$

 ↑
 Factor out 16, a perfect square.

(b) $\sqrt[3]{16} = \sqrt[3]{8 \cdot 2} = \sqrt[3]{8} \cdot \sqrt[3]{2} = \sqrt[3]{2^3} \cdot \sqrt[3]{2} = 2\sqrt[3]{2}$

 ↑ ↑
 Factor (2a)
 out 8,
 a perfect cube.

(c) $\sqrt[3]{-16x^4} = \sqrt[3]{-8 \cdot 2 \cdot x^3 \cdot x} = \sqrt[3]{(-8x^3)(2x)}$

 ↑ ↑
 Factor perfect Combine perfect
 cubes inside radical. cubes.

$$= \sqrt[3]{(-2x)^3 \cdot 2x} = \sqrt[3]{(-2x)^3} \cdot \sqrt[3]{2x} = -2x\sqrt[3]{2x}$$

 ↑
 (2a)

(d) $\sqrt[3]{\dfrac{8x^5}{27}} = \sqrt[3]{\dfrac{2^3 x^3 x^2}{3^3}} = \sqrt[3]{\left(\dfrac{2x}{3}\right)^3 \cdot x^2} = \sqrt[3]{\left(\dfrac{2x}{3}\right)^3} \cdot \sqrt[3]{x^2} = \dfrac{2x}{3}\sqrt[3]{x^2}$ ∎

✏ **N O W W O R K P R O B L E M S 11 A N D 17.**

Two or more radicals can be combined, provided that they have the same index and the same radicand. Such radicals are called **like radicals**.

| EXAMPLE 4 | **Combining Like Radicals** |

(a) $\;-8\sqrt{12} + \sqrt{3} = -8\sqrt{4\cdot3} + \sqrt{3}$

$\qquad\qquad\qquad = -8\cdot\sqrt{4}\,\sqrt{3} + \sqrt{3}$

$\qquad\qquad\qquad = -16\sqrt{3} + \sqrt{3} = -15\sqrt{3}$

(b) $\;\sqrt[3]{8x^4} + \sqrt[3]{-x} + 4\sqrt[3]{27x} = \sqrt[3]{2^3 x^3 x} + \sqrt[3]{-1\cdot x} + 4\sqrt[3]{3^3 x}$

$\qquad\qquad\qquad\qquad\qquad = \sqrt[3]{(2x)^3}\cdot\sqrt[3]{x} + \sqrt[3]{-1}\cdot\sqrt[3]{x} + 4\sqrt[3]{3^3}\cdot\sqrt[3]{x}$

$\qquad\qquad\qquad\qquad\qquad = 2x\sqrt[3]{x} - 1\cdot\sqrt[3]{x} + 12\sqrt[3]{x}$

$\qquad\qquad\qquad\qquad\qquad = (2x + 11)\sqrt[3]{x}$ ∎

NOW WORK PROBLEM 33.

3 **Rationalize Denominators**

When radicals occur in quotients, it is customary to rewrite the quotient so that the denominator contains no radicals. This process is referred to as **rationalizing the denominator**.

The idea is to multiply by an appropriate expression so that the new denominator contains no square roots. For example:

If a Denominator Contains the Factor	Multiply by	To Obtain a Denominator Free of Radicals
$\sqrt{3}$	$\sqrt{3}$	$\left(\sqrt{3}\right)^2 = 3$
$\sqrt{3} + 1$	$\sqrt{3} - 1$	$\left(\sqrt{3}\right)^2 - 1^2 = 3 - 1 = 2$
$\sqrt{2} - 3$	$\sqrt{2} + 3$	$\left(\sqrt{2}\right)^2 - 3^2 = 2 - 9 = -7$
$\sqrt{5} - \sqrt{3}$	$\sqrt{5} + \sqrt{3}$	$\left(\sqrt{5}\right)^2 - \left(\sqrt{3}\right)^2 = 5 - 3 = 2$
$\sqrt[3]{4}$	$\sqrt[3]{2}$	$\sqrt[3]{4}\cdot\sqrt[3]{2} = \sqrt[3]{8} = 2$

In rationalizing the denominator of a quotient, be sure to multiply both the numerator and the denominator by the expression.

| EXAMPLE 5 | **Rationalizing Denominators** |

Rationalize the denominator of each expression:

(a) $\dfrac{1}{\sqrt{3}}$ \qquad (b) $\dfrac{5}{4\sqrt{2}}$ \qquad (c) $\dfrac{\sqrt{2}}{\sqrt{3} - \sqrt{2}}$

Solution (a) The denominator contains the factor $\sqrt{3}$, so we multiply the numerator and denominator by $\sqrt{3}$ to obtain

$$\frac{1}{\sqrt{3}} = \frac{1}{\sqrt{3}}\cdot\frac{\sqrt{3}}{\sqrt{3}} = \frac{\sqrt{3}}{\left(\sqrt{3}\right)^2} = \frac{\sqrt{3}}{3}$$

(b) The denominator contains the factor $\sqrt{2}$, so we multiply the numerator and denominator by $\sqrt{2}$ to obtain

$$\frac{5}{4\sqrt{2}} = \frac{5}{4\sqrt{2}}\cdot\frac{\sqrt{2}}{\sqrt{2}} = \frac{5\sqrt{2}}{4\left(\sqrt{2}\right)^2} = \frac{5\sqrt{2}}{4\cdot2} = \frac{5\sqrt{2}}{8}$$

(c) The denominator contains the factor $\sqrt{3} - \sqrt{2}$, so we multiply the numerator and denominator by $\sqrt{3} + \sqrt{2}$ to obtain

$$\frac{\sqrt{2}}{\sqrt{3} - \sqrt{2}} = \frac{\sqrt{2}}{\sqrt{3} - \sqrt{2}} \cdot \frac{\sqrt{3} + \sqrt{2}}{\sqrt{3} + \sqrt{2}} = \frac{\sqrt{2}(\sqrt{3} + \sqrt{2})}{(\sqrt{3})^2 - (\sqrt{2})^2}$$

$$= \frac{\sqrt{2}\,\sqrt{3} + (\sqrt{2})^2}{3 - 2} = \sqrt{6} + 2 \quad \blacksquare$$

NOW WORK PROBLEM 47.

4 Simplify Expressions with Rational Exponents

Radicals are used to define rational exponents.

If a is a real number and $n \geq 2$ is an integer, then

$$a^{1/n} = \sqrt[n]{a} \tag{3}$$

provided that $\sqrt[n]{a}$ exists.

Note that if n is even and $a < 0$, then $\sqrt[n]{a}$ and $a^{1/n}$ do not exist.

EXAMPLE 6	**Writing Expressions Containing Fractional Exponents as Radicals**

(a) $4^{1/2} = \sqrt{4} = 2$ (b) $8^{1/2} = \sqrt{8} = 2\sqrt{2}$

(c) $(-27)^{1/3} = \sqrt[3]{-27} = -3$ (d) $16^{1/3} = \sqrt[3]{16} = 2\sqrt[3]{2}$ \blacksquare

If a is a real number and m and n are integers containing no common factors, with $n \geq 2$, then

$$a^{m/n} = \sqrt[n]{a^m} = (\sqrt[n]{a})^m \tag{4}$$

provided that $\sqrt[n]{a}$ exists.

We have two comments about equation (4):

1. The exponent $\dfrac{m}{n}$ must be in lowest terms and n must be positive.

2. In simplifying the rational expression $a^{m/n}$, either $\sqrt[n]{a^m}$ or $(\sqrt[n]{a})^m$ may be used, the choice depending on which is easier to simplify. Generally, taking the root first, as in $(\sqrt[n]{a})^m$, is easier.

EXAMPLE 7	**Using Equation (4)**

(a) $4^{3/2} = (\sqrt{4})^3 = 2^3 = 8$ (b) $(-8)^{4/3} = (\sqrt[3]{-8})^4 = (-2)^4 = 16$

(c) $(32)^{-2/5} = (\sqrt[5]{32})^{-2} = 2^{-2} = \dfrac{1}{4}$ (d) $25^{6/4} = 25^{3/2} = (\sqrt{25})^3 = 5^3 = 125$ \blacksquare

NOW WORK PROBLEM 55.

It can be shown that the Laws of Exponents hold for rational exponents. The next example illustrates using the Law of Exponents to simplify.

EXAMPLE 8 **Simplifying Expressions Containing Rational Exponents**

Simplify each expression. Express your answer so that only positive exponents occur. Assume that the variables are positive.

(a) $(x^{2/3}y)(x^{-2}y)^{1/2}$ (b) $\left(\dfrac{2x^{1/3}}{y^{2/3}}\right)^{-3}$ (c) $\left(\dfrac{9x^2y^{1/3}}{x^{1/3}y}\right)^{1/2}$

Solution (a) $(x^{2/3}y)(x^{-2}y)^{1/2} = (x^{2/3}y)[(x^{-2})^{1/2}y^{1/2}]$

$$= x^{2/3}yx^{-1}y^{1/2}$$

$$= (x^{2/3} \cdot x^{-1})(y \cdot y^{1/2})$$

$$= x^{-1/3}y^{3/2}$$

$$= \dfrac{y^{3/2}}{x^{1/3}}$$

(b) $\left(\dfrac{2x^{1/3}}{y^{2/3}}\right)^{-3} = \left(\dfrac{y^{2/3}}{2x^{1/3}}\right)^3 = \dfrac{(y^{2/3})^3}{(2x^{1/3})^3} = \dfrac{y^2}{2^3(x^{1/3})^3} = \dfrac{y^2}{8x}$

(c) $\left(\dfrac{9x^2y^{1/3}}{x^{1/3}y}\right)^{1/2} = \left(\dfrac{9x^{2-(1/3)}}{y^{1-(1/3)}}\right)^{1/2} = \left(\dfrac{9x^{5/3}}{y^{2/3}}\right)^{1/2} = \dfrac{9^{1/2}(x^{5/3})^{1/2}}{(y^{2/3})^{1/2}} = \dfrac{3x^{5/6}}{y^{1/3}}$ ∎

──── **NOW WORK PROBLEM 71.**

 The next two examples illustrate some algebra that you will need to know for certain calculus problems.

EXAMPLE 9 **Writing an Expression as a Single Quotient**

Write the following expression as a single quotient in which only positive exponents appear.

$$(x^2 + 1)^{1/2} + x \cdot \frac{1}{2}(x^2 + 1)^{-1/2} \cdot 2x$$

Solution $(x^2 + 1)^{1/2} + x \cdot \dfrac{1}{2}(x^2 + 1)^{-1/2} \cdot 2x = (x^2 + 1)^{1/2} + \dfrac{x^2}{(x^2 + 1)^{1/2}}$

$$= \dfrac{(x^2 + 1)^{1/2}(x^2 + 1)^{1/2} + x^2}{(x^2 + 1)^{1/2}}$$

$$= \dfrac{(x^2 + 1) + x^2}{(x^2 + 1)^{1/2}}$$

$$= \dfrac{2x^2 + 1}{(x^2 + 1)^{1/2}}$$ ∎

──── **NOW WORK PROBLEM 77.**

| EXAMPLE 10 | Factoring an Expression Containing Rational Exponents |

Factor: $\dfrac{4}{3}x^{1/3}(2x + 1) + 2x^{4/3}$

Solution　We begin by looking for factors that are common to the two terms. Notice that 2 and $x^{1/3}$ are common factors.

$$\frac{4}{3}x^{1/3}(2x + 1) + 2x^{4/3} = 2x^{1/3}\left[\frac{2}{3}(2x + 1) + x\right]$$

$$= \frac{2}{3}x^{1/3}(7x + 2)$$ ∎

NOW WORK PROBLEM **89.**

HISTORICAL FEATURE

The radical sign, $\sqrt{}$, was first used in print by Christoff Rudolff in 1525. It is thought to be the manuscript form of the letter *r* (for the Latin word *radix* = *root*), although this is not quite conclusively proved. It took a long time for $\sqrt{}$ to become the standard symbol for a square root and much longer to standardize $\sqrt[3]{}$, $\sqrt[4]{}$, $\sqrt[5]{}$, and so on. The indexes of the root were placed in every conceivable position, with

$$\sqrt[3]{8}, \quad \sqrt{\text{③}8}, \quad \text{and} \quad \underset{3}{\sqrt{8}}$$

all being variants for $\sqrt[3]{8}$. The notation $\sqrt{}\sqrt{16}$ was popular for $\sqrt[4]{16}$. By the 1700s, the index had settled where we now put it.

The bar on top of the present radical symbol, as follows,

$$\sqrt{a^2 + 2ab + b^2}$$

is the last survivor of the **vinculum**, a bar placed atop an expression to indicate what we would now indicate with parentheses. For example,

$$\overline{ab + c} = a(b + c)$$

A.7 Assess Your Understanding

'Are You Prepared?'

Answers are given at the end of these exercises. If you get a wrong answer, read the pages in red.

1. $(-3)^2 = $ _____; $-3^2 = $ _____ (pp. A6–A8)

2. $\sqrt{16} = $ _____; $\sqrt{(-4)^2} = $ _____ (pp. A6–A8)

Concepts and Vocabulary

3. In the symbol $\sqrt[n]{a}$, the integer *n* is called the _____.

4. *True or False:* $\sqrt[5]{-32} = -2$

5. We call $\sqrt[3]{a}$ the _____ of *a*.

6. *True or False:* $\sqrt[4]{(-3)^4} = -3$

Skill Building

In Problems 7–42, simplify each expression. Assume that all variables are positive when they appear.

7. $\sqrt[3]{27}$

8. $\sqrt[4]{16}$

9. $\sqrt[3]{-8}$

10. $\sqrt[3]{-1}$

11. $\sqrt{8}$

12. $\sqrt[3]{54}$

13. $\sqrt[3]{-8x^4}$

14. $\sqrt[4]{48x^5}$

15. $\sqrt[4]{x^{12}y^8}$

16. $\sqrt[5]{x^{10}y^5}$

17. $\sqrt[4]{\dfrac{x^9y^7}{xy^3}}$

18. $\sqrt[3]{\dfrac{3xy^2}{81x^4y^2}}$

19. $\sqrt{36x}$

20. $\sqrt{9x^5}$

21. $\sqrt{3x^2}\,\sqrt{12x}$

22. $\sqrt{5x}\,\sqrt{20x^3}$

23. $\left(\sqrt{5}\,\sqrt[3]{9}\right)^2$

24. $\left(\sqrt[3]{3}\,\sqrt{10}\right)^4$

25. $\left(3\sqrt{6}\right)\left(2\sqrt{2}\right)$

26. $\left(5\sqrt{8}\right)\left(-3\sqrt{3}\right)$

27. $3\sqrt{2} + 4\sqrt{2}$

28. $6\sqrt{5} - 4\sqrt{5}$

29. $-\sqrt{18} + 2\sqrt{8}$

30. $2\sqrt{12} - 3\sqrt{27}$

31. $\left(\sqrt{3} + 3\right)\left(\sqrt{3} - 1\right)$

32. $\left(\sqrt{5} - 2\right)\left(\sqrt{5} + 3\right)$

33. $5\sqrt[3]{2} - 2\sqrt[3]{54}$

34. $9\sqrt[3]{24} - \sqrt[3]{81}$

35. $\left(\sqrt{x} - 1\right)^2$ **36.** $\left(\sqrt{x} + \sqrt{5}\right)^2$ **37.** $\sqrt[3]{16x^4} - \sqrt[3]{2x}$ **38.** $\sqrt[4]{32x} + \sqrt[4]{2x^5}$

39. $\sqrt{8x^3} - 3\sqrt{50x},$ **40.** $3x\sqrt{9y} + 4\sqrt{25y}$

41. $\sqrt[3]{16x^4y} - 3x\sqrt[3]{2xy} + 5\sqrt[3]{-2xy^4}$ **42.** $8xy - \sqrt{25x^2y^2} + \sqrt[3]{8x^3y^3}$

In Problems 43–54, rationalize the denominator of each expression. Assume that all variables are positive when they appear.

43. $\dfrac{1}{\sqrt{2}}$ **44.** $\dfrac{2}{\sqrt{3}}$ **45.** $\dfrac{-\sqrt{3}}{\sqrt{5}}$ **46.** $\dfrac{-\sqrt{3}}{\sqrt{8}}$

47. $\dfrac{\sqrt{3}}{5 - \sqrt{2}}$ **48.** $\dfrac{\sqrt{2}}{\sqrt{7} + 2}$ **49.** $\dfrac{2 - \sqrt{5}}{2 + 3\sqrt{5}}$ **50.** $\dfrac{\sqrt{3} - 1}{2\sqrt{3} + 3}$

51. $\dfrac{5}{\sqrt[3]{2}}$ **52.** $\dfrac{-2}{\sqrt[3]{9}}$ **53.** $\dfrac{\sqrt{x + h} - \sqrt{x}}{\sqrt{x + h} + \sqrt{x}}$ **54.** $\dfrac{\sqrt{x + h} + \sqrt{x - h}}{\sqrt{x + h} - \sqrt{x - h}}$

In Problems 55–66, simplify each expression.

55. $8^{2/3}$ **56.** $4^{3/2}$ **57.** $(-27)^{1/3}$ **58.** $16^{3/4}$ **59.** $16^{3/2}$ **60.** $25^{3/2}$

61. $9^{-3/2}$ **62.** $16^{-3/2}$ **63.** $\left(\dfrac{9}{8}\right)^{3/2}$ **64.** $\left(\dfrac{27}{8}\right)^{2/3}$ **65.** $\left(\dfrac{8}{9}\right)^{-3/2}$ **66.** $\left(\dfrac{8}{27}\right)^{-2/3}$

In Problems 67–74, simplify each expression. Express your answer so that only positive exponents occur. Assume that the variables are positive.

67. $x^{3/4}x^{1/3}x^{-1/2}$ **68.** $x^{2/3}x^{1/2}x^{-1/4}$ **69.** $(x^3y^6)^{1/3}$ **70.** $(x^4y^8)^{3/4}$

71. $(x^2y)^{1/3}(xy^2)^{2/3}$ **72.** $(xy)^{1/4}(x^2y^2)^{1/2}$ **73.** $(16x^2y^{-1/3})^{3/4}$ **74.** $(4x^{-1}y^{1/3})^{3/2}$

Applications and Extensions

In Problems 75–88, expressions that occur in calculus are given. Write each expression as a single quotient in which only positive exponents and/or radicals appear.

75. $\dfrac{x}{(1 + x)^{1/2}} + 2(1 + x)^{1/2}, \quad x > -1$ **76.** $\dfrac{1 + x}{2x^{1/2}} + x^{1/2}, \quad x > 0$

77. $2x(x^2 + 1)^{1/2} + x^2 \cdot \dfrac{1}{2}(x^2 + 1)^{-1/2} \cdot 2x$ **78.** $(x + 1)^{1/3} + x \cdot \dfrac{1}{3}(x + 1)^{-2/3}, \quad x \neq -1$

79. $\sqrt{4x + 3} \cdot \dfrac{1}{2\sqrt{x - 5}} + \sqrt{x - 5} \cdot \dfrac{1}{5\sqrt{4x + 3}}, \quad x > 5$ **80.** $\dfrac{\sqrt[3]{8x + 1}}{3\sqrt[3]{(x - 2)^2}} + \dfrac{\sqrt[3]{x - 2}}{24\sqrt[3]{(8x + 1)^2}}, \quad x \neq 2, x \neq -\dfrac{1}{8}$

81. $\dfrac{\sqrt{1 + x} - x \cdot \dfrac{1}{2\sqrt{1 + x}}}{1 + x}, \quad x > -1$ **82.** $\dfrac{\sqrt{x^2 + 1} - x \cdot \dfrac{2x}{2\sqrt{x^2 + 1}}}{x^2 + 1}$

83. $\dfrac{(x + 4)^{1/2} - 2x(x + 4)^{-1/2}}{x + 4}, \quad x > -4$ **84.** $\dfrac{(9 - x^2)^{1/2} + x^2(9 - x^2)^{-1/2}}{9 - x^2}, \quad -3 < x < 3$

85. $\dfrac{\dfrac{x^2}{(x^2 - 1)^{1/2}} - (x^2 - 1)^{1/2}}{x^2}, \quad x < -1 \ \text{ or } \ x > 1$ **86.** $\dfrac{(x^2 + 4)^{1/2} - x^2(x^2 + 4)^{-1/2}}{x^2 + 4}$

87. $\dfrac{\dfrac{1 + x^2}{2\sqrt{x}} - 2x\sqrt{x}}{(1 + x^2)^2}, \quad x > 0$ **88.** $\dfrac{2x(1 - x^2)^{1/3} + \dfrac{2}{3}x^3(1 - x^2)^{-2/3}}{(1 - x^2)^{2/3}}, \quad x \neq -1, x \neq 1$

In Problems 89–98, expressions that occur in calculus are given. Factor each expression. Express your answer so that only positive exponents occur.

89. $(x + 1)^{3/2} + x \cdot \dfrac{3}{2}(x + 1)^{1/2}, \quad x \geq -1$ **90.** $(x^2 + 4)^{4/3} + x \cdot \dfrac{4}{3}(x^2 + 4)^{1/3} \cdot 2x$

91. $6x^{1/2}(x^2 + x) - 8x^{3/2} - 8x^{1/2}, \quad x \ge 0$

92. $6x^{1/2}(2x + 3) + x^{3/2} \cdot 8, \quad x \ge 0$

93. $3(x^2 + 4)^{4/3} + x \cdot 4(x^2 + 4)^{1/3} \cdot 2x$

94. $2x(3x + 4)^{4/3} + x^2 \cdot 4(3x + 4)^{1/3}$

95. $4(3x + 5)^{1/3}(2x + 3)^{3/2} + 3(3x + 5)^{4/3}(2x + 3)^{1/2}, \quad x \ge -\dfrac{3}{2}$

96. $6(6x + 1)^{1/3}(4x - 3)^{3/2} + 6(6x + 1)^{4/3}(4x - 3)^{1/2}, \quad x \ge \dfrac{3}{4}$

97. $3x^{-1/2} + \dfrac{3}{2}x^{1/2}, \quad x > 0$

98. $8x^{1/3} - 4x^{-2/3}, \quad x \ne 0$

In Problems 99–106, use a calculator to approximate each radical. Round your answer to two decimal places.

99. $\sqrt{2}$

100. $\sqrt{7}$

101. $\sqrt[3]{4}$

102. $\sqrt[3]{-5}$

103. $\dfrac{2 + \sqrt{3}}{3 - \sqrt{5}}$

104. $\dfrac{\sqrt{5} - 2}{\sqrt{2} + 4}$

105. $\dfrac{3\sqrt[3]{5} - \sqrt{2}}{\sqrt{3}}$

106. $\dfrac{2\sqrt{3} - \sqrt[3]{4}}{\sqrt{2}}$

107. Calculating the Amount of Gasoline in a Tank A Shell station stores its gasoline in underground tanks that are right circular cylinders lying on their sides. See the illustration. The volume V of gasoline in the tank (in gallons) is given by the formula

$$V = 40h^2 \sqrt{\dfrac{96}{h} - 0.608}$$

where h is the height of the gasoline (in inches) as measured on a depth stick.

(a) If $h = 12$ inches, how many gallons of gasoline are in the tank?

(b) If $h = 1$ inch, how many gallons of gasoline are in the tank?

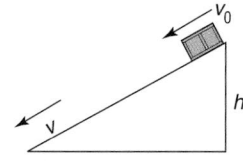

108. Inclined Planes The final velocity v of an object in feet per second (ft/sec) after it slides down a frictionless inclined plane of height h feet is

$$v = \sqrt{64h + v_0^2}$$

where v_0 is the initial velocity (in ft/sec) of the object.

(a) What is the final velocity v of an object that slides down a frictionless inclined plane of height 4 feet? Assume that the initial velocity is 0.

(b) What is the final velocity v of an object that slides down a frictionless inclined plane of height 16 feet? Assume that the initial velocity is 0.

(c) What is the final velocity v of an object that slides down a frictionless inclined plane of height 2 feet with an initial velocity of 4 ft/sec?

Problems 109–112 require the following information.

Period of a Pendulum *The period T, in seconds, of a pendulum of length l, in feet, may be approximated using the formula*

$$T = 2\pi \sqrt{\dfrac{l}{32}}$$

In Problems 109–112, express your answer both as a square root and as a decimal.

109. Find the period T of a pendulum whose length is 64 feet.

110. Find the period T of a pendulum whose length is 16 feet.

111. Find the period T of a pendulum whose length is 8 inches.

112. Find the period T of a pendulum whose length is 4 inches.

Discussion and Writing

113. Give an example to show that $\sqrt{a^2}$ is not equal to a. Use it to explain why $\sqrt{a^2} = |a|$.

'Are you Prepared?' Answers

1. $9; -9$

2. $4; 4$

A.8 Solving Equations

PREPARING FOR THIS SECTION *Before getting started, review the following:*

- Factoring Polynomials (Appendix A, Section A.4, pp. A29–A35)
- Zero-Product Property (Appendix A, Section A.1, p. A2)
- Rational Expressions (Appendix A, Section A.6, pp. A41–A49)

Now work the 'Are You Prepared?' problems on page A64.

OBJECTIVES
1. Solve Linear Equations
2. Solve Rational Equations
3. Solve Equations by Factoring

An **equation in one variable** is a statement in which two expressions, at least one containing the variable, are equal. The expressions are called the **sides** of the equation. Since an equation is a statement, it may be true or false, depending on the value of the variable. Unless otherwise restricted, the admissible values of the variable are those in the domain of the variable. Those admissible values of the variable, if any, that result in a true statement are called **solutions**, or **roots**, of the equation. To **solve an equation** means to find all the solutions of the equation.

For example, the following are all equations in one variable, x:

$$x + 5 = 9, \qquad x^2 + 5x = 2x - 2, \qquad \frac{x^2 - 4}{x + 1} = 0, \qquad \sqrt{x^2 + 9} = 5$$

The first of these statements, $x + 5 = 9$, is true when $x = 4$ and false for any other choice of x. Thus, 4 is a solution of the equation $x + 5 = 9$. We also say that 4 **satisfies** the equation $x + 5 = 9$, because, when we substitute 4 for x, a true statement results.

Sometimes an equation will have more than one solution. For example, the equation

$$\frac{x^2 - 4}{x + 1} = 0$$

has $x = -2$ and $x = 2$ as solutions.

Usually, we will write the solution of an equation in set notation. This set is called the **solution set** of the equation. For example, the solution set of the equation $x^2 - 9 = 0$ is $\{-3, 3\}$.

Some equations have no real solution. For example, $x^2 + 9 = 5$ has no real solution, because there is no real number whose square when added to 9 equals 5.

An equation that is satisfied for every choice of the variable for which both sides are defined is called an **identity**. For example, the equation

$$3x + 5 = x + 3 + 2x + 2$$

is an identity, because this statement is true for any real number x.

One method for solving equations requires that a series of *equivalent equations* be developed from the original equation until an obvious solution results.

Two or more equations that have precisely the same solutions are called **equivalent equations**.

For example, all the following equations are equivalent, because each has only the solution $x = 5$.

$$2x + 3 = 13$$
$$2x = 10$$
$$x = 5$$

The question, though, is "How do I obtain an equivalent equation?" In general, there are five ways to do so.

Procedures That Result in Equivalent Equations

1. Interchange the two sides of the equation:

Replace $\quad 3 = x \quad$ by $\quad x = 3$

2. Simplify the sides of the equation by combining like terms, eliminating parentheses, and so on:

Replace $\quad x + 2 + 6 = 2x + 3(x + 1)$

by $\quad\quad x + 8 = 5x + 3$

3. Add or subtract the same expression on both sides of the equation:

Replace $\quad\quad\quad 3x - 5 = 4$

by $\quad (3x - 5) + 5 = 4 + 5$

4. Multiply or divide both sides of the equation by the same nonzero expression:

Replace $\quad \dfrac{3x}{x - 1} = \dfrac{6}{x - 1} \quad\quad x \neq 1$

by $\quad \dfrac{3x}{x - 1} \cdot (x - 1) = \dfrac{6}{x - 1} \cdot (x - 1)$

5. If one side of the equation is 0 and the other side can be factored, we may use the Zero-Product Property and set each factor equal to 0.

Replace $\quad\quad\quad x(x - 3) = 0$

by $\quad x = 0 \quad$ or $\quad x - 3 = 0$

WARNING
Squaring both sides of an equation does not necessarily lead to an equivalent equation. ■

Whenever it is possible to solve an equation in your head, do so. For example,

The solution of $2x = 8$ is $x = 4$.

The solution of $3x - 15 = 0$ is $x = 5$.

NOW WORK PROBLEM 9.

We now introduce specific types of equations that can be solved algebraically to obtain exact solutions. We start with **linear equations**.

1 Solve Linear Equations

Linear equations are equations such as

$$3x + 12 = 0 \quad\quad \frac{3}{4}x - \frac{1}{5} = 0 \quad\quad 0.62x - 0.3 = 0$$

A **linear equation in one variable** is equivalent to an equation of the form

$$ax + b = 0$$

where a and b are real numbers and $a \neq 0$.

Sometimes a linear equation is called a **first-degree equation**, because the left side is a polynomial in x of degree 1.

| EXAMPLE 1 | **Solving a Linear Equation** |

Solve the equation: $3(x - 2) = 5(x - 1)$

Solution

$$3(x - 2) = 5(x - 1)$$

$3x - 6 = 5x - 5$ — Use the Distributive Property.

$3x - 6 - 5x = 5x - 5 - 5x$ — Subtract 5x from each side.

$-2x - 6 = -5$ — Simplify.

$-2x - 6 + 6 = -5 + 6$ — Add 6 to each side.

$-2x = 1$ — Simplify.

$\dfrac{-2x}{-2} = \dfrac{1}{-2}$ — Divide each side by −2.

$x = -\dfrac{1}{2}$ — Simplify.

✔ **CHECK:** Let $x = -\dfrac{1}{2}$ in the expression in x on the left side of the equation and simplify. Let $x = -\dfrac{1}{2}$ in the expression in x on the right side of the equation and simplify. If the two expressions are equal, the solution checks.

$$3(x - 2) = 3\left(-\frac{1}{2} - 2\right) = 3\left(-\frac{5}{2}\right) = -\frac{15}{2}$$

$$5(x - 1) = 5\left(-\frac{1}{2} - 1\right) = 5\left(-\frac{3}{2}\right) = -\frac{15}{2}$$

Since the two expressions are equal, the solution $x = -\dfrac{1}{2}$ checks.

The solution set is $\left\{-\dfrac{1}{2}\right\}$. ■

NOW WORK PROBLEM 19.

The next example illustrates the solution of an equation that does not appear to be linear, but leads to a linear equation upon simplification.

| EXAMPLE 2 | **Solving an Equation That Leads to a Linear Equation** |

Solve the equation: $(2x - 1)(x - 1) = (x - 5)(2x - 5)$

Solution

$$(2x - 1)(x - 1) = (x - 5)(2x - 5)$$

$2x^2 - 3x + 1 = 2x^2 - 15x + 25$ — Multiply and combine like terms.

$2x^2 - 3x + 1 - 2x^2 = 2x^2 - 15x + 25 - 2x^2$ — Subtract $2x^2$ from each side.

$-3x + 1 = -15x + 25$ — Simplify.

$-3x + 1 - 1 = -15x + 25 - 1$ — Subtract 1 from each side.

$-3x = -15x + 24$ — Simplify.

$-3x + 15x = -15x + 24 + 15x$ — Add 15x to each side.

$12x = 24$ — Simplify.

$\dfrac{12x}{12} = \dfrac{24}{12}$ — Divide each side by 12.

$x = 2$ — Simplify.

✔ CHECK: $(2x - 1)(x - 1) = (2 \cdot 2 - 1)(2 - 1) = (3)(1) = 3$

$\qquad\qquad (x - 5)(2x - 5) = (2 - 5)(2 \cdot 2 - 5) = (-3)(-1) = 3$

Since the two expressions are equal, the solution checks. The solution set is $\{\, 2 \,\}$. ∎

 NOW WORK PROBLEM **29**.

2 Solve Rational Equations

We now introduce another type of equation, the *rational equation*. A **rational equation** is an equation that contains a rational expression. Examples of rational equations are

$$\frac{3}{x + 1} = \frac{2}{x - 1} + 7 \quad \text{and} \quad \frac{x - 5}{x - 4} = \frac{3}{x + 2}$$

To solve a rational equation, multiply both sides of the equation by the least common multiple of the denominators of the rational expressions that make up the rational equation.

EXAMPLE 3 **Solving a Rational Equation**

Solve the equation: $\dfrac{3}{x - 2} = \dfrac{1}{x - 1} + \dfrac{7}{(x - 1)(x - 2)}$

Solution First, we note that the domain of the variable is $\{x \mid x \neq 1, x \neq 2\}$. We clear the equation of rational expressions by multiplying both sides by the least common multiple of the denominators of the three rational expressions, $(x - 1)(x - 2)$.

$$\frac{3}{x - 2} = \frac{1}{x - 1} + \frac{7}{(x - 1)(x - 2)}$$

$$(x - 1)(x - 2)\,\frac{3}{x - 2} = (x - 1)(x - 2)\left[\frac{1}{x - 1} + \frac{7}{(x - 1)(x - 2)}\right]$$

 Multiply both sides by $(x - 1)(x - 2)$. Cancel on the left.

$$3x - 3 = (x - 1)(x - 2)\,\frac{1}{x - 1} + (x - 1)(x - 2)\,\frac{7}{(x - 1)(x - 2)}$$

 Use the Distributive Property on each side; cancel on the right.

$$3x - 3 = (x - 2) + 7$$

 Rewrite the equation

$$3x - 3 = x + 5$$

 Combine like terms.

$$2x = 8$$

 Add 3 to each side; subtract x from each side.

$$x = 4$$

 Divide by 2.

✔ CHECK: $\dfrac{3}{x - 2} = \dfrac{3}{4 - 2} = \dfrac{3}{2}$

$$\frac{1}{x - 1} + \frac{7}{(x - 1)(x - 2)} = \frac{1}{4 - 1} + \frac{7}{(4 - 1)(4 - 2)} = \frac{1}{3} + \frac{7}{3 \cdot 2} = \frac{2}{6} + \frac{7}{6} = \frac{9}{6} = \frac{3}{2}$$

Since the two expressions are equal, the solution $x = 4$ checks.

The solution set is $\{\, 4 \,\}$. ∎

 NOW WORK PROBLEM **39**.

3 Solve Equations By Factoring

Many equations can be solved by factoring and then using the Zero-Product Property.

| EXAMPLE 4 | **Solving Equations by Factoring** |

Solve the equations: (a) $x^2 = 4x$ (b) $x^3 - x^2 - 4x + 4 = 0$

Solution (a) We begin by collecting all terms on one side. This results in 0 on one side and an expression to be factored on the other.

$$x^2 = 4x$$
$$x^2 - 4x = 0$$
$$x(x - 4) = 0 \qquad \text{Factor.}$$
$$x = 0 \quad \text{or} \quad x - 4 = 0 \qquad \text{Apply the Zero-Product Property.}$$
$$x = 4$$

The solution set is $\{0, 4\}$.

✔ **CHECK:** $x = 0$: $0^2 = 4 \cdot 0$ So 0 is a solution.

 $x = 4$: $4^2 = 4 \cdot 4$ So 4 is a solution.

(b) We group the terms of $x^3 - x^2 - 4x + 4 = 0$ as follows:

$$(x^3 - x^2) - (4x - 4) = 0$$

Factor out x^2 from the first grouping and 4 from the second.

$$x^2(x - 1) - 4(x - 1) = 0$$

This reveals the common factor $(x - 1)$, so we have

$$(x^2 - 4)(x - 1) = 0$$
$$(x - 2)(x + 2)(x - 1) = 0 \qquad \text{Factor again.}$$
$$x - 2 = 0 \quad \text{or} \quad x + 2 = 0 \quad \text{or} \quad x - 1 = 0 \qquad \text{Set each factor equal to 0.}$$
$$x = 2 \qquad\qquad x = -2 \qquad\qquad x = 1 \qquad \text{Solve.}$$

The solution set is $\{-2, 1, 2\}$.

✔ **CHECK:** $x = -2$: $(-2)^3 - (-2)^2 - 4(-2) + 4 = -8 - 4 + 8 + 4 = 0$ −2 is a solution.

 $x = 1$: $1^3 - 1^2 - 4(1) + 4 = 1 - 1 - 4 + 4 = 0$ 1 is a solution.

 $x = 2$: $2^3 - 2^2 - 4(2) + 4 = 8 - 4 - 8 + 4 = 0$ 2 is a solution. ∎

NOW WORK PROBLEM 43.

A.8 Assess Your Understanding

'Are You Prepared?'

Answers are given at the end of these exercises. If you get a wrong answer, read the pages listed in red.

1. Find the least common denominator of

$\dfrac{3}{x^2 - 4}$ and $\dfrac{5}{x^2 - 3x + 2}$. (pp. A41–A49)

2. The solution set of the equation $(x - 3)(3x + 5) = 0$ is _____. (p. A2)

3. Factor $2x^2 - 10x$. (pp. A29–A35)

4. Factor $x^3 + 3x^2 - 4x - 12$. (pp. A29–A35)

Concepts and Vocabulary

5. Two equations that have the same solution set are called _____.

6. An equation that is satisfied for every choice of the variable for which both sides are defined is called a(n) _____.

7. *True or False:* The solution of the equation $3x - 8 = 0$ is $\frac{3}{8}$.

8. *True or False:* Some equations have no solution.

Skill Building

In Problems 9–46, solve each equation.

9. $3x = 21$

10. $3x = -24$

11. $5x + 15 = 0$

12. $3x + 18 = 0$

13. $2x - 3 = 5$

14. $3x + 4 = -8$

15. $\frac{1}{3}x = \frac{5}{12}$

16. $\frac{2}{3}x = \frac{9}{2}$

17. $6 - x = 2x + 9$

18. $3 - 2x = 2 - x$

19. $2(3 + 2x) = 3(x - 4)$

20. $3(2 - x) = 2x - 1$

21. $8x - (2x + 1) = 3x - 10$

22. $5 - (2x - 1) = 10$

23. $\frac{1}{2}x - 4 = \frac{3}{4}x$

24. $1 - \frac{1}{2}x = 5$

25. $0.9t = 0.4 + 0.1t$

26. $0.9t = 1 + t$

27. $\frac{2}{y} + \frac{4}{y} = 3$

28. $\frac{4}{y} - 5 = \frac{5}{2y}$

29. $(x + 7)(x - 1) = (x + 1)^2$

30. $(x + 2)(x - 3) = (x - 3)^2$

31. $z(z^2 + 1) = 3 + z^3$

32. $w(4 - w^2) = 8 - w^3$

33. $x^2 = 9x$

34. $x^3 = x^2$

35. $t^3 - 9t^2 = 0$

36. $4z^3 - 8z^2 = 0$

37. $\frac{3}{2x - 3} = \frac{2}{x + 5}$

38. $\frac{-2}{x + 4} = \frac{-3}{x + 1}$

39. $\frac{2}{x - 2} = \frac{3}{x + 5} + \frac{10}{(x + 5)(x - 2)}$

40. $\frac{1}{2x + 3} + \frac{1}{x - 1} = \frac{1}{(2x + 3)(x - 1)}$

41. $x^3 + x^2 - 20x = 0$

42. $x^3 + 6x^2 - 7x = 0$

43. $x^3 + x^2 - x - 1 = 0$

44. $x^3 + 4x^2 - x - 4 = 0$

45. $2x^3 + 4 = x^2 + 8x$

46. $3x^3 + 4x^2 = 27x + 36$

Problems 47–52, list some formulas that occur in applications. Solve each formula for the indicated variable.

47. Electricity $\frac{1}{R} = \frac{1}{R_1} + \frac{1}{R_2}$ for R

48. Finance $A = P(1 + rt)$ for r

49. Mechanics $F = \frac{mv^2}{R}$ for R

50. Chemistry $PV = nRT$ for T

51. Mathematics $S = \frac{a}{1 - r}$ for r

52. Mechanics $v = -gt + v_0$ for t

Discussion and Writing

53. Which of the following pairs of equations are equivalent? Explain.
(a) $x^2 = 9;\ \ x = 3$
(b) $x = \sqrt{9};\ \ x = 3$
(c) $(x - 1)(x - 2) = (x - 1)^2;\ \ x - 2 = x - 1$

54. The equation
$$\frac{5}{x + 3} + 3 = \frac{8 + x}{x + 3}$$
has no solution, yet when we go through the process of solving it we obtain $x = -3$. Write a brief paragraph to explain what causes this to happen.

55. Make up an equation that has no solution and give it to a fellow student to solve. Ask the fellow student to write a critique of your equation.

'Are You Prepared?' Answers

1. $(x - 2)(x + 2)(x - 1)$

2. $\left\{-\frac{5}{3}, 3\right\}$

3. $2x(x - 5)$

4. $(x + 2)(x - 2)(x + 3)$

A.9 Interval Notation; Solving Inequalities

PREPARING FOR THIS SECTION *Before getting started, review the following:*

• Inequalities (Appendix A, Section A.1, p.A9)

✎ Now work the 'Are You Prepared?' problems on page A71.

OBJECTIVES
1 Use Interval Notation
2 Use Properties of Inequalities
3 Solve Linear Inequalities
4 Solve Combined Inequalities

Suppose that a and b are two real numbers and $a < b$. We shall use the notation $a < x < b$ to mean that x is a number *between* a and b. Thus, the expression $a < x < b$ is equivalent to the two inequalities $a < x$ and $x < b$. Similarly, the expression $a \leq x \leq b$ is equivalent to the two inequalities $a \leq x$ and $x \leq b$. The remaining two possibilities, $a \leq x < b$ and $a < x \leq b$, are defined similarly.

Although it is acceptable to write $3 \geq x \geq 2$, it is preferable to reverse the inequality symbols and write instead $2 \leq x \leq 3$ so that, as you read from left to right, the values go from smaller to larger.

A statement such as $2 \leq x \leq 1$ is false because there is no number x for which $2 \leq x$ and $x \leq 1$. Finally, we never mix inequality symbols, as in $2 \leq x \geq 3$.

1 Use Interval Notation

Let a and b represent two real numbers with $a < b$:

A **closed interval**, denoted by **[a, b]**, consists of all real numbers x for which $a \leq x \leq b$.
An **open interval**, denoted by **(a, b)**, consists of all real numbers x for which $a < x < b$.
The **half-open**, or **half-closed**, **intervals** are **(a, b]**, consisting of all real numbers x for which $a < x \leq b$, and **[a, b)**, consisting of all real numbers x for which $a \leq x < b$.

In each of these definitions, a is called the **left endpoint** and b the **right endpoint** of the interval.

The symbol ∞ (read as "infinity") is not a real number but a notational device used to indicate unboundedness in the positive direction. The symbol $-\infty$ (read as "minus infinity" or "negative infinity") also is not a real number, but a notational device used to indicate unboundedness in the negative direction. Using the symbols ∞ and $-\infty$, we can define five other kinds of intervals:

[a, ∞)	consists of all real numbers x for which $x \geq a$
(a, ∞)	consists of all real numbers x for which $x > a$
($-\infty, a$]	consists of all real numbers x for which $x \leq a$
($-\infty, a$)	consists of all real numbers x for which $x < a$
($-\infty, \infty$)	consists of all real numbers x

Note that ∞ and $-\infty$ are never included as endpoints since they are not real numbers.

Table 1 summarizes interval notation, corresponding inequality notation, and their graphs.

Table 1

Interval	Inequality	Graph
The open interval (a, b)	$a < x < b$	
The closed interval $[a, b]$	$a \leq x \leq b$	
The half-open interval $[a, b)$	$a \leq x < b$	
The half-open interval $(a, b]$	$a < x \leq b$	
The interval $[a, \infty)$	$x \geq a$	
The interval (a, ∞)	$x > a$	
The interval $(-\infty, a]$	$x \leq a$	
The interval $(-\infty, a)$	$x < a$	
The interval $(-\infty, \infty)$	All real numbers	

EXAMPLE 1 **Writing Inequalities Using Interval Notation**

Write each inequality using interval notation.

(a) $1 \leq x \leq 3$ (b) $-4 < x < 0$ (c) $x > 5$ (d) $x \leq 1$

Solution (a) $1 \leq x \leq 3$ describes all numbers x between 1 and 3, inclusive. In interval notation, we write $[1, 3]$.

(b) In interval notation, $-4 < x < 0$ is written $(-4, 0)$.

(c) $x > 5$ consists of all numbers x greater than 5. In interval notation, we write $(5, \infty)$.

(d) In interval notation, $x \leq 1$ is written $(-\infty, 1]$. ∎

EXAMPLE 2 **Writing Intervals Using Inequality Notation**

Write each interval as an inequality involving x.

(a) $[1, 4)$ (b) $(2, \infty)$ (c) $[2, 3]$ (d) $(-\infty, -3]$

Solution (a) $[1, 4)$ consists of all numbers x for which $1 \leq x < 4$.

(b) $(2, \infty)$ consists of all numbers x for which $x > 2$.

(c) $[2, 3]$ consists of all numbers x for which $2 \leq x \leq 3$.

(d) $(-\infty, -3]$ consists of all numbers x for which $x \leq -3$. ∎

NOW WORK PROBLEMS 11, 23, AND 31.

2 Use Properties of Inequalities

The product of two positive real numbers is positive, the product of two negative real numbers is positive, and the product of 0 and 0 is 0. For any real number a, the value of a^2 is 0 or positive; that is, a^2 is nonnegative. This is called the **nonnegative property**.

In Words
The square of a real number is never negative.

Nonnegative Property

For any real number a,

$$a^2 \geq 0 \qquad\qquad\qquad \textbf{(1)}$$

If we add the same number to both sides of an inequality, we obtain an equivalent inequality. For example, since $3 < 5$, then $3 + 4 < 5 + 4$ or $7 < 9$. This is called the **addition property** of inequalities.

Addition Property of Inequalities

For real numbers a, b, and c,

$$\text{if } a < b, \text{ then } a + c < b + c \qquad \textbf{(2a)}$$
$$\text{if } a > b, \text{ then } a + c > b + c \qquad \textbf{(2b)}$$

The addition property states that the sense, or direction, of an inequality remains unchanged if the same number is added to each side. Figure 22 illustrates the addition property (2a). In Figure 22(a), we see that a lies to the left of b. If c is positive, then $a + c$ and $b + c$ each lie c units to the right of a and b, respectively. Consequently, $a + c$ must lie to the left of $b + c$; that is, $a + c < b + c$. Figure 22(b) illustrates the situation if c is negative.

Figure 22

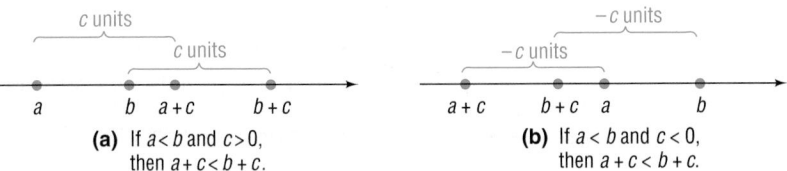

(a) If $a < b$ and $c > 0$,
then $a + c < b + c$.

(b) If $a < b$ and $c < 0$,
then $a + c < b + c$.

DRAW AN ILLUSTRATION SIMILAR TO FIGURE 22 THAT ILLUSTRATES THE ADDITION PROPERTY (2b).

EXAMPLE 3 | **Addition Property of Inequalities**

(a) If $x < -5$, then $x + 5 < -5 + 5$ or $x + 5 < 0$.
(b) If $x > 2$, then $x + (-2) > 2 + (-2)$ or $x - 2 > 0$. ■

NOW WORK PROBLEM 39.

We will use two examples to arrive at our next property.

EXAMPLE 4 | **Multiplying an Inequality by a Positive Number**

Express as an inequality the result of multiplying each side of the inequality $3 < 7$ by 2.

Solution | We begin with

$$3 < 7$$

Multiplying each side by 2 yields the numbers 6 and 14, so we have

$$6 < 14$$ ■

| EXAMPLE 5 | **Multiplying an Inequality by a Negative Number** |

Express as an inequality the result of multiplying each side of the inequality $9 > 2$ by -4.

Solution We begin with

$$9 > 2$$

Multiplying each side by -4 yields the numbers -36 and -8, so we have

$$-36 < -8$$ ∎

Note that the effect of multiplying both sides of $9 > 2$ by the negative number -4 is that the direction of the inequality symbol is reversed.

Examples 4 and 5 illustrate the following general **multiplication properties** for inequalities:

Multiplication Properties for Inequalities

For real numbers a, b, and c,

In Words

Multiplying by a negative number reverses the direction of the inequality.

$$
\begin{array}{ll}
\text{if } a < b \text{ and if } c > 0, \text{ then } ac < bc & \\
\text{if } a < b \text{ and if } c < 0, \text{ then } ac > bc & \textbf{(3a)} \\
\text{if } a > b \text{ and if } c > 0, \text{ then } ac > bc & \\
\text{if } a > b \text{ and if } c < 0, \text{ then } ac < bc & \textbf{(3b)}
\end{array}
$$

The multiplication properties state that the sense, or direction, of an inequality *remains the same* if each side is multiplied by a *positive* real number, whereas the direction is *reversed* if each side is multiplied by a *negative* real number.

| EXAMPLE 6 | **Multiplication Property of Inequalities** |

(a) If $2x < 6$, then $\dfrac{1}{2}(2x) < \dfrac{1}{2}(6)$ or $x < 3$.

(b) If $\dfrac{x}{-3} > 12$, then $-3\left(\dfrac{x}{-3}\right) < -3(12)$ or $x < -36$.

(c) If $-4x > -8$, then $\dfrac{-4x}{-4} < \dfrac{-8}{-4}$ or $x < 2$.

(d) If $-x < 8$, then $(-1)(-x) > (-1)(8)$ or $x > -8$. ∎

✎━━━ **NOW WORK PROBLEM 45.**

3 Solve Linear Inequalities

An **inequality in one variable** is a statement involving two expressions, at least one containing the variable, separated by one of the inequality symbols, $<$, \le, $>$, or \ge. To **solve an inequality** means to find all values of the variable for which the statement is true. These values are called **solutions** of the inequality.

For example, the following are all inequalities involving one variable, x:

$$x + 5 < 8 \qquad 2x - 3 \ge 4 \qquad x^2 - 1 \le 3 \qquad \frac{x + 1}{x - 2} > 0$$

Two inequalities having exactly the same solution set are called **equivalent inequalities**. As with equations, one method for solving an inequality is to replace it by a series of equivalent inequalities until an inequality with an obvious solution, such as $x < 3$, is obtained. We obtain equivalent inequalities by applying some of the same operations as those used to find equivalent equations. The addition property and the multiplication properties form the basis for the following procedures.

Procedures That Leave the Inequality Symbol Unchanged

1. Simplify both sides of the inequality by combining like terms and eliminating parentheses:

 Replace $\quad x + 2 + 6 > 2x + 5(x + 1)$

 by $\qquad x + 8 > 7x + 5$

2. Add or subtract the same expression on both sides of the inequality:

 Replace $\qquad 3x - 5 < 4$

 by $\quad (3x - 5) + 5 < 4 + 5$

3. Multiply or divide both sides of the inequality by the same *positive* expression:

 Replace $\quad 4x > 16$ by $\dfrac{4x}{4} > \dfrac{16}{4}$

Procedures That Reverse the Sense or Direction of the Inequality Symbol

1. Interchange the two sides of the inequality:

 Replace $\quad 3 < x$ by $x > 3$

2. Multiply or divide both sides of the inequality by the same *negative* expression:

 Replace $\quad -2x > 6$ by $\dfrac{-2x}{-2} < \dfrac{6}{-2}$

As the examples that follow illustrate, we solve inequalities using many of the same steps that we would use to solve equations. In writing the solution of an inequality, we may use either set notation or interval notation, whichever is more convenient.

EXAMPLE 7	**Solving an Inequality**

Solve the inequality $4x + 7 \geq 2x - 3$, and graph the solution set.

Solution

$$4x + 7 \geq 2x - 3$$

$$4x + 7 - 7 \geq 2x - 3 - 7 \qquad \text{Subtract 7 from both sides.}$$

$$4x \geq 2x - 10 \qquad \text{Simplify.}$$

$$4x - 2x \geq 2x - 10 - 2x \qquad \text{Subtract 2x from both sides.}$$

$$2x \geq -10 \qquad \text{Simplify.}$$

$$\frac{2x}{2} \geq \frac{-10}{2} \qquad \text{Divide both sides by 2. (The direction of the inequality symbol is unchanged.)}$$

$$x \geq -5 \qquad \text{Simplify.}$$

Figure 23

The solution set is $\{x \mid x \geq -5\}$ or, using interval notation, all numbers in the interval $[-5, \infty)$.

See Figure 23 for the graph of the solution set. ■

NOW WORK PROBLEM 53.

101. A car that averages 25 miles per gallon has a tank that holds 20 gallons of gasoline. After a trip that covered at least 300 miles, the car ran out of gasoline. What is the range of the amount of gasoline (in gallons) that was in the tank at the start of the trip?

102. Repeat Problem 101 if the same car runs out of gasoline after a trip of no more than 250 miles.

103. Arithmetic Mean If $a < b$, show that $a < \dfrac{a + b}{2} < b$. The number $\dfrac{a + b}{2}$ is called the **arithmetic mean** of a and b.

104. Refer to Problem 103. Show that the arithmetic mean of a and b is equidistant from a and b.

105. Geometric Mean If $0 < a < b$, show that $a < \sqrt{ab} < b$. The number \sqrt{ab} is called the **geometric mean** of a and b.

106. Refer to Problems 103 and 105. Show that the geometric mean of a and b is less than the arithmetic mean of a and b.

107. Harmonic Mean For $0 < a < b$, let h be defined by

$$\frac{1}{h} = \frac{1}{2}\left(\frac{1}{a} + \frac{1}{b}\right)$$

Show that $a < h < b$. The number h is called the **harmonic mean** of a and b.

108. Refer to Problems 103, 105, and 107. Show that the harmonic mean of a and b equals the geometric mean squared divided by the arithmetic mean.

Discussion and Writing

109. Make up an inequality that has no solution. Make up one that has exactly one solution.

110. How would you explain to a fellow student the underlying reason for the multiplication properties for inequalities (page A69); that is, the sense or direction of an inequality remains the same if each side is multiplied by a positive real number, while the direction is reversed if each side is multiplied by a negative real number.

111. The inequality $x^2 + 1 < -5$ has no solution. Explain why.

112. Do you prefer to use inequality notation or interval notation to express the solution to an inequality? Give your reasons. Are there particular circumstances when you prefer one to the other? Cite examples.

'Are You Prepared?' Answers

1.

2.

A.10 Problem Solving: Applications of Linear Equations

> **OBJECTIVES**
> 1 Translate Verbal Descriptions into Mathematical Expressions
> 2 Solve Interest Problems
> 3 Solve Mixture Problems
> 4 Solve Uniform Motion Problems
> 5 Solve Constant Rate Job Problems

Applied (word) problems do not come in the form "Solve the equation . . ." Instead, they supply information using words, a verbal description of the real problem. So, to solve applied problems, we must be able to translate the verbal description into the language of mathematics. We do this by using variables to represent unknown quantities and then finding relationships (such as equations) that involve these variables. The process of doing all this is called **mathematical modeling.**

Any solution to the mathematical problem must be checked against the mathematical problem, the verbal description, and the real problem. See Figure 25 for an illustration of the **modeling process.**

Figure 25

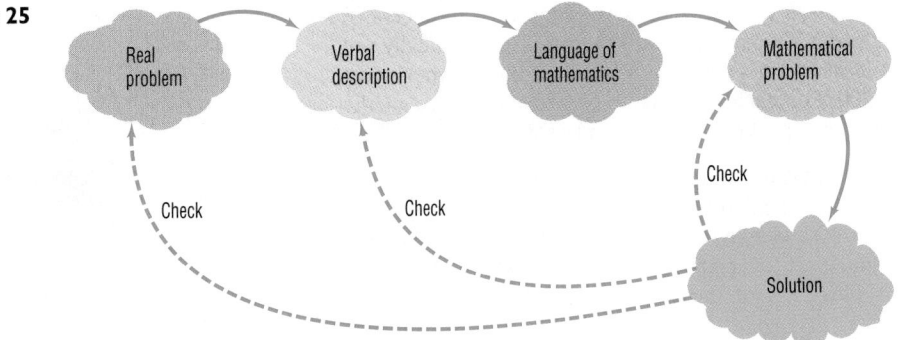

74. $0 < \dfrac{3x + 2}{2} < 4$

75. $1 < 1 - \dfrac{1}{2}x < 4$

76. $0 < 1 - \dfrac{1}{3}x < 1$

77. $(x + 2)(x - 3) > (x - 1)(x + 1)$

78. $(x - 1)(x + 1) > (x - 3)(x + 4)$

79. $x(4x + 3) \le (2x + 1)^2$

80. $x(9x - 5) \le (3x - 1)^2$

81. $\dfrac{1}{2} \le \dfrac{x + 1}{3} < \dfrac{3}{4}$

82. $\dfrac{1}{3} < \dfrac{x + 1}{2} \le \dfrac{2}{3}$

83. $-3 < x + 5 < 2x$

84. $2 < x - 3 < 2x$

85. $x + 2 < 2x - 1 < 5x$

86. $2x - 1 < 3x + 5 < 5x - 7$

Applications and Extensions

87. A young adult may be defined as someone older than 21, but less than 30 years of age. Express this statement using inequalities.

88. Middle-aged may be defined as being 40 or more and less than 60. Express this statement using inequalities.

89. Life Expectancy Metropolitan Life Insurance Co. reported that an average 25-year-old male in 2000 could expect to live at least 50.6 more years, and an average 25-year-old female in 2000 could expect to live at least 55.4 more years.

(a) To what age can an average 25-year-old male expect to live? Express your answer as an inequality.

(b) To what age can an average 25-year-old female expect to live? Express your answer as an inequality.

(c) Who can expect to live longer, a male or a female? By how many years?

90. General Chemistry For a certain ideal gas, the volume V (in cubic centimeters) equals 20 times the temperature T in kelvins (K). If the temperature varies from 353 to 393 K, inclusive, what is the corresponding range of the volume of the gas?

91. Real Estate A real estate agent agrees to sell a large apartment complex according to the following commission schedule: $45,000 plus 25% of the selling price in excess of $900,000. Assuming that the complex will sell at some price between $900,000 and $1,100,000, inclusive, over what range does the agent's commission vary? How does the commission vary as a percent of selling price?

92. Sales Commission A used car salesperson is paid a commission of $25 plus 40% of the selling price in excess of owner's cost. The owner claims that used cars typically sell for at least owner's cost plus $70 and at most owner's cost plus $300. For each sale made, over what range can the salesperson expect the commission to vary?

93. Federal Tax Withholding The percentage method of withholding for federal income tax (2004) states that a single person whose weekly wages, after subtracting withholding allowances, are over $592, but not over $1317, shall have $74.35 plus 25% of the excess over $592 withheld. Over

what range does the amount withheld vary if the weekly wages vary from $600 to $700, inclusive?

SOURCE: *Employer's Tax Guide.* Department of the Treasury, Internal Revenue Service, 2004.

94. Federal Tax Withholding Rework Problem 93 if the weekly wages vary from $800 to $900, inclusive.

95. Electricity Rates Commonwealth Edison Company's summer charge for electricity is 8.275¢ per kilowatt-hour. In addition, each monthly bill contains a customer charge of $7.13. If last summer's bills ranged from a low of $65.96 to a high of $217.02, over what range did usage vary (in kilowatt-hours)?

SOURCE: Commonwealth Edison Co., Chicago, Illinois, 2005.

96. Water Bills The Village of Oak Lawn charges homeowners $27.18 per quarter-year plus $1.90 per 1000 gallons for water usage in excess of 12,000 gallons. In 2005, one homeowner's quarterly bill ranged from a high of $43.33 to a low of $30.03. Over what range did water usage vary?

SOURCE: Village of Oak Lawn, Illinois, 2005.

97. Markup of a New Car The markup over dealer's cost of a new car ranges from 12% to 18%. If the sticker price is $8800, over what range will the dealer's cost vary?

98. IQ Tests A standard intelligence test has an average score of 100. According to statistical theory, of the people who take the test, the 2.5% with the highest scores will have scores of more than 1.96σ above the average, where σ (sigma, a number called the *standard deviation*) depends on the nature of the test. If $\sigma = 12$ for this test and there is (in principle) no upper limit to the score possible on the test, write the interval of possible test scores of the people in the top 2.5%.

99. Computing Grades In your Economics 101 class, you have scores of 68, 82, 87, and 89 on the first four of five tests. To get a grade of B, the average of the first five test scores must be greater than or equal to 80 and less than 90. Solve an inequality to find the range of the score that you need on the last test to get a B.

What do I need to get a B?

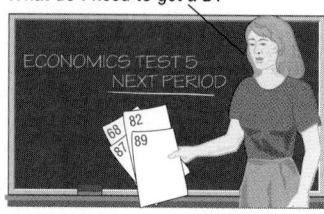

100. Computing Grades Repeat Problem 99 if the fifth test counts double.

In Problems 6–9, determine if the statement is True or False if $a < b$ and $c < 0$.

6. $a + c < b + c$ **7.** $a - c < b - c$ **8.** $ac > bc$ **9.** $\dfrac{a}{c} < \dfrac{b}{c}$

10. *True or False*: The square of any real number is always nonnegative.

Skill Building

In Problems 11–16, express the graph shown in color using interval notation. Also express each graph as an inequality involving x.

11.

12.

13.

14.

15.

16.

In Problems 17–22, an inequality is given. Write the inequality obtained by:
(a) Adding 3 to each side of the given inequality. (b) Subtracting 5 from each side of the given inequality.
(c) Multiplying each side of the given inequality by 3. (d) Multiplying each side of the given inequality by -2.

17. $3 < 5$ **18.** $2 > 1$ **19.** $4 > -3$

20. $-3 > -5$ **21.** $2x + 1 < 2$ **22.** $1 - 2x > 5$

In Problems 23–30, write each inequality using interval notation, and illustrate each inequality using the real number line.

23. $0 \le x \le 4$ **24.** $-1 < x < 5$ **25.** $4 \le x < 6$ **26.** $-2 < x < 0$

27. $x \ge 4$ **28.** $x \le 5$ **29.** $x < -4$ **30.** $x > 1$

In Problems 31–38, write each interval as an inequality involving x, and illustrate each inequality using the real number line.

31. $[2, 5]$ **32.** $(1, 2)$ **33.** $(-3, -2)$ **34.** $[0, 1)$

35. $[4, \infty)$ **36.** $(-\infty, 2]$ **37.** $(-\infty, -3)$ **38.** $(-8, \infty)$

In Problems 39–52, fill in the blank with the correct inequality symbol.

39. If $x < 5$, then $x - 5$ _____ 0. **40.** If $x < -4$, then $x + 4$ _____ 0.

41. If $x > -4$, then $x + 4$ _____ 0. **42.** If $x > 6$, then $x - 6$ _____ 0.

43. If $x \ge -4$, then $3x$ _____ -12. **44.** If $x \le 3$, then $2x$ _____ 6.

45. If $x > 6$, then $-2x$ _____ -12. **46.** If $x > -2$, then $-4x$ _____ 8.

47. If $x \ge 5$, then $-4x$ _____ -20. **48.** If $x \le -4$, then $-3x$ _____ 12.

49. If $2x < 6$, then x _____ 3. **50.** If $3x \le 12$, then x _____ 4.

51. If $-\dfrac{1}{2}x \le 3$, then x _____ -6. **52.** If $-\dfrac{1}{4}x > 1$, then x _____ -4.

In Problems 53–86, solve each inequality. Express your answer using set notation or interval notation. Graph the solution set.

53. $x + 1 < 5$ **54.** $x - 6 < 1$ **55.** $1 - 2x \le 3$

56. $2 - 3x \le 5$ **57.** $3x - 7 > 2$ **58.** $2x + 5 > 1$

59. $3x - 1 \ge 3 + x$ **60.** $2x - 2 \ge 3 + x$ **61.** $-2(x + 3) < 8$

62. $-3(1 - x) < 12$ **63.** $4 - 3(1 - x) \le 3$ **64.** $8 - 4(2 - x) \le -2x$

65. $\dfrac{1}{2}(x - 4) > x + 8$ **66.** $3x + 4 > \dfrac{1}{3}(x - 2)$ **67.** $\dfrac{x}{2} \ge 1 - \dfrac{x}{4}$

68. $\dfrac{x}{3} \ge 2 + \dfrac{x}{6}$ **69.** $0 \le 2x - 6 \le 4$ **70.** $4 \le 2x + 2 \le 10$

71. $-5 \le 4 - 3x \le 2$ **72.** $-3 \le 3 - 2x \le 9$ **73.** $-3 < \dfrac{2x - 1}{4} < 0$

4 Solve Combined Inequalities

Now let's look at how to solve combined inequalities.

EXAMPLE 8

Solving Combined Inequalities

Solve the inequality $-5 < 3x - 2 < 1$ and draw a graph to illustrate the solution.

Solution Recall that the inequality

$$-5 < 3x - 2 < 1$$

is equivalent to the two inequalities

$$-5 < 3x - 2 \quad \text{and} \quad 3x - 2 < 1$$

We will solve each of these inequalities separately.

$$-5 < 3x - 2 \qquad\qquad\qquad\qquad\qquad 3x - 2 < 1$$
$$-5 + 2 < 3x - 2 + 2 \quad \text{Add 2 to both sides.} \quad 3x - 2 + 2 < 1 + 2$$
$$-3 < 3x \qquad\qquad\qquad\qquad\qquad 3x < 3$$
$$\frac{-3}{3} < \frac{3x}{3} \qquad \text{Divide both sides by 3.} \quad \frac{3x}{3} < \frac{3}{3}$$
$$-1 < x \qquad\qquad \text{Simplify.} \qquad\qquad x < 1$$

The solution set of the original pair of inequalities consists of all x for which

$$-1 < x \quad \text{and} \quad x < 1$$

This may be written more compactly as $\{x | -1 < x < 1\}$. In interval notation, the solution is $(-1, 1)$.

See Figure 24 for the graph of the solution set. ∎

Figure 24

We observe in the solution of Example 8 that the two inequalities that were solved required exactly the same steps. A shortcut to solving the original inequality is to deal with the two inequalities at the same time, as follows:

$$-5 < \qquad 3x - 2 < 1$$
$$-5 + 2 < 3x - 2 + 2 < 1 + 2 \quad \text{Add 2 to each part.}$$
$$-3 < \qquad 3x \qquad < 3 \quad \text{Simplify.}$$
$$\frac{-3}{3} < \qquad \frac{3x}{3} \qquad < \frac{3}{3} \quad \text{Divide each part by 3.}$$
$$-1 < \qquad x \qquad < 1 \quad \text{Simplify.}$$

NOW WORK PROBLEM 73.

A.9 Assess Your Understanding

'Are You Prepared?'

Answers are given at the end of these exercises. If you get a wrong answer, read the pages listed in red.

1. Graph the inequality: $x \geq -2$. (p. A5)

2. Graph the inequality: $x < 1$. (p. A5)

Concepts and Vocabulary

3. If each side of an inequality is multiplied by a(n) _____ number, then the direction of the inequality symbol is reversed.

4. A(n) _____ _____, denoted $[a, b]$, consists of all real numbers x for which $a \leq x \leq b$.

5. The _____ _____ state that the sense, or direction, of an inequality remains the same if each side is multiplied by a positive number, while the direction is reversed if each side is multiplied by a negative number.

1 Translate Verbal Descriptions into Mathematical Expressions

Let's look at a few examples that will help you to translate certain words into mathematical symbols.

EXAMPLE 1 | **Translating Verbal Descriptions into Mathematical Expressions**

(a) For uniform motion, the velocity of an object equals the distance traveled divided by the time required.

 Translation: If v is the velocity, s the distance, and t the time, then $v = \dfrac{s}{t}$.

(b) Let x denote a number.

 The number 5 times as large as x is $5x$.
 The number 3 less than x is $x - 3$.
 The number that exceeds x by 4 is $x + 4$.
 The number that, when added to x, gives 5 is $5 - x$. ∎

 ✏️ NOW WORK PROBLEM 7.

Always check the units used to measure the variables of an applied problem. In Example 1(a), if v is measured in miles per hour, the distance s must be expressed in miles and the time t must be expressed in hours. It is a good practice to check units to be sure that they are consistent and make sense.

The steps to follow for setting up applied problems are given below.

Steps for Setting Up Applied Problems

STEP 1: Read the problem carefully, perhaps two or three times. Pay particular attention to the question being asked to identify what you are looking for. If you can, determine realistic possibilities for the answer.

STEP 2: Assign a letter (variable) to represent what you are looking for and, if necessary, express any remaining unknown quantities in terms of this variable.

STEP 3: Make a list of all the known facts, and translate them into mathematical expressions. These may take the form of an equation (or, later, an inequality) involving the variable. If possible, draw an appropriately labeled diagram to assist you. Sometimes a table or chart helps.

STEP 4: Solve the equation for the variable, and then answer the question.

STEP 5: Check the answer with the facts in the problem. If it agrees, congratulations! If it does not agree, try again.

2 Solve Interest Problems

Interest is money paid for the use of money. The total amount borrowed (whether by an individual from a bank in the form of a loan or by a bank from an individual in the form of a savings account) is called the **principal.** The **rate of interest,** expressed as a percent, is the amount charged for the use of the principal for a given period of time, usually on a yearly (that is, per annum) basis.

Simple Interest Formula

If a principal of P dollars is borrowed for a period of t years at a per annum interest rate r, expressed as a decimal, the interest I charged is

$$I = Prt \qquad (1)$$

Interest charged according to formula (1) is called **simple interest.** When using formula (1), be sure to express r as a decimal.

| EXAMPLE 2 | Finance: Computing Interest on a Loan |

Suppose that Juanita borrows $500 for 6 months at the simple interest rate of 9% per annum. What is the interest that Juanita will be charged on this loan? How much does Juanita owe after 6 months?

Solution The rate of interest is given per annum, so the actual time that the money is borrowed must be expressed in years. The interest charged would be the principal, $500, times the rate of interest (9% = 0.09) times the time in years, $\frac{1}{2}$:

$$\text{Interest charged} = I = Prt = (500)(0.09)\left(\frac{1}{2}\right) = \$22.50$$

After 6 months, Juanita will owe what she borrowed plus the interest:
$$\$500 + \$22.50 = \$522.50 \qquad \blacksquare$$

| EXAMPLE 3 | Financial Planning |

Candy has $70,000 to invest and requires an overall rate of return of 9%. She can invest in a safe, government-insured certificate of deposit, but it only pays 8%. To obtain 9%, she agrees to invest some of her money in noninsured corporate bonds paying 12%. How much should be placed in each investment to achieve her goal?

Solution **STEP 1:** The question is asking for two dollar amounts: the principal to invest in the corporate bonds and the principal to invest in the certificate of deposit.

STEP 2: We let x represent the amount (in dollars) to be invested in the bonds. Then $70,000 - x$ is the amount that will be invested in the certificate. (Do you see why?)

STEP 3: We set up a table:

	Principal ($)	Rate	Time (yr)	Interest ($)
Bonds	x	12% = 0.12	1	0.12x
Certificate	70,000 − x	8% = 0.08	1	0.08(70,000 − x)
Total	70,000	9% = 0.09	1	0.09(70,000) = 6300

Since the total interest from the investments is equal to $0.09(70,000) = 6300$, we must have the equation

$$0.12x + 0.08(70,000 - x) = 6300$$

(Note that the units are consistent: the unit is dollars on each side.)

STEP 4: $0.12x + 5600 - 0.08x = 6300$
$$0.04x = 700$$
$$x = 17,500$$

Candy should place $17,500 in the bonds and $70,000 - \$17,500 = \$52,500$ in the certificate.

STEP 5: The interest on the bonds after 1 year is $0.12(\$17,500) = \2100; the interest on the certificate after 1 year is $0.08(\$52,500) = \4200. The total annual interest is $6300, the required amount. $\qquad \blacksquare$

— NOW WORK PROBLEM **17.**

3 Mixture Problems

Oil refineries sometimes produce gasoline that is a blend of two or more types of fuel; bakeries occasionally blend two or more types of flour for their bread. These problems are referred to as **mixture problems** because they combine two or more quantities to form a mixture.

| EXAMPLE 4 | Blending Coffees |

The manager of a Starbucks store decides to experiment with a new blend of coffee. She will mix some B grade Colombian coffee that sells for $5 per pound with some A grade Arabica coffee that sells for $10 per pound to get 100 pounds of the new blend. The selling price of the new blend is to be $7 per pound, and there is to be no difference in revenue from selling the new blend versus selling the other types. How many pounds of the B grade Colombian and A grade Arabica coffees are required?

Solution Let x represent the number of pounds of the B grade Colombian coffee. Then $100 - x$ equals the number of pounds of the A grade Arabica coffee. See Figure 26.

Figure 26

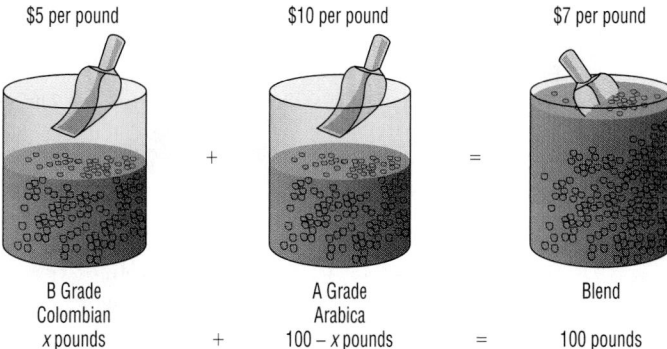

$5 per pound · $10 per pound · $7 per pound

B Grade Colombian x pounds + A Grade Arabica $100 - x$ pounds = Blend 100 pounds

Since there is to be no difference in revenue between selling the A and B grades separately versus the blend, we have

$$\left\{\begin{array}{c}\text{Price per pound} \\ \text{of B grade}\end{array}\right\}\left\{\begin{array}{c}\text{\# Pounds} \\ \text{B grade}\end{array}\right\} + \left\{\begin{array}{c}\text{Price per pound} \\ \text{of A grade}\end{array}\right\}\left\{\begin{array}{c}\text{\# Pounds} \\ \text{A grade}\end{array}\right\} = \left\{\begin{array}{c}\text{Price per pound} \\ \text{of blend}\end{array}\right\}\left\{\begin{array}{c}\text{\# Pounds} \\ \text{blend}\end{array}\right\}$$

$$\$5 \quad \cdot \quad x \quad + \quad \$10 \quad \cdot \, (100 - x) \quad = \quad \$7 \quad \cdot \quad 100$$

We have the equation

$$5x + 10(100 - x) = 700$$
$$5x + 1000 - 10x = 700$$
$$-5x = -300$$
$$x = 60$$

The manager should blend 60 pounds of B grade Colombian coffee with $100 - 60 = 40$ pounds of A grade Arabica coffee to get the desired blend.

✔ CHECK: The 60 pounds of B grade coffee would sell for $(\$5)(60) = \300, and the 40 pounds of A grade coffee would sell for $(\$10)(40) = \400; the total revenue, $700, equals the revenue obtained from selling the blend, as desired. ∎

━━━━━━ NOW WORK PROBLEM 21.

4 Solve Uniform Motion Problems

Objects that move at a constant velocity are said to be in **uniform motion.** When the average velocity of an object is known, it can be interpreted as its constant velocity. For example, a bicyclist traveling at an average velocity of 25 miles per hour is in uniform motion.

> **Uniform Motion Formula**
>
> If an object moves at an average velocity v,* the distance s covered in time t is given by the formula
>
> $$s = vt \qquad \qquad \text{(2)}$$
>
> That is, Distance = Velocity · Time.

| EXAMPLE 5 | **Physics: Uniform Motion** |

Tanya, who is a long-distance runner, runs at an average velocity of 8 miles per hour (mi/hr). Two hours after Tanya leaves your house, you leave in your Honda and follow the same route. If your average velocity is 40 mi/hr, how long will it be before you catch up to Tanya? How far will each of you be from your home?

Solution Refer to Figure 27. We use t to represent the time (in hours) that it takes the Honda to catch up to Tanya. When this occurs, the total time elapsed for Tanya is $t + 2$ hours. Set up the following table:

Figure 27

	Velocity mi/hr	Time hr	Distance mi
Tanya	8	$t + 2$	$8(t + 2)$
Honda	40	t	$40t$

Since the distance traveled is the same, we are led to the following equation:

$$8(t + 2) = 40t$$
$$8t + 16 = 40t$$
$$32t = 16$$
$$t = \frac{1}{2} \text{ hour}$$

It will take the Honda $\frac{1}{2}$ hour to catch up to Tanya. Each of you will have gone 20 miles.

✔ **CHECK:** In 2.5 hours, Tanya travels a distance of $(2.5)(8) = 20$ miles. In $\frac{1}{2}$ hour, the Honda travels a distance of $\left(\frac{1}{2}\right)(40) = 20$ miles. ∎

=== **NOW WORK PROBLEM 27.**

5 Solve Constant Rate Job Problems

This section involves jobs that are performed at a **constant rate.** Our assumption is that, if a job can be done in t units of time, $\frac{1}{t}$ of the job is done in 1 unit of time. Let's look at an example.

*Technically, velocity has both magnitude (speed) and direction. Since all velocities in this section are in the positive direction, we make no distinction between speed and velocity.

| EXAMPLE 6 | **Working Together to Do a Job** |

At 10 AM Danny is asked by his father to weed the garden. From past experience, Danny knows that this will take him 4 hours, working alone. His older brother, Mike, when it is his turn to do this job, requires 6 hours. Since Mike wants to go golfing with Danny and has a reservation for 1 PM, he agrees to help Danny. Assuming no gain or loss of efficiency, when will they finish if they work together? Can they make the golf date?

Solution We set up Table 2. In 1 hour, Danny does $\frac{1}{4}$ of the job, and in 1 hour, Mike does $\frac{1}{6}$ of the job. Let t be the time (in hours) that it takes them to do the job together. In 1 hour, then, $\frac{1}{t}$ of the job is completed. We reason as follows:

$$\left(\begin{array}{c}\text{Part done by Danny}\\\text{in 1 hour}\end{array}\right) + \left(\begin{array}{c}\text{Part done by Mike}\\\text{in 1 hour}\end{array}\right) = \left(\begin{array}{c}\text{Part done together}\\\text{in 1 hour}\end{array}\right)$$

Table 2

	Hours to Do Job	Part of Job Done in 1 Hour
Danny	4	$\frac{1}{4}$
Mike	6	$\frac{1}{6}$
Together	t	$\frac{1}{t}$

From Table 2,

$$\frac{1}{4} + \frac{1}{6} = \frac{1}{t}$$

$$\frac{3 + 2}{12} = \frac{1}{t}$$

$$\frac{5}{12} = \frac{1}{t}$$

$$5t = 12$$

$$t = \frac{12}{5}$$

Working together, the job can be done in $\frac{12}{5}$ hours, or 2 hours, 24 minutes. They should make the golf date, since they will finish at 12:24 PM. ■

NOW WORK PROBLEM 29.

A.10 Assess Your Understanding

Concepts and Vocabulary

1. Using variables to represent unknown quantities and then finding relationships that involve these variables is referred to as _____ _____.

2. The money paid for the use of money is _____.

3. Objects that move at a constant velocity are said to be in _____ _____.

4. *True or False:* The amount charged for the use of principal for a given period of time is called the rate of interest.

5. *True or False:* If an object moves at an average velocity v, the distance s covered in time t is given by the formula $s = vt$.

6. Suppose that you want to mix two coffees to obtain 100 pounds of the blend. If x represents the number of pounds of coffee A, write an algebraic expression that represents the number of pounds of coffee B.

Applications and Extensions

In Problems 7–16, translate each sentence into a mathematical equation. Be sure to identify the meaning of all symbols.

7. Geometry The area of a circle is the product of the number π and the square of the radius.

8. Geometry The circumference of a circle is the product of the number π and twice the radius.

9. **Geometry** The area of a square is the square of the length of a side.

10. **Geometry** The perimeter of a square is four times the length of a side.

11. **Physics** Force equals the product of mass and acceleration.

12. **Physics** Pressure is force per unit area.

13. **Physics** Work equals force times distance.

14. **Physics** Kinetic energy is one-half the product of the mass and the square of the velocity.

15. **Business** The total variable cost of manufacturing x dishwashers is $150 per dishwasher times the number of dishwashers manufactured.

16. **Business** The total revenue derived from selling x dishwashers is $250 per dishwasher times the number of dishwashers sold.

17. **Financial Planning** Betsy, a recent retiree, requires $6000 per year in extra income. She has $50,000 to invest and can invest in B-rated bonds paying 15% per year or in a certificate of deposit (CD) paying 7% per year. How much money should be invested in each to realize exactly $6000 in interest per year?

18. **Financial Planning** After 2 years, Betsy (see Problem 17) finds that she will now require $7000 per year. Assuming that the remaining information is the same, how should the money be reinvested?

19. **Banking** A bank loaned out $12,000, part of it at the rate of 8% per year and the rest at the rate of 18% per year. If the interest received totaled $1000, how much was loaned at 8%?

20. **Banking** Wendy, a loan officer at a bank, has $1,000,000 to lend and is required to obtain an average return of 18% per year. If she can lend at the rate of 19% or at the rate of 16%, how much can she lend at the 16% rate and still meet her requirement?

21. **Blending Teas** The manager of a store that specializes in selling tea decides to experiment with a new blend. She will mix some Earl Grey tea that sells for $5 per pound with some Orange Pekoe tea that sells for $3 per pound to get 100 pounds of the new blend. The selling price of the new blend is to be $4.50 per pound, and there is to be no difference in revenue from selling the new blend versus selling the other types. How many pounds of the Earl Grey tea and Orange Pekoe tea are required?

22. **Business: Blending Coffee** A coffee manufacturer wants to market a new blend of coffee that sells for $3.90 per pound by mixing two coffees that sell for $2.75 and $5 per pound, respectively. What amounts of each coffee should be blended to obtain the desired mixture?
[**Hint:** Assume that the total weight of the desired blend is 100 pounds.]

23. **Business: Mixing Nuts** A nut store normally sells cashews for $4.00 per pound and peanuts for $1.50 per pound. But at the end of the month the peanuts had not sold well, so, in order to sell 60 pounds of peanuts, the manager decided to mix the 60 pounds of peanuts with some cashews and sell the mixture for $2.50 per pound. How many pounds of cashews should be mixed with the peanuts to ensure no change in the profit?

24. **Business: Mixing Candy** A candy store sells boxes of candy containing caramels and cremes. Each box sells for $12.50 and holds 30 pieces of candy (all pieces are the same size). If the caramels cost $0.25 to produce and the cremes cost $0.45 to produce, how many of each should be in a box to make a profit of $3?

25. **Physics: Uniform Motion** A motorboat can maintain a constant speed of 16 miles per hour relative to the water. The boat makes a trip upstream to a certain point in 20 minutes; the return trip takes 15 minutes. What is the speed of the current? (See the figure.)

26. **Physics: Uniform Motion** A motorboat heads upstream on a river that has a current of 3 miles per hour. The trip upstream takes 5 hours, and the return trip takes 2.5 hours. What is the speed of the motorboat? (Assume that the motorboat maintains a constant speed relative to the water.)

27. **A Biathlon** Suppose that you have entered an 87-mile biathlon that consists of a run and a bicycle race. During your run, your average velocity is 6 miles per hour, and during your bicycle race, your average velocity is 25 miles per hour. You finish the race in 5 hours. What is the distance of the run? What is the distance of the bicycle race?

28. **Cyclists** Two cyclists leave a city at the same time, one going east and the other going west. The westbound cyclist bikes at 5 mph faster than the eastbound cyclist. After 6 hours they are 246 miles apart, how fast is each cyclist riding?

29. **Working Together on a Job** Trent can deliver his newspapers in 30 minutes. It takes Lois 20 minutes to do the same route. How long would it take them to deliver the newspapers if they work together?

30. **Working Together on a Job** Patrice, by himself, can paint four rooms in 10 hours. If he hires April to help, they can do the same job together in 6 hours. If he lets April work alone, how long will it take her to paint four rooms?

31. **Enclosing a Garden** A gardener has 46 feet of fencing to be used to enclose a rectangular garden that has a border 2 feet wide surrounding it (see the figure).
(a) If the length of the garden is to be twice its width, what will be the dimensions of the garden?
(b) What is the area of the garden?
(c) If the length and width of the garden are to be the same, what would be the dimensions of the garden?
(d) What would be the area of the square garden?

32. Construction A pond is enclosed by a wooden deck that is 3 feet wide. The fence surrounding the deck is 100 feet long.

(a) If the pond is square, what are its dimensions?

(b) If the pond is rectangular and the length of the pond is to be three times its width, what are its dimensions?

(c) If the pond is circular, what is its diameter?

(d) Which pond has the most area?

33. Football A tight end can run the 100-yard dash in 12 seconds. A defensive back can do it in 10 seconds. The tight end catches a pass at his own 20-yard line with the defensive back at the 15-yard line. (See the figure.) If no other players are nearby, at what yard line will the defensive back catch up to the tight end?

[**Hint**: At time $t = 0$, the defensive back is 5 yards behind the tight end.]

34. Computing Business Expense Therese, an outside salesperson, uses her car for both business and pleasure. Last year, she traveled 30,000 miles, using 900 gallons of gasoline. Her car gets 40 miles per gallon on the highway and 25 in the city. She can deduct all highway travel, but no city travel, on her taxes. How many miles should Therese be allowed as a business expense?

35. Mixing Water and Antifreeze How much water should be added to 1 gallon of pure antifreeze to obtain a solution that is 60% antifreeze?

36. Mixing Water and Antifreeze The cooling system of a certain foreign-made car has a capacity of 15 liters. If the system is filled with a mixture that is 40% antifreeze, how much of this mixture should be drained and replaced by pure antifreeze so that the system is filled with a solution that is 60% antifreeze?

37. Chemistry: Salt Solutions How much water must be evaporated from 32 ounces of a 4% salt solution to make a 6% salt solution?

38. Chemistry: Salt Solutions How much water must be evaporated from 240 gallons of a 3% salt solution to produce a 5% salt solution?

39. Purity of Gold The purity of gold is measured in karats, with pure gold being 24 karats. Other purities of gold are expressed as proportional parts of pure gold. Thus, 18-karat gold is $\frac{18}{24}$, or 75% pure gold; 12-karat gold is $\frac{12}{24}$, or 50% pure gold; and so

on. How much 12-karat gold should be mixed with pure gold to obtain 60 grams of 16-karat gold?

40. Chemistry: Sugar Molecules A sugar molecule has twice as many atoms of hydrogen as it does oxygen and one more atom of carbon than oxygen. If a sugar molecule has a total of 45 atoms, how many are oxygen? How many are hydrogen?

41. Running a Race Mike can run the mile in 6 minutes, and Dan can run the mile in 9 minutes. If Mike gives Dan a head start of 1 minute, how far from the start will Mike pass Dan? (See the figure.) How long does it take?

42. Range of an Airplane An air rescue plane averages 300 miles per hour in still air. It carries enough fuel for 5 hours of flying time. If, upon takeoff, it encounters a head wind of 30 mi/hr, how far can it fly and return safely? (Assume that the wind remains constant.)

43. Emptying Oil Tankers An oil tanker can be emptied by the main pump in 4 hours. An auxiliary pump can empty the tanker in 9 hours. If the main pump is started at 9 AM, when should the auxiliary pump be started so that the tanker is emptied by noon?

44. Cement Mix A 20-pound bag of Economy brand cement mix contains 25% cement and 75% sand. How much pure cement must be added to produce a cement mix that is 40% cement?

45. Emptying a Tub A bathroom tub will fill in 15 minutes with both faucets open and the stopper in place. With both faucets closed and the stopper removed, the tub will empty in 20 minutes. How long will it take for the tub to fill if both faucets are open and the stopper is removed?

46. Using Two Pumps A 5-horsepower (hp) pump can empty a pool in 5 hours. A smaller, 2-hp pump empties the same pool in 8 hours. The pumps are used together to begin emptying this pool. After 2 hours, the 2-hp pump breaks down. How long will it take the larger pump to empty the pool?

47. Comparing Olympic Heroes In the 1984 Olympics, Carl Lewis of the United States won the gold medal in the 100-meter race with a time of 9.99 seconds. In the 1896 Olympics, Thomas Burke, also of the United States, won the gold medal in the 100-meter race in 12.0 seconds. If they ran in the same race repeating their respective times, by how many meters would Lewis beat Burke?

Discussion and Writing

48. Critical Thinking You are the manager of a clothing store and have just purchased 100 dress shirts for $20.00 each. After 1 month of selling the shirts at the regular price, you plan to have a sale giving 40% off the original selling price. However, you still want to make a profit of $4 on each shirt at the sale price. What should you price the shirts at initially to ensure this? If, instead of 40% off at the sale, you give 50% off, by how much is your profit reduced?

49. Critical Thinking Make up a word problem that requires solving a linear equation as part of its solution. Exchange problems with a friend. Write a critique of your friend's problem.

50. Critical Thinking Without solving, explain what is wrong with the following mixture problem: How many liters of 25% ethanol should be added to 20 liters of 48% ethanol to obtain a solution of 58% ethanol? Now go through an algebraic solution. What happens?

51. Computing Average Speed In going from Chicago to Atlanta, a car averages 45 miles per hour, and in going from Atlanta to Miami, it averages 55 miles per hour. If Atlanta is halfway between Chicago and Miami, what is the average speed from Chicago to Miami? Discuss an intuitive solution. Write a paragraph defending your intuitive solution. Then solve the problem algebraically. Is your intuitive solution the same as the algebraic one? If not, find the flaw.

52. Speed of a Plane On a recent flight from Phoenix to Kansas City, a distance of 919 nautical miles, the plane arrived 20 minutes early. On leaving the aircraft, I asked the captain, "What was our tail wind?" He replied, "I don't know, but our ground speed was 550 knots." How can you determine if enough information is provided to find the tail wind? If possible, find the tail wind. (1 knot = 1 nautical mile per hour)

A.11 Complex Numbers

OBJECTIVES 1 Add, Subtract, Multiply, and Divide Complex Numbers

One property of a real number is that its square is nonnegative (greater than or equal to 0). For example, there is no real number x for which

$$x^2 = -1$$

To remedy this situation, we introduce a number called the **imaginary unit**, which we denote by i and whose square is -1; that is,

$$i^2 = -1$$

This should not surprise you. If our universe were to consist only of integers, there would be no number x for which $2x = 1$. This unfortunate circumstance was remedied by introducing numbers such as $\frac{1}{2}$ and $\frac{2}{3}$, the *rational numbers*. If our universe were to consist only of rational numbers, there would be no number x whose square equals 2. That is, there would be no number x for which $x^2 = 2$. To remedy this, we introduced numbers such as $\sqrt{2}$ and $\sqrt[3]{5}$, the *irrational numbers*. The *real numbers*, you will recall, consist of the rational numbers and the irrational numbers. Now, if our universe were to consist only of real numbers, there would be no number x whose square is -1. To remedy this, we introduce a number i, whose square is -1.

In the progression outlined, each time that we encountered a situation that was unsuitable, we introduced a new number system to remedy the situation. And each new number system contained the earlier number system as a subset. The number system that results from introducing the number i is called the **complex number system**.

Complex numbers are numbers of the form $a + bi$, where a and b are real numbers. The real number a is called the **real part** of the number $a + bi$; the real number b is called the **imaginary part** of $a + bi$.

For example, the complex number $-5 + 6i$ has the real part -5 and the imaginary part 6.

When a complex number is written in the form $a + bi$, where a and b are real numbers, we say it is in **standard form**. However, if the imaginary part of a complex number is negative, such as in the complex number $3 + (-2)i$, we agree to write it instead in the form $3 - 2i$.

Also, the complex number $a + 0i$ is usually written merely as a. This serves to remind us that the real numbers are a subset of the complex numbers. The complex number $0 + bi$ is usually written as bi. Sometimes the complex number bi is called a **pure imaginary number**.

1 Add, Subtract, Multiply, and Divide Complex Numbers

Equality, addition, subtraction, and multiplication of complex numbers are defined so as to preserve the familiar rules of algebra for real numbers. Thus, two complex numbers are equal if and only if their real parts are equal and their imaginary parts are equal. That is,

Equality of Complex Numbers

$$a + bi = c + di \quad \text{if and only if} \quad a = c \text{ and } b = d \qquad \textbf{(1)}$$

Two complex numbers are added by forming the complex number whose real part is the sum of the real parts and whose imaginary part is the sum of the imaginary parts. That is,

Sum of Complex Numbers

$$(a + bi) + (c + di) = (a + c) + (b + d)i \qquad \textbf{(2)}$$

To subtract two complex numbers, we use this rule:

Difference of Complex Numbers

$$(a + bi) - (c + di) = (a - c) + (b - d)i \qquad \textbf{(3)}$$

EXAMPLE 1	**Adding and Subtracting Complex Numbers**

(a) $(3 + 5i) + (-2 + 3i) = [3 + (-2)] + (5 + 3)i = 1 + 8i$
(b) $(6 + 4i) - (3 + 6i) = (6 - 3) + (4 - 6)i = 3 + (-2)i = 3 - 2i$ ∎

NOW WORK PROBLEM **11**.

Products of complex numbers are calculated as illustrated in Example 2.

EXAMPLE 2	**Multiplying Complex Numbers**

$$\begin{aligned}
(5 + 3i) \cdot (2 + 7i) &= 5 \cdot (2 + 7i) + 3i(2 + 7i) &&\text{Distributive Property}\\
&= 10 + 35i + 6i + 21i^2 &&\text{Distributive Property}\\
&= 10 + 41i + 21(-1) &&i^2 = -1\\
&= -11 + 41i
\end{aligned}$$

∎

Based on the procedure of Example 2, we define the **product** of two complex numbers by the following formula:

Product of Complex Numbers

$$(a + bi) \cdot (c + di) = (ac - bd) + (ad + bc)i \qquad (4)$$

Do not bother to memorize formula (4). Instead, whenever it is necessary to multiply two complex numbers, follow the usual rules for multiplying two binomials, as in Example 2, remembering that $i^2 = -1$. For example,

$$(2i)(2i) = 4i^2 = -4$$
$$(2 + i)(1 - i) = 2 - 2i + i - i^2 = 3 - i$$

> **NOW WORK PROBLEM 17.**

Algebraic properties for addition and multiplication, such as the Commutative, Associative, and Distributive Properties, hold for complex numbers. However, the property that every nonzero complex number has a multiplicative inverse, or reciprocal, requires a closer look.

Conjugates

If $z = a + bi$ is a complex number, then its **conjugate**, denoted by \bar{z}, is defined as

$$\bar{z} = \overline{a + bi} = a - bi$$

For example, $\overline{2 + 3i} = 2 - 3i$ and $\overline{-6 - 2i} = -6 + 2i$.

EXAMPLE 3 | **Multiplying a Complex Number by Its Conjugate**

Find the product of the complex number $z = 3 + 4i$ and its conjugate \bar{z}.

Solution Since $\bar{z} = 3 - 4i$, we have

$$z\bar{z} = (3 + 4i)(3 - 4i) = 9 - 12i + 12i - 16i^2 = 9 + 16 = 25 \qquad \blacksquare$$

The result obtained in Example 3 has an important generalization.

Theorem The product of a complex number and its conjugate is a nonnegative real number. That is, if $z = a + bi$, then

$$z\bar{z} = a^2 + b^2 \qquad (5)$$

Proof If $z = a + bi$, then

$$z\bar{z} = (a + bi)(a - bi) = a^2 - (bi)^2 = a^2 - b^2i^2 = a^2 + b^2 \qquad \blacksquare$$

To express the reciprocal of a nonzero complex number z in standard form, multiply the numerator and denominator of $\frac{1}{z}$ by its conjugate \bar{z}. That is, if $z = a + bi$ is a nonzero complex number, then

$$\frac{1}{a + bi} = \frac{1}{z} = \frac{1}{z} \cdot \frac{\bar{z}}{\bar{z}} = \frac{\bar{z}}{z\bar{z}} = \frac{a - bi}{a^2 + b^2} = \frac{a}{a^2 + b^2} - \frac{b}{a^2 + b^2}i$$

$$\text{Use (5).}$$

| EXAMPLE 4 | **Writing the Reciprocal of a Complex Number in Standard Form** |

Write $\dfrac{1}{3 + 4i}$ in standard form $a + bi$; that is, find the reciprocal of $3 + 4i$.

Solution The idea is to multiply the numerator and denominator by the conjugate of $3 + 4i$, that is, the complex number $3 - 4i$. The result is

$$\frac{1}{3 + 4i} = \frac{1}{3 + 4i} \cdot \frac{3 - 4i}{3 - 4i} = \frac{3 - 4i}{9 + 16} = \frac{3}{25} - \frac{4}{25}i \qquad \blacksquare$$

To express the quotient of two complex numbers in standard form, we multiply the numerator and denominator of the quotient by the conjugate of the denominator.

| EXAMPLE 5 | **Writing the Quotient of Complex Numbers in Standard Form** |

Write each of the following in standard form.

(a) $\dfrac{1 + 4i}{5 - 12i}$ 　　　　　　(b) $\dfrac{2 - 3i}{4 - 3i}$

Solution (a) $\dfrac{1 + 4i}{5 - 12i} = \dfrac{1 + 4i}{5 - 12i} \cdot \dfrac{5 + 12i}{5 + 12i} = \dfrac{5 + 12i + 20i + 48i^2}{25 + 144}$

$$= \frac{-43 + 32i}{169} = \frac{-43}{169} + \frac{32}{169}i$$

(b) $\dfrac{2 - 3i}{4 - 3i} = \dfrac{2 - 3i}{4 - 3i} \cdot \dfrac{4 + 3i}{4 + 3i} = \dfrac{8 + 6i - 12i - 9i^2}{16 + 9}$

$$= \frac{17 - 6i}{25} = \frac{17}{25} - \frac{6}{25}i \qquad \blacksquare$$

━━━ **NOW WORK PROBLEM 25.**

| EXAMPLE 6 | **Writing Other Expressions in Standard Form** |

If $z = 2 - 3i$ and $w = 5 + 2i$, write each of the following expressions in standard form.

(a) $\dfrac{z}{w}$ 　　　　　(b) $\overline{z + w}$ 　　　　　(c) $z + \overline{z}$

Solution (a) $\dfrac{z}{w} = \dfrac{z \cdot \overline{w}}{w \cdot \overline{w}} = \dfrac{(2 - 3i)(5 - 2i)}{(5 + 2i)(5 - 2i)} = \dfrac{10 - 4i - 15i + 6i^2}{25 + 4}$

$$= \frac{4 - 19i}{29} = \frac{4}{29} - \frac{19}{29}i$$

(b) $\overline{z + w} = \overline{(2 - 3i) + (5 + 2i)} = \overline{7 - i} = 7 + i$

(c) $z + \overline{z} = (2 - 3i) + (2 + 3i) = 4 \qquad \blacksquare$

The conjugate of a complex number has certain general properties that we shall find useful later.

For a real number $a = a + 0i$, the conjugate is $\overline{a} = \overline{a + 0i} = a - 0i = a$. That is,

Theorem The conjugate of a real number is the real number itself.

Other properties that are direct consequences of the definition of the conjugate are given next. In each statement, z and w represent complex numbers.

Theorem

The conjugate of the conjugate of a complex number is the complex number itself.

$$\overline{(\overline{z})} = z \tag{6}$$

The conjugate of the sum of two complex numbers equals the sum of their conjugates.

$$\overline{z + w} = \overline{z} + \overline{w} \tag{7}$$

The conjugate of the product of two complex numbers equals the product of their conjugates.

$$\overline{z \cdot w} = \overline{z} \cdot \overline{w} \tag{8}$$

We leave the proofs of equations (6), (7), and (8) as exercises. See Problems 56–58.

Powers of i

The powers of i follow a pattern that is useful to know.

$$i^1 = i \qquad\qquad i^5 = i^4 \cdot i = 1 \cdot i = i$$
$$i^2 = -1 \qquad\qquad i^6 = i^4 \cdot i^2 = -1$$
$$i^3 = i^2 \cdot i = -i \qquad\qquad i^7 = i^4 \cdot i^3 = -i$$
$$i^4 = i^2 \cdot i^2 = (-1)(-1) = 1 \qquad i^8 = i^4 \cdot i^4 = 1$$

And so on. The powers of i repeat with every fourth power.

EXAMPLE 7

Evaluating Powers of i

(a) $i^{27} = i^{24} \cdot i^3 = (i^4)^6 \cdot i^3 = 1^6 \cdot i^3 = -i$

(b) $i^{101} = i^{100} \cdot i^1 = (i^4)^{25} \cdot i = 1^{25} \cdot i = i$ ∎

EXAMPLE 8

Writing the Power of a Complex Number in Standard Form

Write $(2 + i)^3$ in standard form.

Solution

We use the special product formula for $(x + a)^3$.

$$(x + a)^3 = x^3 + 3ax^2 + 3a^2x + a^3$$

Using this special product formula,

$$(2 + i)^3 = 2^3 + 3 \cdot i \cdot 2^2 + 3 \cdot i^2 \cdot 2 + i^3$$
$$= 8 + 12i + 6(-1) + (-i)$$
$$= 2 + 11i \qquad ∎$$

NOW WORK PROBLEMS 31 AND 39.

In the complex number system, we can discuss square roots of negative numbers.

If N is a positive real number, we define the **principal square root of $-N$**, denoted by $\sqrt{-N}$, as

$$\sqrt{-N} = \sqrt{N}\,i$$

where i is the imaginary unit and $i^2 = -1$.

EXAMPLE 9 **Evaluating the Square Root of a Negative Number**

(a) $\sqrt{-1} = \sqrt{1}i = i$ (b) $\sqrt{-4} = \sqrt{4}i = 2i$

(c) $\sqrt{-8} = \sqrt{8}i = 2\sqrt{2}i$

NOW WORK PROBLEM **47.**

A.11 Assess Your Understanding

Concepts and Vocabulary

1. *True or False:* $i^2 + 1 = 0$

2. $(2 + i)(2 - i) = $ _____.

3. In the complex number $5 + 2i$, the number 5 is called the _____ part; the number 2 is called the _____ part; the number i is called the _____ _____.

4. *True or False:* The conjugate of $2 + 5i$ is $-2 - 5i$.

5. *True or False:* All real numbers are complex numbers.

6. *True or False:* If $z = a + bi$, then $z\bar{z} = a^2 - b^2$.

Skill Building

In Problems 7–44, write each expression in the standard form $a + bi$.

7. $(2 - 3i) + (6 + 8i)$
8. $(4 + 5i) + (-8 + 2i)$
9. $(-3 + 2i) - (4 - 4i)$
10. $(3 - 4i) - (-3 - 4i)$

11. $(2 - 5i) - (8 + 6i)$
12. $(-8 + 4i) - (2 - 2i)$
13. $3(2 - 6i)$
14. $-4(2 + 8i)$

15. $2i(2 - 3i)$
16. $3i(-3 + 4i)$
17. $(3 - 4i)(2 + i)$
18. $(5 + 3i)(2 - i)$

19. $(-6 + i)(-6 - i)$
20. $(-3 + i)(3 + i)$
21. $\dfrac{10}{3 - 4i}$
22. $\dfrac{13}{5 - 12i}$

23. $\dfrac{2 + i}{i}$
24. $\dfrac{2 - i}{-2i}$
25. $\dfrac{6 - i}{1 + i}$
26. $\dfrac{2 + 3i}{1 - i}$

27. $\left(\dfrac{1}{2} + \dfrac{\sqrt{3}}{2}i\right)^2$
28. $\left(\dfrac{\sqrt{3}}{2} - \dfrac{1}{2}i\right)^2$
29. $(1 + i)^2$
30. $(1 - i)^2$

31. i^{23}
32. i^{14}
33. i^{-15}
34. i^{-23}

35. $i^6 - 5$
36. $4 + i^3$
37. $6i^3 - 4i^5$
38. $4i^3 - 2i^2 + 1$

39. $(1 + i)^3$
40. $(3i)^4 + 1$
41. $i^7(1 + i^2)$
42. $2i^4(1 + i^2)$

43. $i^6 + i^4 + i^2 + 1$
44. $i^7 + i^5 + i^3 + i$

In Problems 45–50, perform the indicated operations and express your answer in the form $a + bi$.

45. $\sqrt{-4}$
46. $\sqrt{-9}$
47. $\sqrt{-25}$

48. $\sqrt{-64}$
49. $\sqrt{(3 + 4i)(4i - 3)}$
50. $\sqrt{(4 + 3i)(3i - 4)}$

In Problems 51–58, $z = 3 - 4i$ *and* $w = 8 + 3i$. *Write each expression in the standard form* $a + bi$.

51. $z + \bar{z}$
52. $w - \bar{w}$
53. $z\bar{z}$
54. $\bar{z} - \bar{w}$

55. Use $z = a + bi$ to show that $z + \bar{z} = 2a$ and $z - \bar{z} = 2bi$.

56. Use $z = a + bi$ to show that $\bar{\bar{z}} = z$.

57. Use $z = a + bi$ and $w = c + di$ to show that $\overline{z + w} = \bar{z} + \bar{w}$.

58. Use $z = a + bi$ and $w = c + di$ to show that $\overline{z \cdot w} = \bar{z} \cdot \bar{w}$.

Discussion and Writing

59. Explain to a friend how you would add two complex numbers and how you would multiply two complex numbers. Explain any differences in the two explanations.

Appendix B — Graphing Utilities

Outline

B.1 The Viewing Rectangle

Figure 1

$y = 2x$

Figure 2

All graphing utilities, that is, all graphing calculators and all computer software graphing packages, graph equations by plotting points on a screen. The screen itself actually consists of small rectangles, called **pixels.** The more pixels the screen has, the better the resolution. Most graphing calculators have 2048 pixels per square inch; most computer screens have 4096 to 8192 pixels per square inch. When a point to be plotted lies inside a pixel, the pixel is turned on (lights up). The graph of an equation is a collection of pixels. Figure 1 shows how the graph of $y = 2x$ looks on a TI-84 Plus graphing calculator.

The screen of a graphing utility will display the coordinate axes of a rectangular coordinate system. However, you must set the scale on each axis. You must also include the smallest and largest values of x and y that you want included in the graph. This is called **setting the viewing rectangle** or **viewing window.** Figure 2 illustrates a typical viewing window.

To select the viewing window, we must give values to the following expressions:

Xmin:	the smallest value of x
Xmax:	the largest value of x
Xscl:	the number of units per tick mark on the x-axis
Ymin:	the smallest value of y
Ymax:	the largest value of y
Yscl:	the number of units per tick mark on the y-axis

Figure 3 illustrates these settings and their relation to the Cartesian coordinate system.

Figure 3

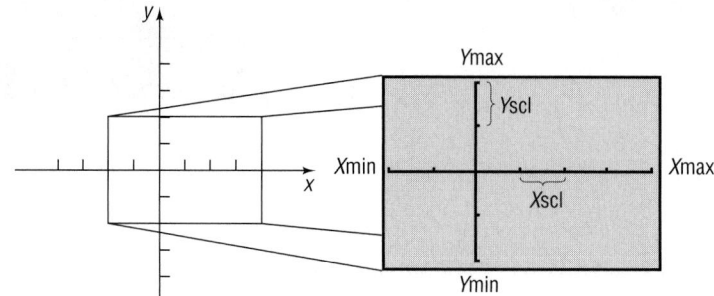

If the scale used on each axis is known, we can determine the minimum and maximum values of x and y shown on the screen by counting the tick marks. Look again at Figure 2. For a scale of 1 on each axis, the minimum and maximum values of x are -10 and 10, respectively; the minimum and maximum values of y are also -10 and 10. If the scale is 2 on each axis, then the minimum and maximum values of x are -20 and 20, respectively; and the minimum and maximum values of y are -20 and 20, respectively.

Conversely, if we know the minimum and maximum values of x and y, we can determine the scales being used by counting the tick marks displayed. We shall follow the practice of showing the minimum and maximum values of x and y in our illustrations so that you will know how the viewing window was set. See Figure 4.

Figure 4

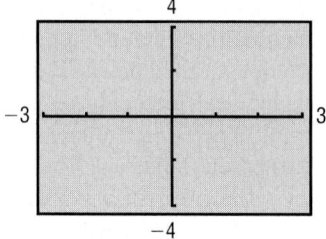

means

$$X\text{min} = -3 \qquad Y\text{min} = -4$$
$$X\text{max} = 3 \qquad Y\text{max} = 4$$
$$X\text{scl} = 1 \qquad Y\text{scl} = 2$$

| EXAMPLE 1 | **Finding the Coordinates of a Point Shown on a Graphing Utility Screen** |

Find the coordinates of the point shown in Figure 5. Assume that the coordinates are integers.

Figure 5

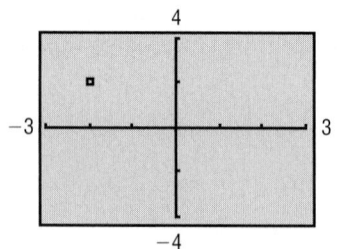

Solution First we note that the viewing window used in Figure 5 is

$$X\text{min} = -3 \qquad Y\text{min} = -4$$
$$X\text{max} = 3 \qquad Y\text{max} = 4$$
$$X\text{scl} = 1 \qquad Y\text{scl} = 2$$

The point shown is 2 tick units to the left on the horizontal axis (scale $= 1$) and 1 tick up on the vertical axis (scale $= 2$). The coordinates of the point shown are $(-2, 2)$. ∎

B.1 Exercises

In Problems 1–4, determine the coordinates of the points shown. Tell in which quadrant each point lies. Assume that the coordinates are integers.

1. **2.** **3.** **4.**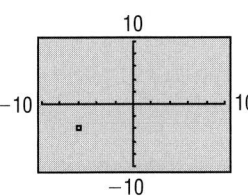

In Problems 5–10, determine the viewing window used.

5. **6.** **7.**

8. **9.** **10.**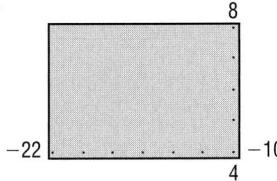

In Problems 11–16, select a setting so that each of the given points will lie within the viewing rectangle.

11. $(-10, 5), (3, -2), (4, -1)$

12. $(5, 0), (6, 8), (-2, -3)$

13. $(40, 20), (-20, -80), (10, 40)$

14. $(-80, 60), (20, -30), (-20, -40)$

15. $(0, 0), (100, 5), (5, 150)$

16. $(0, -1), (100, 50), (-10, 30)$

In Problems 17–20, find the length of the line segment. Assume that the endpoints of each line segment have integer coordinates.

17. **18.** **19.** **20.**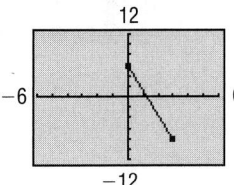

B.2 Using a Graphing Utility to Graph Equations

From Examples 2 and 3 in Foundations, Section F.2, we see that a graph can be obtained by plotting points in a rectangular coordinate system and connecting them. Graphing utilities perform these same steps when graphing an equation. For example, the TI-84 Plus determines 95 evenly spaced input values,[*] starting at Xmin and ending at Xmax, uses the equation to determine the output values, plots these points on the screen, and finally (if in the connected mode) draws a line between consecutive points.

[*]These input values depend on the values of Xmin and Xmax. For example, if Xmin $= -10$ and Xmax $= 10$, then the first input value will be -10 and the next input value will be $-10 + \dfrac{10 - (-10)}{94} = -9.7872$, and so on.

To graph an equation in two variables x and y using a graphing utility requires that the equation be written in the form $y = \{expression\ in\ x\}$. If the original equation is not in this form, replace it by equivalent equations until the form $y = \{expression\ in\ x\}$ is obtained. In general, there are four ways to obtain equivalent equations.

Procedures That Result in Equivalent Equations

1. Interchange the two sides of the equation:

$$\text{Replace} \quad 3x + 5 = y \quad \text{by} \quad y = 3x + 5$$

2. Simplify the sides of the equation by combining like terms, eliminating parentheses, and so on:

$$\text{Replace} \quad (2y + 2) + 6 = 2x + 5(x + 1)$$
$$\text{by} \quad 2y + 8 = 7x + 5$$

3. Add or subtract the same expression on both sides of the equation:

$$\text{Replace} \quad y + 3x - 5 = 4$$
$$\text{by} \quad y + 3x - 5 + 5 = 4 + 5$$

4. Multiply or divide both sides of the equation by the same nonzero expression:

$$\text{Replace} \quad 3y = 6 - 2x$$
$$\text{by} \quad \frac{1}{3} \cdot 3y = \frac{1}{3}(6 - 2x)$$

EXAMPLE 1	**Expressing an Equation in the Form y = {expression in x}**

Solve for y: $2y + 3x - 5 = 4$

Solution We replace the original equation by a succession of equivalent equations.

$$2y + 3x - 5 = 4$$
$$2y + 3x - 5 + 5 = 4 + 5 \qquad \text{Add 5 to both sides.}$$
$$2y + 3x = 9 \qquad \text{Simplify.}$$
$$2y + 3x - 3x = 9 - 3x \qquad \text{Subtract 3x from both sides.}$$
$$2y = 9 - 3x \qquad \text{Simplify.}$$
$$\frac{2y}{2} = \frac{9 - 3x}{2} \qquad \text{Divide both sides by 2.}$$
$$y = \frac{9 - 3x}{2} \qquad \text{Simplify.} \qquad \blacksquare$$

Now we are ready to graph equations using a graphing utility. Most graphing utilities require the following steps:

Steps for Graphing an Equation Using a Graphing Utility

STEP 1: Solve the equation for y in terms of x.

STEP 2: Get into the graphing mode of your graphing utility. The screen will usually display $Y_1 =$ _____, prompting you to enter the expression involving x that you found in Step 1. (Consult your manual for the correct way to enter the expression; for example, $y = x^2$ might be entered as $x^\wedge 2$ or as $x*x$ or as $x\ x^Y 2$).

STEP 3: Select the viewing window. Without prior knowledge about the behavior of the graph of the equation, it is common to select the **standard viewing window**[*] initially. The viewing window is then adjusted based on the graph that appears. In this text the standard viewing window is

$$X\text{min} = -10 \qquad Y\text{min} = -10$$
$$X\text{max} = 10 \qquad Y\text{max} = 10$$
$$X\text{scl} = 1 \qquad Y\text{scl} = 1$$

STEP 4: Graph.

STEP 5: Adjust the viewing window until a complete graph is obtained.

| EXAMPLE 2 | **Graphing an Equation on a Graphing Utility** |

Graph the equation: $6x^2 + 3y = 36$

Solution **STEP 1:** We solve for y in terms of x.

$$6x^2 + 3y = 36$$
$$3y = -6x^2 + 36 \qquad \text{Subtract } 6x^2 \text{ from both sides of the equation.}$$
$$y = -2x^2 + 12 \qquad \text{Divide both sides of the equation by 3 and simplify.}$$

STEP 2: From the $Y_1 =$ _____ screen, enter the expression $-2x^2 + 12$ after the prompt.

STEP 3: Set the viewing window to the standard viewing window.

STEP 4: Graph. The screen should look like Figure 6.

STEP 5: The graph of $y = -2x^2 + 12$ is not complete. The value of Ymax must be increased so that the top portion of the graph is visible. After increasing the value of Ymax to 12, we obtain the graph in Figure 7. The graph is now complete.

Figure 6

Figure 7

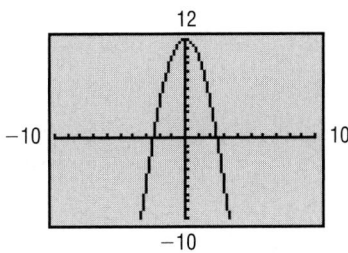

∎

[*]Some graphing utilities have a ZOOM-STANDARD feature that automatically sets the viewing window to the standard viewing window and graphs the equation.

Look again at Figure 7. Although a complete graph is shown, the graph might be improved by adjusting the values of Xmin and Xmax. Figure 8 shows the graph of $y = -2x^2 + 12$ using Xmin $= -4$ and Xmax $= 4$. Do you think this is a better choice for the viewing window?

Figure 8

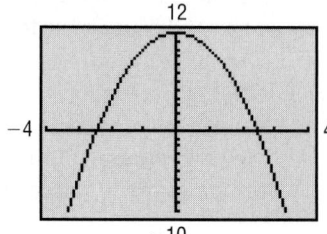

EXAMPLE 3	**Creating a Table and Graphing an Equation**

Create a table and graph the equation: $y = x^3$

Solution Most graphing utilities have the capability of creating a table of values for an equation. (Check your manual to see if your graphing utility has this capability.) Table 1 illustrates a table of values for $y = x^3$ on a TI-84 Plus. See Figure 9 for the graph.

Table 1 **Figure 9**

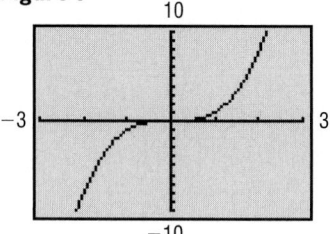

B.2 Exercises

In Problems 1–16, graph each equation using the following viewing windows:

(a) Xmin $= -5$ (b) Xmin $= -10$ (c) Xmin $= -10$ (d) Xmin $= -5$
Xmax $= 5$ Xmax $= 10$ Xmax $= 10$ Xmax $= 5$
Xscl $= 1$ Xscl $= 1$ Xscl $= 2$ Xscl $= 1$
Ymin $= -4$ Ymin $= -8$ Ymin $= -8$ Ymin $= -20$
Ymax $= 4$ Ymax $= 8$ Ymax $= 8$ Ymax $= 20$
Yscl $= 1$ Yscl $= 1$ Yscl $= 2$ Yscl $= 5$

1. $y = x + 2$ **2.** $y = x - 2$ **3.** $y = -x + 2$ **4.** $y = -x - 2$

5. $y = 2x + 2$ **6.** $y = 2x - 2$ **7.** $y = -2x + 2$ **8.** $y = -2x - 2$

9. $y = x^2 + 2$ **10.** $y = x^2 - 2$ **11.** $y = -x^2 + 2$ **12.** $y = -x^2 - 2$

13. $3x + 2y = 6$ **14.** $3x - 2y = 6$ **15.** $-3x + 2y = 6$ **16.** $-3x - 2y = 6$

17–32. *For each of the above equations, create a table, $-3 \leq x \leq 3$, and list points on the graph.*

B.3 Using a Graphing Utility to Locate Intercepts and Check for Symmetry

Value and Zero (or Root)

Most graphing utilities have an eVALUEate feature that, given a value of x, determines the value of y for an equation. We can use this feature to evaluate an equation at $x = 0$ to determine the y-intercept. Most graphing utilities also have a ZERO (or ROOT) feature that can be used to determine the x-intercept(s) of an equation.

| EXAMPLE 1 | Finding Intercepts Using a Graphing Utility |

Use a graphing utility to find the intercepts of the equation $y = x^3 - 8$.

Solution Figure 10(a) shows the graph of $y = x^3 - 8$.

Figure 10

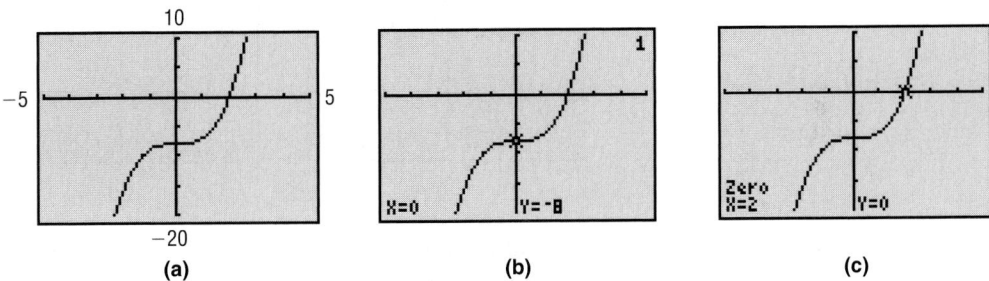

(a) (b) (c)

The eVALUEate feature of a TI-84 Plus graphing calculator accepts as input a value of x and determines the value of y. If we let $x = 0$, we find that the y-intercept is -8. See Figure 10(b).

The ZERO feature of a TI-84 Plus is used to find the x-intercept(s). See Figure 10(c). The x-intercept is 2. ∎

Trace

Most graphing utilities allow you to move from point to point along the graph, displaying on the screen the coordinates of each point. This feature is called TRACE.

| EXAMPLE 2 | Using TRACE to Locate Intercepts |

Graph the equation $y = x^3 - 8$. Use TRACE to locate the intercepts.

Solution Figure 11 shows the graph of $y = x^3 - 8$.

Figure 11

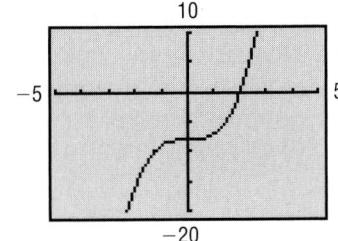

Activate the TRACE feature. As you move the cursor along the graph, you will see the coordinates of each point displayed. When the cursor is on the y-axis, we find that the y-intercept is -8. See Figure 12.

Figure 12

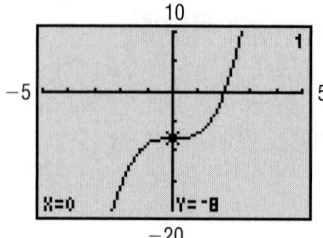

Continue moving the cursor along the graph. Just before you get to the x-axis, the display will look like the one in Figure 13(a). (Due to differences in graphing utilities, your display may be slightly different from the one shown here.)

Figure 13

(a) (b)

In Figure 13(a), the negative value of the y-coordinate indicates that we are still below the x-axis. The next position of the cursor is shown in Figure 13(b). The positive value of the y-coordinate indicates that we are now above the x-axis. This means that between these two points the x-axis was crossed. The x-intercept lies between 1.9148936 and 2.0212766. ∎

EXAMPLE 3

Graphing the Equation $y = \dfrac{1}{x}$

Graph the equation: $y = \dfrac{1}{x}$

With the viewing window set as

$$X\min = -3 \qquad Y\min = -4$$
$$X\max = 3 \qquad Y\max = 4$$
$$X\mathrm{scl} = 1 \qquad Y\mathrm{scl} = 1$$

Figure 14

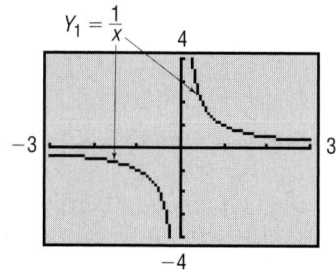

use TRACE to infer information about intercepts and symmetry.

Solution Figure 14 illustrates the graph. We infer from the graph that there are no intercepts; we may also infer that symmetry with respect to the origin is a possibility. The TRACE feature on a graphing utility can provide further evidence of symmetry with respect to the origin. Using TRACE, we observe that for any ordered pair (x, y) the ordered pair $(-x, -y)$ is also a point on the graph. For example, the points $(0.95744681, 1.0444444)$ and $(-0.95744681, -1.0444444)$ both lie on the graph. ∎

B.3 Exercises

In Problems 1–6, use ZERO (or ROOT) to approximate the smaller of the two x-intercepts of each equation. Express the answer rounded to two decimal places.

1. $y = x^2 + 4x + 2$

2. $y = x^2 + 4x - 3$

3. $y = 2x^2 + 4x + 1$

4. $y = 3x^2 + 5x + 1$

5. $y = 2x^2 - 3x - 1$

6. $y = 2x^2 - 4x - 1$

*In Problems 7–14, use ZERO (or ROOT) to approximate the **positive** x-intercepts of each equation. Express each answer rounded to two decimal places.*

7. $y = x^3 + 3.2x^2 - 16.83x - 5.31$

8. $y = x^3 + 3.2x^2 - 7.25x - 6.3$

9. $y = x^4 - 1.4x^3 - 33.71x^2 + 23.94x + 292.41$

10. $y = x^4 + 1.2x^3 - 7.46x^2 - 4.692x + 15.2881$

11. $y = \pi x^3 - (8.88\pi + 1)x^2 - (42.066\pi - 8.88)x + 42.066$

12. $y = \pi x^3 - (5.63\pi + 2)x^2 - (108.392\pi - 11.26)x + 216.784$

13. $y = x^3 + 19.5x^2 - 1021x + 1000.5$

14. $y = x^3 + 14.2x^2 - 4.8x - 12.4$

B.4 Using a Graphing Utility to Solve Equations

For many equations, there are no algebraic techniques that lead to a solution. For such equations, a graphing utility can often be used to investigate possible solutions. When a graphing utility is used to solve an equation, usually *approximate* solutions are obtained. Unless otherwise stated, we shall follow the practice of giving approximate solutions *rounded to two decimal places*.

The ZERO (or ROOT) feature of a graphing utility can be used to find the solutions of an equation when one side of the equation is 0. In using this feature to solve equations, we make use of the fact that the x-intercepts (or zeros) of the graph of an equation are found by letting $y = 0$ and solving the equation for x. Solving an equation for x when one side of the equation is 0 is equivalent to finding where the graph of the corresponding equation crosses or touches the x-axis.

| EXAMPLE 1 | Using ZERO (or ROOT) to Approximate Solutions of an Equation |

Find the solution(s) of the equation $x^2 - 6x + 7 = 0$. Round answers to two decimal places.

Solution The solutions of the equation $x^2 - 6x + 7 = 0$ are the same as the x-intercepts of the graph of $Y_1 = x^2 - 6x + 7$. We begin by graphing the equation. See Figure 15(a).

From the graph there appear to be two x-intercepts (solutions to the equation): one between 1 and 2, the other between 4 and 5.

Figure 15

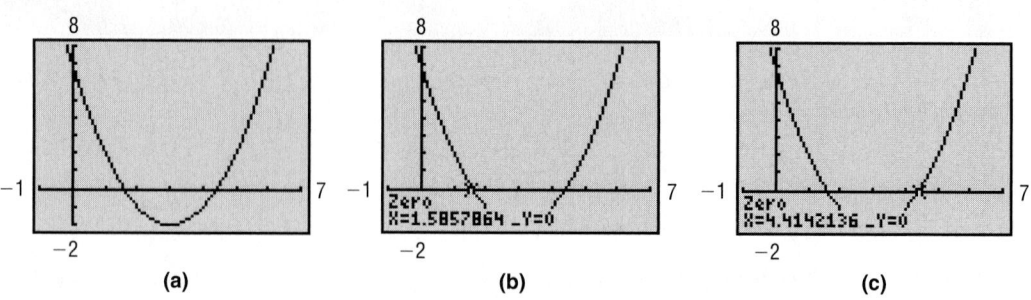

(a) (b) (c)

Using the ZERO (or ROOT) feature of our graphing utility, we determine that the x-intercepts, and so the solutions to the equation, are $x = 1.59$ and $x = 4.41$, rounded to two decimal places. See Figures 15(b) and (c). ∎

A second method for solving equations using a graphing utility involves the INTERSECT feature of the graphing utility. This feature is used most effectively when one side of the equation is not 0.

EXAMPLE 2 **Using INTERSECT to Approximate Solutions of an Equation**

Find the solution(s) to the equation $3(x - 2) = 5(x - 1)$. Round answers to two decimal places.

Solution We begin by graphing each side of the equation as follows: graph $Y_1 = 3(x - 2)$ and $Y_2 = 5(x - 1)$. See Figure 16(a).

At the point of intersection of the graphs, the value of the y-coordinate is the same. We conclude that the x-coordinate of the point of intersection represents the

Figure 16

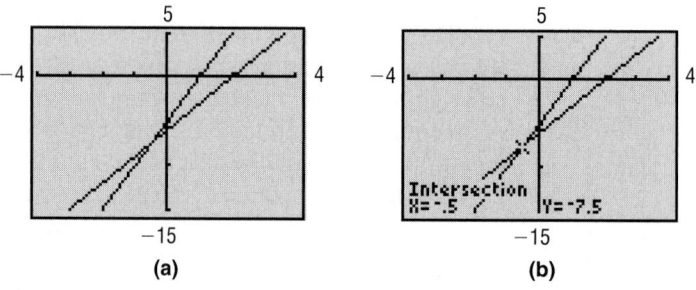

(a) (b)

solution to the equation. Do you see why? The INTERSECT feature on a graphing utility determines the point of intersection of the graphs. Using this feature, we find that the graphs intersect at $(-0.5, -7.5)$. See Figure 16(b). The solution of the equation is therefore $x = -0.5$.

✔ **CHECK:** We can verify our solution by evaluating each side of the equation with -0.5 STOred in x. See Figure 17. Since the left side of the equation equals the right side of the equation, the solution checks. ∎

Figure 17

```
-.5→X
            -.5
3(X-2)
            -7.5
5(X-1)
            -7.5
```

Summary

The steps to follow for approximating solutions of equations are given next.

Steps for Approximating Solutions of Equations Using ZERO (or ROOT)

STEP 1: Write the equation in the form $\{expression\ in\ x\} = 0$.

STEP 2: Graph $Y_1 = \{expression\ in\ x\}$.

Be sure that the graph is complete. That is, be sure that all the intercepts are shown on the screen.

STEP 3: Use ZERO (or ROOT) to determine each x-intercept of the graph.

Steps for Approximating Solutions of Equations Using INTERSECT

STEP 1: Graph $Y_1 = \{expression\ in\ x\ on\ the\ left\ side\ of\ the\ equation\}$.

Graph $Y_2 = \{expression\ in\ x\ on\ the\ right\ side\ of\ the\ equation\}$.

STEP 2: Use INTERSECT to determine each x-coordinate of the point(s) of intersection, if any.

Be sure that the graphs are complete. That is, be sure that all the points of intersection are shown on the screen.

EXAMPLE 3 **Solving a Radical Equation**

Figure 18

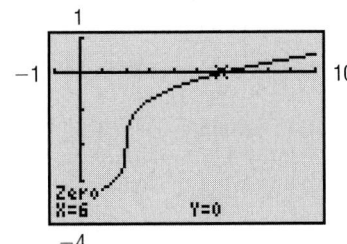

Find the real solutions of the equation $\sqrt[3]{2x - 4} - 2 = 0$.

Solution Figure 18 shows the graph of the equation $Y_1 = \sqrt[3]{2x - 4} - 2$. From the graph, we see one x-intercept near 6. Using ZERO (or ROOT), we find that the x-intercept is 6. The only solution is $x = 6$. ∎

B.5 Square Screens

Figure 19

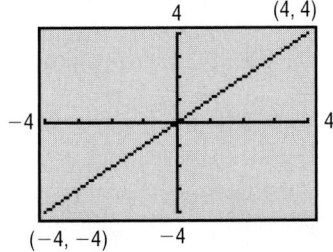

Most graphing utilities have a rectangular screen. Because of this, using the same settings for both x and y will result in a distorted view. For example, Figure 19 shows the graph of the line $y = x$ connecting the points $(-4, -4)$ and $(4, 4)$.

We expect the line to bisect the first and third quadrants, but it doesn't. We need to adjust the selections for Xmin, Xmax, Ymin, and Ymax so that a **square screen** results. On most graphing utilities, this is accomplished by setting the ratio of x to y at $3 : 2$.* For example, if

$$X\text{min} = -6 \qquad Y\text{min} = -4$$
$$X\text{max} = 6 \qquad Y\text{max} = 4$$

*Some graphing utilities have a built-in function that automatically squares the screen. For example, the TI-85 has a ZSQR function that does this. Some graphing utilities require a ratio other than $3 : 2$ to square the screen. For example, the HP 48G requires the ratio of x to y to be $2 : 1$ for a square screen. Consult your manual.

then the ratio of x to y is

$$\frac{X\max - X\min}{Y\max - Y\min} = \frac{6 - (-6)}{4 - (-4)} = \frac{12}{8} = \frac{3}{2}$$

for a ratio of $3:2$, resulting in a square screen.

EXAMPLE 1 **Examples of Viewing Rectangles That Result in Square Screens**

Figure 20

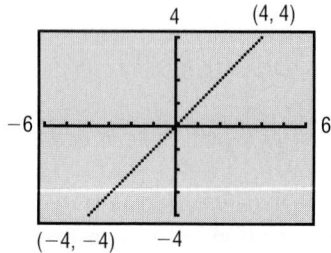

(a) $X\min = -3$
$X\max = 3$
$X\text{scl} = 1$
$Y\min = -2$
$Y\max = 2$
$Y\text{scl} = 1$

(b) $X\min = -6$
$X\max = 6$
$X\text{scl} = 1$
$Y\min = -4$
$Y\max = 4$
$Y\text{scl} = 1$

(c) $X\min = -6$
$X\max = 6$
$X\text{scl} = 1$
$Y\min = -4$
$Y\max = 4$
$Y\text{scl} = 2$ ∎

Figure 20 shows the graph of the line $y = x$ on a square screen using the viewing rectangle given in part (b). Notice that the line now bisects the first and third quadrants. Compare this illustration to Figure 19.

B.5 Exercises

In Problems 1–8, determine which of the given viewing rectangles result in a square screen.

1. $X\min = -3$
$X\max = 3$
$X\text{scl} = 2$
$Y\min = -2$
$Y\max = 2$
$Y\text{scl} = 2$

2. $X\min = -5$
$X\max = 5$
$X\text{scl} = 1$
$Y\min = -4$
$Y\max = 4$
$Y\text{scl} = 1$

3. $X\min = 0$
$X\max = 9$
$X\text{scl} = 3$
$Y\min = -2$
$Y\max = 4$
$Y\text{scl} = 2$

4. $X\min = -6$
$X\max = 6$
$X\text{scl} = 1$
$Y\min = -4$
$Y\max = 4$
$Y\text{scl} = 2$

5. $X\min = -6$
$X\max = 6$
$X\text{scl} = 1$
$Y\min = -2$
$Y\max = 2$
$Y\text{scl} = 0.5$

6. $X\min = -6$
$X\max = 6$
$X\text{scl} = 2$
$Y\min = -4$
$Y\max = 4$
$Y\text{scl} = 1$

7. $X\min = 0$
$X\max = 9$
$X\text{scl} = 1$
$Y\min = -2$
$Y\max = 4$
$Y\text{scl} = 1$

8. $X\min = -6$
$X\max = 6$
$X\text{scl} = 2$
$Y\min = -4$
$Y\max = 4$
$Y\text{scl} = 2$

9. If $X\min = -4$, $X\max = 8$, and $X\text{scl} = 1$, how should $Y\min$, $Y\max$, and $Y\text{scl}$ be selected so that the viewing rectangle contains the point $(4, 8)$ and the screen is square?

10. If $X\min = -6$, $X\max = 12$, and $X\text{scl} = 2$, how should $Y\min$, $Y\max$, and $Y\text{scl}$ be selected so that the viewing rectangle contains the point $(4, 8)$ and the screen is square?

B.6 Using a Graphing Utility to Graph Inequalities

It is easiest to begin with an example.

EXAMPLE 1 **Graphing an Inequality Using a Graphing Utility**

Use a graphing utility to graph: $3x + y - 6 \leq 0$

Solution We begin by graphing the equation $3x + y - 6 = 0$ $(Y_1 = -3x + 6)$. See Figure 21.

Figure 21

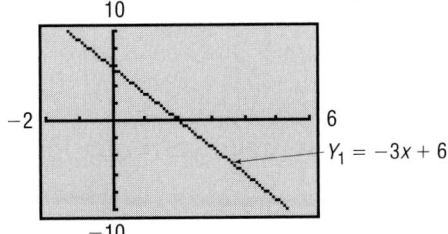

As with graphing by hand, we need to test points selected from each region and determine whether they satisfy the inequality. To test the point $(-1, 2)$, for example, enter $3(-1) + 2 - 6 \leq 0$. See Figure 22(a). The 1 that appears indicates that the statement entered (the inequality) is true. When the point $(5, 5)$ is tested, a 0 appears, indicating that the statement entered is false. Thus, $(-1, 2)$ is a part of the graph of the inequality and $(5, 5)$ is not. Figure 22(b) shows the graph of the inequality on a TI-84 Plus.[*]

Figure 22

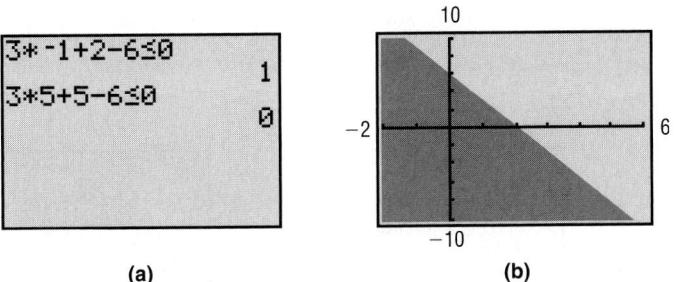

(a) (b)

The steps to follow to graph an inequality using a graphing utility are given next.

Steps for Graphing an Inequality Using a Graphing Utility

STEP 1: Replace the inequality symbol by an equal sign, solve the equation for y, and graph the equation.

STEP 2: In each region, select a test point P and determine if the coordinates of P satisfy the inequality.

 (a) If the test point satisfies the inequality, then so do all the points in the region. Indicate this by using the graphing utility to shade the region.

 (b) If the coordinates of P do not satisfy the inequality, then none of the points in that region do.

[*]Consult your owner's manual for shading techniques.

B.7 Using a Graphing Utility to Solve Systems of Linear Equations

Most graphing utilities have the capability to put the augmented matrix of a system of linear equations in row echelon form. The next example, Example 6 from Section 6.2, demonstrates this feature using a TI-84 Plus graphing calculator.

EXAMPLE 1 **Solving a System of Linear Equations Using a Graphing Utility**

Solve: $\begin{cases} x - y + z = 8 & (1) \\ 2x + 3y - z = -2 & (2) \\ 3x - 2y - 9z = 9 & (3) \end{cases}$

Solution The augmented matrix of the system is

$$\begin{bmatrix} 1 & -1 & 1 & | & 8 \\ 2 & 3 & -1 & | & -2 \\ 3 & -2 & -9 & | & 9 \end{bmatrix}$$

We enter this matrix into our graphing utility and name it A. See Figure 23(a). Using the REF (row echelon form) command on matrix A, we obtain the results shown in Figure 23(b). Since the entire matrix does not fit on the screen, we need to scroll right to see the rest of it. See Figure 23(c).

Figure 23

(a) (b) (c)

The system of equations represented by the matrix in row echelon form is

$$\begin{bmatrix} 1 & -\dfrac{2}{3} & -3 & | & 3 \\ 0 & 1 & \dfrac{15}{13} & | & -\dfrac{24}{13} \\ 0 & 0 & 1 & | & 1 \end{bmatrix} \qquad \begin{cases} x - \dfrac{2}{3}y - 3z = 3 & (1) \\ y + \dfrac{15}{13}z = -\dfrac{24}{13} & (2) \\ z = 1 & (3) \end{cases}$$

Using $z = 1$, we back-substitute to get

$$\begin{cases} x - \dfrac{2}{3}y - 3(1) = 3 \quad \text{(1)} \\ \\ \quad\quad y + \dfrac{15}{13}(1) = -\dfrac{24}{13} \quad \text{(2)} \end{cases} \xrightarrow[\text{Simplify.}]{} \begin{cases} x - \dfrac{2}{3}y = 6 \quad\quad \text{(1)} \\ \\ \quad\quad y = \dfrac{-39}{13} = -3 \quad \text{(2)} \end{cases}$$

Solving the second equation for y, we find that $y = -3$. Back-substituting $y = -3$ into $x - \dfrac{2}{3}y = 6$, we find that $x = 4$. The solution of the system is $x = 4$, $y = -3$, $z = 1$. ■

Notice that the row echelon form of the augmented matrix using the graphing utility differs from the row echelon form in Chapter 10 (p. 725), yet both matrices provide the same solution! This is because the two solutions used different row operations to obtain the row echelon form. In all likelihood, the two solutions parted ways in Step 4 of the algebraic solution, where we avoided introducing fractions by interchanging rows 2 and 3.

Most graphing utilities also have the ability to put a matrix in reduced row echelon form. Figure 24 shows the reduced row echelon form of the augmented matrix from Example 1 using the RREF command on a TI-84 Plus graphing calculator. Using this command, we see that the solution of the system is $x = 4$, $y = -3$, $z = 1$.

Figure 24

Answers

CHAPTER F Foundations: A Prelude to Functions

F.1 Assess Your Understanding (page 6)

5. abscissa; ordinate **6.** quadrants **7.** midpoint **8.** F **9.** F **10.** T

11. **(a)** Quadrant II **(b)** Positive x-axis
(c) Quadrant III **(d)** Quadrant I
(e) Negative y-axis **(f)** Quadrant IV

12. **(a)** Quadrant I **(b)** Quadrant III
(c) Quadrant II **(d)** Quadrant I
(e) Positive y-axis **(f)** Negative x-axis

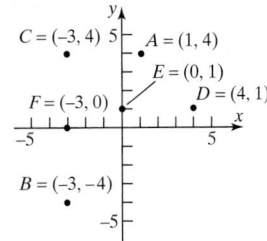

13. The points will be on a vertical line that is 2 units to the right of the y-axis.

14. The points will be on a horizontal line that is 3 units above the x-axis.

15. $\sqrt{5}$
16. $\sqrt{5}$
17. $\sqrt{10}$
18. $\sqrt{10}$
19. $2\sqrt{17}$
20. 5
21. $\sqrt{85}$
22. $\sqrt{29}$
23. $\sqrt{53}$
24. $5\sqrt{5}$
25. $\sqrt{6.89} \approx 2.62$
26. $\sqrt{3.69} \approx 1.92$

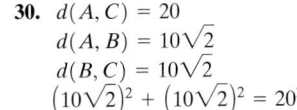

27. $\sqrt{a^2 + b^2}$ **28.** $\sqrt{2}|a|$

29. $d(A, B) = \sqrt{13}$
$d(B, C) = \sqrt{13}$
$d(A, C) = \sqrt{26}$
$\left(\sqrt{13}\right)^2 + \left(\sqrt{13}\right)^2 = \left(\sqrt{26}\right)^2$
Area $= \dfrac{13}{2}$ square units

30. $d(A, C) = 20$
$d(A, B) = 10\sqrt{2}$
$d(B, C) = 10\sqrt{2}$
$\left(10\sqrt{2}\right)^2 + \left(10\sqrt{2}\right)^2 = 20^2$
Area $= 100$ square units

31. $d(A, B) = \sqrt{130}$
$d(B, C) = \sqrt{26}$
$d(A, C) = 2\sqrt{26}$
$\left(\sqrt{26}\right)^2 + \left(2\sqrt{26}\right)^2 = \left(\sqrt{130}\right)^2$
Area $= 26$ square units

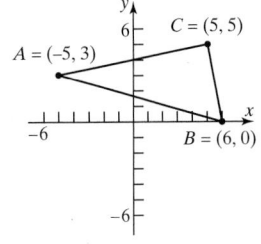

32. $d(A, B) = \sqrt{145}$
$d(A, C) = \sqrt{29}$
$d(B, C) = 2\sqrt{29}$
$\left(\sqrt{29}\right)^2 + \left(2\sqrt{29}\right)^2 = \left(\sqrt{145}\right)^2$
Area $= 29$ square units

33. $d(A, B) = 4$
$d(A, C) = 5$
$d(B, C) = \sqrt{41}$
$4^2 + 5^2 = 16 + 25 = \left(\sqrt{41}\right)^2$
Area $= 10$ square units

34. $d(A, B) = 4$
$d(A, C) = 2\sqrt{5}$
$d(B, C) = 2$
$4^2 + 2^2 = \left(2\sqrt{5}\right)^2$
Area $= 4$ square units

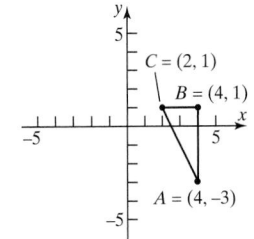

35. $(4, 0)$ **36.** $(0, 2)$ **37.** $\left(\dfrac{3}{2}, 1\right)$ **38.** $\left(3, -\dfrac{1}{2}\right)$ **39.** $(5, -1)$ **40.** $\left(-1, -\dfrac{1}{2}\right)$ **41.** $(1.05, 0.7)$ **42.** $(0.45, 1.7)$ **43.** $\left(\dfrac{a}{2}, \dfrac{b}{2}\right)$ **44.** $\left(\dfrac{a}{2}, \dfrac{a}{2}\right)$

45. $(2, 2); (2, -4)$ **46.** $(13, -3); (-11, -3)$ **47.** $(0, 0); (8, 0)$ **48.** $(0, 1); (0, 7)$ **49.** $\sqrt{17}; 2\sqrt{5}; \sqrt{29}$ **50.** Two triangles are possible. The third vertex is $\left(2\sqrt{3}, 2\right)$ or $\left(-2\sqrt{3}, 2\right)$. **51.** $\left(\dfrac{s}{2}, \dfrac{s}{2}\right)$

52. Let $P_1 = (0, 0)$, $P_2 = (a, 0)$, $P_3 = \left(\dfrac{a}{2}, \dfrac{\sqrt{3}a}{2}\right)$; $d(P_1, P_2) = |a|, d(P_2, P_3) = |a|, d(P_1, P_3) = |a|$, so the points form an equilateral triangle.

Midpoints $P_4 = \left(\dfrac{a}{2}, 0\right)$, $P_5 = \left(\dfrac{3a}{4}, \dfrac{\sqrt{3}a}{4}\right)$, $P_6 = \left(\dfrac{a}{4}, \dfrac{\sqrt{3}a}{4}\right)$

$d(P_4, P_5) = \dfrac{|a|}{2}, d(P_5, P_6) = \dfrac{|a|}{2}, d(P_4, P_6) = \dfrac{|a|}{2}$, so the points form an equilateral triangle.

53. $d(P_1, P_2) = 6; d(P_2, P_3) = 4; d(P_1, P_3) = 2\sqrt{13}$; right triangle
54. $d(P_1, P_2) = \sqrt{53}; d(P_2, P_3) = \sqrt{53}; d(P_1, P_3) = \sqrt{106}$; isosceles right triangle
55. $d(P_1, P_2) = 2\sqrt{17}; d(P_2, P_3) = \sqrt{34}; d(P_1, P_3) = \sqrt{34}$; isosceles right triangle
56. $d(P_1, P_2) = 5\sqrt{5}; d(P_2, P_3) = 10; d(P_1, P_3) = 5$; right triangle
57. $90\sqrt{2} \approx 127.28$ ft
58. $60\sqrt{2} \approx 84.85$ ft
59. **(a)** $(90, 0), (90, 90), (0, 90)$ **(b)** $5\sqrt{2161} \approx 232.43$ ft **(c)** $30\sqrt{149} \approx 366.20$ ft
60. **(a)** $(60, 0), (60, 60), (0, 60)$ **(b)** $40\sqrt{10} \approx 126.49$ ft **(c)** $20\sqrt{185} \approx 272.03$ ft **61.** $d = 50t$ mi. **62.** $d = \sqrt{10{,}000 + 484t^2}$ ft

F.2 Assess Your Understanding (page 15)

3. intercepts **4.** zeros; roots **5.** y-axis **6.** 4 **7.** $(-3, 4)$ **8.** T **9.** F **10.** F **11.** $(0, 0)$ is on the graph. **12.** $(0, 0)$ and $(1, -1)$ are on the graph.
13. $(0, 3)$ is on the graph. **14.** $(0, 1)$ and $(-1, 0)$ are on the graph. **15.** $(0, 2)$ and $\left(\sqrt{2}, \sqrt{2}\right)$ are on the graph. **16.** $(0, 1)$ and $(2, 0)$ are on the graph.
17. $(-2, 0), (0, 2)$ **18.** $(6, 0), (0, -6)$ **19.** $(-4, 0), (0, 8)$ **20.** $(3, 0), (0, -9)$

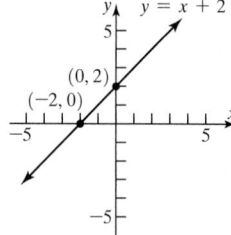

21. $(-1, 0), (1, 0), (0, -1)$ **22.** $(-3, 0), (3, 0), (0, -9)$ **23.** $(-2, 0), (2, 0), (0, 4)$ **24.** $(-1, 0), (1, 0), (0, 1)$

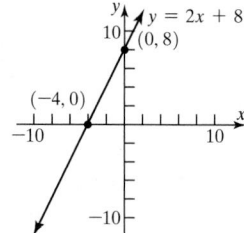

25. $(3, 0), (0, 2)$ **26.** $(2, 0), (0, 5)$ **27.** $(-2, 0), (2, 0), (0, 9)$ **28.** $(-1, 0), (1, 0), (0, 4)$

29.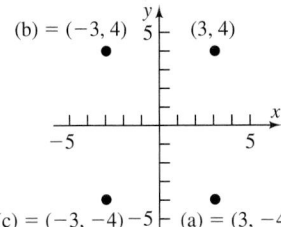

(b) = (−3, 4) (3, 4)

(c) = (−3, −4) (a) = (3, −4)

30.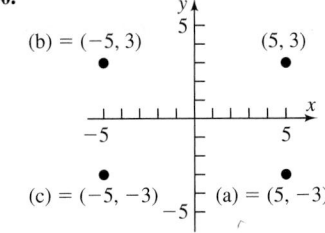

(b) = (−5, 3) (5, 3)

(c) = (−5, −3) (a) = (5, −3)

31.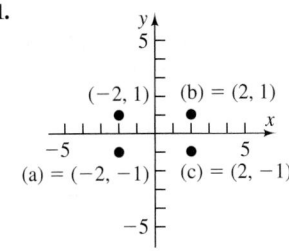

(−2, 1) (b) = (2, 1)

(a) = (−2, −1) (c) = (2, −1)

32.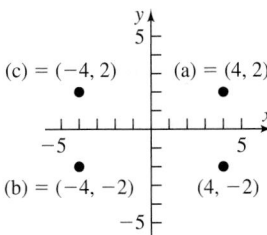

(c) = (−4, 2) (a) = (4, 2)

(b) = (−4, −2) (4, −2)

33.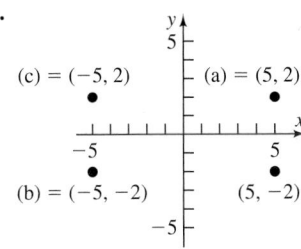

(c) = (−5, 2) (a) = (5, 2)

(b) = (−5, −2) (5, −2)

34.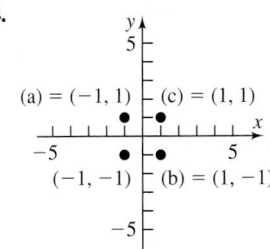

(a) = (−1, 1) (c) = (1, 1)

(−1, −1) (b) = (1, −1)

35.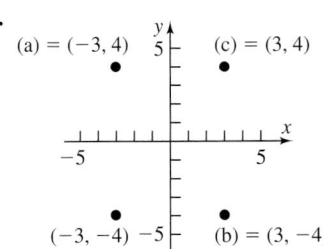

(a) = (−3, 4) (c) = (3, 4)

(−3, −4) (b) = (3, −4)

36.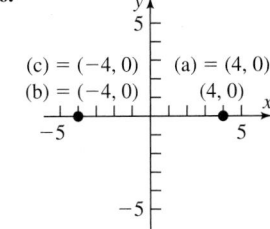

(c) = (−4, 0) (a) = (4, 0)
(b) = (−4, 0) (4, 0)

37.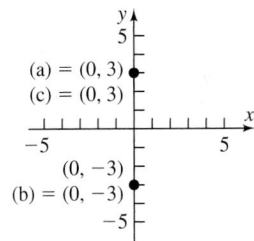

(a) = (0, 3)
(c) = (0, 3)

(0, −3)
(b) = (0, −3)

38.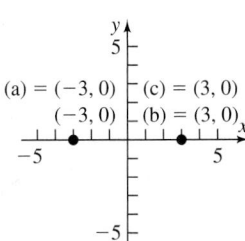

(a) = (−3, 0) (c) = (3, 0)
(−3, 0) (b) = (3, 0)

39. (a) $(−1, 0), (1, 0)$
(b) Symmetric with respect to the x-axis, the y-axis, and the origin

40. (a) $(0, 1)$
(b) No symmetry

41. (a) $\left(-\frac{\pi}{2}, 0\right), (0, 1), \left(\frac{\pi}{2}, 0\right)$
(b) Symmetric with respect to the y-axis

42. (a) $(−2, 0), (0, −3), (2, 0)$
(b) Symmetric with respect to the y-axis

43. (a) $(0, 0)$
(b) Symmetric with respect to the x-axis

44. (a) $(−2, 0), (0, 2), (2, 0), (0, −2)$
(b) Symmetric with respect to the x-axis, the y-axis, and the origin.

45. (a) $(−2, 0), (0, 0), (2, 0)$
(b) Symmetric with respect to the origin

46. (a) $(−4, 0), (0, 0), (4, 0)$
(b) Symmetric with respect to the origin

47. (a) $(−1, 0), (0, −1), (1, 0)$
(b) Symmetric with respect to the y-axis

48. (a) $(0, 0)$
(b) Symmetric with respect to the origin

49. (a) No intercepts
(b) Symmetric with respect to the origin

50. (a) No intercepts
(b) Symmetric with respect to the x-axis

51.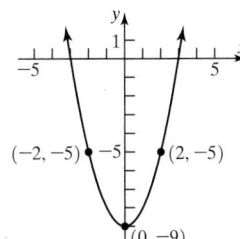

(−2, −5) (2, −5)

(0, −9)

52.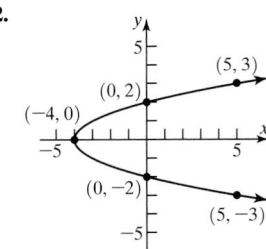

(0, 2) (5, 3)

(−4, 0)

(0, −2)

(5, −3)

53.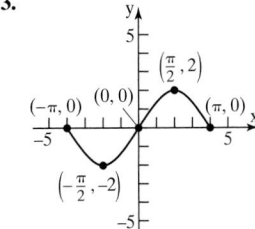

$\left(\frac{\pi}{2}, 2\right)$

(−π, 0) (0, 0) (π, 0)

$\left(-\frac{\pi}{2}, -2\right)$

54.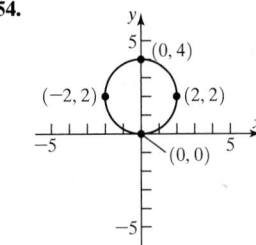

(0, 4)

(−2, 2) (2, 2)

(0, 0)

55. $(-4, 0)$, $(0, -2)$, $(0, 2)$; symmetric with respect to the *x*-axis **56.** $(-9, 0)$, $(0, 3)$, $(0, -3)$; symmetric with respect to the *x*-axis **57.** $(0, 0)$; symmetric with respect to the origin **58.** $(0, 0)$; symmetric with respect to the origin **59.** $(0, 9)$, $(3, 0)$, $(-3, 0)$; symmetric with respect to the *y*-axis

60. $(0, -4)$, $(-2, 0)$, $(2, 0)$; symmetric with respect to the *y*-axis **61.** $(-2, 0)$, $(2, 0)$, $(0, -3)$, $(0, 3)$; symmetric with respect to the *x*-axis, *y*-axis, and origin

62. $(-1, 0)$, $(1, 0)$, $(0, -2)$, $(0, 2)$; symmetric with respect to the *y*-axis, *y*-axis, and origin **63.** $(0, -27)$, $(3, 0)$; no symmetry **64.** $(-1, 0)$, $(1, 0)$, $(0, -1)$; symmetric with respect to the *y*-axis **65.** $(0, -4)$, $(4, 0)$, $(-1, 0)$; no symmetry **66.** $(0, 4)$; symmetric with respect to the *y*-axis **67.** $(0, 0)$; symmetric with respect to the origin **68.** $(2, 0)$, $(-2, 0)$; symmetric with respect to the origin **69.** $(0, 0)$; symmetric with respect to the origin **70.** No intercepts; symmetric with respect to the origin

71.

72.

73.

74.
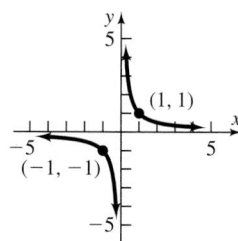

75. $b = 13$ **76.** $b = 2$ **77.** -4 or 1 **78.** $a = -5$ or $a = -1$ **79.** $(-1, -2)$ **80.** -6 **81.** 4 **82.** -2
83. **(a)** $y = \sqrt{x^2}$ and $y = |x|$ have the same graph. **(b)** $\sqrt{x^2} = |x|$ **(c)** $x \geq 0$ for $y = (\sqrt{x})^2$, while x can be any real number for $y = x$.
(d) $y \geq 0$ for $y = \sqrt{x^2}$

F.3 Assess Your Understanding (page 29)

1. undefined; 0 **2.** 3; 2 **3.** $y = b$; *y*-intercept **4.** True **5.** False **6.** True **7.** $m_1 = m_2$; *y*-intercepts; $m_1 m_2 = -1$ **8.** 2 **9.** $-\dfrac{1}{2}$ **10.** False

11. (a) Slope $= \dfrac{1}{2}$

 (b) If *x* increases by 2 units, *y* will increase by 1 unit.

12. (a) Slope $= -\dfrac{1}{2}$

 (b) If *x* increases by 2 units, *y* will decrease by 1 unit.

13. (a) Slope $= -\dfrac{1}{3}$

 (b) If *x* increases by 3 units, *y* will decrease by 1 unit.

14. (a) Slope $= \dfrac{1}{3}$

 (b) If *x* increases by 3 units, *y* will increase by 1 unit.

15. Slope $= -\dfrac{3}{2}$

16. Slope $= -2$

17. Slope $= -\dfrac{1}{2}$

18. Slope $= \dfrac{2}{3}$

19. Slope $= 0$

20. Slope $= 0$

21. Slope undefined

22. Slope undefined

23.

24.

25.

26.

27.

28.

29.

30.

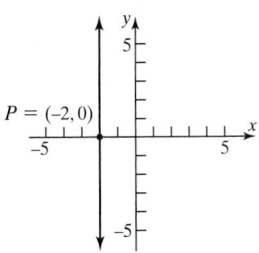

31. $(2,6); (3,10); (4,14)$ **32.** $(-1,5); (0,7); (1,9)$ **33.** $(4,-7); (6,-10); (8,-13)$ **34.** $(0,6); (3,10); (6,14)$ **35.** $(-1,-5); (0,-7); (1,-9)$
36. $(5,0); (6,-1); (7,-2)$ **37.** $x - 2y = 0$ or $y = \frac{1}{2}x$ **38.** $x + 2y = 0$ or $y = -\frac{1}{2}x$ **39.** $x + y = 2$ or $y = -x + 2$ **40.** $x - 3y = -4$ or
$y = \frac{1}{3}x + \frac{4}{3}$ **41.** $2x - y = 3$ or $y = 2x - 3$ **42.** $x + y = 3$ or $y = -x + 3$ **43.** $x + 2y = 5$ or $y = -\frac{1}{2}x + \frac{5}{2}$ **44.** $x - y = -2$ or
$y = x + 2$ **45.** $3x - y = -9$ or $y = 3x + 9$ **46.** $2x - y = 11$ or $y = 2x - 11$ **47.** $2x + 3y = -1$ or $y = -\frac{2}{3}x - \frac{1}{3}$ **48.** $x - 2y = 1$ or
$y = \frac{1}{2}x - \frac{1}{2}$ **49.** $x - 2y = -5$ or $y = \frac{1}{2}x + \frac{5}{2}$ **50.** $x - 5y = -23$ or $y = \frac{1}{5}x + \frac{23}{5}$ **51.** $3x + y = 3$ or $y = -3x + 3$ **52.** $2x + y = -2$ or
$y = -2x - 2$ **53.** $x - 2y = 2$ or $y = \frac{1}{2}x - 1$ **54.** $x - y = -4$ or $y = x + 4$ **55.** $x = 2$; no slope–intercept form **56.** $x = 3$; no slope–intercept
form **57.** $y - 2 = 0$; $y = 2$ **58.** $x - 4 = 0$; $x = 4$ **59.** $2x - y = -4$ or $y = 2x + 4$ **60.** $3x + y = -1$ or $y = -3x - 1$ **61.** $2x - y = 0$ or
$y = 2x$ **62.** $x - 2y = 0$ or $y = \frac{1}{2}x$ **63.** $x = 4$; no slope–intercept form **64.** $y = 2$ **65.** $2x + y = 0$ or $y = -2x$ **66.** $x + 2y = -3$ or
$y = -\frac{1}{2}x - \frac{3}{2}$ **67.** $x - 2y = -3$ or $y = \frac{1}{2}x + \frac{3}{2}$ **68.** $2x + y = 4$ or $y = -2x + 4$ **69.** $y = 4$ **70.** $x = 3$; no slope–intercept form

71. Slope $= 2$; y-intercept $= 3$

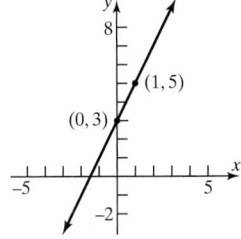

72. Slope $= -3$; y-intercept $= 4$

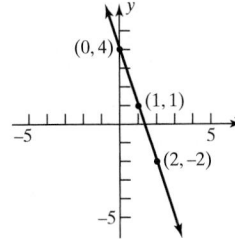

73. $y = 2x - 2$; slope $= 2$; y-intercept $= -2$

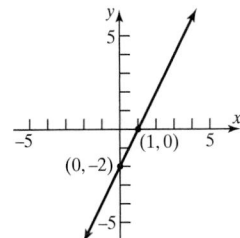

74. $y = -\frac{1}{3}x + 2$; slope $= -\frac{1}{3}$; y-intercept $= 2$

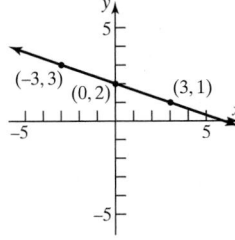

75. Slope $= \frac{1}{2}$; y-intercept $= 2$

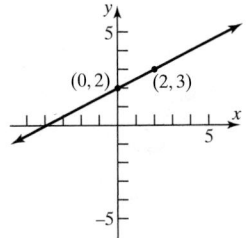

76. Slope $= 2$; y-intercept $= \frac{1}{2}$

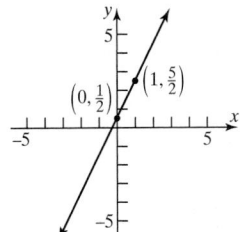

77. Slope $= -\frac{1}{2}$; y-intercept $= 2$

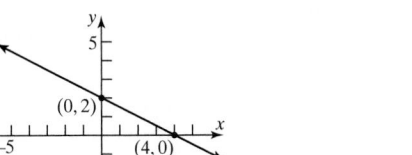

78. Slope $= \frac{1}{3}$; y-intercept $= 2$

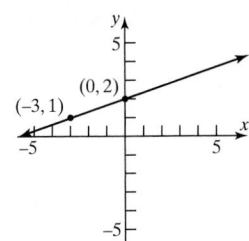

79. Slope $= \frac{2}{3}$; y-intercept $= -2$

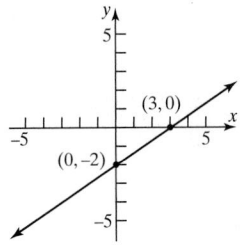

80. Slope $= -\dfrac{3}{2}$; y-intercept $= 3$

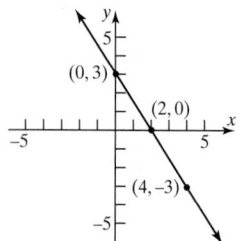

81. Slope $= -1$; y-intercept $= 1$

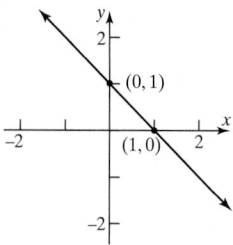

82. Slope $= 1$; y-intercept $= -2$

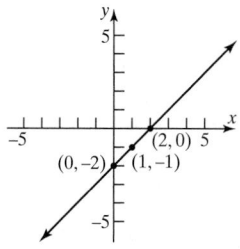

83. Slope undefined; no y-intercept

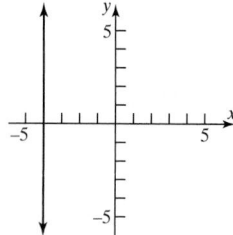

84. Slope $= 0$; y-intercept $= -1$

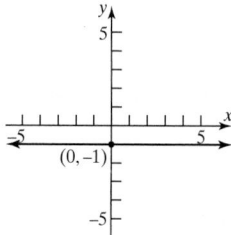

85. Slope $= 0$; y-intercept $= 5$

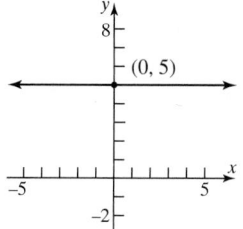

86. Slope undefined; no y-intercept

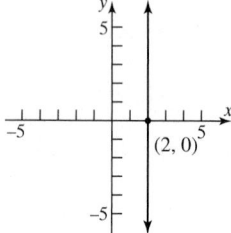

87. Slope $= 1$; y-intercept $= 0$

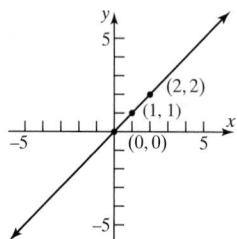

88. Slope $= -1$; y-intercept $= 0$

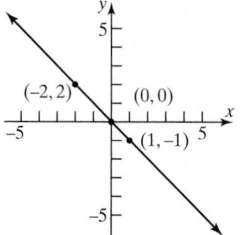

89. Slope $= \dfrac{3}{2}$; y-intercept $= 0$

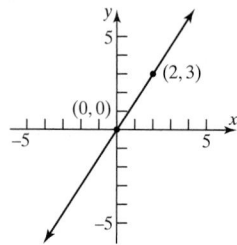

90. Slope $= -\dfrac{3}{2}$; y-intercept $= 0$

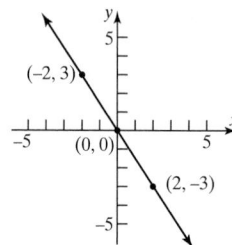

91. **(a)** x-intercept: 3; y-intercept: 2

(b)

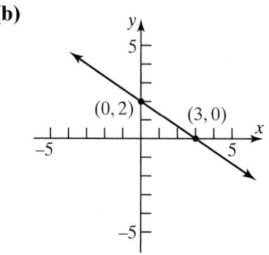

92. **(a)** x-intercept: 2; y-intercept: -3

(b)

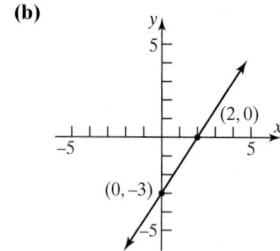

93. **(a)** x-intercept: -10; y-intercept: 8

(b)

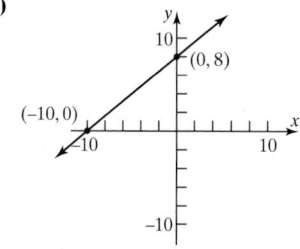

94. **(a)** x-intercept: 4; y-intercept: -6

(b)

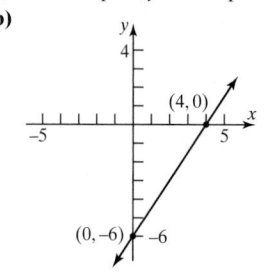

95. **(a)** x-intercept: 3; y-intercept: $\dfrac{21}{2}$

(b)

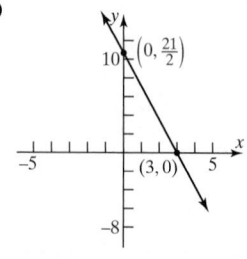

96. **(a)** x-intercept: $\dfrac{18}{5}$; y-intercept: 6

(b)

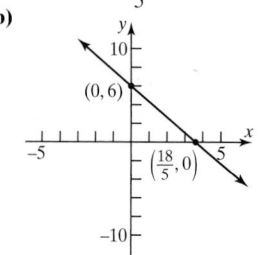

97. **(a)** x-intercept: 2; y-intercept: 3

(b)

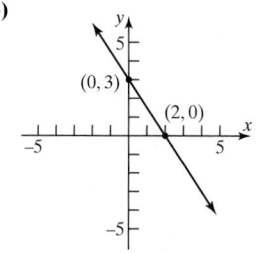

98. (a) x-intercept: 4; y-intercept: -6
(b)

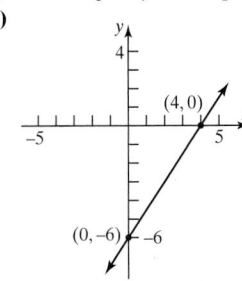

99. (a) x-intercept: 5; y-intercept: -2
(b)

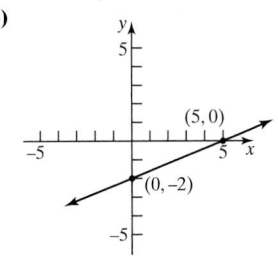

100. (a) x-intercept: -4; y-intercept: 3
(b)

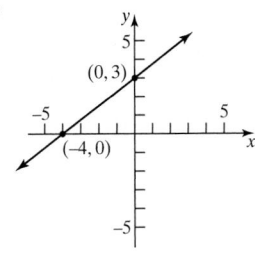

101. $y = 0$ **102.** $x = 0$ **103.** Parallel **104.** Perpendicular **105.** Neither **106.** Neither **107.** $x - y = -2$ or $y = x + 2$ **108.** $x + y = 1$ or

$y = -x + 1$ **109.** $x + 3y = 3$ or $y = -\dfrac{1}{3}x + 1$ **110.** $x + 2y = -2$ or $y = -\dfrac{1}{2}x - 1$

111. $P_1 = (-2, 5)$, $P_2 = (1, 3)$, $m_1 = -\dfrac{2}{3}$; $P_2 = (1, 3)$, $P_3 = (-1, 0)$, $m_2 = \dfrac{3}{2}$; because $m_1 m_2 = -1$, the lines are perpendicular and the points $(-2, 5)$,
$(1, 3)$, and $(-1, 0)$ are the vertices of a right triangle; thus, the points P_1, P_2, and P_3 are the vertices of a right triangle.

112. $P_1 = (1, -1)$, $P_3 = (2, 2)$, $m = 3$; $P_2 = (4, 1)$, $P_4 = (5, 4)$, $m = 3$; $P_3 = (2, 2)$, $P_4 = (5, 4)$, $m = \dfrac{2}{3}$; $P_1 = (1, -1)$, $P_2 = (4, 1)$, $m = \dfrac{2}{3}$; opposite
sides are parallel; the points are the vertices of a parallelogram.

113. $P_1 = (-1, 0)$, $P_2 = (2, 3)$, $m = 1$; $P_3 = (1, -2)$, $P_4 = (4, 1)$, $m = 1$; $P_1 = (-1, 0)$, $P_3 = (1, -2)$, $m = -1$; $P_2 = (2, 3)$, $P_4 = (4, 1)$, $m = -1$;
opposite sides are parallel, and adjacent sides are perpendicular; the points are the vertices of a rectangle.

114. $P_1 = (0, 0)$, $P_2 = (1, 3)$, $m = 3$; $P_3 = (4, 2)$, $P_4 = (3, -1)$, $m = 3$; $P_2 = (1, 3)$, $P_3 = (4, 2)$, $m = -\dfrac{1}{3}$; $P_1 = (0, 0)$, $P_4 = (3, -1)$, $m = -\dfrac{1}{3}$;
opposite sides are parallel; adjacent sides are perpendicular;

$d(P_1, P_2) = \sqrt{10}$; $d(P_2, P_3) = \sqrt{10}$;

$d(P_3, P_4) = \sqrt{10}$; $d(P_1, P_4) = \sqrt{10}$;
all sides have equal lengths; the quadrilateral is a square. **115.** $C = 0.07x + 29$; \$36.70; \$45.10 **116.** $C = 8x + 500$; \$3700; \$6420

117. $C = 0.53x + 1,070,000$ **118.** $S = 0.05x + 375$

119. (a) $C = 0.08275x + 7.58$, $0 \le x \le 400$
(b)

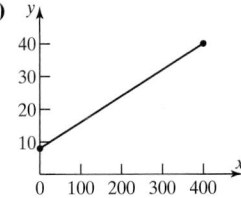

120. (a) $C = 0.0835x + 5.25$, $0 \le x \le 750$
(b)

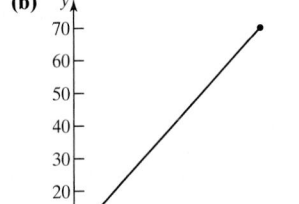

(c) \$15.86 **(d)** \$32.41
(e) Each additional kW-h used adds \$0.08275 to the bill.

(c) \$21.95 **(d)** \$47.00
(e) Each additional kW-h used adds \$0.0835 to the bill.

121. $°C = \dfrac{5}{9}(°F - 32)$; approximately 21.1°C **122. (a)** $K = °C + 273$

(b) $K = \dfrac{5}{9}°F + \dfrac{2297}{9}$

123. (a) $A = \dfrac{1}{5}x + 20,000$

(b) \$80,000 **(c)** Each additional box sold requires
an additional \$0.20 in advertising.

124. Slope from (a, b) to (b, a) is $\dfrac{a - b}{b - a} = -1$. Slope of the line $y = x$ is 1. Since $-1 \cdot 1 = -1$, the line containing the points (a, b) and (b, a) is
perpendicular to the line $y = x$.

The midpoint of (a, b) and $(b, a) = \left(\dfrac{a + b}{2}, \dfrac{b + a}{2}\right)$.

Since $\dfrac{b + a}{2} = \dfrac{a + b}{2}$, the midpoint lies on the line $y = x$.

125. All have the same slope, 2; the lines are parallel.

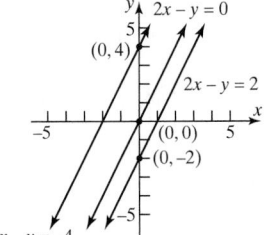

126. Refer to Figure 45. $m_1 m_2 = -1$; $d(A, B) = \sqrt{(m_2 - m_1)^2}$; $d(0, A) = \sqrt{1 + m_2^2}$; $d(0, B) = \sqrt{1 + m_1^2}$. Now show that $[d(0, B)]^2 + [d(0, A)^2] = [d(A, B)]^2$ and apply the converse of the Pythagorean Theorem.
127. (b), (c), (e), (g) **128.** (a), (c), (g) **129.** (c) **130.** (d) **134.** No, consider vertical lines. **135.** No; no **136.** They are the same line.
137. They are the same line. **138.** No **139.** Yes

F.4 Assess Your Understanding (page 36)

3. F **4.** radius **5.** T **6.** F **7.** Center $(2, 1)$; radius 2; $(x - 2)^2 + (y - 1)^2 = 4$ **8.** Center $(1, 2)$; radius $= 2$; $(x - 1)^2 + (y - 2)^2 = 4$
9. Center $\left(\dfrac{5}{2}, 2\right)$; radius $\dfrac{3}{2}$; $\left(x - \dfrac{5}{2}\right)^2 + (y - 2)^2 = \dfrac{9}{4}$ **10.** Center $(1, 2)$; radius $= \sqrt{2}$; $(x - 1)^2 + (y - 2)^2 = 2$

11. $x^2 + y^2 = 4$;
$x^2 + y^2 - 4 = 0$

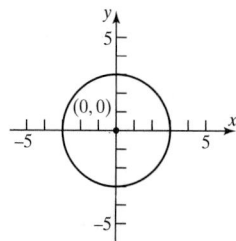

12. $x^2 + y^2 = 9$;
$x^2 + y^2 - 9 = 0$

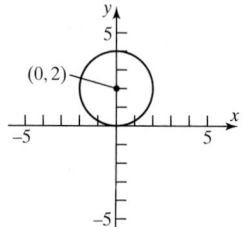

13. $x^2 + (y - 2)^2 = 4$;
$x^2 + y^2 - 4y = 0$

14. $(x - 1)^2 + y^2 = 9$;
$x^2 + y^2 - 2x - 8 = 0$

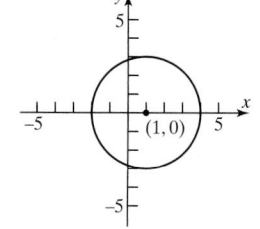

15. $(x - 4)^2 + (y + 3)^2 = 25$;
$x^2 + y^2 - 8x + 6y = 0$

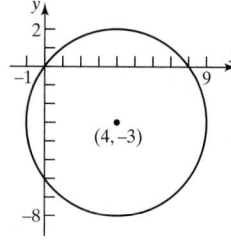

16. $(x - 2)^2 + (y + 3)^2 = 16$;
$x^2 + y^2 - 4x + 6y - 3 = 0$

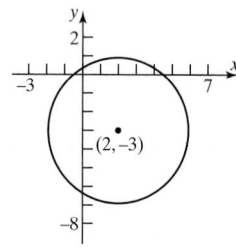

17. $(x + 2)^2 + (y - 1)^2 = 16$;
$x^2 + y^2 + 4x - 2y - 11 = 0$

18. $(x + 5)^2 + (y + 2)^2 = 49$;
$x^2 + y^2 + 10x + 4y - 20 = 0$

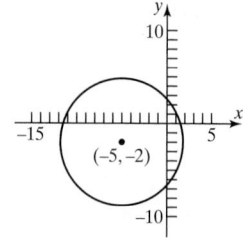

19. $\left(x - \dfrac{1}{2}\right)^2 + y^2 = \dfrac{1}{4}$;
$x^2 + y^2 - x = 0$

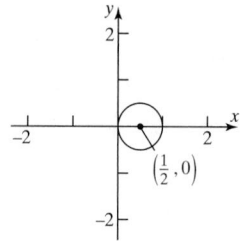

20. $x^2 + \left(y + \dfrac{1}{2}\right)^2 = \dfrac{1}{4}$;
$x^2 + y^2 + y = 0$

21. (a) $(h, k) = (0, 0)$; $r = 2$
(b)

(c) $(\pm 2, 0)$; $(0, \pm 2)$

22. (a) $(h, k) = (0, 1)$; $r = 1$
(b)

(c) $(0, 0)$; $(0, 2)$

23. (a) $(h, k) = (3, 0)$; $r = 2$
(b)

(c) $(1, 0)$; $(5, 0)$

24. (a) $(h, k) = (-1, 1)$; $r = \sqrt{2}$
(b)

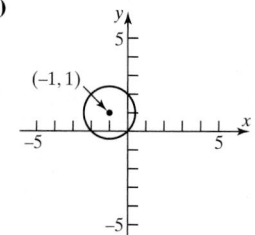

(c) $(0, 0)$; $(0, 2)$; $(-2, 0)$

25. (a) $(h, k) = (1, 2); r = 3$
(b)

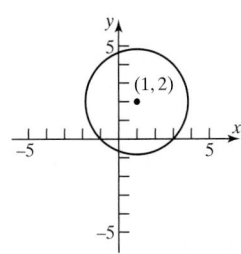

(c) $\left(1 \pm \sqrt{5}, 0\right); \left(0, 2 \pm 2\sqrt{2}\right)$

26. (a) $(h, k) = (-2, -1); r = 5$
(b)

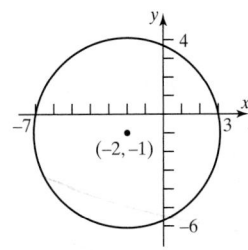

(c) $\left(-2 \pm 2\sqrt{6}, 0\right); \left(0, -1 \pm \sqrt{21}\right)$

27. (a) $(h, k) = (-2, 2); r = 3$
(b)

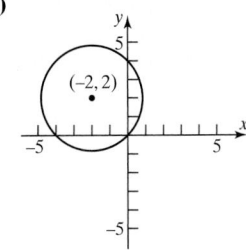

(c) $\left(-2 \pm \sqrt{5}, 0\right); \left(0, 2 \pm \sqrt{5}\right)$

28. (a) $(h, k) = (3, -1); r = 1$
(b)

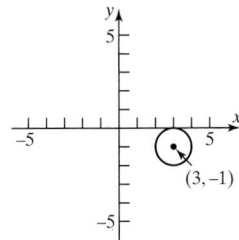

(c) $(3, 0)$

29. (a) $(h, k) = \left(\frac{1}{2}, -1\right); r = \frac{1}{2}$
(b)

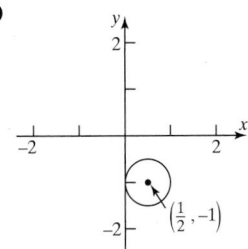

(c) $(0, -1)$

30. (a) $(h, k) = \left(-\frac{1}{2}, -\frac{1}{2}\right); r = 1$
(b)

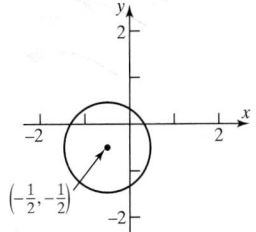

(c) $\left(\frac{-1 \pm \sqrt{3}}{2}, 0\right); \left(0, \frac{-1 \pm \sqrt{3}}{2}\right)$

31. (a) $(h, k) = (3, -2); r = 5$
(b)

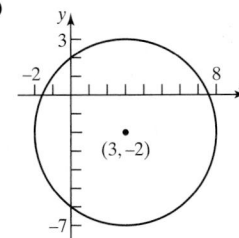

32. (a) $(h, k) = (-2, 0); r = \frac{\sqrt{2}}{2}$
(b)

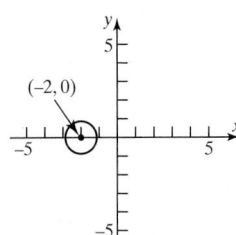

33. (a) $(h, k) = (-2, 0); r = 2$
(b)

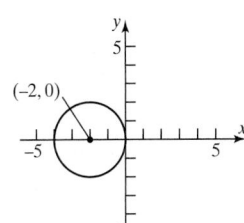

34. (a) $(h, k) = (0, 2); r = 2$
(b)

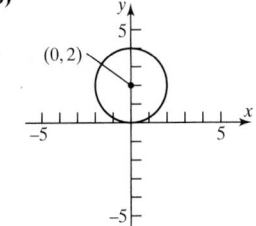

(c) $\left(3 \pm \sqrt{21}, 0\right); (0, -6), (0, 2)$ **(c)** $\left(-2 \pm \frac{\sqrt{2}}{2}, 0\right)$ **(c)** $(0, 0), (-4, 0)$ **(c)** $(0, 0), (0, 4)$

35. $x^2 + y^2 - 13 = 0$ **36.** $x^2 + y^2 - 2x - 19 = 0$ **37.** $x^2 + y^2 - 4x - 6y + 4 = 0$ **38.** $x^2 + y^2 + 6x - 2y + 1 = 0$
39. $x^2 + y^2 + 2x - 6y + 5 = 0$ **40.** $x^2 + y^2 - 4x - 4y + 3 = 0$ **41.** (c) **42.** (d) **43.** (b) **44.** (a) **45.** 18 sq units **46.** $36\pi - 72$ sq units
47. $x^2 + y^2 + 2x + 4y - 4168.16 = 0$
48. (a) $x^2 + (mx + b)^2 = r^2$

$(1 + m^2)x^2 + 2mbx + b^2 - r^2 = 0$

One solution if and only if discriminant $= 0$

$(2mb)^2 - 4(1 + m^2)(b^2 - r^2) = 0$

$-4b^2 + 4r^2 + 4m^2r^2 = 0$

$r^2(1 + m^2) = b^2$

(b) $x = \frac{-2mb}{2(1 + m^2)} = \frac{-2mb}{2b^2/r^2} = -\frac{r^2m}{b}$

$y = m\left(-\frac{r^2m}{b}\right) + b = -\frac{r^2m^2}{b} + b = \frac{-r^2m^2 + b^2}{b} = \frac{r^2}{b}$

(c) Slope of tangent line $= m$

Slope of line joining center to point of tangency $= \frac{r^2/b}{-r^2m/b} = -\frac{1}{m}$

49. $\sqrt{2}x + 4y - 9\sqrt{2} = 0$ **50.** $\sqrt{2}x + 4y - 11\sqrt{2} + 12 = 0$ **51.** $(1, 0)$ **52.** $x + 5y + 13 = 0$ **53.** $y = 2$ **54.** 3 **55.** (b), (c), (e), (g)
56. (b), (e), (g)

CHAPTER 1 Functions and Their Graphs

1.1 Assess Your Understanding (page 51)

5. independent; dependent **6.** range **7.** $[0, 5]$ **8.** $\neq; f, g$ **9.** $(g - f)(x)$ **10.** F **11.** T **12.** T **13.** F **14.** F
15. Function; Domain: {Elvis, Colleen, Kaleigh, Marissa}, Range: {January 8, March 15, September 17} **16.** Not a function **17.** Not a function
18. Function; Domain: {Less than 9th grade, 9th–12th grade, High School Graduate, Some College, College Graduate}; Range:
{\$18,120, \$23,251, \$36,055, \$45,810, \$67,165} **19.** Not a function **20.** Function; Domain: $\{-2, -1, 3, 4\}$, Range: $\{3, 5, 7, 12\}$ **21.** Function;
Domain: $\{1, 2, 3, 4\}$; Range: $\{3\}$ **22.** Function; Domain: $\{0, 1, 2, 3\}$, Range: $\{-2, 3, 7\}$ **23.** Not a function **24.** Not a function **25.** Function;
Domain: $\{-2, -1, 0, 1\}$, Range: $\{0, 1, 4\}$ **26.** Function; Domain: $\{-2, -1, 0, 1\}$, Range: $\{3, 4, 16\}$ **27.** Function **28.** Function **29.** Function
30. Function **31.** Not a function **32.** Not a function **33.** Not a function **34.** Not a function **35.** Function **36.** Function **37.** Not a function
38. Not a function **39. (a)** -4 **(b)** 1 **(c)** -3 **(d)** $3x^2 - 2x - 4$ **(e)** $-3x^2 - 2x + 4$ **(f)** $3x^2 + 8x + 1$ **(g)** $12x^2 + 4x - 4$
(h) $3x^2 + 6xh + 3h^2 + 2x + 2h - 4$ **40. (a)** -1 **(b)** -2 **(c)** -4 **(d)** $-2x^2 - x - 1$ **(e)** $2x^2 - x + 1$ **(f)** $-2x^2 - 3x - 2$
(g) $-8x^2 + 2x - 1$ **(h)** $-2x^2 - 4xh - 2h^2 + x + h - 1$ **41. (a)** 0 **(b)** $\dfrac{1}{2}$ **(c)** $-\dfrac{1}{2}$ **(d)** $\dfrac{-x}{x^2 + 1}$ **(e)** $\dfrac{-x}{x^2 + 1}$ **(f)** $\dfrac{x + 1}{x^2 + 2x + 2}$

(g) $\dfrac{2x}{4x^2 + 1}$ **(h)** $\dfrac{x + h}{x^2 + 2xh + h^2 + 1}$ **42. (a)** $-\dfrac{1}{4}$ **(b)** 0 **(c)** 0 **(d)** $\dfrac{x^2 - 1}{-x + 4}$ **(e)** $\dfrac{-x^2 + 1}{x + 4}$ **(f)** $\dfrac{x^2 + 2x}{x + 5}$ **(g)** $\dfrac{4x^2 - 1}{2x + 4}$

(h) $\dfrac{x^2 + 2xh + h^2 - 1}{x + h + 4}$ **43. (a)** 4 **(b)** 5 **(c)** 5 **(d)** $|x| + 4$ **(e)** $-|x| - 4$ **(f)** $|x + 1| + 4$ **(g)** $2|x| + 4$ **(h)** $|x + h| + 4$ **44. (a)** 0

(b) $\sqrt{2}$ **(c)** 0 **(d)** $\sqrt{x^2 - x}$ **(e)** $-\sqrt{x^2 + x}$ **(f)** $\sqrt{x^2 + 3x + 2}$ **(g)** $\sqrt{4x^2 + 2x}$ **(h)** $\sqrt{x^2 + 2xh + h^2 + x + h}$ **45. (a)** $-\dfrac{1}{5}$

(b) $-\dfrac{3}{2}$ **(c)** $\dfrac{1}{8}$ **(d)** $\dfrac{-2x + 1}{-3x - 5}$ **(e)** $\dfrac{-2x - 1}{3x - 5}$ **(f)** $\dfrac{2x + 3}{3x - 2}$ **(g)** $\dfrac{4x + 1}{6x - 5}$ **(h)** $\dfrac{2x + 2h + 1}{3x + 3h - 5}$ **46. (a)** $\dfrac{3}{4}$ **(b)** $\dfrac{8}{9}$ **(c)** 0 **(d)** $1 - \dfrac{1}{(-x + 2)^2}$

(e) $-1 + \dfrac{1}{(x + 2)^2}$ **(f)** $1 - \dfrac{1}{(x + 3)^2}$ **(g)** $1 - \dfrac{1}{(2x + 2)^2}$ **(h)** $1 - \dfrac{1}{(x + h + 2)^2}$ **47.** All real numbers **48.** All real numbers
49. All real numbers **50.** All real numbers **51.** $\{x | x \neq -4, x \neq 4\}$ **52.** $\{x | x \neq -2, x \neq 2\}$ **53.** $\{x | x \neq 0\}$ **54.** $\{x | x \neq -2, x \neq 0, x \neq 2\}$
55. $\{x | x \geq 4\}$ **56.** $\{x | x \leq 1\}$ **57.** $\{x | x > 9\}$ **58.** $\{x | x > 4\}$ **59.** $\{x | x > 1\}$ **60.** $\{x | x \leq -2\}$ **61. (a)** $(f + g)(x) = 5x + 1$; All real numbers

(b) $(f - g)(x) = x + 7$; All real numbers **(c)** $(f \cdot g)(x) = 6x^2 - x - 12$; All real numbers **(d)** $\left(\dfrac{f}{g}\right)(x) = \dfrac{3x + 4}{2x - 3}; \left\{x \Big| x \neq \dfrac{3}{2}\right\}$ **(e)** 16 **(f)** 11

(g) 10 **(h)** -7 **62. (a)** $(f + g)(x) = 5x - 1$; All real numbers **(b)** $(f - g)(x) = -x + 3$; All real numbers **(c)** $(f \cdot g)(x) = 6x^2 - x - 2$;

All real numbers **(d)** $\left(\dfrac{f}{g}\right)(x) = \dfrac{2x + 1}{3x - 2}; \left\{x \Big| x \neq \dfrac{2}{3}\right\}$ **(e)** 14 **(f)** -1 **(g)** 20 **(h)** 3 **63. (a)** $(f + g)(x) = 2x^2 + x - 1$; All real numbers

(b) $(f - g)(x) = -2x^2 + x - 1$; All real numbers **(c)** $(f \cdot g)(x) = 2x^3 - 2x^2$; All real numbers **(d)** $\left(\dfrac{f}{g}\right)(x) = \dfrac{x - 1}{2x^2}; \{x | x \neq 0\}$ **(e)** 20

(f) -29 **(g)** 8 **(h)** 0 **64. (a)** $(f + g)(x) = 4x^3 + 2x^2 + 4$; All real numbers **(b)** $(f - g)(x) = -4x^3 + 2x^2 + 2$; All real numbers

(c) $(f \cdot g)(x) = 8x^5 + 12x^3 + 2x^2 + 3$; All real numbers **(d)** $\left(\dfrac{f}{g}\right)(x) = \dfrac{2x^2 + 3}{4x^3 + 1}; \left\{x \Big| x \neq -\dfrac{\sqrt[3]{2}}{2}\right\}$ **(e)** 130 **(f)** -222 **(g)** 363 **(h)** 1

65. (a) $(f + g)(x) = \sqrt{x} + 3x - 5; \{x | x \geq 0\}$ **(b)** $(f - g)(x) = \sqrt{x} - 3x + 5; \{x | x \geq 0\}$ **(c)** $(f \cdot g)(x) = 3x\sqrt{x} - 5\sqrt{x}; \{x | x \geq 0\}$
(d) $\left(\dfrac{f}{g}\right)(x) = \dfrac{\sqrt{x}}{3x - 5}; \left\{x \Big| x \geq 0, x \neq \dfrac{5}{3}\right\}$ **(e)** $\sqrt{3} + 4$ **(f)** -5 **(g)** $\sqrt{2}$ **(h)** $-\dfrac{1}{2}$ **66. (a)** $(f + g)(x) = |x| + x$; All real numbers

(b) $(f - g)(x) = |x| - x$; All real numbers **(c)** $(f \cdot g)(x) = x|x|$; All real numbers **(d)** $\left(\dfrac{f}{g}\right)(x) = \dfrac{|x|}{x}; \{x | x \neq 0\}$ **(e)** 6 **(f)** 0 **(g)** 4

(h) 1 **67. (a)** $(f + g)(x) = 1 + \dfrac{2}{x}; \{x | x \neq 0\}$ **(b)** $(f - g)(x) = 1; \{x | x \neq 0\}$ **(c)** $(f \cdot g)(x) = \dfrac{1}{x} + \dfrac{1}{x^2}; \{x | x \neq 0\}$

(d) $\left(\dfrac{f}{g}\right)(x) = x + 1; \{x | x \neq 0\}$ **(e)** $\dfrac{5}{3}$ **(f)** 1 **(g)** $\dfrac{3}{4}$ **(h)** 2 **68. (a)** $(f + g)(x) = \sqrt{x - 1} + \sqrt{4 - x}; \{x | 1 \leq x \leq 4\}$
(b) $(f - g)(x) = \sqrt{x - 1} - \sqrt{4 - x}; \{x | 1 \leq x \leq 4\}$ **(c)** $(f \cdot g)(x) = \sqrt{-x^2 + 5x - 4}; \{x | 1 \leq x \leq 4\}$
(d) $\left(\dfrac{f}{g}\right)(x) = \sqrt{\dfrac{x - 1}{4 - x}}; \{x | 1 \leq x < 4\}$ **(e)** $\sqrt{2} + 1$ **(f)** $\sqrt{3}$ **(g)** $\sqrt{2}$ **(h)** 0 **69. (a)** $(f + g)(x) = \dfrac{6x + 3}{3x - 2}; \left\{x \Big| x \neq \dfrac{2}{3}\right\}$
(b) $(f - g)(x) = \dfrac{-2x + 3}{3x - 2}; \left\{x \Big| x \neq \dfrac{2}{3}\right\}$ **(c)** $(f \cdot g)(x) = \dfrac{8x^2 + 12x}{(3x - 2)^2}; \left\{x \Big| x \neq \dfrac{2}{3}\right\}$ **(d)** $\left(\dfrac{f}{g}\right)(x) = \dfrac{2x + 3}{4x}; \left\{x \Big| x \neq 0; x \neq \dfrac{2}{3}\right\}$ **(e)** $\dfrac{21}{7} = 3$
(f) $-\dfrac{1}{2}$ **(g)** $\dfrac{7}{2}$ **(h)** $\dfrac{5}{4}$ **70. (a)** $(f + g)(x) = \sqrt{x + 1} + \dfrac{2}{x}; \{x | x \geq -1, x \neq 0\}$ **(b)** $(f - g)(x) = \sqrt{x + 1} - \dfrac{2}{x}; \{x | x \geq -1, x \neq 0\}$
(c) $(f \cdot g)(x) = \dfrac{2}{x}\sqrt{x + 1}; \{x | x \geq -1, x \neq 0\}$ **(d)** $\left(\dfrac{f}{g}\right)(x) = \dfrac{x\sqrt{x + 1}}{2}; \{x | x \geq -1, x \neq 0\}$ **(e)** $\dfrac{8}{3}$ **(f)** $\sqrt{5} - \dfrac{1}{2}$ **(g)** $\sqrt{3}$ **(h)** $\dfrac{\sqrt{2}}{2}$

71. $g(x) = 5 - \dfrac{7}{2}x$ **72.** $g(x) = \dfrac{x - 1}{x + 1}$ **73.** 4 **74.** -3 **75.** $2x + h - 1$ **76.** $2x + h + 5$ **77.** $3x^2 + 3xh + h^2$ **78.** $\dfrac{-1}{(x + h + 3)(x + 3)}$

79. $A = -\dfrac{7}{2}$ **80.** $B = 5$ **81.** $A = -4$ **82.** $B = -1$ **83.** $A = 8$; undefined at $x = 3$ **84.** $A = 1, B = 2$ **85.** $A(x) = \dfrac{1}{2}x^2$ **86.** $A = \dfrac{1}{2}x^2$

87. $G(x) = 10x$ **88.** $G(x) = 100 + 10x$ **89. (a)** P is the dependent variable; a is the independent variable. **(b)** $P(20) = 197.34$ million; In 2005, there are 197.34 million people 20 years of age or older. **(c)** $P(0) = 290.580$ million; In 2005, there are 290.580 million people. **90. (a)** N is the dependent variable; r is the independent variable. **(b)** $N(3) = 15.64$ million; In 2005, there are 15.64 million housing units having 3 rooms. **(c)** Housing units cannot have 0 rooms. **91. (a)** 15.1 m, 14.07 m, 12.94 m, 11.72 m **(b)** 1.01 sec, 1.43 sec, 1.75 sec **(c)** 2.02 sec **92. (a)** 7 m, 4.27 m, 1.28 m **(b)** 0.62 sec, 0.88 sec, 1.07 sec **(c)** 1.24 sec **93. (a)** $222 **(b)** $225 **(c)** $220 **(d)** $230 **94. (a)** 1.26 ft^2 **(b)** 1.73 ft^2 **(c)** 1.99 ft^2 **95.** $R(x) = \dfrac{L(x)}{P(x)}$ **96.** $T(x) = V(x) + P(x)$ **97.** $H(x) = P(x) \cdot I(x)$ **98.** $N(x) = I(x) - T(x)$

99. (a) $P(x) = -0.05x^3 + 0.8x^2 + 155x - 500$ **(b)** $P(15) = \$1836.25$ **(c)** When 15 hundred cellphones are sold, the profit is $1836.25.

100. (a) $P(x) = -0.1x^2 + 23x - 400$ **(b)** $P(30) = \$200$ **(c)** When 30 clocks are sold, the profit in $200. **101.** Only $h(x) = 2x$

102. No; f has a domain of all real numbers, while g has a domain of $\{x | x \neq -1\}$.

1.2 Assess Your Understanding (page 59)

3. vertical **4.** 5; -3 **5.** $a = -2$ **6.** F **7.** F **8.** T **9. (a)** $f(0) = 3$; $f(-6) = -3$ **(b)** $f(6) = 0$; $f(11) = 1$ **(c)** Positive **(d)** Negative **(e)** $-3, 6$, and 10 **(f)** $-3 < x < 6$; $10 < x \leq 11$ **(g)** $\{x | -6 \leq x \leq 11\}$ **(h)** $\{y | -3 \leq y \leq 4\}$ **(i)** $-3, 6, 10$ **(j)** 3 **(k)** 3 times **(l)** Once **(m)** 0, 4 **(n)** $-5, 8$ **(o)** $-3, 6, 10$ **10. (a)** $f(0) = 0$; $f(6) = 0$ **(b)** $f(2) = -2$; $f(-2) = 1$ **(c)** Negative **(d)** Positive **(e)** 0, 4, and 6 **(f)** $0 < x < 4$ **(g)** $\{x | -4 \leq x \leq 6\}$ **(h)** $\{y | -2 \leq y \leq 3\}$ **(i)** 0, 4, 6 **(j)** 0 **(k)** Twice **(l)** Once **(m)** 5 **(n)** 2 **(o)** 0, 4, 6 **11.** Not a function **12.** Function **(a)** Domain: all real numbers; Range: $\{y | 0 < y < \infty\}$ **(b)** $(0, 1)$ **(c)** None **13.** Function **(a)** Domain: $\{x | -\pi \leq x \leq \pi\}$; Range: $\{y | -1 \leq y \leq 1\}$ **(b)** $\left(-\dfrac{\pi}{2}, 0\right), \left(\dfrac{\pi}{2}, 0\right), (0, 1)$ **(c)** y-axis **14.** Function **(a)** Domain: $\{x | -\pi \leq x \leq \pi\}$; Range: $\{y | -1 \leq y \leq 1\}$ **(b)** $(-\pi, 0), (\pi, 0), (0, 0)$ **(c)** Origin **15.** Not a function **16.** Not a function **17.** Function **(a)** Domain: $\{x | x > 0\}$; Range: all real numbers **(b)** $(1, 0)$ **(c)** None **18.** Function **(a)** Domain: $\{x | 0 \leq x \leq 4\}$; Range: $\{y | 0 \leq y \leq 3\}$ **(b)** $(0, 0)$ **(c)** None **19.** Function **(a)** Domain: all real numbers; Range: $\{y | y \leq 2\}$ **(b)** $(-3, 0), (3, 0), (0, 2)$ **(c)** y-axis **20.** Function **(a)** Domain: $\{x | x \geq -3\}$; Range: $\{y | y \geq 0\}$ **(b)** $(-3, 0), (0, 2), (2, 0)$ **(c)** None **21.** Function **(a)** Domain: all real numbers; Range: $\{y | y \geq -3\}$ **(b)** $(1, 0), (3, 0), (0, 9)$ **(c)** None

22. Function **(a)** Domain: all real numbers: Range: $\{y | y \leq 5\}$ **(b)** $(-1, 0), (2, 0), (0, 4)$ **(c)** None **23. (a)** Yes **(b)** $f(-2) = 9$; $(-2, 9)$ **(c)** $0, \dfrac{1}{2}$; $(0, -1), \left(\dfrac{1}{2}, -1\right)$ **(d)** All real numbers **(e)** $-\dfrac{1}{2}, 1$ **(f)** -1 **(g)** $-\dfrac{1}{2}, 1$ **24. (a)** No **(b)** $f(-2) = -22$; $(-2, -22)$ **(c)** $-\dfrac{1}{3}, 2$; $\left(-\dfrac{1}{3}, -2\right), (2, -2)$ **(d)** All real numbers **(e)** $0, \dfrac{5}{3}$ **(f)** 0 **(g)** $0, \dfrac{5}{3}$ **25. (a)** No **(b)** $f(4) = -3$; $(4, -3)$ **(c)** 14; $(14, 2)$ **(d)** $\{x | x \neq 6\}$ **(e)** -2 **(f)** $-\dfrac{1}{3}$ **(g)** -2 **26. (a)** Yes **(b)** $f(0) = \dfrac{1}{2}$; $\left(0, \dfrac{1}{2}\right)$ **(c)** $0, \dfrac{1}{2}$; $\left(0, \dfrac{1}{2}\right), \left(\dfrac{1}{2}, \dfrac{1}{2}\right)$ **(d)** $\{x | x \neq -4\}$ **(e)** None **(f)** $\dfrac{1}{2}$ **(g)** None **27. (a)** Yes **(b)** $f(2) = \dfrac{8}{17}$; $\left(2, \dfrac{8}{17}\right)$ **(c)** $-1, 1$; $(-1, 1), (1, 1)$ **(d)** All real numbers **(e)** 0 **(f)** 0 **(g)** 0 **28. (a)** Yes **(b)** $f(4) = 4$; $(4, 4)$ **(c)** -2; $(-2, 1)$ **(d)** $\{x | x \neq 2\}$ **(e)** 0 **(f)** 0 **(g)** 0

29. (a) About 81.07 ft **(b)** About 129.59 ft **(c)** About 26.63 ft **(d)** About 528.13 ft

(e)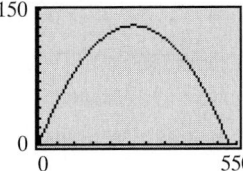

30. (a) $\{x | 0 < x < 1\}$ **(b)**

(f) 115.07 ft and 413.05 ft **(g)** 275 ft; maximum height shown in the table is 131.8 ft **(h)** 264 ft

(c) 0.7 ft; 1.4 ft

31. (a)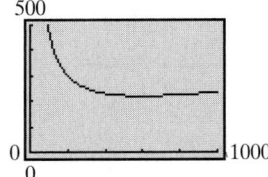

32. (a) $W\left(\dfrac{14{,}110}{5280}\right) \approx 119.84$ pounds

(b)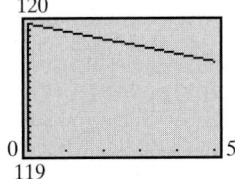

(b)

X	Y1
-50	-625
0	ERROR
50	825
100	470
150	355
200	300
250	269

X=250

(c)

X	Y1
0	120
.5	119.97
1	119.94
1.5	119.91
2	119.88
2.5	119.85
3	119.82

X=0

The weight W varies from 120 pounds to about 119.7 pounds.

(c) 600 mi/hr **(d)** About 0.83 miles

33. (a) 3 **(b)** -2 **(c)** -1 **(d)** 1 **(e)** 2 **(f)** $-\dfrac{1}{3}$ **35.** The x-intercepts can number anywhere from 0 to infinitely many. **36.** There is at most one y-intercept. Yes; $f(x) = 0$, domain $x = a$ **37. (a)** III **(b)** IV **(c)** I **(d)** V **(e)** II **38. (a)** II **(b)** V **(c)** IV **(d)** III **(e)** I

39.

40.

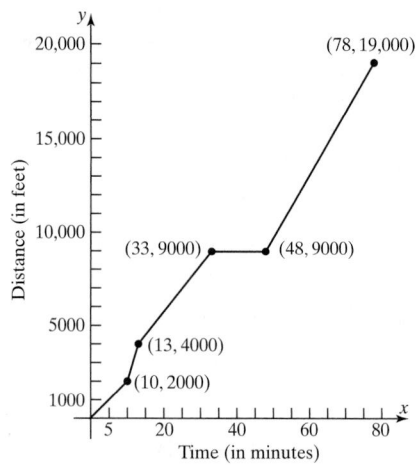

41. (a) 2 hr elapsed during which Kevin was between 0 and 3 mi from home. **(b)** 0.5 hr elapsed during which Kevin was 3 mi from home.
 (c) 0.3 hr elapsed during which Kevin was between 0 and 3 mi from home. **(d)** 0.2 hr elapsed during which Kevin was 0 mi from home.
 (e) 0.9 hr elapsed during which Kevin was between 0 and 2.8 mi from home. **(f)** 0.3 hr elapsed during which Kevin was 2.8 mi from home.
 (g) 1.1 hr elapsed during which Kevin was between 0 and 2.8 mi from home. **(h)** 3 mi **(i)** 2 times
42. (a) $(7, 7.4)$ **(b)** $(4.2, 6)$ **(c)** Increasing from 0 to 30 mi/hr **(d)** 0 mi/hr **(e)** 50 mi/hr **(f)** $(2, 4), (4.2, 6), (7, 7.4), (7.6, 8)$
43. No points whose x-coordinate is 5 or whose y-coordinate is 0 can be on the graph. **44.** Yes. $f(x) = 0$ satisfies the condition that $f(x) = -f(x)$.

1.3 Assess Your Understanding *(page 71)*

6. increasing **7.** even; odd **8.** T **9.** T **10.** F **11.** Yes **12.** No **13.** No **14.** Yes **15.** $(-8, -2); (0, 2); (5, \infty)$ **16.** $(-\infty, -8), (-2, 0), (2, 5)$
17. Yes; 10 **18.** No **19.** $-2, 2; 6, 10$ **20.** $-8, 0, 5; -4, 0, 0$ **21. (a)** $(-2, 0), (0, 3), (2, 0)$ **(b)** Domain: $\{x | -4 \le x \le 4\}$ or $[-4, 4]$; Range:
 $\{y | 0 \le y \le 3\}$ or $[0, 3]$ **(c)** Increasing on $(-2, 0)$ and $(2, 4)$; Decreasing on $(-4, -2)$ and $(0, 2)$ **(d)** Even **22. (a)** $(-1, 0), (0, 2), (1, 0)$
 (b) Domain: $\{x | -3 \le x \le 3\}$ or $[-3, 3]$; Range: $\{y | 0 \le y \le 3\}$ or $[0, 3]$ **(c)** Increasing on $(-1, 0)$ and $(1, 3)$; Decreasing on $(-3, -1)$ and $(0, 1)$
 (d) Even **23. (a)** $(0, 1)$ **(b)** Domain: all real numbers; Range: $\{y | y > 0\}$ or $(0, \infty)$ **(c)** Increasing on $(-\infty, \infty)$ **(d)** Neither **24. (a)** $(1, 0)$
 (b) Domain: $\{x | x > 0\}$ or $(0, \infty)$; Range: all real numbers **(c)** Increasing on $(0, \infty)$ **(d)** Neither **25. (a)** $(-\pi, 0), (0, 0), (\pi, 0)$
 (b) Domain: $\{x | -\pi \le x \le \pi\}$ or $[-\pi, \pi]$; Range: $\{y | -1 \le y \le 1\}$ or $[-1, 1]$ **(c)** Increasing on $\left(-\frac{\pi}{2}, \frac{\pi}{2}\right)$; Decreasing on $\left(-\pi, -\frac{\pi}{2}\right)$ and
 $\left(\frac{\pi}{2}, \pi\right)$ **(d)** Odd **26. (a)** $\left(-\frac{\pi}{2}, 0\right), (0, 1), \left(\frac{\pi}{2}, 0\right)$ **(b)** Domain: $\{x | -\pi \le x \le \pi\}$ or $[-\pi, \pi]$; Range: $\{y | -1 \le y \le 1\}$ or $[-1, 1]$
 (c) Increasing on $(-\pi, 0)$; Decreasing on $(0, \pi)$ **(d)** Even **27. (a)** $\left(0, \frac{1}{2}\right), \left(\frac{1}{2}, 0\right), \left(\frac{5}{2}, 0\right)$ **(b)** Domain: $\{x | -3 \le x \le 3\}$ or $[-3, 3]$; Range:
 $\{y | -1 \le y \le 2\}$ or $[-1, 2]$ **(c)** Increasing on $(2, 3)$; Decreasing on $(-1, 1)$; Constant on $(-3, -1)$ and $(1, 2)$ **(d)** Neither
28. (a) $(-2.3, 0), (0, 1), (3, 0)$ **(b)** Domain: $\{x | -3 \le x \le 3\}$ or $[-3, 3]$; Range: $\{y | -2 \le y \le 2\}$ or $[-2, 2]$ **(c)** Increasing on $(-3, -2)$ and $(0, 2)$;
 Decreasing on $(2, 3)$; Constant on $(-2, 0)$ **(d)** Neither **29. (a)** $0; 3$ **(b)** $-2, 2; 0, 0$ **30. (a)** $0; 2$ **(b)** $-1, 1; 0, 0$ **31. (a)** $\frac{\pi}{2}; 1$
 (b) $-\frac{\pi}{2}; -1$ **32. (a)** $0; 1$ **(b)** $-\pi, \pi; -1, -1$ **33.** Odd **34.** Even **35.** Even **36.** Neither **37.** Odd **38.** Neither **39.** Neither
40. Even **41.** Even **42.** Odd **43.** Odd **44.** Odd
45.

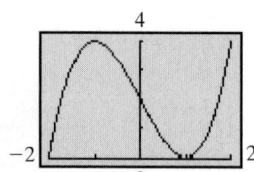

Increasing: $(-2, -1), (1, 2)$
Decreasing: $(-1, 1)$
Local maximum: $(-1, 4)$
Local minimum: $(1, 0)$

46.

Increasing: $(-1, 0), (2, 3)$
Decreasing: $(0, 2)$
Local maximum: $(0, 5)$
Local minimum: $(2, 1)$

47.

Increasing: $(-2, -0.77), (0.77, 2)$
Decreasing: $(-0.77, 0.77)$
Local maximum: $(-0.77, 0.19)$
Local minimum: $(0.77, -0.19)$

48.

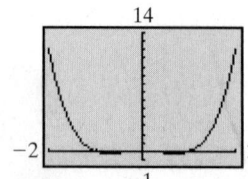

Increasing: $(-0.71, 0), (0.71, 2)$
Decreasing: $(-2, -0.71), (0, 0.71)$
Local maximum: $(0, 0)$
Local minima: $(-0.71, -0.25), (0.71, -0.25)$

49.

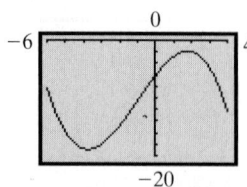

Increasing: $(-3.77, 1.77)$
Decreasing: $(-6, -3.77), (1.77, 4)$
Local maximum: $(1.77, -1.91)$
Local minimum: $(-3.77, -18.89)$

50.

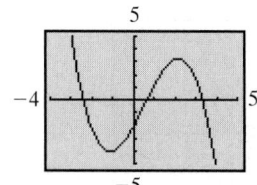

Increasing: $(-1.16, 2.16)$
Decreasing: $(-4, -1.16), (2.16, 5)$
Local maximum: $(2.16, 3.25)$
Local minimum: $(-1.16, -4.05)$

51.

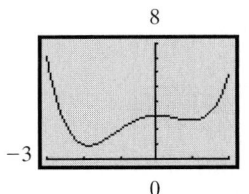

Increasing: $(-1.87, 0), (0.97, 2)$
Decreasing: $(-3, -1.87), (0, 0.97)$
Local maximum: $(0, 3)$
Local minima: $(-1.87, 0.95), (0.97, 2.65)$

52.

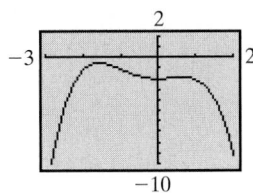

Increasing: $(-3, -1.57), (0, 0.64)$
Decreasing: $(-1.57, 0), (0.64, 2)$
Local maxima: $(-1.57, -0.52), (0.64, -1.87)$
Local minimum: $(0, -2)$

53. (a) -4 **(b)** -8 **(c)** -10 **54. (a)** -4 **(b)** -13 **(c)** -1 **55. (a)** 17 **(b)** -1 **(c)** 11 **56. (a)** -2 **(b)** 0 **(c)** 5 **57. (a)** 5
 (b) 5; the slope of the line joining $(1, f(1))$ and $(3, f(3))$ is 5. **(c)** $y = 5x - 2$ **58. (a)** -4 **(b)** -4; the slope of the line joining $(2, f(2))$ and $(5, f(5))$
 is -4. **(c)** $y = -4x + 1$ **59. (a)** $x - 2$ **(b)** -1; the slope of the line joining $(-2, g(-2))$ and $(1, g(1))$ is -1. **(c)** $y = -x$
60. (a) $x - 1$ **(b)** 1; the slope of the line joining $(-1, g(-1))$ and $(2, g(2))$ is 1. **(c)** $y = x + 3$ **61. (a)** x **(b)** 4; the slope of the line joining $(2, h(2))$
 and $(4, h(4))$ is 4. **(c)** $y = 4x - 8$ **62. (a)** $-2x + 1$ **(b)** -5; the slope of the line joining $(0, h(0))$ and $(3, h(3))$ is -5. **(c)** $y = -5x$

63. (a) $V(x) = x(24 - 2x)^2$
 (b) 972 in.3 **(c)** 160 in.3
 (d) V is largest when $x = 4$.

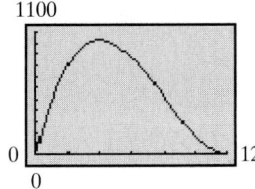

64. (a) $A(x) = x^2 + \dfrac{40}{x}$
 (b) 41 ft^2 **(c)** 24 ft^2
 (d) A is smallest when $x \approx 2.71$ ft.

65. (a)

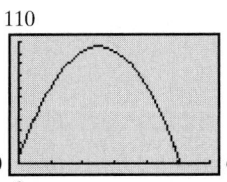

 (b) 2.5 sec
 (c) 106 ft

66. (a)

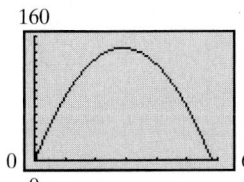

 (b) 2.89 sec
 (c) 144.68 feet
 (d) 156.25 feet

67. (a)

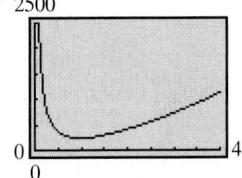

 (b) 10 riding lawn mowers/hr
 (c) \$239/mower

68. (a) 2.16 hr **(b)** 4.47 hr

69. (a), (b), (e)

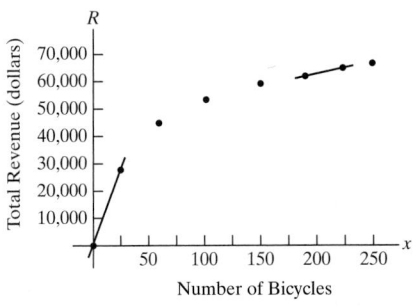

 (c) 1120 dollars/bicycle
 (d) For each additional bicycle sold between 0 and 25 bicycles,
 total revenue increases, on average, by \$1120.
 (f) 75 dollars/bicycle
 (g) For each additional bicycle sold between 190 and 223 bicycles,
 total revenue increases, on average, by \$75.
 (h) The average rate of change is decreasing as the number of
 bicycles increase.

70. (a), (b), (e)

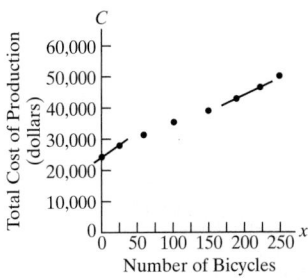

 (c) 150 dollars/bicycle
 (d) For each additional bicycle produced between 0 and 25, total
 cost of production increases, on average, by \$150.
 (f) 113.64 dollars/bicycle
 (g) For each additional bicycle produced between 190 and 223,
 total cost of production increases, on average, by \$113.64.
 (h) The average rate of change appears to remain
 constant as the number of bicycles increase.

71. **(a)** 1 **(b)** 0.5 **(c)** 0.1 **(d)** 0.01 **(e)** 0.001
(f)

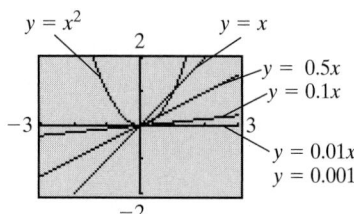

(g) They are getting closer to the tangent line at $(0, 0)$.
(h) They are getting closer to 0.

73. **(a)** 2 **(b)** $2; 2; 2; 2$ **(c)** $y = 2x + 5$
(d)

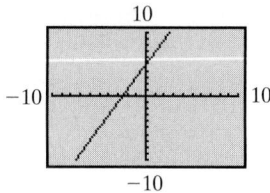

75. **(a)** $2x + h + 2$ **(b)** $4.5; 4.1; 4.01; 4$
(c) $y = 4.01x - 1.01$
(d)

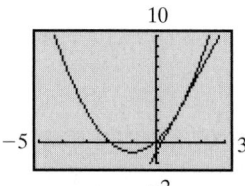

77. **(a)** $4x + 2h - 3$ **(b)** $2; 1.2; 1.02; 1$
(c) $y = 1.02x - 1.02$
(d)

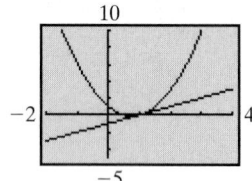

79. **(a)** $-\dfrac{1}{(x + h)x}$ **(b)** $-\dfrac{2}{3}; -\dfrac{10}{11}; -\dfrac{100}{101}; -1$
(c) $y = -\dfrac{100}{101}x + \dfrac{201}{101}$
(d)

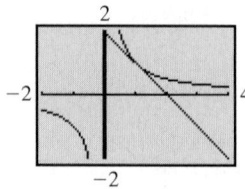

72. **(a)** 3 **(b)** 2.5 **(c)** 2.1 **(d)** 2.01 **(e)** 2.001
(f)

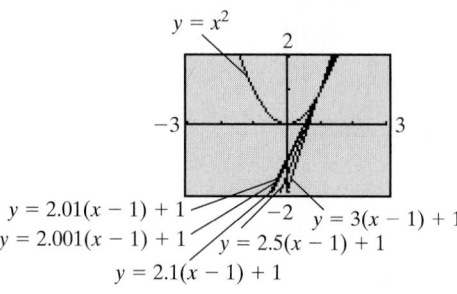

(g) They are getting closer to the tangent line at $(1, 1)$.
(h) They are getting closer to 2.

74. **(a)** -3 **(b)** $-3; -3; -3; -3$ **(c)** $y = -3x + 2$
(d)

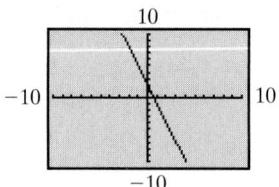

76. **(a)** $4x + 2h + 1$ **(b)** $6; 5.2; 5.02; 5$
(c) $y = 5.02x - 2.02$
(d)

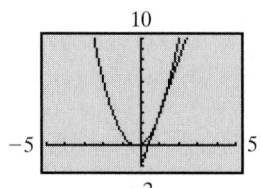

78. **(a)** $-2x - h + 3$ **(b)** $0.5; 0.9; 0.99; 1$
(c) $y = 0.99x - 0.99$
(d)

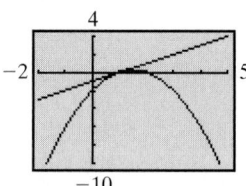

80. **(a)** $\dfrac{-2x - h}{(x^2 + 2xh + h^2)x^2}$ **(b)** $-1.1111; -1.7355; -2$
(c) $y = -1.9704x + 2.9704$
(d)

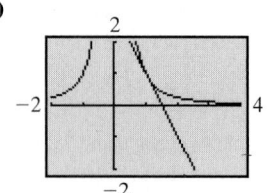

83. At most one **85.** Yes; the function $f(x) = 0$ is both even and odd.

1.4 Assess Your Understanding (page 81)

4. less **5.** piecewise-defined **6.** T **7.** F **8.** F **9.** C **10.** A **11.** E **12.** G **13.** B **14.** D **15.** F **16.** H

17.

18.

19.

20.

21.

22.

23.

24.
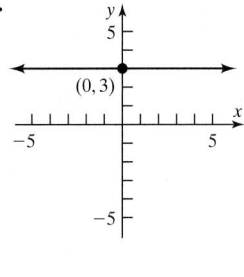

25. (a) 4 **(b)** 2 **(c)** 5 **26. (a)** -1 **(b)** 2 **(c)** 5 **27. (a)** 2 **(b)** 3 **(c)** -4 **28. (a)** 0 **(b)** 0 **(c)** -1

29. (a) All real numbers
 (b) $(0, 1)$
 (c)
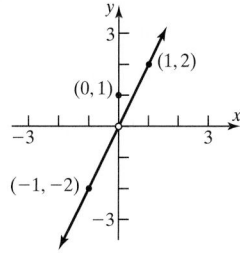
 (d) $\{y \,|\, y \neq 0\}; (-\infty, 0)$ or $(0, \infty)$

30. (a) All real numbers
 (b) $(0, 4)$
 (c)
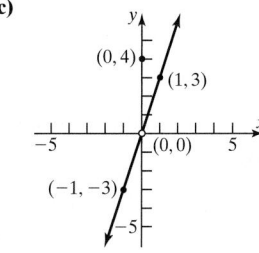
 (d) $\{y \,|\, y \neq 0\}; (-\infty, 0)$ or $(0, \infty)$

31. (a) All real numbers
 (b) $(0, 3)$
 (c)
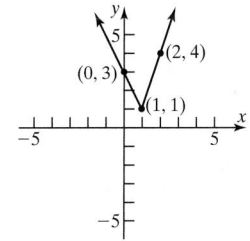
 (d) $\{y \,|\, y \geq 1\}; [1, \infty)$

32. (a) All real numbers
 (b) $(-3, 0), \left(-\dfrac{3}{2}, 0\right), (0, -3)$
 (c)
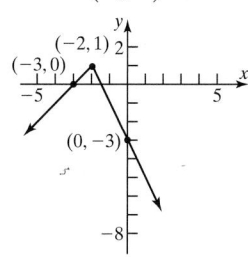
 (d) $\{y \,|\, y \leq 1\}; (-\infty, 1]$

33. (a) $\{x \,|\, x \geq -2\}; [-2, \infty)$
 (b) $(0, 3), (2, 0)$
 (c)
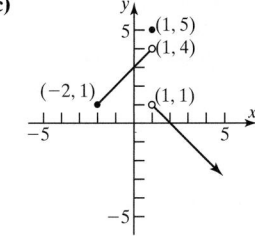
 (d) $\{y \,|\, y < 4, y = 5\}; (-\infty, 4)$ and $\{5\}$

34. (a) $\{x \,|\, x \geq -3\}; [-3, \infty)$
 (b) $\left(-\dfrac{5}{2}, 0\right), (0, -3)$
 (c)
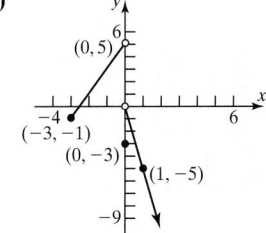
 (d) $\{y \,|\, y < 5\}; (-\infty, 5)$

35. (a) All real numbers
(b) $(-1, 0), (0, 0)$
(c)

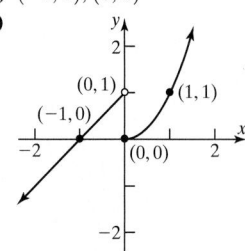

(d) All real numbers

36. (a) All real numbers
(b) $(0, 0)$
(c)

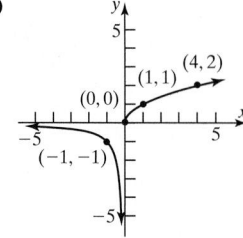

(d) All real numbers

37. (a) $\{x \mid x \geq -2\}; [-2, \infty)$
(b) $(0, 1)$
(c)

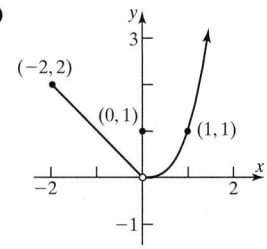

(d) $\{y \mid y > 0\}; (0, \infty)$

38. (a) $\{x \mid x \geq -3\}; [-3, \infty)$
(b) $(-3, 0), (0, 3)$
(c)

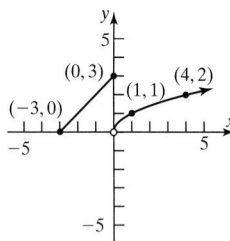

(d) $\{y \mid y \geq 0\}; [0, \infty)$

39. (a) All real numbers
(b) $(x, 0)$ for $0 \leq x < 1$
(c)

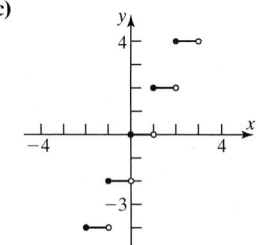

(d) Set of even integers

40. (a) All real numbers
(b) $(x, 0)$ for $0 \leq x < \dfrac{1}{2}$
(c)

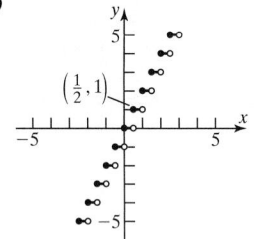

(d) Set of integers

41. $f(x) = \begin{cases} -x & \text{if } -1 \leq x \leq 0 \\ \frac{1}{2}x & \text{if } 0 < x \leq 2 \end{cases}$ (Other answers are possible.)

42. $f(x) = \begin{cases} x & \text{if } -1 \leq x \leq 0 \\ 1 & \text{if } \ 0 < x \leq 2 \end{cases}$ (Other answers are possible.)

43. $f(x) = \begin{cases} -x & \text{if } x \leq 0 \\ -x + 2 & \text{if } 0 < x \leq 2 \end{cases}$ (Other answers are possible.)

44. $f(x) = \begin{cases} 2x + 2 & \text{if } -1 \leq x \leq 0 \\ x & \text{if } \ x > 0 \end{cases}$ (Other answers are possible.)

45. (a) $35 (b) $61 (c) $35.40 **46.** (a) $3 (b) $41 (c) $50 (d) $46

47. (a) $67.43
(b) $477.04

48. (a) $43.62
(b) $166.32

47. (c) $C = \begin{cases} 1.15955x + 9.45 & \text{if } 0 \leq x \leq 50 \\ 0.91025x + 21.915 & \text{if } x > 50 \end{cases}$

48. (c) $C = \begin{cases} 0.9739x + \ \ 6.45 & \text{if } \ 0 \leq x \leq 20 \\ 0.8844x + \ \ 8.24 & \text{if } 20 < x \leq 50 \\ 0.8101x + 11.955 & \text{if } \ x > 50 \end{cases}$

47. (d)

48. (d)

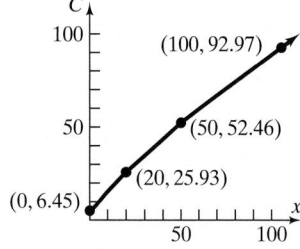

49. For schedule X: $f(x) = \begin{cases} 0.10x & \text{if } \ \ \ \ 0 < x \leq 7300 \\ 730 + 0.15(x - 7300) & \text{if } \ \ \ \ 7300 < x \leq 29{,}700 \\ 4090 + 0.25(x - 29{,}700) & \text{if } \ 29{,}700 < x \leq 71{,}950 \\ 14{,}652.50 + 0.28(x - 71{,}950) & \text{if } \ 71{,}950 < x \leq 150{,}150 \\ 36{,}548.50 + 0.33(x - 150{,}150) & \text{if } 150{,}150 < x \leq 326{,}450 \\ 92{,}727.50 + 0.35(x - 326{,}450) & \text{if } \ \ \ \ \ \ \ \ \ \ \ \ x > 326{,}450 \end{cases}$

50. For schedule Y-1: $f(x) = \begin{cases} 0.10x & \text{if } \ \ \ \ \ 0 < x \leq 14{,}600 \\ 1460 + 0.15(x - 14{,}600) & \text{if } \ \ 14{,}600 < x \leq 59{,}400 \\ 8180 + 0.25(x - 59{,}400) & \text{if } \ \ 59{,}400 < x \leq 119{,}950 \\ 23{,}317.50 + 0.28(x - 119{,}950) & \text{if } 119{,}950 < x \leq 182{,}800 \\ 40{,}915.50 + 0.33(x - 182{,}800) & \text{if } 182{,}800 < x \leq 326{,}450 \\ 88{,}320 + 0.35(x - 326{,}450) & \text{if } \ \ \ \ \ \ \ \ \ \ \ \ x > 326{,}450 \end{cases}$

51. (a)

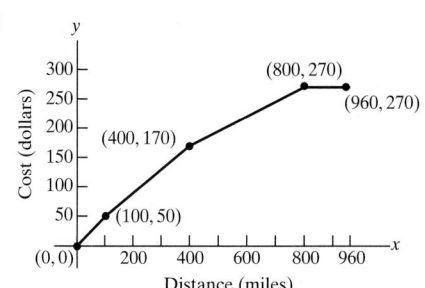

(b) $C = 50 + 0.4(x - 100)$
(c) $C = 170 + 0.25(x - 400)$

52. $C = \begin{cases} 95 & \text{if } x = 7 \\ 119 & \text{if } 7 < x \le 8 \\ 143 & \text{if } 8 < x \le 9 \\ 167 & \text{if } 9 < x \le 10 \\ 190 & \text{if } 10 < x \le 14 \end{cases}$

53. $f(x) = \begin{cases} x & \text{if } 0 \le x < 10 \\ 10 & \text{if } 10 \le x < 500 \\ 30 & \text{if } 500 \le x < 1000 \\ 50 & \text{if } 1000 \le x < 1500 \\ 70 & \text{if } 1500 \le x \end{cases}$

54. $g(x) = \begin{cases} 0.015x & \text{if } 0 \le x \le 1000 \\ 5 + 0.01x & \text{if } x > 1000 \end{cases}$

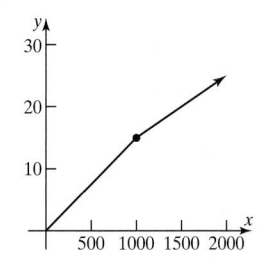

55. (a) 10°C **(b)** 4°C **(c)** −3°C **(d)** −4°C **(e)** The wind chill is equal to the air temperature. **(f)** At wind speed greater than 20 m/sec, the wind chill factor depends only on the air temperature. **56. (a)** −10°C **(b)** −21°C **(c)** −34°C **(d)** −36°C **57.** Each graph is that of $y = x^2$, but shifted vertically. If $y = x^2 + k, k > 0$, the shift is up k units; if $y = x^2 - k, k > 0$, the shift is down k units. **58.** Each graph is that of $y = x^2$, but shifted horizontally. If $y = (x - k)^2, k > 0$, the shift is right k units; if $y = (x + k)^2, k > 0$, the shift is left k units. **59.** Each graph is that of $y = |x|$, but either compressed or stretched. If $y = k|x|$ and $k > 1$, the graph is stretched vertically; if $y = k|x|, 0 < k < 1$, the graph is compressed vertically. **60.** The graph of $y = -f(x)$ is the reflection about the x-axis of the graph of $y = f(x)$. **61.** The graph of $y = f(-x)$ is the reflection about the y-axis of the graph of $y = f(x)$. **62.** Yes. The graph of $y = (x - 1)^3 + 2$ is the graph of $y = x^3$ shifted right 1 unit and up 2 units.
63. They are all U-shaped and open upward. All three go through the points $(-1, 1)$, $(0, 0)$ and $(1, 1)$. As the exponent increases, the steepness of the curve increases (except near $x = 0$). **64.** They all have the same general shape. All three go through the points $(-1, -1)$, $(0, 0)$, and $(1, 1)$. As the exponent increases, the steepness of the curve increases (except near $x = 0$).

1.5 Assess Your Understanding (page 94)

1. horizontal; right **2.** y **3.** $-5; -2; 2$ **4.** T **5.** F **6.** T **7.** B **8.** E **9.** H **10.** D **11.** I **12.** A **13.** L **14.** C **15.** F **16.** J **17.** G
18. K **19.** $y = (x - 4)^3$ **20.** $y = (x + 4)^3$ **21.** $y = x^3 + 4$ **22.** $y = x^3 - 4$ **23.** $y = -x^3$ **24.** $y = -x^3$ **25.** $y = 4x^3$
26. $y = \left(\frac{1}{4}x\right)^3 = \frac{1}{64}x^3$ **27.** (1) $y = \sqrt{x} + 2$; (2) $y = -(\sqrt{x} + 2)$; (3) $y = -(\sqrt{-x} + 2)$
28. (1) $y = -\sqrt{x}$; (2) $y = -\sqrt{x - 3}$; (3) $y = -\sqrt{x - 3} - 2$ **29.** (1) $y = -\sqrt{x}$; (2) $y = -\sqrt{x} + 2$; (3) $y = -\sqrt{x + 3} + 2$ **30.** (1) $y = \sqrt{x} + 2$;
(2) $y = \sqrt{-x} + 2$; (3) $y = \sqrt{-(x + 3)} + 2 = \sqrt{-x - 3} + 2$ **31.** (c) **32.** (d) **33.** (c) **34.** (a)

35.

36.

37.

38.

39.

40.

41.

42.

43.

44.

45.

46.

47.

48.

49.

50.

51.

52.

53.

54.

55.

56.

57.

58.

59.

60.

61.

62.

63.

64.

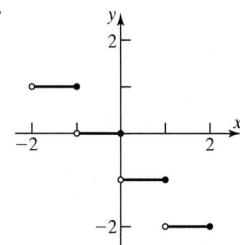

65. (a) $F(x) = f(x) + 3$

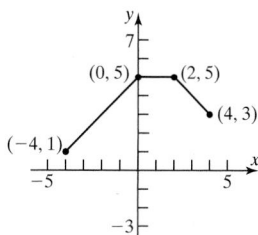

(b) $G(x) = f(x + 2)$

(c) $P(x) = -f(x)$

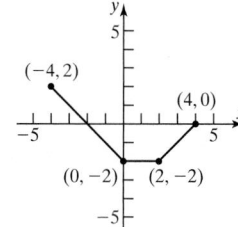

(d) $H(x) = f(x + 1) - 2$

(e) $Q(x) = \dfrac{1}{2}f(x)$

(f) $g(x) = f(-x)$

(g) $h(x) = f(2x)$

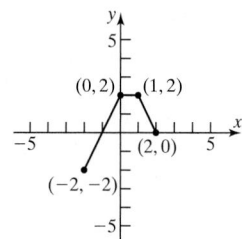

66. (a) $F(x) = f(x) + 3$

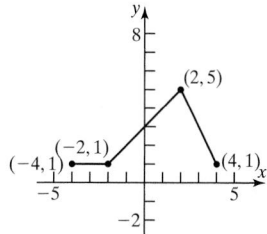

(b) $G(x) = f(x + 2)$

(c) $P(x) = -f(x)$

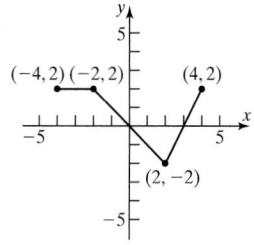

(d) $H(x) = f(x + 1) - 2$

(e) $Q(x) = \dfrac{1}{2}f(x)$

(f) $g(x) = f(-x)$

(g) $h(x) = f(2x)$

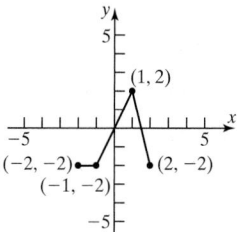

67. (a) $F(x) = f(x) + 3$

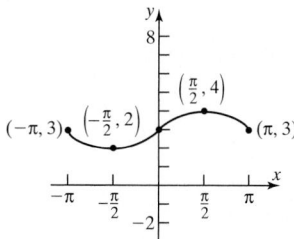

(b) $G(x) = f(x + 2)$

(c) $P(x) = -f(x)$

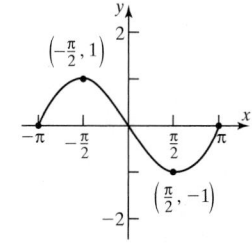

(d) $H(x) = f(x + 1) - 2$

(e) $Q(x) = \frac{1}{2}f(x)$

(f) $g(x) = f(-x)$

(g) $h(x) = f(2x)$

68. (a) $F(x) = f(x) + 3$

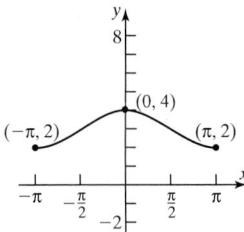

(b) $G(x) = f(x + 2)$

(c) $P(x) = -f(x)$

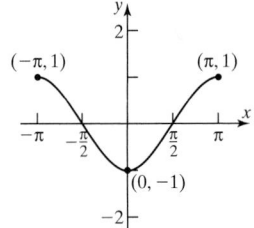

(d) $H(x) = f(x + 1) - 2$

(e) $Q(x) = \frac{1}{2}f(x)$

(f) $g(x) = f(-x)$

(g) $h(x) = f(2x)$

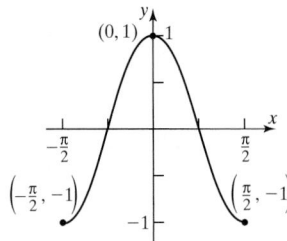

69. (a) -7 and 1 **(b)** -3 and 5 **(c)** -5 and 3 **(d)** -3 and 5 **70. (a)** -12 and -3 **(b)** -5 and 4 **(c)** -8 and 1 **(d)** -1 and 8
71. (a) $(-3, 3)$ **(b)** $(4, 10)$ **(c)** Decreasing on $(-1, 5)$ **(d)** Decreasing on $(-5, 1)$ **72. (a)** $(-4, 5)$ **(b)** $(3, 12)$ **(c)** Decreasing on $(-2, 7)$
(d) Decreasing on $(-7, 2)$
73. (a)

(b)

74. (a)

(b)

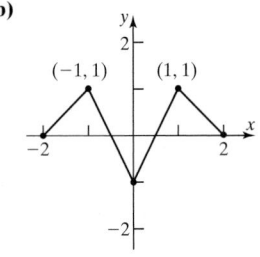

75. $f(x) = (x + 1)^2 - 1$

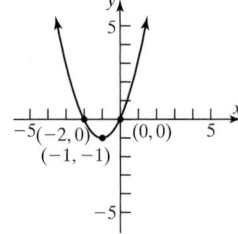

76. $f(x) = (x - 3)^2 - 9$

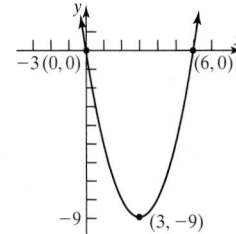

77. $f(x) = (x - 4)^2 - 15$

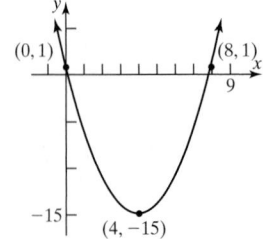

78. $f(x) = (x + 2)^2 - 2$

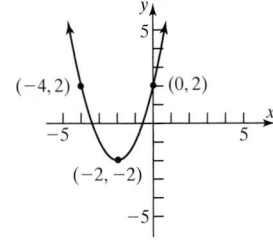

79. $f(x) = \left(x + \dfrac{1}{2}\right)^2 + \dfrac{3}{4}$

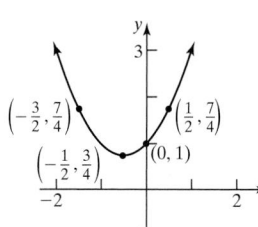

80. $f(x) = \left(x - \dfrac{1}{2}\right)^2 + \dfrac{3}{4}$

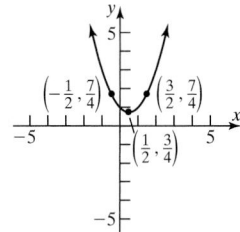

81. $f(x) = 2(x - 3)^2 + 1$

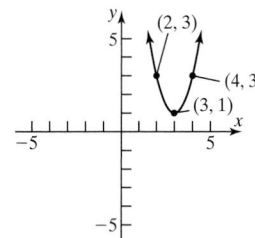

82. $f(x) = 3(x + 1)^2 - 2$

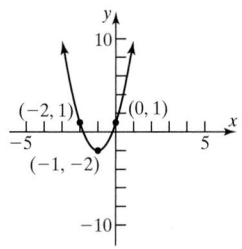

83. $f(x) = -3(x + 2)^2 - 5$

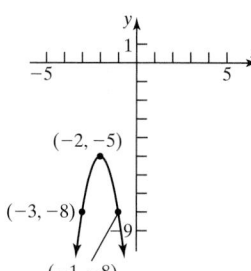

84. $f(x) = -2(x + 3)^2 + 5$

85.

86.

87.

88. **(a)**

(b)

(c)

89. **(a)**

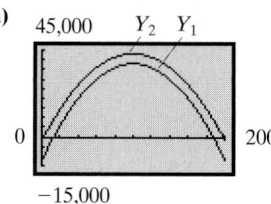

(b) 10% tax
(c) Y_1 is the graph of $p(x)$ shifted down vertically 10,000 units.
 Y_2 is the graph of $p(x)$ vertically compressed by a factor of 0.9.
(d) 10% tax

90. The graph of $y = 4f(x)$ is a vertical stretch by a factor of 4. The graph of $y = f(4x)$ is a horizontal compression by a factor of 4.

1.6 Assess Your Understanding *(page 102)*

1. $V(r) = 2\pi r^3$ **2.** $V(r) = \dfrac{2}{3}\pi r^3$

3. (a) $R(x) = -\dfrac{1}{6}x^2 + 100x$

(b) $13,333.33

(c) 15,000

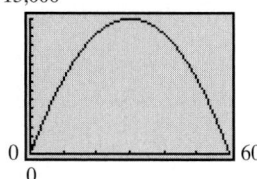

(d) 300; $15,000 **(e)** $50

4. (a) $R(x) = -\dfrac{1}{3}x^2 + 100x$

(b) $6666.67

(c) 10,000

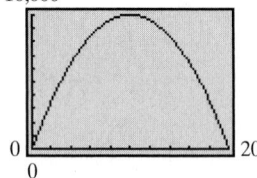

(d) 150; $7500 **(e)** $50

5. (a) $R(x) = -\dfrac{1}{5}x^2 + 20x$

(b) $255

(c) 500

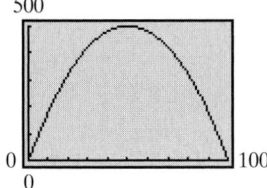

(d) 50; $500 **(e)** $10

6. (a) $R(x) = -\dfrac{1}{20}x^2 + 25x$

(b) $480

(c) 4000

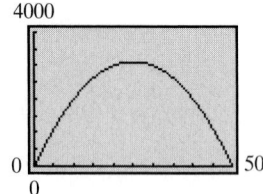

(d) 250; $3125 **(e)** $12.50

7. (a) $A(x) = -x^2 + 200x$

(b) $0 < x < 200$

(c) A is largest when $x = 100$ yd.

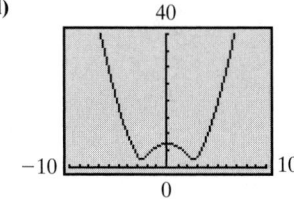

8. (a) $A(x) = -\dfrac{1}{2}x^2 + 1500x$

(b) A is largest when $x = 1500$.

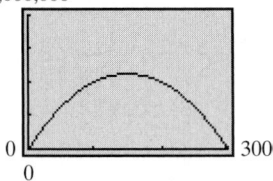

9. (a) $d(x) = \sqrt{x^4 - 15x^2 + 64}$
(b) $d(0) = 8$
(c) $d(1) = \sqrt{50} \approx 7.07$
(d)

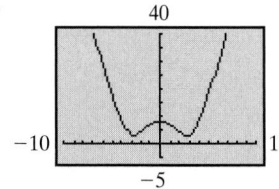

(e) d is smallest when $x \approx -2.74$
or $x \approx 2.74$.

10. (a) $d(x) = \sqrt{x^4 - 13x^2 + 49}$
(b) $d(0) = 7$
(c) $d(-1) = \sqrt{37} \approx 6.08$
(d)

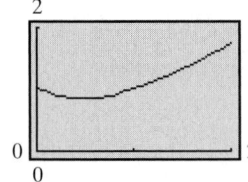

(e) d is smallest when $x \approx -2.55$
or $x \approx 2.55$.

11. (a) $d(x) = \sqrt{x^2 - x + 1}$
(b)

(c) d is smallest when $x = \dfrac{1}{2}$.

12. (a) $d(x) = \sqrt{x^2 + \dfrac{1}{x^2}}$
(b)

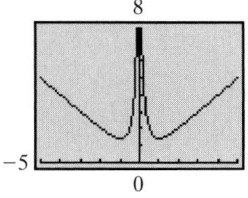

(c) d is smallest when $x = -1$ or $x = 1$.

13. $A(x) = \dfrac{1}{2}x^4$ **14.** $A(x) = \dfrac{9}{2}x - \dfrac{1}{2}x^3$

15. (a) $A(x) = x(16 - x^2)$
(b) Domain: $\{x | 0 < x < 4\}$
(c) The area is largest when $x \approx 2.31$.

16. (a) $A(x) = 2x\sqrt{4 - x^2}$
(b) $p(x) = 4x + 2\sqrt{4 - x^2}$

(c) A is largest when $x \approx 1.41$.

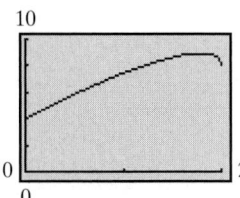

(d) p is largest when $x \approx 1.79$.

17. (a) $A(x) = 4x\sqrt{4 - x^2}$ **(c)** A is largest when $x \approx 1.41$. **(d)** p is largest when $x \approx 1.41$.

 (b) $p(x) = 4x + 4\sqrt{4 - x^2}$

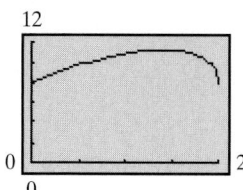

18. (a) $A(r) = 4r^2$ **(b)** $p(r) = 8r$

19. (a) $A(x) = x^2 + \dfrac{25 - 20x + 4x^2}{\pi}$

 (b) Domain: $\{x \mid 0 < x < 2.5\}$

 (c) A is smallest when $x \approx 1.40$ m.

20. (a) $A(x) = \dfrac{\sqrt{3}}{4}x^2 + \dfrac{100 - 60x + 9x^2}{4\pi}$

 (b) $0 < x < \dfrac{10}{3}$

 (c) A is smallest when $x \approx 2.08$ m.

21. (a) $C(x) = x$ **(b)** $A(x) = \dfrac{x^2}{4\pi}$

22. (a) $p(x) = x$ **(b)** $A(x) = \dfrac{1}{16}x^2$

23. (a) $A(r) = 2r^2$ **(b)** $p(r) = 6r$

24. $C(x) = \dfrac{2\sqrt{3}}{3}\pi x$

25. $A(x) = \left(\dfrac{\pi}{3} - \dfrac{\sqrt{3}}{4}\right)x^2$

26. $d(t) = 50t$

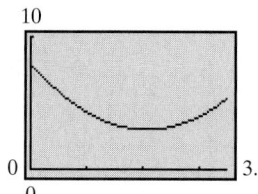

27. (a) $d(t) = \sqrt{2500t^2 - 360t + 13}$

 (b) d is smallest when $t \approx 0.07$ hr.

28. $V(h) = \pi h\left(R^2 - \dfrac{h^2}{4}\right)$ **29.** $V(r) = \dfrac{\pi H(R - r)r^2}{R}$

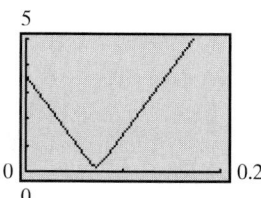

30. (a) $C(x) = 10x + 14\sqrt{x^2 - 10x + 29}, 0 \le x \le 5$

 (b) $C(1) = \$72.61$ **(c)** $C(3) = \$69.60$

 (d)

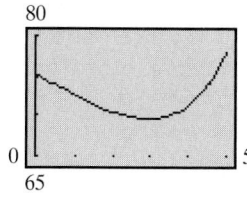

 (e) The least cost occurs when $x \approx 2.96$ mi.

31. (a) $T(x) = \dfrac{12 - x}{5} + \dfrac{\sqrt{x^2 + 4}}{3}$

 (b) $\{x \mid 0 \le x \le 12\}$ **(c)** 3.09 hr

 (d) 3.55 hr

32. $V(h) = \dfrac{\pi}{48}h^3$

Review Exercises *(page 107)*

1. Function; domain $\{-1, 2, 4\}$, range $\{0, 3\}$ **2.** Not a function **3. (a)** 2 **(b)** -2 **(c)** $-\dfrac{3x}{x^2 - 1}$ **(d)** $-\dfrac{3x}{x^2 - 1}$ **(e)** $\dfrac{3(x - 2)}{x^2 - 4x + 3}$

 (f) $\dfrac{6x}{4x^2 - 1}$ **4. (a)** $\dfrac{4}{3}$ **(b)** -4 **(c)** $\dfrac{x^2}{-x + 1}$ **(d)** $-\dfrac{x^2}{x + 1}$ **(e)** $\dfrac{(x - 2)^2}{x - 1}$ **(f)** $\dfrac{4x^2}{2x + 1}$ **5. (a)** 0 **(b)** 0 **(c)** $\sqrt{x^2 - 4}$ **(d)** $-\sqrt{x^2 - 4}$

 (e) $\sqrt{x^2 - 4x}$ **(f)** $2\sqrt{x^2 - 1}$ **6. (a)** 0 **(b)** 0 **(c)** $|x^2 - 4|$ **(d)** $-|x^2 - 4|$ **(e)** $|x^2 - 4x|$ **(f)** $4|x^2 - 1|$ **7. (a)** 0 **(b)** 0 **(c)** $\dfrac{x^2 - 4}{x^2}$

 (d) $-\dfrac{x^2 - 4}{x^2}$ **(e)** $\dfrac{x(x - 4)}{(x - 2)^2}$ **(f)** $\dfrac{x^2 - 1}{x^2}$ **8. (a)** $-\dfrac{8}{5}$ **(b)** $\dfrac{8}{5}$ **(c)** $-\dfrac{x^3}{x^2 - 9}$ **(d)** $-\dfrac{x^3}{x^2 - 9}$ **(e)** $\dfrac{(x - 2)^3}{x^2 - 4x - 5}$ **(f)** $\dfrac{8x^3}{4x^2 - 9}$

9. $\{x \mid x \ne -3, x \ne 3\}$ **10.** $\{x \mid x \ne 2\}$ **11.** $\{x \mid x \le 2\}$ **12.** $\{x \mid x \ge -2\}$ **13.** $\{x \mid x > 0\}$ **14.** $\{x \mid x \ne 0\}$ **15.** $\{x \mid x \ne -3, x \ne 1\}$

16. $\{x \mid x \ne -1, x \ne 4\}$

17. $(f + g)(x) = 2x + 3$; Domain: all real numbers

 $(f - g)(x) = -4x + 1$; Domain: all real numbers

 $(f \cdot g)(x) = -3x^2 + 5x + 2$; Domain: all real numbers

 $\left(\dfrac{f}{g}\right)(x) = \dfrac{2 - x}{3x + 1}$; Domain: $\left\{x \mid x \ne -\dfrac{1}{3}\right\}$

18. $(f + g)(x) = 4x$; Domain: all real numbers

 $(f - g)(x) = -2$; Domain: all real numbers

 $(f \cdot g)(x) = 4x^2 - 1$; Domain: all real numbers

 $\left(\dfrac{f}{g}\right)(x) = \dfrac{2x - 1}{2x + 1}$; Domain: $\left\{x \mid x \ne -\dfrac{1}{2}\right\}$

19. $(f + g)(x) = 3x^2 + 4x + 1$; Domain: all real numbers
$(f - g)(x) = 3x^2 - 2x + 1$; Domain: all real numbers
$(f \cdot g)(x) = 9x^3 + 3x^2 + 3x$; Domain: all real numbers
$\left(\dfrac{f}{g}\right)(x) = \dfrac{3x^2 + x + 1}{3x}$; Domain: $\{x | x \neq 0\}$

20. $(f + g)(x) = 1 + 4x + x^2$; Domain: all real numbers
$(f - g)(x) = -1 + 2x - x^2$; Domain: all real numbers
$(f \cdot g)(x) = 3x + 3x^2 + 3x^3$; Domain: all real numbers
$\left(\dfrac{f}{g}\right)(x) = \dfrac{3x}{1 + x + x^2}$; Domain: all real numbers

21. $(f + g)(x) = \dfrac{x^2 + 2x - 1}{x(x - 1)}$; Domain: $\{x | x \neq 0, 1\}$
$(f - g)(x) = \dfrac{x^2 + 1}{x(x - 1)}$; Domain: $\{x | x \neq 0, 1\}$
$(f \cdot g)(x) = \dfrac{x + 1}{x(x - 1)}$; Domain: $\{x | x \neq 0, 1\}$
$\left(\dfrac{f}{g}\right)(x) = \dfrac{x(x + 1)}{x - 1}$; Domain: $\{x | x \neq 0, 1\}$

22. $(f + g)(x) = \dfrac{4x - 9}{x(x - 3)}$; Domain: $\{x | x \neq 0, 3\}$
$(f - g)(x) = \dfrac{-2x + 9}{x(x - 3)}$; Domain: $\{x | x \neq 0, 3\}$
$(f \cdot g)(x) = \dfrac{3}{x(x - 3)}$; Domain: $\{x | x \neq 0, 3\}$
$\left(\dfrac{f}{g}\right)(x) = \dfrac{x}{3(x - 3)}$; Domain: $\{x | x \neq 0, 3\}$

23. $-4x + 1 - 2h$ **24.** $6x - 2 + 3h$ **25. (a)** Domain: $\{x | -4 \leq x \leq 3\}$; Range: $\{y | -3 \leq y \leq 3\}$ **(b)** $(0, 0)$ **(c)** -1 **(d)** -4
(e) $\{x | 0 < x \leq 3\}$ **(f)**

(g)

(h)

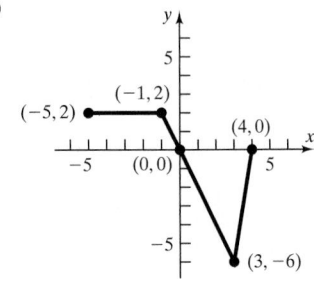

26. (a) Domain: $\{x | -5 \leq x \leq 4\}$; Range: $\{y | -3 \leq y \leq 1\}$ **(b)** 1 **(c)** $(0, 0), (4, 0)$ **(d)** 3
(e) $\{x | -5 \leq x < 0\}$ **(f)**

(g)

(h)

27. (a) Domain: $\{x | -4 \leq x \leq 4\}$ or $[-4, 4]$
Range: $\{y | -3 \leq y \leq 1\}$ or $[-3, 1]$
(b) Increasing on $(-4, -1)$ and $(3, 4)$;
Decreasing on $(-1, 3)$
(c) Local maximum is 1 and occurs at $x = -1$.
Local minimum is -3 and occurs at $x = 3$.
(d) No symmetry
(e) Neither
(f) x-intercepts: $-2, 0, 4$; y-intercept: 0
(g) $-2, 0, 4$

28. (a) Domain: all real numbers or $(-\infty, \infty)$
Range: all real numbers or $(-\infty, \infty)$
(b) Increasing on $(-\infty, -2)$ and $(2, \infty)$;
Decreasing on $(-2, 2)$
(c) Local maximum is 1 and occurs at $x = -2$.
Local minimum is -1 and occurs at $x = 2$.
(d) Origin
(e) Odd
(f) x-intercepts: $-3, 0, 3$; y-intercept: 0
(g) $-3, 0, 3$

29. Odd **30.** Even **31.** Even **32.** Neither **33.** Neither **34.** Neither **35.** Odd **36.** Odd

37.

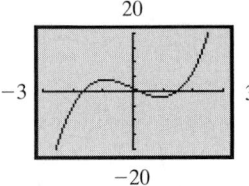

Local maximum: $(-0.91, 4.04)$
Local minimum: $(0.91, -2.04)$
Increasing: $(-3, -0.91)$; $(0.91, 3)$
Decreasing: $(-0.91, 0.91)$

38.

Local maximum: $(1, -3)$
Local minimum: $(-1, -7)$
Increasing: $(-1, 1)$
Decreasing: $(-3, -1)$; $(1, 3)$

39.

Local maximum: $(0.41, 1.53)$
Local minimum: $(-0.34, 0.54)$; $(1.80, -3.56)$
Increasing: $(-0.34, 0.41)$; $(1.80, 3)$
Decreasing: $(-2, -0.34)$; $(0.41, 1.80)$

40.

Local maxima: $(-0.59, 4.62)$; $(2, 3)$
Local minimum: $(0.84, 0.92)$
Increasing: $(-2, -0.59)$; $(0.84, 2)$
Decreasing: $(-0.59, 0.84)$; $(2, 3)$

41. (a) 23 **(b)** 7 **(c)** 47 **42. (a)** 15 **(b)** 3
(c) 57 **43.** -5 **44.** $2(x + 2)$ **45.** $-4x - 5$
46. $x - 1$ **47.** No **48.** Yes **49.** Yes **50.** Yes

51.

52.

53.

54.

55.

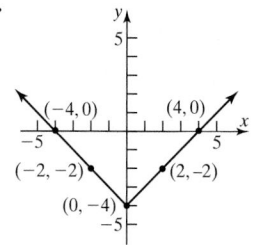

Intercepts: $(-4, 0)$, $(4, 0)$, $(0, -4)$
Domain: all real numbers
Range: $\{y \mid y \geq -4\}$ or $[-4, \infty)$

56.

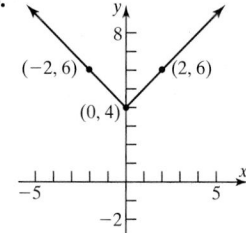

Intercept: $(0, 4)$
Domain: all real numbers
Range: $\{y \mid y \geq 4\}$ or $[4, \infty)$

57.

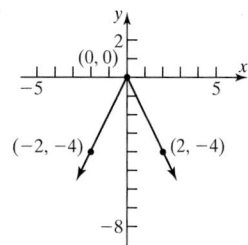

Intercept: $(0, 0)$
Domain: all real numbers
Range: $\{y \mid y \leq 0\}$ or $(-\infty, 0]$

58.

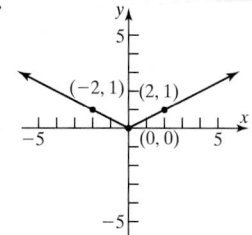

Intercept: $(0, 0)$
Domain: all real numbers
Range: $\{y \mid y \geq 0\}$ or $[0, \infty)$

59.

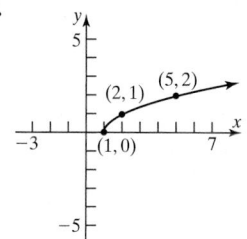

Intercept: $(1, 0)$
Domain: $\{x \mid x \geq 1\}$ or $[1, \infty)$
Range: $\{y \mid y \geq 0\}$ or $[0, \infty)$

60.

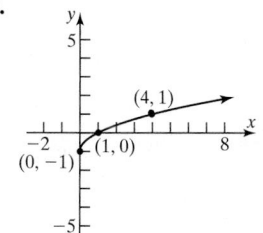

Intercepts: $(0, -1)$, $(1, 0)$
Domain: $\{x \mid x \geq 0\}$ or $[0, \infty)$
Range: $\{y \mid y \geq -1\}$ or $[-1, \infty)$

61.

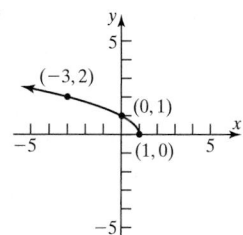

Intercepts: $(0, 1)$, $(1, 0)$
Domain: $\{x \mid x \leq 1\}$ or $(-\infty, 1]$
Range: $\{y \mid y \geq 0\}$ or $[0, \infty)$

62.

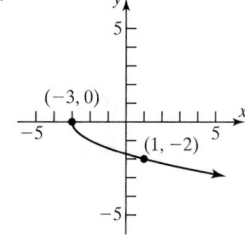

Intercepts: $(-3, 0)$, $\left(0, -\sqrt{3}\right)$
Domain: $\{x \mid x \geq -3\}$ or $[-3, \infty)$
Range: $\{y \mid y \leq 0\}$ or $(-\infty, 0]$

63.

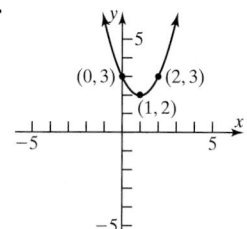

Intercept: $(0, 3)$

Domain: all real numbers

Range: $\{y \mid y \geq 2\}$ or $[2, \infty)$

64.

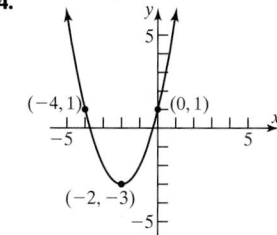

Intercepts: $\left(-2 - \sqrt{3}, 0\right)$,

$\left(-2 + \sqrt{3}, 0\right)$, $(0, 1)$
Domain: all real numbers
Range: $\{y \mid y \geq -3\}$ or $[-3, \infty)$

65.

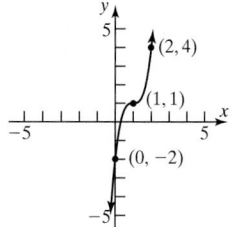

Intercepts: $(0, -2)$,

$\left(1 - \dfrac{\sqrt[3]{9}}{3}, 0\right)$ or about $(0.3, 0)$

Domain: all real numbers
Range: all real numbers

66.

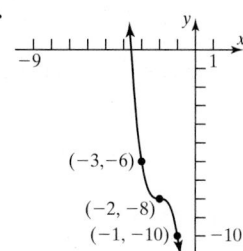

Intercepts: $(0, -24)$,

$\left(-2 - \sqrt[3]{4}, 0\right)$ or about

$(-3.6, 0)$
Domain: all real numbers
Range: all real numbers

67. (a) $\{x|x > -2\}; (-2, \infty)$

(b) $(0, 0)$

(c)

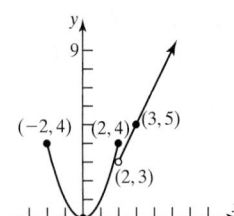

68. (a) $\{x|x > -3\}; (-3, \infty)$

(b) $(0, -1), \left(\dfrac{1}{3}, 0\right)$

(c)

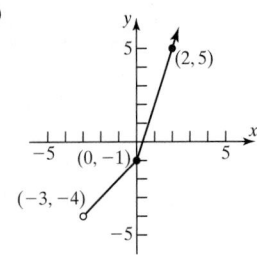

69. (a) $\{x|x \geq -4\}; [-4, \infty)$

(b) $(0, 1)$

(c)

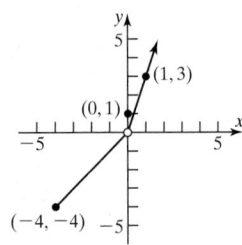

(d) $\{y|y > -6\}; (-6, \infty)$

70. (a) $\{x|x \geq -2\}; [-2, \infty)$

(c)

(d) $\{y|y > -4\}; (-4, \infty)$

(b) $(0, 0)$

(d) $\{y|y \geq 0\}; [0, \infty)$

(d) $\{y|y \geq -4\}; [-4, \infty)$

71. $A = 11$

72. $A = 8$

73. If the radius doubles, the volume of the new sphere is 8 times as large as the original sphere, and the surface area of the new sphere is 4 times as large as the original sphere.

74. (a) $A(x) = (8.5 - 2x)(11 - 2x)$

(b) $0 \leq x < 4.25, 0 < A \leq 93.5$

(c) $A(1) = 58.5$ in.2, $A(1.2) = 52.46$ in.2, $A(1.5) = 44$ in.2

(d)

(e) $A(x) \approx 70$ when $x \approx 0.65$ in.;
$A(x) \approx 50$ when $x \approx 1.28$ in.

75. (a) $A = 2\pi r^2 + \dfrac{200}{r}$ **(b)** 123.22 ft^2 **(c)** 150.53 ft^2

(d) 197.08 ft^2 **(e)** A is smallest when $r \approx 2.52$ ft.

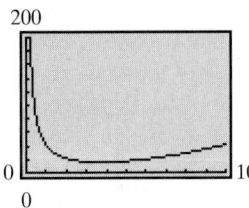

76. 25 square units **77. (a)** 63 **(b)** \$151.90 **78.** $(2, 2)$ **79.** $(2, 3)$ **80.** $\dfrac{20\sqrt{30}}{9} \approx 12.17$ square units

81. (a) $A(x) = 2x^2 + \dfrac{40}{x}$ **(b)** 42 ft^2 **(c)** 28 ft^2

(d)

A is smallest when $x \approx 2.15$ ft.

82. (a) $C(r) = 0.12\pi r^2 + \dfrac{40}{r}$ **(b)** \$16.03 **(c)** \$29.13

(d) The cost is least for $r \approx 3.76$ cm.

Chapter Test *(page 111)*

1. (a) Function; domain: $\{2, 4, 6, 8\}$; range: $\{5, 6, 7, 8\}$ **(b)** Not a function **(c)** Not a function **(d)** Function; domain; all real numbers;

range: $\{y|y \geq 2\}$ **2.** Domain: $\left\{x \Big| x \leq \dfrac{4}{5}\right\}; f(-1) = 3$ **3.** Domain: $\{x|x \neq -2\}; g(-1) = 1$ **4.** Domain: $\{x|x \neq -9, x \neq 4\}; h(-1) = \dfrac{1}{8}$

5. (a) Domain: $\{x|x - 5 \leq x \leq 5\}$; range: $\{y|-3 \leq y \leq 3\}$ **(b)** $(0, 2), (-2, 0),$ and $(2, 0)$ **(c)** $f(1) = 3$ **(d)** $x = -5$ and $x = 3$

(e) $\{x|-5 \leq x < -2$ or $2 < x \leq 5\}; [-5, -2)$ or $(2, 5]$ **(f)** $-2, 2$

6. Local maxima: $f(-0.85) \approx -0.86$

$f(2.35) \approx 15.55$ Local minima: $f(0) = -2$

The function is increasing on the intervals $(-5, -0.85)$ and $(0, 2.35)$ and decreasing on the intervals $(-0.85, 0)$ and $(2.35, 5)$.

7. (a)

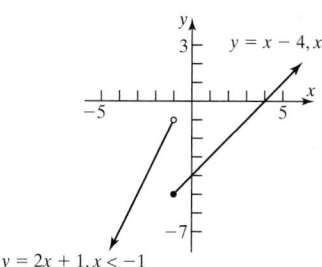

$y = x - 4, x \geq -1$

$y = 2x + 1, x < -1$

(b) $(0, -4), (4, 0)$
(c) $g(-5) = -9$
(d) $g(2) = -2$

8. $\dfrac{f(x) - f(3)}{x - 3} = 3x + 7, x \neq 3$

9. (a) $(f - g) = 2x^2 - 3x + 3$
(b) $(f \cdot g) = 6x^3 - 4x^2 + 3x - 2$
(c) $f(x + h) - f(x) = 4xh + 2h^2$

10. (a)

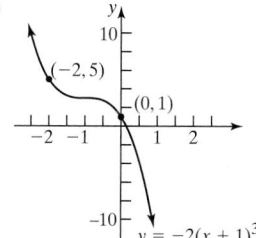

$y = -2(x + 1)^3 + 3$

(b)

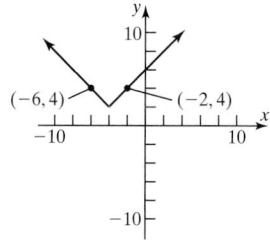

11. (a) 8.67% occurring in 1997 $(x \approx 5)$
(b) The model predicts that the interest rate will be -10.343%. This is not reasonable.

12. (a) $V(x) = \dfrac{x^2}{8} - \dfrac{5x}{4} + \dfrac{\pi x^2}{64}$
(b) 1297.61 ft^3

CHAPTER 2 Linear and Quadratic Functions

2.1 Assess Your Understanding (page 121)

7. slope; y-intercept **8.** $-4; 3$ **9.** positive **10.** True **11.** False **12.** False

13. (a) $m = 2, b = 3$
(b)

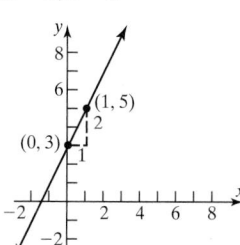

(c) 2 **(d)** Increasing

14. (a) $m = 5, b = -4$
(b)

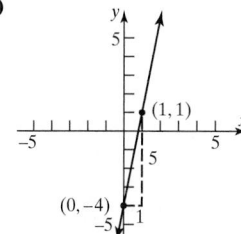

(c) 5 **(d)** Increasing

15. (a) $m = -3, b = 4$
(b)

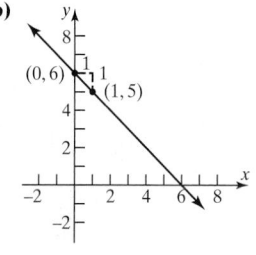

(c) -3 **(d)** Decreasing

16. (a) $m = -1, b = 6$
(b)

(c) -1 **(d)** Decreasing

17. (a) $m = \dfrac{1}{4}, b = -3$
(b)

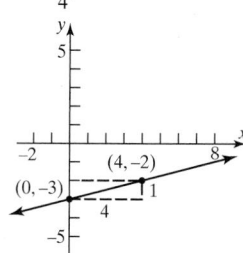

(c) $\dfrac{1}{4}$ **(d)** Increasing

18. (a) $m = -\dfrac{2}{3}, b = 4$
(b)

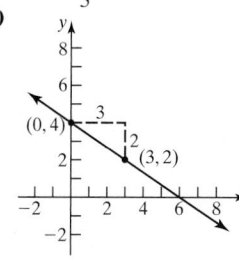

(c) $-\dfrac{2}{3}$ **(d)** Decreasing

19. (a) $m = 0, b = 4$
(b)

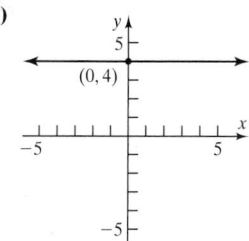

(c) 0 **(d)** Constant

20. (a) $m = 0, b = -2$
(b)

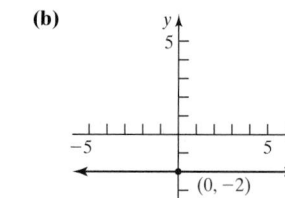

(c) 0 **(d)** Constant

21. (a) 4
(b)

22. (a) -4
(b)

23. (a) 2
(b)

24. (a) 2
(b)

25. (a) 8
(b)

26. (a) 12
(b)

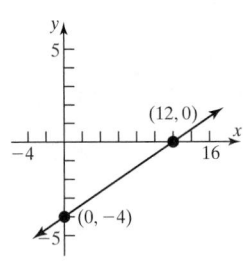

27. Linear; -3 **28.** Nonlinear **29.** Nonlinear **30.** Linear; 4 **31.** Nonlinear **32.** Linear; 0.5 **33.** Linear; 0 **34.** Nonlinear

35. (a) $\dfrac{1}{4}$ **(b)** $\left\{x \middle| x > \dfrac{1}{4}\right\}$ or $\left(\dfrac{1}{4}, \infty\right)$
(c) 1 **(d)** $\{x | x \le 1\}$ or $(-\infty, 1]$
(e)

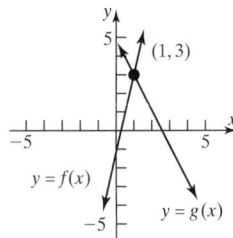

36. (a) $-\dfrac{5}{3}$ **(b)** $\left\{x \middle| x < -\dfrac{5}{3}\right\}$ or $\left(-\infty, -\dfrac{5}{3}\right)$
(c) 2 **(d)** $\{x | x \ge 2\}$ or $[2, \infty)$
(e)

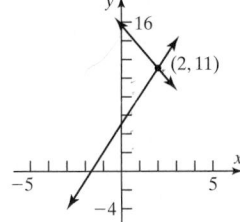

37. (a) 40 **(b)** 88 **(c)** -40 **(d)** $\{x | x > 40\}$ or $(40, \infty)$ **(e)** $\{x | x \le 88\}$ or $(-\infty, 88]$ **(f)** $\{x | -40 < x < 88\}$ or $(-40, 88)$
38. (a) 5 **(b)** -15 **(c)** 15 **(d)** $\{x | x < 5\}$ or $(-\infty, 5)$ **(e)** $\{x | x \ge -15\}$ or $\{-15, \infty)$ **(f)** $\{x | -15 < x < 15\}$ or $(-15, 15)$
39. (a) -4 **(b)** $\{x | x < -4\}$ or $(-\infty, -4)$ **40. (a)** 2 **(b)** $\{x | x \le 2\}$ or $(-\infty, 2]$ **41. (a)** -6 **(b)** $\{x | -6 \le x < 5\}$ or $[-6, 5)$
42. (a) 7 **(b)** $\{x | -4 \le x < 7\}$ or $[-4, 7)$ **43. (a)** $\$45$ **(b)** 180 mi **(c)** 260 mi **44. (a)** $\$24$ **(b)** 64 min **(c)** 144 min
45. (a) $\$778.22$ **(b)** 2006 **(c)** 2012 **46. (a)** $\$671$ billion **(b)** 2008 **(c)** 2013 **47. (a)** $p = \$16, q = 600$ T-shirts **(b)** $0 < p < \$16$
48. (a) $p = \$3; q = 7000$ hot dogs **(b)** $p > \$3$

49. (a) $\{x | 7{,}300 \le x \le 29{,}700\}$ **(b)** $\$2335$
(c) The independent variable is adjusted gross income, x.
 The dependent variable is the tax bill, T.
(d)

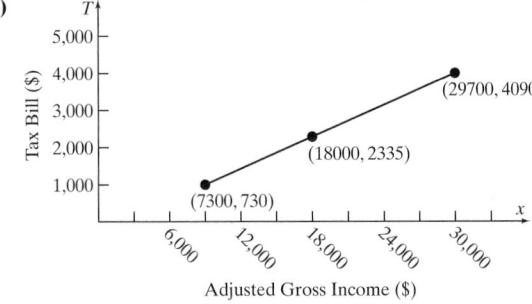

(e) $\$21{,}500$

50. (a) $\{p | p \ge 128\}$ **(b)** $\$7.2$ million
(c)

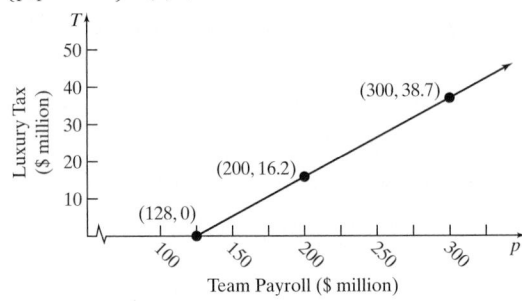

(d) $\$180$ million

51. (a) $x = 5000$ **(b)** $x > 5000$ **52. (a)** $x = 7500$ **(b)** $x > 7500$

53. (a) $V(x) = -1000x + 3000$
(b)

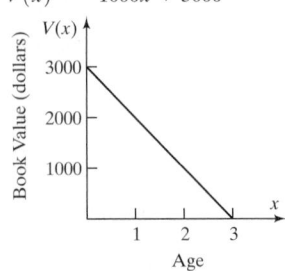

(c) $\$1000$ **(d)** After 1 year

54. (a) $V(x) = -12{,}000x + 120{,}000$
(b)

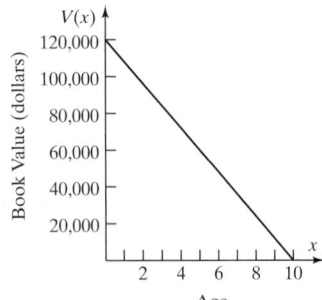

(c) $\$72{,}000$ **(d)** After 4 years

55. (a) $C(x) = 90x + 1800$

(b)

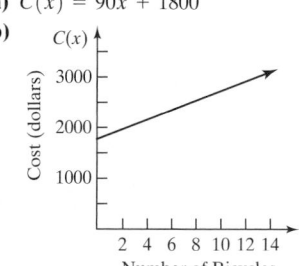

(c) $3060 **(d)** 22 bicycles

56. (a) $1805 **(b)** $C(x) = 90x + 1805$

(c)

(d) $3065 **(e)** 21 bicycles

57. (a) $C(x) = 0.07x + 29$
 (b) $36.70; $45.10
58. (a) $C(x) = 0.05x + 5$
 (b) $10.25; $14
59. (d), (c)
60. (b), (e)

2.2 Assess Your Understanding (page 130)

3. scatter diagram **4.** $y = kx$ **5.** T **6.** T **7.** Linear relation **8.** Nonlinear relation **9.** Linear relation **10.** No relation
11. Nonlinear relation **12.** Nonlinear relation

13. (a)

(c)

(e)

(b) Answers will vary. Using $(4, 6)$ and $(8, 14)$, $y = 2x - 2$. **(d)** $y = 2.0357x - 2.3571$

14. (a)

(c)

(e)

(b) Answers will vary. Using $(5, 2)$ and $(11, 9)$, $y = \dfrac{7}{6}x - \dfrac{23}{6}$. **(d)** $y = 1.1286x - 3.8619$

15. (a)

(c)

(e)

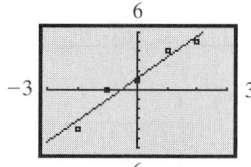

(b) Answers will vary. Using $(-2, -4)$ and $(1, 4)$, $y = \dfrac{8}{3}x + \dfrac{4}{3}$. **(d)** $y = 2.2x + 1.2$

16. (a)

(c)

(e)

(b) Answers will vary. Using $(-2, 7)$ and $(1, 2)$, $y = -\dfrac{5}{3}x + \dfrac{11}{3}$. **(d)** $y = -1.8x + 3.6$

17. (a)

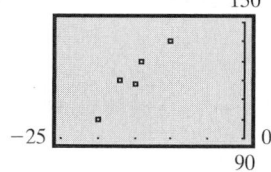

(b) Answers will vary. Using $(-20, 100)$ and $(-15, 118)$, $y = \dfrac{18}{5}x + 172$.

(c)

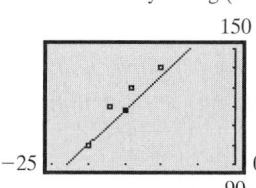

(d) $y = 3.8613x + 180.2920$ **(e)**

18. (a)

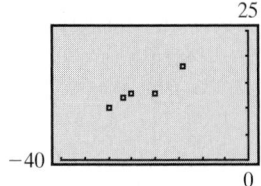

(b) Answers will vary. Using $(-30, 10)$ and $(-14, 18)$, $y = \frac{1}{2}x + 25$.

(c)

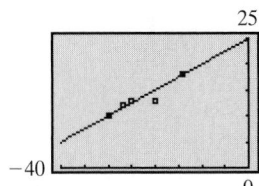

(d) $y = 0.4421x + 23.4559$

(e)

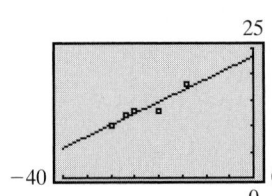

19. $p(b) = 0.00649B$; \$941.05 **20.** $p(b) = 0.00899B$; \$1573.25 **21.** $R(g) = 1.95g$; \$20.48 **22.** $C(A) = 4.75A$; \$16.63 **23.** 5 gallons **24.** 96 ft/sec

25. (a)

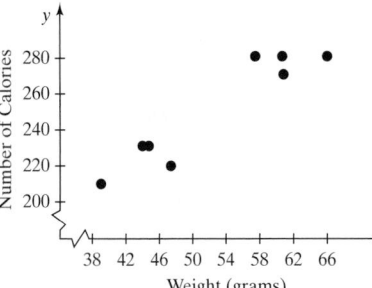

(b) Linear

(c) Answers will vary. Using the points $(39.52, 210)$ and $(66.45, 280)$, $y = 2.599x + 107.288$.

(d)

(e) 269 calories

(f) If the weight of a candy bar is increased by 1 gram, the number of calories will increase by 2.599.

26. (a) No

(b)

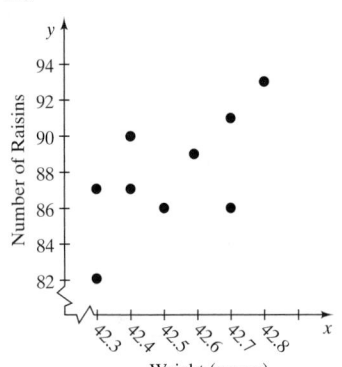

(c) Answers will vary. Using the points $(42.3, 82)$ and $(42.8, 93)$, $y = 22x - 848.6$.

(d)

(e) $N(w) = 22w - 848.6$

(f) Approximately 86 raisins will be in a box weighing 42.5 grams.

(g) If the weight is increased by one gram, the number of raisins will increase by 22 raisins.

27. (a)

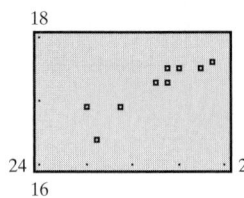

(b) $C(H) = 0.3734H + 7.3268$

(c) If height increases by 1 inch, head circumference increases by about 0.3734 inch.

(d) About 17.0 inches

(e) About 26.98 inches

28. (a)

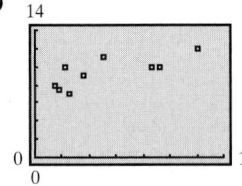

(b) $L(G) = 0.0261G + 7.8738$

(c) If gestation period increases by 1 day, life expectancy increases by about 0.0261 year.

(d) About 10.2 years

29. (a) No

(b)

(c) $D = -1.3355p + 86.1974$

(d) If the price increases \$1, the quantity sold per day decreases by about 1.34 pairs of jeans.

(e) $D(p) = -1.3355p + 86.1974$

(f) $\{p \mid 0 < p \le 64\}$

(g) $D(28) \approx 48.80$; about 49 pairs

30. (a) No

(b)

(c) $S = 2.0667A + 292.8869$

(d) As advertising increases by \$1000, sales increase by about \$2067.

(e) $S(A) = 2.0667A + 292.8869$

(f) $\{A \mid A \ge 0\}$

(g) \$344,554

31.

32.

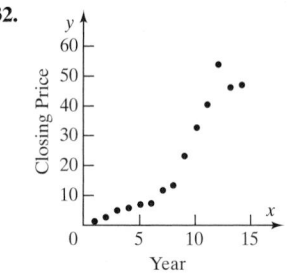

The data do not follow a linear pattern.　　　　The data do not follow a linear pattern.

33. $y = 1.5x + 3.5; r = 1$　**34.** No linear relation　**35.** Average grade is linearly related with time studied.

2.3 Assess Your Understanding (page 142)

7. repeated; multiplicity 2　**8.** discriminant; negative　**9.** F　**10.** T　**11.** Zeros: $0, 9$; x-intercepts: $0, 9$　**12.** Zeros: $0, -4$; x-intercepts: $0, -4$

13. Zeros: $-5, 5$; x-intercepts: $-5, 5$　**14.** Zeros: $-3, 3$; x-intercepts: $-3, 3$　**15.** Zeros: $-3, 2$; x-intercepts: $-3, 2$　**16.** Zeros: $-6, -1$;

x-intercepts: $-6, -1$　**17.** Zeros: $-\dfrac{1}{2}, 3$; x-intercepts: $-\dfrac{1}{2}, 3$　**18.** Zeros: $-1, -\dfrac{2}{3}$; x-intercepts: $-1, -\dfrac{2}{3}$　**19.** Zeros: $-4, 4$;

x-intercepts: $-4, 4$　**20.** Zeros: $-5, 5$; x-intercepts: $-5, 5$　**21.** Zeros: $-6, -2$; x-intercepts: $-6, -2$　**22.** Zeros: $-2, 6$;

x-intercepts: $-2, 6$　**23.** Zero: $\dfrac{3}{2}$; x-intercept: $\dfrac{3}{2}$　**24.** Zero: $\dfrac{4}{5}$; x-intercept: $\dfrac{4}{5}$　**25.** Zeros: $-5, 5$; x-intercepts: $-5, 5$

26. Zeros: $-6, 6$; x-intercepts: $-6, 6$　**27.** Zeros: $-1, 3$; x-intercepts: $-1, 3$　**28.** Zeros: $-3, -1$; x-intercepts: $-3, -1$　**29.** Zeros: $-3, 0$;

x-intercepts: $-3, 0$　**30.** Zeros: $0, \dfrac{4}{3}$; x-intercepts: $0, \dfrac{4}{3}$　**31.** Zeros: $-7, 3$; x-intercepts: $-7, 3$　**32.** Zeros: $3 - \sqrt{22}, 3 + \sqrt{22}$;

x-intercepts: $3 - \sqrt{22}, 3 + \sqrt{22}$　**33.** Zeros: $-\dfrac{1}{4}, \dfrac{3}{4}$; x-intercepts: $-\dfrac{1}{4}, \dfrac{3}{4}$　**34.** Zeros: $-1, \dfrac{1}{3}$; x-intercepts: $-1, \dfrac{1}{3}$

35. Zeros: $\dfrac{-1 - \sqrt{7}}{6}, \dfrac{-1 + \sqrt{7}}{6}$; x-intercepts: $\dfrac{-1 - \sqrt{7}}{6}, \dfrac{-1 + \sqrt{7}}{6}$　**36.** Zeros: $\dfrac{3 - \sqrt{17}}{4}, \dfrac{3 + \sqrt{17}}{4}$; x-intercepts: $\dfrac{3 - \sqrt{17}}{4}, \dfrac{3 + \sqrt{17}}{4}$

37. Zeros: $2 - \sqrt{2}, 2 + \sqrt{2}$; x-intercepts: $2 - \sqrt{2}, 2 + \sqrt{2}$　**38.** Zeros: $-2 - \sqrt{2}, -2 + \sqrt{2}$; x-intercepts: $-2 - \sqrt{2}, -2 + \sqrt{2}$

39. Zeros: $2 - \sqrt{5}, 2 + \sqrt{5}$; x-intercepts: $2 - \sqrt{5}, 2 + \sqrt{5}$　**40.** Zeros: $-3 - 2\sqrt{2}, -3 + 2\sqrt{2}$; x-intercepts: $-3 - 2\sqrt{2}, -3 + 2\sqrt{2}$

41. Zeros: $1, \dfrac{3}{2}$; x-intercepts: $1, \dfrac{3}{2}$　**42.** Zeros: $-\dfrac{3}{2}, -1$; x-intercepts: $-\dfrac{3}{2}, -1$　**43.** No zeros; no x-intercepts

44. No zeros; no x-intercepts　**45.** Zeros: $\dfrac{-1 - \sqrt{5}}{4}, \dfrac{-1 + \sqrt{5}}{4}$; x-intercepts: $\dfrac{-1 - \sqrt{5}}{4}, \dfrac{-1 + \sqrt{5}}{4}$　**46.** Zeros: $\dfrac{-1 - \sqrt{3}}{2}, \dfrac{-1 + \sqrt{3}}{2}$;

x-intercepts: $\dfrac{-1 - \sqrt{3}}{2}, \dfrac{-1 + \sqrt{3}}{2}$　**47.** Zeros: $\dfrac{9 - \sqrt{113}}{8}, \dfrac{9 + \sqrt{113}}{8}$; x-intercepts: $\dfrac{9 - \sqrt{113}}{8}, \dfrac{9 + \sqrt{113}}{8}$

48. No zeros; no x-intercepts　**49.** Zeros: $-\sqrt{5}, \sqrt{5}$; x-intercepts: $-\sqrt{5}, \sqrt{5}$　**50.** Zeros: $-\sqrt{6}, \sqrt{6}$; x-intercepts: $-\sqrt{6}, \sqrt{6}$

51. Zero: $\dfrac{1}{4}$; x-intercept: $\dfrac{1}{4}$　**52.** Zero: $\dfrac{3}{2}$; x-intercept: $\dfrac{3}{2}$　**53.** Zeros: $-\dfrac{3}{5}, \dfrac{5}{2}$; x-intercepts: $-\dfrac{3}{5}, \dfrac{5}{2}$　**54.** Zeros: $-\dfrac{5}{2}, \dfrac{4}{3}$;

x-intercepts: $-\dfrac{5}{2}, \dfrac{4}{3}$　**55.** Zeros: $-\dfrac{1}{2}, \dfrac{2}{3}$; x-intercepts: $-\dfrac{1}{2}, \dfrac{2}{3}$　**56.** Zeros: $-\dfrac{2}{3}, \dfrac{1}{2}$; x-intercepts: $-\dfrac{2}{3}, \dfrac{1}{2}$

57. Zeros: $\dfrac{-\sqrt{2} + 2}{2}, \dfrac{-\sqrt{2} - 2}{2}$; x-intercepts: $\dfrac{-\sqrt{2} + 2}{2}, \dfrac{-\sqrt{2} - 2}{2}$　**58.** Zeros: $\sqrt{2} - 2, \sqrt{2} + 2$; x-intercepts: $\sqrt{2} - 2, \sqrt{2} + 2$

59. Zeros: $\dfrac{-1 - \sqrt{17}}{2}, \dfrac{-1 + \sqrt{17}}{2}$; x-intercepts: $\dfrac{-1 - \sqrt{17}}{2}, \dfrac{-1 + \sqrt{17}}{2}$　**60.** Zeros: $\dfrac{-1 - \sqrt{5}}{2}, \dfrac{-1 + \sqrt{5}}{2}$;

x-intercepts: $\left\{\dfrac{-1 - \sqrt{5}}{2}, \dfrac{-1 + \sqrt{5}}{2}\right\}$　**61.** $\{-6, 0\}$; $(-6, 3), (0, 3)$　**62.** $\{0, 4\}$; $(0, 3), (4, 3)$　**63.** $\left\{-1, -\dfrac{1}{2}\right\}$; $(-1, -1), \left(-\dfrac{1}{2}, \dfrac{1}{2}\right)$

64. $\left\{-\dfrac{2}{3}, 4\right\}$; $\left(-\dfrac{2}{3}, \dfrac{-17}{3}\right), (4, 41)$　**65.** $\{-3, 5\}$; $(-3, 13), (5, 21)$　**66.** $\{-6, 4\}$; $(-6, 3), (4, 33)$　**67.** Zeros: $-2\sqrt{2}, 2\sqrt{2}$; x-intercepts: $-2\sqrt{2}, 2\sqrt{2}$

68. Zeros: $-2, 2$; x-intercepts: $-2, 2$　**69.** Zeros: $-2, -1, 1, 2$; x-intercepts: $-2, -1, 1, 2$　**70.** Zeros: $-\sqrt{5}, \sqrt{5}$; x-intercepts: $-\sqrt{5}, \sqrt{5}$

71. Zeros: $-1, 1$; x-intercepts: $-1, 1$　**72.** Zeros: $-2, 2$; x-intercepts: $-2, 2$　**73.** Zeros: $-2, 1$; x-intercepts: $-2, 1$　**74.** Zeros: $-1, 2$; x-intercepts: $-1, 2$

75. Zeros: $-6, -5$; x-intercepts: $-6, -5$　**76.** Zeros: $-\dfrac{7}{2}, -1$; x-intercepts: $-\dfrac{7}{2}, -1$　**77.** Zero: $-\dfrac{1}{3}$; x-intercept: $-\dfrac{1}{3}$　**78.** Zeros: $-2, 7$;

x-intercepts: $-2, 7$　**79.** Zeros: $-\dfrac{3}{2}, 2$; x-intercepts: $-\dfrac{3}{2}, 2$　**80.** Zeros: $\dfrac{5}{3}, 2$; x-intercepts: $\dfrac{5}{3}, 2$　**81.** Zeros: $0, 16$; x-intercepts: $0, 16$　**82.** Zero: 0;

x-intercept: 0　**83.** Zero: 16; x-intercept: 16　**84.** Zero: 1; x-intercept: 1　**85.** 11 ft by 13 ft　**86.** 17 cm by 18 cm　**87.** 4 ft by 4 ft　**88.** 6 ft by 6 ft

89. (a) The ball strikes the ground after 6 sec.　**(b)** The ball passes the top of the building on its way down after 5 sec.

90. (a) The object will be 15 m above the ground after 0.99 sec and after 3.09 sec.　**(b)** It will strike the ground in 4.08 sec.

(c) It will never reach 100 meters.　**91.** 20 consecutive integers　**92.** 13 sides; no　**93.** $\dfrac{-b + \sqrt{b^2 - 4ac}}{2a} + \dfrac{-b - \sqrt{b^2 - 4ac}}{2a} = \dfrac{-2b}{2a} = -\dfrac{b}{a}$

94. $\left(\dfrac{-b + \sqrt{b^2 - 4ac}}{2a}\right)\left(\dfrac{-b - \sqrt{b^2 - 4ac}}{2a}\right) = \dfrac{b^2 - (b^2 - 4ac)}{4a^2} = \dfrac{4ac}{4a^2} = \dfrac{c}{a}$　**95.** $k = \dfrac{1}{2}$ or $k = -\dfrac{1}{2}$　**96.** $k = 4$ or $k = -4$

97. $ax^2 + bx + c = 0, x = \dfrac{-b \pm \sqrt{b^2 - 4ac}}{2a}$; $ax^2 - bx + c = 0, x = \dfrac{b \pm \sqrt{(-b)^2 - 4ac}}{2a} = \dfrac{b \pm \sqrt{b^2 - 4ac}}{2a} = -\dfrac{-b \pm \sqrt{b^2 - 4ac}}{2a}$

98. For $ax^2 + bx + c = 0$, the solutions are $x_1 = \dfrac{-b + \sqrt{b^2 - 4ac}}{2a}$, $x_2 = \dfrac{-b - \sqrt{b^2 - 4ac}}{2a}$.

For $cx^2 + bx + a = 0$, the solutions are $x_3 = \dfrac{-b + \sqrt{b^2 - 4ac}}{2c}$, $x_4 = \dfrac{-b - \sqrt{b^2 - 4ac}}{2c}$.

$x_1 x_4 = \dfrac{b^2 - (b^2 - 4ac)}{4ac} = 1$ and $x_2 x_3 = \dfrac{b^2 - (b^2 - 4ac)}{4ac} = 1$ **99.** (b) **104.** 0, 1, or 2 times.

2.4 Assess Your Understanding (page 152)

5. parabola **6.** axis or axis of symmetry **7.** $-\dfrac{b}{2a}$ **8.** T **9.** T **10.** T **11.** C **12.** E **13.** F **14.** A **15.** G **16.** B **17.** H **18.** D

19.

20.

21.

22.

23.

24.

25.

26.
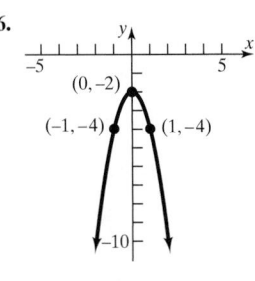

27. $f(x) = (x + 2)^2 - 2$

28. $f(x) = (x - 3)^2 - 10$
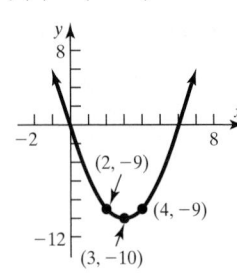

29. $f(x) = 2(x - 1)^2 - 1$
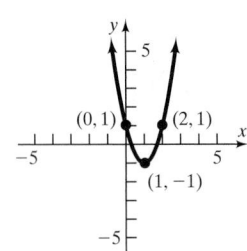

30. $f(x) = 3(x + 1)^2 - 3$
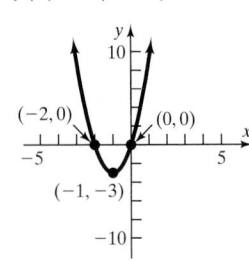

31. $f(x) = -(x + 1)^2 + 1$

32. $f(x) = -2\left(x - \dfrac{3}{2}\right)^2 + \dfrac{13}{2}$
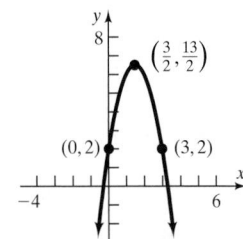

33. $f(x) = \dfrac{1}{2}(x + 1)^2 - \dfrac{3}{2}$
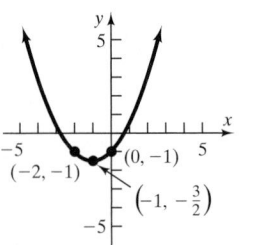

34. $f(x) = \dfrac{2}{3}(x + 1)^2 - \dfrac{5}{3}$

35. (a)

36. (a)

37. (a)

38. (a)
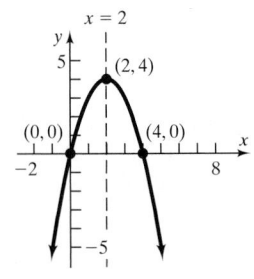

(b) Domain: $(-\infty, \infty)$
Range $[-1, \infty)$
(c) Decreasing on $(-\infty, -1)$
Increasing on $(-1, \infty)$

(b) Domain: $(-\infty, \infty)$
Range $[-4, \infty)$
(c) Decreasing on $(-\infty, 2)$
Increasing on $(2, \infty)$

(b) Domain: $(-\infty, \infty)$
Range $(-\infty, 9]$
(c) Increasing on $(-\infty, -3)$
Decreasing on $(-3, \infty)$

(b) Domain: $(-\infty, \infty)$
Range $(-\infty, 4]$
(c) Increasing on $(-\infty, 2)$
Decreasing on $(2, \infty)$

39. (a)

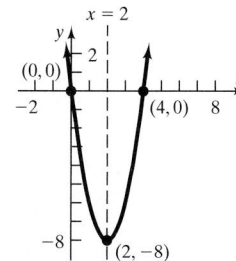

(b) Domain: $(-\infty, \infty)$
Range $[-8, \infty)$
(c) Decreasing on $(-\infty, 2)$
Increasing on $(2, \infty)$

40. (a)

(b) Domain: $(-\infty, \infty)$
Range $[-27, \infty)$
(c) Decreasing on $(-\infty, -3)$
Increasing on $(-3, \infty)$

41. (a)

(b) Domain: $(-\infty, \infty)$
Range $[-9, \infty)$
(c) Decreasing on $(-\infty, -1)$
Increasing on $(-1, \infty)$

42. (a)

(b) Domain: $(-\infty, \infty)$
Range $[-4, \infty)$
(c) Decreasing on $(-\infty, 1)$
Increasing on $(1, \infty)$

43. (a)

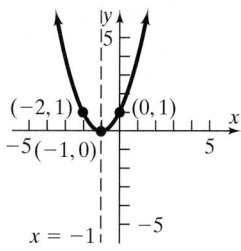

(b) Domain: $(-\infty, \infty)$

Range $[0, \infty)$

(c) Decreasing on $(-\infty, -1)$

Increasing on $(-1, \infty)$

44. (a)

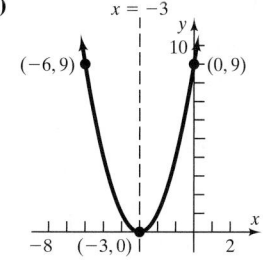

(b) Domain: $(-\infty, \infty)$

Range $[0, \infty)$

(c) Decreasing on $(-\infty, -3)$

Increasing on $(-3, \infty)$

45. (a)

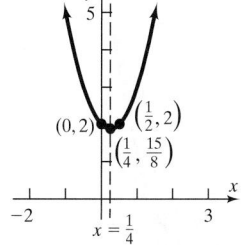

(b) Domain: $(-\infty, \infty)$

Range $\left[\dfrac{15}{8}, \infty\right)$

(c) Decreasing on $\left(-\infty, \dfrac{1}{4}\right)$

Increasing on $\left(\dfrac{1}{4}, \infty\right)$

46. (a)

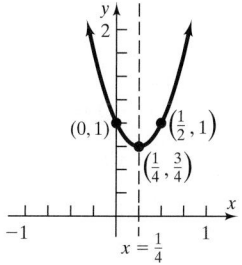

(b) Domain: $(-\infty, \infty)$
Range $\left[\dfrac{3}{4}, \infty\right)$
(c) Decreasing on $\left(-\infty, \dfrac{1}{4}\right)$
Increasing on $\left(\dfrac{1}{4}, \infty\right)$

47. (a)

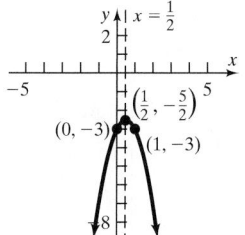

(b) Domain: $(-\infty, \infty)$
Range $\left(-\infty, -\dfrac{5}{2}\right]$
(c) Decreasing on $\left(\dfrac{1}{2}, \infty\right)$
Increasing on $\left(-\infty, \dfrac{1}{2}\right)$

48. (a)

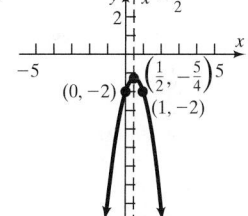

(b) Domain: $(-\infty, \infty)$
Range $\left(-\infty, -\dfrac{5}{4}\right]$
(c) Decreasing on $\left(\dfrac{1}{2}, \infty\right)$
Increasing on $\left(-\infty, \dfrac{1}{2}\right)$

49. (a)

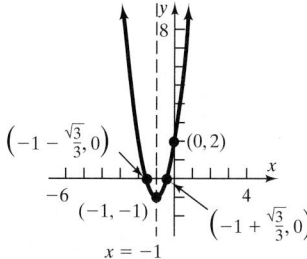

(b) Domain: $(-\infty, \infty)$
Range $[-1, \infty)$
(c) Decreasing on $(-\infty, -1)$

Increasing on $(-1, \infty)$

50. (a)

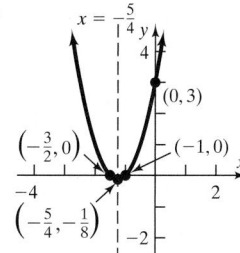

(b) Domain: $(-\infty, \infty)$
Range $\left[-\dfrac{1}{8}, \infty\right)$
(c) Decreasing on $\left(-\infty, -\dfrac{5}{4}\right)$

Increasing on $\left(-\dfrac{5}{4}, \infty\right)$

51. (a)

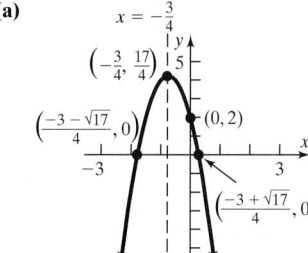

(b) Domain: $(-\infty, \infty)$
Range $\left(-\infty, \dfrac{17}{4}\right]$
(c) Decreasing on $\left(-\dfrac{3}{4}, \infty\right)$

Increasing on $\left(-\infty, -\dfrac{3}{4}\right)$

52. (a)

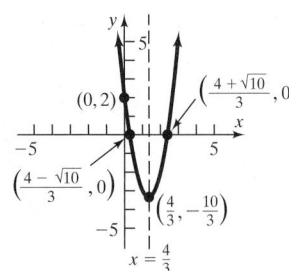

(b) Domain: $(-\infty, \infty)$

Range $\left[-\dfrac{10}{3}, \infty\right)$

(c) Decreasing on $\left(-\infty, \dfrac{4}{3}\right)$

Increasing on $\left(\dfrac{4}{3}, \infty\right)$

53. $f(x) = (x + 1)^2 - 2 = x^2 + 2x - 1$
54. $f(x) = (x - 2)^2 + 1 = x^2 - 4x + 5$
55. $f(x) = -(x + 3)^2 + 5 = -x^2 - 6x - 4$
56. $f(x) = -(x - 2)^2 + 3 = -x^2 + 4x - 1$
57. $f(x) = 2(x - 1)^2 - 3 = 2x^2 - 4x - 1$
58. $f(x) = -2(x + 2)^2 + 6 = -2x^2 - 8x - 2$
59. Minimum value; -18 **60.** Maximum value; 18
61. Minimum value; -21 **62.** Minimum value; -1
63. Maximum value; 21 **64.** Maximum value; 11
65. Maximum value; 13 **66.** Minimum value; -1
67. $a = 6, b = 0, c = 2$ **68.** $a = -3, b = 6, c = 1$

69. (a) (a), (c), (d)

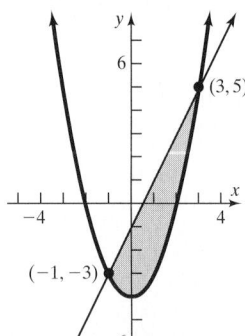

(b) $\{-1, 3\}$

70. (a) (a), (c), (d)

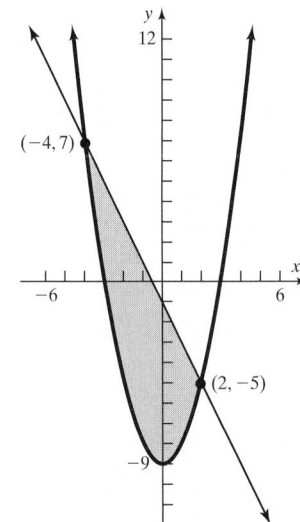

(b) $\{-4, 2\}$

71. (a) (a), (c), (d)

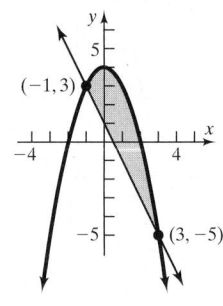

(b) $\{-1, 3\}$

72. (a) (a), (c), (d)

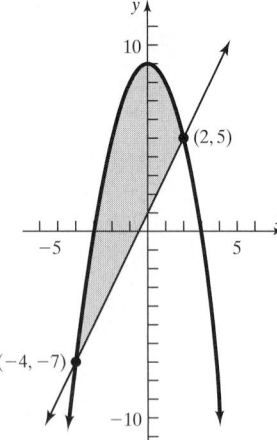

(b) $\{-4, 2\}$

73. (a) (a), (c), (d)

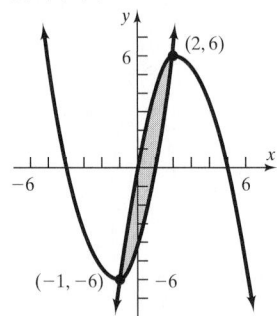

(b) $\{-1, 2\}$

74. (a) (a), (c), (d)

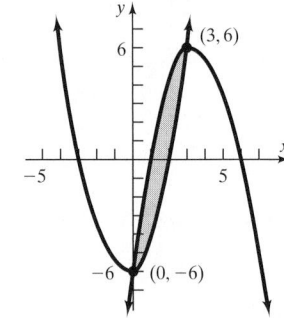

(b) $\{0, 3\}$

75. (a) $a = 1: f(x) = (x + 3)(x - 1) = x^2 + 2x - 3$
$a = 2: f(x) = 2(x + 3)(x - 1) = 2x^2 + 4x - 6$
$a = -2: f(x) = -2(x + 3)(x - 1) = -2x^2 - 4x + 6$
$a = 5: f(x) = 5(x + 3)(x - 1) = 5x^2 + 10x - 15$
(b) The value of a does not affect the x-intercepts, but it changes the y-intercept by a factor of a.

(c) The value of a does not affect the axis of symmetry. It is $x = -1$ for all value of a.
(d) The value of a does not affect the x-coordinate of the vertex. However, the y-coordinate of the vertex is multiplied by a.
(e) The midpoint of the x-intercepts is the x-coordinate of the vertex.

76. (a) $a = 1; f(x) = (x + 5)(x - 3) = x^2 + 2x - 15$
$a = 2; f(x) = 2(x + 5)(x - 3) = 2x^2 + 4x - 30$
$a = -2; f(x) = -2(x + 5)(x - 3) = -2x^2 - 4x + 30$
$a = 5; f(x) = 5(x + 5)(x - 3) = 5x^2 + 10x - 75$
(b) The value of a does not affect the x-intercepts, but it changes the y-intercept by a factor of a.

(c) The value of a does not affect the axis of symmetry. It is $x = -1$ for all values of a.
(d) The value of a does not affect the x-coordinate of the vertex. However, the y-coordinate of the vertex is multiplied by a.
(e) The midpoint of the x-intercepts is the x-coordinate of the vertex.

77. (a) $(-2, -25)$ **(b)** Zeros: $-7, 3$; x-intercepts: $-7, 3$
(c) $-4, 0$; $(-4, -21)$, $(0, -21)$
(d)

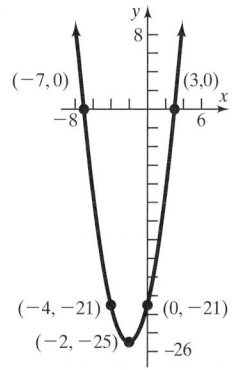

78. (a) $(-1, -9)$
(b) Zeros: $-4, 2$; x-intercepts: $-4, 2$
(c) $-2, 0$; $(-2, -8)$, $(0, -8)$
(d)

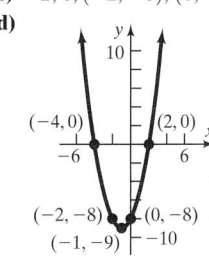

82. If $b^2 - 4ac < 0$ **84.** No

2.5 Assess Your Understanding (page 156)

3. (a) $\{x|x < -2 \text{ or } x > 2\}; (-\infty, -2) \text{ or } (2, \infty)$ **(b)** $\{x|-2 \le x \le 2\}; [-2, 2]$ **4. (a)** $\{x|x < -1 \text{ or } x > 4\}; (-\infty, -1) \text{ or } (4, \infty)$
(b) $\{x|-1 \le x \le 4\}; [-1, 4]$ **5. (a)** $\{x|-2 \le x \le 1\}; [-2, 1]$ **(b)** $\{x|x < -2 \text{ or } x > 1\}; (-\infty, -2) \text{ or } (1, \infty)$
6. (a) $\{x|x < -3 \text{ or } x > 1\}; (-\infty, -3) \text{ or } (1, \infty)$ **(b)** $\{x|-3 \le x \le 1\}; [-3, 1]$ **7.** $\{x|-2 < x < 5\}; (-2, 5)$
8. $\{x|x < -5 \text{ or } x > 2\}; (-\infty, -5) \text{ or } (2, \infty)$ **9.** $\{x|x < 0 \text{ or } x > 4\}; (-\infty, 0) \text{ or } (4, \infty)$ **10.** $\{x|x < -8 \text{ or } x > 0\}; (-\infty, -8) \text{ or } (0, \infty)$
11. $\{x|-3 < x < 3\}; (-3, 3)$ **12.** $\{x|-1 < x < 1\}; (-1, 1)$ **13.** $\{x|x < -4 \text{ or } x > 3\}; (-\infty, -4) \text{ or } (3, \infty)$ **14.** $\{x|-4 < x < -3\}; (-4, -3)$
15. $\left\{x\left|-\dfrac{1}{2} < x < 3\right.\right\}; \left(-\dfrac{1}{2}, 3\right)$ **16.** $\left\{x\left|-\dfrac{2}{3} < x < \dfrac{3}{2}\right.\right\}; \left(-\dfrac{2}{3}, \dfrac{3}{2}\right)$ **17.** $\{x|x < -1 \text{ or } x > 8\}; (-\infty, -1) \text{ or } (8, \infty)$
18. $\{x|x < -5 \text{ or } x > 4\}; (-\infty, -5) \text{ or } (4, \infty)$ **19.** No real solution **20.** No real solution **21.** $\left\{x\left|x < -\dfrac{2}{3} \text{ or } x > \dfrac{3}{2}\right.\right\}; \left(-\infty, -\dfrac{2}{3}\right) \text{ or } \left(\dfrac{3}{2}, \infty\right)$
22. All real numbers; $(-\infty, \infty)$ **23.** $\{x|x \le -4 \text{ or } x \ge 4\}; (-\infty, -4] \text{ or } [4, \infty)$ **24.** $\left\{x\left|0 \le x \le \dfrac{1}{3}\right.\right\}; \left[0, \dfrac{1}{3}\right]$ **25. (a)** $\{-1, 1\}$ **(b)** -1
(c) $\{-1, 4\}$ **(d)** $\{x|x < -1 \text{ or } x > 1\}; (-\infty, -1) \text{ or } (1, \infty)$ **(e)** $\{x|x \le -1\} \text{ or } (-\infty, -1]$ **(f)** $\{x|x < -1 \text{ or } x > 4\}; (-\infty, -1) \text{ or } (4, \infty)$
(g) $\left\{x|x < -\sqrt{2} \text{ or } x > \sqrt{2}\right\}; \left(-\infty, -\sqrt{2}\right) \text{ or } \left(\sqrt{2}, \infty\right)$ **26. (a)** $\{-1, 1\}$ **(b)** $\{1\}$ **(c)** $\{-4, 1\}$
(d) $\{x|x < -1 \text{ or } x > 1\}; (-\infty, -1) \text{ or } (1, \infty)$ **(e)** $\{x|x \ge 1\} \text{ or } [1, \infty)$ **(f)** $\{x|x < -4 \text{ or } x > 1\}; (-\infty, -4) \text{ or } (1, \infty)$
(g) $\left\{x|x < -\sqrt{2} \text{ or } x > \sqrt{2}\right\}; \left(-\infty, -\sqrt{2}\right) \text{ or } \left(\sqrt{2}, \infty\right)$ **27. (a)** $\{-1, 1\}$ **(b)** $-\dfrac{1}{4}$ **(c)** $\{-4, 0\}$ **(d)** $\{x|-1 < x < 1\}; (-1, 1)$
(e) $\left\{x\left|x \le -\dfrac{1}{4}\right.\right\} \text{ or } \left(-\infty, -\dfrac{1}{4}\right]$ **(f)** $\{x|-4 < x < 0\}; (-4, 0)$ **(g)** $\{0\}$ **28. (a)** $\{-2, 2\}$ **(b)** -2 **(c)** $\{-2, 3\}$
(d) $\{x|-2 < x < 2\}; (-2, 2)$ **(e)** $\{x|x \ge -2\}; [-2, \infty)$ **(f)** $\{x|-2 < x < 3\}; (-2, 3)$ **(g)** $\left\{x|-\sqrt{3} \le x \le \sqrt{3}\right\}; \left[-\sqrt{3}, \sqrt{3}\right]$
29. (a) $\{-2, 2\}$ **(b)** $\{-2, 2\}$ **(c)** $\{-2, 2\}$ **(d)** $\{x|x < -2 \text{ or } x > 2\}; (-\infty, -2) \text{ or } (2, \infty)$ **(e)** $\{x|x \le -2 \text{ or } x \ge 2\}; (-\infty, -2] \text{ or } [2, \infty)$
(f) $\{x|x < -2 \text{ or } x > 2\}; (-\infty, -2) \text{ or } (2, \infty)$ **(g)** $\left\{x|x \le -\sqrt{5} \text{ or } x \ge \sqrt{5}\right\}; \left(-\infty, -\sqrt{5}\right] \text{ or } \left[\sqrt{5}, \infty\right)$ **30. (a)** 1 **(b)** $\{-1, 1\}$
(c) $\{0, 1\}$ **(d)** $\{x|x < 1 \text{ or } x > 1\}; (-\infty, 1) \text{ or } (1, \infty)$ **(e)** $\{x|x \le -1 \text{ or } x \ge 1\}; (-\infty, -1] \text{ or } [1, \infty)$
(f) $\{x|x < 0 \text{ or } x > 1\}; (-\infty, 0) \text{ or } (1, \infty)$ **(g)** $\{x|x \le 0 \text{ or } x \ge 2\}; (-\infty, 0] \text{ or } [2, \infty)$ **31. (a)** $\{-1, 2\}$ **(b)** $\{-2, 1\}$ **(c)** 0
(d) $\{x|x < -1 \text{ or } x > 2\}; (-\infty, -1) \text{ or } (2, \infty)$ **(e)** $\{x|-2 \le x \le 1\}; [-2, 1]$ **(f)** $\{x|x < 0\}; (-\infty, 0)$
(g) $\left\{x\left|x < \dfrac{1 - \sqrt{13}}{2} \text{ or } x > \dfrac{1 + \sqrt{13}}{2}\right.\right\}; \left(-\infty, \dfrac{1 - \sqrt{13}}{2}\right) \text{ or } \left(\dfrac{1 + \sqrt{13}}{2}, \infty\right)$ **32. (a)** $\left\{\dfrac{-1 - \sqrt{5}}{2}, \dfrac{-1 + \sqrt{5}}{2}\right\}$ **(b)** $\{-2, 3\}$ **(c)** $\left\{-\dfrac{5}{2}\right\}$
(d) $\left\{\dfrac{-1 - \sqrt{5}}{2} < x < \dfrac{-1 + \sqrt{5}}{2}\right\}; \left(\dfrac{-1 - \sqrt{5}}{2}, \dfrac{-1 + \sqrt{5}}{2}\right)$ **(e)** $\{x|x \le -2 \text{ or } x \ge 3\}; (-\infty, -2] \text{ or } [3, \infty)$ **(f)** $\left\{x\left|x < -\dfrac{5}{2}\right.\right\} \text{ or } \left(-\infty, -\dfrac{5}{2}\right)$
(g) $\{x|-1 \le x \le 0\}; [-1, 0]$ **33. (a)** 5 sec **(b)** The ball is more than 96 feet above the ground from time t between 2 and 3 seconds,
$2 < t < 3$. **34. (a)** 6 sec **(b)** The ball is more than 128 feet above the ground from time t between 2 and 4 seconds, $2 < t < 4$.
35. (a) $\$0, \1000 **(b)** The revenue is more than $\$800,000$ for prices between $\$276.39$ and $\$723.61$, $\$276.39 < p < \723.61.
36. (a) $\$0, \3800 **(b)** The revenue is more than $\$1,200,000$ for prices between $\$800$ and $\$3000$, $\$800 < p < \3000.

2.6 Assess Your Understanding (page 162)

3. $\$500; \$1,000,000$ **4.** $\$1,900; \$1,805,000$ **5.** $40; \$400$ **6.** $20; \$2,000$
7. (a) $\dfrac{625}{16} \approx 39$ ft **(b)** $\dfrac{7025}{32} \approx 219.5$ ft **8. (a)** 156.25 ft **(b)** 78.125 ft **9. (a)** 37.4 years old; 3784
(c) 170 ft **(c)** 312.5 ft **(b)** Decreasing
(d) 220 **(d)** 80

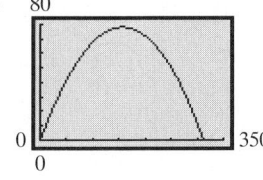

(f) When the height is 100 ft, the projectile
is about 135.7 ft from the cliff. **(f)** When the height is 50 ft, the projectile
has traveled horizontally 62.5 or 250 ft.

10. (a) ≈ 51 years; 10.8%
(b) Increasing
11. (a) 2149 **(b)** 31 years old
(c) The number of murder victims decreases.

12. (a) 4.6% **(b)** 67 years
(c) 20

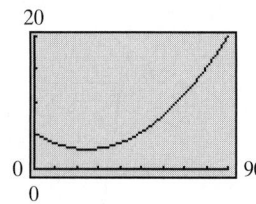

(d) Initially, it declines, but then it increases.

13. (a) 187 or 188; $7031.20 **(b)** $P(x) = -0.2x^2 + 43x - 1750$ **(c)** 107 or 108; $561.20 **14. (a)** 119; $564.06
(b) $P(x) = -0.04x^2 + 8.25x - 250$ **(c)** 103; $175.39 **15. (a)** $R(x) = -\frac{1}{6}x^2 + 100x$ **(b)** $13,333.33 **(c)** 300; $15,000 **(d)** $50

16. (a) $R(x) = -\frac{1}{3}x^2 + 100x$ **(b)** $6666.67 **(c)** 150; $7500 **(d)** $50 **17. (a)** $R(x) = -\frac{1}{5}x^2 + 20x$ **(b)** $255 **(c)** 50; $500 **(d)** $10

18. (a) $R(x) = -\frac{1}{20}x^2 + 25x$ **(b)** $480 **(c)** 250; $3125 **(d)** $12.50 **19. (a)** $A(w) = -w^2 + 200w$ **(b)** A is largest when $w = 100$ yd.
(c) 10,000 sq yd **20. (a)** $A(x) = -x^2 + 1500x$ **(b)** A is largest when $x = 750$ ft. **(c)** 562,500 ft² **21.** 2,000,000 m² **22.** 500,000 m²

23. 18.75 m **24.** Let the point $(0, 25)$ represent the vertex of the arch. Then the equation of the parabola is $h(x) = \dfrac{-25}{(60)^2}x^2 + 25$; $h(10) \approx 24.3$ ft;

$h(20) \approx 22.2$ ft; $h(40) \approx 13.9$ ft. **25.** 3 in. **26.** The width should be approximately 5.6 ft, and the total height of the window should be

approximately 5.6 ft. **27.** $\dfrac{750}{\pi}$ by 375 m **28.** $x = \dfrac{-16}{-6 + \sqrt{3}} \approx 3.75$ ft; total height $= 8 + \dfrac{24}{-6 + \sqrt{3}} - \dfrac{8\sqrt{3}}{-6 + \sqrt{3}} \approx 5.62$ ft **29.** $x = \dfrac{a}{2}$

30. $\left.\begin{array}{c} ah^2 - bh + c = y_0 \\ c = y_1 \\ ah^2 + bh + c = y_2 \end{array}\right\}$ $\left.\begin{array}{c} y_0 + y_2 = 2ah^2 + 2c \\ 4y_1 = 4c \end{array}\right\}$ Area $= \dfrac{h}{3}(2ah^2 + 6c) = \dfrac{h}{3}(y_0 + 4y_1 + y_2)$ **31.** $\dfrac{38}{3}$ **32.** $\dfrac{128}{3}$ **33.** $\dfrac{248}{3}$ **34.** $\dfrac{22}{3}$

35. 25 square units **36.** If x is even, then ax^2 and bx are even and $ax^2 + bx$ is even, which means that $ax^2 + bx + c$ is odd. If x is odd, then ax^2 and bx are odd and $ax^2 + bx$ is even, which means that $ax^2 + bx + c$ is odd. In either case, $f(x)$ is odd.

37. (a) Quadratic, $a < 0$

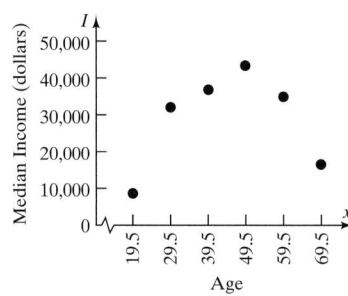

(b) 46.8 yr old
(c) $41,710
(e) 50,000

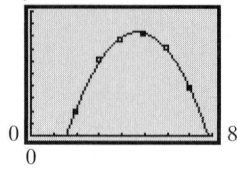

38. (a) Quadratic, $a < 0$

(b) 46.8 yr old
(c) $41,751
(e) 50,000

39. (a) Quadratic, $a < 0$

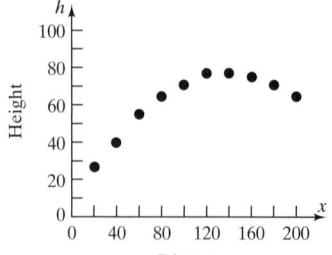

(b) 139.2 ft
(c) 77.4 ft
(e) 80

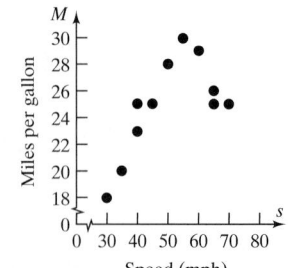

40. (a) Quadratic, $a < 0$

(b) 53.6 mi/hr
(c) 26.4 mi per gal
(e) 32

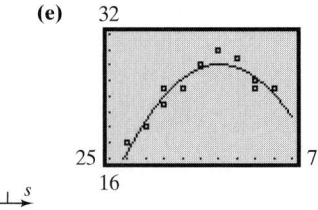

2.7 Assess Your Understanding (page 170)

7. F **8.** T

9. $-2i, 2i$

10. $-3, 3$

11. $-4, 4$

12. $-5i, 5i$
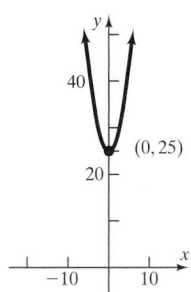

13. $3 - 2i, 3 + 2i$
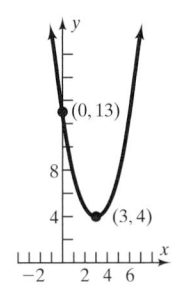

14. $-2 - 2i, -2 + 2i$
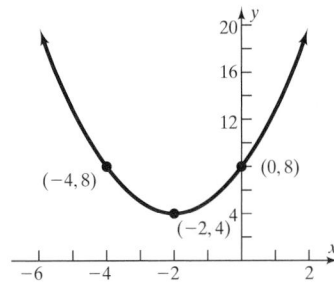

15. $3 - i, 3 + i$
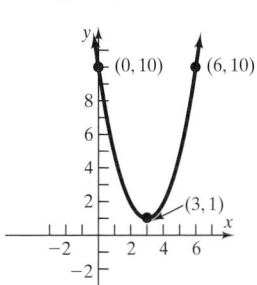

16. $1 - 2i, 1 + 2i$
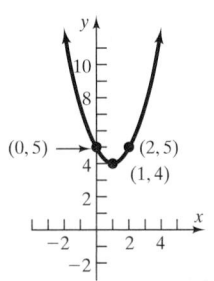

17. $2 - \sqrt{3}, 2 + \sqrt{3}$
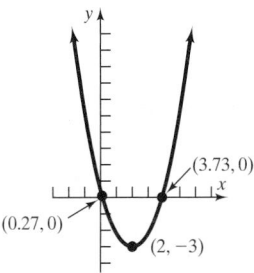

18. $-3 - 2\sqrt{2}, -3 + 2\sqrt{2}$
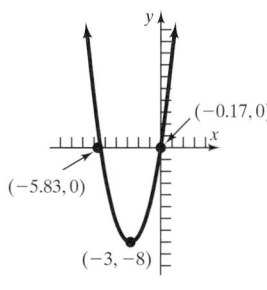

19. $-\dfrac{1}{2} - \dfrac{1}{2}i, -\dfrac{1}{2} + \dfrac{1}{2}i$
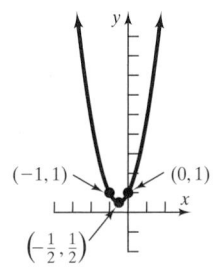

20. $-1 - \dfrac{\sqrt{3}}{3}i, -1 + \dfrac{\sqrt{3}}{3}i$
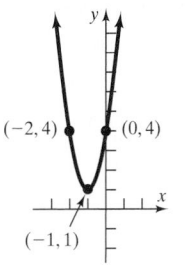

21. $-\dfrac{1}{2} - \dfrac{\sqrt{3}}{2}i, -\dfrac{1}{2} + \dfrac{\sqrt{3}}{2}i$
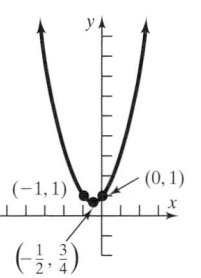

22. $\dfrac{1}{2} - \dfrac{\sqrt{3}}{2}i, \dfrac{1}{2} + \dfrac{\sqrt{3}}{2}i$
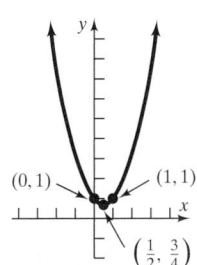

23. $\dfrac{4 - 3\sqrt{2}}{2}, \dfrac{4 + 3\sqrt{2}}{2}$
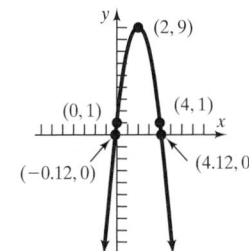

24. $\dfrac{3 - 2\sqrt{3}}{3}, \dfrac{3 + 2\sqrt{3}}{3}$
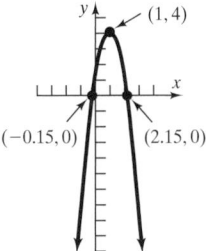

25. Two complex solutions that are conjugates of each other **26.** Two unequal real solutions **27.** Two unequal real solutions **28.** Two complex solutions that are conjugates of each other **29.** A repeated real solution **30.** A repeated real solution **31.** $2 - 3i$ **32.** $4 + i$

2.8 Assess Your Understanding (page 173)

7. $-a; a$ **8.** $-a < u < a$ **9.** $<$ **10.** T **11.** F **12.** F **13. (a)** $\{-9, 3\}$ **(b)** $\{x | -9 \le x \le 3\}; [-9, 3]$ **(c)** $\{x | x < -9 \text{ or } x > 3\}$ $(-\infty, -9)$ or $(3, \infty)$ **14. (a)** $\{0, 4\}$ **(b)** $\{x | 0 \le x \le 4\}; [0, 4]$ **(c)** $\{x | x < 0 \text{ or } x > 4\}; (-\infty, 0)$ or $(4, \infty)$ **15. (a)** $\{-2, 3\}$ **(b)** $\{x | x \le -2 \text{ or } x \ge 3\}; (-\infty, -2]$ or $[3, \infty)$ **(c)** $\{x | -2 < x < 3\}; (-2, 3)$ **16. (a)** $\{-4, 3\}$ **(b)** $\{x | x \le -4 \text{ or } x \ge 3\}; (-\infty, -4]$ or $[3, \infty)$ **(c)** $\{x | -4 < x < 3\}; (-4, 3)$ **17.** $\{-6, 6\}$ **18.** $\{-12, 12\}$ **19.** $\{-4, 1\}$ **20.** $\left\{-\dfrac{1}{3}, 1\right\}$ **21.** $\left\{-1, \dfrac{3}{2}\right\}$ **22.** $\{-1, 2\}$ **23.** $\{-4, 4\}$

24. $\{-1, 1\}$ **25.** $\left\{-\dfrac{1}{2}, \dfrac{1}{2}\right\}$ **26.** $\{-4, 4\}$ **27.** $\left\{-\dfrac{27}{2}, \dfrac{27}{2}\right\}$ **28.** $\{-12, 12\}$ **29.** $\left\{-\dfrac{36}{5}, \dfrac{24}{5}\right\}$ **30.** $\left\{-\dfrac{4}{3}, \dfrac{8}{3}\right\}$ **31.** No real solution
32. No real solution **33.** $\{-3, 3\}$ **34.** $\{-4, 4\}$ **35.** $\{-1, 3\}$ **36.** $\{-4, 3\}$ **37.** $\{-2, -1, 0, 1\}$ **38.** $\{-4, -3, 0, 1\}$

39. $\{x | -6 < x < 6\}; (-6, 6)$

40. $\{x | -9 < x < 9\}; (-9, 9)$

41. $\{x | x < -4 \text{ or } x > 4\}; (-\infty, -4)$ or $(4, \infty)$ **42.** $\{x | x < -1 \text{ or } x > 1\}; (-\infty, -1)$ or $(1, \infty)$

43. $\{x | -4 < x < 4\}; (-4, 4)$

44. $\{x | -5 < x < 5\}; (-5, 5)$

45. $\{x | x < -4 \text{ or } x > 4\}; (-\infty, -4)$ or $(4, \infty)$ **46.** $\{x | x < -3 \text{ or } x > 3\}; (-\infty, -3)$ or $(3, \infty)$

47. $\{x | 1 < x < 3\}; (1, 3)$

48. $\{x | -6 < x < -2\}; (-6, -2)$

49. $\left\{t \middle| -\dfrac{2}{3} \le t \le 2\right\}; \left[-\dfrac{2}{3}, 2\right]$

50. $\{u | -6 \le u \le 1\}; [-6, 1]$

51. $\{x | x \le 1 \text{ or } x \ge 5\}; (-\infty, 1]$ or $[5, \infty)$

52. $\{x | x \le -6 \text{ or } x \ge -2\}; (-\infty, -6]$ or $[-2, \infty)$

53. $\left\{x \middle| -1 < x < \dfrac{3}{2}\right\}; \left(-1, \dfrac{3}{2}\right)$

54. $\{x | -1 < x < 2\}; (-1, 2)$

55. $\{x | x < -1 \text{ or } x > 2\}; (-\infty, -1)$ or $(2, \infty)$ **56.** $\left\{x \middle| x < \dfrac{1}{3} \text{ or } x > 1\right\}; \left(-\infty, \dfrac{1}{3}\right)$ or $(1, \infty)$ **57.** No real solution; \varnothing

58. All real numbers: $(-\infty, \infty)$

59. $|x - 10| < 2; \{x | 8 < x < 12\}$ **60.** $|x + 6| < 3; \{x | -9 < x < -3\}$ **61.** $|2x + 1| > 5; \{x | x < -3 \text{ or } x > 2\}$
62. $|2x - 3| > 1; \{x | x < 1 \text{ or } x > 2\}$ **63.** Between 5.6995 in. and 5.7005 in. **64.** Between 6.1245 in. and 6.1255 in. **65.** IQs less than 71 or greater than 129 **66.** Pregnancies less than 235 days or more than 297 days

Review Exercises (page 176)

1. (a)

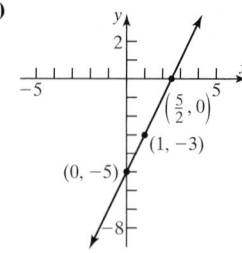

(b) $m = 2; b = -5$
(c) Increasing

2. (a)

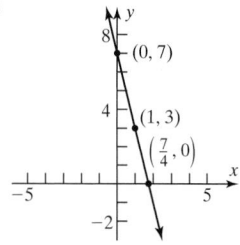

(b) $m = -4; b = 7$
(c) Decreasing

3. (a)

(b) $m = \dfrac{4}{5}; b = -6$
(c) Increasing

4. (a)

(b) $m = -\dfrac{1}{3}; b = 1$
(c) Decreasing

5. (a)

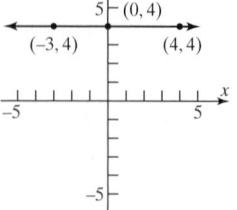

(b) $m = 0, b = 4$
(c) Constant

6. (a)

(b) $m = 0, b = -3$
(c) Constant

7. Zero: -7; y-intercept: 14

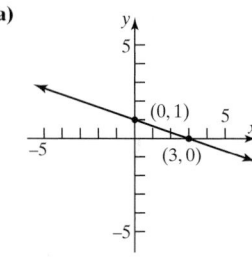

8. Zero: 8; y-intercept: 2

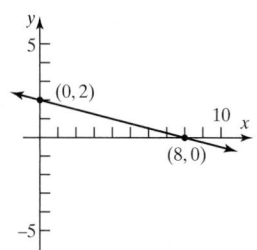

9. Linear; $m = 5$ **10.** Non linear **11.** Zeros: $-9, 8$; x-intercepts: $-9, 8$ **12.** Zeros: $-5, 8$; x-intercepts: $-5, 8$ **13.** Zeros: $-\dfrac{1}{3}, \dfrac{5}{2}$; x-intercepts: $-\dfrac{1}{3}, \dfrac{5}{2}$

14. Zeros: $\dfrac{5}{3}, 4$; x-intercepts: $\dfrac{5}{3}, 4$ **15.** Zeros: $1, 5$; x-intercepts: $1, 5$ **16.** Zeros: $-8, -2$; x-intercepts: $-8, -2$ **17.** Zero: $-\dfrac{1}{3}$; x-intercept: $-\dfrac{1}{3}$

18. Zero: $\dfrac{1}{2}$; x-intercept: $\dfrac{1}{2}$ **19.** Zeros: $\dfrac{2 - \sqrt{6}}{2}, \dfrac{2 + \sqrt{6}}{2}$; x-intercepts: $\dfrac{2 - \sqrt{6}}{2}, \dfrac{2 + \sqrt{6}}{2}$ **20.** Zeros: $\dfrac{-5 - \sqrt{13}}{6}, \dfrac{-5 + \sqrt{13}}{6}$; x-intercepts:

$\dfrac{-5 - \sqrt{13}}{6}, \dfrac{-5 + \sqrt{13}}{6}$ **21.** Zeros: $-\dfrac{1}{2}, 1$; x-intercepts: $-\dfrac{1}{2}, 1$ **22.** Zeros: $-1 - \sqrt{6}, -1 + \sqrt{6}$; x-intercepts: $-1 - \sqrt{6}, -1 + \sqrt{6}$

23. $\{-1, 7\}$ **24.** $\{-8, 2\}$ **25.** $\{-2, 2\}$ **26.** $\{-2, 2\}$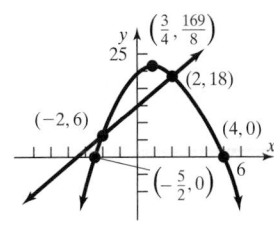

27. Zeros: $-2, -1, 1, 2$; x-intercepts: $-2, -1, 1, 2$ **28.** No real zeros; no x-intercepts **29.** Zeros: $-3, 11$; x-intercepts: $-3, 11$ **30.** Zeros: $-\dfrac{15}{2}, -2$;

x-intercepts: $-\dfrac{15}{2}, -2$ **31.** Zero: 25; x-intercept: 25 **32.** Zeros: $-\dfrac{1}{2}, \dfrac{1}{6}$; x-intercepts: $-\dfrac{1}{2}, \dfrac{1}{6}$

33. **34.** **35.** **36.**

37. **38.** **39. (a)** **40. (a)**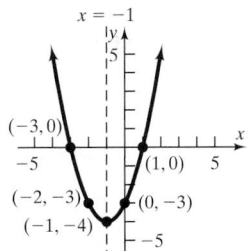

39. (b) Domain: $(-\infty, \infty)$
Range: $[2, \infty)$
(c) Decreasing: $(-\infty, 2)$
Increasing: $(2, \infty)$

40. (b) Domain: $(-\infty, \infty)$
Range: $[-4, \infty)$
(c) Decreasing: $(-\infty, -1)$
Increasing: $(-1, \infty)$

41. (a) **42. (a)** **43. (a)** **44. (a)**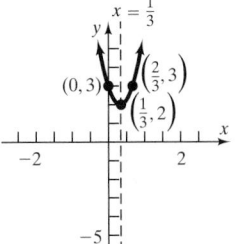

41. (b) Domain: $(-\infty, \infty)$
Range: $[-16, \infty)$

(c) Decreasing: $(-\infty, 0)$

Increasing: $(0, \infty)$

42. (b) Domain: $(-\infty, \infty)$
Range: $(-\infty, 2]$

(c) Increasing: $(-\infty, 0)$

Decreasing: $(0, \infty)$

43. (b) Domain: $(-\infty, \infty)$
Range: $(-\infty, 1]$

(c) Increasing: $\left(-\infty, \dfrac{1}{2}\right)$

Decreasing: $\left(\dfrac{1}{2}, \infty\right)$

44. (b) Domain: $(-\infty, \infty)$
Range: $[2, \infty)$

(c) Decreasing: $\left(-\infty, \dfrac{1}{3}\right)$

Increasing: $\left(\dfrac{1}{3}, \infty\right)$

45. (a)

46. (a)

47. (a)

48. (a)

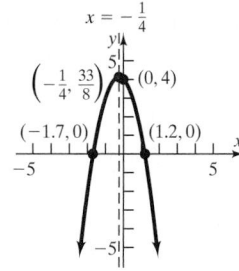

(b) Domain: $(-\infty, \infty)$

Range: $\left[\dfrac{1}{2}, \infty\right)$

(c) Decreasing: $\left(-\infty, -\dfrac{1}{3}\right)$

Increasing: $\left(-\dfrac{1}{3}, \infty\right)$

(b) Domain: $(-\infty, \infty)$

Range: $\left(-\infty, \dfrac{3}{4}\right]$

(c) Increasing: $\left(-\infty, \dfrac{1}{2}\right)$

Decreasing: $\left(\dfrac{1}{2}, \infty\right)$

(b) Domain: $(-\infty, \infty)$

Range: $\left[-\dfrac{7}{3}, \infty\right)$

(c) Decreasing: $\left(-\infty, -\dfrac{2}{3}\right)$

Increasing: $\left(-\dfrac{2}{3}, \infty\right)$

(b) Domain: $(-\infty, \infty)$

Range: $\left(-\infty, \dfrac{33}{8}\right]$

(c) Increasing: $\left(-\infty, -\dfrac{1}{4}\right)$

Decreasing: $\left(-\dfrac{1}{4}, \infty\right)$

49. Minimum value; 1 **50.** Minimum value; -3 **51.** Maximum value; 12 **52.** Maximum value; 22 **53.** Maximum value; 16 **54.** Maximum value; 4

55. $\{x|-8 < x < 2\}$; $(-8, -2)$ **56.** $\left\{x\middle|x \le -\dfrac{1}{3} \text{ or } x \ge 1\right\}$; $\left(-\infty, -\dfrac{1}{3}\right]$ or $[1, \infty)$ **57.** $\left\{x\middle|x \le -\dfrac{1}{3} \text{ or } x \ge 5\right\}$; $\left(-\infty, -\dfrac{1}{3}\right]$ or $[5, \infty)$

58. $\left\{x\middle|\dfrac{1}{4} < x < 3\right\}$; $\left(\dfrac{1}{4}, 3\right)$

59. $-2\sqrt{2}i, 2\sqrt{2}i$

60. $-2\sqrt{2}, 2\sqrt{2}$

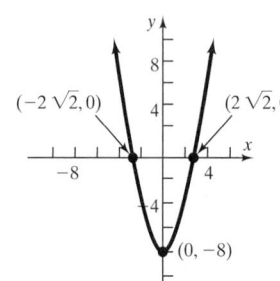

61. $-1 - \sqrt{5}, -1 + \sqrt{5}$

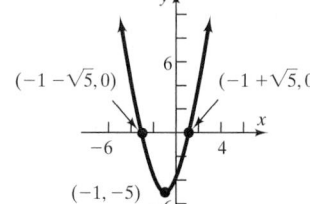

62. $-2 - i, -2 + i$

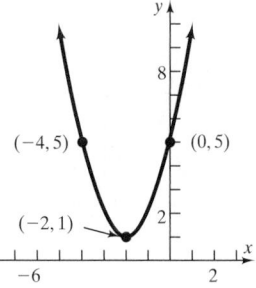

63. $1 - \dfrac{\sqrt{2}}{2}i, 1 + \dfrac{\sqrt{2}}{2}i$

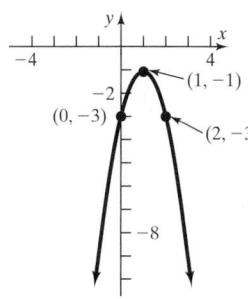

64. $1 - \dfrac{\sqrt{6}}{3}i, 1 + \dfrac{\sqrt{6}}{3}i$

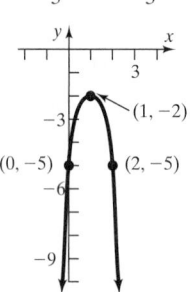

65. $-\dfrac{1}{2} - \dfrac{\sqrt{2}}{2}i, -\dfrac{1}{2} + \dfrac{\sqrt{2}}{2}i$

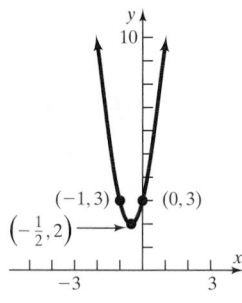

66. $-\dfrac{1}{2} - \dfrac{\sqrt{15}}{10}i, -\dfrac{1}{2} + \dfrac{\sqrt{15}}{10}i$

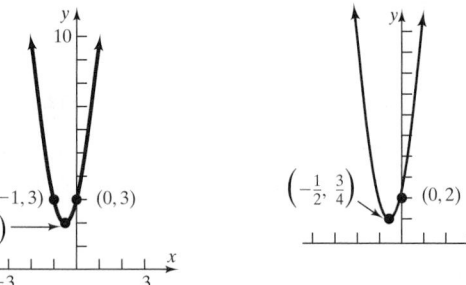

67. $\{-5, 2\}$ **68.** $\left\{-\dfrac{4}{3}, 2\right\}$ **69.** $\left\{-\dfrac{5}{3}, 3\right\}$ **70.** $\{-1, 2\}$

71. $\left\{x\middle|-\dfrac{3}{2} < x < -\dfrac{7}{6}\right\}$; $\left(-\dfrac{3}{2}, -\dfrac{7}{6}\right)$

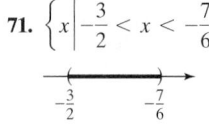

72. $\left\{x\middle|\dfrac{1}{3} < x < \dfrac{2}{3}\right\}$; $\left(\dfrac{1}{3}, \dfrac{2}{3}\right)$

73. $\{x|x \le -2 \text{ or } x \ge 7\}$;

$(-\infty, -2]$ or $[7, \infty)$

74. $\left\{x\middle|x \le -\dfrac{11}{3} \text{ or } x \ge 3\right\}$;

$\left(-\infty, -\dfrac{11}{3}\right]$ or $[3, \infty)$

75. $\left\{x\middle|0 \le x \le \dfrac{4}{3}\right\}$; $\left[0, \dfrac{4}{3}\right]$

76. $\left\{x\middle|-\dfrac{1}{4} \le x \le \dfrac{5}{4}\right\}$; $\left[-\dfrac{1}{4}, \dfrac{5}{4}\right]$

77. $\left\{x\middle|x < -1 \text{ or } x > \dfrac{7}{3}\right\}$;

$(-\infty, -1)$ or $\left(\dfrac{7}{3}, \infty\right)$

78. $\{x|x < -4 \text{ or } x > 5\}$;

$(-\infty, -4)$ or $(5, \infty)$

79. (a) Company A: $C(x) = 0.06x + 7$; Company B: $C(x) = 0.08x$ **(b)** 350 min **(c)** $0 \le x < 350$ **80. (a)** $S(x) = 0.01x + 15{,}000$ **(b)** $25{,}000
(c) $8,500,000 **(d)** $13,500,000 **81.** $p(B) = 0.00657B$; $1083.92 **82.** $R(g) = 2.38g$; $26.66 **83.** 36 senior citizens went on the trip. Each paid
$13.40. **84.** The length of the plasterboard should be cut down to about 6.47 ft. **85.** The legs are 5 cm and 12 cm. **86.** The width is 6 in. and the
length is 8 in. **87. (a)** 63 clubs **(b)** $151.90 **88. (a)** 1992 **(b)** 1901 thousand violent crimes **89.** 50 by 50 ft **90.** 2 by 8 ft **91.** 25 sq units
92. 3.6 ft **93. (a)** **94. (a)** Quadratic, $a < 0$. **(c)** $6408 thousand

(d) 6500

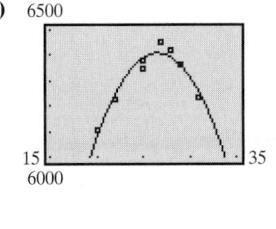

(b) Yes; the two variables appear to
be linearly related.
(c) $y = 1.3902x + 1.114$ **(d)** 38.0 mm

(b) About $26.5 thousand

Chapter Test *(page 179)*

1. (a) $\dfrac{3}{4}$

(b) Decreasing

(c)

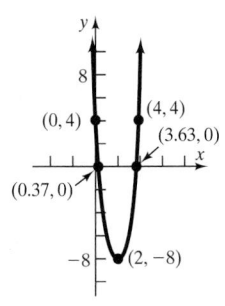

2. $-\dfrac{4}{3}, 2$

3. $\dfrac{2 - \sqrt{6}}{2}, \dfrac{2 + \sqrt{6}}{2}$

4. $\{-1, 3\}$

5. $-3, 0$

6.

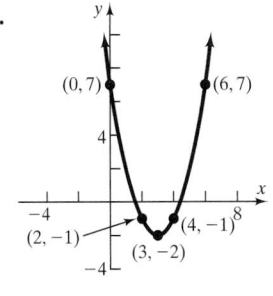

7. (a) Opens up
(b) $(2, -8)$

(c) $x = 2$

(d) x-intercepts: $\dfrac{2(3 - \sqrt{6})}{3}, \dfrac{2(3 + \sqrt{6})}{3}$;
y-intercept: 4

(e)

8. Maximum value; 21
9. $\{x \mid x \le 4 \text{ or } x \ge 6\}$; $(-\infty, 4]$ or $[6, \infty)$

10. $-1 - \dfrac{\sqrt{6}}{2}i, -1 + \dfrac{\sqrt{6}}{2}i$

11. $\left\{-3, \dfrac{7}{3}\right\}$
12. $\{x \mid -11 < x < 5\}$; $(-11, 5)$

13. $\{x \mid x \le -5 \text{ or } x \ge 2\}$;
$(-\infty, -5]$ or $[2, \infty)$

14. (a) $C(m) = 0.15m + 129.50$
(b) $258.50 **(c)** 562 miles
15. (a) 1997, about 8.67%
(b) about -10.347; this doesn't
make sense.

Cumulative Review Exercises *(page 180)*

1. $5\sqrt{2}$; $(1.5, 0.5)$ **2.** $(-2, -1)$ and $(2, 3)$ are on the graph.

3. $\left\{x \mid x \ge -\dfrac{3}{5}\right\}$ or $\left[-\dfrac{3}{5}, \infty\right)$ **4.** $y = -2x + 2$ **5.** $y = -\dfrac{1}{2}x + \dfrac{13}{2}$ **6.** $(x - 2)^2 + (y + 4)^2 = 25$

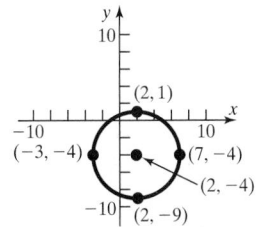

7. Yes, a function **8. (a)** -3 **(b)** $x^2 - 4x - 2$ **(c)** $x^2 + 4x + 1$ **(d)** $-x^2 + 4x - 1$ **(e)** $x^2 - 3$ **(f)** $2x + h - 4$ **9.** $\left\{ z \middle| z \neq \dfrac{7}{6} \right\}$

10. Yes, a function **11. (a)** No **(b)** $-1; (-2, -1)$ is on the graph. **(c)** $-8; (-8, 2)$ is on the graph. **12.** Neither

13. Local maximum is 5.30 and occurs at $x = -1.29$. **14. (a)** -4 **(b)** $\{x \mid x > -4\}$ or $(-4, \infty)$
 Local minimum is -3.30 and occurs at $x = 1.29$.
 Increasing: $(-4, -1.29)$ or $(1.29, 4)$
 Decreasing: $(-1.29, 1.29)$

15. (a) Domain: $\{x \mid -4 \leq x \leq 4\}$; Range: $\{y \mid -1 \leq y \leq 3\}$ **(b)** $(-1, 0), (0, -1), (1, 0)$ **(c)** y-axis **(d)** 1 **(e)** -4 and 4 **(f)** $\{x \mid -1 < x < 1\}$

(g) **(h)** **(i)**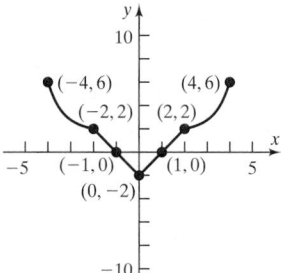

(j) Even **(k)** $(0, 4)$

CHAPTER 3 Polynomial and Rational Functions

3.1 Assess Your Understanding (page 199)

5. smooth; continuous **6.** zero or root **7.** touches **8.** True **9.** False **10.** False **11.** Yes; degree 3 **12.** Yes; degree 4 **13.** Yes; degree 2

14. Yes; degree 1 **15.** No; x is raised to the -1 power. **16.** Yes; degree 2 **17.** No; x is raised to the $\dfrac{3}{2}$ power. **18.** No; x is raised to the $\dfrac{1}{2}$ power.

19. Yes; degree 4 **20.** No; it is the ratio of two polynomials, and the polynomial in the denominator is of positive degree. **21.** Yes; degree 4

22. Yes; degree 5

23. **24.** **25.** **26.**

27. **28.** **29.** **30.**

31. **32.** **33.** **34.**

35. **36.**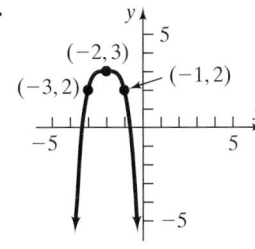

37. $f(x) = x^3 - 3x^2 - x + 3$ for $a = 1$
38. $f(x) = x^3 - 3x^2 - 4x + 12$ for $a = 1$
39. $f(x) = x^3 - x^2 - 12x$ for $a = 1$
40. $f(x) = x^3 + 2x^2 - 8x$ for $a = 1$
41. $f(x) = x^4 - 15x^2 + 10x + 24$ for $a = 1$
42. $f(x) = x^4 - 3x^3 - 15x^2 + 19x + 30$ for $a = 1$
43. $f(x) = x^3 - 5x^2 + 3x + 9$ for $a = 1$
44. $f(x) = x^3 - 12x - 16$ for $a = 1$

45. (a) 7, multiplicity 1; -3, multiplicity 2 **(b)** Graph touches the x-axis at -3 and crosses it at 7. **(c)** Near -3: $f(x) \approx -30(x + 3)^2$; Near 7: $f(x) \approx 300(x - 7)$ **(d)** 2 **(e)** $y = 3x^3$ **46. (a)** -4, multiplicity 1; -3, multiplicity 3 **(b)** Graph crosses the x-axis at -4 and -3. **(c)** Near -4: $f(x) \approx -4(x + 4)$; Near -3: $f(x) \approx 4(x + 3)^2$ **(d)** 3 **(e)** $y = 4x^4$ **47. (a)** 2, multiplicity 3 **(b)** Graph crosses the x-axis at 2 **(c)** Near 2: $f(x) \approx 20(x - 2)^3$ **(d)** 4 **(e)** $y = 4x^5$ **48. (a)** 3, multiplicity 1; -4, multiplicity 3 **(b)** Graph crosses the x-axis at -4 and 3. **(c)** Near -4: $f(x) \approx -14(x + 4)^3$; Near 3: $f(x) \approx 686(x - 3)$ **(d)** 3 **(e)** $y = 2x^4$ **49. (a)** $-\dfrac{1}{2}$, multiplicity 2 **(b)** Graph touches the x-axis at $-\dfrac{1}{2}$. **(c)** Near $-\dfrac{1}{2}$: $f(x) \approx -36.125\left(x + \dfrac{1}{2}\right)^2$ **(d)** 5 **(e)** $y = -2x^6$. **50. (a)** $\dfrac{1}{3}$, multiplicity 2; 1, multiplicity 3 **(b)** Graph touches the x-axis at $\dfrac{1}{3}$ and crosses it at 1. **(c)** Near $\dfrac{1}{3}$: $f(x) \approx -\dfrac{8}{27}\left(x - \dfrac{1}{3}\right)^2$; Near 1: $f(x) \approx \dfrac{4}{9}(x - 1)^3$ **(d)** 4 **(e)** $y = x^5$ **51. (a)** 5, multiplicity 3; -4, multiplicity 2 **(b)** Graph touches the x-axis at -4 and crosses it at 5. **(c)** Near -4: $f(x) \approx -729(x + 4)^2$; Near 5: $f(x) \approx 81(x - 5)^3$ **(d)** 4 **(e)** $y = x^5$ **52. (a)** $-\sqrt{3}$, multiplicity 2; 2, multiplicity 4 **(b)** Graph touches the x-axis at $-\sqrt{3}$ and 2. **(c)** Near $-\sqrt{3}$: $f(x) \approx 194\left(x + \sqrt{3}\right)^2$; Near 2: $f(x) \approx 13.93\left(x - 2\right)^4$ **(d)** 5 **(e)** $y = x^6$ **53. (a)** No real zeros **(b)** Graph neither crosses nor touches the x-axis. **(c)** No real zeros **(d)** 5 **(e)** $y = 3x^6$ **54. (a)** No real zeros **(b)** Graph neither crosses nor touches the x-axis. **(c)** No real zeros **(d)** 5 **(e)** $y = -2x^6$ **55. (a)** 0, multiplicity 2; $-\sqrt{2}$, $\sqrt{2}$, multiplicity 1 **(b)** Graph touches the x-axis at 0 and crosses at $-\sqrt{2}$ and $\sqrt{2}$. **(c)** Near $-\sqrt{2}$: $f(x) \approx 11.31\left(x + \sqrt{2}\right)$; Near 0: $f(x) \approx 4x^2$; Near $\sqrt{2}$: $f(x) \approx -11.31\left(x - \sqrt{2}\right)$ **(d)** 3 **(e)** $y = -2x^4$

56. (a) 0, multiplicity 1; $-\sqrt{3}$, $\sqrt{3}$, multiplicity 1 **(b)** Graph crosses the x-axis at $-\sqrt{3}$, 0, and $\sqrt{3}$. **(c)** Near $-\sqrt{3}$: $f(x) \approx 24\left(x + \sqrt{3}\right)$; Near 0: $f(x) \approx -12x$; Near $\sqrt{3}$: $f(x) \approx 24\left(x - \sqrt{3}\right)$ **(d)** 2 **(e)** $y = 4x^3$ **57.** Could be; zeros: -1, 1, 2; Least degree is 3 **58.** Could be; zeros: -1, 2; Least degree is 4 **59.** Cannot be the graph of a polynomial; gap at $x = -1$ **60.** Can't be; cusp at $x = 0$ **61.** c, e, f **62.** c, e, f **63.** c, e **64.** d, f

65. (a) x-intercept: 1; y-intercept: 1
(b) Touches at 1
(c) $y = x^2$
(d) 1
(e) Near 1: $f(x) = (x - 1)^2$
(f)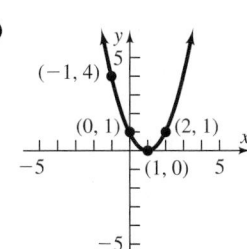

66. (a) x-intercept: 2; y-intercept: -8
(b) Crosses at 2
(c) $y = x^3$
(d) 2
(e) Near 2: $f(x) = (x - 2)^3$
(f)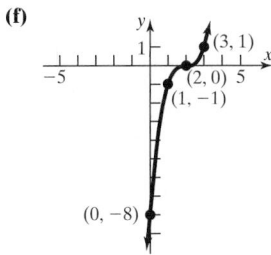

67. (a) x-intercepts: 0, 3; y-intercept: 0
(b) Touches at 0; crosses at 3
(c) $y = x^3$
(d) 2
(e) Near 0: $f(x) \approx -3x^2$; Near 3: $f(x) \approx 9(x - 3)$
(f)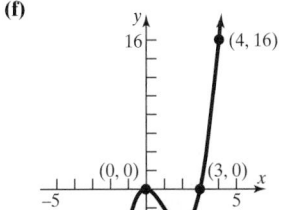

68. (a) x-intercepts: -2, 0; y-intercept: 0
(b) Touches at -2; crosses at 0
(c) $y = x^3$
(d) 2
(e) Near -2: $f(x) \approx -2(x + 2)^2$; Near 0: $f(x) \approx 4x$
(f)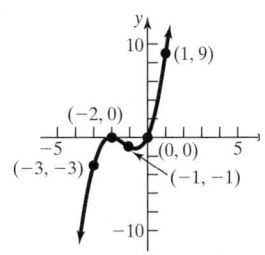

69. (a) x-intercepts: -4, 0; y-intercept: 0
(b) Crosses at -4, 0
(c) $y = 6x^4$
(d) 3
(e) Near -4: $f(x) \approx -384(x + 4)$; Near 0: $f(x) \approx 24x^3$.
(f)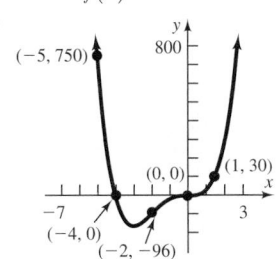

70. (a) x-intercepts: 0, 1; y-intercept: 0
(b) Crosses at 0 and 1
(c) $y = 5x^4$
(d) 3
(e) Near 0: $f(x) \approx -5x$; Near 1: $f(x) \approx 5(x - 1)^3$.
(f)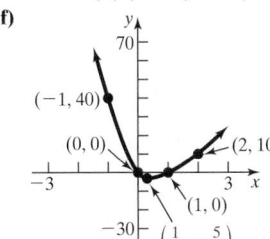

71. (a) x-intercepts: $-2, 0$; y-intercept: 0
(b) Crosses at -2; touches at 0
(c) $y = -4x^3$
(d) 2
(e) Near -2: $f(x) \approx -16(x + 2)$;
Near 0: $f(x) \approx -8x^2$

(f)
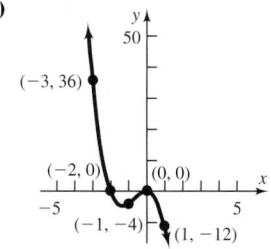

72. (a) x-intercepts: $-4, 0$; y-intercept: 0
(b) Crosses at -4 and 0
(c) $y = -\frac{1}{2}x^4$
(d) 3
(e) Near -4: $f(x) \approx 32(x + 4)$;
Near 0: $f(x) \approx -2x^3$

(f)
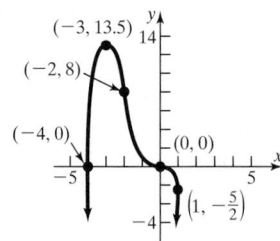

73. (a) x-intercepts: $-4, 1, 2$; y-intercept: 8
(b) Crosses at $-4, 1, 2$
(c) $y = x^3$
(d) 2
(e) Near -4: $f(x) \approx 30(x + 4)$;
Near 1: $f(x) \approx -5(x - 1)$;
Near 2: $f(x) \approx 6(x - 2)$

(f)
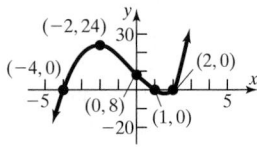

74. (a) x-intercepts: $-4, -1, 3$; y-intercept: -12
(b) Crosses at $-4, -1$, and 3
(c) $y = x^3$
(d) 2
(e) Near -4: $f(x) \approx 21(x + 4)$;
Near -1: $f(x) \approx -12(x + 1)$;
Near 3: $f(x) \approx 28(x - 3)$

(f)
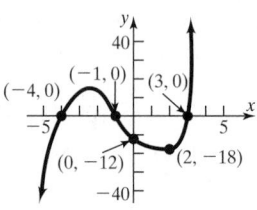

75. $f(x) = 4x - x^3 = -x(x^2 - 4)$
$= -x(x + 2)(x - 2)$
(a) x-intercepts: $-2, 0, 2$; y-intercept: 0
(b) Crosses at $-2, 0$, and 2 **(c)** $y = -x^3$
(d) 2
(e) Near -2: $f(x) \approx -8(x + 2)$; Near 0:
$f(x) \approx 4x$; Near 2: $f(x) \approx -8(x - 2)$

(f)
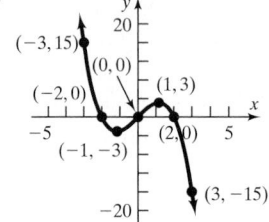

76. $f(x) = x - x^3 = -x(x^2 - 1)$
$= -x(x - 1)(x + 1)$
(a) x-intercepts: $-1, 0, 1$; y-intercept: 0
(b) Crosses at $-1, 0$, and 1 **(c)** $y = -x^3$
(d) 2
(e) Near -1; $f(x) \approx -2(x + 1)$; Near 0:
$f(x) \approx x$; Near 1: $f(x) \approx -2(x - 1)$

(f)
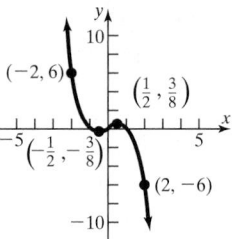

77. (a) x-intercepts: $-2, 0, 2$; y-intercept: 0
(b) Crosses at $-2, 2$; touches at 0
(c) $y = x^4$ **(d)** 3
(e) Near -2: $f(x) \approx -16(x + 2)$;
Near 0: $f(x) \approx -4x^2$;
Near 2: $f(x) \approx 16(x - 2)$

(f)
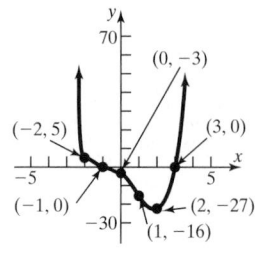

78. (a) x-intercepts: $-4, 0, 3$; y-intercept: 0
(b) Crosses at -4 and 3; touches at 0
(c) $y = x^4$ **(d)** 3
(e) Near -4: $f(x) \approx -112(x + 4)$;
Near 0: $f(x) \approx -12x^2$;
Near 3: $f(x) \approx 63(x - 3)$

(f)
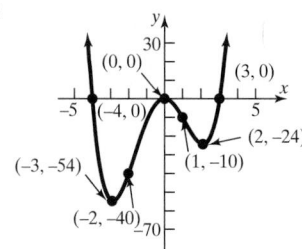

79. (a) x-intercepts: $-2, 2$; y-intercept: 16
(b) Touches at $-2, 2$
(c) $y = x^4$
(d) 3
(e) Near -2: $f(x) \approx 16(x + 2)^2$;
Near 2: $f(x) \approx 16(x - 2)^2$

(f)
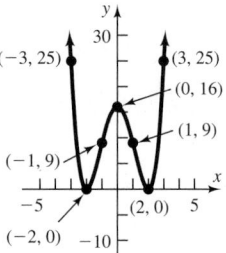

80. (a) x-intercepts: $-1, 3$; y-intercept: -3
(b) Crosses at -1 and 3
(c) $y = x^4$
(d) 3
(e) Near -1: $f(x) \approx -4(x + 1)^3$;
Near 3: $f(x) \approx 64(x - 3)$

(f)

81. (a) x-intercepts: $-1, 1, 3$; y-intercept: -3
(b) Crosses at $-1, 3$; touches at 1
(c) $y = x^4$ **(d)** 3
(e) Near -1: $f(x) \approx -16(x + 1)$;
Near 1: $f(x) \approx -4(x - 1)^2$;
Near 3: $f(x) \approx 16(x - 3)$

(f)
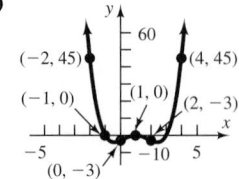

82. (a) x-intercepts: $-1, 1$, and 3; y-intercept: 3
(b) Touches at -1; crosses at 1 and 3
(c) $y = x^4$ **(d)** 3
(e) Near -1: $f(x) \approx 8(x + 1)^2$;
Near 1: $f(x) \approx -8(x - 1)$;
Near 3: $f(x) \approx 32(x - 3)$

(f)
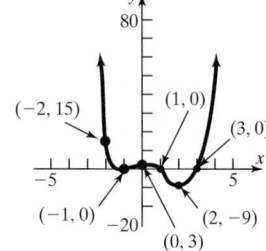

83. (a) x-intercepts: −2, 4; y-intercept: 64
(b) Touches at −2 and 4
(c) $y = x^4$
(d) 3
(e) Near −2: $f(x) \approx 36(x + 2)^2$;
Near 4: $f(x) \approx 36(x - 4)^2$
(f)

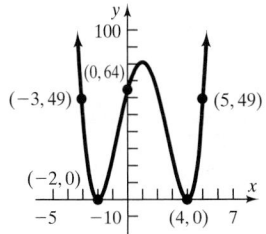

84. (a) x-intercepts: −4, −2, 2; y-intercept: 32
(b) Crosses at −4, −2; touches at 2
(c) $y = x^4$ **(d)** 3
(e) Near −4: $f(x) \approx -72(x + 4)$;
Near −2: $f(x) \approx 32(x + 2)$;
Near 2: $f(x) \approx 24(x + 2)^2$
(f)

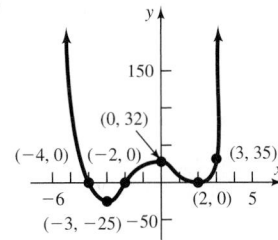

85. (a) x-intercepts: 0, 2; y-intercept: 0
(b) Touches at 0; crosses at 2
(c) $y = x^5$
(d) 4
(e) Near 0: $f(x) \approx -6x^2$;
Near 2: $f(x) \approx 28(x - 2)$
(f)

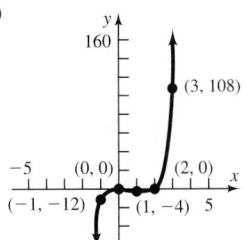

86. (a) x-intercepts: −4, 0; y-intercept: 0
(b) Crosses at −4; touches at 0
(c) $y = x^5$
(d) 4
(e) Near −4: $f(x) \approx 272(x + 4)$;
Near 0: $f(x) \approx 4x^2$
(f)

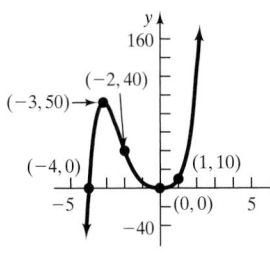

87. (a) x-intercepts: −1, 0, 1; y-intercept: 0
(b) Crosses at 1; touches at −1 and 0
(c) $y = -x^5$ **(d)** 4
(e) Near −1: $f(x) \approx 2(x + 1)^2$;
Near 0: $f(x) \approx x^2$;
Near 1: $f(x) \approx -4(x - 1)$
(f)

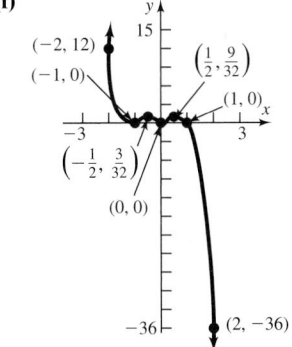

88. (a) x-intercepts: −2, 0, 2, 5; y-intercept: 0
(b) Touches at 0; crosses at −2, 2, and 5
(c) $y = -x^5$ **(d)** 4
(e) Near −2: $f(x) \approx -112(x + 2)$;
Near 0: $f(x) \approx -20x^2$;
Near 2: $f(x) \approx 48(x - 2)$;
Near 5: $f(x) \approx -525(x - 5)$
(f)

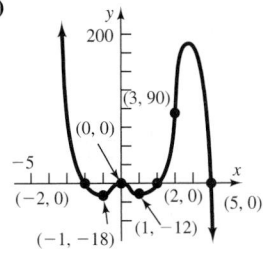

89. (a) Degree 3; $y = x^3$
(b)

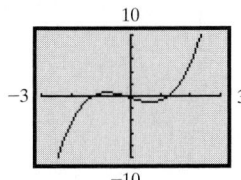

(c) x-intercepts: −1.26, −0.20, 1.26;
y-intercept: −0.31752

(d) Above on (−1.26, −0.20) and (1.26, ∞);
below on (−∞, −1.26) and (−0.20, 1.26)

(e) Local maximum at (−0.80, 0.57);
local minimum at (0.66, −0.99)

(f)

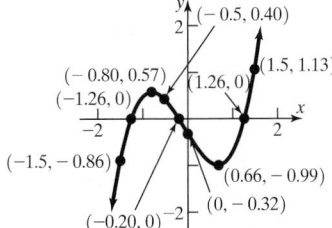

(g) Increasing on (−∞, −0.80) and
(0.66, ∞); decreasing on (−0.80, 0.66)

90. (a) Degree 3; $y = x^3$
(b)

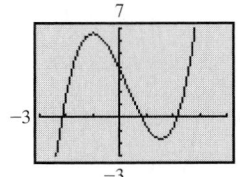

(c) x-intercepts: −2.16, 0.8, 2.16
y-intercept: 3.73248

(d)

Above on (−2.16, 0.8) and (2.16, ∞);
below on (−∞, −2.16) and (0.8, 2.16)

(e) Local maximum at (−1.01, 6.60); local
minimum at (1.54, −1.70)

(f)

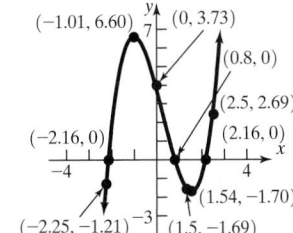

(g) Increasing on (−∞, −1.01) and
(1.54, ∞); decreasing on (−1.01, 1.54)

91. (a) Degree 3; $y = x^3$
(b)

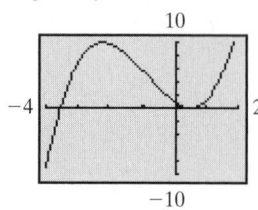

(c) x-intercepts: $-3.56, 0.50$; y-intercept: 0.89
(d)

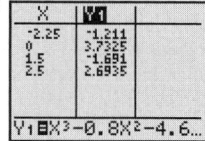

Above on $(-3.56, 0.5)$ and $(0.5, \infty)$; below on $(-\infty, -3.56)$

(e) Local maximum at $(-2.21, 9.91)$; local minimum at $(0.50, 0)$
(f)

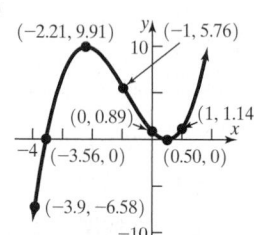

(g) Increasing on $(-\infty, -2.21)$ and $(0.50, \infty)$; decreasing on $(-2.21, 0.50)$

92. (a) Degree 3; $y = x^3$
(b)

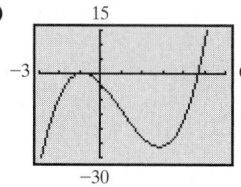

(c) x-intercepts: $-0.9; 4.71$; y-intercept: -3.8151
(d)

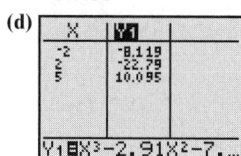

Above on $(4.71, \infty)$; below on $(-\infty, -0.9)$ and $(-0.9, 4.71)$

(e) Local maximum at $(-0.9, 0)$; local minimum at $(2.84, -26.16)$
(f)

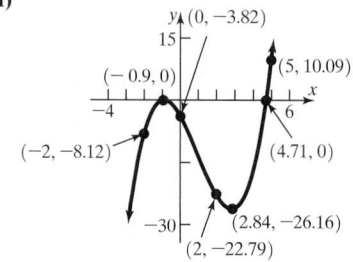

(g) Increasing on $(-\infty, -0.9)$ and $(2.84, \infty)$; decreasing on $(-0.9, 2.84)$

93. (a) Degree 4; $y = x^4$
(b)

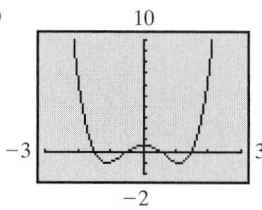

(c) x-intercepts: $-1.5, -0.5, 0.5, 1.5$; y-intercept: 0.5625
(d)

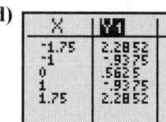

Above on $(-\infty, -1.5)$, $(-0.5, 0.5)$, and $(1.5, \infty)$; below on $(-1.5, -0.5)$ and $(0.5, 1.5)$

(e) Local minima at $(-1.12, -1)$, $(1.12, -1)$; local maximum at $(0, 0.56)$
(f)

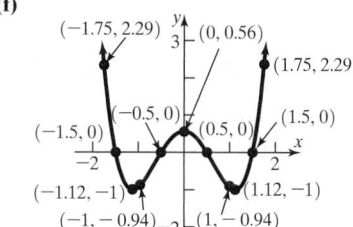

(g) Increasing on $(-1.12, 0)$ and $(1.12, \infty)$; decreasing on $(-\infty, -1.12)$ and $(0, 1.12)$

94. (a) Degree 4; $y = x^4$
(b)

(c) x-intercepts: $-3.90, -1.82, 1.82, 3.90$; y-intercept: 50.2619

(d)

Above on $(-\infty, -3.90)$, $(-1.82, 1.82)$, and $(3.90, \infty)$; below on $(-3.90, -1.82)$ and $(1.82, 3.90)$

(e) Local minima at $(-3.04, -35.30)$, $(3.04, -35.30)$; local maximum at $(0, 50.26)$

(f)

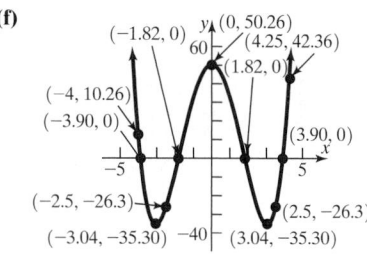

(g) Increasing on $(-3.04, 0)$ and $(3.04, \infty)$; decreasing on $(-\infty, -3.04)$ and $(0, 3.04)$

95. (a) Degree 4; $y = 2x^4$
(b)

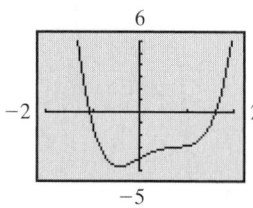

(c) x-intercepts: $-1.07, 1.62$; y-intercept: -4

(d)

Above on $(-\infty, -1.07)$ and $(1.62, \infty)$; below on $(-1.07, 1.62)$

(e) Local minimum at $(-0.42, -4.64)$

(f)

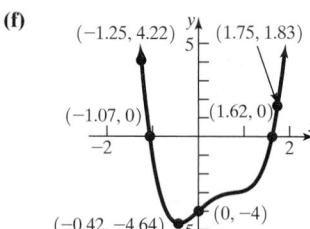

(g) Increasing on $(-0.42, \infty)$; decreasing on $(-\infty, -0.42)$

96. (a) Degree 4; $y = -1.2x^4$

(b)

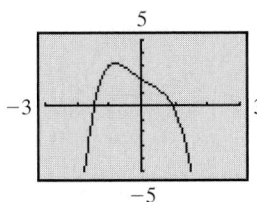

(c) x-intercepts: $-1.47, 0.91$; y-intercept: 2

(d)

Above on $(-1.47, 0.91)$; below on $(-\infty, -1.47)$ and $(0.91, \infty)$

(e) Local maximum at $(-0.81, 3.21)$

(f)

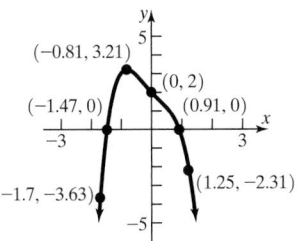

(g) Increasing on $(-\infty, -0.81)$; decreasing on $(-0.81, \infty)$

97. (a) Degree 5; $y = -2x^5$

(b)

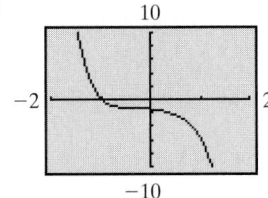

(c) x-intercept: -0.98; y-intercept: $-\sqrt{2}$

(d)

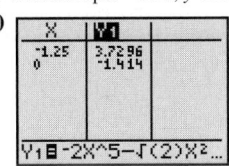

Above on $(-\infty, -0.98)$; below on $(-0.98, \infty)$

(e) None

(f)

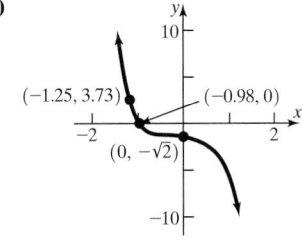

(g) Decreasing on $(-\infty, \infty)$

98. (a) Degree 5; $y = \pi x^5$

(b)

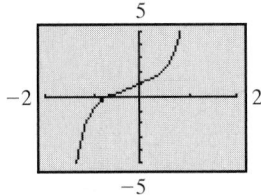

(c) x-intercept: -0.71; y-intercept: 1

(d)

Above on $(-0.71, \infty)$; below on $(-\infty, -0.71)$

(e) None

(f)

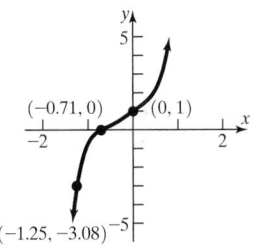

(g) Increasing on $(-\infty, \infty)$

99. (a) Cubic, $a > 0$

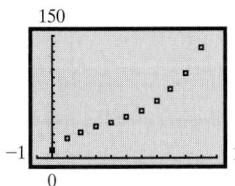

(b) \$7000 per car **(c)** \$20,000 per car
(d) $C(x) = 0.2156x^3 - 2.3473x^2 + 14.3275x + 10.2238$

(e)

(f) About \$171,000
(g) Fixed costs of about \$10,200

100. (a) Cubic, $a > 0$

(b) About \$3.17 per textbook
(c) \$1.85 per textbook
(d) $C(x) = 0.01546x^3 - 0.5951x^2 + 9.1502x + 98.4327$

(e)

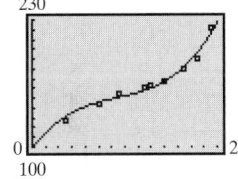

(f) About \$176,000
(g) Fixed costs of about \$98,400

101. (a) Cubic, $a > 0$

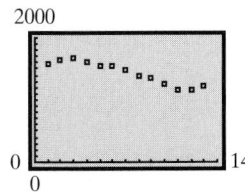

(b) $T(x) = 1.2582x^3 - 28.6146x^2 + 139.5808x + 1453.2098$
(c)

(d) About 1,738,000 thefts

102. (a) Cubic, $a < 0$

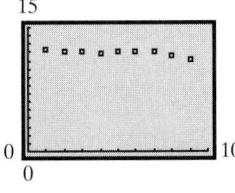

(b) $M(x) = -0.0154x^3 + 0.2130x^2 - 0.8978x + 13.1548$
(c)

(d) About 3600 miles

103. No; yes **107.** $f(x) = \text{int}(x)$; $g(x) = |x|$ **108.** (a)–(d)

3.2 Assess Your Understanding *(page 211)*

5. $y = 1$ **6.** $x = -1$ **7.** Proper **8.** False **9.** True **10.** True **11.** All real numbers except 3; $\{x | x \neq 3\}$ **12.** All real numbers except -3; $\{x | x \neq -3\}$ **13.** All real numbers except 2 and -4; $\{x | x \neq 2, x \neq -4\}$ **14.** All real numbers except -3 and 4; $\{x | x \neq -3, x \neq 4\}$ **15.** All real numbers except $-\frac{1}{2}$ and 3; $\left\{x \,\middle|\, x \neq -\frac{1}{2}, x \neq 3\right\}$ **16.** All real numbers except $\frac{1}{3}$ and -2; $\left\{x \,\middle|\, x \neq \frac{1}{3}, x \neq -2\right\}$ **17.** All real numbers except 2; $\{x | x \neq 2\}$ **18.** All real numbers except -1 and 1; $\{x | x \neq -1, x \neq 1\}$ **19.** All real numbers **20.** All real numbers **21.** All real numbers except -3 and 3; $\{x | x \neq -3, x \neq 3\}$ **22.** All real numbers except -2; $\{x | x \neq -2\}$ **23. (a)** Domain: $\{x | x \neq 2\}$; Range: $\{y | y \neq 1\}$ **(b)** $(0,0)$ **(c)** $y = 1$ **(d)** $x = 2$ **(e)** None **24. (a)** Domain: $\{x | x \neq -1\}$; Range: $\{y | y > 0\}$ **(b)** $(0,2)$ **(c)** $y = 0$ **(d)** $x = -1$ **(e)** None **25. (a)** Domain: $\{x | x \neq 0\}$; Range: all real numbers **(b)** $(-1,0), (1,0)$ **(c)** None **(d)** $x = 0$ **(e)** $y = 2x$ **26. (a)** Domain: $\{x | x \neq 0\}$; Range: $\{y | y \leq -2, y \geq 2\}$ **(b)** None **(c)** None **(d)** $x = 0$ **(e)** $y = -x$ **27. (a)** Domain: $\{x | x \neq -2, x \neq 2\}$; Range: $\{y | y \leq 0, y > 1\}$ **(b)** $(0,0)$ **(c)** $y = 1$ **(d)** $x = -2, x = 2$ **(e)** None **28. (a)** Domain: $\{x | x \neq -1, x \neq 1\}$; Range: all real numbers **(b)** $(0,0)$ **(c)** $y = 0$ **(d)** $x = -1, x = 1$ **(e)** None

29.

30.

31.

32.

33.

34.

35.

36.

37.

38.

39.

40.

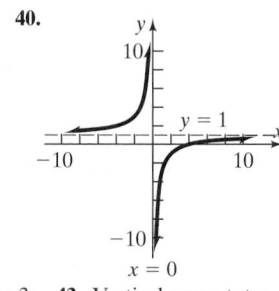

41. Vertical asymptote: $x = -4$ horizontal asymptote: $y = 3$; **42.** Vertical asymptote: $x = 6$ horizontal asymptote: $y = 3$; **43.** Vertical asymptote: $x = 3$; oblique asymptote: $y = x + 5$ **44.** Vertical asymptotes: $x = 2, x = 3$; horizontal asymptote: $y = -1$ **45.** Vertical asymptotes: $x = 1$, $x = -1$ horizontal asymptote: $y = 0$; **46.** Vertical asymptote: $x = 1$ **47.** Vertical asymptote: $x = 0$ horizontal asymptote: $y = 0$; **48.** Vertical asymptotes; $x = 0, x = -2$ horizontal asymptote: $y = 0$; **49.** Vertical asymptote; $x = 0$ oblique asymptote: $y = 3x$; **50.** Vertical asymptotes: $x = -\frac{1}{3}, x = 2$ horizontal asymptote: $y = 2$; **51.** Vertical asymptote: $x = 0$ oblique asymptote: $y = -(x + 1)$; **52.** Vertical asymptotes $x = -1, x = 0$ horizontal asymptote: $y = 0$; **53. (a)** 9.8208 m/sec^2 **(b)** 9.8195 m/sec^2 **(c)** 9.7936 m/sec^2 **(d)** h-axis **54. (a)** 25 **(b)** Approximately 596 **(c)** $y = 2500$

3.3 Assess Your Understanding *(page 227)*

2. in lowest terms **3.** False **4.** False **5.** False **6.** True
7. **1.** Domain: $\{x | x \neq 0, x \neq -4\}$; no y-intercept **2.** R is in lowest terms; x-intercept: -1 **3.** R is in lowest terms; vertical asymptotes: $x = 0, x = -4$ **4.** Horizontal asymptote: $y = 0$, intersected at $(-1, 0)$
5.

Interval	$(-\infty, -4)$	$(-4, -1)$	$(-1, 0)$	$(0, \infty)$
Number Chosen	-5	-2	$-\frac{1}{2}$	1
Value of R	$R(-5) = -\frac{4}{5}$	$R(-2) = \frac{1}{4}$	$R\left(-\frac{1}{2}\right) = -\frac{2}{7}$	$R(1) = \frac{2}{5}$
Location of Graph	Below x-axis	Above x-axis	Below x-axis	Above x-axis
Point on Graph	$\left(-5, -\frac{4}{5}\right)$	$\left(-2, \frac{1}{4}\right)$	$\left(-\frac{1}{2}, -\frac{2}{7}\right)$	$\left(1, \frac{2}{5}\right)$

7.

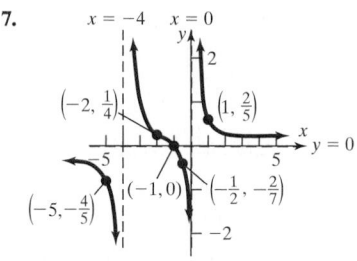

8. 1. Domain $\{x \mid x \neq -2, x \neq 1\}$; y-intercept: 0 **2.** R is in lowest terms; x-intercept: 0 **3.** R is in lowest terms; vertical asymptotes: $x = 1$, $x = -2$
4. Horizontal asymptotes: $y = 0$, intersected at $(0, 0)$
5.

Interval	$(-\infty, -2)$	$(-2, 0)$	$(0, 1)$	$(1, \infty)$
Number Chosen	-3	-1	$\frac{1}{2}$	2
Value of R	$R(-3) = -\frac{3}{4}$	$R(-1) = \frac{1}{2}$	$R(\frac{1}{2}) = -\frac{2}{5}$	$R(2) = \frac{1}{2}$
Location of Graph	Below x-axis	Above x-axis	Below x-axis	Above x-axis
Point on Graph	$\left(-3, -\frac{3}{4}\right)$	$\left(-1, \frac{1}{2}\right)$	$\left(\frac{1}{2}, -\frac{2}{5}\right)$	$\left(2, \frac{1}{2}\right)$

7.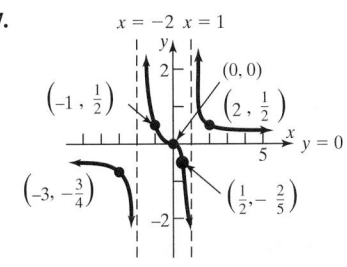

9. 1. $R(x) = \dfrac{3(x+1)}{2(x+2)}$; domain: $\{x \mid x \neq -2\}$; y-intercept: $\dfrac{3}{4}$ **2.** R is in lowest terms; x-intercept: -1 **3.** R is in lowest terms; vertical asymptote: $x = -2$ **4.** Horizontal asymptote: $y = \dfrac{3}{2}$, not intersected

5.

Interval	$(-\infty, -2)$	$(-2, -1)$	$(-1, \infty)$
Number Chosen	-3	$-\frac{3}{2}$	4
Value of R	$R(-3) = 3$	$R\left(-\frac{3}{2}\right) = -\frac{3}{2}$	$R(0) = \frac{3}{4}$
Location of Graph	Above x-axis	Below x-axis	Above x-axis
Point on Graph	$(-3, 3)$	$\left(-\frac{3}{2}, -\frac{3}{2}\right)$	$\left(0, \frac{3}{4}\right)$

7.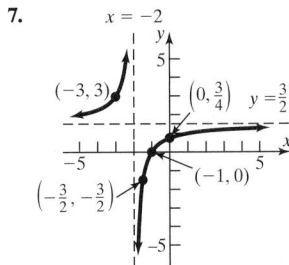

10. 1. $R(x) = \dfrac{2(x+2)}{x-1}$; domain $\{x \mid x \neq 1\}$; y-intercept: -4 **2.** R is in lowest terms; x-intercept: -2 **3.** R is in lowest terms; vertical asymptotes: $x = 1$ **4.** Horizontal asymptotes: $y = 2$, not intersected

5.

Interval	$(-\infty, -2)$	$(-2, 1)$	$(1, \infty)$
Number Chosen	-3	0	2
Value of R	$R(-3) = \frac{1}{2}$	$R(0) = -4$	$R(2) = 8$
Location of Graph	Above x-axis	Below x-axis	Above x-axis
Point on Graph	$\left(-3, \frac{1}{2}\right)$	$(0, -4)$	$(2, 8)$

7.

11. 1. $R(x) = \dfrac{3}{(x+2)(x-2)}$; domain: $\{x \mid x \neq -2, x \neq 2\}$; y-intercept: $-\dfrac{3}{4}$ **2.** R is in lowest terms; no x-intercept **3.** R is in lowest terms; vertical asymptotes: $x = 2$, $x = -2$ **4.** Horizontal asymptote: $y = 0$, not intersected

5.

Interval	$(-\infty, -2)$	$(-2, 2)$	$(2, \infty)$
Number Chosen	-3	0	3
Value of R	$R(-3) = \frac{3}{5}$	$R(0) = -\frac{3}{4}$	$R(3) = \frac{3}{5}$
Location of Graph	Above x-axis	Below x-axis	Above x-axis
Point on Graph	$\left(-3, \frac{3}{5}\right)$	$\left(0, -\frac{3}{4}\right)$	$\left(3, \frac{3}{5}\right)$

7.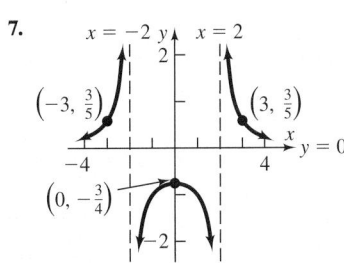

12. 1. $R(x) = \dfrac{6}{(x+2)(x-3)}$; domain $\{x \mid x \neq -2, x \neq 3\}$; y-intercept: -1 **2.** R is in lowest terms; no x-intercept **3.** R is in lowest terms; vertical asymptotes: $x = -2$, $x = 3$ **4.** Horizontal asymptote: $y = 0$, not intersected

5.

Interval	$(-\infty, -2)$	$(-2, 3)$	$(3, \infty)$
Number Chosen	-3	0	4
Value of R	$R(-3) = 1$	$R(0) = -1$	$R(4) = 1$
Location of Graph	Above x-axis	Below x-axis	Above x-axis
Point on Graph	$(-3, 1)$	$(0, -1)$	$(4, 1)$

7.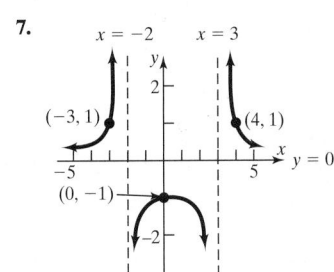

13. 1. $P(x) = \dfrac{x^4 + x^2 + 1}{(x + 1)(x - 1)}$; domain: $\{x \mid x \neq -1, x \neq 1\}$; y-intercept: -1 **2.** P is in lowest terms; no x-intercept **3.** P is in lowest terms; vertical asymptotes: $x = -1, x = 1$ **4.** No horizontal or oblique asymptotes

5.

Interval	$(-\infty, -1)$	$(-1, 1)$	$(1, \infty)$
Number Chosen	-2	0	2
Value of P	$P(-2) = 7$	$P(0) = -1$	$P(2) = 7$
Location of Graph	Above x-axis	Below x-axis	Above x-axis
Point on Graph	$(-2, 7)$	$(0, -1)$	$(2, 7)$

7.

14. 1. $Q(x) = \dfrac{(x^2 + 1)(x + 1)(x - 1)}{(x + 2)(x - 2)}$; domain $\{x \mid x \neq -2, x \neq 2\}$; y-intercept: $\dfrac{1}{4}$ **2.** Q is in lowest terms; x-intercepts: $-1, 1$ **3.** Q is in lowest terms; Vertical asymptotes: $x = -2, x = 2$ **4.** No horizontal or oblique asymptotes

5.

Interval	$(-\infty, -2)$	$(-2, -1)$	$(-1, 1)$	$(1, 2)$	$(2, \infty)$
Number Chosen	-3	$-\frac{3}{2}$	0	$\frac{3}{2}$	3
Value of Q	$Q(-3) = 16$	$Q(-\frac{3}{2}) \approx -2.3$	$Q(0) = \frac{1}{4}$	$Q(\frac{3}{2}) \approx -2.3$	$Q(3) = 16$
Location of Graph	Above x-axis	Below x-axis	Above x-axis	Below x-axis	Above x-axis
Point on Graph	$(-3, 16)$	$(-\frac{3}{2}, -2.3)$	$(0, \frac{1}{4})$	$(\frac{3}{2}, -2.3)$	$(3, 16)$

7.

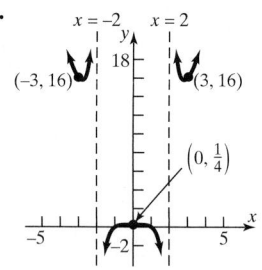

15. 1. $H(x) = \dfrac{(x - 1)(x^2 + x + 1)}{(x + 3)(x - 3)}$; domain: $\{x \mid x \neq -3, x \neq 3\}$; y-intercept: $\dfrac{1}{9}$ **2.** H is in lowest terms; x-intercept: 1 **3.** H is in lowest terms; vertical asymptotes: $x = 3, x = -3$ **4.** Oblique asymptote: $y = x$, intersected at $\left(\dfrac{1}{9}, \dfrac{1}{9}\right)$

5.

Interval	$(-\infty, -3)$	$(-3, 1)$	$(1, 3)$	$(3, \infty)$
Number Chosen	-4	0	2	4
Value of H	$H(-4) \approx -9.3$	$H(0) = \frac{1}{9}$	$H(2) = -1.4$	$H(4) = 9$
Location of Graph	Below x-axis	Above x-axis	Below x-axis	Above x-axis
Point on Graph	$(-4, -9.3)$	$\left(0, \frac{1}{9}\right)$	$(2, -1.4)$	$(4, 9)$

7.

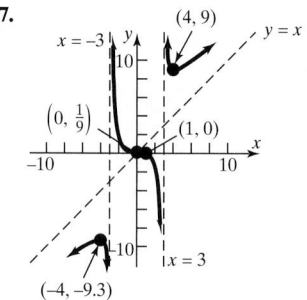

16. 1. $G(x) = \dfrac{(x + 1)(x^2 - x + 1)}{x(x + 2)}$; domain $\{x \mid x \neq -2, x \neq 0\}$; no y-intercept **2.** G is in lowest terms; x-intercept: -1 **3.** G is in lowest terms; vertical asymptotes: $x = 0, x = -2$ **4.** Oblique asymptote: $y = x - 2$, intersected at $(-0.25, -2.25)$

5.

Interval	$(-\infty, -2)$	$(-2, -1)$	$(-1, 0)$	$(0, \infty)$
Number Chosen	-4	-1.5	-0.25	1
Value of G	$G(-4) = -7.875$	$G(-1.5) \approx 3.2$	$G(-0.25) = -2.25$	$G(1) = \frac{2}{3}$
Location of Graph	Below x-axis	Above x-axis	Below x-axis	Above x-axis
Point on Graph	$(-4, -7.875)$	$(-1.5, 3.17)$	$(-0.25, -2.25)$	$(1, \frac{2}{3})$

7.

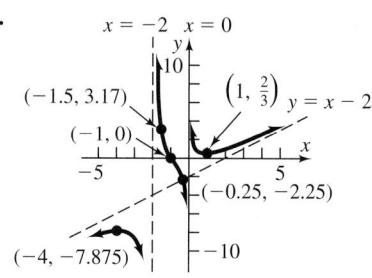

17. 1. $R(x) = \dfrac{x^2}{(x + 3)(x - 2)}$; domain: $\{x \neq -3, x \neq 2\}$; y-intercept: 0 **2.** R is in lowest terms; x-intercept: 0 **3.** R is in lowest terms; vertical asymptotes: $x = 2, x = -3$ **4.** Horizontal asymptote: $y = 1$, intersected at $(6, 1)$

5.

Interval	$(-\infty, -3)$	$(-3, 0)$	$(0, 2)$	$(2, \infty)$
Number Chosen	-6	-1	1	3
Value of R	$R(-6) = 1.5$	$R(-1) = -\frac{1}{6}$	$R(1) = -0.25$	$R(3) = 1.5$
Location of Graph	Above x-axis	Below x-axis	Below x-axis	Above x-axis
Point on Graph	$(-6, 1.5)$	$\left(-1, -\frac{1}{6}\right)$	$(1, -0.25)$	$(3, 1.5)$

7.

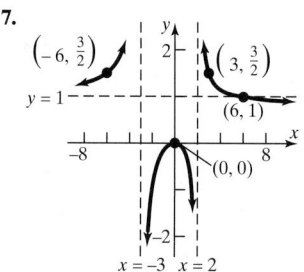

18. 1. $R(x) = \dfrac{(x+4)(x-3)}{(x+2)(x-2)}$; domain $\{x \mid x \ne -2, x \ne 2\}$; y-intercept: 3 **2.** R is in lowest terms; x-intercepts: 3, −4 **3.** R is in lowest terms; vertical asymptotes: $x = -2, x = 2$ **4.** Horizontal asymptote: $y = 1$, intersected at $(8, 1)$

5.

Interval	$(-\infty, -4)$	$(-4, -2)$	$(-2, 2)$	$(2, 3)$	$(3, \infty)$
Number Chosen	−7	−3	0	2.5	8
Value of R	$R(-7) = \frac{2}{3}$	$R(-3) = -1.2$	$R(0) = 3$	$R(2.5) = -1.44$	$R(8) = 1$
Location of Graph	Above x-axis	Below x-axis	Above x-axis	Below x-axis	Above x-axis
Point on Graph	$(-7, \frac{2}{3})$	$(-3, -1.2)$	$(0, 3)$	$(2.5, -1.44)$	$(8, 1)$

7.

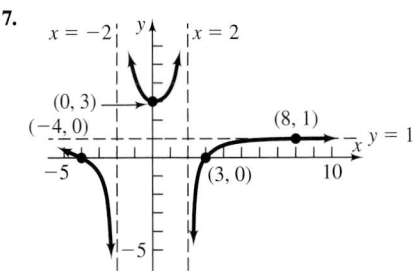

19. 1. $G(x) = \dfrac{x}{(x+2)(x-2)}$; domain: $\{x \mid x \ne -2, x \ne 2\}$; y-intercept: 0 **2.** G is in lowest terms; x-intercept: 0 **3.** G is in lowest terms; vertical asymptotes: $x = -2, x = 2$ **4.** Horizontal asymptote: $y = 0$, intersected at $(0, 0)$

5.

Interval	$(-\infty, -2)$	$(-2, 0)$	$(0, 2)$	$(2, \infty)$
Number Chosen	−3	−1	1	3
Value of G	$G(-3) = -\frac{3}{5}$	$G(-1) = \frac{1}{3}$	$G(1) = -\frac{1}{3}$	$G(3) = \frac{3}{5}$
Location of Graph	Below x-axis	Above x-axis	Below x-axis	Above x-axis
Point on Graph	$\left(-3, -\frac{3}{5}\right)$	$\left(-1, \frac{1}{3}\right)$	$\left(1, -\frac{1}{3}\right)$	$\left(3, \frac{3}{5}\right)$

7.

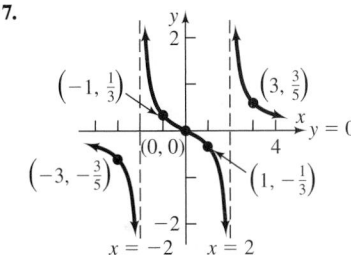

20. 1. $G(x) = \dfrac{3x}{(x+1)(x-1)}$; domain $\{x \mid x \ne -1, x \ne 1\}$; y-intercept: 0 **2.** G is in lowest terms; x-intercept: 0 **3.** G is in lowest terms; vertical asymptotes: $x = -1, x = 1$ **4.** Horizontal asymptote: $y = 0$, intersected at $(0, 0)$

5.

Interval	$(-\infty, -1)$	$(-1, 0)$	$(0, 1)$	$(1, \infty)$
Number Chosen	−2	$-\frac{1}{2}$	$\frac{1}{2}$	2
Value of G	$G(-2) = -2$	$G(-\frac{1}{2}) = 2$	$G(\frac{1}{2}) = -2$	$G(2) = 2$
Location of Graph	Below x-axis	Above x-axis	Below x-axis	Above x-axis
Point on Graph	$(-2, -2)$	$(-\frac{1}{2}, 2)$	$(\frac{1}{2}, -2)$	$(2, 2)$

7.

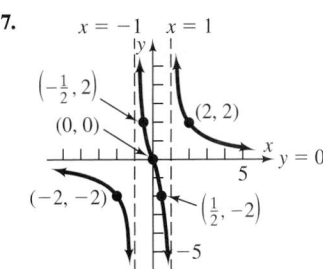

21. 1. $R(x) = \dfrac{3}{(x-1)(x+2)(x-2)}$; domain: $\{x \mid x \ne 1, x \ne -2, x \ne 2\}$; y-intercept: $\dfrac{3}{4}$ **2.** R is in lowest terms; no x-intercept **3.** R is in lowest terms; vertical asymptotes: $x = -2, x = 1, x = 2$ **4.** Horizontal asymptote: $y = 0$, not intersected

5.

Interval	$(-\infty, -2)$	$(-2, 1)$	$(1, 2)$	$(2, \infty)$
Number Chosen	−3	0	1.5	3
Value of R	$R(-3) = -\frac{3}{20}$	$R(0) = \frac{3}{4}$	$R(1.5) = -\frac{24}{7}$	$R(3) = \frac{3}{10}$
Location of Graph	Below x-axis	Above x-axis	Below x-axis	Above x-axis
Point on Graph	$\left(-3, -\frac{3}{20}\right)$	$\left(0, \frac{3}{4}\right)$	$\left(1.5, -\frac{24}{7}\right)$	$\left(3, \frac{3}{10}\right)$

7.

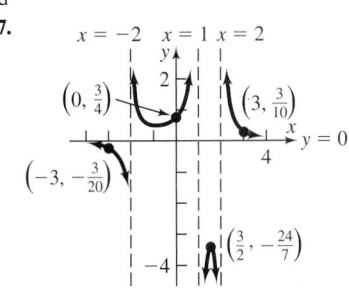

22. 1. $R(x) = \dfrac{-4}{(x+1)(x+3)(x-3)}$; domain $\{x \mid x \neq -1, x \neq -3, x \neq 3\}$; y-intercept: $\dfrac{4}{9}$ **2.** R is in lowest terms; no x-intercept **3.** R is in lowest terms; vertical asymptotes: $x = -3, x = -1, x = 3$ **4.** Horizontal asymptote: $y = 0$, not intersected

5.

Interval	$(-\infty, -3)$	$(-3, -1)$	$(-1, 3)$	$(3, \infty)$
Number Chosen	-4	-2	0	4
Value of R	$R(-4) \approx 0.19$	$R(-2) = -0.8$	$R(0) = \dfrac{4}{9}$	$R(4) \approx -0.11$
Location of Graph	Above x-axis	Below x-axis	Above x-axis	Below x-axis
Point on Graph	$(-4, 0.19)$	$(-2, -0.8)$	$\left(0, \dfrac{4}{9}\right)$	$(4, -0.11)$

7.

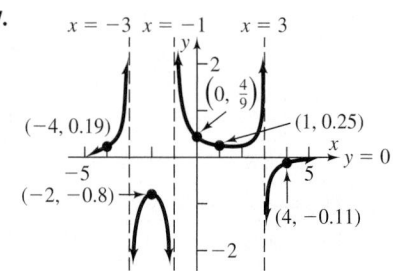

23. 1. $H(x) = \dfrac{(x+1)(x-1)}{(x^2+4)(x+2)(x-2)}$; domain: $\{x \mid x \neq -2, x \neq 2\}$; y-intercept: $\dfrac{1}{16}$ **2.** H is in lowest terms; x-intercepts: $-1, 1$ **3.** H is in lowest terms; vertical asymptotes: $x = -2, x = 2$ **4.** Horizontal asymptote: $y = 0$, intersected at $(-1, 0)$ and $(1, 0)$

5.

Interval	$(-\infty, -2)$	$(-2, -1)$	$(-1, 1)$	$(1, 2)$	$(2, \infty)$
Number Chosen	-3	-1.5	0	1.5	3
Value of H	$H(-3) \approx 0.12$	$H(-1.5) \approx -0.11$	$H(0) = \dfrac{1}{16}$	$H(1.5) \approx -0.11$	$H(3) \approx 0.12$
Location of Graph	Above x-axis	Below x-axis	Above x-axis	Below x-axis	Above x-axis
Point on Graph	$(-3, 0.12)$	$(-1.5, -0.11)$	$\left(0, \dfrac{1}{16}\right)$	$(1.5, -0.11)$	$(3, 0.12)$

7.

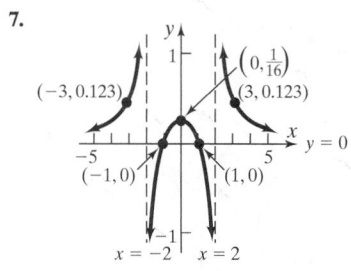

24. 1. $H(x) = \dfrac{x^2 + 4}{(x^2+1)(x+1)(x+1)}$; domain $\{x \mid x \neq -1, x \neq 1\}$; y-intercept: -4 **2.** H is in lowest terms; no x-intercept **3.** H is in lowest terms; vertical asymptotes: $x = -1, x = 1$ **4.** Horizontal asymptote: $y = 0$, not intersected

5.

Interval	$(-\infty, -1)$	$(-1, 1)$	$(1, \infty)$
Number Chosen	-2	0	2
Value of H	$H(-2) = \dfrac{8}{15}$	$H(0) = -4$	$H(2) = \dfrac{8}{15}$
Location of Graph	Above x-axis	Below x-axis	Above x-axis
Point on Graph	$\left(-2, \dfrac{8}{15}\right)$	$(0, -4)$	$\left(2, \dfrac{8}{15}\right)$

7.

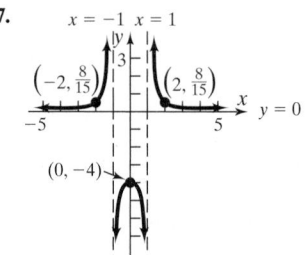

25. 1. $F(x) = \dfrac{(x+1)(x-4)}{x+2}$; domain: $\{x \mid x \neq -2\}$; y-intercept: -2 **2.** F is in lowest terms; x-intercepts: $-1, 4$ **3.** F is in lowest terms; vertical asymptote: $x = -2$ **4.** Oblique asymptote: $y = x - 5$, not intersected

5.

Interval	$(-\infty, -2)$	$(-2, -1)$	$(-1, 4)$	$(4, \infty)$
Number Chosen	-3	-1.5	0	5
Value of F	$F(-3) = -14$	$F(-1.5) = 5.5$	$F(0) = -2$	$F(5) \approx 0.86$
Location of Graph	Below x-axis	Above x-axis	Below x-axis	Above x-axis
Point on Graph	$(-3, -14)$	$(-1.5, 5.5)$	$(0, -2)$	$(5, 0.86)$

7.

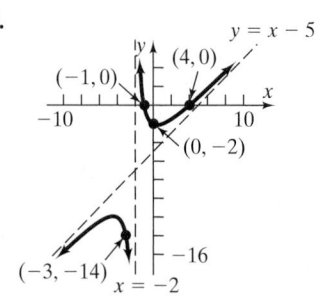

26. 1. $F(x) = \dfrac{(x+2)(x+1)}{x-1}$; domain $\{x \mid x \neq 1\}$; y-intercept: -2 **2.** F is in lowest terms; x-intercepts: $-1, -2$ **3.** F is in lowest terms; vertical asymptote: $x = 1$ **4.** Oblique asymptote: $y = x + 4$, not intersected

24. 1. Domain: $\{x|x \neq 0\}$; no y-intercept **2.** R is in lowest terms; x-intercept: 4 **3.** R is in lowest terms; vertical asymptote: $x = 0$ **4.** Horizontal asymptote: $y = -1$; not intersected

5.

Interval	$(-\infty, 0)$	$(0, 4)$	$(4, \infty)$
Number Chosen	-1	1	5
Value of R	$R(-1) = -5$	$R(1) = 3$	$R(5) = -0.2$
Location of Graph	Below x-axis	Above x-axis	Below x-axis
Point on Graph	$(-1, -5)$	$(1, 3)$	$(5, -0.2)$

7.

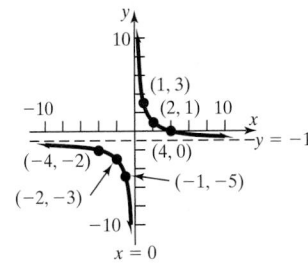

25. 1. Domain: $\{x|x \neq 0, x \neq 2\}$; no y-intercept **2.** H is in lowest terms; x-intercept: -2 **3.** H is in lowest terms; vertical asymptotes: $x = 0, x = 2$
4. Horizontal asymptote: $y = 0$; intersected at $(-2, 0)$

5.

Interval	$(-\infty, -2)$	$(-2, 0)$	$(0, 2)$	$(2, \infty)$
Number Chosen	-3	-1	1	3
Value of H	$H(-3) = -\frac{1}{15}$	$H(-1) = \frac{1}{3}$	$H(1) = -3$	$H(3) = \frac{5}{3}$
Location of Graph	Below x-axis	Above x-axis	Below x-axis	Above x-axis
Point on Graph	$\left(-3, -\frac{1}{15}\right)$	$\left(-1, \frac{1}{3}\right)$	$(1, -3)$	$\left(3, \frac{5}{3}\right)$

7.

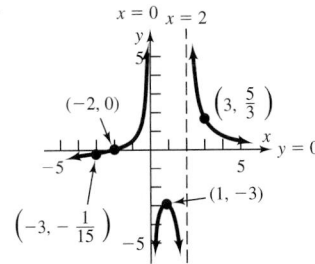

26. 1. $H(x) = \dfrac{x}{(x + 1)(x - 1)}$; domain: $\{x|x \neq -1, x \neq 1\}$; y-intercept: 0 **2.** H is in lowest terms; x-intercept: 0 **3.** H is in lowest terms; vertical asymptotes: $x = -1$ and $x = 1$ **4.** Horizontal asymptote: $y = 0$; intersected at $(0, 0)$

5.

Interval	$(-\infty, -1)$	$(-1, 0)$	$(0, 1)$	$(1, \infty)$
Number Chosen	-2	$-\frac{1}{2}$	$\frac{1}{2}$	2
Value of H	$H(-2) = -\frac{2}{3}$	$H(-\frac{1}{2}) = \frac{2}{3}$	$H(\frac{1}{2}) = -\frac{2}{3}$	$H(2) = \frac{2}{3}$
Location of Graph	Below x-axis	Above x-axis	Below x-axis	Above x-axis
Point on Graph	$\left(-2, -\frac{2}{3}\right)$	$\left(-\frac{1}{2}, \frac{2}{3}\right)$	$\left(\frac{1}{2}, -\frac{2}{3}\right)$	$\left(2, \frac{2}{3}\right)$

7.

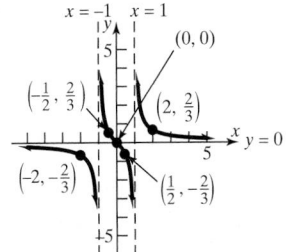

27. 1. $R(x) = \dfrac{(x + 3)(x - 2)}{(x - 3)(x + 2)}$ domain: $\{x|x \neq -2, x \neq 3\}$; y-intercept: 1 **2.** R is in lowest terms; x-intercepts: $-3, 2$ **3.** R is in lowest terms; asymptote: $x = -2, x = 3$ **4.** Horizontal asymptote: $y = 1$; intersected at $(0, 1)$

5.

Interval	$(-\infty, -3)$	$(-3, -2)$	$(-2, 2)$	$(2, 3)$	$(3, \infty)$
Number Chosen	-4	-2.5	0	2.5	4
Value of R	$R(-4) \approx 0.43$	$R(-2.5) \approx -0.82$	$R(0) = 1$	$R(2.5) \approx -1.22$	$R(4) = \frac{7}{3}$
Location of Graph	Above x-axis	Below x-axis	Above x-axis	Below x-axis	Above x-axis
Point on Graph	$(-4, 0.43)$	$(-2.5, -0.82)$	$(0, 1)$	$(2.5, -1.22)$	$\left(4, \frac{7}{3}\right)$

7.

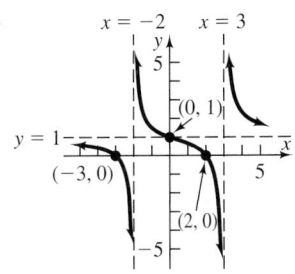

28. 1. $R(x) = \dfrac{(x - 3)^2}{x^2}$; domain: $\{x|x \neq 0\}$; no y-intercept
2. R is in lowest terms; x-intercept: 3;
3. R is in lowest terms; vertical asymptote: $x = 0$
4. Horizontal asymptote: $y = 1$; intersected at $\left(\frac{3}{2}, 1\right)$

5.

Interval	$(-\infty, 0)$	$(0, 3)$	$(3, \infty)$
Number Chosen	-3	1	4
Value of R	$R(-3) = 4$	$R(1) = 4$	$R(4) = \frac{1}{16}$
Location of Graph	Above x-axis	Above x-axis	Above x-axis
Point on Graph	$(-3, 4)$	$(1, 4)$	$\left(4, \frac{1}{16}\right)$

7.

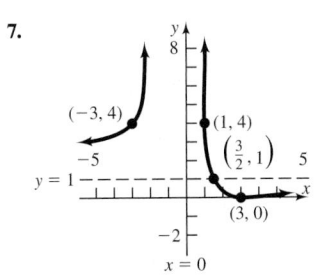

11. (a) x-intercepts: $-4, -2, 0$; y-intercept: 0
(b) Crosses at $-4, -2, 0$
(c) $y = x^3$ **(d)** 2
(e) Near -4: $f(x) \approx 8(x + 4)$;
 Near -2: $f(x) \approx -4(x + 2)$;
 Near 0: $f(x) \approx 8x$
(f)

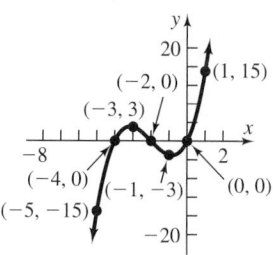

12. (a) x-intercepts: $0, 2, 4$; y-intercept: 0
(b) Crosses at $0, 2$, and 4
(c) $y = x^3$ **(d)** 2
(e) Near 0: $f(x) \approx 8x$;
 Near 2: $f(x) \approx -4(x - 2)$;
 Near 4: $f(x) \approx 8(x - 4)$
(f)

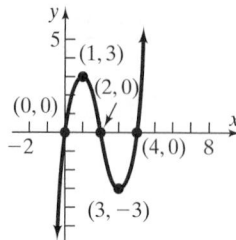

13. (a) x-intercepts: $-4, 2$; y-intercept: 16
(b) Crosses at -4; touches at 2
(c) $y = x^3$ **(d)** 2
(e) Near -4: $f(x) \approx 36(x + 4)$;
 Near 2: $f(x) \approx 6(x - 2)^2$
(f)

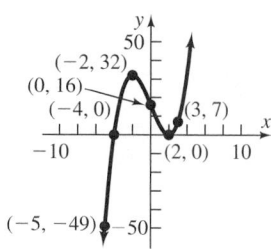

14. (a) x-intercepts: $-4, 2$; y-intercept: -32
(b) Touches at -4; crosses at 2
(c) $y = x^3$ **(d)** 2
(e) Near -4: $f(x) \approx -6(x + 4)^2$;
 Near 2: $f(x) \approx 36(x + 2)$
(f)

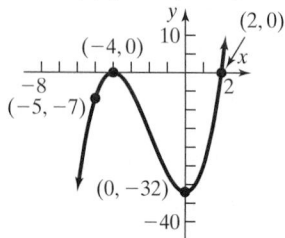

15. (a) $f(x) = -2x^2(x - 2)$;
 x-intercepts: $0, 2$; y-intercept: 0
(b) Touches at 0; crosses at 2
(c) $y = -2x^3$ **(d)** 2
(e) Near 0: $f(x) \approx 4x^2$;
 Near 2: $f(x) \approx -8(x - 2)$
(f)

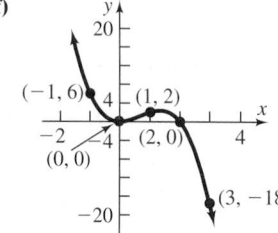

16. (a) $f(x) = -4x(x + 1)(x - 1)$;
 x-intercepts: $-1, 0, 1$; y-intercept: 0
(b) Crosses at $-1, 0, 1$
(c) $y = -4x^3$ **(d)** 2
(e) Near -1: $f(x) \approx -8(x + 1)$;
 Near 0: $f(x) \approx 4x$;
 Near 1: $f(x) \approx -8(x - 1)$
(f)

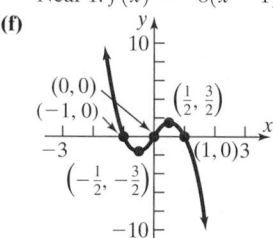

17. (a) x-intercepts: $-3, -1, 1$; y-intercept: 3
(b) Crosses at $-3, -1$; touches at 1 **(c)** $y = x^4$
(d) 3 **(e)** Near -3: $f(x) \approx -32(x + 3)$;
 Near -1: $f(x) \approx 8(x + 1)$; Near 1: $f(x) \approx 8(x - 1)^2$
(f)

18. (a) x-intercepts: $-2, 2, 4$; y-intercept: 32
(b) Crosses at 2 and 4; touches at -2 **(c)** $y = x^4$
(d) 3 **(e)** Near -2: $f(x) \approx 24(x + 2)^2$;
 Near 2: $f(x) = -32(x - 2)$; Near 4: $f(x) \approx 72(x - 4)$
(f)

19. Domain: $\{x \mid x \neq -3, x \neq 3\}$; horizontal asymptote: $y = 0$; vertical asymptotes: $x = -3, x = 3$ **20.** Domain: $\{x \mid x \neq 2\}$; oblique asymptote:
$y = x + 2$; vertical asymptote: $x = 2$ **21.** Domain: $\{x \mid x \neq -2\}$; horizontal asymptote: $y = 1$; vertical asymptote: $x = -2$ **22.** Domain:
$\{x \mid x \neq 1\}$; horizontal asymptote: $y = 1$; vertical asymptote: $x = 1$
23. 1. $R(x) = \dfrac{2(x - 3)}{x}$; domain: $\{x \mid x \neq 0\}$; no y-intercept **2.** R is in lowest terms; x-intercept: 3 **3.** R is in lowest terms; vertical asymptote:
$x = 0$ **4.** Horizontal asymptote: $y = 2$; not intersected

5.

Interval	$(-\infty, 0)$	$(0, 3)$	$(3, \infty)$
Number Chosen	-2	1	4
Value of R	$R(-2) = 5$	$R(1) = -4$	$R(4) = \frac{1}{2}$
Location of Graph	Above x-axis	Below x-axis	Above x-axis
Point on Graph	$(-2, 5)$	$(1, -4)$	$\left(4, \frac{1}{2}\right)$

7.

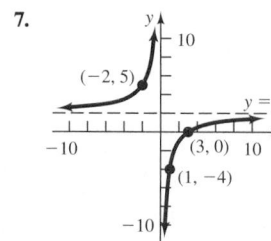

81. 5 **82.** 37 **83.** 2 **84.** 2 **85.** 5 **86.** 13 **87.** $\dfrac{3}{2}$ **88.** $\dfrac{3}{2}$ **89.** $f(0) = -1; f(1) = 10$ **90.** $f(-1) = -6; f(0) = 2$ **91.** $f(-5) = -58;$
$f(-4) = 2$ **92.** $f(-3) = -42; f(-2) = 5$ **93.** $f(1.4) = -0.17536; f(1.5) = 1.40625$ **94.** $f(1.7) = 0.35627; f(1.8) = -1.02112$ **95.** 0.21
96. -0.60 **97.** -4.04 **98.** -2.17 **99.** 1.15 **100.** 0.70 **101.** 2.53 **102.** 2.13 **103.** $k = 5$ **104.** $k = -\dfrac{17}{12}$ **105.** -7 **106.** 1 **107.** If
$f(x) = x^n - c^n$, then $f(c) = c^n - c^n = 0$, so $x - c$ is a factor of f. **108.** If $n \geq 1$ is odd, $(-c)^n = -c^n$, so $f(-c) = (-c)^n + c^n = -c^n + c^n = 0$.
so, $x + c$ is a factor of f when n is odd. **109.** 5 **110.** -3 **111.** 7 in. **112.** 6 cm or 12 cm **113.** All the potential rational zeros are integers. so, r
is either an integer or is not a rational zero (and is therefore irrational).

114. Let $\dfrac{p}{q}$, where p and q have no common factors except 1 and -1, be a solution of the polynomial $f(x) = a_nx^n + a_{n-1}x^{n-1} + \cdots + a_1x + a_0$
whose coefficients are all integers. Then

$$f\left(\frac{p}{q}\right) = a_n\left(\frac{p}{q}\right)^n + a_{n-1}\left(\frac{p}{q}\right)^{n-1} + \cdots + a_1\left(\frac{p}{q}\right) + a_0 = 0$$
$$a_np^n + a_{n-1}p^{n-1}q + \cdots + a_1pq^{n-1} + a_0q^n = 0$$

Because p is a factor of the first n terms of this equation, p must also be a factor of a_0q^n. Since p is not a factor of q (p and q have no common factors except 1 and -1), p must be a factor of a_0. Similarly, q must be a factor of a_n.

116. No, by the Rational Zeros Theorem, $\dfrac{1}{3}$ is not a potential rational zero.

117. No, by the Rational Zeros Theorem, $\dfrac{1}{3}$ is not a potential rational zero.

118. No, by the Rational Zeros Theorem, $\dfrac{3}{5}$ is not a potential rational zero. **119.** No, by the Rational Zeros Theorem, $\dfrac{2}{3}$ is not a potential
rational zero.

3.6 Assess Your Understanding (page 255)

3. one **4.** $3 - 4i$ **5.** True **6.** False **7.** $4 + i$ **8.** $3 - i$ **9.** $-i, 1 - i$ **10.** $2 - i$ **11.** $-i, -2i$ **12.** $-i$ **13.** $-i$ **14.** $2 + i, i$ **15.** $2 - i, -3 + i$
16. $-i, 3 + 2i, -2 - i$ **17.** $f(x) = x^4 - 14x^3 + 77x^2 - 200x + 208; a = 1$ **18.** $f(x) = x^4 - 2x^3 + 6x^2 - 2x + 5; a = 1$
19. $f(x) = x^5 - 4x^4 + 7x^3 - 8x^2 + 6x - 4; a = 1$ **20.** $f(x) = x^6 - 12x^5 + 55x^4 - 120x^3 + 139x^2 - 108x + 85; a = 1$
21. $f(x) = x^4 - 6x^3 + 10x^2 - 6x + 9; a = 1$ **22.** $f(x) = x^5 - 5x^4 + 11x^3 - 13x^2 + 8x - 2; a = 1$ **23.** $-2i, 4$ **24.** $5i, -3$ **25.** $2i, -3, \dfrac{1}{2}$
26. $-3i, -2, \dfrac{1}{3}$ **27.** $3 + 2i, -2, 5$ **28.** $1 - 3i, -1, 6$ **29.** $4i, -\sqrt{11}, \sqrt{11}, -\dfrac{2}{3}$ **30.** $-3i, -3, 4, \dfrac{1}{2}$ **31.** $1, -\dfrac{1}{2} - \dfrac{\sqrt{3}}{2}i, -\dfrac{1}{2} + \dfrac{\sqrt{3}}{2}i;$

$f(x) = (x - 1)\left(x + \dfrac{1}{2} + \dfrac{\sqrt{3}}{2}i\right)\left(x + \dfrac{1}{2} - \dfrac{\sqrt{3}}{2}i\right)$ **32.** $-i, i, -1, 1; f(x) = (x + i)(x - i)(x + 1)(x - 1)$ **33.** $2, 3 - 2i, 3 + 2i;$
$f(x) = (x - 2)(x - 3 + 2i)(x - 3 - 2i)$ **34.** $-5, -4 + i, -4 - i; f(x) = (x + 5)(x + 4 - i)(x + 4 + i)$ **35.** $-i, i, -2i, 2i;$
$f(x) = (x + i)(x - i)(x + 2i)(x - 2i)$ **36.** $-2i, 2i, -3i, 3i; f(x) = (x + 2i)(x - 2i)(x + 3i)(x - 3i)$ **37.** $-5i, 5i, -3, 1;$
$f(x) = (x + 5i)(x - 5i)(x + 3)(x - 1)$ **38.** $-3i, 3i, -7, 4; f(x) = (x + 3i)(x - 3i)(x + 7)(x - 4)$ **39.** $-4, \dfrac{1}{3}, 2 - 3i, 2 + 3i;$
$f(x) = 3(x + 4)\left(x - \dfrac{1}{3}\right)(x - 2 + 3i)(x - 2 - 3i)$ **40.** $-3 - 2i, -3 + 2i, \dfrac{1}{2}, 5; f(x) = 2(x + 3 + 2i)(x + 3 - 2i)\left(x - \dfrac{1}{2}\right)(x - 5)$

41. Zeros that are complex numbers must occur in conjugate pairs; or a polynomial with real coefficients of odd degree must have at least one real zero. **42.** Zeros that are complex numbers must occur in conjugate pairs. **43.** If the remaining zero were a complex number, its conjugate would also be a zero, creating a polynomial of degree 5. **44.** A missing zero is $4 + i$. If the remaining zero were a complex number, its conjugate would also be a zero, creating a polynomial of degree 5.

Review Exercises (page 257)

1. Polynomial of degree 5 **2.** Not a polynomial **3.** Not a polynomial **4.** Polynomial of degree 0

5.

6.

7.

8.

9.

10.
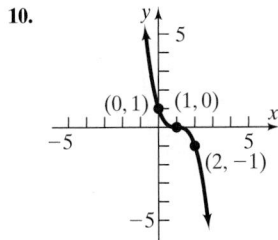

19. $R = f\left(\dfrac{1}{2}\right) = 0$; yes **20.** $R = f\left(-\dfrac{1}{3}\right) = 2$; no **21.** 7; 3 or 1 positive; 2 or 0 negative **22.** 4; 1 positive; 1 negative

23. 6; 2 or 0 positive; 2 or 0 negative **24.** 5; 1 positive; 0 negative **25.** 3; 2 or 0 positive; 1 negative **26.** 3; 1 positive; 2 or 0 negative

27. 4; 2 or 0 positive; 2 or 0 negative **28.** 4; 1 positive; 1 negative **29.** 5; 0 positive; 3 or 1 negative **30.** 5; 5, 3 or 1 positive; 0 negative

31. 6; 1 positive; 1 negative **32.** 6; no positive; no negative **33.** $\pm1, \pm\dfrac{1}{3}$ **34.** $\pm1, \pm3$ **35.** $\pm1, \pm3$ **36.** $\pm1, \pm\dfrac{1}{2}$ **37.** $\pm1, \pm2, \pm\dfrac{1}{4}, \pm\dfrac{1}{2}$

38. $\pm1, \pm2, \pm\dfrac{1}{2}, \pm\dfrac{1}{3}, \pm\dfrac{1}{6}, \pm\dfrac{2}{3}$ **39.** $\pm1, \pm3, \pm9, \pm\dfrac{1}{2}, \pm\dfrac{1}{3}, \pm\dfrac{1}{6}, \pm\dfrac{3}{2}, \pm\dfrac{9}{2}$ **40.** $\pm1, \pm2, \pm3, \pm6, \pm\dfrac{1}{4}, \pm\dfrac{1}{2}, \pm\dfrac{3}{4}, \pm\dfrac{3}{2}$ **41.** $\pm1, \pm2, \pm3, \pm4, \pm6, \pm12, \pm\dfrac{1}{2}, \pm\dfrac{3}{2}$

42. $\pm1, \pm2, \pm3, \pm6, \pm9, \pm18, \pm\dfrac{1}{3}, \pm\dfrac{2}{3}$ **43.** $\pm1, \pm2, \pm4, \pm5, \pm10, \pm20, \pm\dfrac{1}{2}, \pm\dfrac{5}{2}, \pm\dfrac{1}{3}, \pm\dfrac{2}{3}, \pm\dfrac{4}{3}, \pm\dfrac{5}{3}, \pm\dfrac{10}{3}, \pm\dfrac{20}{3}, \pm\dfrac{1}{6}, \pm\dfrac{5}{6}$

44. $\pm1, \pm2, \pm5, \pm10, \pm\dfrac{1}{2}, \pm\dfrac{1}{3}, \pm\dfrac{1}{6}, \pm\dfrac{2}{3}, \pm\dfrac{5}{3}, \pm\dfrac{5}{2}, \pm\dfrac{5}{6}, \pm\dfrac{10}{3}$ **45.** $-3, -1, 2$; $f(x) = (x + 3)(x + 1)(x - 2)$

46. $-4, -5, 1$; $f(x) = (x - 1)(x + 5)(x + 4)$ **47.** $\dfrac{1}{2}$; $f(x) = 2\left(x - \dfrac{1}{2}\right)(x^2 + 1)$ **48.** $-\dfrac{1}{2}$; $f(x) = 2\left(x + \dfrac{1}{2}\right)(x^2 + 1)$

49. $-1, 1$; $f(x) = (x + 1)(x - 1)(x^2 + 2)$ **50.** $2, -2$; $f(x) = (x - 2)(x + 2)(x^2 + 1)$

51. $-\dfrac{1}{2}, \dfrac{1}{2}$; $f(x) = 4\left(x + \dfrac{1}{2}\right)\left(x - \dfrac{1}{2}\right)(x^2 + 2)$ **52.** $\dfrac{1}{2}, -\dfrac{1}{2}$; $f(x) = 4\left(x - \dfrac{1}{2}\right)\left(x + \dfrac{1}{2}\right)(x^2 + 4)$

53. 1, multiplicity 2; $-2, -1$; $f(x) = (x + 2)(x + 1)(x - 1)^2$ **54.** 2, multiplicity 2; $-1, -2$; $f(x) = (x + 1)(x + 2)(x - 2)^2$

55. $-\dfrac{\sqrt{2}}{2}, \dfrac{\sqrt{2}}{2}, 2$; $f(x) = 4\left(x + \dfrac{\sqrt{2}}{2}\right)\left(x - \dfrac{\sqrt{2}}{2}\right)(x - 2)\left(x^2 + \dfrac{1}{2}\right)$ **56.** $-3, -\dfrac{\sqrt{2}}{2}, \dfrac{\sqrt{2}}{2}$; $f(x) = 4(x + 3)\left(x + \dfrac{\sqrt{2}}{2}\right)\left(x - \dfrac{\sqrt{2}}{2}\right)\left(x^2 + \dfrac{1}{2}\right)$

57. $\{-1, 2\}$ **58.** $\left\{-\dfrac{3}{2}\right\}$ **59.** $\left\{\dfrac{2}{3}, -1 + \sqrt{2}, -1 - \sqrt{2}\right\}$ **60.** $\left\{\dfrac{5}{2}\right\}$ **61.** $\left\{\dfrac{1}{3}, \sqrt{5}, -\sqrt{5}\right\}$ **62.** $\left\{-\dfrac{1}{2}, 2, 4\right\}$ **63.** $\{-3, -2\}$ **64.** $\{1\}$

65. $\left\{-\dfrac{1}{3}\right\}$ **66.** $\left\{\dfrac{1}{2}\right\}$ **67.** $\left\{\dfrac{1}{2}, 2, 5\right\}$ **68.** $\left\{-4, -\dfrac{1}{2}, 2\right\}$

69. y-intercept: -6;
x-intercepts: $-3, -1, 2$

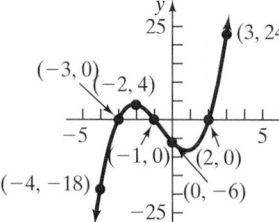

70. y-intercept: -20;
x-intercepts: $-5, -4, 1$

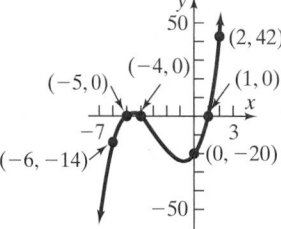

71. y-intercept: -1;
x-intercept: $\dfrac{1}{2}$

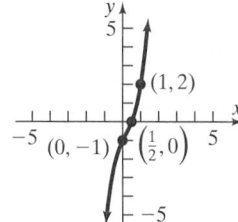

72. y-intercept: 1;
x-intercept: $-\dfrac{1}{2}$

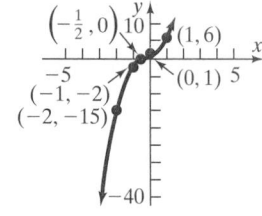

73. y-intercept: -2;
x-intercepts: $-1, 1$

74. y-intercept: -4;
x-intercepts: $-2, 2$

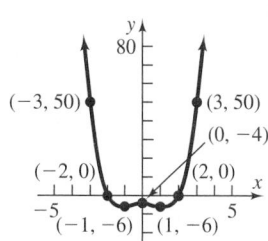

75. y-intercept: -2;
x-intercepts: $-\dfrac{1}{2}, \dfrac{1}{2}$

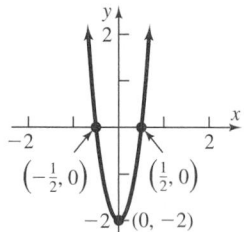

76. y-intercept: -4;
x-intercepts: $-\dfrac{1}{2}, \dfrac{1}{2}$

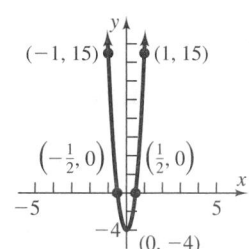

77. y-intercept: 2;
x-intercepts: $-2, -1, 1$

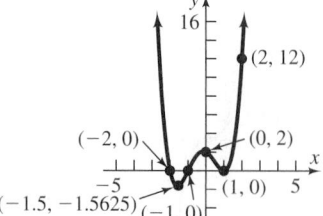

78. y-intercept: 8;
x-intercepts: $-2, -1, 2$

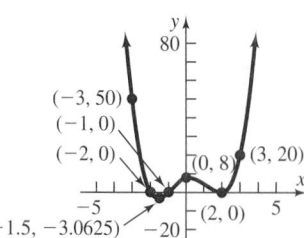

79. y-intercept: 2;
x-intercepts: $-\dfrac{\sqrt{2}}{2}, \dfrac{\sqrt{2}}{2}, 2$

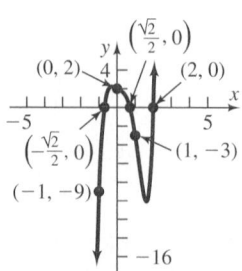

80. y-intercept: -3;
x-intercepts: $-3, -\dfrac{\sqrt{2}}{2}, \dfrac{\sqrt{2}}{2}$

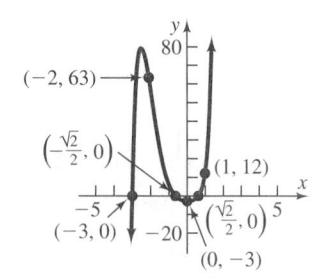

40. $\left\{x\left|-1 < x < \dfrac{2}{3}\text{ or }x > 2\right.\right\}; \left(-1, \dfrac{2}{3}\right)$ or $(2, \infty)$ **41.** $\{x|x > 4\}; (4, \infty)$ **42.** $\{x|0 < x < 1\}; (0, 1)$ **43.** $\{x|x \le -2 \text{ or } x \ge 2\}; (-\infty, -2]$ or $[2, \infty)$ **44.** $\{x|x = 0 \text{ or } x \ge 3\}; 0 \text{ or } [3, \infty)$ **45.** $\{x|x < -4 \text{ or } x \ge 2\}; (-\infty, -4) \text{ or } [2, \infty)$ **46.** $\{x|x < -4 \text{ or } x \ge 1\}; (-\infty, -4)$ or $[1, \infty)$

47.

$f(x) \le g(x)$ if $-1 \le x \le 1$

48.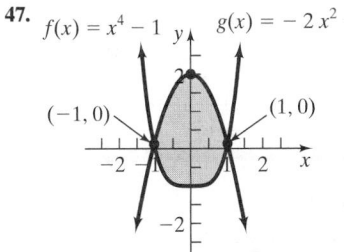

$f(x) \le g(x)$ if $0 \le x \le 1$

49.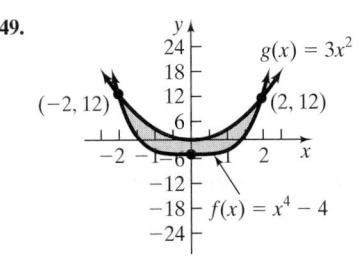

$f(x) \le g(x)$ if $-2 \le x \le 2$

50.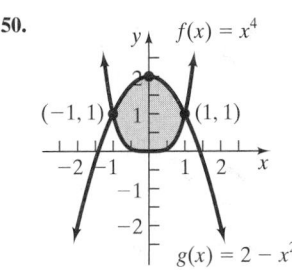

$f(x) \le g(x)$ if $-1 \le x \le 1$

51. Produce at least 250 bicycles
52. Produce at least 300 bicycles

Historical Problems *(page 247)*

1.
$$\left(x - \frac{b}{3}\right)^3 + b\left(x - \frac{b}{3}\right)^2 + c\left(x - \frac{b}{3}\right) + d = 0$$
$$x^3 - bx^2 + \frac{b^2 x}{3} - \frac{b^3}{27} + bx^2 - \frac{2b^2 x}{3} + \frac{b^3}{9} + cx - \frac{bc}{3} + d = 0$$
$$x^3 + \left(c - \frac{b^2}{3}\right)x + \left(\frac{2b^3}{27} - \frac{bc}{3} + d\right) = 0$$
Let $p = c - \dfrac{b^2}{3}$ and $q = \dfrac{2b^3}{27} - \dfrac{bc}{3} + d$. Then $x^3 + px + q = 0$.

2.
$$(H + K)^3 + p(H + K) + q = 0$$
$$H^3 + 3H^2 K + 3HK^2 + K^3 + pH + pK + q = 0$$
Let $3HK = -p$.
$$H^3 - pH - pK + K^3 + pH + pK + q = 0 \qquad H^3 + K^3 = -q$$

3.
$$3HK = -p$$
$$K = -\frac{p}{3H}$$
$$H^3 + \left(-\frac{p}{3H}\right)^3 = -q$$
$$H^3 - \frac{p^3}{27H^3} = -q$$
$$27H^6 - p^3 = -27qH^3$$
$$27H^6 + 27qH^3 - p^3 = 0$$
$$H^3 = \frac{-27q \pm \sqrt{(27q)^2 - 4(27)(-p^3)}}{2 \cdot 27}$$
$$H^3 = \frac{-q}{2} \pm \sqrt{\frac{27^2 q^2}{2^2 (27^2)} + \frac{4(27)p^3}{2^2 (27^2)}}$$
$$H^3 = \frac{-q}{2} \pm \sqrt{\frac{q^2}{4} + \frac{p^3}{27}}$$
$$H = \sqrt[3]{\frac{-q}{2} + \sqrt{\frac{q^2}{4} + \frac{p^3}{27}}}$$
Choose the positive root for now.

4. $H^3 + K^3 = -q$
$$K^3 = -q - H^3$$
$$K^3 = -q - \left[\frac{-q}{2} + \sqrt{\frac{q^2}{4} + \frac{p^3}{27}}\right]$$
$$K^3 = \frac{-q}{2} - \sqrt{\frac{q^2}{4} + \frac{p^3}{27}}$$
$$K = \sqrt[3]{\frac{-q}{2} - \sqrt{\frac{q^2}{4} + \frac{p^3}{27}}}$$

5. $x = H + K$
$$x = \sqrt[3]{\frac{-q}{2} + \sqrt{\frac{q^2}{4} + \frac{p^3}{27}}} + \sqrt[3]{\frac{-q}{2} - \sqrt{\frac{q^2}{4} + \frac{p^3}{27}}}$$ (Note that had we used the negative root in 3 the result would be the same.)

6. $x = 3$ **7.** $x = 2$ **8.** $x = 2$

3.5 Assess your Understanding *(page 248)*

5. Remainder; dividend **6.** $f(c)$ **7.** -4 **8.** False **9.** False **10.** True **11.** $R = f(2) = 8$; no **12.** $R = f(-3) = 161$; no **13.** $R = f(2) = 0$; yes **14.** $R = f(2) = 0$; yes **15.** $R = f(-3) = 0$; yes **16.** $R = f(-3) = 0$; yes **17.** $R = f(-4) = 1$; no **18.** $R = f(-4) = 0$; yes

45. One possibility: $R(x) = \dfrac{x^2}{x^2 - 4}$ **46.** One possibility: $R(x) = -\dfrac{x}{x^2 - 1}$ **47.** One possibility: $R(x) = \dfrac{(x - 1)(x - 3)\left(x^2 + \frac{4}{3}\right)}{(x + 1)^2(x - 2)^2}$

48. One possibility: $R(x) = \dfrac{3(x + 2)(x - 1)^2}{(x + 3)(x - 4)^2}$

49. **(a)** t-axis; $C(t) \to 0$

(b) 0.4

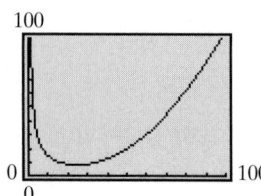

50. **(a)** t-axis; $C(t) \to 0$

(b) 6

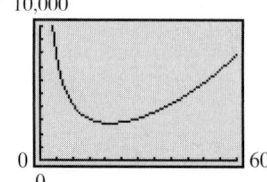

51. **(a)** $\overline{C}(x) = \dfrac{0.2x^3 - 2.3x^2 + 14.3x + 10.2}{x}$

(b) \$9400 **(c)** \approx\$10,933

(d) 60

(c) 0.71 hr after injection

(c) 5 min after injection

(e) 6 **(f)** \$9400

52. **(a)** $\overline{C}(x) = \dfrac{0.015x^3 - 0.595x^2 + 9.15x + 98.43}{x}$

(b) \approx\$11.52 **(c)** \approx\$7.59

(d) 100

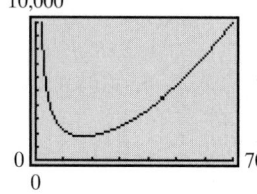

(e) \approx25,058 books **(f)** \approx\$7.59

53. **(a)** $S(x) = 2x^2 + \dfrac{40,000}{x}$

(b) 10,000

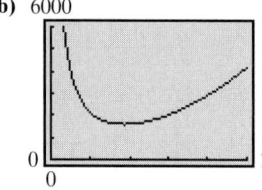

(c) 2784.95 sq in.

(d) 21.54 in. \times 21.54 in. \times 21.54 in.

(e) To minimize the cost of material needed for construction

54. **(a)** $S(x) = 2x^2 + \dfrac{20,000}{x}$

(b) 10,000

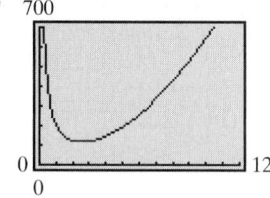

(c) 1754.41 sq in.

(d) 17.10 in. \times 17.10 in. \times 17.10 in.

(e) To minimize the cost of material needed for construction

55. **(a)** $C(r) = 12\pi r^2 + \dfrac{4000}{r}$

(b) 6000

The cost is smallest when $r = 3.76$ cm.

56. **(a)** $A(r) = 2\pi r^2 + \dfrac{200}{r}$

(b) Approximately 123.22 sq ft

(c) Approximately 150.53 sq ft

(d) Approximately 197.08 sq ft

(e) 700

A is smallest when $r = 2.52$ ft.

57. **(a)** $D(p) = \dfrac{429}{p}$ **(b)** 143 **58.** **(a)** $t(s) = \dfrac{1200}{s}$ **(b)** 30 minutes **59.** 450 cm^3 **60.** 24 amperes **61.** 124.76 pounds **62.** 0.012 foot-candle

63. $V = \pi r^2 h$ **64.** $V = \dfrac{\pi}{3}r^2 h$ **65.** $\sqrt[3]{6} \approx 1.82$ in. **66.** $V = \dfrac{kT}{P}$, $k = 5$; 19.375 atmospheres **67.** 2812.5 Joules **68.** 282.2 ft **69.** 384 psi

70. 900 1b **71.** No. Each function is a quotient of polynomials, but it is not written in lowest terms. Each function is undefined for $x = 1$; each graph has a hole at $x = 1$. **72.** All four graphs have a vertical asymptote at $x = 1$; $y = \dfrac{x^2}{x - 1}$ has an oblique asymptote at $y = x + 1$.

3.4 Assess Your Understanding (page 235)

2. False **3.** $\{x | x < -2\}$; $(-\infty, -2)$ **4.** $\{x | x > 5\}$; $(5, \infty)$ **5.** $\{x | x > 4\}$; $(4, \infty)$ **6.** $\{x | x < -8\}$; $(-\infty, -8)$ **7.** $\{x | x \leq -3 \text{ or } 0 \leq x \leq 3\}$; $(-\infty, -3]$ or $[0, 3]$ **8.** $\{x | -1 \leq x \leq 0 \text{ or } x \geq 1\}$; $[-1, 0]$ or $[1, \infty)$ **9.** $\{x | -4 < x < 0 \text{ or } x > 0\}$; $(-4, 0)$ or $(0, \infty)$ **10.** $\{x | x < -5\}$; $(-\infty, -5)$ **11.** $\{x | x \geq 1\}$; $[1, \infty)$ **12.** $\{x | x \geq -2\}$; $[-2, \infty)$ **13.** $\{x | x \leq 1 \text{ or } 2 \leq x \leq 3\}$; $(-\infty, 1]$ or $[2, 3]$ **14.** $\{x | x \leq -3 \text{ or } -2 \leq x \leq -1\}$; $(-\infty, -3]$ or $[-2, -1]$ **15.** $\{x | -1 < x < 0 \text{ or } x > 3\}$; $(-1, 0)$ or $(3, \infty)$ **16.** $\{x | -3 < x < 0 \text{ or } x > 1\}$; $(-3, 0)$ or $(1, \infty)$ **17.** $\{x | x < -1 \text{ or } x > 1\}$; $(-\infty, -1)$ or $(1, \infty)$ **18.** $\{x | -3 < x < 0 \text{ or } 0 < x < 3\}$; $(-3, 0)$ or $(0, 3)$ **19.** $\{x | x < -1 \text{ or } x > 1\}$; $(-\infty, -1)$ or $(1, \infty)$ **20.** $\{x | x > 1\}$; $(1, \infty)$ **21.** $\{x | x < -1 \text{ or } x > 1\}$; $(-\infty, -1)$ or $(1, \infty)$ **22.** $\{x | x < -1 \text{ or } x > 3\}$; $(-\infty, -1)$ or $(3, \infty)$ **23.** $\{x | x \leq -1 \text{ or } 0 < x \leq 1\}$; $(-\infty, -1]$ or $(0, 1]$ **24.** $\{x | x \leq -2 \text{ or } 1 < x \leq 3\}$; $(-\infty, -2]$ or $(1, 3]$ **25.** $\{x | x < -1 \text{ or } x > 1\}$; $(-\infty, -1)$ or $(1, \infty)$ **26.** $\{x | x < -2 \text{ or } x > 2\}$; $(-\infty, -2)$ or $(2, \infty)$ **27.** $\left\{x \middle| x < -\dfrac{2}{3} \text{ or } 0 < x < \dfrac{3}{2}\right\}$; $\left(-\infty, -\dfrac{2}{3}\right)$ or $\left(0, \dfrac{3}{2}\right)$ **28.** $\{x | x < 0 \text{ or } 3 < x < 4\}$; $(-\infty, 0)$ or $(3, 4)$ **29.** $\{x | x < 2\}$; $(-\infty, 2)$ **30.** $\{x | x > 4\}$; $(4, \infty)$ **31.** $\{x | -2 < x \leq 9\}$; $(-2, 9]$ **32.** $\{x | -8 \leq x < -2\}$; $[-8, -2)$ **33.** $\{x | x < 2 \text{ or } 3 < x < 5\}$; $(-\infty, 2)$ or $(3, 5)$ **34.** $\{x | -7 < x < -1 \text{ or } x > 3\}$; $(-7, -1)$ or $(3, \infty)$ **35.** $\{x | x < -3 \text{ or } -1 < x < 1 \text{ or } x > 2\}$; $(-\infty, -3)$ or $(-1, 1)$ or $(2, \infty)$ **36.** $\left\{x \middle| x < -\dfrac{5}{2} \text{ or } -2 < x < -1\right\}$; $\left(-\infty, -\dfrac{5}{2}\right)$ or $(-2, -1)$ **37.** $\{x | x < -5 \text{ or } -4 \leq x \leq -3 \text{ or } x = 0 \text{ or } x > 1\}$; $(-\infty, -5)$ or $[-4, -3]$ or 0 or $(1, \infty)$ **38.** $\{x | x < -1 \text{ or } 0 \leq x < 1 \text{ or } x \geq 2\}$; $(-\infty, -1)$ or $[0, 1)$ or $[2, \infty)$ **39.** $\left\{x \middle| -\dfrac{1}{2} < x < 1 \text{ or } x > 3\right\}$

40. **1.** $f(x) = \dfrac{2x^2 + 9}{x}$; domain $\{x | x \neq 0\}$; no y-intercept **2.** f is in lowest terms; no x-intercepts **3.** f is in lowest terms; vertical asymptote: $x = 0$
4. Oblique asymptote: $y = 2x$, not intersected
5.

Interval	$(-\infty, 0)$	$(0, \infty)$
Number Chosen	-1	1
Value of f	$f(-1) = -11$	$f(1) = 11$
Location of Graph	Below x-axis	Above x-axis
Point on Graph	$(-1, -11)$	$(1, 11)$

7.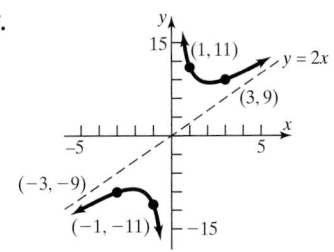

41. **1.** $f(x) = \dfrac{x^3 + 1}{x} = \dfrac{(x + 1)(x^2 - x + 1)}{x}$; domain: $\{x | x \neq 0\}$; no y-intercepts **2.** f is in lowest terms; x-intercept: -1 **3.** f is in lowest terms
vertical asymptote: $x = 0$ **4.** No horizontal or oblique asymptotes
5.

Interval	$(-\infty, -1)$	$(-1, 0)$	$(0, \infty)$
Number Chosen	-2	$-\frac{1}{2}$	1
Value of f	$f(-2) = 3.5$	$f\left(-\frac{1}{2}\right) = -1.75$	$f(1) = 2$
Location of Graph	Above x-axis	Below x-axis	Above x-axis
Point on Graph	$(-2, 3.5)$	$\left(-\frac{1}{2}, -1.75\right)$	$(1, 2)$

7.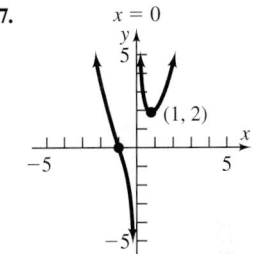

42. **1.** $f(x) = \dfrac{2x^3 + 16}{x} = \dfrac{2(x + 2)(x^2 - 2x + 4)}{x}$; domain $\{x | x \neq 0\}$; no y-intercept **2.** f is in lowest terms; x-intercept: -2 **3.** f is in lowest terms
vertical asymptote: $x = 0$ **4.** No oblique or horizontal asymptotes
5.

Interval	$(-\infty, -2)$	$(-2, 0)$	$(0, \infty)$
Number Chosen	-4	-1	1
Value of f	$f(-4) = 28$	$f(-1) = -14$	$f(1) = 18$
Location of Graph	Above x-axis	Below x-axis	Above x-axis
Point on Graph	$(-3, 16)$	$(-1, -14)$	$(1, 18)$

7.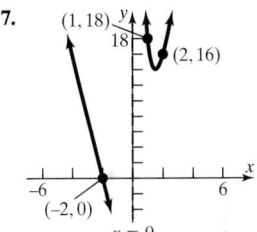

43. **1.** $f(x) = \dfrac{x^4 + 1}{x^3}$; domain: $\{x | x \neq 0\}$; no y-intercepts **2.** f is in lowest terms; no x-intercepts **3.** f is in lowest terms; vertical asymptote: $x = 0$
4. Oblique asymptote: $y = x$, not intersected
5.

Interval	$(-\infty, 0)$	$(0, \infty)$
Number Chosen	-1	1
Value of f	$f(-1) = -2$	$f(1) = 2$
Location of Graph	Below x-axis	Above x-axis
Point on Graph	$(-1, -2)$	$(1, 2)$

7.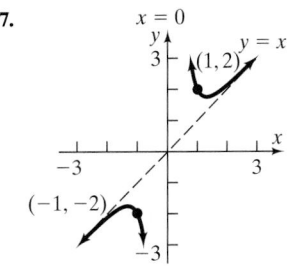

44. **1.** $f(x) = \dfrac{2x^4 + 9}{x^3}$; domain $\{x | x \neq 0\}$; no y-intercept **2.** f is in lowest terms; no x-intercepts **3.** f is in lowest terms; vertical asymptote: $x = 0$
4. Oblique asymptote: $y = 2x$, not intersected
5.

Interval	$(-\infty, 0)$	$(0, \infty)$
Number Chosen	-2	2
Value of f	$f(-2) = -5.125$	$f(2) = 5.125$
Location of Graph	Below x-axis	Above x-axis
Point on Graph	$(-2, -5.125)$	$(2, 5.125)$

7.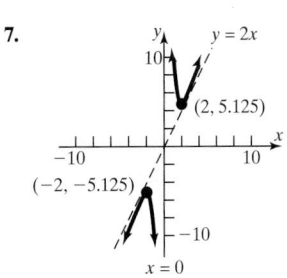

35. 1. $R(x) = \dfrac{(3x + 1)(2x - 3)}{(x - 2)(2x - 3)}$; domain: $\left\{x \mid x \neq \dfrac{3}{2}, x \neq 2\right\}$; y-intercept: $-\dfrac{1}{2}$ **2.** In lowest terms, $R(x) = \dfrac{3x + 1}{x - 2}$; x-intercept: $-\dfrac{1}{3}$

3. Vertical asymptote: $x = 2$; hole at $\left(\dfrac{3}{2}, -11\right)$ **4.** Horizontal asymptote: $y = 3$, not intersected

5.

Interval	$\left(-\infty, -\dfrac{1}{3}\right)$	$\left(-\dfrac{1}{3}, \dfrac{3}{2}\right)$	$\left(\dfrac{3}{2}, 2\right)$	$(2, \infty)$
Number Chosen	-1	0	1.7	6
Value of R	$R(-1) = \dfrac{2}{3}$	$R(0) = -\dfrac{1}{2}$	$R(1.7) \approx -20.3$	$R(6) = 4.75$
Location of Graph	Above x-axis	Below x-axis	Below x-axis	Above x-axis
Point on Graph	$\left(-1, \dfrac{2}{3}\right)$	$\left(0, -\dfrac{1}{2}\right)$	$(1.7, -20.3)$	$(6, 4.75)$

7.

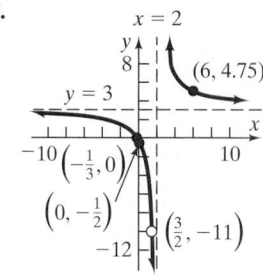

36. 1. $R(x) = \dfrac{(4x + 3)(2x + 5)}{(x - 3)(2x + 5)}$; domain $\left\{x \mid x \neq -\dfrac{5}{2}, x \neq 3\right\}$; y-intercept: -1 **2.** In lowest terms, $R(x) = \dfrac{4x + 3}{x - 3}$; x-intercept: $-\dfrac{3}{4}$

3. Vertical asymptote: $x = 3$, hole at $\left(-\dfrac{5}{2}, \dfrac{14}{11}\right)$ **4.** Horizontal asymptote: $y = 4$

5.

Interval	$(-\infty, -2.5)$	$\left(-2.5, -\dfrac{3}{4}\right)$	$\left(-\dfrac{3}{4}, 3\right)$	$(3, \infty)$
Number Chosen	-3	-1	0	6
Value of R	$R(-3) = \dfrac{3}{2}$	$R(-1) = \dfrac{1}{4}$	$R(0) = -1$	$R(6) = 9$
Location of Graph	Above x-axis	Above x-axis	Below x-axis	Above x-axis
Point on Graph	$\left(-3, \dfrac{3}{2}\right)$	$\left(-1, \dfrac{1}{4}\right)$	$(0, -1)$	$(6, 9)$

7.

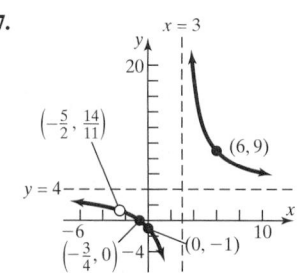

37. 1. $R(x) = \dfrac{(x + 3)(x + 2)}{x + 3}$; domain: $\{x \mid x \neq -3\}$; y-intercept: 2 **2.** In lowest terms, $R(x) = x + 2$; x-intercept: -2 **3.** Vertical asymptote: none; hole at $(-3, -1)$ **4.** Oblique asymptote: $y = x + 2$ intersected at all points except $x = -3$

5.

Interval	$(-\infty, -3)$	$(-3, -2)$	$(-2, \infty)$
Number Chosen	-4	-2.5	0
Value of R	$R(-4) = -2$	$R(-2.5) = -\dfrac{1}{2}$	$R(0) = 2$
Location of Graph	Below x-axis	Below x-axis	Above x-axis
Point on Graph	$(-4, -2)$	$\left(-2.5, -\dfrac{1}{2}\right)$	$(0, 2)$

7.

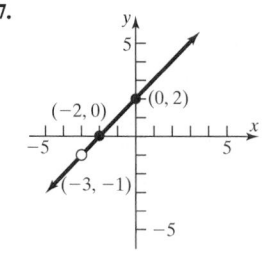

38. 1. $R(x) = \dfrac{(x + 6)(x - 5)}{x + 6}$; domain $\{x \mid x \neq -6\}$; y-intercept: -5 **2.** In lowest terms, $R(x) = x - 5$; x-intercept: 5 **3.** Vertical asymptote: none; hole at $(-6, -11)$ **4.** Oblique asymptote: $y = x - 5$ intersected at all points except $x = -6$

5.

Interval	$(-\infty, -6)$	$(-6, 5)$	$(5, \infty)$
Number Chosen	-7	0	6
Value of R	$R(-7) = -12$	$R(0) = -5$	$R(6) = 1$
Location of Graph	Below x-axis	Below x-axis	Above x-axis
Point on Graph	$(-7, -12)$	$(0, -5)$	$(6, 1)$

7.

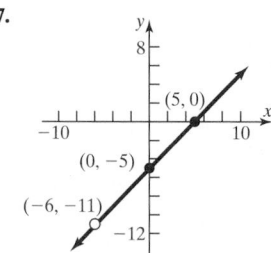

39. 1. $f(x) = \dfrac{x^2 + 1}{x}$; domain: $\{x \mid x \neq 0\}$; no y-intercepts **2.** f is in lowest terms; no x-intercepts **3.** f is in lowest terms; vertical asymptote: $x = 0$ **4.** Oblique asymptotes: $y = x$, not intersected

5.

Interval	$(-\infty, 0)$	$(0, \infty)$
Number Chosen	-1	1
Value of f	$f(-1) = -2$	$f(1) = 2$
Location of Graph	Below x-axis	Above x-axis
Point on Graph	$(-1, -2)$	$(1, 2)$

7.

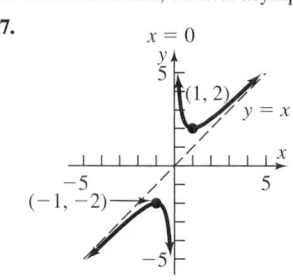

31. 1. Domain: $\{x|x \neq -3\}$; y-intercept: 0 **2.** x-intercepts: $0, 1$ **3.** vertical asymptote: $x = -3$ **4.** Horizontal asymptote: $y = 1$, not intersected

5.

Interval	$(-\infty, -3)$	$(-3, 0)$	$(0, 1)$	$(1, \infty)$
Number Chosen	-4	-1	$\frac{1}{2}$	2
Value of R	$R(-4) = 100$	$R(-1) = -0.5$	$R\left(\frac{1}{2}\right) \approx 0.003$	$R(2) = 0.016$
Location of Graph	Above x-axis	Below x-axis	Above x-axis	Above x-axis
Point on Graph	$(-4, 100)$	$(-1, -0.5)$	$\left(\frac{1}{2}, 0.003\right)$	$(2, 0.016)$

7.

$x = -3$

See enlarged view at right.

Enlarged view

32. 1. Domain $\{x|x \neq 0, x \neq 4\}$; no y-intercept
2. x-intercepts: $1, -2, 3$
3. R is in lowest terms; vertical asymptotes: $x = 0$, $x = 4$
4. Horizontal asymptote: $y = 1$, intersected at $(3.186, 1), (0.314, 1)$

5.

Interval	$(-\infty, -2)$	$(-2, 0)$	$(0, 1)$	$(1, 3)$	$(3, 4)$	$(4, \infty)$
Number Chosen	-3	-1	$\frac{1}{2}$	2	3.186	6
Value of R	$R(-3) \approx 0.16$	$R(-1) = -0.32$	$R(\frac{1}{2}) \approx 0.51$	$R(2) = -0.5$	$R(3.186) \approx 1$	$R(6) = 5$
Location of Graph	Above x-axis	Below x-axis	Above x-axis	Below x-axis	Above x-axis	Above x-axis
Point on Graph	$(-3, 0.16)$	$(-1, -0.32)$	$(\frac{1}{2}, 0.51)$	$(2, -0.5)$	$(3.186, 1)$	$(6, 5)$

7.

$x = 0$ $x = 4$

33. 1. $R(x) = \dfrac{(x + 4)(x - 3)}{(x + 2)(x - 3)}$; domain: $\{x|x \neq -2, x \neq 3\}$; y-intercept: 2
2. In lowest terms, $R(x) = \dfrac{x + 4}{x + 2}$; x-intercept: -4
3. Vertical asymptote: $x = -2$; hole at $\left(3, \dfrac{7}{5}\right)$
4. Horizontal asymptote: $y = 1$, not intersected

5.

Interval	$(-\infty, -4)$	$(-4, -2)$	$(-2, 3)$	$(3, \infty)$
Number Chosen	-5	-3	0	4
Value of R	$R(-5) = \frac{1}{3}$	$R(-3) = -1$	$R(0) = 2$	$R(4) = \frac{4}{3}$
Location of Graph	Above x-axis	Below x-axis	Above x-axis	Above x-axis
Point on Graph	$\left(-5, \frac{1}{3}\right)$	$(-3, -1)$	$(0, 2)$	$\left(4, \frac{4}{3}\right)$

7.

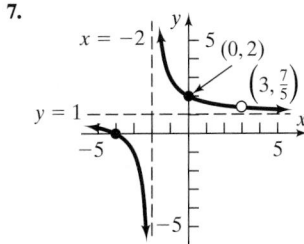

34. 1. $R(x) = \dfrac{(x + 5)(x - 2)}{(x + 5)(x + 3)}$; domain $\{x|x \neq -3, x \neq -5\}$; y-intercept: $-\dfrac{2}{3}$ **2.** In lowest terms, $R(x) = \dfrac{x - 2}{x + 3}$; x-intercept: 2 **3.** Vertical asymptote: $x = -3$; hole at $(-5, 3.5)$ **4.** Horizontal asymptote: $y = 1$, not intersected

5.

Interval	$(-\infty, -5)$	$(-5, -3)$	$(-3, 2)$	$(2, \infty)$
Number Chosen	-6	-4	0	3
Value of R	$R(-6) = \frac{8}{3}$	$R(-4) = 6$	$R(0) = -\frac{2}{3}$	$R(3) = \frac{1}{6}$
Location of Graph	Above x-axis	Above x-axis	Below x-axis	Above x-axis
Point on Graph	$(-6, \frac{8}{3})$	$(-4, 6)$	$(0, -\frac{2}{3})$	$(3, \frac{1}{6})$

7.

5.

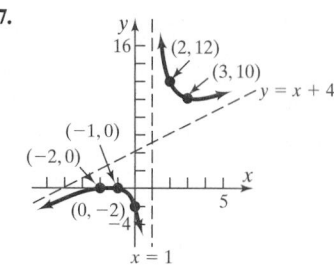

7.

Interval	$(-\infty, -2)$	$(-2, -1)$	$(-1, 1)$	$(1, \infty)$
Number Chosen	-3	-1.5	0	2
Value of F	$F(-3) = -0.5$	$F(-1.5) = 0.1$	$F(0) = -2$	$F(2) = 12$
Location of Graph	Below x-axis	Above x-axis	Below x-axis	Above x-axis
Point on Graph	$(-3, -0.5)$	$(-1.5, 0.1)$	$(0, -2)$	$(2, 12)$

27. 1. $R(x) = \dfrac{(x + 4)(x - 3)}{x - 4}$ domain: $\{x \mid x \neq 4\}$; y-intercept: 3 **2.** R is in lowest terms; x-intercepts: $-4, 3$ **3.** R is in lowest terms; vertical asymptote: $x = 4$ **4.** Oblique asymptote: $y = x + 5$, not intersected

5.

7.

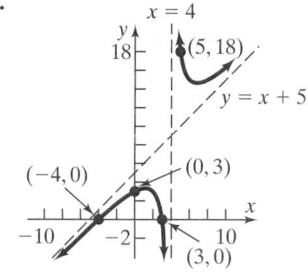

Interval	$(-\infty, -4)$	$(-4, 3)$	$(3, 4)$	$(4, \infty)$
Number Chosen	-5	0	3.5	5
Value of R	$R(-5) = -\frac{8}{9}$	$R(0) = 3$	$R(3.5) = -7.5$	$R(5) = 18$
Location of Graph	Below x-axis	Above x-axis	Below x-axis	Above x-axis
Point on Graph	$(-5, -\frac{8}{9})$	$(0, 3)$	$(3.5, -7.5)$	$(5, 18)$

28. 1. $R(x) = \dfrac{(x + 3)(x - 4)}{x + 5}$; domain $\{x \mid x \neq -5\}$; y-intercept: $-\dfrac{12}{5}$ **2.** R is in lowest terms; x-intercepts: $-3, 4$ **3.** R is in lowest terms; vertical asymptote: $x = -5$ **4.** Oblique asymptote: $y = x - 6$, not intersected

5.

7.

Interval	$(-\infty, -5)$	$(-5, -3)$	$(-3, 4)$	$(4, \infty)$
Number Chosen	-7	-4	0	5
Value of R	$R(-7) = -22$	$R(-4) = 8$	$R(0) = -2.4$	$R(5) = 0.8$
Location of Graph	Below x-axis	Above x-axis	Below x-axis	Above x-axis
Point on Graph	$(-7, -22)$	$(-4, 8)$	$(0, -2.4)$	$(5, -0.8)$

29. 1. $F(x) = \dfrac{(x + 4)(x - 3)}{x + 2}$; domain: $\{x \mid x \neq -2\}$; y-intercept: -6 **2.** F is in lowest terms; x-intercepts: $-4, 3$ **3.** F is in lowest terms; vertical asymptote: $x = -2$ **4.** Oblique asymptote: $y = x - 1$, not intersected

5.

7.

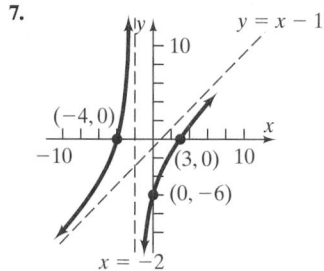

Interval	$(-\infty, -4)$	$(-4, -2)$	$(-2, 3)$	$(3, \infty)$
Number Chosen	-5	-3	0	4
Value of F	$F(-5) = -\frac{8}{3}$	$F(-3) = 6$	$F(0) = -6$	$F(4) = \frac{4}{3}$
Location of Graph	Below x-axis	Above x-axis	Below x-axis	Above x-axis
Point on Graph	$\left(-5, -\frac{8}{3}\right)$	$(-3, 6)$	$(0, -6)$	$\left(4, \frac{4}{3}\right)$

30. 1. $G(x) = \dfrac{(x + 3)(x - 4)}{x + 1}$; domain $\{x \mid x \neq -1\}$; y-intercept: -12 **2.** G is in lowest terms; x-intercepts: $-3, 4$ **3.** G is in lowest terms; vertical asymptote: $x = -1$ **4.** Oblique asymptote: $y = x - 2$, not intersected

5.

7.

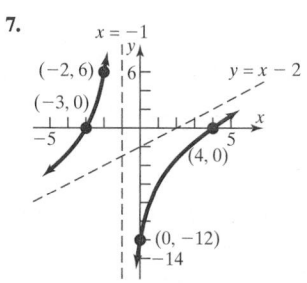

Interval	$(-\infty, -3)$	$(-3, -1)$	$(-1, 4)$	$(4, \infty)$
Number Chosen	-4	-2	0	5
Value of G	$G(-4) = -\frac{8}{3}$	$G(-2) = 6$	$G(0) = -12$	$G(5) = \frac{4}{3}$
Location of Graph	Below x-axis	Above x-axis	Below x-axis	Above x-axis
Point on Graph	$(-4, -\frac{8}{3})$	$(-2, 6)$	$(0, -12)$	$(5, \frac{4}{3})$

29. 1. $F(x) = \dfrac{x^3}{(x+2)(x-2)}$; domain: $\{x|x \neq -2, x \neq 2\}$; y-intercept: 0 **2.** F is in lowest terms; x-intercept: 0 **3.** F is in lowest terms; vertical asymptotes: $x = -2, x = 2$ **4.** Oblique asymptote: $y = x$; intersected at $(0, 0)$

5.

Interval	$(-\infty, -2)$	$(-2, 0)$	$(0, 2)$	$(2, \infty)$
Number Chosen	-3	-1	1	2
Value of F	$F(-3) = -\frac{27}{5}$	$F(-1) = \frac{1}{3}$	$F(1) = -\frac{1}{3}$	$F(3) = \frac{27}{5}$
Location of Graph	Below x-axis	Above x-axis	Below x-axis	Above x-axis
Point on Graph	$\left(-3, -\frac{27}{5}\right)$	$\left(-1, \frac{1}{3}\right)$	$\left(1, -\frac{1}{3}\right)$	$\left(3, \frac{27}{5}\right)$

7.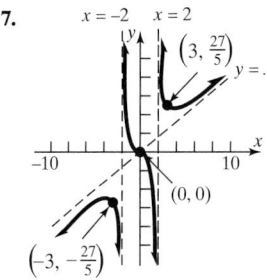

30. 1. Domain: $\{x|x \neq 1\}$; y-intercept: 0 **2.** F is in lowest terms; x-intercept: 0 **3.** F is in lowest terms; vertical asymptote: $x = 1$ **4.** Oblique asymptote: $y = 3x + 6$; intersected at $\left(\frac{2}{3}, 8\right)$

5.

Interval	$(-\infty, 0)$	$(0, 1)$	$(1, \infty)$
Number Chosen	-1	$\frac{1}{2}$	2
Value of F	$F(-1) = -\frac{3}{4}$	$F\left(\frac{1}{2}\right) = 1.5$	$F(2) = 24$
Location of Graph	Below x-axis	Above x-axis	Above x-axis
Point on Graph	$\left(-1, -\frac{3}{4}\right)$	$\left(\frac{1}{2}, 1.5\right)$	$(2, 24)$

7.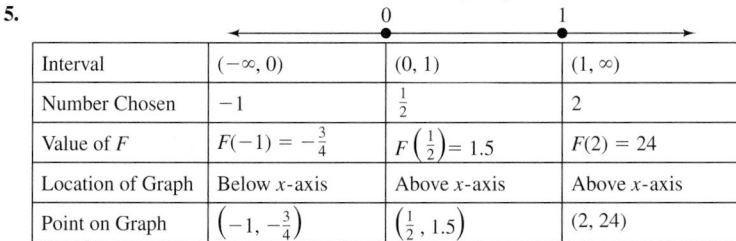

31. 1. Domain: $\{x|x \neq 1\}$; y-intercept: 0 **2.** R is in lowest terms; x-intercept: 0 **3.** R is in lowest terms; vertical asymptote: $x = 1$ **4.** No oblique or horizontal asymptote

5.

Interval	$(-\infty, 0)$	$(0, 1)$	$(1, \infty)$
Number Chosen	-2	$\frac{1}{2}$	2
Value of R	$R(-2) \approx \frac{32}{9}$	$R\left(\frac{1}{2}\right) = \frac{1}{2}$	$R(2) = 32$
Location of Graph	Below x-axis	Above x-axis	Above x-axis
Point on Graph	$\left(-2, \frac{32}{9}\right)$	$\left(\frac{1}{2}, \frac{1}{2}\right)$	$(2, 32)$

7.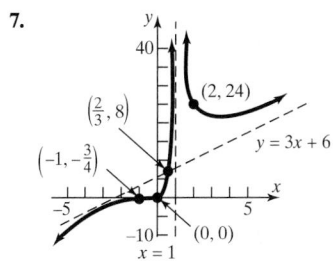

32. 1. $R(x) = \dfrac{x^4}{(x+3)(x-3)}$; domain: $\{x|x \neq -3, x \neq 3\}$; y-intercept: 0 **2.** R is in lowest terms; x-intercept: 0
3. R is in lowest terms; vertical asymptotes: $x = 3$ and $x = -3$ **4.** No oblique or horizontal asymptote.

5.

Interval	$(-\infty, -3)$	$(-3, 0)$	$(0, 3)$	$(3, \infty)$
Number Chosen	-4	-1	1	4
Value of R	$R(-4) \approx 37.57$	$R(-1) = -\frac{1}{8}$	$R(1) = -\frac{1}{8}$	$R(4) = 37$
Location of Graph	Above x-axis	Below x-axis	Below x-axis	Above x-axis
Point on Graph	$\left(-4, 37.57\right)$	$\left(-1, -\frac{1}{8}\right)$	$\left(1, -\frac{1}{8}\right)$	$\left(4, 37.57\right)$

7.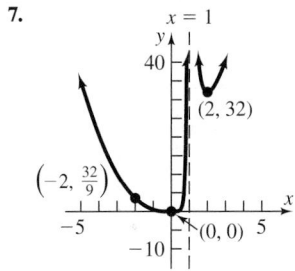

33. 1. $G(x) = \dfrac{(x+2)(x-2)}{(x+1)(x-2)}$; domain: $\{x|x \neq -1, x \neq 2\}$; y-intercept: 2 **2.** In lowest terms, $G(x) = \dfrac{x+2}{x+1}$; x-intercept: -2

3. Vertical asymptote: $x = -1$; hole at $\left(2, \frac{4}{3}\right)$ **4.** Horizontal asymptote: $y = 1$, not intersected

5.

Interval	$(-\infty, -2)$	$(-2, -1)$	$(-1, 2)$	$(2, \infty)$
Number Chosen	-3	-1.5	0	3
Value of G	$G(-3) = \frac{1}{2}$	$G(-1.5) = -1$	$G(0) = 2$	$G(3) = 1.25$
Location of Graph	Above x-axis	Below x-axis	Above x-axis	Above x-axis
Point on Graph	$\left(-3, \frac{1}{2}\right)$	$(-1.5, -1)$	$(0, 2)$	$(3, 1.25)$

7.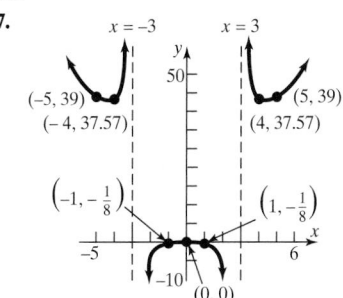

34. 1. $F(x) = \dfrac{(x-1)^2}{(x+1)(x-1)}$; domain: $\{x|x \neq -1, x \neq 1\}$; y-intercept: -1 **2.** In lowest terms, $F(x) = \dfrac{x-1}{x+1}$; no x-intercepts
3. Vertical asymptote: $x = -1$; hole at $(1, 0)$ **4.** Horizontal asymptote: $y = 1$, not intersected

5.

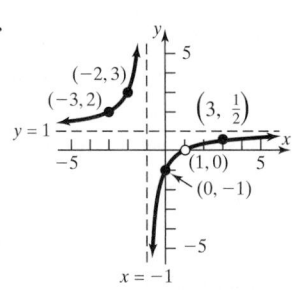

	$(-\infty, -1)$	$(-1, 1)$	$(1, \infty)$
Interval	$(-\infty, -1)$	$(-1, 1)$	$(1, \infty)$
Number Chosen	-2	0	3
Value of F	$F(-2) = 3$	$F(0) = -1$	$F(3) = \frac{1}{2}$
Location of Graph	Above x-axis	Below x-axis	Above x-axis
Point on Graph	$(-2, 3)$	$\left(0, -1\right)$	$\left(3, \frac{1}{2}\right)$

7.

35. $\{x \mid -\infty < x < -2 \text{ or } -1 < x < 2\}$; $(-\infty, -2)$ or $(-1, 2)$ **36.** $\{x \mid -4 \le x \le -1 \text{ or } x \ge 1\}$; $[-4, -1]$ or $[1, \infty)$ **37.** $\{x \mid -3 < x \le 3\}$; $(-3, 3]$

38. $\left\{x \mid x < \frac{1}{3} \text{ or } x > 1\right\}$; $\left(-\infty, \frac{1}{3}\right)$ or $(1, \infty)$ **39.** $\{x \mid x < 1 \text{ or } x > 2\}$; $(-\infty, 1)$ or $(2, \infty)$ **40.** $\left\{x \mid -\frac{5}{2} < x \le -\frac{7}{6}\right\}$; $\left(-\frac{5}{2}, -\frac{7}{6}\right]$

41. $\{x \mid 1 \le x \le 2 \text{ or } x > 3\}$; $[1, 2]$ or $(3, \infty)$ **42.** $\{x \mid x \le -1 \text{ or } 0 < x < 5\}$; $(-\infty, -1]$ or $(0, 5)$ **43.** $\{x \mid x < -4 \text{ or } 2 < x < 4 \text{ or } x > 6\}$;
$(-\infty, -4)$ or $(2, 4)$ or $(6, \infty)$ **44.** $\{x \mid x < -5 \text{ or } -4 < x \le -2 \text{ or } 0 \le x \le 1\}$; $(-\infty, -5)$ or $(-4, -2]$ or $[0, 1]$ **45.** $q(x) = 8x^2 + 5x + 6$;
$R = 10$; g is not a factor of f. **46.** $q(x) = 2x^2 + 12x + 19$; $R = 43$; g is not a factor of f. **47.** $q(x) = x^3 - 4x^2 + 8x - 1$; $R = 0$;
g is a factor of f. **48.** $q(x) = x^3 - x^2 + 2$; $R = 0$; g is a factor of f. **49.** $f(4) = 47{,}105$ **50.** $f(-2) = 204$ **51.** 4, 2, or 0 positive; 2 or 0 negative

52. 1 positive; 2 or 0 negative **53.** $\pm1, \pm3, \pm\frac{1}{2}, \pm\frac{3}{2}, \pm\frac{1}{3}, \pm\frac{1}{4}, \pm\frac{3}{4}, \pm\frac{1}{6}, \pm\frac{1}{12}$ **54.** $\pm1, \pm\frac{1}{2}, \pm\frac{1}{3}, \pm\frac{1}{6}$ **55.** $-2, 1, 4$; $f(x) = (x + 2)(x - 1)(x - 4)$

56. $-1, 4, -2$; $f(x) = (x + 1)(x - 4)(x + 2)$ **57.** $\frac{1}{2}$, multiplicity 2; -2; $f(x) = 4\left(x - \frac{1}{2}\right)^2(x + 2)$ **58.** $-\frac{1}{2}$, multiplicity 2; 2;

$f(x) = 4(x - 2)\left(x + \frac{1}{2}\right)^2$ **59.** 2, multiplicity 2; $f(x) = (x - 2)^2(x^2 + 5)$ **60.** -3, multiplicity 2; $f(x) = (x + 3)^2(x^2 + 2)$ **61.** $\{-3, 2\}$

62. $\{-3, 2\}$ **63.** $\left\{-3, -1, -\frac{1}{2}, 1\right\}$ **64.** $\left\{-3, -2, -\frac{1}{2}, 2\right\}$ **65.** 5 **66.** 11 **67.** $\frac{37}{2}$ **68.** $\frac{17}{3}$ **69.** $f(0) = -1$; $f(1) = 1$ **70.** $f(1) = -2, f(2) = 9$

71. $f(0) = -1$; $f(1) = 1$ **72.** $f(1) = -3, f(2) = 62$ **73.** 1.52 **74.** 1.33 **75.** 0.93 **76.** 1.14 **77.** $4 - i$ **78.** $3 - 4i$ **79.** $-i, 1 - i$ **80.** $1 - i$

81. $-2, 1, 4$; $f(x) = (x + 2)(x - 1)(x - 4)$ **82.** $-2, -1, 4$; $f(x) = (x + 2)(x + 1)(x - 4)$ **83.** -2; $\frac{1}{2}$ (multiplicity 2); $f(x) = 4(x + 2)\left(x - \frac{1}{2}\right)^2$

84. $-\frac{1}{2}$ (multiplicity 2), 2; $f(x) = 4\left(x + \frac{1}{2}\right)^2(x - 2)$ **85.** 2 (multiplicity 2), $-\sqrt{5}i, \sqrt{5}i$; $f(x) = \left(x + \sqrt{5}i\right)\left(x - \sqrt{5}i\right)(x - 2)^2$

86. -3 (multiplicity 2), $-\sqrt{2}i, \sqrt{2}i$; $f(x) = (x + 3)^2\left(x + \sqrt{2}i\right)\left(x - \sqrt{2}i\right)$

87. $-3, 2, -\frac{\sqrt{2}}{2}i, \frac{\sqrt{2}}{2}i$; $f(x) = 2(x + 3)(x - 2)\left(x + \frac{\sqrt{2}}{2}i\right)\left(x - \frac{\sqrt{2}}{2}i\right)$

88. $-3, 2, -\frac{\sqrt{3}}{3}i, \frac{\sqrt{3}}{3}i$; $f(x) = 3(x + 3)(x - 2)\left(x + \frac{\sqrt{3}}{3}i\right)\left(x - \frac{\sqrt{3}}{3}i\right)$ **89.** 199.9 pounds **90.** 14.7 ohms

91. (a) $A(r) = 2\pi r^2 + \dfrac{500}{r}$
(b) 223.22 cm^2
(c) 257.08 cm^2
(d)

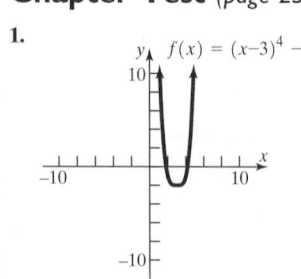

A is smallest when $r \approx 3.41$ cm.

92. (a)

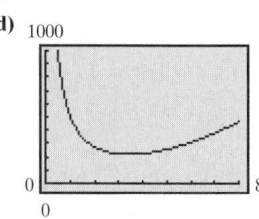

(b) $A(t) = -212.01t^3 + 2429.13t^2$
$+ 59{,}568.85t + 130{,}003.14$
(c) 797,003

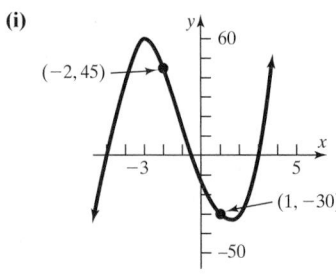

95. (a) Even **(b)** Positive **(c)** Even **(d)** The graph touches the x-axis at $x = 0$, but does not cross it there. **(e)** 8

Chapter Test (page 259)

1.

2. (a) 3 **(b)** Every zero of g lies between -15 and 15.
(c) $\dfrac{p}{q}$: $\pm\frac{1}{2}, \pm1, \pm\frac{3}{2}, \pm\frac{5}{2}, \pm3, \pm5, \pm\frac{15}{2}, \pm15$
(d) $-5, -\frac{1}{2}, 3$; $g(x) = (x + 5)(2x + 1)(x - 3)$
(e) y-intercept: -15 x-intercepts: $-5, -\frac{1}{2}, 3$
(f) Crosses at $-5, -\frac{1}{2}, 3$ **(g)** $y = 2x^3$
(h) Near -5: $g(x) \approx 72(x + 5)$
Near $-\frac{1}{2}$: $g(x) \approx -\frac{63}{4}(2x + 1)$
Near 3: $g(x) \approx 56(x - 3)$
(i)

3. $4, -5i, 5i$ **4.** $\left\{1, \dfrac{5 - \sqrt{61}}{6}, \dfrac{5 + \sqrt{61}}{6}\right\}$ **5.** Domain: $\{x \mid x \neq -10, x \neq 4\}$; asymptotes: $x = -10, y = 2$

6. Domain: $\{x \mid x \neq -1\}$; asymptotes: $x = -1, y = x + 1$

7.

8. Answers may vary. One possibility is $f(x) = x^4 - 4x^3 - 2x^2 + 20x$.

9. Answers may vary. One possibility is $r(x) = \dfrac{2(x - 9)(x - 1)}{(x - 4)(x - 9)}$.

10. $f(0) = 8; f(4) = -36$

Since $f(0) = 8 > 0$ and $f(4) = -36 < 0$, the Intermediate Value Theorem guarantees that there is at least one real zero between 0 and 4.

11. $\{x \mid x < 3 \text{ or } x > 8\}$; $(-\infty, 3)$ or $(8, \infty)$

Cumulative Review (page 261)

1. $\sqrt{26}$

2. $\{x \mid x \leq 0 \text{ or } x \geq 1\}$; $(-\infty, 0]$ or $[1, \infty)$

3. $\{x \mid -1 < x < 4\}$; $(-1, 4)$

4. $f(x) = -3x + 1$

5. $y = 2x - 1$

6.

7. Not a function; 3 has two images **8.** $\{0, 2, 4\}$ **9.** $\left\{x \mid x \geq \dfrac{3}{2}\right\}$; $\left[\dfrac{3}{2}, \infty\right)$

10. Center: $(-2, 1)$; radius: 3

11. x-intercepts: $-3, 0, 3$; y-intercept: 0; symmetric with respect to the origin

12. $y = -\dfrac{2}{3}x + \dfrac{17}{3}$

13. Not a function; it fails the Vertical Line Test.

14. **(a)** 22 **(b)** $x^2 - 5x - 2$ **(c)** $-x^2 - 5x + 2$ **(d)** $9x^2 + 15x - 2$ **(e)** $2x + h + 5$

15. **(a)** $\{x \mid x \neq 1\}$ **(b)** No, $(2, 7)$ is on the graph. **(c)** $4; (3, 4)$ is on the graph. **(d)** $\dfrac{7}{4}; \left(\dfrac{7}{4}, 9\right)$ is on the graph.

16.

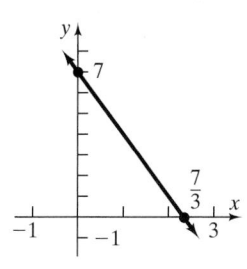

17.

18. $x + 4; m_{\text{sec}} = 6$

19. **(a)** x-intercepts: $-5, -1, 5$; y-intercept: -3
(b) No symmetry
(c) Neither
(d) Increasing: $(-\infty, -3)$ and $(2, \infty)$; decreasing: $(-3, 2)$
(e) Local maximum is 5 and occurs at $x = -3$.
(f) Local minimum is -6 and occurs at $x = 2$.

20. Odd

21. **(a)** Domain: $\{x \mid -3 < x\}$ or $(-3, \infty)$
(b) x-intercept: $-\dfrac{1}{2}$; y-intercept: 1
(c)

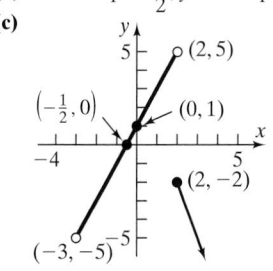

(d) Range: $\{y \mid y < 5\}$ or $(-\infty, 5)$

22.

23. **(a)** $(f + g)(x) = x^2 - 9x - 6$; domain: all real numbers
(b) $\left(\dfrac{f}{g}\right)(x) = \dfrac{x^2 - 5x + 1}{-4x - 7}$; domain: $\left\{x \mid x \neq -\dfrac{7}{4}\right\}$

24. **(a)** $R(x) = -\dfrac{1}{10}x^2 + 150x$
(b) \$14,000 **(c)** 750; \$56,250 **(d)** \$75

CHAPTER 4 Exponential and Logarithmic Functions

4.1 Assess Your Understanding (page 268)

4. $(g \circ f)(x)$ **5.** F **6.** F **7. (a)** -1 **(b)** -1 **(c)** 8 **(d)** 0 **(e)** 8 **(f)** -7 **8. (a)** 5 **(b)** 11 **(c)** 0 **(d)** 0 **(e)** 1 **(f)** 7 **9. (a)** 4 **(b)** 5
(c) -1 **(d)** -2 **10. (a)** 3 **(b)** 4 **(c)** 1 **(d)** -2 **11. (a)** 98 **(b)** 49 **(c)** 4 **(d)** 4 **12. (a)** 95 **(b)** 127 **(c)** 17 **(d)** 1 **13. (a)** 97
(b) $-\dfrac{163}{2}$ **(c)** 1 **(d)** $-\dfrac{3}{2}$ **14. (a)** 4418 **(b)** -191 **(c)** 8 **(d)** -2 **15. (a)** $2\sqrt{2}$ **(b)** $2\sqrt{2}$ **(c)** 1 **(d)** 0 **16. (a)** $\sqrt{13}$ **(b)** $3\sqrt{3}$
(c) $\sqrt{\sqrt{2}+1}$ **(d)** 0 **17. (a)** $\dfrac{1}{17}$ **(b)** $\dfrac{1}{5}$ **(c)** 1 **(d)** $\dfrac{1}{2}$ **18. (a)** $\dfrac{11}{6}$ **(b)** $\dfrac{3}{2}$ **(c)** 1 **(d)** $\dfrac{12}{17}$ **19. (a)** $\dfrac{3}{\sqrt[3]{4}+1}$ **(b)** 1 **(c)** $\dfrac{6}{5}$ **(d)** 0

20. (a) $\dfrac{2\sqrt{10}}{25}$ **(b)** $\dfrac{2}{2\sqrt{2}+1}$ **(c)** 1 **(d)** $\dfrac{2}{3}$ **21.** $\{x|x \neq 0, x \neq 2\}$ **22.** $\left\{x \mid x \neq 0, x \neq \dfrac{2}{3}\right\}$ **23.** $\{x|x \neq -4, x \neq 0\}$ **24.** $\left\{x \mid x \neq -\dfrac{2}{3}, x \neq 0\right\}$

25. $\left\{x \mid x \geq -\dfrac{3}{2}\right\}$ **26.** $\{x|x \leq 1\}$ **27.** $\{x|x \geq 1\}$ **28.** $\{x|x \geq 2\}$ **29. (a)** $(f \circ g)(x) = 6x + 3$; all real numbers **(b)** $(g \circ f)(x) = 6x + 9$;
all real numbers **(c)** $(f \circ f)(x) = 4x + 9$; all real numbers **(d)** $(g \circ g)(x) = 9x$; all real numbers **30. (a)** $(f \circ g)(x) = -2x + 4$; all real
numbers **(b)** $(g \circ f)(x) = -2x - 4$; all real numbers **(c)** $(f \circ f)(x) = x$; all real numbers **(d)** $(g \circ g)(x) = 4x - 12$; all real numbers

31. (a) $(f \circ g)(x) = 3x^2 + 1$; all real numbers **(b)** $(g \circ f)(x) = 9x^2 + 6x + 1$; all real numbers **(c)** $(f \circ f)(x) = 9x + 4$; all real numbers
(d) $(g \circ g)(x) = x^4$; all real numbers **32. (a)** $(f \circ g)(x) = x^2 + 5$; all real numbers **(b)** $(g \circ f)(x) = x^2 + 2x + 5$; all real numbers
(c) $(f \circ f)(x) = x + 2$; all real numbers **(d)** $(g \circ g)(x) = x^4 + 8x^2 + 20$; all real numbers **33. (a)** $(f \circ g)(x) = x^4 + 8x^2 + 16$; all real
numbers **(b)** $(g \circ f)(x) = x^4 + 4$; all real numbers **(c)** $(f \circ f)(x) = x^4$; all real numbers **(d)** $(g \circ g)(x) = x^4 + 8x^2 + 20$; all real numbers

34. (a) $(f \circ g)(x) = 4x^4 + 12x^2 + 10$; all real numbers **(b)** $(g \circ f)(x) = 2x^4 + 4x^2 + 5$; all real numbers **(c)** $(f \circ f)(x) = x^4 + 2x^2 + 2$;
all real numbers **(d)** $(g \circ g)(x) = 8x^4 + 24x^2 + 21$; all real numbers **35. (a)** $(f \circ g)(x) = \dfrac{3x}{2-x}$; $\{x|x \neq 0, x \neq 2\}$
(b) $(g \circ f)(x) = \dfrac{2(x-1)}{3}$; $\{x|x \neq 1\}$ **(c)** $(f \circ f)(x) = \dfrac{3(x-1)}{4-x}$; $\{x|x \neq 1, x \neq 4\}$ **(d)** $(g \circ g)(x) = x$; $\{x|x \neq 0\}$

36. (a) $(f \circ g)(x) = \dfrac{x}{-2 + 3x}$; $\left\{x \mid x \neq 0, x \neq \dfrac{2}{3}\right\}$ **(b)** $(g \circ f)(x) = -2(x + 3)$; $\{x|x \neq -3\}$ **(c)** $(f \circ f)(x) = \dfrac{x+3}{3x+10}$;
$\left\{x \mid x \neq -\dfrac{10}{3}, x \neq -3\right\}$ **(d)** $(g \circ g)(x) = x$; $\{x|x \neq 0\}$ **37. (a)** $(f \circ g)(x) = \dfrac{4}{4+x}$; $\{x|x \neq -4, x \neq 0\}$ **(b)** $(g \circ f)(x) = \dfrac{-4(x-1)}{x}$;
$\{x|x \neq 0, x \neq 1\}$ **(c)** $(f \circ f)(x) = x$; $\{x|x \neq 1\}$ **(d)** $(g \circ g)(x) = x$; $\{x|x \neq 0\}$ **38. (a)** $(f \circ g)(x) = \dfrac{2}{2+3x}$; $\left\{x \mid x \neq 0, x \neq -\dfrac{2}{3}\right\}$
(b) $(g \circ f)(x) = \dfrac{2(x+3)}{x}$; $\{x|x \neq -3, x \neq 0\}$ **(c)** $(f \circ f)(x) = \dfrac{x}{4x+9}$; $\left\{x \mid x \neq -3, x \neq -\dfrac{9}{4}\right\}$ **(d)** $(g \circ g)(x) = x$; $\{x|x \neq 0\}$

39. (a) $(f \circ g)(x) = \sqrt{2x+3}$; $\left\{x \mid x \geq -\dfrac{3}{2}\right\}$ **(b)** $(g \circ f)(x) = 2\sqrt{x} + 3$; $\{x|x \geq 0\}$ **(c)** $(f \circ f)(x) = \sqrt[4]{x}$; $\{x|x \geq 0\}$
(d) $(g \circ g)(x) = 4x + 9$; all real numbers **40. (a)** $(f \circ g)(x) = \sqrt{-2x-1}$; $\left\{x \mid x \leq -\dfrac{1}{2}\right\}$ **(b)** $(g \circ f)(x) = 1 - 2\sqrt{x-2}$; $\{x|x \geq 2\}$
(c) $(f \circ f)(x) = \sqrt{\sqrt{x-2}-2}$; $\{x|x \geq 6\}$ **(d)** $(g \circ g)(x) = -1 + 4x$; all real numbers **41. (a)** $(f \circ g)(x) = x$; $\{x|x \geq 1\}$
(b) $(g \circ f)(x) = |x|$; all real numbers **(c)** $(f \circ f)(x) = x^4 + 2x^2 + 2$; all real numbers **(d)** $(g \circ g)(x) = \sqrt{\sqrt{x-1}-1}$; $\{x|x \geq 2\}$
42. (a) $(f \circ g)(x) = x + 2$; $\{x|x \geq 2\}$ **(b)** $(g \circ f)(x) = \sqrt{x^2 + 2}$; all real numbers **(c)** $(f \circ f)(x) = x^4 + 8x^2 + 20$; all real numbers
(d) $(g \circ g)(x) = \sqrt{\sqrt{x-2}-2}$; $\{x|x \geq 6\}$ **43. (a)** $(f \circ g)(x) = acx + ad + b$; all real numbers **(b)** $(g \circ f)(x) = acx + bc + d$; all real
numbers **(c)** $(f \circ f)(x) = a^2 x + ab + b$; all real numbers **(d)** $(g \circ g)(x) = c^2 x + cd + d$; all real numbers

44. (a) $(f \circ g)(x) = \dfrac{amx + b}{cmx + d}$; $\left\{x \mid x \neq -\dfrac{d}{cm}\right\}$ **(b)** $(g \circ f)(x) = \dfrac{m(ax+b)}{cx+d}$; $\left\{x \mid x \neq -\dfrac{d}{c}\right\}$ **(c)** $(f \circ f)(x) = \dfrac{(a^2 + bc)x + ab + bd}{(ac + cd)x + bc + d^2}$;
$\left\{x \mid x \neq -\dfrac{d}{c}; x \neq -\dfrac{bc + d^2}{ac + cd}\right\}$ **(d)** $(g \circ g)(x) = m^2 x$; all real numbers

45. $(f \circ g)(x) = f(g(x)) = f\left(\dfrac{1}{2}x\right) = 2\left(\dfrac{1}{2}x\right) = x$; $(g \circ f)(x) = g(f(x)) = g(2x) = \dfrac{1}{2}(2x) = x$

46. $(f \circ g)(x) = f(g(x)) = f\left(\dfrac{1}{4}x\right) = 4\left(\dfrac{1}{4}x\right) = x$; $(g \circ f)(x) = g(f(x)) = g(4x) = \dfrac{1}{4}(4x) = x$

47. $(f \circ g)(x) = f(g(x)) = f(\sqrt[3]{x}) = (\sqrt[3]{x})^3 = x$; $(g \circ f)(x) = g(f(x)) = g(x^3) = \sqrt[3]{x^3} = x$
48. $(f \circ g)(x) = f(g(x)) = f(x - 5) = x - 5 + 5 = x$; $(g \circ f)(x) = g(f(x)) = g(x + 5) = x + 5 - 5 = x$
49. $(f \circ g)(x) = f(g(x)) = f\left(\dfrac{1}{2}(x + 6)\right) = 2\left[\dfrac{1}{2}(x + 6)\right] - 6 = x + 6 - 6 = x$;

$(g \circ f)(x) = g(f(x)) = g(2x - 6) = \dfrac{1}{2}(2x - 6 + 6) = x$

50. $(f \circ g)(x) = f(g(x)) = f\left(\dfrac{1}{3}(4 - x)\right) = 4 - 3\left[\dfrac{1}{3}(4 - x)\right] = 4 - (4 - x) = x$;

$(g \circ f)(x) = g(f(x)) = g(4 - 3x) = \dfrac{1}{3}[4 - (4 - 3x)] = \dfrac{1}{3}(3x) = x$

51. $(f \circ g)(x) = f(g(x)) = f\left(\dfrac{1}{a}(x - b)\right) = a\left[\dfrac{1}{a}(x - b)\right] + b = x$; $(g \circ f)(x) = g(f(x)) = g(ax + b) = \dfrac{1}{a}(ax + b - b) = x$

52. $(f \circ g)(x) = f(g(x)) = f\left(\dfrac{1}{x}\right) = \dfrac{1}{\frac{1}{x}} = x$; $(g \circ f)(x) = g(f(x)) = g\left(\dfrac{1}{x}\right) = \dfrac{1}{\frac{1}{x}} = x$

53. $f(x) = x^4$; $g(x) = 2x + 3$ (Other answers are possible.)

54. $f(x) = x^3; g(x) = 1 + x^2$ (Other answers are possible.) **55.** $f(x) = \sqrt{x}; g(x) = x^2 + 1$ (Other answers are possible.)
56. $f(x) = \sqrt{x}; g(x) = 1 - x^2$ (Other answers are possible.) **57.** $f(x) = |x|; g(x) = 2x + 1$ (Other answers are possible.)
58. $f(x) = |x|; g(x) = 2x^2 + 3$ (Other answers are possible.) **59.** $(f \circ g)(x) = 11; (g \circ f)(x) = 2$ **60.** $(f \circ f)(x) = x$ **61.** $-3, 3$ **62.** $-5, 5$

63. $S(t) = \dfrac{16}{9}\pi t^6$ **64.** $V(t) = \dfrac{32}{81}\pi t^9$ **65.** $C(t) = 15{,}000 + 800{,}000t - 40{,}000t^2$ **66.** $A(t) = 40{,}000\pi t$

67. $C(p) = \dfrac{2\sqrt{100 - p}}{25} + 600, 0 \le p \le 100$ **68.** $C(p) = \dfrac{\sqrt{1000 - 5p}}{10} + 400, 0 \le p \le 200$ **69.** $V(r) = 2\pi r^3$ **70.** $V(r) = \dfrac{2}{3}\pi r^3$

71. (a) $f(x) = 0.8282x$ **(b)** $g(x) = 139.5866x$ **(c)** $g(f(x)) = g(0.8282x) = 115.6056x$ **(d)** 115,605.62 yen

72. (a) $K(C(F)) = \dfrac{5F - 2297}{9}$ **(b)** -210.8 k **73.** f and g are odd functions, so $f(-x) = -f(x)$ and $g(-x) = -g(x)$.
Then $(f \circ g)(-x) = f(g(-x)) = f(-g(x)) = -f(g(x)) = -(f \circ g)(x)$. So $f \circ g$ is also odd. **74.** f is an odd function, so $f(-x) = -f(x)$. g is
an even function, so $g(-x) = g(x)$. Then $(f \circ g)(-x) = f(g(-x)) = f(g(x)) = (f \circ g)(x)$. So $f \circ g$ is even.
Also, $(g \circ f)(-x) = g(f(-x)) = g(-f(x)) = g(f(x)) = (g \circ f)(x)$, so $g \circ f$ is even.

4.2 Assess Your Understanding (page 280)

4. One-to-one **5.** $y = x$ **6.** $[4, \infty)$ **7.** F **8.** T **9.** One-to-one **10.** One-to-one **11.** Not one-to-one **12.** Not one-to-one
13. Not one-to-one **14.** One-to-one **15.** One-to-one **16.** One-to-one **17.** One-to-one **18.** One-to-one **19.** Not one-to-one
20. Not one-to-one **21.** One-to-one **22.** Not one-to-one
23.

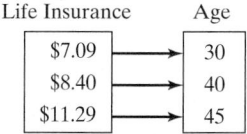

Annual Rainfall	Location
460.00	Mt Waialeale, Hawaii
202.01	Monrovia, Liberia
196.46	Pago Pago, American Samoa
191.02	Moulmein, Burma
182.87	Lae, Papua New Guinea

Domain: {460.00, 202.01, 196.46, 191.02; 182.87}
Range: {Mt. Waialeale, Monrovia, Pago Pago, Moulmein, Lae}

24.

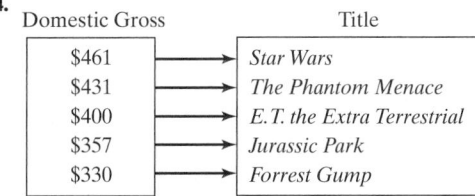

Domestic Gross	Title
$461	Star Wars
$431	The Phantom Menace
$400	E.T. the Extra Terrestrial
$357	Jurassic Park
$330	Forrest Gump

Domain: {461, 431, 400, 357, 330}
Range: {Star Wars, The Phantom Menace, E.T., Jurassic Park, Forrest Gump}

25.

Monthly Cost of Life Insurance	Age
$7.09	30
$8.40	40
$11.29	45

Domain: {$7.09, $8.04, $11.29}
Range: {30, 40, 45}

26.

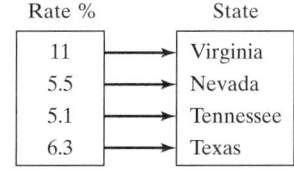

Unemployment Rate %	State
11	Virginia
5.5	Nevada
5.1	Tennessee
6.3	Texas

Domain: {11%, 5.5%, 5.1%, 6.3%}
Range: {Virginia, Nevada, Tennessee, Texas}

27. $\{(5, -3), (9, -2), (2, -1), (11, 0), (-5, 1)\}$
Domain: $\{5, 9, 2, 11, -5\}$
Range: $\{-3, -2, -1, 0, 1\}$
28. $\{(2, -2), (6, -1), (8, 0), (-3, 1), (9, 2)\}$
Domain: $\{2, 6, 8, -3, 9\}$
Range: $\{-2, -1, 0, 1, 2\}$

29. $\{(1, -2), (2, -3), (0, -10), (9, 1), (4, 2)\}$
Domain: $\{1, 2, 0, 9, 4\}$
Range: $\{-2, -3, -10, 1, 2\}$
30. $\{(-8, -2), (-1, -1), (0, 0), (1, 1), (8, 2)\}$
Domain: $\{-8, -1, 0, 1, 8\}$
Range: $\{-2, -1, 0, 1, 2\}$

31.

32.

33.

34.

35.

36.
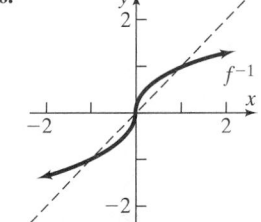

37. $f(g(x)) = f\left(\dfrac{1}{3}(x-4)\right) = 3\left[\dfrac{1}{3}(x-4)\right] + 4$
$= (x-4) + 4 = x;$
$g(f(x)) = g(3x+4) = \dfrac{1}{3}[(3x+4) - 4] = \dfrac{1}{3}(3x) = x$

38. $f(g(x)) = f\left(-\dfrac{1}{2}(x-3)\right) = 3 - 2\left(\dfrac{-1}{2}(x-3)\right)$
$= 3 + (x-3) = x$
$g(f(x)) = g(3 - 2x) = -\dfrac{1}{2}[(3-2x) - 3]$
$= -\dfrac{1}{2}(-2x) = x$

39. $f(g(x)) = f\left(\dfrac{x}{4} + 2\right) = 4\left[\dfrac{x}{4} + 2\right] - 8 = (x+8) - 8 = x;$
$g(f(x)) = g(4x-8) = \dfrac{4x-8}{4} + 2 = (x-2) + 2 = x$

40. $f(g(x)) = f\left(\dfrac{1}{2}x - 3\right) = 2\left(\dfrac{1}{2}x - 3\right) + 6$
$= (x-6) + 6 = x$
$g(f(x)) = g(2x+6) = \dfrac{1}{2}(2x+6) - 3$
$= (x+3) - 3 = x$

41. $f(g(x)) = f\left(\sqrt[3]{x+8}\right) = \left(\sqrt[3]{x+8}\right)^3 - 8 = (x+8) - 8 = x;$
$g(f(x)) = g(x^3 - 8) = \sqrt[3]{(x^3 - 8) + 8} = \sqrt[3]{x^3} = x$

42. $f(g(x)) = f\left(\sqrt{x} + 2\right) = \left(\sqrt{x} + 2 - 2\right)^2 = \left(\sqrt{x}\right)^2 = x$
$g(f(x)) = g((x-2)^2) = \sqrt{(x-2)^2} + 2 = x - 2 + 2 = x$

43. $f(g(x)) = f\left(\dfrac{1}{x}\right) = \dfrac{1}{\left(\dfrac{1}{x}\right)} = x; g(f(x)) = g\left(\dfrac{1}{x}\right) = \dfrac{1}{\left(\dfrac{1}{x}\right)} = x$

44. $f(g(x)) = f(x) = x$
$g(f(x)) = g(x) = x$

45. $f(g(x)) = f\left(\dfrac{4x-3}{2-x}\right) = \dfrac{2\left(\dfrac{4x-3}{2-x}\right) + 3}{\dfrac{4x-3}{2-x} + 4}$
$= \dfrac{2(4x-3) + 3(2-x)}{4x-3 + 4(2-x)} = \dfrac{5x}{5} = x;$
$g(f(x)) = g\left(\dfrac{2x+3}{x+4}\right) = \dfrac{4\left(\dfrac{2x+3}{x+4}\right) - 3}{2 - \dfrac{2x+3}{x+4}}$
$= \dfrac{4(2x+3) - 3(x+4)}{2(x+4) - (2x+3)} = \dfrac{5x}{5} = x$

46. $f(g(x)) = f\left(\dfrac{3x+5}{1-2x}\right) = \dfrac{\dfrac{3x+5}{1-2x} - 5}{2\left(\dfrac{3x+5}{1-2x}\right) + 3}$
$= \dfrac{3x+5 - 5(1-2x)}{2(3x+5) + 3(1-2x)} = \dfrac{13x}{13} = x$
$g(f(x)) = g\left(\dfrac{x-5}{2x+3}\right) = \dfrac{3\left(\dfrac{x-5}{2x+3}\right) + 5}{1 - 2\left(\dfrac{x-5}{2x+3}\right)}$
$= \dfrac{3(x-5) + 5(2x+3)}{(2x+3) - 2(x-5)} = \dfrac{13x}{13} = x$

47. $f^{-1}(x) = \dfrac{1}{3}x$
$f(f^{-1}(x)) = f\left(\dfrac{1}{3}x\right) = 3\left(\dfrac{1}{3}x\right) = x$
$f^{-1}(f(x)) = f^{-1}(3x) = \dfrac{1}{3}(3x) = x$
Domain f = range f^{-1} = all real numbers
Range f = domain f^{-1} = all real numbers

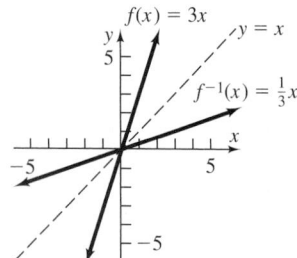

48. $f^{-1}(x) = -\dfrac{1}{4}x$
$f(f^{-1}(x)) = f\left(-\dfrac{1}{4}x\right) = -4\left(-\dfrac{1}{4}x\right) = x$
$f^{-1}(f(x)) = f^{-1}(-4x) = -\dfrac{1}{4}(-4x) = x$
Domain f = range f^{-1} = all real numbers
Range f = domain f^{-1} = all real numbers

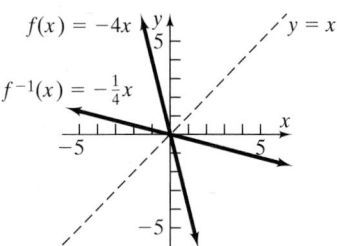

49. $f^{-1}(x) = \dfrac{x}{4} - \dfrac{1}{2}$
$f(f^{-1}(x)) = f\left(\dfrac{x}{4} - \dfrac{1}{2}\right) = 4\left(\dfrac{x}{4} - \dfrac{1}{2}\right) + 2 = (x-2) + 2 = x$
$f^{-1}(f(x)) = f^{-1}(4x+2) = \dfrac{4x+2}{4} - \dfrac{1}{2} = \left(x + \dfrac{1}{2}\right) - \dfrac{1}{2} = x$
Domain f = range f^{-1} = all real numbers
Range f = domain f^{-1} = all real numbers

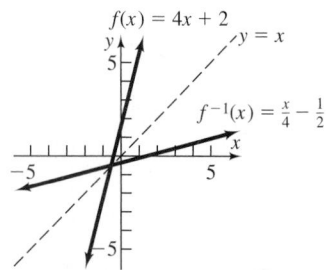

50. $f^{-1}(x) = \dfrac{1-x}{3}$
$f(f^{-1}(x)) = f\left(\dfrac{1-x}{3}\right) = 1 - 3\left(\dfrac{1-x}{3}\right) = 1 - (1-x) = x$
$f^{-1}(f(x)) = f^{-1}(1-3x) = \dfrac{1 - (1-3x)}{3} = \dfrac{3x}{3} = x$
Domain f = range f^{-1} = all real numbers
Range f = domain f^{-1} = all real numbers

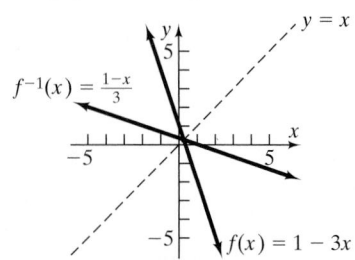

51. $f^{-1}(x) = \sqrt[3]{x+1}$
$f(f^{-1}(x)) = f(\sqrt[3]{x+1}) = (\sqrt[3]{x+1})^3 - 1 = x$
$f^{-1}(f(x)) = f^{-1}(x^3 - 1) = \sqrt[3]{(x^3 - 1) + 1} = x$
Domain f = range f^{-1} = all real numbers
Range f = domain f^{-1} = all real numbers

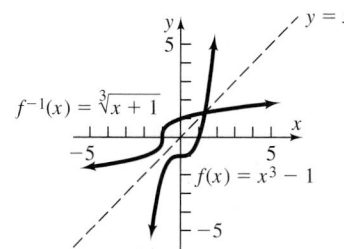

52. $f^{-1}(x) = (x - 1)^{1/3}$
$f(f^{-1}(x)) = f((x-1)^{1/3}) = ((x-1)^{1/3})^3 + 1$
$\qquad = x - 1 + 1 = x$
$f^{-1}(f(x)) = f^{-1}(x^3 + 1) = (x^3 + 1 - 1)^{1/3} = (x^3)^{1/3} = x$
Domain f = range f^{-1} = all real numbers
Range f = domain f^{-1} = all real numbers

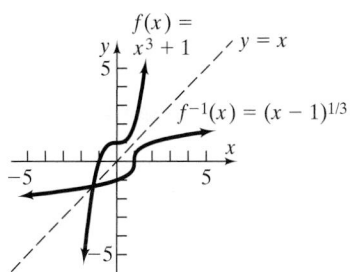

53. $f^{-1}(x) = \sqrt{x - 4}, x \geq 4$
$f(f^{-1}(x)) = f(\sqrt{x-4}) = (\sqrt{x-4})^2 + 4 = x$
$f^{-1}(f(x)) = f^{-1}(x^2 + 4) = \sqrt{(x^2 + 4) - 4} = \sqrt{x^2} = x, x \geq 0$
Domain f = range f^{-1} = $\{x | x \geq 0\}$ or $[0, \infty)$
Range f = domain f^{-1} = $\{x | x \geq 4\}$ or $[4, \infty)$

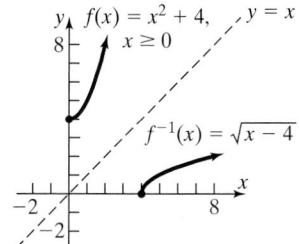

54. $f^{-1}(x) = \sqrt{x - 9}, x \geq 9$
$f(f^{-1}(x)) = f(\sqrt{x-9}) = (\sqrt{x-9})^2 + 9 = x - 9 + 9 = x$
$f^{-1}(f(x)) = f^{-1}(x^2 + 9) = \sqrt{x^2 + 9 - 9} = \sqrt{x^2} = x, x \geq 0$
Domain f = range f^{-1} = $\{x | x \geq 0\}$ or $[0, \infty)$
Range f = domain f^{-1} = $\{x | x \geq 9\}$ or $[9, \infty)$

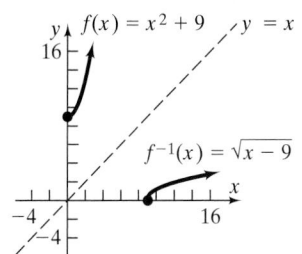

55. $f^{-1}(x) = \dfrac{4}{x}$
$f(f^{-1}(x)) = f\left(\dfrac{4}{x}\right) = \dfrac{4}{\left(\dfrac{4}{x}\right)} = x$

$f^{-1}(f(x)) = f^{-1}\left(\dfrac{4}{x}\right) = \dfrac{4}{\left(\dfrac{4}{x}\right)} = x$

Domain f = range f^{-1} = all real numbers except 0
Range f = domain f^{-1} = all real numbers except 0

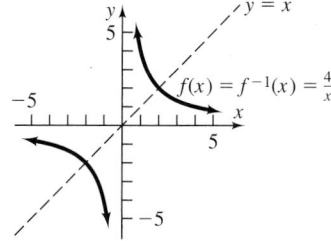

56. $f^{-1}(x) = -\dfrac{3}{x}$
$f(f^{-1}(x)) = f\left(-\dfrac{3}{x}\right) = -\dfrac{3}{\left(-\dfrac{3}{x}\right)} = x$

$f^{-1}(f(x)) = f^{-1}\left(-\dfrac{3}{x}\right) = -\dfrac{3}{\left(-\dfrac{3}{x}\right)} = x$

Domain f = range f^{-1} = all real numbers except 0
Range f = domain f^{-1} = all real numbers except 0

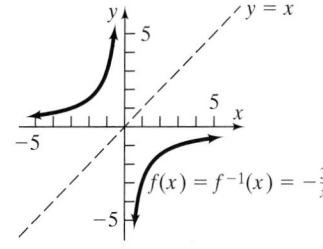

57. $f^{-1}(x) = \dfrac{2x + 1}{x}$
$f(f^{-1}(x)) = f\left(\dfrac{2x+1}{x}\right) = \dfrac{1}{\dfrac{2x+1}{x} - 2} = \dfrac{x}{(2x+1) - 2x} = x$

$f^{-1}(f(x)) = f^{-1}\left(\dfrac{1}{x-2}\right) = \dfrac{2\left(\dfrac{1}{x-2}\right) + 1}{\dfrac{1}{x-2}} = \dfrac{2 + (x - 2)}{1} = x$

Domain f = range f^{-1} = all real numbers except 2
Range f = domain f^{-1} = all real numbers except 0

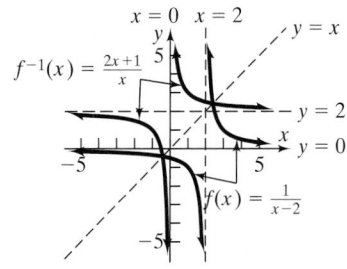

58. $f^{-1}(x) = \dfrac{4}{x} - 2$

$$f(f^{-1}(x)) = f\left(\dfrac{4}{x} - 2\right) = \dfrac{4}{\left(\dfrac{4}{x} - 2\right) + 2} = \dfrac{4}{\left(\dfrac{4}{x}\right)} = x$$

$$f^{-1}(f(x)) = f^{-1}\left(\dfrac{4}{x + 2}\right) = \dfrac{4}{\left(\dfrac{4}{x + 2}\right)} - 2 = x + 2 - 2 = x$$

Domain f = range f^{-1} = all real numbers except -2

Range f = domain f^{-1} = all real numbers except 0

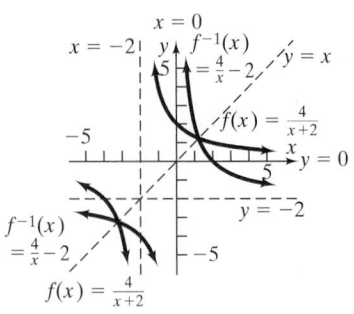

59. $f^{-1}(x) = \dfrac{2 - 3x}{x}$

$$f(f^{-1}(x)) = f\left(\dfrac{2 - 3x}{x}\right) = \dfrac{2}{3 + \dfrac{2 - 3x}{x}} = \dfrac{2x}{3x + 2 - 3x} = \dfrac{2x}{2} = x$$

$$f^{-1}(f(x)) = f^{-1}\left(\dfrac{2}{3 + x}\right) = \dfrac{2 - 3\left(\dfrac{2}{3 + x}\right)}{\dfrac{2}{3 + x}} = \dfrac{2(3 + x) - 3 \cdot 2}{2} = \dfrac{2x}{2} = x$$

Domain f = all real numbers except -3
Range f = domain f^{-1} = all real numbers except 0

60. $f^{-1}(x) = 2 - \dfrac{4}{x}$

$$f(f^{-1}(x)) = f\left(2 - \dfrac{4}{x}\right) = \dfrac{4}{2 - \left(2 - \dfrac{4}{x}\right)} = \dfrac{4}{\left(\dfrac{4}{x}\right)} = x$$

$$f^{-1}(f(x)) = f^{-1}\left(\dfrac{4}{2 - x}\right) = 2 - \dfrac{4}{\left(\dfrac{4}{2 - x}\right)}$$

$$= 2 - (2 - x) = x$$

Domain f = all real numbers except 2
Range f = domain f^{-1} = all real numbers except 0

61. $f^{-1}(x) = \dfrac{-2x}{x - 3}$

$$f(f^{-1}(x)) = f\left(\dfrac{-2x}{x - 3}\right) = \dfrac{3\left(\dfrac{-2x}{x - 3}\right)}{\dfrac{-2x}{x - 3} + 2} = \dfrac{3(-2x)}{-2x + 2(x - 3)} = \dfrac{-6x}{-6} = x$$

$$f^{-1}(f(x)) = f^{-1}\left(\dfrac{3x}{x + 2}\right) = \dfrac{-2\left(\dfrac{3x}{x + 2}\right)}{\dfrac{3x}{x + 2} - 3} = \dfrac{-2(3x)}{3x - 3(x + 2)} = \dfrac{-6x}{-6} = x$$

Domain f = all real numbers except -2
Range f = domain f^{-1} = all real numbers except 3

62. $f^{-1}(x) = \dfrac{x}{x + 2}$

$$f(f^{-1}(x)) = \dfrac{-2\left(\dfrac{x}{x + 2}\right)}{\dfrac{x}{x + 2} - 1} = \dfrac{-2x}{x - (x + 2)} = \dfrac{-2x}{-2} = x$$

$$f^{-1}(f(x)) = \dfrac{\dfrac{-2x}{x - 1}}{\dfrac{-2x}{x - 1} + 2} = \dfrac{-2x}{-2x + 2(x - 1)} = \dfrac{-2x}{-2} = x$$

Domain f = all real numbers except 1
Range f = domain f^{-1} = all real numbers except -2

63. $f^{-1}(x) = \dfrac{x}{3x - 2}$

$$f(f^{-1}(x)) = f\left(\dfrac{x}{3x - 2}\right) = \dfrac{2\left(\dfrac{x}{3x - 2}\right)}{3\left(\dfrac{x}{3x - 2}\right) - 1} = \dfrac{2x}{3x - (3x - 2)} = \dfrac{2x}{2} = x$$

$$f^{-1}(f(x)) = f^{-1}\left(\dfrac{2x}{3x - 1}\right) = \dfrac{\dfrac{2x}{3x - 1}}{3\left(\dfrac{2x}{3x - 1}\right) - 2} = \dfrac{2x}{6x - 2(3x - 1)} = \dfrac{2x}{2} = x$$

Domain f = all real numbers except $\dfrac{1}{3}$
Range f = domain f^{-1} = all real numbers except $\dfrac{2}{3}$

64. $f^{-1}(x) = \dfrac{-1}{x + 3}$

$$f(f^{-1}(x)) = f\left(\dfrac{-1}{x + 3}\right) = \dfrac{3\left(\dfrac{-1}{x + 3}\right) + 1}{-\left(\dfrac{-1}{x + 3}\right)}$$

$$= \dfrac{-3 + (x + 3)}{1} = x$$

$$f^{-1}(f(x)) = f^{-1}\left(\dfrac{3x + 1}{-x}\right) = \dfrac{-1}{\dfrac{3x + 1}{-x} + 3}$$

$$= \dfrac{x}{3x + 1 - 3x} = x$$

Domain f = all real numbers except 0
Range f = domain f^{-1} = all real numbers except -3

65. $f^{-1}(x) = \dfrac{3x + 4}{2x - 3}$

$$f(f^{-1}(x)) = f\left(\dfrac{3x + 4}{2x - 3}\right) = \dfrac{3\left(\dfrac{3x + 4}{2x - 3}\right) + 4}{2\left(\dfrac{3x + 4}{2x - 3}\right) - 3}$$

$$= \dfrac{3(3x + 4) + 4(2x - 3)}{2(3x + 4) - 3(2x - 3)}$$

$$= \dfrac{17x}{17} = x$$

$$f^{-1}(f(x)) = f^{-1}\left(\dfrac{3x + 4}{2x - 3}\right) = \dfrac{3\left(\dfrac{3x + 4}{2x - 3}\right) + 4}{2\left(\dfrac{3x + 4}{2x - 3}\right) - 3}$$

$$= \dfrac{3(3x + 4) + 4(2x - 3)}{2(3x + 4) - 3(2x - 3)} = \dfrac{17x}{17} = x$$

Domain f = all real numbers except $\dfrac{3}{2}$

Range f = domain f^{-1} = all real numbers except $\dfrac{3}{2}$

66. $f^{-1}(x) = \dfrac{3 + 4x}{2 - x}$

$f(f^{-1}(x)) = f\left(\dfrac{3 + 4x}{2 - x}\right) = \dfrac{2\left(\dfrac{3 + 4x}{2 - x}\right) - 3}{\left(\dfrac{3 + 4x}{2 - x}\right) + 4}$

$= \dfrac{2(3 + 4x) - 3(2 - x)}{(3 + 4x) + 4(2 - x)} = \dfrac{11x}{11} = x$

$f^{-1}(f(x)) = f^{-1}\left(\dfrac{2x - 3}{x + 4}\right) = \dfrac{3 + 4\left(\dfrac{2x - 3}{x + 4}\right)}{2 - \left(\dfrac{2x - 3}{x + 4}\right)}$

$= \dfrac{3(x + 4) + 4(2x - 3)}{2(x + 4) - (2x - 3)} = \dfrac{11x}{11} = x$

Domain f = all real numbers except -4

Range f = domain f^{-1} = all real numbers except 2

67. $f^{-1}(x) = \dfrac{-2x + 3}{x - 2}$

$f(f^{-1}(x)) = f\left(\dfrac{-2x + 3}{x - 2}\right) = \dfrac{2\left(\dfrac{-2x + 3}{x - 2}\right) + 3}{\dfrac{-2x + 3}{x - 2} + 2}$

$= \dfrac{2(-2x + 3) + 3(x - 2)}{-2x + 3 + 2(x - 2)} = \dfrac{-x}{-1} = x$

$f^{-1}(f(x)) = f^{-1}\left(\dfrac{2x + 3}{x + 2}\right) = \dfrac{2\left(\dfrac{-2x + 3}{x - 2}\right) + 3}{\dfrac{-2x + 3}{x - 2} + 2}$

$= \dfrac{2(-2x + 3) + 3(x - 2)}{-2x + 3 + 2(x - 2)} = \dfrac{-x}{-1} = x$

Domain f = all real numbers except -2

Range f = domain f^{-1} = all real numbers except 2

68. $f^{-1}(x) = \dfrac{2x - 4}{x + 3}$

$f(f^{-1}(x)) = f\left(\dfrac{2x - 4}{x + 3}\right) = \dfrac{-3\left(\dfrac{2x - 4}{x + 3}\right) - 4}{\left(\dfrac{2x - 4}{x + 3}\right) - 2}$

$= \dfrac{-3(2x - 4) - 4(x + 3)}{(2x - 4) - 2(x + 3)} = \dfrac{-10x}{-10} = x$

$f^{-1}(f(x)) = f^{-1}\left(\dfrac{-3x - 4}{x - 2}\right) = \dfrac{2\left(\dfrac{-3x - 4}{x - 2}\right) - 4}{\left(\dfrac{-3x - 4}{x - 2}\right) + 3}$

$= \dfrac{2(-3x - 4) - 4(x - 2)}{(-3x - 4) + 3(x - 2)} = \dfrac{-10x}{-10} = x$

Domain f = all real numbers except 2

Range f = domain f^{-1} = all real numbers except -3

69. $f^{-1}(x) = \dfrac{2}{\sqrt{1 - 2x}}$

$f(f^{-1}(x)) = f\left(\dfrac{2}{\sqrt{x - 2x}}\right) = \dfrac{\dfrac{4}{1 - 2x} - 4}{2 \cdot \dfrac{4}{1 - 2x}} = \dfrac{4 - 4(1 - 2x)}{2 \cdot 4}$

$= \dfrac{8x}{8} = x$

$f^{-1}(f(x)) = f^{-1}\left(\dfrac{x^2 - 4}{2x^2}\right) = \dfrac{2}{\sqrt{1 - 2\left(\dfrac{x^2 - 4}{2x^2}\right)}} = \dfrac{2}{\sqrt{\dfrac{4}{x^2}}} = \sqrt{x^2}$

$= x$, since $x > 0$

Domain $f = \{x | x > 0\}$ or $(0, \infty)$

Range f = domain $f^{-1} = \left\{x \middle| x < \dfrac{1}{2}\right\}$ or $\left(-\infty, \dfrac{1}{2}\right)$

70. $f^{-1}(x) = \sqrt{\dfrac{3}{3x - 1}}$

$f(f^{-1}(x)) = f\left(\sqrt{\dfrac{3}{3x - 1}}\right) = \dfrac{\dfrac{3}{3x - 1} + 3}{3\left(\dfrac{3}{3x - 1}\right)}$

$= \dfrac{3 + 3(3x - 1)}{9}$

$= \dfrac{9x}{9} = x$

$f^{-1}(f(x)) = f^{-1}\left(\dfrac{x^2 + 3}{3x^2}\right) = \sqrt{\dfrac{3}{3\left(\dfrac{x^2 + 3}{3x^2}\right) - 1}}$

$= \sqrt{\dfrac{3x^2}{(x^2 + 3) - x^2}} = \sqrt{\dfrac{3x^2}{3}}$

$= \sqrt{x^2} = x$, since $x > 0$

Domain $f = \{x | x > 0\}$

Range f = domain $f^{-1} = \left\{x \middle| x > \dfrac{1}{3}\right\}$

71. (a) 0 **(b)** 2 **(c)** 0 **(d)** 1 **72. (a)** $\dfrac{1}{2}$ **(b)** 0 **(c)** 1 **(d)** 0 **73.** 7 **74.** -5 **75.** Domain of f^{-1}: $[-2, \infty)$; range of f^{-1}: $[5, \infty)$. **76.** Domain of f^{-1}: $[5, \infty)$; range of f^{-1}: $[0, \infty)$ **77.** Domain of g^{-1}: $[0, \infty)$; range of g^{-1}: all real numbers **78.** Domain of g^{-1}: $(0, 8)$; range of g^{-1}: $[0, 15]$ **79.** Increasing on the interval $(f(0), f(5))$ **80.** Decreasing on the interval $(f(0), f(5))$ **81.** $f^{-1}(x) = \dfrac{1}{m}(x - b), m \neq 0$ **82.** $f^{-1}(x) = \sqrt{r^2 - x^2}$, $0 \leq x \leq r$ **83.** Quadrant I **84.** Quadrant IV **85.** Possible answer: $f(x) = |x|, x \geq 0$, is one-to-one; $f^{-1}(x) = x, x \geq 0$ **86.** Possible answer: $f(x) = x^4, x \geq 0$, is one-to-one; $f^{-1}(x) = x^{1/4}, x \geq 0$

87. (a) $r(d) = \dfrac{d + 90.39}{6.97}$

(b) $r(d(r)) = \dfrac{6.97r - 90.39 + 90.39}{6.97} = \dfrac{6.97r}{6.97} = r$

$d(r(d)) = 6.97\left(\dfrac{d + 90.39}{6.97}\right) - 90.39 = d + 90.39 - 90.39 = d$

(c) 56 miles per hour

88. (a) $C(H) = \dfrac{H + 10.53}{2.15}$

(b) $H(C(H)) = 2.15\left(\dfrac{H + 10.53}{2.15}\right) - 10.53 = H + 10.53 - 10.53 = H$

$C(H(C)) = \dfrac{2.15C - 10.53 + 10.53}{2.15} = \dfrac{2.15C}{2.15} = C$

(c) 17 inches

89. (a) 77.6 kg

(c) $h(W(h)) = \dfrac{50 + 2.3(h - 60) + 88}{2.3} = \dfrac{2.3h}{2.3} = h$

$W(h(W)) = 50 + 2.3\left(\dfrac{W + 88}{2.3} - 60\right)$

$= 50 + W + 88 - 138 = W$

(b) $h(W) = \dfrac{W - 50}{2.3} + 60 = \dfrac{W + 88}{2.3}$

(d) 73 inches

90. (a) $C(F) = \frac{5}{9}(F - 32)$ **(b)** $C(F(C)) = \frac{5}{9}\left[\frac{9}{5}C + 32 - 32\right] = \frac{5}{9} \cdot \frac{9}{5}C = C$ **(c)** 21.1°C

$F(C(F)) = \frac{9}{5}\left[\frac{5}{9}(F - 32)\right] + 32 = F - 32 + 32 = F$

91. (a) $\{g | 29{,}700 \le g \le 71{,}950\}$ **92. (a)** $\{g | 14{,}600 \le g \le 59{,}400\}$ **93. (a)** t represents time, so $t \ge 0$.

(b) $\{T | 4090 \le T \le 15{,}920\}$ **(b)** $\{T | 1460 \le T \le 8180\}$

(c) $g(T) = \dfrac{T - 4090}{0.28} + 29{,}700$ **(c)** $g(T) = \dfrac{T - 1460}{0.15} + 14{,}600$ **(b)** $t(H) = \sqrt{\dfrac{H - 100}{-4.9}} = \sqrt{\dfrac{100 - H}{4.9}}$

Domain: $\{T | 4090 \le T \le 15{,}920\}$ Domain: $\{T | 1460 \le T \le 8180\}$ **(c)** 2.02 seconds

Range: $\{g | 29{,}700 \le g \le 71{,}950\}$ Range: $\{g | 14{,}600 \le g \le 59{,}400\}$ **94. (a)** $l(T) = \left(\dfrac{T}{2\pi}\right)^2 \cdot (32.2)$ **(b)** 7.34 feet

95. $f^{-1}(x) = \dfrac{-dx + b}{cx - a}; f = f^{-1}$ if $a = -d$ **96.** Yes, the graph must be symmetric about the line $y = x$; $y = \dfrac{1}{x}$ is an example.

98. Possible answer: $y = \begin{cases} \dfrac{1}{x}, & x < 0 \\ x, & x \ge 0 \end{cases}$ is neither an increasing nor a decreasing function on its domain, but is one-to-one. **99.** Yes, if the domain is

$\{x | x \ge 0\}$. **100.** No **101.** On the line $y = x$; no; no **102.** The number of cars manufactured for $800,000

4.3 Assess Your Understanding (page 295)

6. $\left(-1, \dfrac{1}{a}\right), (0, 1), (1, a)$ **7.** 1 **8.** 4 **9.** F **10.** F **11. (a)** 11.212 **(b)** 11.587 **(c)** 11.664 **(d)** 11.665 **12. (a)** 15.426 **(b)** 16.189 **(c)** 16.241

(d) 16.242 **13. (a)** 8.815 **(b)** 8.821 **(c)** 8.824 **(d)** 8.825 **14. (a)** 6.498 **(b)** 6.543 **(c)** 6.580 **(d)** 6.581 **15. (a)** 21.217 **(b)** 22.217

(c) 22.440 **(d)** 22.459 **16. (a)** 21.738 **(b)** 22.884 **(c)** 23.119 **(d)** 23.141 **17.** 3.320 **18.** 0.273 **19.** 0.427 **20.** 8.166 **21.** Not exponential

22. Not exponential **23.** Exponential; $a = 4$ **24.** Exponential; $a = \dfrac{3}{2}$ **25.** Exponential; $a = 2$ **26.** Not exponential **27.** Not exponential

28. Exponential; $a = \dfrac{1}{2}$ **29.** B **30.** F **31.** D **32.** H **33.** A **34.** C **35.** E **36.** G

37.

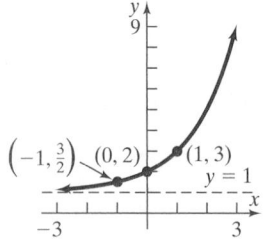

Domain: All real numbers
Range: $\{y | y > 1\}$ or $(1, \infty)$
Horizontal asymptote: $y = 1$

38.

Domain: All real numbers
Range: $\{y | y > -2\}$ or $(-2, \infty)$
Horizontal asymptote: $y = -2$

39.

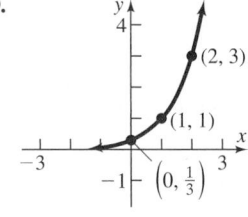

Domain: All real numbers
Range: $\{y | y > 0\}$ or $(0, \infty)$
Horizontal asymptote: $y = 0$

40.

Domain: All real numbers
Range: $\{y | y > 0\}$ or $(0, \infty)$
Horizontal asymptote: $y = 0$

41.

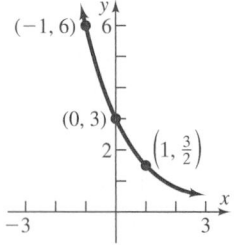

Domain: All real numbers
Range: $\{y | y > 0\}$ or $(0, \infty)$
Horizontal asymptote: $y = 0$

42.

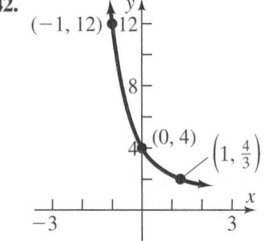

Domain: All real numbers
Range: $\{y | y > 0\}$ or $(0, \infty)$
Horizontal asymptote: $y = 0$

43.

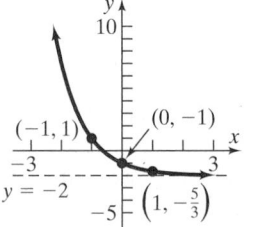

Domain: All real numbers
Range: $\{y | y > -2\}$ or $(-2, \infty)$
Horizontal asymptote: $y = -2$

44.

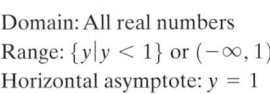

Domain: All real numbers
Range: $\{y | y < 1\}$ or $(-\infty, 1)$
Horizontal asymptote: $y = 1$

45.

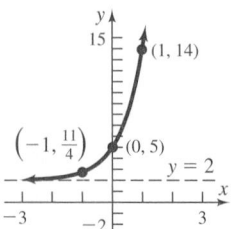

Domain: All real numbers
Range: $\{y | y > 2\}$ or $(2, \infty)$
Horizontal asymptote: $y = 2$

46.

Domain: All real numbers
Range: $\{y | y < 1\}$ or $(-\infty, 1)$
Horizontal asymptote: $y = 1$

47.

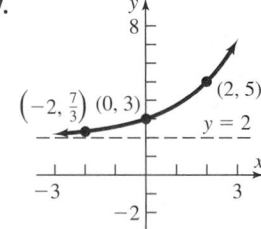

Domain: All real numbers
Range: $\{y | y > 2\}$ or $(2, \infty)$
Horizontal asymptote: $y = 2$

48.

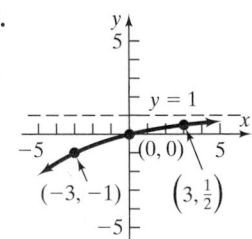

Domain: All real numbers
Range: $\{y | y < 1\}$ or $(-\infty, 1)$
Horizontal asymptote: $y = 1$

49.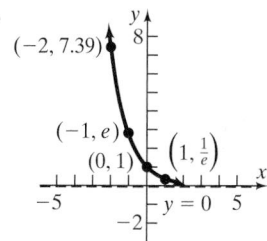

Domain: All real numbers
Range: $\{y|y > 0\}$ or $(0, \infty)$
Horizontal asymptote: $y = 0$

50.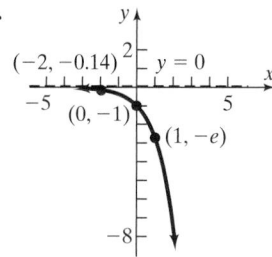

Domain: All real numbers
Range: $\{y|y < 0\}$ or $(-\infty, 0)$
Horizontal asymptote: $y = 0$

51.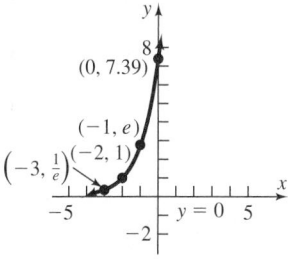

Domain: All real numbers
Range: $\{y|y > 0\}$ or $(0, \infty)$
Horizontal asymptote: $y = 0$

52.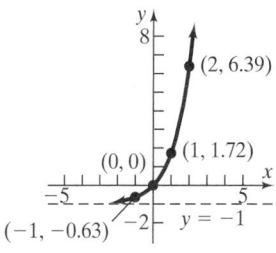

Domain: All real numbers
Range: $\{y|y > -1\}$ or $(-1, \infty)$
Horizontal asymptote: $y = -1$

53.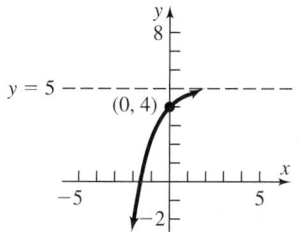

Domain: All real numbers
Range: $\{y|y < 5\}$ or $(-\infty, 5)$
Horizontal asymptote: $y = 5$

54.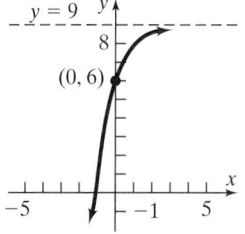

Domain: All real numbers
Range: $\{y|y < 9\}$ or $(-\infty, 9)$
Horizontal asymptote: $y = 9$

55.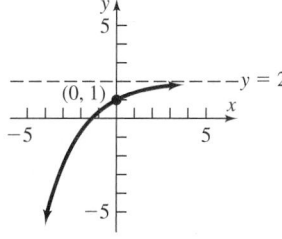

Domain: All real numbers
Range: $\{y|y < 2\}$ or $(-\infty, 2)$
Horizontal asymptote: $y = 2$

56.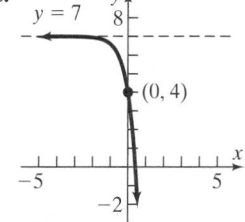

Domain: All real numbers
Range: $\{y|y < 7\}$ or $(-\infty, 7)$
Horizontal asymptote: $y = 7$

57. $\{3\}$ **58.** $\{-6\}$ **59.** $\{-4\}$ **60.** $\{-4\}$ **61.** $\{2\}$ **62.** $\{3\}$ **63.** $\left\{\dfrac{3}{2}\right\}$ **64.** $\{-4\}$ **65.** $\{-\sqrt{2}, 0, \sqrt{2}\}$ **66.** $\left\{0, \dfrac{1}{2}\right\}$ **67.** $\{6\}$ **68.** $\{6\}$

69. $\{-1, 7\}$ **70.** $\{2, 4\}$ **71.** $\{-4, 2\}$ **72.** $\left\{-1, -\dfrac{1}{3}\right\}$ **73.** $\{-4\}$ **74.** $\left\{\dfrac{1}{2}\right\}$ **75.** $\{1, 2\}$ **76.** $\{-6, 2\}$ **77.** (a) $16; (4, 16)$ (b) $-4; \left(-4, \dfrac{1}{16}\right)$

78. (a) $81; (4, 81)$ (b) $-2; \left(-2, \dfrac{1}{9}\right)$ **79.** (a) $\dfrac{9}{4}; \left(-1, \dfrac{9}{4}\right)$ (b) $3; (3, 66)$ **80.** (a) $-\dfrac{14}{5}; \left(-1, -\dfrac{14}{5}\right)$ (b) $3; (3, 122)$ **81.** (a) $10; (-2, 10)$

(b) $3; \left(3, -\dfrac{13}{8}\right)$ **82.** (a) $-5; (-1, -5)$ (b) $-3; (-3, -53)$ **83.** $\dfrac{1}{49}$ **84.** $\dfrac{1}{9}$ **85.** $\dfrac{1}{4}$ **86.** $\dfrac{1}{27}$ **87.** $f(x) = 3^x$ **88.** $f(x) = 5^x$

89. $f(x) = -6^x$ **90.** $f(x) = -e^x$

91.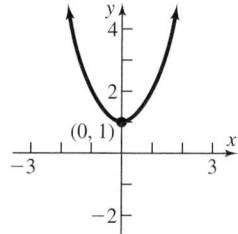

Domain: $(-\infty, \infty)$
Range: $[1, \infty)$
Intercept: $(0, 1)$

92.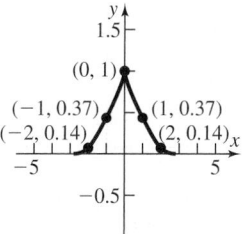

Domain: $(-\infty, \infty)$
Range: $(0, 1]$
Intercept: $(0, 1)$

93.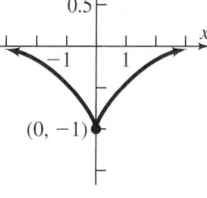

Domain: $(-\infty, \infty)$
Range: $[-1, 0)$
Intercept: $(0, -1)$

94.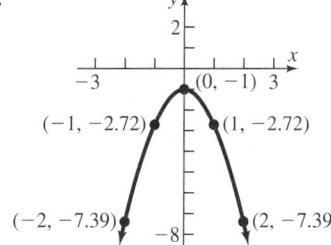

Domain: $(-\infty, \infty)$
Range: $(-\infty, -1]$
Intercept: $(0, -1)$

95. (a) 74% (b) 47% **96.** (a) 568.68 mm Hg (b) 178.27 mm Hg **97.** (a) \$12,123 (b) \$6443 **98.** (a) 34.99 sq mm (b) 3.02 sq mm
99. 3.35 mg; 0.45 mg **100.** 362 students

101. (a) 0.63 (b) 0.98 (c) 1
(d)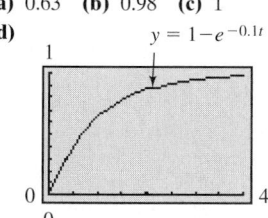

(e) About 7 min

102. (a) 0.895 (b) 0.989 (c) 1
(d)

(e) About 6 min

103. (a) 0.0516 (b) 0.0888
104. (a) 0.16 (b) 0.03
105. (a) 70.95% (b) 72.62% (c) 100%
106. (a) 84 words (b) 153 words

107. (a) 5.41 amp, 7.59 amp, 10.38 amp **(b)** 12 amp **(d)** 3.34 amp, 5.31 amp, 9.44 amp **(e)** 24 amp

(c), (f)

108. (a) 0.06 amp; 0.0364 amp; 0.0134 amp **(d)** 0.12 amp; 0.0728 amp; 0.0268 amp

(b) 0.06 amp **(e)** 0.12 amp

(c)

(f)

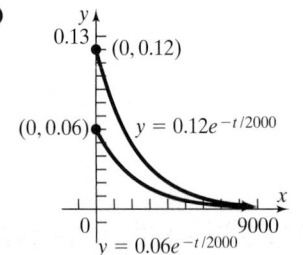

109. $n = 4$: 2.7083; $n = 6$: 2.7181;
$n = 8$: 2.7182788; $n = 10$: 2.7182818

110. 2.71826332

Final Denominator	Value of Expression
$1 + 1$	2.5
$2 + 2$	2.8
$3 + 3$	2.7
$4 + 4$	2.721649485
$5 + 5$	2.717770035
$6 + 6$	2.718348855

$e \approx 2.718281828$

111. $\dfrac{f(x + h) - f(x)}{h} = \dfrac{a^{x+h} - a^x}{h} = \dfrac{a^x a^h - a^x}{h} = \dfrac{a^x(a^h - 1)}{h}$ **112.** $f(A + B) = a^{A+B} = a^A \cdot a^B = f(A) \cdot f(B)$

113. $f(-x) = a^{-x} = \dfrac{1}{a^x} = \dfrac{1}{f(x)}$ **114.** $f(\alpha x) = a^{\alpha x} = (a^x)^\alpha = [f(x)]^\alpha$

115. (a) $f(-x) = \dfrac{1}{2}(e^{-x} - e^{-(-x)}) = \dfrac{1}{2}(e^{-x} - e^x)$

$= -\dfrac{1}{2}(e^x + e^{-x}) = -f(x)$

116. (a) $f(-x) = \dfrac{1}{2}(e^{-x} + e^{-(-x)}) = \dfrac{1}{2}(e^{-x} + e^x)$

$= \dfrac{1}{2}(e^x + e^{-x}) = f(x)$

(b)

(b)

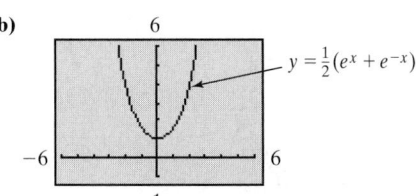

117. $f(1) = 5, f(2) = 17, f(3) = 257, f(4) = 65{,}537$

$f(5) = 4{,}294{,}967{,}297 = 641 \times 6{,}700{,}417$

118. 59 minutes

(c) $(\cosh x)^2 - (\sinh x)^2 = \left[\dfrac{1}{2}(e^x + e^{-x})\right]^2 - \left[\dfrac{1}{2}(e^x - e^{-x})\right]^2$

$= \dfrac{1}{4}[e^{2x} + 2 + e^{-2x} - e^{2x} + 2 - e^{-2x}]$

$= \dfrac{1}{4}(4) = 1$

4.4 Assess Your Understanding (page 308)

4. $\{x | x > 0\}$ or $(0, \infty)$ **5.** $\left(\dfrac{1}{a}, -1\right)$, $(1, 0)$, $(a, 1)$ **6.** 1 **7.** F **8.** T **9.** $2 = \log_3 9$ **10.** $\log_4 16 = 2$ **11.** $2 = \log_a 1.6$ **12.** $\log_a 2.1 = 3$

13. $x = \log_2 7.2$ **14.** $\log_3 4.6 = x$ **15.** $x = \ln 8$ **16.** $\ln M = 2.2$ **17.** $2^3 = 8$ **18.** $3^{-2} = \dfrac{1}{9}$ **19.** $a^6 = 3$ **20.** $b^2 = 4$ **21.** $3^x = 2$ **22.** $2^x = 6$

23. $e^x = 4$ **24.** $e^4 = x$ **25.** 0 **26.** 1 **27.** 2 **28.** -2 **29.** -4 **30.** -2 **31.** $\dfrac{1}{2}$ **32.** $\dfrac{2}{3}$ **33.** 4 **34.** 4 **35.** $\dfrac{1}{2}$ **36.** 3 **37.** $\left\{x | x > 3\right\}$;

$(3, \infty)$ **38.** $\{x | x > 1\}$; $(1, \infty)$ **39.** All real numbers except 0; $\{x | x \neq 0\}$ **40.** $\{x | x > 0\}$; $(0, \infty)$ **41.** $\{x | x > 10\}$; $(10, \infty)$

42. $\left\{x | x > -\dfrac{3}{2}\right\}$; $\left(-\dfrac{3}{2}, \infty\right)$ **43.** $\{x | x > -1\}$; $(-1, \infty)$ **44.** $\{x | x > 5\}$; $(5, \infty)$ **45.** $\{x | x < -1 \text{ or } x > 0\}$; $(-\infty, -1)$ or $(0, \infty)$

46. $\{x | x < 0 \text{ or } x > 1\}$; $(-\infty, 0)$ or $(1, \infty)$ **47.** $\{x | x \geq 1\}$; $[1, \infty)$ **48.** $\{x | x > 0, x \neq 1\}$ **49.** 0.511 **50.** 0.536 **51.** 30.099 **52.** 4.055

53. 2.303 **54.** 0.434 **55.** -53.991 **56.** 1.110 **57.** $\sqrt{2}$ **58.** $a = 2^{1/4} \approx 1.189$

59.

60.

61.

62.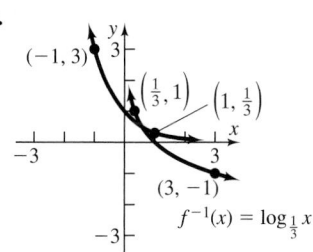

63. B
64. F
65. D
66. H
67. A
68. C
69. E
70. G

71. $x = -4$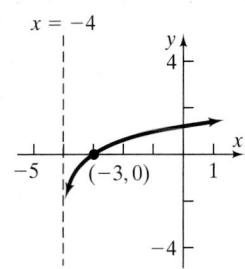

Domain: $(-4, \infty)$
Range: $(-\infty, \infty)$
Vertical asymptote: $x = -4$

72.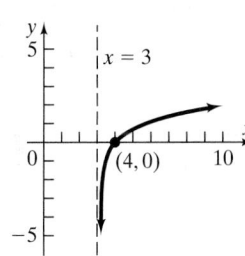

Domain: $(3, \infty)$
Range: $(-\infty, \infty)$
Vertical asymptote: $x = 3$

73.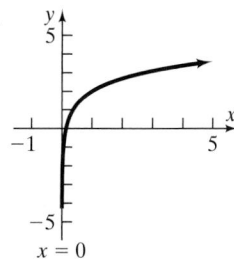

Domain: $(0, \infty)$
Range: $(-\infty, \infty)$
Vertical asymptote: $x = 0$

74.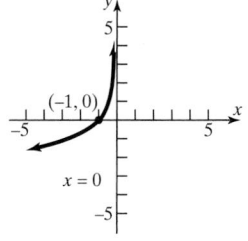

Domain: $(-\infty, 0)$
Range: $(-\infty, \infty)$
Vertical asymptote: $x = 0$

75.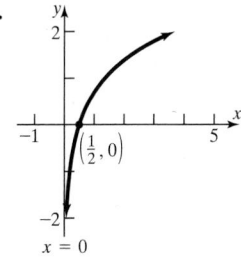

Domain: $(0, \infty)$
Range: $(-\infty, \infty)$
Vertical asymptote: $x = 0$

76.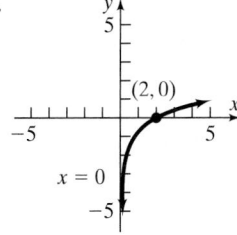

Domain: $(0, \infty)$
Range: $(-\infty, \infty)$
Vertical asymptote: $x = 0$

77.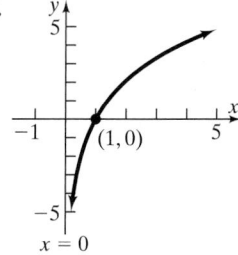

Domain: $(0, \infty)$
Range: $(-\infty, \infty)$
Vertical asymptote: $x = 0$

78.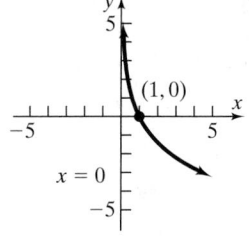

Domain: $(0, \infty)$
Range: $(-\infty, \infty)$
Vertical asymptote: $x = 0$

79.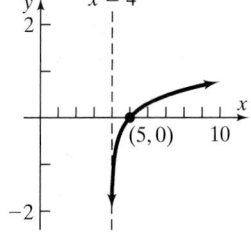

Domain: $(4, \infty)$
Range: $(-\infty, \infty)$
Vertical asymptote: $x = 4$

80.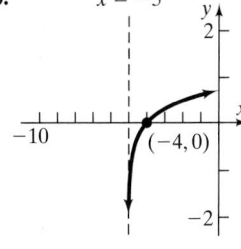

Domain: $(-5, \infty)$
Range: $(-\infty, \infty)$
Vertical asymptote: $x = -5$

81.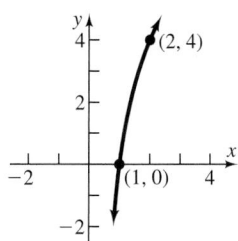

Domain: $(0, \infty)$
Range: $(-\infty, \infty)$
Vertical asymptote: $x = 0$

82.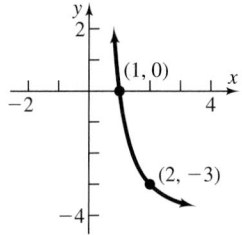

Domain: $(0, \infty)$
Range: $(-\infty, \infty)$
Vertical asymptote: $x = 0$

83.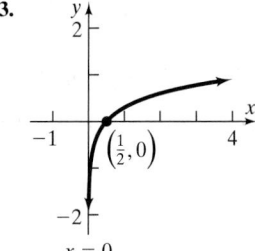

Domain: $(0, \infty)$
Range: $(-\infty, \infty)$
Vertical asymptote: $x = 0$

84.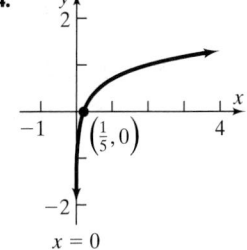

Domain: $(0, \infty)$
Range: $(-\infty, \infty)$
Vertical asymptote: $x = 0$

85.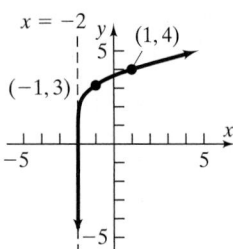

Domain: $(-2, \infty)$
Range: $(-\infty, \infty)$
Vertical asymptote: $x = -2$

86.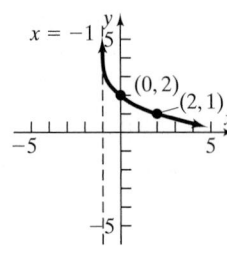

Domain: $(-1, \infty)$
Range: $(-\infty, \infty)$
Vertical asymptote: $x = -1$

87. $\{9\}$

88. $\{125\}$

89. $\left\{\dfrac{7}{2}\right\}$

90. $\left\{\dfrac{11}{3}\right\}$

91. $\{2\}$

92. $\left\{\dfrac{1}{2}\right\}$

93. $\{5\}$

94. $\{-4\}$

95. $\{3\}$

96. $\{4\}$

97. $\{2\}$

98. $\left\{-\dfrac{1}{5}\right\}$

99. $\left\{\dfrac{\ln 10}{3}\right\}$

100. $\left\{\dfrac{\ln 3}{2}\right\}$

101. $\left\{\dfrac{\ln 8 - 5}{2}\right\}$

102. $\left\{\dfrac{1 - \ln 13}{2}\right\}$

103. $\left\{-2\sqrt{2}, 2\sqrt{2}\right\}$

104. $\left\{\dfrac{-1 + \sqrt{85}}{2}, \dfrac{-1 - \sqrt{85}}{2}\right\}$

105. $\{-1\}$

106. $\{-1\}$

107. (a) $\left\{x \middle| x > -\dfrac{1}{2}\right\}; \left(-\dfrac{1}{2}, \infty\right)$ **(b)** $4; (40, 4)$ **(c)** $4; (4, 2)$ **108. (a)** $\{x | x > -1\}; (-1, \infty)$ **(b)** $0; (7, 0)$ **(c)** $3; (3, -1)$

109.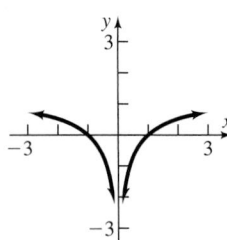

Domain: $\{x | x \neq 0\}$
Range: $(-\infty, \infty)$
Intercepts: $(-1, 0), (1, 0)$

110.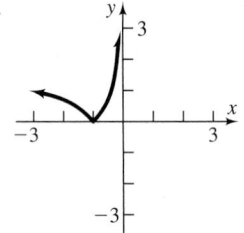

Domain: $\{x | x < 0\}$
Range: $\{y | y \geq 0\}$
Intercept: $(-1, 0)$

111.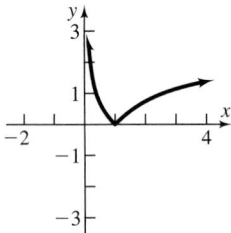

Domain: $\{x | x > 0\}$
Range: $\{y | y \geq 0\}$
Intercept: $(1, 0)$

112.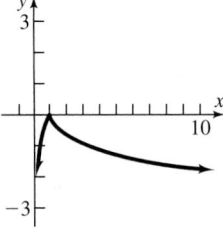

Domain: $\{x | x > 0\}$
Range: $\{y | y \leq 0\}$
Intercept: $(1, 0)$

113. $f^{-1}(x) = \dfrac{\ln x - 3}{2}$ **114.** $f^{-1}(x) = \log_3(x - 4) + 5$ **115.** $f^{-1}(x) = \dfrac{1}{5} \cdot 2^{x+3}$ **116.** $f^{-1}(x) = 5^{x-4} - 1$

117. (a)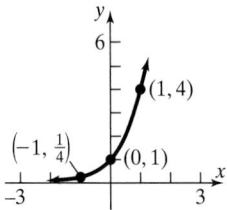

Domain: $(-\infty, \infty)$
Range: $(0, \infty)$
Horizontal asymptote: $y = 0$

(b) $f^{-1}(x) = \log_4 x$

(c)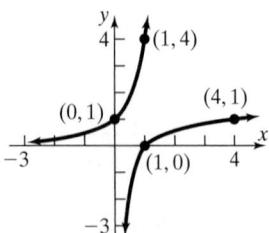

Domain of f^{-1} = range of f = $(0, \infty)$
Range of f^{-1} = domain of f = $(-\infty, \infty)$
Vertical asymptote of f^{-1}: $x = 0$

118. (a)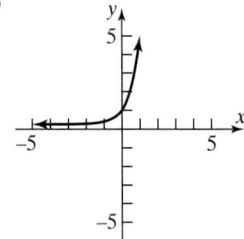

Domain: $(-\infty, \infty)$
Range: $(0, \infty)$
Horizontal asymptote: $y = 0$

(b) $f^{-1}(x) = \log_5 x$

(c)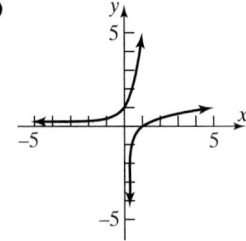

Domain of f^{-1} = range of f = $(0, \infty)$
Domain of f^{-1} = domain of f = $(-\infty, \infty)$
Vertical asymptote of f^{-1}: $x = 0$

119. (a)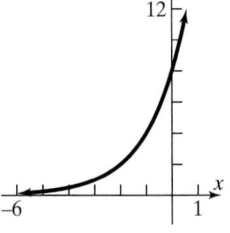

Domain: $(-\infty, \infty)$
Range: $(0, \infty)$
Horizontal asymptote: $y = 0$

(b) $f^{-1}(x) = \log_2 x - 3$

(c)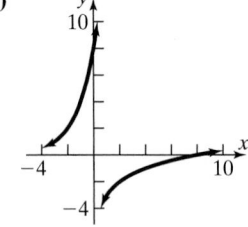

Domain of f^{-1} = range of f = $(0, \infty)$
Range of f^{-1} = domain of f = $(-\infty, \infty)$
Vertical asymptote of f^{-1}: $x = 0$

120. (a)

(c)

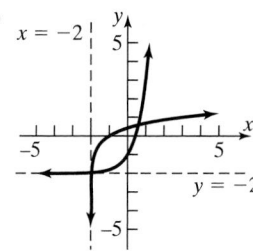

121. (a) 1 **(b)** 2 **(c)** 3 **(d)** It increases **(e)** 0.000316
(f) 3.981×10^{-8} **122. (a)** 0.4283 **(b)** 0.7782 **(c)** 0.5504
123. (a) 5.97 km **(b)** 0.90 km **124. (a)** 1.98 days or slightly less
than 2 days **(b)** 6.58 days or about 6 days and 14 hr
125. (a) 6.93 min **(b)** 16.09 min **(c)** No, since $F(t)$ can never
equal 1 **126. (a)** 4.62 min or about 4 min and 37 sec
(b) 10.73 min or about 10 min and 44 sec **127.** $h \approx 2.29$,
so the time between injections is about 2 hr, 17 min.
128. 3.99 days or about 4 days

Domain: $(-\infty, \infty)$
Range: $(-2, \infty)$
Horizontal asymptote: $y = -2$
(b) $f^{-1}(x) = \log_3(x + 2)$

Domain of f^{-1} = range of $f = (-2, \infty)$
Range of f^{-1} = domain of $f = (-\infty, \infty)$
Vertical asymptote: $x = -2$

129. 0.2695 sec
0.8959 sec

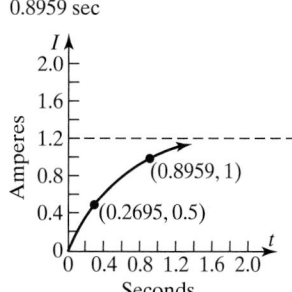

130. (a) 0.0211 **(b)** 38 words **(c)** 54 words **(d)** 109.13 min
131. 50 decibels (dB)
132. 90 dB
133. 110 dB
134. 80 dB **135.** 8.1 **136.** 6.9
137. (a) $k = 11.22$ **(b)** 6.73% **(c)** 0.41% **(d)** 0.14%
138. Because $y = \log_1 x$ means $1^y = 1 = x$, which cannot be true for $x \neq 1$

4.5 Assess Your Understanding (page 318)

1. Sum **2.** 7 **3.** $r \log_a M$ **4.** F **5.** F **6.** T **7.** 71 **8.** -13 **9.** -4 **10.** $\sqrt{2}$ **11.** 7 **12.** 8 **13.** 1 **14.** 2 **15.** 1 **16.** 1 **17.** 2 **18.** 2 **19.** $\dfrac{5}{4}$

20. 42 **21.** 4 **22.** 3 **23.** $a + b$ **24.** $a - b$ **25.** $b - a$ **26.** $-a$ **27.** $3a$ **28.** $3b$ **29.** $\dfrac{1}{5}(a + b)$ **30.** $\dfrac{1}{4}(a - b)$ **31.** $2 + \log_5 x$ **32.** $\log_3 x - 2$

33. $3 \log_2 z$ **34.** $5 \log_7 x$ **35.** $1 + \ln x$ **36.** $1 - \ln x$ **37.** $\ln x + x$ **38.** $\ln x - x$ **39.** $2 \log_a u + 3 \log_a v$ **40.** $\log_2 a - 2 \log_2 b$

41. $2 \ln x + \dfrac{1}{2}\ln(1 - x)$ **42.** $\ln x + \dfrac{1}{2}\ln(1 + x^2)$ **43.** $3 \log_2 x - \log_2(x - 3)$ **44.** $\dfrac{1}{3}\log_5(x^2 + 1) - \log_5(x + 1) - \log_5(x - 1)$

45. $\log x + \log(x + 2) - 2\log(x + 3)$ **46.** $3 \log x + \dfrac{1}{2}\log(x + 1) - 2\log(x - 2)$ **47.** $\dfrac{1}{3}\ln(x - 2) + \dfrac{1}{3}\ln(x + 1) - \dfrac{2}{3}\ln(x + 4)$

48. $\dfrac{4}{3}\ln(x - 4) - \dfrac{2}{3}\ln(x + 1) - \dfrac{2}{3}\ln(x - 1)$ **49.** $\ln 5 + \ln x + \dfrac{1}{2}\ln(1 + 3x) - 3\ln(x - 4)$ **50.** $\ln 5 + 2 \ln x + \dfrac{1}{3}\ln(1 - x) - \ln 4 - 2\ln(x + 1)$

51. $\log_5 u^3 v^4$ **52.** $\log_3\left(\dfrac{u^2}{v}\right)$ **53.** $-\dfrac{5}{2}\log_3 x$ **54.** $\log_2\left(\dfrac{1}{x^3}\right)$ **55.** $\log_4\left[\dfrac{x - 1}{(x + 1)^4}\right]$ **56.** $\log\left(\dfrac{x + 2}{x + 1}\right)$ **57.** $-2\ln(x - 1)$

58. $\log\left[\dfrac{x^2 + 2x - 3}{(x - 2)(x^2 + 7x + 6)}\right]$ **59.** $\log_2[x(3x - 2)^4]$ **60.** $\log_3 x^9$ **61.** $\log_a\left(\dfrac{25x^6}{\sqrt{2x + 3}}\right)$ **62.** $\log\left(\sqrt[3]{x^3 + 1}\sqrt{x^2 + 1}\right)$

63. $\log_2\left[\dfrac{(x + 1)^2}{(x + 3)(x - 1)}\right]$ **64.** $\log_5\left[\dfrac{(3x + 1)^3}{x(2x - 1)^2}\right]$ **65.** 2.771 **66.** 1.796 **67.** -3.880 **68.** -3.907 **69.** 5.615 **70.** 2.584 **71.** 0.874 **72.** 0.303

73. $y = \dfrac{\log x}{\log 4}$

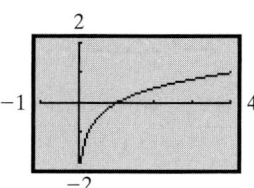

74. $y = \dfrac{\log x}{\log 5}$

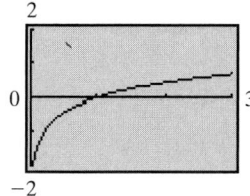

75. $y = \dfrac{\log(x + 2)}{\log 2}$

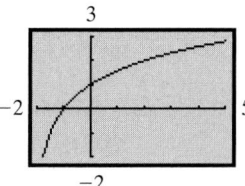

76. $y = \dfrac{\log(x - 3)}{\log 4}$

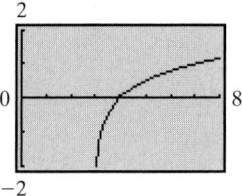

77. $y = \dfrac{\log(x + 1)}{\log(x - 1)}$

78. $y = \dfrac{\log(x - 2)}{\log(x + 2)}$

79. (a) $(f \circ g)(x) = x$; $\{x | x \text{ is any real number}\}$ or $(-\infty, \infty)$
(b) $(g \circ f)(x) = x$; $\{x | x > 0\}$ or $(0, \infty)$ **(c)** 5
(d) $(f \circ h)(x) = \ln x^2$ **(e)** 2
80. (a) $(f \circ g)(x) = x$; $\{x | x \text{ is any real number}\}$ or $(-\infty, \infty)$
(b) $(g \circ f)(x) = x$; $\{x | x > 0\}$ or $(0, \infty)$ **(c)** 3
(d) $(f \circ h)(x) = \log_2(4x)$ **(e)** 5

81. $y = Cx$ **82.** $y = x + C$ **83.** $y = Cx(x + 1)$ **84.** $y = \dfrac{Cx^2}{x + 1}$ **85.** $y = Ce^{3x}$ **86.** $y = Ce^{-2x}$ **87.** $y = Ce^{-4x} + 3$ **88.** $y = Ce^{5x} - 4$

89. $y = \dfrac{\sqrt[3]{C}(2x + 1)^{1/6}}{(x + 4)^{1/9}}$ **90.** $y = \dfrac{\sqrt{C}(x^2 + 1)^{1/6}}{x^{1/4}}$ **91.** 3 **92.** 3 **93.** 1 **94.** $n!$

95. $\log_a\!\left(x + \sqrt{x^2 - 1}\right) + \log_a\!\left(x - \sqrt{x^2 - 1}\right) = \log_a\!\left[\left(x + \sqrt{x^2 - 1}\right)\left(x - \sqrt{x^2 - 1}\right)\right] = \log_a[x^2 - (x^2 - 1)] = \log_a 1 = 0$

96. $\log_a\!\left(\sqrt{x} + \sqrt{x - 1}\right) + \log_a\!\left(\sqrt{x} - \sqrt{x - 1}\right)$

$\qquad = \log_a\!\left[\left(\sqrt{x} + \sqrt{x - 1}\right)\left(\sqrt{x} - \sqrt{x - 1}\right)\right]$

$\qquad = \log_a[x - (x - 1)]$

$\qquad = \log_a 1$

$\qquad = 0$

97. $\ln(1 + e^{2x}) = \ln[e^{2x}(e^{-2x} + 1)] = \ln e^{2x} + \ln(e^{-2x} + 1) = 2x + \ln(1 + e^{-2x})$

98. $\dfrac{f(x + h) - f(x)}{h} = \dfrac{\log_a(x + h) - \log_a(x)}{h}$

$\qquad = \dfrac{1}{h}\log_a\!\left(\dfrac{x + h}{x}\right) = \log_a\!\left(1 + \dfrac{h}{x}\right)^{1/h},\ h \neq 0$

99. $y = f(x) = \log_a x;\ a^y = x$ implies $a^y = \left(\dfrac{1}{a}\right)^{-y} = x$, so $-y = \log_{1/a} x = -f(x)$.

100. $f(AB) = \log_a(AB) = \log_a(A) + \log_a(B) = f(A) + f(B)$ **101.** $f(x) = \log_a x;\ f\!\left(\dfrac{1}{x}\right) = \log_a \dfrac{1}{x} = \log_a 1 - \log_a x = -f(x)$

102. $f(x^\alpha) = \log_a(x^\alpha) = \alpha \log_a(x) = \alpha f(x)$ **103.** $\log_a \dfrac{M}{N} = \log_a(M \cdot N^{-1}) = \log_a M + \log_a N^{-1} = \log_a M - \log_a N$,

\qquad since $a^{\log_a N^{-1}} = N^{-1}$ implies $a^{-\log_a N^{-1}} = N$; i.e., $\log_a N = -\log_a N^{-1}$.

104. $\log_a \dfrac{1}{N} = \log_a 1 - \log_a N = -\log_a N$

4.6 Assess Your Understanding *(page 323)*

5. $\{6\}$ **6.** $\{0\}$ **7.** $\{16\}$ **8.** $\left\{\dfrac{1}{3}\right\}$ **9.** $\{8\}$ **10.** $\left\{\dfrac{1}{3}\right\}$ **11.** $\{3\}$ **12.** $\{5\}$ **13.** $\{5\}$ **14.** $\{4\}$ **15.** $\left\{-1 + \sqrt{1 + e^4}\right\} \approx \{6.456\}$

16. $\left\{\dfrac{1}{e^2 - 1}\right\} \approx \{0.157\}$ **17.** $\{-3\}$ **18.** $\{-3\}$ **19.** $\left\{\dfrac{\ln 3}{\ln 2}\right\} \approx \{1.585\}$ **20.** 0 **21.** 0 **22.** 1 **23.** $\dfrac{\ln 10}{\ln 2} \approx 3.322$ **24.** $\log_3 14 = \dfrac{\ln 14}{\ln 3} \approx 2.402$

25. $-\dfrac{\ln 1.2}{\ln 8} \approx -0.088$ **26.** $\left\{\log_2\!\left(\dfrac{2}{3}\right)\right\} = \left\{\dfrac{\ln\!\left(\dfrac{2}{3}\right)}{\ln 2}\right\} \approx \{-0.585\}$ **27.** $\left\{\dfrac{\ln 3}{2\ln 3 + \ln 4}\right\} \approx \{0.307\}$ **28.** $\left\{\dfrac{\ln(2.5)}{\ln(50)}\right\} \approx \{0.234\}$

29. $\left\{\dfrac{\ln 7}{\ln 0.6 + \ln 7}\right\} \approx \{1.356\}$ **30.** $\left\{\dfrac{\ln\!\left(\dfrac{4}{3}\right)}{\ln\!\left(\dfrac{20}{3}\right)}\right\} \approx 0.152$ **31.** 0 **32.** $\left\{\dfrac{\ln(0.51)}{\ln(9.6333)}\right\} \approx \{-0.297\}$ **33.** $\left\{\dfrac{\ln \pi}{1 + \ln \pi}\right\} \approx \{0.534\}$

34. $\left\{\dfrac{3}{\ln \pi - 1}\right\} \approx \{20.728\}$ **35.** $\left\{\dfrac{\ln 1.6}{3 \ln 2}\right\} \approx \{0.226\}$ **36.** $\left\{\dfrac{\ln\!\left(\dfrac{2}{3}\right)}{0.2 \ln 4}\right\} \approx \{-1.462\}$ **37.** $\left\{\log_4\!\left(-2 + \sqrt{7}\right)\right\} \approx \{-0.315\}$

38. $\left\{\log_3\!\left(\dfrac{3 - \sqrt{5}}{2}\right), \log_3\!\left(\dfrac{3 + \sqrt{5}}{2}\right)\right\} \approx \{-0.876, 0.876\}$ **39.** $\{\log_5 4\} \approx \{0.861\}$ **40.** $\{\log_6 3\} \approx \{0.631\}$ **41.** No real solution

42. No real solution **43.** $\{\log_4 5\} \approx \{1.161\}$ **44.** $\{\log_3 7\} \approx \{1.771\}$ **45.** $\left\{\dfrac{9}{2}\right\}$ **46.** $\{4\}$ **47.** $\{2\}$ **48.** $\{67\}$ **49.** $\{1\}$

50. $\left\{\dfrac{26 + 8\sqrt{10}}{9}, \dfrac{26 - 8\sqrt{10}}{9}\right\} \approx \{5.7,\ 0.078\}$ **51.** $\{16\}$ **52.** $\{81\}$ **53.** $\left\{-1, \dfrac{2}{3}\right\}$ **54.** $\left\{4, \dfrac{1}{4}\right\}$ **55.** $\{0\}$

56. $\left\{\ln\!\left(3 + 2\sqrt{2}\right), \ln\!\left(3 - \sqrt{2}\right)\right\} \approx \{1.763, -1.763\}$ **57.** $\left\{\ln\!\left(2 + \sqrt{5}\right)\right\} \approx \{1.444\}$ **58.** $\left\{\ln\!\left(-2 + \sqrt{5}\right)\right\} \approx \{-1.444\}$

59. $\left\{\dfrac{1}{\dfrac{1}{10^{\log 5}} + \log 3}\right\}$ **60.** $\left\{\dfrac{3}{\dfrac{1}{10^{\log 2}} + \log 6}\right\}$ **61. (a)** $\{5\}, (5, 3)$ **(b)** $\{5\}, (5, 4)$ **(c)** $\{1\}$, yes, at $(1, 2)$ **(d)** $\{5\}$ **(e)** $\left\{-\dfrac{1}{11}\right\}$

62. (a) $\{4\}; (4, 2)$ **(b)** $\{28\}; (28, 3)$ **(c)** No real solution; no intersection **(d)** $\{4\}$ **(e)** $\left\{\dfrac{7}{4}\right\}$

63. (a)

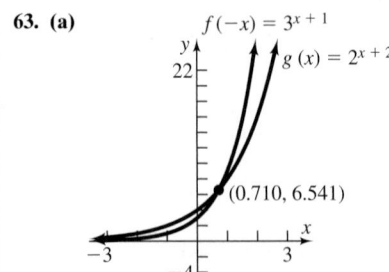

(b) $(0.710, 6.541)$
(c) $\{x \mid x > 0.710\}$ or $(0.710, \infty)$

64. (a)

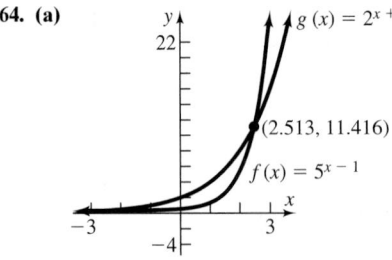

(b) $(2.513, 11.416)$
(c) $\{x \mid x > 2.513\}$ or $(2.513, \infty)$

65. (a), (b)

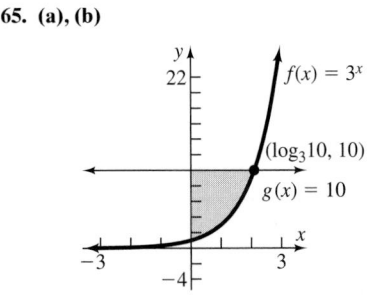

(c) $(\log_3 10, 10)$

66. (a), (b)

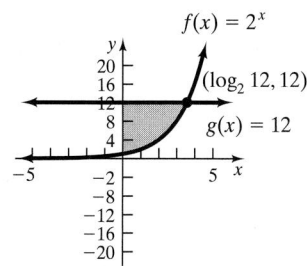

(c) $(\log_2 12, 12)$

67. (a), (b)

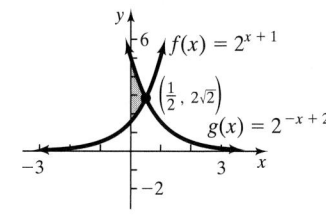

(c) $\left(\dfrac{1}{2}, 2\sqrt{2}\right)$

68. (a), (b)

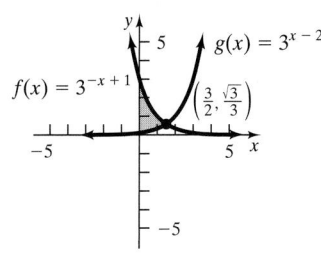

(c) $\left(\dfrac{3}{2}, \dfrac{\sqrt{3}}{3}\right)$

69. (a)

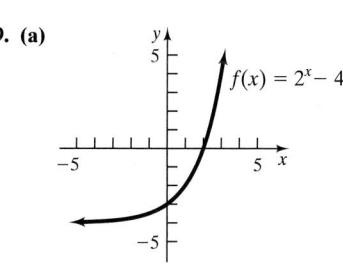

(b) 2 **(c)** $\{x \mid x < 2\}$ or $(-\infty, 2)$

70. (a)

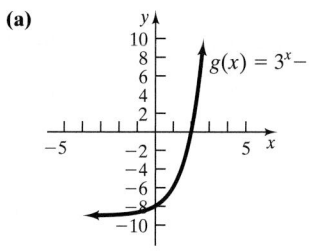

(b) 2 **(c)** $\{x \mid x < 2\}$ or $(-\infty, 2)$

71. $\{2.79\}$ **72.** $\{12.15\}$ **73.** $\{-0.57\}$
74. $\{-1.98, 0.45\}$ **75.** $\{-0.70\}$
76. $\{1.86, 4.54\}$ **77.** $\{0.57\}$ **78.** $\{1.16\}$
79. $\{0.39, 1.00\}$ **80.** $\{0.65\}$ **81.** $\{1.32\}$
82. $\{0.05, 1.48\}$ **83.** $\{1.31\}$ **84.** $\{0.57\}$
85. (a) Around the middle of the year 2006
 (b) In the beginning of the year 2021
86. (a) In the beginning of the year 2016
 (b) In the beginning of the year 2025

87. (a) After 2.4 years **(b)** After 6.5 years **(c)** After 10 years **88. (a)** After 1.4 years **(b)** After 5 years **(c)** After 13 years

4.7 Assess Your Understanding (page 332)

3. $108.29 **4.** $59.83 **5.** $609.50 **6.** $358.84 **7.** $697.09 **8.** $789.24 **9.** $12.46 **10.** $49.35 **11.** $125.23 **12.** $156.83 **13.** $88.72 **14.** $59.14
15. $860.72 **16.** $626.61 **17.** $554.09 **18.** $266.08 **19.** $59.71 **20.** $654.98 **21.** $361.93 **22.** $886.92 **23.** 5.35% **24.** 6.82% **25.** 26%
26. 7.18% **27.** $6\frac{1}{4}\%$ compounded annually **28.** 9% compounded quarterly **29.** 9% compounded monthly **30.** 7.9% compounded daily
31. 104.32 mo (about 8.7 yr); 103.97 mo (about 8.66 yr) **32.** 6.96 yr or 83.52 mo; 6.93 yr or 83.18 mo **33.** 61.02 mo; 60.82 mo **34.** 5.62 yr or
67.43 mo; 5.60 yr or 67.15 mo **35.** 15.27 yr or 15 yr, 3 mo **36.** 16.62 yr or 16 yr, 7 mo **37.** $104,335 **38.** $215.48 **39.** $12,910.62 **40.** $2955.39
41. About $30.17 per share or $3017 **42.** 15.47% **43.** 9.35% **44.** $2315.97 **45.** Not quite. Jim will have $1057.60. The second bank gives a better
deal, since Jim will have $1060.62 after 1 yr. **46.** $1021.60 **47.** Will has $11,632.73; Henry has $10,947.89. **48.** It is better to give $1000 now for
investment to earn $1349.86 in 3 yr. **49. (a)** Interest is $30,000 **(b)** Interest is $38,613.59 **(c)** Interest is $37,752.73. Simple interest at 12% is
best. **50. (a)** 4.34% **(b)** 4.34% **51. (a)** $1364.62 **(b)** $1353.35 **52.** $10,810.76 **53.** $4631.93 **54.** 9.07%

55. (a) 6.1 yr
(b) 18.45 yr

(c) $mP = P\left(1 + \dfrac{r}{n}\right)^{nt}$

$m = \left(1 + \dfrac{r}{n}\right)^{nt}$

$\ln m = \ln\left(1 + \dfrac{r}{n}\right)^{nt} = nt \ln\left(1 + \dfrac{r}{n}\right)$

$t = \dfrac{\ln m}{n \ln\left(1 + \dfrac{r}{n}\right)}$

56. (a) 20.79 yr **(b)** 7.74%
(c) $A = Pe^{rt}$
 $\ln A = \ln P + rt$
 $rt = \ln A - \ln P$
 $t = \dfrac{\ln A - \ln P}{r}$

4.8 Assess Your Understanding (page 343)

1. (a) 500 insects **(b)** $0.02 = 2\%$ **(c)** About 611 insects **(d)** After about 23.5 days **(e)** After about 34.7 days **2. (a)** 1000 **(b)** $0.01 = 1\%$
(c) About 1041 **(d)** After about 53.1 h **(e)** After about 69.3 h **3. (a)** $-0.0244 = -2.44\%$ **(b)** About 391.7 g **(c)** After about 9.1 yr
(d) 28.4 yr **4. (a)** $-0.087 = -8.7\%$ **(b)** About 45.7 g **(c)** After about 4.1 days **(d)** About 7.97 days **5.** 5832; 3.9 days **6.** About 5243;
about 7.85 hours later **7.** 25,198 **8.** 711,111 **9.** 9.797 g **10.** 9.999999467 g; 9.999994668 g **11.** 9727 yr ago **12.** 2882 yr old
13. (a) 5:18 PM **(b)** About 14.3 min **(c)** The temperature of the pizza approaches 70°F. **14. (a)** 45.41°F **(b)** About 16.2 minutes
(c) About 7.26 min **(d)** The temperature approaches 38°F. **15.** 18.63°C; 25.1°C **16.** 45.69°F; 28.46 min **17.** 7.34 kg; 76.6 h **18.** 1.25 V
19. 26.6 days **20.** At about 9:42 PM **21. (a)** 90% **(b)** 12.86% **(c)** 85.77% **(d)** 1996 **(e)** About 5.6 yr **22. (a)** 90% **(b)** 20% **(c)** 47.3%
(d) After 8.44 months **(e)** After 3.7 months **23. (a)** 1000 g **(b)** 43.9% **(c)** 30 g **(d)** 616.6 g **(e)** After 9.85 h **(f)** About 7.9 h
24. (a) 500 eagles **(b)** 16.2% **(c)** Almost 10 eagles **(d)** After 29.8 years **(e)** After 27.3 years
25. (a) 9.23×10^{-3}, or about 0 **(d)** 57.91°, 43.99°, 30.07°
(b) 0.81, or about 1
(c) 5.01, or about 5

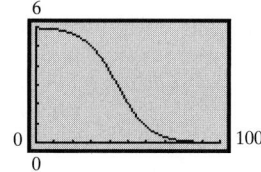

4.9 Assess Your Understanding *(page 349)*

1. (a)

2. (a)

3. (a)

4. (a)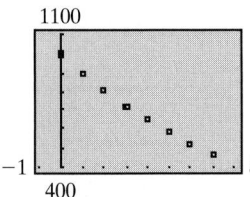

(b) $y = 0.0903(1.3384)^x$
(c) $N(t) = 0.0903e^{0.2915t}$
(d)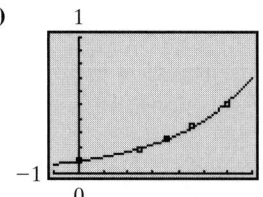
(e) 0.69
(f) After about 7.26 h

(b) $y = 0.0339(1.9474)^x$
(c) $N(t) = 0.0339e^{0.6665t}$
(d)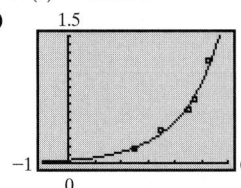
(e) 1.85
(f) 6.19 h

(b) $y = 100.326(0.8769)^x$
(c) $A = 100.326e^{-0.1314t}$
(d)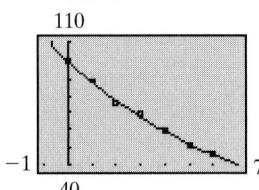
(e) 5.3 weeks **(f)** 0.14 g
(g) After about 12.3 weeks

(b) $y = 998.91(0.8976)^t$
(c) $y = 998.91e^{-0.1081t}$
(d)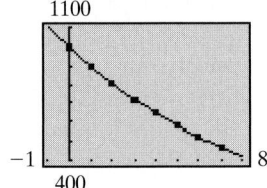
(e) 6.41 days **(f)** 115 g
(g) After about 14.9 days

5. (a)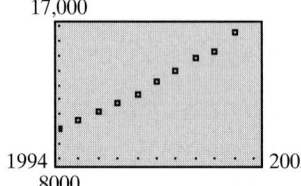

(b) $y = 2.2908 \times 10^{-44}(1.056554737)^x$
(c) 5.66%
(d) $44,230.54
(e) In 2023

7. (a)

6. (a)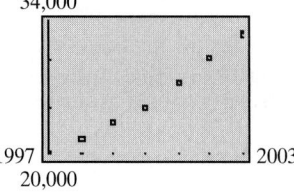

(b) $y = 1.4014 \times 10^{-67}(1.0855)^x$
(c) 8.55%
(d) $131,466.24
(e) In 2014

(b) $y = 32,741.02 - 6070.96 \ln x$
(c)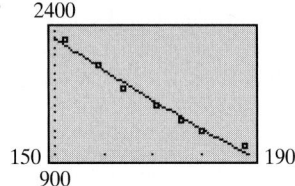

(d) Approximately 168 computers

8. (a)

9. (a)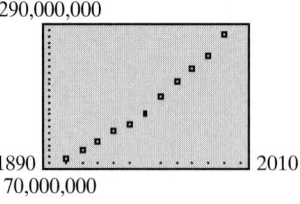

10. (a) $x = 1$ corresponds to 1993.
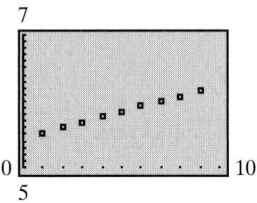

(b) $y = -11,850.72 + 2688.5 \ln x$

(b) $y = \dfrac{799,475,916.5}{1 + 1.56344 \times 10^{14}e^{-0.0160x}}$

(b) $y = \dfrac{10.05267}{1 + 0.8439e^{-0.0320x}}$

(c)

(c)

(c)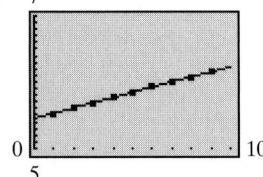

(d) Approximately 152 computers

(d) 799,475,917
(e) Approximately 279,809,184
(f) 2011

(d) 10.053 billion people
(e) Approximately 6.38 billion people
(f) 2013

11. (a) 12,500,000

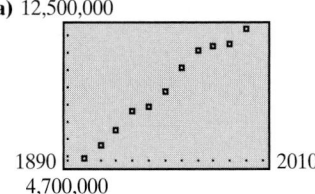

1890 ⌐ 2010
4,700,000

(b) $y = \dfrac{14,471,245.24}{1 + 3.860 \times 10^{20}e^{-0.0246x}}$

(c) 12,500,000

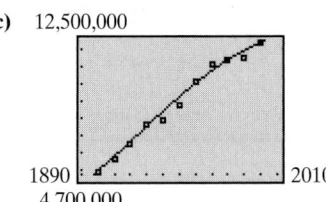

1890 ⌐ 2010
4,700,000

(d) 14,471,245 **(e)** Approximately 12,811,429

12. (a) 13,000,000

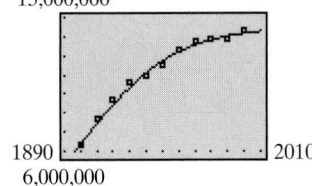

1890 ⌐ 2010
6,000,000

(b) $y = \dfrac{12,529,186.49}{1 + 6.0614 \times 10^{28}e^{-0.0349x}}$

(c) 13,000,000

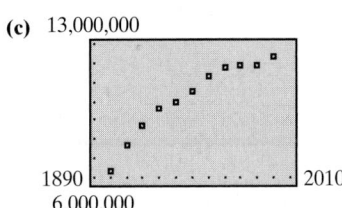

1890 ⌐ 2010
6,000,000

(d) 12,529,186 **(e)** Approximately 12,274,358

Review Exercises *(page 355)*

1. (a) -26 **(b)** -241 **(c)** 16 **(d)** -1 **2. (a)** -1 **(b)** 37 **(c)** 4 **(d)** 5 **3. (a)** $\sqrt{11}$ **(b)** 1 **(c)** $\sqrt{\sqrt{6}+2}$ **(d)** 19 **4. (a)** -5 **(b)** $\sqrt{15}$

(c) -6626 **(d)** $\sqrt{4-\sqrt{5}}$ **5. (a)** e^4 **(b)** $3e^{-2}-2$ **(c)** e^{e^4} **(d)** -17 **6. (a)** $\dfrac{2}{73}$ **(b)** $\dfrac{2}{3}$ **(c)** $\dfrac{2178}{1097}$ **(d)** -9

7. $(f \circ g)(x) = 1 - 3x$, all real numbers; $(g \circ f)(x) = 7 - 3x$, all real numbers;
$(f \circ f)(x) = x$, all real numbers; $(g \circ g)(x) = 9x + 4$, all real numbers

8. $(f \circ g)(x) = 4x + 1$, all real numbers; $(g \circ f)(x) = 4x - 1$, all real numbers;
$(f \circ f)(x) = 4x - 3$, all real numbers; $(g \circ g)(x) = 4x + 3$, all real numbers

9. $(f \circ g)(x) = 27x^2 + 3|x| + 1$, all real numbers; $(g \circ f)(x) = 3|3x^2 + x + 1|$, all real numbers;
$(f \circ f)(x) = 3(3x^2 + x + 1)^2 + 3x^2 + x + 2$, all real numbers; $(g \circ g)(x) = 9|x|$, all real numbers

10. $(f \circ g)(x) = \sqrt{3(1 + x + x^2)}$, all real numbers; $(g \circ f)(x) = 1 + \sqrt{3x} + 3x$, $\{x | x \geq 0\}$;
$(f \circ f)(x) = \sqrt{3\sqrt{3x}}$, $\{x | x \geq 0\}$; $(g \circ g)(x) = 3 + 3x + 4x^2 + 2x^3 + x^4$, all real numbers

11. $(f \circ g)(x) = \dfrac{1 + x}{1 - x}$, $\{x | x \neq 0, x \neq 1\}$; $(g \circ f)(x) = \dfrac{x - 1}{x + 1}$, $\{x | x \neq -1, x \neq 1\}$; $(f \circ f)(x) = x$, $\{x | x \neq 1\}$; $(g \circ g)(x) = x$, $\{x | x \neq 0\}$

12. $(f \circ g)(x) = \sqrt{\dfrac{3 - 3x}{x}}$, $\{x | 0 < x \leq 1\}$; $(g \circ f)(x) = \dfrac{3}{\sqrt{x - 3}}$, $\{x | x > 3\}$;

$(f \circ f)(x) = \sqrt{\sqrt{x - 3} - 3}$, $\{x | x \geq 12\}$; $(g \circ g)(x) = x$, $\{x | x \neq 0\}$

13. (a) One-to-one **(b)** $\{(2, 1), (5, 3), (8, 5), (10, 6)\}$ **14. (a)** One-to-one **(b)** $\{(4, -1), (2, 0), (5, 1), (7, 3)\}$

15.

16.

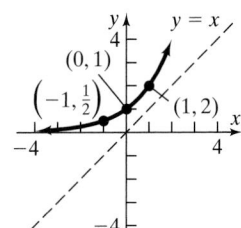

17. $f^{-1}(x) = \dfrac{2x + 3}{5x - 2}$

$f(f^{-1}(x)) = \dfrac{2\left(\dfrac{2x + 3}{5x - 2}\right) + 3}{5\left(\dfrac{2x + 3}{5x - 2}\right) - 2} = x$

$f^{-1}(f(x)) = \dfrac{2\left(\dfrac{2x + 3}{5x - 2}\right) + 3}{5\left(\dfrac{2x + 3}{5x - 2}\right) - 2} = x$

Domain of f = range of f^{-1} = all real numbers except $\dfrac{2}{5}$

Range of f = domain of f^{-1} = all real numbers except $\dfrac{2}{5}$

18. $f^{-1}(x) = \dfrac{2 - 3x}{1 + x}$

$f(f^{-1}(x)) = \dfrac{2 - \dfrac{2 - 3x}{1 + x}}{3 + \dfrac{2 - 3x}{1 + x}} = x$

$f^{-1}(f(x)) = \dfrac{2 - 3\left(\dfrac{2 - x}{3 + x}\right)}{1 + \left(\dfrac{2 - x}{3 + x}\right)} = x$

Domain of f = range of f^{-1} = all real numbers except -3

Range of f = domain of f^{-1} = all real numbers except -1

19. $f^{-1}(x) = \dfrac{x + 1}{x}$

$f(f^{-1}(x)) = \dfrac{1}{\dfrac{x + 1}{x} - 1} = x$

$f^{-1}(f(x)) = \dfrac{\dfrac{1}{x - 1} + 1}{\dfrac{1}{x - 1}} = x$

Domain of f = range of f^{-1} = all real numbers except 1

Range of f = domain of f^{-1} = all real numbers except 0

20. $f^{-1}(x) = x^2 + 2, x \geq 0$
$f(f^{-1}(x)) = \sqrt{x^2 + 2 - 2} = |x| = x, x \geq 0$
$f^{-1}(f(x)) = (\sqrt{x - 2})^2 + 2 = x$
Domain of f = range of f^{-1} = $[2, \infty)$
Range of f = domain of f^{-1} = $[0, \infty)$

21. $f^{-1}(x) = \dfrac{27}{x^3}$
$f(f^{-1}(x)) = \dfrac{3}{\left(\dfrac{27}{x^3}\right)^{1/3}} = x$

22. $f^{-1}(x) = (x - 1)^3$
$f(f^{-1}(x)) = ((x - 1)^3)^{1/3} + 1 = x$
$f^{-1}(f(x)) = (x^{1/3} + 1 - 1)^3 = x$
Domain of f = range of f^{-1} = $(-\infty, \infty)$
Range of f = domain of f^{-1} = $(-\infty, \infty)$

$f^{-1}(f(x)) = \dfrac{27}{\left(\dfrac{3}{x^{1/3}}\right)^3} = x$
Domain of f = range of f^{-1} = all real numbers except 0
Range of f = domain of f^{-1} = all real numbers except 0

23. (a) 81 **(b)** 2 **(c)** $\dfrac{1}{9}$ **(d)** -3 **24. (a)** 3 **(b)** 4 **(c)** $\dfrac{1}{81}$ **(d)** -5 **25.** $\log_5 z = 2$ **26.** $\log_a m = 5$ **27.** $5^{13} = u$ **28.** $a^3 = 4$

29. $\left\{x \middle| x > \dfrac{2}{3}\right\}; \left(\dfrac{2}{3}, \infty\right)$ **30.** $\left\{x \middle| x > -\dfrac{1}{2}\right\}; \left(-\dfrac{1}{2}, \infty\right)$ **31.** $\{x|x < 1 \text{ or } x > 2\}; (-\infty, 1) \text{ or } (2, \infty)$ **32.** $\{x|x < -3 \text{ or } x > 3\}; (-\infty, -3) \text{ or }$
$(3, \infty)$ **33.** -3 **34.** 4 **35.** $\sqrt{2}$ **36.** 0.1 **37.** 0.4 **38.** $\sqrt{3}$ **39.** $\log_3 u + 2 \log_3 v - \log_3 w$ **40.** $8 \log_2 a + 2 \log_2 b$

41. $2 \log x + \dfrac{1}{2} \log(x^3 + 1)$ **42.** $2 \log_5(x + 1) - 2 \log_5 x$ **43.** $\ln x + \dfrac{1}{3}\ln(x^2 + 1) - \ln(x - 3)$ **44.** $2 \ln(2x + 3) - 2 \ln(x - 1) - 2 \ln(x - 2)$

45. $\dfrac{25}{4} \log_4 x$ **46.** $\dfrac{13}{6} \log_3 x$ **47.** $-2 \ln(x + 1)$ **48.** $\log\left(\dfrac{x - 3}{x + 4}\right)$ **49.** $\log\left(\dfrac{4x^3}{[(x + 3)(x - 2)]^{1/2}}\right)$ **50.** $\ln\left[\dfrac{16\sqrt{x^2 + 1}}{\sqrt{x(x - 4)}}\right]$ **51.** 2.124 **52.** 4.392

53.

54.

55.
Domain: $(-\infty, \infty)$
Range: $(0, \infty)$
Horizontal asymptote: $y = 0$

56.
Domain: $(-\infty, \infty)$
Range: $(-\infty, 3)$
Horizontal asymptote: $y = 3$

57.
Domain: $(-\infty, \infty)$
Range: $(0, \infty)$
Horizontal asymptote: $y = 0$

58.
Domain: $(-\infty, \infty)$
Range: $(1, \infty)$
Horizontal asymptote: $y = 1$

59.
Domain: $(-\infty, \infty)$
Range: $(-\infty, 1)$
Horizontal asymptote: $y = 1$

60.
Domain: $(0, \infty)$
Range: $(-\infty, \infty)$
Vertical asymptote: $x = 0$

61.
Domain: $(-3, \infty)$
Range: $(-\infty, \infty)$
Vertical asymptote: $x = -3$

62.
Domain: $(-\infty, \infty)$
Range: $(0, \infty)$
Horizontal asymptote: $y = 0$

63.
Domain: $(-\infty, \infty)$
Range: $(-\infty, 3)$
Horizontal asymptote: $y = 3$

64.
Domain: $(-\infty, 0)$
Range: $(-\infty, \infty)$
Vertical asymptote: $x = 0$

65. $\left\{\dfrac{1}{4}\right\}$ **66.** $\left\{-\dfrac{16}{9}\right\}$ **67.** $\left\{\dfrac{-1 - \sqrt{3}}{2}, \dfrac{-1 + \sqrt{3}}{2}\right\}$ **68.** $\left\{\dfrac{1 - \sqrt{3}}{2}, \dfrac{1 + \sqrt{3}}{2}\right\}$ **69.** $\left\{\dfrac{1}{4}\right\}$ **70.** $\left\{\dfrac{1}{8}\right\}$ **71.** $\left\{\dfrac{2 \ln 3}{\ln 5 - \ln 3} \approx 4.301\right\}$

72. $\left\{\dfrac{2(\ln 7 + \ln 5)}{\ln 7 - \ln 5} \approx 21.133\right\}$ **73.** $\left\{\dfrac{12}{5}\right\}$ **74.** $\{-2, 6\}$ **75.** $\{83\}$ **76.** $\left\{-\dfrac{1}{2}\right\}$ **77.** $\left\{\dfrac{1}{2}, -3\right\}$ **78.** $\{1\}$ **79.** $\{-1\}$ **80.** $\{3, 4\}$

81. $\{1 - \ln 5 \approx -0.609\}$ **82.** $\left\{\dfrac{1 - \ln 4}{2} \approx -0.193\right\}$

83. (a) (a), (g)

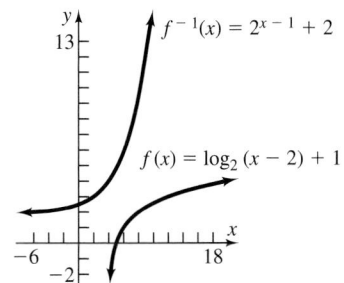

(b) $\{3\}, (6, 3)$ **(c)** $\{10\}, (10, 4)$

(d) $\dfrac{5}{2}$ **(e)** $\left\{x \mid x > \dfrac{5}{2}\right\}$ or $\left(\dfrac{5}{2}, \infty\right)$

(f) $f^{-1}(x) = 2^{x-1} + 2$

84. (a) (a), (g)

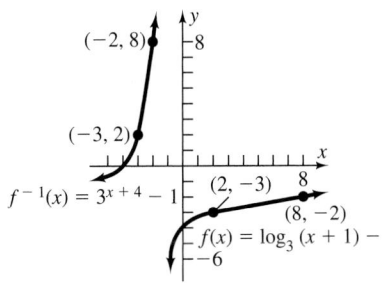

(b) $\{-2\}, (8, -2)$ **(c)** $\{2\}, (2, -3)$

(d) 80 **(e)** $\{x \mid -1 < x < 80\}$, or $(-1, 80)$

(f) $f^{-1}(x) = 3^{x+4} - 1$

85. 3229.5 m
86. 1482 m
87. (a) 37.3 W **(b)** 6.9 dB
88. (a) 11.77 **(b)** 9.56 in.
89. (a) 9.85 yr **(b)** 4.27 yr
90. $20,398.87; 4.04; 17.5 yr
91. $41,668.97
92. (a) 10.436% **(b)** $32,249.24
93. 24,203 yr ago
94. 55.22 min or 55 min and 13 sec
95. 6,835,600,129
96. 7.204 g; 0.519 g

97. (a) 0.3 **(b)** 0.8

(c)

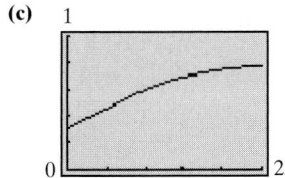

(d) In 2023

98. (a) 165.1

(b) $y = 165.73(0.9951)^x$

(c) 165.1

(d) Approximately 83 sec

99. (a) 10

(c) 15

(b) Wind chill $= 18.921 - 7.096 \ln$ (wind speed) **(d)** About 47 people; 50 people

100. (a) 50

(c) 50

(b) $C = \dfrac{46.93}{1 + 21.273 e^{-0.7306t}}$

(d) Approximately $-3°$F
(e) 2.4 days; during the 10th hour of the 3rd day
(f) 9.5 days

Chapter Test (page 359)

1. (a) $f \circ g = \dfrac{2x + 7}{2x + 3}$; domain: $\left\{x \mid x \neq -\dfrac{3}{2}\right\}$ **(b)** $(g \circ f)(-2) = 5$ **(c)** $(f \circ g)(-2) = -3$

2. (a) The function is not one-to-one. **(b)** The function is one-to-one.

3. $f^{-1}(x) = \dfrac{2 + 5x}{3x}$

	Domain	Range
f	$x \neq \dfrac{5}{3}$	$y \neq 0$
f^{-1}	$x \neq 0$	$y \neq \dfrac{5}{3}$

4. The point $(-5, 3)$ must be on the graph of f^{-1}. **5.** $x = 5$ **6.** $b = 4$ **7.** $x = 625$ **8.** $e^3 + 2 \approx 22.086$ **9.** $\log 20 \approx 1.301$

10. $\log_3 21 = \dfrac{\ln 21}{\ln 3} \approx 2.771$ **11.** $\ln 133 \approx 4.890$

12. Domain: $\{x|-\infty < x < \infty\}$
Range: $\{y|y > -2\}$
Asymptote: $y = -2$

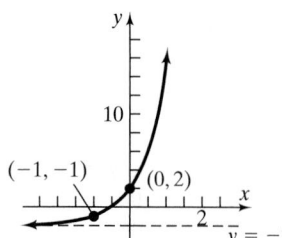

13. Domain: $\{x|x > 2\}$
Range: $\{y|-\infty < y < \infty\}$
Asymptote: $x = 2$

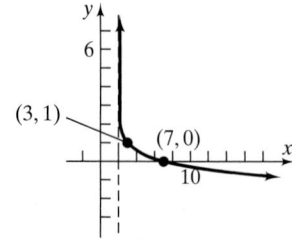

14. $x = 1$ **15.** $x = 91$ **16.** $x = -\ln 2 \approx -0.693$

17. $\dfrac{1 - \sqrt{13}}{2} \approx -1.303,\ \dfrac{1 + \sqrt{13}}{2} \approx 2.303$

18. $x = \dfrac{3 \ln 7}{1 - \ln 7} \approx -6.172$

19. $x = 2\sqrt{6} \approx 4.899$

20. $2 + 3 \log_2 x - \log_2(x - 6) - \log_2(x + 3)$

21. About 250.39 days

22. **(a)** \$35,298 **(b)** \$21,409

23. **(a)** About 83 decibles
(b) The pain threshold will be exceeded if 31,623 people shouted at the same time.

24. $y = \dfrac{213}{1 + 205.86e^{-0.3564t}}$; 197 million U.S. cell phone subscribers

Cumulative Review (page 360)

1. Yes; no **2.** **(a)** 10 **(b)** $2x^2 + 3x + 1$ **(c)** $2x^2 + 4xh + 2h^2 - 3x - 3h + 1$ **3.** $\left(\dfrac{1}{2}, \dfrac{\sqrt{3}}{2}\right)$ is on the graph **4.** -26

5.

6. **(a)**

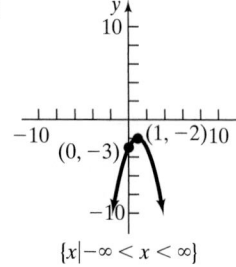

(b) $\{x|-\infty < x < \infty\}$

7. $f(x) = 2(x - 4)^2 - 8 = 2x^2 - 16x + 24$

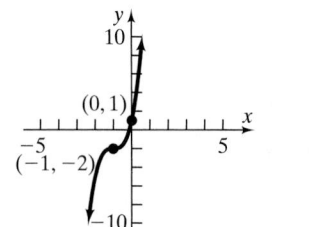

9. $f(g(x)) = \dfrac{4}{(x - 3)^2} + 2$; domain: $\{x|x \neq 3\}$; 3

10. **(a)** Zeros: $-4, -\dfrac{1}{4}, 2$

(b) x-intercepts: $-4, -\dfrac{1}{4}, 2$; y-intercept: -8

(c) Local maximum of 60.75 occurs at $x = -2.5$;
Local minimum of -25 occurs at $x = 1$

(d)

11. **(a)** (a), (c)

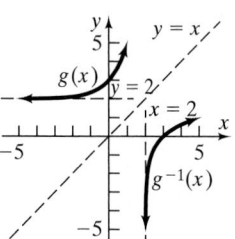

Domain g = range $g^{-1} = (-\infty, \infty)$
Range g = domain $g^{-1} = (2, \infty)$
(b) $g^{-1}(x) = \log_3(x - 2)$

12. $-\dfrac{3}{2}$ **13.** 2

14. **(a)** -1
(b) $\{x|x > -1\}$ or $(-1, \infty)$
(c) 25

15. **(a)**

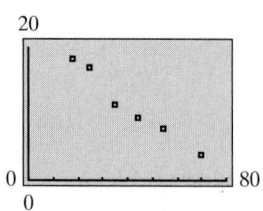

(b) Answers will vary.

CHAPTER 5 Analytic Geometry

5.2 Assess Your Understanding (page 369)

6. parabola **7.** paraboloid of revolution **8.** T **9.** T **10.** T **11.** B **12.** G **13.** E **14.** D **15.** H **16.** A **17.** C **18.** F

19. $y^2 = 16x.$

20. $x^2 = 8y$

21. $x^2 = -12y$

22. $y^2 = -16x$

23. $y^2 = -8x$

24. $x^2 = -4y$

25. $x^2 = 2y$

26. $y^2 = 2x$

27. $x^2 = \dfrac{4}{3}y$

28. $y^2 = \dfrac{9}{2}x$

29. $(x-2)^2 = -8(y+3)$

30. $(y+2)^2 = 8(x-4)$

31. $(y+2)^2 = 4(x+1)$

32. $(x-3)^2 = -8y$

33. $(x+3)^2 = 4(y-3)$

34. $(y-4)^2 = 12(x+1)$

35. $(y+2)^2 = -8(x+1)$

36. $(x+4)^2 = 12(y-1)$

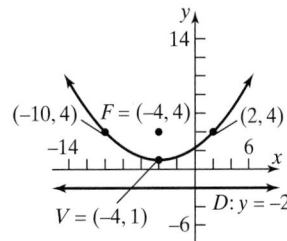

37. Vertex: $(0,0)$; focus: $(0,1)$; directrix: $y = -1$

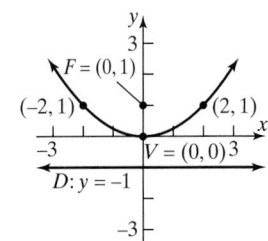

38. Vertex: $(0, 0)$; focus: $(2, 0)$; directrix: $x = -2$

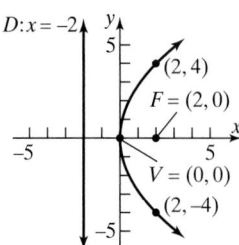

39. Vertex: $(0, 0)$; focus: $(-4, 0)$; directrix: $x = 4$

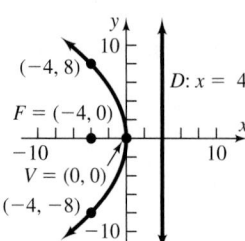

40. Vertex: $(0, 0)$; focus: $(0, -1)$; directrix: $y = 1$

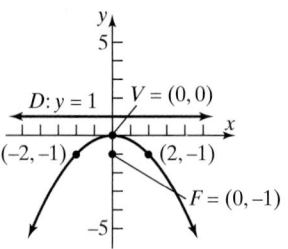

41. Vertex: $(-1, 2)$; focus: $(1, 2)$; directrix: $x = -3$

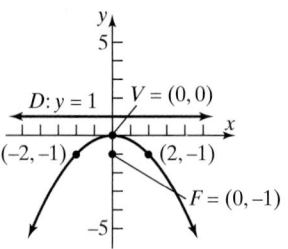

42. Vertex: $(-4, -2)$; focus: $(-4, 2)$; directrix: $y = -6$

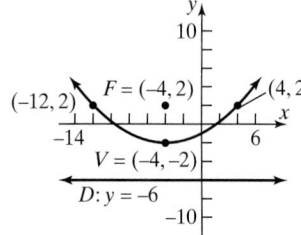

43. Vertex: $(3, -1)$; focus: $\left(3, -\dfrac{5}{4}\right)$; directrix: $y = -\dfrac{3}{4}$

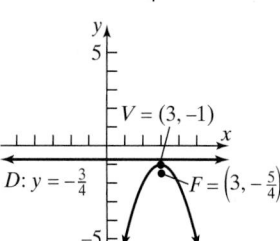

44. Vertex: $(2, -1)$; focus: $(1, -1)$; directrix: $x = 3$

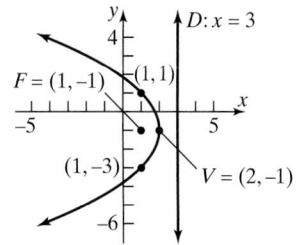

45. Vertex: $(2, -3)$; focus: $(4, -3)$; directrix: $x = 0$

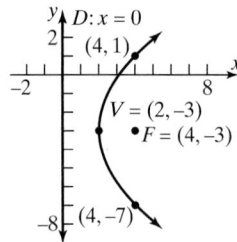

46. Vertex: $(2, 3)$; focus: $(2, 4)$; directrix: $y = 2$

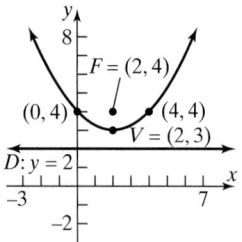

47. Vertex: $(0, 2)$; focus: $(-1, 2)$; directrix: $x = 1$

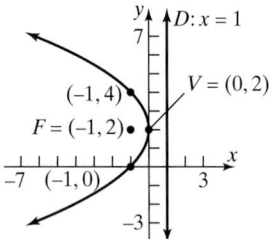

48. Vertex: $(-3, -2)$; focus: $(-3, -1)$; directrix: $y = -3$

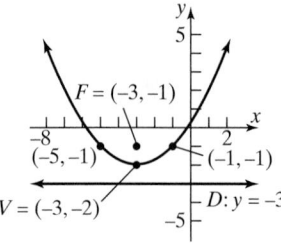

49. Vertex: $(-4, -2)$; focus: $(-4, -1)$; directrix: $y = -3$

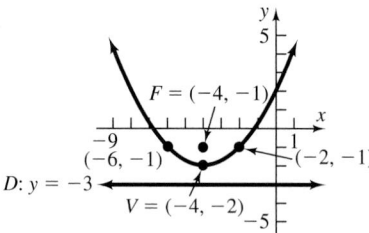

50. Vertex: $(0, 1)$; focus: $(2, 1)$; directrix: $x = -2$

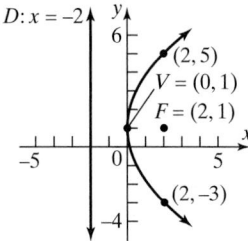

51. Vertex: $(-1, -1)$; focus: $\left(-\dfrac{3}{4}, -1\right)$; directrix: $x = -\dfrac{5}{4}$

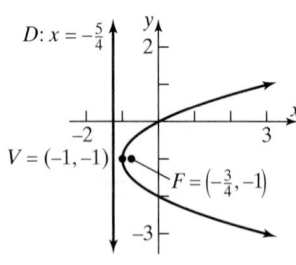

52. Vertex: $(2, -2)$; focus: $\left(2, -\dfrac{3}{2}\right)$; directrix: $y = -\dfrac{5}{2}$

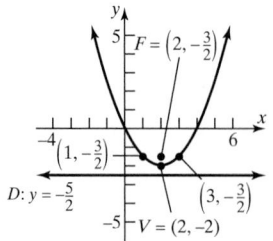

53. Vertex: $(2, -8)$; focus: $\left(2, -\dfrac{31}{4}\right)$; directrix: $y = -\dfrac{33}{4}$

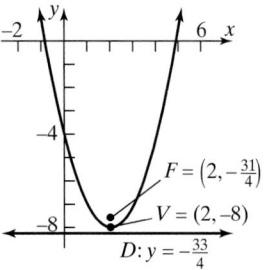

54. Vertex: $(37, -6)$; focus: $\left(\dfrac{147}{4}, -6\right)$;

directrix: $x = \dfrac{149}{4}$

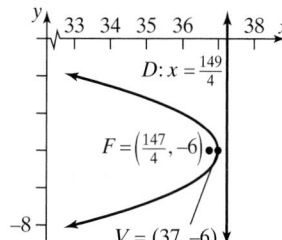

55. $(y - 1)^2 = x$ **56.** $(x - 1)^2 = -(y - 2)$ **57.** $(y - 1)^2 = -(x - 2)$ **58.** $x^2 = 4(y + 1)$

59. $x^2 = 4(y - 1)$ **60.** $(x - 1)^2 = \dfrac{1}{2}(y + 1)$ **61.** $y^2 = \dfrac{1}{2}(x + 2)$ **62.** $y^2 = -(x - 1)$

63. 1.5625 ft from the base of the dish, along the axis of symmetry

64. 1.125 ft or 13.5 in. from the base of the dish **65.** 1 in. from the vertex

66. $4\sqrt{2}$ in. **67.** 20 ft **68.** 15.625 ft **69.** 0.78125 ft **70.** $8\sqrt{2}$ ft

71. 4.17 ft from the base along the axis of symmetry

72. 0.0278 in. from the base of the mirror

73. 24.31 ft, 18.75 ft, 7.64 ft **74.** 27.78 ft

75. $Ax^2 + Ey = 0$, $A \neq 0$, $E \neq 0$

$Ax^2 = -Ey$

$x^2 = -\dfrac{E}{A}y$

This is the equation of a parabola with vertex at $(0, 0)$ and axis of symmetry the y-axis.
The focus is $\left(0, -\dfrac{E}{4A}\right)$; the directrix is the line $y = \dfrac{E}{4A}$. The parabola opens up if $-\dfrac{E}{A} > 0$
and down if $-\dfrac{E}{A} < 0$.

76. $Cy^2 + Dx = 0$, $C \neq 0$, $D \neq 0$

$Cy^2 = -Dx$

$y^2 = -\dfrac{D}{C}x$

This is the equation of a parabola with vertex at $(0, 0)$ and axis of symmetry the x-axis.
The focus is $\left(-\dfrac{D}{4C}, 0\right)$; the directrix is the line $x = \dfrac{D}{4C}$. The parabola opens to the right if
$-\dfrac{D}{C} > 0$ and to the left if $-\dfrac{D}{C} < 0$.

77. $Ax^2 + Dx + Ey + F = 0$, $A \neq 0$

$Ax^2 + Dx = -Ey - F$

$x^2 + \dfrac{D}{A}x = -\dfrac{E}{A}y - \dfrac{F}{A}$

$\left(x + \dfrac{D}{2A}\right)^2 = -\dfrac{E}{A}y - \dfrac{F}{A} + \dfrac{D^2}{4A^2}$

$\left(x + \dfrac{D}{2A}\right)^2 = -\dfrac{E}{A}y + \dfrac{D^2 - 4AF}{4A^2}$

(a) If $E \neq 0$, then the equation may be written as $\left(x + \dfrac{D}{2A}\right)^2 = -\dfrac{E}{A}\left(y - \dfrac{D^2 - 4AF}{4AE}\right)$

This is the equation of a parabola with vertex at $\left(-\dfrac{D}{2A}, \dfrac{D^2 - 4AF}{4AE}\right)$
and axis of symmetry parallel to the y-axis.

(b)–(d) If $E = 0$, the graph of the equation contains no points if $D^2 - 4AF < 0$,
is a single vertical line if $D^2 - 4AF = 0$, and is two vertical lines
if $D^2 - 4AF > 0$.

78. $Cy^2 + Dx + Ey + F = 0$, $C \neq 0$

$Cy^2 + Ey = -Dx - F$

$y^2 + \dfrac{E}{C}y = -\dfrac{D}{C}x - \dfrac{F}{C}$

$\left(y + \dfrac{E}{2C}\right)^2 = -\dfrac{D}{C}x - \dfrac{F}{C} + \dfrac{E^2}{4C^2}$

$\left(y + \dfrac{E}{2C}\right)^2 = -\dfrac{D}{C}x + \dfrac{E^2 - 4CF}{4C^2}$

(a) If $D \neq 0$, then the equation may be written as $\left(y + \dfrac{E}{2C}\right)^2 = -\dfrac{D}{C}\left(x - \dfrac{E^2 - 4CF}{4CD}\right)$.

This is the equation of a parabola with vertex at $\left(\dfrac{E^2 - 4CF}{4CD}, -\dfrac{E}{2C}\right)$
and axis of symmetry parallel to the x-axis.

(b)–(d) If $D = 0$, the graph of the equation contains no points if $E^2 - 4CF < 0$,
is a single horizontal line if $E^2 - 4CF = 0$, and is two horizontal lines
if $E^2 - 4CF > 0$.

5.3 Assess Your Understanding (page 379)

7. ellipse **8.** major **9.** $(0, -5)$; $(0, 5)$ **10.** F **11.** T **12.** T **13.** C **14.** D **15.** B **16.** A

17. Vertices: $(-5, 0)$, $(5, 0)$
Foci: $\left(-\sqrt{21}, 0\right)$, $\left(\sqrt{21}, 0\right)$

18. Vertices: $(-3, 0)$, $(3, 0)$
Foci: $\left(-\sqrt{5}, 0\right)$, $\left(\sqrt{5}, 0\right)$

19. Vertices: $(0, -5)$, $(0, 5)$
Foci: $(0, -4)$, $(0, 4)$

20. Vertices: $(0, -4)$, $(0, 4)$
Foci: $\left(0, -\sqrt{15}\right)$, $\left(0, \sqrt{15}\right)$

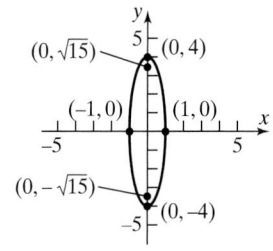

21. $\dfrac{x^2}{4} + \dfrac{y^2}{16} = 1$
Vertices: $(0, -4)$, $(0, 4)$
Foci: $\left(0, -2\sqrt{3}\right)$, $\left(0, 2\sqrt{3}\right)$

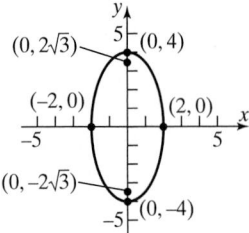

22. $\dfrac{x^2}{18} + \dfrac{y^2}{2} = 1$
Vertices: $\left(-3\sqrt{2}, 0\right)$, $\left(3\sqrt{2}, 0\right)$
Foci: $(-4, 0)$, $(4, 0)$

23. $\dfrac{x^2}{8} + \dfrac{y^2}{2} = 1$
Vertices: $\left(-2\sqrt{2}, 0\right)$, $\left(2\sqrt{2}, 0\right)$
Foci: $\left(-\sqrt{6}, 0\right)$, $\left(\sqrt{6}, 0\right)$

24. $\dfrac{x^2}{4} + \dfrac{y^2}{9} = 1$
Vertices: $(0, -3)$, $(0, 3)$
Foci: $\left(0, -\sqrt{5}\right)$, $\left(0, \sqrt{5}\right)$

25. $\dfrac{x^2}{16} + \dfrac{y^2}{16} = 1$
Vertices: $(-4, 0)$, $(4, 0)$, $(0, -4)$, $(0, 4)$
Focus: $(0, 0)$

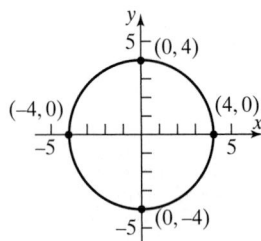

26. $\dfrac{x^2}{4} + \dfrac{y^2}{4} = 1$
Vertices: $(-2, 0)$, $(2, 0)$, $(0, -2)$, $(0, 2)$
Focus: $(0, 0)$

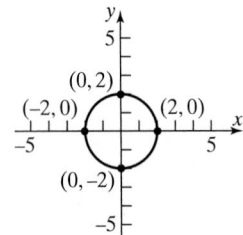

27. $\dfrac{x^2}{25} + \dfrac{y^2}{16} = 1$

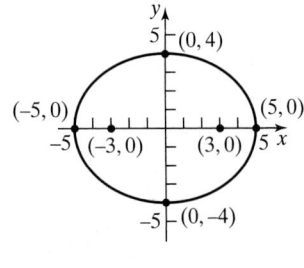

28. $\dfrac{x^2}{9} + \dfrac{y^2}{8} = 1$

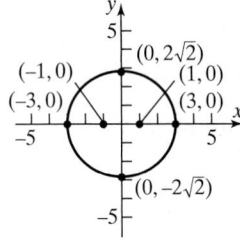

29. $\dfrac{x^2}{9} + \dfrac{y^2}{25} = 1$

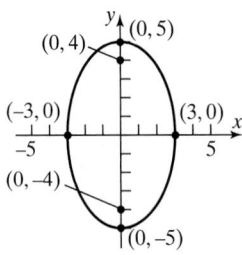

30. $\dfrac{x^2}{3} + \dfrac{y^2}{4} = 1$

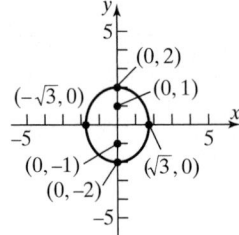

31. $\dfrac{x^2}{9} + \dfrac{y^2}{5} = 1$

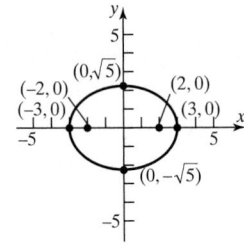

32. $\dfrac{x^2}{12} + \dfrac{y^2}{16} = 1$

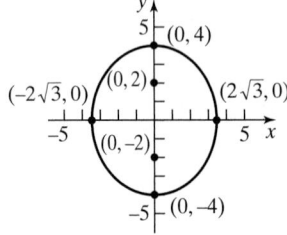

33. $\dfrac{x^2}{25} + \dfrac{y^2}{9} = 1$

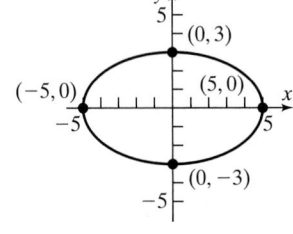

34. $\dfrac{x^2}{48} + \dfrac{y^2}{64} = 1$

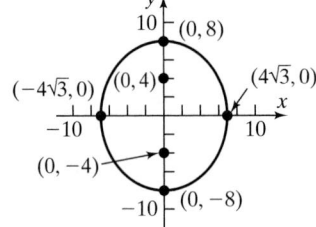

35. $\dfrac{x^2}{4} + \dfrac{y^2}{13} = 1$

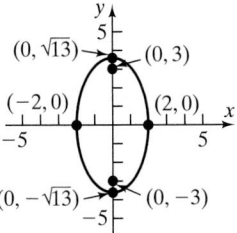

36. $\dfrac{x^2}{16} + y^2 = 1$

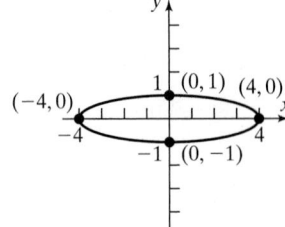

37. $x^2 + \dfrac{y^2}{16} = 1$

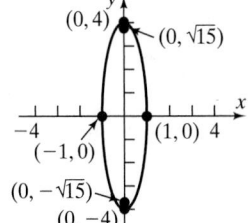

38. $\dfrac{x^2}{25} + \dfrac{y^2}{21} = 1$

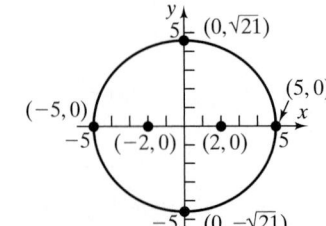

39. $\dfrac{(x + 1)^2}{4} + (y - 1)^2 = 1$

40. $(x + 1)^2 + \dfrac{(y + 1)^2}{4} = 1$

41. $(x - 1)^2 + \dfrac{y^2}{4} = 1$

42. $\dfrac{x^2}{4} + (y - 1)^2 = 1$

43. Center: $(3, -1)$; vertices: $(3, -4), (3, 2)$;

foci: $\left(3, -1 - \sqrt{5}\right), \left(3, -1 + \sqrt{5}\right)$

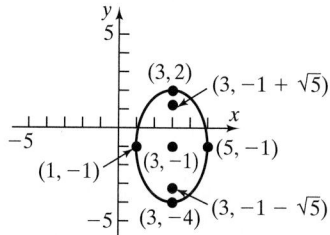

44. Center: $(-4, -2)$;

vertices: $(-7, -2), (-1, -2)$;

foci: $\left(-4 - \sqrt{5}, -2\right), \left(-4 + \sqrt{5}, -2\right)$

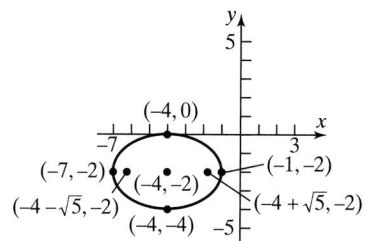

45. $\dfrac{(x + 5)^2}{16} + \dfrac{(y - 4)^2}{4} = 1$

Center: $(-5, 4)$; vertices: $(-9, 4), (-1, 4)$;

foci: $\left(-5 - 2\sqrt{3}, 4\right), \left(-5 + 2\sqrt{3}, 4\right)$

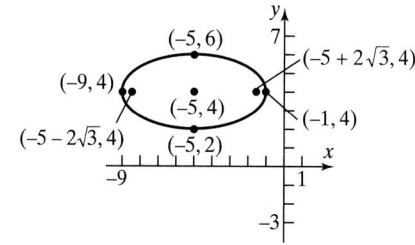

46. $\dfrac{(x - 3)^2}{2} + \dfrac{(y + 2)^2}{18} = 1$

Center: $(3, -2)$;

vertices: $\left(3, -2 - 3\sqrt{2}\right), \left(3, -2 + 3\sqrt{2}\right)$;

foci: $(3, -6), (3, 2)$

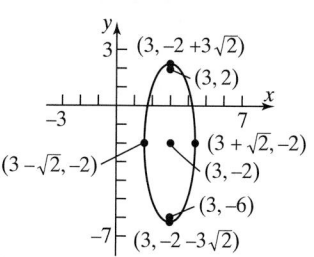

47. $\dfrac{(x + 2)^2}{4} + (y - 1)^2 = 1$

Center: $(-2, 1)$; vertices: $(-4, 1), (0, 1)$;

foci: $\left(-2 - \sqrt{3}, 1\right), \left(-2 + \sqrt{3}, 1\right)$

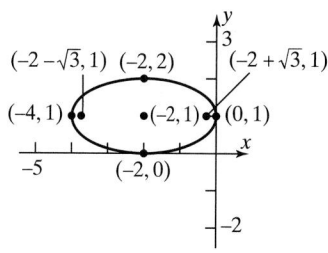

48. $\dfrac{x^2}{3} + (y - 2)^2 = 1$

Center: $(0, 2)$;

vertices: $\left(-\sqrt{3}, 2\right), \left(\sqrt{3}, 2\right)$;

foci: $\left(-\sqrt{2}, 2\right), \left(\sqrt{2}, 2\right)$

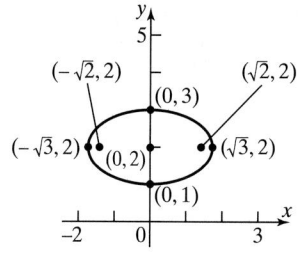

49. $\dfrac{(x - 2)^2}{3} + \dfrac{(y + 1)^2}{2} = 1$

Center: $(2, -1)$; vertices: $\left(2 - \sqrt{3}, -1\right)$;

$\left(2 + \sqrt{3}, -1\right)$; foci: $(1, -1), (3, -1)$

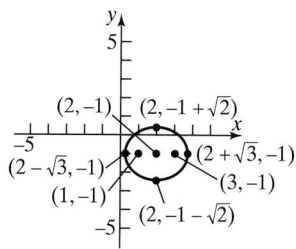

50. $\dfrac{(x + 1)^2}{3} + \dfrac{(y - 1)^2}{4} = 1$

Center: $(-1, 1)$;

vertices: $(-1, -1), (-1, 3)$;

foci: $(-1, 0), (-1, 2)$

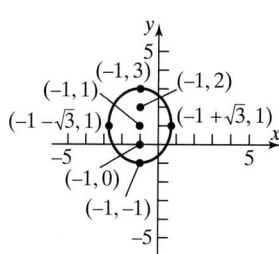

51. $\dfrac{(x - 1)^2}{4} + \dfrac{(y + 2)^2}{9} = 1$

Center: $(1, -2)$; vertices: $(1, -5), (1, 1)$;

foci: $\left(1, -2 - \sqrt{5}\right), \left(1, -2 + \sqrt{5}\right)$

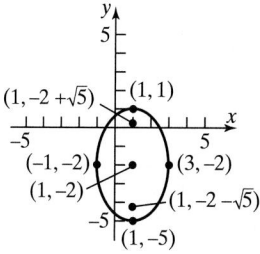

52. $\dfrac{(x + 3)^2}{9} + (y - 1)^2 = 1$

Center: $(-3, 1)$; vertices: $(-6, 1), (0, 1)$;

foci: $\left(-3 - 2\sqrt{2}, 1\right), \left(-3 + 2\sqrt{2}, 1\right)$

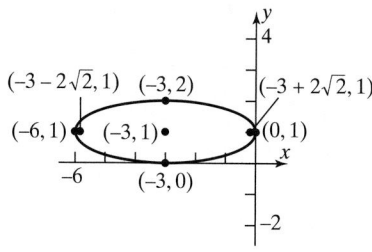

53. $x^2 + \dfrac{(y + 2)^2}{4} = 1$

Center: $(0, -2)$; vertices: $(0, -4), (0, 0)$;

foci: $\left(0, -2 - \sqrt{3}\right), \left(0, -2 + \sqrt{3}\right)$

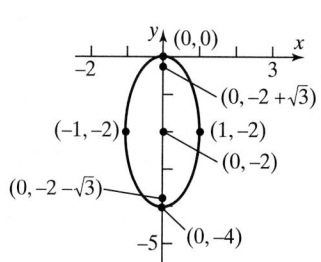

54. $(x - 1)^2 + \dfrac{y^2}{9} = 1$

Center: $(1, 0)$; vertices: $(1, -3), (1, 3)$;

foci: $\left(1, -2\sqrt{2}\right), \left(1, 2\sqrt{2}\right)$

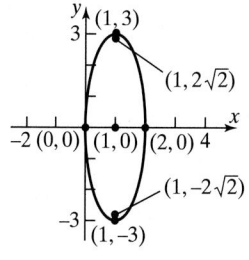

55. $\dfrac{(x-2)^2}{25} + \dfrac{(y+2)^2}{21} = 1$

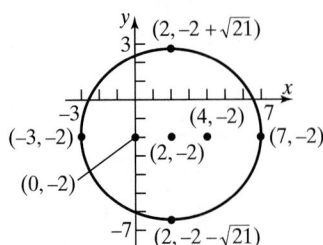

56. $\dfrac{(x+3)^2}{3} + \dfrac{(y-1)^2}{4} = 1$

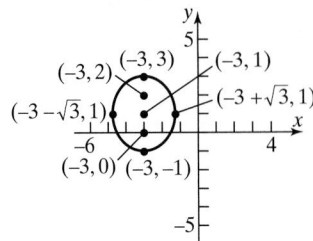

57. $\dfrac{(x-4)^2}{5} + \dfrac{(y-6)^2}{9} = 1$

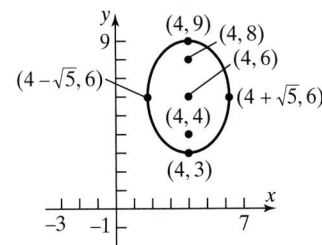

58. $\dfrac{(x+1)^2}{9} + \dfrac{(y-2)^2}{5} = 1$

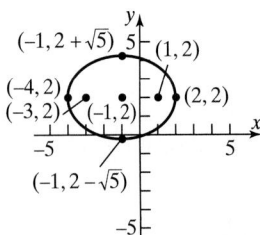

59. $\dfrac{(x-2)^2}{16} + \dfrac{(y-1)^2}{7} = 1$

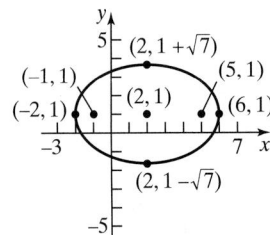

60. $\dfrac{(x-2)^2}{5} + \dfrac{(y-2)^2}{9} = 1$

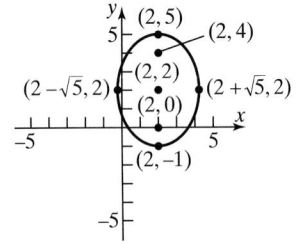

61. $\dfrac{(x-1)^2}{10} + (y-2)^2 = 1$

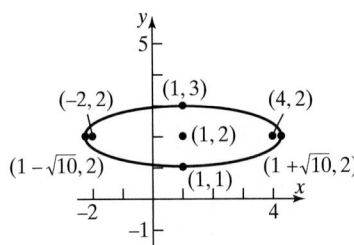

62. $(x-1)^2 + \dfrac{(y-2)^2}{5} = 1$

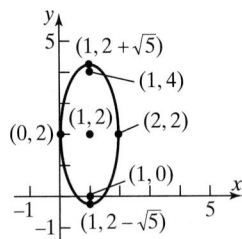

63. $\dfrac{(x-1)^2}{9} + (y-2)^2 = 1$

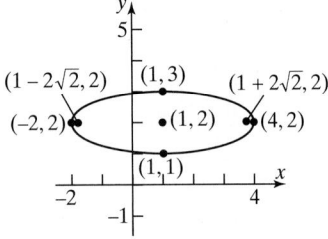

64. $(x-1)^2 + \dfrac{(y-2)^2}{4} = 1$

65.

66.

67.

68.

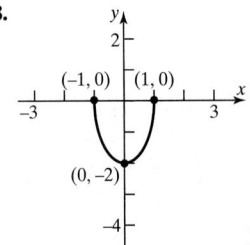

69. $\dfrac{x^2}{100} + \dfrac{y^2}{36} = 1$ **70.** 5 ft from center of bridge, distance is 2.57 ft; 10 ft from center, distance is 4.55 ft; 15 ft from center, distance is 12 ft **71.** 43.3 ft **72.** 112 ft; 25.2 ft **73.** 24.65 ft, 21.65 ft, 13.82 ft

74. 16.67 ft **75.** 0 ft, 12.99 ft, 15 ft, 12.99 ft, 0 ft **76.** 73.69 ft

77. 91.5 million mi; $\dfrac{x^2}{(93)^2} + \dfrac{y^2}{8646.75} = 1$ **78.** 155.5 million mi; $\dfrac{x^2}{(142)^2} + \dfrac{y^2}{19,981.75} = 1$

79. Perihelion: 460.6 million mi; mean distance: 483.8 million mi; $\dfrac{x^2}{(483.8)^2} + \dfrac{y^2}{233,524.2} = 1$

80. Aphelion: 6346 million mi; mean distance: 5448.5 million mi; $\dfrac{x^2}{(5448.5)^2} + \dfrac{y^2}{28,880,646} = 1$

81. 30 ft **82.** $10\sqrt{7}$ ft

83. (a) $Ax^2 + Cy^2 + F = 0$ If A and C are of the same sign and F is of opposite sign, then the equation takes the form
$Ax^2 + Cy^2 = -F$ $\dfrac{x^2}{\left(-\dfrac{F}{A}\right)} + \dfrac{y^2}{\left(-\dfrac{F}{C}\right)} = 1$, where $-\dfrac{F}{A}$ and $-\dfrac{F}{C}$ are positive. This is the equation of an ellipse with center at $(0,0)$.

(b) If $A = C$, the equation may be written as $x^2 + y^2 = -\dfrac{F}{A}$.

This is the equation of a circle with center at $(0,0)$ and radius equal to $\sqrt{-\dfrac{F}{A}}$.

84. $Ax^2 + Cy^2 + Dx + Ey + F = 0$ $A \neq 0, C \neq 0$
$Ax^2 + Dx + Cy^2 + Ey = -F$
$A\left(x^2 + \dfrac{D}{A}x\right) + C\left(y^2 + \dfrac{E}{C}y\right) = -F$
$A\left(x + \dfrac{D}{2A}\right)^2 + C\left(y + \dfrac{E}{2C}\right)^2 = -F + \dfrac{D^2}{4A} + \dfrac{E^2}{4C}$

(a) If $\dfrac{D^2}{4A} + \dfrac{E^2}{4C} - F$ is of the same sign as A (and C), this is the equation of an ellipse with center at $\left(\dfrac{-D}{2A}, \dfrac{-E}{2C}\right)$.

(b) If $\dfrac{D^2}{4A} + \dfrac{E^2}{4C} - F = 0$, the graph is the single point $\left(\dfrac{-D}{2A}, \dfrac{-E}{2C}\right)$.

(c) If $\dfrac{D^2}{4A} + \dfrac{E^2}{4C} - F$ is of the opposite sign to A (and C), the graph contains no points since, in this case, the left side has a sign opposite that of the right side.

5.4 Assess Your Understanding (page 392)

7. hyperbola **8.** transverse axis **9.** $3x = -2y, 3x = 2y$ **10.** F **11.** T **12.** F **13.** B **14.** C **15.** A **16.** D

17. $x^2 - \dfrac{y^2}{8} = 1$

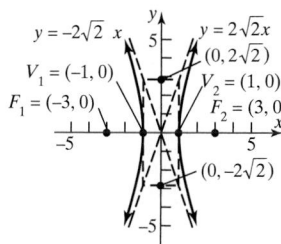

18. $\dfrac{y^2}{9} - \dfrac{x^2}{16} = 1$

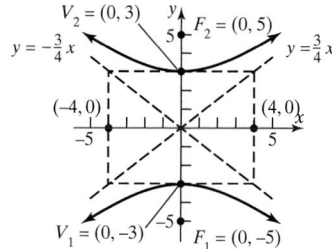

19. $\dfrac{y^2}{16} - \dfrac{x^2}{20} = 1$

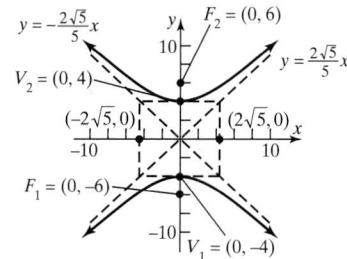

20. $\dfrac{x^2}{4} - \dfrac{y^2}{5} = 1$

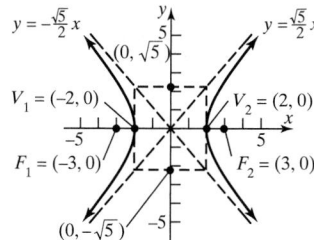

21. $\dfrac{x^2}{9} - \dfrac{y^2}{16} = 1$

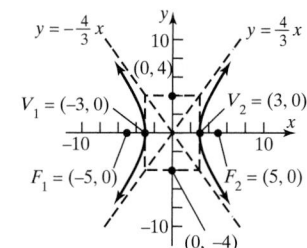

22. $\dfrac{y^2}{4} - \dfrac{x^2}{32} = 1$

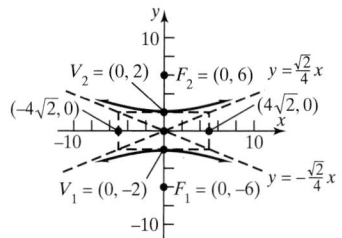

23. $\dfrac{y^2}{36} - \dfrac{x^2}{9} = 1$

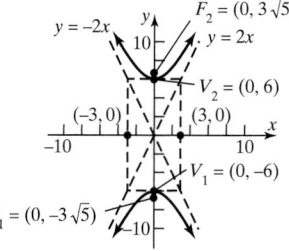

24. $\dfrac{x^2}{16} - \dfrac{y^2}{64} = 1$

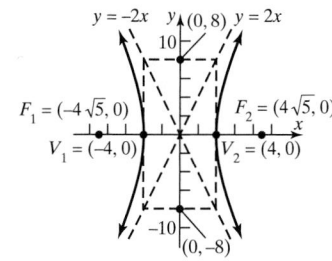

25. $\dfrac{x^2}{8} - \dfrac{y^2}{8} = 1$

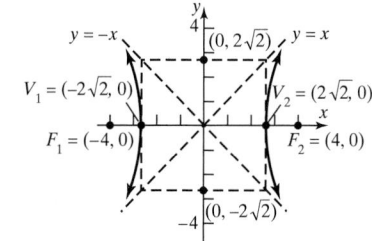

26. $\dfrac{y^2}{2} - \dfrac{x^2}{2} = 1$

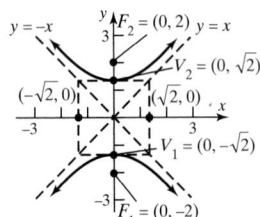

27. $\dfrac{x^2}{25} - \dfrac{y^2}{9} = 1$

Center: $(0, 0)$

Transverse axis: x-axis

Vertices: $(-5, 0)$, $(5, 0)$

Foci: $\left(-\sqrt{34}, 0\right)$, $\left(\sqrt{34}, 0\right)$

Asymptotes: $y = \pm\dfrac{3}{5}x$

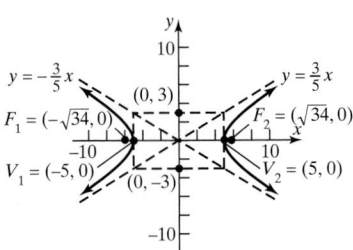

28. $\dfrac{y^2}{16} - \dfrac{x^2}{4} = 1$

Center: $(0, 0)$

Transverse axis: y-axis

Vertices: $(0, -4)$, $(0, 4)$

Foci: $\left(0, -2\sqrt{5}\right)$, $\left(0, 2\sqrt{5}\right)$

Asymptotes: $y = \pm 2x$

29. $\dfrac{x^2}{4} - \dfrac{y^2}{16} = 1$

Center: $(0, 0)$

Transverse axis: x-axis

Vertices: $(-2, 0)$, $(2, 0)$

Foci: $\left(-2\sqrt{5}, 0\right)$, $\left(2\sqrt{5}, 0\right)$

Asymptotes: $y = \pm 2x$

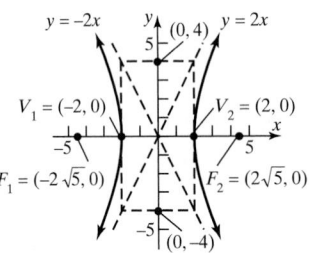

30. $\dfrac{y^2}{4} - \dfrac{x^2}{16} = 1$

Center: $(0, 0)$

Transverse axis: y-axis

Vertices: $(0, -2)$, $(0, 2)$

Foci: $\left(0, -2\sqrt{5}\right)$, $\left(0, 2\sqrt{5}\right)$

Asymptotes: $y = \pm\dfrac{1}{2}x$

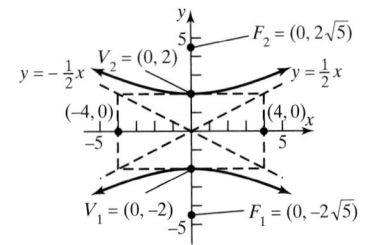

31. $\dfrac{y^2}{9} - x^2 = 1$

Center: $(0, 0)$

Transverse axis: y-axis

Vertices: $(0, -3)$, $(0, 3)$

Foci: $\left(0, -\sqrt{10}\right)$, $\left(0, \sqrt{10}\right)$

Asymptotes: $y = \pm 3x$

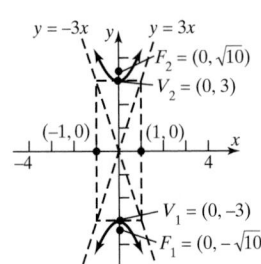

32. $\dfrac{x^2}{4} - \dfrac{y^2}{4} = 1$

Center: $(0, 0)$

Transverse axis: x-axis

Vertices: $(-2, 0)$, $(2, 0)$

Foci: $\left(-2\sqrt{2}, 0\right)$, $\left(2\sqrt{2}, 0\right)$

Asymptotes: $y = \pm x$

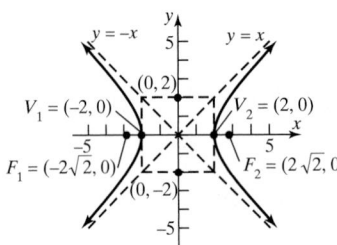

33. $\dfrac{y^2}{25} - \dfrac{x^2}{25} = 1$

Center: $(0, 0)$

Transverse axis: y-axis

Vertices: $(0, -5)$, $(0, 5)$

Foci: $\left(0, -5\sqrt{2}\right)$, $\left(0, 5\sqrt{2}\right)$

Asymptotes: $y = \pm x$

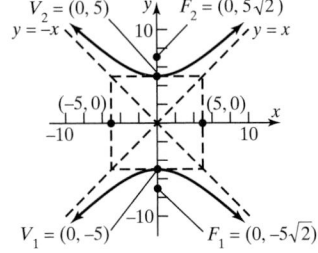

34. $\dfrac{x^2}{2} - \dfrac{y^2}{4} = 1$

Center: $(0, 0)$

Transverse axis: x-axis

Vertices: $\left(-\sqrt{2}, 0\right)$, $\left(\sqrt{2}, 0\right)$

Foci: $\left(-\sqrt{6}, 0\right)$, $\left(\sqrt{6}, 0\right)$

Asymptotes: $y = \pm\sqrt{2}x$

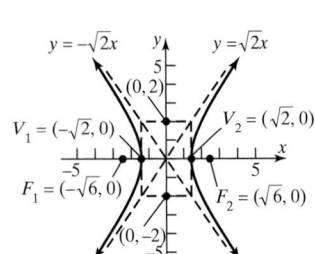

35. $x^2 - y^2 = 1$

36. $y^2 - x^2 = 1$

37. $\dfrac{y^2}{36} - \dfrac{x^2}{9} = 1$

38. $\dfrac{x^2}{4} - \dfrac{y^2}{16} = 1$

39. $\dfrac{(x-4)^2}{4} - \dfrac{(y+1)^2}{5} = 1$

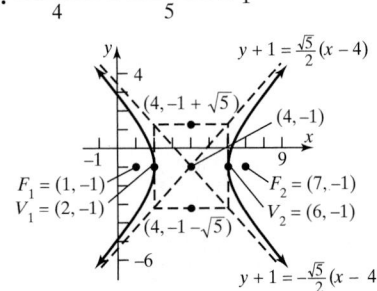

40. $\dfrac{(y-1)^2}{9} - \dfrac{(x+3)^2}{16} = 1$

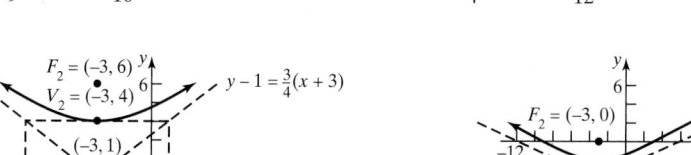

41. $\dfrac{(y+4)^2}{4} - \dfrac{(x+3)^2}{12} = 1$

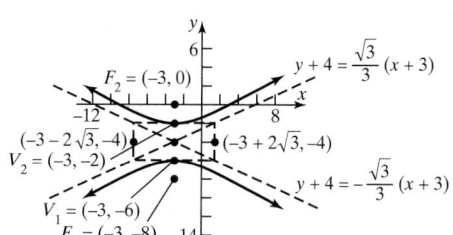

42. $(x-1)^2 - \dfrac{(y-4)^2}{8} = 1$

43. $(x-5)^2 - \dfrac{(y-7)^2}{3} = 1$

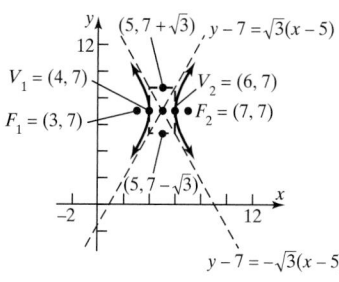

44. $(y-3)^2 - \dfrac{(x+4)^2}{8} = 1$

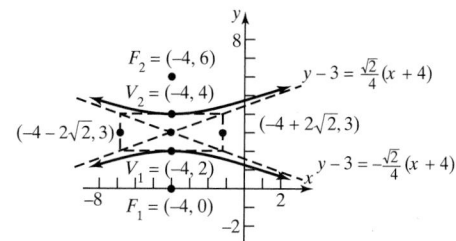

45. $\dfrac{(x-1)^2}{4} - \dfrac{(y+1)^2}{9} = 1$

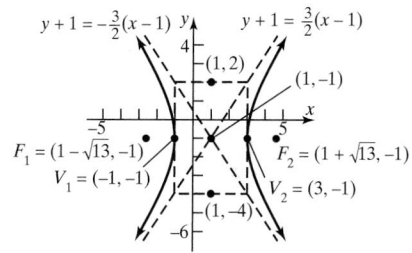

46. $\dfrac{(y+1)^2}{4} - \dfrac{9(x-1)^2}{16} = 1$

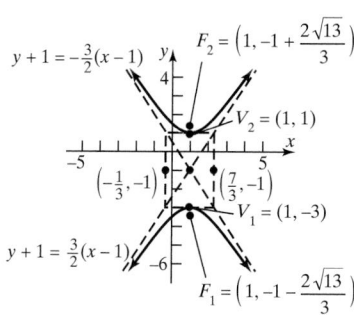

47. $\dfrac{(x-2)^2}{4} - \dfrac{(y+3)^2}{9} = 1$
Center: $(2, -3)$
Transverse axis: parallel to x-axis
Vertices: $(0, -3)$, $(4, -3)$
Foci: $\left(2 - \sqrt{13}, -3\right), \left(2 + \sqrt{13}, -3\right)$
Asymptotes: $y + 3 = \pm\dfrac{3}{2}(x - 2)$

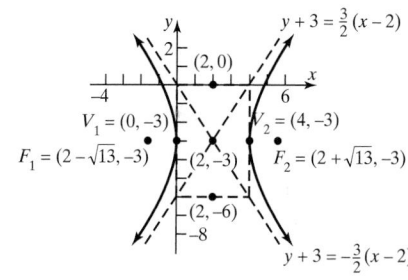

48. $\dfrac{(y+3)^2}{4} - \dfrac{(x-2)^2}{9} = 1$
Center: $(2, -3)$
Transverse axis: parallel to y-axis
Vertices: $(2, -5)$, $(2, -1)$
Foci: $\left(2, -3 - \sqrt{13}\right), \left(2, -3 + \sqrt{13}\right)$
Asymptotes: $y + 3 = \pm\dfrac{2}{3}(x - 2)$

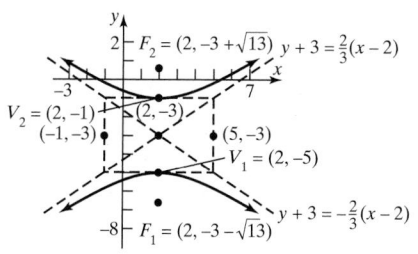

49. $\dfrac{(y-2)^2}{4} - (x+2)^2 = 1$
Center: $(-2, 2)$
Transverse axis: parallel to y-axis
Vertices: $(-2, 0)$, $(-2, 4)$
Foci: $\left(-2, 2 - \sqrt{5}\right), \left(-2, 2 + \sqrt{5}\right)$
Asymptotes: $y - 2 = \pm 2(x + 2)$

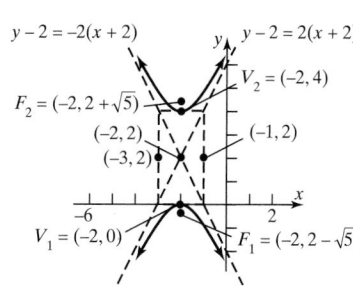

50. $\dfrac{(x+4)^2}{9} - (y-3)^2 = 1$
Center: $(-4, 3)$
Transverse axis: parallel to x-axis
Vertices: $(-7, 3)$, $(-1, 3)$
Foci: $\left(-4 - \sqrt{10}, 3\right), \left(-4 + \sqrt{10}, 3\right)$
Asymptotes: $y - 3 = \pm\dfrac{1}{3}(x + 4)$

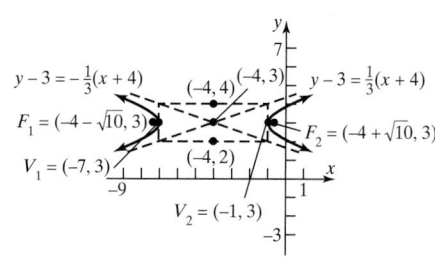

51. $\dfrac{(x+1)^2}{4} - \dfrac{(y+2)^2}{4} = 1$
Center: $(-1, -2)$
Transverse axis: parallel to x-axis
Vertices: $(-3, -2), (1, -2)$
Foci: $\left(-1 - 2\sqrt{2}, -2\right), \left(-1 + 2\sqrt{2}, -2\right)$
Asymptotes: $y + 2 = \pm(x + 1)$

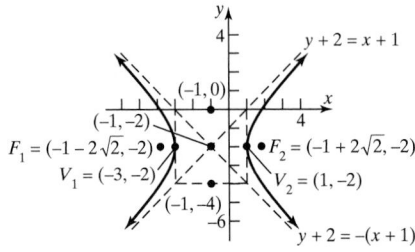

52. $\dfrac{(y-3)^2}{4} - \dfrac{(x+2)^2}{4} = 1$
Center: $(-2, 3)$
Transverse axis: parallel to y-axis
Vertices: $(-2, 1), (-2, 5)$
Foci: $\left(-2, 3 - 2\sqrt{2}\right), \left(-2, 3 + 2\sqrt{2}\right)$
Asymptotes: $y - 3 = \pm(x + 2)$

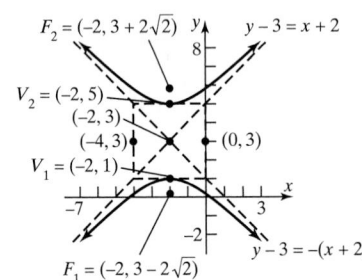

53. $(x-1)^2 - (y+1)^2 = 1$
Center: $(1, -1)$
Transverse axis: parallel to x-axis
Vertices: $(0, -1), (2, -1)$
Foci: $\left(1 - \sqrt{2}, -1\right), \left(1 + \sqrt{2}, -1\right)$
Asymptotes: $y + 1 = \pm(x - 1)$

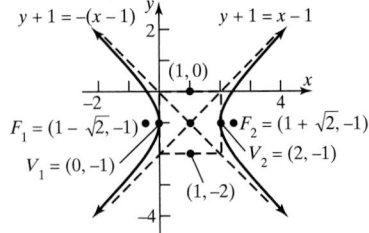

54. $(y-2)^2 - (x-2)^2 = 1$
Center: $(2, 2)$
Transverse axis: parallel to y-axis
Vertices: $(2, 1), (2, 3)$
Foci: $\left(2, 2 - \sqrt{2}\right), \left(2, 2 + \sqrt{2}\right)$
Asymptotes: $y - 2 = \pm(x - 2)$

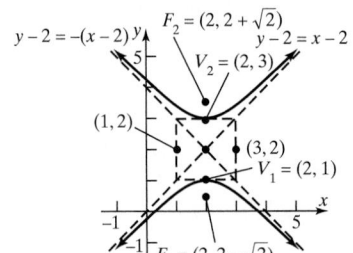

55. $\dfrac{(y-2)^2}{4} - (x+1)^2 = 1$
Center: $(-1, 2)$
Transverse axis: parallel to y-axis
Vertices: $(-1, 0), (-1, 4)$
Foci: $\left(-1, 2 - \sqrt{5}\right), \left(-1, 2 + \sqrt{5}\right)$
Asymptotes: $y - 2 = \pm 2(x + 1)$

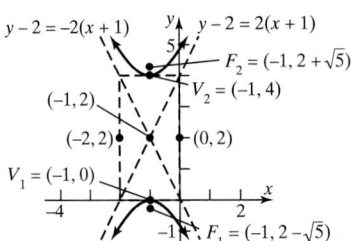

56. $(x+1)^2 - \dfrac{(y-2)^2}{2} = 1$
Center: $(-1, 2)$
Transverse axis: parallel to x-axis
Vertices: $(-2, 2), (0, 2)$
Foci: $\left(-1 - \sqrt{3}, 2\right), \left(-1 + \sqrt{3}, 2\right)$
Asymptotes: $y - 2 = \pm\sqrt{2}(x + 1)$

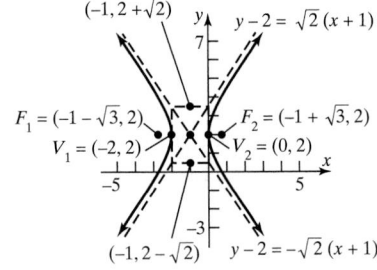

57. $\dfrac{(x-3)^2}{4} - \dfrac{(y+2)^2}{16} = 1$
Center: $(3, -2)$
Transverse axis: parallel to x-axis
Vertices: $(1, -2), (5, -2)$
Foci: $\left(3 - 2\sqrt{5}, -2\right), \left(3 + 2\sqrt{5}, -2\right)$
Asymptotes: $y + 2 = \pm 2(x - 3)$

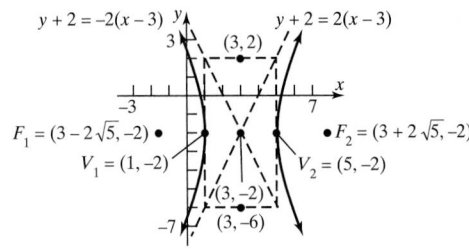

58. $\dfrac{(y+2)^2}{2} - \dfrac{(x-1)^2}{4} = 1$
Center: $(1, -2)$
Transverse axis: parallel to y-axis
Vertices: $\left(1, -2 - \sqrt{2}\right), \left(1, -2 + \sqrt{2}\right)$
Foci: $\left(1, -2 - \sqrt{6}\right), \left(1, -2 + \sqrt{6}\right)$
Asymptotes: $y + 2 = \pm\dfrac{\sqrt{2}}{2}(x - 1)$

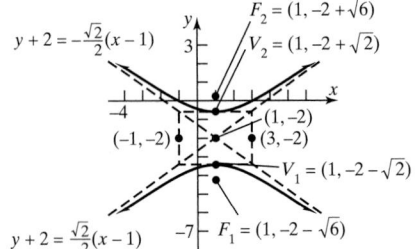

59. $\dfrac{(y-1)^2}{4} - (x+2)^2 = 1$
Center: $(-2, 1)$
Transverse axis: parallel to y-axis
Vertices: $(-2, -1), (-2, 3)$
Foci: $\left(-2, 1 - \sqrt{5}\right), \left(-2, 1 + \sqrt{5}\right)$
Asymptotes: $y - 1 = \pm 2(x + 2)$

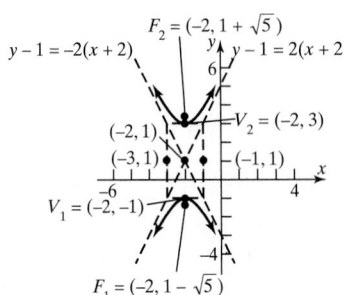

60. $\dfrac{(x+4)^2}{9} - \dfrac{(y+1)^2}{3} = 1$

Center: $(-4, -1)$

Transverse axis: parallel to x-axis

Vertices: $(-7, -1), (-1, -1)$

Foci: $\left(-4 - 2\sqrt{3}, -1\right), \left(-4 + 2\sqrt{3}, -1\right)$

Asymptotes: $y + 1 = \pm\dfrac{\sqrt{3}}{3}(x + 4)$

61. **62.** **63.** **64.**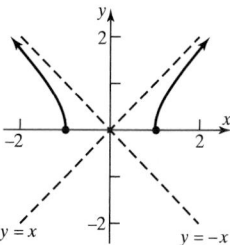

65. The fireworks display is 50,138 feet north of the person at point A. **66.** The lightning strikes 5236 feet north of the person at point A.

67. (a) $y = \pm x$ **68.** 700 feet **71.** $\dfrac{x^2}{4} - y^2 = 1$:

(b) $\dfrac{x^2}{100} - \dfrac{y^2}{100} = 1$ **70.** $\sqrt{2}$ asymptotes $y = \pm\dfrac{1}{2}x$; $y^2 - \dfrac{x^2}{4} = 1$:

asymptotes $y = \pm\dfrac{1}{2}x$

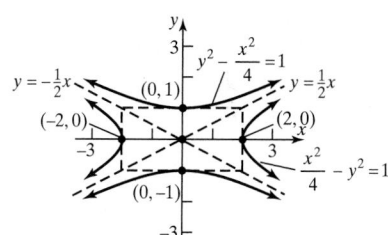

72. $\dfrac{y^2}{a^2} - \dfrac{x^2}{b^2} = 1$

$\dfrac{y^2}{a^2} = 1 + \dfrac{x^2}{b^2}$

$y^2 = a^2\left(1 + \dfrac{x^2}{b^2}\right)$

$y^2 = \dfrac{a^2 x^2}{b^2}\left(\dfrac{b^2}{x^2} + 1\right)$

$y = \pm\dfrac{ax}{b}\sqrt{\dfrac{b^2}{x^2} + 1}$

As $x \to -\infty$ or as $x \to \infty$, the term $\dfrac{b^2}{x^2}$ gets close to 0, so the expression under the radical gets closer to 1.

Thus, the graph of the hyperbola gets closer to the lines $y = -\dfrac{a}{b}x$ and $y = \dfrac{a}{b}x$.

The lines are asymptotes of the hyperbola.

73. $Ax^2 + Cy^2 + F = 0$

$Ax^2 + Cy^2 = -F$

If A and C are of opposite sign and $F \neq 0$, this equation may be written as $\dfrac{x^2}{\left(-\dfrac{F}{A}\right)} + \dfrac{y^2}{\left(-\dfrac{F}{C}\right)} = 1$,

where $-\dfrac{F}{A}$ and $-\dfrac{F}{C}$ are opposite in sign. This is the equation of a hyperbola with center $(0, 0)$.

The transverse axis is the x-axis if $-\dfrac{F}{A} > 0$; the transverse axis is the y-axis if $-\dfrac{F}{A} < 0$.

74. $Ax^2 + Cy^2 + Dx + Ey + F = 0$

where A and C are of opposite sign

$Ax^2 + Dx + Cy^2 + Ey = -F$

$A\left(x^2 + \dfrac{D}{A}x\right) + C\left(y^2 + \dfrac{E}{C}y\right) = -F$

$A\left(x + \dfrac{D}{2A}\right)^2 + C\left(y + \dfrac{E}{2C}\right)^2 = -F + \dfrac{D^2}{4A} + \dfrac{E^2}{4C}$

(a) If $\dfrac{D^2}{4A} + \dfrac{E^2}{4C} - F \neq 0$, this is the equation of hyperbola with

center at $\left(-\dfrac{D}{2A}, -\dfrac{E}{2C}\right)$.

(b) If $\dfrac{D^2}{4A} + \dfrac{E^2}{4C} - F = 0$, the graph is intersecting lines through the

point $\left(-\dfrac{D}{2A}, -\dfrac{E}{2C}\right)$ with slopes $\pm\sqrt{\left|\dfrac{A}{C}\right|}$.

Review Exercises (page 395)

1. Parabola; vertex $(0, 0)$, focus $(-4, 0)$, directrix $x = 4$ **2.** Parabola; vertex $(0, 0)$, focus $\left(0, \dfrac{1}{64}\right)$, directrix $y = -\dfrac{1}{64}$

3. Hyperbola; center $(0, 0)$, vertices $(5, 0)$ and $(-5, 0)$, foci $\left(\sqrt{26}, 0\right)$ and $\left(-\sqrt{26}, 0\right)$, asymptotes $y = \dfrac{1}{5}x$ and $y = -\dfrac{1}{5}x$

4. Hyperbola; center $(0, 0)$, vertices $(0, 5)$ and $(0, -5)$, foci $\left(0, \sqrt{26}\right)$ and $\left(0, -\sqrt{26}\right)$, asymptotes $y = 5x$ and $y = -5x$

5. Ellipse; center $(0, 0)$, vertices $(0, 5)$ and $(0, -5)$, foci $(0, 3)$ and $(0, -3)$

6. Ellipse; center $(0, 0)$, vertices $(0, 4)$ and $(0, -4)$, foci $\left(0, \sqrt{7}\right)$ and $\left(0, -\sqrt{7}\right)$

7. $x^2 = -4(y - 1)$: Parabola; vertex $(0, 1)$, focus $(0, 0)$, directrix $y = 2$

8. $\dfrac{y^2}{3} - \dfrac{x^2}{9} = 1$: Hyperbola; center $(0,0)$, vertices $\left(0, \sqrt{3}\right)$ and $\left(0, -\sqrt{3}\right)$, foci $\left(0, 2\sqrt{3}\right)$ and $\left(0, -2\sqrt{3}\right)$, asymptotes $y = \dfrac{\sqrt{3}}{3}x$, $y = -\dfrac{\sqrt{3}}{3}x$

9. $\dfrac{x^2}{2} - \dfrac{y^2}{8} = 1$: Hyperbola; center $(0,0)$, vertices $\left(\sqrt{2}, 0\right)$ and $\left(-\sqrt{2}, 0\right)$, foci $\left(\sqrt{10}, 0\right)$ and $\left(-\sqrt{10}, 0\right)$, asymptotes $y = 2x$ and $y = -2x$

10. $\dfrac{x^2}{4} + \dfrac{y^2}{9} = 1$: Ellipse; center $(0,0)$, vertices $(0,3)$ and $(0,-3)$, foci $\left(0, \sqrt{5}\right)$ and $\left(0, -\sqrt{5}\right)$

11. $(x - 2)^2 = 2(y + 2)$: Parabola; vertex $(2, -2)$, focus $\left(2, -\dfrac{3}{2}\right)$, directrix $y = -\dfrac{5}{2}$

12. $(y - 1)^2 = \dfrac{1}{2}x$: Parabola; vertex $(0, 1)$, focus $\left(\dfrac{1}{8}, 1\right)$, directrix $x = -\dfrac{1}{8}$

13. $\dfrac{(y - 2)^2}{4} - (x - 1)^2 = 1$: Hyperbola; center $(1, 2)$, vertices $(1, 4)$ and $(1, 0)$, foci $\left(1, 2 + \sqrt{5}\right)$ and $\left(1, 2 - \sqrt{5}\right)$, asymptotes $y - 2 = \pm 2(x - 1)$

14. $(x + 1)^2 + \dfrac{(y - 2)^2}{4} = 1$: Ellipse; center $(-1, 2)$, vertices $(-1, 0)$ and $(-1, 4)$, foci $\left(-1, 2 + \sqrt{3}\right)$ and $\left(-1, 2 - \sqrt{3}\right)$

15. $\dfrac{(x - 2)^2}{9} + \dfrac{(y - 1)^2}{4} = 1$: Ellipse; center $(2, 1)$, vertices $(5, 1)$ and $(-1, 1)$, foci $\left(2 + \sqrt{5}, 1\right)$ and $\left(2 - \sqrt{5}, 1\right)$

16. $\dfrac{(x - 2)^2}{9} + \dfrac{(y + 1)^2}{4} = 1$: Ellipse; center $(2, -1)$, vertices $(5, -1)$ and $(-1, -1)$, foci $\left(2 + \sqrt{5}, -1\right)$ and $\left(2 - \sqrt{5}, -1\right)$

17. $(x - 2)^2 = -4(y + 1)$: Parabola; vertex $(2, -1)$, focus $(2, -2)$, directrix $y = 0$

18. $(y - 2)^2 = -\dfrac{3}{4}(x + 1)$: Parabola; vertex $(-1, 2)$, focus $\left(-\dfrac{19}{16}, 2\right)$, directrix $x = -\dfrac{13}{16}$

19. $\dfrac{(x - 1)^2}{4} + \dfrac{(y + 1)^2}{9} = 1$: Ellipse; center $(1, -1)$, vertices $(1, 2)$ and $(1, -4)$, foci $\left(1, -1 + \sqrt{5}\right)$ and $\left(1, -1 - \sqrt{5}\right)$

20. $(x - 1)^2 - (y + 1)^2 = 1$: Hyperbola; center $(1, -1)$, vertices $(0, -1)$ and $(2, -1)$, foci $\left(1 + \sqrt{2}, -1\right)$ and $\left(1 - \sqrt{2}, -1\right)$, asymptotes $y + 1 = x - 1$ and $y + 1 = -(x - 1)$

21. $y^2 = -8x$

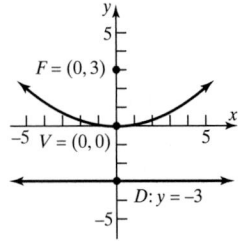

22. $\dfrac{x^2}{16} + \dfrac{y^2}{25} = 1$

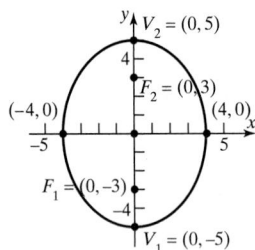

23. $\dfrac{y^2}{4} - \dfrac{x^2}{12} = 1$

24. $x^2 = 12y$

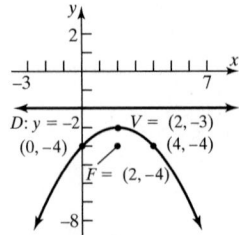

25. $\dfrac{x^2}{16} + \dfrac{y^2}{7} = 1$

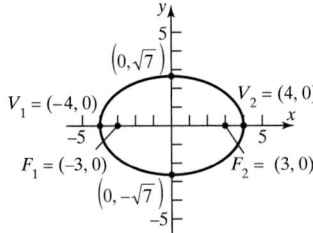

26. $\dfrac{x^2}{4} - \dfrac{y^2}{12} = 1$

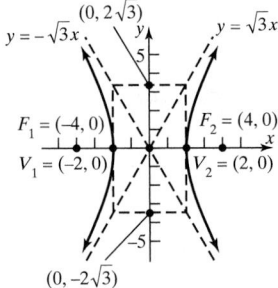

27. $(x - 2)^2 = -4(y + 3)$

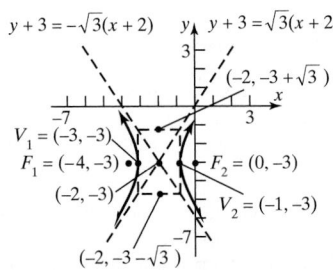

28. $\dfrac{(x + 1)^2}{9} + \dfrac{(y - 2)^2}{8} = 1$

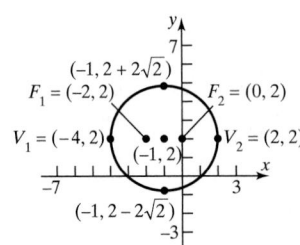

29. $(x + 2)^2 - \dfrac{(y + 3)^2}{3} = 1$

30. $(x - 3)^2 = -4(y - 7)$

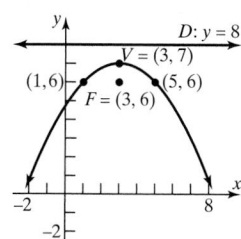

31. $\dfrac{(x + 4)^2}{16} + \dfrac{(y - 5)^2}{25} = 1$

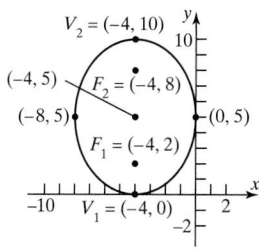

32. $\dfrac{(x - 1)^2}{16} - \dfrac{(y - 3)^2}{20} = 1$

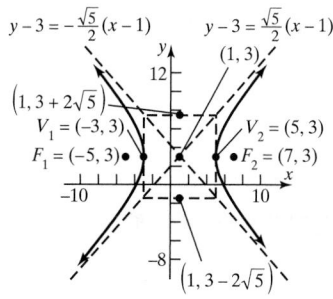

33. $\dfrac{(x + 1)^2}{9} - \dfrac{(y - 2)^2}{7} = 1$

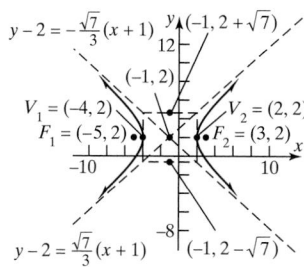

34. $(y + 2)^2 - \dfrac{(x - 4)^2}{15} = 1$

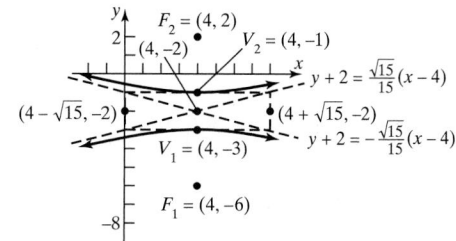

35. $\dfrac{(x - 3)^2}{9} - \dfrac{(y - 1)^2}{4} = 1$

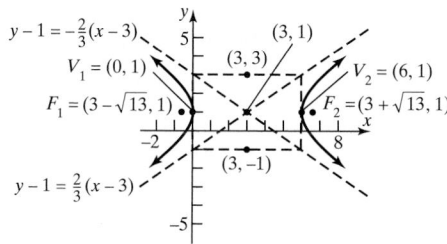

36. $\dfrac{(y - 2)^2}{4} - \dfrac{(x - 4)^2}{1} = 1$

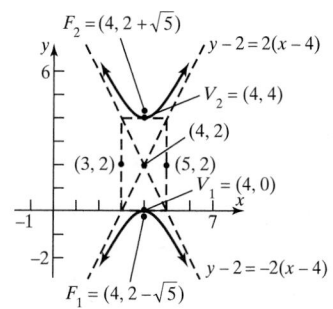

37. $\dfrac{x^2}{5} - \dfrac{y^2}{4} = 1$ **38.** $\dfrac{x^2}{20} + \dfrac{y^2}{4} = 1$

39. The ellipse $\dfrac{x^2}{16} + \dfrac{y^2}{7} = 1$

40. The hyperbola $\dfrac{x^2}{16} - \dfrac{y^2}{9} = 1$

41. $\dfrac{1}{4}$ ft or 3 in.

42. 19.44 ft, 17.78 ft, 11.11 ft

43. 19.72 ft, 18.86 ft, 14.91 ft

44. The foci are located 8.78 ft in from each wall.

45. 450 ft

Chapter Test *(page 396)*

1. Hyperbola; center: $(-1, 0)$; vertices: $(-3, 0)$ and $(1, 0)$; foci: $\left(-1 - \sqrt{13}, 0\right)$ and $\left(-1 + \sqrt{13}, 0\right)$; asymptotes: $y = -\dfrac{3}{2}(x + 1)$ and $y = \dfrac{3}{2}(x + 1)$

2. Parabola; vertex: $\left(1, -\dfrac{1}{2}\right)$; focus: $\left(1, \dfrac{3}{2}\right)$; directrix: $y = -\dfrac{5}{2}$

3. Ellipse; center: $(-1, 1)$; foci: $\left(-1 - \sqrt{3}, 1\right)$ and $\left(-1 + \sqrt{3}, 1\right)$; vertices: $(-4, 1)$ and $(2, 1)$

4. $(x + 1)^2 = 6(y - 3)$

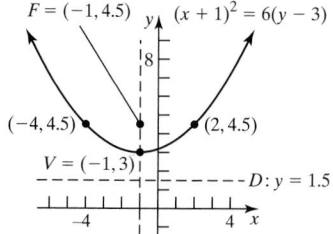

5. $\dfrac{x^2}{7} + \dfrac{y^2}{16} = 1$

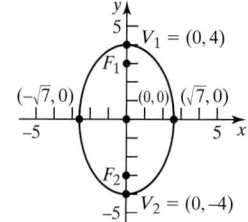

6. $\dfrac{(y - 2)^2}{4} - \dfrac{(x - 2)^2}{8} = 1$

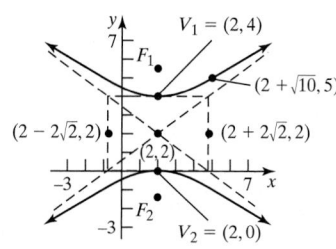

7. The microphone should be located $\dfrac{2}{3}$ feet from the base of the reflector, along its axis of symmetry.

Cumulative Review (page 397)

1. $-6x + 5 - 3h$ **2. (a)** Domain: $(-\infty, \infty)$; range: $(2, \infty)$ **(b)** $y = \log_3(x - 2)$; domain: $(2, \infty)$; range: $(-\infty, \infty)$

3. $x = 2$ or $x = -\dfrac{1}{3}$ or $x = -5$ **4.** $-3 \le x \le 2$ or $[-3, 2]$ **5. (a)** $y = 2x - 2$ **(b)** $(x - 2)^2 + y^2 = 4$ **(c)** $\dfrac{x^2}{9} + \dfrac{y^2}{4} = 1$

(d) $y = 2(x - 1)^2$ **(e)** $\dfrac{y^2}{1} - \dfrac{x^2}{3} = 1$ **(f)** $y = 4^x$ **6. (a)** 18 **(b)** $(2, 18]$

CHAPTER 6 Systems of Equations and Inequalities

6.1 Assess Your Understanding (page 411)

3. inconsistent **4.** consistent **5.** F **6.** T

7. $\begin{cases} 2(2) - (-1) = 5 \\ 5(2) + 2(-1) = 8 \end{cases}$ **8.** $\begin{cases} 3(-2) + 2(4) = 2 \\ (-2) - 7(4) = -30 \end{cases}$ **9.** $\begin{cases} 3(2) - 4\left(\dfrac{1}{2}\right) = 4 \\ \dfrac{1}{2}(2) - 3\left(\dfrac{1}{2}\right) = -\dfrac{1}{2} \end{cases}$ **10.** $\begin{cases} 2\left(-\dfrac{1}{2}\right) + \dfrac{1}{2}(2) = 0 \\ 3\left(-\dfrac{1}{2}\right) - 4(2) = -\dfrac{19}{2} \end{cases}$ **11.** $\begin{cases} 4 - 1 = 3 \\ \dfrac{1}{2}(4) + 1 = 3 \end{cases}$

12. $\begin{cases} (-2) - (-5) = 3 \\ -3(-2) + (-5) = 1 \end{cases}$ **13.** $\begin{cases} 3(1) + 3(-1) + 2(2) = 4 \\ 1 - (-1) - 2 = 0 \\ 2(-1) - 3(2) = -8 \end{cases}$ **14.** $\begin{cases} 4(2) \quad\ - 1 = 7 \\ 8(2) + 5(-3) - 1 = 0 \\ -2 - (-3) + 5(1) = 6 \end{cases}$ **15.** $\begin{cases} 3(2) + 3(-2) + 2(2) = 4 \\ 2 - 3(-2) + 2 = 10 \\ 5(2) - 2(-2) - 3(2) = 8 \end{cases}$

16. $\begin{cases} 4(4) \quad\ - 5(2) = 6 \\ 5(-3) - 2 = -17 \\ -4 - 6(-3) + 5(2) = 24 \end{cases}$ **17.** $x = 6, y = 2$ **18.** $x = 1, y = 2$ **19.** $x = 3, y = 2$ **20.** $x = -1, y = 2$ **21.** $x = 8, y = -4$

22. $x = -\dfrac{13}{4}, y = 2$ **23.** $x = \dfrac{1}{3}, y = -\dfrac{1}{6}$ **24.** $x = -\dfrac{5}{3}, y = 1$ **25.** Inconsistent **26.** Inconsistent **27.** $x = 1, y = 2$ **28.** $x = 1, y = -\dfrac{4}{3}$

29. $x = 4 - 2y$, y is any real number **30.** $y = 3x - 7$, x is any real number **31.** $x = 1, y = 1$ **32.** $x = \dfrac{1}{5}, y = \dfrac{3}{10}$ **33.** $x = \dfrac{3}{2}, y = 1$

34. $x = 2, y = -3$ **35.** $x = 4, y = 3$ **36.** $x = 12, y = 6$ **37.** $x = \dfrac{4}{3}, y = \dfrac{1}{5}$ **38.** $x = \dfrac{1}{2}, y = 2$ **39.** $x = \dfrac{1}{5}, y = \dfrac{1}{3}$ **40.** $x = 4, y = 3$

41. $x = 8, y = 2, z = 0$ **42.** $x = -3, y = 2, z = 1$ **43.** $x = 2, y = -1, z = 1$ **44.** $x = \dfrac{56}{13}, y = -\dfrac{7}{13}, z = \dfrac{35}{13}$ **45.** Inconsistent

46. Inconsistent **47.** $x = 5z - 2, y = 4z - 3$, where z is any real number **48.** $x = \dfrac{4}{7}y + \dfrac{2}{7}$ and $z = -\dfrac{13}{7}y + \dfrac{4}{7}$, where y is any real number

49. Inconsistent **50.** Inconsistent **51.** $x = 1, y = 3, z = -2$ **52.** $x = 1, y = 3, z = -2$ **53.** $x = -3, y = \dfrac{1}{2}, z = 1$ **54.** $x = 3, y = -\dfrac{8}{3}, z = \dfrac{1}{9}$

55. Length 30 ft; width 15 ft **56.** 775 m by 725 m **57.** Cheeseburger \$1.55; shake \$0.85 **58.** 110 adults; 215 seniors **59.** 22.5 lb

60. (a) \$90,000 in AA bonds, \$60,000 in Bank Certificate **(b)** \$130,000 in AA bonds, \$20,000 in Bank Certificate

61. Average wind speed 25 mph; average airspeed 175 mph **62.** 30 mph **63.** 80 \$25 sets and 120 \$45 sets

64. A single hot dog: \$1; a single soft drink: \$0.50 **65.** \$5.56 **66.** Pamela's average speed: 4 mph; the speed of the current: 1 mph

67. Mix 50 mg of first compound with 75 mg of second. **68.** First powder: 30 units; second powder: 15 units

69. $a = \dfrac{4}{3}, b = -\dfrac{5}{3}, c = 1$ **70.** $a = 3, b = -1, c = -6$ **71.** $y = 9000, r = 0.06$ **72.** $y = 1200, r = 0.05$ **73.** $I_1 = \dfrac{10}{71}, I_2 = \dfrac{65}{71}, I_3 = \dfrac{55}{71}$

74. $I_1 = \dfrac{11}{13}, I_2 = \dfrac{6}{13}, I_3 = \dfrac{17}{13}$ **75.** 100 orchestra, 210 main, and 190 balcony seats **76.** 110 adults; 220 children; 75 senior citizens

77. 1.5 chicken, 1 corn, 2 milk **78.** She should place \$8000 in Treasury bills, \$7000 in Treasury bonds, and \$5000 in corporate bonds.

79. If $x =$ price of hamburgers, $y =$ price of fries, $z =$ price of colas, then $x = 2.75 - z$, $y = \dfrac{41}{60} + \dfrac{1}{3}z$, \$0.60 $\le z \le$ \$0.90.

There is not sufficient information:

x	\$2.13	\$2.01	\$1.86
y	\$0.89	\$0.93	\$0.98
z	\$0.62	\$0.74	\$0.89

80. Yes; hamburgers are \$1.95, fries are \$0.95, cola is \$0.80.
81. It will take Beth 30 hr, Bill 24 hr, and Edie 40 hr.

6.2 Assess Your Understanding (page 426)

1. matrix **2.** augmented **3.** T **4.** T

5. $\begin{bmatrix} 1 & -5 & | & 5 \\ 4 & 3 & | & 6 \end{bmatrix}$ **6.** $\begin{bmatrix} 3 & 4 & | & 7 \\ 4 & -2 & | & 5 \end{bmatrix}$ **7.** $\begin{bmatrix} 2 & 3 & | & 6 \\ 4 & -6 & | & -2 \end{bmatrix}$ **8.** $\begin{bmatrix} 9 & -1 & | & 0 \\ 3 & -1 & | & 4 \end{bmatrix}$ **9.** $\begin{bmatrix} 0.01 & -0.03 & | & 0.06 \\ 0.13 & 0.10 & | & 0.20 \end{bmatrix}$ **10.** $\begin{bmatrix} \dfrac{4}{3} & -\dfrac{3}{2} & | & \dfrac{3}{4} \\ \dfrac{1}{4} & \dfrac{1}{3} & | & \dfrac{2}{3} \end{bmatrix}$

11. $\begin{bmatrix} 1 & -1 & 1 & | & 10 \\ 3 & 3 & 0 & | & 5 \\ 1 & 1 & 2 & | & 2 \end{bmatrix}$ **12.** $\begin{bmatrix} 5 & -1 & -1 & | & 0 \\ 1 & 1 & 0 & | & 5 \\ 2 & 0 & -3 & | & 2 \end{bmatrix}$ **13.** $\begin{bmatrix} 1 & 1 & -1 & | & 2 \\ 3 & -2 & 0 & | & 2 \\ 5 & 3 & -1 & | & 1 \end{bmatrix}$ **14.** $\begin{bmatrix} 2 & 3 & -4 & | & 0 \\ 1 & 0 & -5 & | & -2 \\ 1 & 2 & -3 & | & -2 \end{bmatrix}$ **15.** $\begin{bmatrix} 1 & -1 & -1 & | & 10 \\ 2 & 1 & 2 & | & -1 \\ -3 & 4 & 0 & | & 5 \\ 4 & -5 & 1 & | & 0 \end{bmatrix}$

16. $\begin{bmatrix} 1 & -1 & 2 & -1 & | & 5 \\ 1 & 3 & -4 & 2 & | & 2 \\ 3 & -1 & -5 & -1 & | & -1 \end{bmatrix}$ **17.** $\begin{bmatrix} 1 & -3 & | & -2 \\ 0 & 1 & | & 9 \end{bmatrix}$ **18.** $\begin{bmatrix} 1 & -3 & | & -3 \\ 0 & 1 & | & 2 \end{bmatrix}$ **19. (a)** $\begin{bmatrix} 1 & -3 & 4 & | & 3 \\ 0 & 1 & -2 & | & 0 \\ -3 & 3 & 4 & | & 6 \end{bmatrix}$ **(b)** $\begin{bmatrix} 1 & -3 & 4 & | & 3 \\ 2 & -5 & 6 & | & 6 \\ 0 & -6 & 16 & | & 15 \end{bmatrix}$

20. (a) $\begin{bmatrix} 1 & -3 & 3 & | & -5 \\ 0 & 1 & -9 & | & 5 \\ -3 & -2 & 4 & | & 6 \end{bmatrix}$ **(b)** $\begin{bmatrix} 1 & -3 & 3 & | & -5 \\ 2 & -5 & -3 & | & -5 \\ 0 & -11 & 13 & | & -9 \end{bmatrix}$ **21. (a)** $\begin{bmatrix} 1 & -3 & 2 & | & -6 \\ 0 & 1 & -1 & | & 8 \\ -3 & -6 & 4 & | & 6 \end{bmatrix}$ **(b)** $\begin{bmatrix} 1 & -3 & 2 & | & -6 \\ 2 & -5 & 3 & | & -4 \\ 0 & -15 & 10 & | & -12 \end{bmatrix}$

22. (a) $\begin{bmatrix} 1 & -3 & -4 & | & -6 \\ 0 & 1 & 14 & | & 6 \\ -3 & 1 & 4 & | & 6 \end{bmatrix}$ **(b)** $\begin{bmatrix} 1 & -3 & -4 & | & -6 \\ 2 & -5 & 6 & | & -6 \\ 0 & -8 & -8 & | & -12 \end{bmatrix}$ **23. (a)** $\begin{bmatrix} 1 & -3 & 1 & | & -2 \\ 0 & 1 & 4 & | & 2 \\ -3 & 1 & 4 & | & 6 \end{bmatrix}$ **(b)** $\begin{bmatrix} 1 & -3 & 1 & | & -2 \\ 2 & -5 & 6 & | & -2 \\ 0 & -8 & 7 & | & 0 \end{bmatrix}$

24. (a) $\begin{bmatrix} 1 & -3 & -1 & | & 2 \\ 0 & 1 & 4 & | & 2 \\ -3 & -6 & 4 & | & 6 \end{bmatrix}$ **(b)** $\begin{bmatrix} 1 & -3 & -1 & | & 2 \\ 2 & -5 & 2 & | & 6 \\ 0 & -15 & 1 & | & 12 \end{bmatrix}$

25. $\begin{cases} x = 5 \\ y = -1 \end{cases}$ Consistent; $x = 5$, $y = -1$ **26.** $\begin{cases} x = -4 \\ y = 0 \end{cases}$ Consistent; $x = -4$, $y = 0$ **27.** $\begin{cases} x = 1 \\ y = 2 \\ 0 = 3 \end{cases}$ Inconsistent **28.** $\begin{cases} x = 0 \\ y = 0 \\ 0 = 2 \end{cases}$ Inconsistent

29. $\begin{cases} x + 2z = -1 \\ y - 4z = -2 \\ 0 = 0 \end{cases}$ Consistent: $x = -1 - 2z$, $y = -2 + 4z$, z is any real number **30.** $\begin{cases} x + 4z = 4 \\ y + 3z = 2 \\ 0 = 0 \end{cases}$ Consistent: $x = 4 - 4z$, $y = 2 - 3z$, z is any real number **31.** $\begin{cases} x_1 = 1 \\ x_2 + x_4 = 2 \\ x_3 + 2x_4 = 3 \end{cases}$ Consistent: $x_1 = 1$, $x_2 = 2 - x_4$, $x_3 = 3 - 2x_4$, x_4 is any real number **32.** $\begin{cases} x_1 = 1 \\ x_2 + 2x_4 = 2 \\ x_3 + 3x_4 = 0 \end{cases}$ Consistent: $x_1 = 1$, $x_2 = 2 - 2x_4$, $x_3 = -3x_4$, x_4 is any real number

33. $\begin{cases} x_1 + 4x_4 = 2 \\ x_2 + x_3 + 3x_4 = 3 \\ 0 = 0 \end{cases}$ Consistent: $x_1 = 2 - 4x_4$, $x_2 = 3 - x_3 - 3x_4$, x_3, x_4 are any real numbers **34.** $\begin{cases} x_1 = 1 \\ x_2 = 2 \\ x_3 + 2x_4 = 3 \end{cases}$ Consistent: $x_1 = 1$, $x_2 = 2$, $x_3 = 3 - 2x_4$, x_4 is any real number **35.** $\begin{cases} x_1 + x_4 = -2 \\ x_2 + 2x_4 = 2 \\ x_3 - x_4 = 0 \end{cases}$ Consistent: $x_1 = -2 - x_4$, $x_2 = 2 - 2x_4$, $x_3 = x_4$, x_4 is any real number **36.** $\begin{cases} x_1 = 1 \\ x_2 = 2 \\ x_3 = 3 \\ x_4 = 0 \end{cases}$ Consistent: $x_1 = 1$, $x_2 = 2$, $x_3 = 3$, $x_4 = 0$

37. $x = 6$, $y = 2$ **38.** $x = 1$, $y = 2$ **39.** $x = \frac{1}{2}$, $y = \frac{3}{4}$ **40.** $x = \frac{1}{3}$, $y = \frac{2}{3}$ **41.** $x = 4 - 2y$, y is any real number

42. $y = 3x - 7$, x is any real number **43.** $x = \frac{3}{2}$, $y = 1$ **44.** $x = 2$, $y = -3$ **45.** $x = \frac{4}{3}$, $y = \frac{1}{5}$ **46.** $x = \frac{1}{2}$, $y = 2$

47. $x = 8$, $y = 2$, $z = 0$ **48.** $x = -3$, $y = 2$, $z = 1$ **49.** $x = 2$, $y = -1$, $z = 1$ **50.** $x = \frac{56}{13}$, $y = -\frac{7}{13}$, $z = \frac{35}{13}$

51. Inconsistent **52.** Inconsistent **53.** $x = 5z - 2$, $y = 4z - 3$, where z is any real number

54. $x = -\frac{4}{13}z + \frac{6}{13}$, $y = -\frac{7}{13}z + \frac{4}{13}$, where z is any real number **55.** Inconsistent **56.** Inconsistent **57.** $x = 1$, $y = 3$, $z = -2$

58. $x = 1$, $y = 3$, $z = -2$ **59.** $x = -3$, $y = \frac{1}{2}$, $z = 1$ **60.** $x = 3$, $y = -\frac{8}{3}$, $z = \frac{1}{9}$ **61.** $x = \frac{1}{3}$, $y = \frac{2}{3}$, $z = 1$ **62.** $x = \frac{1}{3}$, $y = \frac{2}{3}$, $z = 1$

63. $x = 1$, $y = 2$, $z = 0$, $w = 1$ **64.** $x = 2$, $y = 1$, $z = 0$, $w = 1$ **65.** $y = 0$, $z = 1 - x$, x is any real number **66.** Inconsistent

67. $x = 2$, $y = z - 3$, z is any real number **68.** $x = 1 + \frac{4}{3}z$, $y = 2 - \frac{5}{3}z$, where z is any real number **69.** $x = \frac{13}{9}$, $y = \frac{7}{18}$, $z = \frac{19}{18}$

70. Inconsistent **71.** $x = \frac{7}{5} - \frac{3}{5}z - \frac{2}{5}w$, $y = -\frac{8}{5} + \frac{7}{5}z + \frac{13}{5}w$, where z and w are any real numbers

72. $x = -3 - w$, $y = -7 - 4w$, $z = 4 - w$, where w is any real number **73.** $y = -2x^2 + x + 3$ **74.** $y = x^2 - 4x + 2$

75. $f(x) = 3x^3 - 4x^2 + 5$ **76.** $f(x) = x^3 - 2x^2 + 6$ **77.** 1.5 salmon steak, 2 baked eggs, 1 acorn squash **78.** 1 serving of pork chop, 2 servings of corn on the cob, 2 servings of 2% milk **79.** $4000 in Treasury bills, $4000 in Treasury bonds, $2000 in corporate bonds **80.** $12,000 in Treasury bills, $2000 in Treasury bonds, $6000 in corporate bonds **81.** 8 Deltas, 5 Betas, 10 Sigmas **82.** 6 cases of orange juice, 20 cases of grapefruit juice, 12 cases of tomato juice

83. $I_1 = \frac{44}{23}$, $I_2 = 2$, $I_3 = \frac{16}{23}$, $I_4 = \frac{28}{23}$ **84.** $I_1 = 3.5$, $I_2 = 2.5$, $I_3 = 1$

85. (a)

Amount Invested At		
7%	**9%**	**11%**
0	10,000	10,000
1000	8000	11,000
2000	6000	12,000
3000	4000	13,000
4000	2000	14,000
5000	0	15,000

(b)

Amount Invested At		
7%	**9%**	**11%**
12,500	12,500	0
14,500	8500	2000
16,500	4500	4000
18,750	0	6250

(c) All the money invested at 7% provides $2100, more than what is required.

86. (a) Investing all the money at 7% yields more than $1500. **(b)**

7%	9%	11%
12,500	12,500	0
15,500	6500	3000
18,750	0	6250

(c)

7%	9%	11%
0	12,500	12,500
1000	10,500	13,500
6250	0	18,750

87.

First Liquid	Second Liquid	Third Liquid
50 mg	75 mg	0 mg
36 mg	76 mg	8 mg
22 mg	77 mg	16 mg
8 mg	78 mg	24 mg

88.

First Powder	Second Powder	Third Powder
30 units	15 units	0 units
20 units	14 units	8 units
10 units	13 units	16 units
0 units	12 units	24 units

6.3 Assess Your Understanding (page 438)

1. determinants **2.** $ad - bc$ **3.** F **4.** F **5.** 2 **6.** 7 **7.** 22 **8.** 28 **9.** -2 **10.** -2 **11.** 10 **12.** -169 **13.** -26 **14.** -119 **15.** $x = 6, y = 2$
16. $x = \dfrac{11}{3}, y = \dfrac{2}{3}$ **17.** $x = 3, y = 2$ **18.** $x = -1, y = 2$ **19.** $x = 8, y = -4$ **20.** $x = -\dfrac{13}{4}, y = 2$ **21.** $x = 4, y = -2$ **22.** $x = 2, y = 3$
23. Not applicable **24.** Not applicable **25.** $x = \dfrac{1}{2}, y = \dfrac{3}{4}$ **26.** $x = \dfrac{1}{3}, y = \dfrac{2}{3}$ **27.** $x = \dfrac{1}{10}, y = \dfrac{2}{5}$ **28.** $x = \dfrac{1}{5}, y = \dfrac{3}{10}$ **29.** $x = \dfrac{3}{2}, y = 1$
30. $x = 2, y = -3$ **31.** $x = \dfrac{4}{3}, y = \dfrac{1}{5}$ **32.** $x = \dfrac{1}{2}, y = 2$ **33.** $x = 1, y = 3, z = -2$ **34.** $x = 1, y = 3, z = -2$ **35.** $x = -3, y = \dfrac{1}{2}, z = 1$
36. $x = 3, y = -\dfrac{8}{3}, z = \dfrac{1}{9}$ **37.** Not applicable **38.** Not applicable **39.** $x = 0, y = 0, z = 0$ **40.** $x = 0, y = 0, z = 0$
41. Not applicable **42.** Not applicable **43.** -5 **44.** 1 or -1 **45.** $\dfrac{13}{11}$ **46.** $-\dfrac{7}{6}$ **47.** 0 or -9 **48.** 0 or $-\dfrac{1}{2}$ **49.** -4
50. 8 **51.** 12 **52.** 4 **53.** 8 **54.** 4 **55.** 8 **56.** 12
57. $(y_1 - y_2)x - (x_1 - x_2)y + (x_1 y_2 - x_2 y_1) = 0$
$$(y_1 - y_2)x + (x_2 - x_1)y = x_2 y_1 - x_1 y_2$$
$$(x_2 - x_1)y - (x_2 - x_1)y_1 = (y_2 - y_1)x + x_2 y_1 - x_1 y_2 - (x_2 - x_1)y_1$$
$$(x_2 - x_1)(y - y_1) = (y_2 - y_1)x - (y_2 - y_1)x_1$$
$$y - y_1 = \frac{y_2 - y_1}{x_2 - x_1}(x - x_1)$$

58. From Problem 57 and substitution, we have $y_1 - y_2 = \dfrac{y_3 - y_2}{x_3 - x_2}(x_1 - x_2)$. The line containing (x_2, y_2) and (x_3, y_3) has the form
$y - y_2 = \dfrac{y_3 - y_2}{x_3 - x_2}(x - x_2)$. Any other point on this line can be substituted for (x, y) to get a true statement, and any point that
makes this a true statement is on the line.

59. $\begin{vmatrix} x^2 & x & 1 \\ y^2 & y & 1 \\ z^2 & z & 1 \end{vmatrix} = x^2 \begin{vmatrix} y & 1 \\ z & 1 \end{vmatrix} - x \begin{vmatrix} y^2 & 1 \\ z^2 & 1 \end{vmatrix} + \begin{vmatrix} y^2 & y \\ z^2 & z \end{vmatrix}$
$$= x^2(y - z) - x(y^2 - z^2) + yz(y - z)$$
$$= (y - z)[x^2 - x(y + z) + yz]$$
$$= (y - z)[(x^2 - xy) - (xz - yz)]$$
$$= (y - z)[x(x - y) - z(x - y)]$$
$$= (y - z)(x - y)(x - z)$$

60. If $a = 0$, we have
$$by = s$$
$$cx + dy = t$$
Thus, $y = \dfrac{s}{b}$ and $x = \dfrac{t - dy}{c}$
$$= \dfrac{tb - ds}{bc}$$
Using Cramer's Rule, we get
$$x = \dfrac{sd - tb}{-bc} = \dfrac{tb - sd}{bc}$$
$$y = \dfrac{-sc}{-bc} = \dfrac{s}{b}$$

If $b = 0$, we have
$$ax = s$$
$$cx + dy = t$$
Since $D = ad \neq 0$, then
$a \neq 0$ and $d \neq 0$.
Thus, $x = \dfrac{s}{a}$ and $y = \dfrac{t - cx}{d}$
$$= \dfrac{ta - cs}{ad}$$
Using Cramer's Rule, we get
$$x = \dfrac{sd}{ad} = \dfrac{s}{a}$$
$$y = \dfrac{at - cs}{ad}$$

If $c = 0$, we have
$$ax + by = s$$
$$dy = t$$
Since $D = ad \neq 0$, then
$a \neq 0$ and $d \neq 0$.
Thus, $y = \dfrac{t}{d}$ and $x = \dfrac{s - by}{a}$
$$= \dfrac{sd - bt}{ad}$$
Using Cramer's Rule, we get
$$x = \dfrac{sd - bt}{ad}$$
$$y = \dfrac{at}{ad} = \dfrac{t}{d}$$

If $d = 0$, we have
$$ax + by = s$$
$$cx = t$$
Since $D = -bc \neq 0$, then
$b \neq 0$ and $c \neq 0$.
Thus, $x = \dfrac{t}{c}$ and $y = \dfrac{s - ax}{b}$
$$= \dfrac{sc - at}{bc}$$
Using Cramer's Rule, we get
$$x = \dfrac{-tb}{-bc} = \dfrac{t}{c}$$
$$y = \dfrac{at - sc}{-bc} = \dfrac{sc - at}{bc}$$

61. $\begin{vmatrix} a_{13} & a_{12} & a_{11} \\ a_{23} & a_{22} & a_{21} \\ a_{33} & a_{32} & a_{31} \end{vmatrix} = a_{13}(a_{22}a_{31} - a_{32}a_{21}) - a_{12}(a_{23}a_{31} - a_{33}a_{21}) + a_{11}(a_{23}a_{32} - a_{33}a_{22})$

$$= -[a_{11}(a_{22}a_{33} - a_{32}a_{23}) - a_{12}(a_{21}a_{33} - a_{31}a_{23}) + a_{13}(a_{21}a_{32} - a_{31}a_{22})] = -\begin{vmatrix} a_{11} & a_{12} & a_{13} \\ a_{21} & a_{22} & a_{23} \\ a_{31} & a_{32} & a_{33} \end{vmatrix}$$

62. $\begin{vmatrix} a_{11} & a_{12} & a_{13} \\ ka_{21} & ka_{22} & ka_{23} \\ a_{31} & a_{32} & a_{33} \end{vmatrix} = -ka_{21}(a_{12}a_{33} - a_{32}a_{13}) + ka_{22}(a_{11}a_{33} - a_{31}a_{13}) - ka_{23}(a_{11}a_{32} - a_{31}a_{12})$

$$= k[-a_{21}(a_{12}a_{33} - a_{32}a_{13}) + a_{22}(a_{11}a_{33} - a_{31}a_{13}) - a_{23}(a_{11}a_{32} - a_{31}a_{12})] = k\begin{vmatrix} a_{11} & a_{12} & a_{13} \\ a_{21} & a_{22} & a_{23} \\ a_{31} & a_{32} & a_{33} \end{vmatrix}$$

63. $\begin{vmatrix} a_{11} & a_{12} & a_{11} \\ a_{21} & a_{22} & a_{21} \\ a_{31} & a_{32} & a_{31} \end{vmatrix} = a_{11}(a_{22}a_{31} - a_{32}a_{21}) - a_{12}(a_{21}a_{31} - a_{31}a_{21}) + a_{11}(a_{21}a_{32} - a_{31}a_{22})$

$$= a_{11}a_{22}a_{31} - a_{11}a_{32}a_{21} - a_{12}(0) + a_{11}a_{21}a_{32} - a_{11}a_{31}a_{22} = 0$$

64. $\begin{vmatrix} a_{11} + ka_{21} & a_{12} + ka_{22} & a_{13} + ka_{23} \\ a_{21} & a_{22} & a_{23} \\ a_{31} & a_{32} & a_{33} \end{vmatrix} = (a_{11} + ka_{21})(a_{22}a_{33} - a_{32}a_{23}) - (a_{12} + ka_{22})(a_{21}a_{33} - a_{31}a_{23}) + (a_{13} + ka_{23})(a_{21}a_{32} - a_{31}a_{22})$

$$= a_{11}a_{22}a_{33} - a_{11}a_{32}a_{23} + ka_{21}a_{22}a_{33} - ka_{21}a_{32}a_{23} - a_{12}a_{21}a_{33} + a_{12}a_{31}a_{23}$$
$$- ka_{22}a_{21}a_{33} + ka_{22}a_{31}a_{23} + a_{13}a_{21}a_{32} - a_{13}a_{31}a_{22} + ka_{23}a_{21}a_{32} - ka_{23}a_{31}a_{22}$$
$$= a_{11}a_{22}a_{33} - a_{11}a_{32}a_{23} - a_{12}a_{21}a_{33} + a_{12}a_{31}a_{23} + a_{13}a_{21}a_{32} - a_{13}a_{31}a_{22}$$
$$= a_{11}(a_{22}a_{33} - a_{32}a_{23}) - a_{12}(a_{21}a_{33} - a_{31}a_{23}) + a_{13}(a_{21}a_{32} - a_{31}a_{22})$$
$$= \begin{vmatrix} a_{11} & a_{12} & a_{13} \\ a_{21} & a_{22} & a_{23} \\ a_{31} & a_{32} & a_{33} \end{vmatrix}$$

Historical Problems (page 453)

1. (a) $2 - 5i \longleftrightarrow \begin{bmatrix} 2 & -5 \\ 5 & 2 \end{bmatrix}$, $1 + 3i \longleftrightarrow \begin{bmatrix} 1 & 3 \\ -3 & 1 \end{bmatrix}$ **(b)** $\begin{bmatrix} 2 & -5 \\ 5 & 2 \end{bmatrix}\begin{bmatrix} 1 & 3 \\ -3 & 1 \end{bmatrix} = \begin{bmatrix} 17 & 1 \\ -1 & 17 \end{bmatrix}$ **(c)** $17 + i$ **(d)** $17 + i$

2. (a) $x = k(ar + bs) + l(cr + ds) = r(ka + lc) + s(kb + ld)$ **(b)** $A = \begin{bmatrix} ka + lc & kb + ld \\ ma + nc & mb + nd \end{bmatrix}$
$y = m(ar + bs) + n(cr + ds) = r(ma + nc) + s(mb + nd)$

6.4 Assess Your Understanding (page 453)

1. inverse **2.** square **3.** identity **4.** F **5.** F **6.** F

7. $\begin{bmatrix} 4 & 4 & -5 \\ -1 & 5 & 4 \end{bmatrix}$ **8.** $\begin{bmatrix} -4 & 2 & -5 \\ 3 & -1 & 8 \end{bmatrix}$ **9.** $\begin{bmatrix} 0 & 12 & -20 \\ 4 & 8 & 24 \end{bmatrix}$ **10.** $\begin{bmatrix} -12 & -3 & 0 \\ 6 & -9 & 6 \end{bmatrix}$ **11.** $\begin{bmatrix} -8 & 7 & -15 \\ 7 & 0 & 22 \end{bmatrix}$ **12.** $\begin{bmatrix} 16 & 10 & -10 \\ -6 & 16 & 4 \end{bmatrix}$

13. $\begin{bmatrix} 28 & -9 \\ 4 & 23 \end{bmatrix}$ **14.** $\begin{bmatrix} 22 & 6 \\ 14 & -2 \end{bmatrix}$ **15.** $\begin{bmatrix} 1 & 14 & -14 \\ 2 & 22 & -18 \\ 3 & 0 & 28 \end{bmatrix}$ **16.** $\begin{bmatrix} 14 & 7 & -2 \\ 20 & 12 & -4 \\ -14 & 7 & -6 \end{bmatrix}$ **17.** $\begin{bmatrix} 15 & 21 & -16 \\ 22 & 34 & -22 \\ -11 & 7 & 22 \end{bmatrix}$ **18.** $\begin{bmatrix} 50 & -3 \\ 18 & 21 \end{bmatrix}$ **19.** $\begin{bmatrix} 25 & -9 \\ 4 & 20 \end{bmatrix}$

20. $\begin{bmatrix} 6 & 14 & -14 \\ 2 & 27 & -18 \\ 3 & 0 & 33 \end{bmatrix}$ **21.** $\begin{bmatrix} -13 & 7 & -12 \\ -18 & 10 & -14 \\ 17 & -7 & 34 \end{bmatrix}$ **22.** $\begin{bmatrix} 50 & -3 \\ 18 & 21 \end{bmatrix}$ **23.** $\begin{bmatrix} -2 & 4 & 2 & 8 \\ 2 & 1 & 4 & 6 \end{bmatrix}$ **24.** $\begin{bmatrix} -22 & 29 & 8 & -1 \\ -10 & 17 & 6 & -1 \end{bmatrix}$ **25.** $\begin{bmatrix} 5 & 14 \\ 9 & 16 \end{bmatrix}$

26. $\begin{bmatrix} -1 & 2 & -1 \\ 0 & -12 & 3 \\ 15 & 30 & 0 \end{bmatrix}$ **27.** $\begin{bmatrix} 9 & 2 \\ 34 & 13 \\ 47 & 20 \end{bmatrix}$ **28.** $\begin{bmatrix} 6 & 32 \\ 1 & 3 \\ -2 & -4 \end{bmatrix}$ **29.** $\begin{bmatrix} 1 & -1 \\ -1 & 2 \end{bmatrix}$ **30.** $\begin{bmatrix} 1 & 1 \\ 2 & 3 \end{bmatrix}$ **31.** $\begin{bmatrix} 1 & -\frac{5}{2} \\ -1 & 3 \end{bmatrix}$ **32.** $\begin{bmatrix} -1 & -\frac{1}{2} \\ -3 & -2 \end{bmatrix}$

33. $\begin{bmatrix} 1 & -\frac{1}{a} \\ -1 & \frac{2}{a} \end{bmatrix}$ **34.** $\begin{bmatrix} -\frac{2}{b} & \frac{3}{b} \\ 1 & -1 \end{bmatrix}$ **35.** $\begin{bmatrix} 3 & -3 & 1 \\ -2 & 2 & -1 \\ -4 & 5 & -2 \end{bmatrix}$ **36.** $\begin{bmatrix} 3 & -2 & -4 \\ 3 & -2 & -5 \\ -1 & 1 & 2 \end{bmatrix}$ **37.** $\begin{bmatrix} -\frac{5}{7} & \frac{1}{7} & \frac{3}{7} \\ \frac{9}{7} & \frac{1}{7} & -\frac{4}{7} \\ \frac{3}{7} & -\frac{2}{7} & \frac{1}{7} \end{bmatrix}$ **38.** $\begin{bmatrix} \frac{3}{7} & -\frac{4}{7} & \frac{1}{7} \\ \frac{1}{7} & \frac{1}{7} & -\frac{2}{7} \\ -\frac{5}{7} & \frac{9}{7} & \frac{3}{7} \end{bmatrix}$

39. $x = 3, y = 2$ **40.** $x = 12, y = 28$ **41.** $x = -5, y = 10$ **42.** $x = 9, y = 23$ **43.** $x = 2, y = -1$ **44.** $x = -7, y = -28$

45. $x = \frac{1}{2}, y = 2$ **46.** $x = -\frac{1}{2}, y = 3$ **47.** $x = -2, y = 1$ **48.** $x = 2, y = 1$ **49.** $x = \frac{2}{a}, y = \frac{3}{a}$ **50.** $x = \frac{2}{b}, y = 4$

51. $x = -2, y = 3, z = 5$ **52.** $x = 4, y = -2, z = 1$ **53.** $x = \frac{1}{2}, y = -\frac{1}{2}, z = 1$ **54.** $x = 1, y = -1, z = \frac{1}{2}$

55. $x = -\frac{34}{7}, y = \frac{85}{7}, z = \frac{12}{7}$ **56.** $x = \frac{8}{7}, y = \frac{5}{7}, z = \frac{17}{7}$ **57.** $x = \frac{1}{3}, y = 1, z = \frac{2}{3}$ **58.** $x = 1, y = -1, z = 1$

59. $\begin{bmatrix} 4 & 2 & | & 1 & 0 \\ 2 & 1 & | & 0 & 1 \end{bmatrix} \rightarrow \begin{bmatrix} 1 & \frac{1}{2} & | & \frac{1}{4} & 0 \\ 2 & 1 & | & 0 & 1 \end{bmatrix} \rightarrow \begin{bmatrix} 1 & \frac{1}{2} & | & \frac{1}{4} & 0 \\ 0 & 0 & | & -\frac{1}{2} & 1 \end{bmatrix}$ **60.** $\begin{bmatrix} -3 & \frac{1}{2} & | & 1 & 0 \\ 6 & -1 & | & 0 & 1 \end{bmatrix} \rightarrow \begin{bmatrix} 1 & -\frac{1}{6} & | & -\frac{1}{3} & 0 \\ 6 & -1 & | & 0 & 1 \end{bmatrix} \rightarrow \begin{bmatrix} 1 & -\frac{1}{6} & | & -\frac{1}{3} & 0 \\ 0 & 0 & | & 2 & 1 \end{bmatrix}$

61. $\begin{bmatrix} 15 & 3 & | & 1 & 0 \\ 10 & 2 & | & 0 & 1 \end{bmatrix} \rightarrow \begin{bmatrix} 1 & \frac{1}{5} & | & \frac{1}{15} & 0 \\ 10 & 2 & | & 0 & 1 \end{bmatrix} \rightarrow \begin{bmatrix} 1 & \frac{1}{5} & | & \frac{1}{15} & 0 \\ 0 & 0 & | & -\frac{2}{3} & 1 \end{bmatrix}$ **62.** $\begin{bmatrix} -3 & 0 & | & 1 & 0 \\ 4 & 0 & | & 0 & 1 \end{bmatrix} \rightarrow \begin{bmatrix} 1 & 0 & | & -\frac{1}{3} & 0 \\ 4 & 0 & | & 0 & 1 \end{bmatrix} \rightarrow \begin{bmatrix} 1 & 0 & | & -\frac{1}{3} & 0 \\ 0 & 0 & | & \frac{4}{3} & 1 \end{bmatrix}$

63. $\begin{bmatrix} -3 & 1 & -1 & | & 1 & 0 & 0 \\ 1 & -4 & -7 & | & 0 & 1 & 0 \\ 1 & 2 & 5 & | & 0 & 0 & 1 \end{bmatrix} \rightarrow \begin{bmatrix} 1 & 2 & 5 & | & 0 & 0 & 1 \\ 1 & -4 & -7 & | & 0 & 1 & 0 \\ -3 & 1 & -1 & | & 1 & 0 & 0 \end{bmatrix} \rightarrow \begin{bmatrix} 1 & 2 & 5 & | & 0 & 0 & 1 \\ 0 & -6 & -12 & | & 0 & 1 & -1 \\ 0 & 7 & 14 & | & 1 & 0 & 3 \end{bmatrix} \rightarrow \begin{bmatrix} 1 & 2 & 5 & | & 0 & 0 & 1 \\ 0 & 1 & 2 & | & 0 & -\frac{1}{6} & \frac{1}{6} \\ 0 & 1 & 2 & | & \frac{1}{7} & 0 & \frac{3}{7} \end{bmatrix}$

$\rightarrow \begin{bmatrix} 1 & 2 & 5 & | & 0 & 0 & 1 \\ 0 & 1 & 2 & | & 0 & -\frac{1}{6} & \frac{1}{6} \\ 0 & 0 & 0 & | & \frac{1}{7} & \frac{1}{6} & \frac{11}{42} \end{bmatrix}$

64. $\begin{bmatrix} 1 & 1 & -3 & | & 1 & 0 & 0 \\ 2 & -4 & 1 & | & 0 & 1 & 0 \\ -5 & 7 & 1 & | & 0 & 0 & 1 \end{bmatrix} \rightarrow \begin{bmatrix} 1 & 1 & -3 & | & 1 & 0 & 0 \\ 0 & -6 & 7 & | & -2 & 1 & 0 \\ -5 & 7 & 1 & | & 0 & 0 & 1 \end{bmatrix} \rightarrow \begin{bmatrix} 1 & 1 & -3 & | & 1 & 0 & 0 \\ 0 & 1 & -\frac{7}{6} & | & \frac{1}{3} & -\frac{1}{6} & 0 \\ -5 & 7 & 1 & | & 0 & 0 & 1 \end{bmatrix} \rightarrow \begin{bmatrix} 1 & 1 & -3 & | & 1 & 0 & 0 \\ 0 & 1 & -\frac{7}{6} & | & \frac{1}{3} & -\frac{1}{6} & 0 \\ 0 & 12 & -14 & | & 5 & 0 & 1 \end{bmatrix}$

$\rightarrow \begin{bmatrix} 1 & 1 & -3 & | & 1 & 0 & 0 \\ 0 & 1 & -\frac{7}{6} & | & \frac{1}{3} & -\frac{1}{6} & 0 \\ 0 & 0 & 0 & | & 1 & 2 & 1 \end{bmatrix}$

65. $\begin{bmatrix} 0.01 & 0.05 & -0.01 \\ 0.01 & -0.02 & 0.01 \\ -0.02 & 0.01 & 0.03 \end{bmatrix}$ **66.** $\begin{bmatrix} 0.26 & -0.29 & -0.2 \\ -1.21 & 1.63 & 1.2 \\ -1.84 & 2.53 & 1.8 \end{bmatrix}$ **67.** $\begin{bmatrix} 0.02 & -0.04 & -0.01 & 0.01 \\ -0.02 & 0.05 & 0.03 & -0.03 \\ 0.02 & 0.01 & -0.04 & 0.00 \\ -0.02 & 0.06 & 0.07 & 0.06 \end{bmatrix}$ **68.** $\begin{bmatrix} 0.01 & 0.04 & 0.00 & 0.03 \\ 0.02 & -0.02 & 0.01 & 0.01 \\ -0.04 & 0.02 & 0.04 & 0.06 \\ 0.05 & -0.02 & 0.00 & -0.09 \end{bmatrix}$

69. $x = 4.57, y = -6.44, z = -24.07$ **70.** $x = 4.56, y = -6.06, z = -22.55$ **71.** $x = -1.19, y = 2.46, z = 8.27$
72. $x = -2.05, y = 3.88, z = 13.36$

73. (a) $\begin{bmatrix} 500 & 350 & 400 \\ 700 & 500 & 850 \end{bmatrix}; \begin{bmatrix} 500 & 700 \\ 350 & 500 \\ 400 & 850 \end{bmatrix}$ **(b)** $\begin{bmatrix} 15 \\ 8 \\ 3 \end{bmatrix}$ **(c)** $\begin{bmatrix} 11,500 \\ 17,050 \end{bmatrix}$ **(d)** $[0.10 \quad 0.05]$ **(e)** \$2002.50

74. (a) January sales $\begin{bmatrix} 400 & 250 & 50 \\ 450 & 200 & 140 \end{bmatrix}$; February sales $\begin{bmatrix} 350 & 100 & 30 \\ 350 & 300 & 100 \end{bmatrix}$ **(b)** $\begin{bmatrix} 750 & 350 & 80 \\ 800 & 500 & 240 \end{bmatrix}$ **(c)** $\begin{bmatrix} 100 \\ 150 \\ 200 \end{bmatrix}$ **(d)** $\begin{bmatrix} 143,500 \\ 203,000 \end{bmatrix}$

75. If $D = ad - bc \neq 0$, then $a \neq 0$ and $d \neq 0$, or $b \neq 0$ and $c \neq 0$. Assuming the former,

$\begin{bmatrix} a & b & | & 1 & 0 \\ c & d & | & 0 & 1 \end{bmatrix} \rightarrow \begin{bmatrix} 1 & \frac{b}{a} & | & \frac{1}{a} & 0 \\ c & d & | & 0 & 1 \end{bmatrix} \rightarrow \begin{bmatrix} 1 & \frac{b}{a} & | & \frac{1}{a} & 0 \\ 0 & \frac{D}{a} & | & -\frac{c}{a} & 1 \end{bmatrix} \rightarrow \begin{bmatrix} 1 & \frac{b}{a} & | & \frac{1}{a} & 0 \\ 0 & 1 & | & -\frac{c}{D} & \frac{a}{D} \end{bmatrix} \rightarrow \begin{bmatrix} 1 & 0 & | & \frac{d}{D} & -\frac{b}{d} \\ 0 & 1 & | & -\frac{c}{D} & \frac{a}{D} \end{bmatrix}$

$\qquad R_1 = \frac{1}{a}r_1 \qquad\qquad R_2 = -cr_1 + r_2 \qquad\qquad R_2 = \frac{a}{D}r_2 \qquad\qquad R_1 = -\frac{b}{a}r_2 + r_1$

6.5 Assess Your Understanding (page 461)

5. Proper **6.** Proper **7.** Improper; $1 + \dfrac{9}{x^2 - 4}$ **8.** Improper; $3 + \dfrac{1}{x^2 - 1}$ **9.** Improper; $5x + \dfrac{22x - 1}{x^2 - 4}$ **10.** Improper; $3x + \dfrac{x^2 - 24x - 2}{x^3 + 8}$

11. Improper; $1 + \dfrac{-2(x - 6)}{(x + 4)(x - 3)}$ **12.** Improper; $2x + \dfrac{6x}{x^2 + 1}$ **13.** $\dfrac{-4}{x} + \dfrac{4}{x - 1}$ **14.** $\dfrac{2}{x + 2} + \dfrac{1}{x - 1}$ **15.** $\dfrac{1}{x} + \dfrac{-x}{x^2 + 1}$

16. $\dfrac{\frac{1}{5}}{x + 1} + \dfrac{-\frac{1}{5}x + \frac{1}{5}}{x^2 + 4}$ **17.** $\dfrac{-1}{x - 1} + \dfrac{2}{x - 2}$ **18.** $\dfrac{1}{x + 2} + \dfrac{2}{x - 4}$ **19.** $\dfrac{\frac{1}{4}}{x + 1} + \dfrac{\frac{3}{4}}{x - 1} + \dfrac{\frac{1}{2}}{(x - 1)^2}$ **20.** $\dfrac{-\frac{3}{4}}{x} + \dfrac{-\frac{1}{2}}{x^2} + \dfrac{\frac{3}{4}}{x - 2}$

21. $\dfrac{\frac{1}{12}}{x - 2} + \dfrac{-\frac{1}{12}(x + 4)}{x^2 + 2x + 4}$ **22.** $\dfrac{2}{x - 1} + \dfrac{-2x - 2}{x^2 + x + 1}$ **23.** $\dfrac{\frac{1}{4}}{x - 1} + \dfrac{\frac{1}{4}}{(x - 1)^2} + \dfrac{-\frac{1}{4}}{x + 1} + \dfrac{\frac{1}{4}}{(x + 1)^2}$ **24.** $\dfrac{\frac{1}{2}}{x} + \dfrac{\frac{1}{4}}{x^2} + \dfrac{\frac{3}{4}}{(x - 2)^2} - \dfrac{\frac{1}{2}}{(x - 2)}$

25. $\dfrac{-5}{x + 2} + \dfrac{5}{x + 1} + \dfrac{-4}{(x + 1)^2}$ **26.** $\dfrac{\frac{2}{9}}{x + 2} + \dfrac{\frac{7}{9}}{x - 1} + \dfrac{\frac{2}{3}}{(x - 1)^2}$ **27.** $\dfrac{\frac{1}{4}}{x} + \dfrac{1}{x^2} + \dfrac{-\frac{1}{4}(x + 4)}{x^2 + 4}$ **28.** $\dfrac{\frac{14}{3}}{x - 1} + \dfrac{4}{(x - 1)^2} + \dfrac{-\frac{14}{3}x + \frac{4}{3}}{x^2 + 2}$

29. $\dfrac{\frac{2}{3}}{x + 1} + \dfrac{\frac{1}{3}(x + 1)}{x^2 + 2x + 4}$ **30.** $\dfrac{-6}{x} + \dfrac{7x + 7}{x^2 + 3x + 3}$ **31.** $\dfrac{\frac{2}{7}}{3x - 2} + \dfrac{\frac{1}{7}}{2x + 1}$ **32.** $\dfrac{-\frac{1}{7}}{2x + 3} + \dfrac{\frac{2}{7}}{4x - 1}$ **33.** $\dfrac{\frac{3}{4}}{x + 3} + \dfrac{\frac{1}{4}}{x - 1}$

34. $\dfrac{-3}{x + 1} + \dfrac{2}{x + 2} + \dfrac{2}{x + 3}$ **35.** $\dfrac{1}{x^2 + 4} + \dfrac{2x - 1}{(x^2 + 4)^2}$ **36.** $\dfrac{x}{x^2 + 16} + \dfrac{-16x + 1}{(x^2 + 16)^2}$ **37.** $\dfrac{-1}{x} + \dfrac{2}{x - 3} + \dfrac{-1}{x + 1}$ **38.** $\dfrac{-1}{x^2} + \dfrac{-1}{x^4} + \dfrac{1}{x - 1}$

39. $\dfrac{4}{x-2}+\dfrac{-3}{x-1}+\dfrac{-1}{(x-1)^2}$ **40.** $\dfrac{\frac{3}{8}}{x-1}+\dfrac{\frac{1}{2}}{(x-1)^2}+\dfrac{\frac{5}{8}}{x+3}$ **41.** $\dfrac{x}{(x^2+16)^2}+\dfrac{-16x}{(x^2+16)^3}$ **42.** $\dfrac{1}{(x^2+4)^2}+\dfrac{-4}{(x^2+4)^3}$

43. $\dfrac{-\frac{8}{7}}{2x+1}+\dfrac{\frac{4}{7}}{x-3}$ **44.** $\dfrac{\frac{4}{5}}{2x-1}+\dfrac{\frac{8}{5}}{x+2}$ **45.** $\dfrac{-\frac{2}{9}}{x}+\dfrac{-\frac{1}{3}}{x^2}+\dfrac{\frac{1}{6}}{x-3}+\dfrac{\frac{1}{18}}{x+3}$ **46.** $\dfrac{\frac{13}{24}}{x-2}+\dfrac{-\frac{13}{24}}{x+2}+\dfrac{-\frac{7}{6}}{x^2+2}$

Historical Problem (page 467)

$x = 6$ units, $y = 8$ units

6.6 Assess Your Understanding (page 467)

5.

6.

7.

8.

9.

10.

11.

12.

13.

14.

15.

16.

17.

18.

19.

20.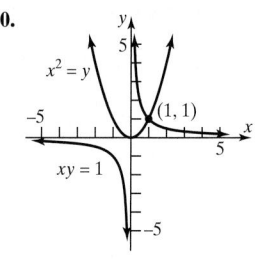

21. No points of intersection

22.

23.

24.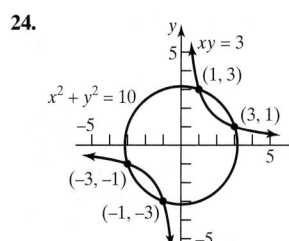

25. $x = 1, y = 4; x = -1, y = -4; x = 2\sqrt{2}, y = \sqrt{2}; x = -2\sqrt{2}, y = -\sqrt{2}$ **26.** $x = 5, y = 2$ **27.** $x = 0, y = 1; x = -\dfrac{2}{3}, y = -\dfrac{1}{3}$

28. $x = -5, y = -\dfrac{3}{2}$ **29.** $x = 0, y = -1; x = \dfrac{5}{2}, y = -\dfrac{7}{2}$ **30.** $x = -2, y = -2; x = -\sqrt{2}, y = -2\sqrt{2}; x = \sqrt{2}, y = 2\sqrt{2}; x = 2, y = 2$

31. $x = 2, y = \dfrac{1}{3}; x = \dfrac{1}{2}, y = \dfrac{4}{3}$ **32.** $x = -2, y = 0; x = 7, y = 6$ **33.** $x = 3, y = 2; x = 3, y = -2; x = -3, y = 2; x = -3, y = -2$

34. $x = -1, y = -2; x = -1, y = 2; x = 1, y = -2; x = 1, y = 2$ **35.** $x = \dfrac{1}{2}, y = \dfrac{3}{2}; x = \dfrac{1}{2}, y = -\dfrac{3}{2}; x = -\dfrac{1}{2}, y = \dfrac{3}{2}; x = -\dfrac{1}{2}, y = -\dfrac{3}{2}$

36. $x = -2\sqrt{2}, y = \sqrt{3}; x = 2\sqrt{2}, y = \sqrt{3}; x = 2\sqrt{2}, y = -\sqrt{3}; x = -2\sqrt{2}, y = -\sqrt{3}$ **37.** $x = \sqrt{2}, y = 2\sqrt{2}; x = -\sqrt{2}, y = -2\sqrt{2}$

38. $x = -5\sqrt{3}, y = \sqrt{3}; x = 5\sqrt{3}, y = -\sqrt{3}$ **39.** No real solution exists. **40.** No real solution exists.

41. $x = \dfrac{8}{3}, y = \dfrac{2\sqrt{10}}{3}; x = -\dfrac{8}{3}, y = \dfrac{2\sqrt{10}}{3}; x = \dfrac{8}{3}, y = -\dfrac{2\sqrt{10}}{3}; x = -\dfrac{8}{3}, y = -\dfrac{2\sqrt{10}}{3}$

42. $x = \dfrac{-\sqrt{2}}{2}, y = \dfrac{-\sqrt{6}}{3}; x = \dfrac{-\sqrt{2}}{2}, y = \dfrac{\sqrt{6}}{3}; x = \dfrac{\sqrt{2}}{2}, y = \dfrac{-\sqrt{6}}{3}; x = \dfrac{\sqrt{2}}{2}, y = \dfrac{\sqrt{6}}{3}$

43. $x = 1, y = \dfrac{1}{2}; x = -1, y = \dfrac{1}{2}; x = 1, y = -\dfrac{1}{2}; x = -1, y = -\dfrac{1}{2}$ **44.** $x = -2, y = -\sqrt{2}; x = -2, y = \sqrt{2}; x = 2, y = -\sqrt{2}; x = 2, y = \sqrt{2}$

45. No real solution exists. **46.** $x = -\sqrt[4]{\dfrac{2}{5}}, y = -\sqrt[4]{\dfrac{2}{3}}; x = -\sqrt[4]{\dfrac{2}{5}}, y = \sqrt[4]{\dfrac{2}{3}}; x = \sqrt[4]{\dfrac{2}{5}}, y = -\sqrt[4]{\dfrac{2}{3}}; x = \sqrt[4]{\dfrac{2}{5}}, y = \sqrt[4]{\dfrac{2}{3}}$

47. $x = \sqrt{3}, y = \sqrt{3}; x = -\sqrt{3}, y = -\sqrt{3}; x = 2, y = 1; x = -2, y = -1$ **48.** $x = -2, y = 2; x = 3, y = -3$

49. $x = 0, y = -2; x = 0, y = 1; x = 2, y = -1$ **50.** $x = 2, y = 1$ **51.** $x = 2, y = 8$ **52.** $x = \dfrac{1}{2}, y = \dfrac{1}{16}$ **53.** $x = 81, y = 3$ **54.** $x = 32, y = 2$

55.

56.

57.

58.

59.

60.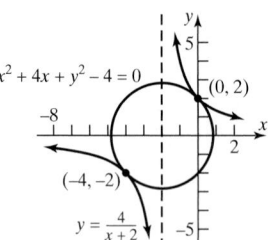

61. $x = 0.48, y = 0.62$ **62.** $x = 0.65, y = 0.52$ **63.** $x = -1.65, y = -0.89$ **64.** $x = -1.37, y = 2.14$
65. $x = 0.58, y = 1.86; x = 1.81, y = 1.05; x = 0.58, y = -1.86; x = 1.81, y = -1.05$
66. $x = 0.64, y = 1.55; x = -0.64, y = -1.55; x = 1.55, y = 0.64; x = -1.55, y = -0.64$ **67.** $x = 2.35, y = 0.85$
68. $x = 1.90, y = 0.64; x = 0.14, y = -2.00$ **69.** 3 and 1; -3 and -1 **70.** 2 and 5 **71.** 2 and 2; -2 and -2 **72.** 2 and 5; -2 and -5
73. $\dfrac{1}{2}$ and $\dfrac{1}{3}$ **74.** $\dfrac{1}{2}$ and -1 **75.** 5 **76.** $\dfrac{1}{7}$ **77.** 5 in. by 3 in. **78.** 4 ft, 6 ft **79.** 2 cm and 4 cm **80.** 8 cm **81.** tortoise: 7 m/hr, hare: $7\dfrac{1}{2}$ m/hr
82. 10.02 ft **83.** 12 cm by 18 cm **84.** 16.57 cm by 13.03 cm **85.** $x = 60$ ft; $y = 30$ ft **86.** It is not possible.
87. $l = \dfrac{P + \sqrt{P^2 - 16A}}{4}; w = \dfrac{P - \sqrt{P^2 - 16A}}{4}$ **88.** $b = P - \dfrac{4h^2 + P^2}{2P}, l = \dfrac{4h^2 + P^2}{4P}$ **89.** $y = 4x - 4$
90. $y = -\dfrac{1}{3}x + \dfrac{10}{3}$ **91.** $y = 2x + 1$ **92.** $y = 4x + 9$ **93.** $y = -\dfrac{1}{3}x + \dfrac{7}{3}$ **94.** $y = \dfrac{3}{2}x + \dfrac{7}{2}$ **95.** $y = 2x - 3$
96. $y = \dfrac{1}{3}x + \dfrac{7}{3}$ **97.** $r_1 = \dfrac{-b + \sqrt{b^2 - 4ac}}{2a}; r_2 = \dfrac{-b - \sqrt{b^2 - 4ac}}{2a}$ **99. (a)** 4.274 ft by 4.274 ft or 0.093 ft by 0.093 ft

6.7 Assess Your Understanding (page 477)

7. satisfied **8.** half-plane **9.** F **10.** T

11.

12.

13.

14.

15.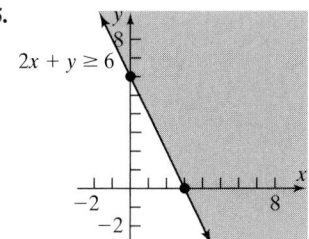
$2x + y \ge 6$

16.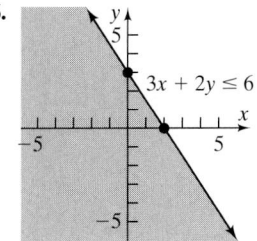
$3x + 2y \le 6$

17.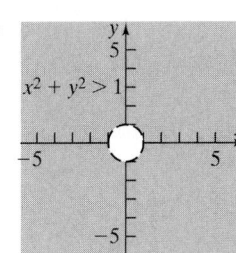
$x^2 + y^2 > 1$

18.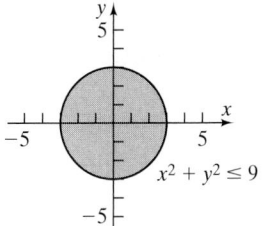
$x^2 + y^2 \le 9$

19.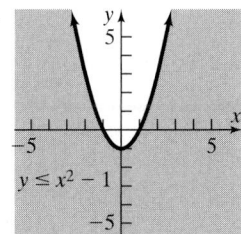
$y \le x^2 - 1$

20.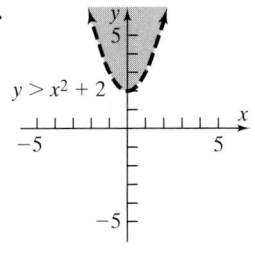
$y > x^2 + 2$

21.
$xy \ge 4$

22.
$xy \le 1$

23.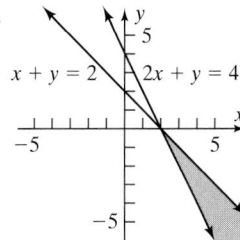
$x + y = 2$ $2x + y = 4$

24.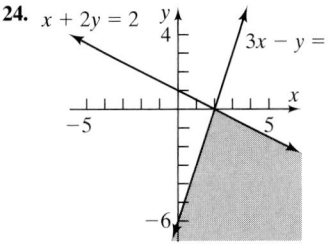
$x + 2y = 2$ $3x - y = 6$

25.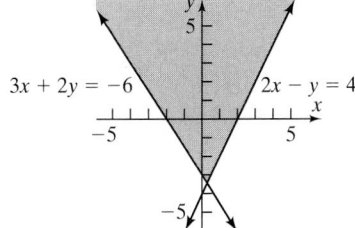
$3x + 2y = -6$ $2x - y = 4$

26.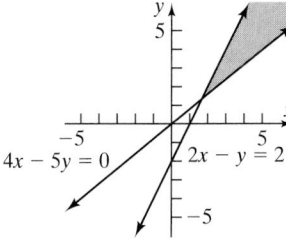
$4x - 5y = 0$ $2x - y = 2$

27.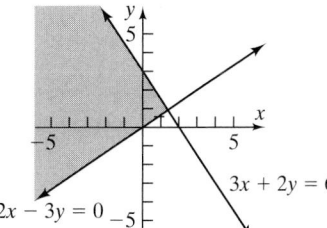
$2x - 3y = 0$ $3x + 2y = 6$

28.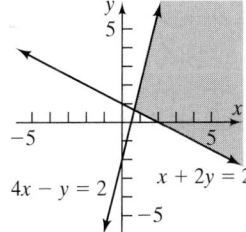
$4x - y = 2$ $x + 2y = 2$

29.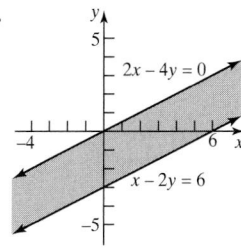
$2x - 4y = 0$ $x - 2y = 6$

30.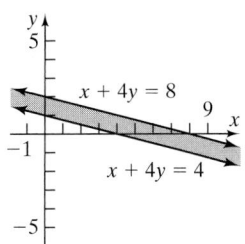
$x + 4y = 8$ $x + 4y = 4$

31.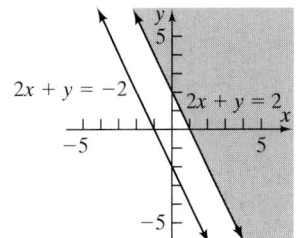
$2x + y = -2$ $2x + y = 2$

32.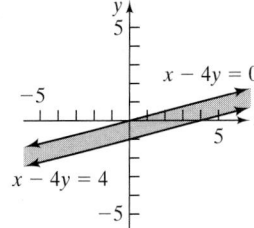
$x - 4y = 0$ $x - 4y = 4$

33. No solution
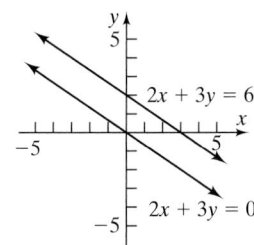
$2x + 3y = 6$ $2x + 3y = 0$

34.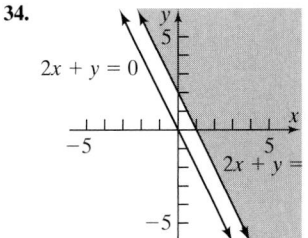
$2x + y = 0$ $2x + y = 2$

35.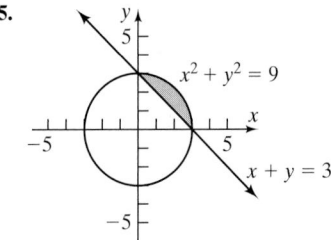
$x^2 + y^2 = 9$ $x + y = 3$

36.

37.

38.

39.

40.

41.

42.
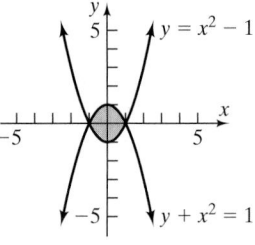

43. Bounded; corner points
$(0, 0), (3, 0), (2, 2), (0, 3)$

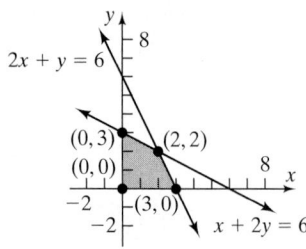

44. Unbounded; corner points
$(0, 4), (4, 0)$

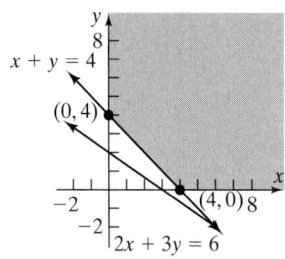

45. Unbounded; corner points
$(2, 0), (0, 4)$

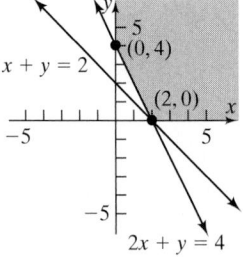

46. Bounded; corner points
$(0, 0), (0, 2), (1, 0)$

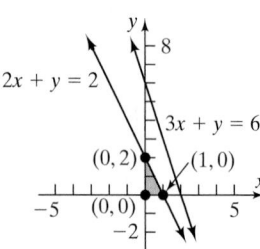

47. Bounded; corner points $(2, 0), (4, 0),$
$\left(\dfrac{24}{7}, \dfrac{12}{7}\right), (0, 4), (0, 2)$

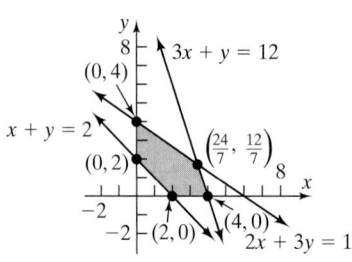

48. Bounded; corner points
$(0, 2), (0, 3), (1, 1)$

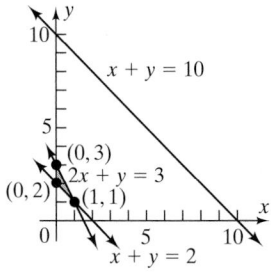

49. Bounded; corner points $(2, 0), (5, 0),$
$(2, 6), (0, 8), (0, 2)$

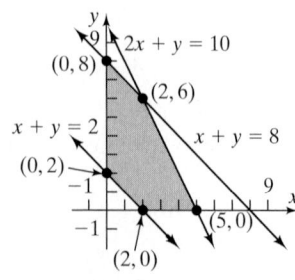

50. Bounded; corner points $(2, 0), (0, 2),$
$(0, 8), (8, 0)$

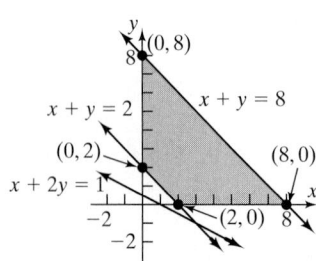

51. Bounded; corner points
$(1, 0), (10, 0), (0, 5), \left(0, \dfrac{1}{2}\right)$

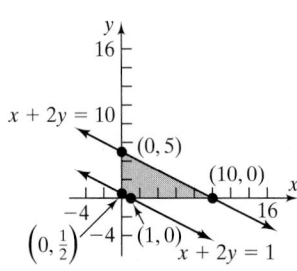

52. Bounded, corner points
$(0, 2), (0, 5), (6, 2), (8, 0), (2, 0)$

53. $\begin{cases} x \le 4 \\ x + y \le 6 \\ x \ge 0 \\ y \ge 0 \end{cases}$ **54.** $\begin{cases} y \le 5 \\ x + y \ge 2 \\ x \le 6 \\ x \ge 0, y \ge 0 \end{cases}$

55. $\begin{cases} x \le 20 \\ y \ge 15 \\ x + y \le 50 \\ x - y \le 0 \\ x \ge 0 \end{cases}$ **56.** $\begin{cases} y \le 6 \\ x \le 5 \\ 3x + 4y \ge 12 \\ y \ge 2x - 8 \\ x \ge 0 \end{cases}$

57. (a) $\begin{cases} x + y \le 50{,}000 \\ x \ge 35{,}000 \\ y \le 10{,}000 \\ x \ge 0 \\ y \ge 0 \end{cases}$

(b)

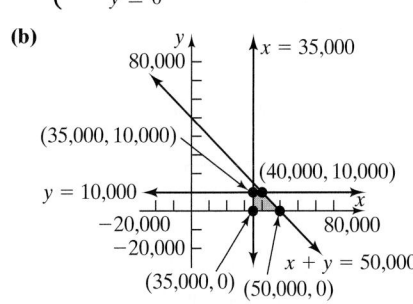

58. (a) $\begin{cases} x \ge 0 \\ y \ge 0 \\ 2x + 3y \le 80 \\ 3x + 4y \le 120 \end{cases}$ **(b)**

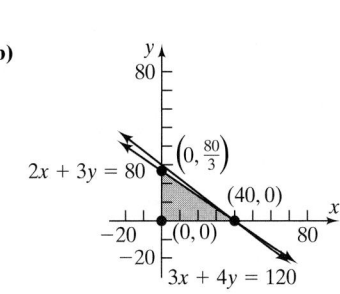

59. (a) $\begin{cases} x \ge 0 \\ y \ge 0 \\ x + 2y \le 300 \\ 3x + 2y \le 480 \end{cases}$ **(b)**

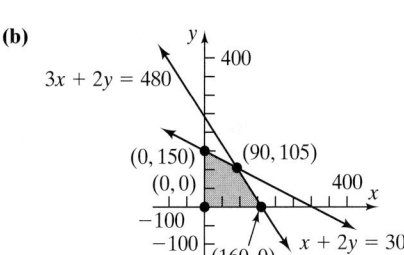

60. (a) $\begin{cases} 4x + 3y \le 960 \\ 2x + 3y \le 720 \\ x \ge 0 \\ y \ge 0 \end{cases}$ **(b)**

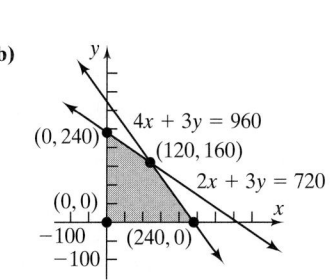

61. (a) $\begin{cases} 3x + 2y \le 160 \\ 2x + 3y \le 150 \\ x \ge 0 \\ y \ge 0 \end{cases}$ **(b)**

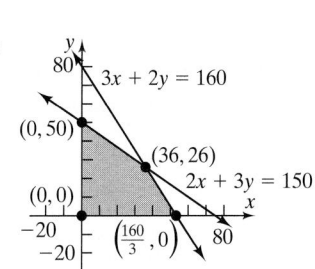

6.8 Assess Your Understanding (page 484)

1. objective function **2.** T **3.** Maximum value is 11; minimum value is 3. **4.** Maximum value is 28; minimum value is 8.
5. Maximum value is 65; minimum value is 4. **6.** Maximum value is 56; minimum value is 3. **7.** Maximum value is 67; minimum value is 20.
8. Maximum value is 65; minimum value is 15. **9.** The maximum value of z is 12, and it occurs at the point $(6, 0)$.
10. Maximum value is 26, and it occurs at the point $(5, 7)$. **11.** The minimum value of z is 4, and it occurs at the point $(2, 0)$.
12. The minimum value of z is 8, and it occurs at the point $(0, 2)$. **13.** The maximum value of z is 20, and it occurs at the point $(0, 4)$.
14. The maximum value of z is 28, and it occurs at the point $(2, 6)$. **15.** The minimum value of z is 8, and it occurs at the point $(0, 2)$.
16. The minimum value of z is $\dfrac{15}{2}$, and it occurs at the point $\left(\dfrac{3}{2}, \dfrac{3}{2}\right)$. **17.** The maximum value of z is 50, and it occurs at the point $(10, 0)$.
18. The maximum value of z is 36, and it occurs at the point $(0, 9)$. **19.** 8 downhill, 24 cross-country; $1760; $1920
20. 24 acres of soybeans and 12 acres of wheat for a maximum profit of $5520; $7200 **21.** 30 acres of soybeans and 10 acres of corn
22. 15 oz of supplement A and 0 oz of supplement B or 7.5 oz of supplement A and 11.25 oz of supplement B, minimum cost $22.50
23. $\dfrac{1}{2}$ hr on machine 1; $5\dfrac{1}{4}$ hr on machine 2 **24.** 15 newer trees and 10 older trees; minimum cost $425 **25.** 100 lb of ground beef and 50 lb of pork
26. (a) $10,000 in a junk bond and $10,000 in Treasury bills **(b)** $12,000 in junk bonds and $8000 in Treasury bills
27. 10 racing skates, 15 figure skates **28.** $20,000 in a AAA bond and $30,000 in a CD **29.** 2 metal samples, 4 plastic samples; $34
30. 15 cans of Gourmet Dog and 25 cans of Chow Hound **31. (a)** 10 first class, 120 coach **(b)** 15 first class, 120 coach
32. 5 oz of supplement A and 1 oz of supplement B; minimum cost $0.38

Review Exercises *(page 487)*

1. $x = 2, y = -1$ **2.** $x = \dfrac{11}{23}, y = \dfrac{8}{23}$ **3.** $x = 2, y = \dfrac{1}{2}$ **4.** $x = -\dfrac{1}{2}, y = 1$ **5.** $x = 2, y = -1$ **6.** $x = 0, y = \dfrac{5}{3}$ **7.** $x = \dfrac{11}{5}, y = -\dfrac{3}{5}$

8. $x = -\dfrac{1}{2}, y = -\dfrac{1}{2}$ **9.** Inconsistent **10.** Inconsistent **11.** $x = 2, y = 3$ **12.** $x = 84, y = -63$ **13.** Inconsistent **14.** $y = -\dfrac{2}{5}x + 2, x$ is any

real number **15.** $x = -1, y = 2, z = -3$ **16.** $x = -1, y = 2, z = 7$ **17.** $x = \dfrac{7}{4}z + \dfrac{39}{4}, y = \dfrac{9}{8}z + \dfrac{69}{8}, z$ is any real number **18.** Inconsistent

19. $\begin{cases} 3x + 2y = 8 \\ x + 4y = -1 \end{cases}$ **20.** $\begin{cases} x + 2y + 5z = -2 \\ 5x \quad - 3z = 8 \\ 2x - y \quad = 0 \end{cases}$ **21.** $\begin{bmatrix} 4 & -4 \\ 3 & 9 \\ 4 & 4 \end{bmatrix}$ **22.** $\begin{bmatrix} -2 & 4 \\ 1 & -1 \\ -6 & 0 \end{bmatrix}$ **23.** $\begin{bmatrix} 6 & 0 \\ 12 & 24 \\ -6 & 12 \end{bmatrix}$ **24.** $\begin{bmatrix} -16 & 12 & 0 \\ -4 & -4 & 8 \end{bmatrix}$ **25.** $\begin{bmatrix} 4 & -3 & 0 \\ 12 & -2 & -8 \\ -2 & 5 & -4 \end{bmatrix}$

26. $\begin{bmatrix} -2 & -12 \\ 5 & 0 \end{bmatrix}$ **27.** $\begin{bmatrix} 8 & -13 & 8 \\ 9 & 2 & -10 \\ 22 & -13 & -4 \end{bmatrix}$ **28.** $\begin{bmatrix} 9 & -31 \\ -6 & -3 \end{bmatrix}$ **29.** $\begin{bmatrix} \dfrac{1}{2} & -1 \\ -\dfrac{1}{6} & \dfrac{2}{3} \end{bmatrix}$ **30.** $\begin{bmatrix} -\dfrac{1}{2} & \dfrac{1}{2} \\ -\dfrac{1}{4} & \dfrac{3}{4} \end{bmatrix}$ **31.** $\begin{bmatrix} \dfrac{5}{7} & \dfrac{9}{7} & \dfrac{3}{7} \\ \dfrac{1}{7} & \dfrac{1}{7} & -\dfrac{2}{7} \\ \dfrac{3}{7} & -\dfrac{4}{7} & \dfrac{1}{7} \end{bmatrix}$ **32.** $\begin{bmatrix} \dfrac{3}{7} & \dfrac{1}{7} & -\dfrac{5}{7} \\ -\dfrac{4}{7} & \dfrac{1}{7} & \dfrac{9}{7} \\ \dfrac{1}{7} & \dfrac{2}{7} & \dfrac{3}{7} \end{bmatrix}$

33. Singular **34.** Singular **35.** $x = \dfrac{2}{5}, y = \dfrac{1}{10}$ **36.** $x = 1, y = \dfrac{3}{2}$ **37.** $x = 9, y = \dfrac{13}{3}, z = \dfrac{13}{3}$ **38.** Inconsistent **39.** $x = -\dfrac{1}{2}, y = -\dfrac{2}{3}, z = -\dfrac{3}{4}$

40. $x = \dfrac{1}{3}, y = \dfrac{4}{5}, z = -\dfrac{1}{15}$ **41.** $z = -1, x = y + 1, y$ is any real number **42.** $y = z, x = -\dfrac{1}{2}z, z$ is any real number

43. $x = 4, y = 2, z = 3, t = -2$ **44.** $x = -\dfrac{17}{15}, y = -\dfrac{1}{5}, z = \dfrac{22}{15}, t = -\dfrac{2}{15}$ **45.** 5 **46.** -12 **47.** 108 **48.** -19 **49.** -100 **50.** 11

51. $x = 2, y = -1$ **52.** $x = 0, y = \dfrac{5}{3}$ **53.** $x = 2, y = 3$ **54.** $x = 0, y = -3$ **55.** $x = -1, y = 2, z = -3$ **56.** $x = \dfrac{37}{9}, y = -\dfrac{19}{6}, z = \dfrac{13}{18}$

57. 16 **58.** -8 **59.** $\dfrac{-\dfrac{3}{2}}{x} + \dfrac{\dfrac{3}{2}}{x - 4}$ **60.** $\dfrac{\dfrac{2}{5}}{x + 2} + \dfrac{\dfrac{3}{5}}{x - 3}$ **61.** $\dfrac{-3}{x - 1} + \dfrac{3}{x} + \dfrac{4}{x^2}$ **62.** $\dfrac{4}{x - 2} + \dfrac{-2}{(x - 2)^2} + \dfrac{-4}{x - 1}$ **63.** $\dfrac{-\dfrac{1}{10}}{x + 1} + \dfrac{\dfrac{1}{10}x + \dfrac{9}{10}}{x^2 + 9}$

64. $\dfrac{\dfrac{6}{5}}{x - 2} + \dfrac{-\dfrac{6}{5}x + \dfrac{3}{5}}{x^2 + 1}$ **65.** $\dfrac{x}{x^2 + 4} + \dfrac{-4x}{(x^2 + 4)^2}$ **66.** $\dfrac{x}{x^2 + 16} + \dfrac{-16x + 1}{(x^2 + 16)^2}$ **67.** $\dfrac{\dfrac{1}{2}}{x^2 + 1} + \dfrac{\dfrac{1}{4}}{x - 1} + \dfrac{-\dfrac{1}{4}}{x + 1}$ **68.** $\dfrac{-\dfrac{4}{5}}{x^2 + 4} + \dfrac{\dfrac{2}{5}}{x + 1} + \dfrac{\dfrac{2}{5}}{x - 1}$

69. $x = -\dfrac{2}{5}, y = -\dfrac{11}{5}; x = -2, y = 1$ **70.** $x = 2, y = -2\sqrt{3}; x = 2, y = 2\sqrt{3}; x = -4, y = 0$ **71.** $x = 2\sqrt{2}, y = \sqrt{2}; x = -2\sqrt{2}, y = -\sqrt{2}$

72. $x = -\sqrt{3}, y = -2\sqrt{2}; x = -\sqrt{3}, y = 2\sqrt{2}; x = \sqrt{3}, y = -2\sqrt{2}; x = \sqrt{3}, y = 2\sqrt{2}$ **73.** $x = 0, y = 0; x = -3, y = 3; x = 3, y = 3$

74. $x = 0, y = -3; x = 0, y = 3$ **75.** $x = \sqrt{2}, y = -\sqrt{2}; x = -\sqrt{2}, y = \sqrt{2}; x = \dfrac{4}{3}\sqrt{2}, y = -\dfrac{2}{3}\sqrt{2}; x = -\dfrac{4}{3}\sqrt{2}, y = \dfrac{2}{3}\sqrt{2}$

76. $x = -\dfrac{5}{21}\sqrt{42}, y = -\dfrac{2}{7}\sqrt{42}; x = \dfrac{5}{21}\sqrt{42}, y = \dfrac{2}{7}\sqrt{42}; x = -2, y = 1; x = 2, y = -1$ **77.** $x = 1, y = -1$

78. $x = -2, y = 0; x = 1, y = 0; x = -1, y = 2$

79. **80.** **81.** **82.**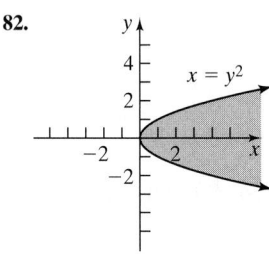

83. Unbounded **84.** Unbounded **85.** Bounded

86. Unbounded

87. Bounded

88. Bounded

89.

90.

91.

92.

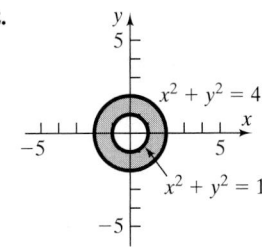

93. The maximum value is 32 when $x = 0$ and $y = 8$. **94.** The maximum value is 20 when $x = 2$ and $y = 4$.

95. The minimum value is 3 when $x = 1$ and $y = 0$. **96.** The minimum value is 4 when $x = 0$ and $y = 4$.

97. 10 **98.** A is any real number, $A \neq 10$. **99.** $y = -\dfrac{1}{3}x^2 - \dfrac{2}{3}x + 1$ **100.** $x^2 + 2x + y^2 + 2y - 3 = 0$

101. 70 pounds of \$3 coffee and 30 pounds of \$6 coffee **102.** 466.25 acres for corn and 533.75 acres for soybeans

103. 1 small, 5 medium, 2 large

104. (a)
$$\begin{aligned} x &\geq 0 \\ y &\geq 0 \\ 4x + 3y &\leq 960 \\ 2x + 3y &\leq 576 \end{aligned}$$

(b)

105. Speedboat: 36.67 km/hr; Aguarico River: 3.33 km/hr

106. 29.69 miles per hour

107. Bruce: 4 hours; Bryce: 2 hours; Marty: 8 hours

108. 12 mermaids and 12 dancing girls to have a profit of \$660; molding has 18 excess hours assigned to it.

109. 35 gasoline engines, 15 diesel engines; 15 gasoline engines, 0 diesel engines

Chapter Test *(page 491)*

1. $x = 3, y = -1$ **2.** Inconsistent. **3.** $x = -z + \dfrac{18}{7}, y = z - \dfrac{17}{7}$, where z can be any real number. **4.** $x = \dfrac{1}{3}, y = -2, z = 0$

5. $\begin{bmatrix} 4 & -5 & 1 & | & 0 \\ -2 & -1 & 6 & | & -19 \\ 1 & 5 & -5 & | & 10 \end{bmatrix}$ **6.** $\begin{cases} 3x + 2y + 4z = -6 \\ 1x + 0y + 8z = 2 \\ -2x + 1y + 3z = -11 \end{cases}$ or $\begin{cases} 3x + 2y + 4z = -6 \\ x + 8z = 2 \\ -2x + y + 3z = -11 \end{cases}$ **7.** $\begin{bmatrix} 6 & 4 \\ 1 & -11 \\ 5 & 12 \end{bmatrix}$ **8.** $\begin{bmatrix} -11 & -19 \\ -3 & 5 \\ 6 & -22 \end{bmatrix}$

9. The operation cannot be performed. **10.** $\begin{bmatrix} 16 & 17 \\ 3 & -10 \end{bmatrix}$ **11.** $\begin{bmatrix} 2 & -1 \\ -\dfrac{5}{2} & \dfrac{3}{2} \end{bmatrix}$ **12.** $B^{-1} = \begin{bmatrix} 3 & 3 & -4 \\ -2 & -2 & 3 \\ -4 & -5 & 7 \end{bmatrix}$ **13.** $x = \dfrac{1}{2}, y = 3$

14. The system is dependent and therefore has an infinite number of solutions. Any ordered pair satisfying the equation $x = \dfrac{1}{4}y - 7$ or $y = -4x + 28$ is a solution to the system. **15.** $x = 1, y = -2, z = 0$ **16.** Inconsistent **17.** -29 **18.** -12 **19.** $x = -2, y = -5$

20. $x = 1, y = -1, z = 4$ **21.** $(1, -3)$ and $(1, 3)$ **22.** $(3, 4)$ and $(1, 2)$

23.

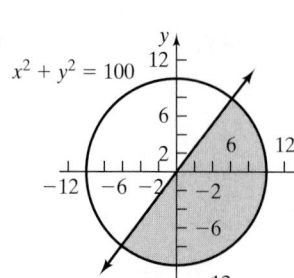

$x^2 + y^2 = 100$

$4x - 3y = 0$

24. $\dfrac{3}{x + 3} + \dfrac{-2}{(x + 3)^2}$

25. $\dfrac{-\dfrac{1}{3}}{x} + \dfrac{\dfrac{1}{3}x}{(x^2 + 3)} + \dfrac{5x}{(x^2 + 3)^2}$

26. The graph is unbounded. The corner points are $(4, 2)$ and $(8, 0)$.

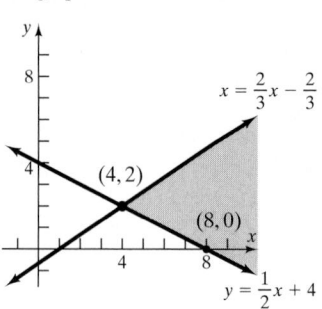

$x = \dfrac{2}{3}x - \dfrac{2}{3}$

$(4, 2)$

$(8, 0)$

$y = \dfrac{1}{2}x + 4$

27. The maximum value of z is 64, and it occurs at the point $(0, 8)$. **28.** Flare jeans cost \$24.50, camisoles cost \$8.50, and T-shirts cost \$6.00.

Cumulative Review *(page 493)*

1. $\left\{ 0, \dfrac{1}{2} \right\}$ **2.** $\{5\}$ **3.** $\left\{ -1, -\dfrac{1}{2}, 3 \right\}$ **4.** $\{-2\}$ **5.** $\left\{ \dfrac{5}{2} \right\}$ **6.** $\left\{ \dfrac{1}{\ln 3} \right\}$ **7.** Odd; symmetric with respect to origin

8. Center: $(1, -2)$; radius $= 4$

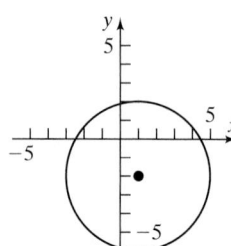

9. Domain: all real numbers
Range: $\{y | y > 1\}$
Horizontal asymptote: $y = 1$

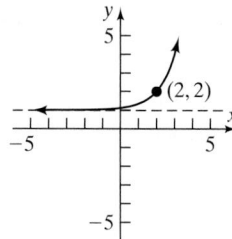

$(2, 2)$

10. $f^{-1}(x) = \dfrac{5}{x} - 2$

Domain of f: $\{x | x \neq -2\}$
Range of f: $\{y | y \neq 0\}$
Domain of f^{-1}: $\{x | x \neq 0\}$
Range of f^{-1}: $\{y | y \neq -2\}$

11. (a)

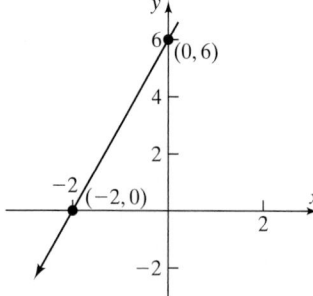

$(0, 6)$

$(-2, 0)$

(b)

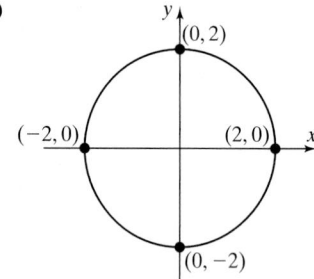

$(0, 2)$

$(-2, 0)$ $(2, 0)$

$(0, -2)$

(c)

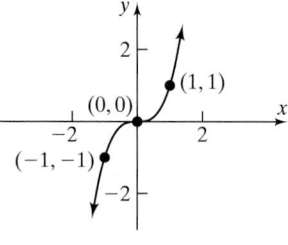

$(1, 1)$

$(0, 0)$

$(-1, -1)$

(d)

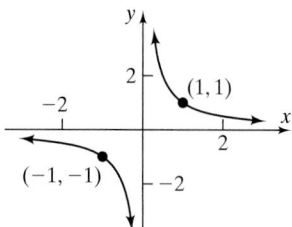

$(1, 1)$

$(-1, -1)$

(e)

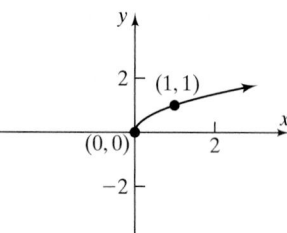

$(1, 1)$

$(0, 0)$

(f)

(g)

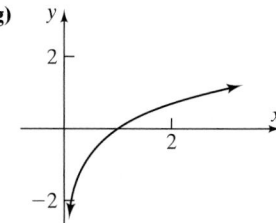

12. (a) -2.28 **(b)** Local maximum of 7 at $x = -1$; local minimum of 3 at $x = 1$ **(c)** $(-\infty, -1), (1, \infty)$

CHAPTER 7 Sequences; Induction; the Binomial Theorem

7.1 Assess Your Understanding (page 502)

3. sequence **4.** $3; 15$ **5.** 20 **6.** T **7.** T **8.** T **9.** 3,628,800 **10.** 362,880 **11.** 504 **12.** 132 **13.** 1260 **14.** 806,400

15. $s_1 = 1, s_2 = 2, s_3 = 3, s_4 = 4, s_5 = 5$ **16.** $s_1 = 2, s_2 = 5, s_3 = 10, s_4 = 17, s_5 = 26$ **17.** $a_1 = \dfrac{1}{3}, a_2 = \dfrac{1}{2}, a_3 = \dfrac{3}{5}, a_4 = \dfrac{2}{3}, a_5 = \dfrac{5}{7}$

18. $b_1 = \dfrac{3}{2}, b_2 = \dfrac{5}{4}, b_3 = \dfrac{7}{6}, b_4 = \dfrac{9}{8}, b_5 = \dfrac{11}{10}$ **19.** $c_1 = 1, c_2 = -4, c_3 = 9, c_4 = -16, c_5 = 25$ **20.** $d_1 = 1, d_2 = -\dfrac{2}{3}, d_3 = \dfrac{3}{5}, d_4 = -\dfrac{4}{7}, d_5 = \dfrac{5}{9}$

21. $s_1 = \dfrac{1}{2}, s_2 = \dfrac{2}{5}, s_3 = \dfrac{2}{7}, s_4 = \dfrac{8}{41}, s_5 = \dfrac{8}{61}$ **22.** $s_1 = \dfrac{4}{3}, s_2 = \dfrac{16}{9}, s_3 = \dfrac{64}{27}, s_4 = \dfrac{256}{81}, s_5 = \dfrac{1024}{243}$ **23.** $t_1 = -\dfrac{1}{6}, t_2 = \dfrac{1}{12}, t_3 = -\dfrac{1}{20}, t_4 = \dfrac{1}{30}, t_5 = -\dfrac{1}{42}$

24. $a_1 = 3, a_2 = \dfrac{9}{2}, a_3 = 9, a_4 = \dfrac{81}{4}, a_5 = \dfrac{243}{5}$ **25.** $b_1 = \dfrac{1}{e}, b_2 = \dfrac{2}{e^2}, b_3 = \dfrac{3}{e^3}, b_4 = \dfrac{4}{e^4}, b_5 = \dfrac{5}{e^5}$ **26.** $c_1 = \dfrac{1}{2}, c_2 = 1, c_3 = \dfrac{9}{8}, c_4 = 1, c_5 = \dfrac{25}{32}$

27. $a_n = \dfrac{n}{n+1}$ **28.** $a_n = \dfrac{1}{n(n+1)}$ **29.** $a_n = \dfrac{1}{2^{n-1}}$ **30.** $a_n = \left(\dfrac{2}{3}\right)^n$ **31.** $a_n = (-1)^{n+1}$ **32.** $a_n = \begin{cases} n & \text{if } n \text{ is odd} \\ \dfrac{1}{n} & \text{if } n \text{ is even} \end{cases}$ or $a_n = n^{(-1)^{n-1}}$

33. $a_n = (-1)^{n+1}n$ **34.** $a_n = (-1)^{n-1}\cdot 2n$ **35.** $a_1 = 2, a_2 = 5, a_3 = 8, a_4 = 11, a_5 = 14$ **36.** $a_1 = 3, a_2 = 1, a_3 = 3, a_4 = 1, a_5 = 3$

37. $a_1 = -2, a_2 = 0, a_3 = 3, a_4 = 7, a_5 = 12$ **38.** $a_1 = 1, a_2 = 1, a_3 = 2, a_4 = 2, a_5 = 3$ **39.** $a_1 = 5, a_2 = 10, a_3 = 20, a_4 = 40, a_5 = 80$

40. $a_1 = 2, a_2 = -2, a_3 = 2, a_4 = -2, a_5 = 2$ **41.** $a_1 = 3, a_2 = \dfrac{3}{2}, a_3 = \dfrac{1}{2}, a_4 = \dfrac{1}{8}, a_5 = \dfrac{1}{40}$ **42.** $a_1 = -2, a_2 = -4, a_3 = -9, a_4 = -23, a_5 = -64$

43. $a_1 = 1, a_2 = 2, a_3 = 2, a_4 = 4, a_5 = 8$ **44.** $a_1 = -1, a_2 = 1, a_3 = 2, a_4 = 9, a_5 = 47$ **45.** $a_1 = A, a_2 = A + d, a_3 = A + 2d, a_4 = A + 3d, a_5 = A + 4d$

46. $a_1 = A, a_2 = rA, a_3 = r^2A, a_4 = r^3A, a_5 = r^4A$

47. $a_1 = \sqrt{2}, a_2 = \sqrt{2 + \sqrt{2}}, a_3 = \sqrt{2 + \sqrt{2 + \sqrt{2}}},$
$a_4 = \sqrt{2 + \sqrt{2 + \sqrt{2 + \sqrt{2}}}},$
$a_5 = \sqrt{2 + \sqrt{2 + \sqrt{2 + \sqrt{2 + \sqrt{2}}}}}$

48. $a_1 = \sqrt{2}, a_2 = \sqrt{\dfrac{\sqrt{2}}{2}}, a_3 = \sqrt{\dfrac{\sqrt{\frac{\sqrt{2}}{2}}}{2}}, a_4 = \sqrt{\dfrac{\sqrt{\frac{\sqrt{\frac{\sqrt{2}}{2}}}{2}}}{2}}, a_5 = \sqrt{\dfrac{\sqrt{\frac{\sqrt{\frac{\sqrt{\frac{\sqrt{2}}{2}}}{2}}}{2}}}{2}}$

49. $3 + 4 + \cdots + (n+2)$ **50.** $3 + 5 + 7 + \cdots + (2n+1)$ **51.** $\dfrac{1}{2} + 2 + \dfrac{9}{2} + \cdots + \dfrac{n^2}{2}$ **52.** $4 + 9 + 16 + \cdots + (n+1)^2$

53. $1 + \dfrac{1}{3} + \dfrac{1}{9} + \cdots + \dfrac{1}{3^n}$ **54.** $1 + \dfrac{3}{2} + \dfrac{9}{4} + \cdots + \left(\dfrac{3}{2}\right)^n$ **55.** $\dfrac{1}{3} + \dfrac{1}{9} + \cdots + \dfrac{1}{3^n}$ **56.** $1 + 3 + 5 + \cdots + (2n-1)$

57. $\ln 2 - \ln 3 + \ln 4 - \cdots + (-1)^n \ln n$ **58.** $8 - 16 + 32 - \cdots + (-1)^{n+1}2^n$ **59.** $\displaystyle\sum_{k=1}^{20} k$ **60.** $\displaystyle\sum_{k=1}^{8} k^3$ **61.** $\displaystyle\sum_{k=1}^{13} \dfrac{k}{k+1}$ **62.** $\displaystyle\sum_{k=1}^{12} (2k-1)$

63. $\displaystyle\sum_{k=0}^{6} (-1)^k \left(\dfrac{1}{3^k}\right)$ **64.** $\displaystyle\sum_{k=1}^{11} (-1)^{k-1} \left(\dfrac{2}{3}\right)^k$ **65.** $\displaystyle\sum_{k=1}^{n} \dfrac{3^k}{k}$ **66.** $\displaystyle\sum_{k=1}^{n} \dfrac{k}{e^k}$ **67.** $\displaystyle\sum_{k=0}^{n} (a + kd)$ or $\displaystyle\sum_{k=1}^{n+1} [a + (k-1)d]$ **68.** $\displaystyle\sum_{k=0}^{n-1} ar^k$ or $\displaystyle\sum_{k=1}^{n} ar^{k-1}$ **69.** 50

70. 160 **71.** 21 **72.** -10 **73.** 90 **74.** 21 **75.** 26 **76.** 10 **77.** 42 **78.** 60 **79.** 96 **80.** 44 **81.** \$2930 **82.** 2162 **83.** \$18,058.03

84. 231 tons **85.** 21 pairs

86. (a) $u_1 = \dfrac{(1 + \sqrt{5}) - (1 - \sqrt{5})}{2\sqrt{5}} = \dfrac{2\sqrt{5}}{2\sqrt{5}} = 1$;

$u_2 = \dfrac{(1 + \sqrt{5})^2 - (1 - \sqrt{5})^2}{2^2\sqrt{5}}$

$= \dfrac{(1 + 2\sqrt{5} + 5) - (1 - 2\sqrt{5} + 5)}{4\sqrt{5}} = \dfrac{4\sqrt{5}}{4\sqrt{5}} = 1$

(b) Set $A = 1 + \sqrt{5}, B = 1 - \sqrt{5}$.

$u_{n+2} = \dfrac{A^{n+2} - B^{n+2}}{2^{n+2}\sqrt{5}}$

$= \dfrac{A^2 \cdot A^n - B^2 \cdot B^n}{2^{n+2}\sqrt{5}}$

$= \dfrac{(1 + \sqrt{5})^2 \cdot A^n - (1 - \sqrt{5})^2 B^n}{2^{n+2}\sqrt{5}}$

$= \dfrac{(6 + 2\sqrt{5}) \cdot A^n - (6 - 2\sqrt{5}) B^n}{2^{n+2}\sqrt{5}}$

$= \dfrac{(3 + \sqrt{5}) A^n - (3 - \sqrt{5}) B^n}{2^{n+1}\sqrt{5}}$

$= \dfrac{(1 + \sqrt{5}) A^n - (1 - \sqrt{5}) B^n + 2A^n - 2B^n}{2^{n+1}\sqrt{5}}$

$= \dfrac{A^{n+1} - B^{n+1}}{2^{n+1}\sqrt{5}} + \dfrac{A^n - B^n}{2^n\sqrt{5}}$

$= u_{n+1} + u_n$ **(c)** This is a Fibonacci sequence since $u_1 = 1 = u_2$ and $u_{n+2} = u_{n+1} + u_n$.

87. Fibonacci sequence
88. (a) $1, 1, 2, 3, 5, 8, 13, 21, 34, 55$
(b) $1, 2, \dfrac{3}{2}, \dfrac{5}{3}, \dfrac{8}{5}, \dfrac{13}{8}, \dfrac{21}{13}, \dfrac{34}{21}, \dfrac{55}{34}, \dfrac{89}{55}$ **(c)** $\dfrac{1 + \sqrt{5}}{2}$
(d) $1, \dfrac{1}{2}, \dfrac{2}{3}, \dfrac{3}{5}, \dfrac{5}{8}, \dfrac{8}{13}, \dfrac{13}{21}, \dfrac{21}{34}, \dfrac{34}{55}, \dfrac{55}{89}$ **(e)** $\dfrac{2}{1 + \sqrt{5}}$
89. (a) 3.630170833 **(b)** 3.669060828 **(c)** 3.669296668 **(d)** 12
90. (a) -0.824 **(b)** 0.1602688 **(c)** 0.0907179533 **(d)** 17
91. $2S = \underbrace{(1 + n) + (1 + n) + \cdots + (n + 1)}_{n \text{ terms}} = n(n + 1)$
$2S = n(n + 1)$
$S = \dfrac{1}{2}n(n + 1)$

7.2 Assess Your Understanding (page 509)

1. arithmetic **2.** F **3.** $s_n - s_{n-1} = (n + 4) - [(n - 1) + 4] = n + 4 - (n + 3) = n + 4 - n - 3 = 1$, a constant; $d = 1$; $s_1 = 5, s_2 = 6, s_3 = 7, s_4 = 8$

4. $s_n - s_{n-1} = (n - 5) - [(n - 1) - 5] = n - 5 - (n - 6) = n - 5 - n + 6 = 1$, a constant; $d = 1$; $s_1 = -4, s_2 = -3, s_3 = -2, s_4 = -1$

5. $a_n - a_{n-1} = (2n - 5) - [2(n - 1) - 5] = 2n - 5 - (2n - 2 - 5) = 2n - 5 - (2n - 7) = 2n - 5 - 2n + 7 = 2$, a constant; $d = 2$; $a_1 = -3, a_2 = -1, a_3 = 1, a_4 = 3$

6. $b_n - b_{n-1} = (3n + 1) - [3(n - 1) + 1] = 3n + 1 - (3n - 3 + 1) = 3n + 1 - (3n - 2) = 3n + 1 - 3n + 2 = 3$, a constant; $d = 3$; $b_1 = 4, b_2 = 7, b_3 = 10, b_4 = 13$

7. $c_n - c_{n-1} = (6 - 2n) - [6 - 2(n - 1)] = 6 - 2n - (6 - 2n + 2) = 6 - 2n - (8 - 2n) = 6 - 2n - 8 + 2n = -2$, a constant; $d = -2$; $c_1 = 4, c_2 = 2, c_3 = 0, c_4 = -2$

8. $d_n - d_{n-1} = (4 - 2n) - [4 - 2(n - 1)] = 4 - 2n - (4 - 2n + 2) = 4 - 2n - (6 - 2n) = 4 - 2n - 6 + 2n = -2$, a constant; $d = -2$; $d_1 = 2, d_2 = 0, d_3 = -2, d_4 = -4$

9. $t_n - t_{n-1} = \left(\dfrac{1}{2} - \dfrac{1}{3}n\right) - \left[\dfrac{1}{2} - \dfrac{1}{3}(n - 1)\right] = \dfrac{1}{2} - \dfrac{1}{3}n - \left(\dfrac{1}{2} - \dfrac{1}{3}n + \dfrac{1}{3}\right) = \dfrac{1}{2} - \dfrac{1}{3}n - \left(\dfrac{5}{6} - \dfrac{1}{3}n\right) = \dfrac{1}{2} - \dfrac{1}{3}n - \dfrac{5}{6} + \dfrac{1}{3}n = -\dfrac{1}{3}$, a constant; $d = -\dfrac{1}{3}$; $t_1 = \dfrac{1}{6}, t_2 = -\dfrac{1}{6}, t_3 = -\dfrac{1}{2}, t_4 = -\dfrac{5}{6}$

10. $t_n - t_{n-1} = \left(\dfrac{2}{3} + \dfrac{n}{4}\right) - \left(\dfrac{2}{3} + \dfrac{n - 1}{4}\right) = \dfrac{2}{3} + \dfrac{n}{4} - \left(\dfrac{2}{3} + \dfrac{n}{4} - \dfrac{1}{4}\right) = \dfrac{2}{3} + \dfrac{n}{4} - \left(\dfrac{5}{12} + \dfrac{n}{4}\right) = \dfrac{2}{3} + \dfrac{n}{4} - \dfrac{5}{12} - \dfrac{n}{4} = \dfrac{1}{4}$, a constant; $d = \dfrac{1}{4}$; $t_1 = \dfrac{11}{12}, t_2 = \dfrac{7}{6}, t_3 = \dfrac{17}{12}, t_4 = \dfrac{5}{3}$

11. $s_n - s_{n-1} = \ln 3^n - \ln 3^{n-1} = n \ln 3 - (n - 1) \ln 3 = n \ln 3 - (n \ln 3 - \ln 3) = n \ln 3 - n \ln 3 + \ln 3 = \ln 3$, a constant; $d = \ln 3$; $s_1 = \ln 3, s_2 = 2 \ln 3, s_3 = 3 \ln 3, s_4 = 4 \ln 3$

12. $s_n - s_{n-1} = e^{\ln n} - e^{\ln(n-1)} = n - (n - 1) = n - n + 1 = 1$, a constant; $d = 1$; $s_1 = 1, s_2 = 2, s_3 = 3, s_4 = 4$ **13.** $a_n = 3n - 1$; $a_5 = 14$

14. $a_n = 4n - 6$; $a_5 = 14$ **15.** $a_n = 8 - 3n$; $a_5 = -7$ **16.** $a_n = 8 - 2n$; $a_5 = -2$ **17.** $a_n = \dfrac{1}{2}(n - 1)$; $a_5 = 2$ **18.** $a_n = \dfrac{4}{3} - \dfrac{1}{3}n$; $a_5 = -\dfrac{1}{3}$

19. $a_n = \sqrt{2}n$; $a_5 = 5\sqrt{2}$ **20.** $a_n = (n - 1)\pi$; $a_5 = 4\pi$ **21.** $a_{12} = 24$ **22.** $a_8 = 13$ **23.** $a_{10} = -26$ **24.** $a_9 = -35$ **25.** $a_8 = a + 7b$

26. $a_7 = 14\sqrt{5}$ **27.** $a_1 = -13; d = 3; a_n = a_{n-1} + 3; a_n = -16 + 3n$ **28.** $a_1 = -3; d = 2; a_n = a_{n-1} + 2; a_n = -5 + 2n$

29. $a_1 = -53; d = 6; a_n = a_{n-1} + 6; a_n = -59 + 6n$ **30.** $a_1 = 74; d = -10; a_n = a_{n-1} - 10; a_n = 84 - 10n$

31. $a_1 = 28; d = -2; a_n = a_{n-1} - 2; a_n = 30 - 2n$ **32.** $a_1 = -18; d = 4; a_n = a_{n-1} + 4; a_n = -22 + 4n$

33. $a_1 = 25; d = -2; a_n = a_{n-1} - 2; a_n = 27 - 2n$ **34.** $a_1 = -40; d = 4; a_n = a_{n-1} + 4; a_n = -44 + 4n$ **35.** n^2 **36.** $n(n + 1)$ **37.** $\dfrac{n}{2}(9 + 5n)$

38. $n(2n - 3)$ **39.** 1260 **40.** 900 **41.** 324 **42.** 301

43.
```
sum(seq(3.45n+4.
12,n,1,20,1))
          806.9
```

44.
```
sum(seq(2.67n-1.
23,n,1,25,1))
            837
■
```

45.
```
sum(seq(2.4n+.4,
n,1,15,1))
            294
```

46.
```
sum(seq(3.5+1.9n
,n,1,15,1))
          280.5
■
```

47.
```
sum(seq(2.58n+2.
32,n,1,25,1))
          896.5
```

48.
```
sum(seq(0.52+3.1
9n,n,1,25,1))
        1049.75
■
```

49. $-\dfrac{3}{2}$

50. 1
51. 1185 seats **52.** 2160 seats
53. 210 of beige and 190 blue
54. **(a)** 42 bricks **(b)** 2130 bricks
55. 30 rows **56.** 8 yr

Historical Problems (page 517)

1. $1\dfrac{2}{3}$ loaves, $10\dfrac{5}{6}$ loaves, 20 loaves, $29\dfrac{1}{6}$ loaves, $38\dfrac{1}{3}$ loaves **2.** **(a)** 1 person **(b)** 2401 kittens **(c)** 2800

7.3 Assess Your Understanding (page 518)

3. geometric **4.** $\dfrac{a}{1 - r}$ **5.** annuity **6.** True **7.** False **8.** True **9.** $r = 3; s_1 = 3, s_2 = 9, s_3 = 27, s_4 = 81$

10. $r = -5; s_1 = -5, s_2 = 25, s_3 = -125, s_4 = 625$ **11.** $r = \dfrac{1}{2}; a_1 = -\dfrac{3}{2}, a_2 = -\dfrac{3}{4}, a_3 = -\dfrac{3}{8}, a_4 = -\dfrac{3}{16}$

12. $r = \dfrac{5}{2}; b_1 = \dfrac{5}{2}, b_2 = \dfrac{25}{4}, b_3 = \dfrac{125}{8}, b_4 = \dfrac{625}{16}$ **13.** $r = 2; c_1 = \dfrac{1}{4}, c_2 = \dfrac{1}{2}, c_3 = 1, c_4 = 2$ **14.** $r = 3; d_1 = \dfrac{1}{3}, d_2 = 1, d_3 = 3, d_4 = 9$

15. $r = 2^{1/3}; e_1 = 2^{1/3}, e_2 = 2^{2/3}, e_3 = 2, e_4 = 2^{4/3}$ **16.** $r = 9; f_1 = 9, f_2 = 81, f_3 = 729, f_4 = 6561$ **17.** $r = \dfrac{3}{2}; t_1 = \dfrac{1}{2}, t_2 = \dfrac{3}{4}, t_3 = \dfrac{9}{8}, t_4 = \dfrac{27}{16}$

18. $r = \dfrac{2}{3}; u_1 = 2, u_2 = \dfrac{4}{3}, u_3 = \dfrac{8}{9}, u_4 = \dfrac{16}{27}$ **19.** Arithmetic; $d = 1$ **20.** Arithmetic; $d = 2$ **21.** Neither **22.** Neither **23.** Arithmetic; $d = -\dfrac{2}{3}$

24. Arithmetic; $d = -\dfrac{3}{4}$ **25.** Neither **26.** Arithmetic; $d = 2$ **27.** Geometric; $r = \dfrac{2}{3}$ **28.** Geometric; $r = \dfrac{5}{4}$ **29.** Geometric; $r = 2$

30. Neither **31.** Geometric; $r = 3^{1/2}$ **32.** Geometric; $r = -1$ **33.** $a_5 = 162; a_n = 2 \cdot 3^{n-1}$ **34.** $a_5 = -512; a_n = -2 \cdot 4^{n-1}$

35. $a_5 = 5; a_n = 5 \cdot (-1)^{n-1}$ **36.** $a_5 = 96; a_n = 6 \cdot (-2)^{n-1}$ **37.** $a_5 = 0; a_n = 0$ **38.** $a_5 = \dfrac{1}{81}; a_n = \left(-\dfrac{1}{3}\right)^{n-1}$ **39.** $a_5 = 4\sqrt{2}; a_n = \left(\sqrt{2}\right)^n$

40. $a_5 = 0; a_n = 0$ **41.** $a_7 = \dfrac{1}{64}$ **42.** $a_8 = 2187$ **43.** $a_9 = 1$ **44.** $a_{10} = 512$ **45.** $a_8 = 0.00000004$ **46.** $a_7 = 100{,}000$ **47.** $-\dfrac{1}{4}(1 - 2^n)$

48. $-\dfrac{1}{6}(1 - 3^n)$ **49.** $2\left[1 - \left(\dfrac{2}{3}\right)^n\right]$ **50.** $-2(1 - 3^n)$ **51.** $1 - 2^n$ **52.** $5\left[1 - \left(\dfrac{3}{5}\right)^n\right]$

53.
```
(1/4)sum(seq(2^n
,n,0,14,1))
        8191.75
```
54.
```
sum(seq(3^n/9,n,
1,15,1))
      2391484.333
■
```
55.
```
sum(seq((2/3)^n,
n,1,15,1))
       1.995432683
```
56.
```
sum(seq(4*3^(n-1
),n,1,15,1))
         28697812
■
```

57.
```
-1sum(seq(2^n,n,
0,14,1))
          -32767
```
58.
```
sum(seq(2*(3/5)^
n,n,0,15,1))
       4.998589445
■
```
59. $\dfrac{3}{2}$ **60.** 6 **61.** 16 **62.** 9 **63.** $\dfrac{8}{5}$ **64.** $\dfrac{4}{7}$ **65.** $\dfrac{20}{3}$ **66.** 12

67. $\dfrac{18}{5}$ **68.** $\dfrac{8}{3}$ **69.** -4 **70.** 2 **71.** \$21,879.11 **72.** \$6655.58

73. (a) 0.775 ft **(b)** 8th **(c)** 15.88 ft **(d)** 20 ft

74. (a) 15.36 ft **(b)** $30(0.8)^n$ **(c)** 19 times **(d)** 270 ft

75. \$349,496.41 **76.** \$16,712.73 **77.** \$96,885.98 **78.** \$66,438.85 **79.** \$305.10 **80.** \$312.44 **81.** 1.845×10^{19} **82.** $\dfrac{1}{3}$ **83.** 10 **84.** 20

85. \$72.67 per share **86.** \$39.64 per share **87.** December 20, 2001 (111 days); \$9999.92 **88.** Option A results in a higher salary in year 5 (\$25,250 versus \$24,761); Option B results in a higher 5-year total (\$116,801 versus \$112,742) **89.** Option B results in more money (\$524.287 versus \$500,500). **90.** Option 2 results in the most: \$16,038,304; Option 1 results in the least: \$14,700,000. **91.** Yes. A constant sequence is both arithmetic and geometric. For example, 3, 3, 3, \ldots is an arithmetic sequence with $a_1 = 3$ and $d = 0$ and is a geometric sequence with $a_1 = 3$ and $r = 1$.

7.4 Assess Your Understanding (page 523)

1. (I) $n = 1: 2(1) = 2$ and $1(1 + 1) = 2$

(II) If $2 + 4 + 6 + \cdots + 2k = k(k + 1)$, then $2 + 4 + 6 + \cdots + 2k + 2(k + 1) = (2 + 4 + 6 + \cdots + 2k) + 2(k + 1)$
$= k(k + 1) + 2(k + 1) = k^2 + 3k + 2 = (k + 1)(k + 2) = (k + 1)[(k + 1) + 1]$.

2. (I) $n = 1: 4(1) - 3 = 1$ and $1(2 - 1) = 1$

(II) If $1 + 5 + 9 + \cdots + (4k - 3) = k(2k - 1)$, then $1 + 5 + 9 + \cdots + (4k - 3) + [4(k + 1) - 3]$
$= [1 + 5 + 9 + \cdots + (4k - 3)] + (4k + 1) = k(2k - 1) + (4k + 1) = 2k^2 + 3k + 1 = (k + 1)(2k + 1)$
$= (k + 1)[2(k + 1) - 1]$.

3. (I) $n = 1: 1 + 2 = 3$ and $\dfrac{1}{2}(1)(1 + 5) = \dfrac{1}{2}(6) = 3$

(II) If $3 + 4 + 5 + \cdots + (k + 2) = \dfrac{1}{2}k(k + 5)$, then $3 + 4 + 5 + \cdots + (k + 2) + [(k + 1) + 2]$

$= [3 + 4 + 5 + \cdots + (k + 2)] + (k + 3) = \dfrac{1}{2}k(k + 5) + k + 3 = \dfrac{1}{2}(k^2 + 7k + 6) = \dfrac{1}{2}(k + 1)(k + 6) = \dfrac{1}{2}(k + 1)[(k + 1) + 5]$.

4. (I) $n = 1: 2(1) + 1 = 3$ and $1(1 + 2) = 3$

(II) If $3 + 5 + 7 + \cdots + (2k + 1) = k(k + 2)$, then $3 + 5 + 7 + \cdots + (2k + 1) + [2(k + 1) + 1]$
$= [3 + 5 + 7 + \cdots + (2k + 1)] + (2k + 3) = k(k + 2) + (2k + 3) = k^2 + 4k + 3 = (k + 1)(k + 3)$
$= (k + 1)[(k + 1) + 2]$.

5. (I) $n = 1: 3(1) - 1 = 2$ and $\dfrac{1}{2}(1)[3(1) + 1] = \dfrac{1}{2}(4) = 2$

(II) If $2 + 5 + 8 + \cdots + (3k - 1) = \dfrac{1}{2}k(3k + 1)$, then $2 + 5 + 8 + \cdots + (3k - 1) + [3(k + 1) - 1]$

$= [2 + 5 + 8 + \cdots + (3k - 1)] + (3k + 2) = \dfrac{1}{2}k(3k + 1) + (3k + 2) = \dfrac{1}{2}(3k^2 + 7k + 4) = \dfrac{1}{2}(k + 1)(3k + 4)$

$= \dfrac{1}{2}(k + 1)[3(k + 1) + 1]$.

6. (I) $n = 1: 3(1) - 2 = 1$ and $\dfrac{1}{2}(1)[3(1) - 1] = 1$

(II) If $1 + 4 + 7 + \cdots + (3k - 2) = \dfrac{1}{2}k(3k - 1)$, then $1 + 4 + 7 + \cdots + (3k - 2) + [3(k + 1) - 2]$

$= [1 + 4 + 7 + \cdots + (3k - 2)] + (3k + 1) = \dfrac{1}{2}k(3k - 1) + (3k + 1) = \dfrac{1}{2}(3k^2 + 5k + 2) = \dfrac{1}{2}(k + 1)(3k + 2)$

$= \dfrac{1}{2}(k + 1)[3(k + 1) - 1]$.

7. (I) $n = 1: 2^{1-1} = 1$ and $2^1 - 1 = 1$

(II) If $1 + 2 + 2^2 + \cdots + 2^{k-1} = 2^k - 1$, then $1 + 2 + 2^2 + \cdots + 2^{k-1} + 2^{(k+1)-1} = (1 + 2 + 2^2 + \cdots + 2^{k-1}) + 2^k$
$= 2^k - 1 + 2^k = 2(2^k) - 1 = 2^{k+1} - 1$.

8. (I) $n = 1: 3^{1-1} = 1$ and $\frac{1}{2}(3^1 - 1) = 1$

(II) If $1 + 3 + 3^2 + \cdots + 3^{k-1} = \frac{1}{2}(3^k - 1)$, then $1 + 3 + 3^2 + \cdots + 3^{k-1} + 3^{(k+1)-1} = (1 + 3 + 3^2 + \cdots + 3^{k-1}) + 3^k$

$$= \frac{1}{2}(3^k - 1) + 3^k = \frac{1}{2}[3^k - 1 + 2(3^k)] = \frac{1}{2}[3(3^k) - 1] = \frac{1}{2}(3^{k+1} - 1).$$

9. (I) $n = 1: 4^{1-1} = 1$ and $\frac{1}{3}(4^1 - 1) = \frac{1}{3}(3) = 1$

(II) If $1 + 4 + 4^2 + \cdots + 4^{k-1} = \frac{1}{3}(4^k - 1)$, then $1 + 4 + 4^2 + \cdots + 4^{k-1} + 4^{(k+1)-1} = (1 + 4 + 4^2 + \cdots + 4^{k-1}) + 4^k$

$$= \frac{1}{3}(4^k - 1) + 4^k = \frac{1}{3}[4^k - 1 + 3(4^k)] = \frac{1}{3}[4(4^k) - 1] = \frac{1}{3}(4^{k+1} - 1).$$

10. (I) $n = 1: 5^{1-1} = 1$ and $\frac{1}{4}(5^1 - 1) = \frac{1}{4}(4) = 1$

(II) If $1 + 5 + 5^2 + \cdots + 5^{k-1} = \frac{1}{4}(5^k - 1)$, then $1 + 5 + 5^2 + \cdots + 5^{k-1} + 5^{(k+1)-1} = (1 + 5 + 5^2 + \cdots + 5^{k-1}) + 5^k$

$$= \frac{1}{4}(5^k - 1) + 5^k = \frac{1}{4}[5^k - 1 + 4(5^k)] = \frac{1}{4}[5(5^k) - 1] = \frac{1}{4}(5^{k+1} - 1).$$

11. (I) $n = 1: \frac{1}{1 \cdot 2} = \frac{1}{2}$ and $\frac{1}{1+1} = \frac{1}{2}$

(II) If $\frac{1}{1 \cdot 2} + \frac{1}{2 \cdot 3} + \frac{1}{3 \cdot 4} + \cdots + \frac{1}{k(k+1)} = \frac{k}{k+1}$, then $\frac{1}{1 \cdot 2} + \frac{1}{2 \cdot 3} + \frac{1}{3 \cdot 4} + \cdots + \frac{1}{k(k+1)} + \frac{1}{(k+1)[(k+1)+1]}$

$$= \left[\frac{1}{1 \cdot 2} + \frac{1}{2 \cdot 3} + \frac{1}{3 \cdot 4} + \cdots + \frac{1}{k(k+1)}\right] + \frac{1}{(k+1)(k+2)} = \frac{k}{k+1} + \frac{1}{(k+1)(k+2)} = \frac{k(k+2) + 1}{(k+1)(k+2)}$$

$$= \frac{k^2 + 2k + 1}{(k+1)(k+2)} = \frac{(k+1)^2}{(k+1)(k+2)} = \frac{k+1}{k+2} = \frac{k+1}{(k+1)+1}.$$

12. (I) $n = 1: \frac{1}{(2 \cdot 1 - 1)(2 \cdot 1 + 1)} = \frac{1}{3}$ and $\frac{1}{2 \cdot 1 + 1} = \frac{1}{3}$

(II) If $\frac{1}{1 \cdot 3} + \frac{1}{3 \cdot 5} + \frac{1}{5 \cdot 7} + \cdots + \frac{1}{(2k-1)(2k+1)} = \frac{k}{2k+1}$, then $\frac{1}{1 \cdot 3} + \frac{1}{3 \cdot 5} + \frac{1}{5 \cdot 7} + \cdots + \frac{1}{(2k-1)(2k+1)}$

$$+ \frac{1}{[2(k+1)-1][2(k+1)+1]} = \left[\frac{1}{1 \cdot 3} + \frac{1}{3 \cdot 5} + \frac{1}{5 \cdot 7} + \cdots + \frac{1}{(2k-1)(2k+1)}\right] + \frac{1}{(2k+1)(2k+3)}$$

$$= \frac{k}{2k+1} + \frac{1}{(2k+1)(2k+3)} = \frac{k(2k+3) + 1}{(2k+1)(2k+3)} = \frac{2k^2 + 3k + 1}{(2k+1)(2k+3)} = \frac{(2k+1)(k+1)}{(2k+1)(2k+3)} = \frac{k+1}{2k+3} = \frac{k+1}{2(k+1)+1}.$$

13. (I) $n = 1: 1^2 = 1$ and $\frac{1}{6} \cdot 1 \cdot 2 \cdot 3 = 1$

(II) If $1^2 + 2^2 + 3^2 + \cdots + k^2 = \frac{1}{6}k(k+1)(2k+1)$, then $1^2 + 2^2 + 3^2 + \cdots + k^2 + (k+1)^2$

$$= (1^2 + 2^2 + 3^2 + \cdots + k^2) + (k+1)^2 = \frac{1}{6}k(k+1)(2k+1) + (k+1)^2 = \frac{1}{6}(2k^3 + 9k^2 + 13k + 6)$$

$$= \frac{1}{6}(k+1)(k+2)(2k+3) = \frac{1}{6}(k+1)[(k+1)+1][2(k+1)+1].$$

14. (I) $n = 1: 1^3 = 1$ and $\frac{1}{4} \cdot 1^2(1+1)^2 = 1$

(II) If $1^3 + 2^3 + 3^3 + \cdots + k^3 = \frac{1}{4}k^2(k+1)^2$, then $1^3 + 2^3 + 3^3 + \cdots + k^3 + (k+1)^3 = (1^3 + 2^3 + 3^3 + \cdots + k^3) + (k+1)^3$

$$= \frac{1}{4}k^2(k+1)^2 + (k+1)^3 = (k+1)^2\left(\frac{1}{4}k^2 + k + 1\right) = \frac{1}{4}(k+1)^2(k^2 + 4k + 4) = \frac{1}{4}(k+1)^2(k+2)^2 = \frac{1}{4}(k+1)^2[(k+1)+1]^2.$$

15. (I) $n = 1: 5 - 1 = 4$ and $\frac{1}{2}(1)(9 - 1) = \frac{1}{2} \cdot 8 = 4$

(II) If $4 + 3 + 2 + \cdots + (5 - k) = \frac{1}{2}k(9 - k)$, then $4 + 3 + 2 + \cdots + (5 - k) + [5 - (k+1)]$

$$= [4 + 3 + 2 + \cdots + (5 - k)] + 4 - k = \frac{1}{2}k(9 - k) + 4 - k = \frac{1}{2}(9k - k^2 + 8 - 2k) = \frac{1}{2}(-k^2 + 7k + 8)$$

$$= \frac{1}{2}(k+1)(8 - k) = \frac{1}{2}(k+1)[9 - (k+1)].$$

16. (I) $n = 1: -(1 + 1) = -2$ and $-\frac{1}{2}(1)(1 + 3) = -2$

(II) If $-2 - 3 - 4 - \cdots - (k+1) = -\frac{1}{2}k(k+3)$, then $-2 - 3 - 4 - \cdots - (k+1) - [(k+1)+1]$

$$= [-2 - 3 - 4 - \cdots - (k+1)] - (k+2) = -\frac{1}{2}k(k+3) - (k+2) = -\frac{1}{2}(k^2 + 3k + 2k + 4) = -\frac{1}{2}(k^2 + 5k + 4)$$

$$= -\frac{1}{2}(k+1)(k+4) = -\frac{1}{2}(k+1)[(k+1)+3].$$

17. (I) $n = 1: 1 \cdot (1 + 1) = 2$ and $\frac{1}{3} \cdot 1 \cdot 2 \cdot 3 = 2$

(II) If $1 \cdot 2 + 2 \cdot 3 + 3 \cdot 4 + \cdots + k(k+1) = \frac{1}{3}k(k+1)(k+2)$, then $1 \cdot 2 + 2 \cdot 3 + 3 \cdot 4 + \cdots + k(k+1)$

$+ (k + 1)[(k + 1) + 1] = [1 \cdot 2 + 2 \cdot 3 + 3 \cdot 4 + \cdots + k(k + 1)] + (k + 1)(k + 2)$

$= \frac{1}{3}k(k + 1)(k + 2) + \frac{1}{3} \cdot 3(k + 1)(k + 2) = \frac{1}{3}(k + 1)(k + 2)(k + 3) = \frac{1}{3}(k + 1)[(k + 1) + 1][(k + 1) + 2].$

18. (I) $n = 1$: $(2 \cdot 1 - 1)(2 \cdot 1) = 2$ and $\frac{1}{3} \cdot 1 \cdot (1 + 1)(4 \cdot 1 - 1) = 2$

(II) If $1 \cdot 2 + 3 \cdot 4 + 5 \cdot 6 + \cdots + (2k - 1)(2k) = \frac{1}{3}k(k + 1)(4k - 1)$, then

$1 \cdot 2 + 3 \cdot 4 + 5 \cdot 6 + \cdots + (2k - 1)(2k) + [2(k + 1) - 1][2(k + 1)] = [1 \cdot 2 + 3 \cdot 4 + 5 \cdot 6 + \cdots + (2k - 1)(2k)]$

$+ (2k + 1)(2k + 2) = \frac{1}{3}k(k + 1)(4k - 1) + (2k + 1)(2k + 2) = \frac{1}{3}(4k^3 + 3k^2 - k) + (4k^2 + 6k + 2)$

$= \frac{1}{3}(4k^3 + 3k^2 - k + 12k^2 + 18k + 6) = \frac{1}{3}(4k^3 + 15k^2 + 17k + 6) = \frac{1}{3}(k^2 + 3k + 2)(4k + 3)$

$= \frac{1}{3}(k + 1)(k + 2)(4k + 3) = \frac{1}{3}(k + 1)[(k + 1) + 1][4(k + 1) - 1].$

19. (I) $n = 1$: $1^2 + 1 = 2$, which is divisible by 2.
(II) If $k^2 + k$ is divisible by 2, then $(k + 1)^2 + (k + 1) = k^2 + 2k + 1 + k + 1 = (k^2 + k) + 2k + 2$. Since $k^2 + k$ is divisible by 2 and $2k + 2$ is divisible by 2, $(k + 1)^2 + (k + 1)$ is divisible by 2.

20. (I) $n = 1$: $1^3 + 2 \cdot 1 = 3$, which is divisible by 3.
(II) If $k^3 + 2k$ is divisible by 3, then $(k + 1)^3 + 2(k + 1) = k^3 + 3k^2 + 3k + 1 + 2k + 2 = (k^3 + 2k) + 3k^2 + 3k + 3$. Since $(k^3 + 2k)$ is divisible by 3 and $3k^2 + 3k + 3$ is divisible by 3, $(k + 1)^3 + 2(k + 1)$ is divisible by 3.

21. (I) $n = 1$: $1^2 - 1 + 2 = 2$ which is divisible by 2.
(II) If $k^2 - k + 2$ is divisible by 2, then $(k + 1)^2 - (k + 1) + 2 = k^2 + 2k + 1 - k - 1 + 2 = (k^2 - k + 2) + 2k$. Since $k^2 - k + 2$ is divisible by 2 and $2k$ is divisible by 2, $(k + 1)^2 - (k + 1) + 2$ is divisible by 2.

22. (I) $n = 1$: $1(1 + 1)(1 + 2) = 6$, which is divisible by 6.
(II) If $k(k + 1)(k + 2)$ is divisible by 6, then $(k + 1)[(k + 1) + 1][(k + 1) + 2] = (k + 1)(k + 2)(k + 3)$
$= (k + 1)(k + 2)k + (k + 1)(k + 2)3 = k(k + 1)(k + 2) + 3(k + 1)(k + 2)$. Now, $k(k + 1)(k + 2)$ is divisible by 6, and since $k + 1$ and $k + 2$ are consecutive integers, either $k + 1$ or $k + 2$ is even (divisible by 2).
In either case, $3(k + 1)(k + 2)$ is divisible by 6, and so $(k + 1)[(k + 1) + 1][(k + 1) + 2]$ is divisible by 6.

23. (I) $n = 1$: If $x > 1$, then $x^1 = x > 1$.
(II) Assume, for an arbitrary natural number k, that if $x > 1$ then $x^k > 1$. Multiply both sides of the inequality $x^k > 1$ by x. If $x > 1$, then $x^{k+1} > x > 1$.

24. (I) $n = 1$: If $0 < x < 1$, then $x^1 = x$ and $0 < x^1 < 1$.
(II) Assume, for an arbitrary natural number k, that if $0 < x < 1$ then $0 < x^k < 1$. Multiply all sides of the inequalities $0 < x^k < 1$ by x. If $0 < x < 1$, then $0 < x^k x < x < 1$ and $0 < x^{k+1} < 1$.

25. (I) $n = 1$: $a - b$ is a factor of $a^1 - b^1 = a - b$.
(II) If $a - b$ is a factor of $a^k - b^k$, then $a^{k+1} - b^{k+1} = a(a^k - b^k) + b^k(a - b)$.
Since $a - b$ is a factor of $a^k - b^k$ and $a - b$ is a factor of $a - b$, then $a - b$ is a factor of $a^{k+1} - b^{k+1}$.

26. (I) $n = 1$: $a^{2 \cdot 1 + 1} + b^{2 \cdot 1 + 1} = a^3 + b^3 = (a + b)(a^2 - ab + b^2)$, so $a + b$ is a factor of $a^3 + b^3$.
(II) If $a + b$ is a factor of $a^{2k+1} + b^{2k+1}$, then $a^{2(k+1)+1} + b^{2(k+1)+1} = a^{2k+3} + b^{2k+3} = a^{2k+1}a^2 + b^{2k+1}b^2$
$= a^{2k+1}a^2 + b^{2k+1}a^2 - b^{2k+1}a^2 + b^{2k+1}b^2 = a^2(a^{2k+1} + b^{2k+1}) - b^{2k+1}(a^2 - b^2) = a^2(a^{2k+1} + b^{2k+1}) - b^{2k+1}(a + b)(a - b)$.
Since $a + b$ is a factor of $a^{2k+1} + b^{2k+1}$ and $a + b$ is a factor of $b^{2k+1}(a + b)(a - b)$, then $a + b$ is a factor of $a^{2(k+1)+1} + b^{2(k+1)+1}$.

27. $n = 1$: $1^2 - 1 + 41 = 41$, which is a prime number.
$n = 41$: $41^2 - 41 + 41 = 1681 = 41^2$, which is not prime.

28. If $2 + 4 + 6 + \cdots + 2k = k^2 + k + 2$, then $2 + 4 + 6 + \cdots + 2k + 2(k + 1)$
$= (2 + 4 + 6 + \cdots + 2k) + 2k + 2 = k^2 + k + 2 + 2k + 2 = k^2 + 3k + 4 = (k^2 + 2k + 1) + (k + 1) + 2$
$= (k + 1)^2 + (k + 1) + 2$.
But $2 \cdot 1 = 2$ and $1^2 + 1 + 2 = 4$. The fact is that $2 + 4 + 6 + \cdots + 2n = n^2 + n$, not $n^2 + n + 2$ (Problem 1).

29. (I) $n = 1$: $ar^{1-1} = a \cdot 1 = a$ and $a \cdot \frac{1 - r^1}{1 - r} = a$, because $r \neq 1$.

(II) If $a + ar + ar^2 + \cdots + ar^{k-1} = a \cdot \frac{1 - r^k}{1 - r}$, then $a + ar + ar^2 + \cdots + ar^{k-1} + ar^{(k+1)-1} = (a + ar + ar^2 + \cdots + ar^{k-1}) + ar^k$

$= a \cdot \frac{1 - r^k}{1 - r} + ar^k = \frac{a(1 - r^k) + ar^k(1 - r)}{1 - r} = \frac{a - ar^k + ar^k - ar^{k+1}}{1 - r} = a \cdot \frac{1 - r^{k+1}}{1 - r}.$

30. (I) $n = 1$: $[a + (1 - 1)d] = a$ and $1 \cdot a + d\frac{1 \cdot (1 - 1)}{2} = a$

(II) If $a + (a + d) + (a + 2d) + \cdots + [a + (k - 1)d] = ka + d\frac{k(k - 1)}{2}$, then

$a + (a + d) + (a + 2d) + \cdots + [a + (k - 1)d] + [a + ((k + 1) - 1)d] = ka + d\frac{k(k - 1)}{2} + a + kd$

$= (k + 1)a + d\frac{k(k - 1) + 2k}{2} = (k + 1)a + d\frac{(k + 1)(k)}{2} = (k + 1)a + d\frac{(k + 1)[(k + 1) - 1]}{2}.$

31. (I) $n = 4$: The number of diagonals in a convex polygon of 4 sides is 2 and $\frac{1}{2} \cdot 4 \cdot (4 - 3) = 2$.

(II) If the number of diagonals in a convex polygon of k sides is $\frac{1}{2}k(k - 3)$, then that of $(k + 1)$ sides is increased by $(k + 1) - 2 = k - 1$. Thus,

the number of diagonals in a convex polygon of $(k + 1)$ sides is $\frac{1}{2}k(k - 3) + (k - 1)$

$$= \frac{1}{2}[k^2 - 3k + 2k - 2] = \frac{1}{2}[k^2 - k - 2] = \frac{1}{2}(k + 1)(k - 2) = \frac{1}{2}(k + 1)[(k + 1) - 3].$$

32. (I) $n = 3$: The sum of the angles of a triangle is $(3 - 2) \cdot 180° = 180°$.

(II) Assume for some $k \geq 3$ that the sum of the angles of a convex polygon of k sides is $(k - 2) \cdot 180°$. A convex polygon of $k + 1$ sides consists of a convex polygon of k sides plus a triangle (see the illustration). The sum of the angles is

$(k - 2) \cdot 180° + 180° = (k - 1) \cdot 180° = [(k + 1) - 2] \cdot 180°$.

k sides

$k + 1$ sides

7.5 Assess Your Understanding (page 529)

1. Pascal triangle **2.** 15 **3.** False **4.** Binomial Theorem **5.** 10 **6.** 35 **7.** 21 **8.** 36 **9.** 50 **10.** 4950 **11.** 1 **12.** 1
13. $\approx 1.8664 \times 10^{15}$ **14.** $\approx 4.1918 \times 10^{15}$ **15.** $\approx 1.4834 \times 10^{13}$ **16.** $\approx 1.7673 \times 10^{10}$ **17.** $x^5 + 5x^4 + 10x^3 + 10x^2 + 5x + 1$
18. $x^5 - 5x^4 + 10x^3 - 10x^2 + 5x - 1$ **19.** $x^6 - 12x^5 + 60x^4 - 160x^3 + 240x^2 - 192x + 64$ **20.** $x^5 + 15x^4 + 90x^3 + 270x^2 + 405x + 243$
21. $81x^4 + 108x^3 + 54x^2 + 12x + 1$ **22.** $32x^5 + 240x^4 + 720x^3 + 1080x^2 + 810x + 243$ **23.** $x^{10} + 5x^8y^2 + 10x^6y^4 + 10x^4y^6 + 5x^2y^8 + y^{10}$
24. $x^{12} - 6x^{10}y^2 + 15x^8y^4 - 20x^6y^6 + 15x^4y^8 - 6x^2y^{10} + y^{12}$ **25.** $x^3 + 6\sqrt{2}x^{5/2} + 30x^2 + 40\sqrt{2}x^{3/2} + 60x + 24\sqrt{2}x^{1/2} + 8$
26. $x^2 - 4\sqrt{3}x^{3/2} + 18x - 12\sqrt{3}x^{1/2} + 9$ **27.** $a^5x^5 + 5a^4bx^4y + 10a^3b^2x^3y^2 + 10a^2b^3x^2y^3 + 5ab^4xy^4 + b^5y^5$
28. $a^4x^4 - 4a^3bx^3y + 6a^2b^2x^2y^2 - 4ab^3xy^3 + b^4y^4$ **29.** 17,010 **30.** −262,440 **31.** −101,376 **32.** 1760 **33.** 41,472 **34.** −314,928 **35.** $2835x^3$
36. $189x^5$ **37.** $314,928x^7$ **38.** $48,384x^3$ **39.** 495 **40.** −84 **41.** 3360 **42.** 252 **43.** 1.00501 **44.** 0.98806

45. $\dbinom{n}{n-1} = \dfrac{n!}{(n-1)![n-(n-1)]!} = \dfrac{n!}{(n-1)!\,1!} = \dfrac{n \cdot (n-1)!}{(n-1)!} = n; \dbinom{n}{n} = \dfrac{n!}{n!\,(n-n)!} = \dfrac{n!}{n!\,0!} = \dfrac{n!}{n!} = 1$

46. $\dbinom{n}{n-j} = \dfrac{n!}{(n-j)![n-(n-j)]!} = \dfrac{n!}{(n-j)!\,j!} = \dfrac{n!}{j!\,(n-j)!} = \dbinom{n}{j}$

47. $2^n = (1 + 1)^n = \dbinom{n}{0}1^n + \dbinom{n}{1}(1)^{n-1}(1) + \cdots + \dbinom{n}{n}1^n = \dbinom{n}{0} + \dbinom{n}{1} + \cdots + \dbinom{n}{n}$

48. $0 = 0^n = (1 - 1)^n = \dbinom{n}{0}1^n + \dbinom{n}{1}(1)^{n-1}(-1) + \dbinom{n}{2}(1)^{n-2}(-1)^2 + \cdots + \dbinom{n}{n}(-1)^n = \dbinom{n}{0} - \dbinom{n}{1} + \dbinom{n}{2} - \cdots + (-1)^n\dbinom{n}{n}$

49. 1 **50.** $12! = 479,001,600 \approx 479,013,972.4$
$20! \approx 2.432902008 \times 10^{18} \approx 2.43292403 \times 10^{18}$
$25! \approx 1.551121004 \times 10^{25} \approx 1.551129917 \times 10^{25}$

Review Exercises (page 532)

1. $a_1 = -\dfrac{4}{3}, a_2 = \dfrac{5}{4}, a_3 = -\dfrac{6}{5}, a_4 = \dfrac{7}{6}, a_5 = -\dfrac{8}{7}$ **2.** $b_1 = 5, b_2 = -7, b_3 = 9, b_4 = -11, b_5 = 13$ **3.** $c_1 = 2, c_2 = 1, c_3 = \dfrac{8}{9}, c_4 = 1, c_5 = \dfrac{32}{25}$

4. $d_1 = e, d_2 = \dfrac{e^2}{2}, d_3 = \dfrac{e^3}{3}, d_4 = \dfrac{e^4}{4}, d_5 = \dfrac{e^5}{5}$ **5.** $a_1 = 3, a_2 = 2, a_3 = \dfrac{4}{3}, a_4 = \dfrac{8}{9}, a_5 = \dfrac{16}{27}$ **6.** $a_1 = 4, a_2 = -1, a_3 = \dfrac{1}{4}, a_4 = -\dfrac{1}{16}, a_5 = \dfrac{1}{64}$

7. $a_1 = 2, a_2 = 0, a_3 = 2, a_4 = 0, a_5 = 2$ **8.** $a_1 = -3, a_2 = 1, a_3 = 5, a_4 = 9, a_5 = 13$ **9.** $6 + 10 + 14 + 18 = 48$ **10.** $2 + (-1) + (-6) = -5$

11. $\displaystyle\sum_{k=1}^{13}(-1)^{k+1}\dfrac{1}{k}$ **12.** $\displaystyle\sum_{k=0}^{n}\dfrac{2^{k+1}}{3^k}$ **13.** Arithmetic; $d = 1; S_n = \dfrac{n}{2}(n + 11)$ **14.** Arithmetic; $d = 4; S_n = n(2n + 5)$ **15.** Neither **16.** Neither

17. Geometric; $r = 8; S_n = \dfrac{8}{7}(8^n - 1)$ **18.** Geometric; $r = 9; S_n = \dfrac{9}{8}(9^n - 1)$ **19.** Arithmetic; $d = 4; S_n = 2n(n - 1)$

20. Arithmetic; $d = -4; S_n = n(3 - 2n)$ **21.** Geometric; $r = \dfrac{1}{2}; S_n = 6\left[1 - \left(\dfrac{1}{2}\right)^n\right]$ **22.** Geometric; $r = -\dfrac{1}{3}; S_n = \dfrac{15}{4}\left[1 - \left(-\dfrac{1}{3}\right)^n\right]$

23. Neither **24.** Neither **25.** 115 **26.** 50 **27.** 75 **28.** −18 **29.** $\dfrac{1093}{2187} \approx 0.49977$ **30.** 682 **31.** 35 **32.** −13 **33.** $\dfrac{1}{10^{10}}$ **34.** 1024 **35.** $9\sqrt{2}$

36. $16\sqrt{2}$ **37.** $\{a_n\} = \{5n - 4\}$ **38.** $\{a_n\} = \{4 - 3n\}$ **39.** $\{a_n\} = \{n - 10\}$ **40.** $\{a_n\} = \{2n + 6\}$ **41.** $\dfrac{9}{2}$ **42.** 4 **43.** $\dfrac{4}{3}$ **44.** $\dfrac{18}{5}$ **45.** 8 **46.** $\dfrac{12}{7}$

47. (I) $n = 1: 3 \cdot 1 = 3$ and $\dfrac{3 \cdot 1}{2}(1 + 1) = 3$

(II) If $3 + 6 + 9 + \cdots + 3k = \dfrac{3k}{2}(k + 1)$, then $3 + 6 + 9 + \cdots + 3k + 3(k + 1) = (3 + 6 + 9 + \cdots + 3k) + (3k + 3)$

$$= \dfrac{3k}{2}(k + 1) + (3k + 3) = \dfrac{3k^2}{2} + \dfrac{3k}{2} + \dfrac{6k}{2} + \dfrac{6}{2} = \dfrac{3}{2}(k^2 + 3k + 2) = \dfrac{3}{2}(k + 1)(k + 2) = \dfrac{3(k + 1)}{2}[(k + 1) + 1].$$

48. (I) $n = 1: (4 \cdot 1 - 2) = 2$ and $2 \cdot 1^2 = 2$

(II) If $2 + 6 + 10 + \cdots + (4k - 2) = 2k^2$, then $2 + 6 + 10 + \cdots + (4k - 2) + [4(k + 1) - 2]$
$= [2 + 6 + 10 + \cdots + (4k - 2)] + (4k + 2) = 2k^2 + (4k + 2) = 2(k + 1)^2$.

49. (I) $n = 1: 2 \cdot 3^{1-1} = 2$ and $3^1 - 1 = 2$

(II) If $2 + 6 + 18 + \cdots + 2 \cdot 3^{k-1} = 3^k - 1$, then $2 + 6 + 18 + \cdots + 2 \cdot 3^{k-1} + 2 \cdot 3^{(k+1)-1} = (2 + 6 + 18 + \cdots + 2 \cdot 3^{k-1}) + 2 \cdot 3^k$
$= 3^k - 1 + 2 \cdot 3^k = 3 \cdot 3^k - 1 = 3^{k+1} - 1$.

50. (I) $n = 1: 3 \cdot 2^{1-1} = 3$ and $3(2^1 - 1) = 3$

(II) If $3 + 6 + 12 + \cdots + 3 \cdot 2^{k-1} = 3(2^k - 1)$, then $3 + 6 + 12 + \cdots + 3 \cdot 2^{k-1} + 3 \cdot 2^{(k+1)-1}$

$= (3 + 6 + 12 + \cdots + 3 \cdot 2^{k-1}) + 3 \cdot 2^k = 3(2^k - 1) + 3 \cdot 2^k = 3(2 \cdot 2^k - 1) = 3(2^{k+1} - 1)$.

51. (I) $n = 1: (3 \cdot 1 - 2)^2 = 1$ and $\frac{1}{2} \cdot 1 \cdot [6(1)^2 - 3(1) - 1] = 1$

(II) If $1^2 + 4^2 + 7^2 + \cdots + (3k - 2)^2 = \frac{1}{2}k(6k^2 - 3k - 1)$, then $1^2 + 4^2 + 7^2 + \cdots + (3k - 2)^2 + [3(k + 1) - 2]^2$

$= [1^2 + 4^2 + 7^2 + \cdots + (3k - 2)^2] + (3k + 1)^2 = \frac{1}{2}k(6k^2 - 3k - 1) + (3k + 1)^2 = \frac{1}{2}(6k^3 - 3k^2 - k) + (9k^2 + 6k + 1)$

$= \frac{1}{2}(6k^3 + 15k^2 + 11k + 2) = \frac{1}{2}(k + 1)(6k^2 + 9k + 2) = \frac{1}{2}(k + 1)[6(k + 1)^2 - 3(k + 1) - 1]$.

52. (II) $n = 1: 1 \cdot (1 + 2) = 3$ and $\frac{1}{6}(1 + 1)(2 \cdot 1 + 7) = 3$

(II) If $1 \cdot 3 + 2 \cdot 4 + 3 \cdot 5 + \cdots + k(k + 2) = \frac{k}{6}(k + 1)(2k + 7)$, then $1 \cdot 3 + 2 \cdot 4 + 3 \cdot 5 + \cdots + k(k + 2)$

$+ (k + 1)[(k + 1) + 2] = [1 \cdot 3 + 2 \cdot 4 + 3 \cdot 5 + \cdots + k(k + 2)] + (k + 1)(k + 3)$

$= \frac{k}{6}(k + 1)(2k + 7) + (k + 1)(k + 3) = (k + 1)\left[\frac{k}{6}(2k + 7) + k + 3\right] = \frac{1}{6}(k + 1)(2k^2 + 13k + 18)$

$= \frac{1}{6}(k + 1)(k + 2)(2k + 9) = \frac{k + 1}{6}[(k + 1) + 1][2(k + 1) + 7]$.

53. 10 **54.** 28 **55.** $x^5 + 10x^4 + 40x^3 + 80x^2 + 80x + 32$ **56.** $x^4 - 12x^3 + 54x^2 - 108x + 81$ **57.** $32x^5 + 240x^4 + 720x^3 + 1080x^2 + 810x + 243$

58. $81x^4 - 432x^3 + 864x^2 - 768x + 256$ **59.** 144 **60.** $-13{,}608$ **61.** 84 **62.** 1792 **63.** (a) 8 bricks (b) 1100 bricks **64.** 360 tiles

65. (a) $20\left(\frac{3}{4}\right)^3 = \frac{135}{16}$ ft (b) $20\left(\frac{3}{4}\right)^n$ ft (c) 13 times (d) 140 ft **66.** \$151{,}873.77 **67.** \$244{,}129.08 **68.** \$23{,}397.17

Chapter Test *(page 533)*

1. $0, \frac{3}{10}, \frac{8}{11}, \frac{5}{4}, \frac{24}{13}$ **2.** $4, 14, 44, 134, 404$ **3.** $2 - \frac{3}{4} + \frac{4}{9} = \frac{61}{36}$ **4.** $-\frac{1}{3} - \frac{14}{9} - \frac{73}{27} - \frac{308}{81} = -\frac{680}{81}$ **5.** $\displaystyle\sum_{k=1}^{10}(-1)^k\left(\frac{k + 1}{k + 4}\right)$ **6.** Neither

7. Geometric; $r = 4$; $S_n = \frac{2}{3}(1 - 4^n)$ **8.** Arithmetic: $d = -8$; $S_n = n(2 - 4n)$ **9.** Arithmetic; $d = -\frac{1}{2}$; $S_n = \frac{n}{4}(27 - n)$

10. Geometric; $r = \frac{2}{5}$; $S_n = \frac{125}{3}\left(1 - \left(\frac{2}{5}\right)^n\right)$ **11.** Neither **12.** $\frac{1024}{5}$ **13.** $243m^5 + 810m^4 + 1080m^3 + 720m^2 + 240m + 32$

14. First we show that the statement holds for $n = 1$. $\left(1 + \frac{1}{1}\right) = 1 + 1 = 2$. The equality is true for $n = 1$, so Condition I holds. Next we assume that

$\left(1 + \frac{1}{1}\right)\left(1 + \frac{1}{2}\right)\left(1 + \frac{1}{3}\right)\cdots\left(1 + \frac{1}{n}\right) = n + 1$ is true for some k, and we determine whether the formula then holds for $k + 1$. We assume that

$\left(1 + \frac{1}{1}\right)\left(1 + \frac{1}{2}\right)\left(1 + \frac{1}{3}\right)\cdots\left(1 + \frac{1}{k}\right) = k + 1$. Now we need to show that $\left(1 + \frac{1}{1}\right)\left(1 + \frac{1}{2}\right)\left(1 + \frac{1}{3}\right)\cdots\left(1 + \frac{1}{k}\right)\left(1 + \frac{1}{k + 1}\right)$

$= (k + 1) + 1 = k + 2$. We do this as follows:

$\left(1 + \frac{1}{1}\right)\left(1 + \frac{1}{2}\right)\left(1 + \frac{1}{3}\right)\cdots\left(1 + \frac{1}{k}\right)\left(1 + \frac{1}{k + 1}\right) = \left[\left(1 + \frac{1}{1}\right)\left(1 + \frac{1}{2}\right)\left(1 + \frac{1}{3}\right)\cdots\left(1 + \frac{1}{k}\right)\right]\left(1 + \frac{1}{k + 1}\right) = (k + 1)$

$\left(1 + \frac{1}{k + 1}\right)$(induction assumption) $= (k + 1) \cdot 1 + (k + 1) \cdot \frac{1}{k + 1} = k + 1 + 1 = k + 2$

Condition II also holds. Thus, the formula holds true for all natural numbers.

15. After 10 years, the Durango will be worth \$6103.11. **16.** The weightlifter will have lifted a total of 8000 pounds after 5 sets.

Cumulative Review *(page 534)*

1. $-3, 3, -3i, 3i$

2. (a)

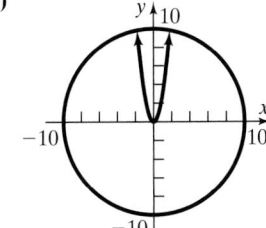

(b) $\left(\sqrt{\frac{-1 + \sqrt{3601}}{18}}, \frac{-1 + \sqrt{3601}}{6}\right), \left(-\sqrt{\frac{-1 + \sqrt{3601}}{18}}, \frac{-1 + \sqrt{3601}}{6}\right)$

(c) The circle and the parabola intersect at

$\left(\sqrt{\frac{-1 + \sqrt{3601}}{18}}, \frac{-1 + \sqrt{3601}}{6}\right), \left(-\sqrt{\frac{-1 + \sqrt{3601}}{18}}, \frac{-1 + \sqrt{3601}}{6}\right)$.

3. $\left\{\ln\left(\frac{5}{2}\right)\right\}$ **4.** $y = 5x - 10$ **5.** $x^2 + y^2 + 2x - 4y - 20 = 0$ **6.** (a) 5 (b) 13 (c) $\frac{6x + 3}{2x - 1}$ (d) $\left\{x \mid x \neq \frac{1}{2}\right\}$ (e) $\frac{7x - 2}{x - 2}$ (f) $\{x \mid x \neq 2\}$

(g) $g^{-1}(x) = \frac{1}{2}(x - 1)$; all reals (h) $f^{-1}(x) = \frac{2x}{x - 3}$; $\{x \mid x \neq 3\}$ **7.** $\frac{x^2}{7} + \frac{y^2}{16} = 1$ **8.** $(x + 1)^2 = 4(y - 2)$

CHAPTER 8 Counting and Probability

8.1 Assess Your Understanding *(page 540)*

1. union **2.** intersection **3.** T **4.** T **5.** $\{1, 3, 5, 6, 7, 9\}$ **6.** $\{1, 2, 3, 4, 5, 6, 7, 8, 9\}$ **7.** $\{1, 5, 7\}$ **8.** $\{1, 9\}$ **9.** $\{1, 6, 9\}$ **10.** $\{1, 6, 9\}$
11. $\{1, 2, 4, 5, 6, 7, 8, 9\}$ **12.** $\{1, 2, 3, 4, 5, 6, 7, 8, 9\}$ **13.** $\{1, 2, 4, 5, 6, 7, 8, 9\}$ **14.** $\{1\}$ **15.** $\{0, 2, 6, 7, 8\}$ **16.** $\{0, 2, 5, 7, 8, 9\}$
17. $\{0, 1, 2, 3, 5, 6, 7, 8, 9\}$ **18.** $\{0, 5, 9\}$ **19.** $\{0, 1, 2, 3, 5, 6, 7, 8, 9\}$ **20.** $\{0, 5, 9\}$ **21.** $\{0, 1, 2, 3, 4, 6, 7, 8\}$ **22.** $\{2, 7, 8\}$ **23.** $\{0\}$
24. $\{0, 1, 2, 3, 5, 6, 7, 8, 9\}$ **25.** $\varnothing, \{a\}, \{b\}, \{c\}, \{d\}, \{a, b\}, \{a, c\}, \{a, d\}, \{b, c\}, \{b, d\}, \{c, d\}, \{a, b, c\}, \{b, c, d\}, \{a, c, d\}, \{a, b, d\}, \{a, b, c, d\}$
26. $\varnothing, \{a\}, \{b\}, \{c\}, \{d\}, \{e\}, \{a, b\}, \{a, c\}, \{a, d\}, \{a, e\}, \{b, c\}, \{b, d\}, \{b, e\}, \{c, d\}, \{c, e\}, \{d, e\}, \{a, b, c\}, \{a, b, d\}, \{a, b, e\}, \{a, c, d\},$
 $\{a, c, e\}, \{a, d, e\}, \{b, c, d\}, \{b, c, e\}, \{b, d, e\}, \{c, d, e\}, \{a, b, c, d\}, \{a, b, c, e\}, \{a, b, d, e\}, \{a, c, d, e\}, \{b, c, d, e\}, \{a, b, c, d, e\}$ **27.** 25
28. 25 **29.** 40 **30.** 50 **31.** 25 **32.** 20 **33.** 37 **34.** 8 **35.** 18 **36.** 31 **37.** 5 **38.** 52 **39.** 175; 125 **40.** 550 **41. (a)** 15 **(b)** 15 **(c)** 15
 (d) 25 **(e)** 40 **42.** 8 **43. (a)** 11,291 thousand **(b)** 72,503 thousand **44. (a)** 23,640 thousand **(b)** 85,677 thousand

8.2 Assess Your Understanding *(page 549)*

3. permutation **4.** combination **5.** T **6.** T **7.** 30 **8.** 42 **9.** 24 **10.** 40,320 **11.** 1 **12.** 1 **13.** 1680 **14.** 336 **15.** 28 **16.** 28 **17.** 35
18. 15 **19.** 1 **20.** 18 **21.** 10,400,600 **22.** 48,620 **23.** $\{abc, abd, abe, acb, acd, ace, adb, adc, ade, aeb, aec, aed, bac, bad, bae, bca, bcd, bce, bda,$
 $bdc, bde, bea, bec, bed, cab, cad, cae, cba, cbd, cbe, cda, cdb, cde, cea, ceb, ced, dab, dac, dae, dba, dbc, dbe, dca, dcb, dce, dea, deb, dec, eab, eac,$
 $ead, eba, ebc, ebd, eca, ecb, ecd, eda, edb, edc\}$; 60 **24.** $\{ab, ba, ac, ca, ad, da, ae, ea, bc, cb, bd, db, be, eb, cd, dc, ce, ec, de, ed\}$; 20 **25.** $\{123, 124,$
 $132, 134, 142, 143, 213, 214, 231, 234, 241, 243, 312, 314, 321, 324, 341, 342, 412, 413, 421, 423, 431, 432\}$; 24 **26.** $\{123, 132, 213, 231, 312, 321, 124, 142,$
 $214, 241, 412, 421, 125, 152, 215, 251, 512, 521, 126, 162, 216, 261, 612, 621, 134, 143, 314, 341, 413, 431, 135, 153, 315, 351, 513, 531, 136, 163, 316, 361,$
 $613, 631, 145, 154, 415, 451, 514, 541, 146, 164, 416, 461, 614, 641, 156, 165, 516, 561, 615, 651, 234, 243, 324, 342, 423, 432, 235, 253, 325, 352, 523, 532,$
 $236, 263, 326, 362, 623, 632, 245, 254, 425, 452, 524, 542, 246, 264, 426, 462, 624, 642, 256, 265, 526, 562, 625, 652, 345, 354, 435, 453, 534, 543, 346, 364,$
 $436, 463, 634, 643, 356, 365, 536, 563, 635, 653, 456, 465, 546, 564, 645, 654\}$; 120 **27.** $\{abc, abd, abe, acd, ace, ade, bcd, bce, bde, cde\}$; 10
28. $\{ab, ac, ad, ae, bc, bd, be, cd, ce, de\}$; 10 **29.** $\{123, 124, 134, 234\}$; 4 **30.** $\{123, 124, 125, 126, 134, 135, 136, 145, 146, 156, 234, 235, 236, 245, 246,$
 $256, 345, 346, 356, 456\}$; 20 **31.** 15 **32.** 15 **33.** 16 **34.** 25 **35.** 8 **36.** 1000 (allowing numbers with initial zeros such as 012) **37.** 24
38. 120 **39.** 60 **40.** 360 **41.** 18,278 **42.** 12,338,352 **43.** 35 **44.** 56 **45.** 1024 **46.** 1024 **47.** 9000 **48.** 80,000 **49.** 120 **50. (a)** 6,760,000
 (b) 3,407,040 **(c)** 3,276,000 **51.** 480 **52.** 117,600 (assuming a combination has no repeated numbers) **53.** 132,860 **54.** 6,302,555,018,760
55. 336 **56.** 5,209,344 **57.** 90,720 **58.** 4,989,600 **59. (a)** 63 **(b)** 35 **(c)** 1 **60. (a)** 3003 **(b)** 20,475 **(c)** 16,653 **61.** 1.157×10^{76}
62. 302,400 **63.** 362,880 **64.** 40,320 **65.** 660 **66.** 70 **67.** 15 **68.** 126

Historical Problems *(page 558)*

1. (a) $\{AAAA, AAAB, AABA, AABB, ABAA, ABAB, ABBA, ABBB, BAAA, BAAB, BABA, BABB, BBAA, BBAB, BBBA, BBBB\}$
 (b) $P(A \text{ wins}) = \dfrac{C(4, 2) + C(4, 3) + C(4, 4)}{2^4} = \dfrac{6 + 4 + 1}{16} = \dfrac{11}{16}$; $P(B \text{ wins}) = \dfrac{C(4, 3) + C(4, 4)}{2^4} = \dfrac{4 + 1}{16} = \dfrac{5}{16}$
2. (a) $\$\dfrac{3}{2} = \1.50 **(b)** $\$\dfrac{1}{2} = \0.50

8.3 Assess Your Understanding *(page 558)*

1. equally likely **2.** complement **3.** F **4.** T **5.** 0, 0.01, 0.35, 1 **6.** $\dfrac{1}{2}, \dfrac{3}{4}, \dfrac{2}{3}, 0$ **7.** Probability model **8.** Probability model

9. Not a probability model **10.** Not a probability model **11.** $S = \{HH, HT, TH, TT\}$; $P(HH) = \dfrac{1}{4}, P(HT) = \dfrac{1}{4}, P(TH) = \dfrac{1}{4}, P(TT) = \dfrac{1}{4}$

12. $S = \{HH, HT, TH, TT\}$; $P(HH) = \dfrac{1}{4}, P(HT) = \dfrac{1}{4}, P(TH) = \dfrac{1}{4}, P(TT) = \dfrac{1}{4}$

13. $S = \{HH1, HH2, HH3, HH4, HH5, HH6, HT1, HT2, HT3, HT4, HT5, HT6, TH1, TH2, TH3, TH4, TH5, TH6, TT1, TT2, TT3, TT4, TT5, TT6\}$; each
 outcome has the probability of $\dfrac{1}{24}$.

14. $S = \{H1H, H2H, H3H, H4H, H5H, H6H, H1T, H2T, H3T, H4T, H5T, H6T, T1H, T2H, T3H, T4H, T5H, T6H, T1T, T2T, T3T, T4T, T5T, T6T\}$; each
 outcome has the probability of $\dfrac{1}{24}$.

15. $S = \{HHH, HHT, HTH, HTT, THH, THT, TTH, TTT\}$; each outcome has the probability of $\dfrac{1}{8}$.

16. $S = \{HHH, HTH, HHT, HTT, THH, TTH, THT, TTT\}$; each outcome has the probability of $\dfrac{1}{8}$.

17. $S = \{$ 1 Yellow, 1 Red, 1 Green, 2 Yellow, 2 Red, 2 Green, 3 Yellow, 3 Red, 3 Green, 4 Yellow, 4 Red, 4 Green$\}$; each outcome has the probability of
 $\dfrac{1}{12}$; thus, $P(2 \text{ Red}) + P(4 \text{ Red}) = \dfrac{1}{12} + \dfrac{1}{12} = \dfrac{1}{6}$.

18. $S = \{$Forward Yellow, Forward Green, Forward Red, Backward Yellow, Backward Green, Backward Red$\}$; each outcome has the probability of $\dfrac{1}{6}$;
 thus, $P(\text{Forward Yellow}) + P(\text{Forward Green}) = \dfrac{1}{6} + \dfrac{1}{6} = \dfrac{1}{3}$.

19. $S = \{$1 Yellow Forward, 1 Yellow Backward, 1 Red Forward, 1 Red Backward, 1 Green Forward, 1 Green Backward, 2 Yellow Forward, 2 Yellow
 Backward, 2 Red Forward, 2 Red Backward, 2 Green Forward, 2 Green Backward, 3 Yellow Forward, 3 Yellow Backward, 3 Red Forward, 3 Red
 Backward, 3 Green Forward, 3 Green Backward, 4 Yellow Forward, 4 Yellow Backward, 4 Red Forward, 4 Red Backward, 4 Green Forward, 4 Green
 Backward$\}$; each outcome has the probability of $\dfrac{1}{24}$; thus, $P(1 \text{ Red Backward}) + P(1 \text{ Green Backward}) = \dfrac{1}{24} + \dfrac{1}{24} = \dfrac{1}{12}$.

20. $S = \{$Yellow 1 Forward, Yellow 1 Backward, Yellow 2 Forward, Yellow 2 Backward, Yellow 3 Forward, Yellow 3 Backward, Yellow 4 Forward, Yellow 4 Backward, Green 1 Forward, Green 1 Backward, Green 2 Forward, Green 2 Backward, Green 3 Forward, Green 3 Backward, Green 4 Forward, Green 4 Backward, Red 1 Forward, Red 1 Backward, Red 2 Forward, Red 2 Backward, Red 3 Forward, Red 3 Backward, Red 4 Forward, Red 4 Backward$\}$; each outcome has the probability of $\frac{1}{24}$; thus, $P(\text{Yellow 2 Forward}) + P(\text{Yellow 4 Forward}) = \frac{1}{24} + \frac{1}{24} = \frac{1}{12}$.

21. $S = \{$11 Red, 11 Yellow, 11 Green, 12 Red, 12 Yellow, 12 Green, 13 Red, 13 Yellow, 13 Green, 14 Red, 14 Yellow, 14 Green, 21 Red, 21 Yellow, 21 Green, 22 Red, 22 Yellow, 22 Green, 23 Red, 23 Yellow, 23 Green, 24 Red, 24 Yellow, 24 Green, 31 Red, 31 Yellow, 31 Green, 32 Red, 32 Yellow, 32 Green, 33 Red, 33 Yellow, 33 Green, 34 Red, 34 Yellow, 34 Green, 41 Red, 41 Yellow, 41 Green, 42 Red, 42 Yellow, 42 Green, 43 Red, 43 Yellow, 43 Green, 44 Red, 44 Yellow, 44 Green$\}$; each outcome has the probability of $\frac{1}{48}$; thus, $E = \{$22 Red, 22 Green, 24 Red, 24 Green$\}$; $P(E) = \frac{n(E)}{n(S)} = \frac{4}{48} = \frac{1}{12}$.

22. $S = \{$Forward 11, Forward 12, Forward 13, Forward 14, Forward 21, Forward 22, Forward 23, Forward 24, Forward 31, Forward 32, Forward 33, Forward 34, Forward 41, Forward 42, Forward 43, Forward 44, Backward 11, Backward 12, Backward 13, Backward 14, Backward 21, Backward 22, Backward 23, Backward 24, Backward 31, Backward 32, Backward 33, Backward 34, Backward 41, Backward 42, Backward 43, Backward 44$\}$; each outcome has the probability of $\frac{1}{32}$; thus, $P(\text{Forward 12}) + P(\text{Forward 14}) + P(\text{Forward 32}) + P(\text{Forward 34}) = \frac{1}{32} + \frac{1}{32} + \frac{1}{32} + \frac{1}{32} = \frac{1}{8}$.

23. A, B, C, F **24.** A **25.** B **26.** F **27.** $P(H) = \frac{4}{5}; P(T) = \frac{1}{5}$ **28.** $P(H) = \frac{1}{3}; P(T) = \frac{2}{3}$

29. $P(1) = P(3) = P(5) = \frac{2}{9}; P(2) = P(4) = P(6) = \frac{1}{9}$ **30.** $P(1) = P(2) = P(3) = P(4) = P(5) = \frac{1}{5}; P(6) = 0$ **31.** $\frac{3}{10}$ **32.** $\frac{2}{5}$ **33.** $\frac{1}{2}$

34. $\frac{1}{2}$ **35.** $\frac{1}{6}$ **36.** $\frac{7}{30}$ **37.** $\frac{1}{8}$ **38.** $\frac{1}{8}$ **39.** $\frac{1}{4}$ **40.** $\frac{3}{8}$ **41.** $\frac{1}{6}$ **42.** $\frac{1}{18}$ **43.** $\frac{1}{18}$ **44.** $\frac{1}{36}$ **45.** 0.55 **46.** 0.10 **47.** 0.70 **48.** 0 **49.** 0.30 **50.** 0.5

51. 0.735 **52.** 0.961 **53.** 0.7 **54.** 0.96 **55.** $\frac{17}{20}$ **56.** $\frac{3}{5}$ **57.** $\frac{11}{20}$ **58.** $\frac{3}{5}$ **59.** $\frac{1}{2}$ **60.** $\frac{1}{10}$ **61.** $\frac{3}{10}$ **62.** $\frac{13}{20}$ **63.** $\frac{2}{5}$ **64.** $\frac{3}{5}$ **65. (a)** 0.57 **(b)** 0.95

(c) 0.83 **(d)** 0.38 **(e)** 0.29 **(f)** 0.05 **(g)** 0.78 **(h)** 0.71 **66. (a)** 0.45 **(b)** 0.75 **(c)** 0.90 **67. (a)** $\frac{25}{33}$ **(b)** $\frac{25}{33}$ **68. (a)** $\frac{7}{13}$ **(b)** $\frac{11}{13}$

69. 0.167 **70.** 0.814 **71.** 0.000033069

Review Exercises (page 562)

1. \emptyset, {Dave}, {Joanne}, {Erica}, {Dave, Joanne}, {Dave, Erica}, {Joanne, Erica}, {Dave, Joanne, Erica}
2. \emptyset, {Green}, {Blue}, {Red}, {Green, Blue}, {Green, Red}, {Blue, Red}, {Green, Blue, Red}
3. {1, 3, 5, 6, 7, 8} **4.** {2, 3, 5, 6, 7, 8, 9} **5.** {3, 7} **6.** {3, 5, 7} **7.** {1, 2, 4, 6, 8, 9} **8.** {1, 4} **9.** {1, 2, 4, 5, 6, 9} **10.** {2, 4, 9} **11.** 17 **12.** 24
13. 29 **14.** 34 **15.** 7 **16.** 45 **17.** 25 **18.** 7 **19.** 336 **20.** 210 **21.** 56 **22.** 35 **23.** 60 **24.** 120 **25.** 128 **26.** 64 **27.** 3024 **28.** 24 **29.** 1680
30. 120 **31.** 91 **32. (a)** 14,400 **(b)** 14,400 **33.** 1,600,000 **34.** 60 **35.** 216,000 **36.** 256 (allowing numbers with initial zeros, such as 011) **37.** 1260
38. 12,600 **39. (a)** 381,024 **(b)** 1260 **40. (a)** 280 **(b)** 280 **(c)** 645 **41. (a)** $8.634628387 \times 10^{45}$ **(b)** 0.6531 **(c)** 0.3469 **42. (a)** 0.29 **(b)** 0.71

43. (a) 0.058 **(b)** 0.942 **44.** 0.15; 0.45 **45.** $\frac{4}{9}$ **46.** $\frac{1}{24}$ **47.** 0.2; 0.26 **48. (a)** 0.68 **(b)** 0.58 **(c)** 0.32

Chapter Test (page 564)

1. {4} **2.** {0, 1, 3, 4, 5, 7, 9} **3.** {1, 9} **4.** {3, 5, 7} **5.** {0, 2, 4, 6, 8} **6.** {1, 2, 3, 5, 6, 7, 8, 9} **7.** $n(\text{physics}) = 22$ **8.** $n(\text{none of the three}) = 3$
9. $n(\text{only biology } and \text{ chemistry}) = 8$ **10.** $n(\text{physics } or \text{ chemistry}) = 45$ **11.** 5040 **12.** 151,200 **13.** 462 **14.** There are 54,264 ways to choose 6 colors from the 21 available colors. **15.** There are 840 distinct arrangements of the letters in the word REDEEMED. **16.** There are 56 different exacta bets for an 8-horse race. **17.** There are 155,480,000 possible license plates using the new format. **18. (a)** 0.95 **(b)** 0.30 **19. (a)** 0.25
(b) 0.55 **20.** 0.19 **21.** $P(\text{win on \$1 play}) = \frac{1}{120,526,770} \approx 0.0000000083$ **22.** $P(\text{exactly 2 fours}) = \frac{1250}{7776} \approx 0.1608$

Cumulative Review (page 565)

1. $\left\{ \frac{1}{3} - \frac{\sqrt{2}}{3}i, \frac{1}{3} + \frac{\sqrt{2}}{3}i \right\}$

2.

3.

4. $\{x | 3.99 \le x \le 4.01\}$ or $[3.99, 4.01]$

5. $\left\{ -\frac{1}{2} + \frac{\sqrt{7}}{2}i, -\frac{1}{2} - \frac{\sqrt{7}}{2}i, -\frac{1}{5}, 3 \right\}$

6.
Domain: all real numbers
Range: $\{y | y > 5\}$
Horizontal asymptote: $y = 5$

7. 2 **8.** $\left\{ \frac{8}{3} \right\}$ **9.** $x = 2, y = -5, z = 3$ **10.** 125; 700

APPENDIX A Review

A.1 Assess Your Understanding (page A9)

1. variable **2.** origin **3.** strict **4.** base; exponent or power **5.** 5 **6.** F **7.** F **8.** F **9.** 4 **10.** -3 **11.** -28 **12.** -2 **13.** $\dfrac{4}{5}$ **14.** $-\dfrac{1}{5}$ **15.** 0

16. $-\dfrac{7}{3}$ **17.** $x = 0$ **18.** $x = 0$ **19.** $x = 3$ **20.** None **21.** None **22.** $x = 1, x = -1$ **23.** $x = 1, x = 0, x = -1$ **24.** $x = 0$ **25.** $\{x \mid x \neq 5\}$

26. $\{x \mid x \neq -4\}$ **27.** $\{x \mid x \neq -4\}$ **28.** $\{x \mid x \neq 6\}$ **29.** [number line] **30.** [number line] **31.** $>$

32. $<$ **33.** $>$ **34.** $<$ **35.** $>$ **36.** $>$ **37.** $=$ **38.** $>$ **39.** $<$ **40.** $=$ **41.** $x > 0$ **42.** $z < 0$ **43.** $x < 2$ **44.** $y > -5$ **45.** $x \leq 1$

46. $x \geq 2$ **47.** [number line -2] **48.** [number line 4] **49.** [number line -1] **50.** [number line 7] **51.** 1 **52.** 5 **53.** 5 **54.** 1 **55.** 1

56. -1 **57.** 22 **58.** 5 **59.** 2 **60.** 13 **61.** 1 **62.** 3 **63.** 2 **64.** 3 **65.** 6 **66.** 2 **67.** 16 **68.** -16 **69.** $\dfrac{1}{16}$ **70.** $-\dfrac{1}{16}$ **71.** $\dfrac{1}{9}$ **72.** 4 **73.** 9

74. 8 **75.** 5 **76.** 6 **77.** 4 **78.** 3 **79.** $\dfrac{1}{64x^6}$ **80.** $-\dfrac{1}{4x^2}$ **81.** $\dfrac{x^4}{y^2}$ **82.** $\dfrac{y^3}{x^3}$ **83.** $\dfrac{1}{x^3 y}$ **84.** $\dfrac{1}{x^3 y}$ **85.** $-\dfrac{8x^3 z}{9y}$ **86.** $\dfrac{1}{2x^6 y^2 z}$ **87.** $\dfrac{16x^2}{9y^2}$ **88.** $\dfrac{216x^6}{125y^6}$

89. $A = lw$ **90.** $P = 2(l + w)$ **91.** $C = \pi d$ **92.** $A = \dfrac{1}{2}bh$ **93.** $A = \dfrac{\sqrt{3}}{4}x^2$ **94.** $P = 3x$ **95.** $V = \dfrac{4}{3}\pi r^3$ **96.** $S = 4\pi r^2$ **97.** $V = x^3$

98. $S = 6x^2$ **99. (a)** $2 \leq 5$ **(b)** $6 > 5$ **100. (a)** $6 \leq 8$ **(b)** $11 > 8$ **101. (a)** Yes **(b)** No **102. (a)** $1.6 \geq 1.5$ **(b)** $1.4 < 1.5$

103. No; $\dfrac{1}{3}$ is larger; $0.000333\ldots$ **104.** No; $\dfrac{2}{3}$ is larger; $0.000666\ldots$ **105.** No **106.** 3.15 or 3.16

A.2 Assess Your Understanding (page A17)

1. right; hypotenuse **2.** $A = \dfrac{1}{2}bh$ **3.** $C = 2\pi r$ **4.** similar **5.** T **6.** T **7.** F **8.** T **9.** T **10.** F **11.** 13 **12.** 10 **13.** 26 **14.** 5 **15.** 25

16. 50 **17.** Right triangle; 5 **18.** Right triangle; 10 **19.** Not a right triangle **20.** Not a right triangle **21.** Right triangle; 25 **22.** Right triangle; 26 **23.** Not a right triangle **24.** Not a right triangle **25.** 8 in^2 **26.** 36 cm^2 **27.** 4 in^2 **28.** 18 cm^2 **29.** $A = 25\pi$ m^2; $C = 10\pi$ m

30. $A = 4\pi$ ft^2; $C = 4\pi$ ft **31.** $V = 224$ ft^3; $S = 232$ ft^2 **32.** $V = 288$ in^3; $S = 280$ in^2 **33.** $V = \dfrac{256}{3}\pi$ cm^3; $S = 64\pi$ cm^2

34. $V = 36\pi$ ft^3; $S = 36\pi$ ft^2 **35.** $V = 648\pi$ in^3; $S = 306\pi$ in^2 **36.** $V = 576\pi$ in^3; $S = 272\pi$ in^2 **37.** π square units

38. $4 - \pi$ square units **39.** 2π square units **40.** $2\pi - 4$ square units **41.** 4 units **42.** 8 units **43.** 67.5 units **44.** 40 units

45. About 16.8 ft **46.** About 1.6 revolutions **47.** 64 ft^2 **48.** 12 ft^2 **49.** $24 + 2\pi \approx 30.28$ ft^2; $16 + 2\pi \approx 22.28$ ft^2

50. $69\pi \approx 216.77$ ft^2; $26\pi \approx 81.68$ ft **51.** About 5.477 mi **52.** About 3.000 mi **53.** From 100 ft: 12.247 mi; From 150 ft: 15.000 mi

54. $a^2 + b^2 = (m^2 - n^2)^2 + (2mn)^2 = m^4 - 2m^2n^2 + n^4 + 4m^2n^2 = m^4 + 2m^2n^2 + n^4 = (m^2 + n^2)^2 = c^2$

A.3 Assess Your Understanding (page A28)

1. 4; 3 **2.** $x^4 - 16$ **3.** $x^3 - 8$ **4.** F **5.** T **6.** F **7.** Monomial; variable: x; coefficient: 2; degree: 3 **8.** Monomial; variable: x; coefficient: -4;
degree: 2 **9.** Not a monomial **10.** Not a monomial **11.** Monomial; variables: x, y; coefficient: -2; degree: 3 **12.** Monomial; variables: x, y;
coefficient: 5; degree: 5 **13.** Not a monomial **14.** Not a monomial **15.** Not a monomial **16.** Not a monomial **17.** Yes; 2 **18.** Yes; 1 **19.** Yes; 0
20. Yes; 0 **21.** No **22.** No **23.** Yes; 3 **24.** Yes; 2 **25.** No **26.** No **27.** $x^2 + 7x + 2$ **28.** $x^3 + 4x^2 - 4x + 6$ **29.** $x^3 - 4x^2 + 9x + 7$
30. $-x^3 + 4x^2 - 4x - 9$ **31.** $6x^5 + 5x^4 + 3x^2 + x$ **32.** $10x^5 + 3x^3 - 10x^2 + 6$ **33.** $7x^2 - x - 7$ **34.** $-7x^2 - 3x$ **35.** $-2x^3 + 18x^2 - 18$
36. $8x^3 - 24x^2 - 48x + 4$ **37.** $2x^2 - 4x + 6$ **38.** $-2x^2 + x - 6$ **39.** $15y^2 - 27y + 30$ **40.** $-4y^3 + 4y^2 + 4y + 12$ **41.** $x^3 + x^2 - 4x$
42. $4x^5 - 4x^3 + 8x^2$ **43.** $-8x^5 - 10x^2$ **44.** $15x^4 - 20x^3$ **45.** $x^3 + 3x^2 - 2x - 4$ **46.** $2x^3 - x^2 - x - 3$ **47.** $x^2 + 6x + 8$ **48.** $x^2 + 8x + 15$
49. $2x^2 + 9x + 10$ **50.** $6x^2 + 5x + 1$ **51.** $x^2 - 2x - 8$ **52.** $x^2 + 2x - 8$ **53.** $x^2 - 5x + 6$ **54.** $x^2 - 6x + 5$ **55.** $2x^2 - x - 6$
56. $6x^2 - 10x - 4$ **57.** $-2x^2 + 11x - 12$ **58.** $-3x^2 - 4x - 1$ **59.** $2x^2 + 8x + 8$ **60.** $2x^2 - 3x - 9$ **61.** $x^2 - xy - 2y^2$ **62.** $2x^2 + xy - 3y^2$
63. $-6x^2 - 13xy - 6y^2$ **64.** $-2x^2 + 7xy - 3y^2$ **65.** $x^2 - 49$ **66.** $x^2 - 1$ **67.** $4x^2 - 9$ **68.** $9x^2 - 4$ **69.** $x^2 + 8x + 16$ **70.** $x^2 + 10x + 25$
71. $x^2 - 8x + 16$ **72.** $x^2 - 10x + 25$ **73.** $9x^2 - 16$ **74.** $25x^2 - 9$ **75.** $4x^2 - 12x + 9$ **76.** $9x^2 - 24x + 16$ **77.** $x^2 - y^2$ **78.** $x^2 - 9y^2$
79. $9x^2 - y^2$ **80.** $9x^2 - 16y^2$ **81.** $x^2 + 2xy + y^2$ **82.** $x^2 - 2xy + y^2$ **83.** $x^2 - 4xy + 4y^2$ **84.** $4x^2 + 12xy + 9y^2$ **85.** $x^3 - 6x^2 + 12x - 8$
86. $x^3 + 3x^2 + 3x + 1$ **87.** $8x^3 + 12x^2 + 6x + 1$ **88.** $27x^3 - 54x^2 + 36x - 8$ **89.** $4x^2 - 11x + 23$; remainder -45 **90.** $3x^2 - 7x + 15$;
remainder -32 **91.** $4x - 3$; remainder $x + 1$ **92.** $3x - 1$; remainder $x - 2$ **93.** $5x^2 - 13$; remainder $x + 27$ **94.** $5x^2 - 11$; remainder
$x + 20$ **95.** $2x^2$; remainder $-x^2 + x + 1$ **96.** x^2; remainder $x - 2$ **97.** $x^2 - 2x + \dfrac{1}{2}$; remainder $\dfrac{5}{2}x + \dfrac{1}{2}$ **98.** $x^2 - \dfrac{2}{3}x - \dfrac{1}{9}$; remainder
$\dfrac{16}{9}x - \dfrac{17}{9}$ **99.** $-4x^2 - 3x - 3$; remainder -7 **100.** $-3x^3 - 3x^2 - 3x - 5$; remainder -6 **101.** $x^2 - x - 1$; remainder $2x + 2$

102. $x^2 + x - 1$; remainder $-2x + 2$ **103.** $x^2 + ax + a^2$; remainder 0 **104.** $x^4 + ax^3 + a^2x^2 + a^3x + a^4$; remainder 0

A.4 Assess Your Understanding (page A36)

1. $3x(x - 2)(x + 2)$ **2.** Prime **3.** T **4.** F **5.** $3(x + 2)$ **6.** $7(x - 2)$ **7.** $a(x^2 + 1)$ **8.** $a(x - 1)$ **9.** $x(x^2 + x + 1)$ **10.** $x(x^2 - x + 1)$
11. $2x(x - 1)$ **12.** $3x(x - 1)$ **13.** $3xy(x - 2y + 4)$ **14.** $12xy(5x - 4y + 6x^2)$ **15.** $(x + 1)(x - 1)$ **16.** $(x + 2)(x - 2)$
17. $(2x + 1)(2x - 1)$ **18.** $(3x + 1)(3x - 1)$ **19.** $(x + 4)(x - 4)$ **20.** $(x + 5)(x - 5)$ **21.** $(5x + 2)(5x - 2)$ **22.** $9(2x + 1)(2x - 1)$
23. $(x + 1)^2$ **24.** $(x - 2)^2$ **25.** $(x + 2)^2$ **26.** $(x - 1)^2$ **27.** $(x - 5)^2$ **28.** $(x + 5)^2$ **29.** $(2x + 1)^2$ **30.** $(3x + 1)^2$ **31.** $(4x + 1)^2$
32. $(5x + 1)^2$ **33.** $(x - 3)(x^2 + 3x + 9)$ **34.** $(x + 5)(x^2 - 5x + 25)$ **35.** $(x + 3)(x^2 - 3x + 9)$ **36.** $-(2x - 3)(4x^2 + 6x + 9)$ or
$(3 - 2x)(9 + 6x + 4x^2)$ **37.** $(2x + 3)(4x^2 - 6x + 9)$ **38.** $-(3x - 4)(9x^2 + 12x + 16)$ or $(4 - 3x)(16 + 12x + 9x^2)$ **39.** $(x + 2)(x + 3)$
40. $(x + 2)(x + 4)$ **41.** $(x + 6)(x + 1)$ **42.** $(x + 1)(x + 8)$ **43.** $(x + 5)(x + 2)$ **44.** $(x + 1)(x + 10)$ **45.** $(x - 8)(x - 2)$

46. $(x - 16)(x - 1)$ **47.** $(x - 8)(x + 1)$ **48.** $(x - 4)(x + 2)$ **49.** $(x + 8)(x - 1)$ **50.** $(x - 2)(x + 4)$ **51.** $(x + 2)(2x + 3)$
52. $(x - 1)(3x + 2)$ **53.** $(x - 2)(2x + 1)$ **54.** $(x + 2)(3x - 1)$ **55.** $(2x + 3)(3x + 2)$ **56.** $(3x - 2)(3x + 1)$ **57.** $(3x + 1)(x + 1)$
58. $(x + 1)(2x + 1)$ **59.** $(z + 1)(2z + 3)$ **60.** $(2z + 1)(3z + 1)$ **61.** $(x + 2)(3x - 4)$ **62.** $(x + 2)(3x + 4)$ **63.** $(x - 2)(3x + 4)$
64. $(x - 2)(3x - 4)$ **65.** $(x + 4)(3x + 2)$ **66.** $(x - 4)(3x - 2)$ **67.** $(x + 4)(3x - 2)$ **68.** $(x - 4)(3x + 2)$ **69.** $(x + 6)(x - 6)$
70. $(x - 3)(x + 3)$ **71.** $2(1 + 2x)(1 - 2x)$ **72.** $3(1 - 3x)(1 + 3x)$ **73.** $(x + 2)(x + 5)$ **74.** $(x + 1)(x + 4)$ **75.** $(x - 7)(x - 3)$
76. $(x - 4)(x - 2)$ **77.** $4(x^2 - 2x + 8)$ **78.** $3(x^2 - 4x + 5)$ **79.** Prime **80.** $(x + 6)^2$ **81.** $-(x - 5)(x + 3)$ **82.** Prime
83. $3(x + 2)(x - 6)$ **84.** $x(x - 2)(x + 10)$ **85.** $y^2(y + 5)(y + 6)$ **86.** $3y(y - 8)(y + 2)$ **87.** $(2x + 3)^2$ **88.** $(3x - 2)^2$
89. $2(3x + 1)(x + 1)$ **90.** $2(x + 1)(4x - 1)$ **91.** $(x - 3)(x + 3)(x^2 + 9)$ **92.** $(x - 1)(x + 1)(x^2 + 1)$ **93.** $(x - 1)^2(x^2 + x + 1)^2$
94. $(x + 1)^2(x^2 - x + 1)^2$ **95.** $x^5(x - 1)(x + 1)$ **96.** $x^5(x - 1)(x^2 + x + 1)$ **97.** $(4x + 3)^2$ **98.** $(3x - 4)^2$ **99.** $-(4x - 5)(4x + 1)$
100. $-(x - 1)(16x + 5)$ **101.** $(2y - 5)(2y - 3)$ **102.** $(3y - 1)(3y + 4)$ **103.** $-(3x - 1)(3x + 1)(x^2 + 1)$ **104.** $-2(2x - 1)(2x + 1)(x^2 + 2)$
105. $(x + 3)(x - 6)$ **106.** $(x + 5)(3x - 7)$ **107.** $(x + 2)(x - 3)$ **108.** $(x - 3)(x - 1)$ **109.** $(3x - 5)(9x^2 - 3x + 7)$
110. $5x(25x^2 + 15x + 3)$ **111.** $(x + 5)(3x + 11)$ **112.** $(x - 3)(7x - 16)$ **113.** $(x - 1)(x + 1)(x + 2)$ **114.** $(x - 3)(x - 1)(x + 1)$
115. $(x - 1)(x + 1)(x^2 - x + 1)$ **116.** $(x + 1)^2(x^2 - x + 1)$ **117.** $25; (x + 5)^2$ **118.** $49; (p + 7)^2$ **119.** $9; (y - 3)^2$ **120.** $4; (x - 2)^2$
121. $\frac{1}{16}; \left(x - \frac{1}{4}\right)^2$ **122.** $\frac{1}{36}; \left(x + \frac{1}{6}\right)^2$ **123.** $2(3x + 4)(9x + 13)$ **124.** $(2x + 1)(30x - 19)$ **125.** $2x(3x + 5)$ **126.** $x^2(32x - 9)$
127. $5(x + 3)(x - 2)^2(x + 1)$ **128.** $6(x + 5)^3(x - 1)(x + 1)$ **129.** $3(4x - 3)(4x - 1)$ **130.** $3x^2(3x + 4)(5x + 4)$
131. $6(3x - 5)(2x + 1)^2(5x - 4)$ **132.** $2(4x + 5)^2(5x + 1)(50x + 31)$ **133.** The possibilities are $(x \pm 1)(x \pm 4) = x^2 \pm 5x + 4$ or
$(x \pm 2)(x \pm 2) = x^2 \pm 4x + 4$, none of which equals $x^2 + 4$. **134.** The possibilities are $(x \pm 1)^2 = x^2 \pm 2x + 1$, neither of which equals
$x^2 + x + 1$.

A.5 Assess Your Understanding (page A41)

1. quotient; divisor; remainder **2.** $-3\overline{)2\,0\,-5\,1}$ **3.** T **4.** T **5.** $x^2 + x + 4$; remainder 12 **6.** $x^2 + x - 4$; remainder 5 **7.** $3x^2 + 11x + 32$;
remainder 99 **8.** $-4x^2 + 10x - 21$; remainder 43 **9.** $x^4 - 3x^3 + 5x^2 - 15x + 46$; remainder -138 **10.** $x^3 + 2x^2 + 5x + 10$; remainder 22
11. $4x^5 + 4x^4 + x^3 + x^2 + 2x + 2$; remainder 7 **12.** $x^4 - x^3 + 6x^2 - 6x + 6$; remainder -16 **13.** $0.1x^2 - 0.11x + 0.321$; remainder -0.3531
14. $0.1x - 0.21$; remainder 0.241 **15.** $x^4 + x^3 + x^2 + x + 1$; remainder 0 **16.** $x^4 - x^3 + x^2 - x + 1$; remainder 0 **17.** No **18.** No **19.** Yes
20. Yes **21.** Yes **22.** Yes **23.** No **24.** Yes **25.** Yes **26.** No **27.** -9

A.6 Assess Your Understanding (page A50)

1. lowest terms **2.** least common multiple **3.** T **4.** F **5.** $\frac{3}{x - 3}$ **6.** $\frac{x}{3}$ **7.** $\frac{x}{3}$ **8.** $\frac{5x + 8}{x}$ **9.** $\frac{4x}{2x - 1}$ **10.** $\frac{(x + 2)^2}{(x - 4)(x + 4)}$ **11.** $\frac{y + 5}{2(y + 1)}$
12. $\frac{y - 1}{y + 1}$ **13.** $\frac{x + 5}{x - 1}$ **14.** $-\frac{x}{x + 2}$ **15.** $-(x + 7)$ **16.** $-(x + 3)$ **17.** $\frac{3}{5x(x - 2)}$ **18.** $\frac{3x}{4(3x + 5)}$ **19.** $\frac{2x}{x + 4}$ **20.** $\frac{6(x + 1)}{x(2x - 1)}$ **21.** $\frac{8}{3x}$
22. $\frac{3}{5x}$ **23.** $\frac{x - 3}{x + 7}$ **24.** $\frac{(x - 5)(x - 2)(x + 3)}{(x - 3)(x - 1)(x + 5)}$ **25.** $\frac{4x}{(x - 2)(x - 3)}$ **26.** $\frac{3(x - 4)}{5x}$ **27.** $\frac{4}{5(x - 1)}$ **28.** $\frac{3}{x - 2}$ **29.** $-\frac{(x - 4)^2}{4x}$ **30.** $-\frac{9x^3}{(x - 3)^2}$
31. $\frac{(x + 3)^2}{(x - 3)^2}$ **32.** $\frac{(x + 1)(x + 2)}{(x - 2)(x - 1)}$ **33.** $\frac{(x - 4)(x + 3)}{(x - 1)(2x + 1)}$ **34.** $\frac{(2x - 3)(4x + 1)}{(3x - 1)(4x - 1)}$ **35.** $\frac{x + 5}{2}$ **36.** $-\frac{3}{x}$ **37.** $\frac{(x - 2)(x + 2)}{2x - 3}$ **38.** $\frac{3(x^2 - 3)}{2x - 1}$
39. $\frac{3x - 2}{x - 3}$ **40.** $\frac{3x - 1}{3x + 2}$ **41.** $\frac{x + 9}{2x - 1}$ **42.** $\frac{4x - 5}{3x + 4}$ **43.** $\frac{4 - x}{x - 2}$ **44.** $\frac{x + 6}{x - 1}$ **45.** $\frac{2(x + 5)}{(x - 1)(x + 2)}$ **46.** $-\frac{3x + 35}{(x - 5)(x + 5)}$ **47.** $\frac{3x^2 - 2x - 3}{(x + 1)(x - 1)}$
48. $\frac{x(5x + 1)}{(x - 4)(x + 3)}$ **49.** $\frac{-(11x + 2)}{(x + 2)(x - 2)}$ **50.** $-\frac{2}{(x - 1)(x + 1)}$ **51.** $\frac{2(x^2 - 2)}{x(x - 2)(x + 2)}$ **52.** $\frac{x^4 + x^3 - x^2 + x - 1}{x^3(x^2 + 1)}$ **53.** $(x - 2)(x + 2)(x + 1)$
54. $(x - 4)^2(x + 3)$ **55.** $x(x - 1)(x + 1)$ **56.** $3(x - 3)(x + 3)(2x + 5)$ **57.** $x^3(2x - 1)^2$ **58.** $x(x - 3)(x + 3)$
59. $x(x - 1)^2(x + 1)(x^2 + x + 1)$ **60.** $x^2(x + 2)^3$ **61.** $\frac{5x}{(x - 6)(x - 1)(x + 4)}$ **62.** $\frac{x^2 + 7x - 1}{(x - 3)(x + 8)}$ **63.** $\frac{2(2x^2 + 5x - 2)}{(x - 2)(x + 2)(x + 3)}$
64. $\frac{3x^2 - 4x + 4}{(x - 1)^2}$ **65.** $\frac{5x + 1}{(x - 1)^2(x + 1)^2}$ **66.** $\frac{-2(2x + 7)}{(x - 1)^2(x + 2)^2}$ **67.** $\frac{-x^2 + 3x + 13}{(x - 2)(x + 1)(x + 4)}$ **68.** $\frac{x^2 - 6x + 11}{(x + 7)(x + 1)^2}$ **69.** $\frac{x^3 - 2x^2 + 4x + 3}{x^2(x + 1)(x - 1)}$
70. $\frac{3x^3 - 5x^2 + 2x + 1}{x^2(x - 1)^2}$ **71.** $\frac{-1}{x(x + h)}$ **72.** $\frac{-(2x + h)}{x^2(x + h)^2}$ **73.** $\frac{x + 1}{x - 1}$ **74.** $\frac{4x^2 + 1}{3x^2 - 1}$ **75.** $\frac{(x - 1)(x + 1)}{2x(2x + 1)}$ **76.** $\frac{x}{(x + 1)^2}$ **77.** $\frac{2(5x - 1)}{(x - 2)(x + 1)^2}$
78. $\frac{-(5x - 4)}{(x - 2)(x + 1)(x + 3)}$ **79.** $\frac{-2x(x^2 - 2)}{(x + 2)(x^2 - x - 3)}$ **80.** $\frac{(x + 3)(x^2 - x - 15)}{x(4x^2 + 5x + 3)}$ **81.** $\frac{-1}{x - 1}$ **82.** $\frac{1}{x}$ **83.** $\frac{19}{(3x - 5)^2}$ **84.** $\frac{13}{(5x - 2)^2}$
85. $\frac{x^2 - 1}{(x^2 + 1)^2}$ **86.** $\frac{x^2 + 4}{(x^2 - 4)^2}$ **87.** $\frac{x(3x + 2)}{(3x + 1)^2}$ **88.** $\frac{x^2(4x - 15)}{(2x - 5)^2}$ **89.** $-\frac{3x^2 + 8x - 3}{(x^2 + 1)^2}$ **90.** $-\frac{2(x^2 - 5x - 9)}{(x^2 + 9)^2}$ **91.** $f = \frac{R_1 \cdot R_2}{(n - 1)(R_1 + R_2)}; \frac{2}{15}$ m
92. $R = \frac{R_1 R_2 R_3}{R_1 R_2 + R_1 R_3 + R_2 R_3}; \frac{20}{11}$ ohms

A.7 Assess Your Understanding (page A57)

3. index **4.** T **5.** cube root **6.** F **7.** 3 **8.** 2 **9.** -2 **10.** -1 **11.** $2\sqrt{2}$ **12.** $3\sqrt[3]{2}$ **13.** $-2x\sqrt[3]{x}$ **14.** $2x\sqrt[4]{3x}$ **15.** x^3y^2 **16.** x^2y **17.** x^2y
18. $\frac{1}{3x}$ **19.** $6\sqrt{x}$ **20.** $3x^2\sqrt{x}$ **21.** $6x\sqrt{x}$ **22.** $10x^2$ **23.** $15\sqrt[3]{3}$ **24.** $300\sqrt[3]{3}$ **25.** $12\sqrt{3}$ **26.** $-30\sqrt{6}$ **27.** $7\sqrt{2}$ **28.** $2\sqrt{5}$ **29.** $\sqrt{2}$
30. $-5\sqrt{3}$ **31.** $2\sqrt{3}$ **32.** $-1 + \sqrt{5}$ **33.** $-\sqrt[3]{2}$ **34.** $15\sqrt[3]{3}$ **35.** $x - 2\sqrt{x} + 1$ **36.** $x + 2\sqrt{5x} + 5$ **37.** $(2x - 1)\sqrt[3]{2x}$ **38.** $(x + 2)\sqrt[4]{2x}$
39. $(2x - 15)\sqrt{2x}$ **40.** $(9x + 20)\sqrt{y}$ **41.** $-(x + 5y)\sqrt[3]{2xy}$ **42.** $5xy$ **43.** $\frac{\sqrt{2}}{2}$ **44.** $\frac{2\sqrt{3}}{3}$ **45.** $-\frac{\sqrt{15}}{5}$ **46.** $-\frac{\sqrt{6}}{4}$ **47.** $\frac{(5 + \sqrt{2})\sqrt{3}}{23}$

48. $\dfrac{(\sqrt{7}-2)\sqrt{2}}{3}$ **49.** $-\dfrac{8\sqrt{5}-19}{41}$ **50.** $\dfrac{9-5\sqrt{3}}{3}$ **51.** $\dfrac{5\sqrt[3]{4}}{2}$ **52.** $-\dfrac{2\sqrt[3]{3}}{3}$ **53.** $\dfrac{2x+h-2\sqrt{x}\sqrt{x+h}}{h}$ **54.** $\dfrac{x+\sqrt{x+h}\sqrt{x-h}}{h}$

55. 4 **56.** 8 **57.** -3 **58.** 8 **59.** 64 **60.** 125 **61.** $\dfrac{1}{27}$ **62.** $\dfrac{1}{64}$ **63.** $\dfrac{27\sqrt{2}}{32}$ **64.** $\dfrac{9}{4}$ **65.** $\dfrac{27\sqrt{2}}{32}$ **66.** $\dfrac{9}{4}$ **67.** $x^{7/12}$ **68.** $x^{11/12}$ **69.** xy^2 **70.** x^3y^6

71. $x^{4/3}y^{5/3}$ **72.** $x^{5/4}y^{5/4}$ **73.** $\dfrac{8x^{3/2}}{y^{1/4}}$ **74.** $\dfrac{8y^{1/2}}{x^{3/2}}$ **75.** $\dfrac{3x+2}{(1+x)^{1/2}}$ **76.** $\dfrac{3x+1}{2x^{1/2}}$ **77.** $\dfrac{x(3x^2+2)}{(x^2+1)^{1/2}}$ **78.** $\dfrac{4x+3}{3(x+1)^{2/3}}$ **79.** $\dfrac{22x+5}{10\sqrt{x}-5\sqrt{4x+3}}$

80. $\dfrac{65x+6}{24\sqrt[3]{(x-2)^2}\sqrt[3]{(8x+1)^2}}$ **81.** $\dfrac{2+x}{2(1+x)^{3/2}}$ **82.** $\dfrac{1}{(x^2+1)^{3/2}}$ **83.** $\dfrac{4-x}{(x+4)^{3/2}}$ **84.** $\dfrac{9}{(9-x^2)^{3/2}}$ **85.** $\dfrac{1}{x^2(x^2-1)^{1/2}}$ **86.** $\dfrac{4}{(x^2+4)^{3/2}}$

87. $\dfrac{1-3x^2}{2\sqrt{x}(1+x^2)^2}$ **88.** $\dfrac{2x(3-2x^2)}{3(1-x^2)^{4/3}}$ **89.** $\dfrac{1}{2}(5x+2)(x+1)^{1/2}$ **90.** $\dfrac{1}{3}(x^2+4)^{1/3}(11x^2+12)$ **91.** $2x^{1/2}(3x-4)(x+1)$ **92.** $2x^{1/2}(10x+9)$

93. $(x^2+4)^{1/3}(11x^2+12)$ **94.** $2x(3x+4)^{1/3}(5x+4)$ **95.** $(3x+5)^{1/3}(2x+3)^{1/2}(17x+27)$ **96.** $12(6x+1)^{1/3}(4x-3)^{1/2}(5x-1)$

97. $\dfrac{3(x+2)}{2x^{1/2}}$ **98.** $\dfrac{4(2x-1)}{x^{2/3}}$ **99.** 1.41 **100.** 2.65 **101.** 1.59 **102.** -1.71 **103.** 4.89 **104.** 0.04 **105.** 2.15 **106.** 1.33 **107.** (a) 15,660.4 gal

(b) 390.7 gal **108.** (a) 16 ft/sec (b) 32 ft/sec (c) 12 ft/sec **109.** $2\sqrt{2}\pi\approx8.89$ sec **110.** $\pi\sqrt{2}\approx4.44$ sec **111.** $\dfrac{\pi\sqrt{3}}{6}\approx0.91$ sec

112. $\dfrac{\pi\sqrt{6}}{12}\approx0.64$ sec

A.8 Assess Your Understanding *(page A64)*

5. equivalent equations **6.** identity **7.** F **8.** T **9.** $\{7\}$ **10.** $\{-8\}$ **11.** $\{-3\}$ **12.** $\{-6\}$ **13.** $\{4\}$ **14.** $\{-4\}$ **15.** $\left\{\dfrac{5}{4}\right\}$ **16.** $\left\{\dfrac{27}{4}\right\}$

17. $\{-1\}$ **18.** $\{1\}$ **19.** $\{-18\}$ **20.** $\left\{\dfrac{7}{5}\right\}$ **21.** $\{-3\}$ **22.** $\{-2\}$ **23.** $\{-16\}$ **24.** $\{-8\}$ **25.** $\{0.5\}$ **26.** $\{-10\}$ **27.** $\{2\}$ **28.** $\left\{\dfrac{3}{10}\right\}$ **29.** $\{2\}$

30. $\{3\}$ **31.** $\{3\}$ **32.** $\{2\}$ **33.** $\{0,9\}$ **34.** $\{0,1\}$ **35.** $\{0,9\}$ **36.** $\{0,2\}$ **37.** $\{21\}$ **38.** $\{-10\}$ **39.** $\{6\}$ **40.** $\left\{-\dfrac{1}{3}\right\}$ **41.** $\{-5,0,4\}$

42. $\{-7,0,1\}$ **43.** $\{-1,1\}$ **44.** $\{-4,-1,1\}$ **45.** $\left\{-2,\dfrac{1}{2},2\right\}$ **46.** $\left\{-3,-\dfrac{4}{3},3\right\}$ **47.** $R=\dfrac{R_1R_2}{R_1+R_2}$ **48.** $r=\dfrac{A-P}{Pt}$ **49.** $R=\dfrac{mv^2}{F}$

50. $T=\dfrac{PV}{nR}$ **51.** $r=\dfrac{S-a}{S}$ **52.** $t=\dfrac{v_0-v}{g}$

A.9 Assess Your Understanding *(page A71)*

3. negative **4.** closed interval **5.** multiplication properties **6.** T **7.** T **8.** T **9.** F **10.** T **11.** $[0,2];0\le x\le2$ **12.** $[2,\infty);x\ge2$
13. $(-1,2);-1<x<2$ **14.** $(-\infty,0];x\le0$ **15.** $[0,3);0\le x<3$ **16.** $(-1,1];-1<x\le1$ **17.** (a) $6<8$ (b) $-2<0$ (c) $9<15$
(d) $-6>-10$ **18.** (a) $5>4$ (b) $-3>-4$ (c) $6>3$ (d) $-4<-2$ **19.** (a) $7>0$ (b) $-1>-8$ (c) $12>-9$ (d) $-8<6$
20. (a) $0>-2$ (b) $-8>-10$ (c) $-9>-15$ (d) $6<10$ **21.** (a) $2x+4<5$ (b) $2x-4<-3$ (c) $6x+3<6$ (d) $-4x-2>-4$
22. (a) $4-2x>8$ (b) $-4-2x>0$ (c) $3-6x>15$ (d) $-2+4x<-10$

23. $[0,4]$

24. $(-1,5)$

25. $[4,6)$

26. $(-2,0)$

27. $[4,\infty)$

28. $(-\infty,5]$

29. $(-\infty,-4)$

30. $(1,\infty)$

31. $2\le x\le5$

32. $1<x<2$

33. $-3<x<-2$

34. $0\le x<1$

35. $x\ge4$

36. $x\le2$

37. $x<-3$

38. $x>-8$

39. $<$ **40.** $<$ **41.** $>$ **42.** $>$ **43.** \ge **44.** \le **45.** $<$ **46.** $<$ **47.** \le **48.** \ge **49.** $<$ **50.** \le **51.** \ge **52.** $<$

53. $\{x|x<4\};(-\infty,4)$

54. $\{x|x<7\};(-\infty,7)$

55. $\{x|x\ge-1\};[-1,\infty)$

56. $\{x|x\ge-1\};[-1,\infty)$

57. $\{x|x>3\};(3,\infty)$

58. $\{x|x>-2\};(-2,\infty)$

59. $\{x|x\ge2\};[2,\infty)$

60. $\{x|x\ge5\};[5,\infty)$

61. $\{x|x>-7\};(-7,\infty)$

62. $\{x|x<5\};(-\infty,5)$

63. $\left\{x\left|x\le\dfrac{2}{3}\right.\right\};\left(-\infty,\dfrac{2}{3}\right]$

64. $\{x|x\le0\};(-\infty,0]$

65. $\{x|x < -20\}$; $(-\infty, -20)$

66. $\left\{x\left|x > -\dfrac{7}{4}\right.\right\}$; $\left(-\dfrac{7}{4}, \infty\right)$

67. $\left\{x\left|x \geq \dfrac{4}{3}\right.\right\}$; $\left[\dfrac{4}{3}, \infty\right)$

68. $\{x|x \geq 12\}$; $[12, \infty)$

69. $\{x|3 \leq x \leq 5\}$; $[3, 5]$

70. $\{x|1 \leq x \leq 4\}$; $[1, 4]$

71. $\left\{x\left|\dfrac{2}{3} \leq x \leq 3\right.\right\}$; $\left[\dfrac{2}{3}, 3\right]$

72. $\{x|-3 \leq x \leq 3\}$; $[-3, 3]$

73. $\left\{x\left|-\dfrac{11}{2} < x < \dfrac{1}{2}\right.\right\}$; $\left(-\dfrac{11}{2}, \dfrac{1}{2}\right)$

74. $\left\{x\left|-\dfrac{2}{3} < x < 2\right.\right\}$; $\left(-\dfrac{2}{3}, 2\right)$

75. $\{x|-6 < x < 0\}$; $(-6, 0)$

76. $\{x|0 < x < 3\}$; $(0, 3)$

77. $\{x|x < -5\}$; $(-\infty, -5)$

78. $\{x|x < 11\}$; $(-\infty, 11)$

79. $\{x|x \geq -1\}$; $[-1, \infty)$

80. $\{x|x \leq 1\}$; $(-\infty, 1]$

81. $\left\{x\left|\dfrac{1}{2} \leq x < \dfrac{5}{4}\right.\right\}$; $\left[\dfrac{1}{2}, \dfrac{5}{4}\right)$

82. $\left\{x\left|-\dfrac{1}{3} < x \leq \dfrac{1}{3}\right.\right\}$; $\left(-\dfrac{1}{3}, \dfrac{1}{3}\right]$

83. $\{x|x > 5\}$; $(5, \infty)$

84. $\{x|x > 5\}$; $(5, \infty)$

85. $\{x|x > 3\}$; $(3, \infty)$

86. $\{x|x > 6\}$; $(6, \infty)$

87. $21 < \text{age} < 30$

88. $40 \leq \text{age} < 60$

89. (a) Male ≥ 75.6 **(b)** Female ≥ 80.4 **(c)** A female can expect to live at least 4.8 years longer. **90.** The volume of the gas ranges from 7060 to 7860 cc, inclusive. **91.** The agent's commission ranges from $45,000 to $95,000, inclusive. As a percent of selling price, the commission ranges from 5% to approximately 8.6%, inclusive. **92.** The commission will vary from $53 to $145, inclusive. **93.** The amount withheld varies from $76.35 to $101.35, inclusive. **94.** The amount withheld varies from $126.35 to $151.35, inclusive. **95.** The usage varies from approximately 710.94 to 2536.44 kilowatt-hours, inclusive. **96.** The water usage varied from 13,500 gal to approximately 20,500 gal. **97.** The dealer's cost varies from $7457.63 to $7857.14, inclusive. **98.** The people in the top 2.5% have test scores in the interval $(123.52, \infty)$. **99.** You need at least a 74 on the last test. **100.** You need at least a 77 on the last test. **101.** The amount of gasoline ranged from 12 to 20 gal, inclusive. **102.** There were 10 gal or less (and at least 0 gal) of gasoline at the start of the trip.

103. $\dfrac{a+b}{2} - a = \dfrac{a+b-2a}{2} = \dfrac{b-a}{2} > 0$; therefore, $a < \dfrac{a+b}{2}$.

$b - \dfrac{a+b}{2} = \dfrac{2b-a-b}{2} = \dfrac{b-a}{2} > 0$; therefore, $b > \dfrac{a+b}{2}$.

104. $\dfrac{a+b}{2} - a = \dfrac{a+b-2a}{2} = \dfrac{b-a}{2}$

$b - \dfrac{a+b}{2} = \dfrac{2b-a-b}{2} = \dfrac{b-a}{2}$;

thus, $\dfrac{a+b}{2}$ is equidistant from a and b.

105. $\left(\sqrt{ab}\right)^2 - a^2 = ab - a^2 = a(b-a) > 0$; thus, $\left(\sqrt{ab}\right)^2 > a^2$ and $\sqrt{ab} > a$.

$b^2 - \left(\sqrt{ab}\right)^2 = b^2 - ab = b(b-a) > 0$; thus $b^2 > \left(\sqrt{ab}\right)^2$ and $b > \sqrt{ab}$.

106. $\dfrac{a+b}{2} - \sqrt{ab} = \dfrac{1}{2}\left[a - 2\sqrt{ab} + b\right] = \dfrac{1}{2}\left(\sqrt{a} - \sqrt{b}\right)^2 > 0$ since $a \neq b$; thus, $\sqrt{ab} < \dfrac{a+b}{2}$.

107. $0 < a < b$ so $0 < \dfrac{1}{b} < \dfrac{1}{a}$; $\dfrac{1}{b} < \dfrac{1}{2}\left(\dfrac{1}{b} + \dfrac{1}{a}\right) < \dfrac{1}{a}$ by Problem 129, so $\dfrac{1}{b} < \dfrac{1}{h} < \dfrac{1}{a}$; therefore $a < h < b$.

108. $\dfrac{1}{h} = \dfrac{1}{2}\left(\dfrac{1}{a} + \dfrac{1}{b}\right)$

$2abh\left(\dfrac{1}{h}\right) = 2abh\left(\dfrac{1}{2}\right)\left(\dfrac{1}{a} + \dfrac{1}{b}\right)$

$2ab = abh\left(\dfrac{1}{a} + \dfrac{1}{b}\right)$

$2ab = bh + ah$

$2ab = (b + a)h$

$\dfrac{2ab}{b+a} = h$

$\dfrac{ab}{\dfrac{a+b}{2}} = h$

$\dfrac{\left(\sqrt{ab}\right)^2}{\dfrac{a+b}{2}} = h$

A.10 Assess Your Understanding (page A79)

1. mathematical modeling **2.** interest **3.** uniform motion **4.** False **5.** True **6.** $100 - x$ **7.** $A = \pi r^2; r =$ radius, $A =$ area
8. $C = 2\pi r; r =$ radius, $C =$ circumference **9.** $A = s^2; A =$ area, $s =$ length of a side **10.** $P = 4s; s =$ length of a side, $P =$ perimeter
11. $F = ma; F =$ force, $m =$ mass, $a =$ acceleration **12.** $P = \dfrac{F}{A}; P =$ pressure, $F =$ force, $A =$ area

13. $W = Fd; W =$ work, $F =$ force, $d =$ distance **14.** $K = \dfrac{1}{2}mv^2; K =$ kinetic energy, $m =$ mass, $v =$ velocity

15. $C = 150x; C =$ total variable cost, $x =$ number of dishwashers **16.** $R = 250x; R =$ total revenue, $x =$ number of dishwashers
17. Invest \$31,250 in bonds and \$18,750 in CDs. **18.** Invest \$43,750 in bonds and \$6250 in CDs. **19.** \$11,600 was loaned out at 8%.
20. She can lend \$333,333.33 at 16%. **21.** Mix 75 lb of Earl Grey tea with 25 lb of Orange Pekoe tea. **22.** 49 lb of coffee I should be mixed with
51 lb of coffee II. **23.** Mix 40 lb of cashews with the peanuts. **24.** Each box should contain 20 caramels and 10 creams. **25.** The speed of the
current is 2.286 mi/hr. **26.** The speed of the boat is 9 mi/hr. **27.** Run: 12 miles; Bicycle: 75 miles **28.** Bicycle 1: 18 mph; Bicycle 2: 23 mph
29. Working together, it takes 12 min. **30.** April would take 15 hr. **31. (a)** The dimensions are 10 ft by 5 ft. **(b)** The area is 50 sq ft.
(c) The dimensions would be 7.5 ft by 7.5 ft. **(d)** The area would be 56.25 sq ft. **32. (a)** Its dimensions are 19 ft by 19 ft. **(b)** Its dimensions
are 28.5 ft by 9.5 ft. **(c)** The diameter is approximately 25.83 ft. **(d)** The circular one has the most area. **33.** The defensive back catches up to
the tight end at the tight end's 45 yd line. **34.** Therese should be allowed 20,000 mi as a business expense. **35.** Add $\dfrac{2}{3}$ gal of water.
36. 5 L should be drained. **37.** Evaporate 10.67 oz of water. **38.** Evaporate 96 oz of water. **39.** 40 g of 12-karat gold should be mixed with 20 g of
pure gold. **40.** There are 11 atoms of oxygen and 22 atoms of hydrogen. **41.** Mike passes Dan $\dfrac{1}{3}$ mi from the start, 2 min from the time Mike
started to race. **42.** It can fly 742.5 mi. **43.** Start the auxiliary pump at 9:45 A.M.. **44.** 5 lb of pure cement must be added. **45.** The tub will fill
in 1 hr. **46.** It will take 1 hr and 45 min more for the 5-hp pump to empty the pool. **47.** Lewis would beat Burke by 16.75 m. **48.** Set the
original price at \$40. At 50% off, there will be no profit. **51.** The average speed is 49.5 mi/hr. **52.** The tail wind was 91.47 knots.

A.11 Assess Your Understanding (page A87)

1. T **2.** 5 **3.** real; imaginary; imaginary unit **4.** F **5.** T **6.** F **7.** $8 + 5i$ **8.** $-4 + 7i$ **9.** $-7 + 6i$ **10.** 6 **11.** $-6 - 11i$ **12.** $-10 + 6i$
13. $6 - 18i$ **14.** $-8 - 32i$ **15.** $6 + 4i$ **16.** $-12 - 9i$ **17.** $10 - 5i$ **18.** $13 + i$ **19.** 37 **20.** -10 **21.** $\dfrac{6}{5} + \dfrac{8}{5}i$ **22.** $\dfrac{5}{13} + \dfrac{12}{13}i$ **23.** $1 - 2i$
24. $\dfrac{1}{2} + i$ **25.** $\dfrac{5}{2} - \dfrac{7}{2}i$ **26.** $-\dfrac{1}{2} + \dfrac{5}{2}i$ **27.** $-\dfrac{1}{2} + \dfrac{\sqrt{3}}{2}i$ **28.** $\dfrac{1}{2} - \dfrac{\sqrt{3}}{2}i$ **29.** $2i$ **30.** $-2i$ **31.** $-i$ **32.** -1 **33.** i **34.** i **35.** -6 **36.** $4 - i$
37. $-10i$ **38.** $3 - 4i$ **39.** $-2 + 2i$ **40.** 82 **41.** 0 **42.** 0 **43.** 0 **44.** 0 **45.** $2i$ **46.** $3i$ **47.** $5i$ **48.** $8i$ **49.** $5i$ **50.** $5i$ **51.** 6 **52.** $6i$ **53.** 25
54. $-5 + 7i$
55. $z + \overline{z} = a + bi + a - bi = 2a + (b - b)i = 2a$
$z - \overline{z} = a + bi - a - bi = 0 + (b + b)i = 2bi$
56. $z = a + bi, \overline{z} = a - bi, \overline{\overline{z}} = \overline{(a - bi)} = a + bi = z$
57. $\overline{z + w} = \overline{(a + bi) + (c + di)} = \overline{(a + c) + (b + d)i} = (a + c) - (b + d)i$
$\overline{z} + \overline{w} = (a - bi) + (c - di) = (a + c) + (-b - d)i = (a + c) - (b + d)i$
58. $\overline{z \cdot w} = \overline{(a + bi)(c + di)} = \overline{(ac - bd) + (ad + bc)i} = (ac - bd) - (ad + bc)i$
$\overline{z} \cdot \overline{w} = (a - bi)(c - di) = [ac - (-b)(-d)] + [a(-d) + (-b)(c)]i = (ac - bd) + (-ad - bc)i = (ac - bd) - (ad + bc)i$

APPENDIX B Graphing Utilities

B.1 Exercises (page B3)

1. $(-1, 4)$; II **2.** $(3, 4)$; 1 **3.** $(3, 1)$; I **4.** $(-6, -4)$; III **5.** Xmin $= -6$, Xmax $= 6$, Xscl $= 2$, Ymin $= -4$, Ymax $= 4$, Yscl $= 2$
6. Xmin $= -3$, Xmax $= 3$, Xscl $= 1$, Ymin $= -2$, Ymax $= 2$, Yscl $= 1$ **7.** Xmin $= -6$, Xmax $= 6$, Xscl $= 2$, Ymin $= -1$, Ymax $= 3$, Yscl $= 1$
8. Xmin $= -9$, Xmax $= 9$, Xscl $= 3$, Ymin $= -12$, Ymax $= 4$, Yscl $= 4$ **9.** Xmin $= 3$, Xmax $= 9$, Xscl $= 1$, Ymin $= 2$, Ymax $= 10$, Yscl $= 2$
10. Xmin $= -22$, Xmax $= -10$, Xscl $= 2$, Ymin $= 4$, Ymax $= 8$, Yscl $= 1$ **11.** Xmin $= -11$, Xmax $= 5$, Xscl $= 1$, Ymin $= -3$, Ymax $= 6$,
Yscl $= 1$ **12.** Xmin $= -3$, Xmax $= 7$, Xscl $= 1$, Ymin $= -4$, Ymax $= 9$, Yscl $= 1$ **13.** Xmin $= -30$, Xmax $= 50$, Xscl $= 10$, Ymin $= -90$,
Ymax $= 50$, Yscl $= 10$ **14.** Xmin $= -90$, Xmax $= 30$, Xscl $= 10$, Ymin $= -50$, Ymax $= 70$, Yscl $= 10$ **15.** Xmin $= -10$, Xmax $= 110$,
Xscl $= 10$, Ymin $= -10$, Ymax $= 160$, Yscl $= 10$ **16.** Xmin $= -20$, Xmax $= 110$, Xscl $= 10$, Ymin $= -10$, Ymax $= 60$, Yscl $= 10$
17. $4\sqrt{10}$ **18.** $\sqrt{149}$ **19.** $2\sqrt{65}$ **20.** $\sqrt{205}$

B.2 Exercises (page B6)

1. (a)

(b)

(c)

(d)

2. (a) **(b)** **(c)** **(d)**

3. (a) **(b)** **(c)** **(d)**

4. (a) **(b)** **(c)** **(d)**

5. (a) **(b)** **(c)** **(d)**

6. (a) **(b)** **(c)** **(d)**

7. (a) **(b)** **(c)** **(d)**

8. (a) **(b)** **(c)** **(d)**

9. (a) **(b)** **(c)** **(d)**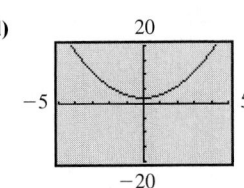

AN126 ANSWERS Section B.2

10. (a) (b) (c) (d)

11. (a) (b) (c) (d)

12. (a) (b) (c) (d)

13. (a) (b) (c) (d)

14. (a) (b) (c) (d)

15. (a) (b) (c) (d)

16. (a) (b) (c) (d)

17. Y1=X+2

18. Y1=X−2

19. Y1=−X+2

20. Y1=−X−2

21. Y1=2X+2

22. Y1=2X−2

23. Y1=−2X+2

24. Y1=−2X−2

25.

X	Y1
-3	11
-2	6
-1	3
0	2
1	3
2	6
3	11

$Y_1=X^2+2$

26.

X	Y1
-3	7
-2	2
-1	-1
0	-2
1	-1
2	2
3	7

$Y_1=X^2-2$

27.

X	Y1
-3	-7
-2	-2
-1	1
0	2
1	1
2	-2
3	-7

$Y_1=-X^2+2$

28.

X	Y1
-3	-11
-2	-6
-1	-3
0	-2
1	-3
2	-6
3	-11

$Y_1=-X^2-2$

29.

X	Y1
-3	7.5
-2	6
-1	4.5
0	3
1	1.5
2	0
3	-1.5

$Y_1=-(3/2)X+3$

30.

X	Y1
-3	-7.5
-2	-6
-1	-4.5
0	-3
1	-1.5
2	0
3	1.5

$Y_1=(3/2)X-3$

31.

X	Y1
-3	-1.5
-2	0
-1	1.5
0	3
1	4.5
2	6
3	7.5

$Y_1=(3/2)X+3$

32.

X	Y1
-3	1.5
-2	0
-1	-1.5
0	-3
1	-4.5
2	-6
3	-7.5

$Y_1=-(3/2)X-3$

B.3 Exercises *(page B9)*

1. -3.41 **2.** -4.65 **3.** -1.71 **4.** -1.43 **5.** -0.28 **6.** -0.22 **7.** 3.00 **8.** 2.00 **9.** 4.50 **10.** 1.70 **11.** $0.32, 12.30$ **12.** $0.64, 13.60$ **13.** $1.00, 23.00$ **14.** 1.07

B.5 Exercises *(page B12)*

1. Yes **2.** No **3.** Yes **4.** Yes **5.** No **6.** Yes **7.** Yes **8.** Yes **9.** Answers may vary. A possible answer is $Y\min = 4$, $Y\max = 12$, and $Y\mathrm{scl} = 1$. **10.** Answers may vary. A possible answer is $Y\min = -2$, $Y\max = 10$, and $Y\mathrm{scl} = 2$.

Index

Chapter Test Prep Video CD t/a
College Algebra: Concepts Through Functions
Michael Sullivan & Michael Sullivan, III
ISBN 0-13-243175-0
CD License Agreement
© 2007 Pearson Education, Inc.
Pearson Prentice Hall
Pearson Education, Inc.
Upper Saddle River, NJ 07458
All rights reserved.
Pearson Prentice Hall™ is a trademark of Pearson Education, Inc.

READ THIS LICENSE CAREFULLY BEFORE OPENING THIS PACKAGE. BY OPENING THIS PACKAGE, YOU ARE AGREEING TO THE TERMS AND CONDITIONS OF THIS LICENSE. IF YOU DO NOT AGREE, DO NOT OPEN THE PACKAGE. PROMPTLY RETURN THE UNOPENED PACKAGE AND ALL ACCOMPANYING ITEMS TO THE PLACE YOU OBTAINED THEM. THESE TERMS APPLY TO ALL LICENSED SOFTWARE ON THE DISK EXCEPT THAT THE TERMS FOR USE OF ANY SHAREWARE OR FREEWARE ON THE DISKETTES ARE AS SET FORTH IN THE ELECTRONIC LICENSE LOCATED ON THE DISK:

1. **GRANT OF LICENSE and OWNERSHIP:** The enclosed CD-ROM ("Software") is licensed, not sold, to you by Pearson Education, Inc. publishing as Pearson Prentice Hall ("We" or the "Company") in consideration of your adoption of the accompanying Company textbooks and/or other materials, and your agreement to these terms. You own only the disk(s) but we and/or our licensors own the Software itself. This license allows instructors and students enrolled in the course using the Company textbook that accompanies this Software (the "Course") to use and display the enclosed copy of the Software on up to one computer of an educational institution, for academic use only, so long as you comply with the terms of this Agreement. You may make one copy for back up only. We reserve any rights not granted to you.

2. **USE RESTRICTIONS:** You may not sell or license copies of the Software or the Documentation to others. You may not transfer, distribute or make available the Software or the Documentation, except to instructors and students in your school who are users of the adopted Company textbook that accompanies this Software in connection with the course for which the textbook was adopted. You may not reverse engineer, disassemble, decompile, modify, adapt, translate or create derivative works based on the Software or the Documentation. You may be held legally responsible for any copying or copyright infringement that is caused by your failure to abide by the terms of these restrictions.

3. **TERMINATION:** This license is effective until terminated. This license will terminate automatically without notice from the Company if you fail to comply with any provisions or limitations of this license. Upon termination, you shall destroy the Documentation and all copies of the Software. All provisions of this Agreement as to limitation and disclaimer of warranties, limitation of liability, remedies or damages, and our ownership rights shall survive termination.

4. **DISCLAIMER OF WARRANTY: THE COMPANY AND ITS LICENSORS MAKE NO WARRANTIES ABOUT THE SOFTWARE, WHICH IS PROVIDED "AS-IS." IF THE DISK IS DEFECTIVE IN MATERIALS OR WORKMANSHIP, YOUR ONLY REMEDY IS TO RETURN IT TO THE COMPANY WITHIN 30 DAYS FOR REPLACEMENT UNLESS THE COMPANY DETERMINES IN GOOD FAITH THAT THE DISK HAS BEEN MISUSED OR IMPROPERLY INSTALLED, REPAIRED, ALTERED OR DAMAGED. THE COMPANY DISCLAIMS ALL WARRANTIES, EXPRESS OR IMPLIED, INCLUDING WITHOUT LIMITATION, THE IMPLIED WARRANTIES OF MERCHANTABILITY AND FITNESS FOR A PARTICULAR PURPOSE. THE COMPANY DOES NOT WARRANT, GUARANTEE OR MAKE ANY REPRESENTATION REGARDING THE ACCURACY, RELIABILITY, CURRENTNESS, USE, OR RESULTS OF USE, OF THE SOFTWARE.**

5. **LIMITATION OF REMEDIES AND DAMAGES: IN NO EVENT, SHALL THE COMPANY OR ITS EMPLOYEES, AGENTS, LICENSORS OR CONTRACTORS BE LIABLE FOR ANY INCIDENTAL, INDIRECT, SPECIAL OR CONSEQUENTIAL DAMAGES ARISING OUT OF OR IN CONNECTION WITH THIS LICENSE OR THE SOFTWARE, INCLUDING, WITHOUT LIMITATION, LOSS OF USE, LOSS OF DATA, LOSS OF INCOME OR PROFIT, OR OTHER LOSSES SUSTAINED AS A RESULT OF INJURY TO ANY PERSON, OR LOSS OF OR DAMAGE TO PROPERTY, OR CLAIMS OF THIRD PARTIES, EVEN IF THE COMPANY OR AN AUTHORIZED REPRESENTATIVE OF THE COMPANY HAS BEEN ADVISED OF THE POSSIBILITY OF SUCH DAMAGES.** SOME JURISDICTIONS DO NOT ALLOW THE LIMITATION OF DAMAGES IN CERTAIN CIRCUMSTANCES, SO THE ABOVE LIMITATIONS MAY NOT ALWAYS APPLY.

6. **GENERAL:** THIS AGREEMENT SHALL BE CONSTRUED IN ACCORDANCE WITH THE LAWS OF THE UNITED STATES OF AMERICA AND THE STATE OF NEW YORK, APPLICABLE TO CONTRACTS MADE IN NEW YORK, EXCLUDING THE STATE'S LAWS AND POLICIES ON CONFLICTS OF LAW, AND SHALL BENEFIT THE COMPANY, ITS AFFILIATES AND ASSIGNEES. This Agreement is the complete and exclusive statement of the agreement between you and the Company and supersedes all proposals, prior agreements, oral or written, and any other communications between you and the company or any of its representatives relating to the subject matter. If you are a U.S. Government user, this Software is licensed with "restricted rights" as set forth in subparagraphs (a)-(d) of the Commercial Computer-Restricted Rights clause at FAR 52.227-19 or in subparagraphs (c)(1)(ii) of the Rights in Technical Data and Computer Software clause at DFARS 252.227-7013, and similar clauses, as applicable. Should you have any questions concerning this agreement or if you wish to contact the Company for any reason, please contact in writing: Pearson Education, Inc., One Lake Street, Upper Saddle River, New Jersey 07458 "AS IS" LICENSE AGREEMENT AND LIMITED *********************************

System Requirements

-Windows	-Macintosh
Pentium II 300 MHz processor	PowerPC G3 233 MHz or better
Windows 98, NT, 2000, ME, or XP	Mac OS 9.x or 10.x
64 MB RAM (128 MB RAM required for Windows XP)	64 MB RAM
4.3 MB available hard drive space (optional-for minimum QuickTime installation)	10 MB available hard drive space for Mac OS 9, 19 MB on OS X (optional-if QuickTime installation is needed)
800 x 600 resolution	800 x 600 resolution
8x or faster CD-ROM drive	8x or faster CD-ROM drive
QuickTime 6.x	QuickTime 6.x
Sound card	

Support Information

If you are having problems with this software, call (800) 677-6337 between 8:00 a.m. and 8:00 p.m. EST, Monday through Friday, and 5:00 p.m. through 12:00 a.m. EST on Sundays. You can also get support by filling out the web form located at: http://247.prenhall.com/mediaform

Our technical staff will need to know certain things about your system in order to help us solve your problems more quickly and efficiently. If possible, please be at your computer when you call for support. You should have the following information ready:

- Textbook ISBN
- CD-ROM ISBN
- corresponding product and title
- computer make and model
- Operating System (Windows or Macintosh) and Version
- RAM available
- hard disk space available
- Sound card? Yes or No
- printer make and model
- network connection
- detailed description of the problem, including the exact wording of any error messages.

NOTE: Pearson does not support and/or assist with the following:

- third-party software (i.e. Microsoft including Microsoft Office suite, Apple, Borland, etc.)
- homework assistance
- Textbooks and CD-ROMs purchased used are not supported and are non-replaceable. To purchase a new CD-ROM, contact Pearson Individual Order Copies at 1-800-282-0693.

PERMUTATIONS/COMBINATIONS

$$0! = 1 \qquad 1! = 1$$

$$n! = n(n-1) \cdots (3)(2)(1)$$

$$P(n,r) = \frac{n!}{(n-r)!}$$

$$C(n,r) = \binom{n}{r} = \frac{n!}{(n-r)! \, r!}$$

GEOMETRIC SERIES

$$\text{If } |r| < 1, \quad a + ar + ar^2 + \cdots = \sum_{k=1}^{\infty} ar^{k-1} = \frac{a}{1-r}$$

GEOMETRIC SEQUENCE

$$a + ar + ar^2 + \cdots + ar^{n-1} = a \cdot \frac{1 - r^n}{1 - r}$$

ARITHMETIC SEQUENCE

$$a + (a+d) + (a+2d) + \cdots + [a + (n-1)d]$$
$$= \frac{n}{2}\big(2a + (n-1)d\big)$$
$$= \frac{n}{2}(a + an)$$

BINOMIAL THEOREM

$$(a+b)^n = a^n + \binom{n}{1} ba^{n-1} + \binom{n}{2} b^2 a^{n-2} + \cdots + \binom{n}{n-1} b^{n-1} a + b^n$$

PROPERTIES OF LOGARITHMS

$$\log_a (MN) = \log_a M + \log_a N$$

$$\log_a \left(\frac{M}{N}\right) = \log_a M - \log_a N$$

$$\log_a M^r = r \log_a M$$

$$\log_a M = \frac{\log M}{\log a} = \frac{\ln M}{\ln a}$$

CONICS

Parabola

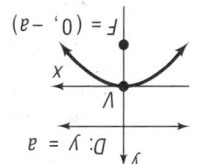

$y^2 = 4ax$ D: $x = -a$, F = (a, 0), V

$y^2 = -4ax$ D: $x = a$, F = (-a, 0), V

$x^2 = 4ay$ D: $y = -a$, F = (0, a), V

$x^2 = -4ay$ D: $y = a$, F = (0, -a), V

Ellipse

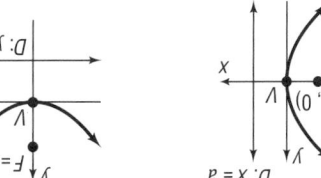

$$\frac{x^2}{a^2} + \frac{y^2}{b^2} = 1, \quad a > b, \quad c^2 = a^2 - b^2$$

$$\frac{x^2}{b^2} + \frac{y^2}{a^2} = 1, \quad a > b, \quad c^2 = a^2 - b^2$$

Hyperbola

$$\frac{x^2}{a^2} - \frac{y^2}{b^2} = 1, \quad c^2 = a^2 + b^2 \qquad \text{Asymptotes: } y = \frac{b}{a}x, \quad y = -\frac{b}{a}x$$

$$\frac{y^2}{a^2} - \frac{x^2}{b^2} = 1, \quad c^2 = a^2 + b^2 \qquad \text{Asymptotes: } y = \frac{a}{b}x, \quad y = -\frac{a}{b}x$$

FORMULAS/EQUATIONS

Distance Formula If $P_1 = (x_1, y_1)$ and $P_2 = (x_2, y_2)$, the distance from P_1 to P_2 is

$$d(P_1, P_2) = \sqrt{(x_2 - x_1)^2 + (y_2 - y_1)^2}$$

Standard Equation of a Circle The standard equation of a circle of radius r with center at (h, k) is

$$(x - h)^2 + (y - k)^2 = r^2$$

Slope Formula The slope m of the line containing the points $P_1 = (x_1, y_1)$ and $P_2 = (x_2, y_2)$ is

$$m = \frac{y_2 - y_1}{x_2 - x_1} \qquad \text{if } x_1 \neq x_2$$

$$m \text{ is undefined} \qquad \text{if } x_1 = x_2$$

Point–Slope Equation of a Line The equation of a line with slope m containing the point (x_1, y_1) is

$$y - y_1 = m(x - x_1)$$

Slope–Intercept Equation of a Line The equation of a line with slope m and y-intercept b is

$$y = mx + b$$

Quadratic Formula The solutions of the equation $ax^2 + bx + c = 0, a \neq 0$, are

$$x = \frac{-b \pm \sqrt{b^2 - 4ac}}{2a}$$

If $b^2 - 4ac > 0$, there are two unequal real solutions.
If $b^2 - 4ac = 0$, there is a repeated real solution.
If $b^2 - 4ac < 0$, there are two complex solutions that are not real.

GEOMETRY FORMULAS

Circle $r = $ Radius, $A = $ Area, $C = $ Circumference

$$A = \pi r^2 \qquad C = 2\pi r$$

Triangle $b = $ Base, $h = $ Altitude (Height), $A = $ area

$$A = \tfrac{1}{2} b h$$

Rectangle $l = $ Length, $w = $ Width, $A = $ area, $P = $ perimeter

$$A = lw \qquad P = 2l + 2w$$

Rectangular Box $l = $ Length, $w = $ Width, $h = $ Height, $V = $ Volume, $S = $ Surface area

$$V = lwh \qquad S = 2lw + 2lh + 2wh$$

Sphere $r = $ Radius, $V = $ Volume, $S = $ Surface area

$$V = \tfrac{4}{3} \pi r^3 \qquad S = 4\pi r^2$$

Right Circular Cylinder $r = $ Radius, $h = $ Height, $V = $ Volume, $S = $ Surface area

$$V = \pi r^2 h \qquad S = 2\pi r^2 + 2\pi r h$$